Get the Most Out of Your Math for Elementary Teachers MyMathLab® Course

Pearson

Beckmann Skills Review MyMathLab Course

MyMathLab has helped millions of students succeed in their math courses. The Beckmann Skills Review MyMathLab course is uniquely tailored to work in tandem with the Beckmann text.

The **Skills Review MyMathLab** provides review and skill development that complements the text, helping students brush up on skills needed to be successful in class. The MyMathLab course doesn't mirror the problems from the text, but instead covers basic skills needed prior to class, eliminating the need to spend valuable class time re-teaching basics that students should already know. This enables students to have a richer experience in the classroom while working through the book activities and problems.

Prepare Students for the Classroom with Tools for Success

The MyMathLab course includes a wealth of resources that not only helps students learn the concepts, but puts the concepts in the context of teaching in their own classroom.

IMAP (Integrating Math and Pedagogy) Videos give students insight into children's reasoning, and **Classroom Videos** show real teachers with their students or discussing classroom strategies.

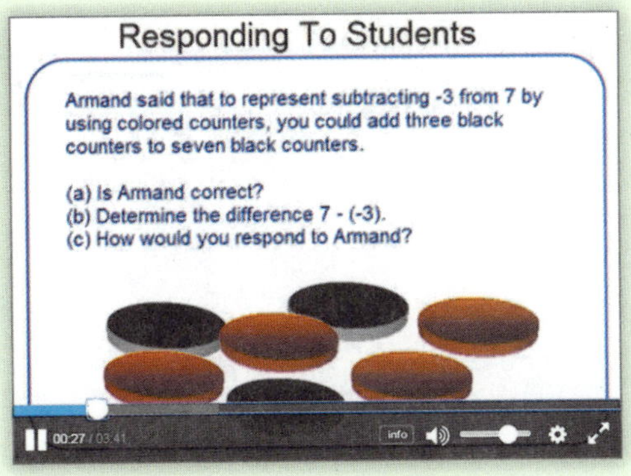

Responding to Students Videos show common questions or misconceptions that children have about mathematical topics and approaches to addressing these questions.

www.mymathlab.com

MATHEMATICS
for Elementary Teachers
with Activities

5th EDITION

Sybilla Beckmann
University of Georgia

Pearson

Director, Portfolio Management: Anne Kelly
Senior Courseware Portfolio Manager: Marnie Greenhut
Courseware Portfolio Management Assistant: Stacey Miller
Content Producer: Kathleen A. Manley
Managing Producer: Karen Wernholm
Development Editor: Elka Block
Media Producer: Nicholas Sweeny
Manager, Content Development: Robert Carroll
Product Marketing Manager: Alicia Frankel
Marketing Assistant: Hanna Lafferty
Senior Author Support/Technology Specialist: Joe Vetere
Program Design Lead: Barbara Atkinson
Production Coordination, Composition, Photo Research, Permission Clearance, and Illustrations: Cenveo® Publisher Services
Cover Design: Barbara T. Atkinson
Cover Image: The Davis Group, Inc./Lauri Kaihlanen

Acknowledgements of third party content appear on page IA-8, which constitutes an extension of this copyright page.

Library of Congress Cataloging-in-Publication Data

Names: Beckmann, Sybilla.
Title: Mathematics for elementary teachers, with activities / Sybilla Beckmann, University of Georgia.
Description: Fifth edition. | New York, N.Y. : Pearson, [2018]
Identifiers: LCCN 2016035793| ISBN 9780134392790 | ISBN 0134392795
Subjects: LCSH: Mathematics--Study and teaching (Elementary)--Activity programs.
Classification: LCC QA135.6 .B43 2018 | DDC 372.7/044--dc23 LC record available at https://lccn.loc.gov/2016035793

Pearson

ISBN 13: 978-0-13-439279-0
ISBN 10: 0-13-439279-5

6 2020

To Will, Joey, and Arianna

CONTENTS

Throughout the text, each section concludes with Practice Problems, Answers to Practice Problems, and Problems.

FOREWORD

by Roger Howe, Ph.D., Yale University

We owe a debt of gratitude to Sybilla Beckmann for this book.

Mathematics educators commonly hear that teachers need a "deep understanding" of the mathematics they teach. In this text, this pronouncement is not mere piety, it is the guiding spirit.

With the 1989 publication of its *Curriculum and Evaluation Standards for School Mathematics*, the National Council of Teachers of Mathematics (NCTM) initiated a new era of ferment and debate about mathematics education. The NCTM *Standards* achieved widespread acceptance in the mathematics education community. Many states created or rewrote their standards for mathematics education to conform to the NCTM *Standards*, and the National Science Foundation funded large-scale curriculum development projects to create mathematics programs consistent with the *Standards*' vision.

But this rush of activity largely ignored a major lesson from the 1960s' "New Math" era of mathematics education reform: in order to enable curricular reform, it is vital to raise the level of teachers' capabilities in the classroom. In 1999, the publication of *Knowing and Teaching Elementary Mathematics* by Liping Ma finally focused attention on teachers' understanding of mathematics principles they were teaching. Ma adapted interview questions (originally developed by Deborah Ball) to compare the basic mathematics understanding of American teachers and Chinese teachers. The differences were dramatic. Where American teachers' understanding was foggy, the Chinese teachers' comprehension was crystal clear. This vivid evidence showed that the difference in Asian and American students'achievement, revealed in many international comparisons, correlated to a difference in the mathematical knowledge of the teaching corps.

The Mathematical Education of Teachers, published by the Conference Board on the Mathematical Sciences (CBMS), was one response to Ma's work. Its first recommendation gave official voice to the dictum: "Prospective teachers need mathematics courses that develop a deep understanding of the mathematics they will teach." This report provided welcome focus on the problem, but the daunting task of creating courses to fulfill this recommendation remained.

Sybilla Beckmann has risen admirably to that challenge. Keeping mathematical principles firmly in mind while listening attentively to her students and addressing the needs of the classroom, she has written a text that links mathematical principles to their day-to-day uses. For example, in Chapter 4, Multiplication, the first section is devoted to the meaning of multiplication. First, it is defined through grouping: $A \times B$ means "the number of objects in A groups of B objects each." Beckmann then analyzes other common situations where multiplication arises to show that the definition applies to each. The section problems, then, do not simply provide practice in multiplication; they require students to show how the definition applies.

Subsequent sections continue to connect applications of multiplication (e.g., finding areas, finding volumes) to the definition. This both extends students' understanding of the definition and unites varied applications under a common roof. Reciprocally, the applications are used to develop the key properties of multiplication, strengthening the link between principle and practice. In the next chapter, the definition of multiplication is revisited and adapted to include fraction multiplication as well as whole numbers. Rather than emphasizing the procedure for multiplying fractions, this text focuses on how the procedure follows from the definitions.

Here and throughout the book, students are taught not merely specific mathematics, but the coherence of mathematics and the need for careful definitions as a basis for reasoning. By inculcating the point of view that mathematics makes sense and is based on precise language and careful reasoning, this book conveys far more than knowledge of specific mathematical topics: it can transmit some of the spirit of doing mathematics and create teachers who can share that spirit with their students. I hope the book will be used widely, with that goal in mind.

ABOUT THE AUTHOR

Sybilla Beckmann is the Josiah Meigs Distinguished Teaching Professor of Mathematics at the University of Georgia. She received her PhD in mathematics from the University of Pennsylvania and taught at Yale University as a J. W. Gibbs Instructor of Mathematics. Her early research was on Arithmetic Geometry, but her current research is in mathematical cognition, the mathematical education of teachers, and mathematics content for students at all levels, but especially for PreK through the middle grades. She has developed mathematics courses for prospective elementary and middle grades teachers at the University of Georgia and wrote this book for such courses. She is interested in helping college faculty learn to teach mathematics courses for elementary and middle grades teachers, and she works with graduate students and postdoctoral fellows toward that end. A member of the writing teams for the *Common Core State Standards for Mathematics* and for NCTM's *Curriculum Focal Points for Prekindergarten Through Grade 8 Mathematics,* she has worked on the development of several state mathematics standards. She was a member of the Committee on Early Childhood Mathematics of the National Research Council and coauthor of its report *Mathematics Learning in Early Childhood: Paths Toward Excellence and Equity.* She has also been a member on numerous other national panels and committees working to improve mathematics education. Several years ago she taught an average 6th grade mathematics class every day at a local public school in order to better understand school mathematics teaching. Sybilla has won a number of awards including the Louise Hay Award for Contributions to Mathematics Education from the Association for Women in Mathematics and the Mary P Dolciani Award from the Mathematical Association of America.

Sybilla enjoys playing piano, swimming, dancing, and traveling with her family. She and her husband, Will Kazez, live in Athens, Georgia. They look forward to visits from their two children who are now away working and in graduate school.

Follow Sybilla on Twitter @SybillaBeckmann

PREFACE

Introduction

I wrote *Mathematics for Elementary School Teachers with Activities* to help future elementary school teachers develop a deep understanding of the mathematics they will teach. People commonly think that elementary school mathematics is simple and that it shouldn't require college-level study to teach it. But to teach mathematics well **teachers must know more than just *how* to carry out basic mathematical procedures; they must be able to explain *why*** mathematics works the way it does. Knowing why requires a much deeper understanding than knowing how. By learning to explain why mathematics works the way it does, teachers will learn to make sense of mathematics. I hope they will carry this "sense of making sense" into their own future classrooms.

Because I believe in deep understanding, this book focuses on *explaining why.* Prospective elementary school teachers will learn to explain why the standard procedures and formulas of elementary mathematics are valid, why nonstandard methods can also be valid, and why other seemingly plausible ways of reasoning are not correct. The book emphasizes key concepts and principles, and it guides prospective teachers in giving explanations that draw on these key concepts and principles. In this way, teachers will come to organize their knowledge around the key concepts and principles of mathematics so they will be able to help their students do likewise.

Common Core State Standards for Mathematics

Now that most states have adopted the *Common Core State Standards for Mathematics* (CCSSM), we have the opportunity to join together to work toward its successful implementation. Part of CCSSM's success will rest on teachers' own understanding of the mathematics they teach; this understanding must become deeper and stronger than has been common.

To develop this deeper understanding, teachers must study the mathematics they will teach in an especially active way, by engaging in mathematical practices. The eight **CCSSM Standards for Mathematical Practice** outline and summarize some of the processes that are important in mathematics. These Standards for Mathematical Practice ask students to reason, construct, and critique arguments, and to make sense of mathematics. They ask students to look for structure and to apply mathematics. They ask students to monitor and evaluate their progress and to persevere. These standards apply not only to K–12 students, but to all of us who study and practice mathematics, including all of us who teach mathematics.

Throughout this book, the Class Activities, Problems and the text itself have been designed to foster ongoing active engagement in mathematical practices while studying mathematics content. At the beginning of each chapter, I briefly describe ways to engage in a few of the practices that are especially suited to the material in that chapter.

This book is centered on the mathematical content of prekindergarten through grade 8. It addresses almost all of the K–8 **CCSSM Standards for Mathematical Content** from a teacher's perspective, with a focus on how ideas develop and connect and on powerful ways of representing and reasoning about the ideas. Each section is labeled with the grade levels at which there are directly related CCSSM Standards for Mathematical Content. The development of the material goes beyond what is expected of elementary school students. For example, in third grade, students learn that shapes can have the same perimeter but different area. Section 12.8 explores that idea beyond the third grade level, by considering all possible shapes that have a fixed perimeter, including circles, and by asking what the full range of possible areas is for all those shapes.

The chapters are designed to help prospective teachers study how mathematical ideas develop across grade levels. For example, in third grade, students learn about areas of rectangles (Section 12.1), and they find areas of shapes by using the additivity principle (Section 12.2). In sixth grade, students use the additivity principle to explain area formulas, such as those for triangles and parallelograms (Sections 12.3, 12.4); in seventh grade they can apply the principle to see where the area formula for circles comes from (Section 12.6); and in eighth grade they can use the principle to explain the Pythagorean theorem (Section 12.9). Thus teachers can see how a simple but powerful idea introduced in third grade leads to important mathematics across grade levels.

Content and Organization

The book is **organized around the operations** instead of around the different types of numbers. In my view, there are two key advantages to focusing on the operations. The first is a more advanced, unified perspective, which emphasizes that a given operation (addition, subtraction, multiplication, or division) retains its meaning across all the different types of numbers. Prospective teachers who have already studied numbers and operations in the traditional way for years will find that method enables them to take a broader view and to consider a different perspective. A second advantage is that fractions, decimals, and percents—traditional weak spots—can be studied repeatedly throughout a course, rather than only at the end. The repeated coverage of fractions, decimals, and percents allows students to gradually become used to reasoning with these numbers, so they aren't overwhelmed when they get to multiplication and division of fractions and decimals.

A section on **solving problems and explaining solutions** appears near the beginning of Chapter 2 on fractions. Fractions provide an especially rich domain for explanations and for challenging problems. The section helps students think about how explanations in math can be different from explanations in other fields of knowledge.

Arithmetic in bases other than 10 is not explicitly discussed, but other bases are implied in several activities and problems. For example, Class Activities 3M Regrouping in Base 12, and 3N Regrouping in Base 60, and as well as Practice exercise 6, and Problems 7–11 in Section 3.3 also involve the idea of other bases. These activities and problems allow students to grapple with the significance of the base in place value without getting bogged down in the mechanics of arithmetic in other bases.

Visual representations, including number lines, double number lines, strip diagrams (also known as tape diagrams), and base-ten drawings are used repeatedly throughout the book and help prospective teachers learn to explain and make sense of mathematical ideas, solution methods, and standard notation. Chapter 9 introduces U.S. teachers to the impressive strip-diagram method for solving algebra word problems, which is used in grades 3–6 in Singapore, where children get the top math scores in the world. The text shows how reasoning about strip diagrams leads to standard algebraic techniques.

Activities

Class Activities were written as part of the text and are central and integral to full comprehension. All good teachers of mathematics know mathematics is not a spectator sport. We must actively think through mathematical ideas to make sense of them for ourselves. When students work on problems in the class activities—first on their own, then in a pair or a small group, and then within a whole class discussion—they have a chance to think through the mathematical ideas several times. By discussing mathematical ideas and explaining their solution methods to each other, students can deepen and extend their thinking. As every mathematics teacher knows, students really learn mathematics when they have to explain it to someone else.

A number of activities and problems offer opportunities to **critique reasoning**. For example, in Class Activities 2Q, 3O, 7N, 12Q, 14R, 15E, 16B students investigate common errors in comparing fractions, adding fractions, distinguishing proportional relationships from those that are not, determining perimeter, working with similar shapes, displaying data, and probability. Since most misconceptions have a certain plausibility about them, it is important to understand what makes them mathematically incorrect. By examining what makes misconceptions incorrect, teachers deepen their understanding of key concepts and principles, and they develop their sense of valid mathematical reasoning. I also hope that, by studying and analyzing these misconceptions, teachers will be able to explain to their students why an erroneous method is wrong, instead of just saying "You can't do it that way."

Other activities and problems **examine calculation methods that are nonstandard but nevertheless correct**. For example, in Class Activities 2P, 2S, 3L, 3P, 4K, 6H, and 6O, teachers explore ways to compare fractions, calculate with percents, subtract whole numbers, add and subtract mixed numbers, multiply mentally, divide whole numbers, and divide fractions in nonstandard but logically valid ways. When explaining why nonstandard methods are correct, teachers have further opportunities to draw on key concepts and principles and to see how these concepts and principles underlie calculation methods. By examining nonstandard methods, teachers also learn there can be more than one correct way to solve a problem. They see how valid logical reasoning, not convention or authority, determines whether a method is correct. I hope that by having studied and analyzed a variety of valid solution methods, teachers will be prepared to value their students' creative mathematical activity. Surely a student who has found an unusual but correct solution method will be discouraged if told the method is incorrect. Such a judgment also conveys to the student entirely the wrong message about what mathematics is.

Studying nonstandard methods of calculation provides valuable opportunities, but the common methods deserve to be studied and appreciated. These methods are remarkably clever and make highly efficient use of underlying principles. Because of these methods, we know that a wide range of problems can always be solved straightforwardly. The common methods are major human achievements and part of the world's heritage; like all mathematics, they are especially wonderful because they cross boundaries of culture and language.

I believe *Mathematics for Elementary School Teachers with Activities* is an excellent fit for the recommendations of the Conference Board of the Mathematical Sciences regarding the mathematical preparation of teachers. I also believe that the book helps prepare teachers to teach in accordance with the principles and standards of the National Council of Teachers of Mathematics (NCTM).

Pedagogical Features

Practice Exercises give students the opportunity to try out problems. Solutions appear in the text immediately after the Practice Exercises, providing students with many examples of the kinds of good explanations they should learn to write. By attempting the Practice Exercises themselves and checking their solutions against the solutions provided, students will be better prepared to provide good explanations in their homework.

Problems are opportunities for students to explain the mathematics they have learned, without being given an answer at the end of the text. Problems are typically assigned as homework. Solutions appear in the *Instructor's Solutions Manual.*

* **Problems with a purple asterisk are** more challenging, involve an extended investigation, or are designed to extend students' thinking beyond the central areas of study.

The **core** icon denotes central material. These problems and activities are highly recommended for mastery of the material.

UPDATED! CCSS The **Common Core** icon indicates that the section addresses the Common Core State Standards for Mathematics at the given grade level. The treatment of the topic goes beyond what students at that grade level are expected to do, because teachers need to know how mathematical ideas develop and progress. New to this edition are CCSS references in the margin that address these specific standards. Common Core standards have also been noted on every activity. These icons and the Common Core material have been designed to not be a distraction for states that have not adopted the CCSSM.

UPDATED! **FROM THE FIELD** boxes have been expanded to include more detail on children's literature that is related to the mathematical content of the section, as well as an added focus on related research. Also included are **IMAP** video clips of children working on math problems, resources to help future teachers prepare lessons, and more!

Section Summary and Study Items are provided at the end of each section to help students organize their thinking and focus on key ideas as they study. Each chapter ends with a Chapter Summary pulling together the ideas from each section.

Content Changes for the Fifth Edition

This edition has been enhanced extensively:

- Chapter 1 introduces the term "place value parts" to highlight base-ten structure, as recommended by Professors Roger Howe and Karen Fuson. New base-ten drawings better show the structure of repeated bundling in groups of ten.
- Based on research, Chapter 2 has an improved discussion in the text and new activities that use measurement to highlight how fractions are numbers. This "measurement sense of fraction" continues to be developed in revised fraction multiplication and division activities in Chapters 5 and 6 and in revised activities on developing equations for proportional relationships in Chapter 7.
- Chapter 3 extends the discussion of addition and subtraction on number lines by including the case of subtraction as a difference or unknown addend.
- Based on research, Chapter 4 has a revised introduction to multiplication to better highlight how multiplication is about coordinating numbers and sizes of equal groups of quantities. New color-coordinated math drawings continue into Chapters 5, 6, and 7 to highlight connections across multiplication, division, and proportional relationships, and across whole numbers and fractions.
- Chapter 4 clarifies that there are different written methods for implementing the standard multiplication algorithm (e.g., the partial-products method). Revisions to the activities promote more active reasoning about arrays to develop the standard multiplication algorithm.
- Chapter 5 has an improved discussion and revised activities on how the definition of multiplication in Chapter 4 continues to apply to the case of fractions.
- Chapter 6 clarifies that there are different written methods for implementing the standard division algorithm (e.g., a version of the scaffold method). The Class Activity on interpreting the standard division algorithm in terms of bundled toothpicks has been improved. It now starts with simpler examples.
- Based on research, Chapter 6 highlights cases of equivalent division problems in fraction (and decimal) division.
- Based on research, Chapter 7 develops different ways of reasoning about multiplication and division with quantities to solve ratio and proportion problems. There is a new section

on proportional relationships, which includes ways of reasoning about strip diagrams and variable parts to develop equations, including for lines through the origin in a coordinate plane. The section on inversely proportional relationships now puts more emphasis on reasoning about multiplication and division with quantities.

- In Chapter 9, some Class Activities and problems were revised and moved for a more coherent and streamlined development. The section on sequences now includes graphs and equations for arithmetic sequences as part of the development of linear relationships, which started in Chapter 7 with proportional relationships. The sections on functions and linear functions now build more explicitly on proportional relationships and arithmetic sequences.
- Chapter 10 now starts with a discussion of lines and angles, including more angle terminology to facilitate clear communication, and a new Class Activity on folding paper to create angles. Class Activities on categories of shapes are revised for a better focus on relationships among categories.
- Chapter 11 now better highlights how measurement concepts underlie the process of measurement.
- Chapter 12 includes more opportunities for problem solving, including with shearing. The section on circles has an improved discussion of the relationship between the diameter and circumference of circles that extends ideas about proportional relationships developed in Chapter 7. The section on the Pythagorean theorem now includes a discussion of its converse.
- In Chapter 13 revised art clarifies the locations of bases in solid shapes.
- In Chapter 14 revisions to the art, text, Class Activities, and problems offer better opportunities to explore what various transformations do to points, lines, line segments, angles, and shapes. New and revised Class Activities and problems provide opportunities to use a compass and straightedge to construct the results of transformations and to construct lines and angles with desired properties. There are additional problems about applying congruence.
- Chapter 14 makes stronger connections to ratio and proportional relationships discussed in Chapter 7, including a new section on dilations and similarity that builds on ideas about proportional relationships in Chapter 7, and a more explicit discussion of how similarity applies to lines and their slope.
- Based on research, Chapter 15 includes revised text, Class Activities, and problems about reasoning to draw inferences from random samples.
- In Chapter 15, examples have been updated using more current data. Revisions to the text, Class Activities, and problems provide better opportunities for comparing distributions.
- Chapter 16 has a better exposition of probabilities of events in uniform probability models.

I am enthusiastic about these changes and am sure students and professors will be as well.

Online Learning

UPDATED! Skills Review MyMathLab® Online Course (access code required)

MyMathLab comes from a **trusted partner** with educational expertise and an eye on the future. Knowing that you are using a Pearson product means knowing that you are using quality content. That means that our eTexts are accurate and our assessment tools work. Whether you are just getting started with MyMathLab, or have a question along the way, we're here to help you learn about our technologies and how to incorporate them into your course.

Modules are organized by topic and not correlated to the textbook, as the intent is primarily for students to brush up on skills where they lack confidence and not as a primary means of assessment. Students can reinforce their skills on math content necessary for successful completion of Class Activities and Problems so instructors can make productive use of class time and not have to reteach the basics.

- In addition to the traditional skills review problems there are a variety of problem types.
    - Problems using Integrated Mathematics and Pedagogy (IMAP) videos test students' understanding of concepts and content in the context of children's reasoning processes.
    - Problems using the eManipulatives have students interact with the eManipulative and then answer assigned questions.
    - Problems assessing understanding of both the new Common Core videos and the new Demonstration videos.
- Integrated Mathematics and Pedagogy (IMAP) videos show how elementary and middle school students work through problems. These videos provide great insight into student thinking.
- An IMAP implementation guide and correlation provide additional guidance for integrating the IMAP videos into your course.
- NEW! Common Core videos help students see content in the context of the Common Core State Standards for Mathematics.
- NEW! Demonstration videos show the author presenting various concepts to help make them more concrete and understandable for students.
- NEW! GeoGebra interactives and worksheets enable students to manipulate GeoGebra applets and answer questions to solidfy understanding.
- NEW! Summaries of children's literature related to mathematics by Kirsten Keels and the author expand on the content in the From the Field boxes in the text.
- NEW! Summaries of research on children's mathematical thinking and learning by I. Burak Ölmez, Eric Siy, and the author expand on the content in the From the Field boxes in the text.
- A multimedia textbook with links to Class Activities, IMAP videos, and all new videos and animations.
- An Additional Activity Bank consisting of activities not included in the 5th edition of the text as well as additional activities from the author's own collection.
- An Image Resource Library that contains all art from the text and is available for instructors to use in their own presentations and handouts.

Resources for Success

Michael Matthews of University of Nebraska at Omaha

Instructor Resources

To assist in teaching using an inquiry-based approach, a variety of instructor resources are available.

Annotated Instructor's Edition

ISBN 10: 0134423348 ISBN 13: 9780134423340

The Annotated Instructor's Edition contains answers to selected problems right on the page in addition to Teaching Tips from the author pointing out problematic areas for students or additional activities and other resources instructors can find in MyMathLab.

The following resources can be downloaded from www.pearsonhighered.com or in MyMathLab.

Active Teachers, Active Learners videos

These videos help instructors enrich their classroom by expanding their own knowledge of teaching using an inquiry-based approach. Classroom video of author Sybilla Beckmann and her students discovering various concepts in class, along with optional voiceover commentary from Beckmann, can help any instructor thinking about an inquiry-based approach. Footage from instructors' classrooms at other universities is also included to provide a view into other styles and methods of teaching an inquiry-based course. Instructors speak about making the switch to teaching using this method, and students give feedback about their experience in the course.

These videos are the ideal resource for instructors who are considering switching to a more inquiry-based approach from a more traditional lecture style, teaching an inquiry-based course for the first time, training adjuncts or graduate assistants who will be teaching the course, or for instructors who seek new ideas to integrated into their existing inquiry-based classes.

Instructor's Resource Manual

Written by the author, this manual includes general advice on teaching the course, advice for struggling students, and sample syllabi, as well as support and ideas for each chapter.

Instructor's Solutions Manual

Michael Matthews of University of Nebraska at Omaha

This solutions manual contains worked-out solutions to all problems in the text.

Instructor's Testing Manual

Written by the author, this manual includes guidance on assessment, including sample test problems.

www.mymathlab.com

ACKNOWLEDGMENTS

In writing this book, I have benefited from much help and advice, for which I am deeply grateful. First and foremost, I thank Andrew Izsák and our *Teachers' Multiplicative Reasoning Research Group*, including Torrey Kulow, I. Burak Ölmez, Dean Stevenson, Eun Jung, and Eun Kyung Kang. Our research has significantly informed a number of revisions to this edition. The research has been generously funded by the National Science Foundation (Grant No. DRL-1420307), the Spencer Foundation, and the University of Georgia. The opinions expressed in this book are those of the author and do not necessarily reflect the views of NSF, the Spencer Foundation, or the University of Georgia.

Thanks also to Kirsten Keels, I. Burak Ölmez, and Eric Siy for their excellent work finding and summarizing children's literature, current research on children's mathematical thinking and learning, and interesting data sets. For helpful comments and conversations I'm grateful to Michael Berglund, Brian Bonsignore, Theresa Brons, Diana Chang, Chris Drupieski, Brad Findell, Susan Friel, Karen Fuson, Sayonita GhoshHajra, Natalie Hobson, Karen Heinz, Jacob Hicks, Roger Howe, Lauren Huckaba, Rick Hudson, Niles Johnson, Sheri Johnson, Joey Kazez, Jeremy Kilpatrick, Anne Marie Marshall, Michael Matthews, Besty McNeil, Doris Mohr, Stacy Musgrave, Tom Needham, Hans Parshall, Rachel Roberts, Christopher Sabino, Aysa Sahin, Bolanle Salaam, John Samons, Lauri Semarne, Andrew Talian, Maren Turbow, Jonathan Williams, Josh Woods, and Matt Zawodniak. I thank my students for generously sharing their thinking with me and for teaching me many interesting and surprising solution methods.

I would also like to express my appreciation to the following reviewers and class testers for many helpful comments on this edition and earlier drafts:

* George Avrunin, University of Massachusetts-Amherst
Gerald Beer, California State University, Los Angeles
* Kayla Blyman, University of Kentucky
Barbara Britton, Eastern Michigan University
Michaele F. Chappell, Middle Tennessee State University
Christopher Danielson, Normandale Community College
Maria Diamantis, Southern Connecticut State University
* Jim Dudley, University of New Mexico
Ron English, Niagara County Community College
Larry Feldman, Indiana University of Pennsylvania
Victor Ferdinand, Ohio State University
Richard Francis, Southeast Missouri State University
Shannon Guerrero, Northern Arizona University
* Kimberly Haughee, Helena College University of Montana
* Josh Himmelsbach, University of Maryland
Rick A. Hudson, University of Southern Indiana
Gail Johnston, Iowa State University
Judy Kidd, James Madison University
Michael Edward Matthews, University of Nebraska at Omaha
* Kelly McCormick, University of Southern Maine
*Betsy McNeal, Ohio State University
Doris J. Mohr, University of Southern Indiana
Michelle Moravec, McLennan Community College
Michael Nakamaye, University of New Mexico
Shirley Pereira, Grossmont College
Kevin Peterson, Columbus State University
* Nick Pilewski, Ohio University
Cheryl Roddick, San Jose State University
David Ruszkiewicz, Milwaukee Area Technical College
Ayse A. Sahin, DePaul University
William A. Sargeant, University of Wisconsin – Whitewater
Dan Schultz-Ela, Mesa State College
* Sonya Sherrod, Texas Tech University
Julia Shew, Columbus State Community College
Darla Shields, Slippery Rock University
Wendy Hagerman Smith, Radford University
* Bill Smith, Rowan University
Ginger Warfield, University of Washington
David C. Wilson, Buffalo State College
* Rebecca Wong, West Valley College
* Xiaoxia Xie, Auburn University
Carol Yin, LaGrange College

The Active Teachers, Active Learners videos were a huge undertaking that could not have been possible without the contributions of my colleagues. Few people feel comfortable being videotaped in their classrooms, so I would like to thank the following contributors and their students for taking that chance and helping make the videos a reality: Sarah Donaldson, Covenant College; Michael Edward Matthews, University of Nebraska at Omaha; Janice Rech, University of Nebraska at Omaha; and Bridgette Stevens, University of Northern Iowa.

I am very grateful for the support and excellent work of everyone at Pearson who has contributed to this project. I especially want to thank my editor, Marnie Greenhut, for her enthusiastic support, as well as Joe Vetere, Kathy Manley, Nick Sweeny, Alicia Frankel, and Barbara Atkinson for invaluable assistance. It is only through their dedicated hard work that this book has become a reality.

I thank my family—Will, Joey, and Arianna Kazez, and my parents Martin and Gloria Beckmann—for their patience, support, and encouragement. My children long ago gave up asking, "Mom, are you still working on that book?" and are now convinced that books take forever to write.

Sybilla Beckmann
Athens, Georgia

1

Numbers and the Base-Ten System

What are numbers and where do they come from? What humans consider to be numbers has evolved over the course of history, and the way children learn about numbers parallels this development. When did humans first become aware of numbers? The answer is uncertain, but it is at least many tens of thousands of years ago. Some scholars believe that numbers date back to the beginning of human existence, citing as a basis for their views the primitive understanding of numbers observed in some animals. (See [20] for a fascinating account of this and also of the human mind's capacity to comprehend numbers.)

In this chapter, we discuss elementary ideas about numbers, which reveal surprising intricacies that we are scarcely aware of as adults. We will study the base-ten system—a remarkably powerful and efficient system for writing numbers and a major achievement in human history. Not only does the base-ten system allow us to express arbitrarily large numbers and arbitrarily small numbers—as well as everything in between—but also it enables us to quickly compare numbers and assess the ballpark size of a number. The base-ten system is familiar to adults, but its slick compactness hides its inner workings. We will examine with care those inner workings of the base-ten system that children must grasp to make sense of numbers.

In this chapter, we focus on the following topics and practices within the *Common Core State Standards for Mathematics (CCSSM)*.

Standards for Mathematical Content in the CCSSM

In the domain of *Counting and Cardinality* (Kindergarten) young children learn to say and write small counting numbers and to count collections of things. In the domain of *Numbers and Operations in Base Ten* (Kindergarten through Grade 5), students learn to use the powerful base-ten system. This system starts with the idea of making groups of ten and gradually extends this idea to the greater and to the smaller place values of decimals.

Standards for Mathematical Practice in the CCSSM

Opportunities to engage in all eight of the Standards for Mathematical Practice described in the CCSSM occur throughout the study of counting and the base-ten system. The following standards are especially appropriate for emphasis while studying counting and the base-ten system:

- **2 Reason abstractly and quantitatively.** Students engage in this practice when they make sense of number words and symbols by viewing numbers as representing quantities and when they use numbers to describe quantities.
- **5 Use appropriate tools strategically.** The base-ten system represents numbers in a very compact, abstract way. By reasoning with appropriate tools, such as drawings of tens and ones or number lines that show decimals, students learn to make sense of the powerful base-ten system.
- **7 Look for and make use of structure.** The base-ten system has a uniform structure, which creates symmetry and patterns. Students engage in this practice when they seek to understand how increasingly greater base-ten units can always be created and how the structure of the base-ten system allows us to compare numbers and find numbers in between numbers.

1.1 The Counting Numbers

CCSS Pre-K, Common Core State Standards Grades K–4

What are numbers and why do we have them? We use numbers to tell us "how many" or "how much," in order to communicate specific, detailed information about collections of things and about quantities of stuff. Although there are many different kinds of numbers (e.g., fractions, decimals, and negative numbers), the most basic numbers, and the starting point for young children, are the **counting numbers**—the numbers 1, 2, 3, 4, 5, 6, ... (The dots indicate that the list continues without end.)

counting numbers

How Is It Different to View the Counting Numbers as a List and for Describing Set Size?

There are two distinctly different ways to think about the counting numbers. Connecting these two views is a major mathematical idea that very young children must grasp before they can do school arithmetic.

CLASS ACTIVITY

1A The Counting Numbers as a List, p. CA-1

One way to think about the counting numbers is as a list. The list of counting numbers starts with 1 and continues, with every number having a unique successor. Except for the number 1, every number in the list has a unique predecessor. So the list of counting numbers is an *ordered* list. Every counting number appears exactly once in this ordered list. The ordering of the list of counting numbers is important because of the second way of thinking about the counting numbers.

The second way to think about the counting numbers is as "telling how many." In other words, a counting number describes how many things are in a set[1]. The number of things in a set is called the **cardinality of a set**. Think for a moment about how surprisingly abstract the notion of cardinality is. We use the number 3 to quantify a limitless variety of collections—3 cats, 3 toy dinosaurs, 3 jumps, 3 claps, and so on. The number 3 is the abstract, common aspect that all examples of sets of 3 things share.

cordinality of a set

For sets of up to about 3, 4, or 5 objects, we can usually recognize the number of objects in the set immediately, without counting the objects one by one. The process of immediate recognition of the exact number of objects in a set is called *subitizing* and is discussed further in [68]. But in general, we must count the objects in a set to determine how many there are. The process of counting the objects in a set connects the "list view" of the counting numbers with the "cardinality view." As adults, this connection is so familiar that we are usually not even aware of it. But for young children, this connection is not obvious, and grasping it is an important milestone (see [68]).

How Do Children Connect Counting to Cardinality?

CLASS ACTIVITY

1B Connecting Counting Numbers as a List with Cardinality, p. CA-2

one-to-one correspondence

When we count a set of objects one by one, we make a **one-to-one correspondence** between an initial portion of the list of counting numbers and the set. For example, when a child counts a set of 5 blocks, the child makes a one-to-one correspondence between the list 1, 2, 3, 4, 5 and the set of blocks. This means that each block is paired with exactly one number and each number is paired with exactly one block, as indicated in Figure 1.1. Such a one-to-one correspondence connects the "list" view of the counting numbers with cardinality. However, there is another critical piece of understanding this connection relies on, that adults typically take for granted but is not obvious to young children: *The last number we say when we count a set of objects tells us the total number of objects in the set*, as indicated in Figure 1.2. It is for this reason that the order of the counting numbers is so important, unlike the order of the letters of the alphabet, for example.

CCSS K.CC.4

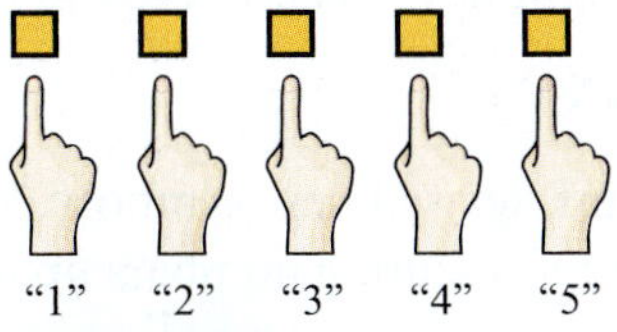

Figure 1.1 Counting 5 blocks makes a one-to-one correspondence between the list 1, 2, 3, 4, 5 and the blocks.

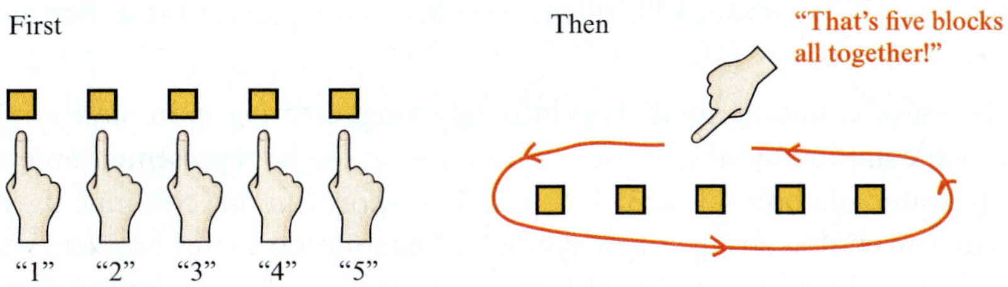

Figure 1.2 When we count objects, the last number we say tells us the total number of objects.

[1]A **set** is a collection of distinct "things." These things can include concepts and ideas, such as the concept of an infinitely long straight line, and imaginary things, such as heffalumps.

The connection between the list and cardinality views of the counting numbers is especially important in understanding that numbers *later* in the list correspond with *larger* quantities and that numbers *earlier* in the list correspond with *smaller* quantities. In particular, starting at any counting number, the next number in the list describes the size of a set that has one more object in it, and the previous number in the list describes the size of a set that has one less object in it.

What Are the Origins of the Base-Ten System for Representing Counting Numbers?

Let's think about the list of counting numbers and the symbols we use to represent these numbers. The symbols for the first nine counting numbers 1, 2, 3, 4, 5, 6, 7, 8, 9 have been passed along to us by tradition and could have been different. Instead of the symbol 4, we could be using a completely different symbol. In fact, one way to represent 4 is simply with 4 tally marks. So why don't we just use tally marks to represent counting numbers?

Suppose a shepherd living thousands of years ago used tally marks to keep track of his sheep. If his tally marks were not organized, he likely had a hard time comparing the number of sheep he had on different days, as shown in Figure 1.3. But if he grouped his tally marks, comparing the number of sheep was easier.

Figure 1.3 A shepherd's tally of sheep.

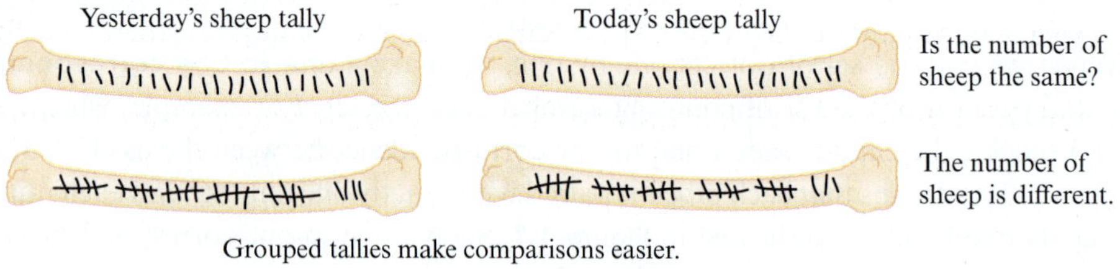

As people began to live in cities and engage in trade, they needed to work with larger numbers. But tally marks are cumbersome to write in large quantities. Instead of writing tally marks, it's more efficient to write a single symbol that represents a group of tally marks, such as the Roman numeral V, which represents 5. Recording 50 sheep might have been written as:

VVVVVVVVVV

However, it's hard to read all those Vs, so the Romans devised new symbols: X for 10, L for 50, C for 100, and M for 1000. These symbols are fine for representing numbers up to a few thousand, but what about representing 10,000? Once again,

MMMMMMMMMM

is difficult to read. What about 100,000 or 1,000,000? To represent these, one might want to create yet more new symbols.

Because the list of counting numbers is infinitely long, creating more new symbols is a problem. How can each counting number in this infinitely long list be represented uniquely? Starting with our Hindu–Arabic symbols—1, 2, 3, 4, 5, 6, 7, 8, 9—how can one continue the list without resorting to creating an endless string of new symbols? The solution to this problem was not obvious and was a significant achievement in the history of human thought. The **base-ten system**, or **decimal system**, is the ingenious system we use today to write (and say) counting numbers without resorting to creating more and more new symbols. The base-ten system requires using only ten distinct symbols—the **digits** 0, 1, 2, 3, 4, 5, 6, 7, 8, 9. The key innovation of the base-ten system is that rather than using new symbols to represent larger and larger numbers, it uses *place value*.

base-ten system

decimal system

digit

What Is Place Value in the Base-Ten System?

place value

Place value means that the quantity that a digit in a number represents depends—in a very specific way—on the position of the digit in the number.

Do Class Activity 1C before you read on.

IMAP *Watch Zenaida discuss the place value of 32, 120, and 316.*

CLASS ACTIVITY

1C How Many Are There?, p. CA-4

How did you and your classmates organize the toothpicks in the Class Activity 1C? If you made bundles of 10 toothpicks and then made 10 bundles of 10 to make a bundle of 100, then you began to reinvent place value and the base-ten system. Place value for the base-ten system works by creating larger and larger units by *repeatedly bundling them in groups of ten.*

Ten plays a special role in the base-ten system, but its importance is not obvious to children. Teachers must repeatedly draw children's attention to the role that ten plays. For example, a young child might be able to count that there are 14 beads in a collection, such as the collection shown in Figure 1.4, but the child may not realize that 14 actually stands for 1 ten and 4 ones. With the perspective presented in Figure 1.4, the numbers 10, 11, 12, 13, 14 are just the counting numbers that follow 9. Young children begin to learn about place value and the base-ten system when they (1) learn to organize collections of between 10 and 19 small objects into one group of ten and some ones, as in Figure 1.5, and (2) understand that the digit 1 in 14 does not stand for "one" but instead stands for *1 group of ten*. Some teachers like to help young students learn the meaning of the digit 1 in the numbers 10 to 19 with the aid of cards, such as the ones shown in Figure 1.6. These cards show how a number such as 17 is made up of 1 ten and 7 ones.

Figure 1.4 The important role that ten plays is not obvious even when counting more than 10 items.

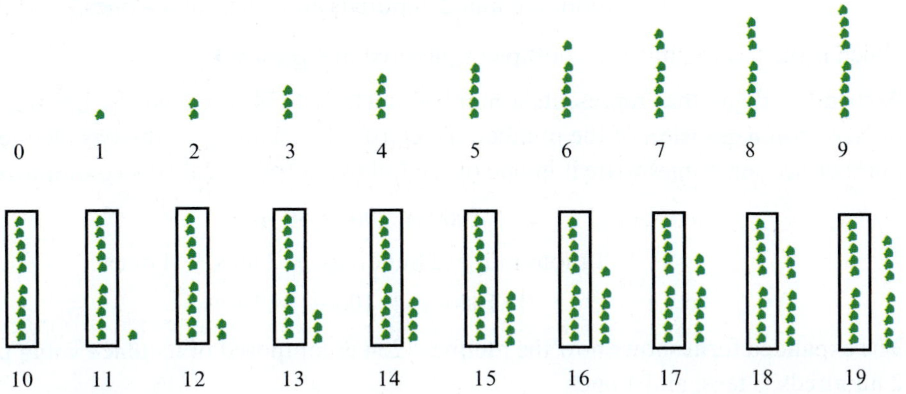

Figure 1.5 Organizing collections of between 1 and 19 objects according to the base-ten system.

Children extend their understanding of the base-ten system in two ways: (1) viewing a group of ten as a unit in its own right, and (2) understanding that a two-digit number such as 37 stands for 3 tens and 7 ones and can be represented with bundled objects and simple drawings like those in Figure 1.7.

Just as 10 ones are grouped to make a new unit of ten, 10 tens are grouped to make a new unit of one-hundred and 10 hundreds are grouped to make a new unit of one-thousand, as indicated in

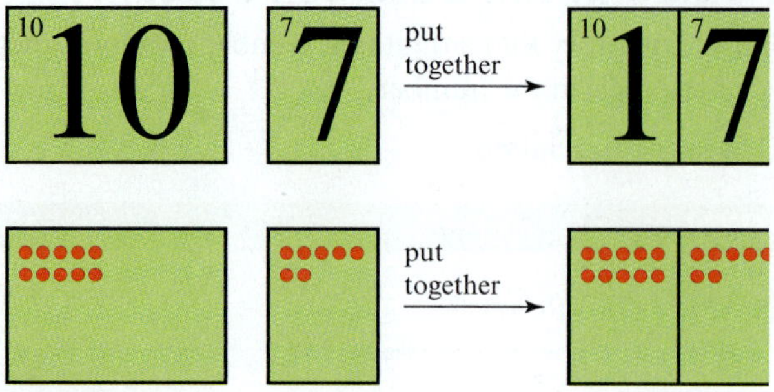

Figure 1.6 Cards that can be put together and taken apart to show numbers between 10 and 19 as tens and ones.

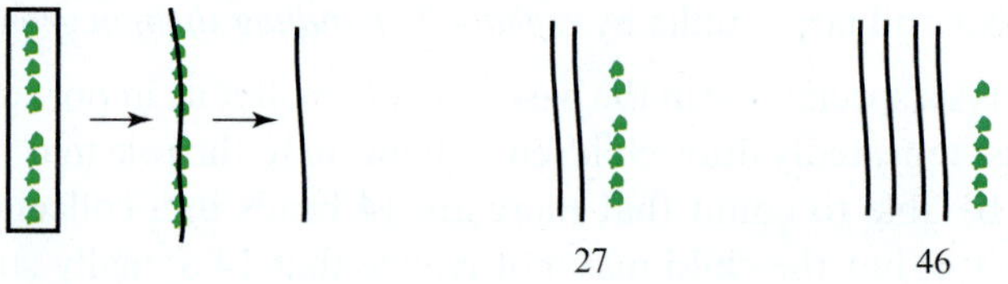

CCSS
1.NBT.2a
2.NBT.1a
4.NBT.1

Figure 1.7 Math drawings for representing two-digit numbers as tens and ones.

base-ten units

Figure 1.8. Continuing in this way, 10 thousands are grouped to make a new unit of ten-thousand, and so on. Thus, arbitrarily large **base-ten units** can be made by grouping 10 of the previously made units. These increasingly large units are represented in successive places to the left in a number. *The value of each place in a number is ten times the value of the place to its immediate right*, as indicated in Figure 1.9, which shows the standard names (used in the United States) of the place values up to the billions place.

In general, a string of digits, such as 1234, stands for the total amount that all its places taken together represent. Within the string, each digit stands for that many of that place's value. In the number 1234, the 1 stands for 1 thousand, the 2 stands for 2 hundreds, the 3 stands for 3 tens, and the 4 stands for 4 ones, giving

1 thousand and 2 hundreds and 3 tens and 4 ones,

which is the total number of toothpicks pictured in Figure 1.10.

base-ten representation
base-ten expansion
expanded form

A string of digits that represents a number, such as 1234, is called the **base-ten representation** or **base-ten expansion** of the number. To clarify the meaning of the base-ten representation of a number, we sometimes write it in one of the following forms, called **expanded form**:

$$1000 + 200 + 30 + 4$$

$$1 \text{ thousand} + 2 \text{ hundreds} + 3 \text{ tens} + 4 \text{ ones}$$

$$1 \cdot 1000 + 2 \cdot 100 + 3 \cdot 10 + 4$$

place value parts

The expanded form shows how the number 1234 is composed of its **place value parts**, 1 thousand, 2 hundreds, 3 tens, and 4 ones.

Notice that the way the toothpicks in Figure 1.10 are organized corresponds with the base-ten representation for the total number of toothpicks being depicted. Notice also how compactly this rather large number of toothpicks is represented by the short string 1234. Now imagine showing ten times as many toothpicks as are in the bundle of 1000. This would be a lot of toothpicks, yet writing this number as 10,000 only takes 5 digits!

Although the base-ten system is highly efficient and practical, children have difficulty learning what written numbers represent because they must keep the place values in mind. Because these values are not shown explicitly, even interpreting written numbers requires a certain level of abstract thinking.

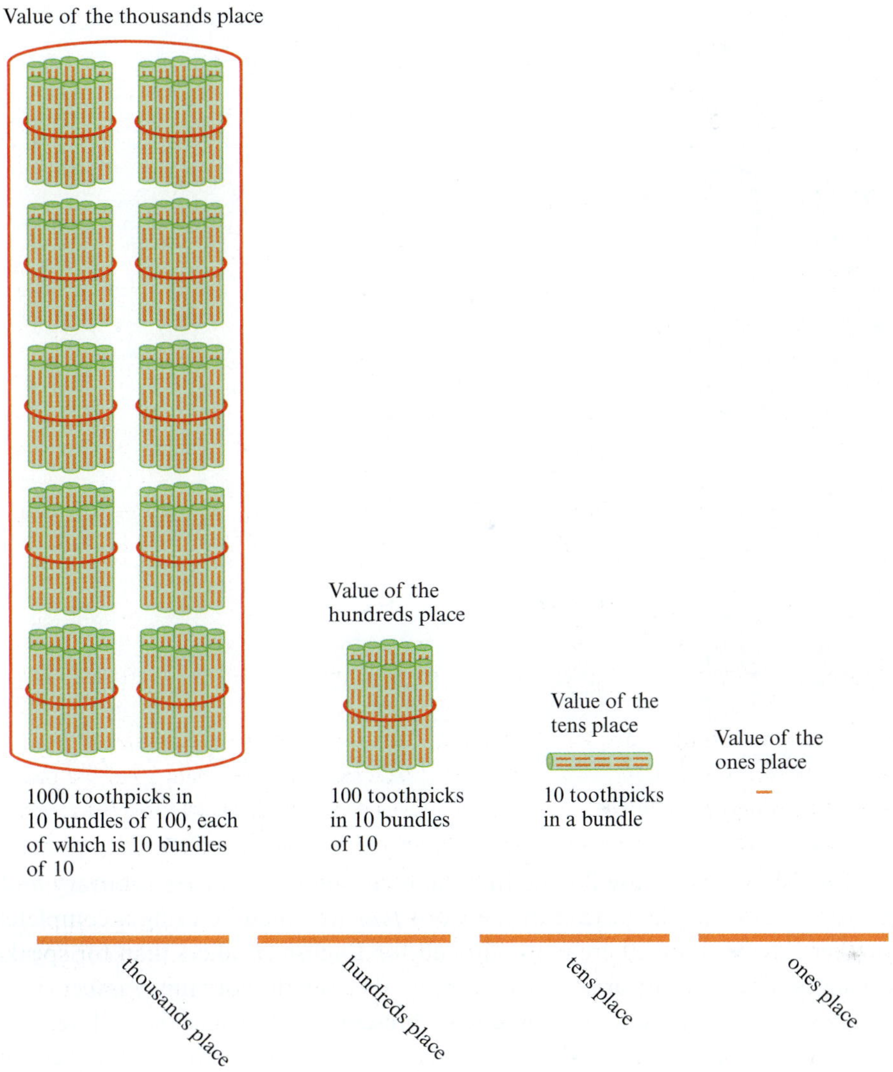

Figure 1.8 Bundled toothpicks representing base-ten units and values of places in the base-ten system.

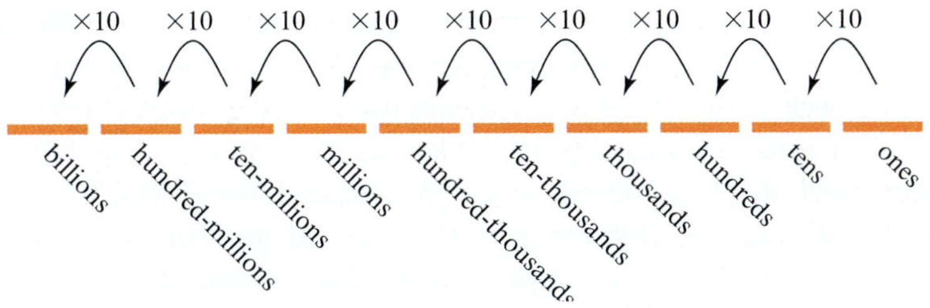

Figure 1.9 An illustration of how each place's value is ten times the value of the place to its right.

In summary, by using place value, the base-ten system allows us to write any counting number, no matter how large, using only the ten digits 0 through 9. The key idea of place value is to create larger and larger units, by making the value of each new place ten times the value of the place to its right. By using place value, every counting number can be expressed in a unique way as a string of digits.

What Is Difficult about Counting Number Words?

Before young children learn to write the symbols for the counting numbers, they learn to say the number words. Unfortunately, some of the words used to say the counting numbers in English do not correspond well with the base-ten representations of these numbers. This makes the early learning of numbers more difficult for children who speak English than for children who speak other languages.

Figure 1.10

Representing 1234 with bundled toothpicks to show how it is composed of its place value parts.

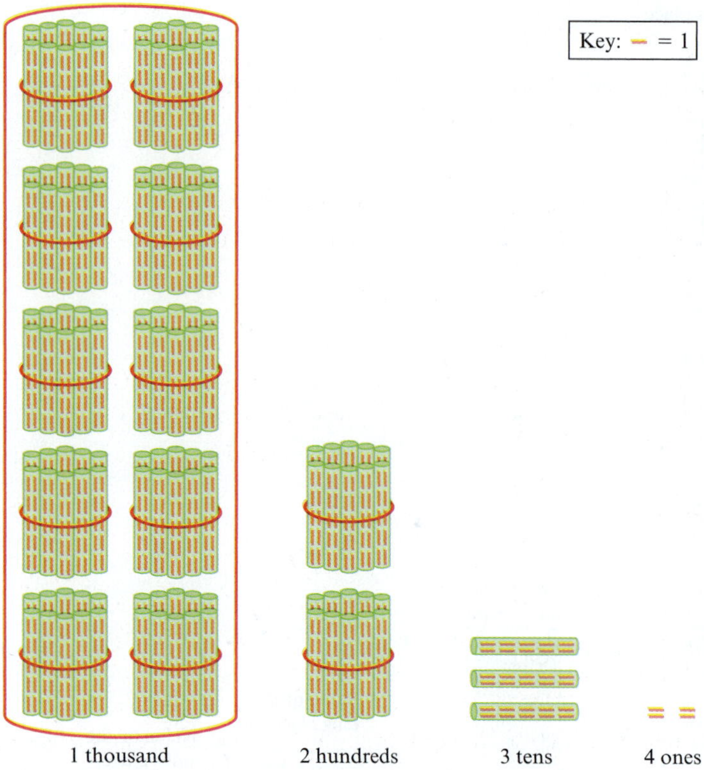

The English words we use for the first ten counting numbers are arbitrary and could have been different. For example, instead of the word *four* we could be using a completely different word. Numbers greater than 10 are more difficult for English speakers than for speakers of some other languages. The difficulty arises because the way we say the counting numbers from 11 to 19 in English does not correspond to their base-ten representations. Notice that "eleven" does not sound like "one ten and one," which is what 11 stands for; nor does "twelve" sound very much like "one ten and two," which is what 12 stands for. To add to the confusion, "thirteen, fourteen, . . . , nineteen" sound like the reverses of "one ten and three, one ten and four, . . . , one ten and nine," which is what 13, 14, . . . , 19 represent. From 20 onward, most of the English words for counting numbers do correspond fairly well to their base-ten representations. For example, "twenty" sounds roughly like "two tens," "sixty-three" sounds very much like "six tens and three," and "two-hundred eighty-four" sounds very much like "2 hundreds and eight tens and four," which is what 284 represents. Note, however, that it's easy for children to confuse decade numbers with teen numbers because their pronunciation is so similar. For example, "sixty" and "sixteen" sound similar.

Why Do We Need the Whole Numbers?

whole numbers

The **whole numbers** are the counting numbers together with zero:

$$0, 1, 2, 3, 4, 5, \ldots$$

The notion of zero may seem natural to us today, but our early ancestors struggled to discover and make sense of zero. Although humans have always been acquainted with the notion of "having none," as in having no sheep or having no food to eat, the concept of 0 as a number was introduced far later than the counting numbers—not until sometime before 800 A.D. (See [10].) Even today, the notion of 0 is difficult for many children to grasp. This difficulty is not surprising: Although the counting numbers can be represented nicely by sets of objects, you have to show *no* objects in order to represent the number 0 in a similar fashion. But how does one *show* no objects? We might use a picture as in Figure 1.11.

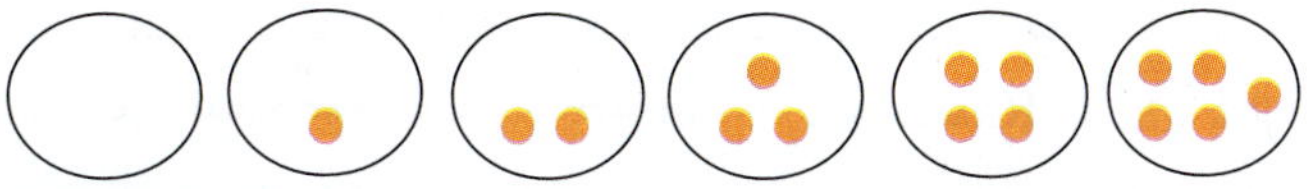

Figure 1.11 Representing the whole numbers.

Notice that place value requires the use of 0. To write three-hundred, for example, we must show that the 3 is in the hundreds place and then show that there are no tens and no ones. We do this by writing 300.

What Ideas Lead to Number Lines?

We can use counting numbers to count events as well as the number of physical things in a collection. For example, children might count how many times they have hopped or jumped or how many steps they have taken. Many children's games involve moving a game piece along a path. If the path is labeled with successive counting numbers, like the one in **Figure 1.12**, we can call the path a **number path**. Number paths are informal precursors to the concepts of distance and length, and to the mathematical concept of a number line.

number path

Figure 1.12 A number path.

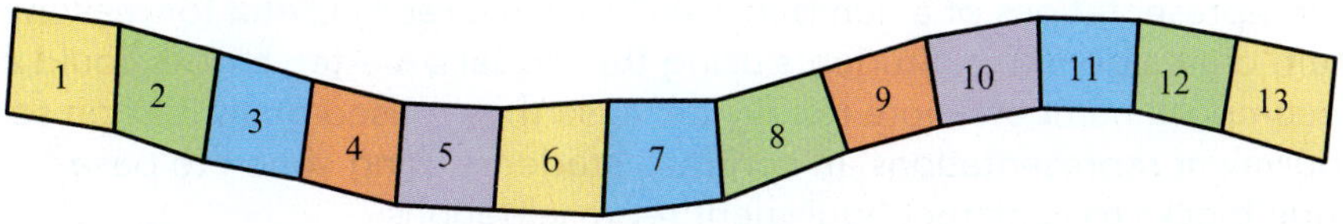

number line

unit (number line)

scale (number line)

A **number line** is a line on which we have chosen one location as 0, and another location, typically to the right of 0, as 1. Number lines stretch infinitely far in both directions, although in practice we can only show a small portion of a number line (and that portion may or may not include 0 and 1). The distance from 0 to 1 is called a **unit**, and the choice of a unit is called the **scale** of the number line. Once choices for the locations of 0 and 1 have been made, each counting number is represented by the point on the number line located that many units to the right of 0.

Number lines are an important way to represent numbers because they allow the concept of number to be expanded to decimals, fractions, and negative numbers and they unify different kinds of numbers and present them as a coherent whole.

Although number paths and number lines are similar, there is a critical distinction between them. This distinction makes using number paths with the youngest children better than using number lines. Number paths clearly show distinct "steps" along the path that children can count, just as they might count their own steps, hops, or jumps. In contrast, to interpret a number line correctly, we must rely on the ideas of length and distance from 0, as indicated in **Figure 1.13**. Instead of

CCSS
2.MD.6

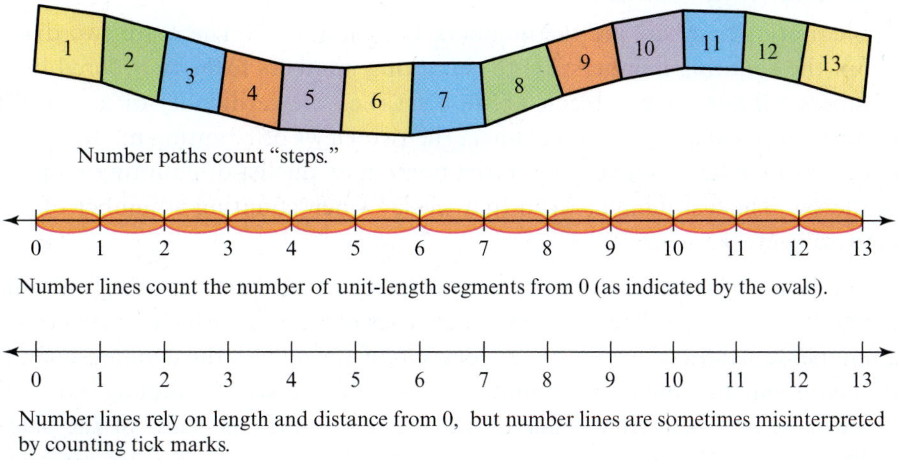

Figure 1.13 Number paths versus number lines: counting steps versus length.

FROM THE FIELD **Research**

Burris, J. T. (2013). Virtual place value. *Teaching Children Mathematics, 20*(4), 228–236.

This study investigated how four third-grade classes engaged with base-ten blocks to build and identify quantities and to write corresponding numbers during a unit on place value. Two classes used virtual base-ten blocks; the other two used concrete base-ten blocks. Both the virtual and the concrete base-ten blocks could be grouped, ungrouped, and regrouped into units, tens, hundreds, and thousands. The students' reasoning was analyzed in terms of a conceptual framework of counting stages. At the most basic stage students are able to count by ones but are unable to view a number such as 32 as 3 tens and 2 ones. At the most advanced stage, students are able to count by tens and ones and can move fluidly between ways of thinking about a number. For example, they can view the 3 in 32 both as thirty and as 3 tens. The study found that in both groups, most students were at the most advanced stage. A difference between the two groups was in how efficiently students could create equivalent representations of a number, a skill that is directly useful for reasoning about multi-digit algorithms. Students using the virtual base-ten blocks could compose and decompose numbers more readily because they reused quantities on screen to create equivalent representations. In contrast, students using concrete base-ten blocks had to trade blocks to construct equivalent representations.

IMAP *Watch Maryann's students discuss place value.*

focusing on length, young children tend to count "tick marks" along a number line. The habit of counting tick marks instead of attending to length can lead to omitting 0 and to misinterpretations about locations of fractions on number lines. We will examine some of these errors when we study fractions in Chapter 2.

SECTION SUMMARY AND STUDY ITEMS

Section 1.1 The Counting Numbers

The counting numbers are the numbers 1, 2, 3, 4, ... There are two distinct ways to think about the counting numbers: (1) the counting numbers form an ordered list, and (2) counting numbers tell how many objects are in a set (i.e., the cardinality of a set). When we count the number of objects in a set, we connect the two views of counting numbers by making a one-to-one correspondence between an initial portion of the list of counting numbers and the objects in the set. The last number word that we say when counting a number of objects tells us how many objects there are.

In the base-ten system, every counting number can be written using only the ten symbols (or digits) 0, 1, 2, . . . , 9. The base-ten system uses place value, which means that the value a digit in a number represents depends on the location of the digit in the number. Each place has the value of a base-ten unit, and base-ten units are created by repeated bundling by ten. Each base-ten unit is the value of a place. The value of a place is ten times the value of the place to its immediate right.

Unfortunately, the way we say the English names of the counting numbers 11 through 19 does not correspond to the way we write these numbers.

The counting numbers can be displayed on number paths, which are informal precursors to number lines. The whole numbers, 0, 1, 2, 3, . . . , can be displayed on number lines.

Key Skills and Understandings

1. Describe the two views of the counting numbers—as a list and as used for cardinality. Discuss the connections between the list and cardinality views of the counting numbers.
2. Explain what it means for the base-ten system to use place value. Discuss what problem the development of the base-ten system solved.
3. Describe base-ten units and explain how adjacent place values are related in the base-ten system.
4. Describe and make math drawings to represent a given counting number in terms of bundled objects in a way that fits with the base-ten representation for that number of objects.
5. Describe how to represent whole numbers on a number line and discuss the difference between a number path (as described in the text) and a number line.

Practice Exercises for Section 1.1

1. If a young child can correctly say the number word list "one, two, three, four, five," will the child necessarily be able to determine how many bears are in a collection of 5 toy bears that are lined up in a row? Discuss why or why not.
2. If a young child can correctly say the number word list "one, two, three, four, five" and point one by one to each bear in a collection of 5 toy bears while saying the number words, does the child necessarily understand that there are 5 bears in the collection? Discuss why or why not.
3. What problem in the history of mathematics did the development of the base-ten system solve?
4. Make a math drawing showing how to organize 19 objects in a way that fits with the structure of the base-ten system.
5. Describe how to organize 100 toothpicks in a way that fits with the structure of the base-ten system. Explain how your organization reflects the structure of the base-ten system and how it fits with the way we write the number 100.

Answers to Practice Exercises for Section 1.1

1. No, the child might not be able to determine that there are 5 bears in the collection because the child might not be able to make a one-to-one correspondence between the number words 1, 2, 3, 4, 5 and the bears. For example, the child might point twice to one of the bears and count two numbers for that bear, or the child might skip over a bear while counting. See also the next practice exercise and its answer.
2. No, the child might not understand that the last number word that is said while counting the bears tells how many bears there are in the collection.
3. The base-ten system solved the problem of having to invent more and more new symbols to stand for larger and larger numbers. By using the base-ten system, and place value, every counting number can be written using only the ten digits 0, 1, . . . , 9.
4. See **Figure 1.5**.
5. First, bundle all the toothpicks into bundles of 10. Then gather those 10 bundles of 10 into a single bundle. This repeated bundling in groups of 10 is the basis of the base-ten system. The 1 in 100 stands for this 1 large bundle of 10 bundles of 10.

PROBLEMS FOR SECTION 1.1

1. In your own words, discuss the connection between the counting numbers as a list and the counting numbers as they are used to describe how many objects are in sets. Include a discussion of what you will need to attend to if you are teaching young children who are learning to count.

2. If you give a child in kindergarten or first grade a bunch of beads or other small objects and ask the child to show you what the 3 in 35 stands for, the child might show you 3 of the beads. You might be tempted to respond that the 3 really stands for "thirty" and not 3. Of course it's true that the 3 does stand for thirty, but is there a better way you could respond, so as to draw attention to the base-ten system? How could you organize the beads to make your point?

3. For each of the following collections of small objects, draw a simple picture and write a brief description for how to organize the objects in a way that corresponds to the way we use the base-ten system to write the number for that many objects.

a. 47 beads

b. 328 toothpicks

c. 1000 toothpicks

4. In your own words, describe how you can use collections of objects (such as toothpicks or Popsicle sticks) to show the values of some base-ten units. Discuss also how the values of adjacent places in base-ten representations of numbers are related.

5. In your own words, discuss the beginning ideas of place value and the base-ten system that young children who can count beyond ten must begin to learn. Include a discussion of some of the hurdles faced by English speakers.

6. Children sometimes mistakenly read the number 1001 as "one hundred one." Why do you think a child would make such a mistake? Make a math drawing showing how to represent 1001 with bundled objects.

7. Explain why the bagged and loose toothpicks pictured in Figure 1.14 are not organized in a way that fits with the structure of the base-ten system. Describe how to alter the appearance of these bagged and loose toothpicks so that the same total number of toothpicks are organized in a way that is compatible with the base-ten system.

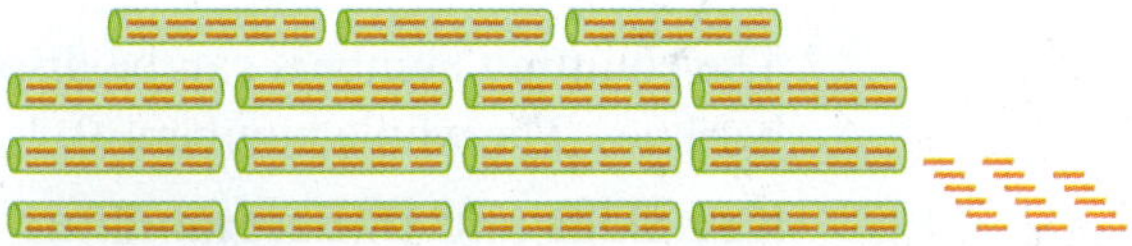

Figure 1.14 Why are these not organized according to the base-ten system?

8. Describe key features of the base-ten system. Compared to more primitive ways of writing numbers, what is one advantage and one disadvantage of the base-ten system?

***9.** Draw number lines like the ones in Figure 1.15.

a. Plot 900 on the first number line and explain your reasoning.

b. Plot 250 on the second number line and explain your reasoning.

c. Plot 6200 on the third number line and explain your reasoning.

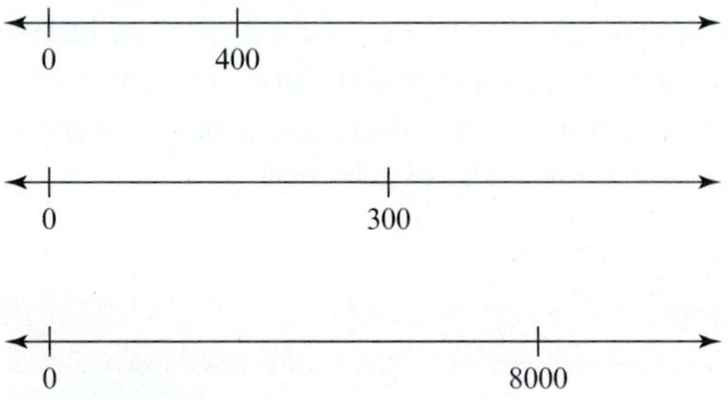

Figure 1.15 Number lines.

***10.** The students in Ms. Caven's class have a large poster showing a million dots. Now, the students would really like to see a billion of something. Think of at least two different ways that you might attempt to show a billion of something and discuss whether your methods would be feasible. Be specific and back up your explanations with calculations.

1.2 Decimals and Negative Numbers

CCSS Common Core State Standards Grades 4, 5, 6

The counting numbers are the most basic kinds of numbers, followed by the whole numbers. However, many situations, both practical and theoretical, require other numbers, such as fractions, decimals, and negative numbers. Although fractions, decimals, and negative numbers may appear to be different, they become unified when they are represented on a number line.

What Are the Origins of Decimals and Negative Numbers?

Why do we have decimals (and fractions) and negative numbers? How did these numbers arise?

In ancient times, a farmer filling bags with grain might have had only enough grain to fill the last bag half full. When trading goods, the farmer needed a way to express partial quantities. In modern times, we buy gasoline by the gallon (or liter), but we don't always buy a whole number of gallons. So we need a precise way to describe numbers that are in between whole numbers. Both fractions and decimals arise by creating new units that are less than 1, but decimals are created by extending the base-ten system.

The introduction of negative numbers came relatively late in human development. Although the ancient Babylonians may have had the concept of negative numbers around 2000 B.C., negative numbers were not always accepted by mathematicians even as late as the sixteenth century A.D. ([10]). The difficulty lies in interpreting the meaning of negative numbers. How can negative numbers be represented? This problem may seem perplexing at first, but in fact there are understandable interpretations of negative numbers such as temperature below zero or elevations below sea level or ground level.

How Do Decimals Extend the Base-Ten System?

The essential structure of the base-ten system is that the value of each place is ten times the value of the place to its right. So, moving to the *left* across the places in the base-ten system, the value of the places are successively *multiplied* by 10. Likewise, moving to the *right*, the values of the places are successively *divided* by 10, as indicated in Figure 1.16.

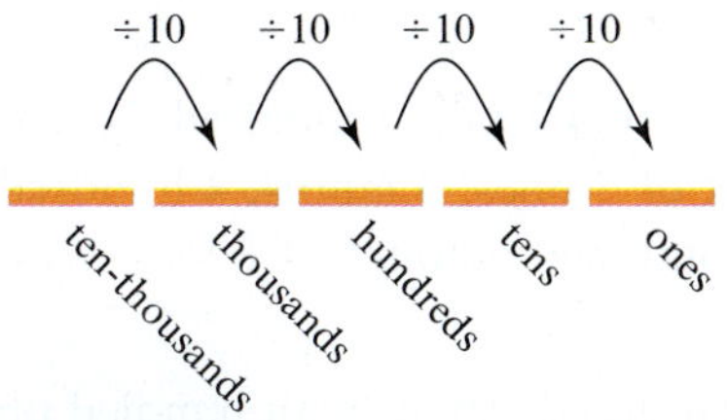

Figure 1.16 Moving to the right, the values of places in the base-ten system are divided by 10.

decimal point

Using the base-ten system, decimals are created by establishing places to the right of the ones place. We indicate the location of the ones place by placing a **decimal point** (.) to its right. Starting at the ones place, we divide the unit 1 into 10 equal pieces to create a new unit, a *tenth*. Tenths are recorded in the place to the right of the ones place. Then we divide a tenth into 10 equal pieces to create a new unit, a hundredth. Hundredths are recorded in the place to the right of the tenths place. Figure 1.17 depicts the process of dividing by 10 to create smaller and smaller place values to the right of the ones place. This process continues without end, as shown in Figure 1.18.

decimal notation

decimal

decimal number

decimal representation

decimal expansion

When we use the base-ten system to represent a number as a string of digits, possibly including a decimal point, and possibly having infinitely many nonzero digits to the right of the decimal point, we say the number is in **decimal notation** and call it a **decimal** or a **decimal number**. We may also refer to the string of digits representing the number as a **decimal representation** or **decimal expansion** of the number.

Figure 1.17

Tenths are created by dividing 1 into 10 equal pieces; hundredths are created by dividing a tenth into 10 equal pieces.

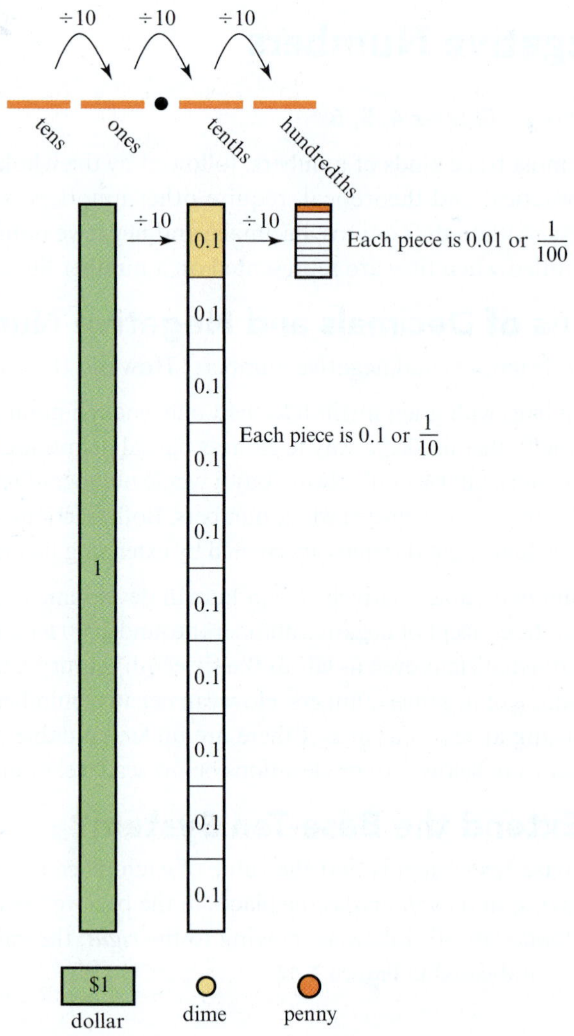

CCSS
5.NBT.1

Just as 2345 stands for the combined amount of 2 thousands, 3 hundreds, 4 tens, and 5 ones, the decimal 2.345 stands for the combined amount of 2 ones, 3 tenths, 4 hundredths, and 5 thousandths and the decimal 23.45 stands for the combined amount of 2 tens, 3 ones, 4 tenths, and 5 hundredths.

Using the base-ten structure, we can represent (some) decimals with bundled objects in the same way that we represent whole numbers with bundled objects, as Class Activity 1D shows. The only difference is that a single object must be allowed to represent a place value that is less than 1. Although this may seem surprising at first, it is a common idea. After all, a penny represents $0.01.

IMAP *Watch Megan and Donna represent decimal numbers with blocks.*

CLASS ACTIVITY

1D Representing Decimals with Bundled Objects, p. CA-6

How Do Decimals as Lengths Develop into Decimals on Number Lines?

A good way to represent positive decimals is as lengths; this way of representing decimals leads naturally to placing decimals on number lines. Figure 1.18 indicates how to represent the decimals 1.2, 1.23, and 1.234 as lengths. The meter, which is the main unit of length used in the metric system, is a natural unit to use when representing decimals as lengths because the metric system was designed to be compatible with the base-ten system.

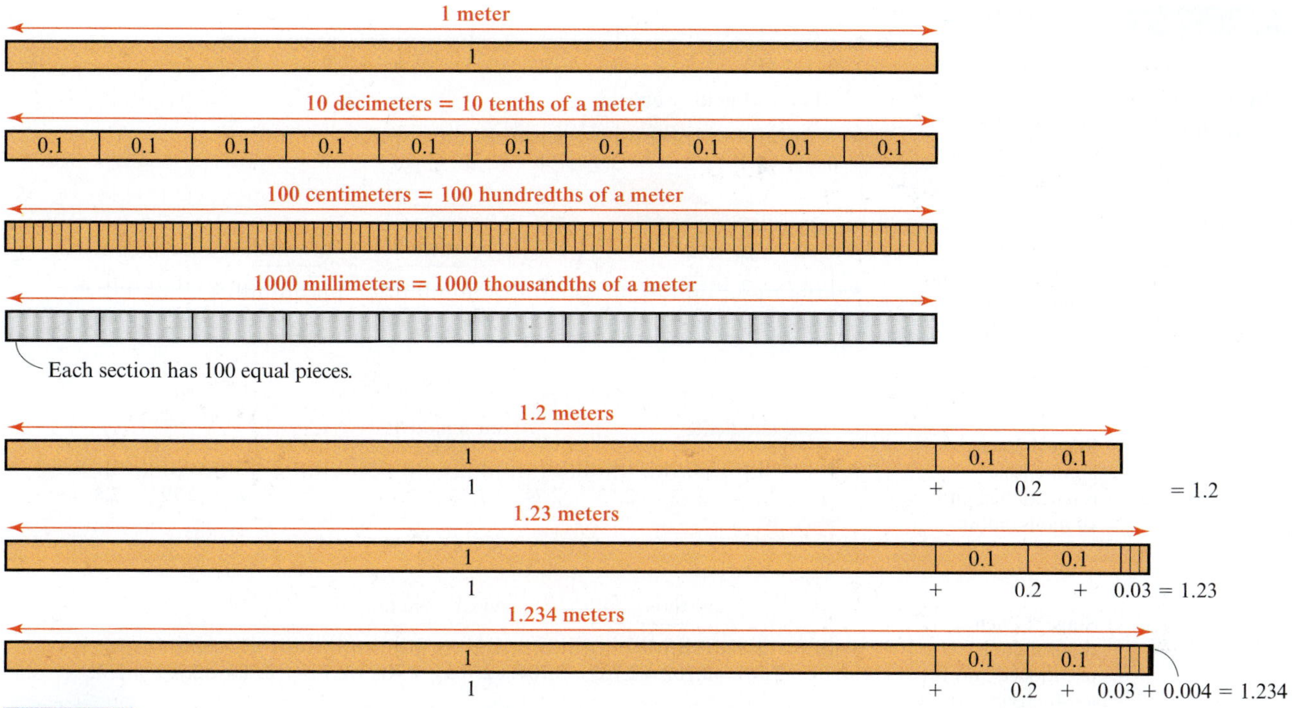

Figure 1.18 Representing decimals as lengths using metric length units.

CLASS ACTIVITY

1E Representing Decimals as Lengths, p. CA-7

If you did Class Activity 1E you used strips of paper, which are not very durable. For instruction, you might want to use a more durable material, such as lengths of plastic tubing (see [84]).

To connect lengths with number lines, imagine representing a positive decimal as a length by using strips of paper (or pieces of plastic tubing) as in Class Activity 1E. Now imagine placing the left end of the length of paper at 0. Then the right end of the length of paper lands on the point on the number line that the length represents, as in Figure 1.19. In this way, we can view number lines as related to lengths, and we can view points on number lines in terms of their distances from 0. More generally, a positive number N is located to the right of 0 at a distance of N units away from 0.

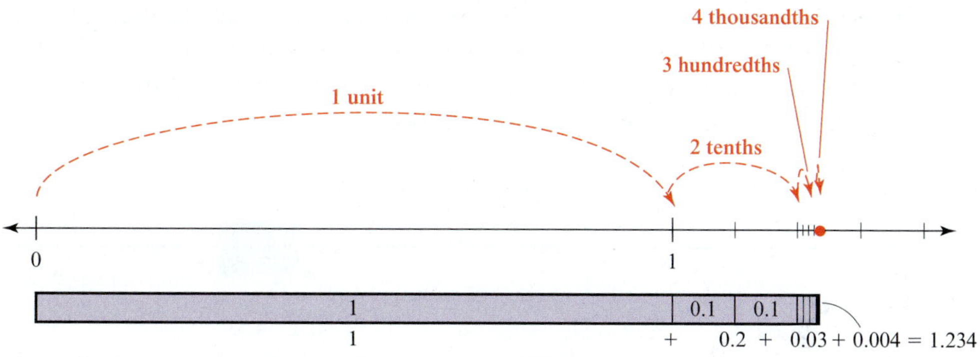

Figure 1.19 Plotting decimals on a number line.

One way to think about decimals is as "filling in" the locations on the number line between the whole numbers. You can think of plotting decimals as points on the number line in successive stages according to the structure of the base-ten system. At the first stage, the whole numbers are placed on a number line so that consecutive whole numbers are one unit apart. (See Figure 1.20).

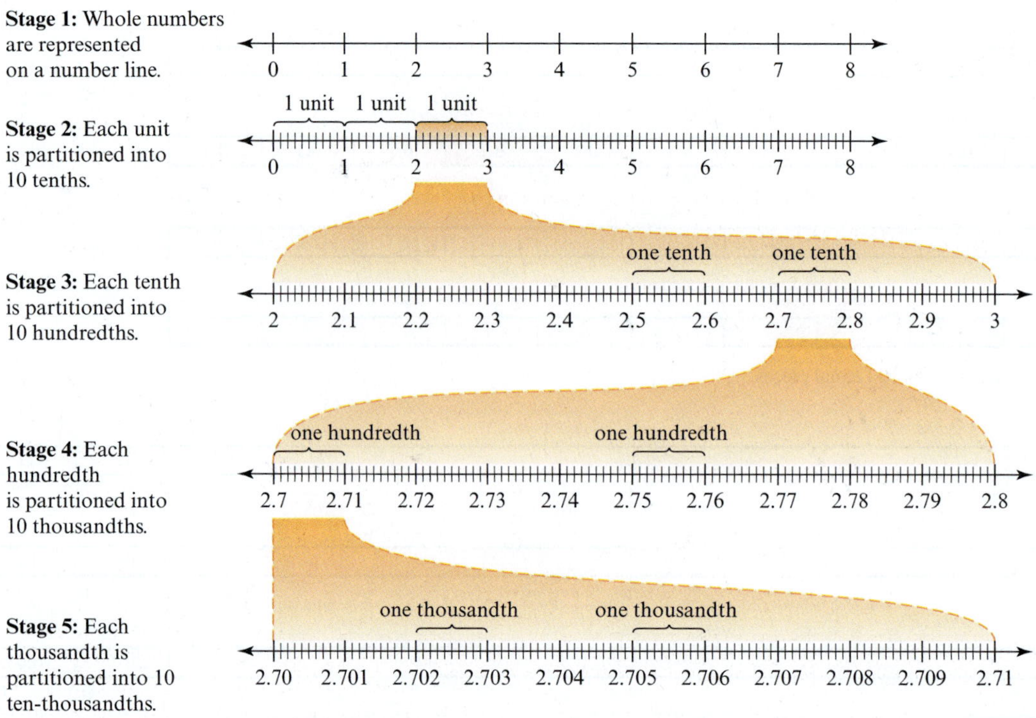

Figure 1.20 Decimal numbers "fill in" number lines.

At the second stage, the decimals that have entries in the tenths place, but no smaller place, are spaced equally between the whole numbers, breaking each interval between consecutive whole numbers into 10 smaller intervals each one-tenth unit long. See the Stage 2 number line in Figure 1.20. Notice that, although the interval between consecutive whole numbers is broken into 10 intervals, there are only 9 tick marks for decimal numbers in the interval, one for each of the 9 nonzero entries, 1 through 9, that go in the tenths places.

We can think of the stages as continuing indefinitely. At each stage in the process of filling in the number line, we plot new decimals. The tick marks for these new decimals should be shorter than the tick marks of the decimal numbers plotted at the previous stage. We use shorter tick marks to distinguish among the stages and to show the structure of the base-ten system.

The digits in a decimal are like an address. When we read a decimal from left to right, we get more and more detailed information about where the decimal is located on a number line. The left-most digit specifies a "big neighborhood" in which the number is located. The next digit to the right narrows the location of the decimal to a smaller neighborhood of the number line. Subsequent digits to the right specify ever more narrow neighborhoods in which the decimal is located, as indicated in Figure 1.21. When we read a decimal from left to right, it's almost like specifying a geographic location by giving the country, state, county, zip code, street, and street number, except that decimals can have infinitely more detailed locations.

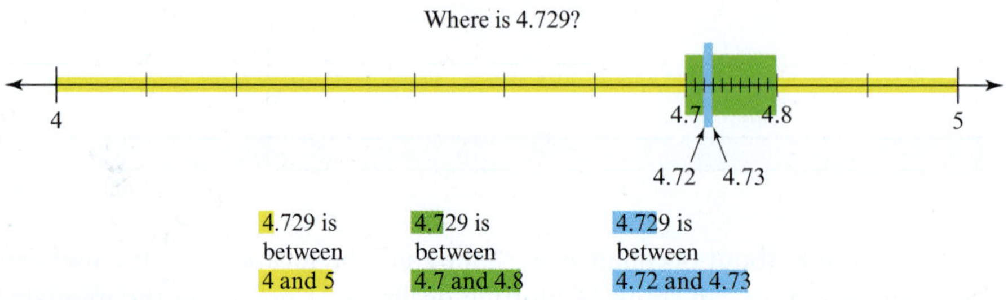

Figure 1.21 Digits to the right in a decimal describe the decimal's location on a number line with ever greater specificity.

Figure 1.22

Zooming in on the location of 1.738.

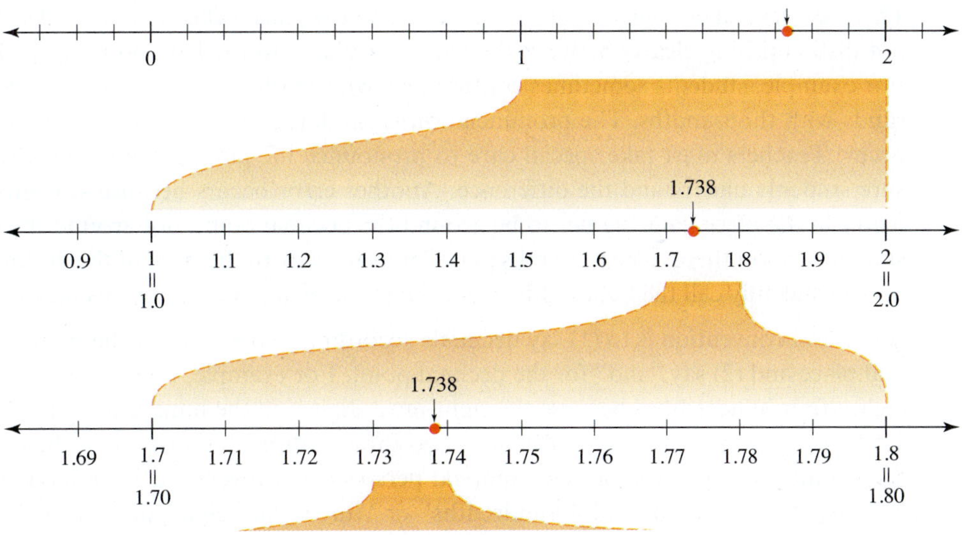

By "zooming in" on narrower and narrower portions of the number line, as in **Figure 1.22**, we can see in greater detail where a decimal is located.

When plotting decimals on number lines (or when comparing, adding, or subtracting decimals), it is often useful to append zeros to the right-most nonzero digit to express explicitly that the values in these smaller places are zero. For example, 1.78, 1.780, 1.7800, 1.78000, and so on, all stand for the same number. These representations show explicitly that the number 1.78 has 0 thousandths, 0 ten-thousandths, and 0 hundred-thousandths. Similarly, we may append zeros to the left of the left-most nonzero digit in a number to express explicitly that the values in these larger places are zero. For example, instead of writing .58, we may write 0.58, which perhaps makes the decimal point more clearly visible.

CLASS ACTIVITY

1F Zooming In on Number Lines, p. CA-8

IMAP
Watch Vanessa write 10 hundredths.

CLASS ACTIVITY

1G Numbers Plotted on Number Lines, p. CA-11

What Is Difficult about Decimal Words?

The names for the values of the places to the right of the ones place are symmetrically related to the names of the values of the places to the left of the ones place, as shown in **Figure 1.23**.

Figure 1.23

Symmetry in the place value names is around the ones place.

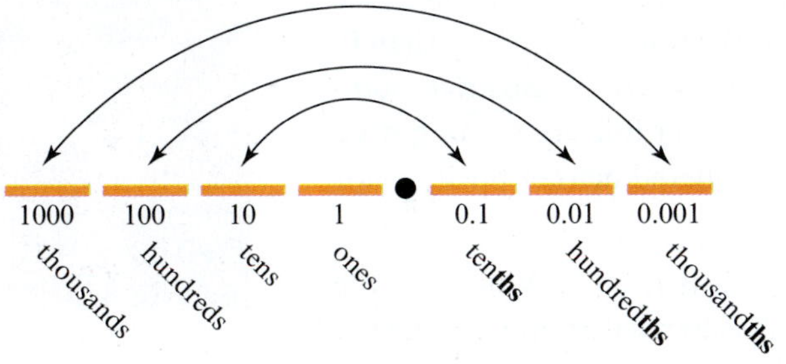

There are several common errors associated with the place value names for decimals. One error is not distinguishing clearly between the values of places to the left and right of the decimal point. For example, students sometimes confuse tens with ten**ths** or hundreds with hundred**ths** or thousands with thousand**ths**. The pronunciation is similar, so it's easy to see how this confusion can occur! Teachers must take special care to pronounce the place value names clearly and to make sure students understand the difference. Another error occurs because students expect the symmetry in the place value names to be around the decimal point, not around the ones place. Some students expect there to be a "oneths place" immediately to the right of the decimal point, and they may mistakenly call the hundredths place the tenths place because of this misunderstanding.

A cultural convention is to (1) say decimals according to the value of the right-most nonzero decimal place and (2) say "and" for the decimal point. For example, we usually say 3.84 as "three and eighty-four hundredths" because the right-most digit is in the hundredths place. Similarly, we say 1.592 as "one and five-hundred ninety-two thousandths" because the right-most digit is in the thousandths place. From a mathematical perspective, however, it is perfectly acceptable to say 3.84 as "3 and 8 tenths and 4 hundredths" or "three point eight four." In fact, we can't use the usual cultural conventions when saying decimals that have infinitely many digits to the right of the decimal point. For example, the number pi, which is 3.1415 . . . must be read as "three point one four one five . . ." because there is no right-most nonzero digit in this number! Furthermore, the conventional way of saying decimals is logical, but the reason for this will not be immediately obvious to students who are just learning about decimals and place value. We will explain why when we discuss adding and subtracting fractions in Chapter 3.

What Are Negative Numbers and Where Are They on Number Lines?

negative
minus sign
integers

For any number N, its **negative** is also a number and is denoted $-N$. The symbol $-$ is called a **minus sign**. For example, the negative of 4 is -4, which can be read *negative four* or *minus four.* The set of numbers consisting of 0, the counting numbers, and the negatives of the counting numbers, is called the **integers**.

$$\ldots, -5, -4, -3, -2, -1, 0, 1, 2, 3, 4, 5, \ldots$$

CCSS 6.NS.5

We can think of a negative number, $-N$, as the "opposite" of N. Negative numbers are commonly used to denote amounts owed, temperatures below zero, and even for locations below ground or below sea level. For example, we could use -100 to represent owing 100 dollars. The temperature $-4°$ Celsius stands for 4 degrees below 0° Celsius. An altitude of -50 feet means 50 feet below sea level. In some places, negative numbers are even used to indicate floor levels. The photo at the right shows a floor directory in a French department store. Floor 0 is ground level and Floor -1 is one flight below ground level (the basement).

positive numbers
negative numbers

On a number line, we display the negative numbers in the same way as we display 0 and the numbers greater than 0. To the right of 0 on the number line are the **positive numbers**. To the left of 0 on the number line are the **negative numbers**. The number 0 is considered neither positive nor negative.

Given a positive number N that is located N units to the right of zero, its negative, $-N$ is located N units to the *left* of 0 on the number line. Therefore positive and negative numbers are placed symmetrically with respect to 0 on the number line, as seen in **Figure 1.24**. Note that this symmetry on the *number line* should not be confused with the symmetry of *place values* with respect to the ones place. More generally, for any number M, positive or negative, its **negative** or **opposite**, $-M$, is the number on the opposite side of the number line that is the same distance from 0. For example, $-(-2) = 2$. The distance of M from 0 is called the **absolute value** of M and is denoted $|M|$. So for example, $|-2| = 2$.

negative, opposite

absolute value

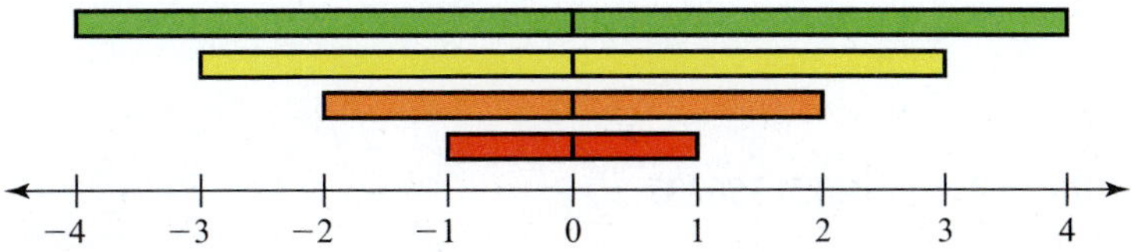

CCSS 6.NS.6a

Figure 1.24 N and $-N$ are the same distance from 0, creating symmetry in the number line.

Figure 1.25 shows how we use distance to plot the negative number −2.41 on a number line. We start at 0, then we move to the left: first 2 units, then another 4 tenths, and then another 1 hundredth. In this way, the number −2.41 is plotted 2.41 units away from 0, to the left of 0.

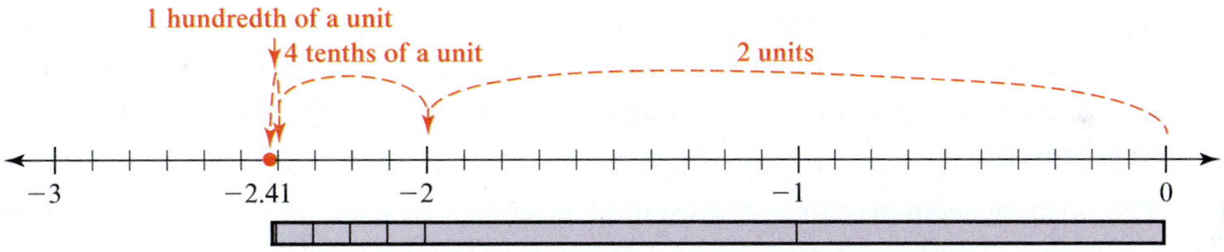

Figure 1.25 Plotting −2.41.

CLASS ACTIVITY

1H Negative Numbers on Number Lines, p. CA-13

Can Decimals with Infinitely Many Nonzero Entries Be Located on Number Lines?

Some decimals extend infinitely far to the right. For example, it turns out that the decimal representation of $\sqrt{2}$ (the positive number that when multiplied with itself is 2) is

$$\sqrt{2} = 1.41421356237\ldots$$

which goes on forever and never ends. *Every* decimal, even a decimal with infinitely many nonzero entries to the right of the decimal point, has a definite location on a number line. However, such a decimal will never fall *exactly* on a tick mark, no matter what the scale of the number line. **Figure 1.26** shows where $\sqrt{2}$ is located on number lines of various scales. The fact that a decimal like $\sqrt{2}$ has a location on the number line may seem like a murky idea; in fact, it is a very subtle point, whose details were not fully worked out by mathematicians until the late nineteenth century.

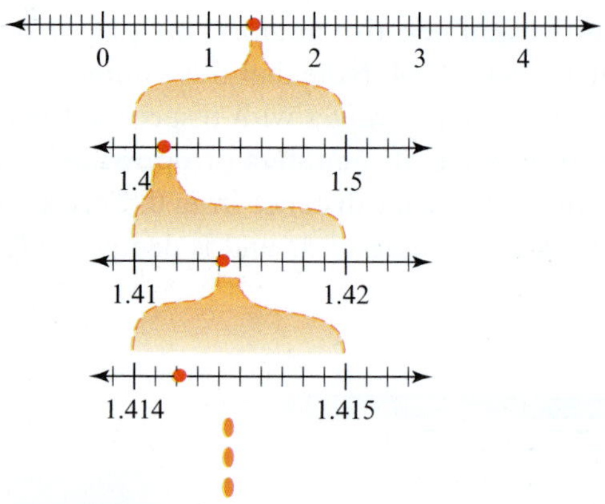

Figure 1.26 Plotting $\sqrt{2} = 1.41421356237\ldots$ on a number line.

SECTION SUMMARY AND STUDY ITEMS

Section 1.2 Decimals and Negative Numbers

The decimals extend the base-ten system and are formed by creating places to the right of the ones place while preserving the essential structure of the base-ten system: the value of each place is ten times the value of the place to its immediate right. Just as the counting numbers can be represented with bundled objects, finite (positive) decimals can also be represented with bundled objects, provided that a single object stands for the value of the lowest place one needs to represent.

The positive decimals can be represented as lengths in a way that fits with the structure of the base-ten system. This way of representing decimals fits nicely with the metric system (meters, decimeters, centimeters, and millimeters.) Decimals can also be represented on number lines in a way that is compatible with viewing positive decimals as lengths: the location of a positive decimal is its distance from 0. We can also think of the decimals as filling in the number line: First, we plot whole numbers, then we plot tenths, then hundredths, and so on.

The negative numbers are the numbers to the left of zero on the number line.

Key Skills and Understandings

1. Describe and draw rough pictures to represent a given positive decimal in terms of bundled objects in a way that fits with and shows the structure of the base-ten system. (Take care to state the meaning of one of the objects.)
2. Describe and draw rough pictures to represent a given positive decimal as a length in a way that fits with and shows the structure of the base-ten system.
3. Discuss how decimals fill in a number line, label tick marks on number lines, and plot numbers on number lines.
4. Show how to zoom in on portions of number lines to see the portions in greater detail.
5. View negative numbers as owed amounts, and plot a given negative number on a number line.

Practice Exercises for Section 1.2

1. Describe three ways discussed in the text to represent a decimal such as 1.234. For each way, show how to represent 1.234.
2. Describe how to represent 0.0278 with bundles of small objects in a way that fits with and shows the structure of the base-ten system. In this case, what does one small object represent?
3. You have bundles of toothpicks like the ones shown in Figure 1.27 and you want to use these bundled toothpicks to represent a decimal. List at least three decimals that you could use these bundles to represent and explain your answer in each case.

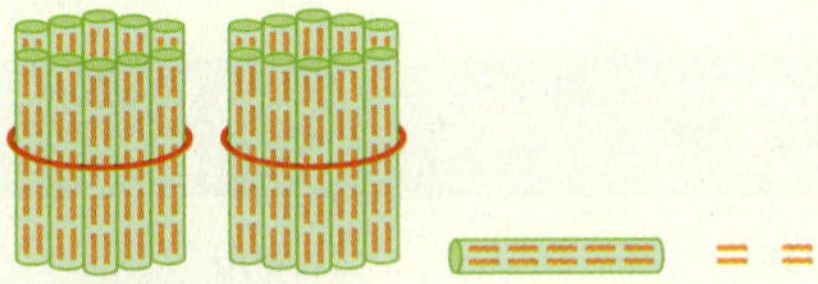

Figure 1.27 Which decimals can this represent?

4. Describe and make drawings showing how to represent 1.369 and 1.07 as lengths by using strips of paper in a way that fits with and shows the structure of the base-ten system.
5. Draw a number line on which long tick marks are whole numbers and on which 0.003 can be plotted (in its approximate location). Now show how to "zoom in" on smaller and smaller portions of the number line (as in Figure 1.22 and Class Activity 1F, part 1) until you have zoomed in to a portion of the number line in which the long tick marks are thousandths and so that 0.003 can be plotted on each number line. Label the long tick marks on each number line, and plot 0.003 on each number line (in its approximate location).
6. Draw number lines like the ones in Figure 1.28. Label the unlabeled tick marks on your three number lines in three different ways. In each case, your labeling should fit with the structure of the base-ten system and the fact that the tick marks at the ends of the number lines are longer than the other tick marks.

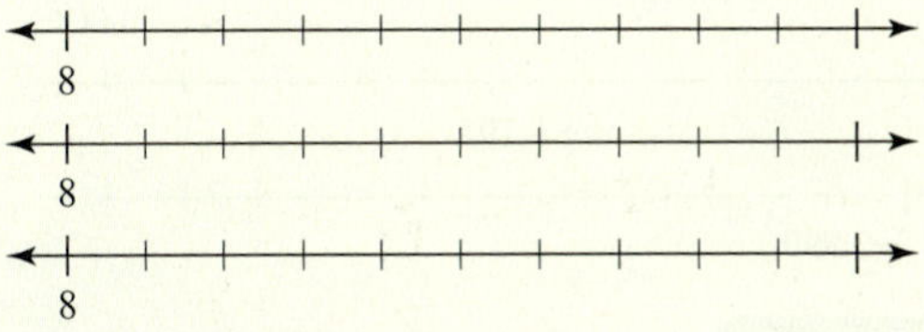

Figure 1.28 Labeling tick marks in three different ways.

7. Draw number lines like the ones in Figure 1.29. Label the unlabeled tick marks on your three number lines in three different ways. In each case, your labeling should fit with the structure of the base-ten system and the fact that the tick marks at the ends of the number lines are longer than the other tick marks.

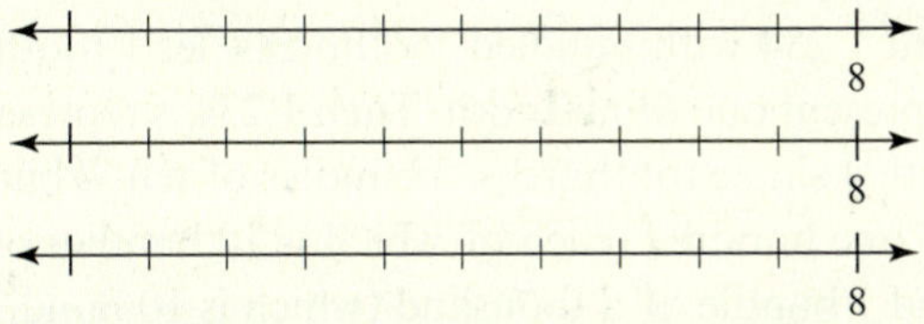

Figure 1.29 Labeling tick marks in three different ways.

8. Draw a number line like the one in Figure 1.30 for the exercises that follow. For each exercise, label all the tick marks on the number line. The number to be plotted need not land on a tick mark.
   a. Plot 28.369 on a number line on which the long tick marks are tenths.
   b. Plot 1.0601 on a number line on which the long tick marks are thousandths.
   c. Plot 14.8577 on a number line on which the long tick marks are whole numbers.

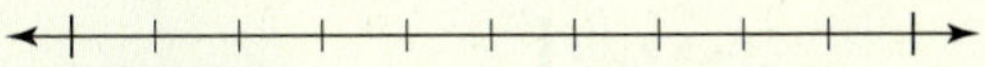

Figure 1.30 A number line.

9. Draw number lines like the ones in Figure 1.31. Label the tick marks on your number lines with appropriate decimals.

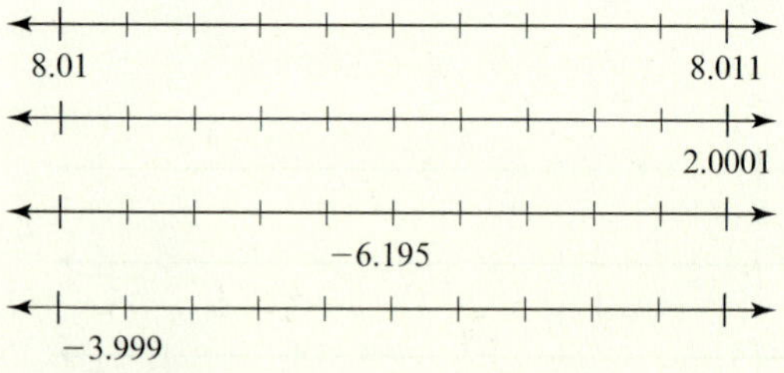

Figure 1.31 Labeling tick marks.

10. Describe where the negative numbers are located on the number line. Then describe how to plot the numbers −2 and −2.41 on a number line on which 0 and 1 have been plotted.

11. Draw a number line like the one in Figure 1.30 for the exercises that follow. For each exercise, label all the tick marks on the number line. The number to be plotted need not land on a tick mark.

a. Plot −7.65 on a number line on which the long tick marks are whole numbers.

b. Plot −0.0118 on a number line on which the long tick marks are hundredths.

c. Plot −1.584 on a number line on which the long tick marks are tenths.

***12.** Give examples of decimal numbers that cannot be represented with bundles of toothpicks—even if you had as many toothpicks as you wanted.

Answers to Practice Exercises for Section 1.2

1. The decimal 1.234 can be represented with bundled objects, as a length, or on a number line. To represent 1.234 with bundled toothpicks, let 1 toothpick represent one thousandth. Then 1.234 is represented with 4 single toothpicks, 3 bundles of ten, 2 bundles of one hundred (each of which is 10 bundles of 10) and 1 bundle of a thousand (which is 10 bundles of 100). See Figure 1.18 representing 1.234 as a length.

2. Represent 0.0278 as 2 bundles of 100 objects (each of which is 10 bundles of 10), 7 bundles of 10 objects, and 8 individual objects. In this case, each individual object must represent one ten-thousandth, since 0.0278 is 2 hundredths and 7 thousandths and 8 ten-thousandths.

3. If one toothpick represents 1, then Figure 1.27 represents 214. The following table shows several other possibilities:

1 Toothpick Represents	Figure 1.27 Represents
100	21,400
10	2140
1	214
0.1	21.4
0.01	2.14
0.001	0.214

4. Let a long strip of paper (e.g., 1 meter long) represent 1 unit of length, as at the top of Figure 1.32. Make another 1-unit-long strip, and subdivide it into 10 strips of equal length. Each of these strips is then 0.1 unit long. Make another 0.1-unit-long

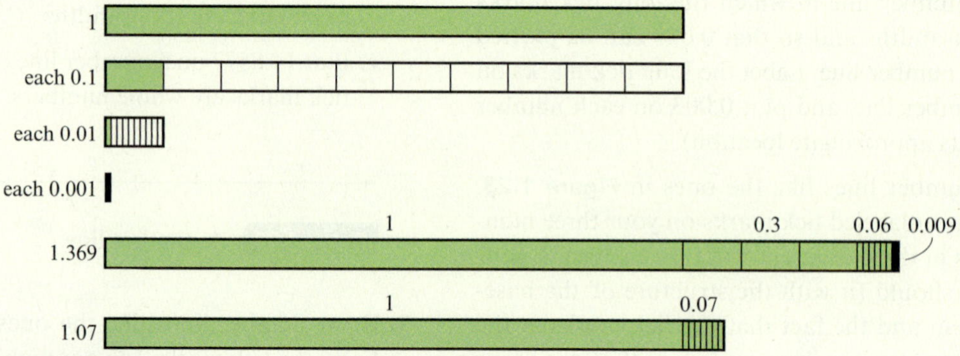

Figure 1.32 Representing decimals as lengths.

strip, and subdivide this strip into 10 strips of equal length. Each of these smaller strips is then 0.01 unit long. Make another 0.01-unit-long strip, and subdivide this strip into 10 strips of equal length. Each of these tiny strips is then 0.001 unit long.

To represent 1.369 as a length, make a long strip by laying the 1-unit-long strip next to 3 of the 0.1-unit-long strips, 6 of the 0.01-unit-long strips, and 9 of the 0.001-unit-long strips, as in the middle of Figure 1.32.

To represent 1.07 as a length, make a long strip by laying the 1-unit-long strip next to 7 of the 0.01-unit-long strips. (Don't use any 0.1-unit-long strips or 0.001-unit-long strips.) See the bottom of Figure 1.32.

5. See Figure 1.33.

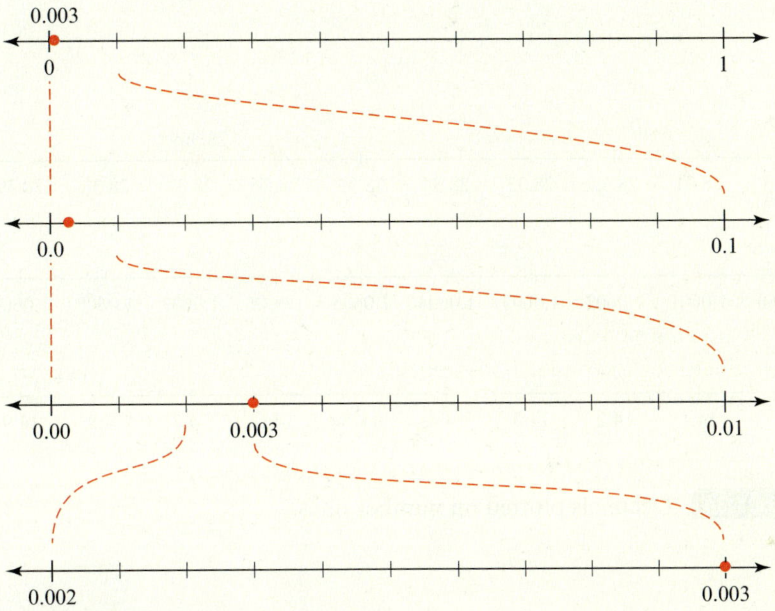

Figure 1.33 Zooming in on 0.003.

6. See Figure 1.34 for one way to label the tick marks. We can think of the second number line as "zoomed in" on the left portion of the first, and we can think of the third number line as "zoomed in" on the left portion of the second.

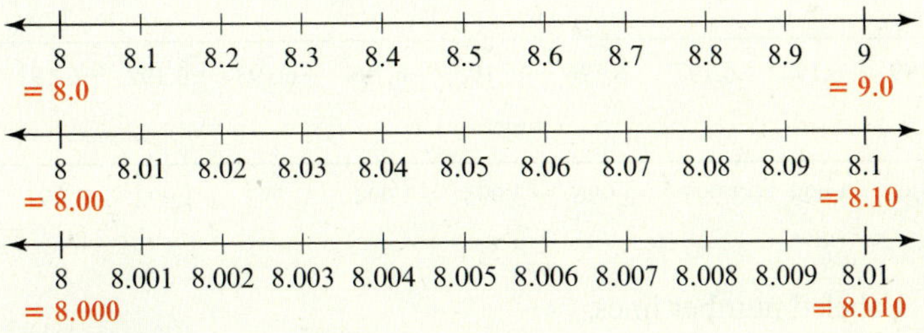

Figure 1.34 Labeled number lines.

7. See Figure 1.35 for one way to label the tick marks. We can think of the second number line as "zoomed in" on the right portion of the first and the third number line as "zoomed in" on the second.

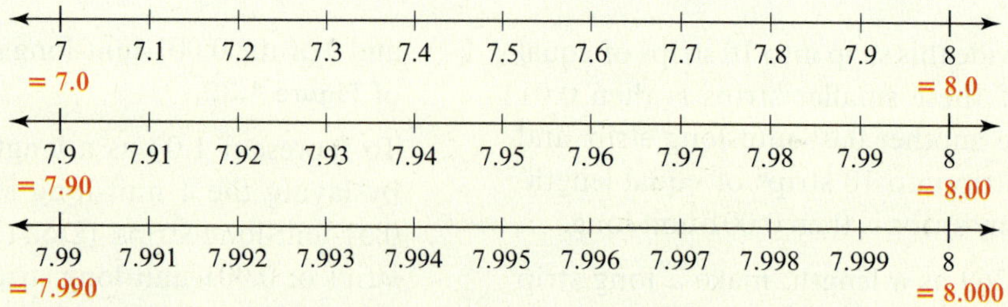

Figure 1.35 Labeled number lines.

8. See **Figure 1.36**.

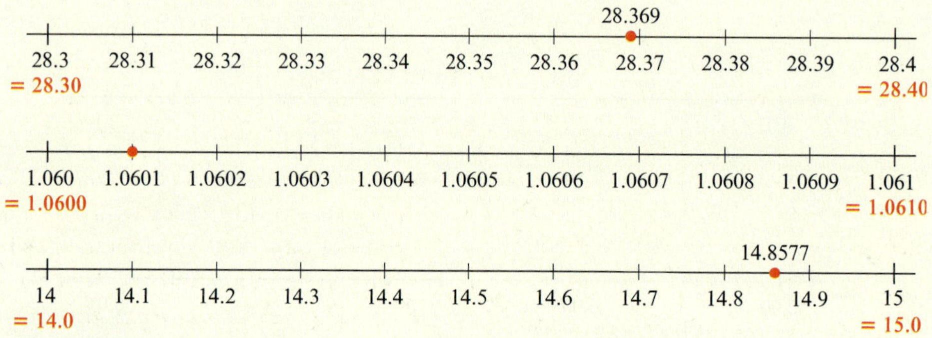

Figure 1.36 Decimals plotted on number lines.

9. See **Figure 1.37**.

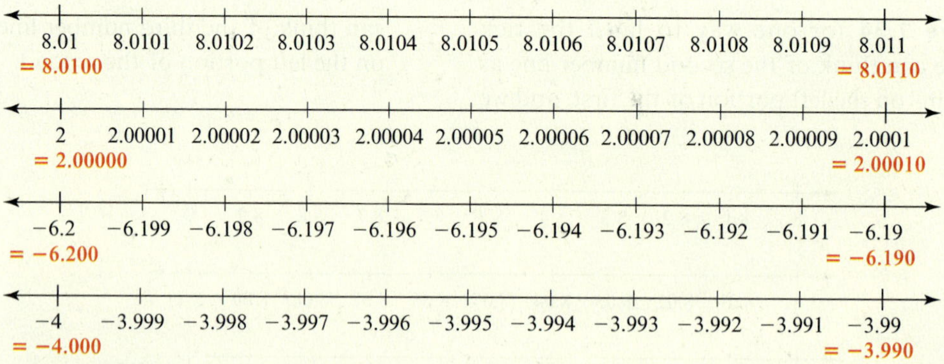

Figure 1.37 Labeled number lines.

10. The negative numbers are located to the left of 0 on the number line. The number -2 is located 2 units from 0 to the left of 0 (and recall that the distance of 1 unit is the distance between 0 and 1). See **Figure 1.25** for how to locate -2.41.

11. See Figure 1.38.

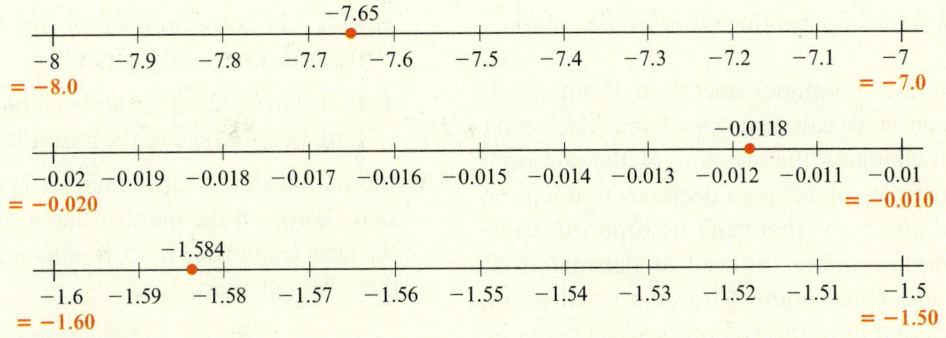

Figure 1.38 Negative numbers plotted on number lines.

12. Realistically, we would be hard-pressed to represent numbers with more than 4 nonzero digits with toothpicks. Even 999 would be difficult to represent. However, there are some numbers whose expanded form *can't* be represented by toothpicks (in the manner described in the text)—even if you had as many toothpicks as you wanted. For example, consider

$$0.333333333\ldots$$

where the 3s go on forever. Which place's value would you pick to be represented by 1 toothpick? If you did pick such a place, you would have to represent the values of the places to the right by tenths of a toothpick, hundredths of a toothpick, thousandths of a toothpick, and so on, forever, in order to represent this number.

As an aside, here is a surprising fact: We *can* represent 0.333333333 . . . with toothpicks, but in a different way (not by bundling so as to show the places in the decimal number). Namely, we can represent 0.333333333 . . . by one third of a toothpick, because it so happens that one third = 0.333333333 . . . (which you can see by dividing 1 by 3).

PROBLEMS FOR SECTION 1.2

1. Suppose you want to show how the structure of the base-ten system remains the same to the left and right of the decimal point. You have bundles of toothpicks like the ones shown in Figure 1.39 and you want to use these bundled toothpicks to represent a decimal. List at least three decimals that you could use these bundles to represent and explain your answer in each case.

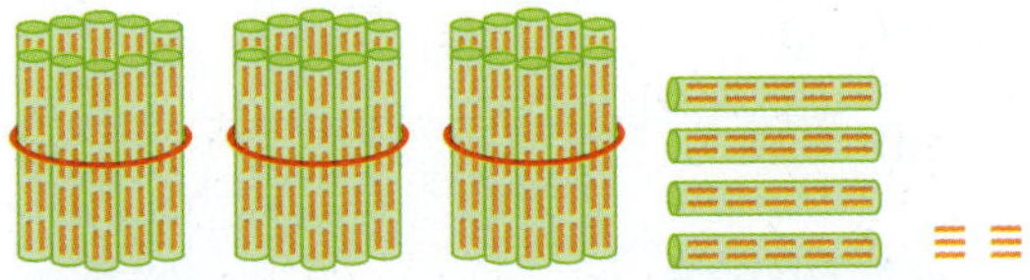

Figure 1.39 Which decimals can these bundled toothpicks represent?

2. Make math drawings of small bundled objects to show how to represent the accompanying decimals. Your drawings should correspond to the decimal representation of the numbers. In each case, list two other decimals that your drawing could represent. Explain how to interpret your drawings as representing these decimals.

a. 0.26

b. 13.4

c. 1.28

d. 0.000032

3. Describe and make drawings showing how to represent 1.438 and 0.804 as lengths by using strips of paper in a way that fits with and shows the structure of the base-ten system.

4. Jerome says that the unlabeled tick mark on the number line in Figure 1.40 should be 7.10. Why might Jerome think this? Explain to Jerome why he is not correct. Describe how you could help Jerome understand the correct answer.

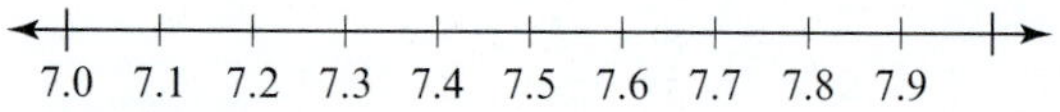

Figure 1.40 How to label the unlabeled tick mark.

5. Students are sometimes uncertain about which zeros in decimals can be dropped and which can't (without changing the meaning of the number). Give examples of zeros in decimals that can be dropped and zeros that can't be dropped. Consider whole numbers as well as decimals that require a decimal point. Choose a small set of examples that nevertheless covers all the types of cases where zeros can be dropped and where they can't.

6. Draw a number line on which the tick marks are whole numbers. Now show how to zoom in on smaller and smaller portions of the number line (as in Figure 1.22 and Class Activity 1F, part 1) until you have zoomed in to a portion of the number line in which the long tick marks are thousandths. (You may choose which portions of the number line to zoom in on, but of course, your example should be different from others you have seen.)

7. Draw a number line on which the long tick marks are whole numbers and on which 7.0028 can be plotted (in its approximate location). Show how to zoom in on smaller and smaller portions of the number line (as in Figure 1.22 and Class Activity 1F, part 1) until you have zoomed in to a portion of the number line in which the long tick marks are thousandths and so that 7.0028 can be plotted on each number line. Label the long tick marks on each number line, and plot 7.0028 on each number line (in its approximate location).

8. Use a number line like the one in Figure 1.41 for the problems that follow. For each problem, label all the tick marks on the number line. The number to be plotted need not land on a tick mark.

Figure 1.41 A number line.

a. Plot 13.58 on a number line on which the long tick marks are whole numbers.

b. Plot 0.193 on a number line on which the long tick marks are tenths.

c. Plot 26.9999 on a number line on which the long tick marks are thousandths.

d. Plot 2.379 on a number line on which the long tick marks are tenths.

e. Plot 7.148 on a number line on which the long tick marks are whole numbers.

f. Plot 9.075132 on a number line on which the long tick marks are thousandths.

9. Cierral plots the decimal number 7.001 in the location shown on the number line in Figure 1.42. Is the label legitimate? If so, how should Cierral label the other tick marks?

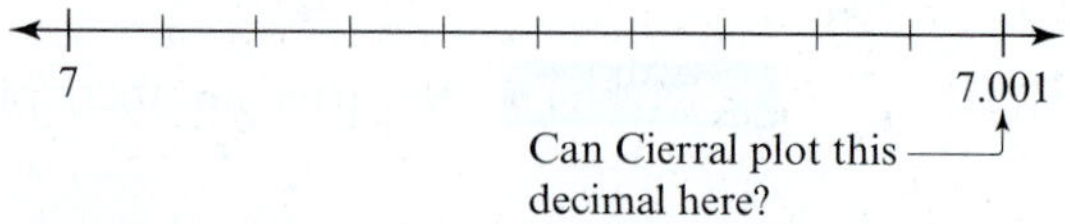

Figure 1.42 How to label the tick marks.

10. Juan plots the decimal number 9.999 in the location shown on the number line in Figure 1.43. Is this label legitimate? If so, how should Juan label the other tick marks?

Figure 1.43 How to label the tick marks.

11. For each number line in Figure 1.44 (a)–(d), draw three copies of the line. Use your number lines to show three different ways to label the tick marks in the original line. In each case, your labeling should fit with the structure of the base-ten system and the fact that the tick marks at the ends of the number lines are longer than the other tick marks.

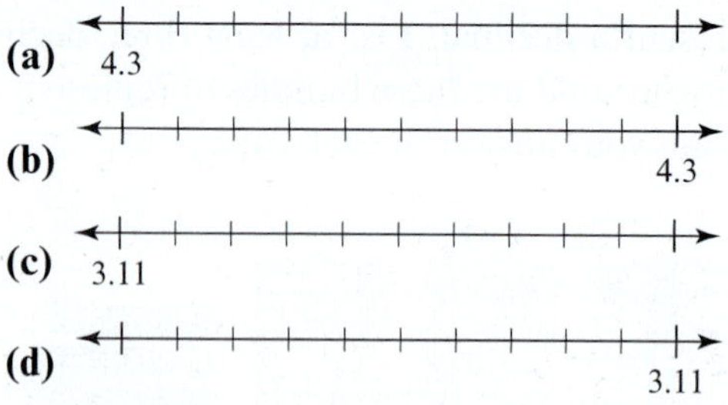

Figure 1.44 How to label the tick marks.

12. Using −1, −2, and −1.68 as examples, describe in your own words where negative numbers are located on a number line, based on the locations of 0 and 1.

13. Students sometimes get confused about the locations of decimals and negative numbers on number lines. Draw a number line on which you have plotted 0, 1, and -1. Then give at least three examples of numbers that are between 0 and 1 and at least three examples of numbers that are between 0 and -1. Plot all your examples on your number line. Pick examples that give students some sense of the variety of numbers that are between 0 and 1 and between 0 and -1. Why do you think your choice of examples will be good for students to see?

14. Draw a number line like the one in Figure 1.41 for the problems that follow. For each problem, label all the tick marks on the number line. The number to be plotted need not land on a tick mark.

a. Plot -4.3 on a number line on which the long tick marks are whole numbers.

b. Plot -0.28 on a number line on which the long tick marks are whole numbers.

c. Plot -0.28 on a number line on which the long tick marks are tenths.

d. Plot -6.193 on a number line on which the long tick marks are tenths.

e. Plot -6.193 on a number line on which the long tick marks are hundredths.

***15.** Explain why it is the case that whenever N is a negative number, $|N| = -N$.

1.3 Reasoning to Compare Numbers in Base Ten

CCSS Common Core State Standards Grades 2, 4, 5, 6

If Timothy Elementary School raised \$1023 for a fundraiser and Barrow Elementary School raised \$789, which school raised more money? Of course you know right away that Timothy Elementary raised more, but why can we compare numbers the way we do? Our way for comparing numbers relies on the nature of the base-ten system, which we will investigate in this section. We also discuss the concepts of "greater than" and "less than" and the idea of comparing numbers by viewing them as quantities and as locations on number lines.

How Can We Compare Nonnegative Numbers by Viewing Them as Amounts?

If A and B are nonnegative numbers such that A represents a larger quantity than B, then we say that A is **greater than** B, and we write

greater than

$$A > B$$

less than

Similarly, if A represents a smaller quantity than B, we say that A is **less than** B, and we write

$$A < B$$

CCSS
1.NBT.3
2.NBT.4

A good way to remember which symbol to use is to notice that the "wide side" faces the larger number. Some teachers have their students draw alligator teeth on the symbols, as in Figure 1.45, to help them: Alligators are hungry and always want to eat the larger amount.

inequality

A statement of the form $A > B$ or $A < B$ is called an **inequality**. An example of an inequality is $7 > 3$ or $3 < 7$.

Figure 1.45
Alligator teeth on a "less than" symbol.

Why Can We Compare Numbers the Way We Do?

Think for a moment about how you compare two positive numbers to decide which one is greater. For example, how do you know that 584,397 is less than 600,214? As adults, we are used to comparing numbers, so we usually don't stop to think about why it works. But how do we know that we can compare numbers by starting at the left-most place? The answer lies in the nature of the base-ten system.

CCSS 4.NBT.2

In the base-ten system, the value of a place is always greater than the largest number that can be made from all the lower places to its right.[2] We can see this by thinking about place values in terms of bundling base-ten units. Think about counting a collection of toothpicks one by one and organizing them into bundles to show how many you have counted so far:

- 9 toothpicks "fill up" the ones place; one more toothpick makes 1 bundle of ten in the next place to the left.
- 99 toothpicks "fill up" the tens and ones places; one more toothpick makes 1 bundle of a hundred in the next place to the left.
- 999 toothpicks "fill up" the hundreds, tens, and ones places; one more toothpick makes 1 bundle of a thousand in the next place to the left.

So even if you make the largest possible number that uses only the hundreds, tens, and ones places, this number—999—is still smaller than the smallest number you can make using a nonzero entry in the thousands place—namely, 1000. So the thousands place counts more than the lower hundreds place, tens place, and ones place combined.

CCSS 5.NBT.3b

Because the structure of the base-ten system is uniform, the situation is the same for decimals as for whole numbers. For example, even if we make the largest possible number that uses only the hundredths and thousandths places—namely, 0.099—this number is still less than the smallest number we can make using a nonzero entry in the tenths place—namely, 0.1. (See Figure 1.46.)

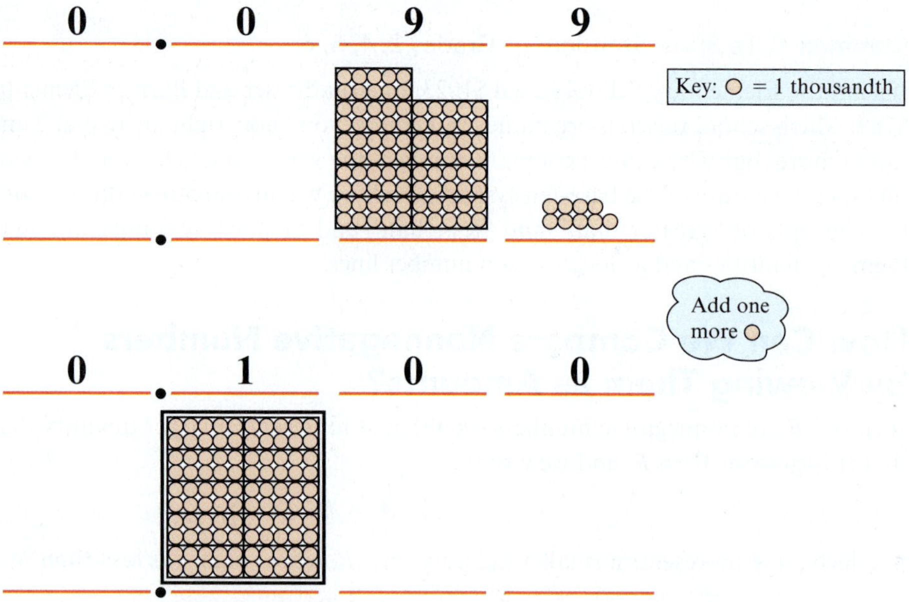

Figure 1.46 Higher place values count more than the largest number that can be made from lower place values.

In general, *the value of a place is greater than the largest number made with places of smaller values.* Therefore, to compare the sizes of numbers in base-ten, we compare like places, starting at the place of largest value in which at least one of the numbers has a nonzero entry.

For example,

$$1234 > 789$$

because 1234 has a 1 in the thousands place and 789 has 0 in the thousands place (even though this zero is usually not written).

$$\begin{array}{c} 1234 \\ \updownarrow \\ 0789 \end{array}$$

[2]See the technical exception on page 29.

Similarly,

$$1.2378 < 1.24$$

because both numbers have a 1 in the ones place and a 2 in the tenths place, but 1.24 has a larger digit, namely, 4, in the hundredths place than does 1.2378, which only has a 3.

$$\begin{array}{c} 1.2\,3\,7\,8 \\ \updownarrow\,\updownarrow\,\updownarrow \\ 1.2\,4 \end{array}$$

Notice that determining the greater number is similar to putting words in alphabetical order, except that for numbers in base ten, we compare values in *like places*, whereas for words, we compare letters starting from the *beginning* of each word. Maybe this is why a decimal such as 1.01 might at first glance look as if it is less than 0.998.

IMAP *Watch Megan and Donna compare decimals and see a common misconception about ordering decimals.*

CLASS ACTIVITY

1I Critique Reasoning About Comparing Decimals, p. CA-14

1J Finding Decimals Between Decimals, p. CA-15

A Technical Exception to the Rule for Comparing Numbers in Base Ten There is a rare exception to the preceding method for determining which of two numbers in base-ten is greater. The exception occurs when one of the numbers is a decimal that has an infinitely repeating 9, such as

$$37.569999999999\ldots$$

In this case, the preceding method would lead us to say that

$$37.57 > 37.569999999999\ldots$$

However, this is *not true*. In fact, as we will see in Section 8.6,

$$37.57 = 37.569999999999\ldots$$

How Can Number Lines Help Us to Compare Numbers?

It is especially easy to compare numbers that have been plotted on a number line. When we use a number line, we can compare negative numbers as well as positive ones.

CCSS 6.NS.7a

If A and B are numbers plotted on a number line, then

$A < B$ provided that A is to the left of B on the number line.

$A > B$ provided that A is to the right of B on the number line.

These interpretations of "greater than" and "less than" are consistent with the interpretation we developed previously for positive numbers, where we viewed positive numbers as representing quantities. Why? If a positive number A represents a larger quantity than another positive number B, then A will be plotted farther from 0 than B. Since both are plotted to the right of 0, A will be to the right of B. Likewise, if A represents a smaller quantity than B, then we will plot A closer to 0 than B. Since both are to the right of 0, A will be to the left of B as shown in Figure 1.47.

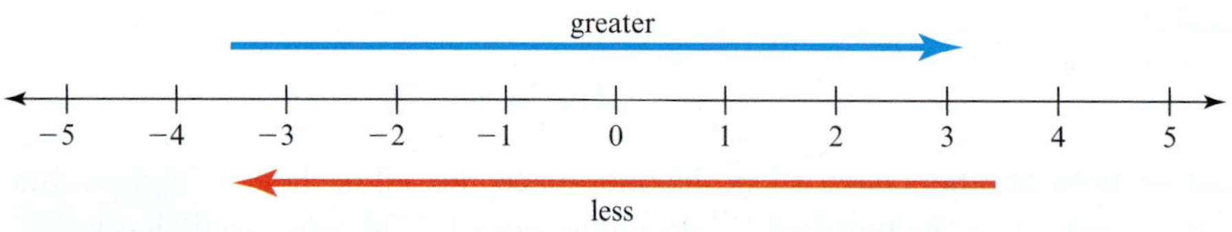

Figure 1.47 Going to the right on the number line, numbers become greater; going to the left, numbers become less.

With the number line interpretation of $>$ and $<$, we can also compare negative numbers. For example, we plot -4 to the left of -3, as in **Figure 1.48**; therefore, $-4 < -3$. By thinking of a number and its negative as opposites that are the same distance from 0, we can see why the statement

$$-4 < -3$$

is consistent with the statement

$$4 > 3$$

The numbers 4 and -4 are farther from 0 than are 3 and -3, as **Figure 1.48** shows. But 4 and 3 are to the *right* of 0, making 4 *greater than* 3. In contrast, -4 and -3 are to the *left* of 0, making -4 *less than* -3.

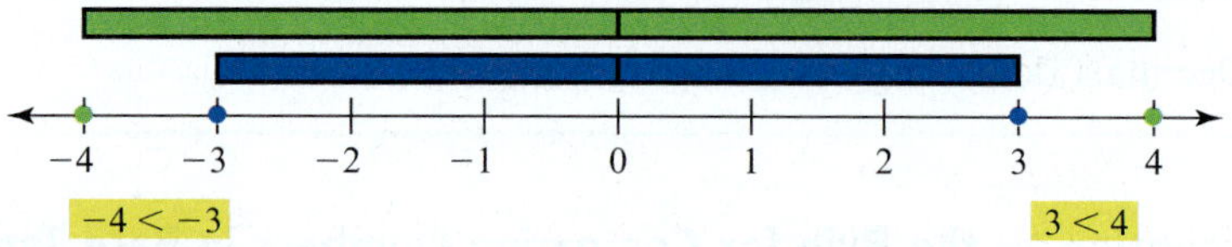

Figure 1.48 Showing that $-4 < -3$ is consistent with $3 < 4$.

CCSS
6.NS.7b

How Can We Use Real-World Contexts to Compare Negative Numbers?

In addition to explaining why $-4 < -3$ by using a number line, we can explain why $-4 < -3$ by viewing the negative numbers in real-world contexts such as locations below ground level or as amounts owed. Think of -4 as owing \$4 and -3 as owing \$3. If you owe \$4, you have less than if you owe \$3; therefore,

$$-4 < -3$$

CLASS ACTIVITY

1K Decimals Between Decimals on Number Lines, p. CA-16

1L "Greater Than" and "Less Than" with Negative Numbers, p. CA-17

SECTION SUMMARY AND STUDY ITEMS

Section 1.3 Reasoning to Compare Numbers in Base Ten

The value of a place is greater than the largest number made from places of lower value. For this reason, we compare numbers in the base-ten system by looking first at the place of greatest value, then moving to places of lower value if the numbers agree in the place of greatest value. On number lines, numbers become greater as one moves to the right. Given any two distinct numbers, there is always another number in between. (In fact, there are infinitely many numbers in between.) Taking the negative of two numbers reverses the comparison between them.

Key Skills and Understandings

1. Given any two numbers in base ten (including negative numbers), determine which is greater, and put any collection of numbers in order from least to greatest (or vice versa).

2. Explain the rationale for comparing positive numbers in base ten by first comparing the place of greatest value (and then moving to places of lower value).
3. Draw rough pictures of bundled objects to show that one number is greater than another.
4. Describe and draw rough pictures representing numbers as lengths and showing that one number is greater than another.
5. Use a number line to demonstrate that one number is greater than another (including negative numbers).
6. Describe negative numbers as owed amounts to explain why one negative number is greater than another.
7. Given two distinct numbers in base ten (positive or negative or 0), find another number in between them.

Practice Exercises for Section 1.3

1. Which is greater, 0.01 or 0.0099999999999?
2. Make a math drawing that shows bundled objects representing 1.2 and 0.89 and demonstrates that 1.2 > 0.89.
3. Describe and make math drawings showing how to represent 1.2 and 0.89 as lengths by using strips of paper that are 1 unit long, 0.1 unit long, 0.01 unit long, and 0.001 unit long as in Class Activity 1E. Use these representations to explain why 1.2 > 0.89.
4. Use a number line to show that 1.2 > 0.89.
5. Use a number line to show that −1.2 < −0.89.
6. Explain why −1.2 < −0.89, without using a number line.
7. Find a number between 1.4142133 and 1.41422, and plot all three numbers visibly and distinctly on a number line in which the tick marks fit with the structure of the base-ten system. The numbers need not land on tick marks.
8. Give two different decimals that are between 3.456 and 3.457.
9. Children who have heard of a googol, which is the number that is written as a 1 followed by one hundred zeros, will often think it's the largest number. Is it? Is there a largest number?
10. The smallest whole number that is greater than zero is 1. Is there a smallest decimal that is greater than zero?

Answers to Practice Exercises for Section 1.3

1. The number 0.01 is greater because it has a 1 in the hundredths place, whereas 0.0099999999999 has a 0 in the hundredths place and all higher places.
2. See Figure 1.49.

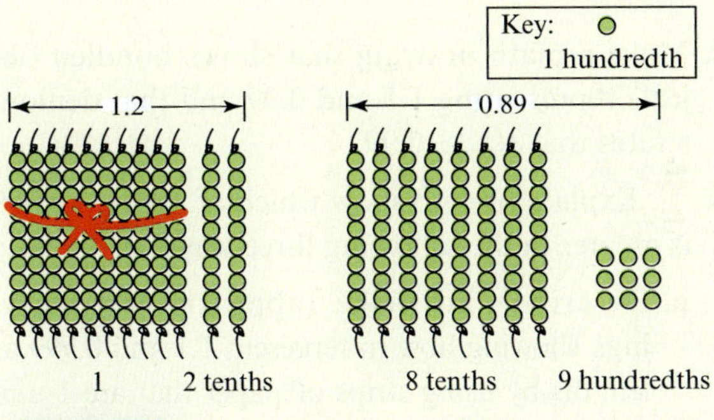

Figure 1.49 1.2 > 0.89.

3. Let a long strip of paper (e.g., 1 meter long) represent 1 unit of length, as at the top of Figure 1.50. Make another 1-unit-long strip, and subdivide it

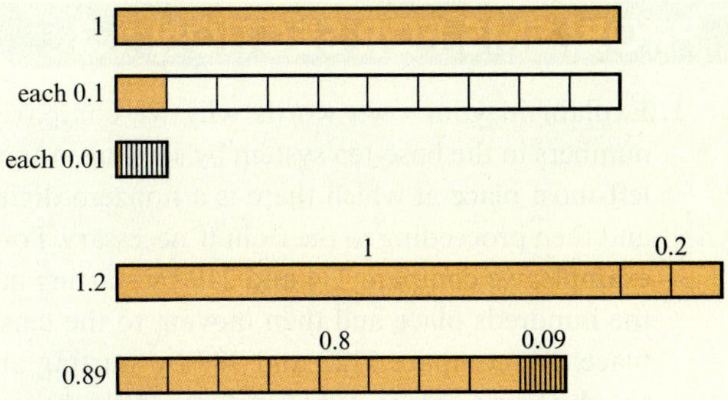

Figure 1.50 Representing decimal numbers as lengths.

into 10 strips of equal length. Each of these strips is then 0.1 unit long. Make another 0.1-unit-long strip, and subdivide this strip into 10 strips of equal length. Each of these smaller strips is then 0.01 unit long.

See the bottom of Figure 1.50 for how to represent 1.2 and 0.89 as lengths. Since the 1.2-unit-long strip is longer than the 0.89-unit-long strip, we conclude that $1.2 > 0.89$.

4. See Figure 1.51.

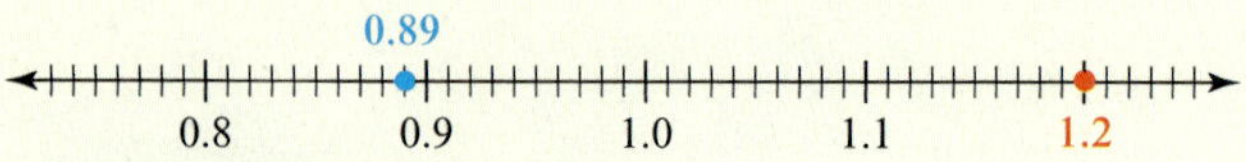

Figure 1.51 $1.2 > 0.89$ because 1.2 is to the right of 0.89.

5. See Figure 1.52.

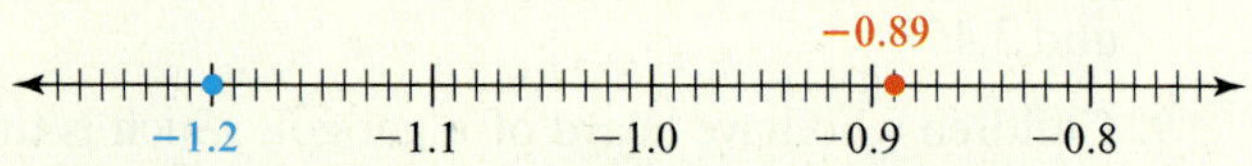

Figure 1.52 $-1.2 < -0.89$ because -1.2 is to the left of -0.89.

6. To explain why $-1.2 < -0.89$, consider having -1.2 as owing \$1.20 and having -0.89 as owing \$0.89. If you owe \$1.20, then you have less than if you owe only \$0.89. Therefore, having -1.2 means you have less than if you have -0.89. So, $-1.2 < -0.89$.

7. The number 1.414215 is one example of a number that is in between 1.4142133 and 1.41422. There are many other examples. See Figure 1.53.

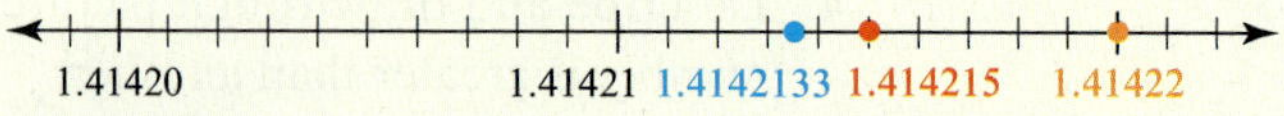

Figure 1.53 The number 1.414215 is between 1.4142133 and 1.41422.

8. The numbers 3.4563 and 3.4567 both lie in between 3.456 and 3.457.

9. The number a googol plus one, for example, is greater than a googol. There is no largest number because, no matter what number you choose as candidate for the largest number, that number plus one is a larger number.

10. No, there is no smallest decimal that is greater than zero. Consider the following list of numbers:

0.1
0.01
0.001
0.0001
0.00001
0.000001

Imagine the list continuing forever. The numbers in this list get smaller and smaller, getting ever closer to 0 without ever reaching 0. No matter what decimal you choose that is greater than 0, it will have a nonzero entry somewhere. Based on where the first nonzero entry in your chosen number is, you can pick a decimal from the preceding list that is even smaller than your chosen number. Therefore, there can be no smallest decimal that is greater than zero.

PROBLEMS FOR SECTION 1.3

1. Explain in your own words why we compare numbers in the base-ten system by starting at the left-most place at which there is a nonzero digit and then proceeding to the right if necessary. For example, we compare 234 and 219 by starting at the hundreds place and then moving to the tens place. We compare 1122 and 987 by starting at the thousands place. What is it about the base-ten system that allows us to compare numbers this way?

2. Make a math drawing that shows bundled objects representing 1.1 and 0.99 and that demonstrates that $1.1 > 0.99$.

3. Explain how to show which of 1.1 and 0.999 is greater in the following three ways:

a. Describe and make (approximate) drawings showing how to represent 1.1 and 0.999 as lengths by using strips of paper that are 1 unit long, 0.1 unit long, 0.01 unit long, and 0.001 unit long as in Class Activity 1E. Use these representations to show which of 1.1 and 0.999 is greater.

b. Use a carefully drawn number line to show which of 1.1 and 0.999 is greater. You may wish to "zoom in" on portions of the number line.

c. Make math drawings showing how to represent 1.1 and 0.999 with bundled objects in a way that fits with the structure of the base-ten system. Use these representations to show which of 1.1 and 0.999 is greater.

4. Some students have difficulty comparing decimals with 0 or comparing decimals that have zeros in some places (see [88] and [83]). Plot each of the given pairs or triples of decimals on a carefully drawn number line to show which of the pair or triple is greatest. You may "zoom in" on portions of your number lines in order to show locations more clearly.

a. Compare 0.6 and 0.

b. Compare 0.00 and 0.7.

c. Compare 3.00, 3.0, and 3.

d. Compare 3.7777 and 3.77.

5. Mary is labeling tick marks on a number line. Starting at a tick mark labeled 4.90, she labels the next tick marks to the right with 4.91, 4.92, 4.93, . . . , 4.99. Then she labels the next tick mark to the right of 4.99 with 4.100. Mary then says that 4.100 is greater than 4.99.

a. Discuss what error Mary is making. Why might Mary make this error?

b. Describe how you might help Mary understand her error and how to correct it.

6. Mark says that 0.178 is greater than 0.25. Why might Mark think this? Explain in at least two different ways why Mark's statement is not correct.

7. Find a number between 3.24 and 3.241, if there is one. If there is no number between them, explain why not.

8. Is there more than one decimal between 8.45 and 8.47? If so, find two such decimals.

9. a. Find a number between 3.8 and 3.9, and plot all three numbers visibly and distinctly on a number line like the one in Figure 1.54. Your labeling of the tick marks should fit with the structure of the base-ten system.

b. Describe how to use money to find a number between 3.8 and 3.9.

Figure 1.54 A number line with long and short tick marks.

10. For each of the following pairs of numbers, find a decimal between the two numbers, and plot all three numbers visibly and distinctly on a number line like the one in Figure 1.54. Label all the longer tick marks. Your labeling of the tick marks should fit with the structure of the base-ten system.

a. The numbers 2.981 and 2.982

b. The numbers 13 and 12.9999

c. The numbers 13 and 13.0001

11. Explain in two different ways why $-8 < -5$.

12. Explain in two different ways why $-3.25 < -1.4$.

13. Some students confuse decimals and negative numbers and mix up the locations of 0 and 1 when comparing decimals (see [83]).

a. Plot 0, 1, 0.6, -0.7, and -0.06 on a carefully drawn number line. You may "zoom in" on portions of the number line in order to show them more clearly.

b. Put 0, 1, 0.6, -0.7, and -0.06 in order from least to greatest.

14. For each of the following pairs of numbers, find a decimal between the two numbers, and plot all three numbers visibly and distinctly on a number line like the one in Figure 1.54. Label all the longer tick marks. Your labeling of the tick marks should fit with the structure of the base-ten system.

a. The numbers -1.5 and -1.6

b. The numbers -34.9714 and -34.9835

c. The numbers -7.834 and -7.83561

***15.** Describe an infinite list of decimals, all of which are greater than 3.514, but get closer and closer to 3.514.

*16. The smallest integer that is greater than 2 is 3. Is there a smallest decimal that is greater than 2? Explain why or why not in your own words.

*17. On the number line in Figure 1.55, assume that $-A$ is negative and B is positive. Where can 0 be located on that number line if $A < B$? Explain your answer.

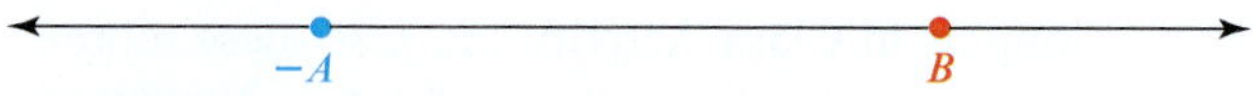

Figure 1.55 Points $-A$ and B on a number line.

*18. On the number line in Figure 1.55, assume that $-A$ is negative and B is positive. Where can 0 be located on that number line if $A > B$? Explain your answer.

1.4 Reasoning about Rounding

CCSS Common Core State Standards Grades 3, 4, 5

Why do we round numbers? Sometimes we round because we don't know a quantity exactly, and sometimes we round because we want to convey the approximate size of a quantity. For example, a company might know that it spent $174,586.74 on some equipment, but in a report to management, an employee might describe the expenditure as $175,000. This rounded number is more quickly and easily grasped and is close enough to the actual expenditure for the discussion at hand. We also round numbers to estimate the result of a calculation. For example, to estimate 5.9×8.2 we can round 5.9 to 6 and 8.2 to 8, estimating 5.9×8.2 to be about 6×8, which is 48.

How Can We Use Place Value Understanding to Round Numbers?

round

CCSS 3.NBT.1

To **round** a number in base ten means to find a nearby number that has fewer (or no more) nonzero digits. When we round a number, we ignore the "detail" that the values in the lower places provide and we only focus on the more significant portion of the number. We can round to the nearest 100, to the nearest 10, to the nearest 1, to the nearest tenth, to the nearest hundredth, and so on. For any place value, we can round to that place's value. To round a given number to the nearest 100, we must find the number that is closest to our given number and has only zeros in places smaller than the hundreds place. To round a given number to the nearest tenth, we must find the number that is closest to our given number and has only zeros in places smaller than the tenths place. In general, to round a given number to a given place's value, we must find the number that is closest to our given number and that has only zeros in the smaller places.

To understand rounding, think about zooming out on a number line so that you can see the tick marks that have the desired place value. When rounding a number to the nearest hundred, we zoom out to a number line with tick marks representing hundreds as in Figure 1.56. Then we ask which tick mark the number is closest to. The number A in the figure is closer to 3600 than to 3700, so A rounded to the nearest hundred is 3600. The number B in the figure is closer to 3700 than to 3600, so B rounded to the nearest hundred is 3700.

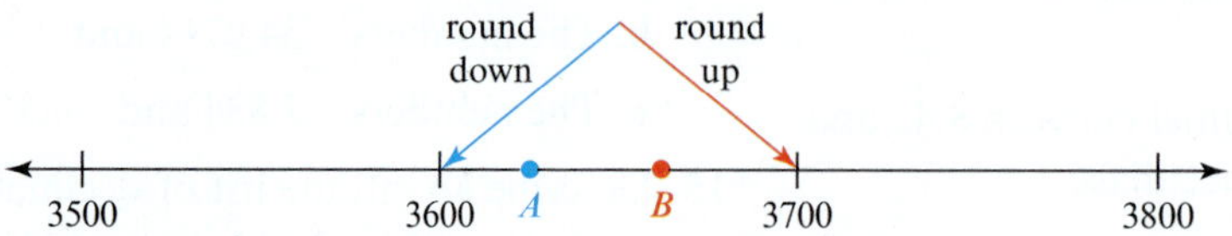

Figure 1.56 Rounding to the nearest hundred on a number line labeled with tick marks representing hundreds.

Some teachers like to show their students a "mountain" like the one in Figure 1.56. The numbers on the right part of an interval "roll" toward the higher tick mark and numbers on the left part of an interval "roll" toward the lower tick mark.

A number line perspective can help us explain why the usual method for rounding a number makes sense. To round 14,378 to the nearest thousand, we first observe that the number lies between these two thousands: 14,000 and 15,000. Which thousand is 14,378 closer to? To answer this we look at the digit in the next lower place, which is the hundreds place. See Figure 1.57. Because the digit is 3, the number 14,378 is closer to 14,000 than to 15,000. Therefore 14,378 rounded to the nearest thousand is 14,000.

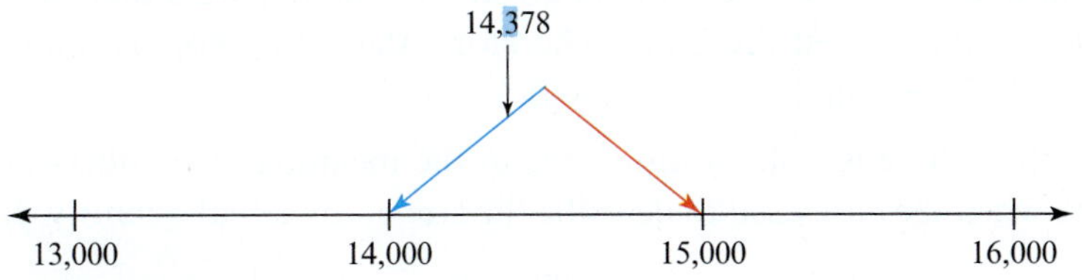

Figure 1.57 Rounding 14,378 to the nearest thousand by thinking about placement on a number line labeled with tick marks representing thousands.

CCSS 5.NBT.4

In general, by looking to the value of the place immediately to the right of the place we are rounding to, we can determine whether to round up or down: a digit less than 5 means the number is closer to the smaller value, whereas a digit greater than 5 means the number is closer to the larger value. Figure 1.58 shows how 1.382 rounded to the nearest tenth is 1.4 because the 8 in the hundredths place means the number is closer to 1.4 than to 1.3.

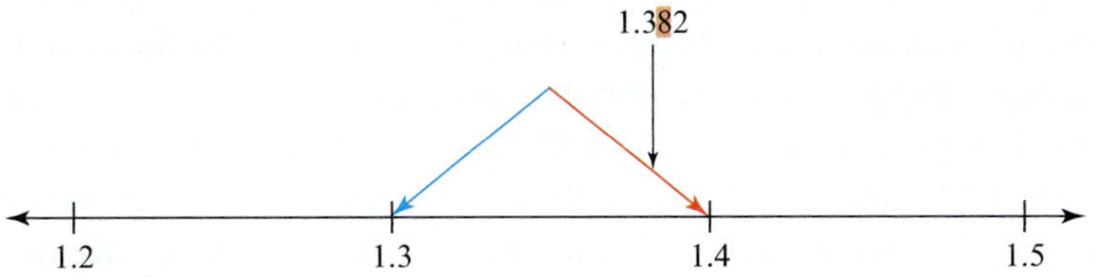

Figure 1.58 Rounding 1.382 to the nearest tenth by thinking about placement on a number line labeled with tick marks representing tenths.

What do we do about numbers that are exactly halfway in between two tick marks? That is, what do we do when the digit is 5? For example, how do we round 14,500 when we are rounding to the nearest thousand? How do we round 1.35 when we are rounding to the nearest tenth? We need a convention to break ties like these. The most common convention, and the one you should use unless otherwise specified, is to *round up* when there is a tie. So when rounding 14,500 to the nearest thousand, round to 15,000. When rounding 1.35 to the nearest tenth, round to 1.4.

CLASS ACTIVITY

1M Explaining Rounding, p. CA-18

1N Rounding with Number Lines, p. CA-19

1O Can We Round This Way?, p. CA-20

CCSS
SMP6

What Is the Significance of Rounding When Working with Numbers That Represent Actual Quantities?

When we use a number to describe the size of an actual quantity, such as a distance or a population, we generally assume that this number has been rounded. Furthermore, we assume that *the way a number is written indicates the rounding that has taken place.* For example, when the distance between two locations is described as 6.2 miles, we assume from the presence of a digit in the tenths place, but no smaller place, that the actual distance has been rounded to the nearest tenth of a mile. Therefore, the actual distance could be anywhere between 6.15 and 6.25 miles (but less than 6.25 miles). Similarly, if the population of a city is described as 93,000, we assume from the nonzero entry in the thousands place and the zeros in all smaller value places that the actual population has been rounded to the nearest thousand. Therefore, the actual population could be anywhere between 92,500 and 93,500 (but less than 93,500).

Because of rounding, there is a slight difference in the meaning of numbers in the abstract and decimal numbers when they are used to describe the size of an actual quantity. In the abstract,

$$8.00 = 8$$

However, when we write 8 for the weight of an object in kilograms, it has a different meaning than when we write 8.00 for the weight of an object in kilograms. Writing 8 indicates that the actual weight has been rounded to the nearest one, whereas writing 8.00 means that the actual weight has been rounded to the nearest hundredth. If the weight is reported as 8.00 kilograms, then the actual weight could be anywhere between 7.995 and 8.005 kilograms (but less than 8.005 kilograms). If, however, the weight is reported as 8 kilograms, then the actual weight could be anywhere between 7.5 and 8.5 kilograms (but less than 8.5 kilograms). Therefore, writing 8.00 conveys that the weight is known much more accurately than if the weight is reported as 8 kilograms.

When you solve a problem that involves real or realistic quantities, round your answer so that it does not appear to be more accurate than it actually is. Your answer cannot be any more accurate than the numbers you started with, so round your answer to fit the rounding of your initial numbers. For example, suppose that the population of a city is given as 1.6 million people. After some calculations, you project that the city will have 1.95039107199 million people in 10 years. Although the decimal 1.95039107199 may be the exact answer to your calculations, do not report your answer this way, because it makes your answer appear to be more accurate than it actually is. You must assume that the initial number 1.6 is rounded to the nearest tenth. Therefore, you should also round the answer, 1.95039107199, to the nearest tenth and report the projected population in 10 years as 2.0 million.

SECTION SUMMARY AND STUDY ITEMS

Section 1.4 Reasoning about Rounding

To round a number to a given place's value means to find the closest number that has zeros in all smaller places. When we work with actual quantities, the way the number is written generally indicates the rounding that has taken place and, therefore, the precision with which the quantity is known.

Key Skills and Understandings

1. Given any number, round it to a given place.
2. Use an appropriate number line to explain how to round a given number to a given place.
3. Recognize that numbers representing actual quantities have generally been rounded.

Practice Exercises for Section 1.4

1. Round 6.248 to the nearest tenth. Explain why you round the number the way you do. Use a number line to support your explanation.
2. Round 39,995 to the nearest ten. Explain why you round the number the way you do. Use a number line to support your explanation.
3. Round 173.465 to the nearest hundred, to the nearest ten, to the nearest one, to the nearest tenth, and to the nearest hundredth.
4. The distance between two cities is described as 1500 miles. Should you assume that this is the exact distance between the cities? If not, what could the distance between the cities possibly be?

Answers to Practice Exercises for Section 1.4

1. You are rounding to the nearest tenth, so on a number line you must find tick marks representing tenths that 6.248 lies between. The number 6.248 is between the tenths 6.2 and 6.3. But because of the 4 in the hundredths place, 6.248 is less than 6.25, which is halfway between 6.2 and 6.3. Therefore, 6.248 is closer to 6.2 than to 6.3, so 6.248 rounded to the nearest tenth is 6.2. See Figure 1.59.

Figure 1.59 Rounding 6.248 to the nearest tenth.

2. You are rounding to the nearest ten, so on a number line you must find tick marks representing tens that 39,995 lies between. The number 39,995 lies exactly halfway between 39,990 and 40,000, as shown in Figure 1.60. By the "with 5 round up" convention, round up to 40,000.

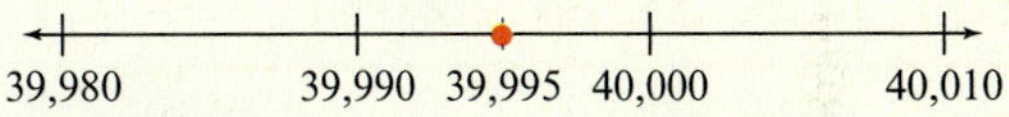

Figure 1.60 Rounding 39,995 to the nearest ten.

3. To the nearest hundred: 200

   To the nearest ten: 170

   To the nearest one: 173

   To the nearest tenth: 173.5

   To the nearest hundredth: 173.47
4. Because the reported distance has zeros in the tens and ones places, assume that the distance has been rounded to the nearest hundred. Therefore, the distance between the two cities is probably not 1500 miles; it could be anywhere in the range between 1450 miles and 1550 (but less than 1550).

PROBLEMS FOR SECTION 1.4

1. Round 2.1349 to the nearest hundredth. Explain in your own words the reasoning behind why we round the number the way we do. Use a number line to support your explanation.
2. Round 27,003 to the nearest hundred. Explain in your own words the reasoning behind why we round the number the way we do. Use a number line to support your explanation.
3. Round 9995.2 to the nearest ten. Explain in your own words the reasoning behind why we round the number the way we do. Use a number line to support your explanation.
4. Adam has made up his own method of rounding. Starting at the right-most place in a number, he keeps rounding to the value of the next place to the left until he reaches the place

to which the number was to be rounded. For example, Adam would use the following steps to round 11.3524 to the nearest tenth:

$$11.3524 \rightarrow 11.352 \rightarrow 11.35 \rightarrow 11.4$$

Is Adam's method a valid way to round? Explain why or why not.

5. The label on a snack food package says that one serving of the snack food contains 0 grams of trans fat. Does this mean that there is no trans fat in one serving of the snack food? If not, what amounts of trans fat in a serving of the snack food are possible? Discuss these questions in light of the information presented in this section.

6. The weight of an object is reported as 12,000 pounds. Should you assume that this is the exact weight of the object? If not, what could the weight of the object possibly be? Explain.

7. In a report, a population is given as 2700. Should you assume that this is the exact population? If not, what can you say about the exact population? Explain.

CHAPTER SUMMARY

Section 1.1 The Counting Numbers Page 2

- The counting numbers are the numbers 1, 2, 3, 4, ... There are two distinct ways to think about the counting numbers: on the one hand, the counting numbers form an ordered list; on the other hand, counting numbers tell how many objects are in a set (i.e., the cardinality of a set). When we count the number of objects in a set, we connect the two views of counting numbers by making a one-to-one correspondence between an initial portion of the list of counting numbers and the objects in the set. The last number word that we say when we count a number of objects tells us how many objects there are. Page 2
- In the base-ten system, every counting number can be written using only the ten symbols 0, 1, 2, . . . , 9. The base-ten system uses place value, which means that the value that a digit in a number represents depends on the location of the digit in the number. Each place has the value of a base-ten unit, and base-ten units are created by repeated bundling by ten. Each base-ten unit is the value of a place. The value of a place is ten times the value of the place to its immediate right. Page 5
- Unfortunately, the way we say the English names of the counting numbers 11 through 19 does not fit well with the way we write these numbers. Page 7
- The counting numbers can be displayed on number paths, which are informal precursors to number lines. The whole numbers, 0, 1, 2, 3, . . . , can be displayed on number lines. Page 9

Key Skills and Understandings

1. Describe the two views of the counting numbers—as a list and as used for cardinality. Discuss the connections between the list and cardinality views of the counting numbers. Page 2
2. Explain what it means for the base-ten system to use place value. Discuss what problem the development of the base-ten system solved. Page 4
3. Describe base-ten units and describe how adjacent place values are related in the base-ten system. Page 5
4. Describe and make math drawings to represent a given counting number in terms of bundled objects in a way that fits with the base-ten representation for that number of objects. Page 6
5. Describe how to represent whole numbers on a number line and discuss the difference between a number path (as described in the text) and a number line. Page 9

Section 1.2 Decimals and Negative Numbers

Page 13

- The decimals extend the base-ten system and are formed by creating places to the right of the ones place while preserving the essential structure of the base-ten system: the value of each place is ten times the value of the place to its immediate right. Just as the counting numbers can be represented with bundled objects, finite (positive) decimals can also be represented with bundled objects, provided that a single object stands for the value of the lowest place one needs to represent. Page 13
- The positive decimals can be represented as lengths in a way that fits with the structure of the base-ten system. This way of representing decimals fits nicely with the metric system (meters, decimeters, centimeters, and millimeters). Decimals can also be represented on number lines in a way that is compatible with viewing positive decimals as lengths: the location of a positive decimal is its distance from 0. We can also think of the decimals as filling in the number line: First, we plot whole numbers, then we plot tenths, then hundredths, and so on. Page 14
- The negative numbers are the numbers to the left of zero on the number line. Page 18

Key Skills and Understandings

1. Describe and draw rough pictures to represent a given positive decimal in terms of bundled objects in a way that fits with and shows the structure of the base-ten system. (Take care to state the meaning of one of the objects.) Page 13
2. Describe and draw rough pictures to represent a given positive decimal as a length in a way that fits with and shows the structure of the base-ten system. Page 15
3. Discuss how decimals fill in a number line, label tick marks on number lines, and plot numbers on number lines. Page 15
4. Show how to zoom in on portions of number lines to see the portions in greater detail. Page 16
5. View negative numbers as owed amounts, and plot a given negative number on a number line. Page 18

Section 1.3 Reasoning to Compare Numbers in Base Ten

Page 27

- The value of a place is greater than the largest number made from places of smaller values. For this reason, we compare numbers in the base-ten system by looking first at the place of greatest value, then moving to places of smaller values if the numbers agree in the place of greatest value. On number lines, numbers become greater as one moves to the right. Given any two distinct numbers, there is always another number in between. (In fact, there are infinitely many numbers in between.) Taking the negative of two numbers reverses the comparison between them. Page 27

Key Skills and Understandings

1. Given any two numbers in base ten (including negative numbers), determine which is greater, and put any collection of numbers in order from least to greatest (or vice versa). Page 27
2. Explain the rationale for comparing positive numbers in base ten by first comparing the place of greatest value (and then moving to places of lower value). Page 27
3. Draw rough pictures of bundled objects to show that one number is greater than another. Page 28
4. Describe and draw rough pictures representing numbers as lengths and showing that one number is greater than another. Page 29

5. Use a number line to demonstrate that one number is greater than another (including negative numbers).	Page 29
6. Describe negative numbers as owed amounts to explain why one negative number is greater than another.	Page 30
7. Given two distinct numbers in base ten (positive or negative or 0), find another number in between them.	Page 31
Section 1.4 Reasoning about Rounding	Page 34
▪ To round a number to a given place's value means to find the closest number that has zeros in all smaller places. When we work with actual quantities, the way the number is written generally indicates the rounding that has taken place and, therefore, the precision with which the quantity is known.	Page 34
Key Skills and Understandings	
1. Given any number, round it to a given place.	Page 34
2. Use an appropriate number line to explain how to round a given number to a given place.	Page 34
3. Recognize that numbers representing actual quantities have generally been rounded.	Page 36

2

Fractions and Problem Solving

We use fractions and percent when we need to describe a portion or a part of something. Fractions and percent arise in mathematics, science, and daily life. The study of fractions and percent offers ongoing opportunities to engage in challenging problem solving and mathematical reasoning.

In this chapter, we focus on several topics and practices within the *Common Core State Standards for Mathematics (CCSSM)*.

Standards for Mathematical Content in the CCSSM

In the domain of *Number and Operations—Fractions* (Grades 3–5), students develop an understanding of fractions as numbers that can be described as numbers of parts and can be represented on a number line. They explain equivalence of fractions, and they compare fractions by reasoning about the sizes and the number of parts. In the domain of *Ratios and Proportional Relationships* (Grades 6, 7), students work with percent as fractions with denominator 100, and they solve basic percent problems.

Standards for Mathematical Practice in the CCSSM

Opportunities to engage in all eight of the Standards for Mathematical Practice described in the Common Core State Standards occur throughout the study of fractions and percent.

Solving problems is the heart of mathematics and is an essential part of every topic. The following standards are especially appropriate while studying fractions and percent and when highlighting problem solving.

- **1 Make sense of problems and persevere in solving them.** Students engage in this practice when they persist in trying different approaches to solving problems and when they seek to learn and use new ideas and new ways of thinking.
- **3 Construct viable arguments and critique the reasoning of others.** Students engage in this practice when they use the definition of fraction to explain why an amount can be described with a certain fraction, when they use reasoning about sizes of parts to compare fractions, and when they explore a proposed way of comparing fractions to determine if it is or is not valid.
- **5 Use appropriate tools strategically.** Students engage in this practice when they use math drawings, number lines, and tables as thinking aids during the problem-solving process and to communicate and explain logical lines of reasoning.

2.1 Solving Problems and Explaining Solutions

CCSS Common Core State Standards All grades

Fractions are fertile ground for problem solving and reasoning. Before we study fractions, let's think about how to solve problems and explain solutions. We will examine some simple but sensible guidelines for solving problems and think about what qualifies as a good explanation *in mathematics*. A main theme of this book is explaining why: Why are the familiar procedures and formulas of elementary mathematics valid? Why is a student's response incorrect? Why is a different way of carrying out a calculation often perfectly correct? We will be seeking mathematical answers to these questions.

What Is the Role of Problem Solving?

Mathematics exists to solve problems. With mathematics, we can solve a vast variety of problems in technology, science, business and finance, medicine, daily life, and mathematics itself. The potential uses of mathematics are limited only by human ingenuity. Solving problems is not only the most important *end* of mathematics; it is also a *means* for learning mathematics. Mathematicians have long known that good problems can deepen our thinking about mathematics, guide us to new ways of using mathematical techniques, help us recognize connections between topics in mathematics, and force us to confront mathematical misconceptions we may hold. By working on good problems, we learn mathematics better.

STANDARDS ON PROBLEM SOLVING

- The *Principles and Standards for School Mathematics* of the National Council of Teachers of Mathematics (NCTM) [64] includes this Process Standard:
- **Problem Solving** Instructional programs from prekindergarten through grade 12 should enable all students to
  - build new mathematical knowledge through problem solving;
  - solve problems that arise in mathematics and in other contexts;

- apply and adapt a variety of appropriate strategies to solve problems;
- monitor and reflect on the process of mathematical problem solving.

(From Principles and Standards for School Mathematics. Copyright © by National Council of Teachers of Mathematics. Used by permission of National Council of Teachers of Mathematics.)

- The *Common Core State Standards for Mathematics (CCSSM)* includes this Standard for Mathematical Practice:
  - **1 Make sense of problems and persevere in solving them.** Mathematically proficient students start by explaining to themselves the meaning of a problem and looking for entry points to its solution. . . . [They] plan a solution pathway rather than simply jumping into a solution attempt. . . . They monitor and evaluate their progress and change course if necessary. . . . Mathematically proficient students check their answers to problems using a different method, and they continually ask themselves, "Does this make sense?" (From Common Core Standards for Mathematical Practice. Published by Common Core Standards Initiative.)

The Process of Problem Solving In 1945, the mathematician George Polya presented a four-step guideline for solving problems in his book *How to Solve It* [72]. These four steps are simple and sensible, and they have become a framework for thinking about problem solving.

Polya's Steps

1. Understand the problem.
2. Devise a plan.
3. Carry out the plan.
4. Look back.

(From *How to Solve it: A New Aspect of Mathematical Method* by George Polya. Published by Princeton University Press, © 1945.)

Teachers and researchers have used Polya's steps to develop more detailed lists of prompts or questions to help students monitor their thinking during problem solving.

Polya's first step, *understand the problem*, is the most important. It may seem obvious that if you don't understand a problem, you won't be able to solve it, but it is easy to rush into a problem and try to do "something like we did in class" before you think about what the problem is asking. So *slow down* and read problems carefully. In some cases, drawing a diagram or a simple math picture can help you understand the problem.

In the midst of solving a problem (Polya's steps 2 and 3), it's important to monitor and reflect on the process of problem solving. Think about what information you know, what information you are looking for, and how to relate those pieces of information. If you get stuck, think about whether you have seen similar problems. Consider whether you could adapt or modify the reasoning you used for another problem to the problem at hand. Be willing to try another approach to solving the problem.

It's important to persevere when attempting to solve a problem. Students sometimes think they can solve a problem only if they've seen one just like it before, but this is not true. Your common sense and natural thinking abilities are powerful tools that will serve you well if you use them. By persevering, you will develop these thinking abilities.

Polya's fourth step, *look back*, gives you an opportunity to catch mistakes. Check to see if your answer is plausible. For example, if the problem was to find the height of a telephone pole, then answers such as 2.3 feet or 513 yards are unlikely—look for a mistake somewhere. Looking back also gives you an opportunity to make connections: Have you seen this type of answer before? What did you learn from this problem? Could you use these ideas in some other way? Is there another way to solve the problem? When you look back, you have an opportunity to learn from your own work.

IMAP *Watch Elise talk about problem solving with her students.*

Solving Problems for Yourself Students sometimes wonder why they need to solve problems themselves: Why can't the teacher just *show* us how to solve the problem? Of course, teachers do show solutions to many problems. However, sometimes teachers should step back and *guide* their students, helping them to use fundamental concepts and principles to *figure out* how to solve a problem. Why? Because the process of grappling with a problem can help students understand the underlying concepts and principles. Teachers who are too quick to tell students how to solve problems may actually rob them of valuable learning experiences. The process of making sense of things for yourself is the essence of education—use it in mathematics and do not underestimate its power.

How Can We Use Strip Diagrams and Other Math Drawings?

math drawings

Visual representations can often help us make sense of a problem, formulate a solution strategy, and explain a line of reasoning. Simple drawings that show relationships between quantities and are quick and easy to make can be especially helpful. We call such drawings **math drawings**. Math drawings should be as simple as possible and include only those details that are relevant to solving the problem.

strip diagram
tape diagram

One type of math drawing is the **strip diagram**, also called a **tape diagram**. Strip diagrams use lengths of rectangular strips to represent quantities, as shown on the left in Figure 2.1. Because strip diagrams use lengths, they connect readily with number lines. Strip diagrams can help students formulate equations, including algebraic equations. Strip diagrams are used throughout this book.

Figure 2.1
A strip diagram and a drawing that is not a math drawing.

Problem: In an aquarium, $\frac{3}{4}$ of the fish are goldfish. There are 18 other fish. How many goldfish are there?

Math drawing (a strip diagram):

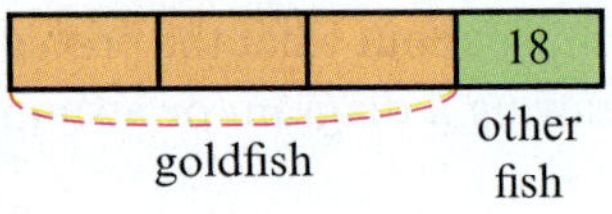

3 parts,
18 goldfish in each part

$3 \times 18 = 54$ goldfish

Not a math drawing:

Why Should We Explain Solutions?

In mathematics, we want more than just answers to problems. We want to know why a method works. Why does it give the correct answer to the problem? As a teacher, you will need to explain why the mathematics you are teaching works the way it does. However, there is an even more compelling reason for providing explanations. When you try to explain something to someone else, you clarify your own thinking and you learn more yourself. When you try to explain a solution, you may find that you don't understand it as well as you thought. The exercise of explaining is valuable because it provides an opportunity to learn more, to uncover an error, or to clear up a misconception. Even if you understood the solution well, you will understand it better after explaining it.

Communicating about mathematics gives both children and adults an opportunity to make sense of mathematics. According to the NCTM [64, p. 56] (From Principles and Standards for School Mathematics. Copyright © by National Council of Teachers of Mathematics. Used by permission of National Council of Teachers of Mathematics.),

> *From children's earliest experiences with mathematics*, it is important to help them understand that assertions should always have reasons. Questions such as "Why do you think it is true?" and "Does anyone think the answer is different, and why do you think so?" help students see that statements need to be supported or refuted by evidence.

When we communicate about mathematics in order to explain and convince, we must use *reasoning*. Logical reasoning is the essence of mathematics. In mathematics, everything but the fundamental starting assumptions has a reason, and the whole structure of mathematics is built up by reasoning.

STANDARDS ON REASONING, PROOF, AND COMMUNICATION

- The *Principles and Standards for School Mathematics* of the National Council of Teachers of Mathematics (NCTM) [64] includes these Process Standards:
- **Communication** Instructional programs from prekindergarten through grade 12 should enable all students to
  - organize and consolidate their mathematical thinking through communication;
  - communicate their mathematical thinking coherently and clearly to peers, teachers, and others;
  - analyze and evaluate the mathematical thinking and strategies of others;
  - use the language of mathematics to express mathematical ideas precisely.
- **Reasoning and Proof** Instructional programs from prekindergarten through grade 12 should enable all students to
  - recognize reasoning and proof as fundamental aspects of mathematics;
  - make and investigate mathematical conjectures;
  - develop and evaluate mathematical arguments and proofs;
  - select and use various types of reasoning and methods of proof.

  (From Principles and Standards for School Mathematics. Copyright © by National Council of Teachers of Mathematics. Used by permission of National Council of Teachers of Mathematics.)
- The *Common Core State Standards for Mathematics (CCSSM)* includes this Standard for Mathematical Practice:
  - **3 Construct viable arguments and critique the reasoning of others.** Mathematically proficient students understand and use stated assumptions, definitions, and previously established results in constructing arguments. . . . They justify their conclusions, communicate them to others, and respond to the arguments of others. . . . Mathematically proficient students are also able to compare the effectiveness of two plausible arguments, distinguish correct logic or reasoning from that which is flawed, and—if there is a flaw in the argument—explain what it is. . . . (From Common Core Standards for Mathematical Practice. Published by Common Core Standards Initiative.)

How Are Explanations in Mathematics Different from Other Explanations? In mathematics, we seek particular kinds of explanations: those using logical reasoning and based on initial assumptions that are either explicitly stated or assumed to be understood by the reader or listener. Explanations can vary according to different areas of knowledge. There are many different kinds of explanations—even of the same phenomenon. For example, consider this question: Why are there seasons? (See Figure 2.2.)

Figure 2.2

Why are there seasons?

The simplest answer is "because that's just the way it is." Every year, we observe the passing of the seasons, and we expect to see the cycle of spring, summer, fall, and winter continue indefinitely. The cycle of seasons is an observed fact that has been documented since humans began to keep records. We could stop here, but when we ask why there are seasons, we are searching for a deeper explanation.

A poetic explanation for the seasons might refer to the cycles of birth, death, and rebirth around us. In our experiences, nothing remains unchanged forever, and many things are parts of a cycle. The cycle of the seasons is one of the many cycles that we observe.

Most cultures have stories that explain why we have seasons. The ancient Greeks, for example, explained the seasons with the story of Persephone and her mother, Demeter, who tends the earth. When Pluto, god of the underworld, stole Persephone, forcing her to become his bride, Demeter was heartbroken. Pluto and Demeter arranged a compromise, and Persephone could stay with her mother for half a year and return to Pluto in the underworld for the remaining half of every year. When Persephone is in the underworld, Demeter is sad and does not tend the earth. Leaves fall from the trees, flowers die, and it is fall and winter. When Persephone returns, Demeter is happy again and tends the earth. Leaves grow on the trees, flowers bloom, and it is spring and summer. This is a beautiful story, but we can still ask for another kind of explanation.

Modern scientists explain the reason for the seasons by the tilt of the earth's axis relative to the plane in which the earth travels around the sun. When the northern hemisphere is tilted toward the sun, it is summer there; when it is tilted away from the sun, it is winter. Perhaps this settles the matter, but a seeker could still ask for more. Why are the earth and sun positioned the way they are? Why does the earth revolve around the sun and not fly off alone into space? These questions can lead again to poetry, or to the spiritual, or to further physical theories. Maybe they lead to an endless cycle of questions.

How Do We Write Good Mathematical Explanations? While an oral explanation helps you develop your solution to a problem, written explanations push you to polish, refine, and clarify your ideas. This is as true in mathematical writing as in any other kind of writing, and it is true at all levels. You should write explanations of your solutions to problems, and your students should write explanations of their solutions, too. Some elementary school teachers have successfully integrated mathematics and writing in their classrooms and use writing to help their students develop their understanding of mathematics.

Like any kind of writing, it takes work and practice to write good mathematical explanations. When you solve a problem, do not attempt to write the final draft of your solution right from the start. Use scratch paper to work on the problem and collect your ideas. Then, write your solution as part of the *looking back* stage of problem solving. Think of your explanations as an essay. As with any essay that aims to convince, what counts is not only factual correctness but also persuasiveness, explanatory power, and clarity of expression. In mathematics, we persuade by giving a thorough, logical argument, in which chains of logical deductions are strung together connecting the starting assumptions to the desired conclusion.

CHARACTERISTICS OF GOOD EXPLANATIONS IN MATHEMATICS

1. The explanation is factually correct, or nearly so, with only minor, inconsequential flaws.
2. The explanation addresses the specific question or problem that was posed. It is focused, detailed, and precise. Key points are emphasized. There are no irrelevant or distracting points.
3. The explanation is clear, convincing, and logical. A clear and convincing explanation is characterized by the following:
   - **a.** The explanation could be used to teach another (college) student, possibly even one who is not in the class.
   - **b.** The explanation could be used to convince a skeptic.
   - **c.** The explanation does not require the reader to make a leap of faith.
   - **d.** If applicable, supporting math drawings, diagrams, and equations are used appropriately and as needed.
   - **e.** The explanation is coherent.
   - **f.** Clear, complete sentences are used.

Good mathematical explanations are thorough. They should not have gaps that require leaps of faith. On the other hand, a good explanation should not belabor points that are well known to the audience or not central to the explanation. For example, if your solution contains the calculation $356 \div 7$, a college-level explanation need not describe how the calculation is carried out, except when it is necessary for the solution. Unless your instructor tells you otherwise, assume that you are writing your explanations for your classmates.

The box above lists characteristics of good mathematical explanations. When you write an explanation, check whether it has these characteristics. The more you work at writing explanations, and the more you ponder and analyze what makes good explanations, the better you will write explanations, and the better you will understand the mathematics involved. Note that the solutions to the practice exercises in each section provide you with many examples of the kinds of explanations you should learn to write.

Now let's stretch our thinking as we solve problems and explain solutions with fractions!

2.2 Defining and Reasoning About Fractions

CCSS Common Core State Standards Grades 3 through 6

What does a fraction such as $\frac{3}{4}$ mean? There are actually several different ways to define fractions. In this section we will adopt a definition of fractions, and we will use this definition repeatedly as we solve problems and reason about fractions. We will see how to represent fractions with math drawings and on number lines, and we will examine some common difficulties in understanding fractions.

Definition of Fractions

How Do We Define Fractions?

Fractions occur naturally whenever we want to consider a portion of an object or quantity. Consider how fractions are used in the following ordinary situations: $\frac{1}{8}$ of a pizza, $\frac{2}{3}$ of the houses in the neighborhood, $\frac{9}{10}$ of the profit, $\frac{5}{4}$ of a cup of water.

whole
unit amount
reference amount

All these examples use the word *of*, and all the fractions represent a portion *of* some object, collection of objects, or quantity. To define what we mean by a fraction $\frac{A}{B}$, we start by specifying a **whole** or **unit amount** or **reference amount**. So suppose there is an object (such as a sandwich), a collection (such as a group of people or a bunch of bagels), or other quantity (such as a quantity of water or of money). Let's call this object, collection, or quantity our whole or unit amount or reference amount. What do we mean by $\frac{A}{B}$ of the whole we are considering? We will define $\frac{A}{B}$ in essentially the same way as the *Common Core State Standards. Note:* Throughout this chapter, we will work only with fractions $\frac{A}{B}$ for which A and B are nonnegative numbers and B is not zero.

unit fraction $\frac{1}{B}$

In order to define $\frac{A}{B}$, we define the **unit fraction $\frac{1}{B}$** first. If the whole we are considering can be partitioned (divided) into B parts of equal size, then the amount formed by 1 of those parts is what we mean by $\frac{1}{B}$ of our whole. In other words, an amount is $\frac{1}{B}$ of our whole, if B copies of it joined together are the same size as the whole.

fraction $\frac{A}{B}$

Then the **fraction $\frac{A}{B}$** of our whole is the amount formed by A parts (or copies of parts), each of size $\frac{1}{B}$ of the whole. See Figure 2.3.

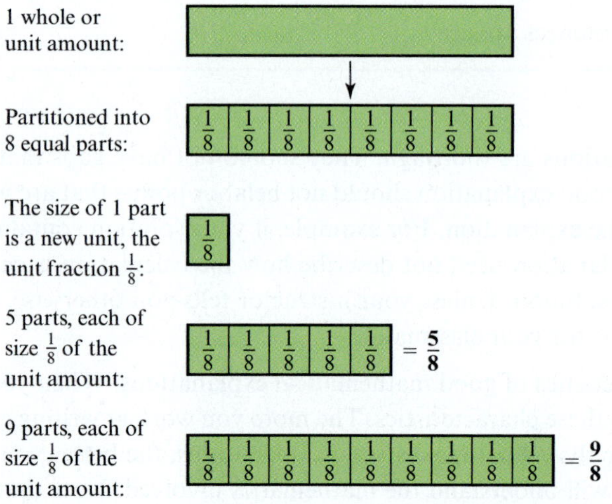

Figure 2.3
Fractions tell us how many of what size parts there are.

numerator
denominator

CCSS
3.NF.1

If $\frac{A}{B}$ is a fraction, then A is called the **numerator** and B is called the **denominator**. These names fit because the word *numerator* comes from *number*, and the numerator tells us the number of parts. The word *denominator* comes from *name* and is related to *denomination*, which tells what type something is. For example, in money, the denominator of a bill tells us what type of bill it is—in other words, what value it has. The denominator of a fraction tells us how many equal parts are in 1 whole. This determines the *size* of each part, which we denote by the unit fraction $\frac{1}{B}$. So a fraction $\frac{A}{B}$ tells us how many (A) of what size parts $\left(\frac{1}{B}\right)$. It is similar to base-ten notation because a digit in a place tells us how many there are of that place's unit.

Figure 2.4 highlights the three key components in the definition of fractions. Sometimes we focus on the details that tell us the exact amount a fraction specifies. For example, $\frac{5}{8}$ of a pizza is the amount in 5 parts, each of size $\frac{1}{8}$ of a pizza. But sometimes we focus on the idea that a fraction is a number, just like 2 or 3, that tells us how much of a unit amount there is. For example, we can imagine having 2 pizzas or 3 pizzas or $\frac{5}{8}$ pizzas. The numbers 2, 3, and $\frac{5}{8}$ all tell us *how many pizzas* we are considering.

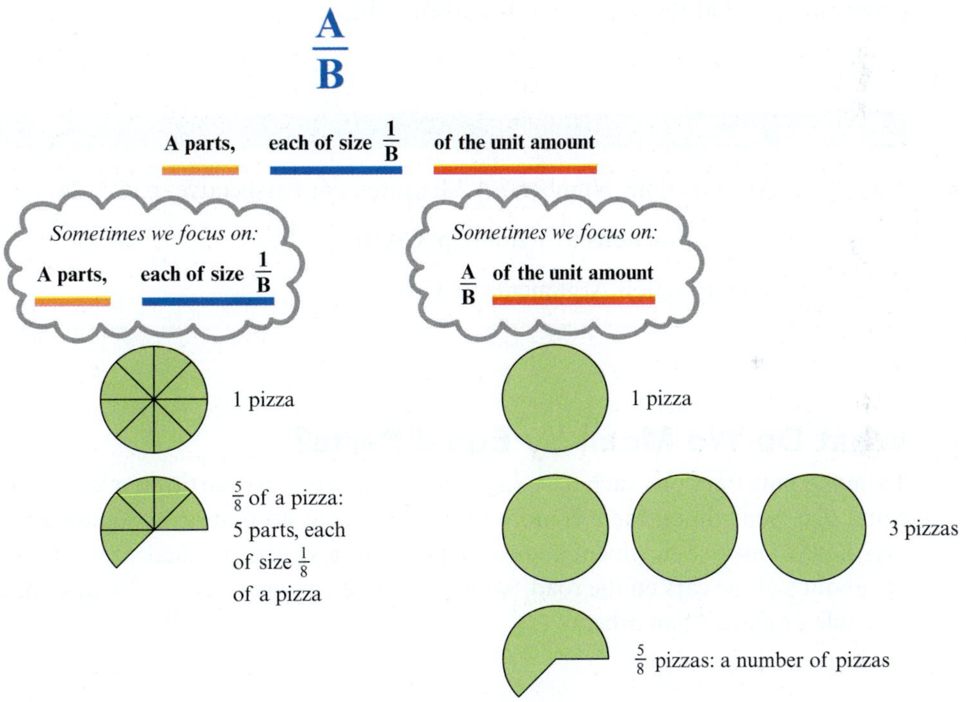

Figure 2.4 Three key components in the definition of fractions.

IMAP
Watch Jacky use fractions strips to compare fractions and a practicing teacher explain Jacky's thinking.

CLASS ACTIVITY

2A Getting Familiar with Our Definition of Fractions, p. CA-21

2B Using Our Fraction Definition to Solve Problems, p. CA-23

Why Are Fractions Numbers?

Fractions always specify how much of a unit amount (or reference amount) there is, just as all nonnegative numbers do. We can think of fractions, and all nonnegative numbers, as the result of measuring by a unit amount. For example, if you have 3 liters of juice, the 3 tells you how many of the unit amount "1 liter" you have. Similarly, if you have $\frac{1}{8}$ of a liter of juice, the $\frac{1}{8}$ tells you how many (i.e., how much) of the unit amount "1 liter" you have. The unit amount for a fraction is what the fraction is *of*—the amount that measures as 1. For example, the unit amount for $\frac{2}{5}$ in "$\frac{2}{5}$ of Hogwart's land" is all of Hogwart's land, or $\frac{5}{5}$ of Hogwart's land (*not* 5). The unit amount for $\frac{3}{4}$ in "$\frac{3}{4}$ of a meter" is 1 meter or $\frac{4}{4}$ meter (*not* 4). The unit amount for a fraction $\frac{A}{B}$ that does not refer to a physical quantity is just the number 1 or $\frac{B}{B}$ (*not* the denominator B).

Attending closely to a fraction's unit amount is critically important for (1) solving problems, (2) explaining arithmetic with fractions, and (3) understanding that fractions are numbers in the same way that 2 and 3 are numbers. You can use language that emphasizes how much of a unit amount a fraction specifies. For example, you could describe a piece of pizza as "$\frac{1}{8}$ of a pizza" rather than as "1 out of 8 pieces of pizza."

The next Class Activities ask you to attend to the relationship between a fraction and its unit amount in different kinds of situations. In Class Activity 2C, you will look into how a fraction is "measured by" its unit amount and see how we can use fractions to compare separate amounts. In Class Activity 2D, you will see that a single quantity can be described with *different* fractions depending on what the unit amount is taken to be.

CLASS ACTIVITY

2C Why Are Fractions Numbers? A Measurement Perspective, p. CA-24

2D Relating Fractions to Wholes, p. CA-26

2E Critiquing Fraction Arguments, p. CA-27

What Do We Mean by Equal Parts?

To make unit fractions such as $\frac{1}{2}$, $\frac{1}{3}$, $\frac{1}{4}$, and so on, we must partition a unit amount into parts of equal size. Sometimes, there is more than one way to do that, as in Figure 2.5. Other times, you may have to make a judgment about what constitutes parts of equal size. For example, when talking about $\frac{2}{3}$ of the cars on the road, we treat each car as equal, even though some cars may be more valuable or bigger than others.

Figure 2.5
Different ways to make 3 parts of equal size.

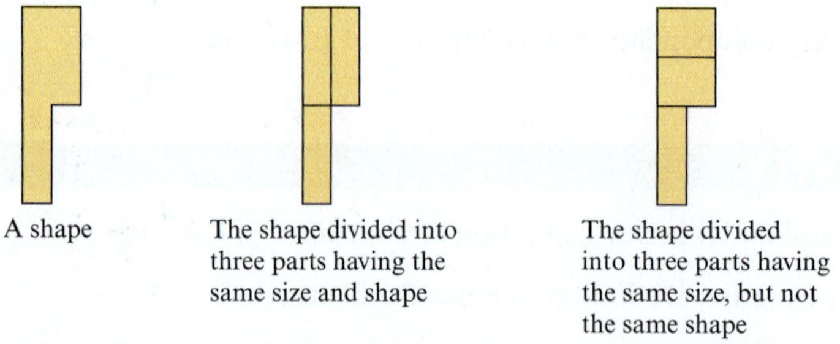

We can avoid ambiguity about equal parts by focusing on lengths of fraction strips and lengths of intervals on number lines. Coming up, we'll see how to use fraction strips to place fractions on number lines.

How Can We Interpret Fractions as Lengths and as Numbers on Number Lines?

As with decimals, a good way to represent fractions is with lengths. When a whole (or unit amount) is represented with a strip of paper, a drawing of a strip, or a rod, then length is salient and it's natural to make equal parts by dividing lengthwise as in Figure 2.3.

CCSS 3.NF.2

As with decimals, we focus on length when we place fractions on number lines. If A and B are whole numbers and B is not zero, then the fraction $\frac{A}{B}$ is located on a number line according to the rule for locating numbers in base-ten on number lines given in Section 1.2. Thus, $\frac{A}{B}$ is located to the right of zero at a distance of $\frac{A}{B}$ units from 0. Therefore, $\frac{A}{B}$ is at the right end of a line segment whose left end is at 0 and whose length is $\frac{A}{B}$ of one unit.

For example, where is $\frac{5}{3}$ on a number line? The fraction $\frac{5}{3}$ is at the right end of a line segment whose left end is at 0 and whose length is $\frac{5}{3}$ of a unit. To construct such a line segment, use the meaning of $\frac{5}{3}$: Divide the line segment from 0 to 1 into 3 equal parts, and make a line segment consisting of 5 copies of those parts. Place this five-part line segment on the number line so that its left endpoint is at 0, as shown in Figure 2.6. Then the right endpoint of the five-part line segment is at the fraction $\frac{5}{3}$.

Figure 2.6

Plotting $\frac{5}{3}$ on a number line.

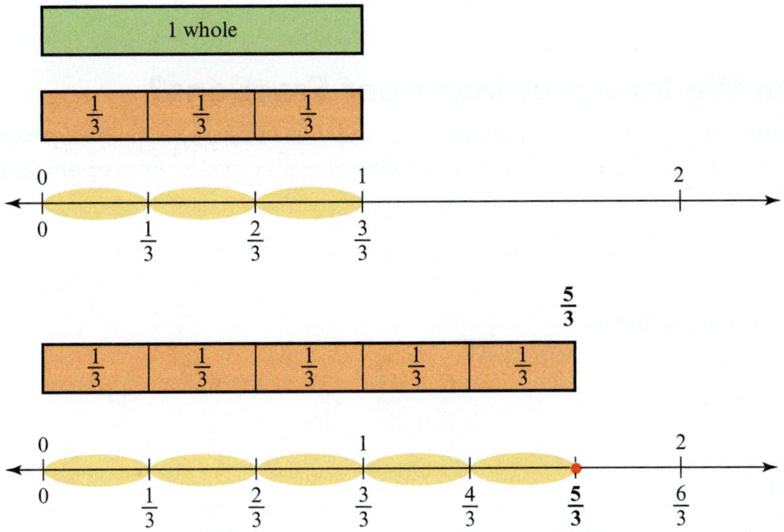

To emphasize the important point that a number is plotted on a number line according to its *distance* from 0, some teachers find it helpful to circle segments on the number line as shown in Figure 2.6. Highlighting the segments helps students focus on their length, and understand that the segment between 0 and $\frac{5}{3}$ consists of 5 pieces, each of which is $\frac{1}{3}$ of a unit long. When students pay attention to the lengths of the segments, they avoid some of the common number line errors that are due to counting tick marks without attending to length and distance. Class Activity 2F examines some of these errors.

CLASS ACTIVITY

2F Critique Fraction Locations on Number Lines, p. CA-29

CLASS ACTIVITY

2G Fractions on Number Lines, p. CA-30

FROM THE FIELD **Research**

Shaughnessy, M. M. (2011). Identify fractions and decimals on a number line. *Teaching Children Mathematics, 17*(7), 428–434.

In order to examine students' difficulties with labeling fractions and decimals on a number line, the author interviewed students in an urban school district in California. She found that more students successfully labeled decimals on number lines than they did fractions and that more students successfully labeled points when an interval was equally partitioned than when it was not. The author found four common errors. Some students used unconventional notation to label points on the number line. Some students would redefine the unit of the number line and base their representation on a different unit. For example, a student who saw twenty ticks between zero and two (ten between each unit) labeled the eighth tick as 8/20. Some students just counted numbers of tick marks to determine numerators and denominators and did not attend to the distances between tick marks. Some students counted tick marks and turned that count into a fraction, for example by plotting 1/6 at the sixth tick mark. The author recommends that teachers should consider these misconceptions when planning tasks.

How Can We Interpret Improper Fractions?

proper fraction

improper fraction

A fraction $\frac{A}{B}$ (where A and B are whole numbers and B is not zero) is called a **proper fraction** if the numerator A is smaller than the denominator B. Otherwise, it's called an **improper fraction**. For example,

$$\frac{2}{5}, \quad \frac{3}{8}, \quad \frac{15}{16}$$

are proper fractions, whereas

$$\frac{6}{5}, \quad \frac{11}{8}, \quad \frac{27}{16}$$

are improper fractions.

Students sometimes think that improper fractions don't make sense because "we don't have enough pieces." For example, if 3 pieces make the whole, how can you consider the 5 pieces in $\frac{5}{3}$? Students must be able to think about making more of those pieces.

Number lines are especially good for showing improper fractions. When we show improper fractions with pieces of an object, we must take care to specify what unit amount the fraction refers to. For example, does the shaded region in Figure 2.7 represent $\frac{5}{3}$ or $\frac{5}{6}$? We can't say unless we know what the fraction is supposed to be *of*. The shaded region is $\frac{5}{3}$ of the three-piece rectangle, but it

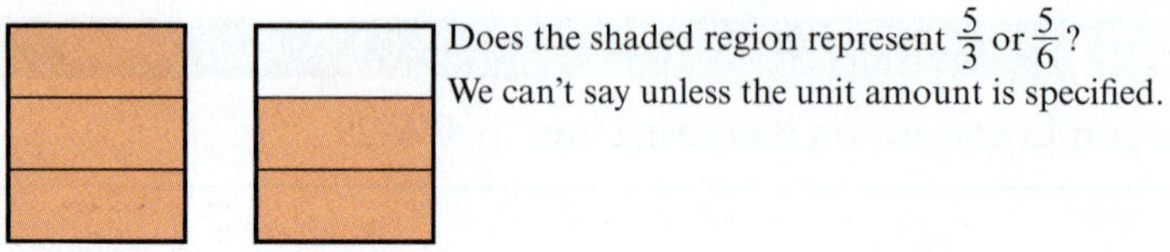

Figure 2.7
Showing $\frac{5}{3}$ with pieces of rectangles.

is also $\frac{5}{6}$ of the two rectangles. In contrast, on number lines, the whole is always the line segment between 0 and 1, so no ambiguity arises. On a number line, proper and improper fractions are on an equal footing. See Figure 2.8.

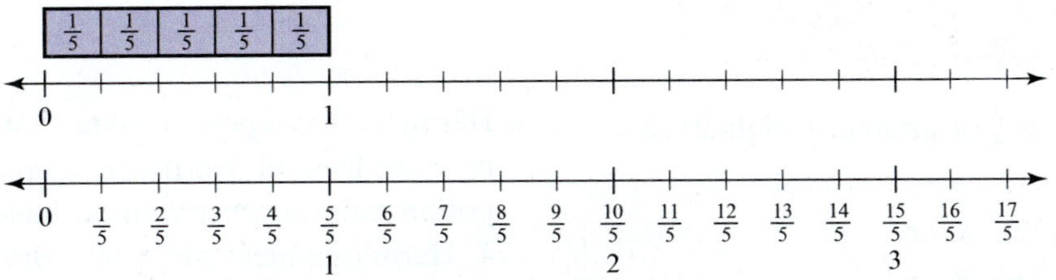

Figure 2.8 Proper and improper fractions on a number line.

CLASS ACTIVITY

2H Improper Fraction Problem Solving with Pattern Tiles, p. CA-31

Preview: Decimal Representations of Fractions

Because fractions can be plotted on number lines, and because every location on a number line corresponds to a number in base ten, it must be possible to represent fractions in base ten as decimals. In fact, to write a fraction $\frac{A}{B}$ as a decimal number we divide A by B as follows:

$$\frac{A}{B} = A \div B$$

We will see why it makes sense to write a fraction in decimal notation by dividing in Chapter 6.

So,

$$\frac{5}{16} = 5 \div 16 = 0.3125 \qquad \frac{1}{12} = 1 \div 12 = 0.083333\ldots \qquad \frac{2}{7} = 2 \div 7 = 0.285714\ldots$$

SECTION SUMMARY AND STUDY ITEMS

Section 2.2 Defining and Reasoning About Fractions

Our definition of fraction is as follows: If A and B are whole numbers, and B is not zero, and if a whole or unit amount can be partitioned into B equal parts, then $\frac{1}{B}$ of the whole is the amount formed by 1 part, and $\frac{A}{B}$ of the whole is the amount formed by A parts (or copies of parts), each of size $\frac{1}{B}$ of the whole.

When working with fractions, pay close attention to the *whole* or *unit amount* namely, the object, collection, or quantity that the fraction is *of.* Even though fractions are expressed in terms of a pair of numbers, a fraction represents a single number, and as such, it can be plotted on a number line.

Key Skills and Understandings

1. Find fractional amounts of an object, collection, or quantity, and justify your reasoning.
2. In problems, determine the whole or unit amount associated with a fraction appearing in the problem.
3. Use fractions to compare quantities.
4. Plot fractions, including improper fractions, on number lines and explain why the location fits with the definition of fraction.

Practice Exercises for Section 2.2

1. This rectangle of x's is $\frac{3}{4}$ of another (original) rectangle of x's.

x x x x
x x x x
x x x x
x x x x
x x x x
x x x x

Show the original rectangle.

2. This rectangle of x's is $\frac{8}{3}$ of another (original) rectangle of x's.

x x x x x x x x x x x x x x x x
x x x x x x x x x x x x x x x x
x x x x x x x x x x x x x x x x

Show the original rectangle.

3. Simone got $\frac{3}{4}$ of a full bar of chocolate. Hank was supposed to get $\frac{1}{4}$ of the full chocolate bar, but he has to get his share from Simone. What fraction of Simone's chocolate should Hank get? Make a math drawing that helps you solve this problem. Use this drawing to help you explain your solution. Use our definition of fraction to describe the fractions in this problem, drawing attention to the unit amount that each fraction is *of*. What are the different unit amounts in this problem?

4. If one serving of juice gives you $\frac{3}{2}$ of your daily value of vitamin C, how much of your daily value of vitamin C will you get in $\frac{2}{3}$ of a serving of juice? Make a math drawing that helps you solve this problem. Use this drawing to help you explain your solution. Use our definition of fraction to describe the fractions in this problem, drawing attention to the unit amount that each fraction is *of*. What are the different unit amounts in this problem?

5. Show $\frac{1}{8}$ of the combined amount in the 3 pies in **Figure 2.9** and explain why your answer is correct.

Figure 2.9 Three pies.

6. Hermione has a potion recipe that calls for 4 drams of snake liver oil. Hermione wants to make $\frac{2}{3}$ of the potion recipe. Rather than calculate $\frac{2}{3}$ of the number 4, Hermione measures $\frac{2}{3}$ of a dram of snake liver oil 4 times and uses that amount of snake liver oil in her potion. Use drawings and our definition of fraction to explain why Hermione's method is valid.

7. Divide the shaded shape in **Figure 2.10** into 4 equal-sized parts in two different ways: one where all parts have the same area and shape, and one where all parts have the same area, but some parts do not have the same shape.

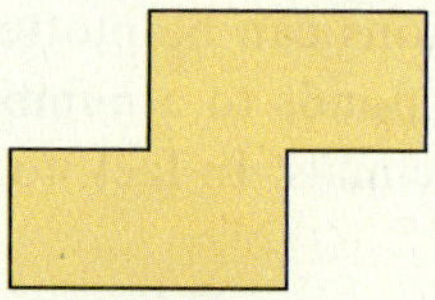

Figure 2.10 Dividing a shape into 4 equal parts.

8. Explain why we can use the fractions in (a) and (b) to describe the shaded region in **Figure 2.11**. How can two different numbers describe the same shaded region? How must we interpret the shaded region in each case?

a. $\frac{5}{6}$

b. $\frac{5}{3}$

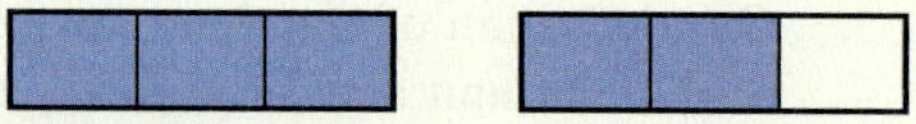

Figure 2.11 What fraction describes the shaded region?

9. Starting with a number line on which 0 and 1 have already been plotted, explain where to plot $\frac{5}{8}$ and explain why that location fits with the definition of fraction.

10. Plot 0, 1, and $\frac{8}{7}$ on a number line like the one in **Figure 2.12** in such a way that each number falls on a tick mark. Lengthen the tick marks of whole numbers.

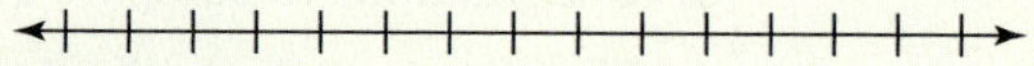

Figure 2.12 A number line for plotting fractions.

11. Place equally spaced tick marks on the number line in Figure 2.13 so that you can plot $\frac{1}{7}$ on a tick mark. Then plot $\frac{1}{7}$. Explain your reasoning.

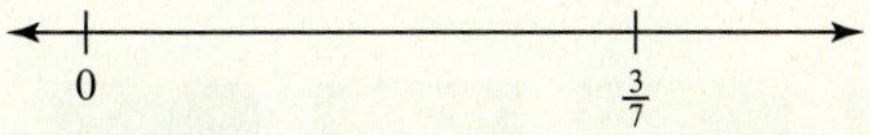

Figure 2.13 Plot $\frac{1}{7}$.

12. Place equally spaced tick marks on the number line in Figure 2.14 so that you can plot 1 on a tick mark. Then plot 1. Explain your reasoning.

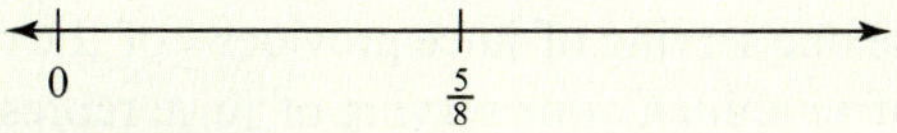

Figure 2.14 Plot 1.

13. Plot $\frac{11}{8}$ on a number line like the one in Figure 2.15.

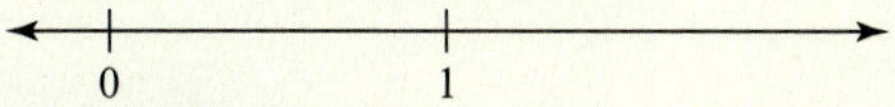

Figure 2.15 A number line.

Answers to Practice Exercises for Section 2.2

1. Because the rectangle shown in the practice exercise is $\frac{3}{4}$ of another rectangle, it must consist of 3 parts, each of which is one fourth of the other rectangle. So divide the given rectangle into 3 equal parts:

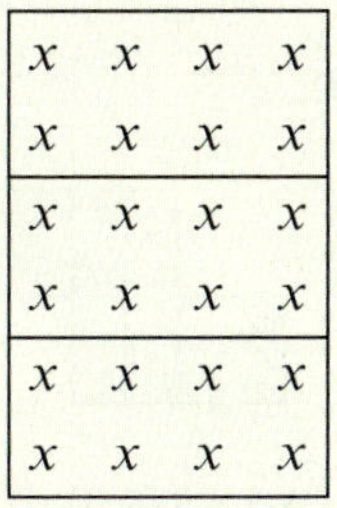

The original rectangle that we are looking for must consist of 4 of these parts because the parts are fourths of the original rectangle:

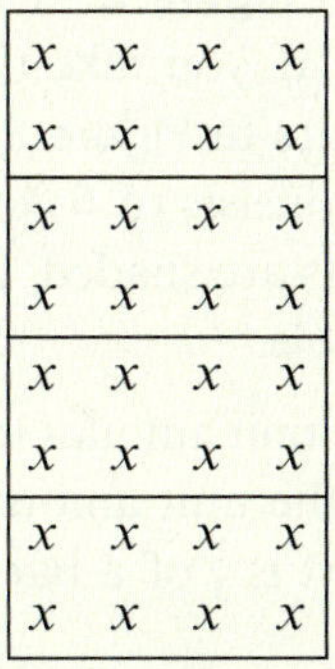

2. Because the rectangle shown in the practice exercise is $\frac{8}{3}$ of another rectangle, it must consist of 8 parts, each of which is $\frac{1}{3}$ of the other rectangle. So the rectangle shown in the practice exercise must consist of 8 equal parts:

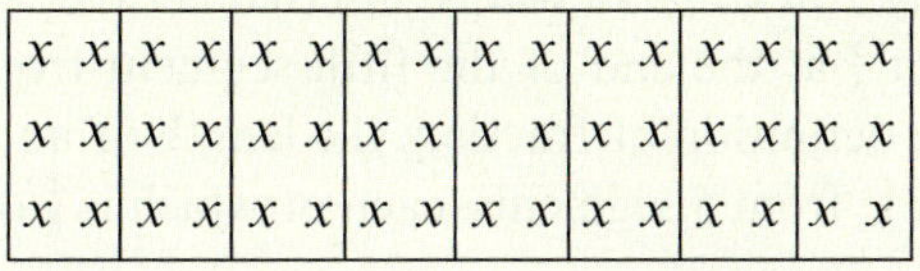

The original rectangle must consist of 3 of these parts:

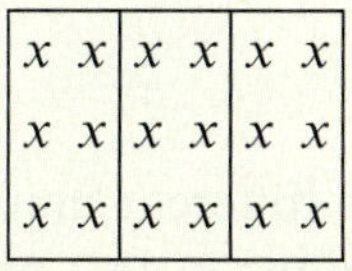

3. As Figure 2.16 shows, we can divide the full chocolate bar into 4 equal parts, and Simone has 3 of those 4 parts. One of Simone's 3 parts should go to Hank. Therefore, Hank should receive $\frac{1}{3}$ of Simone's chocolate.

Figure 2.16 Different ways to describe the same amount.

The 1 part that is to go to Hank is $\frac{1}{4}$ of the full bar of chocolate because it is 1 out of 4 equal parts making the full chocolate bar. But Hank's 1 part is also 1 out of 3 equal parts that make Simone's chocolate, so it is $\frac{1}{3}$ of Simone's chocolate. In the problem, the full bar of chocolate is one unit amount, but Simone's chocolate is another unit amount.

4. Because one serving of juice provides $\frac{3}{2}$ of the daily value of vitamin C, one serving of juice represents 3 parts, each of which is $\frac{1}{2}$ of the daily value of vitamin C. The daily value of vitamin C is represented in 2 of those parts, as shown in Figure 2.17. Those 2 parts are each $\frac{1}{3}$ of a serving of juice. So if you drink $\frac{2}{3}$ of a serving of juice, you will get the full daily value of vitamin C.

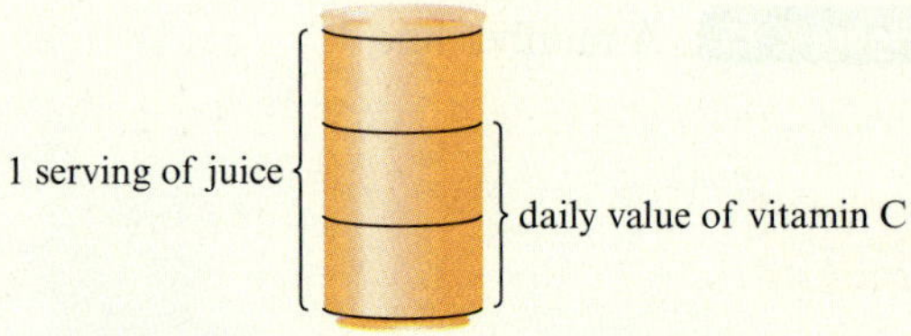

Figure 2.17 One serving of juice provides $\frac{3}{2}$ of the daily value of vitamin C.

There are two different unit amounts associated with the fractions in this problem: a serving of juice and the daily value of vitamin C.

5. If you shade $\frac{1}{8}$ of each pie individually, as in Figure 2.18, then collectively, the 3 shaded parts, which are labeled 1, are $\frac{1}{8}$ of the 3 pies combined. Why?

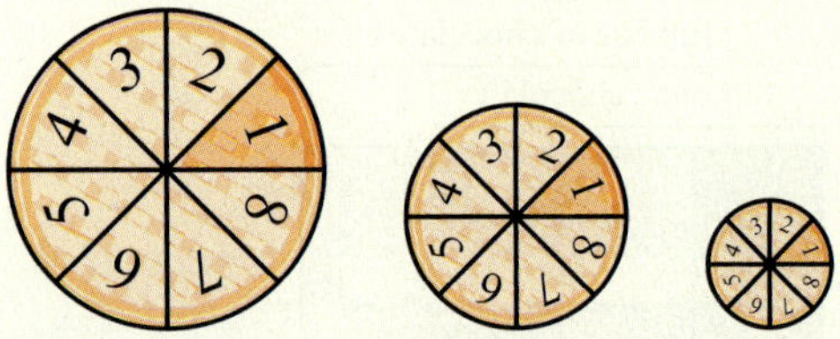

Figure 2.18 $\frac{1}{8}$ of three pies.

Because when 8 copies of the 3 shaded parts are joined, they form the whole three pies.

6. If we represent the 4 drams of snake liver oil with drawings of 4 jars, as at the top of Figure 2.19, then the shaded regions in the middle of Figure 2.19 represent $\frac{1}{3}$ of the 4 drams of snake liver oil, because 3 copies of these shaded regions make the full 4 drams. Therefore, 2 copies of the shaded regions in the middle of Figure 2.19 form $\frac{2}{3}$ of the 4 drams of snake liver oil, as shown at the bottom of Figure 2.19.

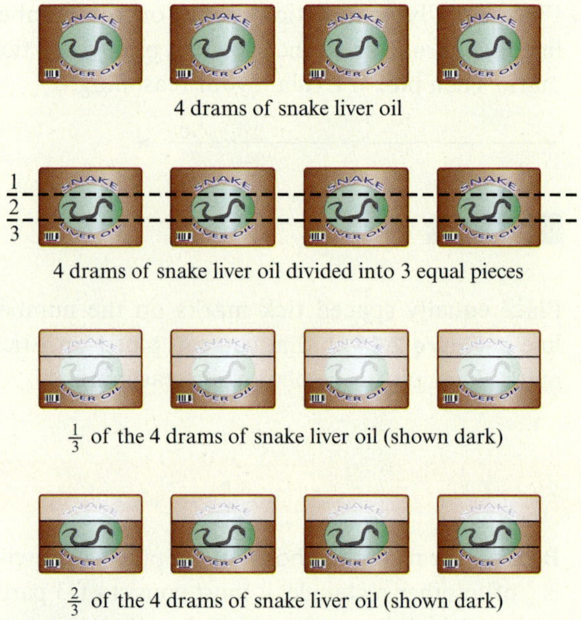

Figure 2.19 Snake liver oil.

7. See Figure 2.20.

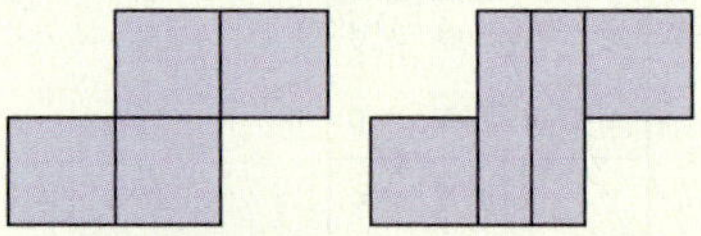

Figure 2.20 Two ways to divide the shape into 4 equal parts.

8. We can use different numbers to describe the shaded part in Figure 2.11 by choosing different unit amounts. If you take the unit amount to be the full two bars in Figure 2.11, then because the unit amount consists of 6 equal parts and because 5 of those parts are shaded, the shaded part is $\frac{5}{6}$ of the chosen whole.

   If you take the unit amount to be 1 bar (of 3 parts), then because the unit amount consists of 3 equal parts, each part is $\frac{1}{3}$ of a bar. So 5 parts are $\frac{5}{3}$ of a bar.

9. First, divide the line segment between 0 and 1 into 8 pieces of equal length. According to the definition of unit fractions, each of these segments is $\frac{1}{8}$ of a unit long. Use tick marks to show where these segments begin and end, but note that it turns out only 7 tick marks will be inserted between 0 and 1. Plot $\frac{5}{8}$ at the end of the fifth segment from 0. By the definition of fraction, the length of an interval made from 5 segments, each of which is $\frac{1}{8}$ of a unit

long, is $\frac{5}{8}$ of a unit. So the location of $\frac{5}{8}$ fits with the definition of fraction.

10. To plot $\frac{8}{7}$, we need to work with sevenths. Therefore, we must divide the 1-unit segment from 0 to 1 into 7 equal parts, as shown in Figure 2.21.

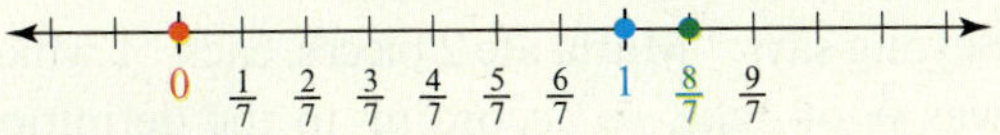

Figure 2.21 Plotting 0, 1, and $\frac{8}{7}$.

11. On a number line on which adjacent tick marks are $\frac{1}{7}$ apart, the point $\frac{3}{7}$ will be the third tick mark to the right of 0, and $\frac{1}{7}$ is the first tick mark, as shown in Figure 2.22.

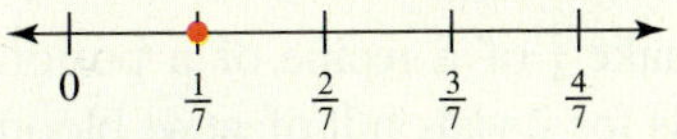

Figure 2.22 Adjacent tick marks are $\frac{1}{7}$ apart.

12. On a number line on which adjacent tick marks are $\frac{1}{8}$ apart, $\frac{5}{8}$ will be at the fifth tick mark to the right of 0 and $1 = \frac{8}{8}$ will be at the eighth tick mark to the right of 0, as shown in Figure 2.23.

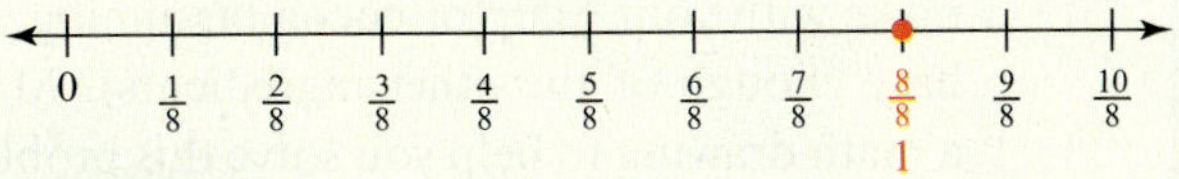

Figure 2.23 Adjacent tick marks are $\frac{1}{8}$ apart.

13. To plot $\frac{11}{8}$ on a number line, divide the 1-unit segment from 0 to 1 into 8 equal parts, and measure off 11 of those parts, as shown in Figure 2.24.

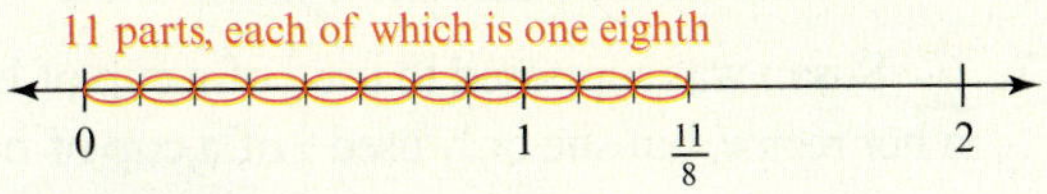

Figure 2.24 Plotting $\frac{11}{8}$.

PROBLEMS FOR SECTION 2.2

1. Michael says that the dark marbles in Figure 2.25 can't represent $\frac{1}{3}$ because "there are 5 marbles and 5 is more than 1, but $\frac{1}{3}$ is supposed to be less than 1." In a short paragraph, discuss the situation.

Figure 2.25 Dark and light marbles.

2. When Jean was asked to say what the 3 in the fraction $\frac{2}{3}$ means, Jean said that the 3 is the whole. Explain why it is not completely correct to say that "3 is the whole." What is a better way to say what the 3 in the fraction $\frac{2}{3}$ means?

3. This rectangle of x's is $\frac{4}{5}$ of another (original) rectangle of x's.

x x x x
x x x x
x x x x
x x x x
x x x x

Show the original rectangle. Explain why your solution is correct.

4. This rectangle of x's is $\frac{6}{5}$ of another (original) rectangle of x's.

x x x x x
x x x x x
x x x x x
x x x x x
x x x x x
x x x x x

Show the original rectangle. Explain why your solution is correct.

5. Kaitlyn gave $\frac{1}{2}$ of her candy bar to Arianna. Arianna gave $\frac{1}{3}$ of the candy she got from Kaitlyn to Cameron.

a. What fraction of a candy bar did Cameron get? Make a math drawing to help you solve this problem and explain your solution. Use our definition of fraction in your explanation and attend to the unit amount that each fraction is *of*.

b. Describe the different unit amounts that occur in part (a). Discuss how one amount can be described with two different fractions depending on what the unit amount is taken to be.

6. You were supposed to use $\frac{2}{3}$ of a cup of cocoa to make a batch of cookies, but you only have $\frac{1}{3}$ of a cup of cocoa.

a. What fraction of the cookie recipe can you make with your $\frac{1}{3}$ cup of cocoa (assuming you have enough of the other ingredients)? Make a math drawing to help you solve this problem and explain your solution. Use our definition of fraction in your explanation and attend to the unit amount that each fraction is *of*.

b. Describe the different unit amounts that occur in part (a). Discuss how one amount can be described with two different fractions depending on what the unit amount is taken to be.

7. Susan was supposed to use $\frac{5}{4}$ of a cup of butter in her recipe, but she only used $\frac{3}{4}$ of a cup of butter.

a. What fraction of the butter that she should have used did Susan actually use? Make a math drawing to help you solve this problem and explain your solution. Use our definition of fraction in your explanation and attend to the unit amount that each fraction is *of*.

b. Describe the different unit amounts that occur in part (a). Discuss how one amount can be described with two different fractions depending on what the unit amount is taken to be.

8. If $\frac{3}{4}$ of a cup of a snack food gives you your daily value of calcium, then what fraction of your daily value of calcium is in 1 cup of the snack food? Make a math drawing to help you solve this problem and explain your solution.

9. Make up a word problem or situation where *one* object (or collection, or quantity) is *both* $\frac{1}{2}$ of something and $\frac{1}{3}$ of something else.

10. *Cake problem:* Figure 2.26 represents two cakes of different sizes. Each cake was divided into 12 equal pieces. Marla ate one piece from each cake. What fraction of the total amount of cake (in the two cakes combined) did Marla eat?

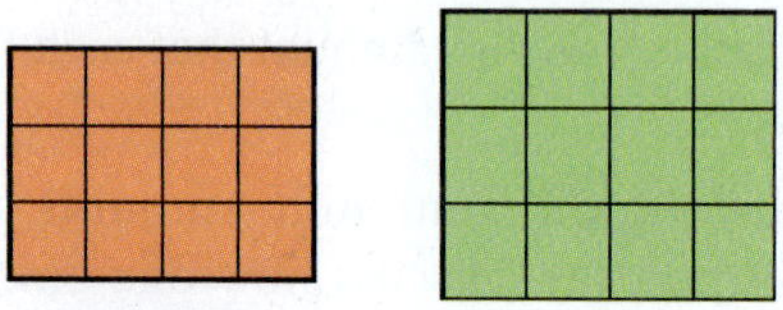

Figure 2.26 Two cakes, each divided into 12 equal pieces.

a. Ben says: "Marla ate $\frac{2}{24}$ of the cake because she ate 2 pieces out of a total of 24 pieces." Discuss Ben's reasoning. Is it correct? Explain why or why not. If the two cakes were the same size instead of different sizes, would Ben's reasoning be correct in that case? Explain.

b. Seyong says: "Marla ate 2 pieces, each of which was $\frac{1}{12}$ of cake, so according to the definition of fraction, Marla ate $\frac{2}{12}$ of the cake." Discuss Seyong's reasoning. Is it correct? Explain why or why not. What if the two cakes each weighed 1 pound? In that case, could Seyong's reasoning be used to make a correct statement? Explain.

c. Explain how to solve the cake problem with valid reasoning that uses our definition of fraction.

11. Harry wants to make $\frac{3}{4}$ of a recipe of a potion. The full recipe calls for 2 vials full of newt blood. Instead of calculating $\frac{3}{4}$ of 2, Harry measures $\frac{3}{4}$ of a vial of newt blood 2 times and uses this amount to make $\frac{3}{4}$ of his potion recipe. Is Harry's method valid? Using our definition of fractions, explain why or why not.

12. Figure 2.27 shows a picture of three pies. Draw a copy of the three pies. Shade $\frac{3}{8}$ of the combined amount in the three pies, and use our definition of fractions to explain why your answer is correct.

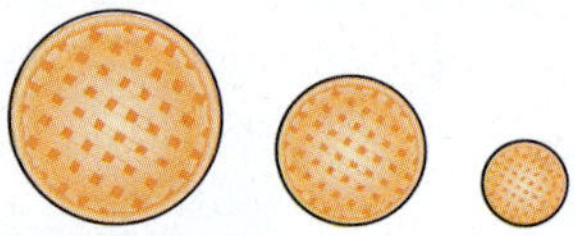

Figure 2.27 Three pies.

13. Discuss why it can be confusing to show an improper fraction such as $\frac{7}{3}$ with pieces of an object. What is another way to show the fraction $\frac{7}{3}$?

14. Draw a number line like the one in Figure 2.28 and plot $\frac{1}{3}$ and $\frac{2}{3}$ on your number line. In your own words, explain why those locations for these fractions fit with our definition of fraction.

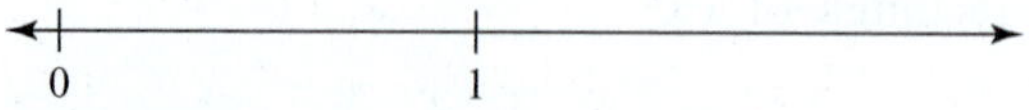

Figure 2.28 A number line.

15. Draw a number line like the one in Figure 2.28 and plot $\frac{5}{3}$ on your number line. In your own words, explain why this location fits with our definition of fraction.

16. Draw a number line like the one in Figure 2.29. Then plot 0, 1, $\frac{4}{3}$, and $\frac{11}{3}$ on your number line in such a way that each number falls on a tick mark. Lengthen the tick marks of whole numbers.

Figure 2.29 A number line.

17. Give two different fractions that you can legitimately use to describe the shaded region in Figure 2.30. For each fraction, explain why you can use that fraction to describe the shaded region. Write an unambiguous question about the shaded region in Figure 2.30 that can be answered by naming a fraction. Explain why your question is not ambiguous.

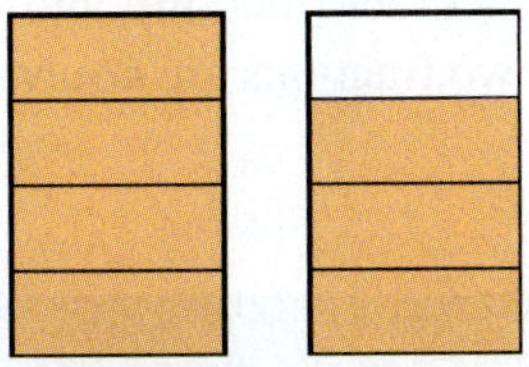

Figure 2.30 Describe the shaded part with a fraction.

18. Marquez says that the shaded region in Figure 2.31 represents the fraction $\frac{9}{12}$. Carmina says the shaded region in Figure 2.31 represents the fraction $\frac{9}{4}$. Explain why each of the two students' answers can be considered correct. Then write an unambiguous question about the shaded region in Figure 2.31 that can be answered by naming a fraction. Explain why your question is not ambiguous.

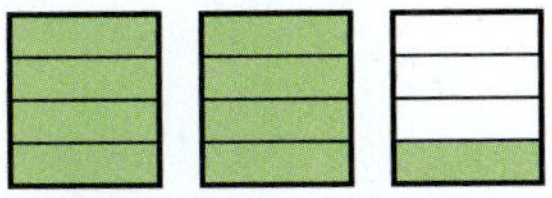

Figure 2.31 A shaded region.

***19.** NanHe made a design that used hexagons, rhombuses, and triangles like the ones shown in Figure 2.32. NanHe counted how many of each shape she used in her design and determined that $\frac{4}{11}$ of the shapes she used were hexagons, $\frac{5}{11}$ were rhombuses, and $\frac{2}{11}$ were triangles. You may combine your answers to parts (a) and (b).

a. Even though $\frac{4}{11}$ of the shapes NanHe used were hexagons, does this mean that the hexagons in NanHe's design take up $\frac{4}{11}$ of the area of her design? If not, what fraction of the area of NanHe's design do the hexagons take up?

b. Write a short paragraph describing how the notion of "equal parts" relates to your answer in part (a) of this problem.

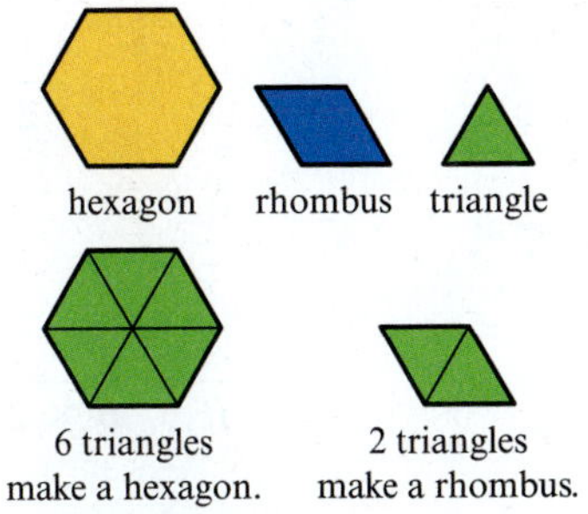

Figure 2.32 Pattern blocks.

***20.** A county has two elementary schools, both of which have an after-school program for the convenience of working parents. In school A, $\frac{1}{3}$ of the children attend the after-school program; in school B, $\frac{2}{3}$ of the children attend the after-school program. There are a total of 1500 children in the two schools combined.

a. If school A has 600 students and school B has 900 students, then what is the fraction of elementary school students in the county who attend the after-school program? What if it's the other way around, and school A has 900 students and school B has 600 students?

b. Using the data in the problem statement, make up two more examples of numbers of students in each school. (These don't have to be entirely realistic numbers.) For each of your examples, determine the fraction of elementary school children in the county who attend the after-school program.

c. Jamie says that if $\frac{1}{3}$ of the children in school A attend the after-school program, and if $\frac{2}{3}$ of the children in the school B attend the after-school program, then $\frac{1}{2}$ of the children from the two schools combined attend the after-school program. Jamie arrives at this answer because she says $\frac{1}{2}$ is halfway between $\frac{1}{3}$ and $\frac{2}{3}$. Is Jamie's answer of $\frac{1}{2}$ *always* correct? Is Jamie's answer of $\frac{1}{2}$ *ever* correct, and if so, under what circumstances?

2.3 Reasoning About Equivalent Fractions

CCSS Common Core State Standards Grades 4, 5

When we write a whole number in base-ten notation, there is only one way to do so (without the use of a decimal point). The only way to write 1234 in base ten (without a decimal point) is 1234. But the case of fractions is completely different because every fraction is equal to *infinitely many* other fractions. For example,

$$\frac{1}{2} = \frac{2}{4} = \frac{3}{6} = \frac{4}{8} = \frac{5}{10} = \cdots$$

In general,

$$\frac{A}{B} = \frac{A \times 2}{B \times 2} = \frac{A \times 3}{B \times 3} = \frac{A \times 4}{B \times 4} = \frac{A \times 5}{B \times 5} = \cdots$$

equivalent

In this section, we will see why every fraction is equal to infinitely many other fractions, and we will study some consequences of this fact. First some terminology: we say two fractions are **equivalent** if they represent the same number, in other words, if they are equal.

CLASS ACTIVITY

2I Explaining Equivalent Fractions, p. CA-32

2J Critique Fraction Equivalence Reasoning, p. CA-33

Why Is Every Fraction Equal to Infinitely Many Other Fractions?

Equivalent Fractions

As we consider why every fraction is equal to infinitely many other fractions, let's start with an example. Why is

$$\frac{3}{4}$$

of a rod the same amount of the rod as

$$\frac{3 \times 5}{4 \times 5} = \frac{15}{20}$$

CCSS 4.NF.1

of the same rod? We can use our definition of fraction to explain this equivalence. To form $\frac{3}{4}$ of the rod, first divide the rod into 4 equal parts. Then $\frac{3}{4}$ of the rod consists of 3 parts. Divide each of the 4 equal parts into 5 small, equal parts, as shown in Figure 2.33. Now the rod consists of a total of 4×5 or 20 small parts. The original 3 parts representing $\frac{3}{4}$ of the rod have each been subdivided into 5 equal parts; therefore, these 3 parts have become 3×5 or 15 smaller parts, as shown in the shaded portions of Figure 2.33. So, 3 of the original $\frac{1}{4}$ parts of the rod is the same amount of the rod as 3×5 of the smaller $\frac{1}{20}$ parts of the rod. Therefore,

$$\frac{3}{4} = \frac{3 \times 5}{4 \times 5} = \frac{15}{20}$$

Figure 2.33

Subdivide each part into 5 parts.

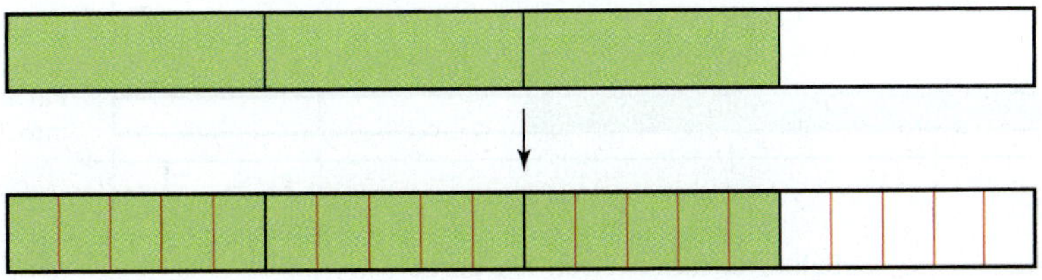

	Number of parts shaded	Size of each part
$\frac{3}{4}$	3	$\frac{1}{4}$
$\frac{3 \cdot 5}{4 \cdot 5}$	$3 \cdot 5$	$\frac{1}{4 \cdot 5}$
$\frac{15}{20}$	15	$\frac{1}{20}$

When showing this equation to students, you might prefer to condense it as such:

$$\overset{\times 5}{\frac{3}{4} = \frac{15}{20}}_{\times 5}$$

There wasn't anything special about the numbers 3, 4, and 5 used here—the same reasoning will apply to other counting numbers substituted for these numbers. Therefore, in general, if N is any counting number, $\frac{A}{B}$ of a whole is the same amount of the whole as $\frac{A \times N}{B \times N}$ of the whole. In other words,

$$\frac{A}{B} = \frac{A \times N}{B \times N}$$

CCSS 5.NF.5b

Another way to explain why $\frac{A}{B} = \frac{A \times N}{B \times N}$ is to multiply by 1 in the form of $\frac{N}{N}$:

$$\begin{aligned} \frac{A}{B} &= \frac{A}{B} \times 1 \\ &= \frac{A}{B} \times \frac{N}{N} \\ &= \frac{A \times N}{B \times N} \end{aligned}$$

This explanation for why $\frac{A}{B} = \frac{A \times N}{B \times N}$ relies on the more advanced notion of fraction multiplication, whereas the previous explanation relies only on the definition of fraction.

Why Use Equivalent Fractions?

Because every fraction is equal to infinitely many other fractions, you have a lot of flexibility when you work with fractions. When you start with a fraction, say, $\frac{3}{4}$, you know that it is always equal to many other fractions with larger denominators, for example,

$$\frac{3}{4} = \frac{6}{8} = \frac{9}{12} = \frac{12}{16} = \frac{15}{20} = \cdots$$

common denominators

When working with two fractions simultaneously, it is often desirable to give them **common denominators**, which just means the *same* denominators. Why?

When we give fractions common denominators, we describe the fractions in terms of *like parts*. The fractions $\frac{2}{3}$ and $\frac{3}{5}$ are in terms of thirds and fifths, respectively. We can write these two fractions with the common denominator $3 \times 5 = 15$:

$$\frac{2}{3} = \frac{2 \times 5}{3 \times 5} = \frac{10}{15}$$
$$\frac{3}{5} = \frac{3 \times 3}{5 \times 3} = \frac{9}{15}$$

As you see in **Figure 2.34**, when we give $\frac{2}{3}$ and $\frac{3}{5}$ common denominators, we subdivide the thirds of $\frac{2}{3}$ and the fifths of $\frac{3}{5}$ into fifteenths, so that both fractions are described with like parts, namely fifteenths.

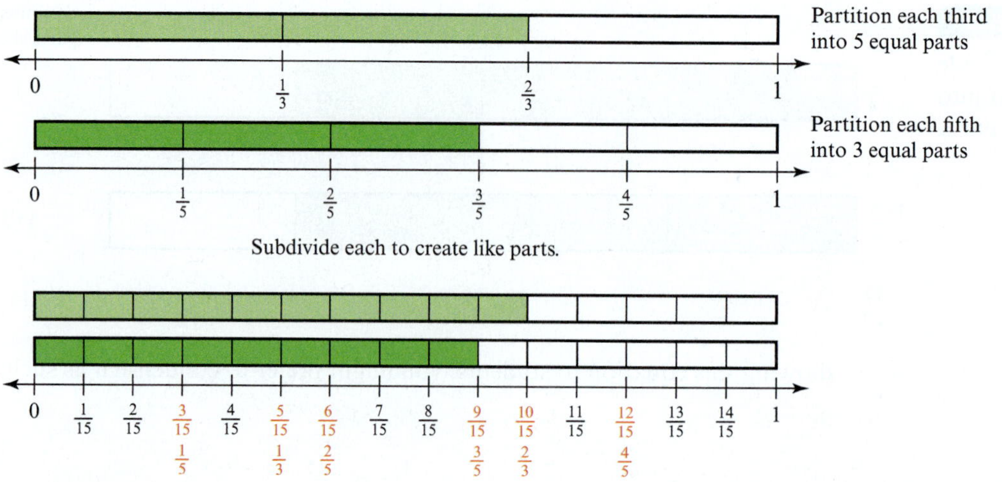

Figure 2.34 Common denominators give fractions like parts.

The fractions $\frac{2}{3}$ and $\frac{3}{5}$ have many other common denominators as well, such as 30 and 45, but 15 is the smallest one.

For any two fractions, multiplying the denominators always produces a common denominator; however, it may not be the least common denominator. The number 24 is a common denominator for the fractions $\frac{3}{8}$ and $\frac{5}{6}$, whereas $8 \times 6 = 48$. We will discuss least common denominators again in Section 8.5 when we discuss least common multiples.

When you solve problems, don't just assume you will need to find common denominators. We use common denominators when we want to work with parts of the *same size*. In some problem-solving situations, you might need fractions with the *same number* of parts. Then you should find equivalent fractions with *common numerators*. When solving problems, look for equivalent fractions that help you reason about the situation.

CLASS ACTIVITY

2K Interpreting and Using Common Denominators, p. CA-34

2L Solving Problems by Using Equivalent Fractions, p. CA-35

2M Problem Solving with Fractions on Number Lines, p. CA-36

2N Measuring One Quantity with Another, p. CA-37

How Can we Interpret Simplifying Fractions?

Every fraction is equal to infinitely many other fractions, but in a collection of fractions that are equal to each other, there is one that is the simplest.

simplest form A fraction $\frac{A}{B}$ (where A and B are whole numbers and B is not zero) is said to be in **simplest form** (or in lowest terms) if there is no whole number other than 1 that divides both A and B evenly.

The fraction

$$\frac{3}{4}$$

is in simplest form because no whole number other than 1 divides both 3 and 4 evenly. The fraction

$$\frac{30}{35}$$

is not in simplest form because the number 5 divides both 30 and 35 evenly. Notice, however, that we can put the fraction $\frac{30}{35}$ in simplest form as follows:

$$\frac{30}{35} = \frac{6 \times 5}{7 \times 5} = \frac{6}{7} \tag{2.1}$$

This equation is the reverse of the equation that gives $\frac{6}{7}$ the denominator 35:

$$\frac{6}{7} = \frac{6 \times 5}{7 \times 5} = \frac{30}{35}$$

When showing Equation 2.1 to students, you might prefer to condense it as

$$\frac{30}{35} \overset{\div 5}{=} \frac{6}{7} \quad (\div 5 \text{ on numerator and denominator})$$

simplify To **simplify** a fraction means to write it as an equivalent fraction by dividing the numerator and denominator by the same number, as in the example just given.

If you think in terms of rods, simplifying a fraction is like joining together small rod pieces to make larger rod pieces, as shown in **Figure 2.35**, with the example

$$\frac{18}{24} \overset{\div 6}{=} \frac{3}{4} \quad (\div 6 \text{ on numerator and denominator})$$

Figure 2.35

Simplifying.

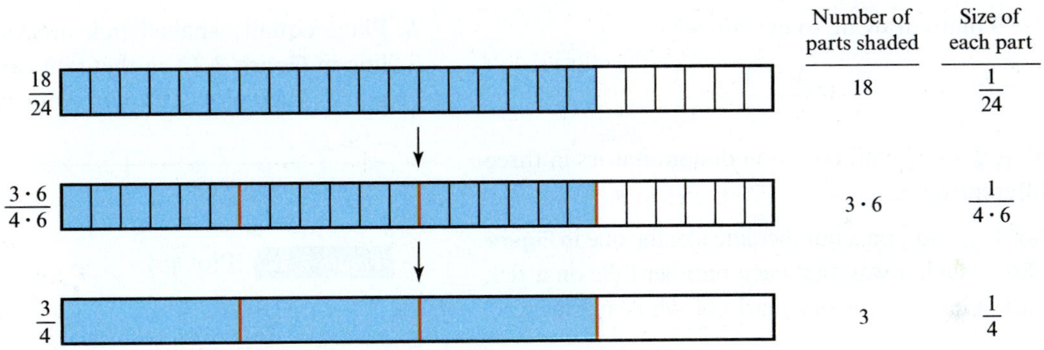

In other words, simplifying a fraction reverses the procedure of subdividing rod pieces that was demonstrated earlier in **Figure 2.33**.

Notice that we can sometimes use several steps when simplifying fractions. For example:

$$\frac{18}{24} \overset{\div 2}{=} \frac{9}{12} \overset{\div 3}{=} \frac{3}{4} \quad (\div 2, \text{ then } \div 3 \text{ on numerator and denominator})$$

We will discuss simplifying fractions again in Section 8.5 when we discuss greatest common factors.

SECTION SUMMARY AND STUDY ITEMS

Section 2.3 Reasoning About Equivalent Fractions

Any fraction $\frac{A}{B}$ is equivalent to infinitely many other fractions via this relationship:

$$\frac{A}{B} = \frac{A \times N}{B \times N}$$

We often call fractions that are equal and represent the same number *equivalent*.

We use equivalent fractions to give two fractions common denominators. When we give fractions common denominators, we express the fractions in terms of like parts. We can simplify fractions by dividing the numerator and denominator by the same number. The simplest form of a fraction is an equivalent fraction that has the smallest possible numerator and denominator.

Key Skills and Understandings

1. Given a fraction, use our definition of fraction, math drawings, and number lines to explain why multiplying the numerator and denominator by the same counting number produces an equivalent fraction. In the process, attend to the number of parts and the size of the parts.
2. Explain that giving fractions common denominators expresses the fractions in terms of like parts.
3. Solve problems in which you will need to make equivalent fractions and justify your solutions.
4. Simplify fractions, and explain the process in terms of math drawings.
5. Given two fractions on a number line, place equally spaced tick marks to plot other fractions on the number line as well.

Practice Exercises for Section 2.3

1. Use a math drawing to explain why

$$\frac{4}{5} = \frac{4 \times 3}{5 \times 3} = \frac{12}{15}$$

2. Write $\frac{3}{8}$ and $\frac{5}{6}$ with common denominators in three different ways.

3. Plot 1, $\frac{5}{3}$, and $\frac{7}{4}$ on a number line like the one in **Figure 2.36** in such a way that each number falls on a tick mark. Lengthen the tick marks of whole numbers.

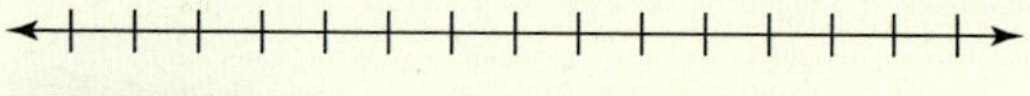

Figure 2.36 A number line for plotting fractions.

4. Plot 1, 0.9, and $\frac{5}{4}$ on a number line like the one in **Figure 2.36** in such a way that each number falls on a tick mark. Lengthen the tick marks of whole numbers.

5. The line segment in **Figure 2.37** has length $\frac{9}{8}$ unit. Explain how to subdivide the line segment and possibly add pieces to the line segment to create a line segment of length $\frac{3}{4}$ unit. Do this without first creating a segment of length 1 unit. Explain how you know the resulting line segment is the correct length.

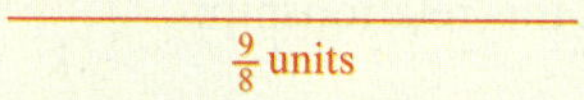

Figure 2.37 Create a line segment of length $\frac{3}{4}$ units given that this line segment has length $\frac{9}{8}$ units.

6. Place equally spaced tick marks on the number line in **Figure 2.38** so that you can plot $\frac{2}{3}$ on a tick mark. Then plot $\frac{2}{3}$. Explain your reasoning.

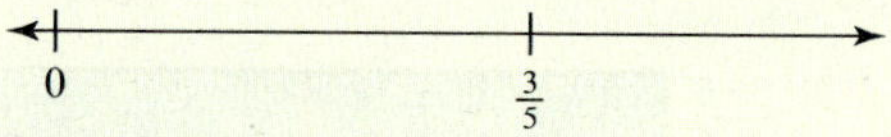

Figure 2.38 Plot $\frac{2}{3}$.

7. Plot $\frac{29}{3}$ on a number line like the one in **Figure 2.39**.

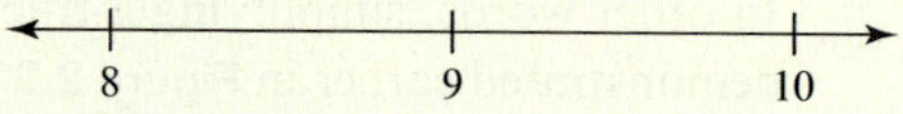

Figure 2.39 A number line.

8. Kelsey has $\frac{3}{5}$ of a chocolate bar. She wants to give some of her chocolate bar to Janelle. What fraction of her chocolate bar should Kelsey give Janelle so that Kelsey will be left with $\frac{1}{2}$ of the original chocolate bar?

a. Make math drawings to help you solve this problem. Explain your answer, attending carefully to the unit amount that each fraction is *of*.

b. In solving this problem, how do $\frac{3}{5}$ and $\frac{1}{2}$ each appear as equivalent fractions?

9. Ken ordered $\frac{3}{4}$ of a ton of gravel. Ken wants $\frac{1}{4}$ of his order of gravel delivered now and $\frac{3}{4}$ delivered later. What fraction of a ton of gravel should Ken get delivered now?

a. Make math drawings to help you solve this problem. Explain your answer, attending carefully to the unit amount that each fraction is *of*.

b. In solving the problem, how does $\frac{3}{4}$ appear as an equivalent fraction?

10. Use the math drawing in **Figure 2.40** to help you explain the equation simplifying $\frac{10}{15}$:

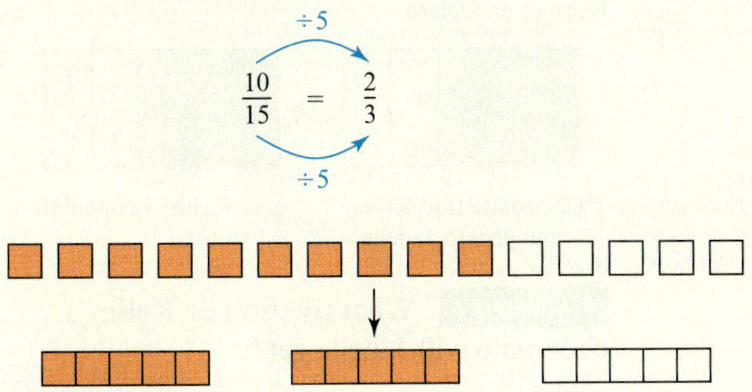

Figure 2.40 Explaining why $\frac{10}{15} = \frac{2}{3}$.

11. Put the following fractions in simplest form:

$$\frac{45}{72}, \frac{24}{36}, \frac{56}{88}$$

12. The strips in **Figure 2.41** show the relative amounts of money that Annie and Aria have saved. Write two sentences, one in which you use a fraction to describe how Aria's savings compare with Annie's and another in which you use a fraction to describe how Annie's savings compare with Aria's. For each fraction, state what its unit amount is.

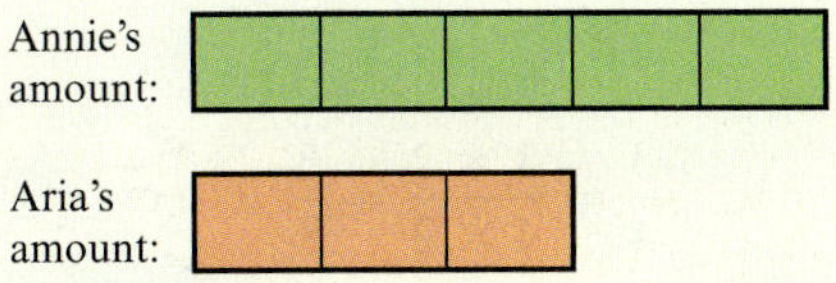

Figure 2.41 Annie's and Aria's savings.

Answers to Practice Exercises for Section 2.3

1. If you have a rod that is divided into 5 equal parts, and 4 are shown shaded, then if you divide each of the 5 parts into 3 smaller parts, the shaded amount will now consist of 4×3 small parts, and the whole rod will consist of 5×3 small parts. Thus, the shaded part of the rod can be described both as $\frac{4}{5}$ of the rod and as

$$\frac{4 \times 3}{5 \times 3} = \frac{12}{15}$$

of the rod. Therefore $\frac{4}{5}$ and $\frac{12}{15}$ represent the same number.

2. There are infinitely many ways to write $\frac{3}{8}$ and $\frac{5}{6}$ with common denominators. Here are two:

$$\frac{3}{8} = \frac{3 \times 6}{8 \times 6} = \frac{18}{48}$$
$$\frac{5}{6} = \frac{5 \times 8}{6 \times 8} = \frac{40}{48}$$

$$\frac{3}{8} = \frac{3 \times 3}{8 \times 3} = \frac{9}{24}$$
$$\frac{5}{6} = \frac{5 \times 4}{6 \times 4} = \frac{20}{24}$$

3. Common denominators will make like sized parts.

$$\frac{5}{3} = \frac{5 \times 4}{3 \times 4} = \frac{20}{12}$$
$$\frac{7}{4} = \frac{7 \times 3}{4 \times 3} = \frac{21}{12}$$

$$1 = \frac{12}{12}$$

Therefore $\frac{5}{3}, \frac{7}{4}$ and 1 can be plotted as shown in **Figure 2.42**.

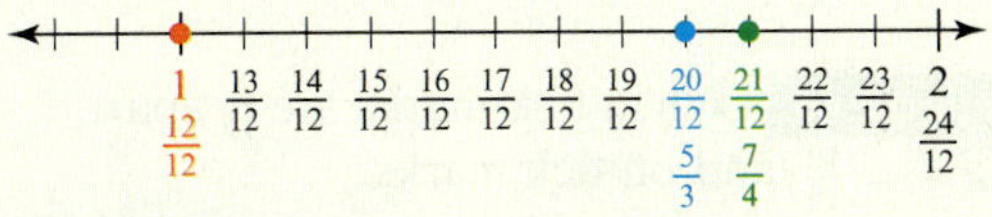

Figure 2.42 Plotting 1, $\frac{5}{3}$, and $\frac{7}{4}$.

4. Common denominators will make like sized parts.

$$0.9 = \frac{9}{10} = \frac{9 \times 2}{10 \times 2} = \frac{18}{20}$$
$$\frac{5}{4} = \frac{5 \times 5}{4 \times 5} = \frac{25}{20}$$

$$1 = \frac{20}{20}$$

Therefore 1, 0.9, and $\frac{5}{4}$ can be plotted as shown in Figure 2.43.

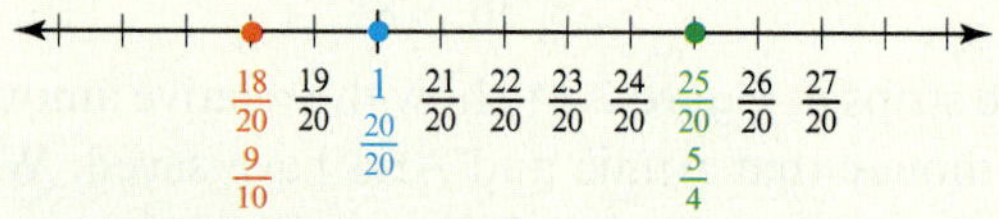

Figure 2.43 Plotting 1, 0.9, and $\frac{5}{4}$.

5. Because the original line segment is $\frac{9}{8}$ units long, it consists of 9 equal parts, each of which is $\frac{1}{8}$ of a unit long.

Since

$$\frac{3}{4} = \frac{3 \times 2}{4 \times 2} = \frac{6}{8}$$

we should create a line segment consisting of 6 of the $\frac{1}{8}$-unit-long parts, as shown in Figure 2.44.

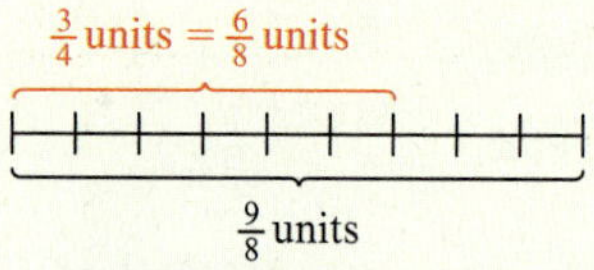

Figure 2.44 From $\frac{9}{8}$ units to $\frac{3}{4}$ units.

6. $\frac{3}{5}$ and $\frac{2}{3}$ can both be expressed in terms of fifteenths. So, if we plot these fractions on a number line on which adjacent tick marks are $\frac{1}{15}$ apart, they will land on tick marks. The fraction $\frac{3}{5}$ will be at the ninth tick mark to the right of 0, and $\frac{2}{3}$ will be at the tenth tick mark to the right of 0, as shown in Figure 2.45.

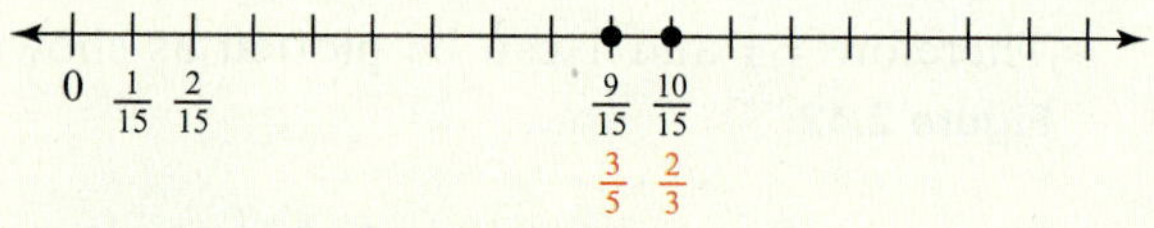

Figure 2.45 When tick marks are $\frac{1}{15}$ apart, $\frac{3}{5} = \frac{9}{15}$ and $\frac{2}{3} = \frac{10}{15}$ land on tick marks.

7. The fraction $\frac{29}{3}$ is located 29 one-thirds to the right of 0. Because $8 = \frac{8}{1} = \frac{24}{3}$ and $9 = \frac{9}{1} = \frac{27}{3}$ and $10 = \frac{10}{1} = \frac{30}{3}$, it follows that $\frac{29}{3}$ is between 9 and 10 as shown in Figure 2.46.

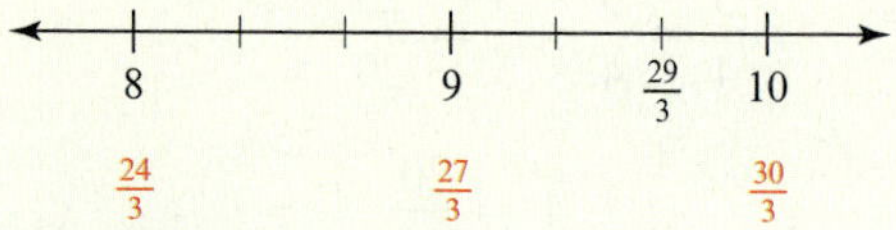

Figure 2.46 Plotting $\frac{29}{3}$.

8. a. See Figure 2.47. If the original chocolate bar is divided into 10 equal parts, then the amount of chocolate that Kelsey has is $\frac{6}{10}$ of the original chocolate. The amount that Kelsey wants to keep is $\frac{5}{10}$ of the original chocolate; so Kelsey should give 1 part out of her 6 parts to Janelle. Therefore, Janelle should get $\frac{1}{6}$ of Kelsey's chocolate.

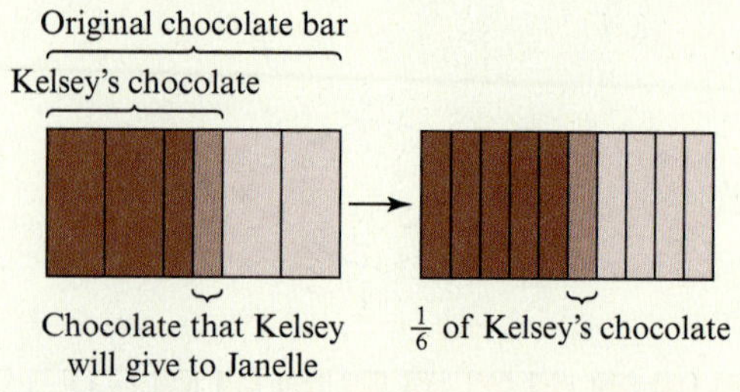

Figure 2.47 What fraction of Kelsey's chocolate will Janelle get?

b. In solving this problem, $\frac{3}{5}$ becomes

$$\frac{3 \times 2}{5 \times 2} = \frac{6}{10}$$

and $\frac{1}{2}$ becomes

$$\frac{1 \times 5}{2 \times 5} = \frac{5}{10}$$

9. a. In Figure 2.48, 1 ton of gravel is represented by a rectangle, and Ken's order is represented by $\frac{3}{4}$ of the rectangle. Then Ken's order is divided into 4 equal parts and of those parts one is darkly shaded. This darkly shaded amount is $\frac{3}{16}$ of the original rectangle representing 1 ton. Therefore, $\frac{1}{4}$ of Ken's order is $\frac{3}{16}$ of a ton.

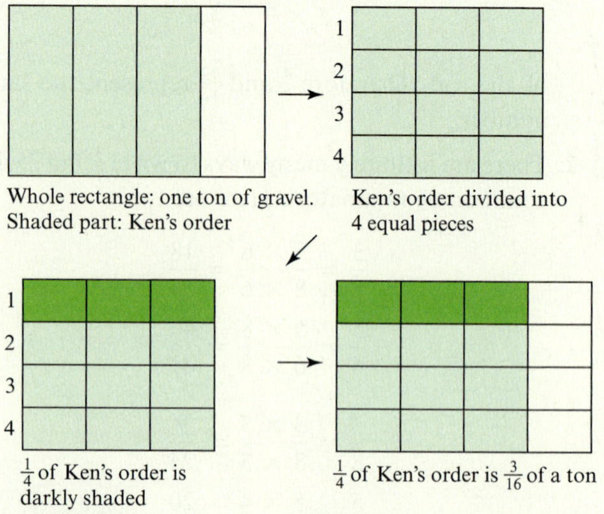

Figure 2.48 What fraction of a ton should Ken get now?

b. In solving this problem, the $\frac{3}{4}$ of a ton of gravel appears as

$$\frac{3 \times 4}{4 \times 4} = \frac{12}{16}$$

of a ton of gravel.

10. If we look at the collection of squares at the top of Figure 2.40, we see that 10 out of 15 of these squares are shaded. Therefore, $\frac{10}{15}$ of the squares are shaded. The squares can be put in groups of 5. Viewing the squares in groups, we see that 2×5 out of 3×5 squares are shaded, so we can also say that $\frac{2 \times 5}{3 \times 5}$ of the squares are shaded. The squares form 3 groups, and 2 of those groups are shaded. So $\frac{2}{3}$ of the squares are shaded. Therefore, we have three ways of writing the fraction of squares that are shaded, and these three ways must all be equal:

$$\frac{10}{15} = \frac{2 \times 5}{3 \times 5} = \frac{2}{3}$$

11.

$$\frac{45}{72} = \frac{5 \times 9}{8 \times 9} = \frac{5}{8}$$

$$\frac{24}{36} = \frac{2 \times 12}{3 \times 12} = \frac{2}{3}$$

$$\frac{56}{88} = \frac{7 \times 8}{11 \times 8} = \frac{7}{11}$$

Notice that we could also simplify each of these fractions in several steps. For example,

$$\frac{45}{72} = \frac{15 \times 3}{24 \times 3} = \frac{15}{24} = \frac{5 \times 3}{8 \times 3} = \frac{5}{8}$$

12. Aria's savings are $\frac{3}{5}$ as much as Annie's savings. Annie's savings are $\frac{5}{3}$ as much as Aria's savings. You can also say that Annie's savings are $1\frac{2}{3}$ times as much as Aria's savings. The unit amount for $\frac{3}{5}$ is Annie's savings. The unit amount for $\frac{5}{3}$ (or $1\frac{2}{3}$) is Aria's savings.

PROBLEMS FOR SECTION 2.3

1. Use a math drawing to explain in your own words why multiplying the numerator and denominator of a fraction by the same number results in the same number (equivalent fraction). Give a general explanation, but illustrate with

$$\frac{3}{4} = \frac{3 \times 3}{4 \times 3}$$

Students may find it confusing that we *multiply* the number yet we *divide* the parts in the math drawing. Attend carefully to this point and to the number and size of the parts.

2. Explain in two different ways why

$$\frac{2}{3} = \frac{2 \times 4}{3 \times 4}$$

3. Using the fractions $\frac{2}{3}$ and $\frac{3}{4}$, describe how to give two fractions common denominators. In terms of math drawings, what are you doing when you give fractions common denominators? What stays the same and what changes?

4. Using a math drawing, explain why dividing both the numerator and denominator of

$$\frac{6}{8}$$

by 2 produces the same number (an equivalent fraction). Discuss how to see division by 2 in both the numerator and denominator in terms of your math drawing. Attend carefully to points that might be difficult for students.

5. Using a math drawing with the example

$$\frac{6}{9}$$

describe how to simplify a fraction, and explain why the procedure makes sense. In terms of the math drawing, what are you doing when you simplify the fraction? What changes and what stays the same?

6. Draw a number line like the one in Figure 2.49. Then plot 0, $\frac{3}{8}$, and $\frac{1}{3}$ on your number line in such a way that each number falls on a tick mark. Lengthen the tick marks of whole numbers.

Figure 2.49 Number line.

7. Draw a number line like the one in Figure 2.49. Then plot $\frac{1}{2}$, $\frac{3}{5}$, and $\frac{5}{8}$ on your number line in such a way that each number falls on a tick mark. Lengthen the tick marks of whole numbers (if there are any).

8. Draw a number line like the one in Figure 2.49. Then plot 0, 0.3, and $\frac{3}{5}$ on your number line in such a way that each number falls on a tick mark. Lengthen the tick marks of whole numbers.

9. Draw a number line like the one in Figure 2.49. Then plot 1, $\frac{5}{6}$, and $\frac{8}{7}$ on your number line in such a way that each number falls on a tick mark. Lengthen the tick marks of whole numbers.

10. Draw a number line like the one in Figure 2.50. Place equally spaced tick marks on your number line so that you can plot $\frac{1}{3}$ on a tick mark without first plotting 1. Then plot $\frac{1}{3}$. Explain your reasoning.

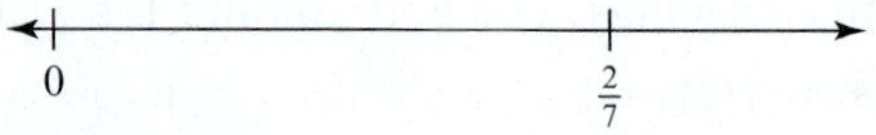

Figure 2.50 Plot $\frac{1}{3}$.

11. Draw a number line like the one in Figure 2.51. Place equally spaced tick marks on your number line so that you can plot $\frac{1}{2}$ on a tick mark without first plotting 1. Then plot $\frac{1}{2}$. Explain your reasoning.

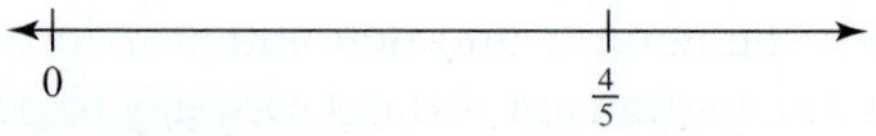

Figure 2.51 Plot $\frac{1}{2}$.

12. Draw a number line like the one in Figure 2.52. Place equally spaced tick marks on your number line so that you can plot $\frac{5}{12}$ on a tick mark without first plotting 1. Then plot $\frac{5}{12}$. Explain your reasoning.

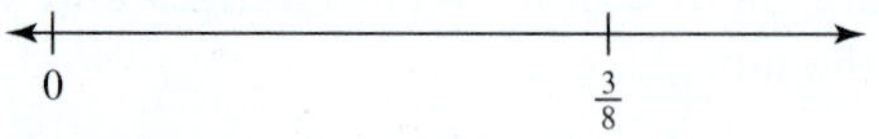

Figure 2.52 Plot $\frac{5}{12}$.

13. Draw a number line like the one in Figure 2.53. Plot $\frac{47}{6}$ on your number line. Explain your reasoning.

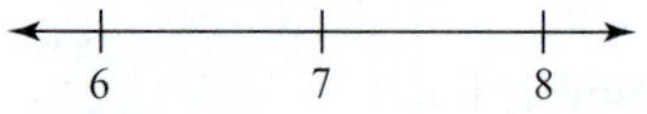

Figure 2.53 Plot $\frac{47}{6}$.

14. Erin says the tick mark indicated in Figure 2.54 should be labeled 2.2. Is Erin right or not? If not, why not, and how can she label the tick mark properly?

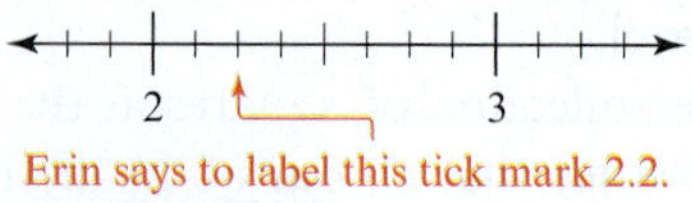

Figure 2.54 How to label the tick mark?

15. First you poured $\frac{3}{4}$ of a cup of water into an empty bowl. Then you scooped out $\frac{1}{3}$ of a cup of water. How much water is left in the bowl?

a. Make math drawings to help you solve this problem. Explain your answer, attending carefully to the unit amount that each fraction is *of*.

b. In solving the problem, how do $\frac{3}{4}$ and $\frac{1}{3}$ each appear as equivalent fractions?

16. First you poured $\frac{3}{4}$ of a cup of water into an empty bowl. Then you scooped out $\frac{1}{2}$ of the water in the bowl. How much water is left in the bowl?

a. Make math drawings to help you solve this problem. Explain your answer, attending carefully to the unit amount that each fraction is *of*.

b. In solving the problem, how does $\frac{3}{4}$ appear as an equivalent fraction?

17. You want to make a recipe that calls for $\frac{2}{3}$ of a cup of flour, but you only have $\frac{1}{2}$ of a cup of flour left. Assuming you have enough of the other ingredients, what fraction of the recipe can you make?

a. Make math drawings to help you solve this problem. Explain your answer, attending carefully to the unit amount that each fraction is *of*.

b. In solving the problem, how do $\frac{2}{3}$ and $\frac{1}{2}$ each appear as equivalent fractions?

18. Ken got $\frac{2}{3}$ of a ton of gravel even though he ordered $\frac{3}{4}$ of a ton of gravel. What fraction of his order did Ken get?

a. Make math drawings to help you solve this problem. Explain your answer, attending carefully to the unit amount that each fraction is *of*.

b. In solving this problem, how do $\frac{2}{3}$ and $\frac{3}{4}$ each appear as equivalent fractions?

19. One serving of SnackOs is $\frac{2}{3}$ of a cup. Jane wants to eat $\frac{1}{3}$ of a serving of SnackOs. What fraction of a cup of SnackOs should Jane eat?

a. Make math drawings to help you solve this problem. Explain your answer, attending carefully to the unit amount that each fraction is *of*.

b. In solving the problem, how does $\frac{2}{3}$ appear as an equivalent fraction?

20. So far, Benny has run $\frac{1}{2}$ of a mile, but that is only $\frac{4}{5}$ of his total running distance. What is his total running distance? Explain how to reason about equivalent fractions and the number line in Figure 2.55 to solve the problem.

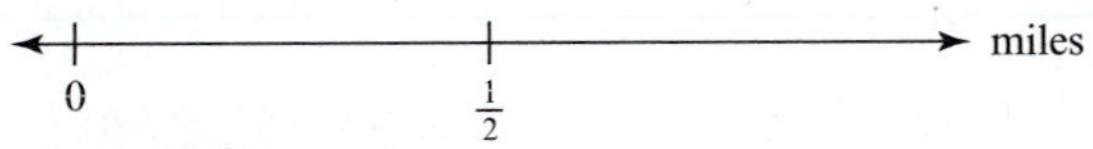

Figure 2.55 Reasoning about a number line to solve a problem.

21. So far, Mahelia has run $\frac{3}{2}$ miles, but that is only $\frac{2}{5}$ of her total running distance. What is her total running distance? Explain how to reason about equivalent fractions and the number line in Figure 2.56 to solve the problem.

Figure 2.56 Reasoning about a number line to solve a problem.

22. a. See Figure 2.57. How many of Strip A does it take to make Strip B exactly? How much of Strip B does it take to make Strip A exactly? Explain your answers using our definition of fraction.

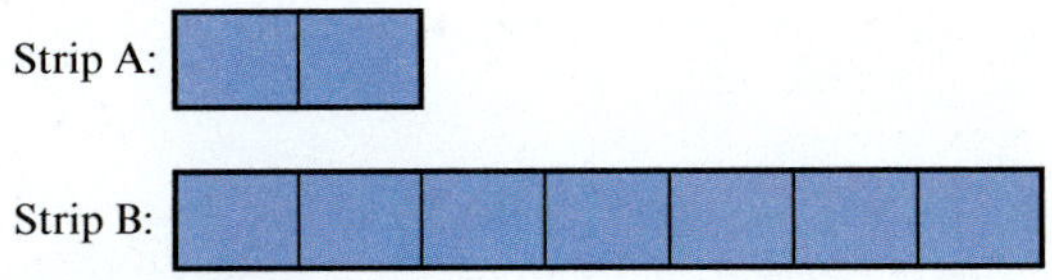

Figure 2.57 Measuring one strip by another strip.

b. See Figure 2.58. A candy company puts small candies on strips of different lengths. How many of Candy Strip C does it take to make Candy Strip D exactly? How much of Candy Strip D does it take to make Candy Strip C? Explain how to answer each question with two equivalent fractions using our definition of fraction.

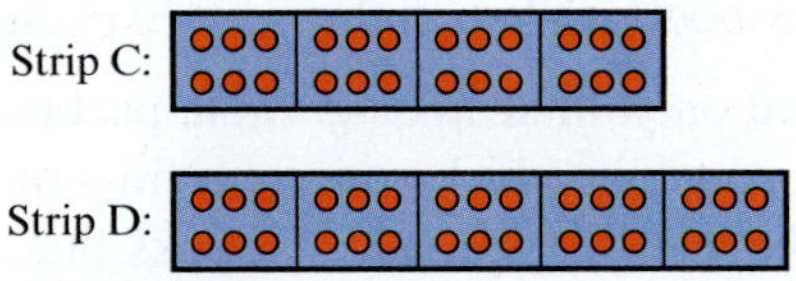

Figure 2.58 Measuring one candy strip by another.

c. See Figure 2.59. How many of Candy Strip E does it take to make Candy Strip F exactly? How much of Candy Strip F does it take to make Candy Strip E exactly? Explain how to answer each question with two equivalent fractions using our definition of fraction.

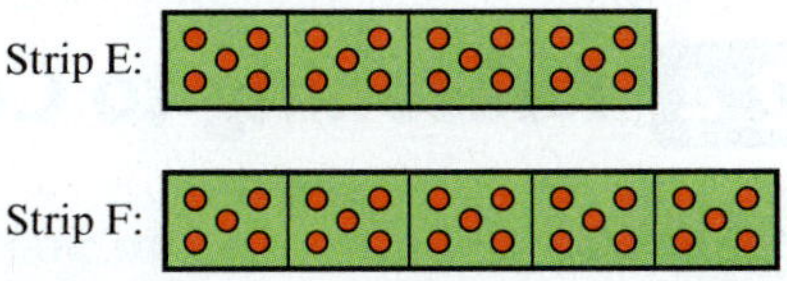

Figure 2.59 Measuring one candy strip by another.

d. Even if you did not do so in parts b and c, explain how to think about the candy strips so that some of your answers in part b are the same as some of your answers in part c.

23. The strips in Figure 2.60 show the relative amounts of books that Rachel and Leah have read so far this school year. (Note that each rectangle represents some fixed number of books, but this number may be greater than 1).

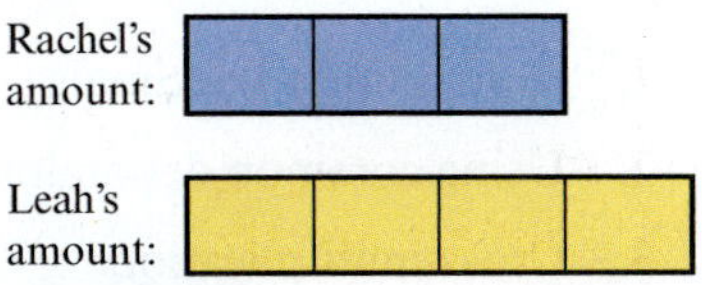

Figure 2.60 Amounts of books Rachel and Leah have read this year.

a. Write a sentence in which you use a fraction to describe how the amount of books that Rachel has read compares with the amount of books Leah has read. Explain briefly.

b. Write a sentence in which you use a fraction to describe how the amount of books that Leah has read compares with the amount of books Rachel has read. Explain briefly.

24. a. Draw a strip diagram like the one in Figure 2.60 to show that Adam has read $\frac{5}{6}$ as many books as Joseph this year. Explain briefly.

b. Based on your strip diagram in part (a), write a sentence in which you use a fraction to describe how the amount of books that Joseph has read compares with the amount of books that Adam has read.

***25.** Becky moves into an apartment with two friends on August 1. Becky's friends have been in the apartment since July 1. The electric bill comes every two months, and the next one will be for the electricity used in July and August. The bill is not broken down by month. What fraction of the July/August electric bill should Becky pay, and what fraction should her two friends pay, if they want to divide the bill fairly?

a. Solve Becky's electric bill problem if the apartment has a large communal area that is frequently used by all the friends, the friends typically eat meals together, and not much electricity is used in the separate sleeping areas. Explain your reasoning.

b. Solve Becky's electric bill problem if the friends spend most of their time in their separate rooms, don't eat meals together, and don't spend much time together in their communal area. Explain your reasoning.

2.4 Reasoning to Compare Fractions

CCSS Common Core State Standards Grades 3, 4, 5

Given two numbers in base ten, we can determine which one is greater by comparing their digits. However, comparing fractions is more complicated, because every fraction is equal to infinitely many other fractions. Unlike whole numbers, we can't always tell just by comparing digits whether the fractions are equal or whether one is greater than the other. You will probably recognize that a familiar fraction such as $\frac{1}{2}$ is equal to $\frac{3}{6}$ or $\frac{50}{100}$, but can you tell right away, just by looking, that

$$\frac{411}{885} \text{ and } \frac{548}{1180}$$

are equal? It is not obvious.

In this section we will reason about numbers and sizes of parts to compare fractions. We start by examining four general methods for determining whether two fractions are equal, or if not, which one is greater:

1. Converting to decimals
2. Using common denominators
3. Cross-multiplication
4. Using common numerators

How Can We Use Decimals to Compare Fractions?

We can convert every fraction to a decimal number by dividing the denominator into the numerator. Therefore, we can compare two fractions simply by converting both fractions to decimals and comparing the decimals.

Is $\frac{17}{35}$ equal to $\frac{43}{87}$, or if not, which is greater?

$$\frac{17}{35} = 17 \div 35 = 0.4857\ldots$$
$$\frac{43}{87} = 43 \div 87 = 0.4942\ldots$$

Since

$$0.4942\ldots > 0.4857\ldots$$

it follows that

$$\frac{43}{87} > \frac{17}{35}$$

Why Can We Compare Fractions by Using Common Denominators?

CCSS 3.NF.3d

When two fractions have the same denominator, we can determine whether they are equal, or if not, which is greater, by comparing the numerators.

For example, which is greater, $\frac{4}{7}$ or $\frac{3}{7}$? Both fractions are described in terms of like parts—sevenths—so we only have to see which one has more parts (see Figure 2.61). Four sevenths is more than 3 sevenths, which we can record as

$$\frac{4}{7} > \frac{3}{7}$$

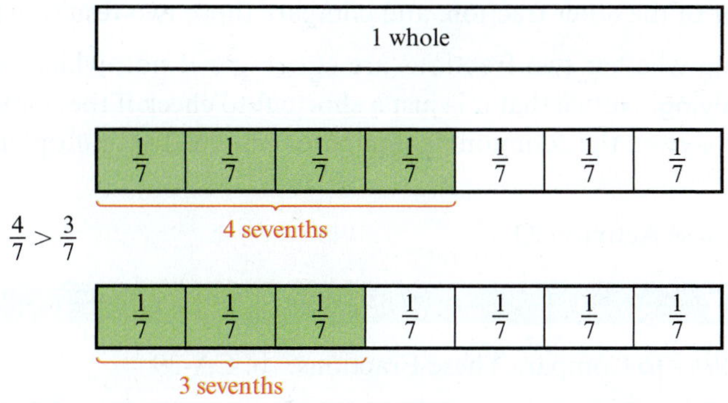

Figure 2.61 Comparing fractions that have the same denominators.

CCSS 4.NF.2

But what if you want to compare two fractions that don't have the same denominator? How can you find out if they are equal; or if they are not, how can you find out which fraction is greater? You can give the two fractions a common denominator. It can be any common denominator—it does not have to be the least one. A common denominator that always works is obtained by multiplying the two denominators.

For example, to compare $\frac{3}{4}$ and $\frac{5}{6}$, we can give the two fractions the common denominator $4 \times 6 = 24$ to compare them:

$$\frac{3}{4} = \frac{3 \times 6}{4 \times 6} = \frac{18}{24}$$
$$\frac{5}{6} = \frac{5 \times 4}{6 \times 4} = \frac{20}{24}$$

Because

$$\frac{20}{24} > \frac{18}{24}$$

it follows that

$$\frac{5}{6} > \frac{3}{4}$$

We would reach the same conclusion by using the common denominator 12 instead of 24.

What are we really doing when we give two fractions common denominators in order to compare the sizes of the fractions? As we saw in Section 2.3, when we give fractions common denominators, we create *like parts*. Once the fractions are described in terms of like parts, we only have to see how many of those parts each fraction is made of.

Why Can We Compare Fractions by Cross-Multiplying?

When we compared $\frac{3}{4}$ and $\frac{5}{6}$ by giving them the common denominators $4 \times 6 = 24$, we saw that

$$\frac{3}{4} = \frac{3 \times 6}{4 \times 6} = \frac{18}{24}$$

and

$$\frac{5}{6} = \frac{5 \times 4}{6 \times 4} = \frac{20}{24}$$

Notice that to compare the fractions all we needed to compare was

$$3 \times 6 \quad \text{and} \quad 5 \times 4$$

because those were the numerators when we gave $\frac{3}{4}$ and $\frac{5}{6}$ a common denominator by multiplying the two denominators. So, all we really had to do was to multiply the numerator of each fraction $\frac{3}{4}$ and $\frac{5}{6}$ by the denominator of the other fraction, and compare those two resulting numbers.

cross-multiplying

This method for checking whether two fractions are equal—or if not, which one is greater—is often called **cross-multiplying**. Notice that it is just a shortcut to check if the numerators are equal when the two fractions are given the common denominator obtained by multiplying the two original denominators.

Before you read on, do Class Activity 2O.

CLASS ACTIVITY

2O What Is Another Way to Compare These Fractions?, p. CA-39

Why Can We Compare Fractions by Using Common Numerators?

CCSS
3.NF.3d
4.NF.2

Which fraction is greater,

$$\frac{5}{8} \quad \text{or} \quad \frac{5}{9}?$$

Both fractions represent 5 parts, but $\frac{5}{8}$ is 5 *eighths*, whereas $\frac{5}{9}$ is 5 *ninths*. If an object is divided into 8 equal parts, then each part is larger than if the object is divided into 9 equal parts—fewer parts making up the same whole means that each part has to be larger. So eighths are bigger than ninths and therefore,

$$\frac{5}{8} > \frac{5}{9}$$

as shown in **Figure 2.62**. Notice that to make a valid comparison, we use the same whole for both fractions.

Figure 2.62
Comparing fractions that have the same numerator.

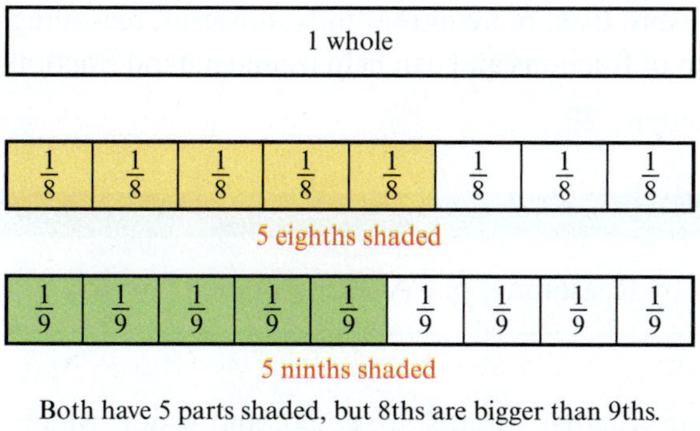

The reasoning we just used shows us that when two fractions have the same numerator, the fraction with the greater denominator is *less than* the fraction with the smaller denominator, as we see on the number lines in Figure 2.63.

Figure 2.63
Fractions with the same numerator and increasing denominators become smaller.

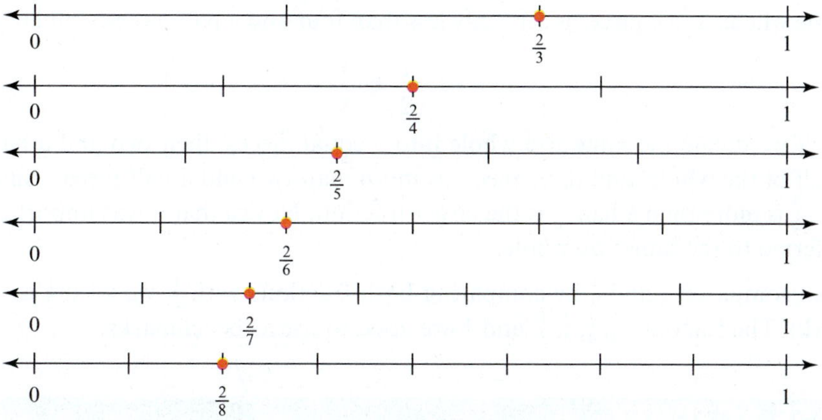

If we want to compare two fractions and the fractions don't already have a common numerator, then we can give the fractions a common numerator to compare them. For example, which fraction is greater,

$$\frac{4}{235} \quad \text{or} \quad \frac{6}{301}?$$

We can give the two fractions the common numerator of 12 to compare them:

$$\frac{4}{235} = \frac{4 \times 3}{235 \times 3} = \frac{12}{705}$$

$$\frac{6}{301} = \frac{6 \times 2}{301 \times 2} = \frac{12}{602}$$

Since 705 is greater than 602,

$$\frac{12}{705} < \frac{12}{602}$$

because 12 pieces of an object that has been divided into 705 equal pieces is a smaller amount than 12 pieces of the object if it has been divided into 602 equal pieces. Therefore, $\frac{4}{235}$ is less than $\frac{6}{301}$.

How Can We Reason About Benchmarks to Compare Fractions?

The general methods just described for comparing fractions are useful because they are efficient and they always work, but sometimes it is possible to use other reasoning to compare fractions. Why would we want to use other reasoning when we already have several good methods? In some cases, other reasoning can be more efficient. Even more importantly, however, reasoning in other ways can help us think about the definition of fractions and can help us understand fractions better.

Before you read on, do Class Activity 2P.

CLASS ACTIVITY

2P Comparing Fractions by Reasoning, p. CA-39

CCSS 4.NF.2

Consider the case of determining whether $\frac{3}{5}$ and $\frac{4}{9}$ are equal, and if not, which is larger. We can reason that $\frac{4}{9}$ is less than $\frac{1}{2}$,

$$\frac{4}{9} < \frac{1}{2}$$

because if we divide a whole into 9 equal pieces, then it would take four and a half pieces to make half of the whole, and 4 pieces is certainly less than four and a half pieces. Similarly, we can say that

$$\frac{1}{2} < \frac{3}{5}$$

because if we divide the same size whole into 5 equal pieces, then two and a half of them would make half of the whole, and three pieces is more than two and a half pieces. Since $\frac{4}{9}$ is less than a half, and $\frac{3}{5}$ is more than a half, $\frac{3}{5}$ is the larger fraction. Notice that it was important that both fractions referred to the same size whole.

benchmark (or landmark)

When we compared $\frac{3}{5}$ and $\frac{4}{9}$ by comparing both fractions with $\frac{1}{2}$, we used $\frac{1}{2}$ as a **benchmark (or landmark)**. The fractions $\frac{1}{2}$, $\frac{1}{4}$, $\frac{3}{4}$, $\frac{1}{3}$, and 1 are good to use as benchmarks.

IMAP *Watch Ally, Jacky, and Jace compare and convert fractions.*

CLASS ACTIVITY

2Q Can We Reason This Way?, p. CA-39

SECTION SUMMARY AND STUDY ITEMS

Section 2.4 Reasoning to Compare Fractions

We can compare fractions by converting them to decimals, by giving the fractions common denominators, by giving the fractions common numerators, or by cross-multiplying the fractions. Other reasoning can also be used to compare fractions: for example, comparing each fraction to a benchmark, such as $\frac{1}{2}$ or 1.

Key Skills and Understandings

1. Compare fractions by giving them common denominators, and use math drawings and our definition of fraction to explain the rationale for this method of comparison.
2. Compare fractions using cross-multiplication, and explain that this method can be viewed as a shortcut for comparing fractions by giving them a common denominator.
3. Compare fractions that have the same numerator, and use math drawings and our definition of fraction to explain the rationale for this method of comparison.

4. Compare fractions by comparing them to benchmarks such as $\frac{1}{2}$ and 1. Compare fractions by reasoning about the number of parts and the sizes of the parts.
5. Recognize that reasoning about fraction comparison requires the fractions to have the same size whole.

Practice Exercises for Section 2.4

1. Compare the sizes of the following pairs of fractions in two ways: by giving the fractions common denominators and by cross-multiplying. How are these two methods related?

$$\frac{2}{3} \text{ and } \frac{3}{5}$$
$$\frac{8}{13} \text{ and } \frac{13}{21}$$
$$\frac{15}{20} \text{ and } \frac{6}{8}$$

2. Complete the following to make true statements about comparing fractions:

a. If two fractions have the same denominators, then the one with __________ is greater.

b. If two fractions have the same numerator, then the one with __________ is greater.

Use our definition of fraction to explain why your answers make sense.

3. For each of the following pairs of fractions, determine which is larger. Use reasoning other than finding common denominators, cross-multiplying, or converting to decimals to explain your answers.

a. $\frac{6}{11}$ versus $\frac{6}{13}$

b. $\frac{5}{8}$ versus $\frac{7}{12}$

c. $\frac{5}{21}$ versus $\frac{7}{24}$

d. $\frac{21}{22}$ versus $\frac{56}{57}$

e. $\frac{97}{100}$ versus $\frac{35}{38}$

4. Explain why

$$\frac{A}{B} > \frac{C}{D} \quad \text{exactly when} \quad A \times D > C \times B$$

5. Without using a calculator, order the following numbers from largest to smallest: $\frac{41}{20}$, 2, $\frac{11}{5}$, −2, −2.3, $-\frac{19}{10}$. Explain your reasoning.

6. Find a fraction between $\frac{2}{11}$ and $\frac{3}{11}$ whose numerator and denominator are whole numbers.

7. Find a decimal between $\frac{21}{34}$ and $\frac{34}{55}$, and plot all three numbers visibly and distinctly on a number line. Be sure the labeling fits with the structure of the base-ten system. The numbers need not land on tick marks.

Answers to Practice Exercises for Section 2.4

1.

Fractions	Common Denominator	Cross-Multiplying	Conclusion
$\frac{2}{3}$ $\frac{3}{5}$	$\frac{2 \times 5}{3 \times 5} > \frac{3 \times 3}{5 \times 3}$	$2 \times 5 > 3 \times 3$	$\frac{2}{3} > \frac{3}{5}$
$\frac{8}{13}$ $\frac{13}{21}$	$\frac{8 \times 21}{13 \times 21} < \frac{13 \times 13}{21 \times 13}$	$8 \times 21 < 13 \times 13$	$\frac{8}{13} < \frac{13}{21}$
$\frac{15}{20}$ $\frac{6}{8}$	$\frac{15 \times 8}{20 \times 8} = \frac{6 \times 20}{8 \times 20}$	$15 \times 8 = 6 \times 20$	$\frac{15}{20} = \frac{6}{8}$

Notice that to compare $\frac{15}{20}$ and $\frac{6}{8}$ with the common denominator method, you could choose a smaller common denominator than 20×8. For instance, you could choose 40 as the common denominator. In this case, you compare the numerators 2×15 and 5×6 rather than comparing 8×15 and 20×6, which is what you compare with the cross-multiplying method. See the discussion in the text about comparing the two methods.

2. a. If two fractions have the same denominators, then the one with the greater numerator is greater. This is because having the same denominators means the fractions are in terms of like parts and the numerators tell us how many parts of each fraction we are considering.

b. If two fractions have the same numerator, then the one with the smaller denominator is greater. See the discussion in the text to explain why.

3. a. $\frac{6}{11} > \frac{6}{13}$. If you divide a whole into 11 pieces, then each piece will be larger than if you divide that same size whole into 13 pieces. So, 6 pieces of a whole that is divided into 11 pieces will be more than 6 pieces of a same size whole that is divided into 13 pieces.

b. $\frac{5}{8}$ is one piece more than half $(\frac{4}{8})$, and $\frac{7}{12}$ is one piece more than half $(\frac{6}{12})$. However, when a whole is divided into 8 pieces, each piece is larger than when an identical whole is divided into 12 pieces—fewer pieces making up the same amount means each piece must be larger. The eighths pieces are larger than the twelfths pieces, and since each fraction, $\frac{5}{8}$ and $\frac{7}{12}$, is one piece more than one half, it follows that $\frac{5}{8} > \frac{7}{12}$.

c. Notice that both fractions are close to $\frac{1}{4}$. The fractions $\frac{1}{4}$ and $\frac{5}{20}$ are equal; therefore, $\frac{5}{21}$ is a little less than $\frac{1}{4}$. (See the reasoning in the previous answer.) The fractions $\frac{1}{4}$ and $\frac{6}{24}$ are equal; therefore, $\frac{7}{24}$ is a little bigger than $\frac{1}{4}$. (See the reasoning in the previous answer.) So, $\frac{5}{21} < \frac{7}{24}$.

d. Notice that each fraction is "one piece less than 1 whole." The fraction $\frac{21}{22}$ is $\frac{1}{22}$ less than a whole, and $\frac{56}{57}$ is $\frac{1}{57}$ less than a whole. But $\frac{1}{22}$ is bigger than $\frac{1}{57}$ because, if you divide a whole into 22 pieces, each piece will be bigger than if you divide an identical whole into 57 pieces. Therefore, $\frac{21}{22} < \frac{56}{57}$ because $\frac{21}{22}$ is a bigger piece away from a whole than is $\frac{56}{57}$.

e. $\frac{97}{100} > \frac{35}{38}$. The same reasoning used in the answer to the previous part applies here, too. This time, each fraction is 3 pieces away from 1 whole.

4. To compare two fractions $\frac{A}{B}$ and $\frac{C}{D}$, we can give them the common denominator $B \times D = D \times B$. Then

$$\frac{A}{B} = \frac{A \times D}{B \times D}$$

and

$$\frac{C}{D} = \frac{C \times B}{D \times B}$$

Since the fractions $\frac{A \times D}{B \times D}$ and $\frac{C \times B}{D \times B}$ have the same denominator, the fraction $\frac{A \times D}{B \times D}$ is greater than $\frac{C \times B}{D \times B}$ exactly when the numerator $A \times D$ is greater than the numerator $C \times B$. This in turn means that the original fraction $\frac{A}{B}$ is greater than $\frac{C}{D}$ exactly when $A \times D$ is greater than $C \times B$. In other words,

$$\frac{A}{B} > \frac{C}{D}$$

exactly when

$$A \times D > C \times B$$

5. Every positive number is greater than every negative number, so we know that $\frac{41}{20}$, 2, and $\frac{11}{5}$ are greater than -2, -2.3, and $-\frac{19}{10}$. We can give $\frac{41}{20}$, 2, and $\frac{11}{5}$ the common denominator of 20 by letting

$$2 = \frac{2}{1} = \frac{2 \times 20}{1 \times 20} = \frac{40}{20}$$

and

$$\frac{11}{5} = \frac{11 \times 4}{5 \times 4} = \frac{44}{20}$$

Comparing the numerators, we see that

$$\frac{11}{5} > \frac{41}{20} > 2$$

Since $\frac{19}{10} = 1.9$, and since

$$2.3 > 2 > 1.9$$

it follows that

$$-1.9 > -2 > -2.3$$

Therefore,

$$\frac{11}{5} > \frac{41}{20} > 2 > -1.9 > -2 > -2.3$$

6. Since $\frac{2}{11} = \frac{4}{22}$ and $\frac{3}{11} = \frac{6}{22}$, the fraction $\frac{5}{22}$ is one example of a fraction between $\frac{2}{11}$ and $\frac{3}{11}$.

7. The decimal representation of $\frac{21}{34}$ is 0.61764 . . . and the decimal representation of $\frac{34}{55}$ is 0.61818 . . .; so, 0.618 is one example of a decimal that is in between $\frac{21}{34}$ and $\frac{34}{55}$. There are many other examples. See **Figure 2.64**.

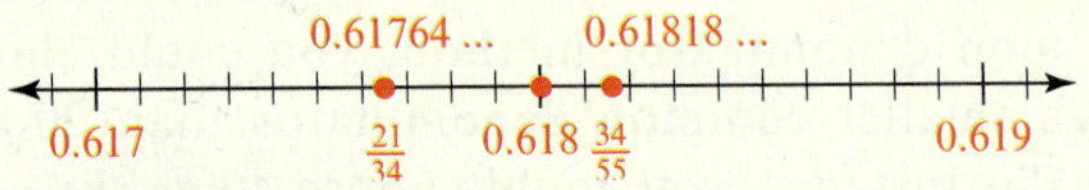

Figure 2.64 The number 0.618 is between $\frac{21}{34}$ and $\frac{34}{55}$.

PROBLEMS FOR SECTION 2.4

1. Julie says that the drawing in Figure 2.65 shows that

$$\frac{5}{8} > \frac{3}{4}$$

by comparing the areas. Critique Julie's reasoning.

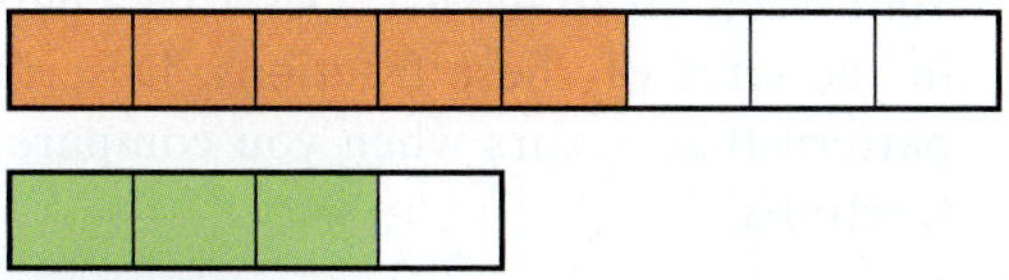

Figure 2.65 Is $\frac{5}{8} > \frac{3}{4}$?

2. In your own words, explain in detail why we can determine which of two fractions is greater by giving the two fractions common denominators. What is the rationale behind this method? What are we really doing when we give the fractions common denominators?

3. In your own words, explain in detail why we can determine which of two fractions is greater by using the cross-multiplying method. What is the rationale behind this method? What are we really doing when we cross-multiply in order to compare fractions?

4. In your own words, explain in detail why it is the case that if two fractions have the same numerator, the fraction with the smaller denominator is greater.

5. Which fraction is greater, $\frac{35}{109}$ or $\frac{36}{104}$? To explain your answer, use our definition of fraction along with reasoning other than finding common denominators, common numerators, cross-multiplying, or converting to decimals.

6. Which fraction is greater, $\frac{15}{31}$ or $\frac{23}{47}$? To explain your answer, use our definition of fraction along with reasoning other than finding common denominators, common numerators, cross-multiplying, or converting to decimals.

7. Which fraction is greater, $\frac{19}{20}$ or $\frac{27}{28}$? To explain your answer, use our definition of fraction along with reasoning other than finding common denominators, common numerators, cross-multiplying, or converting to decimals.

8. Without using a calculator, order the following numbers from smallest to largest: $\frac{11}{10}$, 1.2, $\frac{9}{8}$, -1.1, -0.98, -1. Explain your reasoning.

9. Find a number between $\frac{23}{84}$ and $\frac{29}{98}$ and plot all three numbers visibly and distinctly on a number line like the one in Figure 2.66, which has a set of longer tick marks and a set of shorter tick marks. The labeling of the number line should fit with the structure of the base-ten system. Label all the longer tick marks. The numbers need not land on tick marks.

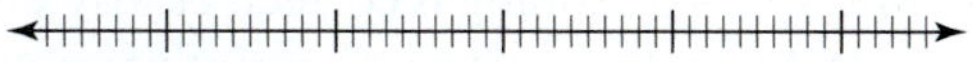

Figure 2.66 Number line.

10. Find a number between $\frac{78}{134}$ and $\frac{124}{213}$ and plot all three numbers visibly and distinctly on a number line like the one in Figure 2.66, which has a set of longer tick marks and a set of shorter tick marks. The labeling of the number line should fit with the structure of the base-ten system. Label all the longer tick marks. The numbers need not land on tick marks.

11. Find two different fractions between $\frac{2}{5}$ and $\frac{3}{5}$ whose numerators and denominators are whole numbers, and plot all four fractions on a number line.

12. Give two different methods for solving the following problem: Find two different fractions in between $\frac{3}{5}$ and $\frac{2}{3}$ whose numerators and denominators are whole numbers.

13. Give two different methods for solving the following problem: Find two different fractions between $\frac{5}{7}$ and $\frac{6}{7}$ whose numerators and denominators are whole numbers.

14. A student says that $\frac{1}{5}$ is halfway between $\frac{1}{4}$ and $\frac{1}{6}$. Use a carefully drawn number line to show that this is not correct. What fraction is halfway between $\frac{1}{4}$ and $\frac{1}{6}$? Explain your reasoning.

15. Sam has a method for comparing fractions: He just looks at the denominator. Sam says the fraction with the larger denominator is smaller because, if there are more pieces, each piece is smaller. Discuss Sam's ideas.

16. Minju says that fractions that use bigger numbers are greater than fractions that use smaller numbers. Make up two problems for Minju to help her reconsider her ideas. For each problem, explain how to solve it, and explain why you chose that problem for Minju.

17. Malcolm says that

$$\frac{8}{11} > \frac{7}{10}$$

because

$$8 > 7 \text{ and } 11 > 10$$

Discuss Malcolm's reasoning. Even though it is true that $\frac{8}{11} > \frac{7}{10}$, is Malcolm's reasoning correct? If Malcolm's reasoning is correct, clearly explain why. If Malcolm's reasoning is not correct, give Malcolm two examples that show why not.

***18.** You may combine your answers to all three parts of this problem.

a. Is it valid to compare

$$\frac{30}{70} \text{ and } \frac{20}{50}$$

by "cancelling" the 0s and comparing

$$\frac{3}{7} \text{ and } \frac{2}{5}$$

instead? Explain your answer.

b. Is it valid to compare

$$\frac{15}{25} \text{ and } \frac{105}{205}$$

by "cancelling" the 5s and comparing

$$\frac{1}{2} \text{ and } \frac{10}{20}$$

instead? Explain your answer.

c. Write a paragraph discussing the distinction between your answer in (a) and your answer in (b).

***19.** Consider the following list of fractions:

$$\frac{1}{1}, \frac{2}{1}, \frac{3}{2}, \frac{5}{3}, \frac{8}{5}, \ldots$$

You do not have to explain your answers to the following parts:

a. Describe a pattern in the list of fractions. Use your description to find the next 5 entries in the list after $\frac{8}{5}$. You will now have the first 10 entries in the list of fractions.

b. Use either the cross-multiplying method or the common denominator method to compare the sizes of the 1st, 3rd, 5th, 7th, and 9th fractions in the list. Describe a pattern in the sizes of these fractions. Describe a pattern that occurs when you compare the fractions.

c. Use either the cross-multiplying method or the common denominator method to compare the sizes of the 2nd, 4th, 6th, 8th, and 10th fractions in the list. Describe a pattern in the sizes of these fractions. Describe a pattern that occurs when you compare the fractions.

d. Convert the 10 fractions on your list to decimals, and plot them on a number line. Zoom in on portions of your number line so that you can show clearly where each decimal number is plotted relative to the others.

e. If you could find more and more entries in the list of fractions, and plot them on a number line, in what region of the number line would they be located? Do you think these numbers would get closer and closer to a particular number?

***20.** Suppose you start with a proper fraction and you add 1 to both the numerator and the denominator. For example, if you started with $\frac{2}{3}$, then you'd get a new fraction $\frac{2+1}{3+1} = \frac{3}{4}$.

a. Give at least 5 examples of proper fractions $\frac{A}{B}$. In each example, compare the sizes of $\frac{A}{B}$ and $\frac{A+1}{B+1}$. What do you notice? Make sure you are working with proper fractions (where the numerator is less than the denominator).

b. Frank says that if $\frac{A}{B}$ is a proper fraction, then $\frac{A+1}{B+1}$ is always greater than $\frac{A}{B}$ because $\frac{A+1}{B+1}$ has more parts. Regardless of whether Frank's conclusion is correct, discuss whether Frank's reasoning is valid. Did Frank give a convincing explanation that $\frac{A+1}{B+1}$ is greater than $\frac{A}{B}$? If not, what objections could you make to Frank's reasoning?

c. Explain the phenomenon you discovered in part (a). Compare the sizes of $\frac{A}{B}$ and $\frac{A+1}{B+1}$. Explain why the fraction you say is larger really *is* larger.

2.5 Reasoning About Percent

CCSS Common Core State Standards Grades 6 and 7

Almost any time we open a newspaper, walk into a store, or listen to a sports broadcast, we encounter percentages. We can think of percentages as special kinds of fractions—namely, ones that have denominator 100 (and whose numerators need not be whole numbers). In this section, we will study the basic types of percent problems and a variety of methods for solving these problems.

How Do We Define Percent?

percent The word **percent**, which is usually represented by the symbol %, means "of each hundred" or "out of a hundred" (per = "of each" or "for each" or "out of" and cent = "hundred.") So 35% means 35 out of 100, or $\frac{35}{100}$, and in general,

$$P\% = \frac{P}{100}$$

When we work with percentages, we can apply the definition of fraction. For example, if 35% of the trees at an arboretum are evergreen, then this means that the fraction of trees at the arboretum that are evergreen is $\frac{35}{100}$. According to the definition of $\frac{35}{100}$, the trees can be divided into 100 equal parts so that 35 of those parts consist of evergreen trees, as shown in Figure 2.67.

Figure 2.67 An arboretum where 35% of the trees are evergreen.

On the other hand, the fraction of trees at the arboretum that are evergreen can also be expressed as $\frac{\text{\# evergreens}}{\text{\# trees}}$. So we have two ways of expressing the fraction of trees at the arboretum that are evergreen, which therefore must be equal:

$$\frac{35}{100} = \frac{\text{\# evergreens}}{\text{\# trees}}$$

In general, when we write a given fraction as a percent, we express that fraction as an equivalent fraction with denominator 100.

Why use percents when we could use ordinary fractions? By using the denominator 100, it becomes easy to compare fractional amounts of different quantities. For example, if the fraction of trees that are evergreen at one arboretum is $\frac{160}{500}$ and the fraction of trees that are evergreen at another arboretum is $\frac{200}{800}$, we have to do some calculating to tell which arboretum has the greater fraction of trees that are evergreen. On the other hand, if we are told that 32% of the trees at one arboretum are evergreens whereas 25% of the trees at another arboretum are evergreens, then we know immediately that the first arboretum has a greater fraction of trees that are evergreen.

How Can We Use Math Drawings to Reason About Percentages?

Percentages that we encounter frequently are shown in Table 2.1, along with their equivalent fractions expressed in simplest form. These percentages are also shown in Figure 2.68 as the shaded portion of each math drawing.

TABLE 2.1 Common percentages expressed as fractions

$25\% = \frac{25}{100} = \frac{1}{4}$,	$50\% = \frac{50}{100} = \frac{1}{2}$,	$75\% = \frac{75}{100} = \frac{3}{4}$
$10\% = \frac{10}{100} = \frac{1}{10}$,	$20\% = \frac{20}{100} = \frac{1}{5}$,	$5\% = \frac{5}{100} = \frac{1}{20}$

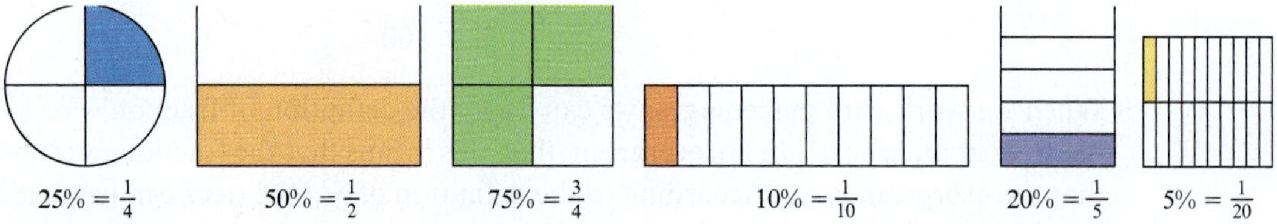

Figure 2.68 Math drawings of percentages.

If you are familiar with the fraction representations of the percentages in Table 2.1, you can easily extend them to nearby percentages and fractions, as follows:

55% is a little more than $\frac{1}{2}$ and is halfway between 50% and 60%,

or

15% is $\frac{1}{4}$ minus 10%.

Some of the class activities, exercises, and problems ask you to use math drawings to solve percent problems. Although making drawings is usually not an efficient way of solving percent problems, drawings can help you develop a better understanding of percents. Just as it would not be reasonable to pull out bundled toothpicks whenever we work with numbers in base ten, it would not be reasonable to do all percent calculations with drawings. However, both physical objects and drawings are valuable learning and thinking aids.

CLASS ACTIVITY

2R Math Drawings, Percentages, and Fractions, p. CA-40

How Can We Organize Our Thinking About Basic Percent Problems?

A percent expresses a fraction as an equivalent fraction that has denominator 100. So a basic percent problem involves a statement that there are two equivalent ways to express a fraction, namely

$$P\% = \frac{\text{portion}}{\text{whole amount}}$$

or in other words that

$$\frac{P}{100} = \frac{\text{portion}}{\text{whole amount}} \tag{2.2}$$

Another formulation of this equation is

$P\%$ of the whole amount is the portion

or

the portion is $P\%$ of the whole amount

A helpful way to organize the information in a basic percent problem is with a *percent table*—namely, a table of this form:

$$\begin{aligned} P\% &\rightarrow \text{portion} \\ 100\% &\rightarrow \text{whole amount} \end{aligned}$$

CCSS
6.RP.3c

There are three basic kinds of percent problems: one for each of the cases where one of P (the percent), the "portion," or the "whole amount" is unknown and to be determined, and the other two amounts are known. To solve such a problem, we must solve Equation 2.2 for the unknown amount. In this section, we consider three different (but interrelated) ways to solve Equation 2.2:

1. using standard techniques of algebra, such as cross-multiplying Equation 2.2 to obtain the equation

$$P \cdot (\text{whole amount}) = 100 \cdot (\text{portion})$$

and then solving this equation;
2. using a percent table and either reasoning about benchmark percentages, "going through 1%," or "going through 1";
3. making equivalent fractions (without cross-multiplying).

Using Algebra to Solve Percent Problems Consider this basic percent problem in which the "portion" is unknown (and to be calculated) and the percent, P, and the whole amount are known:

Susie must pay 6% tax on her purchase of \$44. How much tax must Susie pay?

To solve this problem, we must solve the equation

$$\frac{6}{100} = \frac{\text{tax}}{\$44}$$

Solving algebraically, such as by multiplying both sides of the equation by 44, we find that

$$\text{tax} = 0.06 \cdot \$44 = \$2.64$$

Basic percent problems can always be solved algebraically by setting up and solving an equation.

Next, we'll examine some other ways of reasoning to solve percent problems.

How Can We Reason About Percent Tables to Solve Percent Problems?

By reasoning about percent tables, we can develop a better feel for percentages while solving a variety of problems. Class Activity 2S will help you reason about percent tables and benchmark percentages.

CLASS ACTIVITY

2S Reasoning About Percent Tables to Solve "Portion Unknown" Percent Problems, p. CA-41

Percent tables allow us to reason flexibly with percentages. They also allow for a general method, which we can call "going through 1%." Part 3 of Class Activity 2S demonstrated this method. To calculate 6% of \$40 by going through 1%, we can reason as follows. Since 100% is \$40, we can find 1% by dividing by 100. So, 1% is \$0.40. Then, 6% is 6 times as much, which is $6 \cdot \$0.40 = \$2.40$. We can summarize this reasoning in the percent table in Table 2.2.

TABLE 2.2 Using a percent table to calculate 6% of \$40

100%	→	\$40
1%	→	$\$40 \div 100 = \$0.40$
6%	→	$6 \cdot \$0.40 = \$2.40$

We can use the method of going through 1% with a percent table to solve other kinds of problems, such as the next "whole unknown" percent problem:

> A store gave \$15,000 to schools in the community. This \$15,000 represents 3% of the store's annual profit. What was the store's annual profit?

The portion (\$15,000) and the percent that it represents (3%) are both known, and the whole 100% (the annual profit) is to be determined. To solve the problem, we can reason that if 3% of the store's annual profit is \$15,000, then to find 1% we should divide by 3. So 1% of the store's annual profit is $\$15{,}000 \div 3 = \$5{,}000$. Then 100% of the store's annual profit is 100 times as much, which is $100 \cdot \$5{,}000 \cdot \$500{,}000$. This reasoning is recorded in Table 2.3.

TABLE 2.3 Calculating 100% of an amount if 3% of the amount is \$15,000

3%	→	\$15,000
1%	→	$\$15{,}000 \div 3 = \$5{,}000$
100%	→	$100 \cdot \$5{,}000 = \$500{,}000$

Instead of going through 1 *percent* it is sometimes easier to solve a problem by going through the *amount* 1. Consider this problem:

> 3 is what percent of 8?

In this problem, the whole (8) and the portion (3) are known and the percent is to be determined. Table 2.4 corresponds to this reasoning: if 100% represents 8, then dividing by 4, we see that 25% represents 2. Dividing that by 2 we see that 12.5% represents 1. Multiplying by 3, we find that 3 is 37.5% of 8.

TABLE 2.4 Using a percent table to calculate what percent 3 is of 8 by going through 1

100%	→	8
100% ÷ 4 = 25%	→	2
25% ÷ 2 = 12.5%	→	1
3 · 12.5% = 37.5%	→	3

CLASS ACTIVITY

2T Reasoning About Percent Tables, p. CA-43

How Can We Reason About Equivalent Fractions to Solve Percent Problems?

Sometimes it's easy to solve a percent problem by thinking in terms of equivalent fractions. For example, how much is a 6% tax on a purchase of \$50? We can solve this problem mentally by thinking that a 6% tax means \$6 in tax for every \$100. Since \$50 is half as much as \$100, we pay half as much tax, namely, \$3. Formulated in terms of equivalent fractions, this reasoning corresponds to solving

$$\frac{6}{100} = \frac{\text{tax}}{50}$$

by dividing the numerator and denominator of $\frac{6}{100}$ by 2 to create an equivalent fraction with denominator 50:

$$\frac{6}{100} \overset{\div 2}{=} \frac{\text{tax}}{50}$$

Thus, tax = 6 ÷ 2 = 3, and so the tax is \$3.

Or consider this "percent unknown" problem:

> In a class of 25 students, 3 are absent. What percent of the class is absent?

To find this percent, we must find the fraction that has denominator 100 and is equivalent to

$$\frac{3}{25}$$

We can do this by multiplying the numerator and denominator by 4:

$$\frac{3}{25} \overset{\cdot 4}{=} \frac{12}{100} = 12\%$$

CLASS ACTIVITY

2U Percent Problem Solving, p. CA-44

FROM THE FIELD Children's Literature

- Murphy, S. J. (2003). *The grizzly gazette.* New York, NY: HarperCollins. Corey wants to be the next camp mascot, but she faces some steep competition! She needs the majority vote to win so she uses her knowledge of percentages to interpret the daily poll results in her camp's newspaper.
- Pallotta, J. (2001). *Twizzlers: percentages book.* New York, NY: Scholastic. In this picture book, a class uses Twizzlers to explore the relationship among percentages, decimals, and fractions.
- Gifford, S. (2003). *Piece=part=portion: fractions=decimals=percents.* Berkeley, CA: Tricycle Press. This book uses real-world objects to illustrate fractional parts and shows the equivalent decimals and percentages related to those fractions.

SECTION SUMMARY AND STUDY ITEMS

Section 2.5 Reasoning About Percent

Percents are special fractions—namely, fractions that have denominator 100. A basic percent problem states that a *portion* is $P\%$ of a *whole amount,* which states that two fractions are equivalent:

$$\frac{P}{100} = \frac{\text{portion}}{\text{whole amount}}$$

Key Skills and Understandings

1. Use several methods to solve basic percent problems in which two of the percent P, the portion, and the whole amount are known, and one is unknown.
2. Know how to reason with the following to solve basic percent problems: working with equivalent fractions; using a percent table; going through 1%; going through 1; and using math drawings, benchmark fractions, and mental calculation if appropriate.

Practice Exercises for Section 2.5

1. For each of the shapes shown in Figure 2.69, determine what percent of the shape is shaded. Give your answer rounded to the nearest multiple of 5 (i.e., 5, 10, 15, 20, . . .).

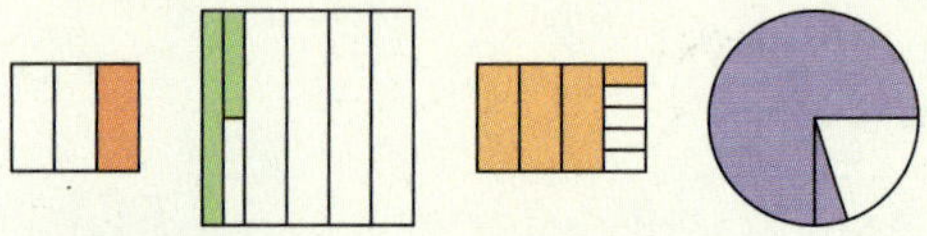

Figure 2.69 What percent of each shape is shaded?

2. A restaurant server received a $7.00 tip on a meal he served. If this tip represents 20% of the cost of the meal, then how much did the meal cost? Solve this problem with the aid of a math drawing and a percent table.
3. If $12.3 million is 75% of the budget, then how much is the full budget? Solve the problem with the aid of a math drawing and a percent table. Explain your reasoning.

4. There were 4800 gallons of water in a tank. Some of the water was drained out, leaving 65% of the original amount of water in the tank. How many gallons of water are in the tank? Solve the problem with the aid of a math drawing and a percent table. Explain your reasoning.

5. If your daily value of vitamin C is 60 milligrams, then how many milligrams is 95% of your daily value of vitamin C? Solve the problem with the aid of a math drawing and a percent table. Explain your reasoning.

6. George was given 9 grams of medicine, but the full dose that he was supposed to receive is 20 grams. What percent of his full dose did George receive? First, solve the problem with the aid of a math drawing and a percent table. Explain your reasoning. Then show how to solve the problem by making equivalent fractions (without cross-multiplying).

7. **a.** 1260 is 150% of what number?

   b. What percent of 760 is 950?

   c. The fraction $\frac{3}{7}$ is what percent of $\frac{4}{7}$?

8. Show how to use a percent table and "going through 1%" or "going through 1" to solve the following problems:

   a. What percent of 200 is 15?

   b. 68 is 4% of what number?

   c. What percent of 80 is 3?

   d. 375 is 125% of what number?

9. Show how to solve the following problems by working with equivalent fractions (without cross-multiplying):

   a. A dress cost \$75. Now the dress is being sold at a \$6 discount. What percent is the discount?

   b. Of the rabbits on a rabbit farm, 35% are white. If there are 70 white rabbits on the rabbit farm, how many rabbits in all are on the farm?

   c. There are 25 children in a class. If 16% of the children in the class get free lunch, how many children get free lunch?

10. Billy has 20% more marbles than Sammy. If Billy has 12 more marbles than Sammy, then how many marbles does Billy have? Explain your solution.

11. There are 200 marbles in a bucket. Of the 200 marbles, 80% have swirled colors and 20% are solid colors. How many swirled marbles must be removed so that 75% of the remaining marbles are swirled?

Answers to Practice Exercises for Section 2.5

1. See **Figure 2.70**.

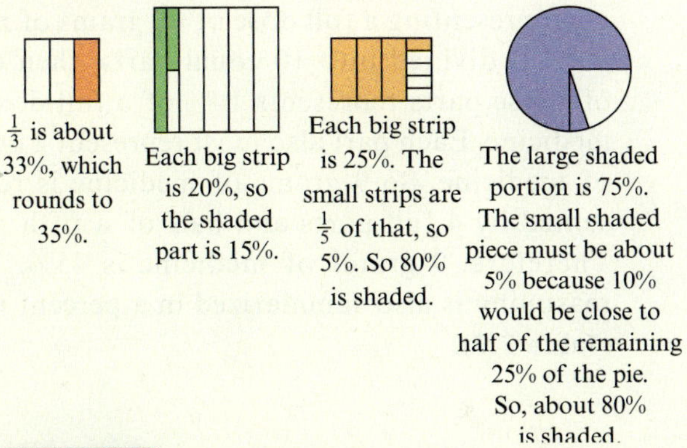

Figure 2.70 Percent shaded.

2. Twenty percent is equal to $\frac{1}{5}$. If \$7.00 represents one-fifth of the cost of the meal, then the meal must cost 5 times \$7.00, which is \$35.00. See **Figure 2.71** for the corresponding math drawing and **Table 2.5**, for a percent table.

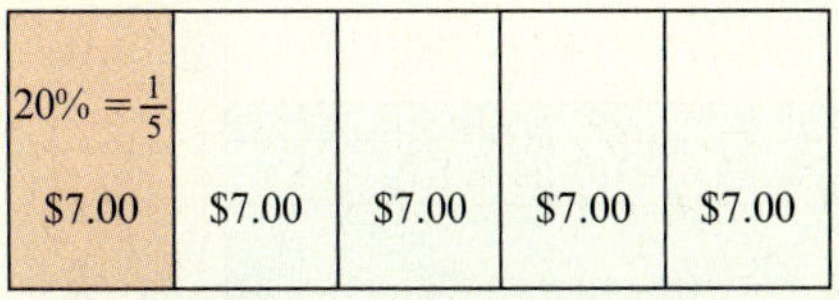

Figure 2.71 Calculating 100% from 20%.

TABLE 2.5 Using a percent table

20%	→	\$7.00
100%	→	$5 \cdot \$7.00 = \$35.00$

3. Because

$$75\% = \frac{3}{4}$$

we can think of the \$12.3 million that make up 75% of the budget as distributed equally among the 3 parts of $\frac{3}{4}$, as shown in **Figure 2.72**. Each of those

3 parts must, therefore, contain \$4.1 million. The full budget is made of 4 of those parts. Thus, the full budget must be 4 • \$4.1 = \$16.4 million. This reasoning is also summarized in a percent table (Table 2.6).

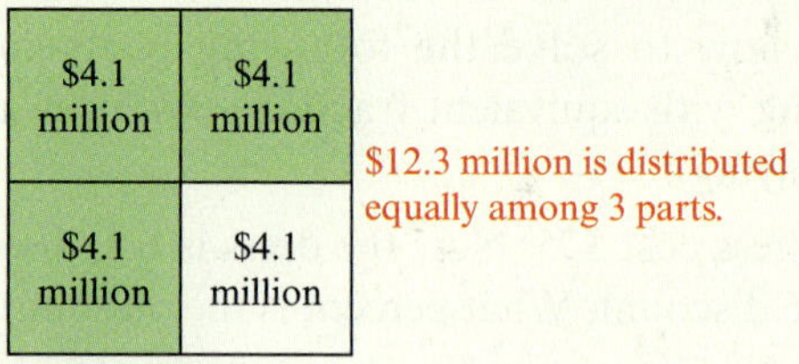

Figure 2.72 Calculating the full budget.

TABLE 2.6 Using a percent table

75%	→	\$12.3 million
25%	→	\$12.3 ÷ 3 = \$4.1 million
100%	→	4 • \$4.1 = \$16.4 million

4. As Figure 2.73 shows, we can think of 65% as 50% + 10% + 5%. Now, 50% of 4800 is $\frac{1}{2}$ of 4800, which is 2400. Since 10% of 4800 is $\frac{1}{10}$ of 4800, which is 480, it follows that 5% is half of 480, which is 240. Therefore, 65% of 4800 is 2400 + 480 + 240, which is 3120, so 3120 gallons of water are left in the tank. This reasoning is also summarized in a percent table (Table 2.7).

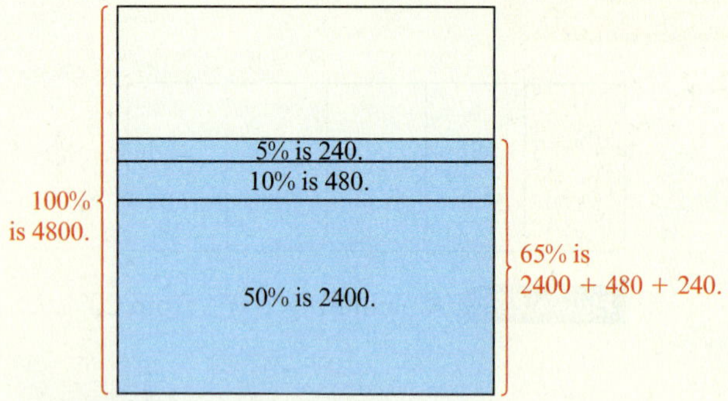

Figure 2.73 Calculating 65% of 4800.

TABLE 2.7 Using a percent table

100%	→	4800
50%	→	4800 ÷ 2 = 2400
10%	→	4800 ÷ 10 = 480
5%	→	480 ÷ 2 = 240
65%	→	2400 + 480 + 240 = 3120

5. As Figure 2.74 shows, 95% is 5% less than 100%. The 10 vertical strips in Figure 2.74 are each 10%, and half of one of these vertical strips is 5%. Because 10% of 60 milligrams is $\frac{1}{10}$ of 60 milligrams, which is 6 milligrams, 5% of 60 milligrams is half of 6 milligrams, which is 3 milligrams. So, 95% of 60 milligrams is 3 milligrams less than 60 milligrams, which is 57 milligrams. This reasoning is also summarized in a percent table (Table 2.8).

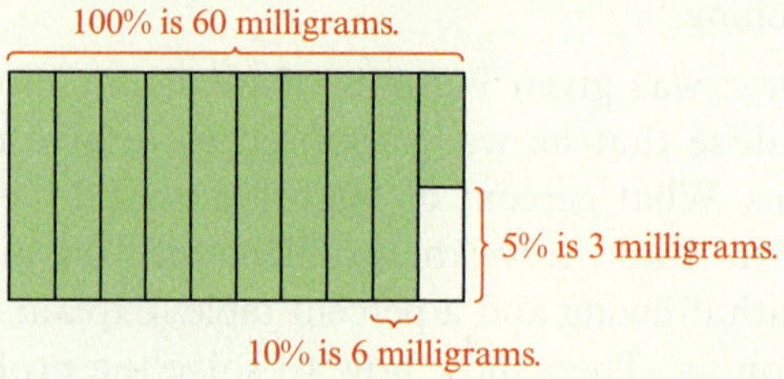

Figure 2.74 What is 95% of 60 milligrams?

TABLE 2.8 Using a percent table

100%	→	60 mg
10%	→	6 mg
5%	→	3 mg
95%	→	60 − 3 = 57 mg

6. One way to determine what percent 9 grams is of 20 grams is shown in Figure 2.75. If a rectangle representing a full dose of 20 grams of medicine is divided into 10 equal parts, then each of those parts represents 10% of a full dose of medicine. Each part also must represent 2 grams of medicine. So 9 grams of medicine is represented by 4 full parts and half of a fifth part. Therefore, 9 grams of medicine is 45%. This reasoning is also summarized in a percent table (Table 2.9).

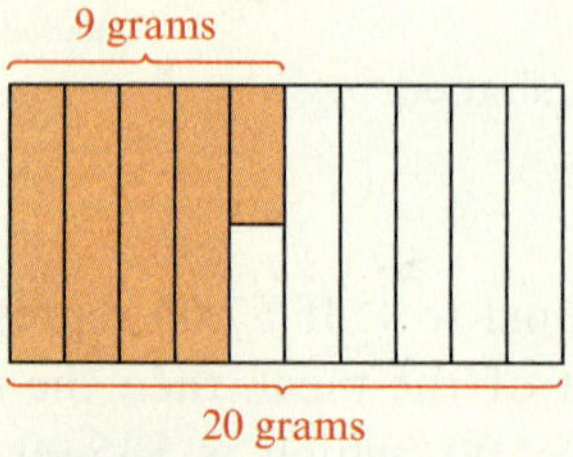

Figure 2.75 Nine grams is what percent of 20 grams?

TABLE 2.9 Using a percent table

100%	→	20 grams
10%	→	2 grams
5%	→	1 gram
45%	→	4 · 2 + 1 = 9 grams

We can also solve the problem by working with equivalent fractions:

$$\frac{9}{20} = \frac{45}{100} = 45\% \quad (\cdot 5 \text{ in numerator and denominator})$$

7. a. If 150% of a number is 1260, then to find 50% of the number, first divide by 3. So 50% of the number is 420. Therefore, 100% of the number is twice as much, which is 840. So the number is 840. The following percent table records this line of reasoning:

150% → 1260

50% → 420

100% → 840

b. Using equivalent fractions:

$$\frac{950}{760} = \frac{95}{76} = \frac{5 \cdot 19}{4 \cdot 19} = \frac{5}{4} = \frac{125}{100} = 125\%$$

So 950 is 125% of 760.

c. Write a percent table:

$100\% \rightarrow \frac{4}{7}$

$25\% \rightarrow \frac{1}{7}$

$75\% \rightarrow \frac{3}{7}$

So $\frac{3}{7}$ is 75% of $\frac{4}{7}$.

8. a.

$100\% \rightarrow 200$

$100\% \div 200 = \frac{1}{2}\% \rightarrow 1$

$15 \cdot \frac{1}{2}\% = 7.5\% \rightarrow 15$

So 15 is 7.5% of 200.

b.

$4\% \rightarrow 68$

$1\% \rightarrow 68 \div 4 = 17$

$100\% \rightarrow 17 \cdot 100 = 1700$

So 68 is 4% of 1700.

c.

$100\% \rightarrow 80$

$100\% \div 80 = 1.25\% \rightarrow 1$

$3 \cdot 1.25\% = 3.75\% \rightarrow 3$

So 3 is 3.75% of 80.

d.

$125\% \rightarrow 375$

$1\% \rightarrow 375 \div 125 = 3$

$100\% \rightarrow 100 \cdot 3 = 300$

So 125% of 300 is 375.

9. a.

$$\frac{6}{75} = \frac{2}{25} = \frac{8}{100} = 8\% \quad (\div 3,\ \text{then} \cdot 4 \text{ in numerator and denominator})$$

So 6 is 8% of 75, and the discount on the dress is 8%.

b.

$$35\% = \frac{35}{100} = \frac{70}{200} \quad (\cdot 2 \text{ in numerator and denominator})$$

So 70 is 35% of 200; therefore, there are 200 rabbits on the farm.

c.

$$16\% = \frac{16}{100} = \frac{4}{25} \quad (\div 4 \text{ in numerator and denominator})$$

So 16% of 25 is 4, and 4 children get free lunch.

10. Billy has 20% more marbles than Sammy and this is 12 marbles; this means that 20% of Sammy's marbles is 12 marbles. Therefore, 100% of Sammy's marbles is 5 times as much, which is 5 · 12 = 60 marbles, as summarized in the next percent table and as shown in Figure 2.76.

20% of Sammy's marbles → 12

100% of Sammy's marbles → 5 · 12 = 60

Since Sammy has 60 marbles and Billy has 12 more, Billy has 60 + 12 = 72 marbles.

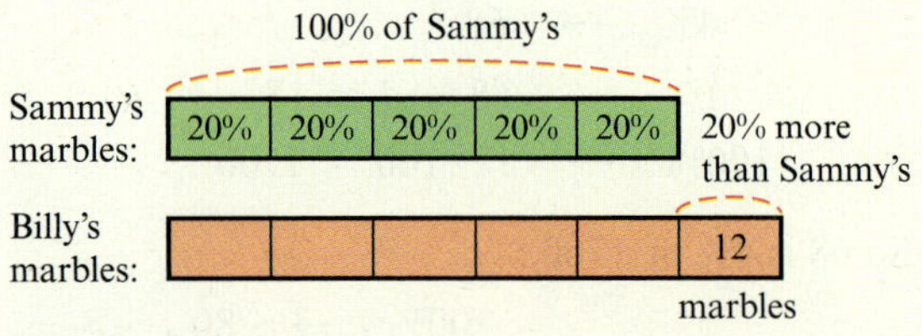

Figure 2.76 Billy has 20% more marbles than Sammy and this is 12 marbles.

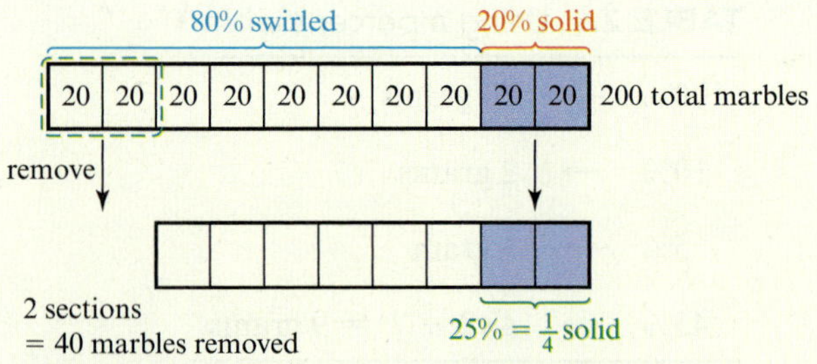

Figure 2.77 Removing swirled marbles so that the solid marbles become 25%.

11. Because 80% of the 200 marbles are swirled and 20% are solid colors, 160 marbles are swirled and 40 are solid colors. We want these 40 solid-color marbles to be 25% of the remaining marbles. So, if 25% of the remaining marbles is 40, then 100% of them is 4 times as much, namely, 160.

Since $200 - 160 = 40$, we must remove 40 swirled marbles.

Another way to solve the problem is shown in Figure 2.77.

The 200 marbles are shown in 10 sections, with the swirled marbles in 8 sections (80%) and the solid marbles in 2 sections (20%). Those 2 sections of solid marbles must become 25% of the marbles. Since 25% is $\frac{1}{4}$, the figure shows that 2 sections of swirled marbles must be removed. Since each section has 20 marbles $(200 \div 10 = 20)$, the 2 sections to be removed consist of 40 marbles. So, 40 swirled marbles must be removed.

PROBLEMS FOR SECTION 2.5

1. If your daily value of carbohydrates is 300 grams, then how many grams is 95% of your daily value of carbohydrates? Solve the problem with the aid of a math drawing or a percent table, or both, explaining your reasoning.

2. A road crew ordered $\frac{5}{2}$ tons of gravel, but they only received $\frac{3}{2}$ tons of gravel. What percent of their order did the road crew receive? Solve the problem with the aid of a math drawing or a percent table, or both, explaining your reasoning.

3. If your daily value of dietary fiber is 25 grams and if you only ate 80% of your daily value of dietary fiber, then how many grams of dietary fiber did you eat? Solve the problem with the aid of a math drawing or a percent table, or both, explaining your reasoning.

4. If 36,000 people make up 15% of a population, then what is the total population? Solve the problem with the aid of a math drawing or a percent table, or both, explaining your reasoning.

5. If 180 milligrams of potassium constitute 5% of your daily value of potassium, then how many milligrams is your full daily value of potassium? Solve the problem with the aid of a math drawing or a percent table, or both, explaining your reasoning.

6. In Happy Valley, the average rainfall in July is 5 inches, but this year only 3.5 inches of rain fell in July. What percent of the average July rainfall did Happy Valley receive this year? Solve the problem with the aid of a math drawing or a percent table, or both, explaining your reasoning.

7. If a $\frac{3}{4}$-cup serving of cereal provides your full daily value of vitamin B6, then what percentage of your daily value of vitamin B6 will you receive in $\frac{1}{2}$ of a cup of the cereal? Solve the problem with the aid of a math drawing or a percent table, or both, explaining your reasoning.

8. a. Mentally determine what percent 225 is of 250. Use a percent table to help you explain why your method makes sense.

b. Show how to determine what percent 225 is of 250 by finding equivalent fractions (without cross-multiplying).

c. Mentally determine what percent 960 is of 1600. Use a percent table to help you explain why your method makes sense.

d. Show how to determine what percent 960 is of 1600 by finding equivalent fractions (without cross-multiplying).

9. Show how to use a percent table and either "going through 1%" or "going through 1" to solve the following problems:

a. 690 is 23% of what number?

b. What percent of 40 is 3?

c. 630 is 9% of what number?

d. 12% of what number is 480?

10. If a $\frac{2}{3}$-cup serving of cereal provides your full daily value of folic acid, then what percent of your daily value of folic acid will you receive in $\frac{1}{2}$ of a cup of the cereal? Solve the problem with the aid of a math drawing or a percent table, or both, explaining your reasoning.

11. The Biggo Corporation hopes that 95% of its 4600 employees will participate in the charity fund drive. How many employees does the Biggo Corporation hope will participate in the fund drive? Solve the problem with the aid of a math drawing or a percent table, or both, explaining your reasoning.

12. A company has bought 3.4 acres of land out of the 4 acres of land that it plans to buy. What percent of the land has the company already bought? Solve the problem with a math drawing or a percent table, or both, explaining your reasoning.

13. The mayor says that $3.6 million has been spent and that this represents 75% of the money allocated for a project. What was the total amount of money that was allocated for the project? Solve the problem with a math drawing or a percent table, or both, explaining your reasoning.

14. Sixty percent of a city's population of 84,000 live within 5 miles of the library. How many people live within 5 miles of the library? Solve the problem with a math drawing or a percent table, or both, explaining your reasoning.

15. Frank ran 80% as far as Denise. How far did Denise run as a percentage of Frank's running distance? Draw a picture or diagram to help you solve the problem. Use your picture to help explain your answer.

16. GrandMart sells 115% as much soda as BigMart. How much soda does BigMart sell, calculated as a percentage of GrandMart's soda sales? Explain your answer.

17. Connie and Benton paid for identical plane tickets, but Benton spent more than Connie (and Connie's ticket was not free).

a. If Connie spent 75% as much as Benton, then did Benton spend 125% as much as Connie? If not, then what percentage of Connie's ticket price did Benton spend? Explain your answer.

b. If Benton spent 125% as much as Connie, then did Connie spend 75% as much as Benton? If not, then what percentage of Benton's ticket price did Connie spend? Explain your answer.

18. At a newsstand, 75% of the items sold are newspapers. Does this mean that 75% of the newsstand's income comes from selling newspapers? Why or why not? Write a paragraph to explain. Include examples to illustrate your points.

19. A company produces two types of handbags and sells a total of 500 handbags per day. Of the 500 bags sold per day, 30% are style A and 70% are style B. Suppose the company begins to sell additional style A bags every day, but does not sell any additional style B bags. How many more style A bags would the company have to sell per day so that 50% of their handbag sales are style A bags? Explain your solution.

20. A county has two elementary schools, school A and school B. At school A, 30% of the children speak Spanish at home. At school B, 20% of the children speak Spanish at home. Use this information about the schools in all parts of this problem.

a. Tom says that 50% of the elementary school children in the county speak Spanish at home. Tom found 50% by adding 30% and 20%. Is Tom correct? Explain why or why not.

b. DeShun says that 25% of the elementary school children in the county speak Spanish at home. DeShun found 25% by averaging 30% and 20%.

i. If there are 400 children in school A and 100 children in school B, is DeShun correct that 25% of the elementary school children in the county speak Spanish at home? If not, what percent is it?

ii. If there are 100 children in school A and 400 children in school B, is DeShun correct that 25% of the elementary school children in the county speak Spanish at home? If not, what percent is it?

iii. Find circumstances under which DeShun is correct that 25% of the elementary school children in the county speak Spanish at home.

*21. Brand A soup is 20% solids and 80% water. Use this information about brand A soup in all parts of this problem.

a. If a can of brand A soup is mixed with a full can of water, what percent of the mixture is water? Explain your answer.

b. The manufacturer of brand A soup has found a way to remove half of the water in the soup to make a more concentrated soup. What percent of this concentrated soup is water? Explain your answer.

c. What percent of the water in the soup would the manufacturer of brand A soup have to remove so that the remaining, concentrated soup becomes 50% solids? Explain your answer.

*22. At a store, there is a display of 300 cans of beans. Of the 300 cans, 60% are brand A and 40% are brand B. How many cans of brand B beans must be removed so that 75% of the remaining cans are brand A? Explain your solution.

*23. Suppose that 200 pounds of freshly picked cucumbers are 99% water by weight. After several days, some of the water from the cucumbers has evaporated, and the cucumbers are now 98% water. How much do the cucumbers weigh now? Explain your solution.

CHAPTER SUMMARY

Section 2.2 Defining and Reasoning About Fractions

Page 48

- Our definition of fraction is as follows: If A and B are whole numbers, and B is not zero, and if a whole or unit amount can be partitioned into B equal parts, then $\frac{1}{B}$ of the whole is the amount formed by 1 part, and $\frac{A}{B}$ of the whole is the amount formed by A parts (or copies of parts), each of size $\frac{1}{B}$ of the whole. Page 48
- When working with fractions, pay close attention to the *whole,* or *unit amount,* namely, the object, collection, or quantity that the fraction is *of.* Even though fractions are expressed in terms of a pair of numbers, a fraction represents a single number, and as such, it can be plotted on a number line. Page 49

Key Skills and Understandings

1. Find fractional amounts of an object, collection, or quantity, and justify your reasoning. Page 48
2. In problems, determine the whole associated with a fraction appearing in the problem. Page 49
3. Use fractions to compare quantities. Page 49
4. Plot fractions, including improper fractions, on number lines and explain why the location fits with the definition of fraction. Page 51

Section 2.3 Reasoning About Equivalent Fractions

Page 60

- Any fraction $\frac{A}{B}$ is equivalent to infinitely many other fractions via this relationship: Page 60

$$\frac{A}{B} = \frac{A \times N}{B \times N}$$

- We often call fractions that are equal and represent the same number *equivalent.* Page 60
- We use equivalent fractions to give two fractions common denominators. When we give fractions common denominators, we express the fractions in terms of like parts. Page 61

▪ We can simplify fractions by dividing the numerator and denominator by the same number. The simplest form of a fraction is an equivalent fraction that has the smallest possible numerator and denominator.	Page 62
Key Skills and Understandings	
1. Given a fraction, use our definition of fraction, math drawings, and number lines to explain why multiplying the numerator and denominator by the same counting number produces an equivalent fraction. In the process, attend to the number of parts and the size of the parts.	Page 60
2. Explain that giving fractions common denominators expresses the fractions in terms of like parts.	Page 61
3. Solve problems in which you will need to make equivalent fractions, and justify your solutions.	Page 62
4. Simplify fractions, and explain the process in terms of math drawings.	Page 62
5. Given two fractions on a number line, place equally spaced tick marks to plot other fractions on the number line as well.	Page 64
Section 2.4 Reasoning to Compare Fractions	Page 70
▪ We can compare fractions by converting them to decimals, by giving the fractions common denominators, by giving the fractions common numerators, or by cross-multiplying the fractions. Other reasoning can also be used to compare fractions: for example, comparing each fraction to a benchmark, such as $\frac{1}{2}$ or 1.	Page 70
Key Skills and Understandings	
1. Compare fractions by giving them common denominators, and use math drawings and our definition of fraction to explain the rationale for this method of comparison.	Page 71
2. Compare fractions using cross-multiplication, and explain that this method can be viewed as a shortcut for comparing fractions by giving them a common denominator.	Page 72
3. Compare fractions that have the same numerator, and use math drawings and our definition of fraction to explain the rationale for this method of comparison.	Page 72
4. Compare fractions by comparing them to benchmarks such as $\frac{1}{2}$ and 1. Compare fractions by reasoning about the number of parts and the sizes of the parts.	Page 73
5. Recognize that reasoning about fraction comparison requires the fractions to have the same size whole.	Page 74
Section 2.5 Reasoning About Percent	Page 79
▪ Percents are special fractions—namely, fractions that have denominator 100. A basic percent problem states that a *portion* is $P\%$ of a *whole amount,* which states that two fractions are equivalent: $$\frac{P}{100} = \frac{\text{portion}}{\text{whole amount}}$$	Page 79
Key Skills and Understandings	
1. Use several methods to solve basic percent problems in which two of the percent P, the portion, and the whole amount are known, and one is unknown.	Page 80
2. Know how to reason with the following to solve basic percent problems: working with equivalent fractions; using a percent table; going through 1%; going through 1; and using math drawings, benchmark fractions, and mental calculation if appropriate.	Page 80

3

Addition and Subtraction

In this chapter, we start by considering what addition and subtraction mean and what kinds of problems addition and subtraction solve. The types of problems we can solve with addition and subtraction are surprisingly rich and varied.

We'll also see how the commutative and associative properties of addition are building blocks for addition and subtraction methods. These properties allow us the flexibility of taking apart and rearranging addition and subtraction problems to make them easier to solve. In particular, the properties of addition underlie methods young children learn to become fluent with the facts of single-digit addition and subtraction.

We'll then study the efficient general methods for adding and subtracting in the base-ten system and with fractions and analyze how these methods work. What are we really doing when we "carry" in an addition problem? Why do we add and subtract fractions in the way we do rather than by just adding the numerators and adding the denominators? We'll study logical explanations for why the standard processes work. Finally, we'll see why we add and subtract negative numbers the way we do.

We focus on the following topics and practices within the *Common Core State Standards for Mathematics.*

Standards for Mathematical Content in the CCSSM

In the domain of *Operations and Algebraic Thinking* (K–Grade 5) students learn what kinds of problems addition and subtraction solve and increasingly sophisticated ways of thinking and reasoning about addition and subtraction with small numbers. In the domain of *Number and Operations in Base Ten* (K–Grade 5), students learn to compose and decompose numbers according to place value and to make sense of methods for adding and subtracting multidigit numbers. In the domain of *Number and Operations—Fractions* (Grades 3–5), students extend their understanding of addition and subtraction to fractions.

Standards for Mathematical Practice in the CCSSM

Opportunities to engage in all eight of the Standards for Mathematical Practice described in the *Common Core State Standards* occur throughout the study of addition and subtraction, but the following standards may be especially appropriate for emphasis:

- **3 Construct viable arguments and critique the reasoning of others.** Students engage in this practice when they make sense of methods of addition and subtraction and when they explain their methods with the aid of math drawings.
- **4 Model with mathematics.** Students engage in this practice when they use addition or subtraction to solve problems and when they examine situations critically to determine whether addition or subtraction apply.
- **7 Look for and make use of structure.** Students engage in this practice when they apply properties of addition or decompose and recompose numbers in order to add or subtract. For example, to add $8 + 7$ a student might break 7 into $2 + 5$, combine the 2 with 8 to make a 10, and then add on the remaining 5 to make 15.

(From Common Core Standards for Mathematical Practice. Published by Common Core Standards Initiative.)

3.1 Interpretations of Addition and Subtraction

CCSS Common Core State Standards Grades K, 1, 2

The most familiar ways to think of addition and subtraction are as *combining* and *taking away*. But as we'll see, there is a surprising variety of types of problems in which we add or subtract. We can also view addition and subtraction on number lines, which will allow us to understand addition and subtraction of negative numbers.

What Are the First Ways of Thinking About Addition and Subraction?

sum

If A and B are non-negative numbers, then we can define the **sum**

$$A + B$$

basic definition of addition

as the total number of units (or objects) we will have if we start with A units (or objects) and then get B more units (or objects). The numbers A and B in a sum are called **terms, addends,** or **summands**. We can represent the sum

$$3 + 4$$

as the total number of ducks that will be in the pond if there were 3 ducks and 4 more flew in.

difference

If A and B are non-negative numbers such that A is greater than or equal to B, then we can define the **difference**

$$A - B$$

as the total number of units (or objects) we will have if we start with A units (or objects) and take away B of those units (or objects). The numbers A and B in a difference can be called **terms**. The number A is sometimes called the **minuend**, and the number B is sometimes called the **subtrahend**. We can represent

basic definition of subtraction

$$7 - 2$$

as the number of ducks that are left in the pond if there were 7 ducks and 2 flew away.

How Are Addition and Subtraction Related?

Every statement about subtraction corresponds to a statement about addition. Namely, to say that

$$A - B = C$$

is equivalent to saying that

$$C + B = A$$

Why? We can use the equation $A - B = C$ to represent the situation where we have A apples, we give away B apples, and we are left with C apples. Reversing the process, we can start with the C apples we have left. If we now get back the B apples, we will then have the A apples we originally started with. This latter scenario corresponds to the equation

$$C + B = A$$

Notice that the strip diagram in Figure 3.1 fits naturally with the equation $A - B = C$, as well as with the equation $C + B = A$ (or $B + C = A$), showing us visually that the two equations are equivalent.

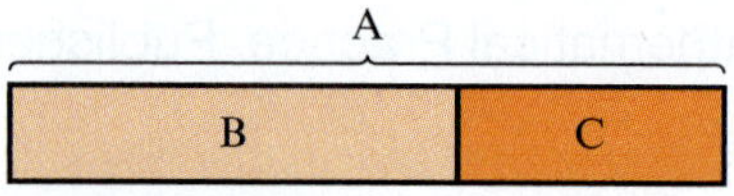

Figure 3.1 Representing $A - B = C$ and $C + B = A$ with a strip diagram.

We will see that the relationship between addition and subtraction gives rise to problems that can be solved by addition, but are not "add to" problems, and problems that can be solved by subtraction, but are not "take from" problems.

CLASS ACTIVITY

3A Relating Addition and Subtraction—The Shopkeeper's Method of Making Change, p. CA-45

CCSS Table 1

What Are the Different Types of Addition and Subtraction Word Problems?

We defined addition in terms of "adding to," and we defined subtraction in terms of "taking from." These interpretations of addition and subtraction are the most basic and simple interpretations, but they allow for more variety than you might realize, as you will see in Class Activity 3B.

CLASS ACTIVITY

3B Writing Add To and Take From Problems, p. CA-46

CCSS
K.OA
1.OA.1
2.OA.1

What Are Add to and Take from Problems?

These addition and subtraction problems involve change over time and are therefore sometimes called "change problems." The Add To and Take From problems each have three subtypes: Result Unknown, Change Unknown, and Start Unknown. See Figures 3.2 and 3.3 for examples.

Figure 3.2
Three types of Add To problems.

Add To Problems

Result Unknown Rachel had 23 CDs. She got 18 more CDs. How many does she have now?	23 + 18 = □	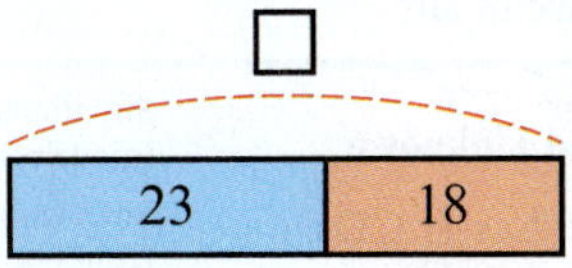
Change Unknown Rachel had 23 CDs. After she got some more CDs, she had 41 CDs. How many CDs did Rachel get?	Situation equation (models the situation): 23 + □ = 41 Solution equation (could be used to solve): 41 − 23 = □	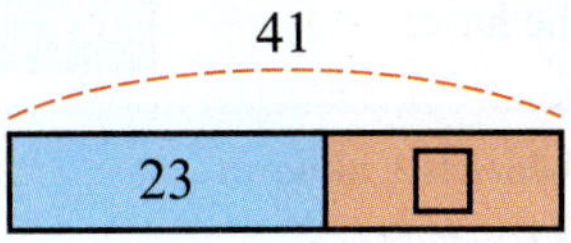
Start Unknown Rachel had some CDs. After she got 18 more, she had 41 CDs. How many CDs did Rachel have before?	Situation equation (models the situation): □ + 18 = 41 Solution equation (could be used to solve): 41 − 18 = □	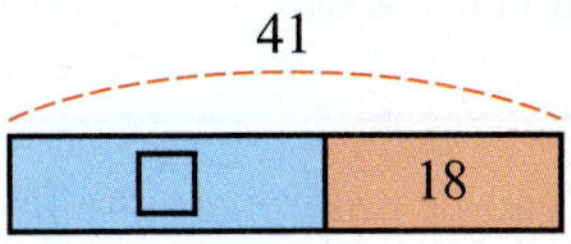

Figure 3.3
Three types of Take From problems.

Take From Problems

Result Unknown Rachel had 41 CDs. She gave away 23. How many does she have now?	41 − 23 = □	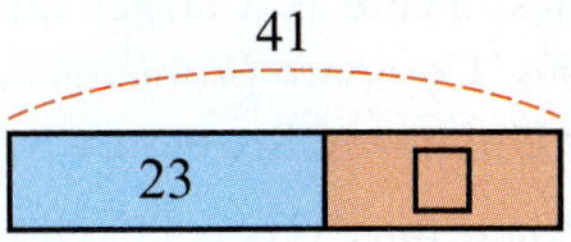
Change Unknown Rachel had 41 CDs. After she gave some away, she had 18 left. How many did she give away?	Situation equation (models the situation): 41 − □ = 18 Solution equation (could be used to solve): 41 − 18 = □	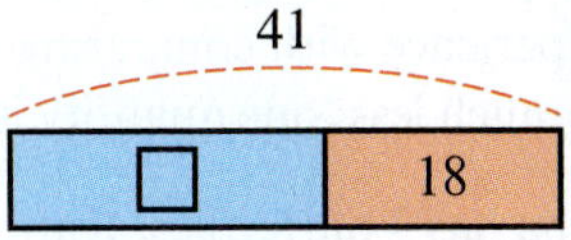
Start Unknown Rachel had some CDs. After she gave 23 away, she had 18 left. How many CDs did Rachel have before?	Situation equation (models the situation): □ − 23 = 18 Solution equation (could be used to solve): 23 + 18 = □	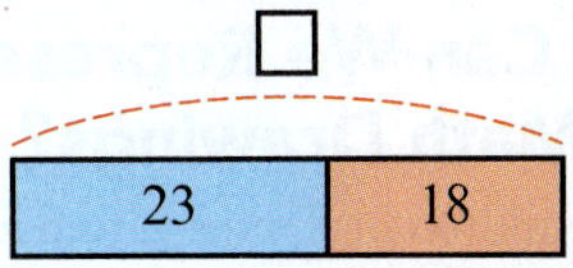

CCSS
K.OA.3, K.OA.4
1.OA.1
2.OA.1

What Are Put Together/Take Apart Problems?

These addition and subtraction problems involve two distinct parts that make a whole but do not involve change over time. These problems are sometimes called "part-part-whole" problems. Put Together/Take Apart problems have three subtypes: Total Unknown, Addend Unknown (each

Figure 3.4

Three types of Put Together/Take Apart problems.

Put Together/Take Apart Problems

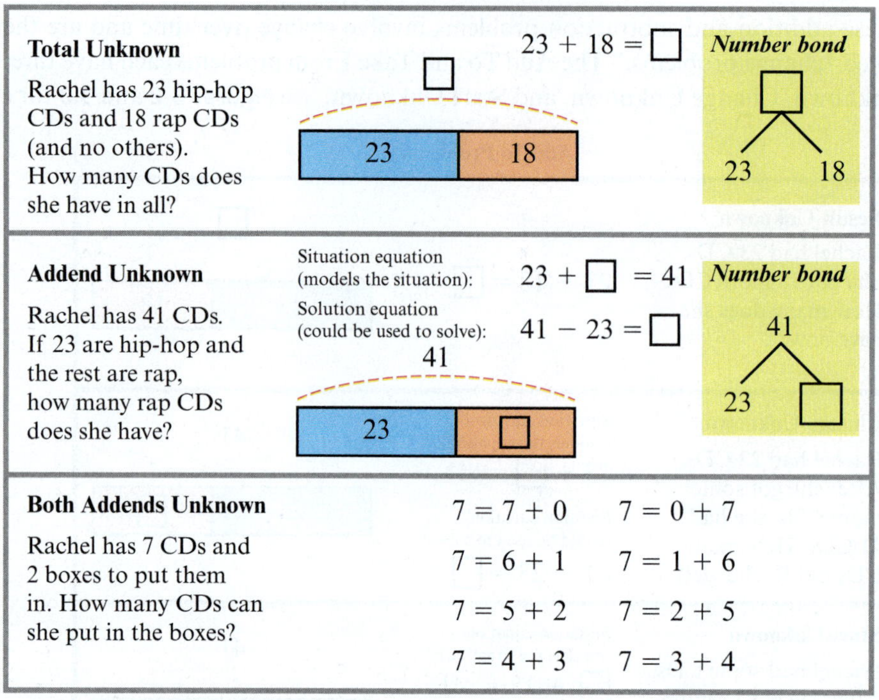

addend can be unknown) and Both Addends Unknown. Both Addends Unknown problems are a special case and are used mainly with whole numbers 10 and under as a foundation for important strategies that will be discussed in the next section. See Figure 3.4 for some examples.

CCSS

1.OA.1
2.OA.1

What Are Compare Problems?

Compare problems are addition and subtraction problems that involve the comparison of two quantities. There is a larger quantity, a smaller quantity, and the difference between the two quantities. Compare problems have three subtypes: Difference Unknown, Bigger Unknown, and Smaller Unknown. See Figure 3.5 for some examples.

Each of these three types of Compare problems can be formulated in a "more" version and in a "fewer" version. Later in this section, we'll see how this language can make the problem harder or easier.

The complex wording of compare problems can be difficult for children to grasp. They will first need experience with comparing quantities without having to decide specifically how much more, or how much less, one quantity is than another, as in the next problems:

Anna has 7 butterflies. Katie has 9 butterflies. Who has more butterflies?

Anna has 7 butterflies. Katie has 9 butterflies. Who has fewer butterflies? (or: Who has less?)

How Can We Represent Problems with Equations and Math Drawings?

When children in Kindergarten and Grade 1 start learning some of the different types of addition and subtraction word problems, they first use only small numbers, and they typically solve the problems by representing the situation with small objects, their fingers, or math drawings. For example, young children could draw dots or other small marks to represent objects. They can also use number bond diagrams, such as those shown in Figure 3.4, which they can continue to use with larger numbers later on. Strip diagrams, such as those in Figures 3.2–3.4, use lengths to show relationships between quantities. They can eventually be related to number lines.

Each type of addition and subtraction problem can be formulated mathematically by one or more equations. Because Add To and Take From problems involve change over time, there is a single equation

Figure 3.5 Three types of Compare problems.

Compare Problems

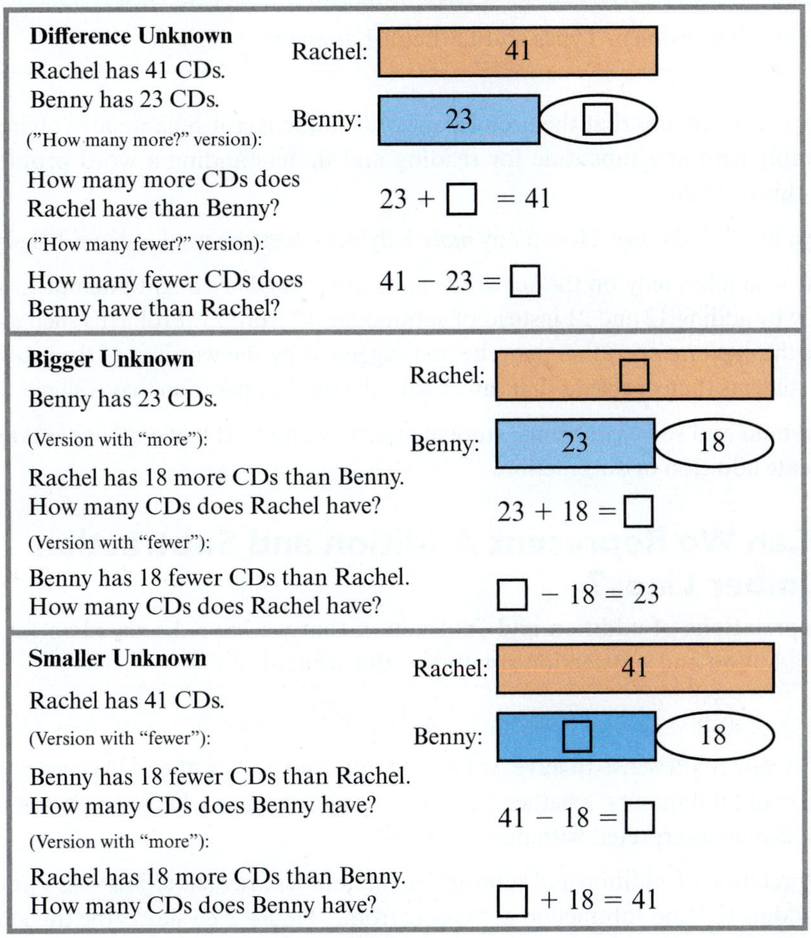

that fits most naturally with the situation of the problem, a "situation equation" (see Figures 3.2 and 3.3). But addition equations can always be reformulated in terms of subtraction, and subtraction equations can always be reformulated in terms of addition because of the connection between addition and subtraction. Such reformulated equations may be helpful for solving the problem. For example, to solve the Take From, Start Unknown problem in Figure 3.3, students might undo the "situation equation" (? − 23 = 18) and find the "solution equation" (23 + 18 = ?). Observe also how number bond diagrams and strip diagrams can help to reveal the different equations that can be associated with a single situation.

Put Together/Take Apart, Both Addends Unknown problems are different from the other types of problems because they have multiple solutions and they give rise to collections of equations, as shown in Figure 3.4. Such equations help children understand that the equal sign does not just mean "becomes" but that it tells us the left and right sides stand for the same amount. This broader understanding of equations and the equal sign is essential for algebra as well as for important strategies that will be discussed in the next section.

CLASS ACTIVITY

3C Writing Put Together/Take Apart and Compare Problems, p. CA-47

Why Can't We Rely on Keywords Alone?

In Class Activity 3D, observe how attending only to keywords in a problem might lead students to solve a problem incorrectly.

CLASS ACTIVITY

3D Identifying Problem Types and Difficult Language, p. CA-48

It's important to realize that the use of keywords alone is not reliable for solving word problems. There simply isn't any substitute for reading and understanding a word problem! For example, consider this problem:

> Tanya has 12 ladybugs. How many *more* ladybugs does she need to have 21 ladybugs *altogether*?

A student who relies only on the keywords *more* and *altogether* might attempt to solve this problem incorrectly by adding 12 and 21 instead of subtracting 12 from 21. Problems such as this one, which is solved by the *opposite* operation than the one suggested by the wording of the problem, are more difficult for students than problems that can be solved with the operation that *is* suggested by the wording.

To understand and solve problems, students must attend to the overall context as well as to words that indicate addition or subtraction.

CCSS 2.MD.6

How Can We Represent Addition and Subtraction on Number Lines?

The interpretations of addition and subtraction that we have discussed so far encompass most ordinary addition and subtraction situations. But what about

$$\sqrt{2} + \sqrt{3}?$$

It doesn't really make sense to have $\sqrt{2}$ of an apple and get another $\sqrt{3}$ of an apple. Addition and subtraction of all numbers, whether positive or negative, and whether represented by decimals or fractions, can be interpreted with number lines.

Our interpretation of addition and subtraction on number lines should fit with our definition of addition as "adding to" and subtraction as "taking from." Figure 3.6a shows the transition from addition as combining individual things, to increasing length, to the result of movement along a number line. Figure 3.6b shows the transition from subtraction as taking from, to decreasing length, to the result of movement along a number line. Figure 3.7 shows another way to think about subtraction on a

Figure 3.6a

From addition as combining to addition as moving along a number line.

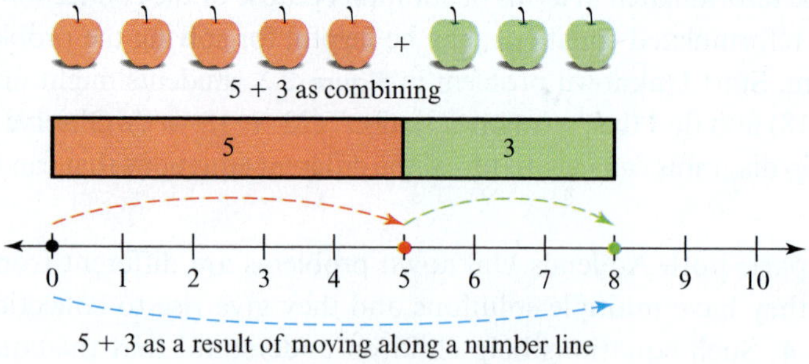

Figure 3.6b

From subtraction as taking from to subtraction as moving along a number line.

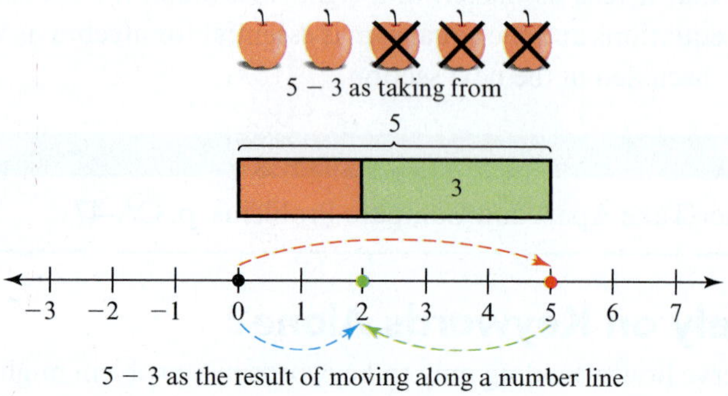

FROM THE FIELD Children's Literature

Shaskan, T. S. (2009). *If you were a plus sign*. Mankato, MN: Picture Window Books. In this story, the author explains how the plus sign is used in mathematics. Students will discover that addition allows you to find total amounts and to join amounts together.

Murphy, S. J. (2002). *Safari park*. New York, NY: HarperCollins. Grandpa takes his family to a new amusement park and buys each of his grandchildren 20 tickets for rides. Challenge your class to use tape diagrams to model how each grandchild spent her or his tickets and to figure out different ways to use extra tickets.

Pallotta, J. (2002). *The Hershey's Kisses subtraction book*. New York, NY: Scholastic. This book introduces students to subtraction by focusing on Hershey's Kisses at a circus carnival. Mathematical models for subtraction are included on every page, and students can explore the relationship between addition and subtraction throughout the book.

Cleary, B. P. (2006). *The action of subtraction*. Minneapolis, MN: Millbrook Press. The colorful illustrations in this book model subtraction in memorable ways, and students are able to compare how amounts change before and after subtraction.

Murphy, S. J. (1997). *Elevator magic*. New York, NY: HarperCollins. A young boy operates the elevator for his mother as she runs a few errands. He uses subtraction facts to help him locate the correct floor for each errand.

IMAP *Watch Tonya's class model strategies for solving a problem.*

number line, namely as a difference, or unknown addend. With this view, we can think of $5 - 3 = ?$ as $3 + ? = 5$. To get from 3 to 5 we need to move 2 more units to the right, so $5 - 3 = 2$.

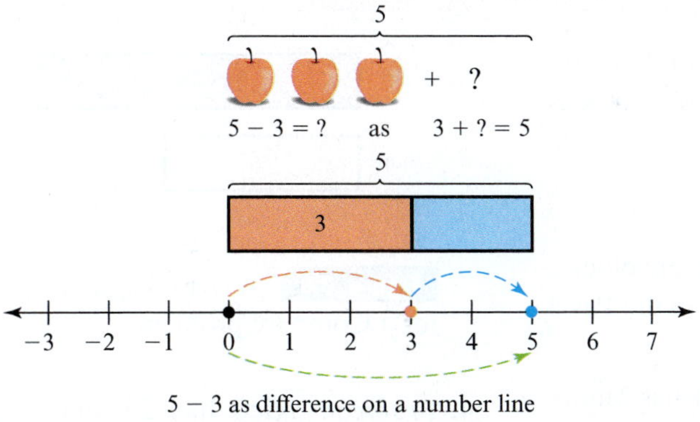

Figure 3.7 From subtraction as unknown addend to subtraction as difference on a number line.

SECTION SUMMARY AND STUDY ITEMS

Section 3.1 Interpretations of Addition and Subtraction

The most basic way to interpret addition is as Adding To, and the most basic way to interpret subtraction is as Taking From. But we also use addition and subtraction to solve problems that arise in Put Together/Take Apart situations and in Compare situations. For all these types of problems, there are subtypes depending on which quantity is unknown. If we view addition and subtraction in terms of movement along a number line, we can make sense of addition and subtraction when any numbers are involved.

Key Skills and Understandings

1. Write Add To, Take From, Put Together/Take Apart, and Compare problems of all subtypes and write equations and make math drawings to represent the problems.
2. Recognize that keywords alone are not effective for solving problems and that problems whose keywords indicate the opposite operation of a solution are difficult.
3. Explain how to use a number line to add and subtract numbers (for nonnegative numbers only).

Practice Exercises for Section 3.1

1. Write a Put Together/Take Apart, Addend Unknown problem and write an equation that fits naturally with the problem.
2. Write an Add To, Start Unknown problem that fits naturally with the equation

$$? + 7 = 15$$

3. Write two versions of a Compare, Smaller Unknown problem. Use "more" in one version and use "fewer" in the other and identify which is the harder problem. Draw a strip diagram and write an addition equation and a subtraction equation that fit with the problems.
4. Write a Take From Change Unknown problem that fits naturally with the equation

$$23 - ? = 9$$

Draw a strip diagram from the problem.

5. Write a word problem that a student who relies only on keywords might solve incorrectly by subtracting 31 − 28, but that is solved correctly by adding 31 + 28. Discuss why the student might solve the problem incorrectly.

Answers to Practice Exercises for Section 3.1

1. Pavel has 14 blocks in all. Eight of his blocks are red and the rest are blue. How many blue blocks does Pavel have?

$$8 + ? = 14$$

2. Pavel had some blocks. After he got 7 more blocks, he had 15 blocks in all. How many blocks did Pavel have at first?
3. *"More" version:* Becky has 23 rocks, which is 9 more than Sam has. How many rocks does Sam have?

   "Fewer" version: Becky has 23 rocks. Sam has 9 fewer rocks than Becky. How many rocks does Sam have?

   See Figure 3.8 for a strip diagram. The two equations 23 − 9 = ? and ? + 9 = 23 fit with both versions of the problem. The subtraction equation fits more closely with the "fewer" version and the addition equation fits more closely with the "more" version. The "more" version is harder because "more" indicates addition, but the problem is solved by subtracting 9 from 23.

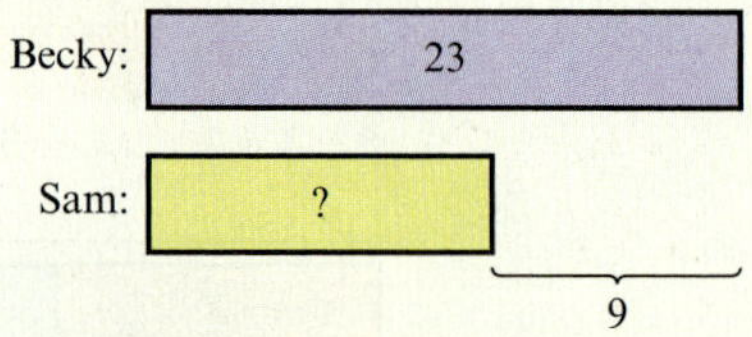

Figure 3.8 A strip diagram for a Compare problem.

4. *Problem:* Nico has 23 snap-together building blocks. After he used some of his blocks to build a robot, he had 9 blocks left. How many blocks did Nico use to build his robot?

   See Figure 3.9 for a strip diagram.

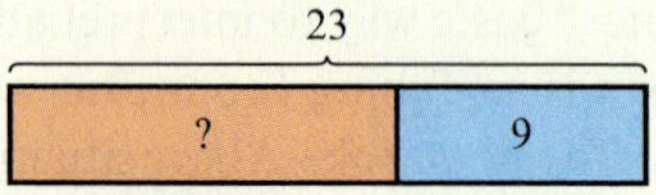

Figure 3.9 Strip diagram for a Take From problem.

5. *Problem:* Katie had some pencils. After she gave away 28 pencils, she had 31 pencils left. How many pencils did Katie have at first?

The words "gave away" and "left" in the problem indicate subtraction and may lead a student who is not thinking carefully about the problem statement to simply subtract 28 from 31. In fact, the subtraction equation, $? - 28 = 31$, fits with the problem. But this does not mean that 28 should be subtracted from 31 to solve the problem.

PROBLEMS FOR SECTION 3.1

1. a. Write an Add To problem that fits naturally with the equation

$$? + 17 = 42$$

Draw a strip diagram for the problem.

b. Write an Add To problem that fits naturally with the equation

$$17 + ? = 42$$

Draw a strip diagram for the problem.

c. Write a Put Together/Take Apart problem that fits naturally with the equation

$$? + 17 = 42 \quad \text{or} \quad 17 + ? = 42$$

Draw a strip diagram or number bond for the problem.

2. Write a Put Together/Take Apart, Both Addends Unknown problem for the number 6. Write equations that fit with the solutions of the problem.

3. a. Write two versions of a Compare, Difference Unknown problem that involves the numbers 17 and 42. Use "more" in one version and "fewer" in the other.

b. Draw a strip diagram, and write an addition equation and a subtraction equation that fit with the problems.

c. Identify the version of the problem you wrote in part (a) that is harder for students. Explain why it is harder.

4. a. Write two versions of a Compare, Bigger Unknown problem that involves the numbers 17 and 42. Use "more" in one version and "fewer" in the other.

b. Draw a strip diagram, and write an addition equation and a subtraction equation that fit with the problems.

c. Identify the version of the problem you wrote in part (a) that is harder for students. Explain why it is harder.

5. Identify the type and subtype of each of the following problems.

a. Shawn has 15 marbles, which is 7 more marbles than Kyle has. How many marbles does Kyle have?

b. Tiffany has 12 blocks, 5 of which are cubes and the rest cylinders. How many blocks are cylinders?

c. Peter had some carrots. After he ate 3 of them, he had 14 carrots left. How many carrots did Peter have before?

d. In a bag of 17 marbles, 9 marbles belong to Kelly and the rest belong to Shauntay. How many marbles belong to Shauntay?

6. Write an Add To, a Put Together/Take Apart, and a Compare problem, each of which can be solved by subtracting $32 - 17$.

7. Write a Take From, a Put Together/Take Apart, and a Compare problem, each of which can be solved by adding $32 + 17$.

8. a. Write an Add To problem that can be solved by subtracting numbers in the problem. Show how a student could use a strip diagram to see why the problem can be solved by subtracting.

b. Write a Take From problem that can be solved by adding numbers in the problem. Show how a student could use a strip diagram to see why the problem can be solved by adding.

3.2 The Commutative and Associative Properties of Addition, Mental Math, and Single-Digit Facts

CCSS Common Core State Standards Grades K, 1, 2

In this section we'll study two fundamental properties of addition, the commutative and associative properties, and we'll see how these properties underlie mental methods of addition that both children and adults can use to solve problems in flexible and creative ways. The commutative and associative properties of addition also underlie the increasingly advanced methods that young children learn as they travel along the path to fluency with single-digit addition and subtraction facts.

The commutative and associative properties of addition, together with the other properties of arithmetic (which include the commutative and associative properties of multiplication and the distributive property) form the building blocks of all of arithmetic. Ultimately, every calculation strategy, whether a mental method of calculation or a standard algorithm, relies on these properties. These properties allow us to take numbers apart, to break arithmetic problems into pieces that are easier to solve, and to put the pieces back together. The strategy of decomposing into simpler pieces, analyzing the pieces, and then putting them back together is important at every level of mathematics and in all branches of mathematics.

First, let's look at how we use parentheses to group numbers in mathematical expressions.

What Is the Role of Parentheses in Expressions with Three or More Terms?

A sum

$$A + B + C$$

with 3 terms means the sum of $A + B$ and C. In other words, according to the meaning of $A + B + C$, we add from left to right. Likewise, if there are 4 or more terms in a sum, the sum stands for the result obtained by adding from left to right.

But what if we want to indicate the sum of A with $B + C$? We can show this with parentheses. In mathematical expressions, parentheses are used to group numbers and operations (+, −, ×, ÷). To indicate the sum of A with $B + C$, we write

$$A + (B + C)$$

So,

$$\begin{aligned} 17 + (18 + 2) &= 17 + 20 \\ &= 37 \end{aligned}$$

Notice that we can express the meaning of the sum

$$A + B + C$$

by using parentheses:

$$(A + B) + C$$

CLASS ACTIVITY

3E Mental Math, p. CA-49

Why Does the Associative Property of Addition Make Sense?

How can you make the problem

$$7384 + 999 + 1$$

easy to solve mentally? Rather than adding from left to right, it is easier to first add 999 and 1 to make 1000, and then add 7384 and 1000 to get 8384. In other words, rather than adding from left to right, calculating

$$(7384 + 999) + 1$$

according to the meaning of the sum $7384 + 999 + 1$, it is easier to group the 999 with the 1 and calculate

$$7384 + (999 + 1)$$

instead. Why is it legitimate to switch the way the numbers in the sum are grouped? Because, according to the associative property of addition, both ways of calculating the sum give equal results; in other words,

$$(7384 + 999) + 1 = 7384 + (999 + 1)$$

associative property of addition

The **associative property of addition** tells us that when we add any three numbers, it doesn't matter whether we add the first two and then add the third, or whether we add the first number to the sum of the second and the third—either way we will always get the same answer. In other words, the associative property of addition says that if A, B, and C are any three numbers, then

$$(A + B) + C = A + (B + C)$$

That is, the sum of $A + B$ and C is equal to the sum of A and $B + C$.

It might help to understand the term *associative property* by remembering that to *associate* with someone means to keep company with them. In the equation

$$(7384 + 999) + 1 = 7384 + (999 + 1)$$

the number 999 can either *associate* with 7384, or it can *associate* with 1.

We assume that the associative property of arithmetic is true for all numbers, but we can use an example to see why this property of addition makes sense. Suppose there is a group of 2 marbles, a second group of 3 marbles, and a third group of 4 marbles, as at the top of Figure 3.10. If we put the first and second groups together, we can describe the total number of marbles as

$$(2 + 3) + 4$$

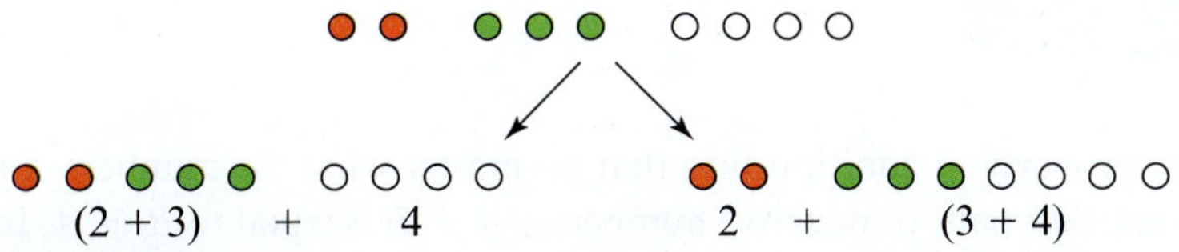

Figure 3.10 Demonstrating the associative property of addition.

But if we put the second and third groups of marbles together, we can describe the total number of marbles as

$$2 + (3 + 4)$$

Since the total number of marbles is the same no matter how we describe it, these two expressions for the total must be equal. In other words,

$$(2 + 3) + 4 = 2 + (3 + 4)$$

The same kind of relationship would hold if other numbers of marbles replaced the numbers 2, 3, and 4. The associative property of addition states that this kind of relationship always holds, no matter what numbers are involved, even if these numbers are fractions, decimals, or negative numbers.

The associative property of addition allows us to group sums of numbers any way we like. For example, to add

$$0.01 + 2.47 + 3.97 + 0.02 + 0.01$$

we can group the first two terms and the last three terms to make the addition easy:

$$(0.01 + 2.47) + (3.97 + 0.02 + 0.01) = 2.48 + 4.00 = 6.48$$

Why Does the Commutative Property of Addition Make Sense?

How can a young child calculate $2 + 9$? The child might think about the problem as $9 + 2$ and count on 2 from 9, by saying "ten, eleven." In replacing $2 + 9$ with $9 + 2$, the child has used the *commutative property of addition* (even if she does not know that term).

commutative property of addition

The **commutative property of addition** states that for all real numbers A and B,

$$A + B = B + A$$

To explain why the commutative property of addition makes sense, suppose that you have 3 green apples and 2 red apples. Then you have

$$3 + 2$$

apples in all. But if you think about the 2 red apples first, and then the 3 green apples, the total is

$$2 + 3$$

apples. Either way it's the same number of apples in all, as shown in Figure 3.11, and therefore

$$3 + 2 = 2 + 3$$

Figure 3.11 The commutative property of addition: $3 + 2 = 2 + 3$.

The commutative property of addition says that no matter what the numbers A and B are—even if they are fractions, decimals, or negative numbers—$A + B$ is equal to $B + A$. In other words,

$$A + B = B + A$$

for all numbers A and B.

In this setting, the word *commute* means *to change places with*. In going from $3 + 2$ to $2 + 3$, the 2 and the 3 change places. Students sometimes get confused between the commutative and associative properties. Notice that the commutative property involves changing the order of terms; in other words, the commutative property involves *rearranging* terms. The associative property involves *regrouping* terms by changing the location of parentheses.

Combined, the commutative and associative properties of addition allow for great flexibility in calculating sums: You can rearrange the terms any way you like, and you can combine the terms with each other any way you like. This flexibility is valuable not only for mental calculations, but it is also an essential skill in algebra.

How Do We Interpret the Equal Sign Correctly?

equation
equal sign, =

An **equation** is a mathematical statement saying that two numbers or expressions are equal, meaning that they stand for the same amount. We use the **equal sign**, **=**, to express this equality. The equal sign means "*is equal to,*" or in other words "*is the same amount as.*"

CCSS
1.OA.7

The most common way that we use the equal sign is to show the result of a calculation, as in

$$7 + 5 = 12$$

Therefore, students often interpret the equal sign as meaning "*calculate the answer*" rather than "*is the same amount as.*"

If students take a "*calculate the answer*" view of the equal sign, then an equation such as

$$7 + 5 = 5 + 7$$

which illustrates the commutative property of addition, won't make sense. The equations we use to describe the properties of arithmetic require an "*is the same amount as*" view of the equal sign.

Notice that the equations for the commutative and associative properties arise by seeing one amount in two different ways.

CCSS
1.OA.3
1.OA.6
2.OA.2

What Are Children's Learning Paths for Single-Digit Facts and How Do They Rely on Properties of Addition?

According to the *Common Core State Standards*, by the end of second grade, children should know the single-digit addition facts

$$\begin{array}{cccc} 1+1=2 & 1+2=3 & \cdots & 1+9=10 \\ \vdots & \vdots & & \vdots \\ 9+1=10 & 9+2=11 & \cdots & 9+9=18 \end{array}$$

and the associated single-digit subtraction facts from $2 - 1 = 1$ to $18 - 9 = 9$. The *Common Core State Standards* support a learning path in which children develop fluency with these addition and subtraction facts by reasoning in progressively sophisticated ways. This learning path is outlined in Class Activities 3F and 3G. Importantly, the learning path is *not mere rote memorization* but relies instead on reasoning with the commutative and associative properties of addition, even though children may not be aware that they are using these properties.

CLASS ACTIVITY

3F Children's Learning Paths for Single-Digit Addition, p. CA-50

3G Children's Learning Paths for Single-Digit Subtraction, p. CA-52

How Can Children Use The Commutative Property in Single-Digit Addition? As you saw in Class Activity 3F, the commutative property of addition is important for fluency with single-digit addition problems because it allows children to replace addition problems of the form

$$\text{smaller number} + \text{larger number}$$

with the problem

$$\text{larger number} + \text{smaller number}$$

For example, instead of solving $2 + 7 = ?$, a child can solve $7 + 2 = ?$ if the child understands the commutative property of addition (even if she doesn't know the term). If the child doesn't yet have quick recall of either $2 + 7$ or $7 + 2$, then the child can "count on" from 7 as described in Class Activity 3F. The commutative property of addition lightens the load of learning all the single-digit addition facts: It reduces these facts to knowing the "doubles," $1 + 1, 2 + 2, 3 + 3, \ldots, 9 + 9$, and the facts of the form (larger number) + (smaller number).

How Can Children Use the Associative Property in Derived Fact Methods? Children use the associative property of addition (whether or not they know this term) when they use the more advanced *derived fact methods* described in Class Activity 3F to solve single-digit addition problems. Especially important is the **make-a-ten method**. To use this method to calculate $8 + 7$, a child breaks 7 into $2 + 5$, joins the 2 with the 8 to make a 10, then adds the remaining 5 to the 10 to get the solution, 15. The dot picture in Figure 3.12 illustrates this method. The following equations show that this method uses the associative property of addition:

make-a-ten method

$$\begin{aligned} 8 + 7 &= 8 + (2 + 5) \\ &= (8 + 2) + 5 \\ &= 10 + 5 = 15 \end{aligned}$$

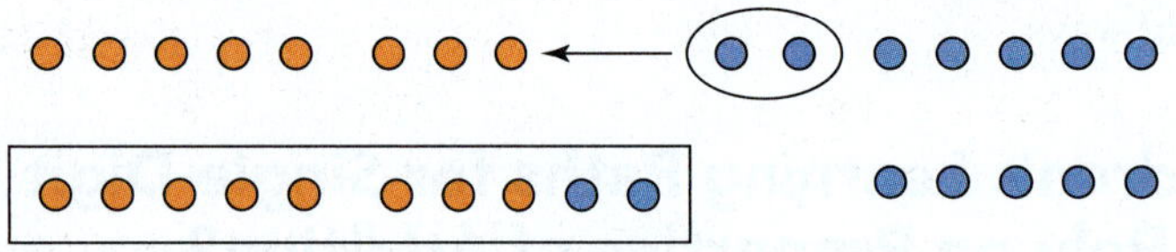

Figure 3.12 The make-a-ten method.

The associative property was used in going from the first line of the equation to the second line: the 2 goes from associating with the 5 to make 7 to associating with the 8 to make a 10. So a child who knows that 8 needs 2 more to make a 10 and who knows that 7 breaks into $2 + 5$ can derive that $8 + 7 = 15$.

The make-a-ten method draws children's attention to 10 and helps reinforce the understanding of the teen numbers 11 through 19 as a 10 and some ones. When children use the *subtract from ten method* described in Class Activity 3G to calculate $13 - 8$, they view 13 as $10 + 3$ and subtract the 8 from the 10, leaving 2 to add to 3 to make 5. The understanding of teen numbers as a 10 and some ones is critically important to making sense of multidigit addition and subtraction. When children solve a multidigit addition problem such as $28 + 47$, the first step in the common method is

$$\begin{array}{r} {\scriptstyle 1} \\ 28 \\ +\,47 \\ \hline 5 \end{array}$$

To understand this regrouping method, a child must know not only that $8 + 7 = 15$ but also that this 15 is 1 ten and 5 ones.

The derived fact methods help children connect and relate addition facts and derive facts they don't yet know from ones they have already learned. Derived fact methods give children the opportunity to engage in a powerful technique that is used across arithmetic, geometry, and algebra and at all levels of mathematics: the technique of taking apart, analyzing, and putting back together.

CCSS 1.OA.4

How Is Viewing Subtraction Problems as Unknown Addend Problems Helpful for Children? Class Activity 3G described two *unknown addend methods* children can use for subtraction. To solve $12 - 7 = ?$ a child can first view this problem as the unknown addend problem $7 + ? = 12$. To solve this unknown addend problem children can either count on from 7 to 12; or they can add 3 to make 10, then add another 2 to make 12, and then recognize the unknown addend as $3 + 2 = 5$. Although children could solve $12 - 7 = ?$ by counting down 7 from 12, teachers consider this method slow and error-prone. In contrast, using an unknown addend method to subtract makes subtraction as easy for children as addition. Learning unknown addend methods to subtract also helps children understand the link between subtraction and addition.

FROM THE FIELD Research

Betts, P. (2015). Counting on using a number game. *Teaching Children Mathematics, 21*(7), 430–436.

Researchers have documented counting all and counting on as two strategies to count the total number of objects when two sets are joined. In counting all, the child restarts counting from 1 once the sets are joined. In counting on, the child starts counting the joined set from the number of items in the first set. The author documents some instructional games that can help children transition from counting all to counting on, including the *number board game*. On a number board, children have tokens that move upwards based on the roll of a die. When children move their token on the board, they have to coordinate the number on the die with the movement and numbers on the board. Working with four students, the author found that three of the four were able to transition to counting-on strategies by the end of four sessions. Their teachers also saw the students using the strategies in class. The activity helped some students develop strategies of decomposing and subitizing. For example, when one student rolled a five, she would keep track of her movements with two head nods, a pause, two nods, a pause, and one nod. Two other students also demonstrated this strategy. The author recommends such coordinating activities to develop students' own strategies that facilitate counting on.

What Are Some Special Strategies for Multidigit Addition and Subtraction?

Some of the methods that are part of children's learning paths for single-digit addition and subtraction generalize to multidigit situations, thereby providing children and adults with flexible, quick ways to solve addition and subtraction problems.

Make-a-Round-Number Method The make-a-round-number method is just like the make-a-ten method, except that instead of just making a ten, we make other round numbers. The make-a-round-number method uses the associative property of addition to shift one piece of an addend and join the piece with the other addend. To use this method to calculate a sum, look for a nice round number that is close to one of the addends. For example, to add $376 + 199$ mentally, break 376 into $375 + 1$ and join the 1 with 199 to make 200. Therefore, the sum is $375 + 200 = 575$. The following equations show that we used the associative property of addition in applying this method:

$$\begin{aligned} 376 + 199 &= (375 + 1) + 199 \\ &= 375 + (1 + 199) \\ &= 375 + 200 = 575 \end{aligned}$$

Rounding and Compensating Another way to calculate $376 + 199$ is to round and compensate: Suppose we add 200 to 376 instead of adding 199 to 376. This makes 576. But we added 1 too many, so we must take 1 away from 576. Therefore, $376 + 199 = 575$.

We can write corresponding equations as follows:

$$\begin{aligned} 376 + 199 &= 376 + 200 - 1 \\ &= 576 - 1 \\ &= 575 \end{aligned}$$

We can also round and compensate in subtraction problems. Consider the problem 684 − 295. The number 295 is close to 300. So, to calculate

$$684 - 295$$

we can reason as follows:

> Taking 300 objects away from 684 objects leaves 384 objects. But when we took 300 away, we took away 5 more than we should have (because 300 is 5 more than 295). This means that we must *add* 5 to 384 to get the answer, 389.

The following equations correspond to this line of reasoning:

$$\begin{aligned} 684 - 295 &= 684 - 300 + 5 \\ &= 384 + 5 \\ &= 389 \end{aligned}$$

Subtraction Problems as Unknown Addend Problems We saw previously that it can be helpful for children to view a subtraction problem such as $14 - 9 = ?$ as an unknown addend problem, $9 + ? = 14$. Viewing a subtraction problem as an unknown addend problem can also be helpful in solving multidigit subtraction problems mentally. Consider once again the example $684 - 295 = ?$. We can view this subtraction problem as the unknown addend problem $295 + ? = 684$.

With this point of view, we start with 295 and keep adding numbers until we reach 684:

$$\begin{aligned} 295 + 5 &= 300 \\ 300 + 300 &= 600 \\ 600 + 84 &= 684 \end{aligned}$$

or

$$295 + 5 + 300 + 84 = 684$$

All together, starting with 295, we added

$$5 + 300 + 84 = 389$$

to reach 684. Therefore, $684 - 295 = 389$. This is the method that shopkeepers frequently use to make change.

How Can We Organize Strings of Equations So They Communicate Accurately?

To show a line of reasoning for a calculation, it is useful to write strings of equations that correspond to the reasoning. Rather than writing

$$\begin{aligned} 198 + 357 + 2 &= 198 + 2 + 357 \\ 198 + 2 + 357 &= 200 + 357 \\ 200 + 357 &= 557 \end{aligned}$$

it is neater and easier to follow this sequence of steps:

$$\begin{aligned} 198 + 357 + 2 &= 198 + 2 + 357 \\ &= 200 + 357 \\ &= 557 \end{aligned}$$

When several equations are strung together, it is for the purpose of concluding that the *first* expression is equal to the *last* expression. From the previous equations we conclude that 198 + 357 + 2 is equal to 557. This string of equations used the following general property of equality, which we assume holds true:

If A is equal to B and if B is equal to C, then A is also equal to C.

When you use an equal sign, *make sure that the quantities before and after the equal sign really are equal to each other*. Suppose you calculate

$$499 + 165$$

by taking 1 from 165, adding this 1 to 499 to make 500, and then adding on the remaining 164 to get 664. When showing this method of calculation with equations, it is easy to make the following mistake:

Warning: incorrect

$$\begin{aligned} 499 + 1 &= 500 + 164 \\ &= 664 \end{aligned}$$

These equations are incorrect because 499 + 1 *is not equal to* 500 + 164. Instead, you can write the following correct equations:

$$\begin{aligned} 499 + 165 &= 499 + 1 + 164 \\ &= 500 + 164 \\ &= 664 \end{aligned}$$

CLASS ACTIVITY

3H Reasoning to Add and Subtract, p. CA-53

SECTION SUMMARY AND STUDY ITEMS

Section 3.2 The Commutative and Associative Properties of Addition, Mental Math, and Single-Digit Facts

The associative property of addition says that, for all numbers A, B, and C,

$$(A + B) + C = A + (B + C)$$

The commutative property of addition says that, for all real numbers A and B,

$$A + B = B + A$$

We can use these properties to help children learn the basic addition facts. The commutative property allows children to cut the number of facts that must be memorized almost in half. The associative property is used in the important make-a-ten strategy. The commutative and associative properties of addition can also help make some multidigit addition problems easier to carry out mentally.

Key Skills and Understandings

1. State the associative property of addition and explain how to get its equation by viewing one amount in two ways.
2. Give examples to show how to use the associative property of addition to make problems easier to do mentally, including the make-a-ten strategy.
3. State the commutative property of addition and explain how to get its equation by viewing one amount in two ways.

4. Give examples to show how to use the commutative property of addition to make problems easier to do mentally, including the "count on from the larger addend" methods, and explain how it helps children cut down on the memorization of basic addition facts.
5. Describe how to view subtraction problems as unknown addend problems, explain how this can help young children with basic subtraction facts, and explain how this can be applied to other mental math problems.
6. Write correct equations to go along with a mental method of addition or subtraction. Identify where the commutative or associative properties of addition have been used in calculations.
7. Explain how to add or subtract by using methods other than the common addition and subtraction algorithms.

Practice Exercises for Section 3.2

1. State the associative property of addition and show how to use a collection of small objects to illustrate why this property makes sense.
2. State the commutative property of addition and show how to use a collection of small objects to illustrate why this property makes sense.
3. Describe how a child who is learning the single-digit addition facts could use the commutative property of addition (even if the child doesn't know the term *commutative property*).
4. Describe how a child who is learning the single-digit addition facts can use the make-a-ten method to add 7 + 4. Write equations that show this method. Which property of addition does this method use?
5. Describe two ways children who are learning the single-digit addition and subtraction facts can solve subtraction problems by viewing subtraction problems as unknown addend problems.
6. Give an example that demonstrates how you can use the associative property of addition to make a multidigit addition problem easier to do mentally. Describe in words how you can solve the problem mentally. Write equations that show where you used the associative property.
7. The sequence of equations that follows shows a way of using properties of addition to calculate a sum. Say specifically which properties of addition were used, and where.

$$
\begin{aligned}
27 + 89 + 13 &= 27 + (89 + 13) \\
&= 27 + (13 + 89) \\
&= (27 + 13) + 89 \\
&= 40 + 89 \\
&= 129
\end{aligned}
$$

8. For each of the addition problems that follow, write equations that correspond to a mental method for calculating the sum that uses the associative and/or the commutative properties of addition. Say specifically which properties of addition were used, and where.

   a. 993 + 2389

   b. 398 + (76 + 2)

9. Each arithmetic problem in this exercise has a description for how to solve the problem. In each case, write a sequence of equations that correspond to the given description.

   a. *Problem:* 23 + 45

   Solution: 23 plus 40 is 63, and then 5 more makes 68.

   b. *Problem:* 800 − 297

   Solution: 800 − 300 = 500, but subtracting 300 subtracts 3 too many, so we must add 3 back, making 503.

10. Nancy writes the following equations to solve 37 + 14:

$$30 + 10 = 40 + 7 = 47 + 4 = 51$$

   Write correct equations that solve 37 + 14 and that incorporate Nancy's solution strategy.

11. Jim writes the following equations to solve 85 − 15:

$$85 - 10 = 75 - 5 = 70$$

   Write correct equations that solve 85 − 15 and that incorporate Jim's solution strategy.

12. Find ways to solve the following set of addition and subtraction problems *other than* by using the common addition or subtraction algorithms. In each case, explain your reasoning, and also write equations that incorporate your thinking.

a. $786 - 47$

b. $427 + 28$

c. $999 + 999$

d. $1002 - 986$

e. $237 - 40$

Answers to Practice Exercises for Section 3.2

1. $A + (B + C) = (A + B) + C$ for all numbers A, B, C. See Figure 3.10.

2. $A + B = B + A$ for all numbers A, B. See Figure 3.11.

3. A young child uses the commutative property of addition when she "counts on from larger." For example, if the child is asked to add $2 + 9$, the child can use the commutative property to replace this sum with $9 + 2$ and count on 2 from 9, by saying "10, 11" instead of counting on 9 from 2.

4. To add $7 + 4$ with the make-a-ten method, the child breaks 4 into $3 + 1$, joins the 3 with the 7 to make a 10, then joins this 10 with the remaining 1 to make 11. In equations:

$$\begin{aligned} 7 + 4 &= 7 + (3 + 1) \\ &= (7 + 3) + 1 \\ &= 10 + 1 = 11 \end{aligned}$$

The associative property of addition was used at the second equal sign.

5. A child can view a subtraction problem such as $13 - 9 = ?$ as the unknown addend problem $9 + ? = 13$. If the child is still counting on, the child can count up from 9 to 13, saying "10, 11, 12, 13" and using his fingers to keep track of how many numbers he counted on. Since he counted on 4 numbers, he concludes that $9 + 4 = 13$, so $13 - 9 = 4$. If the child is ready for derived fact methods and no longer needs to count on, then the child can think: "Starting at 9, one more is 10, and then 3 more is 13, so that's 1 and 3 which is 4," again concluding that $9 + 4 = 13$, so that $13 - 9 = 4$.

6. To calculate $49 + 37$ mentally, move 1 from 37 to 49, so that the sum becomes $50 + 36$, which is 86. Corresponding equations are:

$$\begin{aligned} 49 + 37 &= 49 + (1 + 36) \\ &= (49 + 1) + 36 \\ &= 50 + 36 \\ &= 86 \end{aligned}$$

The associative property of addition is used at the second equal sign to switch the placement of parentheses from grouping 1 and 36 together, to grouping 49 and 1 together.

7. The associative property of addition was used at the first equal sign to say that $(27 + 89) + 13 = 27 + (89 + 13)$. The commutative property of addition was used at the second equal sign to change $89 + 13$ to $13 + 89$. The associative property of addition was used at the third equal sign to say that

$$27 + (13 + 89) = (27 + 13) + 89.$$

8. a. $993 + 2389 = 993 + (7 + 2382) = (993 + 7) + 2382 = 1000 + 2382 = 3382$. The associative property of addition was used in rewriting $993 + (7 + 2382)$ as $(993 + 7) + 2382$.

b. $398 + (76 + 2) = 398 + (2 + 76) = (398 + 2) + 76 = 400 + 76 = 476$. The commutative property of addition was used to rewrite $76 + 2$ as $2 + 76$. The associative property of addition was used to rewrite $398 + (2 + 76)$ as $(398 + 2) + 76$.

9. a.
$$\begin{aligned} 23 + 45 &= 23 + (40 + 5) \\ &= (23 + 40) + 5 \\ &= 63 + 5 \\ &= 68 \end{aligned}$$

The associative property of addition was used at the second equal sign to group the 40 with the 23 instead of with the 5 (where it was part of 45).

b.
$$\begin{aligned} 800 - 297 &= 800 - 300 + 3 \\ &= 500 + 3 \\ &= 503 \end{aligned}$$

10.
$$\begin{aligned} 37 + 14 &= 30 + 10 + 7 + 4 \\ &= 40 + 7 + 4 \\ &= 47 + 4 \\ &= 51 \end{aligned}$$

11.
$$\begin{aligned} 85 - 15 &= 85 - 10 - 5 \\ &= 75 - 5 \\ &= 70 \end{aligned}$$

12. Here are equations you could write:

a. $786 - 47 = 786 - 50 + 3$
$= 736 + 3$
$= 739$

b. $427 + 28 = 427 + 3 + 25$
$= 430 + 25$
$= 455$

c. $999 + 999 = 999 + 1000 - 1$
$= 1999 - 1$
$= 1998$

d. $986 + 4 = 990$
$990 + 10 = 1000$
$1000 + 2 = 1002$

so,

$$986 + 4 + 10 + 2 = 1002$$

so,

$$1002 - 986 = 4 + 10 + 2 = 16$$

e. $237 - 40 = 240 - 40 - 3$
$= 200 - 3$
$= 197$

PROBLEMS FOR SECTION 3.2

1. Many teachers have a collection of small cubes that can be snapped end-to-end to make "trains" of cubes. The cubes come in various colors.

a. Describe how to use snap-together cubes in different colors to demonstrate the commutative property of addition. Explain how to get the property's equation by viewing one amount in two ways.

b. Describe how to use snap-together cubes in different colors to demonstrate the associative property of addition. Explain how to get the property's equation by viewing one amount in two ways.

2. Discuss the difference between the commutative property of addition and the associative property of addition.

3. Figure 3.13 indicates a make-a-ten method for adding 6 + 8.

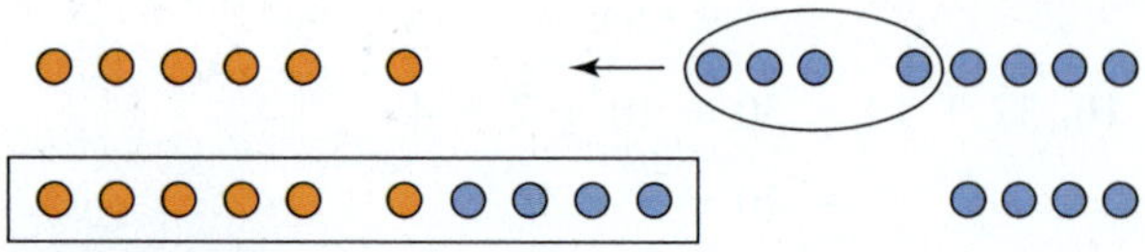

Figure 3.13 A dot picture for 6 + 8.

a. Write equations that correspond to the make-a-ten method for adding 6 + 8 depicted in Figure 3.13. Your equations should make careful and appropriate use of parentheses. Which property of arithmetic do your equations and Figure 3.13 illustrate?

b. Make a math drawing for 7 + 5 that illustrates a make-a-ten method. Write equations that correspond to the strategy indicated in your picture, making careful and appropriate use of parentheses.

4. Tomaslav has learned the following facts well:

- All the sums of whole numbers that add to 10 or less—Tomaslav knows these facts "forwards and backwards." For example, he knows not only that 5 + 2 is 7, but also that 7 decomposes into 5 + 2 or 2 + 5.
- 10 + 1, 10 + 2, 10 + 3, . . . , 10 + 10.
- the *doubles* 1 + 1, 2 + 2, 3 + 3, . . . , 10 + 10.

For each sum in this problem, describe at least three different ways that Tomaslav could use reasoning, together with the facts he knows well, to determine the sum. In each case, write equations that correspond to the strategies you describe. Take care to use parentheses appropriately and as needed.

a. 8 + 7

b. 6 + 7

c. 8 + 9 (Try to find four or five different ways to solve this.)

5. Give an example of an arithmetic problem that can be made easy to solve mentally by using the associative property of addition. Write equations that show your use of the associative property of addition. Your use of this property must genuinely make the problem easier to solve.

6. Give an example of an arithmetic problem that can be made easy to solve mentally by using the commutative property of addition. Write equations that show your use of the commutative property of addition. Your use of this property must genuinely make the problem easier to solve.

7. Describe a way to calculate $304 - 81$ mentally, by using reasoning other than the common subtraction algorithm. Then write a coherent sequence of equations that correspond to your reasoning.

8. To calculate $159 - 73$, a student writes the following equations:

$$160 - 70 = 90 - 3 = 87 - 1 = 86$$

Although the student has a good idea for solving the problem, his equations are not correct. In words, describe the student's solution strategy and discuss why the strategy makes sense; then write a correct sequence of equations that correspond to this solution strategy. Write your equations in the following form:

$$\begin{aligned} 159 - 73 &= \text{some expression} \\ &= \text{some expression} \\ &= \vdots \\ &= 86 \end{aligned}$$

9. To calculate $201 - 88$, a student writes the following equations:

$$88 + 2 = 90 + 10 = 100 + 100 = 200 + 1 = 201$$
$$2 + 10 = 12 + 100 = 112 + 1 = 113$$

Although the student has a good idea for solving the problem, his equations are not correct. In words, describe the student's solution strategy and explain why it makes sense; then write correct equations that correspond to this solution strategy.

10. David and Ashley want to calculate $8.27 − $2.98 by first calculating $8.27 − $3 = $5.27. David says that they must *subtract* $0.02 from $5.27, but Ashley says that they must *add* $0.02 to $5.27.

a. Draw a number line (which need not be perfectly to scale) to help you explain who is right and why. Do not just say which answer is numerically correct; use the number line to help you explain why the answer must be correct.

b. Explain in another way who is right and why.

11. Tylishia says that she can calculate $324 - 197$ by adding 3 to both numbers and calculating $327 - 200$ instead.

a. Draw a number line (which need not be perfectly to scale) to help you explain why Tylishia's method is valid.

b. Explain in another way why Tylishia's method is valid.

c. Could you adapt Tylishia's method to other subtraction problems, such as to the problem $183-49$? If so, give at least two more examples and show how to apply Tylishia's method in each case.

***12.** Is there an *associative property of subtraction*? In other words, is

$$A - (B - C) = (A - B) - C$$

true for all real numbers A, B, C? Explain your answer. If the equation is always true, explain why; if the equation is not always true, find another expression that *is* equal to the expression

$$A - (B - C)$$

and explain why the two expressions are equal.

***13.** Is there a *commutative property of subtraction*? Explain why or why not. In your answer, include a statement of what a commutative property of subtraction would be if there were such a property.

3.3 Why the Standard Algorithms for Addition and Subtraction in Base Ten Work

CCSS Common Core State Standards Grades 1, 2, 3, 4, 5

Have you ever wondered why the methods we use to add and subtract multidigit numbers in base ten work? These methods were developed by mathematicians in ancient times. How did these mathematicians know that their clever methods would yield correct results? To develop the calculation methods, the mathematicians had to understand how to decompose and recompose numbers according to place value and how to apply properties of arithmetic. Today the methods of arithmetic still provide fertile ground for developing a sense of place value and a sense of how to take numbers apart and recombine them suitably. For this reason, even though calculators are ubiquitous, children should learn to make sense of calculation methods and to use them with an understanding of why they work.

What Is an Algorithm?

algorithm

An **algorithm** is a method or a procedure for carrying out a calculation.

When you bake a cake, you probably follow a recipe. A cake recipe is a kind of algorithm: a step-by-step procedure taking various ingredients and turning them into a cake. In the same way, there are step-by-step procedures for adding, subtracting, multiplying, and dividing numbers. Just as there is mystery in how flour, eggs, butter, and sugar can combine to make cake, there is mystery in how the standard algorithms result in correct answers to arithmetic problems. Somehow, we throw numbers in, mix them up in a specific way, and out comes the correct answer to the addition problem. In this book, we won't examine the mysteries of cake baking, but we will uncover the mysteries of the algorithms of arithmetic. In fact, they aren't mysteries at all, but clever, efficient ways of calculating that make complete sense once you know how to look at them.

How Does the Addition Algorithm Develop?

When we use the standard algorithm to add numbers in base ten, such as 149 + 85, we put the numbers one under the other, lining up like places. Then we add column by column, *regrouping* as needed. When adding numbers this way, if the digits in a column add to 10 (or more), we write down the ones digit of the sum and shift the remaining 10 to become a 1 in the next place to the left. Traditionally we write this 1 at the top of the next column to the left, although some educators have found it is better for children to write the 1 at the *bottom* of the next column to the left. This process of shifting a 10 from the sum of the digits in one place to a 1 in the next place to the left is called **regrouping**. Regrouping is also called **trading** or **carrying**.

regrouping

Here are two ways to show regrouping:

$$\begin{array}{r} {\scriptstyle 11} \\ 149 \\ +\ \ 85 \\ \hline 234 \end{array} \qquad \begin{array}{r} 149 \\ +\ \ 85 \\ \hline {\scriptstyle 11} \\ 234 \end{array}$$

Why do we add column by column? Why do we regroup? The answers to these questions lie in the interpretation of addition as combining and in the idea of place value.

IMAP *Watch Gretchen use the base-ten blocks and the hundred chart.*

CLASS ACTIVITY

3I Adding and Subtracting with Base-Ten Math Drawings, p. CA-55

3J Understanding the Standard Addition Algorithm, p. CA-56

3K Understanding the Standard Subtraction Algorithm, p. CA-57

CCSS
2.NBT.7
2.NBT.9

How Do Math Drawings and Bundled Objects Support Understanding of the Addition Algorithm for Whole Numbers? Consider the sum 149 + 85. We can represent this sum as the total number of things when we combine 149 things with 85 things. In order to analyze the addition algorithm, represent the numbers 149 and 85 with bundles, as described in Section 1.1. Represent 149 things as 1 bundle of 100 (which is 10 bundles of ten), 4 bundles of ten, and 9 individual things, as shown in **Figure 3.14**. Similarly, represent 85 things as 8 bundles of ten and 5 individual things. When all these things are combined, how many are there? Adding like bundles, we have the following:

14 individual things (9 plus 5 more)

12 bundles of 10 (4 bundles of ten plus 8 more bundles of ten)

1 bundle of 100

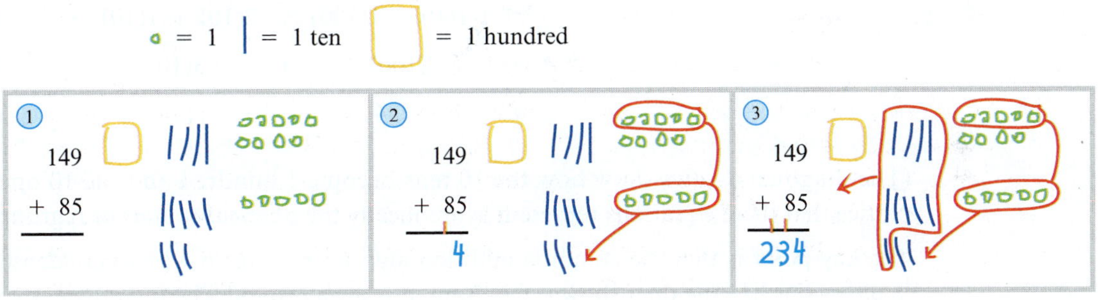

Figure 3.14
Base-ten math drawings for 149 + 85.

Symbolically, we can write this as follows:

$$\begin{array}{r} 1(100) + 4(10) + 9(1) \\ + 8(10) + 5(1) \\ \hline = 1(100) + 12(10) + 14(1) \end{array}$$

So there are

$$1(100) + 12(10) + 14(1)$$

things, but this expression is not the expanded form of a number in ordinary base-ten notation because there are 12 tens and 14 ones. This is where we regroup our bundles. *The regrouping of bundles is the physical representation of the regrouping process in the addition algorithm.*

To regroup the bundled things, convert the 14 individual things into 1 bundle of ten and 4 individuals. This 1 bundle of ten is the small "carried" 1 we write above the 4 in the standard procedure. The small 1 above the tens in the following example really stands for 1 ten:

$$\begin{array}{r} {\scriptstyle 11} \\ 149 \\ +85 \\ \hline 234 \end{array}$$

Likewise, regroup the 12 bundles of 10 things into 1 bundle of a hundred (namely, 10 bundles of ten) and 2 bundles of ten. This 1 bundle of a hundred is the small "carried" 1 we write above the 1 in the standard procedure. That small 1 really stands for 1 hundred. When you collect like bundles, you will have 2 bundles of a hundred, 3 bundles of ten, and 4 individual things.

Here is the same example written symbolically:

$$\begin{array}{rl}
1(100) + 4(10) + 9(1) & \\
+ 8(10) + 5(1) & \\
\hline
= 1(100) + \cancel{12(10)} + \cancel{14(1)} & \\
+ 1(10) + 4(1) & \quad \text{4 ones become 1 ten and 4 ones.} \\
+ 1(100) + 2(10) & \quad \text{12 tens become 1 hundred and 2 tens.} \\
\hline
= 2(100) + 3(10) + 4(1) &
\end{array}$$

We can also write this regrouping in equation form:

$$\begin{aligned}
1(100) + 12(10) + 14(1) &= 1(100) + 10(10) + 2(10) + 10(1) + 4(1) \\
& \qquad\qquad\quad \swarrow \qquad\qquad\qquad \swarrow \\
&= 1(100) + 1(100) + 2(10) + 1(10) + 4(1) \\
&= 2(100) + 3(10) + 4(1) \\
&= 234
\end{aligned}$$

(The diagonal arrows show how the 10 tens become 1 hundred and the 10 ones become 1 ten.) Notice that these equations represent symbolically the *physical actions* of regrouping the objects.

The key point is that the common addition algorithm is just a way to condense the information in equations like the ones shown; therefore, it is a way to quickly and efficiently record the physical action of adding and regrouping actual objects. This example demonstrates why the common addition algorithm gives us correct answers to addition problems.

CCSS
5.NBT.7

How Does the Addition Algorithm for Decimals Develop? When we use the common addition algorithm to add decimals, we use the same ideas as for whole numbers. The first step in the algorithm is to line up like places, by lining up the decimal points.

As we saw in Section 1.2, bundled objects can be used to represent (finite) decimals, as long as the meaning of 1 object is interpreted suitably. To represent the sum 0.834 + 6.7 with things, let 1 thing represent 0.001. Then a bundle of 10 things represents 0.01, a bundle of 100 things (a bundle of 10 tens) represents 0.1, and a bundle of 1000 things (a bundle of 10 hundreds) represents 1. So 0.834 is then represented by

8 bundles of 100, 3 bundles of 10, and 4 individual things,

while 6.7 is represented by

6 bundles of 1000 and 7 bundles of 100.

Notice that we chose 1 thing to represent 0.001 in *both* 0.834 and 6.7. In this way, when we use the bundled things to represent the sum 0.834 + 6.7, thousandths will be added to thousandths, hundredths to hundredths, tenths to tenths, and ones to ones. If 1 thing were to represent 0.001 in 0.834, but 0.1 in 6.7, then we would add thousandths to tenths and hundredths to ones, which wouldn't make any sense. This would be like treating a penny as 0.01 of a dollar in one setting and 0.1 of a dollar in another setting. *The consistent choice for the meaning of 1 thing when representing both decimals is necessary so that like terms will be added*—tens to tens, ones to ones, tenths to tenths, hundredths to hundredths, and so on.

Once like places have been lined up, the addition algorithm proceeds in the same way as if the decimal points weren't there, and the explanation for why the algorithm works is the same as for whole numbers. Students often like to place zeros in lower place values, such as by writing 6.7 as 6.700.

$$\begin{array}{r} \scriptstyle 1 \\ 0.834 \\ +\ 6.700 \\ \hline 7.534 \end{array}$$

How Does the Subtraction Algorithm Develop?

Now let's turn to subtraction. When we subtract numbers in the base-ten system using the standard paper-and-pencil algorithm, we first put the numbers one under the other, lining up like places. Then we subtract column by column, *regrouping* as needed so that we can subtract the numbers in a column without a negative number resulting. When subtracting numbers this way, if the digit at the top of a column is less than the digit below it, we cross out the digit at the top of the next column to the left, changing this digit to one that is 1 less, and we replace the digit at the top of our original column with 10 plus the original digit. (For example, we replace a 2 with 12.) If we can't carry out these steps because there is a 0 in the next column to the left, then we keep moving to the left, crossing out 0s until we come to a digit that is greater than 0. We cross this nonzero digit out and replace it with the digit that is 1 less, we replace all the intervening 0s with 9s, and, as before, we replace the digit at the top of the original column with 10 plus this digit. This process of changing digits so as to be able to subtract in a column is called **regrouping**. Regrouping is also called **trading** or **borrowing**.

regrouping

Regrouping is used in the following examples:

$$\begin{array}{r} 142 \\ -\ \ 83 \\ \hline \end{array} \quad \rightarrow \quad \begin{array}{r} \overset{0}{\cancel{1}}\,\overset{13}{\cancel{4}}\,\overset{12}{\cancel{2}} \\ -\ \ 8\ 3 \\ \hline 5\ 9 \end{array}$$

$$\begin{array}{r} 1002 \\ -\ \ \ 53 \\ \hline \end{array} \quad \rightarrow \quad \begin{array}{r} \overset{0}{\cancel{1}}\,\overset{9}{\cancel{0}}\,\overset{9}{\cancel{0}}\,\overset{12}{\cancel{2}} \\ -\ \ \ 5\ 3 \\ \hline 9\ 4\ 9 \end{array}$$

As with the addition algorithm, we want to understand why this procedure makes sense. We will learn why by interpreting subtraction as taking away and by considering place value.

How Do Math Drawings and Bundled Objects Support Understanding of the Subtraction Algorithm? As with addition, we can model the subtraction algorithm, particularly the regrouping process, with bundles of things. Consider the difference 142 − 83. We can represent 142 with 1 bundle of a hundred, 4 bundles of ten, and 2 individual things. How many things will be left when we take 83 things away? When we try to take 3 individual things away from 142, we first need to do some unbundling. *This unbundling is a physical representation of the regrouping process.* This process is illustrated in Figure 3.15. One of the 4 bundles of ten can be unbundled and added to the individual things. The result is 1 bundle of a hundred, 3 bundles of ten, and 12 individuals. Using equations, we have

$$\begin{aligned} 142 &= 1(100) + 4(10) + 2(1) \\ &= 1(100) + 3(10) + 1(10) + 2(1) \\ &\qquad\qquad\qquad\qquad \searrow \\ &= 1(100) + 3(10) + 10(1) + 2(1) \\ &= 1(100) + 3(10) + 12(1) \end{aligned}$$

Figure 3.15

Base-ten math drawings for 142 − 83.

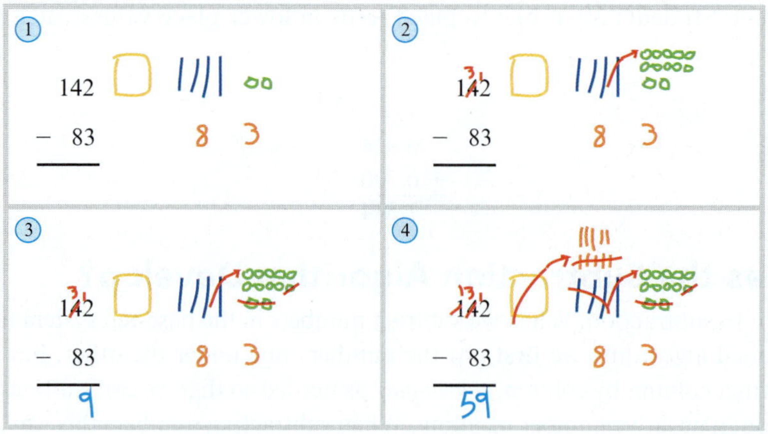

CCSS
2.NBT.7
2.NBT.9

Now there won't be any problem taking 3 individual things away, but what about taking away 8 bundles of ten? There are only 3 bundles of ten. So we must unbundle the 1 bundle of a hundred as 10 bundles of ten and combine these 10 bundles of ten with the 3 bundles of ten to make 13 bundles of ten. We continue from the previous equations to get

$$
\begin{aligned}
142 &= 1(100) + \quad 3(10) + 12(1) \\
&\searrow \\
&= 10(10) + 3(10) + 12(1) \\
&= 13(10) \qquad + 12(1)
\end{aligned}
$$

IMAP

Watch children solving multidigit addition and subtraction problems in these video clips: Andrew, 120 + 96, using a method he invented; Connor, 39 + 25; Estephania, 1000 − 1; Freddie, 400 − 150, 150 + ? = 450; Gilberto, 265 + 537; Gretchen, 70 − 23; Hally, 1000 − 6, 1000 − 1; Johanna, 1000 − 4; Johanna, 29 + 30 + 31; Talecia, 638 + 476.

After regrouping, there is no difficulty taking 3 individual things away from the 12 individual things and 8 bundles of ten away from the 13 bundles of ten. Notice that the physical actions of moving the things correspond exactly to the regrouping process that takes place in the common subtraction algorithm. In expanded form, the subtraction problem 142 − 83 can now be rewritten as

$$
\begin{array}{r}
13(10) + 12(1) \\
\underline{-\,[8(10) + \ \ 3(1)]} \\
= 5(10) + \ \ 9(1) \ = 59
\end{array}
$$

The key point is that the standard subtraction algorithm is just a way to condense the information in equations like those given previously. Therefore, it is a quick and efficient way to record the physical actions of regrouping and then taking away actual objects. This is why the common subtraction algorithm gives us correct answers to subtraction problems.

How Does the Subtraction Algorithm for Decimals Develop? As with the addition algorithm for decimals, we can generally still represent decimal subtraction problems with bundles of things by an appropriate interpretation of 1 thing. Like places (and therefore decimal points) are lined up so that like terms are subtracted—tens from tens, ones from ones, tenths from tenths, hundredths from hudredths, and so on.

CLASS ACTIVITY

3L A Third-Grader's Method of Subtraction, p. CA-58

3M Regrouping in Base 12, p. CA-59

3N Regrouping in Base 60, p. CA-59

SECTION SUMMARY AND STUDY ITEMS

Section 3.3 Why the Standard Algorithms for Addition and Subtraction in Base Ten Work

The standard addition and subtraction algorithms work in terms of objects grouped in ones, tens, hundreds, and so on. Regrouping occurs when we make or break bundles.

Key Skills and Understandings

1. Explain the addition and subtraction algorithms in terms of bundled objects and base-ten math drawings, paying special attention to regrouping.
2. Explain that we line up like places so that we add or subtract hundreds and hundreds, tens and tens, ones and ones, and so on. This is why we line up decimal points when adding and subtracting decimals.

Practice Exercises for Section 3.3

1. Make a base-ten math drawing for 37 + 26 that a child could use side by side with the standard algorithm for adding 37 + 26 to help make sense of the algorithm.
2. Make a base-ten math drawing for 41 − 28 that a child could use side by side with the standard algorithm for subtracting 41 − 28 to help make sense of the algorithm.
3. Write equations with numbers in expanded form showing how to regroup the number 104 so that 69 can be subtracted from it.
4. Why do we line up decimal points before adding or subtracting decimals? What are we really doing?
5. Ellie solves the subtraction problem 2.5 − 0.13 with toothpicks. She represents 2.5 with 2 bundles of 10 toothpicks and 5 individual toothpicks, and she represents 0.13 with 1 bundle of 10 toothpicks and 3 individual toothpicks. Ellie gets the answer 1.2. Is she right? If not, explain why not and discuss how she could use the toothpicks correctly.
6. A store buys action figures in boxes. Each box contains 50 bags, and each bag contains 6 action figures. At the beginning of the month, the store has

   7 unopened boxes, 15 unopened bags, and 3 individual action figures.

   At the end of the month, the store has

   2 unopened boxes, 37 unopened bags, and 5 individual action figures.

   How many action figures did the store sell during the month (assuming they got no additional shipments of action figures)? Write your answer in terms of boxes, bags, and individual action figures. *Work with boxes, bags, and individuals in a sort of expanded form and use regrouping to solve this problem.*

Answers to Practice Exercises for Section 3.3

1. See Figure 3.16.

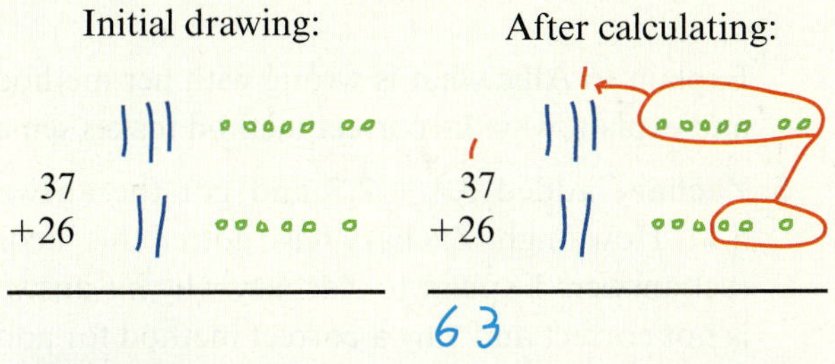

Figure 3.16 A base-ten math drawing for 37 + 26.

2. See Figure 3.17.

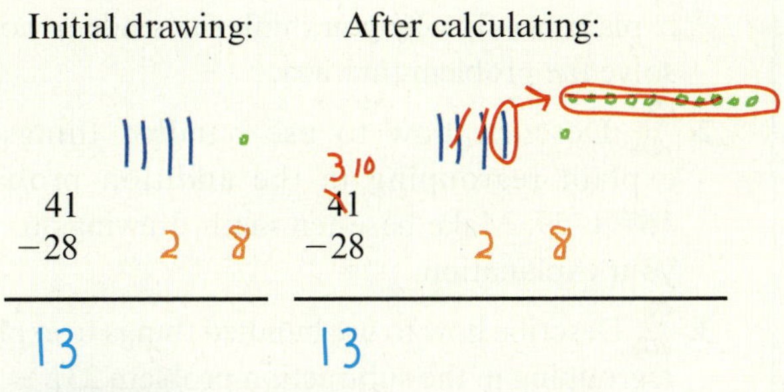

Figure 3.17 A base-ten math drawing for 41 − 28.

3. Here's the regrouping process, with equations in expanded form:

$$\begin{aligned} 104 &= 1(100) + 0(10) + 4(1) \\ &= 10(10) + 0(10) + 4(1) \\ &= 10(10) + 4(1) \\ &= 9(10) + 1(10) + 4(1) \\ &= 9(10) + 10(1) + 4(1) \\ &= 9(10) + 14(1) \end{aligned}$$

4. When we line up decimal points, we are really lining up *like places,* so that we will add or subtract ones with ones, tenths with tenths, hundredths with hundredths, and so on.

5. No, Ellie's answer is not correct—1 toothpick must represent the same amount when representing both 2.5 and 0.13. Ellie should let 1 toothpick represent $\frac{1}{100}$ in both cases. Then 2.5 is represented by 2 bundles of 100 toothpicks and 5 bundles of 10 toothpicks, while 0.13 is represented by 1 bundle of 10 toothpicks and 3 individual toothpicks. Now Ellie should be able to see that she'll need to regroup in order to subtract. It might help Ellie to think in terms of money: 2.5 and 0.13 can be represented by \$2.50 and \$0.13. Ellie's way of using the toothpicks would be like saying that a dime is equal to a penny.

6. We must solve the following problem:

$$\begin{array}{r} 7 \text{ boxes} + 15 \text{ bags} + 3 \text{ individual} \\ -\,(2 \text{ boxes} + 37 \text{ bags} + 5 \text{ individual}) \\ \hline \end{array}$$

We can solve this by first regrouping the 7 boxes, 15 bags, and 3 individual action figures. If we open one of the bags, then there is one less bag, but 6 more individual figures, so there are 7 boxes, 14 bags, and 9 individual action figures.

If we open one of the boxes, then there is one less box, but 50 more bags of action figures, so there are 6 boxes, 64 bags, and 9 individual action figures.

It's still the same number of action figures—they are just arranged in a different way. In equation form we can write this as

$$\begin{aligned} &7 \text{ boxes} + 15 \text{ bags} + 3 \text{ individual} \\ &= 7 \text{ boxes} + 14 \text{ bags} + (6 + 3) \text{ individual} \\ &= 6 \text{ boxes} + (50 + 14) \text{ bags} + 9 \text{ individual} \\ &= 6 \text{ boxes} + 64 \text{ bags} + 9 \text{ individual} \end{aligned}$$

Now we are ready to subtract the 2 boxes, 37 bags, and 5 individual action figures:

$$\begin{array}{r} 6 \text{ boxes} + 64 \text{ bags} + 9 \text{ individual} \\ -\,(2 \text{ boxes} + 37 \text{ bags} + 5 \text{ individual}) \\ \hline 4 \text{ boxes} + 27 \text{ bags} + 4 \text{ individual} \end{array}$$

So a total of 4 boxes, 27 bags, and 4 individual action figures were sold during the month.

PROBLEMS FOR SECTION 3.3

1. Refer to Class Activity 3I, Adding and Subtracting with Base-Ten Math Drawings. Show how students 1, 2, and 3 might solve the addition problem 29 + 46 and how student 4 might solve the subtraction problem 54 − 28. In each case explain briefly why you think the student would solve the problem that way.

2. Describe how to use bundled things to explain regrouping in the addition problem 167 + 59. Make base-ten math drawings to aid your explanation.

3. Describe how to use bundled things to explain regrouping in the subtraction problem 231 − 67. Make math drawings to aid your explanation.

4. Allie solves the subtraction problem 304 − 9 as follows:

$$\begin{array}{r} \overset{2}{\cancel{3}}0\overset{14}{\cancel{4}} \\ -\quad 9 \\ \hline 205 \end{array}$$

Explain to Allie what is wrong with her method, and explain why the correct method makes sense.

5. Zachary added 3.4 + 2.7 and got the answer 5.11. How might Zachary have gotten this incorrect answer? Explain to Zachary why his answer is not correct and why a correct method for adding 3.4 + 2.7 makes sense.

6. To solve 512 − 146, a student writes the following:

$$\begin{array}{rr} 512 & 400 \\ -146 & -30 \\ \hline -4 & 370 \quad 512 - 146 = 366 \\ -30 & -4 \\ \hline 400 & 366 \end{array}$$

Describe the student's solution strategy and discuss why the strategy makes sense. Expanded forms may be helpful to your discussion.

7. *Problem:* Matteo is 4 feet 3 inches tall. Nico is 3 feet 11 inches tall. How much taller is Matteo than Nico?

Sarah solved this problem as follows:

$$\begin{array}{r} \overset{3}{\cancel{4}}\text{ ft } \overset{13}{\cancel{3}}\text{in} \\ -3\text{ ft } 11\text{ in} \\ \hline 2\text{ in} \end{array}$$

So Sarah gave 2 inches as the answer. Is Sarah right? If not, explain what is wrong with her method, and show how to *modify* her method of regrouping to make it correct. (Do not start from scratch.)

8. *Problem:* A container holds 2 quarts and 4 fluid ounces. The container is now filled with 6 fluid ounces of liquid. How much liquid must be added to the container to make it full?

John solved this problem as follows:

$$\begin{array}{r} \overset{1}{\cancel{2}}\text{ q } \overset{14}{\cancel{4}}\text{ fl oz} \\ -\qquad 6\text{ fl oz} \\ \hline 1\text{ q } 8\text{ fl oz} \end{array}$$

So John gave 1 quart and 8 fluid ounces as the answer. Is John right? If not, explain what is wrong with his method, and show how to *modify* his method of regrouping to make it correct. (Do not start from scratch.)

9. On a space shuttle mission, a certain experiment is started 2 days, 14 hours, and 30 minutes into the mission. The experiment takes 1 day, 21 hours, and 47 minutes to run. When will the experiment be completed? Give your answer in days, hours, and minutes into the mission. Work with a sort of expanded form. In other words, work with

$$2(\text{days}) + 14(\text{hours}) + 30(\text{minutes})$$

and

$$1(\text{day}) + 21(\text{hours}) + 47(\text{minutes})$$

and *regroup among days, hours, and minutes* to solve this problem.

10. We can write dates and times in a sort of expanded form. For example, October 4th, 6:53 P.M. can be written as

$$4(\text{days}) + 18(\text{hours}) + 53(\text{minutes})$$

(In some circumstances you might also want to include the month and the year.) How long is it from 3:27 P.M. on October 4 to 7:13 A.M. on October 19? Give your answer in days, hours, and minutes. Work in the type of expanded form previously described, and *regroup among days, hours, and minutes* to solve this problem.

11. Erin wants to figure out how much time elapsed between 9:45 A.M. and 11:30 A.M. Erin does the following:

$$\begin{array}{r} 1\overset{0}{\cancel{1}}:\overset{12}{\cancel{3}}\overset{1}{0} \\ -\quad 9:45 \\ \hline 1:85 \end{array}$$

and says the answer is 1 hour and 85 minutes.

a. Is Erin right? If not, explain what is wrong with her method, and show how to *modify* her method of regrouping to make it correct. (Do not start from scratch.)

b. Solve the problem of how much time elapsed between 9:45 A.M. and 11:30 A.M. in another way than your modification of Erin's method. Explain your method.

12. The standard subtraction algorithm described in the text is not the only correct subtraction algorithm. Some people use the following algorithm instead: Line up the numbers as in the standard algorithm, and subtract column by column, proceeding from right to left. The only difference between this new algorithm and our standard one is in regrouping. To regroup with the new algorithm, move one column to the left and cross out the digit at the *bottom* of the column, replacing it with that digit *plus 1*. (So replace an 8 with a 9, replace a 9 with a 10, etc.) Then, as in the standard algorithm, replace the digit at the top of the original column with the original number plus 10. The following example shows the steps of this new algorithm:

$$\begin{array}{r} 132 \\ -\ 79 \\ \hline \end{array} \;\rightarrow\; \begin{array}{r} 13\overset{12}{\cancel{2}} \\ -\ \overset{8}{\cancel{7}}9 \\ \hline 3 \end{array} \;\rightarrow\; \begin{array}{r} 1\overset{13}{\cancel{3}}\overset{12}{\cancel{2}} \\ -\ \overset{1}{\cancel{0}}\overset{8}{\cancel{7}}9 \\ \hline 53 \end{array}$$

a. Use the new algorithm to solve 524 − 198 and 1003 − 95. Verify that the new algorithm gives correct answers.

b. Explain why it makes sense that this new algorithm gives correct answers to subtraction problems. What is the reasoning behind this new algorithm?

c. What are some advantages and disadvantages of this new algorithm compared with our common one?

13. The standard subtraction algorithm described in the text is not the only correct subtraction algorithm. The next subtraction algorithm is called *adding the complement*. For a 3-digit whole number, N, the *complement* of N is $999 - N$. For example, the complement of 486 is

$$999 - 486 = 513$$

Notice that regrouping is never needed to calculate the complement of a number. To use the adding-the-complement algorithm to subtract a 3-digit whole number, N, from another 3-digit whole number, start by adding the complement of N rather than subtracting N. For example, to solve

$$723 - 486$$

first add the complement of 486:

$$\begin{array}{r} 723 \\ +\ 513 \\ \hline 1236 \end{array}$$

Then cross out the 1 in the thousands column, and add 1 to the resulting number:

$$\not{1}236 \rightarrow 236 + 1 = 237$$

Therefore, according to the *adding-the-complement* algorithm, $723 - 486 = 237$.

a. Use the adding-the-complement algorithm to calculate 301 − 189 and 295 − 178. Verify that you get the correct answer.

b. Explain why the adding-the-complement algorithm gives you the correct answer to any 3-digit subtraction problem. Focus on the relationship between the original problem and the addition problem in adding the complement. For example, how are the problems 723 − 486 and 723 + 513 related? Work with the *complement* relationship, $513 = 999 - 486$, and notice that $999 = 1000 - 1$.

c. What are some advantages and disadvantages of the adding-the-complement subtraction algorithm compared with the common subtraction algorithm described in the text?

***14.** Here's how Mo solved the subtraction problem 635 − 813:

$$\begin{array}{r} 635 \\ -813 \\ \hline -222 \end{array}$$

Mo did this by working from right to left, saying

$$5 - 3 = 2$$
$$3 - 1 = 2$$
$$6 - 8 = -2$$

a. Is Mo's answer right?

b. Solve

$$\begin{array}{r} 6(100) + 3(10) + 5(1) \\ -\ [8(100) + 1(10) + 3(1)] \\ \hline \end{array}$$

by working with expanded forms. Discuss how Mo's work compares with your work in expanded forms.

***15.** After you solve part (b) of this problem, think carefully about your explanation in part (a). You may find that you need to revise it.

a. What is the smallest number of coins you can use to make 43 cents if quarters, dimes, nickels, and pennies are available? Explain why your answer is correct.

b. What is the smallest number of coins you can use to make 43 cents if quarters, dimes, and pennies (but no nickels) are available? Explain why your answer is correct.

3.4 Reasoning About Fraction Addition and Subtraction

CCSS Common Core State Standards Grades 4, 5

Why do the procedures we use to add and subtract fractions make sense? We will answer this question in this section. Because mixed numbers (such as $2\frac{3}{4}$) are defined in terms of addition, we will also study these numbers and the algorithm for converting them to improper fractions. Similarly, because we can always express a finite decimal as a sum of fractions by writing the decimal in expanded form, we will see why we can write finite decimals as fractions. Finally, we will consider fraction addition and subtraction word problems.

IMAP *Watch Felisha use an area to model to add rational numbers.*

CLASS ACTIVITY

3O Why Do We Add and Subtract Fractions the Way We Do?, p. CA-60

CCSS 4.NF.3a,d

Why Do We Add and Subtract Fractions with Like Denominators the Way We Do?

If two fractions have the same denominator, then you can add or subtract these fractions by adding or subtracting the numerators and leaving the denominator unchanged. For example,

$$\frac{2}{7} + \frac{3}{7} = \frac{2+3}{7} = \frac{5}{7} \quad \text{and} \quad \frac{2}{3} + \frac{2}{3} = \frac{2+2}{3} = \frac{4}{3}$$

Why does this make sense? We can interpret the sum

$$\frac{2}{7} + \frac{3}{7}$$

as the total number of cubic yards of wood chips you have if you first get $\frac{2}{7}$ of a cubic yard of wood chips and then get another $\frac{3}{7}$ of a cubic yard of wood chips (see Figure 3.18). The fraction $\frac{2}{7}$ means 2 parts, each of size $\frac{1}{7}$ and the fraction $\frac{3}{7}$ means 3 parts, each of size $\frac{1}{7}$. So if you first get 2 parts and then get 3 more parts, then all together, you have 2 + 3 parts, or 5 parts. What size parts are they? Each of the 5 parts is $\frac{1}{7}$ of a cubic yard, so you have $\frac{5}{7}$ of a cubic yard of wood chips. The reasoning for $\frac{2}{3} + \frac{2}{3}$ is exactly the same. Notice how thinking about fractions according to our definition—as numbers of parts of a certain size—helps us reason about fraction addition.

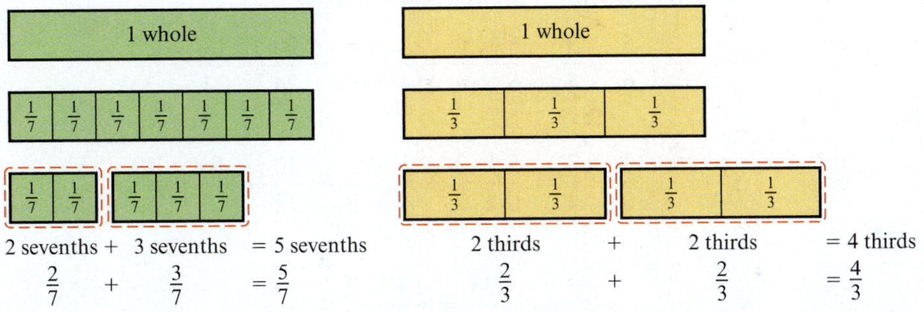

Figure 3.18 Adding fractions with like denominators.

Why Do We Add and Subtract Fractions with Unlike Denominators by Finding Common Denominators?

How do we add or subtract fractions that have different denominators, such as $\frac{5}{6} + \frac{3}{4}$? We can interpret the sum

$$\frac{5}{6} + \frac{3}{4}$$

Figure 3.19

Using a common denominator to add.

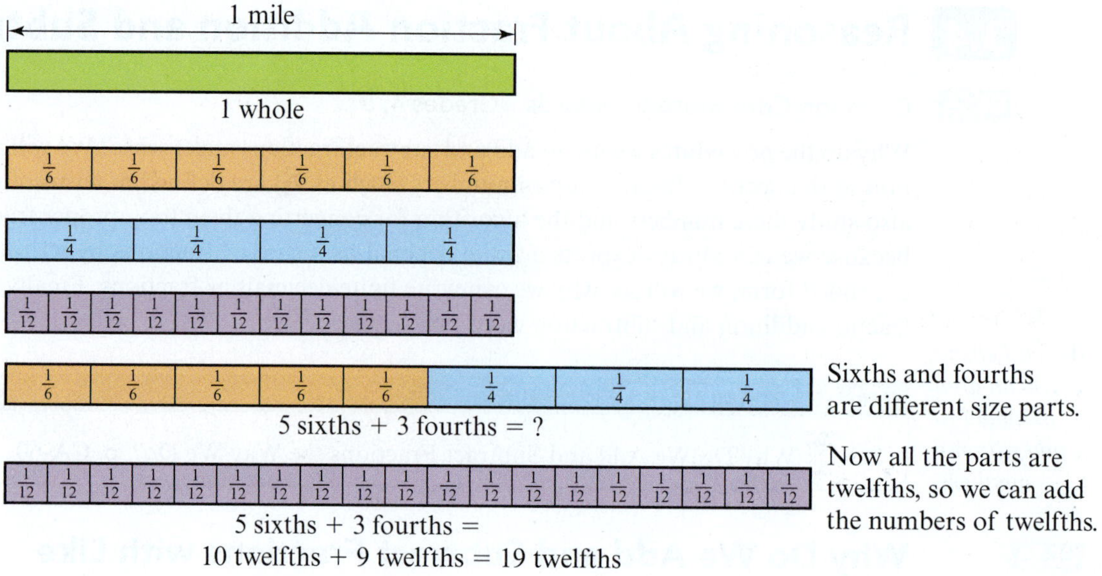

CCSS
5.NF.1
5.NF.2

as the total distance walked if you first walk $\frac{5}{6}$ of a mile and then walk $\frac{3}{4}$ of a mile. (See Figure 3.19.) In this case, the first distance walked is described in terms of *sixths* of a mile and the second distance walked is described in terms of *fourths* of a mile. Sixths of a mile and fourths of a mile are different-size distances, so we can't just add the 5 in $\frac{5}{6}$ to the 3 in $\frac{3}{4}$. We need a common distance with which to describe both $\frac{5}{6}$ of a mile and $\frac{3}{4}$ of a mile. In other words, we need to break *sixths* and *fourths* into *like parts*.

As in comparing fractions, the process of breaking into like parts is achieved numerically by giving the fractions $\frac{5}{6}$ and $\frac{3}{4}$ a common denominator. *Any* common denominator will do; it does not have to be the least one. You can always produce a common denominator by multiplying the two denominators. So, to add

$$\frac{5}{6} \text{ and } \frac{3}{4}$$

we can use the common denominator 6×4, which is 24. If we want to work with smaller numbers, we can use 12, which is a common denominator because $12 = 6 \times 2$ and $12 = 4 \times 3$. Once we have common denominators, the fractions are described in terms of like parts, so we simply add the numerators to determine the total number of parts:

$$\frac{5}{6} + \frac{3}{4} = \frac{5 \times 4}{6 \times 4} + \frac{3 \times 6}{4 \times 6} = \frac{20}{24} + \frac{18}{24} = \frac{38}{24}$$

or

$$\frac{5}{6} + \frac{3}{4} = \frac{5 \times 2}{6 \times 2} + \frac{3 \times 3}{4 \times 3} = \frac{10}{12} + \frac{9}{12} = \frac{19}{12}$$

Since

$$\frac{38}{24} = \frac{19 \times 2}{12 \times 2} = \frac{19}{12}$$

the two ways just shown of adding $\frac{5}{6}$ and $\frac{3}{4}$ produce equal results. When we use the common denominator 12, the resulting sum $\frac{19}{12}$ is in simplest form. If the least common denominator is not used when adding fractions, then the resulting sum will not be in simplest form.

Figure 3.19 shows the addition of $\frac{5}{6}$ and $\frac{3}{4}$ with the common denominator 12, using fraction strips. In this case, each sixth is broken into 2 parts and each fourth is broken into 3 parts to produce twelfths.

Why Can We Express Mixed Numbers as Improper Fractions?

mixed number A mixed number or **mixed fraction** is a number that is written in the form

$$A\frac{B}{C}$$

where A, B, and C are nonzero whole numbers, and $\frac{B}{C}$ is a proper fraction (i.e., the numerator is less than the denominator). So

$$2\frac{3}{4} \quad \text{and} \quad 5\frac{7}{8}$$

are mixed numbers. The mixed number

$$A\frac{B}{C}$$

stands for the sum of its whole number part and its fractional part:

$$A + \frac{B}{C}$$

Since a whole number A can also be written as a fraction, namely, $\frac{A}{1}$, we can write the mixed number $A\frac{B}{C}$ as follows:

$$A\frac{B}{C} = A + \frac{B}{C} = \frac{A}{1} + \frac{B}{C} = \frac{A \times C}{1 \times C} + \frac{B}{C} = \frac{A \times C}{C} + \frac{B}{C} = \frac{A \times C + B}{C}$$

Now we can use the way we add fractions to show that every mixed number can be written as an improper fraction:

IMAP *Watch Ally compare and convert fractions. And, watch Rachel correct her own misconception about converting mixed numbers to improper fractions using an area model.*

$$A\frac{B}{C} = \frac{A \times C + B}{C}$$

So,

$$2\frac{3}{4} = \frac{2 \times 4 + 3}{4} = \frac{11}{4}$$

We sometimes convert mixed numbers to improper fractions in order to add, subtract, multiply, or divide them.

CLASS ACTIVITY

3P Critiquing Mixed Number Addition and Subtraction Methods, p. CA-62

CCSS 4.NF.5

Why Can We Write Finite Decimals as Fractions and How Does That Explain Decimal Names?

Every finite decimal stands for a finite sum of fractions. We can see this representation by writing the decimal in its expanded form—for example,

$$2.7 = 2 + \frac{7}{10}$$

$$32.85 = 30 + 2 + \frac{8}{10} + \frac{5}{100}$$

$$0.491 = \frac{4}{10} + \frac{9}{100} + \frac{1}{1000}$$

By giving these fractions a common denominator, we can write the decimal as a fraction. For example,

$$\begin{aligned} 2.7 &= 2 + \frac{7}{10} \\ &= \frac{20}{10} + \frac{7}{10} \\ &= \frac{27}{10} \end{aligned}$$

or

$$\begin{aligned} 0.491 &= \frac{4}{10} + \frac{9}{100} + \frac{1}{1000} \\ &= \frac{400}{1000} + \frac{90}{1000} + \frac{1}{1000} \\ &= \frac{491}{1000} \end{aligned}$$

It is for this reason that we can read 0.491 as "four-hundred ninety-one thousandths."

Observe that this method for writing decimals as fractions applies only to finite decimals. It does not apply to a decimal such as

$$0.4444444\ldots$$

where the 4s continue forever. In fact, it turns out that some numbers, such as $\sqrt{2} = 1.4142\ldots$ cannot be written as fractions whose numerator and denominator are whole numbers.

When Is Combining Not Adding?

The old admonishment "You can't add apples and oranges" is especially apt when adding and subtracting fractions, because sometimes fractions of quantities that are to be combined refer to *different wholes.*

Whenever we add two numbers, such as

$$3 + 4 \quad \text{or} \quad \frac{2}{3} + \frac{3}{4}$$

CCSS
4.NF.3a
5.NF.2

both summands and the sum refer to the same whole or unit amount. Similarly, whenever we subtract two numbers, such as

$$4 - 3 \quad \text{or} \quad \frac{3}{4} - \frac{2}{3}$$

the minuend, the subtrahend, and the difference all refer to the same whole or unit amount.

For example, we can interpret the sum

$$3 + 4$$

as the total number of apples when 3 apples are combined with 4 apples. But if you interpret 3 + 4 by combining 3 bananas and 4 blocks, as in Figure 3.20, then the best you can say is that you have *7 objects* in all.

Figure 3.20 Combining bananas and blocks.

Similarly, we can interpret the sum

$$\frac{1}{3} + \frac{1}{4}$$

as the fraction of a pie we will have in all if we have $\frac{1}{3}$ of the pie and then get another $\frac{1}{4}$ *of the same pie or an equivalent pie*. The $\frac{1}{3}$, the $\frac{1}{4}$, and the sum $\frac{1}{3} + \frac{1}{4} = \frac{7}{12}$ *all refer to the same pie,* or to copies of an equivalent pie. But suppose we have $\frac{1}{3}$ of a small pie and $\frac{1}{4}$ of a large pie, as in **Figure 3.21**. Does the combined amount represent the sum

$$\frac{1}{3} + \frac{1}{4}?$$

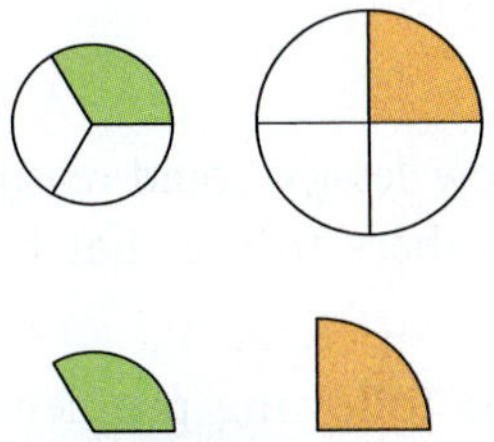

Figure 3.21 Combining $\frac{1}{3}$ of a small pie and $\frac{1}{4}$ of a large pie does not make $\frac{1}{3} + \frac{1}{4}$ of *a pie*.

No, it does not, because the $\frac{1}{3}$ and the $\frac{1}{4}$ refer to *different wholes* that are not equivalent. All we can say is that we have 2 pieces of pie. We can't say that we have $\frac{7}{12}$ of a pie—because which pie would the $\frac{7}{12}$ refer to?

So when working with fraction addition or subtraction word problems, be especially careful that the fractions in question refer to the same underlying wholes. In some cases, fractional amounts that are to be combined or taken away refer to different wholes; therefore, the problem is not a fraction addition or subtraction problem.

CLASS ACTIVITY

3Q Are These Word Problems for $\frac{1}{2} + \frac{1}{3}$?, p. CA-63

3R Are These Word Problems for $\frac{1}{2} - \frac{1}{3}$?, p. CA-64

3S What Fraction Is Shaded?, p. CA-65

3T Addition with Whole Numbers, Decimals, Fractions, and Mixed Numbers: What Are Common Ideas?, p. CA-65

SECTION SUMMARY AND STUDY ITEMS

Section 3.4 Reasoning About Fraction Addition and Subtraction

To add or subtract fractions, we first give the fractions common denominators so as to work with like parts. Once we have like parts, we can simply add or subtract the number of parts. A mixed number stands for the sum of a whole number and a fraction; when we add the whole number part to the fraction part, the result is an improper fraction. When we work with fraction addition and subtraction word problems, we must pay close attention to underlying wholes and to the wording of the problem.

Key Skills and Understandings

1. Describe how to add or subtract fractions, and explain why the process makes sense. In particular, explain why we give the fractions common denominators.
2. Use math drawings and number lines to help explain the logic behind the procedure of turning a mixed number into an improper fraction.
3. Write fraction addition and subtraction word problems.
4. Determine if a given word problem fits a given fraction addition or subtraction problem.
5. Use estimation and a sense of the size of fractions to assess the reasonableness of the answer to a fraction addition or subtraction problem.

Practice Exercises for Section 3.4

1. Using the example $\frac{1}{4} + \frac{5}{6}$, explain why we must give fractions a common denominator in order to add them.

2. When we add or subtract fractions, *must* we use the least common denominator? Are there any advantages to using the least common denominator?

3. The usual procedure for converting a mixed number, such as $7\frac{1}{3}$, into an improper fraction is this:

$$7\frac{1}{3} = \frac{7 \times 3 + 1}{3} = \frac{22}{3}$$

Explain the logic behind this procedure.

4. You showed Tommy the drawing in Figure 3.22 to explain why $2\frac{1}{4} = \frac{9}{4}$, but Tommy says that it shows $\frac{9}{12}$, not $\frac{9}{4}$. What must you clarify?

Figure 3.22 Showing $2\frac{1}{4} = \frac{9}{4}$.

5. We usually call the number 0.43 "forty-three hundredths" and not "four tenths and three hundredths." Why do these two mean the same thing?

6. Jessica says that

$$\frac{1}{2} + \frac{2}{3} = \frac{3}{5}$$

and shows the drawing in Figure 3.23 to prove it.

a. What is the problem with Jessica's reasoning? Why is it not correct? Don't just explain how to do the problem correctly; explain what is faulty with Jessica's reasoning.

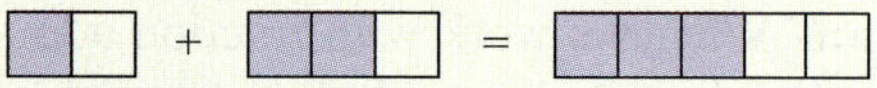

Figure 3.23 Does $\frac{1}{2} + \frac{2}{3}$ equal $\frac{3}{5}$?

b. Explain how Jessica could reason about the sizes of the numbers to see that her answer is not correct.

7. Which of the following problems can be solved by adding $\frac{1}{2} + \frac{1}{3}$? For those problems that can't be solved by adding $\frac{1}{2} + \frac{1}{3}$, solve the problem in another way if there is enough information to do so, or explain why the problem cannot be solved.

a. In Ms. Dock's class, $\frac{1}{2}$ of the boys want pizza for lunch and $\frac{1}{3}$ of the girls want pizza for lunch. What fraction of the children want pizza for lunch?

b. $\frac{1}{2}$ of Jane's shirts have the color red on them somewhere. $\frac{1}{3}$ of Jane's shirts have the color pink on them somewhere. What fraction of Jane's shirts have either red or pink on them somewhere?

c. $\frac{1}{2}$ of Jane's shirts have the color red on them somewhere; of the shirts that belong to Jane and do not have red on them somewhere, $\frac{1}{3}$ have the color pink on them somewhere. What fraction of Jane's shirts have either red or pink on them somewhere?

8. Can the problem that follows be solved by subtracting $\frac{1}{2} - \frac{1}{3}$? If not, explain why not, and solve the problem in a different way if there is enough information to do so.

Problem: There is $\frac{1}{2}$ of a pie left over from yesterday. Pratima eats $\frac{1}{3}$ of the leftover pie. How much pie is left?

9. What is wrong with the following problem? Give two different ways to restate the problem. Explain how to solve your restated problems.

Problem: Liat has $\frac{3}{4}$ cup of juice. Sumin has $\frac{1}{3}$ less. How much juice does Sumin have?

10. For each square in Figure 3.24, determine the fraction of the square that is shaded. Explain your reasoning. You may assume that all lengths that appear to be equal really are equal.

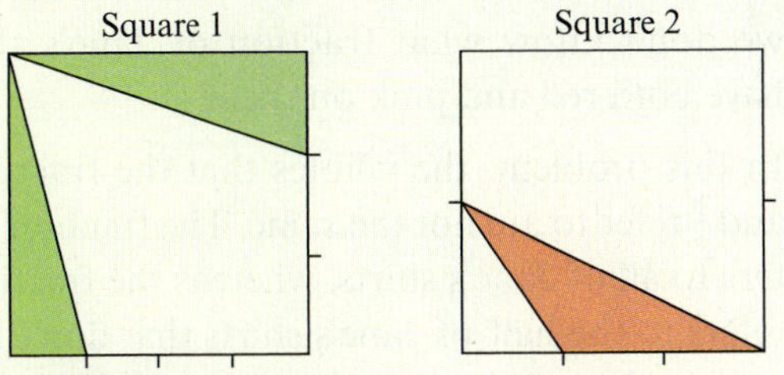

Figure 3.24 What fraction of each square is shaded?

11. Each of the drawings in Figure 3.25 shows two adjacent lots of land, lot A and lot B. In each case, 20% of lot A is shown shaded and 40% of lot B is shown shaded. What percent of the *combined amount* of lot A and lot B is shaded in each case?

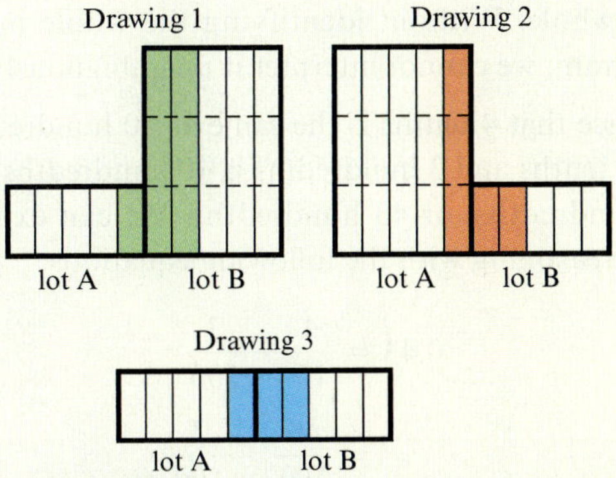

Figure 3.25 What percent of the combined amount of lots A and B is shaded?

Answers to Practice Exercises for Section 3.4

1. When we add $\frac{1}{4}$ and $\frac{5}{6}$, we first give the fractions a common denominator so they have *like parts*. We can interpret $\frac{1}{4} + \frac{5}{6}$ as the total amount of pie you have if you start with $\frac{1}{4}$ of a pie and then get $\frac{5}{6}$ of another equivalent (same size) pie. But fourths and sixths are pieces of different size, so we must express both kinds of pie pieces in terms of like pieces. If we divide each of the 4 fourths of a pie into 3 pieces and each of the 6 sixths of the other pie into 2 pieces, then $\frac{1}{4}$ of the first pie becomes $\frac{3}{12}$ and $\frac{5}{6}$ of the second pie becomes $\frac{10}{12}$. Now both fractions of pie are expressed in terms of twelfths of a pie, as seen in Figure 3.26. Therefore, the total amount of pie you have consists of $3 + 10 = 13$ pieces, and these pieces are twelfths of a pie. Thus, you have $\frac{13}{12}$ of a pie, which is $1\frac{1}{12}$ of a pie.

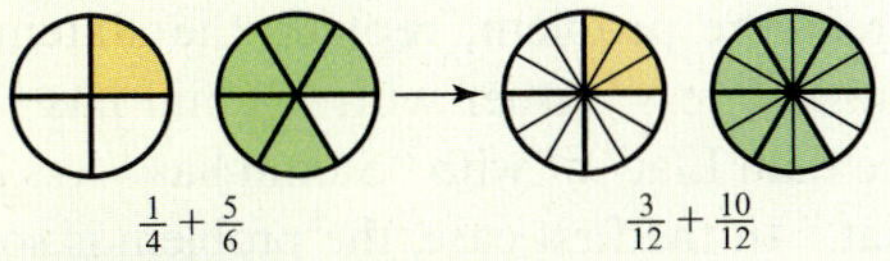

Figure 3.26 Finding common denominators to add $\frac{1}{4}$ and $\frac{5}{6}$.

Numerically, this process of subdividing pieces of pie can be expressed succinctly with the following equations:

$$\frac{1}{4} + \frac{5}{6} = \frac{1 \times 3}{4 \times 3} + \frac{5 \times 2}{6 \times 2} = \frac{3}{12} + \frac{10}{12} = \frac{13}{12} = 1\frac{1}{12}$$

2. To add fractions, we do not need to use the least common denominator. Any common denominator will do. In practice exercise 1, we could have used the common denominator 24 instead of 12 to add $\frac{1}{4} + \frac{5}{6}$. However, the least common denominator, is smaller than other common denominators and therefore may be easier to work with.

3. Remember that $7\frac{1}{3}$ stands for $7 + \frac{1}{3}$, and that $7 = \frac{7}{1}$. To calculate

$$7\frac{1}{3} = \frac{7}{1} + \frac{1}{3}$$

as a fraction, we need to first find a common denominator:

$$\begin{aligned} 7\frac{1}{3} &= 7 + \frac{1}{3} \\ &= \frac{7}{1} + \frac{1}{3} \\ &= \frac{7 \times 3}{1 \times 3} + \frac{1}{3} \\ &= \frac{7 \times 3 + 1}{3} \\ &= \frac{22}{3} \end{aligned}$$

Notice that the next-to-last step shows the procedure for turning a mixed number into an improper fraction. This procedure is really just a shorthand way to give 7 and $\frac{1}{3}$ a common denominator and add them.

4. You must identify the whole in the diagram. Tommy is taking the whole to be the full collection of squares, but to interpret the diagram as representing $2\frac{1}{4}$, we must take a strip of 4 squares to be the whole. Without identifying the whole in the diagram, we cannot interpret it unambiguously.

5. Notice that 4 tenths is the same as 40 hundredths, so 4 tenths and 3 hundredths is 40 hundredths and 3 hundredths, or 43 hundredths. We can express this reasoning with the following equations:

$$\begin{aligned} 0.43 &= \frac{4}{10} + \frac{3}{100} \\ &= \frac{40}{100} + \frac{3}{100} \\ &= \frac{43}{100} \end{aligned}$$

6. a. Jessica's drawing shows that $\frac{1}{2}$ of the 2-block bar combined with $\frac{2}{3}$ of the 3-block bar does indeed make up $\frac{3}{5}$ of a 5-block bar. The problem is that Jessica is using three different *wholes*. The fraction $\frac{1}{2}$ refers to a 2-block whole, the fraction $\frac{2}{3}$ refers to a 3-block whole, and the fraction $\frac{3}{5}$ refers to a 5-block whole. When we add fractions, such as $\frac{1}{2} + \frac{2}{3}$, both fractions in the sum and the sum itself should all refer to the same whole.

b. Jessica could reason that because $\frac{2}{3}$ is greater than $\frac{1}{2}$, the sum $\frac{1}{2} + \frac{2}{3}$ must be greater than 1. Because $\frac{3}{5}$ is less than 1, her answer must be incorrect.

7. None of the problems can be solved by adding $\frac{1}{2} + \frac{1}{3}$.

a. In this problem, the number of *boys* in Ms. Dock's class is the whole associated with the fraction $\frac{1}{2}$, whereas the number of *girls* in Ms. Dock's class is the whole associated with the fraction $\frac{1}{3}$. These two wholes are different. When we add any two numbers, the numbers must refer to the same wholes. There is not enough information to determine what fraction of the children in Ms. Dock's class want pizza for lunch, because we do not know if there is the same number of girls as boys in Ms. Dock's class.

b. In this problem, the fractions $\frac{1}{2}$ and $\frac{1}{3}$ refer to the same whole—namely, Jane's shirts—but some of the shirts that have pink on them may also have red on them somewhere. So, if we start with the $\frac{1}{2}$ of Jane's shirts that have red on them, and if we then want to add on the remaining shirts that have pink on them, we do not know what fraction of Jane's shirts these remaining shirts are. We can't solve this problem because we don't know what fraction of Jane's shirts have both red and pink on them.

c. In this problem, the wholes that the fraction $\frac{1}{2}$ and $\frac{1}{3}$ refer to are not the same. The fraction $\frac{1}{2}$ refers to all of Jane's shirts, whereas the fraction $\frac{1}{3}$ refers to the half of Jane's shirts that don't have red on them. We can solve this problem, however, because the shirts that have pink but not red on them are $\frac{1}{3}$ of $\frac{1}{2}$ of Jane's shirts, which is $\frac{1}{6}$ of Jane's shirts. (You can see this by drawing a diagram.) Therefore, the fraction of Jane's shirts that have either red or pink on them is

$$\frac{1}{2} + \frac{1}{6} = \frac{4}{6} = \frac{2}{3}$$

8. The problem cannot be solved by subtracting $\frac{1}{2} - \frac{1}{3}$ because the fractions $\frac{1}{2}$ and $\frac{1}{3}$ refer to different wholes. The $\frac{1}{2}$ refers to the whole pie, whereas the $\frac{1}{3}$ refers to the pie that is left over. We can solve this problem because when Pratima eats $\frac{1}{3}$ of the $\frac{1}{2}$ pie that is left over, she is eating $\frac{1}{6}$ of the pie (as you can see by drawing a diagram). So we know the following:

$$\frac{1}{2} + \frac{1}{6} = \frac{4}{6} = \frac{2}{3}$$

of the pie has been eaten. Therefore, $\frac{1}{3}$ of the pie is left.

9. The problem is with the statement "Sumin has $\frac{1}{3}$ less." Does this mean that Sumin has $\frac{1}{3}$ *cup* less juice than Liat, or does it mean that the *amount of juice* Sumin has is $\frac{1}{3}$ less than the amount of juice Liat has? It is not clear which meaning is intended. To correct the problem, replace the statement, "Sumin has $\frac{1}{3}$ less" either with "Sumin has $\frac{1}{3}$ cup less juice than Liat" or with "Sumin has $\frac{1}{3}$ less juice than Liat." In the first case, the problem is solved by calculating

$$\frac{3}{4} - \frac{1}{3} = \frac{9}{12} - \frac{4}{12} = \frac{5}{12}$$

so that Sumin has $\frac{5}{12}$ cup of juice. In the second case, the problem is solved by first calculating $\frac{1}{3}$ of $\frac{3}{4}$ cup of juice, which is $\frac{1}{4}$ cup of juice. (There are 3 equal parts in $\frac{3}{4}$, and each part is $\frac{1}{4}$.) Therefore,

Sumin has $\frac{1}{4}$ cup less juice than Liat, so Sumin has $\frac{3}{4} - \frac{1}{4} = \frac{1}{2}$ cup of juice.

10. The shaded region in square 1 consists of two pieces. One piece is $\frac{1}{8}$ of the square, as shown on the left in **Figure 3.27**. The other piece is $\frac{1}{6}$ of the square, as shown on the right in **Figure 3.27**. Therefore, the shaded region in square 1 is $\frac{1}{8} + \frac{1}{6} = \frac{7}{24}$ of the square.

$\frac{1}{8}$ of the square is shaded. $\frac{1}{6}$ of the square is shaded.

Figure 3.27 The original shaded region is $\frac{1}{8} + \frac{1}{6}$ of the square.

We can think of the shaded region in square 2 as the difference between the shaded region on the left in **Figure 3.28** and the shaded region on the right in **Figure 3.28**. Therefore, the shaded region in square 2 is $\frac{1}{4} - \frac{1}{12} = \frac{1}{6}$ of the square.

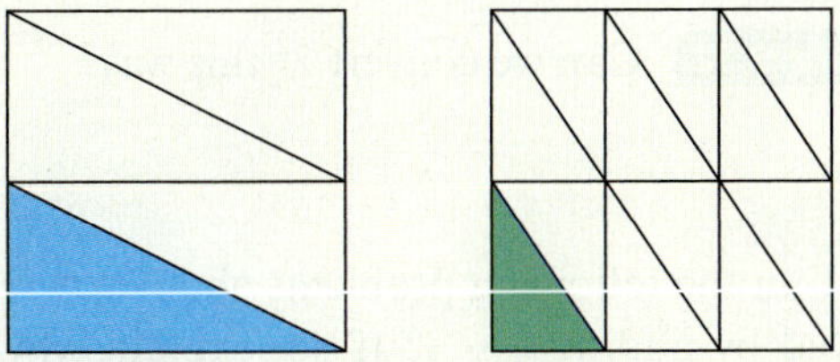

$\frac{1}{4}$ of the square is shaded. $\frac{1}{12}$ of the square is shaded.

Figure 3.28 The original shaded region is $\frac{1}{4} - \frac{1}{12}$ of the square.

11. See **Figure 3.29**.

Drawing 1

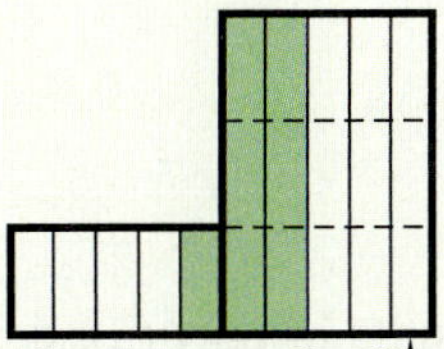

There are 20 small pieces of this size making up the combined lots A and B. Therefore, each such piece is $\frac{1}{20}$, which is 5%. Since 7 are shaded, that means 35% of the combined amount is shaded.

Drawing 2

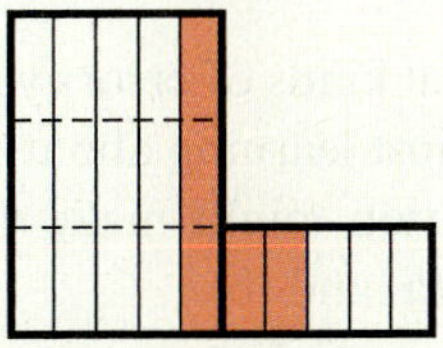

Here, 5 pieces are shaded. Each piece is 5% of the total. Therefore, 25% of the total is shaded.

Drawing 3

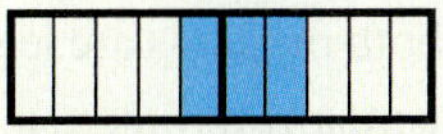

$\frac{3}{10} = \frac{30}{100} = 30\%$ of the combined amount is shaded.

Figure 3.29 Various percents of the combined amount of lots A and B are shaded.

PROBLEMS FOR SECTION 3.4

1. Using the example $\frac{2}{7} + \frac{3}{7}$, explain why we add fractions with like denominators the way we do. What is the logic behind the procedure? Make math drawings to support your explanation.

2. Using the example $\frac{2}{3} + \frac{3}{4}$, explain why we add fractions the way we do. What is the logic behind the procedure? Make math drawings to support your explanation.

3. Using the example $3\frac{5}{6}$, describe the procedure for turning a mixed number into an improper fraction, and explain in your own words why this procedure makes sense. What is the logic behind the procedure? Make math drawings to support your explanation.

4. Use two number lines, one labeled with (fractions and) mixed numbers, the other labeled with (proper and) improper fractions, to help you explain why the procedure for turning mixed numbers into improper fractions described on page 125 makes sense.

5. **a.** For each of the following decimals, show how to write the decimal as a fraction by first putting the decimal in expanded form:

   i. 2.34

   ii. 124.5

   iii. 7.938

   b. Based on your results in part (a), describe a quick way to rewrite a finite decimal as a fraction. Illustrate with the example 2748.963.

6. **a.** Show how to calculate the sum $\frac{2}{5} + 0.25$ and show how to write the answer as a fraction and as a decimal.

   b. Show how to calculate the sum $3.8 + \frac{3}{8}$ and show how to write the answer as a fraction and as a decimal.

   c. Discuss briefly what kinds of errors you think students who are just learning about fraction and decimal addition might make with the problems in parts (a) and (b).

7. Show how to calculate $5\frac{3}{4} + 1\frac{2}{3}$ in two different ways. In each case, express your answer as a mixed number. Explain why both methods are legitimate.

8. Show how to calculate $4\frac{2}{3} - 1\frac{3}{4}$ in two different ways. In each case, express your answer as a mixed number. Explain why both methods are legitimate.

9. Find two *different* positive fractions whose sum is $\frac{1}{11}$.

10. **a.** John says $\frac{2}{3} + \frac{2}{3} = \frac{4}{6}$ and uses the drawing in Figure 3.30 as evidence. Discuss what is wrong with John's reasoning. What underlying misconception does John have? *Don't* just explain how to do the problem correctly; explain why John's reasoning is faulty.

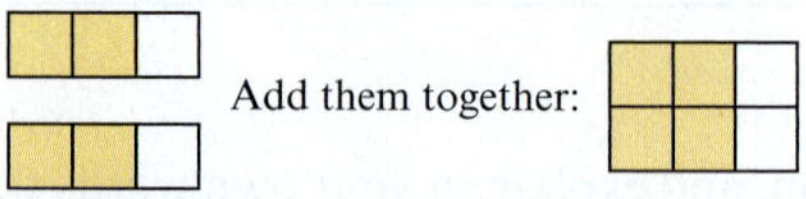

Figure 3.30 Showing how $\frac{2}{3} + \frac{2}{3} = \frac{4}{6}$.

   b. Explain how John could reason about the sizes of the numbers to see that his answer is not correct.

11. Denise says that $\frac{2}{3} - \frac{1}{2} = \frac{1}{3}$ and gives the reasoning indicated in Figure 3.31 to support her answer. Is Denise right? If not, what is wrong with her reasoning and how could you help her understand her mistake and fix it? *Don't* just explain how to solve the problem correctly; explain where Denise's reasoning is flawed.

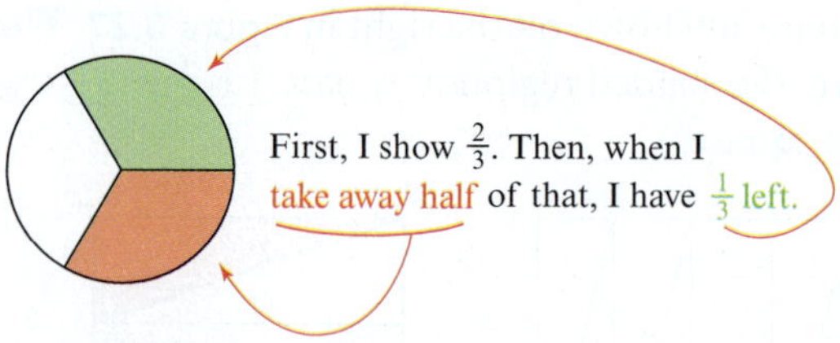

Figure 3.31 Denise's idea for $\frac{2}{3} - \frac{1}{2}$.

12. Arnold says that $2\frac{2}{3} = \frac{4}{5}$, and he uses the drawing in Figure 3.32 to support his conclusion. What is wrong with Arnold's reasoning? Do not just state the correct way to convert $2\frac{2}{3}$; explain why Arnold's reasoning is not valid.

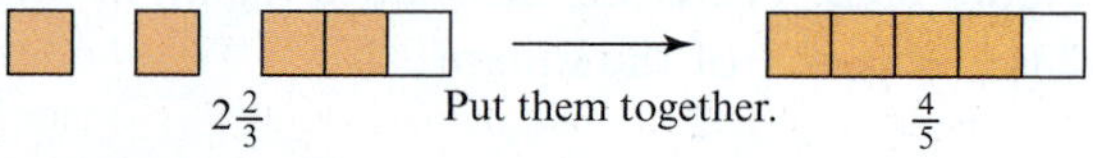

Figure 3.32 Can we convert $2\frac{2}{3}$ this way?

13. Can the following problems about voters be solved by adding $\frac{1}{2} + \frac{1}{3}$? If so, explain why. If not, explain why not. Solve the problems if they can be solved. Write a different problem about voters that can be solved by adding $\frac{1}{2} + \frac{1}{3}$.

   Problem 1 About Voters: In Kneebend County, $\frac{1}{2}$ of the female voters and $\frac{1}{3}$ of the male voters voted for a certain referendum. Altogether, what fraction of the voters voted for the referendum in Kneebend County?

   Problem 2 About Voters: In Kneebend County, $\frac{1}{2}$ of the female voters and $\frac{1}{3}$ of the male voters voted for a certain referendum. There are the same number of women voters as men voters in Kneebend County. Altogether, what fraction of the voters voted for the referendum in Kneebend County?

14. Can the following problems about a bird feeder be solved by subtracting $\frac{3}{4} - \frac{1}{2}$? If so, explain why. If not, explain why not. Solve the problems if they can be solved. Write a different

problem about a bird feeder that can be solved by subtracting $\frac{3}{4} - \frac{1}{2}$.

Problem 1 About a Bird Feeder: A bird feeder was filled with $\frac{3}{4}$ of a full bag of bird seed. The birds ate $\frac{1}{2}$ of what was in the bird feeder. What fraction of a full bag of bird seed did the birds eat?

Problem 2 About a Bird Feeder: A bird feeder was filled with $\frac{3}{4}$ of a full bag of bird seed. The birds ate $\frac{1}{2}$ of what was in the bird feeder. What fraction of a full bag of bird seed is left in the bird feeder?

15. Can the following problems about Sarah's bead collection be solved by adding $\frac{1}{4} + \frac{1}{5}$? If so, explain why. If not, explain why not. Solve the problems if they can be solved. Write a different story problem about Sarah's bead collection that can be solved by adding $\frac{1}{4} + \frac{1}{5}$.

Problem 1 About Sarah's Bead Collection: One-fourth of the beads in Sarah's collection are pink. One-fifth of the beads in Sarah's collection are long. What fraction of the beads in Sarah's collection are either pink or long?

Problem 2 About Sarah's Bead Collection: One-fourth of the beads in Sarah's collection are pink. One-fifth of the beads in Sarah's collection that are not pink are long. What fraction of the beads in Sarah's collection are either pink or long?

16. Can the following problem about Jim's medicine be solved by subtracting $\frac{1}{3} - \frac{1}{4}$? If so, explain why. If not, explain why not and solve the problem if it can be solved. Write a different problem that can be solved by subtracting $\frac{1}{3} - \frac{1}{4}$.

Problem About Jim's Medicine: Jim has a small container filled with 1 tablespoon of medicine. Jim poured out $\frac{1}{3}$ tablespoon of medicine, then he poured out $\frac{1}{4}$ tablespoon of medicine. Now how much medicine is left in the container?

17. a. Write and solve a word problem for $\frac{3}{4} + \frac{2}{3}$.

b. Write and solve a word problem for $\frac{3}{4} - \frac{2}{3}$.

18. a. Write and solve a word problem for $2\frac{1}{2} + 1\frac{1}{3}$.

b. Write and solve a word problem for $2\frac{1}{2} - 1\frac{1}{3}$.

19. a. Write and solve a Compare problem that can be solved by calculating $\frac{3}{4} - \frac{2}{3}$.

b. Write and solve a Compare problem that can be solved by calculating $\frac{2}{3} + 1\frac{1}{2}$.

20. a. Write and solve a Put Together/Take Apart problem for $\frac{1}{2} + ? = \frac{2}{3}$.

b. Write and solve a Put Together/Take Apart problem for $\frac{1}{2} + \frac{2}{3} = ?$

21. For each square in Figure 3.33, determine the fraction of the square that is shaded. Explain your reasoning. You may assume that all lengths that appear to be equal really are equal. Do not use any area formulas.

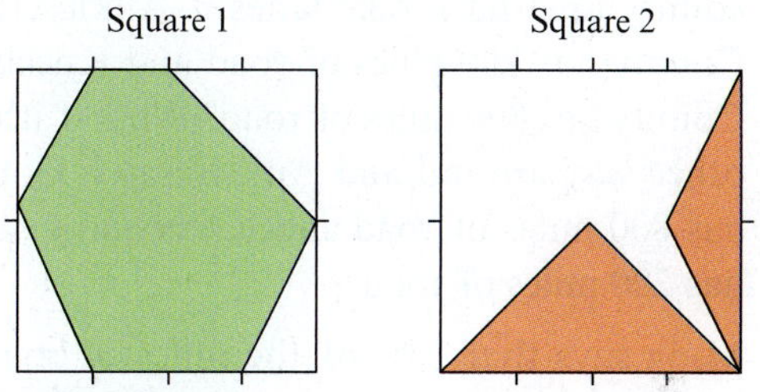

Figure 3.33 What fraction of each square is shaded?

22. For each square in Figure 3.34, determine the fraction of the square that is shaded. Explain your reasoning. You may assume that all lengths that appear to be equal really are equal. Do not use any area formulas.

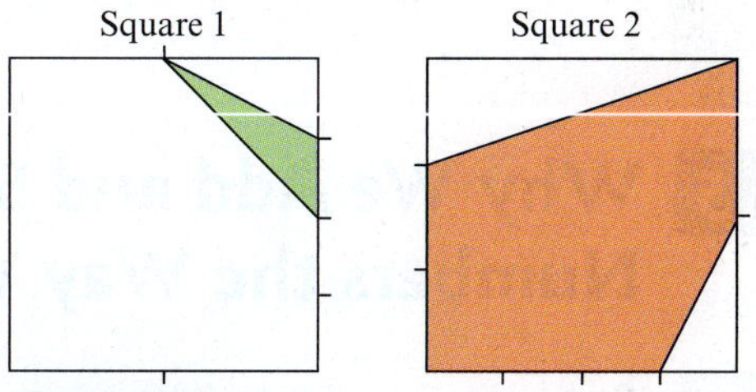

Figure 3.34 What fraction of each square is shaded?

23. Broad Street divides Popperville into an east side and a west side. On the east side of Popperville, 20% of the children qualify for a reduced-price lunch. On the west side of Popperville, 30% of the children qualify for a reduced-price lunch. Is it correct to calculate the percentage of children in all of Popperville who qualify for reduced-price lunch by adding 20% and 30% to get 50%? If the answer is no, why not? Explain in detail and calculate the correct percentage in at least two examples.

24. Anklescratch County and Kneebend County are two adjacent counties. In Anklescratch, 30% of the miles of road have bike lanes, whereas in Kneebend, only 10% of the miles of road have bike lanes.

a. Elliot says that in the two-county area, 40% of the miles of road have bike lanes. He found 40% by adding 30% and 10%. Is Elliot correct or not? Explain.

b. What percent of miles of road in the two-county area have bike lanes if Anklescratch County has 200 miles of road and Kneebend County has 800 miles of road? What if it's the other way around and Anklescratch County has 800 miles of road and Kneebend County has 200 miles of road?

c. Ming says that 20% of the miles of road in the two-county area have bike lanes because 20% is the average of 10% and 30%. Explain why Ming's answer could be either correct or incorrect. Under what circumstances will Ming's answer be correct?

***25.** There are two elementary schools in the town of South Elbow. In the first school, 20% of the children are Hispanic. In the second school, 30% of the children are Hispanic. Write a paragraph about what you can and cannot tell from these data alone. Include examples to support your points.

***26.** Suppose you start with a fraction and you add 1 to both the numerator and the denominator. For example, if you started with $\frac{2}{3}$, then you'd get a new fraction $\frac{2+1}{3+1} = \frac{3}{4}$. Is this procedure of adding 1 to the numerator and the denominator the same as adding the number 1 to the original fraction? (For example, is $\frac{2+1}{3+1}$ equal to $\frac{2}{3} + 1$?) Explain your answer.

***27.** In the first part of the season, the Bluejays play 18 games and win 7, while the Robins play 3 games and win 1. In the second part of the season, the Bluejays play 2 games and win 1, while the Robins play 17 games and win 8.

a. Which team won a larger fraction of its games in the first part of the season?

b. Which team won a larger fraction of its games in the second part of the season?

c. Which team won the larger fraction of its games overall? Is the answer surprising in light of the answers to parts (a) and (b)?

3.5 Why We Add and Subtract with Negative Numbers the Way We Do

CCSS Common Core State Standards Grade 7

Where do the rules for adding and subtracting with negative numbers come from? These rules come from extending arithmetic and its properties to be consistent with arithmetic with positive numbers. In this section, we will use a combination of word problems and properties of arithmetic to show why we add and subtract with negative numbers the way we do. We will see that some of the different types of word problems discussed in Section 3.1 are useful for interpreting addition and subtraction with negative numbers.

If we view negative numbers as the "opposites" of positive numbers, we can interpret negative numbers as temperatures below zero, locations below ground (in a mine or a building, for example), locations below sea level, dollar amounts owed, or negatively charged particles. Using contexts such as these, we can find word problems that fit with negative number arithmetic.

CLASS ACTIVITY

3U Using Word Problems to Find Rules for Adding and Subtracting Negative Numbers, p. CA-66

In this section we will see why it makes sense to interpret

$$A + (-B) \quad \text{as} \quad A - B$$

and

$$A - (-B) \quad \text{as} \quad A + B$$

To do so, we first consider sums such as $(-3) + 3$ and $(-7) + 7$.

CCSS 7.NS.1a

Why Does Adding a Number to Its Negative Result in Zero?

How can we interpret $(-3) + 3$? A word problem for $(-3) + 3 = ?$ follows:

> It was $-3°$ C at dawn. In the meantime, the temperature went up $3°$ C. Now what is the temperature?

Since $-3°$ C means 3 degrees C below 0, the new temperature must be $0°$ C, as we see in Figure 3.35, so $(-3) + 3 = 0$. In general,

$$(-N) + N = 0$$

additive inverses

for any number N. This equation states that $-N$ and N are **additive inverses**—in other words, that $-N$ and N add up to 0.

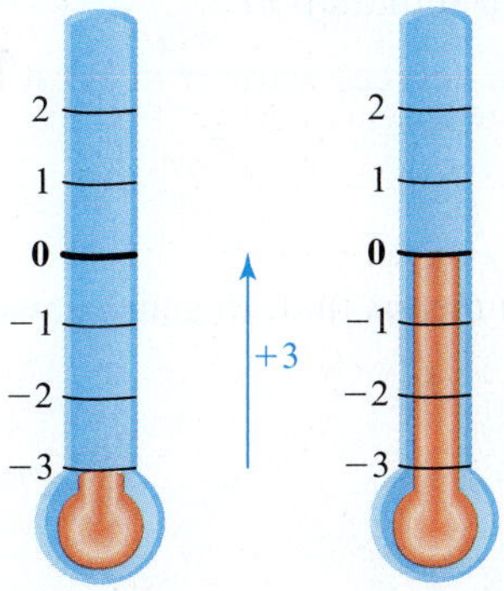

Figure 3.35 Using an increase in temperature to explain why $(-3) + 3 = 0$.

CCSS 7.NS.1c,d

Why Does It Make Sense to Interpret $A + (-B)$ as $A - B$?

How can we interpret the sum $5 + (-3)$? You may know the rule that $5 + (-3) = 5 - 3$, but where does this rule come from and why does it make sense? We can deduce this rule from the associative property of addition and the link between subtraction and addition.

To explain why the sum $5 + (-3)$ must be equal to $5 - 3$, we will first evaluate the sum

$$5 + (-3) + 3 \tag{3.1}$$

in two different ways by applying the associative property of addition. Then we will apply the link between subtraction and addition.

So consider the sum $5 + (-3) + 3$ of Equation 3.1.

According to the associative property of addition, we can associate the -3 in Equation 3.1 with either the 3 or with the 5, and we will get the same end result either way. When we associate the -3 with the 3, we get this:

$$5 + \underbrace{(-3) + 3}_{0} = 5 + 0 = 5$$

because $(-3) + 3 = 0$, as previously discussed. On the other hand, when we associate the -3 with 5 and represent the unknown $5 + (-3)$ as ? we get this:

$$\underbrace{5 + (-3)}_{?} + 3 = ? + 3$$

Whichever way we evaluate $5 + (-3) + 3$ we must get the same answer. Therefore

$$? + 3 = 5 \tag{3.2}$$

But according to the link between addition and subtraction, Equation 3.2 is equivalent to the subtraction equation

$$5 - 3 = ?$$

Since we are using ? to stand for $5 + (-3)$, we conclude that

$$5 + (-3) = 5 - 3$$

The same reasoning applies with other numbers; so, in general, it makes sense to interpret

$$A + (-B) \quad \text{as} \quad A - B$$

Why Does It Make Sense to Interpret $A - (-B)$ as $A + B$?

How do we interpret a difference such as $5 - (-3)$? Instead of interpreting the subtraction as taking away, let's interpet it as comparison (see Section 3.1). A Compare, Difference Unknown word problem for $5 - (-3) = ?$ is as follows:

> The temperature was 5° C in West Lafayette. At the same time it was −3° C in Indianapolis. How much warmer was it in West Lafayette than in Indianapolis?

Since −3° C means 3 degrees C below 0, it is $3 + 5 = 8$ degrees warmer in West Lafayette than in Indianapolis, as we see in Figure 3.36. So

$$5 - (-3) = 5 + 3$$

The same reasoning applies with other numbers, which means that, in general, we can interpret

$$A - (-B) \quad \text{as} \quad A + B$$

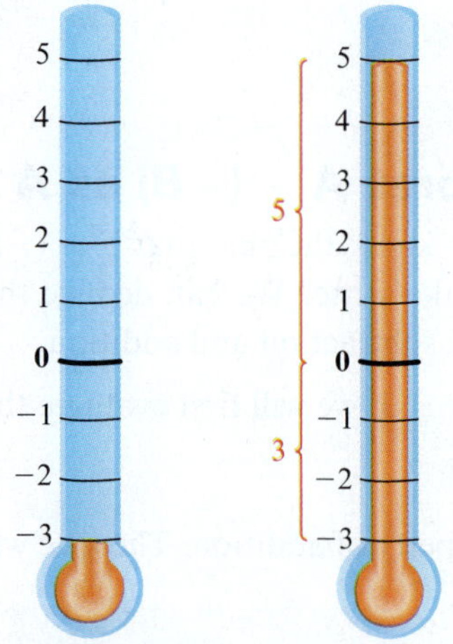

Figure 3.36 Using a comparison of temperatures to explain why $5 - (-3) = 5 + 3$.

How Can We Extend Addition and Subtraction on Number Lines to Negative Numbers?

In Section 3.1 we saw how to represent addition and subtraction on number lines when only positive numbers are involved. How should we interpret addition and subtraction on a number line when negative numbers are involved? For example, where should $5 + (-3)$ be on the number line? Since

$$5 + (-3) = 5 - 3$$

and, in general, since

$$A + (-B) = A - B$$

when we use a number line to add a negative number, we should move to the *left* instead of to the right. Similarly, where should $5 - (-3)$ be on the number line? Since

$$5 - (-3) = 5 + 3$$

and, in general, since

$$A - (-B) = A + B$$

when we use a number line to subtract a negative number, we should move to the *right* instead of to the left.

addition on number lines

These discussions indicate that we should interpret **addition and subtraction on number lines** as follows:

- If A and B are any two numbers, then they are represented by points on a number line. The sum

$$A + B$$

corresponds to the point on the number line that is located as follows:

1. Go to A on the number line.
2. Move a distance of $|B|$ units (the distance that B is away from 0)
   - Move to the right if B is positive.
   - Move to the left if B is negative.
3. The resulting point is the location of $A + B$.

CCSS 7.NS.1b

Where is $2 + (-3)$ on a number line? As shown in **Figure 3.37**, go to 2 and move 3 units to the left. Move left because -3 is negative. You end up at -1; so,

$$2 + (-3) = -1$$

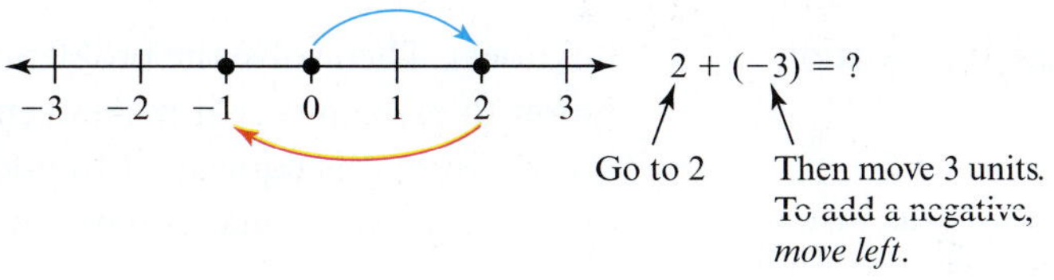

Figure 3.37 Using a number line to show that $2 + (-3) = -1$.

- If A and B are any two numbers, the difference

$$A - B$$

corresponds to the point on the number line that is located as follows:

subtraction on number lines

1. Start at A on the number line.
2. Move a distance of $|B|$ units (the distance that B is away from 0)
   - Move to the *left* if B is positive.
   - Move to the *right* if B is negative.
3. The resulting point is the location of $A - B$.

Where is $(-2) - (-5)$ on a number line? As shown in **Figure 3.38**, go to -2 and move 5 units to the *right*. Move right because -5 is negative and we are subtracting. You end up at 3; so, $(-2) - (-5) = 3$.

Although the combining and taking from interpretations of addition and subtraction fit with our common sense and intuition, they do not work well with certain numbers. The number line interpretation of addition and subtraction, although more abstract, applies to *all* numbers.

Figure 3.38

Using a number line to show that $(-2) - (-5) = 3$.

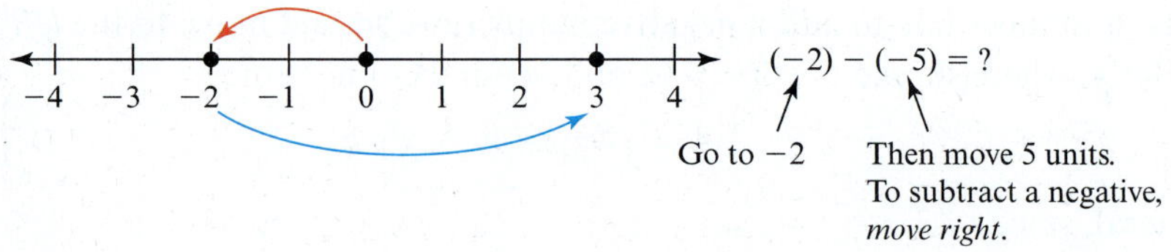

SECTION SUMMARY AND STUDY ITEMS

Section 3.5 Why We Add and Subtract with Negative Numbers the Way We Do

We can use word problems together with properties of addition and the link between addition and subtraction to make sense of the rules for adding and subtracting with negative numbers. Using these rules, we can extend addition and subtraction on number lines to the case of negative numbers.

Key Skills and Understandings

1. Write and solve word problems involving addition and subtraction with negative numbers.
2. Apply properties of arithmetic and the connection between addition and subtraction to negative numbers.
3. Explain how to use a number line to add and subtract numbers, including negative numbers.

Practice Exercises for Section 3.5

1. Write a Compare problem that fits with the equation

$$(-3) - 7 = ?$$

and in which one quantity is -3 and the other quantity unknown and is 7 less than -3.

2. Write a Compare problem that fits with the equation

$$4 - (-3) = ?$$

and in which one quantity is 4, the other quantity is -3, and the difference between the two quantities is unknown. Then solve the problem and use your solution to explain why it makes sense that subtracting -3 from 4 is equivalent to adding 3 to 4, or in other words why it makes sense that

$$4 - (-3) = 4 + 3$$

3. Use a number line to calculate $0 - (-2)$. Briefly explain the method.

4. Use a number line to calculate $(-3) - (-3)$. Briefly explain the method.

Answers to Practice Exercises for Section 3.5

1. *Problem:* The temperature was −3° C in Buffalo. At the same time, it was 7° C colder in Syracuse. What was the temperature in Syracuse?

As we see in Figure 3.39, the temperature that is 7 degrees C colder than −3° C is −10° C.

2. *Problem:* Mary is on floor 4 of a building while Jonathan is on floor −3 of the same building. How many floors does Jonathan have to go up to get to Mary's floor?

For Jonathan to get to Mary's floor, he first needs to go up 3 floors to get to floor 0, and then he needs to go up another 4 floors to get to Mary's floor. Altogether, that means going up $4 + 3$ floors. So based on this solution, we see that subtracting

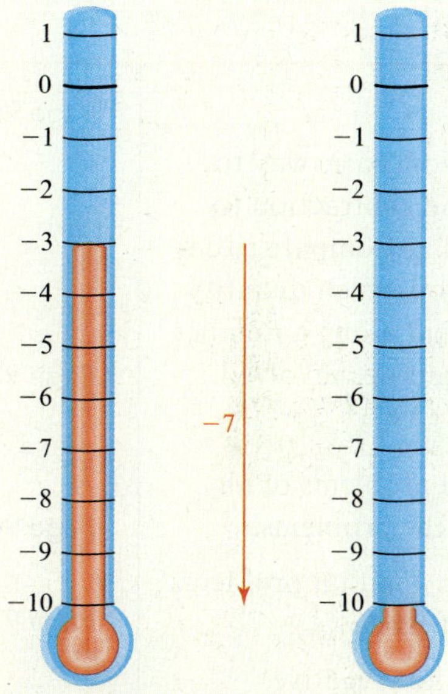

Figure 3.39 Seven degrees colder than −3° C is −10° C, so (−3) − 7 = −10.

−3 from 4 is equivalent to adding 3 to 4 or, in other words, that

$$4 - (-3) = 4 + 3$$

3. See **Figure 3.40**. To find the location of 0 − (−2), go to 0 first. Since we are subtracting, we might think we will move to the left, but since −2 is a negative number, we move 2 units to the *right*, ending at 2.

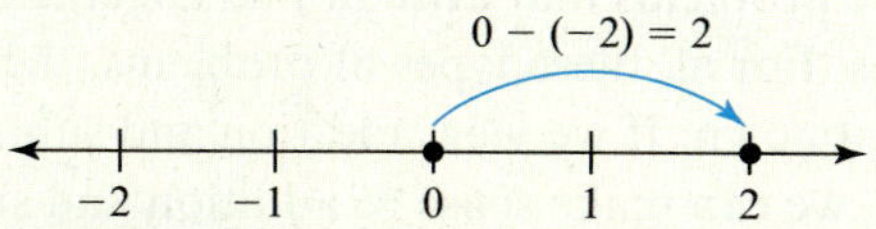

Figure 3.40 0 − (−2) = 2.

4. See **Figure 3.41**. To find the location of (−3) − (−3), go to −3 first. Since we are subtracting, we might think we will move to the left, but since −3 is a negative number, we move 3 units to the *right*, ending at 0.

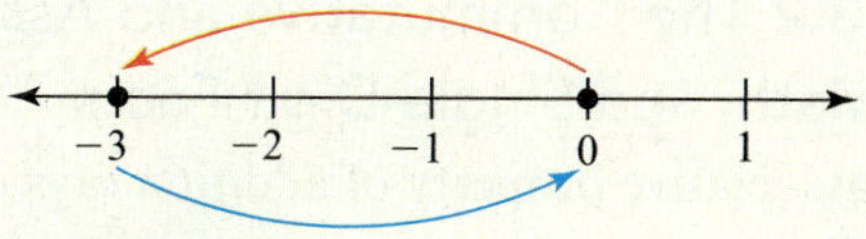

Figure 3.41 (−3) − (−3) = 0.

PROBLEMS FOR SECTION 3.5

1. a. Write an Add To problem that fits naturally with the equation

$$(-4) + 3 = ?$$

Solve the problem, and explain why the solution makes sense.

b. Write a Take From problem that fits naturally with the equation

$$(-4) - 3 = ?$$

Solve the problem, and explain why the solution makes sense.

2. a. Write a Compare problem that fits with the equations

$$(-7) + ? = -1 \text{ and } (-1) - (-7) = ?$$

Solve the problem, and explain why the solution makes sense.

b. Write a Compare problem that fits with the equations

$$(-7) + 6 = ? \text{ and } ? - (-7) = 6$$

Solve the problem, and explain why the solution makes sense.

3. a. Show how to use a number line to calculate −1 + (−2).

b. Show how to use a number line to calculate 1 − (−2).

c. Show how to use a number line to calculate −1 − 2.

d. Show how to use a number line to calculate −1 − (−2).

4. a. For each of these four expressions,

$$-2 + 6, -(2 + 6), -2 - 6, -(2 - 6)$$

draw a number line (with a consistent scale) and plot the expression on the number line. In each case, show clearly why the expression is located where it is based on the rules for plotting sums, differences, and negatives of numbers on number lines.

b. Based on your work in part (a), compare and contrast the four expressions shown there.

CHAPTER SUMMARY

Section 3.1 Interpretations of Addition and Subtraction — Page 93

- The most basic way to interpret addition is as Adding To, and the most basic way to interpret subtraction is as Taking From. But we also use addition and subtraction to solve problems that arise in Put Together/Take Apart situations and in Compare situations. For all these types of problems, there are subtypes depending on which quantity is unknown. If we view addition and subtraction in terms of movement along a number line, we can make sense of addition and subtraction when any numbers are involved. Page 93

Key Skills and Understandings

1. Write Add To, Take From, Put Together/Take Apart, and Compare problems of all subtypes and write equations and make math drawings to represent the problems. Page 94
2. Recognize that keywords alone are not effective for solving problems and that problems whose keywords indicate the opposite operation of a solution are difficult. Page 97
3. Explain how to use a number line to add and subtract numbers (for nonnegative numbers only). Page 98

Section 3.2 The Commutative and Associative Properties of Addition, Mental Math, and Single-Digit Facts — Page 102

- The associative property of addition says that, for all numbers A, B, and C,

$$(A + B) + C = A + (B + C)$$

The commutative property of addition says that, for all numbers A and B,

$$A + B = B + A$$

Page 103

- We can use these properties to help children learn the basic addition facts. The commutative property allows children to cut the number of facts that must be memorized almost in half. The associative property is used in the important make-a-ten strategy. The commutative and associative properties of addition can also help make some multidigit addition problems easier to carry out mentally. Page 105

Key Skills and Understandings

1. State the associative property of addition and explain how to get its equation by viewing one amount in two ways. Page 103
2. Give examples to show how to use the associative property of addition to make problems easier to do mentally, including the make-a-ten strategy. Page 106
3. State the commutative property of addition and explain how to get its equation by viewing one amount in two ways. Page 104
4. Give examples to show how to use the commutative property of addition to make problems easier to do mentally, including the "count on from the larger addend" methods, and explain how it helps children cut down on the memorization of basic addition facts. Page 105
5. Describe how to view subtraction problems as unknown addend problems, explain how this can help young children with basic subtraction facts, and explain how this can be applied to other mental math problems. Page 106
6. Write correct equations to go along with a mental method of addition or subtraction. Identify where the commutative or associative properties of addition have been used in calculations. Page 108
7. Explain how to add or subtract by using methods other than the common addition and subtraction algorithms. Page 107

Section 3.3 Why the Standard Algorithms for Addition and Subtraction in Base Ten Work — Page 114

- The standard addition and subtraction algorithms work in terms of objects grouped in ones, tens, hundreds, and so on. Regrouping occurs when we make or break bundles. — Page 114

Key Skills and Understandings

1. Explain the addition and subtraction algorithms in terms of bundled objects and base-ten math drawings, paying special attention to regrouping. — Page 115
2. Explain that we line up like places so that we add or subtract hundreds and hundreds, tens and tens, ones and ones, and so on. This is why we line up decimal points when adding and subtracting decimals. — Page 116

Section 3.4 Reasoning About Fraction Addition and Subtraction — Page 123

- To add or subtract fractions, we first give the fractions common denominators so as to work with like parts. Once we have like parts, we can simply add or subtract the number of parts. A mixed number stands for the sum of a whole number and a fraction; when we add the whole number part to the fraction part, the result is an improper fraction. When we work with fraction addition and subtraction word problems, we must pay close attention to underlying wholes and to the wording of the problem. — Page 123

Key Skills and Understandings

1. Describe how to add or subtract fractions, and explain why the process makes sense. In particular, explain why we give the fractions common denominators. — Page 123
2. Use math drawings and number lines to help explain the logic behind the procedure of turning a mixed number into an improper fraction. — Page 125
3. Write fraction addition and subtraction word problems. — Page 126
4. Determine if a given word problem fits a given fraction addition or subtraction problem. — Page 127
5. Use estimation and a sense of the size of fractions to assess the reasonableness of the answer to a fraction addition or subtraction problem. — Page 128

Section 3.5 Why We Add and Subtract with Negative Numbers the Way We Do — Page 134

- We can use word problems together with properties of addition and the link between addition and subtraction to make sense of the rules for adding and subtracting with negative numbers. Using these rules, we can extend addition and subtraction on number lines to the case of negative numbers. — Page 134

Key Skills and Understandings

1. Write and solve word problems involving addition and subtraction with negative numbers. — Page 134
2. Apply properties of arithmetic and the connection between addition and subtraction to negative numbers. — Page 135
3. Explain how to use a number line to add and subtract numbers, including negative numbers. — Page 136

4

Multiplication

In this chapter, we will study what multiplication means and what kinds of problems it solves. We examine ways of representing multiplication, the properties of multiplication, and the procedures we use to multiply whole numbers. Because multiplication is so familiar, we may take it for granted, and not stop to question what it means or why it works. As we will see, the procedures we use when we multiply are based on very clever uses of the properties of arithmetic.

We focus on the following topics and practices within the *Common Core State Standards for Mathematics* (*CCSSM*) in this chapter.

Standards for Mathematical Content in the CCSSM

In the domain of *Operations and Algebraic Thinking* (K–Grade 5) students learn what kinds of problems multiplication solves, and they learn to reason about multiplication with small numbers. In the domain of *Number and Operations in Base Ten* (K–Grade 5), students learn to decompose numbers according to place value and to apply properties of multiplication to make sense of methods for multiplying multidigit numbers.

Standards for Mathematical Practice in the CCSSM

Opportunities to engage in all eight of the Standards for Mathematical Practice described in the CCSSM occur throughout the study of multiplication, although the following standards may be especially appropriate for emphasis:

- **2 Reason abstractly and quantitatively.** Students engage in this practice when they connect the definition of multiplication with contexts and math drawings to determine whether a problem can be solved by multiplication.
- **3 Construct viable arguments and critique the reasoning of others.** Students engage in this practice when they make sense of methods of multiplication and when they explain their methods with the aid of math drawings.
- **7 Look for and make use of structure.** Students engage in this practice when they see and apply the structure of the distributive property to multiply—for example, by calculating 6×8 as 5×8 plus 1×8, or by calculating 3×42 as 3×40 plus 3×2.

(From Common Core Standards for Mathematical Practice. Published by Common Core Standards Initiative.)

4.1 Interpretations of Multiplication

CCSS **Common Core State Standards** Grades 3, 4, 7

What is multiplication? In this section we define multiplication and examine some categories of problems that can be solved with multiplication.

How Can We Define Multiplication?

What does multiplication mean? To explain what 3×5 means and why it equals 15 you might use repeated addition: adding 3 fives makes 15. But if we were to define multiplication as repeated addition, it would not generalize to fractions and decimals. So instead, we define multiplication in terms of *equal groups*.

If M and N are nonnegative numbers, then

$$M \cdot N \text{ or } M \times N \text{ or } M * N$$

basic definition of multiplication
product
multiplier
multiplicand

which we read as "M times N," means the number of units (or objects) in M equal groups if there are N units (or objects) in 1 group. The value (result) of $M \cdot N$ is called the **product** of M and N. Notice that the numbers M and N play different roles. To distinguish these roles, we call the number of equal groups (M) the **multiplier** and the number of units in one group (N) the **multiplicand**. It is important for students in the elementary and middle grades to think about multiplication in terms of equal groups, but they do not need to use the terms "multiplier" and "multiplicand."

CCSS 3.OA.1

The choice to write the multiplier *first* and the multiplicand *second* is a convention (which is different in some countries). For consistency and clear communication, we will stick to this convention when we are using the meaning of multiplication.

If P is the product of M and N, then we can write the multiplication equation

$$M \cdot N = P$$

Figure 4.1 summarizes the different roles of the multiplier, the multiplicand, and the product. *Multiplication is about coordinating numbers and sizes of equal groups*. Unlike in addition and subtraction, where all the numbers refer to the same kinds of things (see Figure 4.2), in multiplication the numbers refer to different things. The multiplier M is a number of groups, whereas the multiplicand N and the product P are numbers of *units*. So N and P tell us the *size* of 1 group and M groups, respectively. In contrast, M tells us *how many groups* of size N are in P. As a shorthand, we can describe the product P as the size of M groups, each of size N. Even shorter, we can say that P is M groups of N.

factors
multiple

When $M \cdot N = P$, we can also refer to M and N as **factors** of P or of $M \cdot N$, and we can say that P is a **multiple** of N (and of M). For example, 3 and 5 are factors of 15 and 15 is a multiple of 5 (and of 3) because $3 \cdot 5 = 15$.

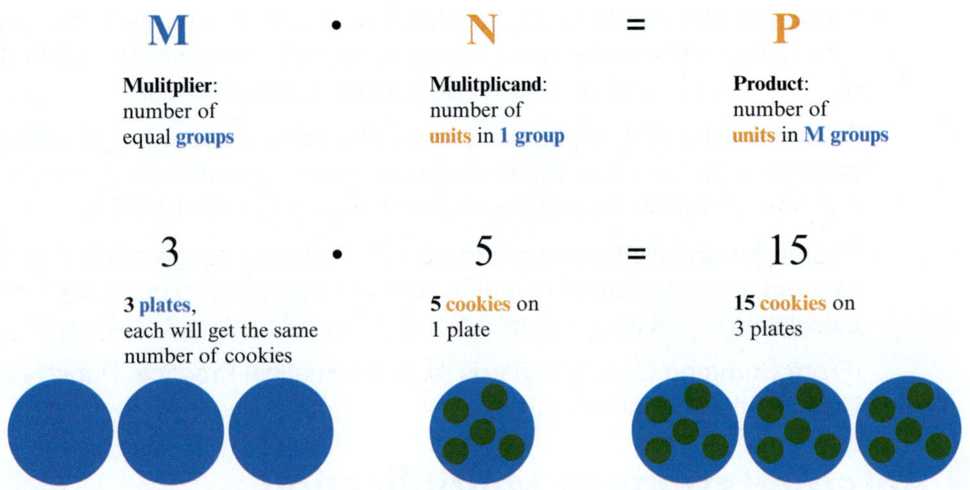

Figure 4.1 Multiplication is about coordinating numbers and sizes of equal groups.

Figure 4.2 Unlike multiplication, in addition (and subtraction) all the numbers must have the same referents.

How Can We Tell If a Problem Is Solved by Multiplication?

If you have a word problem, how can you tell if it is solved by multiplication instead of in some other way, such as by addition or division? This question is an important one for teachers, because students just guess at which operation to use to solve a problem if they don't understand what addition, subtraction, multiplication, and division mean. Multiplication applies to situations that involve equal groups. In some cases, the groups are evident. For example, Figure 4.3 clearly shows 3 equal groups with 4 dots in one group. So according to the definition of multiplication, there are $3 \cdot 4$ dots in all. Similarly, if there are 237 bags and each bag contains 46 potatoes, then the total number of potatoes is $237 \cdot 46$ (which turns out to be 10,902 potatoes). Whenever a collection of objects is arranged into M groups, and there are N objects in each group, then we know that, according to the definition of multiplication, there are $M \cdot N$ objects in all M groups.

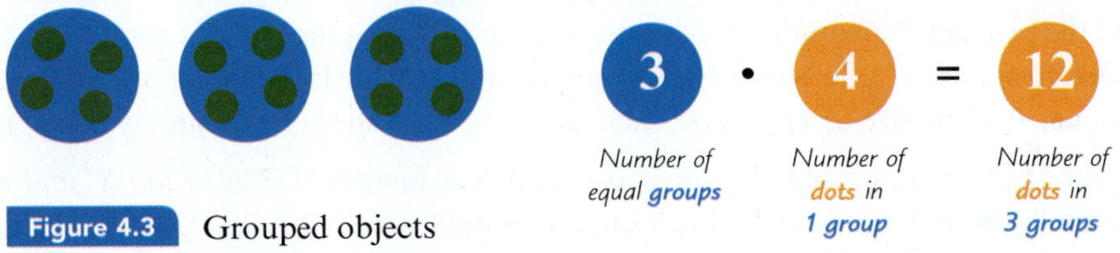

Figure 4.3 Grouped objects showing multiplicative structure.

Next, we will examine types of word problems that can be solved with multiplication. In some cases the groups in these word problems are not obvious, and we must first figure out what the groups are in order to determine that the problem can be solved by multiplication.

CLASS ACTIVITY

4A Showing Multiplicative Structure, p. CA-67

Array Problems In some problems, we want to determine the total number of objects that are in an *array*. A two-dimensional **array** is a rectangular arrangement of things into (horizontal) rows and (vertical) columns, such that each row has the same number of things and each column has the same number of things.

array

CCSS 3.OA.3 Column 1 of Table 2

Array problem: How many cans (viewed from above) are shown in **Figure 4.4**?

Figure 4.4 An array of cans viewed from above.

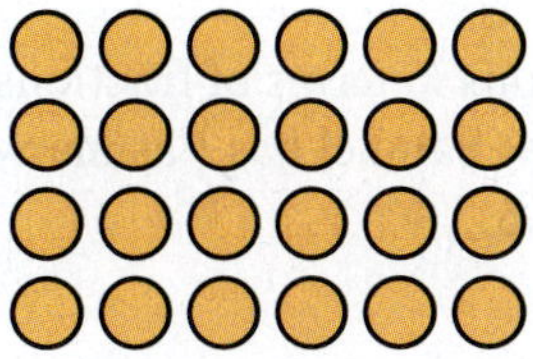

We can multiply to determine how many cans are depicted in the array in **Figure 4.4**. Why? View each *row* in the array as a group, as on the left in **Figure 4.5**. With this view, the array consists of 4 equal groups, and there are 6 cans in one group. Therefore, according to the definition of multiplication, the total number of cans in the array is $4 \cdot 6$.

We can also view each *column* in the array of cans as a group, as shown on the right in **Figure 4.5**. With this perspective, the array consists of 6 equal groups, and there are 4 cans in one group. Therefore, according to the definition of multiplication, the total number of cans in the array is $6 \cdot 4$ (which is the reverse of the previous $4 \cdot 6$).

Figure 4.5 Subdividing an array of cans into groups.

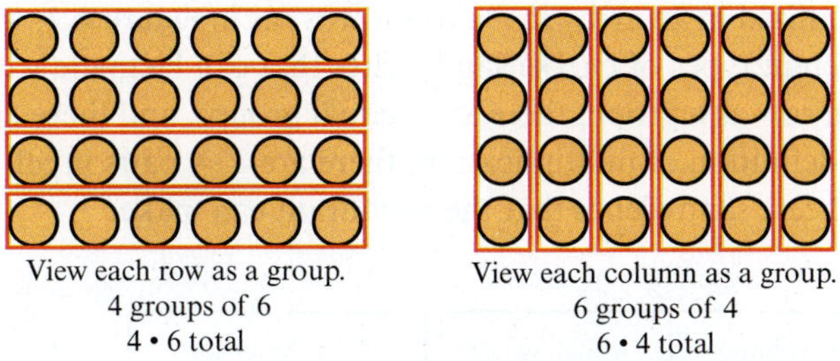

Arrays are especially important because they can be divided into natural groups in two different ways. The ability to see multiplication problems in two different ways is very helpful to students as they learn single-digit multiplication facts.

CCSS 4.OA.1 4.OA.2

Multiplicative Comparison Problems Some problems describe a multiplicative relationship between two quantities.

Multiplicative Comparison problem: Robert and Samuel each have a hermit crab collection. Robert has 3 times as many hermit crabs as Samuel. Samuel has 7 hermit crabs. How many hermit crabs does Robert have?

The wording of this problem indicates that it involves multiplication, and Figure 4.6 shows Robert's hermit crab collection as 3 equal groups, each of the same size as Samuel's collection of 7 hermit crabs. Therefore, according to the definition of multiplication, Robert has 3 · 7 hermit crabs.

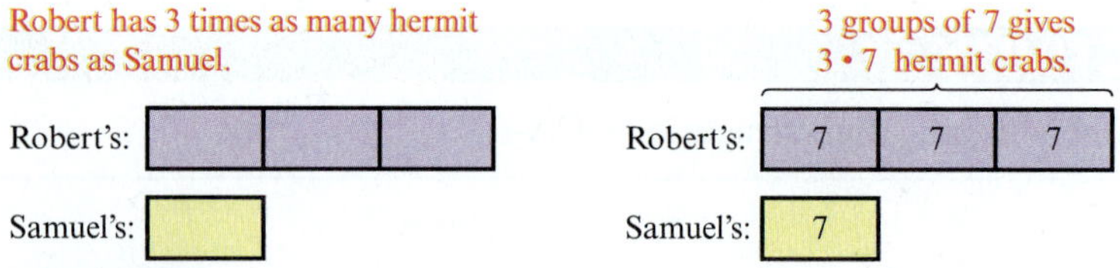

Figure 4.6 Using a strip diagram to show equal groups in a multiplicative comparison problem.

Strip diagrams can be especially helpful for multiplicative comparison problems because the "N times as many as" language of these problems is often difficult for students to grasp. By drawing a strip diagram, students think about size comparisons and the meaning of the phrase "N times as many as."

Notice that we can view multiplicative comparison situations in terms of fractions. In the problem about Robert's and Samuel's hermit crabs, we can say that Samuel has $\frac{1}{3}$ as many hermit crabs as Robert, which we can see from the strip diagram Figure 4.6.

Ordered Pair Problems In some problems, we want to determine how many ordered pairs of things can be made. A pair of things (two things) is an **ordered pair** if one of the things is designated as first and the other is designated second.

ordered pair

CCSS 7.RP.8b

Ordered Pair problem: A restaurant serves cheese sandwiches that are made from a piece of bread and a piece of cheese. There are 3 types of bread to choose from: wheat, white, and rye, and there are 4 types of cheese to choose from: cheddar, provolone, Swiss, and American. How many types of cheese sandwiches can the restaurant make with these choices?

We can multiply to determine how many types of cheese sandwiches the restaurant can make. Why? Each type of cheese sandwich can be considered an ordered pair consisting of a type of bread and a type of cheese. (Note that the type of bread was designated first and the type of cheese second, although it could just as well be the other way around as long as consistency is maintained.) These pairs are organized in an array in a natural way, as in Figure 4.7, where each row of the array shows all the different types of cheese sandwiches that can be made with one particular kind of bread. Viewing each row as a group, there are 3 equal groups, and there are 4 pairs in each group, so according to the definition of multiplication, there are 3 · 4 pairs in all. Therefore, there are 3 · 4 different types of cheese sandwiches that the restaurant can make.

(white, cheddar)	(white, provolone)	(white, Swiss)	(white, American)
(wheat, cheddar)	(wheat, provolone)	(wheat, Swiss)	(wheat, American)
(rye, cheddar)	(rye, provolone)	(rye, Swiss)	(rye, American)

Figure 4.7 Types of cheese sandwiches organized in an array.

Instead of organizing the types of cheese sandwiches into an array, they can also be organized into a tree diagram, as in Figure 4.8(a), or into a list, as in Figure 4.8(b).

tree diagram

A **tree diagram** is a diagram consisting of line segments, called branches, that connect pieces of information. To read a tree diagram, start at the far left and follow branches all the way across

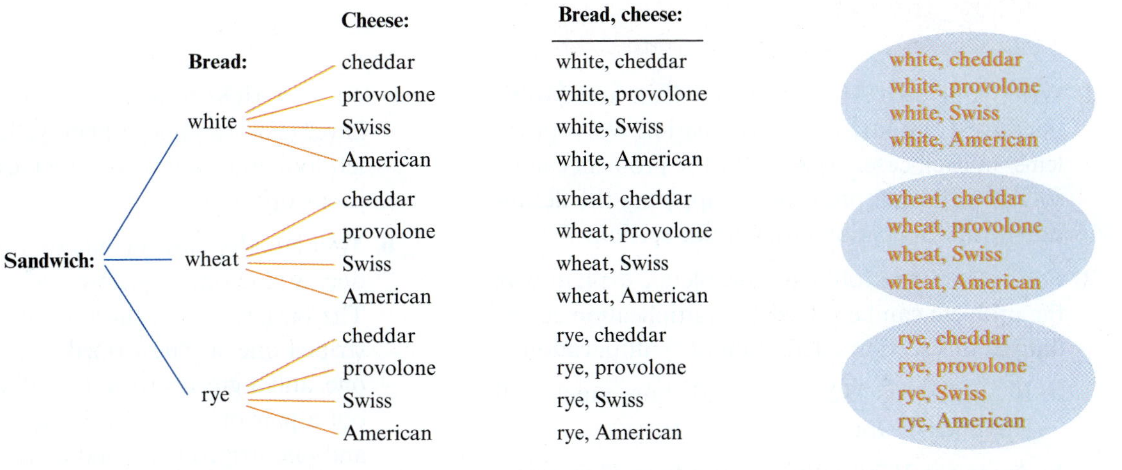

(a) Tree Diagram **(b) Organized List** **(c) 3 groups, 4 pairs in each group $3 \cdot 4$ total pairs**

Figure 4.8 Types of cheese sandwiches organized in (a) a tree diagram, (b) an organized list, and (c) equal groups.

to the right. The tree diagram and the list in Figure 4.8(a), (b) each have a structure consisting of 3 groups of 4, as indicated in Figure 4.8(c). Therefore, as before, we conclude that the total number of types of cheese sandwiches is $3 \cdot 4$.

Ordered Pair problems arise when finding probabilities of compound events (Grade 7 in the *Common Core State Standards*).

CLASS ACTIVITY

4B Writing Multiplication Word Problems, p. CA-69

In this section, we examined some types of word problems that can be solved with multiplication. Problems about areas of rectangles and volumes of boxes (rectangular prisms) are also important types of problems that can be solved with multiplication; we will study these types of problems in Section 4.3. But first, we study the special role that multiplication by 10 plays in the decimal system.

SECTION SUMMARY AND STUDY ITEMS

Section 4.1 Interpretations of Multiplication

For non-negative numbers, the product $M \cdot N$ stands for the number of units in M equal groups of size N units. We can therefore show multiplicative structure in a situation by exhibiting equal groups. We can exhibit equal groups with arrays, organized lists, and tree diagrams.

Key Skills and Understandings

1. Explain why multiplication solves a problem by exhibiting or describing equal groups. Write a corresponding multiplication expression ($M \cdot N$) or equation ($M \cdot N = P$).
2. Write Array, Ordered Pair, and Multiplicative Comparison word problems for a given multiplication problem (such as $4 \cdot 6 = ?$), as well as multiplication word problems for which the equal groups are more evident.

Practice Exercises for Section 4.1

1. Write your own Array, Ordered Pair, and Multiplicative Comparison multiplication word problems. In each case, explain why the problem can be solved with multiplication by applying the definition of multiplication given in this section.

2. For each of the following problems, explain why the problem can be solved by multiplication according to this section's definition of multiplication.

   a. If apples cost \$2 per pound, how much will 5 pounds of apples cost?

   b. There are 1000 milliliters in a liter. How many milliliters are in 4 liters?

   c. If you drive a steady 120 kilometers per hour for 2 hours, how far will you have driven?

3. Explain the difference among the next three problems and the way they are solved.

   a. Your laundry basket contains 4 plain socks: a red one, a blue one, a yellow one, and a green one. The basket also contains 4 striped socks: a red striped one, a blue striped one, a yellow striped one, and a green striped one. If you want to wear a plain sock on your left foot and a striped sock on your right foot, how many options do you have? (For example, a plain yellow sock on your left foot and a red striped sock on your right foot is one option.)

   b. Your laundry basket contains 4 plain socks: a red one, a blue one, a yellow one, and a green one. The basket also contains 4 striped socks: a red striped one, a blue striped one, a yellow striped one, and a green striped one. If you want to pick out a pair of socks consisting of one plain sock and one striped sock, and then wear that pair of socks how many options do you have? (For example, a red striped sock on your left foot and a plain yellow sock on your right foot is one option.)

   c. Your laundry basket contains 4 plain socks: a red one, a blue one, a yellow one, and a green one. The basket also contains 4 striped socks: a red striped one, a blue striped one, a yellow striped one, and a green striped one. If you reach into the laundry basket, pick out a sock, and put it on your left foot and then reach in again, pick out another sock, and put it on your right foot, how many different possible outcomes are there?

Answers to Practice Exercises for Section 4.1

1. See the text for examples of word problems and for explanations for why the problems can be solved with multiplication. Be sure to use correct language in your Multiplicative Comparison problems. Problems should include a phrase of the form "*M* times as many as," although the word *many* can be replaced with *much*, *long*, *tall*, *wide*, or similar words. Be sure also in your explanation for why your problem can be solved by multiplying $A \cdot B$, that you explain how to see *A* equal groups with *B* objects in one group.

2. **a.** If you view each pound as a group, then 5 pounds is 5 groups. Each group contains \$2 worth of apples. So the total cost of the apples is the total dollar value of 5 groups of \$2, which is $5 \cdot 2$ dollars, according to the definition of multiplication in this section.

   b. If you view each of the 4 liters as a group and each 1-liter group contains 1000 milliliters, then 4 liters can be viewed as 4 groups, each of which contains 1000 milliliters. So according to the definition of the multiplication in this section, the total number of milliliters in 4 liters is $4 \cdot 1000$.

   c. If you view each hour as a group, and during each 1-hour time period you travel 120 kilometers, then each 1-hour group "consists of" 120 kilometers. So during 2 hours, you travel a total of "2 groups of 120 kilometers," which is $2 \cdot 120$ kilometers.

3. The discussion here is only a brief outline. Problem 7 asks you to solve the three problems in detail.

   Part (a) is an example of an ordered pair problem. The information can be organized into an array, a tree diagram, or an ordered list, each of which has the structure of 4 groups of 4. So there are $4 \cdot 4 = 16$ options.

   To solve part (b), multiply the solution for part (a) by 2. So there are $2 \cdot 16 = 32$ options.

   For part (c), the possible outcomes can be organized into 8 groups of 7, so there are $8 \cdot 7$, or 56 possible outcomes.

PROBLEMS FOR SECTION 4.1

1. Use this section's definition of multiplication to explain why each of the following problems can be solved by multiplying:
   - **a.** There are 3 feet in a yard. If a rug is 5 yards long, how long is it in feet?
   - **b.** There are 5280 feet in a mile. How long in feet is a 4-mile-long stretch of road?
   - **c.** Will is driving 65 miles per hour. If he continues driving at that speed, how far will he drive in 3 hours?
2. Write an Array word problem for $6 \times 8 = ?$. Explain clearly why the problem can be solved by multiplying 6×8 by using the definition of multiplication in this section. (You may replace the numbers 6 and 8 with different numbers.)
3. Write an Ordered Pair word problem for $6 \cdot 8 = ?$. Explain clearly why the problem can be solved by multiplying $6 \cdot 8$ by using the definition of multiplication in this section.
4. Write a Multiplicative Comparison word problem for $6 \cdot 8 = ?$. Explain clearly why the problem can be solved by multiplying $6 \cdot 8$ by using the definition of multiplication in this section.
5. **a.** Write a Multiplicative Comparison word problem for $3 \cdot 5 = ?$.
   - **b.** Draw a strip diagram for your problem in part (a) and explain how this section's definition of multiplication applies to solve the problem.
   - **c.** Reword your problem in part (a) so that it is about the same situation but you use the fraction $\frac{1}{3}$ in the statement of the problem.
6. Write a Multiplicative Comparison problem in which you describe one quantity as 3 times as much as another quantity but which cannot be solved by multiplying the numbers given in the problem.
7. Solve the three problems that are given in the 3 parts of Practice Exercise 3. Explain the solutions in detail.
8. John, Trey, and Miles want to know how many two-letter secret codes there are that don't have a repeated letter. For example, they want to count BA and AB, but they don't want to count doubles such as ZZ or XX. John says there are $26 + 25$ because you don't want to use the same letter twice; that's why the second number is 25. Trey says he thinks it should be times, not plus: $26 \cdot 25$. Miles says the number is $26 \cdot 26 - 26$ because you need to take away the double letters. Discuss the boys' ideas. Which answers are correct, which are not, and why? Explain your answers clearly and thoroughly, drawing on this section's definition of multiplication.
9. **a.** A 40-member club will elect a president and then elect a vice president. How many possible outcomes are there?
   - **b.** A 40-member club will elect a pair of co-presidents. How many possible outcomes are there?
   - **c.** Are the answers to (a) and (b) the same or different? Explain why they are the same or why they are different.

4.2 Why Multiplying by 10 Is Special in Base Ten

CCSS Common Core State Standards Grade 4

Why is multiplying by 10, by 100, by 1000, and so on, special in the base-ten system? The answer lies in how the values of the places in the base-ten system are related.

CLASS ACTIVITY

4C Multiplying by 10, p. CA-70

Multiplying by 10, 100, 1000, and so on, is special because of the structure of place value in the base-ten system: In the base-ten system, the value of each place is 10 times the value of the place to its immediate right. Consider what happens when we multiply the number 34 by 10. The number 34 stands for 3 tens and 4 ones and can be represented by 3 bundles of 10 toothpicks and 4 individual toothpicks, as shown in **Figure 4.9**. Then $10 \cdot 34$ stands for the total number of toothpicks in 10 groups of 34 toothpicks. As **Figure 4.10** shows, when we form 10 groups of 34 toothpicks, each of the 3 original tens becomes bundled into 1 group of 100 and each of the 4 original individual toothpicks is bundled into 1 group of 10. Therefore, when we multiply 34 by 10, the 3 in the tens place moves one place over to the hundreds place and the 4 in the ones place moves one place over to the tens place. *Notice that this shifting occurs precisely because the value of the hundreds place is 10 times the value of the tens place and the value of the tens place is 10 times the value of the ones place.*

CCSS 4.NBT.1

Figure 4.9 The number 34 represented by 3 bundles of 10 and 4 individual toothpicks.

Figure 4.10 Ten groups of 34 are bundled into 3 hundreds and 4 tens, so the 3 tens become 3 hundreds and the 4 ones become 4 tens.

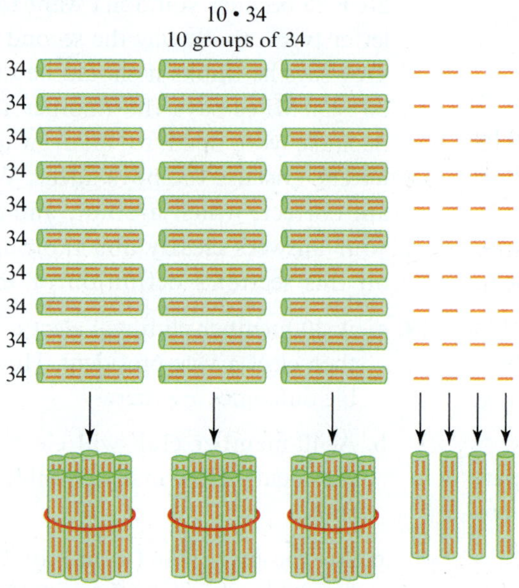

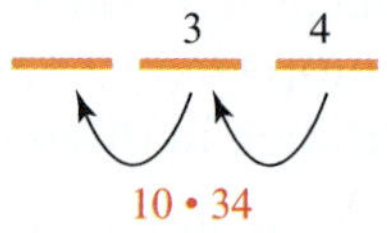

The situation is similar for other numbers.

We can think of

$$10 \cdot 0.034$$

as the total in 10 groups of 0.034. Each group of 0.034 consists of 3 hundredths and 4 thousandths, which can be represented with 4 small objects (each representing one thousandth) and 3 bundles of ten. In fact, if we let 1 toothpick in **Figure 4.10** stand for 1 thousandth, then **Figure 4.10** shows this:

The 4 thousandths move one place to the left to become 4 hundredths because the value of the hundredths place is 10 times the value of the thousandths place.

The 3 hundredths move one place to the left to become 3 tenths because the value of the tenths place is 10 times the value of the hundredths place.

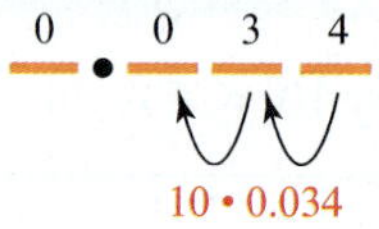

CCSS
5.NBT.1
5.NBT.2

In general, when we multiply a number by 10, all the digits move one place to the left because the value of each place is 10 times the value of the place to its right.

What about multiplying by 100, or 1000, or 10,000, and so on? Because

$$100 = 10 \cdot 10$$

multiplying by 100 has the same effect as multiplying by 10 twice. Therefore, multiplying by 100 moves each digit in the decimal representation of a number *two* places to the left. Similarly, because

$$1000 = 10 \cdot 10 \cdot 10$$

multiplying by 1000 has the same effect as multiplying by 10 three times. Therefore, multiplying by 1000 moves each digit in the base-ten representation of a number *three* places to the left.

SECTION SUMMARY AND STUDY ITEMS

Section 4.2 Why Multiplying by 10 Is Special in Base Ten

Because the value of each place in the base-ten system is 10 times the value of the place to its right, multiplying a number by 10 moves each digit one place to the left.

Key Skills and Understandings

1. Describe multiplication by 10 as moving digits one place to the left.
2. Use math drawings of bundled objects to explain why multiplying a number by 10 moves each digit one place to the left.
3. Explain that because $100 = 10 \cdot 10$, and $1000 = 10 \cdot 10 \cdot 10$, multiplying a number by 100 moves each digit 2 places to the left and multiplying a number by 1000 moves each digit 3 places to the left. Explain similar results for other base-ten place values.

Practice Exercises for Section 4.2

1. a. What are the limitations of the following two statements?

- To multiply a number by 10, put a 0 at the end of the number.
- To multiply a number by 10, move the decimal point one place to the right.

b. What is a better way to describe what multiplying by 10 does to the base-ten representation of a number than either of the statements in part (a)?

2. Using the example $10 \cdot 3.4$ to illustrate, explain why we move the digits in the base-ten representation of a number one place to the left when we multiply by 10.

Answers to Practice Exercises for Section 4.2

1. a. Although the first statement is valid for whole numbers, it is not correct for decimals. For example, $10 \cdot 2.8$ is not equal to 2.80. The second statement is correct, but you wouldn't want to use it with students who are only studying whole number multiplication.

b. A better way to describe what multiplying by 10 does to a number is to say that each digit moves one place to the left.

2. Using bundled toothpicks, we find that the number 3.4 can be represented as shown in Figure 4.9, as long as 1 toothpick represents $\frac{1}{10}$ and a bundle of 10 toothpicks represents 1. Now use the explanation that is given in the text and that accompanies Figure 4.10.

PROBLEMS FOR SECTION 4.2

1. Using the example $10 \cdot 47$ to illustrate, explain in your own words why we move the digits in a number 1 place to the left when we multiply by 10.

2. Mary says that $10 \cdot 3.7 = 3.70$. Why might Mary think this? Explain to Mary why her answer is not correct and why the correct answer is right. If you tell Mary a procedure, be sure to tell Mary why the procedure makes sense.

3. Now that you understand why multiplying a number by 10 shifts the digits 1 place to the left, explain how we can deduce that multiplying by 10,000 shifts the digits 4 places to the left and multiplying by 100,000 shifts the digits 5 places to the left. How should we think about the numbers 10,000 and 100,000 to make these deductions?

*4. **a.** Find the decimal representation of $\frac{1}{37}$ to at least 6 places (or as many as your calculator shows). Notice the repeating pattern.

b. Now find the decimal representations of $\frac{10}{37}$ and of $\frac{26}{37}$ to at least 6 places. Compare the repeating patterns with each other and to the decimal representation of $\frac{1}{37}$. What do you notice?

c. Write $10 \cdot \frac{1}{37} = \frac{10}{37}$ and $100 \cdot \frac{1}{37} = \frac{100}{37}$ as mixed numbers.

d. What happens to the decimal representation of a number when it is multiplied by 10? By 100? Use your answer, and part (c), to explain the relationships you noticed in part (b).

*5. **a.** Find the decimal representation of $\frac{1}{41}$ to at least 10 decimal places. Notice the repeating pattern.

b. Use your answer in part (a) to find the decimal representations of the numbers

$$10 \cdot \frac{1}{41}, \quad 100 \cdot \frac{1}{41}, \quad 1000 \cdot \frac{1}{41},$$
$$10{,}000 \cdot \frac{1}{41}, \quad 100{,}000 \cdot \frac{1}{41}$$

without a calculator.

c. Write the numbers

$$10 \cdot \frac{1}{41} = \frac{10}{41},$$
$$100 \cdot \frac{1}{41} = \frac{100}{41},$$
$$1000 \cdot \frac{1}{41} = \frac{1000}{41},$$
$$10{,}000 \cdot \frac{1}{41} = \frac{10{,}000}{41},$$
$$100{,}000 \cdot \frac{1}{41} = \frac{100{,}000}{41}$$

as mixed numbers.

d. Use your answers in part (b) to find the decimal representations of the fractional parts of the mixed numbers you found in part (c). Do not use your calculator or do long division; use part (b). Explain your reasoning.

4.3 The Commutative and Associative Properties of Multiplication, Areas of Rectangles, and Volumes of Boxes

CCSS Common Core State Standards Grades 3, 4, 5

In Chapter 3, we saw that the commutative and associative properties of addition are important building blocks for addition, which allow us to calculate sums flexibly. In this section we'll study the commutative and associative properties of multiplication. We'll see why these properties make sense and how they allow us to calculate products flexibly.

There also are geometric ways to think about the commutative and associative properties of multiplication. We can view the commutative property of multiplication in terms of areas of rectangles and we can view the associative property of multiplication in terms of volumes of boxes. In order to examine the commutative and associative properties of multiplication from this geometric perspective, we will first analyze why we can multiply to find the area of a rectangle and to find the volume of a box. Thus, we will identify the origins of the area formula for rectangles and the volume formula for boxes.

Finally, in the problems we will see how to apply the reasoning behind finding areas of rectangles and volumes of boxes to estimating numbers of things with multiplication.

Why Does the Commutative Property of Multiplication Make Sense?

commutative property of multiplication

The **commutative property of multiplication** says that for all numbers A and B

$$A \cdot B = B \cdot A$$

For example,

$$84 \cdot 2 = 2 \cdot 84$$

which says that $84 \cdot 2$ and $2 \cdot 84$ are the same amount.

We assume that this property always holds for *any* pair of numbers, but we can explain why this property makes sense for counting numbers.

If you think of the commutative property of multiplication in terms of groupings, it can seem somewhat mysterious: Why do 3 groups of 5 marbles have the same total number of marbles as 5 groups of 3 marbles? Why do 297 baskets of potatoes with 43 potatoes in each basket contain the same total number of potatoes as 43 baskets of potatoes with 297 potatoes in each basket? Of course, we can calculate $3 \cdot 5$ and $5 \cdot 3$ and see that they are equal, and we can calculate $297 \cdot 43$ and $43 \cdot 297$ and see that they are equal, but by simply calculating, we don't see why the commutative property should hold in other situations as well. A conceptual explanation that will show why this property makes sense will help explain its meaning. We can give a conceptual explanation by working with arrays or with areas of rectangles.

Suppose you have 3 groups with 5 marbles in each group. According to the definition of multiplication, this is a total of

$$3 \cdot 5$$

marbles. You can arrange the 3 groups of marbles so that they form the 3 rows of an array, as shown on the left in **Figure 4.11**. But if you now choose the *columns* of the array to be the groups, as on the right in **Figure 4.11**, then there are 5 groups with 3 marbles in each group. Therefore, according to the definition of multiplication, there is a total of

CCSS 3.OA.5

$$5 \cdot 3$$

marbles. But the total number of marbles is the same, either way you count them. Therefore,

$$3 \cdot 5 = 5 \cdot 3$$

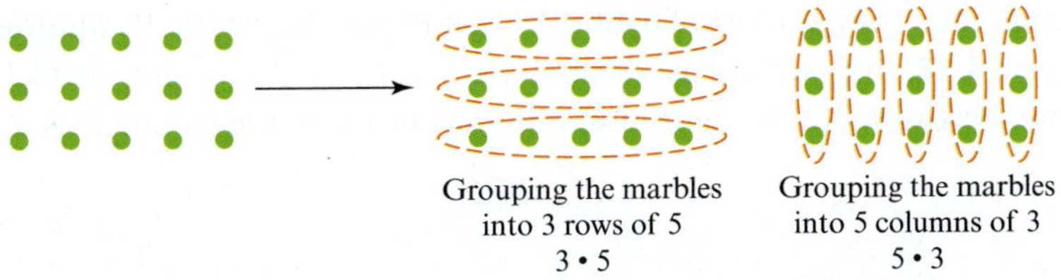

Figure 4.11 Grouping marbles to show that $3 \cdot 5 = 5 \cdot 3$.

By imagining drawings like **Figure 4.11**, we can see why the commutative property of multiplication should hold for counting numbers other than 3 and 5. If A and B are any counting numbers, there will be an A by B array similar to **Figure 4.11** illustrating that

$$A \cdot B = B \cdot A$$

We can also think about the commutative property of multiplication geometrically, in terms of areas of rectangles. In order to do so, let's first examine why we can multiply to find areas of rectangles.

CCSS 3.MD.5

Why Can We Multiply to Find Areas of Rectangles?

You probably remember learning the length times width, $L \cdot W$, formula for areas of rectangles. Why is this formula valid? We can explain this formula by using the definition of multiplication. But first, what is area?

square inch **square foot** **square unit** **area**

Area is usually measured in square units. Depending on what we are describing—the page of a book, the floor of a room, a football field, a parcel of land—we usually measure area in square inches, square feet, square yards, square miles, square centimeters, square meters, or square kilometers. A **square inch**, often written 1 in^2, is the area of a square that is 1 inch wide and 1 inch long, as is the square in Figure 4.12. Similarly, a **square foot**, often written 1 ft^2, is the area of a square that is 1 foot wide and 1 foot long. In general, for any unit of length, a **square unit** is the area of a square that is 1 unit wide and 1 unit long. The **area** of a region, in square units, is the number of 1-unit-by-1-unit squares it takes to cover the region without gaps or overlaps (where squares may be cut apart if necessary).

Figure 4.12

Area in square inches.

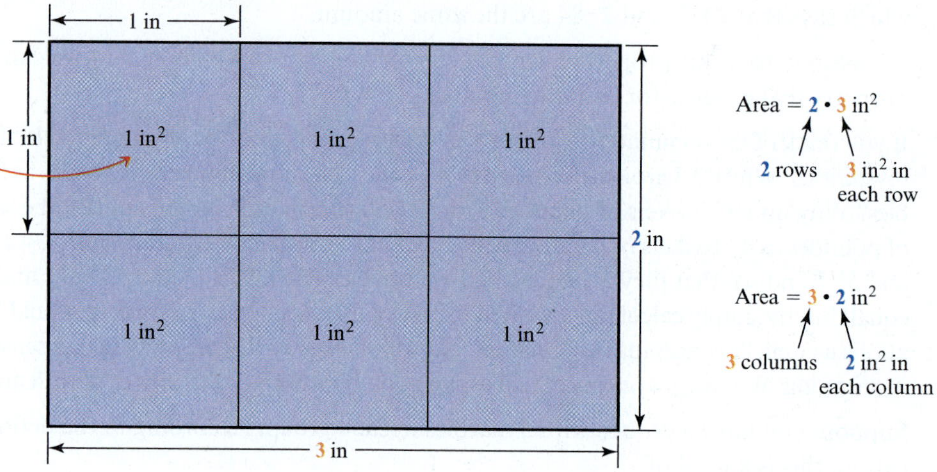

CCSS 3.MD.7a

The rectangle in Figure 4.12 is 3 inches wide and 2 inches long (or high). Why can its area be calculated by multiplying its length times its width? When we subdivide the rectangle into 1-inch-by-1-inch squares, we have 2 rows of these squares, and each row has 3 squares. In other words, the rectangle is made up of 2 groups with 3 squares in each group. Therefore, according to the definition of multiplication, the rectangle is made up of $2 \cdot 3$, or 6 squares. Because each 1-inch-by-1-inch square has area 1 square inch, the area of the whole rectangle is $2 \cdot 3$ square inches. We can use the same reasoning to determine areas of other rectangles.

In general, if L and W are any whole numbers, then a rectangle that is L units long and W units wide can be subdivided into L rows of 1-unit-by-1-unit squares, with W squares in each row, as indicated in Figure 4.13. In other words, the rectangle consists of L groups, with W squares in each group. Therefore, according to the definition of multiplication, the rectangle is made of

$$L \cdot W$$

Figure 4.13

A rectangle that is L units long and W units wide.

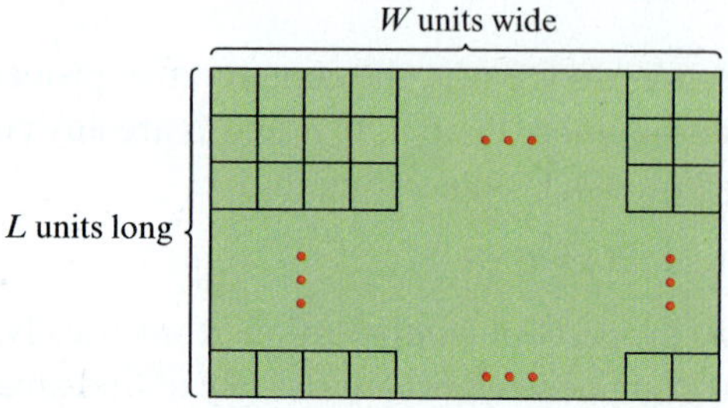

1-unit-by-1-unit squares. Because each 1-unit-by-1-unit square has area 1 square unit, the entire L-unit-by-W-unit rectangle has area

$$L \cdot W \text{ square units}$$

This explanation for why the length-times-width formula is valid applies not only when the length and width of the rectangle are whole numbers, but even when the length and width are fractions or decimals. So, a rectangle that is L units by W units has area $L \cdot W$ square units regardless of what kinds of (positive) numbers L and W are.

How Can We Use Area to Explain the Commutative Property of Multiplication?

Returning to the commutative property of multiplication, we can see why this property makes sense by using areas of rectangles. For example, a rug that is 3 feet by 5 feet can be thought of as being made of 3 rows with 5 squares, each of area 1 square foot, in each row. Therefore, the area of the rug is

$$3 \cdot 5 \text{ square feet}$$

On the other hand, the rug can be thought of as made of 5 columns with 3 squares, each of area 1 square foot, in each column, as shown in Figure 4.14. Therefore, the area of the rug is

$$5 \cdot 3 \text{ square feet}$$

But the area is the same either way you calculate it; therefore,

$$3 \cdot 5 = 5 \cdot 3$$

As before, the same line of reasoning will work when other counting numbers replace 3 and 5.

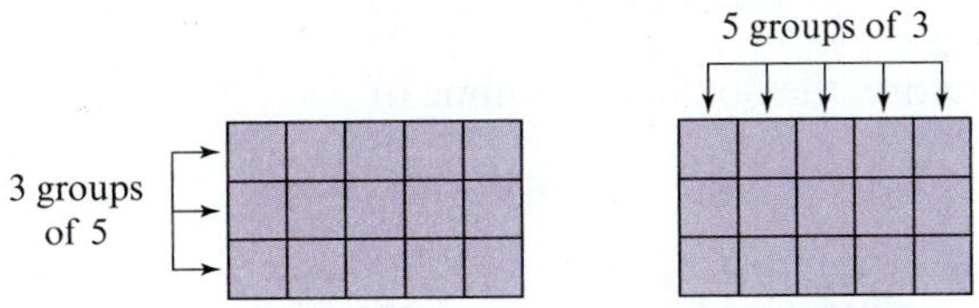

Figure 4.14
Using the area of a rectangle to show that $3 \cdot 5 = 5 \cdot 3$.

CLASS ACTIVITY

4D Explaining the Commutative Property of Multiplication with Arrays and Area, p. CA-71

Why Can We Multiply to Find Volumes of Boxes?

CCSS 5.MD.3

Not only are *areas* naturally related to multiplication, but *volumes* are as well, as we'll see next.

But first, what is volume? Depending on what you want to measure—a dose of liquid medicine, the size of a compost pile, the volume of coal in a mountain—volume can be measured in cubic inches, cubic feet, cubic yards, cubic miles, cubic centimeters, cubic meters, or cubic kilometers. (Volumes of liquids are also commonly measured in liters, milliliters, gallons, quarts, cups, and fluid ounces.)

One cubic centimeter, often written 1 cm^3, is the volume of a cube that is 1 centimeter high, 1 centimeter long, and 1 centimeter wide. A drawing of such a cube is shown in Figure 4.15, along with a cube of volume 1 cubic inch, 1 $inch^3$. One cubic yard, often written 1 yd^3, is the volume of a cube that is 1 yard high, 1 yard long, and 1 yard wide. In general, for any unit of length, 1 **cubic unit**, often written 1 $unit^3$, is the volume of a cube that is 1 unit high, 1 unit long, and 1 unit wide.

cubic unit

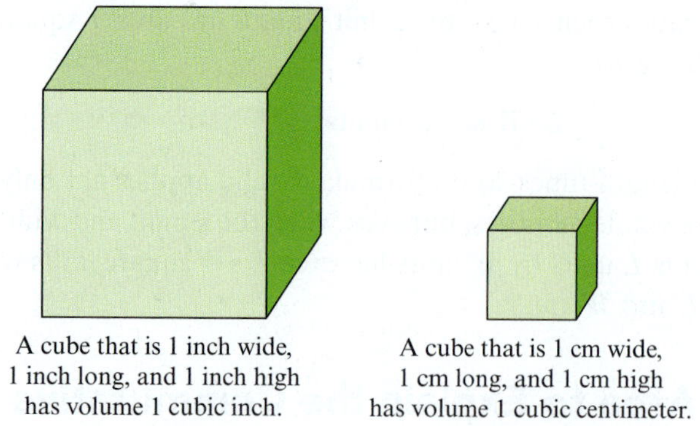

Figure 4.15 Cubes of volume 1 $inch^3$ and 1 cm^3.

We will be working with volumes of boxes and box shapes. (These shapes are also called rectangular prisms.) The **volume**, in cubic units, of a box or box shape is just the number of 1-unit-by-1-unit-by-1-unit cubes that it would take to fill the box or make the box shape (without gaps or overlaps). You may remember a formula for the volume of a box, but if so, assume for a moment that you don't know this formula. We will see how to derive the formula for the volume of a box from the definition of multiplication.

volume

CCSS 5.MD.5a

Suppose you have a box that is 4 inches high, 2 inches long, and 3 inches wide, as pictured in Figure 4.16. What is the volume of this box? If you have a set of building blocks that are all 1 inch high, 1 inch long, and 1 inch wide, then you can use the building blocks to build the box. The number of blocks needed is the volume of the box in cubic inches. We can use multiplication to describe the number of blocks needed by considering the box to be made of 4 horizontal layers, as shown on the right in Figure 4.16. Each layer consists of 2 rows of 3 blocks; so, according to the definition of multiplication, each layer contains $2 \cdot 3$ blocks. There are 4 layers, each containing $2 \cdot 3$ blocks. So, according to the definition of multiplication, there are

$$4 \cdot (2 \cdot 3)$$

blocks making up the box. Therefore, the box has a volume of

$$4 \cdot (2 \cdot 3) = 4 \cdot 6 = 24 \text{ cubic inches}$$

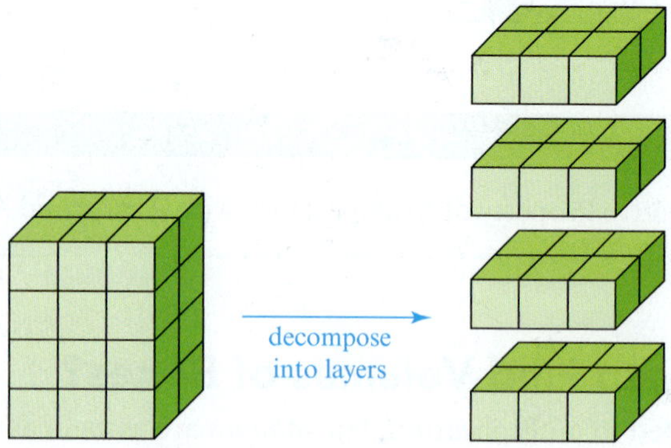

Figure 4.16 A 4-inch-high, 2-inch-long, 3-inch-wide box (scale drawing).

The reasoning used in this example applies generally. Suppose you have a box that is H units high, L units long, and W units wide. If H, L, and W are whole numbers, then as before, you can build such a box out of 1-unit-by-1-unit-by-1-unit cubes. How many cubes does it take? If you consider the box to be made of horizontal layers, then there are H layers. Each layer is made up of L rows of W blocks (or W rows of L blocks) and therefore contains $L \cdot W$ blocks, according to the definition of multiplication. There are H layers with $L \cdot W$ blocks in each layer. Therefore, according to the definition of multiplication, the box is made out of

$$H \cdot (L \cdot W)$$

1-unit-by-1-unit-by-1-unit cubes. Consequently, the box has volume

$$H \cdot (L \cdot W) \text{ cubic units}$$

Notice that the order in which the letters H, L, and W appear and the way the parentheses are placed in the expression $H \cdot (L \cdot W)$ corresponds to the way the box was divided into groups.

As with areas, it turns out that this height-times-length-times-width formula for the volume of a box remains valid even when the height, length, and width of the box are not whole numbers. The volume of a box that is H units high, L units long, and W units wide is always $H \cdot (L \cdot W)$ cubic units.

Next, we examine the use of parentheses in expressions like $H \cdot (L \cdot W)$.

Why Does the Associative Property of Multiplication Make Sense?

Recall that the associative property of addition says that, for all real numbers A, B, and C,

$$(A + B) + C = A + (B + C)$$

As we have seen, the associative property of addition gives us flexibility in calculating sums. Likewise, the associative property of multiplication will give us flexibility in calculating products.

associative property of multiplication

The **associative property of multiplication** says that for all numbers A, B, and C,

$$(A \cdot B) \cdot C = A \cdot (B \cdot C)$$

We assume that this property holds for all numbers, but as we'll see, we can explain why it makes sense for counting numbers.

Why is the associative property valid? Why is $A \cdot (B \cdot C)$ always the same amount as $(A \cdot B) \cdot C$?

When we have a specific example, we can check that both amounts turn out to be the same. For example:

$$(4 \cdot 2) \cdot 3 = 8 \cdot 3 = 24$$
$$4 \cdot (2 \cdot 3) = 4 \cdot 6 = 24$$

Therefore,

$$(4 \cdot 2) \cdot 3 = 4 \cdot (2 \cdot 3)$$

But what if we used different numbers? Why *must* the amounts *always* come out the same? To explain this, we will develop a conceptual explanation for why

$$(4 \cdot 2) \cdot 3 = 4 \cdot (2 \cdot 3)$$

This explanation will be general in the sense that it will also explain why the equation is true if we were to replace the numbers 4, 2, and 3 with other counting numbers. Our explanation is based on calculating the volume of a box in two different ways.

Previously, we decomposed a 4-inch-high, 2-inch-long, 3-inch-wide box shape into 4 groups of blocks, with $2 \cdot 3$ blocks in each group, as shown on the left of Figure 4.17. When we view it that way, we see that the total number of blocks in the box shape is

$$4 \cdot (2 \cdot 3)$$

On the other hand, the box shape can be decomposed into $4 \cdot 2$ groups, with 3 blocks in each group, as shown on the right of Figure 4.17. According to the definition of multiplication, there are

$$(4 \cdot 2) \cdot 3$$

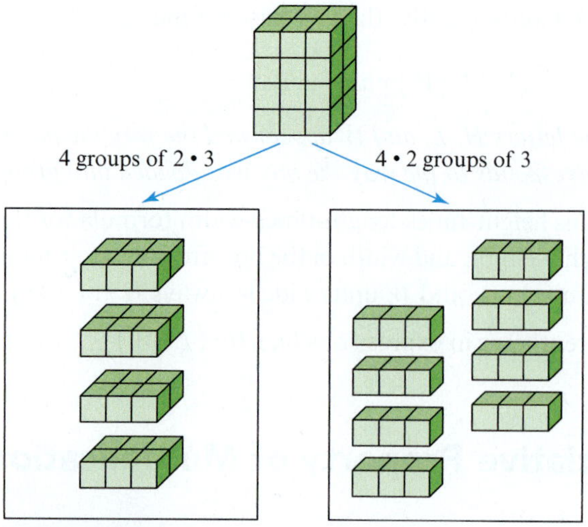

Figure 4.17 Showing why $4 \cdot (2 \cdot 3) = (4 \cdot 2) \cdot 3$.

blocks in the box shape. There are $4 \cdot 2$ groups because the groups of 3 blocks can themselves be arranged into 4 groups of 2. But the total number of blocks in the box shape is the same either way we count them. Therefore,

$$(4 \cdot 2) \cdot 3 = 4 \cdot (2 \cdot 3)$$

which is what we wanted to establish. Notice that the same argument works when the numbers 4, 2, and 3 are replaced with other counting numbers—only the size of the box will change. This reasoning explains why the associative property of multiplication makes sense for all counting numbers.

CLASS ACTIVITY

4E Describing the Volume of a Box with Multiplication and Explaining the Associative Property, p. CA-72

CCSS
SMP2
SMP7

How Do the Associative and Commutative Properties of Multiplication Help Us Calculate Flexibly?

How do we apply the commutative property of multiplication? The commutative property gives us flexibility by allowing us to calculate in different ways. For example, if there are 15 packages of soap and 2 bars of soap in each package, how many bars of soap are there? One way to determine the total number of bars of soap is to count by twos 15 times.

$$2, 4, 6, 8, 10, 12, 14, 16, 18, 20, 22, 24, 26, 28, 30$$

but a quicker way is to calculate

$$15 + 15 = 30$$

Why can we determine the answer either way? Because of the commutative property of multiplication! The first method determines what 15 twos make, in other words, the first method calculates

$$15 \cdot 2$$

The second method calculates what two 15s make, in other words, the second method calculates

$$2 \cdot 15$$

The commutative property of multiplication tells us that we will get the answer either way.

CCSS 3.NBT.3

How do we apply the associative property of multiplication? One important way is by breaking a number apart and reassociating one of the parts. For example, how can we multiply $7 \cdot 80$ mentally? We can first recall the single-digit multiplication fact $7 \cdot 8 = 56$ and then multiply that result by 10 to get the answer, 560. The next equations correspond to this strategy:

$$\begin{aligned} 7 \cdot 80 &= 7 \cdot (8 \cdot 10) \\ &= (7 \cdot 8) \cdot 10 \\ &= 56 \cdot 10 = 560 \end{aligned}$$

The equations show that we broke 80 apart into $8 \cdot 10$ and then applied the associative property to reassociate the 8 with the 7 instead of with the 10 at the second equal sign.

We often apply the associative and commutative properties together. For example, how can we make

$$5 \cdot 41 \cdot 2$$

easy to calculate mentally? Since 10 is easy to multiply with, we could multiply the 5 and the 2 first to make 10, then multiply 10 and 41 to make 410. Teachers sometimes like to record this method this way:

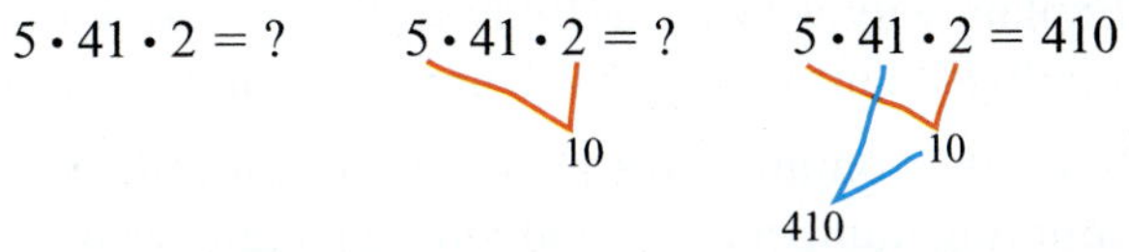

IMAP *Watch a teacher facilitate a child's thinking while he attempts to explain why he thinks that 19,200 tens is another way to say 192,000.*

Notice that even though we may not see it right away, the teacher's method of recording how to calculate $5 \cdot 41 \cdot 2$ actually does involve the commutative and associative properties, as we see from the next equations:

$$\begin{aligned} 5 \cdot 41 \cdot 2 &= 41 \cdot 5 \cdot 2 \\ &= 41 \cdot 10 \\ &= 410 \end{aligned}$$

The commutative property was used at the first equal sign and the associative property was used at the second equal sign—to associate the 5 with the 2 instead of with the 41—as well as to simply drop the parentheses altogether.

CLASS ACTIVITY

4F How Can We Use the Associative and Commutative Properties of Multiplication? p. CA-74

SECTION SUMMARY AND STUDY ITEMS

Section 4.3 The Commutative and Associative Properties of Multiplication, Areas of Rectangles, and Volumes of Boxes

The commutative property of multiplication states that

$$A \cdot B = B \cdot A$$

for all numbers A and B.

The area of a rectangle, in square units, is the number of 1-unit-by-1-unit squares it takes to cover the rectangle without gaps or overlaps. We can multiply $L \cdot W$ to find the area of an L-by-W rectangle because the rectangle can be subdivided into L rows, each of which contains W 1-unit-by-1-unit squares.

We can explain why the commutative property of multiplication makes sense by subdividing an array into groups in two different ways.

The associative property of multiplication states that

$$(A \cdot B) \cdot C = A \cdot (B \cdot C)$$

for all numbers A, B, and C.

The volume of a box, in cubic units, is the number of 1-unit-by-1-unit-by-1-unit cubes it takes to fill or make the box (without gaps or overlaps). We can multiply $H \cdot (L \cdot W)$ to find the volume of a box that is H units high, L units long, and W units wide because the box can be subdivided into H layers, each of which contains L rows of W 1-unit-by-1-unit-by-1-unit cubes.

We can explain why the associative property of multiplication makes sense by subdividing groups of grouped objects in two different ways.

We use the combination of the commutative and associative properties in many mental calculations, such as in calculating $60 \cdot 700$ by recalling that $6 \cdot 7 = 42$ and moving 42 three places to the left to make 42,000.

Key Skills and Understandings

1. Explain why we can multiply to find the area of a rectangle by describing rectangles as subdivided into groups of 1-unit-by-1-unit squares.
2. State the commutative property of multiplication, and explain why it makes sense (for counting numbers) by subdividing rectangles or arrays in two different ways.
3. Explain why we can multiply to find the volume of a box by describing boxes as subdivided into groups of groups of 1-unit-by-1-unit-by-1-unit cubes.
4. State the associative property of multiplication, and explain why it makes sense (for counting numbers) by subdividing boxes two different ways or by subdividing groups of groups of objects in two different ways.
5. Give examples of how to use the associative and commutative properties of multiplication in problems and recognize when these properties have been used.
6. Apply the methods used in explaining the area formula for rectangles and the volume formula for boxes to estimate numbers of items by multiplying.

Practice Exercises for Section 4.3

1. State the commutative property of multiplication.
2. Use our definition of multiplication and a math drawing to give a conceptual explanation for why $2 \cdot 4 = 4 \cdot 2$. Your explanation should be general enough to remain valid if other counting numbers were to replace 2 and 4.
3. Give an example to show how to apply the commutative property of multiplication to make a math problem easier to solve mentally.
4. Use the definition of multiplication to explain why the area of a carpet that is 20 feet long and 12 feet wide is $20 \cdot 12$ square feet.
5. Use the definition of multiplication to explain why a box that is 3 feet wide, 2 feet long, and 4 feet high has volume $4 \cdot (2 \cdot 3)$ cubic feet.
6. One cubic foot of water weighs about 62 pounds. How much will the water weigh in a swimming pool that is 20 feet wide, 30 feet long, and 4 feet deep?
7. State the associative property of multiplication.
8. Give an example of how to use the associative property of multiplication to make a calculation easy to do mentally.
9. Explain how you use the associative property of multiplication when you calculate $7 \cdot 600$ mentally.
10. Use the definition of multiplication and the idea of decomposing a box shape in two different ways to explain why $4 \cdot (2 \cdot 3) = (4 \cdot 2) \cdot 3$.
11. Describe how to use the design in Figure 4.18 to explain why

$$5 \cdot (2 \cdot 2) = (5 \cdot 2) \cdot 2$$

Your explanation should be general, so that it explains why the previous equation is true when you replace the numbers 5, 2, and 2 with other counting numbers, and the design in Figure 4.18 is changed accordingly.

Figure 4.18 A design of spirals.

Answers to Practice Exercises for Section 4.3

1. $A \cdot B = B \cdot A$ for all numbers A, B.

2. See the array in Figure 4.19. It shows that the total number of objects in 2 groups with 4 objects in each group ($2 \cdot 4$ objects) can also be thought of as made out of 4 groups with 2 objects in each group ($4 \cdot 2$ objects). But you have the same number of objects either way you count them; therefore,

$$2 \cdot 4 = 4 \cdot 2$$

Figure 4.19 Using an array to show $2 \cdot 4 = 4 \cdot 2$.

3. If sponges come 2 to a package and if you have 13 packages of sponges, then the total number of sponges you have is $13 \cdot 2$. To calculate $13 \cdot 2$, you can instead calculate $2 \cdot 13$, according to the commutative property of multiplication. Two groups of 13 is just $13 + 13$, which is easy to calculate mentally as 26.

4. Think of a carpet that is 20 feet long and 12 feet wide as made up of 20 rows of squares with 12 squares in each row, each square being 1 foot wide and 1 foot long. In other words, the carpet consists of 20 groups, with 12 squares in each group. Therefore, according to the definition of multiplication, the carpet is made of $20 \cdot 12$ squares. Each square has area 1 square foot, so the area of the carpet is $20 \cdot 12$ square feet.

5. See the explanation associated with Figure 4.16 in the text. Substitute "feet" where the text says "inches."

6. Because the water in the pool is in the shape of a box that is 4 feet high, 20 feet wide, and 30 feet long, it can be thought of as

$$4 \cdot (30 \cdot 20) = 2400$$

1-foot-by-1-foot-by-1-foot cubes of water. Each of those cubes of water weighs 62 pounds, so the water in the pool will weigh

$$2400 \cdot 62 \text{ pounds} = 148{,}800 \text{ pounds}$$

7. $(A \cdot B) \cdot C = A \cdot (B \cdot C)$ for all numbers A, B, C.

8. See the next practice exercise. See also problems 10, 11, 13, and 14 in the Problems section.

9. When we calculate $7 \cdot 600$ mentally, we first calculate $7 \cdot 6$ (a "basic fact") and then we multiply by 100. When we do this, we take the 6 from the 600 and group it with the 7 instead of with 100. In equation form, we can express this re-association as follows:

$$\begin{aligned} 7 \cdot 600 &= 7 \cdot (6 \cdot 100) \\ &= (7 \cdot 6) \cdot 100 \\ &= 42 \cdot 100 \\ &= 4200 \end{aligned}$$

The associative property of multiplication is used at the second equal sign.

10. See Figure 4.17 and the discussion in the text.

11. See Figure 4.20. On the one hand, we can think of the design as being made up of 5 clusters of spirals, with 2 groups of 2 spirals in each cluster. In this arrangement there are $5 \cdot (2 \cdot 2)$ spirals in the design. On the other hand, we can think of the design as made up of $5 \cdot 2$ groups of spirals with 2 spirals in each group. There are $5 \cdot 2$ groups because there are 5 sets of 2 groups. This means there are $(5 \cdot 2) \cdot 2$ spirals in the design. There are the same number of spirals, no matter how they are counted. Therefore,

$$5 \cdot (2 \cdot 2) = (5 \cdot 2) \cdot 2$$

Figure 4.20 Using grouping to explain why $5 \cdot (2 \cdot 2) = (5 \cdot 2) \cdot 2$.

PROBLEMS FOR SECTION 4.3

1. There are 31 envelopes with 3 stickers in each envelope. Seyong calculates the total number of stickers by counting by threes 31 times. Natasha adds 31 + 31 + 31 = 93 instead. Discuss the two calculation methods. Are both legitimate? Explain and relate to the material in this section.

2. Here is Amy's explanation for why the commutative property of multiplication is true for counting numbers:

   Whenever I take two counting numbers and multiply them, I always get the same answer as when I multiply them in the reverse order. For example,

   $$6 \cdot 8 = 48$$
   $$8 \cdot 6 = 48$$
   $$9 \cdot 12 = 108$$
   $$12 \cdot 9 = 108$$
   $$3 \cdot 15 = 45$$
   $$15 \cdot 3 = 45$$

   It always works that way; no matter which numbers I multiply, I will get the same answer either way I multiply them.

   Discuss why Amy's explanation is not complete. Then explain why the commutative property of multiplication is valid by viewing arrays in two ways.

3. Using the definition of multiplication, explain why it makes sense to multiply $4 \cdot 6$ to determine the area of a 4-foot-by-6-foot rug in square feet.

4. Use the definition of multiplication to explain why a box that is 5 inches high, 4 inches wide, and 3 inches long has a volume of

   $$5 \cdot (3 \cdot 4) \text{ cubic inches}$$

   Explain the parentheses in the expression $5 \cdot (3 \cdot 4)$.

5. Figure A in Figure 4.21 shows a 5-unit-high, 3-unit-wide, and 2-unit-long box made out of blocks. Figures B through G of Figure 4.21 show different ways of subdividing the box into natural groups of blocks. For each of these ways of subdividing the box, use the definition of multiplication as we have described it to write an expression for the total number of blocks in the box. Each expression should involve the numbers 5, 3, and 2, the multiplication symbol, and parentheses. Explain why you write your expressions as you do.

Figure 4.21 Different ways to subdivide a box.

6. Write three different expressions for the total number of curlicues in Figure 4.22. Each expression should involve only the following: the numbers 2, 3, 5, and 8; the multiplication symbol $\times$ or $\cdot$; and parentheses. For each expression, use the definition of multiplication as we have described it to explain why your expression represents the total number of curlicues in Figure 4.22.

Figure 4.22 How many curlicues?

7. Suppose you have 60 pennies arranged into 12 stacks with 5 pennies in each stack. The 12 stacks are arranged into 3 rows with 4 stacks in each row, as shown in Figure 4.23. For each of the following expressions, explain how to see the pennies grouped that way. Your explanation of the grouping should not involve rearranging the pennies.

a. $3 \cdot (4 \cdot 5)$

b. $(3 \cdot 4) \cdot 5$

c. $5 \cdot (3 \cdot 4)$

d. $4 \cdot (3 \cdot 5)$

Figure 4.23 Stacks of pennies.

8. To calculate $3 \cdot 80$ mentally, we can just calculate $3 \cdot 8 = 24$ and then put a zero on the end to get the answer, 240. Use Figure 4.24 to help you explain why this method of calculation is valid.

Figure 4.24 Explaining a mental method for calculating $3 \cdot 80$.

9. Write equations to show how the commutative and associative properties of multiplication are involved when you calculate $40 \cdot 800$ mentally by relying on basic multiplication facts (such as $4 \cdot 8$). Write your equations in the form

$$\begin{aligned} 40 \cdot 800 &= \text{some expression} \\ &= \vdots \\ &= \text{some expression} \end{aligned}$$

Indicate specifically where the commutative and associative properties of multiplication are used.

10. Explain how to use the associative property of multiplication to make $16 \cdot 25$ easy to calculate mentally. Write equations that show why your method is valid, and show specifically where you have used the associative property of multiplication. Write your equations in the form

$$\begin{aligned} 16 \cdot 25 &= \text{some expression} \\ &= \vdots \\ &= \text{some expression} \end{aligned}$$

11. Use the associative property of multiplication to make the problem $32 \cdot 0.25$ easy to calculate mentally. Write equations to show your use of the associative property of multiplication. Explain how your solution method is related to solving $32 \cdot 0.25$ by thinking in terms of money.

12. Explain how to make the following product easy to calculate mentally (there are five 2s and five 5s in the product):

$$2 \cdot 2 \cdot 2 \cdot 2 \cdot 2 \cdot 5 \cdot 5 \cdot 5 \cdot 5 \cdot 5$$

13. Julia says that it's easy to multiply a number by 4 because you just "double the double." Explain Julia's idea, and explain why it uses the associative property of multiplication.

14. Carmen says that it's easy to multiply even numbers by 5 because you just take half of the number and put a zero on the end. Write equations that incorporate Carmen's method and that demonstrate why her method is valid. Use the case $5 \cdot 22$ for the sake of concreteness. Write your equations in the following form:

$$\begin{aligned} 5 \cdot 22 &= \text{some expression} \\ &= \text{some expression} \\ &\vdots \\ &= 110 \end{aligned}$$

15. The Browns need new carpet for a room with a rectangular floor that is 35 feet wide and 43 feet long. To save money, they will install the carpet themselves. The carpet comes on a large roll that is 9 feet wide. The carpet store will cut any length of carpet they like, but the Browns must buy the full 9 feet in width.

a. Draw clear, detailed pictures showing two different ways the Browns could lay their carpet.

b. For each way of laying the carpet, find how much carpet the Browns will need to buy from the carpet store. Which way is less expensive for the Browns?

16. If a roll of a certain kind of wrapping paper is unrolled, the wrapping paper forms a rectangle that is 3 feet wide and 20 feet long. The wrapping paper is to be covered with an array of ladybugs by repeating the design shown in Figure 4.25. (The arrows and the "3 inches," "2 inches" are not part of the design; they show the dimensions of a portion of the design.) How many ladybugs will be on the wrapping paper? Explain your reasoning.

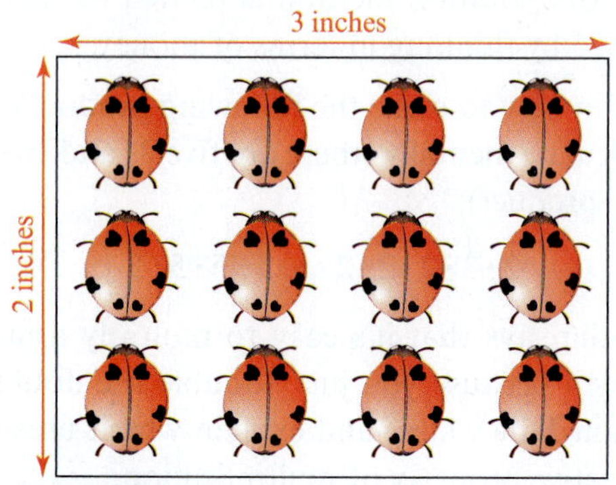

Figure 4.25 A ladybug design for wrapping paper.

17. Ms. Dunn's class wants to estimate the number of blades of grass in a lawn that is shaped roughly like a 30-foot-by-40-foot rectangle. The students cut 1-inch-by-1-inch squares out of sheets of paper and place these square holes over patches of grass. Each student cuts the grass from their 1-square-inch patch and counts the number of blades of grass they cut. Inside, the class calculates the average number of blades of grass they cut. This average is 37. Using these data, determine approximately how many blades of grass are in the lawn. Explain your reasoning.

18. Imagine that you are standing on a sandy beach, like the one in Figure 4.26, gazing off into the distance. How many grains of sand might you be looking at? To solve this problem, make reasonable assumptions about how wide the beach is and how far down the length of the beach you can see. Make a reasonable assumption about how many grains of sand are in a very small piece of beach, and explain why your assumption is reasonable. Based on your assumptions, make a calculation that will give you a fairly good estimate of the number of grains of sand that you can see. Explain your reasoning.

Figure 4.26 How many grains of sand do you see, looking down a sandy beach?

19. A lot of gumballs are in a glass container. The container is shaped like a box with a square base. When you look down on the top of the container, you see about 50 gumballs at the surface. When you look at one side of the container, you see about 60 gumballs up against the glass. You also notice that there are about 9 gumballs against each vertical edge of the container. Given this information, estimate the total number of gumballs in the container. Explain your reasoning.

20. Figure 4.27 shows a grocery store display of cases of soft drinks. The display consists of a large box shape with a "staircase" on top. The display is 7 cases wide and 5 cases deep; it is 4 cases tall in the front and 8 cases tall at the back. How many cases of soft drinks are there in the display? Solve this problem in at least two different ways, and explain your method in each case.

Figure 4.27 A display of cases of soft drinks.

21. Use the facts that

$$1 \text{ mile} = 1760 \text{ yards}$$
$$1 \text{ yard} = 3 \text{ feet}$$
$$1 \text{ foot} = 12 \text{ inches}$$

to calculate the number of inches in a mile. Do this in two different ways to illustrate the associative property of multiplication.

22. A roll of wrapping paper is 30 inches wide. When you unroll the wrapping paper and cut off a portion, you get a rectangular piece of wrapping paper that is 30 inches wide and can have various lengths. The wrapping paper has a design of stars on it. If you were to subdivide the paper into 2-inch-by-2-inch squares, each such square would have 25 stars on it. Use this information about the wrapping paper in parts (a), (b), and (c).

a. How long a piece of the wrapping paper would you need to get at least 1000 stars? Explain your reasoning.

b. How long a piece of the wrapping paper would you need to get at least 1,000,000 stars? Explain your reasoning. Realistically, could this length come from a single roll of wrapping paper? Why or why not?

c. How long a piece of the wrapping paper would you need to get 1,000,000,000 stars? Explain your reasoning. If you had this length of wrapping paper and you wanted to show it to the children in your class, could you roll it out in the school yard? Why or why not?

23. Estimate how many neatly stacked hundred-dollar bills you could fit in a briefcase that is 20 inches long, 11 inches wide, and 2 inches thick on the inside. Describe your method, and explain why it gives a good estimate.

***24.** A cube that is 10 inches wide, 10 inches long, and 10 inches high is made out of smaller cubes that are each 1 inch wide, 1 inch long, and 1 inch high. The large cube is then painted on the outside.

a. How many of the smaller cubes that make up the large cube have paint on them? Explain.

b. How many of the smaller cubes have paint on exactly two sides? Explain.

c. How many of the smaller cubes have paint on exactly three sides? Explain.

***25.** Investigate the following two questions, and explain your conclusions:

a. If you make a rectangular garden that is twice as wide and twice as long as a rectangular garden that you already have, how will the area of the larger garden compare with the area of the original garden? (Will the larger garden be twice as big, 3 times as big, 4 times as big, etc.?)

b. If you make a cardboard box that is twice as wide, twice as tall, and twice as long as a cardboard box that you already have, how will the volume of the larger box compare with the volume of the original box? (Will the larger box be twice as big, 3 times as big, 4 times as big, etc.?)

***26.** The Better Baking Company is introducing a new line of reduced-fat brownies in addition to its regular brownies. The batter for the reduced-fat brownies contains $\frac{1}{3}$ less fat than the batter for the regular brownies. Both types of brownies will be baked in the same size rectangular pan, which is 24 inches wide and 30 inches long. The bakers cut the regular brownies from this pan by dividing the width into 12 equal segments and by dividing the length into 10 equal segments, so that each regular brownie is 2 inches by 3 inches. Each regular brownie contains 6.3 grams of fat. In addition to using a reduced-fat batter, the Better Baking Company would like to further reduce the amount of fat in their new brownies by making these brownies smaller than the regular ones. You have been contacted to help with this task. Present two different ways to divide the length and width of the pan to produce smaller brownies. In each case, calculate the amount of fat in each brownie and explain the basis for your calculation. The brownies should be of a reasonable size, and there should be no waste left over in the pan after cutting the brownies. The length and width of the brownies do not necessarily have to be whole numbers of inches.

4.4 The Distributive Property

CCSS Common Core State Standards Grades 3, 4, 5

The associative and commutative properties apply in situations when we are adding only (as in Chapter 3) or multiplying only (as in Section 4.3). In this section, we introduce a property that applies to addition and multiplication together—namely, the distributive property of multiplication over addition. The distributive property is the most important and computationally powerful tool in all of arithmetic. It allows for tremendous flexibility in performing mental calculations, and, as we will see in Section 4.6, the distributive property is the foundation of the standard multiplication algorithm.

As with the other properties of arithmetic that we have studied, we will explain why the distributive property makes sense as well as discussing how it is useful in solving problems. Similarly to the commutative property of multiplication, the distributive property can be viewed geometrically in terms of arrays (or areas). As with all the other properties of arithmetic, the distributive property relies on counting a collection of things in two different ways.

Before we discuss the distributive property of multiplication over addition, let's review the conventions for interpreting expressions involving both multiplication and addition.

How Do We Interpret Expressions Involving Both Multiplication and Addition?

Expressions that involve both addition and multiplication must be interpreted suitably, according to the conventions developed by mathematicians. This situation is entirely unlike the situation where only addition, or only multiplication, is involved in an expression. In those situations, parentheses can be dropped safely and adjacent numbers can be combined at will. However, when both multiplication and addition are present in an expression, parentheses cannot generally be dropped without changing the value of the expression. For example,

$$7 + 5 \cdot 2$$

is *not equal* to

$$(7 + 5) \cdot 2$$

To properly interpret an expression such as

$$7 + 5 \cdot 2$$

or

$$5 \cdot 17 + 9 \cdot 10^2 - 12 \cdot 94 + 20 \div 5$$

order of operations

we need to use the following conventions on **order of operations**:

- Expressions inside parentheses are always considered first, subject to the following conventions.
- All powers (exponents) are considered first.
- Multiplications and divisions are considered from left to right.
- Additions and subtractions are considered from left to right.

Therefore,

$$\begin{aligned} 7 + 5 \cdot 2 &= 7 + 10 \\ &= 17 \end{aligned}$$

whereas

$$\begin{aligned} (7 + 5) \cdot 2 &= 12 \cdot 2 \\ &= 24 \end{aligned}$$

Similarly,

$$\begin{aligned} &5 \cdot 17 + 9 \cdot 10^2 - 12 \cdot 94 + 20 \div 5 \\ &= 5 \cdot 17 + 9 \cdot 100 - 12 \cdot 94 + 20 \div 5 \\ &= 85 + 900 - 1128 + 4 \\ &= -139 \end{aligned}$$

Note that the order of operations conventions allow us to *interpret* expressions unambiguously, but they do not dictate how to calculate. For example, as we'll see next when we study the distributive property, we could calculate $7 \cdot (10 + 2)$ by calculating $7 \cdot 10 = 70$, calculating $7 \cdot 2 = 14$, and adding 70 and 14 to make 84.

Why Does the Distributive Property Make Sense?

distributive property

The **distributive property** of multiplication over addition says that for all numbers, A, B, and C,

$$A \cdot (B + C) = A \cdot B + A \cdot C$$

Notice the use of parentheses to group the B and C: The expression

$$A \cdot (B + C)$$

means A times the *quantity* $B + C$. The distributive property tells us that $A \cdot (B + C)$ and $A \cdot B + A \cdot C$ always stand for the same amount.

As with the other properties of arithmetic that we have studied, we assume that the distributive property holds for all numbers. However, we can explain why the distributive property makes sense for counting numbers by decomposing arrays of objects or by decomposing rectangles.

CLASS ACTIVITY

4G Explaining the Distributive Property, p. CA-75

Consider an array of dots consisting of 4 horizontal rows, with 7 dots in each row, as shown in Figure 4.28. The coloring decomposes this array into two smaller arrays. There are two ways of expressing the total number of dots in the array based on this decomposition. On the one hand, there are 4 rows with $5 + 2$ dots in each row, and therefore the total number of dots is

$$4 \cdot (5 + 2)$$

Figure 4.28

An illustration of the distributive property: $4 \cdot (5 + 2) = 4 \cdot 5 + 4 \cdot 2$.

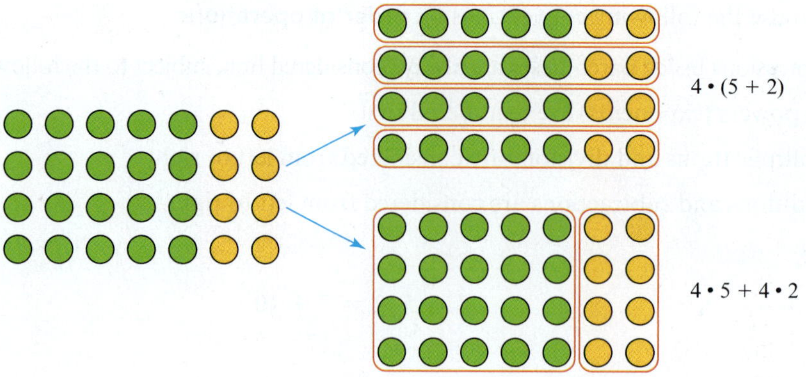

Notice that parentheses are needed to say 4 times the *quantity* 5 plus 2. On the other hand, the coloring decomposes the array of dots into two arrays. The array on the left consists of 4 rows with 5 dots in each row, and thus contains $4 \cdot 5$ dots; the array on the right consists of 4 rows with 2 dots in each row and hence contains $4 \cdot 2$ dots. Therefore, combining these two arrays of dots, there is a total of

$$4 \cdot 5 + 4 \cdot 2$$

dots. But the total number of dots is the same, either way you count them. Therefore,

$$4 \cdot (5 + 2) = 4 \cdot 5 + 4 \cdot 2$$

This explains why the distributive property makes sense *in this case*. But notice that the reasoning is general in the sense that if we were to replace the numbers 4, 5, and 2 with other counting numbers, and if we were to correspondingly adjust the size of array of dots, then our argument would still hold. Therefore, the distributive property makes sense for all counting numbers.

Variations on the Distributive Property

Several useful variations on the distributive property are listed next. We can obtain all these variations from the original distributive property by using the commutative property of multiplication, by using the distributive property repeatedly, or by using the fact that

$$B - C = B + (-C)$$

The Original Distributive Property

$$A \cdot (B + C) = A \cdot B + A \cdot C$$

for all numbers A, B, and C.

Variation 1

$$(A + B) \cdot C = A \cdot C + B \cdot C$$

for all numbers A, B, and C.

Variation 2

$$A \cdot (B + C + D) = A \cdot B + A \cdot C + A \cdot D$$

for all numbers A, B, C, and D.

Variation 3

$$A \cdot (B - C) = A \cdot B - A \cdot C$$

for all real numbers A, B, and C.

Collectively, we will refer to the original distributive property and all its variations simply as "the distributive property."

CCSS
SMP2
SMP7
4.NBT.5

How Does the Distributive Property Help Us Calculate Flexibly?

Just like the other properties of arithmetic that we have studied, we can often apply the distributive property to make mental arithmetic problems easier to solve.

For example, what is an easy way to calculate $41 \cdot 25$ mentally? We can use the following strategy:

40 times 25 is 1000, plus one more 25 is 1025.

The following sequence of equations corresponds to this strategy and uses the first variation on the distributive property at the second equal sign:

$$\begin{aligned} 41 \cdot 25 &= (40 + 1) \cdot 25 \\ &= 40 \cdot 25 + 1 \cdot 25 \\ &= 1000 + 25 \\ &= 1025 \end{aligned}$$

What is an easy way to calculate $39 \cdot 25$ mentally? We can use a strategy similar to the previous one, this time taking away a group of 25:

40 groups of 25 makes 1000; take away a group of 25, and we are left with 975.

The following sequence of equations corresponds to this strategy and uses the third variation on the distributive property (varied by the first variation) at the second equal sign.

$$\begin{aligned} 39 \cdot 25 &= (40 - 1) \cdot 25 \\ &= 40 \cdot 25 - 1 \cdot 25 \\ &= 1000 - 25 \\ &= 975 \end{aligned}$$

CLASS ACTIVITY

4H Applying the Distributive Property to Calculate Flexibly, p. CA-76

4I Critique Multiplication Strategies, p. CA-77

Where Does FOIL Come From?

You may have learned the FOIL method for multiplying expressions of the form

$$(A + B) \cdot (C + D)$$

FOIL **FOIL** stands for *First*, *Outer*, *Inner*, *Last*, in order to remind you that

$$(A + B) \cdot (C + D) = A \cdot C + A \cdot D + B \cdot C + B \cdot D$$

where $A \cdot C$ is *First*, $A \cdot D$ is *Outer*, $B \cdot C$ is *Inner*, and $B \cdot D$ is *Last*.

Why is FOIL valid? We can derive FOIL by using the distributive property several times. Hence, FOIL is an extension of the distributive property. To derive FOIL, let's first treat $C + D$ as a single entity; think of $C + D$ as c. Then, by the distributive property,

$$(A + B) \cdot c = A \cdot c + B \cdot c$$

Therefore,

$$(A + B) \cdot \overbrace{(C + D)}^{c} = A \cdot \overbrace{(C + D)}^{c} + B \cdot \overbrace{(C + D)}^{c}$$

Now we can apply the distributive property again, this time to $A \cdot (C + D)$ and to $B \cdot (C + D)$. Stringing all these equations together, we have

$$\begin{aligned}(A + B) \cdot (C + D) &= A \cdot (C + D) + B \cdot (C + D) \\ &= A \cdot C + A \cdot D + B \cdot C + B \cdot D\end{aligned}$$

thereby proving that FOIL is valid.

We can also see why FOIL makes sense by subdividing arrays or rectangles. Figure 4.29 shows a rectangle made of $10 + 3$ rows with $10 + 4$ small squares in each row. According to the definition of multiplication, the total number of small squares in the rectangle is

$$(10 + 3) \cdot (10 + 4)$$

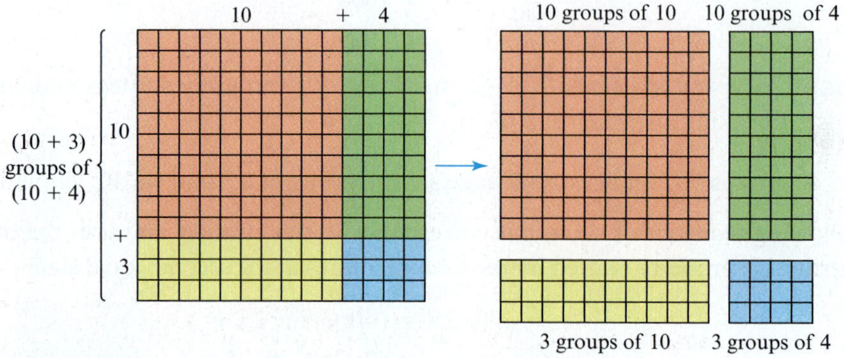

Figure 4.29 Subdividing a rectangle to show that $(10 + 3) \cdot (10 + 4) = 10 \cdot 10 + 10 \cdot 4 + 3 \cdot 10 + 3 \cdot 4$.

As we see by the coloring in Figure 4.29, we can subdivide the rectangle into 4 natural parts: 10 groups of 10 small squares, 10 groups of 4 small squares, 3 groups of 10 small squares, and 3 groups of 4 small squares. Therefore, the total number of small squares in the rectangle is

$$10 \cdot 10 + 10 \cdot 4 + 3 \cdot 10 + 3 \cdot 4$$

But the total number of small squares is the same either way you count them. Therefore,

$$(10 + 3) \cdot (10 + 4) = 10 \cdot 10 + 10 \cdot 4 + 3 \cdot 10 + 3 \cdot 4$$

which is the FOIL equation. The same reasoning will apply when the numbers 10, 3, 10, and 4 are replaced with other numbers, and the rectangle is changed accordingly. Thus, we see why FOIL must be valid for all counting numbers.

How Can We Extend the Distributive Property?

How do we multiply an expression of the form

$$(A + B) \cdot (C + D + E)?$$

Using our previous reasoning, let's first treat $C + D + E$ as a single entity; think of $C + D + E$ as c. Then, by the distributive property,

$$(A + B) \cdot c = A \cdot c + B \cdot c$$

Therefore,

$$(A + B) \cdot \overbrace{(C + D + E)}^{c} = A \cdot \overbrace{(C + D + E)}^{c} + B \cdot \overbrace{(C + D + E)}^{c}$$

Now we can apply the distributive property again, this time to $A \cdot (C + D + E)$ and to $B \cdot (C + D + E)$. Stringing all these equations together, we have

$$\begin{aligned}(A + B) \cdot (C + D + E) &= A \cdot (C + D + E) + B \cdot (C + D + E)\\ &= A \cdot C + A \cdot D + A \cdot E + B \cdot C + B \cdot D + B \cdot E\end{aligned}$$

Notice that on the right-hand side of the previous equation, each of A and B is multiplied with each of C, D, and E.

extended distributive property

Repeating the reasoning of the previous paragraph with other products, we see that FOIL and the previous equation are both special cases of an **extended distributive property**, which tells us that, to multiply a sum by another sum, we can multiply each number in the first sum by each number in the second sum and add all these products.

SECTION SUMMARY AND STUDY ITEMS

Section 4.4 The Distributive Property

The distributive property states that

$$A \cdot (B + C) = A \cdot B + A \cdot C$$

for all real numbers A, B, and C. We can explain why the distributive property is valid (for counting numbers) by expressing the total number of objects in a subdivided array in two different ways.

We apply the distributive property in many calculations. This property is the single most powerful property of arithmetic. (In Section 4.6, we will see that it is the key idea underlying the multiplication algorithm.)

The convention on order of operations governs how we interpret expressions that involve several operations, such as multiplication and addition.

Key Skills and Understandings

1. State the distributive property, and explain why it makes sense (for counting numbers) by describing the total number of objects in a subdivided array in two different ways. Use simple situations to explain or illustrate the distributive property.
2. Give examples of how to apply the distributive property in problems and recognize when the distributive property has been used.

Practice Exercises for Section 4.4

1. Does the expression

$$3 \cdot 4 + 2$$

have a different meaning than the expression

$$3 \cdot \quad 4 + 2$$

which has a big space between the $\cdot$ and the 4? Explain.

2. Dana and Sandy are working on

$$8 \cdot 5 + 20 \div 4$$

Dana says the answer is 15, but Sandy says the answer is 45. Who's right, who's wrong, and why?

3. Mr. Greene has a bag of plastic spiders to give out. After putting 4 spiders in each of 23 bags, he has 8 spiders left. Write an expression using the numbers 4, 23, and 8; the symbols $\times$ or $\cdot$ and $+$; and parentheses, if needed, for the total number of spiders Mr. Greene had to give out. If you use parentheses, explain why you need them; if you do not use parentheses, explain why you do not need them.

4. State the distributive property.

5. There are 15 goodie bags. Each goodie bag contains 2 pencils, 4 stickers, and 3 wiggly worms. Write two different expressions using the numbers 15, 2, 4,

and 3; the symbols × or · and +; and parentheses, if needed, for the total number of objects in the 15 goodie bags. If you use parentheses in an expression, explain why you need them; if you do not use parentheses, explain why you do not need them. Use your two expressions to illustrate the distributive property (variation 2).

6. Using a specific example, explain why the distributive property makes sense. Even though you use a specific example, your explanation should be general, in the sense that we should be able to see why it will hold true when other numbers replace your numbers.

7. Give an example of how to apply the distributive property to make a calculation easy to carry out mentally. Describe how to carry out this mental calculation. Write equations showing how the calculation strategy used the distributive property.

8. Compute mentally:

$$97{,}346 \cdot 142{,}349 + 2{,}654 \cdot 142{,}349$$

9. Draw an array that shows why

$$20 \cdot 19 = 20 \cdot 20 - 20 \cdot 1$$

Also, use this equation to help you calculate 20 × 19 mentally.

10. Draw a subdivided array to show that

$$(10 + 2) \cdot (10 + 3) = 10 \cdot 10 + 10 \cdot 3 + 2 \cdot 10 + 2 \cdot 3$$

Then write equations that use properties of arithmetic to show why the preceding equation is true.

Answers to Practice Exercises for Section 4.4

1. The expressions

$$3 \cdot 4 + 2$$

and

$$3 \cdot \quad 4 + 2$$

have the same meaning. According to the convention on order of operations, multiplication is considered before addition, no matter how big a space there is following the multiplication symbol. If you want to write 3 times the *quantity* 4 + 2, then you must use parentheses and write

$$3 \cdot (4 + 2)$$

2. Sandy is right that the answer is 45. Dana, like many students, just worked from left to right. She did not use the conventions on order of operations: Consider multiplication and division first, then addition and subtraction. According to the conventions on the order of operations,

$$8 \cdot 5 + 20 \div 4 = 40 + 5 = 45$$

3. Since there are 23 bags with 4 spiders in each bag, the total number of spiders is $23 \cdot 4$. But there are also 8 more spiders, so Mr. Greene has

$$23 \cdot 4 + 8$$

spiders. We do not need parentheses because, according to the convention on order of operations, we can calculate the foregoing expression by multiplying 23 times 4 and then adding 8 to that quantity, which is exactly what we want to express.

4. $A \cdot (B + C) = A \cdot B + A \cdot C$ for all numbers A, B, C.

5. Since there are 15 goodie bags and each goodie bag has a total of 2 + 4 + 3 objects in it, the total number of objects in the goodie bags is

$$15 \cdot (2 + 4 + 3)$$

We need parentheses in this expression to show that 15 multiplies the entire quantity 2 + 4 + 3.

Another expression for the total number of objects in the goodie bags is

$$15 \cdot 2 + 15 \cdot 4 + 15 \cdot 3$$

because there are $15 \cdot 2$ pencils, $15 \cdot 4$ stickers, and $15 \cdot 3$ wiggly worms. Because of the conventions on order of operations, we do not need parentheses here, since we carry out multiplication before addition. Since it's the same total number of objects either way we count them, we have the equation

$$15 \cdot (2 + 4 + 3) = 15 \cdot 2 + 15 \cdot 4 + 15 \cdot 3$$

which illustrates the distributive property (variation 2).

6. See the explanation in the text for why $4 \cdot (5 + 2) = 4 \cdot 5 + 4 \cdot 2$ using the array in Figure 4.28. See also the discussion on why that explanation is a general one. Note in particular that the explanation *does not* rely on evaluating either $4 \cdot (5 + 2)$ or $4 \cdot 5 + 4 \cdot 2$ as equal to the specific number 28. If it did, it would not be a general explanation because we wouldn't be able to tell how the explanation

would work if the numbers 4, 5, and 2 in the example were replaced with other counting numbers.

7. We can calculate $7 \cdot 12$ mentally by first calculating $7 \cdot 10 = 70$, then $7 \cdot 2 = 14$ and then adding $70 + 14 = 84$. This strategy uses the distributive property, as shown by the following equations:

$$\begin{aligned} 7 \cdot 12 &= 7 \cdot (10 + 2) \\ &= 7 \cdot 10 + 7 \cdot 2 \\ &= 70 + 14 = 84 \end{aligned}$$

8. By the distributive property,

$$\begin{aligned} &97{,}346 \cdot 142{,}349 + 2{,}654 \cdot 142{,}349 \\ &\quad = (97{,}346 + 2{,}654) \cdot 142{,}349 \\ &\quad = 100{,}000 \cdot 142{,}349 \\ &\quad = 14{,}234{,}900{,}000 \end{aligned}$$

9. Figure 4.30 shows 20 rows of 20 squares. If you take away 20 rows of 1 square—namely, a vertical strip of squares—then you are left with 20 rows of 19 squares, thus illustrating that

$$\begin{aligned} 20 \cdot 19 &= 20 \cdot 20 - 20 \cdot 1 \\ &= 400 - 20 \\ &= 380 \end{aligned}$$

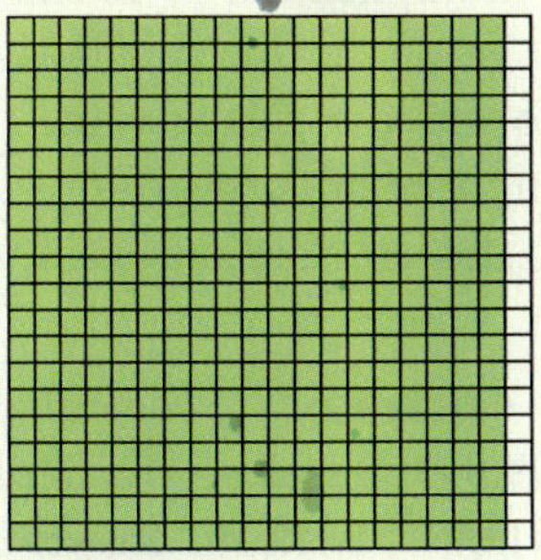

Figure 4.30 An array showing that $20 \cdot 19 = 20 \cdot 20 - 20 \cdot 1$.

10. The array in Figure 4.31 shows $10 + 2$ groups of $10 + 3$ small squares. According to the definition of multiplication, there are a total of $(10 + 2) \cdot (10 + 3)$ small squares in this array. The array is broken into four smaller arrays: 10 groups of 10, 10 groups of 3, 2 groups of 10, and 2 groups of 3. But it's the same number of small squares no matter how you count them; therefore,

$$\begin{aligned} (10 + 2) \cdot (10 + 3) &= 10 \cdot 10 + 10 \cdot 3 \\ &\quad + 2 \cdot 10 + 2 \cdot 3 \end{aligned}$$

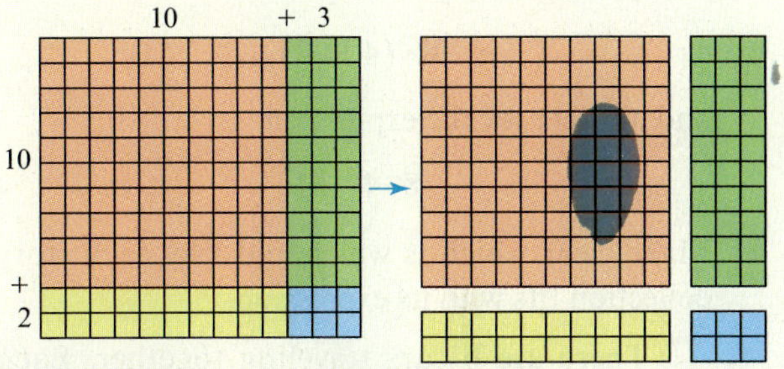

Figure 4.31 Subdividing an array to show that $(10 + 2) \cdot (10 + 3) = 10 \cdot 10 + 10 \cdot 3 + 2 \cdot 10 + 2 \cdot 3$.

We can also explain this equation by using the distributive property:

$$\begin{aligned} (10 + 2) \cdot (10 + 3) &= (10 + 2) \cdot 10 \\ &\quad + (10 + 2) \cdot 3 \\ &= 10 \cdot 10 + 2 \cdot 10 \\ &\quad + 10 \cdot 3 + 2 \cdot 3 \end{aligned}$$

The distributive property was used at both equal signs. In fact, it was used twice at the second equal sign.

PROBLEMS FOR SECTION 4.4

1. Ben and Charles are working on

$$4 + 3 \cdot 2 \cdot 10$$

Ben says the answer is 64. Charles says the answer is 140. Who's right, who's wrong, and why?

2. a. There are 6 cars traveling together. Each car has 2 people in front and 3 in the back. Write an expression using the numbers 6, 2, and 3; the symbols $\times$ or $\cdot$ and $+$; and parentheses, if needed, for the total number of people riding in the 6 cars. If you use parentheses, explain why you need them; if you do not use parentheses, explain why you do not need them.

b. Describe a collection of things whose total number is given by the expression

$$6 \cdot 2 + 3$$

3. The students in Mrs. Black's class are arranged as follows: 7 tables have 4 students sitting at each of them; 2 tables have 3 students sitting at each of them. Write an expression using the numbers 7, 4, 2, and 3; the symbols $\times$ or $\cdot$ and $+$; and parentheses, if needed, for the total number of students in Mrs. Black's class. If you use parentheses, explain why you need them; if you do not use parentheses, explain why you do not need them.

4. Describe one collection of things whose total number is given by the expression

$$8 \cdot (4 + 2)$$

and another for the expression

$$8 \cdot 4 + 2$$

Make clear which is which and why each story collection fits with its expression.

5. There are 6 cars traveling together. Each car has 2 people in front and 3 people in back. Explain how to use this situation to illustrate the distributive property.

6. Draw arrays to help you explain why the equations shown are true, without evaluating any of the products or sums. Explain your answers clearly.

   a. $(5 + 1) \cdot 7 = 5 \cdot 7 + 1 \cdot 7$

   b. $6 \cdot (5 + 2) = 6 \cdot 5 + 6 \cdot 2$

   c. $(10 - 1) \cdot 6 = 10 \cdot 6 - 1 \cdot 6$

7. Explain how to use the distributive property to make $31 \cdot 25$ easy to calculate mentally. Write an equation that corresponds to your strategy. Without drawing all the detail, draw a rough picture of an array that illustrates this calculation strategy.

8. Explain how to calculate $29 \cdot 20$ mentally by using $30 \cdot 20$. Write an equation that uses subtraction and the distributive property and that corresponds to your strategy. Without drawing all the detail, draw a rough picture of an array that illustrates this calculation strategy.

9. Ted thinks that because $10 \cdot 10 = 100$ and $2 \cdot 5 = 10$, he should be able to calculate $12 \cdot 15$ by adding $100 + 10$ to get 110. Explain to Ted in two different ways that, even though his method is not correct, his calculations can be part of a correct way to calculate $12 \cdot 15$.

   a. by drawing an array

   b. by writing equations that use the distributive property

10. Working on the multiplication problem $21 \cdot 34$, SunJae says that he can take 1 from the 21 and put it with the 34 to get $20 \cdot 35$ and that this new multiplication problem, $20 \cdot 35$, should have the same answer as $21 \cdot 34$. Is SunJae right? How might you convince him that his reasoning is or is not correct in a way other than simply showing him the answers to the two multiplication problems?

11. a. Use the distributive property several times to show why

$$(10 + 2) \cdot (10 + 4)$$
$$= 10 \cdot 10 + 10 \cdot 4 + 2 \cdot 10 + 2 \cdot 4$$

   b. Draw an array to show why

$$(10 + 2) \cdot (10 + 4)$$
$$= 10 \cdot 10 + 10 \cdot 4 + 2 \cdot 10 + 2 \cdot 4$$

   c. Relate the steps in your equations in part (a) to your array in part (b).

12. In Section 4.2, we drew pictures of bundled objects to explain why multiplication by 10 shifts the digits in base ten. Now use the distributive property, expanded forms, and place value to explain why multiplying a number by 10 moves each digit one place to the left. Use the example $10 \cdot 12{,}345$ to illustrate.

*13. a. Use an ordinary calculator to calculate $666{,}666{,}666 \cdot 999{,}999{,}999$. Based on the calculator's display, guess how to write the answer in ordinary decimal notation (showing *all* digits).

   b. Now think some more, and determine how to write the product $666{,}666{,}666 \cdot 999{,}999{,}999$ in ordinary base ten notation, showing all its digits, without multiplying longhand or using a calculator or computer.

*14. Without using a calculator or computer and without simply multiplying longhand, calculate the product $9{,}999{,}999{,}999 \cdot 9{,}999{,}999{,}999$. Give the answer in ordinary base-ten notation, showing all its digits. (Both numbers in the product have 10 nines.) Explain your method.

*__15.__ Check the following:

$$11 - 2 = 3 \cdot 3$$
$$1111 - 22 = 33 \cdot 33$$
$$111111 - 222 = 333 \cdot 333$$

Continue to find at least three more in the pattern. Does the pattern continue? Now explain why there is such a pattern. *Hint:* Notice that $1111 - 22 = 11 \cdot 101 - 11 \cdot 2$.

16. Determine which of the following two numbers is greater, and explain your reasoning:

a. $1{,}000{,}000 \cdot (1 + 2 + 3 + 4 + \cdots + 1{,}000{,}001)$

b. $1{,}000{,}001 \cdot (1 + 2 + 3 + 4 + \cdots + 1{,}000{,}000)$

*__17.__ The **square** of a number is just the number times itself. For example, the square of 4 is $4 \cdot 4 = 16$.

a. Find the squares of the numbers 15, 25, 35, 45, ..., 95, 105, 115, 125, ..., 195, 205, and three other whole numbers that end in 5.

b. Find some patterns in the answers to part (a). Specifically, in all of your answers to part (a), what do you notice about the last two digits, and what do you notice about the number formed by deleting the last two digits?

c. Use what you discovered in part (b) to predict the squares of 2,005 and 10,005.

d. Every whole number ending in 5 must be of the form $10A + 5$ for some whole number A. Find the square of $10A + 5$—namely, calculate $(10A + 5) \cdot (10A + 5)$.

e. How does your answer to part (d) explain the pattern you found in part (b)?

*__18.__ The square of a number is just the number times itself. For example, the square of 4 is 16.

a. Calculate the squares of 1, 2, . . . , 9 and many other whole numbers, including 17, 34, 61, 82, 99, 123, 255, 386, and 728. Record the ones digits in each case. What do you notice? Do any of your squares have a ones digit of 7, for example? Are any other digits missing from the ones digits of squares?

b. Which digits can never occur as the ones digit of a square of a whole number? Explain why some digits cannot occur as the ones digit of a square of a whole number.

c. Based on what you've discovered, could the number

139,787,847,234,329,483

be the square of a whole number? Why or why not?

4.5 Properties of Arithmetic, Mental Math, and Single-Digit Multiplication Facts

CCSS **Common Core State Standards** Grades 3, 4, 5

The properties of arithmetic are the building blocks for all of arithmetic. They allow us to break calculations apart into pieces, calculate the pieces, and then put the pieces back together to obtain the full solution. This process of taking apart, analyzing, and putting back together is a powerful technique that is used across all of mathematics: in arithmetic, in geometry, and in algebra—and at all levels. In this section we'll examine how we can combine the properties of arithmetic to take calculations apart. First we'll look at the single-digit multiplication facts and see how students who are learning these facts can build up their knowledge by breaking apart facts they haven't yet mastered into ones they already know. Then we'll examine other mental calculations, including percent calculations. Throughout this section, note how breaking calculations apart is essentially an algebraic way of looking at the calculations.

What Is a Learning Path for Single-Digit Multiplication Facts?

In school, students must develop fluency with the single-digit multiplication facts from $1 \times 1 = 1$ to $9 \times 9 = 81$. Fluency with these facts is essential to succeed in mathematics, but the process of becoming fluent is not merely rote memorization.

When students first begin to learn to multiply, they multiply by skip-counting. For example, to multiply 5 by 7, a student will count by 5s seven times:

5, 10, 15, 20, 25, 30, 35

A visual aid to connect skip-counting with arrays, and thus with multiplication, is shown in Figure 4.32. To skip-count with the aid of an array, students can cover the array with a sheet of paper and then uncover more and more of the array as they count while sliding down the paper.

Figure 4.32 Skip-counting with an array.

CCSS 3.OA.5

It may seem that the single-digit multiplication facts should simply be memorized. However, the *Common Core State Standards* indicate a process of learning multiplication facts by learning relationships among the facts. In this way, students organize the facts so that they know them better, and they learn to reason "algebraically" by applying properties of arithmetic. For example, if a student knows that $7 \times 5 = 35$, and the student knows the commutative property of multiplication, then the student will be able to determine that $5 \times 7 = 35$. The student might then figure out 6×7 by thinking that 6 groups of 7 is one more group of 7 than 5 groups of 7, so 6×7 is 7 more than 35, which is 42. This reasoning is indicated and summarized by the decomposed array and the equations in Figure 4.33.

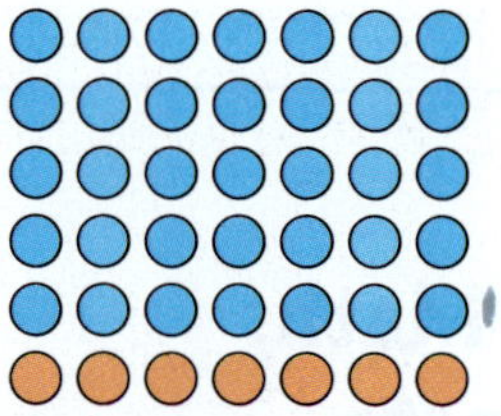

$$6 \times 7 = (5 + 1) \times 7 \quad \text{(distributive property)}$$
$$= (5 \times 7) + (1 \times 7)$$
$$= 35 + 7 = 42$$

Figure 4.33 Deriving 6×7 from 5×7 and 1×7.

CLASS ACTIVITY

4J Using Properties of Arithmetic to Aid the Learning of Basic Multiplication Facts, p. CA-78

How Is Algebra Behind Flexible Calculation Strategies?

Sections 4.3 and 4.4 included examples of how to apply properties of arithmetic as calculation strategies. We used whole numbers in these examples, but the properties of arithmetic apply to all numbers, including percents, for example. In Chapter 2 we reviewed ways of making percent calculations easy to do with the aid of percent tables. Now we'll see how to write equations that fit with the percent table methods, so that we can examine the algebra that is in these methods.

In Chapter 2 we worked with percent calculations by viewing them as problems about equivalent fractions. But we can also think about percent calculations as multiplication problems. Why? Suppose we want to calculate $P\%$ of some quantity Q. Just as "2 of Q" is $2 \cdot Q$ and "3 of Q" is $3 \cdot Q$ so too

$$P\% \text{ of } Q \text{ is } P\% \cdot Q$$

So for example, what is 95% of 460? To calculate mentally, we could reason that 10% of 460 is 46, so 5% of 460 is half of 46, which is 23. Then, thinking of 95% as 100% − 5%, we can calculate 95% of 460 by taking 23 away from 460, which leaves 437. (We can even take 23 away from 460 in two steps by first taking away 20 to make 440 and then taking away another 3 to make 437.) The next percent table summarizes this line of reasoning:

$$100\% \rightarrow 460$$
$$10\% \rightarrow 46$$
$$5\% \rightarrow 23$$
$$95\% = 100\% - 5\% \rightarrow 460 - 23 = 437$$

To show the algebra that is in this mental math strategy, we can write the following equations that fit with the strategy:

IMAP *Watch Javier use number sense to solve how many eggs in six dozen, 6 × 12, then twelve dozen, 12 × 12.*

IMAP *See the following video clips for examples of children solving multiplication problems: Javier, 6 × 12, then 12 × 12 Ally, 6 × 14 Andrew, 7 × 12 Brooke, 15 × 12 Connor, 4 × 25 Estephania, rolls of candies Gilberto, 12 × 15 Jennifer, 7 × 12 Rachel, 45 × 36 Shannon, 4 × 15 Maryann, 251 × 12*

$$\begin{aligned} 95\% \cdot 460 &= (100\% - 5\%) \cdot 460 \\ &= 100\% \cdot 460 - 5\% \cdot 460 \\ &= 460 - \left(\frac{1}{2} \cdot 10\%\right) \cdot 460 \\ &= 460 - \frac{1}{2} \cdot (10\% \cdot 460) \\ &= 460 - \frac{1}{2} \cdot 46 \\ &= 460 - 23 = 437 \end{aligned}$$

Notice that to formulate these equations, we have to think about the calculation in a big-picture way. These equations show that the distributive property was used at the second equal sign and the associative property of multiplication was used at the fourth equal sign.

FROM THE FIELD Research

Flowers, J. M., & Rubenstein, R. N. (2010/2011). Multiplication fact fluency using doubles. *Mathematics Teaching in the Middle School, 16*(5), pp. 296–301.

Many middle school students do not have solid fluency with multiplication facts, which creates a gap in their development and undermines their confidence and disposition for further mathematical learning. The authors have found that students have a natural interest in doubling and they use doubling as part of an approach in which students accumulate new facts by reasoning from familiar facts. Students develop problem solving and reasoning skills, along with confidence. The authors offer a sequence of doubling problem sets as well as a "doubling around the room" activity, in which one student says "1" and each consecutive student doubles the previous number. Once students are familiar with doubling, the authors recommend that students connect doubling tasks to multiplication fact fluency. For example, 3 times 7 is three sevens, which is two groups of 7 and one more 7. To determine 8 times 7, a student could reason that double 7 is 14, that double 14 is 28 (and that is 4 sevens) and that double 28 is 56, which is 8 sevens. The authors also recommend that students make arrays on graph paper and keep an inventory of facts that are known and facts that are still to be learned. The doubling activities provide links to earlier mathematics (the place value structure of our base-ten system, the distributive property, and composing and decomposing) and to later mathematics (probability, exponential growth, proportional reasoning).

Keep in mind one important comment about writing equations that fit with a mental calculation method: We use these equations to show the algebraic structure of the calculations and thus to better understand the nature of the calculations and to go more deeply into their inner workings. In ordinary circumstances, carry out mental calculations in a quick and efficient way—and *mentally*!

CLASS ACTIVITY

4K Solving Arithmetic Problems Mentally, p. CA-80

4L Writing Equations That Correspond to a Method of Calculation, p. CA-81

4M Showing the Algebra in Mental Math, p. CA-82

SECTION SUMMARY AND STUDY ITEMS

Section 4.5 Properties of Arithmetic, Mental Math, and Single-Digit Multiplication Facts

The commutative and associative properties of addition and multiplication and the distributive property are key in explaining calculations, and they link arithmetic and algebra. Given a description of a mental method of calculation, we can write a sequence of equations that correspond to the method and thus show the "algebra" in the mental math. Many of the basic multiplication facts are related to each other via properties of arithmetic. Students can capitalize on such relationships to aid their learning of the basic facts.

Key Skills and Understandings

1. Given a mental method of calculation, write a string of equations that correspond to the method and that show which properties of arithmetic were used (knowingly or not).
2. Use equations and decomposed arrays when describing how basic multiplication facts are related to other basic facts via properties of arithmetic.

Practice Exercises for Section 4.5

1. Write equations that correspond to the following reasoning for determining 5×7:

   I know 2×7 is 14. Then another 2×7 makes 28. And one more 7 makes 35.

   Which properties of arithmetic are used? Explain. Draw an array that corresponds to the reasoning.

2. A child is having difficulty remembering 8×8. Draw two arrays showing how 8×8 is related to other possibly easier facts involving smaller numbers. For each array, write a corresponding equation relating 8×8 to other multiplication facts. Which properties of arithmetic do you use?

3. The string of equations that follows corresponds to a mental method for calculating $45 \cdot 11$. Explain in words why the method of calculation makes sense. Which properties of arithmetic are used, and where are they used?

$$\begin{aligned} 45 \cdot 11 &= 45 \cdot (10 + 1) \\ &= 45 \cdot 10 + 45 \cdot 1 \\ &= 450 + 45 \\ &= 495 \end{aligned}$$

4. Each arithmetic problem in this exercise has a description for solving the problem. In each case, write a string of equations that corresponds to the given description. Identify properties of arithmetic that are used. Write your equations in the following form:

$$\begin{aligned} \text{original} &= \text{some expression} \\ &= \vdots \\ &= \text{some expression} \end{aligned}$$

a. *Problem:* What is $6 \cdot 40$?

Solution: 6 times 4 is 24; then you multiply by 10 and get 240.

b. *Problem:* What is 110% of 62?

Solution: 100% of 62 is 62 and 10% of 62 is 6.2, so all together it's 68.2.

c. *Problem:* Calculate the 7% tax on a purchase of $25.

Solution: 10% of $25 is $2.50, so 5% is $1.25. 1% of $25 is 25 cents, so 2% is 50 cents. This means 7% is $1.25 plus 50 cents, which is $1.75.

d. *Problem:* What is 45% of 300?

Solution: 50% of 300 is 150. 10% of 300 is 30, and half of that is 15, so the answer is 135.

e. *Problem:* Find $\frac{3}{4}$ of 72.

Solution: Half of 72 is 36. Half of 36 is 18. Then, to add 36 and 18, I did 40 plus 14, which is 54.

f. *Problem:* Calculate $59 \cdot 70$.

Solution: 6 times 7 is 42, so 60 times 70 is 4200. Therefore, 59 times 70 is 70 less, which is 4130.

5. For each of the problems that follow, use the distributive property to help make the problem easy to solve mentally. In each case, write a string of equations that correspond to your strategies. Write your equations in the following form:

$$\begin{aligned} \text{original} &= \text{some expression} \\ &= \vdots \\ &= \text{some expression} \end{aligned}$$

a. Calculate 51% of 140.

b. Calculate 95% of 60.

c. Calculate $\frac{5}{8} \cdot 280$.

d. Calculate 99% of 80.

6. For each of the following arithmetic problems, use properties of arithmetic to make the problem easy to solve mentally. Write a string of equations that corresponds to your method. Say which properties of arithmetic you use and where you use them.

a. $25 \cdot 84$

b. $49 \cdot 6$

c. $486 \cdot 5$

7. Joey, a second-grader, used the following method to mentally calculate the number of minutes in a day: First, Joey calculated 25 times 6 by finding 4 times 25 plus two times 25, which is 150. Then he subtracted 6 from 150 to get 144. Then he multiplied 144 by 10 to get the answer: 1440. Write a sequence of equations that corresponds to Joey's method and shows why his method is legitimate. What properties of arithmetic are involved?

Answers to Practice Exercises for Section 4.5

1. The distributive property is used at the second equal sign in the following equations:

$$\begin{aligned} 5 \times 7 &= (2 + 2 + 1) \times 7 \\ &= 2 \times 7 + 2 \times 7 + 1 \times 7 \\ &= 14 + 14 + 7 \\ &= 28 + 7 = 35 \end{aligned}$$

The subdivided array in Figure 4.34 corresponds to the arithmetic.

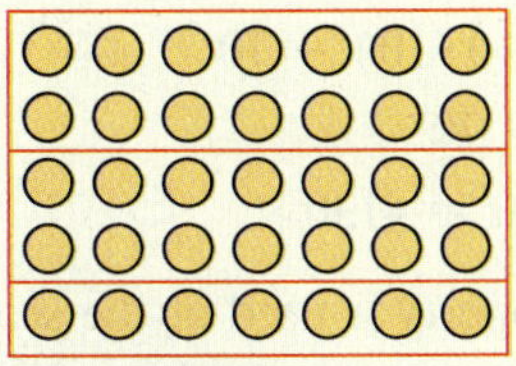

Figure 4.34 Subdividing an array.

2. See Figure 4.35 for the arrays. The array on the left breaks 8 groups of 8 stars into 2 groups of 4×8. The corresponding equation is

$$8 \times 8 = 2 \times (4 \times 8)$$

which uses the associative property of multiplication. We could also write the equation

$$8 \times 8 = 4 \times 8 + 4 \times 8$$

which uses the distributive property. If the student knows the fact $4 \times 8 = 32$, then he might be able to add $32 + 32$ quickly to get the correct answer to 8×8.

The array on the right of Figure 4.35 breaks 8 groups of 8 stars into 5 groups of 8 and 3 groups of 8. The corresponding equation is

$$8 \times 8 = 5 \times 8 + 3 \times 8$$

If the student knows 5×8 and 3×8, then she might be able to add $40 + 24$ quickly to get the correct answer to 8×8.

Figure 4.35 Relating 8×8 to other multiplication facts.

3. The product $45 \cdot 11$ stands for the total number of objects in 45 groups that have 11 objects in each group. These objects can be broken into 45 groups of 10 and another 45 groups of 1. The number of objects in 45 groups of 10 is $45 \cdot 10$, which is 450, and the number of objects in 45 groups of 1 is 45. In all, that makes $450 + 45$, which is 495 objects. The distributive property is used to say that $45 \cdot (10 + 1) = 45 \cdot 10 + 45 \cdot 1$.

4. a.
$$\begin{aligned} 6 \cdot 40 &= 6 \cdot (4 \cdot 10) \\ &= (6 \cdot 4) \cdot 10 \\ &= 24 \cdot 10 \\ &= 240 \end{aligned}$$

The associative property of multiplication is used at the second equal sign in order to group the 4 with the 6 instead of with the 10.

b.
$$\begin{aligned} 110\% \cdot 62 &= (100\% + 10\%) \cdot 62 \\ &= 100\% \cdot 62 + 10\% \cdot 62 \\ &= 62 + 6.2 \\ &= 68.2 \end{aligned}$$

The distributive property is used at the second equal sign.

c.
$$\begin{aligned} 7\% \cdot 25 &= (5\% + 2\%) \cdot 25 \\ &= 5\% \cdot 25 + 2\% \cdot 25 \\ &= \frac{1}{2} \cdot 10\% \cdot 25 + 2 \cdot 1\% \cdot 25 \\ &= \frac{1}{2} \cdot 2.5 + 2 \cdot 0.25 \\ &= 1.25 + 0.50 \\ &= 1.75 \end{aligned}$$

The distributive property is used at the second equal sign. Although parentheses are not shown, the associative property of multiplication is used twice at the fourth equal sign in order to calculate 10% of 25 and 1% of 25 first, and then multiply those by $\frac{1}{2}$ and 2, respectively.

d.
$$\begin{aligned} 45\% \cdot 300 &= (50\% - 5\%) \cdot 300 \\ &= 50\% \cdot 300 - 5\% \cdot 300 \\ &= 150 - \frac{1}{2} \cdot 10\% \cdot 300 \\ &= 150 - \frac{1}{2} \cdot 30 \\ &= 150 - 15 \\ &= 135 \end{aligned}$$

The distributive property is used at the second equal sign. Although parentheses are not shown, the associative property of multiplication is used at the fourth equal sign to calculate 10% of 300 first, and then find half of that.

e.
$$\begin{aligned} \frac{3}{4} \cdot 72 &= \left(\frac{1}{2} + \frac{1}{2} \cdot \frac{1}{2}\right) \cdot 72 \\ &= \frac{1}{2} \cdot 72 + \frac{1}{2} \cdot \frac{1}{2} \cdot 72 \\ &= 36 + \frac{1}{2} \cdot 36 \\ &= 36 + 18 \\ &= 36 + (4 + 14) \\ &= (36 + 4) + 14 \\ &= 40 + 14 \\ &= 54 \end{aligned}$$

The distributive property is used at the second equal sign. Although parentheses are not shown, the associative property of multiplication is used at the third equal sign to find half of 72 first. Then we take half of that. The associative property of addition was used at the sixth equal sign to change the placement of the parentheses.

f.
$$\begin{aligned} 59 \cdot 70 &= (60 - 1) \cdot 70 \\ &= 60 \cdot 70 - 1 \cdot 70 \\ &= (6 \cdot 10) \cdot (7 \cdot 10) - 70 \\ &= (6 \cdot 7) \cdot (10 \cdot 10) - 70 \\ &= 42 \cdot 100 - 70 \\ &= 4200 - 70 = 4130 \end{aligned}$$

The distributive property is used at the second equal sign. The commutative and associative properties of multiplication are used at the fourth equal sign. Notice that even though the stated solution to the problem did not explicitly mention multiplying 10 by 10 to get 100 and then multiplying 42 by 100, these calculations were used implicitly. So the previous equations actually expand on the stated solution, filling in the implied details.

5. a. We can decompose 51% as 50% plus 1%. 50% of 140 is half of 140, which is 70. 1% of 140 is 1.4. So 51% of 140 is 70 plus 1.4, which is 71.4. Using equations, we have

$$\begin{aligned} 51\% \cdot 140 &= (50\% + 1\%) \cdot 140 \\ &= 50\% \cdot 140 + 1\% \cdot 140 \\ &= 70 + 1.4 = 71.4 \end{aligned}$$

The distributive property was used at the second equal sign.

b. We can decompose 95% as 100% minus 5%. Now 10% of 60 is 6, so 5% of 60 is half of that, which is 3. Therefore, 95% of 60 is 60 − 3, which is 57. Using equations, we have

$$\begin{aligned} 95\% \cdot 60 &= (100\% - 5\%) \cdot 60 \\ &= 100\% \cdot 60 - 5\% \cdot 60 \\ &= 60 - \frac{1}{2} \cdot 10\% \cdot 60 \\ &= 60 - \frac{1}{2} \cdot 6 \\ &= 60 - 3 = 57 \end{aligned}$$

The distributive property was used at the second equal sign.

c. Use the fact that $\frac{5}{8} = \frac{4}{8} + \frac{1}{8} = \frac{1}{2} + \frac{1}{8}$.

$$\begin{aligned} \frac{5}{8} \cdot 280 &= \left(\frac{1}{2} + \frac{1}{8}\right) \cdot 280 \\ &= \frac{1}{2} \cdot 280 + \frac{1}{2} \cdot \frac{1}{2} \cdot \frac{1}{2} \cdot 280 \\ &= 140 + 35 = 175 \end{aligned}$$

The distributive property was used at the second equal sign. The associative property of multiplication could be used at the third equal sign to find $\frac{1}{8}$ of 280 by finding $\frac{1}{2}$ of 280, then $\frac{1}{2}$ of that result, and half of that result.

d.
$$\begin{aligned} 99\% \cdot 80 &= (100\% - 1\%) \cdot 80 \\ &= 100\% \cdot 80 - 1\% \cdot 80 \\ &= 80 - 0.8 \\ &= 79.2 \end{aligned}$$

The distributive property was used at the second equal sign.

6. a.
$$\begin{aligned} 25 \cdot 84 &= 25 \cdot (4 \cdot 21) \\ &= (25 \cdot 4) \cdot 21 \\ &= 100 \cdot 21 \\ &= 2100 \end{aligned}$$

The associative property of multiplication was used to rewrite $25 \cdot (4 \cdot 21)$ as $(25 \cdot 4) \cdot 21$—in other words, to multiply the 4 with the 25 instead of with the 21.

b.
$$\begin{aligned} 49 \cdot 6 &= (50 - 1) \cdot 6 \\ &= 50 \cdot 6 - 1 \cdot 6 \\ &= 300 - 6 \\ &= 294 \end{aligned}$$

The distributive property was used to rewrite $(50 - 1) \cdot 6$ as $50 \cdot 6 - 1 \cdot 6$.

c.
$$\begin{aligned} 486 \cdot 5 &= (243 \cdot 2) \cdot 5 \\ &= 243 \cdot (2 \cdot 5) \\ &= 243 \cdot 10 \\ &= 2430 \end{aligned}$$

The associative property of multiplication was used to rewrite $(243 \cdot 2) \cdot 5$ as $243 \cdot (2 \cdot 5)$.

7. The following sequence of equations shows in detail why Joey's method is valid and how it uses properties of arithmetic:

$$\begin{aligned} 24 \cdot 60 &= 24 \cdot (6 \cdot 10) \\ &= (24 \cdot 6) \cdot 10 \\ &= [(25 - 1) \cdot 6] \cdot 10 \\ &= (25 \cdot 6 - 1 \cdot 6) \cdot 10 \\ &= [6 \cdot 25 - 6] \cdot 10 \\ &= [(4 + 2) \cdot 25 - 6] \cdot 10 \\ &= [4 \cdot 25 + 2 \cdot 25 - 6] \cdot 10 \\ &= [100 + 50 - 6] \cdot 10 \\ &= (150 - 6) \cdot 10 \\ &= 144 \cdot 10 = 1440 \end{aligned}$$

The associative property of multiplication is used at the second equal sign.

The distributive property is used at the fourth and seventh equal signs.

The commutative property of multiplication is used at the fifth equal sign.

PROBLEMS FOR SECTION 4.5

1. Josh consistently remembers that $7 \times 7 = 49$, but he keeps forgetting 7×8.
   a. Explain to Josh how 7×7 and 7×8 are related. Draw an array to help you show this relationship.
   b. Write an equation relating 7×8 to 7×7. Which property of arithmetic does your equation use? Explain.

2. Demarcus knows his 1×, 2×, and 3× multiplication tables. He also knows 4×1, 4×2, 4×3, 4×4, and 4×5.
   a. Describe how the three arrays in Figure 4.36 provide Demarcus with three different ways to determine 4×6 from multiplication facts that he already knows. In each case, write an equation that corresponds to the array and that shows how 4×6 is related to other multiplication facts.

Figure 4.36 Different ways to think about 4×6.

   b. Draw arrays showing two different ways that Demarcus could use the multiplication facts he already knows to determine 4×7. In each case, write an equation that corresponds to the array and that shows how 4×7 is related to other multiplication facts.
   c. Draw arrays showing two different ways that Demarcus could use the multiplication facts he already knows to determine 4×8. In each case, write an equation that corresponds to the array and that shows how 4×8 is related to other multiplication facts.

---

3. Suppose that a student has learned the following basic multiplication facts:
   - The ×1, ×2, ×3, ×4, and ×5 tables—that is,

$1 \times 1 = 1$	$1 \times 2 = 2$	$1 \times 3 = 3$	$1 \times 4 = 4$	$1 \times 5 = 5$
$2 \times 1 = 2$	$2 \times 2 = 4$	$2 \times 3 = 6$	$2 \times 4 = 8$	$2 \times 5 = 10$
$3 \times 1 = 3$	$3 \times 2 = 6$	$3 \times 3 = 9$	$3 \times 4 = 12$	$3 \times 5 = 15$
$4 \times 1 = 4$	$4 \times 2 = 8$	$4 \times 3 = 12$	$4 \times 4 = 16$	$4 \times 5 = 20$
$5 \times 1 = 5$	$5 \times 2 = 10$	$5 \times 3 = 15$	$5 \times 4 = 20$	$5 \times 5 = 25$
$6 \times 1 = 6$	$6 \times 2 = 12$	$6 \times 3 = 18$	$6 \times 4 = 24$	$6 \times 5 = 30$
$7 \times 1 = 7$	$7 \times 2 = 14$	$7 \times 3 = 21$	$7 \times 4 = 28$	$7 \times 5 = 35$
$8 \times 1 = 8$	$8 \times 2 = 16$	$8 \times 3 = 24$	$8 \times 4 = 32$	$8 \times 5 = 40$
$9 \times 1 = 9$	$9 \times 2 = 18$	$9 \times 3 = 27$	$9 \times 4 = 36$	$9 \times 5 = 45$

   - The squares $1 \times 1 = 1, 2 \times 2 = 4, 3 \times 3 = 9, \ldots, 9 \times 9 = 81$.

   For each of the multiplication problems (a) through (c), find at least two different ways that some of the preceding facts, together with properties of arithmetic, could be used to mentally calculate the answer to the problem. Explain your answers, drawing arrays to illustrate how the following problems are related to facts in the given lists:

   a. 6×7
   b. 7×8
   c. 6×8

4. For each of the multiplication problems (a) through (c), describe a way to make the problem easy to solve mentally. Then write equations that correspond to your method of calculation. Write your equations in the following form:

$$\begin{aligned} 24 \cdot 25 &= \text{some expression} \\ &= \text{some expression} \\ &= \vdots \end{aligned}$$

In each case, state which properties of arithmetic you used, and indicate where you used those properties.

a. $24 \cdot 25$

b. $25 \cdot 48$

c. $51 \cdot 6$

5. Suppose that the sales tax where you live is 6%. Compare the total amount of sales tax you would pay if you bought a pair of pants and a shirt at the same time, versus if you first bought the pants and then went back to the store and bought the shirt. Write equations to show why the distributive property is relevant to this problem.

6. Clint and Sue went out to dinner and had a nice meal that cost \$64.82. With a 7% tax of \$4.54, the total came to \$69.36. They want to leave a tip of approximately 15% of the cost of the meal (before the tax). Describe a way that Clint and Sue can mentally figure the tip.

7. Your favorite store is having a 10%-off sale, meaning that the store will take 10% off the price of each item you buy. When the clerk rings up your purchases, she takes 10% off the total (before tax), rather than 10% off each item. Will you get the same discount either way? Is there a property of arithmetic related to this? Explain!

8. The exchanges that follow are taken from *Developing Children's Understanding of the Rational Numbers: A New Model and an Experimental Curriculum* by Joan Moss and Robbie Case [53, p. 135]. "*Experimental S1*" and "*Experimental S3*" are two of the fourth-grade students who participated in an experimental curriculum described in the article.*

Experimenter: What is 65% of 160?

Experimental S1: Fifty percent (of 160) is 80. I figure 10%, which would be 16. Then I divided by 2, which is 8 (5%) then 16 plus 8 um . . . 24. Then I do 80 plus 24, which would be 104.

Experimental S3: Ten percent of 160 is 16; 16 times 6 equals 96. Then I did 5%, and that was 8, so . . . , 96 plus 8 equals 104.

For each of the two students' responses, write strings of equations that correspond to the student's method for calculating 65% of 160. State which properties of arithmetic were used and where. (Be specific.)

Write your string of equations in the following form:

$$\begin{aligned} 65\% \cdot 160 &= \text{some expression} \\ &= \vdots \\ &= \text{some expression} \end{aligned}$$

9. Here is Marco's method for calculating $38 \cdot 60$:

Four times 6 is 24, so 40 times 60 is 2400. Then 2 times 6 is 12, so it's 2400 − 120, which is 2280.

Write equations that incorporate Marco's method and that also show why his method is valid. Write your equations in the following form:

$$\begin{aligned} 38 \cdot 60 &= \text{some expression} \\ &= \text{some expression} \\ &= \vdots \\ &= 2280 \end{aligned}$$

Which properties of arithmetic did Marco use (knowingly or not), and where?

10. Jenny uses the following method to find 28% of 60,000 mentally:

Twenty-five percent is $\frac{1}{4}$, and $\frac{1}{4}$ of 60 is 15, so 25% of 60,000 is 15,000. One percent of 60,000 is 600, and that times 3 is 1800. So the answer is 15,000 + 1,800, which is 16,800.

Write a string of equations that calculates 28% of 60,000 and that incorporates Jenny's ideas. Write your equations in the following form:

$$\begin{aligned} 28\% \cdot 60{,}000 &= \text{some expression} \\ &= \text{some expression} \\ &= \vdots \\ &= 16{,}800 \end{aligned}$$

11. Use properties of arithmetic to calculate 35% of 440 mentally. Describe your strategy in words, and write a string of equations that corresponds

*From Journal for Research in Mathematics Education, Vol. 30, No. 2, 122–147 by Robbie Case and Joan Moss. Copyright © 1999 by National Council for Teachers of Mathematics (NCTM). Used by permission of National Council for Teachers of Mathematics (NCTM).

to your strategy. Indicate which properties of arithmetic you used, and where. Be specific. Write your equations in the following form:

$$\begin{aligned} 35\% \cdot 440 &= \text{some expression} \\ &= \vdots \\ &= \text{some expression} \end{aligned}$$

12. Use the distributive property to make it easy for you to calculate 30% of 240 mentally. Then use the associative property of multiplication to solve the same problem. In each case, explain your strategy in words, and then write equations that correspond to your strategy. Write your equations in the following form:

$$\begin{aligned} 30\% \cdot 240 &= \text{some expression} \\ &= \text{some expression} \\ &= \vdots \end{aligned}$$

13. Tamar calculated $41 \cdot 41$ as follows:

Four 4s is 16, so four 40s is 160 and forty 40s is 1600. Then forty-one 40s is another 40 added on, which is 1640. So forty-one 41s is 41 more, which is 1681.

a. Explain briefly why it makes sense for Tamar to solve the problem the way she does. What is the idea behind her strategy?

b. Write equations that incorporate Tamar's work and that show clearly why Tamar's method calculates the correct answer to $41 \cdot 41$. Which properties of arithmetic did Tamar use (knowingly or not), and where? Be thorough and be specific. Write your equations in the following format:

$$\begin{aligned} 41 \cdot 41 &= \text{some expression} \\ &= \text{some expression} \\ &= \vdots \\ &= 1681 \end{aligned}$$

14. Here is how Nya solved the problem $\frac{3}{4} \cdot 72$:

Half of 72 is 36. Half of 36 is 18. Then, to add 36 and 18, I did 40 plus 14, which is 54.

Write a string of equations that incorporate Nya's ideas. Which properties of arithmetic did Nya use (knowingly or not), and where? Be thorough and be specific. Write your equations in the following format:

$$\begin{aligned} \frac{3}{4} \cdot 72 &= \text{some expression} \\ &= \text{some expression} \\ &= \vdots \\ &= 54 \end{aligned}$$

15. **a.** Lindsay calculates two-fifths of 1260 by using the following strategy: First, she finds half of 1260, which is 630. Then she subtracts one tenth of 1260, which is 126, from 630 and gives the answer, 504. Discuss the ideas behind Lindsay's strategy. Then write a string of equations that incorporate this strategy and that show why the strategy is valid. What property of arithmetic is involved? Write your equations in the following form:

$$\begin{aligned} \frac{2}{5} \cdot 1260 &= \text{some expression} \\ &= \text{some expression} \\ &= \vdots \\ &= 504 \end{aligned}$$

b. Terrell calculates $\frac{2}{5}$ of 1260 in the following way: First he multiplies 1260 by 2 to get 2520. Then he multiplies 2520 by 2 to get 5040 and divides this by 10 to get 504. Discuss the idea behind Terrell's strategy. Then write a string of equations that incorporate Terrell's strategy, and that show why the strategy is valid. Write your equations in the form shown previously.

16. While working on the multiplication problem $38 \cdot 25$, Maria says this:

Four times 25 is 100, so 40 times 25 is 1000. Now take away 2, so the answer is 998.

Is Maria's method correct or not? If it is correct, write equations that incorporate Maria's work and that show why it's correct. If Maria's reasoning is not correct, work with portions that are right to correct Maria's work, and write a string of equations that corresponds to your corrected method for calculating $38 \cdot 25$. Write your equations in the following form:

$$\begin{aligned} 38 \cdot 25 &= \text{some expression} \\ &= \text{some expression} \\ &= \vdots \end{aligned}$$

*17. There is an interesting mental technique for multiplying certain pairs of numbers. The following examples will illustrate how it works:

- To calculate $32 \cdot 28$, notice that the two factors 28 and 32 are both 2 away—in opposite directions—from 30. To calculate $32 \cdot 28$, do the following:

$$\begin{aligned} 30 \cdot 30 - 2 \cdot 2 &= 900 - 4 \\ &= 896 \end{aligned}$$

Notice that you can do this calculation in your head.

- Similarly, to calculate $59 \cdot 61$, notice that both factors are 1 away—in opposite directions—from 60. Then $59 \cdot 61$ is

$$\begin{aligned} 60 \cdot 60 - 1 \cdot 1 &= 3600 - 1 \\ &= 3599 \end{aligned}$$

Once again, notice that you can do this mentally.

a. Use the method just shown to calculate $83 \cdot 77$, $195 \cdot 205$, and one other multiplication problem like this that you make up.

b. Now explain why this method works. *Hint:* A drawing might be helpful. Another approach is to notice that the technique applies to multiplication problems of the form $(A + B) \cdot (A - B)$.

***18.** Try out this next mathematical magic trick. Do the following on a piece of paper:

a. Write the number of days a week you would like to go out (from 1 to 7).

b. Multiply the number by 2.

c. Add 5.

d. Multiply by 50.

e. If you have already had your birthday this year, add 1762 if it is 2012. (Add 1763 if it is 2013, add 1764 if it is 2014, and so on.) If you have not yet had your birthday this year, add 1761 if it is 2012. (Add 1762 if it is 2013, add 1763 if it is 2014, and so on.)

f. Finally, subtract the 4-digit year you were born. You should now have a 3-digit number. If not, try again. If you have a 3-digit number, continue with the following:

The first digit of your answer is your original number (i.e., how many times you want to go out each week). The second two digits are your current age.

Is it magic, or is it math? Explain why the trick works.

4.6 Why the Standard Algorithm for Multiplying Whole Numbers Works

CCSS Common Core State Standards Grades 4, 5

In this section, we draw on much of our work in the chapter—multiplication by 10, the commutative and associative properties of multiplication, and the distributive property—to explain why the standard algorithm (procedure) for multiplying whole numbers is valid.

The standard algorithm for multiplying multiple-digit whole numbers is an efficient paper-and-pencil method of calculation. This method is useful because it converts a multiplication problem with numbers that have several digits to many multiplication problems with 1-digit numbers. Therefore, someone who knows the 1-digit multiplication tables (from $1 \times 1 = 1$ to $9 \times 9 = 81$) can multiply any pair of whole numbers by using the standard algorithm. But why does this clever method give the correct answer to multiplication problems? What makes it work?

Before we continue, let's first make sense of the questions at the end of the previous paragraph. Notice the distinction between what multiplication *means* and the longhand *procedure* for multiplying. If you have 58 bags of widgets and there are 764 widgets in each bag, then you can ask how many widgets you have in all. There is some specific total number of widgets. But what is this number? According to what multiplication means, the total number of widgets is

$$58 \times 764$$

The usual way of writing the steps in the standard algorithm is as follows:

Multiply 8×4, write the 2, carry the 3; multiply 8×6, add the carried 3, write the 1, carry the 5, etc.

Why does this process for calculating 58 × 764 give the actual total number of widgets in 58 bags that have 764 widgets in each bag?

The answer will be explained in this section.

What Are Methods for Writing the Steps of the Standard Multiplication Algorithm?

partial-products method

One especially accessible way of recording the steps of the standard multiplication algorithm is called the **partial-products method**. The partial-products method is almost the same as the common way of writing the standard algorithm, but it shows more steps, and children are better able to illustrate and explain calculations using it. A few examples will illustrate how the partial-products method works. In the examples below, the arrows and the expressions to the right of the arrows have been added to show how to carry out the steps. You do not necessarily have to write these arrows and expressions when you use the method yourself.

Common Method

$$\begin{array}{r} {}^{4} \\ 38 \\ \times\ 6 \\ \hline 228 \end{array}$$

Partial-Products Method

$$\begin{array}{rl} 38 & \\ \times\ 6 & \\ \hline 48 & \leftarrow 6 \times 8 \\ 180 & \leftarrow 6 \times 3 \text{ tens} \\ \hline 228 & \leftarrow \text{add} \end{array}$$

Common Method

$$\begin{array}{r} {}^{2\,1} \\ 274 \\ \times\ \ 3 \\ \hline 822 \end{array}$$

Partial-Products Method

$$\begin{array}{rl} 274 & \\ \times\ 3 & \\ \hline 12 & \leftarrow 3 \times 4 \\ 210 & \leftarrow 3 \times 7 \text{ tens} \\ 600 & \leftarrow 3 \times 2 \text{ hundreds} \\ \hline 822 & \leftarrow \text{add} \end{array}$$

Common Method

$$\begin{array}{r} {}^{1} \\ {}^{1} \\ 45 \\ \times\ 23 \\ \hline 135 \\ 900 \\ \hline 1035 \end{array}$$

Partial-Products Method

$$\begin{array}{rl} 45 & \\ \times\ 23 & \\ \hline 15 & \leftarrow 3 \times 5 \\ 120 & \leftarrow 3 \times 4 \text{ tens} \\ 100 & \leftarrow 2 \text{ tens} \times 5 \\ 800 & \leftarrow 2 \text{ tens} \times 4 \text{ tens} \\ \hline 1035 & \leftarrow \text{add} \end{array}$$

How Can We Relate the Common and Partial-Products Written Methods for the Standard Algorithm?

Compare the common and partial-products methods for writing the steps in calculating 58×764:

Common Method

$$\begin{array}{rl} {}^{32} & \\ {}^{53} & \\ 764 & \\ \times\ 58 & \\ \hline 6{,}112 & \leftarrow 32 + 480 + 5{,}600 \\ & \\ 38{,}200 & \leftarrow 200 + 3{,}000 + 35{,}000 \\ \hline 44{,}312 & \end{array}$$

Partial-Products Method

$$\begin{array}{rl} 764 & \\ \times\ 58 & \\ \hline 32 & \leftarrow 8 \times 4 \\ 480 & \leftarrow 8 \times 6 \text{ tens} \\ 5{,}600 & \leftarrow 8 \times 7 \text{ hundreds} \\ 200 & \leftarrow 5 \text{ tens} \times 4 \\ 3{,}000 & \leftarrow 5 \text{ tens} \times 6 \text{ tens} \\ 35{,}000 & \leftarrow 5 \text{ tens} \times 7 \text{ hundreds} \\ \hline 44{,}312 & \end{array}$$

Notice that the common method condenses the steps in the partial-products method. In this example, three lines of calculations in the partial-products method are condensed to one line in the common method. The 6,112 produced in the common method can be obtained by adding the three lines 8×4, 8×6 tens, and 8×7 hundreds of the partial-products method; the 38,200 can be obtained by adding 5 tens $\times$ 4, 5 tens $\times$ 6 tens, and 5 tens $\times$ 7 hundreds. When you multiply and then add carried numbers in the common method, you are really just combining lines from the partial-products method.

Because the partial-products method is an expanded version of the common method, we can explain why the common method correctly calculates the answers to multiplication problems by explaining why the partial-products method correctly calculates the answers to multiplication problems. We can also use the partial-products method to explain why we put zeros in some of the lines of the common method.

Why Do We Place Extra Zeros on Some Lines When We Use the Common Method to Record the Steps of the Standard Algorithm?

We can use the partial-products method to explain why we put an initial zero in the ones column of the second line in the common method. In the previous example, this is the line produced by 5×4, 5×6, and 5×7. As the partial-products method reveals, that line is actually produced by *multiplying by 50 rather than by 5*. By placing the initial zero in the ones place, all digits are moved one place to the left, which has the same effect as multiplying by 10. The net effect is that this second line is now actually 50 times 764 rather than 5 times 764.

What if the multiplication problem was 258×764 instead of 58×764? Using the common method, there would be a third line produced by 2×4, 2×6, and 2×7. Before beginning the calculations for this third line, you would put down two zeros. Why two zeros? Because this line should really be produced by multiplying by 200 instead of by 2. By placing zeros in the ones and tens places, all digits are moved two places to the left, which has the same effect as multiplying by 100. The net effect is that the third line is then 200 times 764 instead of 2 times 764.

Why Does the Standard Multiplication Algorithm Produce Correct Answers?

CLASS ACTIVITY

4N How Can We Develop and Understand the Standard Multiplication Algorithm? p. CA-83

CCSS 4.NBT.5 5.NBT.5

We will now see why the standard multiplication algorithm calculates correct answers to multiplication problems. We will use the example 58×764 to illustrate. Think of this problem as representing the total number of widgets in 58 rows that each contain 764 widgets. The key to explaining why the standard algorithm gives the correct answer to 58×764 lies in observing that each step in the algorithm counts the number of widgets in a portion of the 58 rows of 764 widgets.

Figure 4.37 indicates 58 rows with 764 widgets in each row. Therefore, according to the definition of multiplication, Figure 4.37 represents a total of 58×764 widgets.

The widget collection in Figure 4.37 has been decomposed into portions. How are these portions determined? The portions come from decomposing the numbers 58 and 764 into their place value parts. Notice that the number of widgets in each portion corresponds to a line in the partial-products method. For instance, the line 5600, which comes from 8×700, is the number of widgets in the piece in the lower left-hand corner, which represents 8 groups of 700 widgets. Similarly, the line 200, which comes from 50×4, is the number of widgets in the piece at the top right-hand corner,

Figure 4.37

Subdividing 58 rows of 764 widgets.

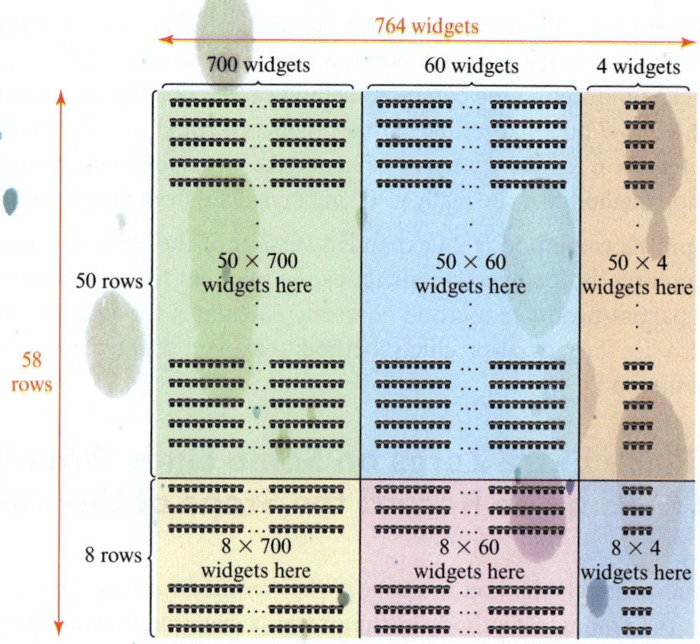

which represents 50 groups of 4 widgets. It's the same for all the other lines in the partial-products method—each line corresponds to a portion of the collection of widgets. When you add up the lines in the method, you get the total number of widgets. Therefore, the partial-products method correctly calculates the total number of widgets; in other words, it correctly calculates 58 × 764. Since the common method is just a condensation of the partial-products method, it too must correctly calculate answers to multiplication problems.

We can also write equations that show how the reasoning we just used is based on place value and the distributive property:

$$\begin{aligned} 58 \times 764 &= (50 + 8) \times (700 + 60 + 4) \\ &= 50 \times (700 + 60 + 4) + 8 \times (700 + 60 + 4) \\ &= 50 \times 700 + 50 \times 60 + 50 \times 4 + 8 \times 700 + 8 \times 60 + 8 \times 4 \\ &= 35{,}000 + 3{,}000 + 200 + 5{,}600 + 480 + 32 \\ &= 44{,}312 \end{aligned}$$

As the preceding equations show, the six lines produced by the partial-products method are exactly the six products

$$50 \times 700, \quad 50 \times 60, \quad 50 \times 4, \quad 8 \times 700, \quad 8 \times 60, \quad 8 \times 4$$

produced by using the distributive property several times, after decomposing 58 and 764 into their place value parts.

SECTION SUMMARY AND STUDY ITEMS

Section 4.6 Why the Standard Algorithm for Multiplying Whole Numbers Works

The partial-products written method for the standard algorithm is closely related to the common method but shows more detail. We can explain the validity of the standard multiplication algorithm by subdividing arrays according to place value and by writing numbers in expanded form and applying the distributive property.

Key Skills and Understandings

1. Relate a multiplication problem to an array, and subdivide the array so that the pieces correspond to the lines in the partial-products method. Use the subdivision to explain the validity of the partial-products method. Show the portions of the array that correspond to the lines in the common method.
2. Solve a multiplication problem by writing equations that use expanded forms and the distributive property. Relate the equations to the lines in the partial-products method. Use the relationship to explain why the partial-products method calculates the correct answer to the multiplication problem.
3. Understand that the standard multiplication algorithm can be explained in terms of the definition of multiplication, place value, and properties of arithmetic.

Practice Exercises for Section 4.6

1. Draw an array on graph paper, and use your array to explain why the partial-products method calculates the answer to 24×35. If graph paper is not available, draw a rectangle to represent your array rather than drawing 24 rows with 35 items in each row.
2. Relate the partial-products method and array you drew for the previous exercise to the steps in the common method for calculating 24×35.
3. Solve the multiplication problem 24×35 by writing equations that use expanded forms and the distributive property. Relate your equations to the steps in the partial-products method for calculating 24×35.
4. Relate your equations from the previous exercise to the steps in the common method for writing the standard algorithm for calculating 24×35.
5. Cameron wants to calculate 23×23. She says,

$$20 \times 20 = 400 \text{ and } 3 \times 3 = 9$$

so

$$23 \times 23 = 400 + 9 = 409$$

Critique, Cameron's reasoning and compare her work with the steps in the partial-products method for calculating 23×23.

Answers to Practice Exercises for Section 4.6

1. Figure 4.38 shows an array consisting of 24 rows of small squares with 35 small squares in each row. According to the meaning of multiplication, there is a total of 24×35 small squares in the array. The shading shows how to subdivide the array into 4 pieces: one with 20 rows of 30, one with 20 rows of 5, one with 4 rows of 30, and one with 4 rows of 5 small squares. The number of small squares in each of these 4 pieces is calculated as follows:

$$20 \times 30 = 600$$
$$20 \times 5 = 100$$
$$4 \times 30 = 120$$
$$4 \times 5 = 20$$

But the number of small squares in these 4 pieces corresponds exactly to the 4 lines we produce when we use the partial-products method to calculate 24×35:

$$\begin{array}{r} 35 \\ \times\ 24 \\ \hline 20 \\ 120 \\ 100 \\ +\ 600 \\ \hline 840 \end{array}$$

So the steps in the partial-products method calculate the number of squares in the four parts that make up a full array of 24 rows of 35 squares. When we add the numbers together at the end, we add up the total number of squares in all four parts, and therefore get the total number of small squares in 24 rows of 35 squares, which is 24 × 35 small squares.

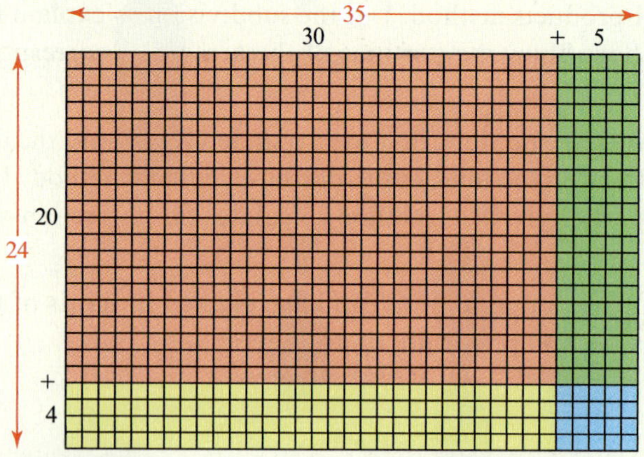

Figure 4.38 An array of 24 rows with 35 small squares in each row subdivided into four parts.

2. The common method for writing the standard algorithm is a condensed version of the partial-products method. The two lines we produce when we multiply 24 × 35 by using the common method each come from combining two of the lines in the partial-products method:

$$\begin{array}{r}
\overset{1}{\overset{2}{3}}5 \\
\times\ 24 \\
\hline
140 \\
700 \\
\hline
840
\end{array}
\qquad
\begin{array}{r}
35 \\
\times\ 24 \\
\hline
\left\{\begin{array}{r} 20 \\ 120 \end{array}\right. \\
\left\{\begin{array}{r} 100 \\ 600 \end{array}\right. \\
\hline
840
\end{array}$$

In terms of the array in Figure 4.38, the line 140 in the standard algorithm comes from the bottom 4 rows, which represent 4 × 35; the line 700 in the common method comes from the top 20 rows, which represent 20 × 35.

3. $24 \times 35 = (20 + 4) \times (30 + 5)$

$$\begin{aligned}
&= 20 \times (30 + 5) + 4 \times (30 + 5) \\
&= 20 \times 30 + 20 \times 5 + 4 \times 30 + 4 \times 5 \\
&= 600 + 100 + 120 + 20 \\
&= 840
\end{aligned}$$

The distributive property was used at the second and third equal signs. The four products, 20 × 30, 20 × 5, 4 × 30, and 4 × 5, that result from the use of expanded forms and the distributive property to calculate 24 × 35 are exactly the four products seen in the partial-products method (see exercise 1).

4. Refer to the equations in the answer to exercise 3. The two terms in the second line of the equations—namely, 4 × (30 + 5) and 20 × (30 + 5)—correspond to the first and second lines, respectively, produced by the common method—namely, 140 and 700 (see the answer to exercise 2). So the common method comes from breaking the bottom factor, 24, into expanded form 20 + 4, and multiplying each of these components by the top factor, 35. The common method calculates 24 × 35 by calculating

$$20 \times 35 + 4 \times 35$$

which uses the distributive property.

5. When we calculate 23 × 23 using the partial-products method, we produce four lines of products:

$$\begin{array}{r}
23 \\
\times\ 23 \\
\hline
9 \\
60 \\
60 \\
+\,400 \\
\hline
529
\end{array}$$

Cameron's calculations produced only two of those lines—the 9 and the 400—so she is missing the two lines of 60. We can see these two missing 60s in Figure 4.39, which shows 23 rows of 23 small squares, subdivided into four portions. The array shows that Cameron has left out 20 × 3 and 3 × 20, which correspond to the darkly shaded portions.

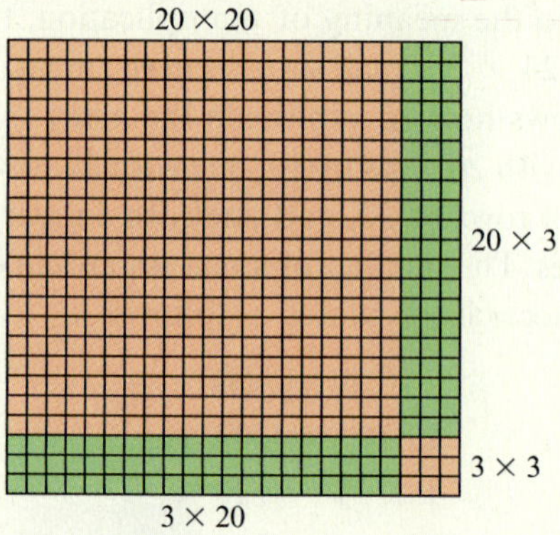

Figure 4.39 23 × 23 = 20 × 20 + 20 × 3 + 3 × 20 + 3 × 3.

PROBLEMS FOR SECTION 4.6

1. Solve the multiplication problem

$$\begin{array}{r} 96 \\ \times\ 8 \\ \hline \end{array}$$

in three different ways: by using the common method for writing the standard algorithm, by using the partial-products method for writing the standard algorithm, and by writing the numbers in expanded forms and using properties of arithmetic. For each of the three methods, discuss how the steps in that method are related to the steps in the other methods.

2. Solve the multiplication problem

$$\begin{array}{r} 84 \\ \times\ 76 \\ \hline \end{array}$$

in three different ways: by using the common method, by using the partial-products method, and by writing the numbers in expanded forms and using properties of arithmetic. Discuss how the steps in the partial-products method are related to the steps in the other methods.

3. Solve the multiplication problem

$$\begin{array}{r} 237 \\ \times\ 43 \\ \hline \end{array}$$

in three different ways: by using the common method, by using the partial-products method, and by writing the numbers in expanded forms and using properties of arithmetic. Discuss how the steps in the partial-products method are related to the steps in the other methods.

4. When we multiply

$$\begin{array}{r} 37 \\ \times\ 26 \\ \hline \end{array}$$

by using the common method, we start the second line by writing a zero:

$$\begin{array}{r} \overset{4}{3}7 \\ \times\ 26 \\ \hline 222 \\ 0 \end{array}$$

Explain why we place this zero in the second line.

5. **a.** Use the partial-products method to calculate 8×24.

   b. Draw an array for 8×24. (You may wish to use graph paper.) Subdivide the array in a natural way so that the parts of the array correspond to the steps in the partial-products method.

   c. Solve 8×24 by writing equations that use expanded forms and the distributive property. Relate your equations to the steps in the partial-products method.

6. **a.** Use the partial-products method to calculate 19×28.

   b. On graph paper, draw an array for 19×28. If graph paper is not available, draw a rectangle to represent the array rather than drawing 19 rows with 28 items in each row. Subdivide the array in a natural way so that the parts of the array correspond to the steps of the partial-products method.

   c. Solve 19×28 by writing equations that use expanded forms and the distributive property. Relate your equations to the steps in the partial-products method.

7. **a.** Use the partial-products and common methods to calculate 27×28.

   b. On graph paper, draw an array for 27×28. If graph paper is not available, draw a rectangle to represent the array rather than drawing 27 rows with 28 items in each row. Subdivide the array in a natural way so that the parts of the array correspond to the steps in the partial-products method.

   c. On the array that you drew for part (b), show the parts that correspond to the steps of the common method.

   d. Solve 27×28 by writing equations that use expanded forms and the distributive property. Relate your equations to the steps in the partial-products method.

8. **a.** Draw an array on graph paper, and use your array to explain why the partial-products method calculates the correct answer to 23×27. If graph paper is not available, draw a rectangle to represent the array rather than drawing 23 rows of 27 items.

   b. Relate the array you drew for part (a) to the steps in the common method for calculating 23×27. Use this relationship to explain why the common method calculates the correct answer to 23×27.

9. Solve the multiplication problem 23 × 27 by writing equations that use expanded forms and the distributive property. Relate your equations to the steps in the partial-products method for calculating 23 × 27. Use this relationship to explain why the partial-products method calculates the correct answer to 23 × 27.

10. a. Use the partial-products and common methods to calculate 43 × 275.

b. Draw a large rectangle to represent an array for 43 × 275. Subdivide the rectangle in a natural way so that the parts of the rectangle correspond to the steps in the partial-products method.

c. On the rectangle that you drew for part (b), show the parts that correspond to the steps in the common method.

d. Solve 43 × 275 by writing equations that use expanded forms and the distributive property. Relate your equations to the steps in the partial-products method.

11. a. Use the common method to calculate

$$\begin{array}{r} 37 \\ \times\ 24 \\ \hline \end{array}$$

b. On graph paper, draw an array for 24 × 37 (or draw a rectangle to represent such an array). Subdivide the array in a natural way so that the parts of the array correspond to the steps of the common method in part (a).

c. Write equations that use the distributive property and that correspond to the steps in the common method in part (a).

d. Use the common method to calculate

$$\begin{array}{r} 24 \\ \times\ 37 \\ \hline \end{array}$$

e. On graph paper, draw an array for 37 × 24 (or draw a rectangle to represent such an array). Subdivide the array in a natural way so that the parts of the array correspond to the steps of the common method in part (d). Compare with part (b).

f. Write equations that use the distributive property and that correspond to the steps in the common method in part (d). Are these the same equations as in part (c)?

g. Other than the distributive property, which you used in parts (c) and (f), which property of arithmetic is relevant to this problem? Explain and discuss.

12. The **lattice method** is a technique that is sometimes used for multiplication. Figure 4.40 shows how to use this method to multiply a 2-digit number by a 2-digit number.

a. Use the lattice method to calculate 38 × 54 and 72 × 83.

b. Use the partial-products method to calculate 38 × 54 and 72 × 83.

c. Explain how the lattice method is related to the partial-products method.

d. Discuss advantages and disadvantages of using the lattice method.

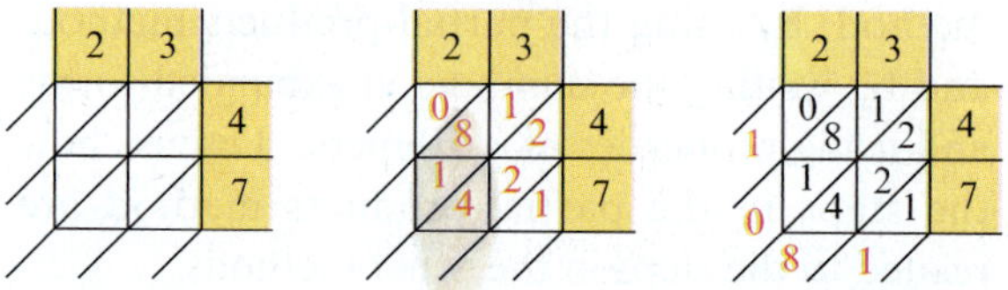

Draw a lattice as shown in the figure. Place the numbers you want to multiply on the top and down the side.

Multiply each digit along the top with each digit along the side. Write the answers in the cells as shown.

Starting at the bottom right, and moving left, then up, add numbers in diagonal strips, carrying as necessary. The answer is 1081.

Figure 4.40 Using the lattice method to show that 23 × 47 = 1081.

13. The following method for multiplying 21 × 23 relies on repeatedly doubling numbers (by adding them to themselves):

$$\begin{array}{rl} 23 & \\ +\ 23 & \\ \hline 46 & 2 \\ +\ 46 & \\ \hline 92 & 4 \\ +\ 92 & \\ \hline 184 & 8 \\ +\ 184 & \\ \hline 368 & 16 \end{array} \qquad 21 = 16 + 4 + 1 \qquad \begin{array}{r} 368 \\ 92 \\ +\ 23 \\ \hline 483 \end{array}$$

The answer, 483, to the multiplication problem 21 × 23, is shown on the bottom of the column on the right.

Notice that this method requires only addition; students do not need to know how to multiply.

a. Discuss and explain the method of repeated doubling. Address the following questions in your discussion:

i. What is the significance of the column of the numbers 2, 4, 8, 16?

ii. What is the significance of writing $21 = 16 + 4 + 1$?

iii. How are the numbers 368, 92, and 23 chosen for the column on the right? Why do we add those numbers?

b. Use the method of repeated doubling to calculate 26×35. Show your work.

c. Use the method of repeated doubling to calculate 37×51. Show your work.

CHAPTER SUMMARY

Section 4.1 Interpretations of Multiplication Page 143

- For non-negative numbers, the product $M \times N$ stands for the number of units in M equal groups of size N units. We can therefore show multiplicative structure in a situation by exhibiting equal groups. We can exhibit equal groups with arrays, organized lists, and tree diagrams. Page 143

Key Skills and Understandings

1. Explain why multiplication solves a problem by exhibiting or describing equal groups. Write a corresponding multiplication expression $(M \cdot N)$ or equation $(M \cdot N = P)$. Page 144
2. Write Array, Ordered Pair, and Multiplicative Comparison word problems for a given multiplication problem (such as $4 \cdot 6 = ?$), as well as multiplication word problems for which the equal groups are more evident. Page 145

Section 4.2 Why Multiplying by 10 Is Special in Base Ten Page 149

- Because the value of each place in the base-ten system is 10 times the value of the place to its right, multiplying a number by 10 moves each digit one place to the left. Page 150

Key Skills and Understandings

1. Describe multiplication by 10 as moving digits one place to the left. Page 150
2. Use math drawings of bundled objects to explain why multiplying a number by 10 moves each digit one place to the left. Page 150
3. Explain that because $100 = 10 \cdot 10$, and $1000 = 10 \cdot 10 \cdot 10$, multiplying a number by 100 moves each digit 2 places to the left and multiplying a number by 1000 moves each digit 3 places to the left. Explain similar results for other base-ten place values. Page 151

Section 4.3 The Commutative and Associative Properties of Multiplication, Areas of Rectangles, and Volumes of Boxes Page 152

- The commutative property of multiplication states that

$$A \cdot B = B \cdot A$$

for all numbers A and B. Page 153

- The area of a rectangle, in square units, is the number of 1-unit-by-1-unit squares it takes to cover the rectangle without gaps or overlaps. We can multiply $L \cdot W$ to find the area of an L-by-W rectangle because the rectangle can be subdivided into L rows, each of which contains W 1-unit-by-1-unit squares. Page 154
- We can explain why the commutative property of multiplication makes sense by subdividing an array into groups in two different ways. Page 155
- The associative property of multiplication states that

$$(A \cdot B) \cdot C = A \cdot (B \cdot C)$$

for all numbers A, B, and C. Page 157

- The volume of a box, in cubic units, is the number of 1-unit-by-1-unit-by-1-unit cubes it takes to fill or make the box (without gaps or overlaps). We can multiply $H \cdot (L \cdot W)$ to find the volume of a box that is H units high, L units long, and W units wide because the box can be subdivided into H layers, each of which contains L rows of W 1-unit-by-1-unit-by-1-unit cubes. Page 155
- We can explain why the associative property of multiplication makes sense by subdividing groups of grouped objects in two different ways. Page 157
- We use the combination of the commutative and associative properties in many mental calculations, such as in calculating $60 \cdot 700$ by recalling that $6 \cdot 7 = 42$ and moving 42 three places to the left to make 42,000. Page 158

Key Skills and Understandings

1. Explain why we can multiply to find the area of a rectangle by describing rectangles as subdivided into groups of 1-unit-by-1-unit squares. Page 154
2. State the commutative property of multiplication, and explain why it makes sense (for counting numbers) by subdividing rectangles or arrays in two different ways. Page 153
3. Explain why we can multiply to find the volume of a box by describing boxes as subdivided into groups of groups of 1-unit-by-1-unit-by-1-unit cubes. Page 155
4. State the associative property of multiplication, and explain why it makes sense (for counting numbers) by subdividing boxes two different ways or by subdividing groups of groups of objects in two different ways. Page 157
5. Give examples of how to use the associative and commutative properties of multiplication in problems and recognize when these properties have been used. Page 158
6. Apply the methods used in explaining the area formula for rectangles and the volume formula for boxes to estimate numbers of items by multiplying. Page 164

Section 4.4 The Distributive Property Page 166

- The distributive property states that

$$A \cdot (B + C) = A \cdot B + A \cdot C$$

for all numbers A, B, and C. We can explain why the distributive property is valid (for counting numbers) by expressing the total number of objects in a subdivided array in two different ways. Page 167

- We apply the distributive property in many calculations. This property is the single most powerful property of arithmetic. (In Section 4.6, we will see that it is the key idea underlying the multiplication algorithms.) Page 169

5

Multiplication of Fractions, Decimals, and Negative Numbers

In this chapter, we will study multiplication of fractions, decimals, and negative numbers. We will also study the related topics of powers and scientific notation. Why are the procedures we use in these topics valid? For example, why do we put the decimal point where we do when we multiply decimals? Why do we multiply fractions by multiplying the numerators and the denominators, even though we don't add fractions by adding the numerators and adding the denominators? We will answer these questions by using the definition of multiplication, as it applies to fractions and decimals. We will also study fraction and decimal word problems and see why we must pay close attention to the wording of these problems.

We focus on the following topics and practices within the *Common Core State Standards for Mathematics* (*CCSSM*) in this chapter.

Standards for Mathematical Content in the CCSSM

In the domain of *Number and Operations—Fractions* (Grades 3–5), students extend multiplication to fractions and they solve problems with and reason about fraction multiplication. In the domain of *Number and Operations in Base Ten* (K–Grade 5), students extend multiplication to decimals, they solve problems with and reason about decimal multiplication, and they work with powers of 10. In the domain of *The Number System* (Grades 6–8), students extend multiplication to negative numbers by extending properties of multiplication to negative numbers.

Standards for Mathematical Practice in the CCSSM

Opportunities to engage in all eight of the Standards for Mathematical Practice described in the CCSSM occur throughout the study of multiplication. The following standards may be especially appropriate for emphasis while studying multiplication of fractions, decimals, and negative numbers:

- **2 Reason abstractly and quantitatively.** Students engage in this practice when they extend the definition of multiplication to fractions and when they analyze a context to determine whether a problem can be solved by multiplying fractions.
- **3 Construct viable arguments and critique the reasoning of others.** Students engage in this practice when they use math drawings to compute products of fractions and to explain their reasoning.
- **7 Look for and make use of structure.** Students engage in this practice when they use the structure of the base-ten system to connect decimal multiplication to whole number multiplication and to reason about placement of the decimal point in a product of decimals.

  (From Common Core Standards for Mathematical Practice. Published by Common Core Standards Initiative.)

5.1 Making Sense of Fraction Multiplication

CCSS Common Core State Standards Grades 4, 5, 6

In this section, we will see how our definition of multiplication extends to the multiplication of fractions and why the standard procedure for multiplying fractions makes sense and agrees with the definition.

How Can We Extend Our Previous Understanding of Multiplication to Fractions?

Just as with whole number multiplication, we need to think about what multiplication with fractions means before we can make sense of fraction multiplication calculations. In Chapter 4 we defined multiplication as follows: If M and N are non-negative numbers, then

$$M \cdot N$$

means the number of units (or objects) in M equal groups if there are N units (or objects) in 1 group. Figure 5.1 summarizes the different roles that M, N, and their product, P, play in a multiplication equation $M \cdot N = P$.

M	•	**N**	=	**P**
Number of equal **groups**		Number of **units** in **1 group**		Number of **units** in **M groups**

Figure 5.1 The definition of multiplication applies to fractions and decimals as well as to whole numbers.

Fraction Multiplication, Part 1

Our definition of multiplication applies not only to whole numbers but also to fractions and decimals. Figure 5.2 provides some examples in which the groups are servings of food. Nutrition labels on food packages often describe quantities of food in terms of servings as well as in terms of standard measurement units such as cups, ounces, grams, or liters. We can imagine preparing 3 servings

CCSS
4.NF.4
5.NF.4a
5.NF.6

of a food (Figure 5.2 (a)), but we can also imagine eating less than a whole serving, such as $\frac{2}{3}$ of a serving or $\frac{1}{2}$ of a serving (Figures 5.2 (b) and (c)).

Notice that to clarify the meaning of multiplication when the multiplier M is a fraction, it helps to change the wording a little. For example, to interpret the multiplier $\frac{2}{3}$ in

$$\frac{2}{3} \cdot 15$$

Fraction Multiplication, Part 2

it makes more sense to say "$\frac{2}{3}$ *of a group*" than to say "$\frac{2}{3}$ *equal groups.*" In this case, we don't need to say "equal" because we already know that all three of the thirds that make the whole group are equal in size based on what $\frac{2}{3}$ means. If we had 4 groups or 3 groups or 2 groups, we would need to make clear that all those separate groups are the same size. Also, because $\frac{2}{3}$ is less than 1, we have less than a whole group, and the $\frac{2}{3}$ tells us *how much of a group* we have, so it makes sense to say

$$\frac{2}{3} \text{ of a group}$$

The wording "$\frac{2}{3}$ of a group" helps us see that for a group of size 15 units, the size of the product $\frac{2}{3} \cdot 15$ is

$$\frac{2}{3} \textit{ of } 15 \text{ units}$$

This is the familiar interpretation of multiplication as "*of*."

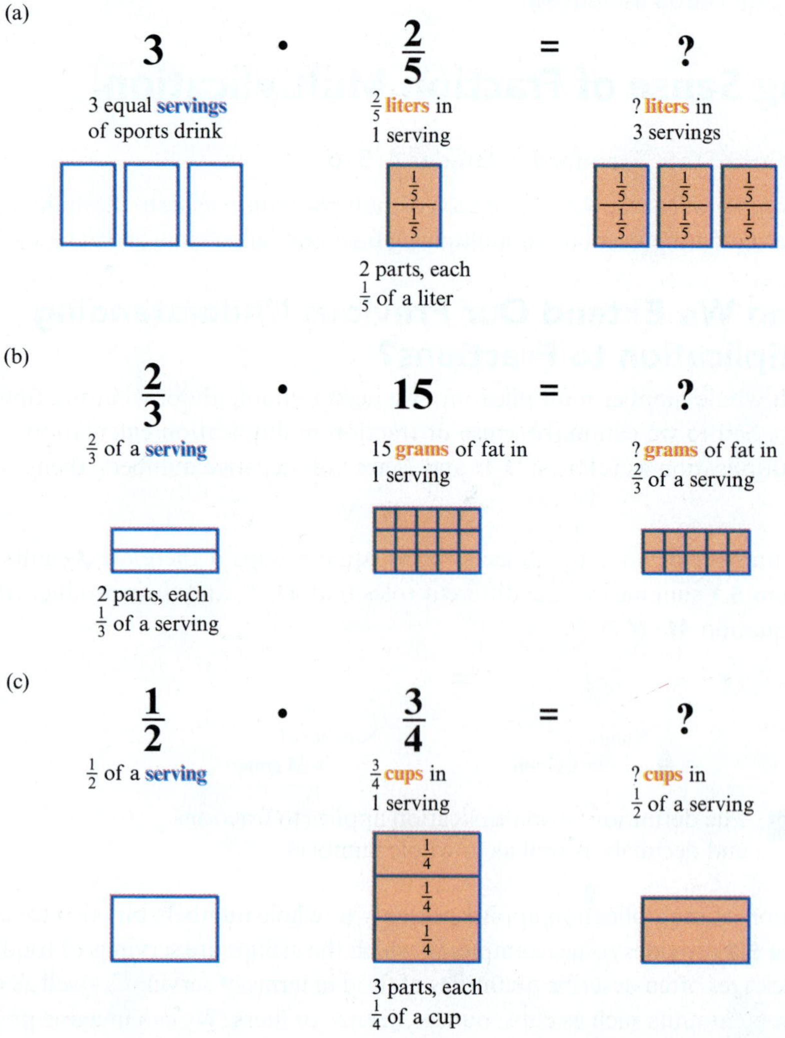

Figure 5.2 Interpreting multiplication with fractions.

CLASS ACTIVITY

5A Extending Multiplication to Fractions Part I, p. CA-86

5B Making Multiplicative Comparisons, p. CA-88

5C Extending Multiplication to Fractions, Part II, p. CA-89

What Is the Procedure for Multiplying Fractions?

The procedure for multiplying fractions is straightforward: Just multiply the numerators, and multiply the denominators. In other words,

$$\frac{A}{B} \cdot \frac{C}{D} = \frac{A \cdot C}{B \cdot D}$$

For example, consider the product

$$\frac{2}{3} \cdot \frac{4}{5} = \frac{2 \cdot 4}{3 \cdot 5} = \frac{8}{15}$$

Notice that the procedure can be used even when whole numbers are involved, because a whole number can always be written as a fraction by "putting it over 1." For example,

$$\frac{2}{3} \cdot 17 = \frac{2}{3} \cdot \frac{17}{1} = \frac{2 \cdot 17}{3 \cdot 1} = \frac{34}{3} = 11\frac{1}{3}$$

One way to multiply mixed numbers is to first convert the mixed numbers to improper fractions. For example,

$$2\frac{3}{4} \cdot 1\frac{2}{3} = \frac{11}{4} \cdot \frac{5}{3} = \frac{55}{12} = 4\frac{7}{12}$$

Why Is the Procedure for Multiplying Fractions Valid?

The procedure for multiplying fractions seems sensible, because we multiply the numerators and the denominators. However, remember that we don't add fractions by simply adding the numerators and adding the denominators. So, why is the procedure of multiplying the numerators and multiplying the denominators valid for fraction multiplication? What is the logic behind it? We will use the definitions of fractions and of multiplication, and logical reasoning to answer those questions.

CLASS ACTIVITY

5D Explaining Why the Procedure for Multiplying Fractions Is Valid, p. CA-91

According to our definition of multiplication, in the equation

$$\frac{2}{7} \cdot \frac{3}{4} = ?$$

the $\frac{2}{7}$ is a number of groups, the $\frac{3}{4}$ is the number of units in 1 group, and we are looking for the number of units in $\frac{2}{7}$ of a group. So we want to know how much of a unit there is in $\frac{2}{7}$ of $\frac{3}{4}$ of a unit.

Let's let a rectangle stand for 1 unit, as in Figure 5.3 (a). Because we will be taking $\frac{2}{7}$ of $\frac{3}{4}$, let's show $\frac{3}{4}$ of the unit next, as in Figures 5.3 (b) and (c). To find $\frac{2}{7}$ of that $\frac{3}{4}$, let's first divide that $\frac{3}{4}$ into 7 equal parts as in Figure 5.3 (d). Then 2 of those parts represent $\frac{2}{7}$ of $\frac{3}{4}$ of the unit, as in Figure 5.3 (e).

Figure 5.3

Finding $\frac{2}{7}$ of $\frac{3}{4}$ of a unit.

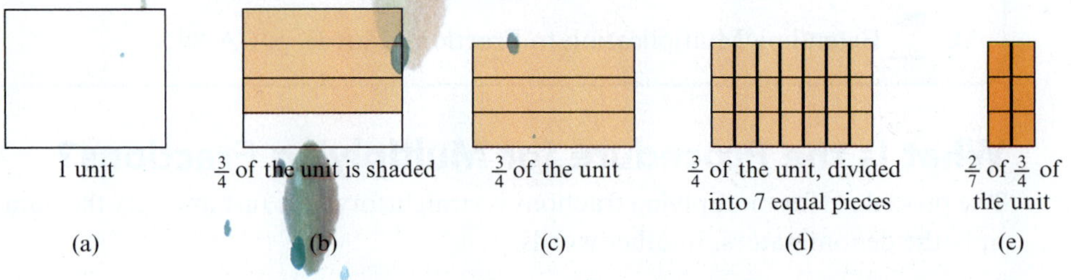

Here is the crucial point: We must identify the $\frac{2}{7}$ of the $\frac{3}{4}$ of the unit *as a fraction of the original 1 unit*. To do so, let's put $\frac{2}{7}$ of the $\frac{3}{4}$ of the unit back inside the original unit, as shown in Figure 5.4. We see that the original unit has been decomposed into 7 groups of 4, or $7 \cdot 4$ small, equal parts. The darkly shaded portion consists of 2 groups of 3, or $2 \cdot 3$ of those small parts. So the $\frac{2}{7}$ of the $\frac{3}{4}$ of the unit is represented by $2 \cdot 3$ parts each of which is $\frac{1}{7 \cdot 4}$ of the unit. This means that $\frac{2}{7}$ of the $\frac{3}{4}$ of the unit is

$$\frac{2 \cdot 3}{7 \cdot 4}$$

of the unit. Therefore,

$$\frac{2}{7} \cdot \frac{3}{4} = \frac{2 \cdot 3}{7 \cdot 4}$$

Figure 5.4

How many units is $\frac{2}{7}$ of $\frac{3}{4}$ of a unit?

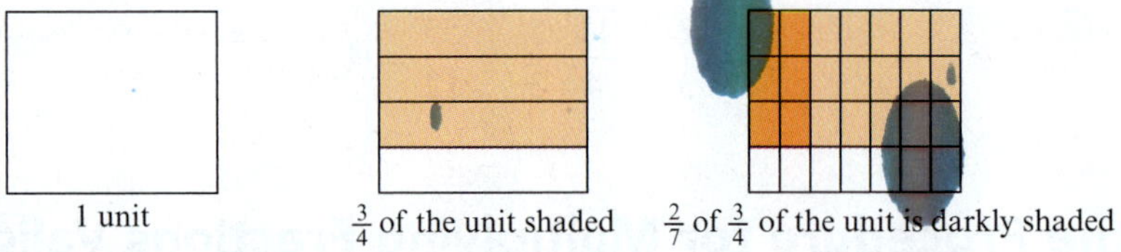

In other words, the procedure for multiplying fractions gives the answer we expect from the definitions of fractions and of multiplication.

There wasn't anything special about the numbers 2, 3, 7, and 4; any other counting numbers could be substituted, and the argument would work in the same way. Therefore, the explanation we developed also explains why we multiply other fractions by multiplying the numerators and multiplying the denominators.

$$\frac{A}{B} \cdot \frac{C}{D} = \frac{A \cdot C}{B \cdot D}$$

Notice that, in order to understand fraction multiplication, we had to work with different reference amounts. In the example, we first took $\frac{3}{4}$ of the original unit. Then we took $\frac{2}{7}$ of this $\frac{3}{4}$ portion. We had to treat the $\frac{3}{4}$ of the unit as a new whole in its own right in order to take $\frac{2}{7}$ of it. In the end, when we found the portion that was $\frac{2}{7}$ of the $\frac{3}{4}$ of the unit, we had to determine what fraction that portion was of the original unit. In other words, we had to go back to the original whole again. As always, when we work with fractions, we must pay close attention to the underlying reference amounts or what the fractions are *of*.

IMAP *Watch Felisha use an area problem to solve a story problem.*

CLASS ACTIVITY

5E When Do We Multiply Fractions?, p. CA-93

5F What Fraction Is Shaded?, p. CA-94

FROM THE FIELD Research

Wyberg, T., Whitney, S. R., Cramer, K. A., Monson, D. S., & Leavitt, S. (2011/2012). Unfolding fraction multiplication. *Mathematics Teaching in the Middle School, 17*(5), pp. 288–294.

The article presents a sequence of activities using number lines and paper folding to help sixth-grade students develop an understanding of fraction multiplication. The authors first asked students to solve a whole number multiplication problem: Joshkin makes $12 an hour. How much does he make in 4 hours of work? They then asked a related fraction times whole number question: How much will Joshkin earn if he works for 3/4 of an hour? Almost all students were able to reason about a number line to solve the problem. They partitioned the $12 into four parts, and then they iterated the $3 intervals three times to find the answer, $9. This activity helped students experience multiplication as a process of partitioning and iterating—not just as repeated addition. Students' understanding of fractions also evolved: the 3/4 changed from being 3 out of 4 equal parts to operating on an amount by dividing it by 4 and then multiplying the result by 3. For a fraction times a fraction, the authors found that a number line was harder. They therefore introduced paper folding as a way for students to model fraction multiplication. Many students were able to develop and explain the fraction multiplication procedure by folding paper. To extend students' thinking, the authors asked students to return to number lines to explain fraction multiplication. This gave students the opportunity to apply the fraction multiplication algorithm and to reason about fraction multiplication in several different ways.

SECTION SUMMARY AND STUDY ITEMS

Section 5.1 Making Sense of Fraction Multiplication

The meaning of $M \cdot N$ as the number of units in M equal groups if there are N units in 1 group applies not only to whole numbers but also to fractions and decimals. To multiply fractions, we multiply the numerators and denominators. We can explain why this procedure makes sense by seeing fraction multiplication as taking a fraction of a fraction.

Key Skills and Understandings

1. Interpret the meaning of multiplication with fractions. Write word problems for a given fraction multiplication problem, and solve the problems by using logic and a math drawing.
2. Use simple word problems and logical reasoning about math drawings to explain why the procedure for multiplying fractions is valid.
3. Given a word problem, determine whether it can be solved by fraction multiplication.

Practice Exercises for Section 5.1

1. What does $\frac{3}{4}\cdot\frac{1}{2}$ mean? Give an example of a word problem that is solved by calculating $\frac{3}{4}\cdot\frac{1}{2}$.
2. Use the definition of multiplication to explain why

$$\frac{2}{7}\cdot\frac{3}{4}=\frac{2\cdot 3}{7\cdot 4}$$

3. Anthony is trying to calculate $\frac{5}{7}\cdot 2$. He makes a math drawing as in **Figure 5.5** and concludes from his drawing that $\frac{5}{7}\cdot 2=\frac{10}{14}$, because there are 14 small pieces and 10 of them are shaded. Critique Anthony's reasoning.

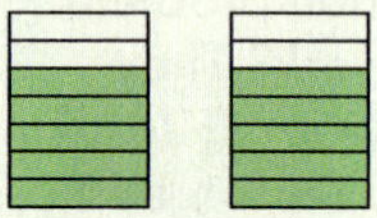

Figure 5.5 Is $\frac{5}{7}\cdot 2=\frac{10}{14}$?

4. Which of the following problems are solved by calculating $\frac{1}{3}\cdot\frac{2}{3}$, and which are not?
   a. A recipe calls for $\frac{2}{3}$ cup of sugar. You want to make $\frac{1}{3}$ of the recipe. How much sugar should you use?
   b. $\frac{2}{3}$ of the cars at a car dealership have power steering. $\frac{1}{3}$ of the cars at the same car dealership have side-mounted airbags. What fraction of the cars at the car dealership have both power steering and side-mounted airbags?
   c. $\frac{2}{3}$ of the cars at a car dealership have power steering. $\frac{1}{3}$ of those cars that have power steering have side-mounted airbags. What fraction of the cars at the car dealership have both power steering and side-mounted airbags?
   d. Ed put $\frac{2}{3}$ of a bag of candies in a batch of cookies that he made. Ed ate $\frac{1}{3}$ of the batch of cookies. How many candies did Ed eat?
   e. Ed put $\frac{2}{3}$ of a bag of candies in a batch of cookies that he made. Ed ate $\frac{1}{3}$ of the batch of cookies. What fraction of a bag of candies did Ed eat?
5. Maryann is calculating $\frac{2}{3}\cdot 6\frac{3}{4}$. Rather than converting the $6\frac{3}{4}$ to an improper fraction, she solves the problem this way. She finds $\frac{2}{3}$ of 6, which is 4, and she finds $\frac{2}{3}$ of $\frac{3}{4}$, which is $\frac{1}{2}$. Then Maryann says the answer is $4\frac{1}{2}$. Write equations that correspond to Maryann's work and that explain why her work is correct. Which property of arithmetic is involved?

Answers to Practice Exercises for Section 5.1

1. $\frac{3}{4}\cdot\frac{1}{2}$ means the number of units in $\frac{3}{4}$ of $\frac{1}{2}$ of a unit. A sample word problem: A recipe calls for $\frac{1}{2}$ pound of sea slugs. You decide to make $\frac{3}{4}$ of the recipe. How many pounds of sea slugs will you need?
2. Review the explanation in the text. Be sure to describe $\frac{2}{7}\cdot\frac{3}{4}$ as "$\frac{2}{7}$ of $\frac{3}{4}$" as well as in terms of the definition of multiplication. Be sure to explain how to see multiplying numerators and denominators.
3. Although Anthony has drawn a good representation of the problem, his reasoning is not completely right. Ten pieces are shaded, and these 10 pieces do represent $\frac{5}{7}$ of 2 units. But Anthony must remember to describe the shaded amount as a number of units, so in terms of *one* rectangle. This amount is $\frac{10}{7}$ of a unit. When drawing pictures such as Anthony's, it's a good idea to also draw 1 unit somewhere as a reminder that this is your reference amount.
4. a. Yes, $\frac{1}{3}\cdot\frac{2}{3}$ is the fraction of a cup of sugar you should use, since you will need $\frac{1}{3}$ of $\frac{2}{3}$ cup of sugar.
   b. No, from the information given, we don't know if $\frac{1}{3}$ of the cars that have power steering have side-mounted airbags.
   c. Yes, $\frac{1}{3}$ of $\frac{2}{3}$ of the cars at the car dealership have both power steering and side-mounted airbags.
   d. No, we can't tell how many candies Ed ate, only what fraction of a bag of candies Ed ate.
   e. Yes, Ed ate $\frac{1}{3}$ of $\frac{2}{3}$ of a bag of candies.
5.

$$\begin{aligned}\frac{2}{3}\cdot 6\frac{3}{4}&=\frac{2}{3}\cdot\left(6+\frac{3}{4}\right)\\&=\frac{2}{3}\cdot 6+\frac{2}{3}\cdot\frac{3}{4}\\&=4+\frac{1}{2}\\&=4\frac{1}{2}\end{aligned}$$

The distributive property was used at the second equal sign.

PROBLEMS FOR SECTION 5.1

1. a. Anita had $\frac{1}{2}$ of a bag of fertilizer left. She used $\frac{3}{4}$ of what was left. What question about Anita's fertilizer can be answered by calculating $\frac{3}{4} \cdot \frac{1}{2}$?

   b. Hermione wants to make 4 batches of a potion. Each batch of potion requires $\frac{2}{3}$ cup of toober pus. What question about the potion will be answered by calculating $4 \cdot \frac{2}{3}$?

   c. There were 6 pieces of pizza left. Tommy ate $\frac{3}{4}$ of them. What question about the pizza can be answered by calculating $\frac{3}{4} \cdot 6$?

2. a. Discuss the meaning of $\frac{1}{2} \cdot \frac{1}{4}$. Include a simple word problem and a math drawing in your discussion.

   b. Discuss the meaning of $\frac{1}{4} \cdot \frac{1}{2}$. Include a simple word problem and a math drawing in your discussion.

   c. Discuss the difference between parts (a) and (b).

3. Paul used $\frac{3}{4}$ cup of butter in the batch of brownies he made. He ate $\frac{1}{6}$ of the batch of brownies. What fraction of a cup of butter did he consume when he ate the brownies? Make math drawings to help you solve this problem. Explain in detail how your drawings help you solve the problem.

4. Which of the following are word problems for $\frac{1}{2} \cdot \frac{1}{3} = ?$, and which are not? Explain your answer in each case.

   a. $\frac{1}{3}$ of the children in a class have black hair. $\frac{1}{2}$ have curly hair. How many children in the class have curly black hair?

   b. $\frac{1}{3}$ of the children in a class have black hair. $\frac{1}{2}$ of the children who have black hair also have curly hair. How many children in the class have curly black hair?

   c. $\frac{1}{3}$ of the children in a class have black hair. $\frac{1}{2}$ of the children who have black hair also have curly hair. What fraction of the children in the class have curly black hair?

   d. $\frac{1}{3}$ of the children in a class have black hair. $\frac{1}{2}$ of the children in the class have curly hair. What fraction of the children in the class have curly black hair?

5. Which of the following are word problems for $\frac{3}{4} \cdot 5 = ?$, and which are not? Explain your answer in each case.

   a. A cake was cut into pieces of equal size. There are 5 pieces of cake left. John gets $\frac{3}{4}$ of them. What fraction of the cake does he get?

   b. A cake was cut into pieces of equal size. There are 5 pieces of cake left. John gets $\frac{3}{4}$ of them. How many pieces of cake does he get?

   c. A cake for a class party was cut into pieces of equal size. There are 5 pieces of cake left. $\frac{3}{4}$ of the class still wants cake. What fraction of the cake will be eaten?

   d. A cake for a class party was cut into pieces of equal size. There are 5 pieces of cake left. $\frac{3}{4}$ of the class still wants cake. How many pieces does each person get?

6. Consider this word problem about baking brownies:

   You are baking brownies for your class. You put white frosting on $\frac{1}{3}$ of the brownies and you put small red hearts on $\frac{1}{4}$ of the brownies. How many brownies have both white frosting and small red hearts on them?

   a. Can the brownie problem be solved? If so, solve the problem; if not, explain why not.

   b. Is the brownie problem a problem for $\frac{1}{4} \cdot \frac{1}{3}$? If so, explain briefly why it is; if not, modify the problem so that it is a problem for $\frac{1}{4} \cdot \frac{1}{3}$.

   c. Is the brownie problem a problem for $\frac{1}{3} + \frac{1}{4}$? If so, explain briefly why it is; if not, modify the problem so that it is a problem for $\frac{1}{3} + \frac{1}{4}$.

7. Explain why it would be easy to interpret the drawing in Figure 5.6 incorrectly as showing that $4 \cdot \frac{3}{5} = \frac{12}{20}$. Explain how to interpret the drawing as representing $4 \cdot \frac{3}{5}$.

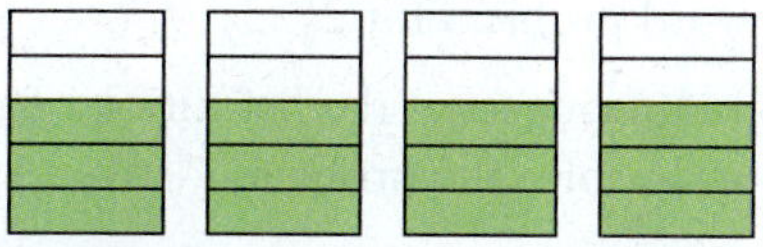

Figure 5.6 A drawing for $4 \cdot \frac{3}{5}$.

8. Write a word problem for

$$\frac{1}{3} \cdot \frac{1}{4} = ?$$

Use the definition of multiplication and math drawings to determine the answer to the multiplication problem. Explain your answer.

9. Write a word problem for

$$2 \cdot \frac{3}{5} = ?$$

Use the definition of multiplication and math drawings to determine the answer to the multiplication problem. Explain your answer.

10. Write a word problem for

$$\frac{2}{3} \cdot 5 = ?$$

Use the definition of multiplication and math drawings to determine the answer to the multiplication problem. Explain your answer.

11. Write a simple word problem for

$$\frac{2}{3} \cdot \frac{4}{5} = ?$$

Use your word problem and math drawings to explain why it makes sense that the answer to the fraction multiplication problem is

$$\frac{2 \cdot 4}{3 \cdot 5}$$

In particular, use your drawings to explain why we multiply the numerators and why we multiply the denominators.

12. One serving of Gooey Gushers provides 12% of the daily value of vitamin C. Tim ate $3\frac{1}{2}$ servings of Gooey Gushers.

a. Calculate $3\frac{1}{2} \cdot 12\%$ without using a calculator. Show your work.

b. What question about the Gooey Gushers can you answer by calculating $3\frac{1}{2} \cdot 12\%$?

13. a. Write a word problem for $2\frac{1}{2} \cdot 3\frac{1}{2} = ?$.

b. Use math drawings and the definition of multiplication to solve the problem.

c. Use the distributive property to calculate $2\frac{1}{2} \cdot 3\frac{1}{2}$ by rewriting this product as $(2 + \frac{1}{2}) \cdot (3 + \frac{1}{2})$.

d. Identify the four terms produced by the distributive property (in part [c]) in a math drawing like the one in part (b).

e. Now write the mixed numbers $2\frac{1}{2}$ and $3\frac{1}{2}$ as improper fractions, and use the standard procedure for multiplying fractions to calculate $2\frac{1}{2} \cdot 3\frac{1}{2}$. How do you see the product of the numerators in your drawing in part (b)? How can you see the product of the denominators in your drawing?

14. Manda says that

$$3\frac{2}{3} \cdot 2\frac{1}{5} = 3 \cdot 2 + \frac{2}{3} \cdot \frac{1}{5}$$

Explain why Manda has made a good attempt, but her answer is not correct. Explain how to work with what Manda has already written and modify it to get the correct answer. In other words, don't just start from scratch and show Manda how to do the problem, but rather take what she has already written, use it, and make it mathematically correct. Which property of arithmetic is relevant to correcting Manda's work? Explain.

15. a. Write an expression that uses both multiplication and addition (or subtraction) to describe the total fraction of Figure 5.7 that is shaded. (For example, $\frac{5}{7} \cdot \frac{2}{9} + \frac{1}{3}$ is an expression that uses both multiplication and addition.) Explain your reasoning. Then determine what fraction of the figure is shaded (in simplest form). You may assume that lengths appearing to be equal really are equal.

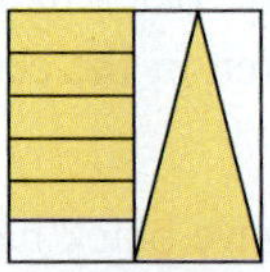

Figure 5.7 What fraction is shaded?

b. Draw a figure in which you shade $\frac{1}{4} \cdot \frac{1}{3} + \frac{3}{5} \cdot \frac{1}{3}$ of the figure. Explain your reasoning, then determine what fraction of the figure is shaded (in simplest form).

16. a. Write an expression that uses both multiplication and addition (or subtraction) to describe the total fraction of Figure 5.8 that is shaded. (For example, $\frac{5}{7} \cdot \frac{2}{9} + \frac{1}{3}$ is an expression that uses both multiplication and addition.) Explain your reasoning. Then determine what fraction of the figure is shaded (in simplest form). You may assume that lengths appearing to be equal really are equal.

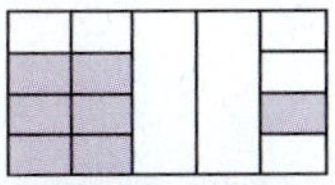

Figure 5.8 What fraction is shaded?

b. Draw a figure in which you shade $\frac{5}{7} - \frac{3}{5} \cdot \frac{1}{7}$ of the figure. Explain your reasoning. Then determine what fraction of the figure is shaded (in simplest form).

***17.** To understand fraction multiplication thoroughly, we must be able to work simultaneously with different reference amounts. Using the example $\frac{3}{5} \cdot \frac{3}{4} = ?$, explain why this is so. What are the different reference amounts that are associated with the fractions in the problem $\frac{3}{5} \cdot \frac{3}{4} = ?$ (including the answer to the problem)?

***18.** To understand fraction multiplication thoroughly, we must understand how to partition a whole number of objects into equal parts. Where do we need to understand how to partition a whole number of objects into equal parts in order to understand $\frac{2}{3} \cdot \frac{2}{5}$ thoroughly? Be specific.

***19.** The liquid in a car's radiator is 75% water and 25% antifreeze. Suppose that 30% of the radiator's liquid is drained out and replaced with pure antifreeze. What percent of the radiator's liquid is now antifreeze? Make math drawings to help you solve this problem. Explain your answer.

***20.** You are holding a yellow flask and a red flask. The yellow flask contains more than 1 cup of yellow paint, and the red flask contains an equal amount of red paint. You pour 1 cup of the red paint in the red flask into the yellow flask and mix thoroughly. You then pour 1 cup of the paint mixture in the yellow flask into the red flask and mix thoroughly.

a. Without doing any calculations, which do you think will be greater, the percentage of yellow paint in the red flask or the percentage of red paint in the yellow flask? Explain your reasoning.

b. Now use calculations to figure out the answer to the problem in part (a).

***21.** Discuss why we must develop an understanding of multiplication that goes beyond seeing it as repeated addition.

***22.** Explain in your own words why multiplication means the same thing whether we are multiplying fractions or whole numbers. Use word problems for $3 \cdot 4$ and $\frac{1}{3} \cdot \frac{1}{4}$ to illustrate.

5.2 Making Sense of Decimal Multiplication

CCSS Common Core State Standards Grades 5, 6

To multiply (finite) decimals, the standard procedure is first to multiply the numbers without the decimal points, and then to place the decimal point in the answer according to a certain rule. Why is this procedure for multiplying decimals valid? The text, class activities, practice exercises, and problems in this section will help you answer this question in a number of different ways.

What Is the Procedure for Multiplying Decimals?

What is the standard procedure for multiplying decimals? We first multiply the numbers without the decimal points. Then we add the number of digits to the right of the decimal points in the numbers we want to multiply, and we put the decimal point in the answer to the product computed without the decimal points that many places from the end. So, to multiply

$$\begin{array}{r} 1.36 \\ \times\ 3.7 \\ \hline \end{array}$$

we first multiply as if there were no decimal points:

$$\begin{array}{r} 136 \\ \times\ 37 \\ \hline 5032 \end{array}$$

Then we add the number of digits to the right of the decimal points in our two original numbers and put the decimal point that many places from the end in our answer. There are 2 digits behind the decimal point in 1.36 and another 1 digit in 3.7, for a total of $1 + 2 = 3$ digits; so the decimal point goes 3 digits from the end:

$$\begin{array}{r} 1.36 \\ \times\ 3.7 \\ \hline 5.032 \end{array}$$

CLASS ACTIVITY

5G Decimal Multiplication Word Problems and Estimation, p. CA-95

How Can We Use Estimation to Determine Where the Decimal Point Goes?

Suppose you have forgotten the rule about where to put the decimal point in the answer to a decimal multiplication problem. One quick way to figure out where to put the decimal point is to think about the sizes of the numbers. For example, 1.36 is between 1 and 2, and 3.7 is between 3 and 4; so 1.36×3.7 must be between $1 \times 3 = 3$ and $2 \times 4 = 8$. So, where will it make sense to put the decimal point in 5032 to get the answer to 1.36×3.7? The numbers 503.2 and 50.32 are far too big. The number 0.5032 is too small. The only number that makes sense is 5.032 because it is between 3 and 8.

Why Is the Rule for Placing the Decimal Point Valid?

CLASS ACTIVITY

5H Explaining Why We Place the Decimal Point Where We Do When We Multiply Decimals, p. CA-96

CCSS 5.NBT.7

How can we explain why we add the number of places behind the decimal points to determine where to place the decimal point in the answer to a decimal multiplication problem? One way is to relate the decimal multiplication problem to the multiplication problem without the decimal points. For example, how are the multiplication problems

$$0.12 \times 6.24$$

and

$$12 \times 624$$

related? As indicated in Figure 5.9, to get from 0.12 to 12, we must multiply by ten 2 times; to get from 6.24 to 624, we must multiply by ten 2 times. Therefore, to get from

$$0.12 \times 6.24$$

to

$$12 \times 624$$

we must multiply by ten 2 times and then another 2 times. We know that

$$12 \times 624 = 7488$$

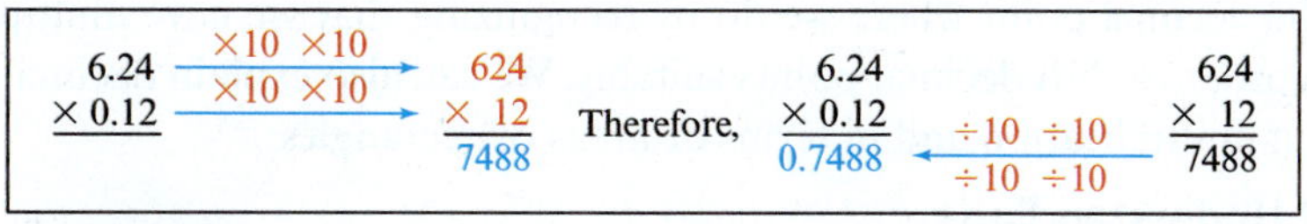

Figure 5.9 Relating 0.12×6.24 and 12×624.

Therefore, to get back to the answer to the original problem, 0.12×6.24, we must divide 7488 by ten 2 times and then another 2 times. When we divide by ten 2 times and then another 2 times, we shift the number $2 + 2$ places to the right. So we have shown that to calculate 0.12×6.24, we must move the answer to 12×624 to the right $2 + 2$ places, or move the decimal point $2 + 2$ places to the left, which is exactly the procedure for decimal multiplication. There wasn't anything special about the numbers we used here—the same line of reasoning applies for any other decimal multiplication problem.

Another way to explain why the standard procedure for multiplying decimals is valid is by writing the decimals as fractions. Then we can use the procedure for multiplying fractions that we have already studied. Every finite decimal can be written as a fraction with a denominator that is a product of 10s. For example,

$$1.25 = \frac{125}{100} = \frac{125}{10 \times 10}$$

$$0.003 = \frac{3}{1000} = \frac{3}{10 \times 10 \times 10}$$

We can use this way of writing the denominators to explain the placement of the decimal point when we multiply decimals.

$$\begin{aligned} 1.25 \times 0.003 &= \frac{125}{10 \times 10} \times \frac{3}{10 \times 10 \times 10} \\ &= \frac{125 \times 3}{(10 \times 10) \times (10 \times 10 \times 10)} \end{aligned}$$

These equations tell us that to calculate the answer to 1.25×0.003, we should calculate the answer to 125×3 and divide by ten 2 times and then another 3 times. When we divide by ten 2 times and then another 3 times, we move the number 2 places and then 3 places to the right, or equivalently, we move the decimal point 2 places and then 3 places to the left. In other words, to calculate the answer to 1.25×0.003, we should calculate the answer to 125×3 and then move the decimal point a total of $2 + 3 = 5$ places to the left, which is exactly the standard decimal multiplication procedure.

CLASS ACTIVITY

5I Decimal Multiplication and Areas of Rectangles, p. CA-97

SECTION SUMMARY AND STUDY ITEMS

Section 5.2 Making Sense of Decimal Multiplication

To multiply decimals, the standard procedure is to multiply the numbers without the decimal points present. Then we add the number of digits to the right of the decimal points in the two original numbers and put the decimal point in the answer that many places from the end. We can explain why we place the decimal point where we do by recognizing that we have multiplied and then divided by tens in order to shift decimal points suitably. We can also explain decimal multiplication in terms of fraction multiplication and in terms of areas of rectangles.

Key Skills and Understandings

1. Explain why we put the decimal point where we do when we multiply decimals.
2. Use estimation to determine where to place the decimal point in a decimal multiplication problem.
3. Describe decimal multiplication in terms of area.
4. Write word problems for a given decimal multiplication problem.

Practice Exercises for Section 5.2

1. The product of 1.35×7.2 is 9.72, but shouldn't the answer have 3 digits to the right of its decimal point? Why doesn't it?
2. Write a word problem for 8.3×4.15.
3. Suppose you multiply a decimal that has 5 digits to the right of its decimal point by a decimal that has 2 digits to the right of its decimal point. Explain why you put the decimal point $5 + 2$ places from the end of the product without the decimal points.
4. Use the definition of multiplication to explain why we can multiply to find the area of a 2.4-unit-by-1.6-unit rectangle.
5. Without multiplying, find the area of a 2.4-unit-by-1.6-unit rectangle. Then use that result to explain why the product 2.4×1.6 has a (nonzero) entry in the hundredths place (but not in lower places).

Answers to Practice Exercises for Section 5.2

1. The rule says to find the product without the decimal points (i.e., 135×72) and then move the decimal point in that answer $2 + 1 = 3$ places to the left. But

$$135 \times 72 = 9720$$

ends in a zero. That's why that third digit doesn't appear in the answer.

2. You bought 8.3 gallons of gas. Gas costs $4.15 per gallon. How much did you pay?
3. Let's work with a particular example to illustrate, say, 1.23456×1.23. To get from 1.23456 to 123456, we must multiply by ten 5 times. To get from 1.23 to 123, we must multiply by ten 2 times. So, to get from

$$1.23456 \times 1.23$$

to

$$123456 \times 123$$

we must multiply by ten 5 times and then another 2 times. Therefore, to get from the answer to

$$123456 \times 123$$

back to the answer to

$$1.23456 \times 1.23$$

we must divide by ten 5 times and then another 2 times.

When we divide the answer to

$$123456 \times 123$$

by ten 5 times and then another 2 times, we move the decimal point to the left 5 + 2 places. The explanation works the same way for any other decimals with 5 digits and 2 digits to the right of their decimal points.

4. Starting with a 1-unit-by-1-unit square, **Figure 5.10** shows 1.6 of the square, then 2 groups of 1.6 of the square, and finally 2.4 groups of the 1.6 of the square. So, according to the definition of multiplication, the 2.4-unit-by-1.6-unit rectangle in **Figure 5.10** consists of 2.4 × 1.6 squares and therefore has area 2.4 × 1.6 square units.

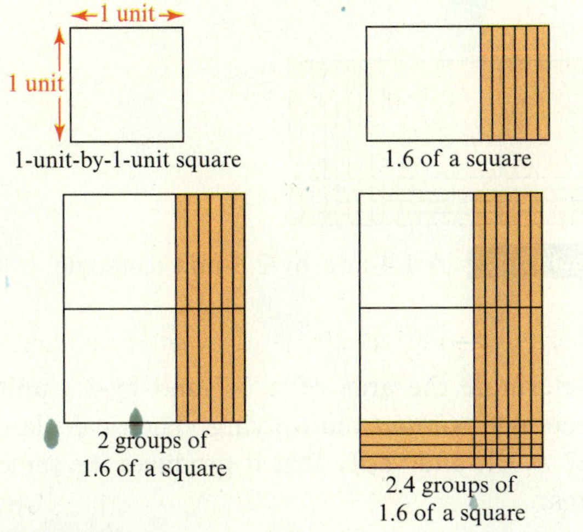

Figure 5.10 Showing 2.4 × 1.6 as 2.4 groups of 1.6 square units.

5. If we take a 2.4-unit-by-1.6-unit rectangle apart, we can recombine it as shown in **Figure 5.11**. We see that the rectangle is made of 3 full 1-unit-by-1-unit squares, $\frac{8}{10}$ of a square, and $\frac{4}{100}$ of a square. Therefore, the total area of the 2.4-unit-by-1.6-unit rectangle is 3.84 square units. The values in the hundredths place come from the small $\frac{1}{10}$-unit-by-$\frac{1}{10}$-unit squares, which have area $\frac{1}{100}$ square units.

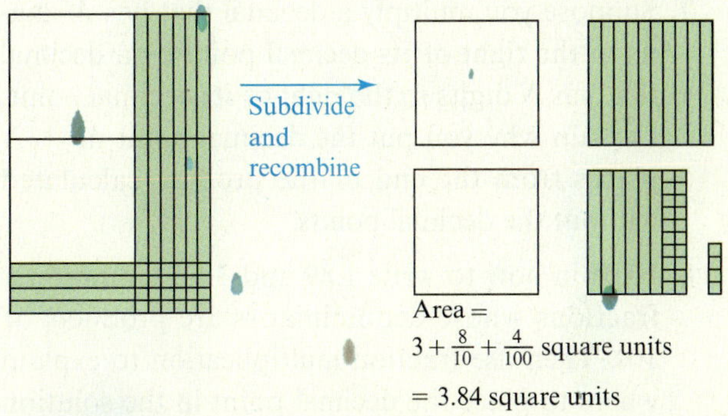

Figure 5.11 Determining the area of a 2.4-unit-by-1.6-unit rectangle by subdividing and recombining.

PROBLEMS FOR SECTION 5.2

1. Write and solve your own word problem for 14.3 × 1.39.

2. Write a word problem for 1.3 × 2.79. Solve your problem without using a calculator.

3. Leah is working on the multiplication problem 2.43 × 0.148. Ignoring the decimal points, Leah multiplies 243 × 148 and gets the answer 35964. But Leah can't remember the rule about where to put the decimal point in this answer to get the correct answer to 2.43 × 0.148. Explain how Leah can use reasoning about the sizes of the numbers to determine where to put the decimal point.

4. Ron used a calculator to determine that

$$0.35 \times 2.4 = 0.84$$

He wants to know why the rule about adding up the number of places to the right of the decimal point doesn't work in this case. Why aren't there 2 + 1 = 3 digits to the right of the decimal point in the answer? Is Ron correct that the rule about adding the number of places to the right of the decimal points is not correct in this case? Explain.

5. When we multiply 0.48 × 3.9, we first multiply as if the decimal points were not there:

```
  3.9
× .48
  312
 1560
 1872
```

Then we place a decimal point in 1872 to get the correct answer. Explain clearly why it makes sense to put the decimal point where we do.

6. Suppose you multiply a decimal that has 2 digits to the right of its decimal point by a decimal that has 3 digits to the right of its decimal point. Explain why you put the decimal point 2 + 3 places from the end of the product calculated without the decimal points.

7. Suppose you multiply a decimal that has M digits to the right of its decimal point by a decimal that has N digits to the right of its decimal point. Explain why you put the decimal point $M + N$ places from the end of the product calculated without the decimal points.

8. Explain how to write 1.89 and 3.57 as improper fractions whose denominators are products of 10s. Then use fraction multiplication to explain where to place the decimal point in the solution to 1.89×3.57. Show how to use the products of 10s in the denominators to explain why we add the number of digits behind the decimal points in 1.89 and 3.57.

9. *A Shampoo Problem:* A bottle contains 25.4 fluid ounces of shampoo. Katie uses 0.25 of the bottle. How much shampoo is left?

a. Is the shampoo problem a word problem for 0.25×25.4, is it a word problem for $25.4 - 0.25$, or is it not a word problem for either of these? Explain.

b. Write a new shampoo word problem for 0.25×25.4 and write a new shampoo word problem for $25.4 - 0.25$. Make clear which is which.

10. Determine the area of the rectangle in Figure 5.12 *without* multiplying. Then calculate 1.4×2.8, and verify that it produces the same area.

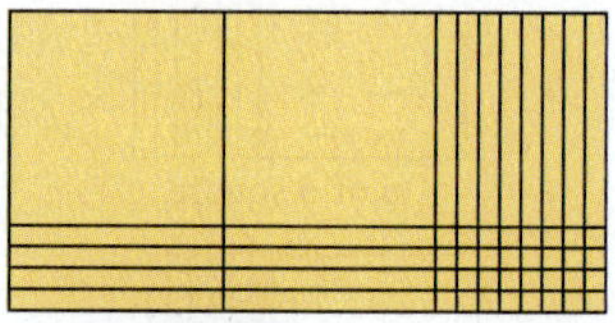

Figure 5.12 A 1.4-unit-by-2.8-unit rectangle.

11. Determine the area of a 1.7-unit-by-3.1-unit rectangle *without* multiplying. Then calculate 1.7×3.1, and verify that it produces the same area.

5.3 Extending Multiplication to Negative Numbers

CCSS Common Core State Standards Grade 7

So far, we have examined multiplication only for numbers that are not negative. What is the meaning of multiplication for negative numbers? In this section, we will examine several different ways to understand multiplication of negative numbers.

CLASS ACTIVITY

5J Using the Distributive property to Explain Multiplication with Negative Numbers (and 0), p. CA-98

What should $3 \cdot -2$ mean? We know that $3 \cdot 2$ stands for the number of units in 3 equal groups if there are 2 units in 1 group. If we try to use the same interpretation for $3 \cdot -2$, then we have the following:

$3 \cdot -2$ is the number of units in 3 equal groups if there are -2 units in 1 group.

What does this mean? One way to interpret this sensibly is to think of "-2 units in 1 group" as meaning "each group *owes* 2 objects." For example, if Lakeisha, Mary, and Jayna each *owe* \$2.00, then we can think of each girl as *having* -2 dollars, and all together, the 3 girls have $3 \cdot -2$ dollars. Notice that with this interpretation, it makes sense to say

$$3 \cdot -2 = -6$$

because all together, the 3 girls collectively owe 6 dollars, as represented by the -6. In general, if A and B are positive numbers, then we can define $A \cdot -B$ to be $-(A \cdot B)$; that is,

$$A \cdot -B = -(A \cdot B) \tag{5.1}$$

This definition is consistent with the interpretation of negative numbers as amounts owed. Equation 5.1 gives us the familiar rule:

A positive times a negative is negative.

What about a negative number times a positive number, such as

$$-2 \cdot 3$$

We could perhaps interpret this product as the total number of objects in 2 owed groups with 3 objects in each group. Similarly, we could interpret a negative number times a negative number, such as

$$-4 \cdot -5$$

as the total number of objects in 4 owed groups, with each group owing 5 objects. But these interpretations seem difficult to grasp.

CCSS 7.NS.2a,c

There is another way we can understand how to multiply negative numbers, which draws upon the properties of arithmetic. We have seen why the distributive property and the commutative and associative properties of addition and multiplication make sense for counting numbers. These properties of arithmetic are fundamental and describe the algebraic structure of numbers and operations. We have seen that these properties of arithmetic underlie mental calculations and standard calculation algorithms. Because the properties of arithmetic are so important and fundamental, we should want them to hold for any number system that is an extension of the counting numbers.

Let us now assume that the properties of arithmetic we have studied hold not only for positive numbers but for negative numbers as well. We will see that this assumption determines how we multiply negative numbers.

How do we multiply a negative number with a positive number, such as

$$-2 \cdot 3$$

According to the commutative property of multiplication,

$$-2 \cdot 3 = 3 \cdot -2$$

which must be equal to -6, according to Equation 5.1 and the previous discussion about how to interpret $3 \cdot -2$. In general, if A and B are any positive numbers, then because we are assuming that the commutative property of multiplication holds, we must define $-A \cdot B$ to be $-(A \cdot B)$. That is,

$$-A \cdot B = -(A \cdot B) \tag{5.2}$$

This equation gives us the following familiar rule:

A negative times a positive is negative.

Finally, how do we multiply a negative number with a negative number, as in

$$-4 \cdot -5$$

We are assuming that the distributive property holds for both positive and negative numbers. Therefore,

$$(-4 \cdot -5) + (4 \cdot -5) = (-4 + 4) \cdot -5$$
$$= 0 \cdot -5$$
$$= 0$$

So

$$(-4 \cdot -5) + (4 \cdot -5) = 0$$

Since $4 \cdot -5 = -20$ (by Equation 5.1), we have

$$(-4 \cdot -5) + (-20) = 0$$

But 20 is the only number which, when added to -20, yields zero. Therefore, $-4 \cdot -5$ must equal 20. By this logic, if A and B are any positive numbers, then if we assume that the distributive property holds, we must define $-A \cdot -B$ to be $A \cdot B$; that is,

$$-A \cdot -B = A \cdot B \qquad (5.3)$$

This equation gives us the following familiar rule:

A negative times a negative is positive.

SECTION SUMMARY AND STUDY ITEMS

Section 5.3 Extending Multiplication to Negative Numbers

We can explain why a positive number times a negative number is negative by considering negative numbers as representing owed amounts. We can use properties of arithmetic to explain the way we multiply with negative numbers.

Key Skills and Understandings

1. Explain why a positive number times a negative number should be negative by considering negative numbers as owed amounts (for example).
2. Use properties of arithmetic to explain the rules for multiplying with negative numbers.

Practice Exercises for Section 5.3

1. Explain why it makes sense that $6 \cdot -7 = -42$ by interpreting negative numbers as amounts owed.
2. Given that $6 \cdot -7 = -42$, use a property of arithmetic to explain why it makes sense that $-7 \cdot 6 = -42$.
3. Given that $6 \cdot -7 = -42$, use a property of arithmetic to explain why it makes sense that $-6 \cdot -7 = 42$.

Answers to Practice Exercises for Section 5.3

1. View $6 \cdot -7$ as the amount 6 people have if each owes $7. See the text for details.
2. Apply the commutative property of multiplication. See the text for details.
3. Apply the distributive property. See the text for details.

PROBLEMS FOR SECTION 5.3

1. Explain the following in your own words.

   a. Explain why it makes sense that $5 \cdot -2 = -10$ by interpreting negative numbers as amounts owed or as negatively charged particles, for example.

   b. Given that $5 \cdot -2 = -10$, use a property of arithmetic to explain why it makes sense that $-2 \cdot 5 = -10$.

   c. Given that $5 \cdot -2 = -10$, use a property of arithmetic to explain why it makes sense that $-5 \cdot -2 = 10$.

2. For each of the following cases, either explain why the case cannot occur or give an example to show how it can.

   a. Two positive numbers whose product is less than both numbers

   b. Two positive numbers whose product is between the two numbers

   c. A positive number and a negative number whose product is less than both numbers

   d. A positive number and a negative number whose product is between the two numbers

   e. A positive number and a negative number whose product is greater than both numbers

3. For each of the following cases, either explain why the case cannot occur or give an example to show how it can.

   a. Two negative numbers whose product is greater than both numbers

   b. Two negative numbers whose product is in between the two numbers

   c. Two negative numbers whose product is less than both numbers

5.4 Powers and Scientific Notation

CCSS Common Core State Standards Grades 5, 6, 8

Many scientific applications require the use of very large or very small numbers: Distances between stars are huge, whereas molecular distances are tiny. These kinds of numbers can be cumbersome to write in ordinary decimal notation. Therefore, a special notation called scientific notation is often used to write such numbers. When a number is in scientific notation, we can see its order of magnitude—how big or small it is—at a glance.

Use a calculator to multiply

$$123{,}456{,}789 \times 987{,}654{,}321$$

How is the answer displayed? The answer is probably displayed in one of the following forms:

1.2193263 17

or

1.2193263 E 17

Both of these displays represent

$$1.2193263 \times 10^{17}$$

which is in scientific notation. Scientific notation involves multiplying by powers of 10, such as 10^{17}, a subject we discuss next.

What Are Powers and Exponents?

A convenient notation for writing an expression such as

$$10 \times 10 \times 10 \times 10 \times 10 \times 10$$

which is 6 tens multiplied together, is

$$10^6$$

So,

$$1{,}000{,}000 = 10 \times 10 \times 10 \times 10 \times 10 \times 10 = 10^6$$

Power of 10
exponent

The expression 10^6 is read "ten to the sixth power" or just "ten to the sixth," and we can refer to an expression like 10^6 as a **power of 10**. The number 6 in 10^6 is called the **exponent** of 10^6.

More generally, if A is any real number and B is any counting number, then A^B stands for B As multiplied together:

$$A^B = \underbrace{A \times A \times \cdots \times A}_{B \text{ times}}$$

For example,

$$2^5 = 2 \times 2 \times 2 \times 2 \times 2 = 32$$

CCSS
5.NBT.2
6.EE.1

As with powers of 10, we read A^B as "A to the Bth power" or "A to the B," and B is called the exponent of A^B.

Table 5.1 shows powers of 10 from 10^1 to 10^{12}. Notice the correlation between the number of zeros and the exponent on the 10 that is visible in Table 5.1. For example, 10,000 is written as a 1 followed by 4 zeros; it also can be written as 10^4, where the exponent on the 10 is 4. Similarly, 1,000,000 is written as a 1 followed by 6 zeros and it can also be written as 10^6.

TABLE 5.1 Powers of 10

Ten	$10 = 10^1$
Hundred	$100 = 10 \times 10 = 10^2$
Thousand	$1000 = 10 \times 10 \times 10 = 10^3$
Ten thousand	$10{,}000 = 10 \times 10 \times 10 \times 10 = 10^4$
Hundred thousand	$100{,}000 = 10 \times 10 \times 10 \times 10 \times 10 = 10^5$
Million	$1{,}000{,}000 = 10 \times 10 \times 10 \times 10 \times 10 \times 10 = 10^6$
Ten million	$10{,}000{,}000 = 10^7$
Hundred million	$100{,}000{,}000 = 10^8$
Billion	$1{,}000{,}000{,}000 = 10^9$
Ten billion	$10{,}000{,}000{,}000 = 10^{10}$
Hundred billion	$100{,}000{,}000{,}000 = 10^{11}$
Trillion	$1{,}000{,}000{,}000{,}000 = 10^{12}$

What about decimal places to the right of the decimal point? As before, there is special notation to write one tenth, one hundredth, one thousandth, and so on, as powers of 10. This time, the powers are *negative*, as in the following:

$$\begin{array}{ccccc} \frac{1}{10} & = & \frac{1}{10^1} & = & 10^{-1} \\ \frac{1}{100} & = & \frac{1}{10^2} & = & 10^{-2} \\ \frac{1}{1000} & = & \frac{1}{10^3} & = & 10^{-3} \\ \frac{1}{10{,}000} & = & \frac{1}{10^4} & = & 10^{-4} \\ \vdots & & \vdots & & \vdots \end{array}$$

Since 1/10 = 0.1 and 1/100 = 0.01 and 1/1000 = 0.001, and so on, the negative exponents fit the following pattern of positive exponents:

$$\begin{array}{lcl} 10{,}000 & = & 10^4 \\ 1{,}000 & = & 10^3 \\ 100 & = & 10^2 \\ 10 & = & 10^1 \\ 1 & = & 10^0 \\ 0.1 & = & 10^{-1} \\ 0.01 & = & 10^{-2} \\ 0.001 & = & 10^{-3} \\ 0.0001 & = & 10^{-4} \end{array}$$

Notice that it makes sense to *define* 10^0 to be the number 1—that clearly fits with the pattern of moving the decimal point one place to the left (moving down the left-hand column of numbers) and lowering the exponent on the 10 by 1 (moving down the column on the right). There is another good reason to define 10^0 to be 1, which you will see at the end of the next activity.

CLASS ACTIVITY

5K Multiplying Powers of 10, p. CA-99

What Is Scientific Notation?

scientific notation A number is in **scientific notation** if it is written as a decimal that has exactly 1 nonzero digit to the left of the decimal point, multiplied by a power of 10. So a number is in scientific notation if it is written in the form

$$\#.\#\#\#\#\# \times 10^{\#}$$

CCSS 8.EE.3

where the # to the left of the decimal point is not zero and where any number of digits can be displayed to the right of the decimal point. Thus, neither

$$121.93263 \times 10^{15}$$

nor

$$12.193263 \times 10^{16}$$

is in scientific notation, but

$$1.2193263 \times 10^{17}$$

is in scientific notation.

Other than 0, every real number can be expressed in scientific notation. To write a number in scientific notation, think about how multiplication by powers of 10 works. For example, how do we express

$$847{,}930{,}000$$

in scientific notation? We need to find an exponent that will make the following equation true:

$$847{,}930{,}000 = 8.4793 \times 10^{?}$$

The decimal point in 8.4793 must be moved 8 places to the right to get 847,930,000; therefore, we should multiply 8.4793 by 10^8. In other words,

$$847{,}930{,}000 = 8.4793 \times 10^{8}$$

How do we write

$$0.0000345$$

in scientific notation? We need to find an exponent that will make the following equation true:

$$0.0000345 = 3.45 \times 10^{?}$$

The decimal point in 3.45 must be moved 5 places to the left to get 0.0000345; therefore, we must multiply 3.45 by

$$0.00001 = \frac{1}{10^5} = 10^{-5}$$

in order to get 0.0000345. So

$$0.0000345 = 3.45 \times 10^{-5}$$

Here are a few more examples:

Ordinary decimal notation	Scientific notation
12	1.2×10^1
123	1.23×10^2
1234	1.234×10^3
12345	1.2345×10^4
1234.5	1.2345×10^3
123.45	1.2345×10^2
12.345	1.2345×10^1
1.2345	1.2345 or 1.2345×10^0
0.12345	1.2345×10^{-1}
0.012345	1.2345×10^{-2}
0.0012345	1.2345×10^{-3}

Although scientific notation is mainly used in scientific settings, it is common to use a variation of scientific notation when discussing large numbers in more common situations, such as when budgets or populations are concerned. For example, an amount of money may be described as $3.5 billion, which means

$$\$3.5 \times (1 \text{ billion})$$

or

$$\$3.5 \times 1{,}000{,}000{,}000$$

Two other ways to express \$3.5 billion are

$$\$3.5 \times 10^9$$

and

$$\$3{,}500{,}000{,}000$$

The form

$$\$3.5 \text{ billion}$$

is probably more quickly and easily grasped by most people than any of the other forms.

By the way, what is a big number—or a small number? Children sometimes ask questions such as the following:

> Is 100 a big number? What about 1000, is that a big number?

The answer is, "It depends." We wouldn't think of 1000 grains of sand as a lot of sand, but we might consider 10 pages of homework to be a lot.

googol Speaking of big numbers, there is one special number name that many children find amusing and fascinating: a **googol**, the name for the number whose decimal representation is a 1 followed by 100 zeros:

$$\underbrace{10000000000\ldots0000000000}_{100 \text{ zeros}}$$

googolplex A googol is so large that, according to current theories in physics, it is even larger than the number of atoms in the universe. Another very large number that has a special name is a **googolplex**—a number whose decimal representation is a 1 followed by a googol zeros:

$$\underbrace{10000000000\ldots0000000000}_{\text{a googol zeros}}$$

It is very difficult to comprehend such a number.

SECTION SUMMARY AND STUDY ITEMS

Section 5.4 Powers and Scientific Notation

Exponents allow us to express repeated multiplication succinctly. Scientific notation is a way to write numbers by using powers of 10. Very large numbers and very small numbers are usually put in scientific notation in order to work with them and grasp them more easily.

Key Skills and Understandings

1. Use exponents to express powers.
2. Explain why we add exponents when we multiply two powers of the same number (such as two powers of 10).
3. Explain why it makes sense that we define 10^0 to be 1.
4. Put numbers that are in ordinary decimal notation in scientific notation, and put numbers that are in scientific notation in ordinary decimal notation.
5. Use scientific notation in calculations.

Practice Exercises for Section 5.4

1. Write the following numbers as powers of 10:
   a. 10,000,000,000,000
   b. 0.1
   c. 0.000001
   d. 1
   e. 10
   f. the number whose decimal representation is a 1 followed by 200 zeros
   g. a googol
   h. a googolplex
2. We call 10^6 a million, we call 10^9 a billion, and we call 10^{12} a trillion. What are 10^{13} and 10^{14} called? If we call 10^{15} a thousand trillion and 10^{18} a million trillion, then what are 10^{19}, 10^{20}, and 10^{21} called?
3. Explain why $10^A \times 10^B = 10^{A+B}$ is always true whenever A and B are counting numbers.
4. Write the following numbers in scientific notation:
   a. 153,293,043,922
   b. 0.00000321
   c. $(2.398 \times 10^{15}) \times (3.52 \times 10^9)$
   d. $(5.9 \times 10^{15}) \times (8.3 \times 10^9)$
5. Write 1.2 trillion in ordinary decimal notation and in scientific notation.
6. The astronomical unit (AU) is used to measure distances. One AU is the average distance from the earth to the sun, which is 92,955,630 miles. We are 2 billion AU from the center of the Milky Way galaxy (our galaxy). How many miles are we from the center of the Milky Way galaxy? Give your answer in scientific notation. Explain how the meaning of multiplication applies to this problem.
7. Sam uses a calculator to multiply 666,666 × 7,777,777. The calculator's answer is displayed as follows:

   5.1851795 E 12

   So Sam writes

   $$666{,}666 \times 7{,}777{,}777 = 5{,}185{,}179{,}500{,}000$$

   Is Sam's answer correct or not? If not, why not?

Answers to Practice Exercises for Section 5.4

1. a. $10{,}000{,}000{,}000{,}000 = 10^{13}$
   b. $0.1 = 10^{-1}$
   c. $0.000001 = 10^{-6}$
   d. $1 = 10^0$
   e. $10 = 10^1$
   f. The number whose decimal representation is a 1 followed by 200 zeros can be written 10^{200}.
   g. A googol can be written 10^{100}.
   h. A googlpex can be written 10^{googol} or $10^{(10^{100})}$.
2. Since 10^{12} is a trillion, 10^{13} is ten trillion and 10^{14} is a hundred trillion. Since 10^{18} is a million trillion, 10^{19} is ten-million trillion, 10^{20} is a hundred-million trillion, and 10^{21} is a billion trillion.
3. The expression $10^A \times 10^B$ stands for A tens multiplied by B tens. If we multiply A tens with B tens, then all together, we multiply $A \times B$ tens. Therefore,

   $$10^A \times 10^B = 10^{(A+B)}$$

   We can also explain this relationship with the following equations:

   $$\begin{aligned} &10^A \times 10^B \\ &= \underbrace{10 \times 10 \times \cdots \times 10}_{A \text{ times}} \times \underbrace{10 \times 10 \times \cdots \times 10}_{B \text{ times}} \\ &= \underbrace{10 \times 10 \times \cdots \times 10 \times 10 \times 10 \times \cdots \times 10}_{A + B \text{ times}} \end{aligned}$$

4. a. $1.53293043922 \times 10^{11}$
   b. 3.21×10^{-6}
   c. 8.44096×10^{24}
   d. 4.897×10^{25}
5. Ordinary decimal notation: 1,200,000,000,000. Scientific notation: 1.2×10^{12}.
6. Since we are 2 billion AU from the center of the galaxy and each AU is 92,955,630 miles, the number of miles from the earth to the center of the galaxy is the total number of objects in 2 billion groups (each AU is a group) when there are

92,955,630 objects in each group (each object is 1 mile). Therefore, according to the meaning of multiplication, we are

$$\begin{aligned} 2 \text{ billion} \times 92{,}955{,}630 &= 2 \times 10^9 \times 92{,}955{,}630 \\ &= 185{,}911{,}260 \times 10^9 \\ &\approx 1.86 \times 10^8 \times 10^9 \\ &= 1.86 \times 10^{17} \end{aligned}$$

miles from the center of the galaxy.

7. No, Sam's answer is not correct. When the calculator displays the answer to 666,666 × 7,777,777 as

5.1851795 E 12

this stands for

$$5.1851795 \times 10^{12}$$

However, the calculator is forced to round its answer because it can display only so many digits on its screen. Therefore, although it is true that

$$5.1851795 \times 10^{12} = 5{,}185{,}179{,}500{,}000$$

this is not the exact answer to 666,666 × 7,777,777. Instead, it is the answer to 666,666 × 7,777,777 rounded to the nearest hundred-thousand.

PROBLEMS FOR SECTION 5.4

1. Write the following numbers as powers of 10:

a. 0.000001

b. 10,000,000

c. the number whose decimal representation is a 1 followed by 50 zeros

d. the number whose decimal representation is a decimal point followed by 10 zeros, followed by a 1

e. the number whose decimal representation is a decimal point followed by 50 zeros, followed by a 1

2. a. The winnings of a lottery were \$250 million. Write 250 million in ordinary decimal notation and in scientific notation.

b. A company's revenues are \$15 billion. Write 15 billion in ordinary decimal notation and in scientific notation.

3. Write the following numbers in scientific notation:

a. 201,348,761,098

b. 0.000000078

c. $(2.4 \times 10^{12}) \times (8.6 \times 10^{11})$

d. $(6.1 \times 10^{13}) \times (9.2 \times 10^{8})$

4. A calculator might display the answer to

555,555 × 6,666,666

as

3.7037 E 12

a. What does the calculator's display mean?

b. What information about the product 555,555 × 6,666,666 can you obtain from the calculator's display? Can you determine the exact value of 555,555 × 6,666,666 in ordinary decimal notation? Why or why not? Can you determine how many digits are in the ordinary decimal representation of the product 555,555 × 6,666,666? Explain.

5. Tanya says that the ones digit of 2^{59} is a 5 because her calculator's display for 2^{59} reads

5.76460752303 E 17

and there is a 5 in the ones place. Is Tanya right? Why or why not?

6. Is 2×10^7 equal to 2^7? Is 1×10^9 equal to 1^9? If not, explain the distinctions between the expressions.

7. Calculate 8 × 123,456,123,456 using a calculator. What pattern do you see in the digits of the answer? Can you guess what the answer to

8 × 123,456,123,456,123,456

will be? What about

8 × 123,456,123,456,123,456,123,456

Use powers of 10 and the distributive property to explain why your guesses must be right.

8. Let's say that you want to write the product 179,234,652 × 437,481,694 as a whole number in ordinary decimal notation, showing all its digits. Find a way to use a calculator that displays no more than 12 digits to help you do this in an efficient way. Do not just multiply longhand. Explain your technique, and explain why it works.

9. Light travels at a speed of about 300,000 kilometers per second.

a. How far does light travel in one day? Give your answer both in scientific notation and in ordinary decimal notation. Explain your work.

b. How far does light travel in one year? Give your answer both in scientific notation and in ordinary decimal notation, rounded to the nearest hundred-billion kilometers. Explain your work. The distance that light travels in one year is called a *light year*.

10. According to scientific theories, the solar system formed between 5 and 6 billion years ago. Light travels 186,282 miles per second. How far has the light from the forming solar system traveled in 5.5 billion years? Give your answer in scientific notation.

***11.** Suppose that a laboratory has 1 gram of a radioactive substance that has a half-life of 100 years. "A half-life of 100 years" means that, no matter what amount of the radioactive substance one starts with, after 100 years, only half of it will be left. So, after the first 100 years, only half a gram would be left, and after another 100 years, only a quarter of a gram would be left.

a. How many hundreds of years will it take until there is less than one hundred millionth of a gram left of the laboratory's radioactive substance? (Give a whole number of hundreds of years.)

b. How many hundreds of years will it take until there is less than one billionth of a gram left of your radioactive substance? (Give a whole number of hundreds of years.)

***12.** Suppose you multiply a 6-digit number by an 8-digit number. How many digits will the product have? The following problem will help you answer this question:

a. When you write a 6-digit number and an 8-digit number in scientific notation, what will the exponents on the 10s be? Explain.

b. Write

$$(1.3 \times 10^5) \times (2.5 \times 10^7)$$

and

$$(7.9 \times 10^5) \times (8.3 \times 10^7)$$

in scientific notation.

c. Suppose you write a 6-digit number and an 8-digit number in scientific notation. If you multiply these numbers and write the product in scientific notation, what will the exponent on the 10 be when the product is expressed in scientific notation? Explain why your answer is in an "either . . . or . . ." form. (*Hint:* See part (b).)

d. Use your answer to part (c) to answer the following question: When you multiply a 6-digit number by an 8-digit number, how many digits will the product have?

CHAPTER SUMMARY

Section 5.1 Making Sense of Fraction Multiplication — Page 197

- The meaning of $M \cdot N$ as the number of units in M equal groups if there are N units in 1 group applies not only to whole numbers but also to fractions and decimals. — Page 197
- To multiply fractions, we multiply the numerators and denominators. We can explain why this procedure makes sense by seeing fraction multiplication as taking a fraction of a fraction. — Page 199

Key Skills and Understandings	
1. Interpret the meaning of multiplication with fractions. Write word problems for a given fraction multiplication problem, and solve the problems by using logic and a math drawing.	Page 197
2. Use simple word problems and logical reasoning about math drawings to explain why the procedure for multiplying fractions is valid.	Page 199
3. Given a word problem, determine whether it can be solved by fraction multiplication.	Page 201
Section 5.2 Making Sense of Decimal Multiplication	Page 205
▪ To multiply decimals, the standard procedure is to multiply the numbers without the decimal points present. Then we add the number of digits to the right of the decimal points in the two original numbers and put the decimal point in the answer that many places from the end. We can explain why we place the decimal point where we do by recognizing that we have multiplied and then divided by tens in order to shift decimal points suitably. We can also explain decimal multiplication in terms of fraction multiplication and in terms of areas of rectangles.	Page 205
Key Skills and Understandings	
1. Explain why we put the decimal point where we do when we multiply decimals.	Page 206
2. Use estimation to determine where to place the decimal point in a decimal multiplication problem.	Page 206
3. Describe decimal multiplication in terms of area.	Page 207
4. Write word problems for a given decimal multiplication problem.	Page 208
Section 5.3 Extending Multiplication to Negative Numbers	Page 210
▪ We can explain why a positive number times a negative number is negative by considering negative numbers as representing owed amounts. We can use properties of arithmetic to explain the way we multiply with negative numbers.	Page 210
Key Skills and Understandings	
1. Explain why a positive number times a negative number should be negative by considering negative numbers as owed amounts (for example).	Page 210
2. Use properties of arithmetic to explain the rules for multiplying with negative numbers.	Page 210
Section 5.4 Powers and Scientific Notation	Page 213
▪ Exponents allow us to express repeated multiplication succinctly	Page 213
▪ Scientific notation is a way to write numbers by using powers of 10. Very large numbers and very small numbers are usually put in scientific notation in order to work with them and grasp them more easily.	Page 213
Key Skills and Understandings	
1. Use exponents to express powers.	Page 214
2. Explain why we add exponents when we multiply two powers of the same number (such as two powers of 10).	Page 215
3. Explain why it makes sense that we define 10^0 to be 1.	Page 215
4. Put numbers that are in ordinary decimal notation in scientific notation, and put numbers that are in scientific notation in ordinary decimal notation.	Page 216
5. Use scientific notation in calculations.	Page 218

6

Division

In this chapter, we will study division of different kinds of numbers: whole numbers, fractions, decimals, and negative numbers. As with multiplication, we will study the meaning of division, and we will analyze why the various division procedures are valid based on the meaning of division. We will also study the crucial link between division and fractions.

We will focus on the following topics and practices within the *Common Core State Standards for Mathematics* (*CCSSM*).

Standards for Mathematical Content in the CCSSM

In the domain of *Operations and Algebraic Thinking* (K–Grade 5), students represent and solve problems about division and they understand the relationship between multiplication and division. In the domain of *Number and Operations in Base Ten* (K–Grade 5), students use place value understanding and properties of operations to perform and become fluent in multidigit division. In the domains of *Number and Operations—Fractions* (Grades 3–5) and *The Number System* (Grades 6–8), students extend their understanding of whole number division to divide fractions by fractions and to divide with negative numbers.

Standards for Mathematical Practice in the CCSSM

Opportunities to engage in all eight of the Standards for Mathematical Practice described in the CCSSM occur throughout the study of division. The following standards may be especially appropriate for emphasis while studying division:

- **4 Model with mathematics.** Students engage in this practice when they solve a real-world problem by formulating a numerical division problem and then solve and interpret the solution to the numerical division problem appropriately to solve the original real-world problem.
- **5 Use appropriate tools strategically.** Students engage in this practice when they use methods for recording multidigit division strategically and when they use strip diagrams, double number lines, tables, or the connection between multiplication and division to reason about and solve fraction division problems.
- **6 Attend to precision.** Students engage in this practice when they think carefully about how to interpret a remainder in a problem-solving situation and when they distinguish between the meaning of $3 \div 4$ and of $\frac{3}{4}$ and yet also recognize the connection between these expressions.

(From Common Core Standards for Mathematical Practice. Published by Common Core Standards Initiative.

6.1 Interpretations of Division

CCSS Common Core State Standards Grades 3, 4, 7

What does division mean? Think about a simple word problem you might give your students to help them understand what $15 \div 3$ means. Does an example like giving out 15 cookies to 3 children come to mind? There is another quite different but equally valid way to think about the meaning of division. In this section, we will discuss these two interpretations of division. Just as every subtraction problem can be rewritten as an addition problem, every division problem can be rewritten as a multiplication problem. The two interpretations of division arise from this connection between division and multiplication.

Have you ever wondered why we are not allowed to divide by zero? Is there a reason behind this "law"? In this section we'll see that the link between division and multiplication gives us a mathematical reason for why we can't divide by zero.

How Can We Define Division?

Division is denoted in three standard ways:

$$A \div B, \quad A/B, \quad \text{and} \quad B\overline{)A}$$

quotient
dividend
divisor

All three are read "A divided by B." In a division problem $A \div B$, the result is called the **quotient**, the number A is called the **dividend**, and the number B is called the **divisor**.

CLASS ACTIVITY

6A What Does Division Mean?, p. CA-100

What does division mean? In the same way that we can think of subtraction as addition with an unknown addend, we can also think of division as multiplication with an unknown factor. Thinking this way,

$$18 \div 6 = ?$$

means

$$? \cdot 6 = 18 \quad \text{or} \quad 6 \cdot ? = 18$$

As Figure 6.1 indicates, because the multiplier and multiplicand play different roles, we get two different types of division, how-many-groups division and how-many-units-in-1-group division.

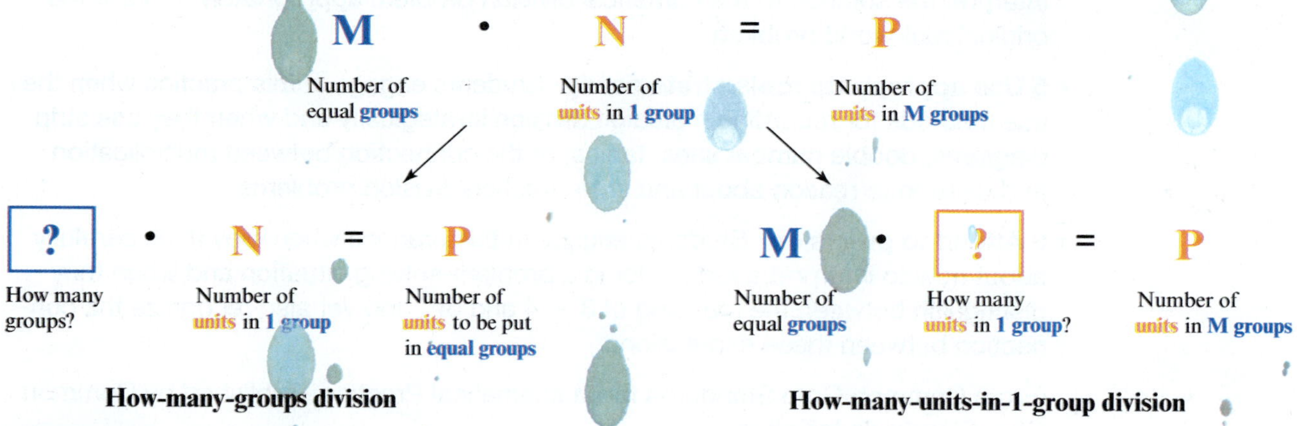

Figure 6.1 Two interpretations of division.

The How-Many-Groups Interpretation If A and B are nonnegative numbers, and B is not 0, then, according to the how-many-groups interpretation of division,

(Exact) division $A \div B$

$A \div B$ means the number of groups that are formed when A units (or objects) are divided into equal groups with B units (or objects) in 1 group.

With this how-many-groups interpretation of division, the problem

$$A \div B = ?$$

is equivalent to the multiplication problem

$$? \cdot B = A$$

with the number of groups unknown, as illustrated in Figure 6.2. In other words, with this interpretation, $A \div B$ means "the number of Bs that are in A." This interpretation of division is sometimes called the *measurement model of division* or the *subtractive model of division.*

Figure 6.2 Interpreting division as how-many-groups.

CCSS 3.OA.2

As an example, consider this word problem for $18 \div 6 = ?$:

There are 18 cookies to be put into packages. Each package will have 6 cookies. How many packages will there be?

Figure 6.3 shows how to interpret this as a how-many-groups problem.

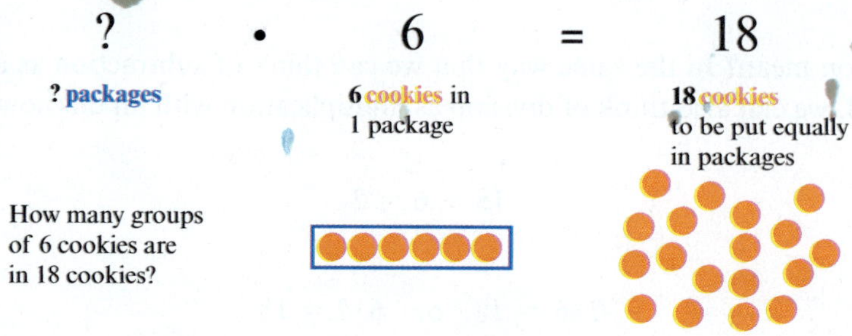

Figure 6.3 Interpreting a word problem as how-many-groups division.

The How-Many-Units-in-1-Group Interpretation If A and B are nonnegative numbers and B is not 0, then, according to the how-many-units-in-1-group interpretation of division,

(Exact) division $A \div B$

> $A \div B$ means the number of units (or objects) that are in 1 group when A units (or objects) are divided equally among B groups.

With this interpretation of division, the problem

$$A \div B = ?$$

is equivalent to the multiplication problem

$$B \cdot ? = A$$

with the number in 1 group unknown, as illustrated in **Figure 6.4**. This interpretation of division is sometimes called the *partitive model of division* or the *sharing model of division*.

Figure 6.4 Interpreting division as how-many-units-in-1-group.

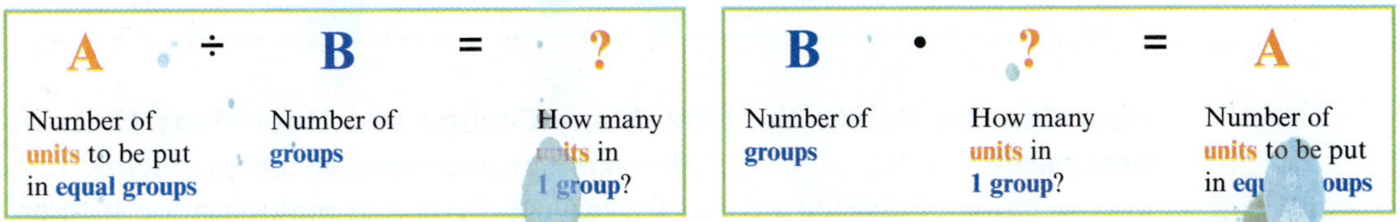

CCSS 3.OA.2

As an example, consider this word problem for $18 \div 6 = ?$:

> There are 18 cookies to be put equally into 6 packages. How many cookies will be in each package?

Figure 6.5 shows how to interpret this as a how-many-units-in-1-group problem.

Figure 6.5 Interpreting a word problem as how-many-units-in-1-group division.

How Can We Distinguish How-Many-Groups from How-Many-Units-in-1-Group Problems?

To determine what kind of division problem a problem is, it helps to reformulate the problem in terms of multiplication. For example, consider this problem:

> Seven cups of snodgrass filled 3 identical containers full. How many cups of snodgrass are needed to fill one container full?

We can reformulate this problem with the equation

$$\underset{\text{Number of groups}}{\underset{\uparrow}{3}} \cdot \underset{\text{Size of 1 group}}{\underset{\uparrow}{(\text{cups of snodgrass in 1 container})}} = \underset{\text{Size of 3 groups}}{\underset{\uparrow}{7\text{ cups}}}$$

which we can solve by dividing $7 \div 3$. Since there are 3 groups and the "cups of snodgrass in 1 container" is the amount in each group, the problem is a how-many-units-in-1-group problem.

Here is another example:

> Beezlebugs cost \$3 per pound. You have \$14 to spend on beezlebugs. How many pounds of beezlebugs can you buy?

The wording "\$3 per pound" means that each pound is worth \$3, so we can think of each pound as a "group of \$3." We can, therefore, reformulate the beezlebug problem with the equation

(pounds of beezlebugs)	·	\$3	=	\$14
↑		↑		↑
Number of groups		Size of 1 group		Size of all the groups

which we can solve by dividing $14 \div 3$. Since the "pounds of beezlebugs" stand for the number of groups and since there are \$3 in each group, the problem is a how-many-groups problem.

CLASS ACTIVITY

6B Division Word Problems, p. CA-101

How Can We Relate the How-Many-Groups and How-Many-Units-in-1-Group Interpretations? Why do we get the same answer to division problems, regardless of which interpretation of division we use? By viewing division as multiplication with an unknown factor, we can explain why a division problem will have the same answer, regardless of whether the how-many-groups or the how-many-units-in-1-group interpretation of division is used. The key lies in the commutative property of multiplication. According to the how-many-groups interpretation, solving $A \div B = ?$ is equivalent to solving $? \cdot B = A$. But according to the commutative property of multiplication, $? \cdot B = B \cdot ?$. Therefore, solving $? \cdot B = A$ is equivalent to solving $B \cdot ? = A$, which is equivalent to solving $A \div B = ?$ with the how-many-units-in-1-group interpretation of division. Thus, we will always get the same answer to a division problem, regardless of which way we interpret division.

What Are Array, Area, and Other Division Word Problems?

CCSS 4.OA.2, Table 2, columns 2 and 3

In Chapter 4 we studied array, ordered pair, and multiplicative comparison word problems, and we saw why areas of rectangles are calculated by multiplying. In each of these cases, it is possible to describe the situation in terms of equal groups and, therefore, in terms of multiplication. By viewing division problems as multiplication problems that have a known product, one known factor, and one unknown factor, we can create division problems that concern arrays, areas, multiplicative comparisons, and ordered pairs.

For example, consider these problems (drawings to accompany the first three problems are shown in **Figure 6.6**):

1. There are 54 cans arranged in an array that has 6 rows. How many cans are in each row?

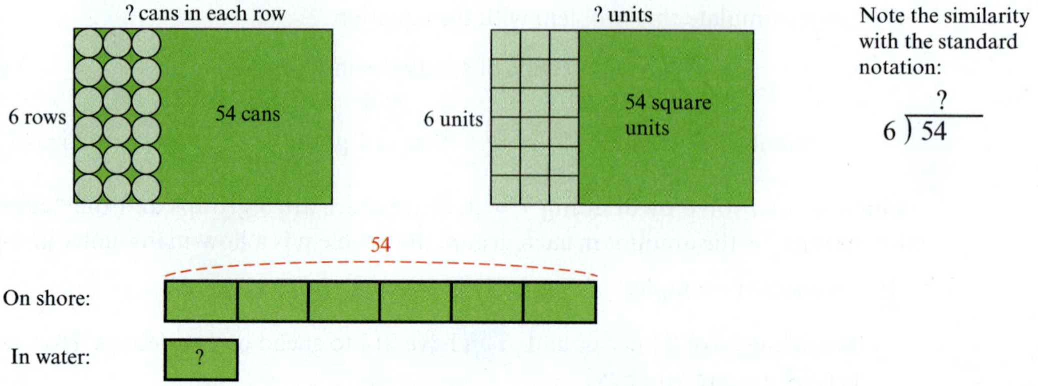

Figure 6.6 Math drawings for array, area, and multiplicative comparison division problems.

2. A rectangular room has area 54 square meters. One side of the room is 6 meters long. How long is an adjacent side?
3. There were 54 penguins on shore, and that was 6 times as many penguins as were in the water. How many penguins were in the water?
4. In a game, when you roll a 6-sided number cube and then pick a card, there are 54 different possible outcomes. How many different cards are there that you can pick?

Each of these problems can be viewed as a problem for

$$6 \times ? = 54 \quad \text{or} \quad ? \times 6 = 54$$

Therefore, each of these problems can be solved by calculating $54 \div 6 = 9$.

Interestingly, notice that math drawings for array and area division problems are similar to the division notation of the form $6\overline{)54}$, as indicated in **Figure 6.6**.

Why Can't We Divide by Zero?

CCSS
7.NS.2b

When we defined division, $A \div B$, we said that the divisor, B, should not be 0. Why not? The next class activity will help you explain this in several ways. One way is to translate a division problem with divisor 0 into an equivalent multiplication problem.

CLASS ACTIVITY

6C Why Can't We Divide by Zero?, p. CA-102

How Do We Divide with Negative Numbers?

So far we have defined division for nonnegative numbers. How can we make sense of division problems like

$$-18 \div 3 = ?$$

or

$$18 \div -3 = ?$$

We could perhaps talk about dividing $18 that is owed among 3 people. Another way is to rewrite a division problem with negative numbers as a multiplication problem. For example, the problem

$$-18 \div 3 = ?$$

is equivalent to

$$? \times 3 = -18$$

IMAP
The following video clips show examples of children solving division problems: Arriel, 18 ÷ 3 Cheyenne, 18 ÷ 6 Myrna, 20 ÷ 4

(if we use the how-many-groups interpretation). Since $-6 \times 3 = -18$, and since no other number times 3 is -18, it follows that

$$-18 \div 3 = -6$$

By considering how multiplication works when negative numbers are involved, we can see that the following rules apply to division problems involving negative numbers:

$$\text{negative} \div \text{positive} = \text{negative}$$

$$\text{positive} \div \text{negative} = \text{negative}$$

$$\text{negative} \div \text{negative} = \text{positive}$$

FROM THE FIELD Children's Literature

Shaskan, T. S. (2009). *If you were a divided-by sign.* Minneapolis, MN: Picture Window Books.

This children's book introduces vocabulary related to division and shows the relationship between repeated subtraction and division. The illustrations also show the equal sharing model of division.

Dodds, D. A. (1999). *The great divide: A mathematical marathon.* Cambridge, MA: Candlewick Press.

The Great Divide is quite a challenging marathon. Each obstacle in the race divides the bicyclists into smaller and smaller groups until only one racer heads toward the finish line.

Hutchins, P. (1986). *The doorbell rang.* New York, NY: Greenwillow Books.

Victoria and Sam have twelve cookies to share between them at tea time. A new friend arrives every time the doorbell rings, though, and they have to figure out how to share the cookies equally among themselves as more people visit their home.

Calvert, P. (2011). *The multiplying menace divides: A math adventure.* Watertown, MA: Charlesbridge.

Rumpelstiltskin teams up with a witch named Matilda to seek revenge on Prince Peter and his kingdom. They plan to use the Great Staff of Product-Quo to divide the kingdom into frogs, but Peter uses mathematical reasoning to outsmart them once again.

Murphy, S. J. (1997). *Divide and ride.* New York, NY: HarperCollins.

Every time 11 best friends want to go on a carnival ride, they divide themselves among the seats and have friends left over. Since they don't want to leave anyone behind, they invite others to go with them on the ride.

Pinczes, E. J. (1995). *A remainder of one.* New York, NY: Houghton Mifflin.

The 25th squadron wants to please the queen by marching in perfectly even rows, but one bug-soldier ends up in a line all on his own during the parade. Even after dividing the squadron into three lines, Joe is still a remainder of one. Help him decide how many rows of soldiers will allow him to march with the rest of the bug-soldiers during the parade.

SECTION SUMMARY AND STUDY ITEMS

Section 6.1 Interpretations of Division

There are two interpretations of division: the how-many-groups interpretation and the how-many-units-in-1-group interpretation. With the how-many-groups interpretation of division, $A \div B$ means the number of Bs that are in A—that is, the number of groups when A units are divided into equal groups with B units in 1 group. With the how-many-units-in-1-group interpretation of division, $A \div B$ means the number of units in 1 group when A units are divided equally among B groups. Every division problem can be reformulated as a multiplication problem. With the how-many-groups interpretation,

$$A \div B = ? \text{ is equivalent to } ? \cdot B = A$$

With the how-many-units-in-1-group interpretation,

$$A \div B = ? \text{ is equivalent to } B \cdot ? = A$$

Key Skills and Understandings

1. Write and recognize whole number division word problems for both interpretations of division.
2. Explain why we can't divide by 0, but why we *can* divide 0 by a nonzero number.
3. Use division to solve problems.

Practice Exercises for Section 6.1

1. For each of the following word problems, write the corresponding numerical division problem and decide which interpretation of division is involved (the how-many-groups or the how-many-units-in-1-group).
   a. There are 5235 tennis balls that are to be put into packages of 3. How many packages of balls can be made?
   b. If 50 fluid ounces of mouthwash costs \$5, then what is the price per fluid ounce of this mouthwash?
   c. If 50 fluid ounces of mouthwash costs \$5, then how much of this mouthwash is worth \$1?
   d. If you have a full 50-fluid-ounce bottle of mouthwash and you use 2 fluid ounces per day, then how many days will this mouthwash last?
   e. If 1 yard is 3 feet, how long is an 84-foot-long stretch of sidewalk in yards?
   f. If 1 gallon is 16 cups, how many gallons are 64 cups of lemonade?
   g. If you drive 315 miles at a constant speed and it takes you 5 hours, then how fast did you go?
2. Write two word problems for $300 \div 12 = ?$, one for each of the two interpretations of division.
3. a. Write a multiplicative comparison word problem for $18 \div 3 = ?$.
   b. Write an area word problem for $18 \div 3 = ?$.
4. Josh says that $0 \div 5$ doesn't make sense because if you have nothing to divide among 5 groups, then you won't be able to put anything in the 5 groups. Which interpretation of division is Josh using? What does Josh's reasoning actually say about $0 \div 5$?
5. Roland says,

   I have 23 pencils to give out to my students, but I have no students. How many pencils should each student get? There's no possible answer because I can't give out 23 pencils to my students if I have no students.

   Write a division problem that corresponds to what Roland said. Which interpretation of division does this use? What does Roland's reasoning say about the answer to this division problem?
6. Katie says,

   I have no candies to give out and I'm going to give each of my friends 0 candies. How many friends do I have? I could have 10 friends, or 50 friends, or 0 friends—there's no way to tell.

   Write a division problem that corresponds to what Katie said. Which interpretation of division does this use? What does Katie's reasoning say about the answer to this division problem?
7. Use the how-many-groups interpretation of division to explain why $1 \div 0$ is not defined.
8. Explain why $0 \div 0$ is not defined (or indeterminate) by rewriting $0 \div 0 = ?$ as a multiplication problem.

Answers to Practice Exercises for Section 6.1

1. a. $5235 \div 3$. This uses the how-many-groups interpretation. 1745 packages can be made.
   b. $5 \div 50$. This uses the how-many-units-in-1-group interpretation. Each ounce of mouthwash represents a group. We want to divide \$5 equally among these groups. Each fluid ounce costs \$0.10.
   c. $50 \div 5$. This uses the how-many-units-in-1-group interpretation. Each dollar represents a group. We want to divide 50 ounces among the 5 groups. \$1 buys 10 fluid ounces.

d. 50 ÷ 2. This uses the how-many-groups interpretation. Each 2 fluid ounces is a group. We want to know how many of these groups are in 50 fluid ounces. The mouthwash will last for 25 days.

e. 84 ÷ 3. This uses the how-many-groups interpretation. Each 3 feet is a group (a yard). We want to know how many of these groups are in 84 feet. There are 28 yards in 84 feet.

f. 64 ÷ 16. This uses the how-many-groups interpretation. Each 16 cups is a group (a gallon). We want to know how many of these groups are in 64 cups. The answer is 4 gallons.

g. 315 ÷ 5 = 63 miles per hour. This uses the how-many-units-in-1-group interpretation. Divide the 315 miles equally among the 5 hours. Each hour represents a group. In each hour, you drove 315 ÷ 5 miles. This means your speed was 63 miles per hour.

2. A good example for the how-many-groups interpretation is "how many feet are in 300 inches?" Because each foot is 12 inches, this problem can be interpreted as "how many 12s are in 300?" An example for the how-many-units-in-1-group interpretation is "300 snozzcumbers will be divided equally among 12 hungry boys. How many snozzcumbers does each hungry boy get?"

3. a. Kaya saved $18 and that is 3 times as much as her little sister, Ana, saved. How much did Ana save?

b. A rectangular poster is to have an area of 18 square feet. If the poster will be 3 feet wide, how long should it be?

4. Josh is using the how-many-units-in-1-group interpretation of division. Josh's statement can be reinterpreted as this: If you have 0 objects and you divide them equally among 5 groups, then there will be 0 objects in each group. Therefore, 5 × 0 = 0 and so 0 ÷ 5 = 0.

5. The division problem that corresponds to what Roland said is this: 23 ÷ 0 = ?. Written in terms of multiplication it is 0 × ? = 23. Roland is using the how-many-units-in-1-group interpretation of division. Each group is represented by a student. Roland wants to divide 23 objects equally among 0 groups. But as Roland says, this is impossible to do. Therefore, 23 ÷ 0 is undefined.

6. The division problem that corresponds to what Katie said is this: 0 ÷ 0 = ?. Written in terms of multiplication it is ? × 0 = 0. Katie is using the how-many-groups interpretation of division. Each group is represented by a friend. There are 0 objects to be distributed equally, with 0 objects in each group. But from this information, there is no way to determine how many groups there are. There could be *any* number of groups—50, or 100, or 1000. So the reason that 0 ÷ 0 is undefined (or indeterminate) is there isn't *one unique answer*.

7. With the how-many-groups interpretation, 1 ÷ 0 means the number of groups there are when 1 object is divided into groups with 0 objects in each group. But if you put 0 objects in each group, then there's no way to distribute the 1 object—it can never be distributed among the groups, no matter how-many-groups there are. In other words, ? × 0 = 1 cannot be solved. Therefore, 1 ÷ 0 is not defined.

8. Any division problem $A \div B = ?$ can be rewritten in terms of multiplication—namely, as either $? \times B = A$ or as $B \times ? = A$. Therefore, 0 ÷ 0 = ? means the same as ? × 0 = 0 or 0 × ? = 0. But *any number* times 0 is 0, so the ? can stand for any number. Since there isn't one unique answer to 0 ÷ 0 = ?, we say that 0 ÷ 0 is undefined (or indeterminate).

PROBLEMS FOR SECTION 6.1

1. For each of the following word problems, write the corresponding numerical division problem, state which interpretation of division is involved (the how-many-groups or the how-many-units-in-1-group) and solve the problem:

a. If 252 rolls are to be put in packages of 12, then how many packages of rolls can be made?

b. If you have 506 stickers to give out equally to 23 children, then how many stickers will each child get?

c. Given that 1 gallon is 8 pints, how many gallons of water are 48 pints of water?

d. If your car used 12 gallons of gasoline to drive 360 miles, then how many miles per gallon did your car get?

e. If you drove 177 miles at a constant speed and if it took you 3 hours, then how fast were you going?

f. Given that 1 foot is 12 inches, how many feet long is an 84-inch-long board?

2. Write two word problems for $63 \div 7 = ?$, one for each of the two interpretations of division. Solve each problem.

3. a. Write an array problem for $21 \div 3 = ?$ and make a math drawing for the problem.

b. Write an area problem for $21 \div 3 = ?$ and make a math drawing for the problem.

4. a. Write a multiplicative comparison problem for $3 \cdot ? = 21$.

b. Draw a strip diagram to accompany your problem in part (a).

c. Discuss why a student might mistakenly attempt to solve your problem in part (a) by multiplying 21 by 3 instead of by dividing 21 by 3. Explain how drawing a strip diagram might be helpful.

d. Reformulate your problem in part (a), but this time use the fraction $\frac{1}{3}$.

5. a. Is $0 \div 5$ defined or not? Write a word problem for $0 \div 5$, and use your problem to discuss whether or not $0 \div 5$ is defined.

b. Is $5 \div 0$ defined or not? Write a word problem for $5 \div 0$, and use your problem to discuss whether or not $5 \div 0$ is defined.

6. a. Is $0 \div 3$ defined or not? Explain your reasoning.

b. Is $3 \div 0$ defined or not? Explain your reasoning.

7. Write and solve one word problem for $35 \div 70 = ?$ and another for $70 \div 35 = ?$. Say which is which.

8. a. Use the definition of powers of 10 to show how to write the following expressions as a single power of 10 (i.e., in the form $10^{\text{something}}$):

i. $10^5 \div 10^2$

ii. $10^6 \div 10^4$

iii. $10^7 \div 10^6$

b. In each of (i), (ii), and (iii), compare the exponents involved. In each case, what is the relationship among the exponents?

c. Explain why it is always true that $10^A \div 10^B = 10^{A-B}$ when A and B are counting numbers and A is greater than B.

6.2 Division and Fractions and Division with Remainder

CCSS Common Core State Standards Grades 4, 5

When we *multiply* two whole numbers, the product is always a whole number. So it's interesting that when we *divide* two whole numbers, the quotient might not be a whole number. Furthermore, when the quotient is not a whole number, we have different options for how to express the answer to the division problem. In this section, we'll study the different kinds of answers we can give to whole number division problems. We'll see that if the division problem arose from a word problem, we should take the context into account when deciding which answer to the numerical division problem to use. We'll also see how the different answers to a division problem are related. Especially important in this relationship is the link between division and fractions, which we examine first.

How Can We Explain the Connection Between Fractions and Division?

What is the relationship between fractions and division? Have you noticed that the same notation is often used for division as well as for fractions? Instead of writing the "divided by" symbol, ÷, we sometimes write $2 \div 5$ as $2/5$. But the expression $2/5$ also stands for the fraction $\frac{2}{5}$, so the notation we use equates fractions with division. However, according to our definitions of fractions and of division, the expressions

$$\frac{2}{5} \quad \text{and} \quad 2 \div 5$$

have the following *different* meanings:

- $\frac{2}{5}$ of a pie is the amount of pie formed by 2 pieces when the pie is divided into 5 equal pieces.
- $2 \div 5$ is the amount of pie one person will receive if 2 (identical) pies are divided equally among 5 people (using the how-many-units-in-1-group interpretation).

Notice the difference: $\frac{2}{5}$ refers to *2 pieces of pie*, whereas $2 \div 5$ refers to dividing *2 pies*. But is it the same amount of pie either way? Before you read on, please do Class Activities 6D, 6E, and 6F.

IMAP *Watch Felisha solve fractions pictorially.*

CLASS ACTIVITY

6D Relating Whole Number Division and Fractions, p. CA-103

6E Using Measurement Ideas to Relate Whole Number Division and Fractions, p. CA-103

6F Relating Whole Number Division and Fraction Multiplication, p. CA-104

CCSS 5.NF.3

Let's return to the case of dividing 2 pies equally among 5 people. To do so, you can divide each pie into 5 equal pieces and give each person 1 piece from each pie, as shown in Figure 6.7. One person's share of pie consists of 2 pieces of pie, and each of those pieces is $\frac{1}{5}$ of a pie. (That is, each piece comes from a pie that has been divided into 5 equal pieces.) So each of the 5 people sharing the 2 pies gets $\frac{2}{5}$ of a pie. Thus, 1 person's share of pie can be described in two ways: as $2 \div 5$ of a pie and as $\frac{2}{5}$ of a pie. Therefore, both ways of describing a person's share of pie must be equal, and we have

$$2 \div 5 = \frac{2}{5}$$

The preceding discussion applies equally well when other whole numbers replace 2 and 5 (except that 5 should not be replaced with 0). So, in general, if A and B are whole numbers and B is not 0, then $A \div B$ really is equal to $\frac{A}{B}$.

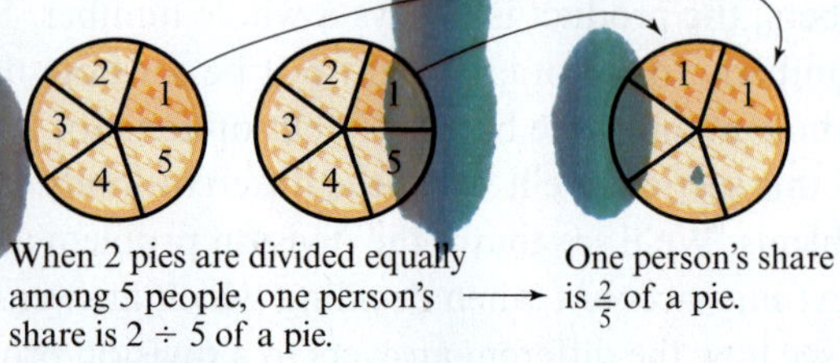

Figure 6.7 Explaining why $2 \div 5 = \frac{2}{5}$ by dividing 2 pies equally among 5 people and determining that 1 person's share is $\frac{2}{5}$ of a pie.

How Are Exact Division and Division with Remainder Related?

Consider the division problem $23 \div 4$. Depending on the context, any one of the following three answers could be most appropriate:

$$23 \div 4 = 5.75$$

$$23 \div 4 = 5\frac{3}{4}$$

$23 \div 4$ has whole number quotient 5, remainder 3

The first two answers fit with division as we have been interpreting it so far—exact division. For example, to divide \$23 equally among 4 people, each person should get $\$23 \div 4 = \$5.75$. Or, if you have 23 cups of flour, and a batch of cookies requires 4 cups of flour, then you can make $23 \div 4 = 5\frac{3}{4}$ batches of cookies (assuming that it is possible to make $\frac{3}{4}$ of a batch of cookies). But if you have 23 pencils to divide equally among 4 children, then it doesn't make sense to give each child 5.75 or $5\frac{3}{4}$ pencils. Instead, it is best to give each child 5 pencils and keep the remaining 3 pencils in reserve. How does this answer, 5, remainder 3, fit with division as we have defined it?

CCSS 4.OA.3

In order to get the third answer, 5, remainder 3, we need to interpret division in a different manner than we have so far. For each of the two main interpretations of division, there is an alternative formulation, which allows for a remainder. In these alternative formulations, we seek a quotient that is a *whole number*.

If A and B are whole numbers, and B is not zero, then

division with remainder, how-many-units-in-1-group

$A \div B$ is the largest whole number of objects that are in each group when A objects are divided equally among B groups. The **remainder** is the number of objects left over (i.e., that can't be placed in a group). This is the **how-many-units-in-1-group** interpretation, with remainder.

division with remainder, how-many-groups

$A \div B$ is the largest whole number of groups that can be made when A objects are divided into groups with B objects in each group. The **remainder** is the number of objects left over (i.e., that can't be placed in a group). This is the **how-many-groups** interpretation, with remainder.

For division with remainder, the connection between division and multiplication is different than it is for exact division. For example,

$23 \div 4$ has whole number quotient 5, remainder 3

corresponds with equations that involve both multiplication and addition:

$$5 \times 4 + 3 = 23 \quad \text{or} \quad 4 \times 5 + 3 = 23$$

Note the contrast with exact division: $23 \div 4 = 5.75$ corresponds with

$$5.75 \times 4 = 23 \quad \text{or} \quad 4 \times 5.75 = 23.$$

How is division with remainder connected with exact division? We'll study this question next.

How Can We Connect Whole-Number-with-Remainder Answers to Mixed Number Answers? How are the whole-number-with-remainder answer and the mixed number answer to a whole number division problem related? Given the division problem $29 \div 6$, the whole-number-with-remainder answer is 4 remainder 5, and its (exact) mixed number answer is $4\frac{5}{6}$. The remainder is 5, the divisor is 6, and the fractional part of the mixed number answer, $4\frac{5}{6}$ is $\frac{5}{6}$, which is in the form

$$\frac{\text{remainder}}{\text{divisor}}$$

This will be the case in general. We can see why by supposing we have 29 pizzas to be divided equally among 6 classrooms. How many pizzas does each classroom get? Since 29 ÷ 6 has whole number quotient 4, remainder 5, each classroom will get 4 pizzas, and there will be 5 pizzas left over. Instead of just giving away the remaining 5 pizzas, we might want to divide these 5 pizzas into 6 equal parts. When 5 pizzas are divided equally among 6 classrooms, how much pizza does each classroom get? Each classroom gets

$$5 \div 6 = \frac{5}{6}$$

of a pizza, according to the work we did previously, in which we equated fractions and division, and as we also see in Figure 6.8.

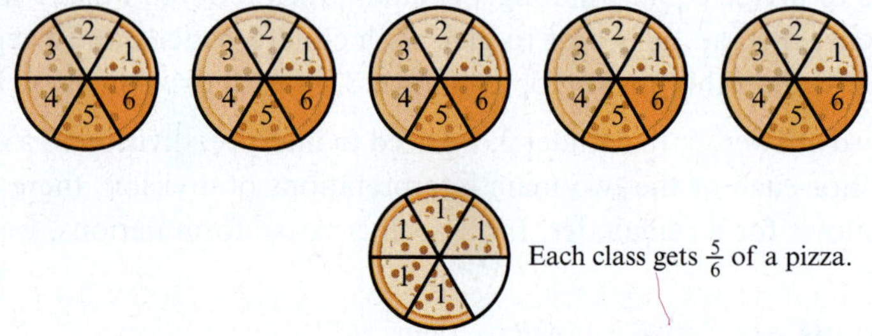

Figure 6.8 Dividing 5 pizzas among 6 classrooms.

So, instead of giving each class 4 pizzas and having 5 pizzas left over, we can give each class $4\frac{5}{6}$ pizzas. In other words,

29 ÷ 6 has whole number quotient 4, remainder 5

is equivalent to

$$29 \div 6 = 4\frac{5}{6}$$

The same reasoning works with other numbers. If A and B are whole numbers, then

$A \div B$ has whole number quotient Q, remainder R

is equivalent to

$$A \div B = Q\frac{R}{B}$$

In Class Activity 6G, you will interpret whole-number-with-remainder and mixed number answers to division problems in the context of word problems. You'll see that you should pick the form of the answer to a numerical division problem suitably in order to best answer a related word problem. Even then, the answer to the word problem is not necessarily identical to the answer to the numerical division problem.

CLASS ACTIVITY

6G What to Do with the Remainder?, p. CA-105

Why Can We Use Division to Convert Improper Fractions to Mixed Numbers? In Chapter 3, we discussed how to turn mixed numbers into improper fractions by viewing a mixed number as the sum of its whole number part and its fractional part. Now we will go the other way around and turn improper fractions into mixed numbers. We do so by viewing an improper fraction in terms of division.

For example, to convert the improper fraction $\frac{27}{4}$ to a mixed number, write $\frac{27}{4}$ in terms of division, and calculate the mixed number answer to the corresponding division problem:

$$\frac{27}{4} = 27 \div 4$$

$27 \div 4$ has whole number quotient 6, remainder 3

Thus,

$$\frac{27}{4} = 27 \div 4 = 6\frac{3}{4}$$

IMAP▶

See the following video clips for examples of children solving division problems: Felisha, 1 shared by 4, then by 3 Fraction kids, 8 shared by 6

Another way to explain why the preceding procedure for turning $\frac{27}{4}$ into a mixed number makes sense is to notice that $\frac{27}{4}$ stands for 27 parts each of which is $\frac{1}{4}$. Each group of 4 parts makes a whole. When we divide 27 by 4, we find that there are 6 groups of 4 and 3 left over. In terms of the fraction $\frac{27}{4}$, the 6 groups of 4 parts make 6 wholes. The 3 remaining parts make $\frac{3}{4}$ of a whole because each of the parts is one fourth.

The same reasoning works with other numbers. If A and B are whole numbers, and if

$A \div B$ has whole number quotient Q, remainder R

then

$$\frac{A}{B} = Q\frac{R}{B}$$

SECTION SUMMARY AND STUDY ITEMS

Section 6.2 Division and Fractions and Division with Remainder

Division and fractions are related by $A \div B = \frac{A}{B}$. Given a whole number division problem, the exact quotient might not be a whole number. In this case, the quotient can be expressed as a mixed number or fraction or as a decimal, or it can be expressed as a whole number with a remainder.

Key Skills and Understandings

1. Explain why it makes sense that the result of dividing $A \div B$ is $\frac{A}{B}$.
2. Write and recognize whole number division word problems that are best answered exactly, with either a decimal or a mixed number, or answered with a whole number and a remainder.
3. In word problems, interpret quotients and remainders appropriately, and recognize the distinction between solving a numerical division problem that is related to a word problem and solving the word problem.
4. Describe how the whole-number-with-remainder answer to a division problem is related to the mixed number answer, and explain why this relationship holds.

Practice Exercises for Section 6.2

1. Use the definition of fractions and the definition of division, from the how-many-units-in-1-group viewpoint, to explain why $3 \div 10 = \frac{3}{10}$. Your explanation should be general, in the sense that you could see why $3 \div 10 = \frac{3}{10}$ would still be true if other numbers were to replace 3 and 10.
2. Describe how the whole number with remainder and mixed number answers to $14 \div 3$ are related. Use a simple word problem to help you explain this relationship.
3. For each of the following word problems, write the corresponding numerical division problem. Also, interpret the meaning of the whole-number-with-remainder answer and the meaning of the mixed number answer to the division problem in terms of the word problem and its answer.
   - **a.** You have 27 pints of soup and several containers that hold 4 pints. How many containers will you need to hold all the soup?
   - **b.** How long will it take you to drive 105 miles at a steady speed of 45 miles per hour?
   - **c.** You drove 135 miles at a steady speed and it took you 2 hours. How fast were you going?
4. If March 5th is a Wednesday, then why do you know right away that March 12th, 19th, 26th, and April 2nd are Wednesdays? Explain this, using mathematics and knowledge of our calendar system.
5. Suppose that today is Friday. What day of the week will it be 36 days from today? What day of the week will it be 52 days from today? What about 74 days from today? Explain how division with a remainder is relevant to these questions.

Answers to Practice Exercises for Section 6.2

1. If there are 3 (identical) pies to be divided equally among 10 people, then according to the definition of division (with the how-many-units-in-1-group interpretation) each person will get $3 \div 10$ of a pie. To divide the pies, you can divide each pie into 10 pieces and give each person 1 piece from each of the 3 pies. One person's share is shown shaded in Figure 6.9. Each person then gets 3 pieces, where each piece is $\frac{1}{10}$ of a pie. Therefore, each person gets $\frac{3}{10}$ of a pie, according to the definition of fractions.

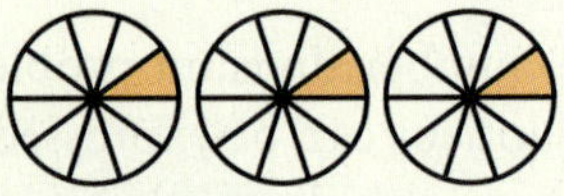

Figure 6.9 The shaded portion is 1 person's share when 3 pies are divided equally among 10 people.

Because each person's share can be described both as $3 \div 10$ of a pie and as $\frac{3}{10}$ of a pie, it follows that $\frac{3}{10} = 3 \div 10$.

2. The whole-number-with-remainder answer to $14 \div 3$ is 4, remainder 2. The mixed number answer is $4\frac{2}{3}$. To connect these two answers, consider an example. If you have 14 cookies to be divided equally among 3 people, then you could give each person 4 cookies and have 2 cookies left over (the remainder 2), or you could take those 2 remaining cookies and divide them equally among the 3 people. You could do this by dividing each cookie into 3 equal parts and giving each person 2 parts. In this way, each person would get $4\frac{2}{3}$ cookies.
3. **a.** $27 \div 4$. The whole-number-with-remainder answer to $27 \div 4$ is 6, remainder 3. This means that you can fill 6 containers completely full and will then have 3 pints remaining. So you will need another container, for a total of 7 containers, to hold all the soup. The mixed number answer to $27 \div 4$ is $6\frac{3}{4}$. This means that you can fill 6 containers completely full and fill the 7th container $\frac{3}{4}$ full. Again, the answer to the word problem is therefore 7.

   b. $105 \div 45$. The whole-number-with-remainder answer to $105 \div 45$ is 2, remainder 15. This means that after 2 full hours of driving, you will still have another 15 miles to go. The mixed number answer to $105 \div 45$ is $2\frac{15}{45} = 2\frac{1}{3}$. This means it will take $2\frac{1}{3}$ hours to drive the 105 miles. Since $\frac{1}{3}$ of an hour is 20 minutes, it will take 2 hours, 20 minutes.

   c. $135 \div 2$. The whole-number-with-remainder answer to $135 \div 2$ is 67, remainder 1. This means you went 67 whole miles in each hour, and another part of a mile in each hour. The whole-number-with-remainder answer to the numerical division problem is really not appropriate or illuminating for solving the word problem. The mixed number answer to $135 \div 2$ is $67\frac{1}{2}$. This means that you drove $67\frac{1}{2}$ miles each hour, in other words, that you were driving at a speed of $67\frac{1}{2}$ miles per hour.

4. Every 7 days after Wednesday is another Wednesday. March 12th, 19th, 26th, and April 2nd are 7, 14, 21, and 28 days after March 5th, and 7, 14, 21, and 28 are multiples of 7, so these days are all Wednesdays, too.

5. Every 7 days after a Friday is another Friday. So 35 days after today (assumed to be a Friday) is another Friday. Therefore, 36 days from today is a Saturday, because it is 1 day after a Friday. Notice that we only needed to find the remainder of 36 when divided by 7 in order to determine the answer. $52 \div 7$ has whole number quotient 7, remainder 3, so 52 days from today will be 3 days after a Friday (because the 7 groups of 7 get us to another Friday). Three days after a Friday is a Monday. $74 \div 7 = 10$, remainder 4, so 74 days from today will be 4 days after a Friday, which is a Tuesday.

PROBLEMS FOR SECTION 6.2

1. Use the definition of fractions and how-many-units-in-1-group division, to explain in your own words why $3 \div 7 = \frac{3}{7}$. Your explanation should be general, in the sense that you could see why $3 \div 7 = \frac{3}{7}$ would still be true if other numbers were to replace 3 and 7.

2. Using our definition of fractions, the idea of measuring one strip by another strip, and how-many-groups division, explain why $5 \div 8 = \frac{5}{8}$ and $8 \div 5 = \frac{8}{5}$. Your explanation should be general in the sense that you could see why the equations would still be true if other numbers were to replace 5 and 8.

3. a. Explain why $11 \div 5 = \frac{1}{5} \cdot 11$ using how-many-units-in-1-group division, our definition of multiplication, and a math drawing.

b. More generally, explain why dividing a number by a natural number N is equivalent to multiplying the number by $\frac{1}{N}$.

4. Describe how to get the mixed number answer to $23 \div 5$ from the whole-number-with-remainder answer. Explain why your method makes sense by interpreting it in terms of a simple word problem.

5. a. Write a simple how-many-units-in-1-group word problem for $17 \div 5$ for which the answer "3, remainder 2" is appropriate. Explain what the answer "3, remainder 2" means in the context of the word problem.

b. Write a simple how-many-groups word problem for $17 \div 5$ for which the answer "3, remainder 2" is appropriate. Explain what the answer "3, remainder 2" means in the context of the word problem.

c. Write a simple how-many-units-in-1-group word problem for $17 \div 5$ for which the answer $3\frac{2}{5}$ is appropriate. Explain what the answer $3\frac{2}{5}$ means in the context of the word problem.

d. Write a simple how-many-groups word problem for $17 \div 5$ for which the answer $3\frac{2}{5}$ is appropriate. Explain what the answer $3\frac{2}{5}$ means in the context of the word problem.

6. Write and solve four different word problems for $21 \div 4$.

a. In the first word problem, the answer should be best expressed as 5, remainder 1. Explain why this is the best answer. Interpret the meaning of "5, remainder 1" in the context of the problem.

b. In the second word problem, the answer should be best expressed as $5\frac{1}{4}$. Explain why this is the best answer. What does the $\frac{1}{4}$ mean in the context of the problem?

c. In the third word problem, the answer should be best expressed as 5.25. Explain why this is the best answer.

d. The answer to the fourth word problem should be 6 (even though $21 \div 4 \neq 6$). Explain why this is the best answer.

7. For each of the problems that follow, write the corresponding numerical division problem and solve the problem. Determine the best form (or forms) of the answer: a mixed number, a decimal, a whole number with remainder, or a whole number that is not equal to the solution of the numerical division problem. Briefly explain your answers.

a. For purposes of maintenance, a 58-mile stretch of road will be divided into 3 equal parts. How many miles of road are in each part?

b. For purposes of maintenance, a 58-mile stretch of road will be divided into sections of 15 miles. Each section of 15 miles will be the responsibility of a particular road crew. How many road crews are needed?

c. You have 75 pencils to give out to a class of 23 children who insist on fair distribution. How many pencils will each child get?

d. You have 7 packs of chips to share equally among 4 hungry people. How many packs of chips will each person get?

8. Explain how to solve the next problems with division, assuming that today is Friday. Which interpretation of division do you use?

a. What day of the week will it be 50 days from today?

b. What day of the week will it be 60 days from today?

c. What day of the week will it be 91 days from today?

d. What day of the week will it be 365 days from today?

9. In your own words, describe a procedure for turning an improper fraction, such as $\frac{19}{4}$, into a mixed number, and explain why this procedure makes sense.

***10.** Halloween (October 31) of 2014 was on a Friday, which was great for kids.

a. How can you use division to determine what day of the week Halloween was on in 2015? (The year 2014 was not a leap year, so Halloween of 2015 was 365 days from Halloween of 2014.)

b. After 2014, when are the next two times that Halloween falls on either a Friday or a Saturday? Again, use mathematics to determine this. Explain your reasoning. (The years 2016, 2020, 2024, etc. are leap years, so they have 366 days instead of 365.)

***11.** Must there be at least one Friday the 13th in every year? Use division to answer this question. (You may answer only for years that aren't leap years.) To get started on solving this problem, answer the following: If January 13th falls on a Monday, then what day of the week will February 13th, March 13th, and so forth, fall on? Use division with remainder to answer these questions. Now consider what will happen if January 13th falls on a Tuesday, a Wednesday, and so on.

***12.** Presidents' Day is the third Monday in February. In 2016, Presidents' Day was on February 15. What is the date of Presidents' Day in 2017? Use mathematics to solve this problem without looking at a calendar. Explain your reasoning clearly.

***13.** I'm thinking of a number. When you divide it by 2, it has remainder 1; when you divide it by 3, it has remainder 1; when you divide by 4, 5, or 6, it always has remainder 1. The number I am thinking of is greater than 1. What could my number be?

***14.** I'm thinking of a number. When you divide it by 12, it has remainder 2; and when you divide it by 16, it also has remainder 2. The number I am thinking of is greater than 2. Find at least three numbers that could be my number. How are these numbers related?

***15.** Three robbers have just acquired a large pile of gold coins. They go to bed, leaving their faithful servant to guard it. In the middle of the night, the first robber gets up, gives 2 gold coins from the pile to the servant as hush money, divides the remaining pile of gold evenly into 3 parts, takes 1 part, forms the remaining 2 parts back into a single pile and goes back to bed. A little later, the second robber gets up, gives 2 gold coins from the remaining pile to the servant as hush money, divides the remaining pile evenly into 3 parts, takes 1 part, forms the remaining 2 parts back into a single pile and goes back to bed. A little later, the third robber gets up and does the very same thing. In the morning, when they count up the gold coins, there are 100 of them left. How many were in the pile originally? Explain your answer.

*16. A year that is not a leap year has 365 days. (Leap years have 366 days and generally occur every four years.) There are 7 days in a week and 52 whole weeks in a year. How many whole weeks are there in 3 years? How many whole weeks are there in 7 years? Is the number of whole weeks in 3 years three times the number of whole weeks in 1 year? Is the number of whole weeks in 7 years seven times the number of whole weeks in 1 year? Explain the discrepancy!

6.3 Why Division Algorithms Work

CCSS Common Core State Standards Grades 4, 5

In Chapter 4, we studied the reasoning and logic of the longhand multiplication procedures. In this section, we will do the same for division. Students sometimes wonder why we should study the division algorithm when we can simply use a calculator to calculate answers to division problems. As we'll see in this section, to understand why the division algorithm works, students must think about the meaning of division and they must reason about place value and properties of arithmetic. When the focus is on reasoning and sense-making, the study of the algorithms of arithmetic can provide students with the opportunity to deepen their understanding of some of the most fundamental ideas in mathematics.

How Can We Reason to Solve Division Problems?

To analyze why division algorithms work, we will first study how to reason to solve division problems without algorithms and without a calculator.

CLASS ACTIVITY

6H Discuss Division Reasoning, p. CA-107

As you saw in Class Activity 6H, we can divide without using a longhand division procedure. For example, suppose that we want to put 110 candies into packages of 8 candies each. How many packages can we make, and how many candies will be left over? To solve this problem, we must calculate $110 \div 8$, which we can do in the following way:

> Ten packages use $10 \cdot 8 = 80$ candies. Three packages use $3 \cdot 8 = 24$ candies. So far, we have used $80 + 24 = 104$ candies in $10 + 3 = 13$ packages. Six more candies make 110, so we can make 13 packages of candies with 6 candies left over.

This equation expresses this reasoning:

$$10 \cdot 8 + 3 \cdot 8 + 6 = 110 \qquad (6.1)$$

According to the distributive property, we can rewrite this equation as

$$13 \cdot 8 + 6 = 110 \qquad (6.2)$$

which shows that $110 \div 8$ has whole number quotient 13, remainder 6.

To solve the problem, we repeatedly added multiples of 8 until we got as close as possible to 110 without going over. Another approach is to start at 110 and repeatedly *subtract* multiples of 8:

> Ten packages will use up $10 \cdot 8 = 80$ candies. Then there will be $110 - 80 = 30$ candies left. Three packages will use $3 \cdot 8 = 24$ candies. Then there will be only $30 - 24 = 6$ candies left. So we can make $10 + 3 = 13$ packages, and 6 candies will be left over.

The equations to express this reasoning are

$$110 - 10 \cdot 8 = 30$$
$$30 - 3 \cdot 8 = 6$$

These equations can be condensed to the single equation

$$110 - 10 \cdot 8 - 3 \cdot 8 = 6 \tag{6.3}$$

According to the distributive property, we can rewrite this equation as

$$110 - 13 \cdot 8 = 6 \tag{6.4}$$

Since 6 is less than 8,

$$110 \div 8 \text{ has whole number quotient 13, remainder 6}$$

CCSS
4.NBT.6
5.NBT.6

As these examples show, to divide, we can repeatedly subtract multiples of the divisor or we can repeatedly add multiples of the divisor.

Why Does the Scaffold Method of Division Work?

As we just saw, we can calculate $110 \div 8$ by repeatedly subtracting multiples of 8. Repeated subtraction is the basis of long division algorithms. In order to understand the standard long division algorithm, we will work with a flexible method called the **scaffold method**. Depending on how the method is used, it can be less efficient than the standard long division algorithm, or it can be a way to write the steps of the standard algorithm. In any case, the scaffold method allows flexibility in calculating and is a stepping-stone to the standard algorithm.

Table 6.1 shows how to calculate $4581 \div 7$ by using the standard longhand algorithm. The steps of the standard algorithm are recorded in two ways, with the common method and with the scaffold method. Both methods arrive at the conclusion that 4581 divided by 7 is 654 with remainder 3.

TABLE 6.1 Standard long division implemented with two methods: the common method, and the scaffold method

Common Method	Scaffold Method	
	4	← How many 7s are in 31?
	50	← How many tens of 7s are in 381?
654	600	← How many hundreds of 7s are in 4581?
7)4581	7)4581	
−42	−4200	← 600 sevens
38	381	← What is left over after subtracting 600 sevens
−35	−350	← 50 sevens
31	31	← What is left over after subtracting another 50 sevens
−28	−28	← 4 sevens
3	3	← What is left over after subtracting another 4 sevens

To use the scaffold method to calculate $4581 \div 7$ and to understand why this method works, let's use the how-many-groups interpretation of division and think of $4581 \div 7$ as the largest whole number of sevens in 4581—or in other words, the largest whole number that we can multiply 7 by without going over 4581. With that in mind, examine the scaffold method shown in the right panel in Table 6.1. To carry out the scaffold method, we begin by asking how many hundreds of sevens are in 4581. We can start with hundreds because 1000 sevens is 7000, which is already greater than

4581. There are 600 sevens in 4581 because $600 \cdot 7 = 4200$, but 700 sevens would be 4900, which is greater than 4581. We subtract the 600 sevens—namely 4200—from 4581, leaving 381. Now we ask how many tens of sevens are in 381. There are 50 sevens in 381 because $50 \cdot 7 = 350$, but $60 \cdot 7 = 420$ is greater than 381. We subtract the 50 sevens—namely, 350—from 381, leaving 31. Finally, we ask how many sevens are in 31. There are 4, leaving 3 as a remainder. To determine the answer to $4581 \div 7$ we add the numbers at the top of the scaffold:

$$600 + 50 + 4 = 654$$

Therefore,

$$4581 \div 7 \text{ has whole number quotient 654, remainder 3}$$

Another way to write the scaffold method is to write the numbers 600, 50, 4 to the *side* of the scaffold instead of at the top, as shown in **Table 6.2**. However, some teachers have found this notation to be confusing for students from other countries, where similar looking notation may be used but the *divisor* is written to the right.

TABLE 6.2 Another way of writing the scaffold method

7)4581	
−4200	600
381	
−350	50
31	
−28	4
3	

In carrying out the scaffold method, we started with 4581, subtracted 600 sevens, subtracted another 50 sevens, subtracted another 4 sevens, and in the end, 3 were left over. In other words,

$$4581 - 600 \cdot 7 - 50 \cdot 7 - 4 \cdot 7 = 3 \qquad (6.5)$$

Notice that all together, starting with 4581, we subtracted a total of 654 sevens and were left with 3. We can record this observation by first rewriting Equation (6.5) as

$$4581 - (600 \cdot 7 + 50 \cdot 7 + 4 \cdot 7) = 3$$

and then applying the distributive property to get

$$4581 - (600 + 50 + 4) \cdot 7 = 3$$

or

$$4581 - 654 \cdot 7 = 3 \qquad (6.6)$$

Equation (6.6) tells us that when we take 654 sevens away from 4581, we are left with 3. Therefore, we can conclude that 654 is the largest whole number of 7s in 4581, and therefore $4581 \div 7$ has whole number quotient 654, remainder 3.

In general, why does the scaffold method of division give correct answers to division problems, based on the meaning of division? When you solve a division problem $A \div B$ by using the scaffold method, you start with the number A, and you repeatedly subtract multiples of B (numbers times B) until a number remains that is less than B. Since you subtracted as many Bs as possible, when you add the total number of Bs that were subtracted, that is the largest whole number of Bs that are in A. What's left over is the remainder. Therefore, the answer provided by the scaffold method to calculate $A \div B$ gives the answer that we expect for $A \div B$ based on the meaning of division.

CLASS ACTIVITY

6I Why the Scaffold Method of Division Works, p. CA-108

We used the how-many-groups interpretation of division to make sense of the scaffold method, but can the how-many-units-in-1-group interpretation also be used? **Table 6.3** shows that the answer is yes.

TABLE 6.3 Using the scaffold method of long division with the how-many-units-in-1-group viewpoint to divide 4581 objects equally among 7 groups

4	←	From 31, how many individuals can we put in each group?
50	←	From 381, how many tens can we put in each group?
600	←	From 4581, how many hundreds can we put in each group?
7)4581		
−4200	←	Put 600 in each group.
381	←	Left over after putting 600 in each group
−350	←	Put another 50 in each group.
31	←	Left over after putting 50 in each group
−28	←	Put another 4 in each group.
3	←	Left over after putting 4 in each group

Using the Scaffold Method Flexibly Some teachers like to introduce long division with the scaffold method, using it as a stepping-stone to the standard long division algorithm. Why? The scaffold method is flexible. It allows students at different levels of computational fluency to carry out long division successfully and in a way that makes sense to them. As **Table 6.4** indicates, students can learn to become more efficient until they are doing the standard algorithm version of the scaffold method (the version shown in **Table 6.1**).

Why Does the Common Method for Implementing the Standard Division Algorithm Work?

What is the logic behind the way we usually write the standard division algorithm? To understand the logic of this algorithm we'll need to focus on several elements: what division means; how to break numbers apart by place value; how to work with a number part by part, thereby applying properties of arithmetic (even if only implicitly); and how the values of adjacent places are related.

Standard Division Algorithm CCSS Grades 4, 5, 6

The common method for implementing the standard division algorithm is nicely interpreted from the how-many-units-in-1-group viewpoint. To do so, consider the number we want to divide (the dividend) as a number of objects bundled into ones, tens, hundreds, thousands, and so on. Consider the case of dividing 4581 objects equally among 7 groups, and think of the objects as bundled into 4 thousands, 5 hundreds, 8 tens, and 1 individual object. At the first step of the method we ask how many hundreds of objects we can put in each group. We can put 6 hundreds in each of the 7 groups, using 42 hundreds, as we see in the first step of the method in **Table 6.5**. Then $45 - 42 = 3$ hundreds remain. We must unbundle these 3 hundreds in order to subdivide them. Unbundled, the 3 hundreds become 30 tens. We can combine these 30 tens with the 8 tens we have in the 4581 objects. So, we now have 38 tens to distribute among the 7 groups. This process of unbundling and combining corresponds to "bringing down" the 8 next to the 3 from $45 - 42$ in the long division process. From the 38 tens we now have, we can give each of the 7 groups 5 tens, leaving $38 - 35 = 3$ tens. We must unbundle these 3 tens in order to subdivide them. Unbundled, the

TABLE 6.4 Using the scaffold method of long division flexibly on the way to the standard algorithm

Hundreds and tens are easy to work with, but this is inefficient:		Fewer steps make this more efficient; 5s and 2s are easy to work with:	Efficient use of the scaffold method is a way to implement the standard algorithm:
8)5072		4	4
−800	100	10	30
4272		20	600
−800	100	100	8)5072
3472		500	−4800
−800	100	8)5072	272
2672		−4000	−240
−800	100	1072	32
1872		−800	−32
−800	100	272	0
1072		−160	
−800	100	112	
272		−80	
−80	10	32	
192		−32	
−80	10	0	
112			
−80	10		
32			
−32	4		
0			

3 tens become 30 ones. We can combine these 30 ones with the 1 one we have in the 4581 objects. So we now have 31 ones to distribute among the 7 groups. Again, the process of unbundling and combining corresponds to "bringing down" the 1 next to the 3 from 38 − 35 in the long division process. From the 31 ones, each group gets another 4 individual objects, leaving 31 − 28 = 3 objects remaining. All together, each of the 7 groups got 654 objects, and 3 objects were left over, as before.

TABLE 6.5 Interpreting the standard division algorithm from the how-many-units-in-1-group viewpoint to divide 4581 objects equally among 7 groups

654		
7)4581		
−42	←	Each group gets 6 hundreds.
38	←	3 hundreds and 8 tens = 38 tens remain.
−35	←	Each group gets another 5 tens.
31	←	3 tens and 1 one = 31 ones remain.
−28	←	Each group gets another 4 ones.
3	←	3 ones remain.

CLASS ACTIVITY

6J Interpreting the Standard Division Algorithm as Dividing Bundled Toothpicks, p. CA-109

Why Do We Calculate Decimal Answers to Whole Number Division Problems the Way We Do?

If a whole number division problem doesn't have a whole number quotient, then we have several options for describing the answer to the division problem: We can describe it as a whole number quotient with a remainder; we can describe it as a mixed number (or fraction); or we can describe it as a decimal. We discussed whole-number-with-remainder answers and mixed number (and fraction) answers to division problems in Section 6.2. To get the decimal answer to a whole number division problem, all we need to do is extend the standard division algorithm into smaller decimal places.

For example, suppose there is \$243 that is to be divided equally among 7 people. In this case, neither a whole-number-with-remainder answer nor a mixed number answer to 243 ÷ 7 is practical. Instead, a decimal answer is most appropriate. Table 6.6 shows how to use the standard division algorithm to calculate this decimal answer to the hundredths place and how to interpret each step in the process in terms of doling out money in stages that fit with the decimal system. Notice that there are 3 cents left over, which cannot be divided further because we do not have denominations smaller than one-hundredth of a dollar. In an abstract setting, or in a context where thousandths, ten-thousandths, and so on make sense, we could continue the division process, dividing the remaining 3 hundredths to as many decimal places as we needed. We would find that

$$243 \div 7 = 34.7142\ldots$$

CLASS ACTIVITY

6K Interpreting the Calculation of Decimal Answers to Whole Number Division Problems in Terms of Money, p. CA-110

6L Critique Reasoning About Decimal Answers to Division Problems, p. CA-110

TABLE 6.6 Calculating the decimal answer to 243 ÷ 7 by viewing the division algorithm as dividing \$243 equally among 7 people in stages that fit with the decimal system: doling out tens, then ones, then dimes, then pennies

34.71	
7)243.000	
−21	← Each person gets 3 tens.
33	← 3 tens and 3 ones = 33 ones remain.
−28	← Each person gets 4 ones.
50	← 5 ones = 50 dimes (tenths) remain.
−49	← Each person gets 7 dimes (tenths).
10	← 1 dime (tenth) = 10 pennies (hundredths) remain.
−7	← Each person gets 1 penny (hundredth).
3	← 3 pennies (hundredths) remain.

CCSS 7.NS.2d

Why Can We Express Fractions as Decimals?

We can use the standard division algorithm to express a fraction whose numerator and denominator are whole numbers as a decimal by equating the fraction with division:

$$\frac{A}{B} = A \div B$$

For example, how can we express the fraction $\frac{5}{16}$ as a decimal? First note that

$$\frac{5}{16} = 5 \div 16$$

Then use the division algorithm to calculate $5 \div 16$ as a decimal:

$$\begin{array}{r} 0.3125 \\ 16\overline{)5.0000} \\ \underline{-48} \\ 20 \\ \underline{-16} \\ 40 \\ \underline{-32} \\ 80 \\ \underline{-80} \\ 0 \end{array}$$

Therefore,

$$\frac{5}{16} = 0.3125$$

Similarly, we can calculate

$$\frac{1}{12} = 1 \div 12 = 0.083333333\ldots$$

The decimal representation of $\frac{1}{12}$ has infinitely many digits to the right of the decimal point, unlike the decimal representation of $\frac{5}{16} = 0.3125$, which has only 4 (nonzero) digits to the right of its decimal point.

By equating fractions with division, and by using the division algorithm, we can express any fraction whose numerator and denominator are whole numbers as a decimal.

How Can We Reason About Math Drawings to Express Fractions as Decimals?

In simple cases we can use math drawings to determine the decimal representations of fractions. These drawings can help us better understand the relationship between decimals and fractions.

Figure 6.10 shows that

$$\frac{1}{4} = 0.25$$

in the following way: Consider a large square (made up of 100 small squares) as representing 1. To show $\frac{1}{4}$ of the large square, we must divide the large square into 4 equal parts. One of those 4 equal parts represents $\frac{1}{4}$ of the large square. Because each vertical strip of 10 small squares represents $\frac{1}{10}$ of the large square and each small square represents $\frac{1}{100}$ of the large square, **Figure 6.10** shows that

$$\frac{1}{4} = 2 \cdot \frac{1}{10} + 5 \cdot \frac{1}{100} = 0.25$$

Figure 6.10

The shaded area is $\frac{1}{4}$ of the large square. Therefore, $\frac{1}{4} = 2 \cdot \frac{1}{10} + 5 \cdot \frac{1}{100} = 0.25$.

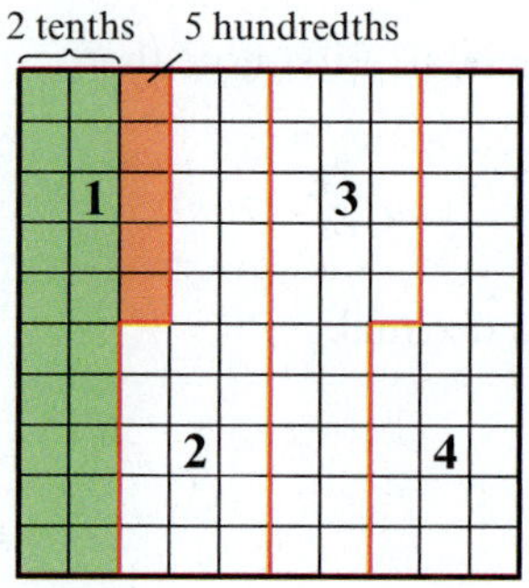

CCSS 5.NBT.6

What Are Issues to Consider when Dividing with Multidigit Divisors?

When we use a division algorithm with a divisor that has two or more digits, we must estimate quotients during the process. Students should be aware that even if they round numbers correctly and make sensible estimates, they may still need to revise their estimates during the division process. For example, when using the standard division algorithm for the division problem $17\overline{)1581}$, it is reasonable to round the 17 to 20 and the 158 to 160 and to estimate the digit in the tens place of the quotient as 8 because $160 \div 20 = 8$:

$$\begin{array}{r} 8 \\ 17\overline{)1581} \\ -136 \\ \hline 22 \end{array}$$

However, the remaining 22 is greater than 17, and so the digit in the tens place of the quotient is actually 9:

$$\begin{array}{r} 93 \\ 17\overline{)1581} \\ -153 \\ \hline 51 \\ -51 \\ \hline 0 \end{array}$$

SECTION SUMMARY AND STUDY ITEMS

Section 6.3 Why Division Algorithms Work

We can solve whole number division problems in a primitive way by repeatedly subtracting the divisor or repeatedly adding the divisor. The scaffold method is a method that works with entire numbers instead of just portions of them. This method allows for flexibility in calculation but it can also be implemented as the standard division algorithm. We can explain why the scaffold method is valid by using the distributive property to “collect up” multiples of the divisor that were subtracted during the division process. We can interpret each step in the common method for implementing the standard division algorithm by viewing numbers as representing objects bundled according to place value. During this method, we repeatedly unbundle amounts and combine them with the amount in the next-lower place.

Key Skills and Understandings

1. Use the scaffold method of division, and interpret the process from either the how-many-groups or the how-many-units-in-1-group point of view.
2. Use the common method for implementing the standard long division algorithm, and interpret the process from the how-many-units-in-1-group point of view. Explain the "bringing down" steps in terms of unbundling the remaining amount and combining it with the amount in the next-lower place.
3. Understand and use nonstandard methods of division.
4. Use the standard division algorithm to write fractions as decimals and to give decimal answers to whole number division problems.

Practice Exercises for Section 6.3

1. Use the scaffold method to calculate $31\overline{)73125}$. Interpret the steps in the scaffold in terms of the following word problem:

> There are 73,125 beads that will be put into bags with 31 beads in each bag. How many bags of beads can be made, and how many beads will be left over?

2. Here is one way to calculate $239 \div 9$:

> Ten nines make 90. Another 10 nines make 180. Five more nines make 225. One more nine makes 234. Another five ones make 239. All together that's 26 nines, with 5 left over. So the answer is 26, remainder 5.

Write a single equation that incorporates this reasoning. Use your equation and the distributive property to write another equation, that shows the answer to $239 \div 9$.

3. Here is one way to calculate $2687 \div 4$:

> Five hundred fours is 2000. That leaves 687. One hundred fours is 400. That leaves 287. Another 50 fours is 200. Now 87 are left. Another 20 fours is 80. Now there are 7 left, and we can get one more 4 out of that with 3 left. All together there were $500 + 100 + 50 + 20 + 1 = 671$ fours in 2687 with 3 left over, so $2687 \div 4 = 671$, remainder 3.

a. Write a scaffold that corresponds to the reasoning in the problem.

b. Calculate $2687 \div 4$, using a scaffold with fewer steps than your scaffold for part (a).

c. Even though your scaffold in part (a) uses more steps than necessary, is it still based on sound reasoning? Explain.

4. Calculate $2950 \div 13$ without using a calculator or a division algorithm.

5. Calculate $1000 \div 27$ without using a calculator or a division algorithm.

6. Show how to use the common method for implementing the standard division algorithm to calculate $4581 \div 7$. Interpret each step in the algorithm in terms of dividing 4581 toothpicks equally among 7 groups, where the toothpicks are bundled according to the decimal system. In particular, be sure to explain how to interpret the "bringing down" steps in terms of the bundles.

7. Show how to use the common method for implementing the standard division algorithm to calculate the decimal answer to $20 \div 11$ to the hundredths place. Interpret each step in terms of dividing \$20 equally among 11 people.

8. Show how to use the standard division algorithm to write the fraction $\frac{5}{8}$ as a decimal.

9. Use the large square in **Figure 6.11** (which is subdivided into 100 small squares) to help you explain why the decimal representation of $\frac{1}{8}$ is 0.125.

10. Marina says that since $300 \div 50 = 6$, that means you can put 300 marbles into 6 groups with 50 marbles in each group. Marina thinks that knowing $300 \div 50 = 6$ should help her calculate $300 \div 55$. At first she tries $300 \div 50 + 300 \div 5$, but that answer seems too big. Is there some other way to use the fact that $300 \div 50 = 6$ in order to calculate $300 \div 55$?

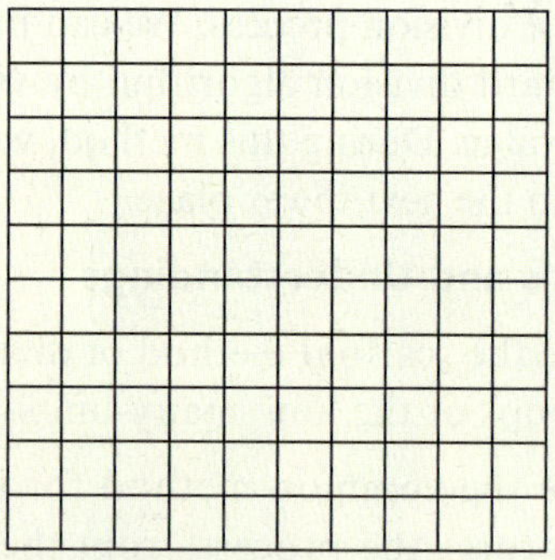

Figure 6.11 A square subdivided into 100 smaller squares.

Answers to Practice Exercises for Section 6.3

1.

$$\begin{array}{r} 8 \\ 50 \\ 300 \\ 2000 \\ 31\overline{)73125} \\ \underline{-62000} \\ 11125 \\ \underline{-9300} \\ 1825 \\ \underline{-1550} \\ 275 \\ \underline{-248} \\ 27 \end{array}$$

so $73125 \div 31$ has whole number quotient 2358, remainder 27

If we make 2000 bags of beads, we will use $2000 \times 31 = 62{,}000$ beads, which will leave 11,125 beads remaining from the original 73,125. If we make another 300 bags of beads, we will use $300 \times 31 = 9300$ beads, which leaves 1825 beads remaining. If we make another 50 bags of beads, we will use 1550 beads, leaving 275 beads. Finally, if we make another 8 bags of beads, we will use 248 beads, leaving only 27 beads, which is not enough for another bag. All together, we made $2000 + 300 + 50 + 8 = 2358$ bags of beads, and 27 beads are left over.

2. $10 \cdot 9 + 10 \cdot 9 + 5 \cdot 9 + 1 \cdot 9 + 5 = 239$

So, by the distributive property,

$$(10 + 10 + 5 + 1) \cdot 9 + 5 = 239$$

Therefore,

$$26 \cdot 9 + 5 = 239$$

which means that $239 \div 9$ has whole number quotient 26, remainder 5.

3. a.

$$\begin{array}{r} 1 \\ 20 \\ 50 \\ 100 \\ 500 \\ 4\overline{)2687} \\ \underline{-2000} \\ 687 \\ \underline{-400} \\ 287 \\ \underline{-200} \\ 87 \\ \underline{-80} \\ 7 \\ \underline{-4} \\ 3 \end{array}$$

So $2687 \div 4$ has whole number quotient 671, remainder 3

b.

$$\begin{array}{r} 1 \\ 70 \\ 600 \\ 4\overline{)2687} \\ \underline{-2400} \\ 287 \\ \underline{-280} \\ 7 \\ \underline{-4} \\ 3 \end{array}$$

So $2687 \div 4$ has whole number quotient 671, remainder 3

c. Even though the scaffold in part (a) uses more steps than necessary, it is still based on sound reasoning. Instead of subtracting the full 600 fours from 2687, the scaffold subtracted the same number of fours in two steps instead of one: first subtracting 500 fours and then subtracting another 100 fours. If we think in terms of putting 2687 cookies into packages of 4, we are first making 500 packages and then making another 100 packages instead of making 600 packages straight away. Similarly, instead of subtracting the full 70 fours from 287, the scaffold subtracted 50 fours and then another 20 fours. All together, we are still finding the same total number of fours in 2687, just in a slightly less efficient way than in the scaffold in part (b).

4. There are many ways you could do this. Here is one way: 100 thirteens is 1300, so 200 thirteens is 2600. Taking 2600 away from 2950 leaves 350. Another 20 thirteens is 260, leaving 90. Five thirteens make 65. Now we have 25 left. We can only get one more 13, and 12 will be left. All together we have $200 + 20 + 5 + 1 = 226$ thirteens and 12 are left, so $2950 \div 13$ has whole number quotient 226, remainder 12.

5. There are many ways you could do this. Here is one of them: $10 \cdot 27 = 270$ and $20 \cdot 27 = 540$, so $30 \cdot 27 = 540 + 270 = 810$. Five 27s must be half of 270, which is 135. Therefore,

$$\begin{aligned} 35 \cdot 27 &= (30 + 5) \cdot 27 \\ &= 30 \cdot 27 + 5 \cdot 27 \\ &= 810 + 135 \\ &= 945. \end{aligned}$$

Two 27s is 54, and adding that to 945 makes 999. Therefore, $37 \cdot 27 = 999$, so $37 \cdot 27 + 1 = 1000$, and this means that $1000 \div 27$ has whole number quotient 37, remainder 1.

6. See **Table 6.5** for a brief sketch.

7. Using the standard algorithm, we find that $20 \div 11 = 1.81\ldots$:

$$\begin{array}{r} 1.81 \\ 11\overline{)20.00} \end{array}$$

−11	←	Each person gets \$1.
90	←	\$9 = 90 dimes remain.
−88	←	Each person gets 8 dimes.
20	←	2 dimes = 20 pennies remain.
−11	←	Each person gets 1 penny.
9	←	9 pennies remain.

If we think in terms of dividing \$20 equally among 11 people, then at the first step we ask how many ones we should give each person. Each of the 11 people will get \$1, leaving \$20 − \$11 = \$9 remaining. If we trade each dollar for 10 dimes, we will have 90 dimes. We can now give each of the 11 people 8 dimes, using a total of 88 dimes, leaving 2 dimes left. If we trade each dime for 10 pennies, we will have 20 pennies. Each person gets 1 penny, leaving $20 - 11 = 9$ pennies. Therefore, each person gets \$1.81, and $20 \div 11 = 1.81\ldots$. The decimal representation of $20 \div 11$ continues to have digits in the thousandths place, the ten-thousandths place, and so on, but we can't use our money interpretation for these places because we don't have denominations smaller than 1 penny.

8. The following long division calculation shows that $\frac{5}{8} = 0.625$:

$$\begin{array}{r} 0.625 \\ 8\overline{)5.000} \\ \underline{-\,48} \\ 20 \\ \underline{-\,16} \\ 40 \\ \underline{-\,40} \\ 0 \end{array}$$

9. By thinking of the large square as representing 1, each small square then represents $\frac{1}{100}$. To divide the large square into 8 equal pieces, first make 8 strips of 10 small squares, as indicated in **Figure 6.12**. Each of the 8 equal pieces into which we want to divide the large square gets one of these strips. Then there are still 20 small squares left to be divided into 8 equal pieces. Make 8 groups of 2 small squares from these 20 small squares. Each of the 8 equal pieces gets 2 of these small squares. That leaves 4 small squares left to be divided into 8 equal pieces. Dividing each small square in half makes 8 half-squares. All together, we obtain $\frac{1}{8}$ of the big square by collecting these various parts: one strip of 10 small squares, 2 small squares, and

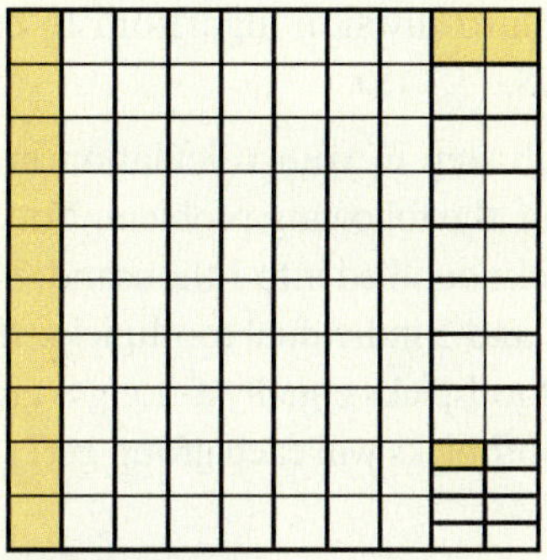

Figure 6.12 Showing $\frac{1}{8} = 0.125$.

$\frac{1}{2}$ of a small square, as shown in Figure 6.12. Notice that since 1 small square is $\frac{1}{100}$ of the large square, $\frac{1}{2}$ of a small square is $\frac{1}{200} = \frac{5}{1000}$ of the large square. Therefore,

$$\frac{1}{8} = 1 \cdot \frac{1}{10} + 2 \cdot \frac{1}{100} + 5 \cdot \frac{1}{1000}$$
$$= 0.125$$

10. Marina is right that 300 ÷ 50 + 300 ÷ 5 won't work—it divides the 300 marbles into groups *twice*, so it gives too many groups. Also, if you get 6 groups when you put 50 marbles in each group, then you must get fewer groups when you put more marbles in each group. If you think about removing one of those 6 groups of marbles, then you could take the 50 marbles in that group and from them, put 5 marbles in each of the remaining 5 groups. Then you'd have 5 groups of 55 marbles, with 25 marbles left over. Therefore, 300 ÷ 55 = 5, remainder 25.

PROBLEMS FOR SECTION 6.3

1. a. Calculate 4215 ÷ 6 and 62,635 ÷ 32 in two ways: with the common method for implementing the standard division algorithm and with the scaffold method.

b. Compare the common method for implementing the standard division algorithm and the scaffold method. How are the methods alike? How are the methods different? What are advantages and disadvantages of each method?

2. a. Use the scaffold method to calculate 793 ÷ 4.

b. Interpret the steps in your scaffold in terms of the following word problem: If you have 793 cookies and you want to put them in packages of 4, how many packages will there be, and how many cookies will be left over?

c. Write a single equation, as in the text and Practice Exercise 2, that incorporates the steps of your scaffold. Use your equation and the distributive property to write another equation, as in the text and Practice Exercise 2, that shows the answer to 793 ÷ 4. Relate this last equation to portions in the scaffold method.

3. a. Use the common method for implementing the standard division algorithm to calculate 1875 ÷ 8.

b. Interpret each step in your calculation in part (a) in terms of the following problem: You have 1875 toothpicks bundled into 1 thousand, 8 hundreds, 7 tens, and 5 individual toothpicks. If you divide these toothpicks equally among 8 groups, how many toothpicks will each group get?

4. a. Use the common method for implementing the standard division algorithm to determine the decimal answer to 2893 ÷ 6 to the hundredths place.

b. Interpret each step in your division calculation in part (a) in terms of dividing $2893 equally among 6 people.

5. Tamarin calculates 834 ÷ 25 in the following way:

> I know that four 25s make 100, so I counted 4 for each of the 8 hundreds. This gives me 32. Then there is one more 25 in 34, but there will be 9 left. So the answer is 33, remainder 9.

a. Explain Tamarin's method in detail, and explain why her method is legitimate. (Do not just state that she gets the correct answer; explain why her method gives the correct answer.) Include equations as part of your explanation.

b. Use Tamarin's method to calculate 781 ÷ 25.

6. Felicia is working on the following problem: There are 730 balls to be put into packages of 3. How many packages can be made, and how many balls will be left over? Here are Felicia's ideas:

> One hundred packages of balls will use 300 balls. After another 100 packages, we will have used up 600 balls. Another 30 packages will use another 90 balls, for a total of 690 balls used. Ten more packages brings us to 720 balls used. Three more packages will bring us to 729 balls used, with 1 ball left over. All together we could make 100 + 100 + 30 + 10 + 3 = 243 packages of balls with 1 ball left over.

Write equations as in the text that correspond to Felicia's work. Explain how your equations show that 730 ÷ 3 has whole number quotient 243, remainder 1.

7. Rodrigo calculates 650 ÷ 15 in the following way:

150	← 10	300	← 20
+ 150	← 10	+ 300	← 20
300	← 20	600	← 40
600	← 40		43 R 5
+ 30	← 2		
630	← 42		
+ 15			
645	← 43		
+ 5	← left		
650			

a. Explain why Rodrigo's method makes sense.

b. Write equations that correspond to Rodrigo's work and that demonstrate that 650 ÷ 15 has whole number quotient 43, remainder 5.

8. Meili calculates 1200 ÷ 45 in the following way:

45	450	← 10	
× 10	+ 450	← 10	
450	900		
	+ 90	← 2	10 + 10 + 2 + 2 + 2
	990		= 26
	+ 90	← 2	26 R 30
	1080		
	+ 90	← 2	
	1170		
	+ 30	← left	
	1200		

a. Explain why Meili's strategy makes sense.

b. Write equations that correspond to Meili's work and that demonstrate that 1200 ÷ 45 has whole number quotient 26, remainder 30.

9. Calculate 623 ÷ 8 without using a calculator or any division algorithm. Show your work. Then briefly describe your reasoning.

10. Calculate 2000 ÷ 75 without using a calculator or any division algorithm. Show your work. Then briefly describe your reasoning.

11. Use some or all of the multiplication facts

$$2 \cdot 35 = 70$$

$$10 \cdot 35 = 350$$

$$20 \cdot 35 = 700$$

repeatedly to calculate 2368 ÷ 35 without using a calculator. Explain your method.

12. Use the two multiplication facts, $30 \cdot 12 = 360$ and $12 \cdot 12 = 144$, to calculate 550 ÷ 12 without the use of a calculator, or a division algorithm. Use both multiplication facts, and explain your method. Give your answer as a whole number with a remainder.

13. Describe how the whole-number-with-remainder and the decimal answer to 23 ÷ 6 are related and explain why this relationship holds.

14. Show how to use division to determine the decimal representation of $\frac{1}{37}$.

15. Describe how to use either dimes and pennies or a subdivided square to determine the tenths and hundredths places in the decimal representation of $\frac{1}{11}$. Explain your reasoning.

16. Describe how to use either dimes and pennies or a subdivided square to determine the tenths and hundredths places in the decimal representation of $\frac{1}{9}$. Explain your reasoning.

17. a. Use the standard division algorithm to determine the decimal representation of $\frac{1}{9}$ to the ten-thousandths place.

b. Interpret the steps to the hundredths place in part (a) in terms of dividing $1 equally among 9 people.

18. Describe how to use either dimes and pennies or a subdivided square to determine the tenths and hundredths places in the decimal representation of $\frac{1}{7}$. Explain your reasoning.

19. a. Use division to determine the decimal representation of $\frac{1}{7}$ to 7 decimal places.

b. Interpret the steps to the hundredths place in part (a) in terms of dividing $1 equally among 7 people.

20. Jessica calculates that 7 ÷ 3 = 2, remainder 1. When Jessica is asked to write her answer as a decimal, she simply puts the remainder 1 behind the decimal point:

$$7 \div 3 = 2.1$$

Is Jessica correct or not? If not, explain why not, and explain to Jessica in a concrete way why the correct answer makes sense.

21. a. Write and solve a simple word problem for 1200 ÷ 30.

b. Use the situation of your word problem in part (a) to help you solve 1200 ÷ 31 without a calculator or a division algorithm by modifying your solution to 1200 ÷ 30.

c. Use the situation of your word problem in part (a) to help you solve 1200 ÷ 29 without a calculator or a division algorithm by modifying your solution to 1200 ÷ 30.

***22. a.** Suppose you want to estimate

$$459 \div 38$$

by rounding 38 up to 40. Both

$$440 \div 40$$

and

$$480 \div 40$$

are easy to calculate mentally. Use reasoning about division to determine which division problem, 440 ÷ 40 or 480 ÷ 40, should give you a better estimate to 459 ÷ 38. Then check your answer by solving the division problems.

b. Suppose you want to estimate

$$459 \div 42$$

by rounding 42 down to 40. As before, both 440 ÷ 40 and 480 ÷ 40 are easy to calculate mentally. Use reasoning about division to determine which division problem, 440 ÷ 40 or 480 ÷ 40, should give you a better estimate to 459 ÷ 42. Then check your answer by solving the division problems.

c. Suppose you want to estimate 632 ÷ 58. What is a good way to round the numbers 632 and 58 so that you get an easy division problem which will give a good estimate to 632 ÷ 58? Explain, drawing on what you learned from parts (a) and (b).

d. Suppose you want to estimate 632 ÷ 62. What is a good way to round the numbers 632 and 62 so that you get an easy division problem which will give a good estimate to 632 ÷ 62? Explain, drawing on what you learned from parts (a) and (b).

***23.** Bob wants to estimate 1893 ÷ 275. He decides to round 1893 up to 2000. Since he rounded 1893 up, Bob says that he'll get a better estimate if he rounds 275 in the opposite direction—namely, down to 250 rather than up to 300. Therefore, Bob says that 1893 ÷ 275 is closer to 8 (which is 2000 ÷ 250) than to 7, because 2000 ÷ 300 is a little less than 7.

This problem will help you investigate whether or not Bob's reasoning is correct.

a. Which of 2000 ÷ 250 and 2000 ÷ 300 gives a better estimate to 1893 ÷ 275? Can Bob's reasoning (just described) be correct?

b. Use the meaning of division (either of the two main interpretations) to explain the following: When estimating the answer to a division problem $A \div B$, if you round A up, you will generally get a better estimate if you also round B up rather than down. Draw diagrams to aid your explanation.

c. Now consider the division problem 1978 ÷ 205. Suppose you round 1978 up to 2000. In this case, will you get a better estimate to 1978 ÷ 205 if you round 205 up to 250 (and compute 2000 ÷ 250) or if you round 205 down to 200 (and compute 2000 ÷ 200)? Reconcile this with your findings in part (b).

***24.** A student calculates 6998 ÷ 7 as follows, and concludes that 6998 ÷ 7 = 1000 − 1, remainder 5, which is 999, remainder 5:

$$\begin{array}{r} -1 \\ 1000 \\ 7\overline{)6998} \\ \underline{-7000} \\ -2 \\ \underline{-(-7)} \\ 5 \end{array}$$

Even though it is not conventional to use negative numbers in a scaffold, explain why the student's method corresponds to legitimate reasoning. Write a simple word problem for 6998 ÷ 7, and use your problem to discuss the reasoning that corresponds to the scaffold.

***25.** When you divide whole numbers using an ordinary calculator, the answer is displayed as a decimal. But what if you want the answer as a whole number with a remainder? Here's a method to determine the remainder with an ordinary calculator, illustrated with the example of 236 ÷ 7.

Use the calculator to divide the whole numbers:

$$236 \div 7 = 33.71428\ldots$$

Subtract the whole number part of the answer:

$$33.71428\ldots - 33 = 0.71428\ldots$$

Multiply the resulting decimal by the divisor (which was 7):

$$0.71428\ldots \cdot 7 = 4.99999 \text{ (or 5)}$$

Round the resulting number to the nearest whole number. This is your remainder. In this example the remainder is 5. So $236 \div 7 = 33$, remainder 5.

a. Using a calculator, solve at least two more division problems with whole numbers, and use the preceding method to find the answer as a whole number with a remainder. Check that your answers are correct.

b. Now solve the same division problems you did in part (a), except this time give your answers as mixed numbers instead of as decimals.

c. Use the mixed number version to help you explain why the calculator method for determining the remainder works.

6.4 Fraction Division from the How-Many-Groups Perspective

CCSS Common Core State Standards Grades 5, 6

Did you know that there are methods for dividing fractions other than the standard "invert and multiply" method that you probably use? In this section, we will examine fraction division from a how-many-groups perspective. This perspective leads naturally to a general method for calculating fraction division: Give the fractions a common denominator and then divide the numerators. By viewing division as multiplication with an unknown factor, we will develop a related method for dividing fractions: Divide the numerators and divide the denominators. These methods are different from the standard "invert and multiply" fraction division method that you are probably most familiar with (which we will examine in the next section).

How Can We Interpret Fraction Division as How-Many-Groups?

Recall that with how-many-groups division, we interpret $A \div B = ?$ as $? \cdot B = A$ (see Figure 6.13). This applies not only to whole numbers but

Figure 6.13 How-many-groups division.

A	÷	B	=	?
Number of units to be put in equal groups		Number of units in 1 group		How many groups?

?	•	B	=	A
How many groups?		Number of units in 1 group		Number of units to be put in equal groups

also to fractions. For example, we can interpret

$$\frac{8}{3} \div \frac{2}{3} = ? \quad \text{as} \quad ? \cdot \frac{2}{3} = \frac{8}{3}$$

So $\frac{8}{3} \div \frac{2}{3} = ?$ asks how many groups of $\frac{2}{3}$ we can make from $\frac{8}{3}$, as in Figure 6.14.

Figure 6.14 Interpreting $\frac{8}{3} \div \frac{2}{3} = ?$ as how-many-groups division.

$$? \quad \cdot \quad \frac{2}{3} \quad = \quad \frac{8}{3}$$

? servings

How many groups of $\frac{2}{3}$ cups are in $\frac{8}{3}$ cups?

$\frac{2}{3}$ cups popcorn is 1 serving

$\frac{8}{3}$ cups is ? servings

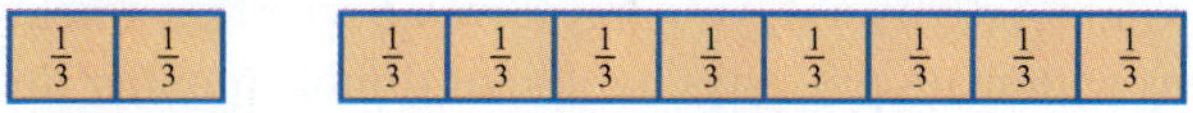

2 parts, each $\frac{1}{3}$ of a cup

8 parts, each $\frac{1}{3}$ of a cup

Figure 6.15 will help you compare two versions of a how-many-groups problem, one with whole numbers and one with fractions.

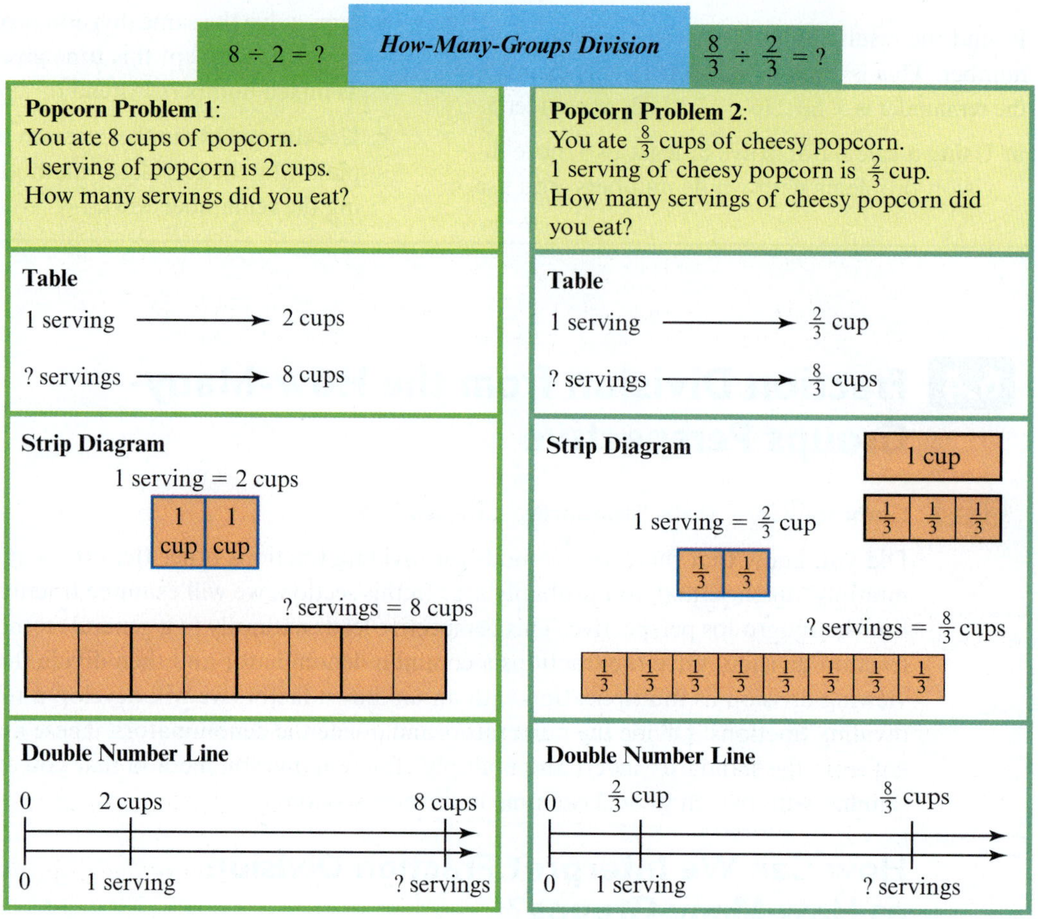

Figure 6.15 How-many-groups division with whole numbers and with fractions.

Observe how the tables, strip diagrams, and double number lines highlight the how-many-groups structure of the problem. Observe also that the structure is the same, whether whole numbers or fractions are involved. A **double number line** is a pair of coordinated number lines that is used to show how quantities are related. Amounts that line up on the two number lines are related. You can use double number lines and tables to help you organize your thinking and solve problems.

double number line

IMAP▶ *Watch Elliott solve 1 divided by 1/3 by using his understanding of division.*

CLASS ACTIVITY

6M How-Many-Groups Fraction Division Problems, p. CA-111

6N Equivalent Division Problems, p. CA-112

Why Can We Divide Fractions by Giving Them Common Denominators and Then Dividing the Numerators?

CCSS 5.NF.7 6.NS.1

In Class Activity 6M, you may have noticed that when you solve how-many-groups division problems with math drawings, tables, or double number lines, it helps to give the two fractions a common denominator. We use common denominators to create like parts for quantities. Once we have like parts, we can relate quantities by reasoning about the numbers of parts of each quantity. In Class Activity 6N you saw that when you have like-sized parts, a fraction division problem is equivalent to a whole number division problem.

In fact, one general method for dividing fractions is to give the fractions common denominators and then divide the numerators. This method turns a fraction division problem into an equivalent whole number division problem. Let's use an example to see why this method is valid. Consider

$$\frac{2}{3} \div \frac{1}{2} = ?$$

and the how-many-groups problem for it:

> How many times will we need to fill a $\frac{1}{2}$ cup measuring cup with water and pour it into a container that holds $\frac{2}{3}$ cup of water in order to fill the container?

Or, said another way,

> How many $\frac{1}{2}$ cups of water are in $\frac{2}{3}$ cup of water?

Figure 6.16

$\frac{2}{3} \div \frac{1}{2} = ?$ How many $\frac{1}{2}$ cups of water are in $\frac{2}{3}$ cup?

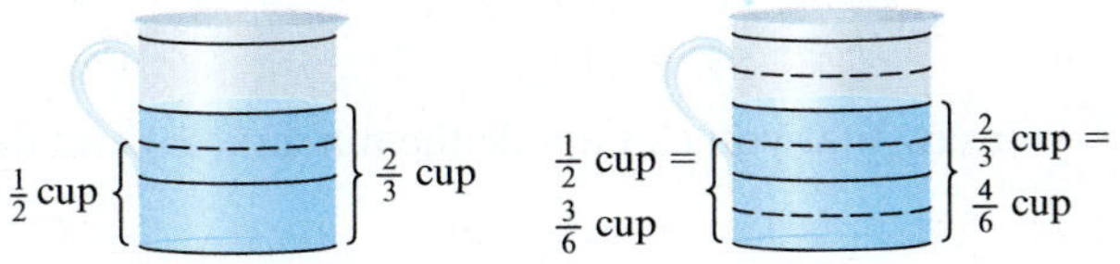

From **Figure 6.16** we can say right away that the answer to this problem is "one and a little more" because one-half cup clearly fits in two-thirds of a cup, but then a little more is still needed to fill the two-thirds of a cup. We can give a more precise answer by giving $\frac{1}{2}$ and $\frac{2}{3}$ a common denominator thereby creating like parts. So we can rephrase the problem "how many $\frac{1}{2}$ cups of water are in $\frac{2}{3}$ cup of water?" as "how many $\frac{3}{6}$ cups are in $\frac{4}{6}$ cups?" or as "how many 3 sixths are in 4 sixths?." Because the sixths are a common unit, this is equivalent to the problem "how many 3s are in 4?" which is the problem $4 \div 3 = ?$, whose answer is $\frac{4}{3} = 1\frac{1}{3}$. See **Figure 6.17**.

Figure 6.17

Using common denominators to interpret $\frac{2}{3} \div \frac{1}{2} = ?$ as an equivalent whole number division problem.

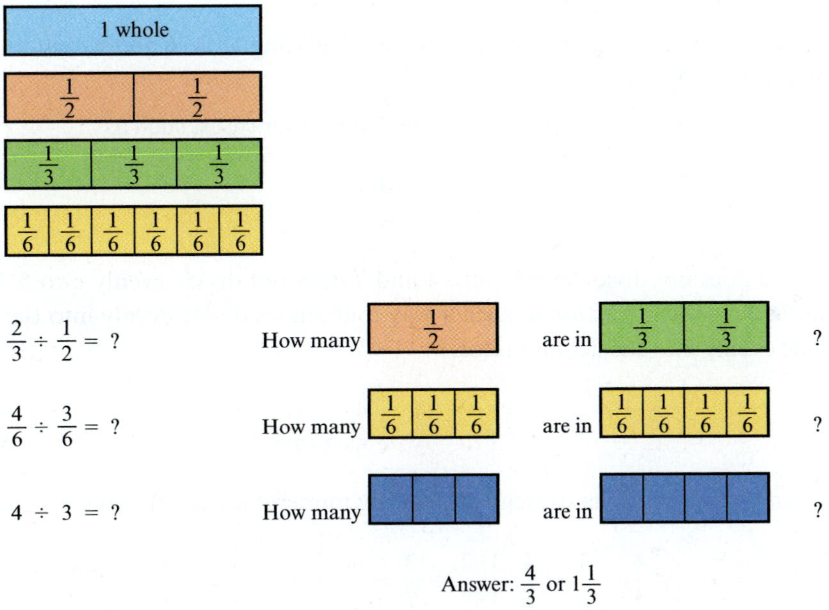

The same line of reasoning will work for any fraction division problem. To divide fractions we can first give them common denominators and then divide the resulting numerators.

For example,

$$\frac{3}{5} \div \frac{1}{4} = \frac{12}{20} \div \frac{5}{20} = 12 \div 5 = \frac{12}{5}$$

Why Can We Divide Fractions by Dividing the Numerators and the Denominators?

CLASS ACTIVITY

6O Dividing Fractions by Dividing the Numerators and Dividing the Denominators, p. CA-113

Think about a quick way to calculate

$$\frac{18}{55} \div \frac{6}{11}$$

If you did Class Activity 6O, you saw that you can divide the numerators and divide the denominators like this:

$$\frac{18}{55} \div \frac{6}{11} = \frac{18 \div 6}{55 \div 11} = \frac{3}{5}$$

We can explain why this method is valid by viewing a division problem as an unknown-factor multiplication problem. In this example, we can view

$$\frac{18}{55} \div \frac{6}{11} = \frac{?}{?}$$

as

$$\frac{?}{?} \cdot \frac{6}{11} = \frac{18}{55} \quad \text{or} \quad \frac{6}{11} \cdot \frac{?}{?} = \frac{18}{55}$$

We now see that we can find the numerator by calculating $18 \div 6$ and we can find the denominator by calculating $55 \div 11$.

The previous example worked well, but what about other cases, such as

$$\frac{4}{5} \div \frac{3}{7}$$

In this case 3 does not divide evenly into 4 and 7 does not divide evenly into 5. However, we can write $\frac{4}{5}$ as an equivalent fraction in such a way that 3 does divide evenly into the numerator and 7 does divide evenly into the denominator:

$$\frac{4}{5} = \frac{4 \cdot 3 \cdot 7}{5 \cdot 3 \cdot 7}$$

Now we can use the previous strategy of dividing numerators and dividing denominators:

$$\frac{4}{5} \div \frac{3}{7} = \frac{4 \cdot 3 \cdot 7}{5 \cdot 3 \cdot 7} \div \frac{3}{7} = \frac{4 \cdot 3 \cdot 7 \div 3}{5 \cdot 3 \cdot 7 \div 7} = \frac{4 \cdot 7}{5 \cdot 3} = \frac{28}{15}$$

You might have recognized the next-to-last step as the result of applying the "invert and multiply" method of fraction division (which we will examine in the next section).

IMAP

See the following video clips for examples of children solving division problems:

Elliot, $1 \div \frac{1}{3}$, then $1\frac{1}{2} \div \frac{1}{3}$

Javier, $9 \div \frac{3}{4}$

Shelby, $12 \div \frac{1}{2}$

Trina, division of fractions

FROM THE FIELD Research

Newton, K. J., & Sands, J. (2012). Why don't we just divide across? *Mathematics Teaching in the Middle School, 17*(6), pp. 340–345.

This study investigated ways that sixth graders in a school for students with language-based learning differences could reason about and make sense of fraction division. After reviewing fraction multiplication, students were asked to solve $\frac{8}{21} \div \frac{2}{3} = ?$. Students developed a "divide across" method, obtaining the solution 4/7 by calculating $8 \div 2 = 4$ and $21 \div 3 = 7$. They found this method easy and intuitive, but then struggled to solve $3/5 \div 1/2 = ?$. However, one student solved the problem by first rewriting 3/5 as 6/10. He then used the divide across method to calculate $3/5 \div 1/2 = 6/10 \div 1/2 = 6/5$. After exploring the divide across method in other examples, students decided that although the method was simple and intuitive, it was often tedious to implement. At that point, students were ready to learn another method. The class explored equivalent division and multiplication problems, such as $24 \div 2$ and $24 \cdot 1/2$. After working through many problems, students found that dividing by a number is equivalent to multiplying by the number's reciprocal, leading to the standard "invert and multiply" procedure for fraction division. Several students, including some who had struggled the most with mathematics, demonstrated flexibility and were able to analyze problems and determine which of the two methods (divide across or invert and multiply) would be most appropriate.

SECTION SUMMARY AND STUDY ITEMS

Section 6.4 Fraction Division from the How-Many-Groups Perspective

We can interpret fraction division from the how-many-groups viewpoint. This perspective leads to a general method for dividing fractions: give the fractions common denominators and then divide the numerators.

Key Skills and Understandings

1. Write and recognize fraction division word problems for the how-many-groups interpretation of division.
2. Solve fraction division word problems with the aid of math drawings, tables, and double number lines. Know how to interpret math drawings appropriately.
3. Explain why we can divide fractions by giving them a common denominator and then dividing the numerators.
4. Explain why we can divide fractions by dividing numerators and dividing denominators.

Practice Exercises for Section 6.4

1. Write a how-many-groups word problem for $\frac{3}{4} \div \frac{2}{3} = ?$ and solve the problem with the aid of a strip diagram, a table, and a double number line.

2. Annie wants to solve the division problem $\frac{3}{4} \div \frac{1}{2} = ?$ by using the following word problem:

I need $\frac{1}{2}$ cup of chocolate chips to make a batch of cookies. How many batches of cookies can I make with $\frac{3}{4}$ of a cup of chocolate chips?

Annie makes a drawing like the one in Figure 6.18. Explain why it would be easy for Annie to misinterpret her drawing as showing that $\frac{3}{4} \div \frac{1}{2} = 1\frac{1}{4}$. How should Annie interpret her drawing so as to conclude that $\frac{3}{4} \div \frac{1}{2} = 1\frac{1}{2}$?

3. Write a simple how-many-groups word problem for $\frac{5}{2} \div \frac{2}{3} = ?$ and use the word problem to help you explain why you can solve the division problem by first giving the fractions a common denominator and then dividing the numerators.

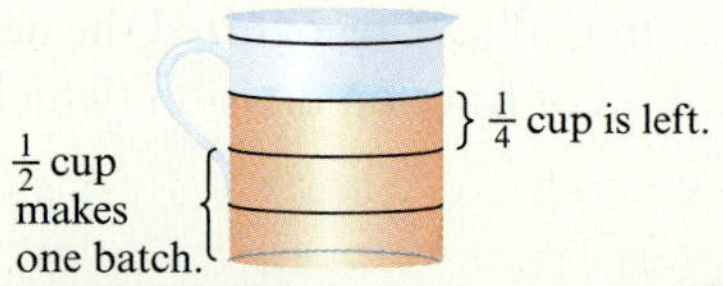

Figure 6.18 How many batches of cookies can be made with $\frac{3}{4}$ cup of chocolate chips if 1 batch requires $\frac{1}{2}$ cup of chocolate chips?

4. Use the fact that we can rewrite the division problem $\frac{7}{11} \div \frac{3}{5} = ?$ as a multiplication problem with an unknown factor to explain why the division problem can be solved by multiplying $\frac{7}{11}$ by $\frac{5}{3}$.

Answers to Practice Exercises for Section 6.4

1. *Word Problem:* If 1 pound of beans fills a bucket $\frac{2}{3}$ full, how much will beans that fill the bucket $\frac{3}{4}$ full weigh?

See Figure 6.19 for a table, strip diagram, and double number line showing that the beans in $\frac{3}{4}$ of a bucket weigh $1\frac{1}{8}$ pounds.

2. Annie's diagram shows that she can make 1 full batch of cookies from her $\frac{3}{4}$ cup of chocolate chips and that $\frac{1}{4}$ cup of chocolate chips will be left over. Because $\frac{1}{4}$ cup of chocolate chips is left over, it would be easy for Annie to misinterpret her picture as showing $\frac{3}{4} \div \frac{1}{2} = 1\frac{1}{4}$. The answer to the problem

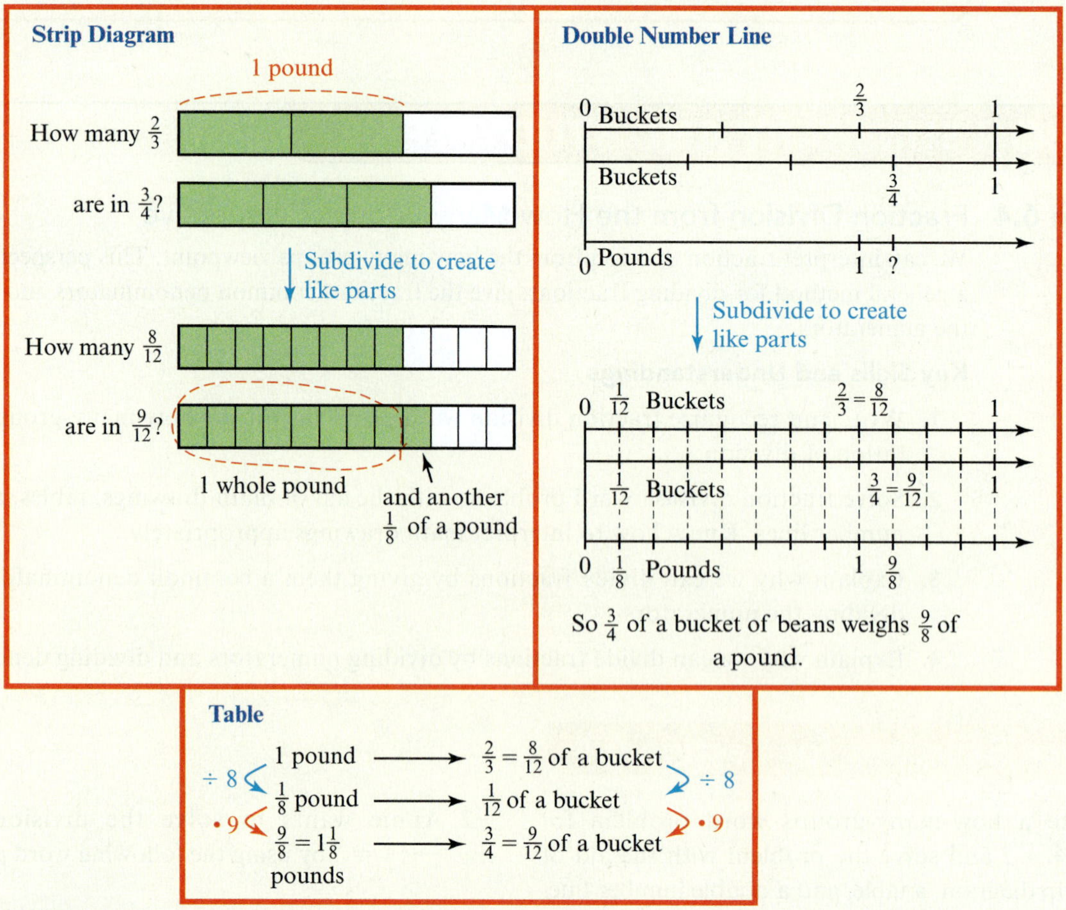

Figure 6.19 Solving fraction division with a strip diagram, a double number line, and a table.

is supposed to be the number of *batches* Annie can make. In terms of batches, the remaining $\frac{1}{4}$ cup of chocolate chips makes $\frac{1}{2}$ of a batch of cookies. We can see this because 2 quarter-cup sections make a full batch, so each quarter-cup section makes $\frac{1}{2}$ of a batch of cookies. Thus, by interpreting the remaining $\frac{1}{4}$ cup of chocolate chips in terms of batches, we see that Annie can make $1\frac{1}{2}$ batches of chocolate chips, thereby showing that $\frac{3}{4} \div \frac{1}{2} = 1\frac{1}{2}$, not $1\frac{1}{4}$.

3. *Word Problem:* One serving of juice is $\frac{2}{3}$ of a cup. If you drink $2\frac{1}{2} = \frac{5}{2}$ cups of juice, how many servings will that be? To solve the problem, we need to find how many $\frac{2}{3}$ are in $\frac{5}{2}$, which, by giving the fractions common denominators, is the same as finding how many $\frac{4}{6}$ are in $\frac{15}{6}$. In other words, to solve the problem we need to find how-many-groups of 4 sixths are in 15 sixths. But to find how-many-groups of 4 sixths are in 15 sixths we just need to find how many 4s are in 15, which is $15 \div 4 = 3\frac{3}{4}$. So to solve the division problem, we first rewrote it in terms of common denominators:

$$\frac{5}{2} \div \frac{2}{3} = \frac{15}{6} \div \frac{4}{6}$$

and then we divided the numerators:

$$15 \div 4 = 3\frac{3}{4}$$

4. Work with the unknown factor multiplication problem

$$? \cdot \frac{3}{5} = \frac{7}{11}$$

Now change $\frac{7}{11}$ into an equivalent fraction so that the multiplication problem will "work." If we change $\frac{7}{11}$ to the equivalent fraction

$$\frac{7 \cdot 3 \cdot 5}{11 \cdot 3 \cdot 5}$$

then to solve

$$? \cdot \frac{3}{5} = \frac{7 \cdot 3 \cdot 5}{11 \cdot 3 \cdot 5}$$

the ? should be

$$\frac{7 \cdot 5}{11 \cdot 3}$$

which is equal to

$$\frac{7}{11} \cdot \frac{5}{3}$$

PROBLEMS FOR SECTION 6.4

1. Explain how to use the math drawings of Figures 6.14 and 6.15 (including the double number line) to solve Popcorn Problem 2 of Figure 6.15.

2. *A Bread Problem:* If 1 loaf of bread requires $1\frac{1}{4}$ cups of flour, then how many loaves of bread can you make with 10 cups of flour? (Assume that you have enough of all other ingredients on hand.)

   Solve the bread problem with the aid of a math drawing, a table, or a double number line. Explain your reasoning.

3. *A Measuring Problem:* You are making a recipe that calls for $\frac{2}{3}$ cup of water, but you can't find your $\frac{1}{3}$ cup measure. You can, however, find your $\frac{1}{4}$ cup measure. How many times should you fill your $\frac{1}{4}$ cup measure to measure $\frac{2}{3}$ cup of water?

   Solve the measuring problem with the aid of a math drawing, a table, or a double number line. Explain your reasoning.

4. Write a how-many-groups word problem for $4 \div \frac{2}{3} = ?$ and solve your problem with the aid of a math drawing, a table, or a double number line. Explain your reasoning.

5. Write a how-many-groups word problem for $5\frac{1}{4} \div 1\frac{3}{4} = ?$ and solve your problem with the aid of a math drawing, a table, or a double number line. Explain your reasoning.

6. Jose and Mark are making cookies for a bake sale. Their recipe calls for $2\frac{1}{4}$ cups of flour for each batch. They have 5 cups of flour. Jose and Mark realize that they can make two batches of cookies and that there will be some flour left. Since the recipe doesn't call for eggs, and since they have plenty of the other ingredients on hand, they decide they can make a fraction of a batch in addition to the two whole batches. But Jose and Mark have a difference of opinion. Jose says that

$$5 \div 2\frac{1}{4} = 2\frac{2}{9}$$

so he says that they can make $2\frac{2}{9}$ batches of cookies. Mark says that two batches of cookies will use up $4\frac{1}{2}$ cups of flour, leaving $\frac{1}{2}$ left, so they should be able to make $2\frac{1}{2}$ batches. Mark draws the picture in Figure 6.20 to explain his thinking to Jose.

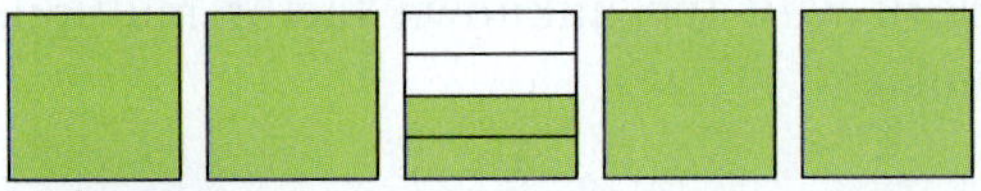

Figure 6.20 Representing $5 \div 2\frac{1}{4}$ by considering how many $2\frac{1}{4}$ cups of flour are in 5 cups of flour.

Discuss the boys' mathematics: What's right, what's not right, and why? If anything is incorrect, how could you modify it to make it correct?

7. Marvin has 11 yards of cloth to make costumes for a play. Each costume requires $1\frac{1}{2}$ yards of cloth.

   a. Solve the following two problems:

      i. How many costumes can Marvin make, and how much cloth will be left over?

      ii. What is $11 \div 1\frac{1}{2}$?

   b. Compare and contrast your answers in part (a).

8. *A Laundry Problem:* You need $\frac{3}{4}$ cup of laundry detergent to wash 1 full load of laundry. How many loads of laundry can you wash with 5 cups of laundry detergent? (Assume that you can wash fractional loads of laundry.)

   Solve the laundry problem with the aid of a math drawing, a table, or a double number line. Explain your reasoning.

9. Write a how-many-groups word problem for $2 \div \frac{3}{4} = ?$ and solve your problem with the aid of a math drawing, a table, or a double number line. Explain your reasoning.

10. Write a how-many-groups word problem for $\frac{1}{3} \div \frac{1}{4} = ?$ and solve your problem with the aid of a math drawing, a table, or a double number line. Explain your reasoning.

11. Write a how-many-groups word problem for $\frac{1}{2} \div \frac{2}{3} = ?$ and solve your problem with the aid of a math drawing, a table, or a double number line. Explain your reasoning.

12. Write a simple how-many-groups word problem for $\frac{1}{3} \div \frac{1}{2} = ?$. Use the word problem together with a math drawing table, or double number line to help you explain in your own words why you can solve the division problem by first giving the fractions a common denominator and then dividing the numerators.

13. a. By rewriting the division problem

$$\frac{18}{33} \div \frac{9}{11} = ?$$

   as a multiplication problem with an unknown factor, explain why it is valid to divide the fractions as follows:

$$\frac{18}{33} \div \frac{9}{11} = \frac{18 \div 9}{33 \div 11} = \frac{2}{3}$$

   b. Give an example of a numerical fraction division problem that can be made easy to solve by using the method of dividing the numerators and dividing the denominators that is demonstrated in part (a).

14. Use the fact that we can rewrite the division problem $\frac{7}{10} \div \frac{2}{3} = ?$ as a multiplication problem with an unknown factor to explain in your own words why the division problem can be solved by multiplying $\frac{7}{10}$ by $\frac{3}{2}$.

15. a. Tyrone says that $\frac{1}{2} \div 5$ doesn't make sense because 5 is bigger than $\frac{1}{2}$ and you can't divide a smaller number by a bigger number. Give Tyrone an example of a sensible word problem for $\frac{1}{2} \div 5$. Solve your problem, and explain your solution.

   b. Kim says that $4 \div \frac{1}{3}$ can't be equal to 12 because when you divide, the answer should be smaller. Kim thinks the answer should be $\frac{1}{12}$ because that is less than 4. Give Kim an example of a word problem for $4 \div \frac{1}{3}$, and explain why it makes sense that the answer really is 12, not $\frac{1}{12}$.

*16. Fraction division word problems involve the simultaneous use of different unit amounts. Solve the paint problem that follows with the aid of a math drawing, a table, or a double number line. Describe how you must work simultaneously with different unit amounts in solving the problem.

   A Paint Problem: You need $\frac{3}{4}$ of a bottle of paint to paint a poster board. You have $3\frac{1}{2}$ bottles of paint. How many poster boards can you paint?

6.5 Fraction Division from the How-Many-Units-in-1-Group Perspective

CCSS Common Core State Standards Grades 5, 6

Have you ever wondered why the standard "invert and multiply" procedure for dividing fractions is valid? In this section we will review this standard procedure and explain why it is valid. To do so, we will use the how-many-units-in-1-group perspective on division. Often how-many-units-in-1-group fraction division word problems don't seem like division problems at all, but rather like proportion problems—which in fact they are. Finally, we'll consider a variety of word problems and pay special attention to distinguishing fraction division word problems from fraction multiplication word problems. In particular, we'll clarify the meaning of dividing *in* half and dividing *by* one half.

How Can We Interpret Fraction Division as How-Many-Units-in-1-Group?

Recall that with how-many-units-in-1-group division, we interpret $A \div B = ?$ as $B \cdot ? = A$ (see Figure 6.21). This applies not only to whole

Figure 6.21 How-many-units-in-1-group division.

A	÷	B	=	?
Number of units to be put in equal groups		Number of groups		How many units in 1 group?

B	•	?	=	A
Number of groups		How many units in 1 group?		Number of units to be put in equal groups

numbers but also to fractions. For example, we can interpret

$$\frac{4}{5} \div \frac{2}{3} = ? \quad \text{as} \quad \frac{2}{3} \cdot ? = \frac{4}{5}$$

So $\frac{4}{5} \div \frac{2}{3} = ?$ asks for the size of 1 whole group when $\frac{2}{3}$ of a group has size $\frac{4}{5}$ units, as in Figure 6.22.

Figure 6.22 Interpreting $\frac{4}{5} \div \frac{2}{3} = ?$ as how-many-units-in-1-group division.

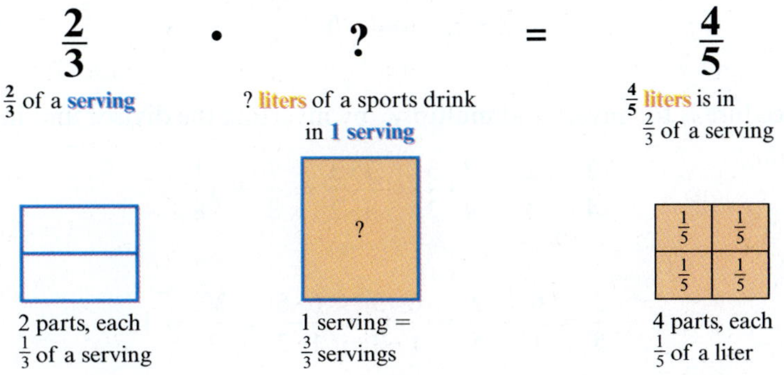

CCSS 5.NF.7 6.NS.1

Figure 6.23 will help you compare two versions of a how-many-units-in-1-group problem, one with whole numbers and one with fractions. Observe how the tables, strip diagrams, and double number lines highlight the how-many-units-in-1-group structure of the problem. Observe also that the structure is the same, whether whole numbers or fractions are involved.

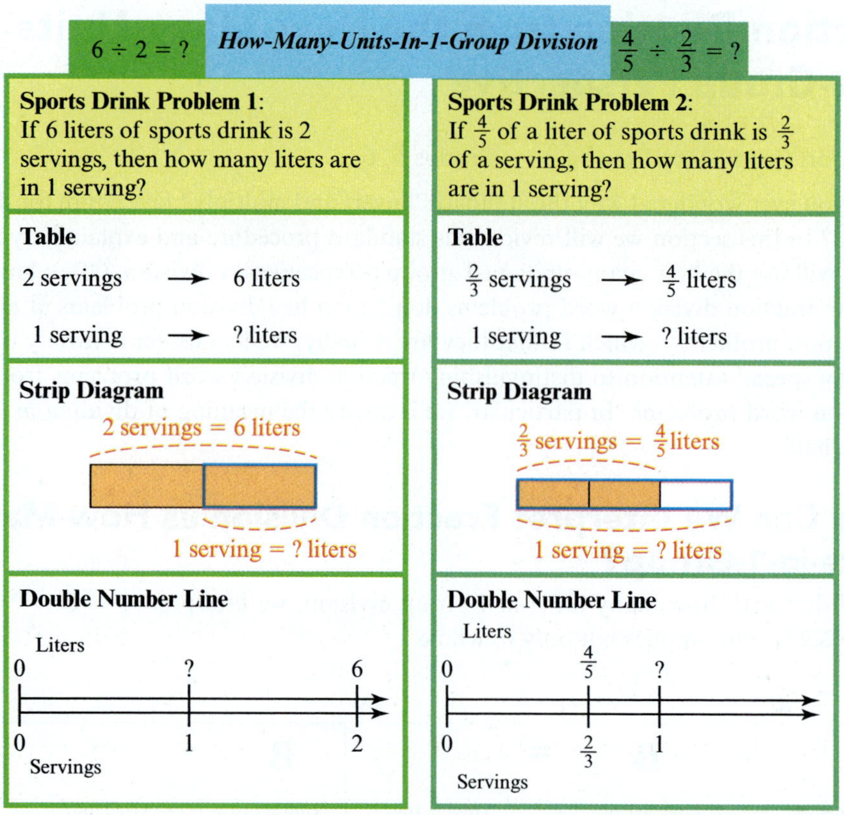

Figure 6.23 How-many-units-in-1-group division with whole numbers and with fractions.

CLASS ACTIVITY

6P How-Many-Units-in-1-Group Fraction Division Problems, p. CA-114

How Does the "Invert and Multiply" or "Multiply by the Reciprocal" Procedure Work?

To divide fractions, such as

$$\frac{3}{4} \div \frac{2}{3} \quad \text{and} \quad 6 \div \frac{2}{5}$$

the standard procedure is to "invert and multiply" by inverting the divisor and multiplying by it:

$$\frac{3}{4} \div \frac{2}{3} = \frac{3}{4} \cdot \frac{3}{2} = \frac{3 \cdot 3}{4 \cdot 2} = \frac{9}{8} = 1\frac{1}{8}$$

and

$$6 \div \frac{2}{5} = \frac{6}{1} \div \frac{2}{5} = \frac{6}{1} \cdot \frac{5}{2} = \frac{6 \cdot 5}{1 \cdot 2} = \frac{30}{2} = 15$$

and, in general,

$$\frac{A}{B} \div \frac{C}{D} = \frac{A}{B} \cdot \frac{D}{C} = \frac{A \cdot D}{B \cdot C}$$

Some teachers describe the process as "flip the factor" to help students remember which fraction to invert.

As an interesting special case, notice that we can apply the fraction division procedure to whole number division because every whole number is equal to a fraction (e.g., $2 = \frac{2}{1}$ and $3 = \frac{3}{1}$). Therefore,

$$2 \div 3 = \frac{2}{1} \div \frac{3}{1} = \frac{2}{1} \cdot \frac{1}{3} = \frac{2 \cdot 1}{1 \cdot 3} = \frac{2}{3}$$

Notice that this result, $2 \div 3 = \frac{2}{3}$, agrees with our finding earlier in this chapter on the link between division and fractions—namely, that $A \div B = \frac{A}{B}$.

reciprocal

Another way to describe the standard method for dividing fractions is in terms of the reciprocal of the divisor. The **reciprocal** of a fraction $\frac{C}{D}$ is the fraction $\frac{D}{C}$. In order to divide fractions, we multiply the dividend by the reciprocal of the divisor.

multiplicative inverse

The reciprocal, $\frac{D}{C}$, of a fraction $\frac{C}{D}$ is the **multiplicative inverse** of the fraction, meaning that a fraction and its reciprocal multiply to 1:

$$\frac{C}{D} \cdot \frac{D}{C} = \frac{C \cdot D}{D \cdot C} = 1$$

It is this property of reciprocals that lies behind the reason that the standard procedure for dividing fractions works.

Next, we'll explain why the standard "invert and multiply" or "multiply by the reciprocal" procedure for dividing fractions is valid.

Why Is the "Invert and Multiply" Procedure Valid?

Why can we divide fractions by multiplying by the reciprocal of the divisor? Let's see how to use the how-many-units-in-1-group interpretation of division to explain.

Consider the following how-many-units-in-1-group problem for $\frac{1}{2} \div \frac{3}{5}$:

> You used $\frac{1}{2}$ can of paint to paint $\frac{3}{5}$ of a wall. How many cans of paint will it take to paint the whole wall?

This is a how-many-units-in-1-group problem because it corresponds with this multiplication equation:

$$\frac{3}{5} \cdot (\text{amount to paint the whole wall}) = \frac{1}{2}$$

as we see in **Figure 6.24**.

Figure 6.24 How-many-units-in-1-group fraction division.

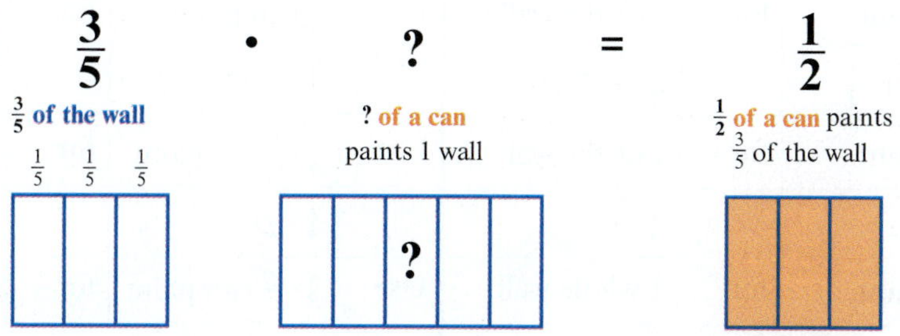

Fraction Division CCSS Grades 5, 6

We will now see why it makes sense to solve this division problem by multiplying $\frac{1}{2}$ by the reciprocal of $\frac{3}{5}$, namely, by $\frac{5}{3}$. Let's focus on the wall to be painted, as shown in **Figure 6.25**. Think of dividing the wall into 5 equal sections, 3 of which you painted with the $\frac{1}{2}$ can of paint. If you used $\frac{1}{2}$ can of paint to paint 3 sections, then each of the 3 sections required $\frac{1}{3}$ of the paint, or $\frac{1}{3} \cdot \frac{1}{2}$ of a can of paint. To determine how much paint you will need for the whole wall, multiply the amount you need for 1 section by 5. So you can determine the amount of paint you need for the whole wall by multiplying the

$\frac{1}{2}$ can of paint by $\frac{1}{3}$ and then multiplying that result by 5, as summarized in **Table 6.7**. But to multiply a number by $\frac{1}{3}$ and then multiply it by 5 is the same as multiplying the number by $\frac{5}{3}$. Therefore, we can determine the number of cans of paint you need for the whole wall by multiplying $\frac{1}{2}$ by $\frac{5}{3}$:

$$\frac{5}{3} \cdot \frac{1}{2} = \frac{5}{6}$$

Figure 6.25

The amount of paint needed for the whole wall is $\frac{5}{3}$ of the $\frac{1}{2}$ can used to cover $\frac{3}{5}$ of the wall.

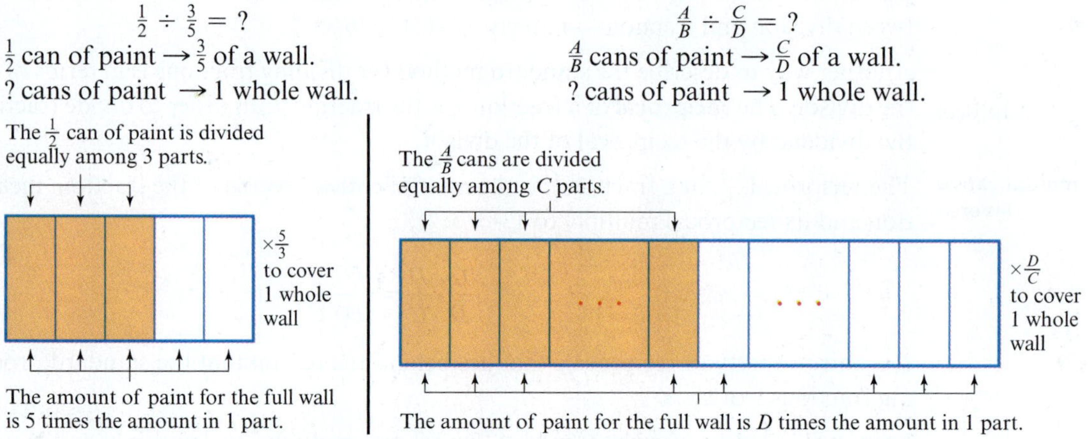

Another way to summarize this reasoning is to observe that the whole wall is $\frac{5}{3}$ of the part painted by $\frac{1}{2}$ can of paint, so the whole will take $\frac{5}{3}$ as much paint, which is $\frac{5}{3} \cdot \frac{1}{2}$ cans of paint.

This is exactly the "invert and multiply" procedure for dividing $\frac{1}{2} \div \frac{3}{5}$. It shows that you will need $\frac{5}{6}$ can of paint for the whole wall.

The preceding argument works when other fractions replace $\frac{1}{2}$ and $\frac{3}{5}$, as summarized on the right in **Figure 6.25** and **Table 6.7**, thereby explaining why

$$\frac{A}{B} \div \frac{C}{D} = \frac{A}{B} \cdot \frac{D}{C}$$

In other words, to divide fractions, multiply the dividend by the reciprocal of the divisor.

TABLE 6.7 To solve $\frac{1}{2} \div \frac{3}{5}$ or $\frac{A}{B} \div \frac{C}{D}$, determine how much paint to use for a whole wall if $\frac{1}{2}$ can of paint covers $\frac{3}{5}$ of the wall or if $\frac{A}{B}$ of a can of paint covers $\frac{C}{D}$ of the wall

$\frac{1}{2} \div \frac{3}{5} = ?$				$\frac{A}{B} \div \frac{C}{D} = ?$			
Use	$\frac{1}{2}$ can paint	for	$\frac{3}{5}$ of the wall.	Use	$\frac{A}{B}$ can paint	for	$\frac{C}{D}$ of the wall.
	$\downarrow \div 3$ or $\cdot \frac{1}{3}$		$\downarrow \div 3$ or $\cdot \frac{1}{3}$		$\downarrow \div C$ or $\cdot \frac{1}{C}$		$\downarrow \div C$ or $\cdot \frac{1}{C}$
Use	$\frac{1}{6}$ can paint	for	$\frac{1}{5}$ of the wall.	Use	$\frac{A}{B} \cdot \frac{1}{C}$ can paint	for	$\frac{1}{D}$ of the wall.
	$\downarrow \cdot 5$		$\downarrow \cdot 5$		$\downarrow \cdot D$		$\downarrow \cdot D$
Use	$\frac{5}{6}$ can paint	for	1 whole wall.	Use	$\frac{A}{B} \cdot \frac{D}{C}$ can paint	for	1 whole wall.
in one step:				**in one step:**			
Use	$\frac{1}{2}$ can paint	for	$\frac{3}{5}$ of the wall.	Use	$\frac{A}{B}$ can paint	for	$\frac{C}{D}$ of the wall.
	$\downarrow \cdot \frac{5}{3}$		$\downarrow \cdot \frac{5}{3}$		$\downarrow \cdot \frac{D}{C}$		$\downarrow \cdot \frac{D}{C}$
Use	$\frac{5}{6}$ can paint	for	1 whole wall.	Use	$\frac{A}{B} \cdot \frac{D}{C}$ can paint	for	1 whole wall.
so $\frac{1}{2} \div \frac{3}{5} = \frac{1}{2} \cdot \frac{5}{3}$				so $\frac{A}{B} \div \frac{C}{D} = \frac{A}{B} \cdot \frac{D}{C}$			

How Is Dividing *by* $\frac{1}{2}$ Different from Dividing *in* $\frac{1}{2}$?

In mathematics, language is used much more precisely and carefully than in everyday conversation. This is one source of difficulty in learning mathematics. For example, consider the following two phrases:

dividing *by* $\frac{1}{2}$

dividing *in* $\frac{1}{2}$

You may feel that these two phrases mean the same thing; however, mathematically, they do not. To divide a number—say, 5—by $\frac{1}{2}$ means to calculate $5 \div \frac{1}{2}$. Remember that we read $A \div B$ as A divided by B. We would divide 5 by $\frac{1}{2}$ if we wanted to know how many half cups of flour are in 5 cups of flour, for example. (Notice that there are 10 half cups of flour in 5 cups of flour, not $2\frac{1}{2}$.)

On the other hand, to divide a number *in* half means to find half of that number. So to divide 5 in half means to find $\frac{1}{2}$ of 5. One half of 5 means $\frac{1}{2} \times 5$. So dividing in $\frac{1}{2}$ is the same as dividing by 2.

When you examine the word problems in Class Activity 6Q, pay close attention to the wording in order to decide if they are division problems or not. When it comes to word problems, there is simply no substitute for careful reading and close attention to meaning!

CLASS ACTIVITY

6Q Are These Division Problems?, p. CA-115

How Can We Make Sense of Complex Fractions?

How can we make sense of fractions such as

$$\frac{\frac{2}{3}}{5}, \quad \frac{\frac{2}{3}}{\frac{4}{5}}, \quad \frac{1+\frac{2}{3}}{1+\frac{4}{5}}, \quad \frac{-5}{6}, \quad \frac{-5}{-6}$$

Fractions such as these, where the numerator, the denominator, or both are themselves fractions or other expressions are often called **complex fractions**. As we saw in Section 6.2, when A and B are counting numbers and B is not zero,

complex fractions

$$\frac{A}{B} = A \div B$$

CCSS 7.NS.3

We can use this connection between fractions and division to *define* complex fractions in terms of division.

So, for example,

$$\frac{\frac{2}{3}}{5} = \frac{2}{3} \div 5 = \frac{2}{3} \div \frac{5}{1} = \frac{2}{3} \cdot \frac{1}{5} = \frac{2}{15}$$

and

$$\frac{-5}{-6} = (-5) \div (-6) = 5 \div 6 = \frac{5}{6}$$

(See Section 6.1 for division with negative numbers.)

SECTION SUMMARY AND STUDY ITEMS

Section 6.5 Fraction Division from the How-Many-Units-in-1-Group Perspective

We can interpret fraction division from the how-many-units-in-1-group viewpoint. The "invert and multiply" or "multiply by the reciprocal" procedure is nicely explained from this perspective.

Key Skills and Understandings

1. Write and recognize fraction division word problems for the how-many-units-in-1-group as well as from the how-many-groups interpretations.
2. Solve division problems with the aid of math drawings, tables, and double number lines as well as numerically. Know how to interpret visual supports appropriately.
3. Use the how-many-units-in-1-group interpretation of division to explain why the "invert and multiply" procedure for dividing fractions is valid.

Practice Exercises for Section 6.5

1. Write a how-many-units-in-1-group word problem for $2 \div \frac{3}{4} = ?$. Use the situation of the problem to help you explain why you can solve the division problem by multiplying 2 by the reciprocal of $\frac{3}{4}$.

2. Which of the following are solved by the division problem $\frac{3}{4} \div \frac{1}{2}$? For those that are, which interpretation of division is used? For those that are not, determine how to solve the problem, if it can be solved.

a. $\frac{3}{4}$ of a bag of jelly worms make $\frac{1}{2}$ cup. How many cups of jelly worms are in 1 bag?

b. $\frac{3}{4}$ of a bag of jelly worms make $\frac{1}{2}$ cup. How many bags of jelly worms does it take to make 1 cup?

c. You have $\frac{3}{4}$ of a bag of jelly worms and a recipe that calls for $\frac{1}{2}$ cup of jelly worms. How many batches of your recipe can you make?

d. You have $\frac{3}{4}$ cup of jelly worms and a recipe that calls for $\frac{1}{2}$ cup of jelly worms. How many batches of your recipe can you make?

e. If $\frac{3}{4}$ pound of candy costs $\frac{1}{2}$ of a dollar, then how many pounds of candy should you be able to buy for 1 dollar?

f. If you have $\frac{3}{4}$ pound of candy and you divide the candy in $\frac{1}{2}$, how much candy will you have in each portion?

g. If $\frac{1}{2}$ pound of candy costs \$1, then how many dollars should you expect to pay for $\frac{3}{4}$ pound of candy?

3. Frank, John, and David earned \$14 together. They want to divide it equally, except that David should only get a half share, since he did half as much work as either Frank or John did (and Frank and John worked equal amounts). Write a division problem to find out how much Frank should get. Which interpretation of division does this problem use?

4. Bill leaves a tip of \$4.50 for a meal. If the tip is 15% of the cost of the meal, then how much did the meal cost? Write a division problem to solve this. Which interpretation of division does this problem use?

5. Compare the arithmetic needed to solve the following problems:

a. What fraction of a $\frac{1}{3}$ cup measure is filled when we pour in $\frac{1}{4}$ cup of water?

b. What is $\frac{1}{4}$ of $\frac{1}{3}$ cup?

c. How much more is $\frac{1}{3}$ cup than $\frac{1}{4}$ cup?

d. If $\frac{1}{4}$ cup of water fills $\frac{1}{3}$ of a plastic container, then how much water will the full container hold?

6. It takes an assembly line $\frac{1}{2}$ hour to produce enough boxes of widgets to fill a truck $\frac{2}{3}$ full. At that rate, how long does it take the assembly line to produce enough boxes of widgets to fill the truck completely full? Explain how to use the double number line in Figure 6.26 to help you solve this problem.

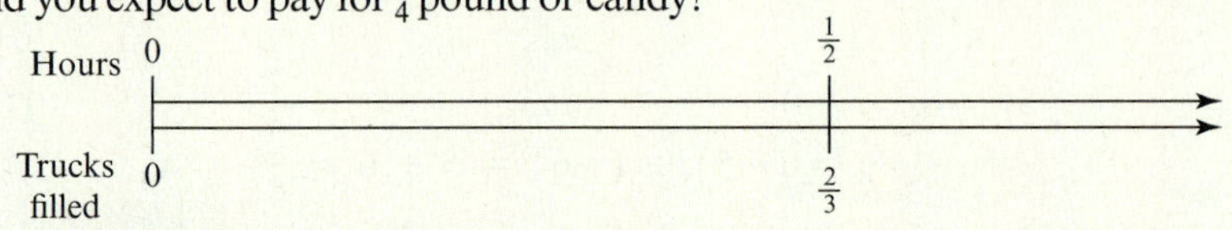

Figure 6.26 A double number line.

Answers to Practice Exercises for Section 6.5

1. If 2 tons of dirt fill a truck $\frac{3}{4}$ full, then how many tons of dirt will be needed to fill the truck completely full?

We can see that this is a how-many-units-in-1-group type of problem because the 2 tons of dirt fill $\frac{3}{4}$ of a group (the truck) and we want to know the number of tons in 1 whole group. Or we could realize that if 6 tons of dirt fill 3 trucks, then $6 \div 3$ tons of dirt fill one truck, and this uses the how-many-units-in-1-group view of division. So the same must be true when we replace 6 with 2 and 3 with $\frac{3}{4}$.

Figure 6.27 shows a truck bed divided into 4 equal parts, with 3 of those parts filled with dirt. Because the 3 parts are filled with 2 tons of dirt, each of the 3 parts must contain $\frac{1}{3}$ of the 2 tons of dirt. To fill the truck completely, 4 parts, each containing $\frac{1}{3}$ of the 2 tons of dirt, are needed. Because "4 parts, each $\frac{1}{3}$" means $\frac{4}{3}$, to fill the truck completely, $\frac{4}{3}$ of the 2 tons of dirt are needed. Therefore, the truck takes $\frac{4}{3} \cdot 2$ tons of dirt to fill it. So $2 \div \frac{3}{4} = \frac{4}{3} \cdot 2$.

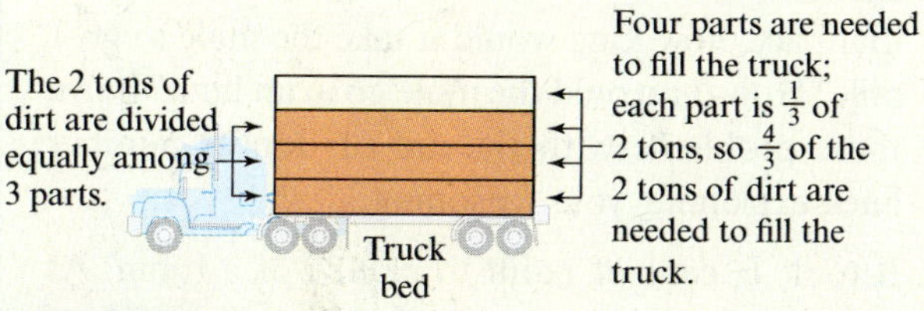

Figure 6.27 Showing why $2 \div \frac{3}{4} = \frac{4}{3} \cdot 2$ by considering how many tons of dirt it takes to fill a truck if 2 tons fills it $\frac{3}{4}$ full.

2. a. This problem can be rephrased as "if $\frac{1}{2}$ cup of jelly worms fill $\frac{3}{4}$ of a bag, then how many cups fill a whole bag?" Therefore, this is a how-many-units-in-1-group division problem illustrating $\frac{1}{2} \div \frac{3}{4}$, not $\frac{3}{4} \div \frac{1}{2}$. Since $\frac{1}{2} \div \frac{3}{4} = \frac{1}{2} \cdot \frac{4}{3} = \frac{2}{3}$, there are $\frac{2}{3}$ cup of jelly worms in a whole bag.

b. This problem is solved by $\frac{3}{4} \div \frac{1}{2}$, according to the how-many-units-in-1-group interpretation. A group is a cup, and each unit is a bag of jelly worms.

c. This problem can't be solved because you don't know how many cups of jelly worms are in $\frac{3}{4}$ of a bag.

d. This problem is solved by $\frac{3}{4} \div \frac{1}{2}$, according to the how-many-groups interpretation. One group consists of $\frac{1}{2}$ cup of jelly worms.

e. This problem is solved by $\frac{3}{4} \div \frac{1}{2}$, according to the how-many-units-in-1-group interpretation. This is because you can think of the problem as saying that $\frac{3}{4}$ pound of candy fills $\frac{1}{2}$ of a group and you want to know how many pounds fills 1 whole group.

f. This problem is solved by $\frac{3}{4} \cdot \frac{1}{2}$, not $\frac{3}{4} \div \frac{1}{2}$. It is dividing *in* half, not dividing *by* half.

g. This problem is solved by $\frac{3}{4} \div \frac{1}{2}$, according to the how-many-groups interpretation because you want to know how many $\frac{1}{2}$ pounds are in $\frac{3}{4}$ pound. One group consists of $\frac{1}{2}$ pound of candy.

3. If we consider Frank and John as each representing 1 group, and David as representing half of a group, then the $14 should be distributed equally among $2\frac{1}{2}$ groups. Therefore, this is a how-many-units-in-1-group division problem. Each group should get

$$14 \div 2\frac{1}{2} = 14 \div \frac{5}{2} = 14 \cdot \frac{2}{5} = \frac{28}{5}$$
$$= 5\frac{3}{5} = 5\frac{6}{10} = 5.60$$

dollars. Therefore, Frank and John should each get $5.60, and David should get half of that, which is $2.80.

4. According to the how-many-units-in-1-group interpretation, the problem is solved by $4.50 ÷ 0.15 because $4.50 fills 0.15 of a group and we want to know how much is in 1 whole group. So the meal cost

$$\$4.50 \div 0.15 = \$4.50 \div \frac{15}{100} = \$4.50 \cdot \frac{100}{15}$$
$$= \frac{\$450}{15} = \$30$$

5. Each problem, except for the first and last, requires different arithmetic to solve it.

a. This is asking, "$\frac{1}{4}$ equals what times $\frac{1}{3}$?" We solve this by calculating $\frac{1}{4} \div \frac{1}{3}$, which is $\frac{3}{4}$. We can also think of this as a division problem with the how-many-groups interpretation because we want to know how many $\frac{1}{3}$ cups are in $\frac{1}{4}$ cup. According to the meaning of division, this is $\frac{1}{4} \div \frac{1}{3}$.

b. This is asking, "What is $\frac{1}{4}$ of $\frac{1}{3}$?" We solve this by calculating $\frac{1}{4} \cdot \frac{1}{3} = \frac{1}{12}$.

c. This is asking, "What is $\frac{1}{3} - \frac{1}{4}$?" The answer is $\frac{1}{12}$, which happens to be the same answer as in part (b), but the arithmetic to solve it is different.

d. Since $\frac{1}{4}$ cup of water fills $\frac{1}{3}$ of a plastic container, the full container will hold 3 times as much water, or $3 \cdot \frac{1}{4} = \frac{3}{4}$ of a cup. We can also think of this as a division problem with the how-many-units-in-1-group interpretation, where $\frac{1}{4}$ cup of water is put into $\frac{1}{3}$ of a group. We want to know how much is in 1 group. According to the meaning of division, it's $\frac{1}{4} \div \frac{1}{3}$, which again is equal to $\frac{3}{4}$.

6. To determine how long it will take to produce enough widgets to fill the truck completely full, we must find the location on the top number line that is above 1, as indicated on the first double number line in Figure 6.28. If we give the bottom number line tick marks at every third, then $\frac{2}{3}$ is the second tick mark and $1 = \frac{3}{3}$ is the third tick mark. The number above $\frac{1}{3}$ must be half of $\frac{1}{2}$, which is $\frac{1}{4}$, as shown on the second double number line in Figure 6.28. So on the top number line, the tick marks are spaced $\frac{1}{4}$ apart. The number above 1 is $\frac{3}{4}$, so it will take $\frac{3}{4}$ of an hour to produce enough widgets to fill the truck completely full.

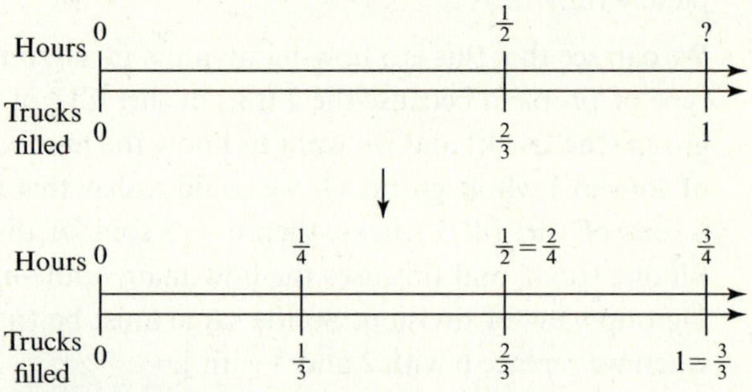

Figure 6.28 Using a double number line to solve a fraction division problem.

PROBLEMS FOR SECTION 6.5

1. Explain how to use the math drawings of Figures 6.22 and 6.23 (including the double number line) to solve Sports Drink Problem 2 of Figure 6.23.

2. Write a how-many-units-in-1-group word problem for $4 \div \frac{1}{3} = ?$ and use your problem and a math drawing, table, or double number line to explain why it makes sense to solve $4 \div \frac{1}{3} = ?$ by "inverting and multiplying"—in other words, by multiplying 4 by $\frac{3}{1}$.

3. Write a how-many-units-in-1-group word problem for $4 \div \frac{2}{3} = ?$ and use your problem and a math drawing, table, or double number line to explain why it makes sense to solve $4 \div \frac{2}{3}$ by "inverting and multiplying"—in other words, by multiplying 4 by $\frac{3}{2}$.

4. Write a how-many-units-in-1-group word problem for $\frac{1}{2} \div \frac{3}{4} = ?$ and use your problem and a math drawing, table, or double number line to explain why it makes sense to solve $\frac{1}{2} \div \frac{3}{4}$ by "inverting and multiplying"—in other words, by multiplying $\frac{1}{2}$ by $\frac{4}{3}$.

5. Write a how-many-units-in-1-group word problem for $1 \div 2\frac{1}{2} = ?$ and use your problem and a math drawing, table, or double number line to explain why it makes sense to solve $1 \div 2\frac{1}{2}$ by "inverting and multiplying."

6. If $1\frac{1}{2}$ cups of a cereal weigh 2 pounds, how much does 1 cup of the cereal weigh? Solve this problem with the aid of a double number line, explaining your reasoning.

7. It took a mule $\frac{2}{3}$ of an hour to go $\frac{4}{5}$ of a mile. At that pace, how long would it take the mule to go 1 mile? How far would the mule go in an hour? Solve these problems with the aid of double number lines, explaining your reasoning.

8. It took $1\frac{1}{3}$ cans of paint to paint $\frac{2}{5}$ of a room. At that rate, how many cans of paint will it take to paint the whole room? What fraction of the room can you paint with 1 can of paint? Solve these problems with the aid of double number lines, explaining your reasoning.

9. Write a word problem for $\frac{3}{4} \cdot \frac{1}{2} = ?$ and another word problem for $\frac{3}{4} \div \frac{1}{2} = ?$. (Make clear which is which.) In each case, use elementary reasoning about the situation to solve your problem. Explain your reasoning.

10. For each of the following problems, determine if it is a problem for $\frac{2}{3} \div \frac{1}{4}$ or for $\frac{1}{4} \div \frac{2}{3}$. If so, say what type of division (how-many-groups or how-many-units-in-1-group) and explain how you can tell. If not, say how the problem could be solved (if it can be solved).

a. You divided $\frac{2}{3}$ of a gallon of paint equally among 4 containers. How much paint is in each container?

b. It takes $\frac{1}{4}$ of a gallon of paint to paint a wall. How many walls (of the same size) can you paint with $\frac{2}{3}$ of a gallon?

c. It took $\frac{2}{3}$ of a gallon of paint to paint a wall. How much paint did it take to paint $\frac{1}{4}$ of the wall?

d. It took $\frac{2}{3}$ of a gallon of paint to paint $\frac{1}{4}$ of a wall. How much paint will it take for the whole wall?

e. After using $\frac{2}{3}$ of a gallon of paint to paint a wall, you had $\frac{1}{4}$ of a gallon of paint left. How much paint did you have at first?

f. It takes $\frac{2}{3}$ of a gallon of paint to paint a wall. How much of the wall can you paint with $\frac{1}{4}$ of a gallon?

11. Sam picked $\frac{1}{2}$ gallon of blueberries. He poured the blueberries into one of his plastic containers and noticed that the berries filled the container $\frac{2}{3}$ full. Solve the following problems by reasoning about a math drawing, a table, or a double number line:

a. How many of Sam's containers will 1 gallon of blueberries fill? (Assume that Sam has a number of containers of the same size.)

b. How many gallons of blueberries does it take to fill Sam's container completely full?

12. A road crew is building a road. So far, $\frac{2}{3}$ of the road has been completed and this portion of the road is $\frac{3}{4}$ of a mile long. Solve the following problems by reasoning about a math drawing, a table, or a double number line:

a. How long will the road be when it is completed?

b. When the road is 1 mile long, what fraction of the road will be completed?

13. Will has mowed $\frac{2}{3}$ of his lawn, and so far it has taken him 45 minutes. For each of the following problems, solve the problem in two ways: (1) by reasoning about the situation and (2) by interpreting the problem as a division problem (say whether it is a how-many-groups or a how-many-units-in-1-group type of problem) and by solving the division problem using standard paper-and-pencil methods. Do not use a calculator. Verify that you get the same answer both ways.

a. How long will it take Will to mow the entire lawn (all together)?

b. What fraction of the lawn can Will mow in 1 hour?

14. Grandma's favorite muffin recipe uses $1\frac{3}{4}$ cups of flour for 1 batch of 12 muffins. For each of the problems (a) through (c), solve the problem in two ways: (1) by reasoning about the situation and (2) by interpreting the problem as a division problem (say whether it is a how-many-groups or a how-many-units-in-1-group type of problem) and by solving the division problem using standard paper-and-pencil methods. Do not use a calculator. Verify that you get the same answer both ways.

a. How many cups of flour are in 1 muffin?

b. How many muffins does 1 cup of flour make?

c. If you have 3 cups of flour, then how many batches of muffins can you make? (Assume that you can make fractional batches of muffins and that you have enough of all the ingredients.)

15. Tran writes the following word problem for $1\frac{3}{4} \div \frac{1}{2} = ?$:

There are $1\frac{3}{4}$ pizzas left. Tran eats half of the pizza that is left. How many pieces of pizza did Tran eat?

Tran shows the diagrams in **Figure 6.29**, saying that they show that

$$1\frac{3}{4} \div \frac{1}{2} = 3\frac{1}{2}$$

a. Even though Tran gets the correct answer to the division problem, explain why his problem is not a correct word problem for $1\frac{3}{4} \div \frac{1}{2} = ?$.

b. Write a correct word problem for $1\frac{3}{4} \div \frac{1}{2} = ?$ that is about $1\frac{3}{4}$ pizzas.

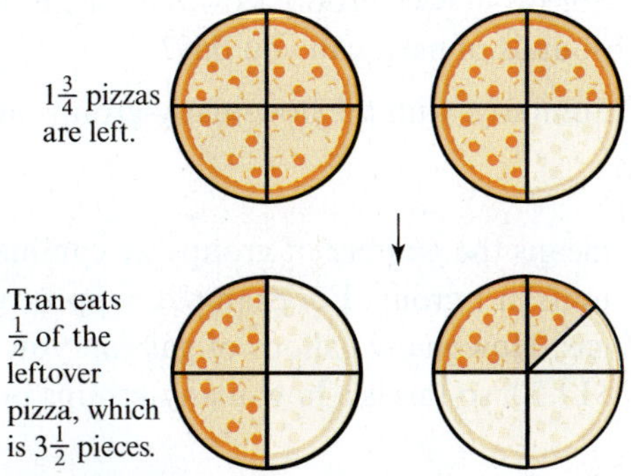

Figure 6.29 Why is this not a word problem for $1\frac{3}{4} \div \frac{1}{2}$?

6.6 Dividing Decimals

CCSS Common Core State Standards Grades 5, 6

CCSS 5.NBT.7 6.NS.3

Why do we divide decimals the way we do? When we divide finite decimals, such as

$$2.35\overline{)3.714}$$

the standard procedure is to move the decimal point in both the divisor (2.35) and the dividend (3.714) the same number of places to the right until the divisor becomes a whole number:

$$235\overline{)371.4}$$

Then a decimal point is placed above the decimal point in the new dividend (371.4):

$$\begin{array}{r} . \\ 235\overline{)371.4} \end{array}$$

From then on, division proceeds just as if we were doing the whole number division problem 3714 ÷ 235, except that a decimal point is left in place for the answer. Why does this procedure of shifting the decimal points make sense? In this section, we will explain in several different ways why we can shift the decimal points as we do. Our technique for shifting decimal points in division problems also allows us to calculate efficiently with large numbers, such as those in the millions, billions, and trillions. But first we consider the two interpretations of division for decimals.

CLASS ACTIVITY

6R Reasoning and Estimation with Decimal Division, p. CA-116

6S Decimal Division, p. CA-117

How Do the Two Interpretations of Division Extend to Decimals?

The two interpretations that we have used for whole numbers and for fractions also apply to decimals.

The How-Many-Groups Interpretation With the how-many-groups interpretation of division, 35 ÷ 7 means the number of groups we can make when we divide 35 objects into groups with 7 objects in each group. For example, if meat costs $7 per pound and you have $35, then how many pounds of meat can you buy?

Similarly, with the how-many-groups interpretation of division,

$$14.5 \div 2.45$$

means the number of groups we can make when we divide 14.5 units into equal groups with 2.45 units in 1 group. For example, suppose gas costs $2.45 per gallon and you have $14.50 to spend on gas. How many gallons of gas can you buy? We want to know how-many-groups of $2.45 are in $14.50, so this is a how-many-groups problem for 14.5 ÷ 2.45.

The How-Many-Units-in-1-Group Interpretation With the how-many-units-in-1-group interpretation of division, 35 ÷ 7 means the number of units in one group if 35 units are divided equally among 7 groups. For example, if 7 identical action figures cost $35, then how much did one action figure cost?

Similarly, with the how-many-units-in-1-group interpretation of division,

$$9.48 \div 5.3$$

means the number of units in one group if 9.48 units are in 5.3 equal groups. You can also think of the 9.48 objects as "filling" 5.3 groups. For example, if you bought 5.3 pounds of peaches for \$9.48, then how much did 1 pound of peaches cost? The \$9.48 is distributed equally among or "fills" 5.3 groups, and we want to know how many dollars are in one group. Or, if you bought 5 pounds of peaches for \$10, then 1 pound of peaches would cost \$10 ÷ 5, and this is a how-many-units-in-1-group division problem. So the same must be true when we replace 10 with 9.48 and 5 with 5.3. Therefore, this is a how-many-units-in-1-group problem for 9.48 ÷ 5.3.

Next we turn to different methods for explaining why we move the decimal point as we do when we divide decimals.

How Can We Use Multiplying and Dividing by the Same Power of 10 to Explain Why We Shift Decimal Points?

One way to explain why it is valid to shift the decimal points in a division problem is as follows: When we shift the decimal points in both the divisor and the dividend the same number of places, we have multiplied and divided by the same power of 10, thereby replacing the original problem with an equivalent problem that can be solved by previous methods. Recall that multiplying a decimal number by a power of 10 shifts the digits in the number and therefore can be thought of as shifting the decimal point. Consider the following example:

$$2.35\overline{)3.714}$$

Because

$$2.35 \times 100 = 235$$
$$3.714 \times 100 = 371.4$$

it follows that

$$\begin{aligned} & \qquad\qquad \times 100 \qquad\qquad\quad \div 100 \\ & \qquad\qquad\quad \downarrow \qquad\qquad\qquad\quad \downarrow \\ 3.714 \div 2.35 &= (3.714 \times 100) \div (2.35 \times 100) \\ &= 371.4 \div 235 \end{aligned}$$

Since we multiplied and divided by the same number—namely, 100—the original problem 3.714 ÷ 2.35 is equivalent to the new problem 371.4 ÷ 235. In other words, the two problems have the identical answer. We can give the same explanation by writing the division problems in fraction form:

$$\begin{aligned} \frac{3.714}{2.35} &= \frac{3.714}{2.35} \times \frac{100}{100} \\ &= \frac{3.714 \times 100}{2.35 \times 100} \\ &= \frac{371.4}{235} \end{aligned}$$

So the shifting of decimal points is really just a way to replace the problem

$$3.714 \div 2.35$$

with the equivalent problem

$$371.4 \div 235$$

How Can We Use Dollars and Cents to Explain Why We Shift Decimal Points?

If the decimals in a division problem have no digits in places lower than the hundredths place, then we can use money to interpret the division problem. For example, we can interpret

$$1.27\overline{)4.5}$$

as follows:

4.5 ÷ 1.27 is the number of pounds of plums we can buy for \$4.50 if plums cost \$1.27 per pound.

(Note that we are using the how-many-groups interpretation of division because we want to know how many groups of \$1.27 are in \$4.50.) If we think in terms of cents rather than dollars, then an equivalent way to rephrase the preceding statement is as follows:

4.5 ÷ 1.27 is the number of pounds of plums we can buy for 450 cents if plums cost 127 cents per pound.

But this last statement also describes the division problem

$$127\overline{)450}$$

Therefore, the problems 4.5 ÷ 1.27 and 450 ÷ 127 are equivalent. In other words, they must have the same answer. We can buy the same number of pounds of plums whether we are thinking in terms of dollars or in terms of cents. So, when we shift the decimal points of 1.27 and 4.5 two places to the right in the division problem $1.27\overline{)4.5}$ to obtain the problem $127\overline{)450}$, we have simply replaced the original problem with a problem that has the same answer and can be solved with the techniques we learned for whole number division.

How Can We Change the Unit to Explain Why We Shift Decimal Points?

A more general way to explain why we can shift the decimal points in a decimal division problem is to think about representing the division problem with bundled objects or with pictures representing bundled objects. For example, consider the division problem

$$0.29\overline{)1.7}$$

We can interpret this division problem as asking, "How many 0.29s are in 1.7?" from a how-many-groups viewpoint. Figure 6.30 represents this division problem if we interpret the smallest cube as $\frac{1}{100}$. But if we interpret the smallest cube as 1 instead of $\frac{1}{100}$, then Figure 6.30 asks "how many 29s are in 170?" which is the division problem

$$29\overline{)170}$$

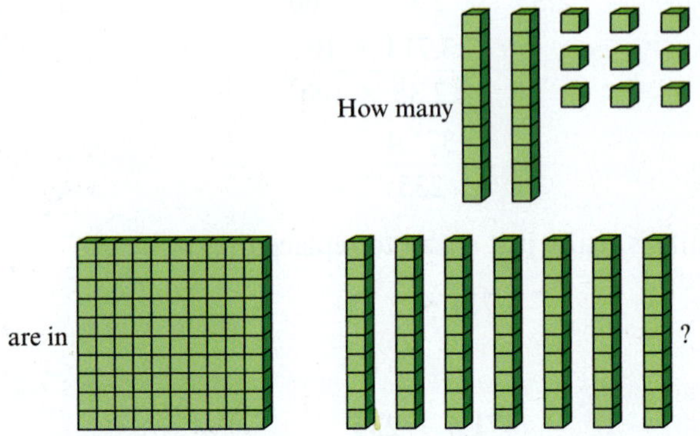

Figure 6.30
Representing 1.7 ÷ 0.29.

The answer must be the same, no matter which way we interpret the figure; therefore, the division problems $1.7 \div 0.29$ and $170 \div 29$ are equivalent problems. Notice that from the diagram we can estimate that the answer is between 5 and 6 because 5 copies of the cubes on the top will be fewer cubes than on the bottom, but 6 copies of the cubes on the top will be more cubes than those on the bottom.

By allowing the smallest cube in Figure 6.30 to represent different values, we see that this figure can represent infinitely many different division problems, as indicated in Table 6.8. Therefore, all of these division problems must have the same answer.

TABLE 6.8 Different ways to interpret Figure 6.30

If the Small Cube Represents	Then Figure 6.30 Represents
⋮	⋮
$\frac{1}{10{,}000}$	$0.017 \div 0.0029$
$\frac{1}{1000}$	$0.17 \div 0.029$
$\frac{1}{100}$	$1.7 \div 0.29$
$\frac{1}{10}$	$17 \div 2.9$
1	$170 \div 29$
10	$1700 \div 290$
100	$17{,}000 \div 2900$
⋮	⋮
1 million	(170 million) $\div$ (29 million)
⋮	⋮

How Can We Use Estimation to Decide Where to Put the Decimal Point?

As a way of checking your work, or as a backup if you forget what to do about the decimal points when dividing decimals, you can often estimate to determine where to put the decimal point in the answer to a decimal division problem.

For example, suppose that you want to calculate

$$0.7\overline{)2.53}$$

but instead you calculate

$$\begin{array}{r} 36.14 \\ 7\overline{)253} \end{array}$$

How should you move the decimal point in 36.14 to get the correct answer to $2.53 \div 0.7$? By shifting the decimal point in 36.14, you can get any of the following numbers:

$$\ldots, 0.03614, 0.3614, 3.614, 36.14, 361.4, 3614, \ldots$$

Which number in this list is a reasonable answer to $2.53 \div 0.7$? You can reason that 2.53 is close to 3 and 0.7 is close to 1, so the answer to $2.53 \div 0.7$ should be close to $3 \div 1$ which is 3. Therefore, the correct answer should be 3.614.

How Can We Divide Numbers in the Millions, Billions, and Trillions?

Earlier, we explained why we can shift the decimal point in both the divisor and the dividend the same number of places to obtain an equivalent division problem. To divide decimals, we shift the decimal points to the right, thereby turning the decimal division problem into a whole number division problem. But we can also shift the decimal points to the left. By doing this, we can replace a division problem involving large numbers with an equivalent division problem that involves smaller numbers. For example, to calculate

$$(170 \text{ million}) \div (29 \text{ million}) = 170{,}000{,}000 \div 29{,}000{,}000$$

we can shift the decimal points (which are not shown, but are to the right of the final 0s in each number) 6 places to the left. We thereby obtain the equivalent problem

$$170 \div 29$$

which doesn't involve so many zeros. In this way, we can make division problems involving numbers in the millions, billions, and trillions easier to solve.

What do we do if a problem mixes trillions, billions, and millions? For example, how much money will each person get if \$1.3 billion is divided equally among 8 million people? If we express 1.3 billion in terms of millions, then both numbers will be in terms of millions. Since 1 billion is 1000 million,

$$1.3 \text{ billion} = 1.3 \times 1000 \text{ million} = 1300 \text{ million}$$

Therefore,

$$(1.3 \text{ billion}) \div (8 \text{ million}) = (1300 \text{ million}) \div (8 \text{ million}) = 1300 \div 8$$

Since $1300 \div 8 = 162.5$, each person will get \$162.50.

SECTION SUMMARY AND STUDY ITEMS

Section 6.6 Dividing Decimals

We can interpret decimal division from the how-many-groups as well as from the how-many-units-in-1-group viewpoint. To divide decimals, we move the decimal points in both the divisor and dividend the same number of places so that the divisor becomes a whole number; then we proceed as with whole number division. There are several ways to explain this process of shifting the decimal points in decimal division.

Key Skills and Understandings

1. Write decimal division word problems for both interpretations of division.
2. Explain in several different ways why we move the decimal points the way we do when we divide decimals, and calculate decimal divisions (without a calculator).
3. Use estimation to determine the location of the decimal point in a decimal division problem.
4. Explain why many division problems are equivalent, such as $6000 \div 2000$, $600 \div 200$, $60 \div 20$, $6 \div 2$, and $0.6 \div 0.2$.

Practice Exercises for Section 6.6

1. Theresa needs to cut a piece of wood 0.33 inches thick, or just a little less thick. Theresa's ruler shows subdivisions of $\frac{1}{32}$ of an inch. How many $\frac{1}{32}$ of an inch thick should Theresa cut her piece of wood? What type of division problem is this?

2. Describe a quick way to mentally calculate $0.11 \div 0.125$ by thinking in terms of fractions.

3. Write a how-many-groups word problem for $0.35 \div 1.45 = ?$.

4. Write a how-many-units-in-1-group word problem for $0.35 \div 1.45 = ?$.

5. Without calculating the answers, give two different explanations for why the two division problems

$$0.65\overline{)4.3} \quad \text{and} \quad 65\overline{)430}$$

must have the same answer.

6. Show how to calculate $2.3 \div 0.008$ without a calculator.

7. Explain how to calculate

$$(4 \text{ billion}) \div (2 \text{ million})$$

mentally.

8. Use the idea of multiplying and dividing by the same number to explain why the problems

$$170{,}000{,}000 \div 29{,}000{,}000$$

and

$$170 \div 29$$

are equivalent.

9. Use Figure 6.30 to explain why the problems

$$170{,}000{,}000 \div 29{,}000{,}000$$

and

$$170 \div 29$$

are equivalent.

Answers to Practice Exercises for Section 6.6

1. $0.33 \div \frac{1}{32} = 10.56$, so Theresa should cut her piece of wood $\frac{10}{32}$ of an inch thick. This uses the how-many-groups interpretation of division, exact division (although in the end we round down because Theresa needs to cut her piece of wood a little less thick than the actual answer). Each $\frac{1}{32}$ of an inch represents one group. We want to know how many of these groups of $\frac{1}{32}$ of an inch are in 0.33 inches.

2. Observe that $0.125 = \frac{1}{8}$. To divide by a fraction, we can "invert and multiply." So, to divide by $\frac{1}{8}$, we multiply by 8. Therefore, $0.11 \div 0.125 = 0.11 \div \frac{1}{8} = 0.11 \times 8 = 0.88$.

3. "If apples cost \$1.45 per pound, then how many pounds of apples can you buy for \$0.35?" is a how-many-groups problem for $0.35 \div 1.45$, because you want to know how many groups of \$1.45 are in \$0.35.

4. "If you paid \$0.35 for 1.45 liters of water, then how much does 1 liter of water cost?" is a how-many-units-in-1-group problem for $0.35 \div 1.45$, because \$0.35 is distributed equally among 1.45 liters and you want to know how many dollars are "in" 1 liter. Or you could notice that if you paid \$6 for 2 liters of water, then 1 liter would cost $\$6 \div 2$, and this uses the how-many-units-in-1-group view of division. So the same must be true when we replace \$6 with \$0.35 and 2 liters with 1.45 liters.

5. *Explanation 1:* When we shift the decimal points in the numbers 0.65 and 4.3 two places to the right, we have really just multiplied and divided the problem $4.3 \div 0.65$ by 100, thereby arriving at an equivalent division problem. In equations,

$$4.3 \div 0.65 = \underset{\times 100 \downarrow}{(4.3 \times 100)} \div \underset{\div 100 \downarrow}{(0.65 \times 100)}$$
$$= 430 \div 65$$

or

$$4.3 \div 0.65 = \frac{4.3}{0.65} = \frac{4.3}{0.65} \times \frac{100}{100}$$
$$= \frac{4.3 \times 100}{0.65 \times 100}$$
$$= \frac{430}{65} = 430 \div 65$$

Explanation 2: Thinking in terms of money, we can interpret $4.3 \div 0.65 = ?$ as asking, "How many groups of \$0.65 are in \$4.30?" If we phrase the question in terms of cents instead of dollars, it becomes "How-many-groups of 65 cents are in 430 cents?", which we can solve by calculating $430 \div 65$. The answer must be the same either way we ask the question. Therefore, $4.3 \div 0.65$ and $430 \div 65$ are equivalent problems.

We can give a third explanation by using a diagram like Figure 6.30, in which we ask how many groups of 6 ten-sticks and 5 small squares are in 4 hundred-squares and 3 ten-sticks. By interpreting the small cube as $\frac{1}{100}$, we find that the diagram represents $4.3 \div 0.65$. By interpreting the small cube as 1, we see that the diagram represents $430 \div 65$. The answer must be the same either way we interpret it, so once again, we conclude that $4.3 \div 0.65$ and $430 \div 65$ are equivalent problems.

6. To calculate $2.3 \div 0.008$, we can move both decimal points 3 places to the right and calculate $2300 \div 8$ instead:

$$\begin{array}{r} 287.5 \\ 8\overline{)2300.0} \\ \underline{-16} \\ 70 \\ \underline{-64} \\ 60 \\ \underline{-56} \\ 40 \\ \underline{-40} \\ 0 \end{array}$$

7. Since 1 billion is 1000 million, 4 billion is 4000 million. So,

$$(4 \text{ billion}) \div (2 \text{ million}) =$$
$$(4000 \text{ million}) \div (2 \text{ million}) = 4000 \div 2 = 2000$$

Mentally, we just think that 4000 million divided by 2 million is the same as 4000 divided by 2.

8. If we divide the numbers 170 million and 29 million in the division problem $170{,}000{,}000 \div 29{,}000{,}000$ by 1 million, then we will have divided and multiplied by 1 million to obtain the new problem $170 \div 29$. Since we divided and multiplied by the same number, the new problem and the old problem are equivalent. In other words, they have the same answer. In equation form,

$$\begin{array}{ccc} (170 \text{ million}) & \div & (29 \text{ million}) \\ \cdot\, 1 \text{ million} & & \div\, 1 \text{ million} \\ \downarrow & & \downarrow \end{array}$$
$$= (170 \cdot 1 \text{ million}) \div (29 \cdot 1 \text{ million})$$
$$= 170 \div 29 = 5.86$$

In fraction form,

$$(170 \text{ million}) \div (29 \text{ million}) = \frac{170 \cdot 1 \text{ million}}{29 \cdot 1 \text{ million}}$$
$$= \frac{170}{29} \cdot \frac{1 \text{ million}}{1 \text{ million}} = \frac{170}{29}$$
$$= 170 \div 29$$

9. As shown in Table 6.8, if we let the small cube in Figure 6.30 represent 1 million, then Figure 6.30 represents

$$170{,}000{,}000 \div 29{,}000{,}000$$

But if we let the small cube represent 1, then Figure 6.30 represents

$$170 \div 29$$

We must get the same answer either way we interpret the diagram; therefore, the two division problems $170{,}000{,}000 \div 29{,}000{,}000$ and $170 \div 29$ are equivalent.

PROBLEMS FOR SECTION 6.6

1. Sue needs to cut a piece of wood 0.4 of an inch thick, or just a little less thick. Sue's ruler shows sixteenths of an inch. How many sixteenths of an inch thick should Sue cut her piece of wood? What type of division problem is this?

2. **a.** Write a how-many-groups word problem for $5.6 \div 1.83 = ?$.

   b. Write a how-many-units-in-1-group word problem for $5.6 \div 1.83 = ?$.

   c. Write a word problem (any type) for $0.75 \div 2.4 = ?$.

3. a. Calculate 28.3 ÷ 0.07 to the hundredths place without a calculator. Show your work.

   b. Describe the standard procedure for determining where to put the decimal point in the answer to 28.3 ÷ 0.07.

   c. Explain in two different ways why the placement of the decimal point that you described in part (b) is valid.

4. a. Calculate 16.8 ÷ 0.35 to the hundredths place without a calculator. Show your work.

   b. Describe the standard procedure for determining where to put the decimal point in the answer to 16.8 ÷ 0.35.

   c. Explain in two different ways why the placement of the decimal point that you described in part (b) is valid.

5. Ramin must calculate 8.42 ÷ 3.6 longhand, but he can't remember what to do about decimal points. Instead, Ramin solves the division problem 842 ÷ 36 longhand and gets the answer 23.38. Ramin knows that he must shift the decimal point in 23.38 somehow to get the correct answer to 8.42 ÷ 3.6. Explain how Ramin could use estimation to determine where to put the decimal point.

6. a. Make a math drawing like Figure 6.30, and use your drawing to help you explain why the division problems 2.15 ÷ 0.36 and 215 ÷ 36 are equivalent.

   b. Use a math drawing to help you explain why the division problems $\frac{7}{8} \div \frac{3}{8}$ and 7 ÷ 3 are equivalent.

   c. Discuss how parts (a) and (b) are related.

7. *A federal debt problem:* If the federal debt is $11 trillion, and if this debt were divided equally among 300 million people, then how much would each person owe? Describe how to estimate mentally the answer to the federal debt problem, and explain briefly why your strategy makes sense.

8. *A tax cut problem:* If a 1.3 trillion dollar tax cut were divided equally among 300 million people over a 10-year period, then how much would each person get each year? Assume that you have only a very simple calculator that cannot use scientific notation and that displays at most 8 digits. Describe how to use such a calculator to solve the tax cut problem, and explain why your solution method is valid.

9. Light travels at about 300,000 kilometers per second. If Pluto is 6 billion kilometers away from us, then how long does light from Pluto take to reach us? Explain why you can calculate the way you do.

10. A newly discovered star is about 75 trillion kilometers away from us. Light travels at about 300,000 kilometers per second. How long does light from the new star take to reach us? Explain why you can calculate the way you do.

11. Light travels at a speed of about 300,000 kilometers per second. The distance that light travels in 1 year is called a *light year*. The star Alpha Centauri is 4.34 light years from earth. How many years would it take a rocket traveling at 6000 kilometers per hour to reach Alpha Centauri? Solve this problem, and explain how you use the meanings of multiplication and division in solving it.

12. Susan has a 5-pound bag of flour and an old recipe of her grandmother's calling for 1 kilogram of flour. She reads on the bag of flour that it weighs 2.26 kilograms. She also reads on the bag of flour that 1 serving of flour is about $\frac{1}{4}$ cup and that there are about 78 servings in the bag of flour.

    a. Based on the information given, how many cups of flour should Susan use in her grandmother's recipe? Solve this problem, and explain how you use the definitions of multiplication and division in solving it.

    b. How can Susan measure this amount of flour as precisely as possible if she has the following measuring containers available: 1 cup, $\frac{1}{2}$ cup, $\frac{1}{3}$ cup, $\frac{1}{4}$ cup measures, 1 tablespoon? Remember that 1 cup = 16 tablespoons.

13. In ordinary language, the term *divide* means "partition and make smaller," as in "Divide and conquer."

    a. In mathematics, does dividing always make smaller? In other words, if you start with a number N and divide it by another number M, is the resulting quotient $N \div M$ necessarily less than N? Explain briefly.

    b. For which positive numbers, M, is $10 \div M$ less than 10? Determine all such positive numbers M. Use the meaning of division (either interpretation) to explain why your answer is correct.

14. When Mary converted a recipe from metric measurements to U.S. customary measurements, she discovered that she needed 8.63 cups of flour. Mary has a 1-cup measure, a $\frac{1}{2}$-cup measure, a $\frac{1}{4}$-cup measure, and a measuring tablespoon, which is $\frac{1}{16}$ cup. How should Mary use her measuring implements to measure the 8.63 cups of flour as accurately and efficiently as possible? Explain your reasoning.

*** 15.** Suppose you need to know how many thirty-secondths ($\frac{1}{32}$) of an inch 0.685 inch is (rounded to the nearest thirty-secondth of an inch). Explain why the following method is a legitimate way to solve this problem:

- Calculate $0.685 \times 32 = 21.92$, and round the result to the nearest whole number, namely, 22.
- Use your result from the previous step, 22, to form the fraction $\frac{22}{32}$. Then 0.685 inch is $\frac{22}{32}$ inch rounded to the nearest thirty-secondth of an inch.

Don't just verify that this method gives the right answer; explain why it works.

16. Sarah is building a carefully crafted cabinet and calculates that she must cut a certain piece of wood 33.33 inches long. Sarah has a standard tape measure that shows subdivisions of one-sixteenth ($\frac{1}{16}$) of an inch. How should Sarah measure 33.33 inches with her tape measure, using the closest sixteenth of an inch? (How many whole inches and how many sixteenths of an inch?) Explain your reasoning.

CHAPTER SUMMARY

Section 6.1 Interpretations of Division — Page 223

- There are two interpretations of division: the how-many-groups interpretation and the how-many-units-in-1-group interpretation. With the how-many-groups interpretation of division, $A \div B$ means the number of Bs that are in A—that is, the number of groups when A units are divided into equal groups with B units in 1 group. With the how-many-units-in-1-group interpretation of division, $A \div B$ means the number of units in 1 group when A objects are divided equally among B groups. — Page 223
- Every division problem can be reformulated as a multiplication problem. With the how-many-groups interpretation,

$$A \div B = ? \text{ is equivalent to } ? \cdot B = A$$

With the how-many-units-in-1-group interpretation,

$$A \div B = ? \text{ is equivalent to } B \cdot ? = A$$

Page 224

Key Skills and Understandings

1. Write and recognize whole number division word problems for both interpretations of division. — Page 223
2. Explain why we can't divide by 0, but why we *can* divide 0 by a nonzero number. — Page 227
3. Use division to solve problems. — Page 229

Section 6.2 Division and Fractions and Division with Remainder — Page 231

- Division and fractions are related by $A \div B = \frac{A}{B}$. Given a whole number division problem, the exact quotient might not be a whole number. In this case, the quotient can be expressed as a mixed number or fraction or as a decimal, or it can be expressed as a whole number with a remainder. — Page 232

Key Skills and Understandings	
1. Explain why it makes sense that the result of dividing $A \div B$ is $\frac{A}{B}$.	Page 232
2. Write and recognize whole number division word problems that are best answered exactly, with either a decimal or a mixed number, or answered with a whole number and a remainder.	Page 233
3. In word problems, interpret quotients and remainders appropriately, and recognize the distinction between solving a numerical division problem that is related to a word problem and solving the word problem.	Page 233
4. Describe how the whole-number-with-remainder answer to a division problem is related to the mixed number answer, and explain why this relationship holds.	Page 233
Section 6.3 Why Division Algorithms Work	Page 239
▪ We can solve whole number division problems in a primitive way by repeatedly subtracting the divisor or repeatedly adding the divisor.	Page 239
▪ The scaffold method is a method that works with entire numbers instead of just portions of them. This method allows for flexibility in calculation but it can also be implemented as the standard division algorithm. We can explain why the scaffold method is valid by using the distributive property to "collect up" multiples of the divisor that were subtracted during the division process.	Page 240
▪ We can interpret each step in the common method for implementing the standard division algorithm by viewing numbers as representing objects bundled according to place value. During this method, we repeatedly unbundle amounts and combine them with the amount in the next-lower place.	Page 242
Key Skills and Understandings	
1. Use the scaffold method of division, and interpret the process from either the how-many-groups or the how-many-units-in-1-group point of view.	Page 240
2. Use the common method for implementing the standard long division algorithm, and interpret the process from the how-many-units-in-1-group point of view. Explain the "bringing down" steps in terms of unbundling the remaining amount and combining it with the amount in the next-lower place.	Page 242
3. Understand and use nonstandard methods of division.	Page 239
4. Use the standard division algorithm to write fractions as decimals and to give decimal answers to whole number division problems.	Page 245
Section 6.4 Fraction Division from the How-Many-Groups Perspective	Page 253
▪ We can interpret fraction division from the how-many-groups viewpoint. This perspective leads to a general method for dividing fractions: give the fractions common denominators and then divide the numerators.	Page 253
Key Skills and Understandings	
1. Write and recognize fraction division word problems for the how-many-groups interpretation of division.	Page 253
2. Solve fraction division word problems with the aid of math drawings, tables, and double number lines. Know how to interpret math drawings appropriately.	Page 254
3. Explain why we can divide fractions by giving them a common denominator and then dividing the numerators.	Page 254
4. Explain why we can divide fractions by dividing numerators and dividing denominators.	Page 256

Section 6.5 Fraction Division from the How-Many-Units-in-1-Group Perspective	Page 261
▪ We can interpret fraction division from the how-many-units-in-1-group viewpoint. The "invert and multiply" or "multiply by the reciprocal" procedure is nicely explained from this perspective.	Page 261
Key Skills and Understandings	
1. Write and recognize fraction division word problems for the how-many-units-in-1-group as well as from the how-many-groups interpretations.	Page 261
2. Solve division problems with the aid of math drawings, tables, and double number lines as well as numerically. Know how to interpret visual supports appropriately.	Page 261
3. Use the how-many-units-in-1-group interpretation of division to explain why the "invert and multiply" procedure for dividing fractions is valid.	Page 263
Section 6.6 Dividing Decimals	Page 270
▪ We can interpret decimal division from the how-many-groups as well as from the how-many-units-in-1-group viewpoint. To divide decimals, we move the decimal points in both the divisor and dividend the same number of places so that the divisor becomes a whole number; then we proceed as with whole number division. There are several ways to explain this process of shifting the decimal points in decimal division.	Page 270
Key Skills and Understandings	
1. Write decimal division word problems for both interpretations of division.	Page 270
2. Explain in several different ways why we move the decimal points the way we do when we divide decimals, and calculate decimal divisions (without a calculator).	Page 271
3. Use estimation to determine the location of the decimal point in a decimal division problem.	Page 273
4. Explain why many division problems are equivalent, such as $6000 \div 2000$, $600 \div 200$, $60 \div 20$, $6 \div 2$, and $0.6 \div 0.2$.	Page 272

7

Ratio and Proportional Relationships

Ratios and proportional relationships are essential to mathematics and to all of science, and they are useful in daily life. They are a foundation for understanding rate of change, slope, linear-relationships, and other relationships in mathematics and science. In this chapter, we will motivate and define the concepts of ratio and proportional relationship. We will then see ways of reasoning to solve proportion problems with the aid of tables, graphs, double number lines, and strip diagrams. We will see how unit rates arise from ratios, how they connect ratios to fractions, and how they are behind the common cross-multiplying method for solving proportions. We will see that graphs of proportional relationships are of a special type and we will develop equations for proportional relationships by reasoning quantitatively. We will distinguish proportional relationships from other kinds of relationships, including inversely proportional relationships. Finally, we will study percent increase and decrease.

We focus on the following topics and practices within the *Common Core State Standards for Mathematics*.

Standards for Mathematical Content in the CCSSM

In the domain of *Ratio and Proportional Relationships* (Grades 6 and 7), students learn ratio concepts, including the concept of unit rate associated with a ratio, and they use ratio language, such as "3 cups flour to 2 cups water" and "3 cups flour for every 2 cups water." They reason about ratios and rates with the aid of tables, double number lines, and strip diagrams to solve problems. They analyze and graph proportional relationships and they distinguish them from other kinds of relationships. They also use proportional relationships to solve multistep percent problems, such as problems involving percent increase and decrease.

Standards for Mathematical Practice in the CCSSM

Opportunities to engage in all eight of the Standards for Mathematical Practice described in the *Common Core State Standards* occur throughout the study of ratio and proportional relationships, although the following standards may be especially appropriate for emphasis:

- **2 Reason abstractly and quantitatively.** Students engage in this practice when they use a ratio to describe a quality that mixtures or other related quantities have in common, and when they recognize that a single ratio can apply to both small and large amounts of the mixture or the related quantities.
- **4 Model with mathematics.** Students engage in this practice when they use ratios and proportional relationships to model situations and when they examine relationships critically to determine if they are proportional or not and why or why not.
- **5 Use appropriate tools strategically.** Students engage in this practice when they make and reason logically about ratio tables, double number lines, and strip diagrams as they solve problems involving ratios and proportional relationships.

  (From Common Core Standards for Mathematical Practice. Published by Common Core Standards Initiative.)

7.1 Motivating and Defining Ratio and Proportional Relationships

CCSS Common Core State Standards Grades 6, 7, and 8

What are ratios and why do we have the concept of ratio? In this section, we'll see some ways that ratios arise and we'll define what ratio means. In the same way that there are two ways to interpret the meaning of division (how-many-groups and how-many-units-in-1-group), there are two ways to interpret the meaning of ratio.

The concept of *ratio* allows us to describe mathematically certain qualities of quantities that are mixed, combined, or related. Ratios can describe very different kinds of qualities, such as the flavor of a juice mixture, the color of a paint mixture, a speed of walking, or the density of a substance. In each case, there can be small or large amounts of the substances that have the same quality and are in the same ratio. A small amount of yellow and blue paint mixed together could be the same shade of green as a larger amount of paint; a ratio of yellow to blue paint specifies that shade of green. You could walk for a short time and distance at the same speed as a longer time and distance; a ratio of distance to time describes that walking speed. A small piece of pine wood has the same density as a large piece of the same kind of pine wood; a ratio of mass to volume describes that density. A mathematical definition of ratio should help us determine if quantities have the same quality, such as the same color or flavor or if the quantities have different qualities.

We can notate ratios in several ways:

CCSS 6.RP.1

- We can use words, as in "7 to 2" or "7 meters in 2 seconds" or "7 meters for every 2 seconds" or "7 parts flour to 2 parts milk."
- We can use a colon, as in 7:2.

Although it is common to write ratios in fraction notation (such as $\frac{7}{2}$), we will keep ratio and fraction notation separate because they have different meanings. In Section 7.3, we will see how to connect ratios and fractions.

How Can We Define Ratio from the Multiple-Batches Perspective?

What does it mean for two quantities to be in a specific ratio? Let's use an example to motivate a definition. Suppose we have a mixture of 3 cups grape juice and 5 cups peach juice. What other amounts of grape and peach juice will make a mixture of the same flavor and color? If we think of the original mixture as *1 batch*, and we make several copies of that batch and combine them, that combination will have the same flavor and color as the original (see Figure 7.1). If we take $\frac{1}{2}$ of the original mixture, or some other portion, that will also have the same flavor and color as the original. In other words, all mixtures made from

$$N \cdot 3 \text{ cups grape juice, } N \cdot 5 \text{ cups peach juice}$$

where N is any positive number of batches, will have the same flavor and color. We can think of the specific flavor and color of all those mixtures as encoded by the single pair 3, 5.

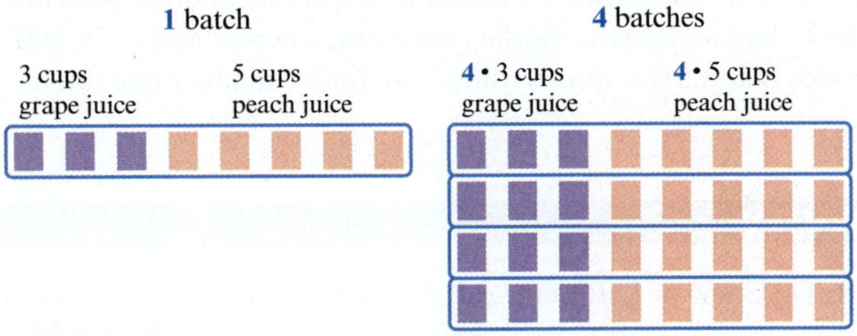

Figure 7.1 A mixture of 4 batches has the same flavor and color as a mixture of 1 batch of grape and peach juice.

In general, from the multiple-batches perspective, if we have A units of one quantity and B units of a second quantity (where A and B are positive numbers) we consider the two together as 1 batch, or 1 composed unit (a unit formed from two or more units). We say that two quantities are in the **ratio** A to B if there are

ratio (multiple-batches)

$$N \cdot A \text{ units of the first quantity, } N \cdot B \text{ units of the second quantity}$$

for some positive number of batches N (where N is the same in $N \cdot A$ and $N \cdot B$). Here we are using our definition of multiplication, so that N is a number of groups (batches), and A and B are numbers of units in 1 group.

With the multiple-batches perspective, we think of 1 batch (or composed unit) as fixed and we *vary the number of batches* to get varying quantities in the same ratio (see Figure 7.2).

With the multiple-batches perspective on ratio, the units of measurement for the quantities can be the same or different. For example, a ratio

$$5 \text{ meters in } 2 \text{ seconds}$$

describes a speed. In this case it makes more sense to think of "5 meters in 2 seconds" as a composed unit than as a batch. The speed 15 meters in 6 seconds is in the same ratio because it is

$$3 \cdot 5 \text{ meters in } 3 \cdot 2 \text{ seconds}$$

The speed 2.5 meters in 1 second is in the same ratio because it is

$$\frac{1}{2} \cdot 5 \text{ meters in } \frac{1}{2} \cdot 2 \text{ seconds}$$

Number of batches	Cups grape	Cups peach
1 batch	1 • 3	1 • 5
2 batches	2 • 3	2 • 5
3 batches	3 • 3	3 • 5
4 batches	4 • 3	4 • 5
N batches	N • 3	N • 5

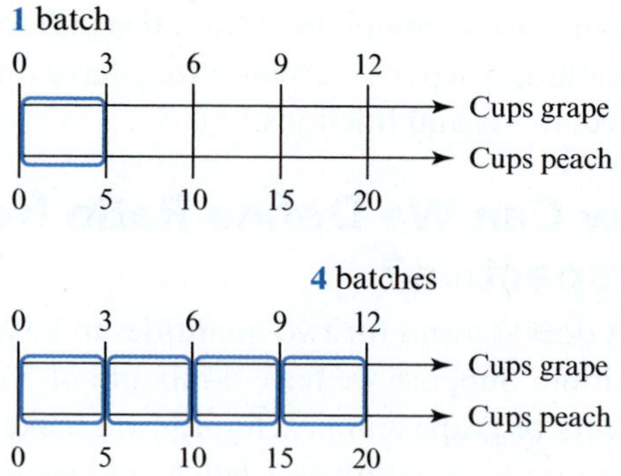

Figure 7.2 A table and double number lines showing varying quantities in the same ratio.

ratio table

When we view a ratio from the multiple-batches perspective, we often display values that are in that ratio in a *ratio table* or on a *double number line*, as in Figure 7.2. A **ratio table** is a coordinated arrangement of entries that are in the same ratio and are displayed horizontally or vertically. A composed unit or a batch is represented by a set of coordinated entries in a ratio table and by a pair of coordinated intervals on a double number line. Double number lines are much like tables, except that they show the relative sizes of the quantities, whereas entries in tables need not be organized by size. Notice that the two number lines in a double number line can have different scales, as in Figure 7.2.

CLASS ACTIVITY

7A Mixtures: The Same or Different? p. CA-118

How Can We Define Ratio from the Variable-Parts Perspective?

Recipes are sometimes described in terms of parts, such as mixing 1 part rice with 3 parts water or mixing 1 part cement with 2 parts sand and 4 parts gravel. In these cases, 1 part could be any size we like; it could be the size of a small container if we only need a little or it could be the size of a large container if we need a lot. This motivates our second definition of ratio. Again, consider the mixture of 3 cups grape juice, 5 cups peach juice, and the question of what other amounts of grape and peach juice will make a mixture of the same flavor and color. If we think of the original mixture as *3 parts* grape juice and *5 parts* peach juice and if we change the size of each part from 1 cup to, say, 4 cups, then the new mixture still follows the same overall recipe and so will have the same flavor and color (see Figure 7.3). In fact, if the size of each part is $\frac{1}{2}$ a cup or any other number of cups, a mixture of

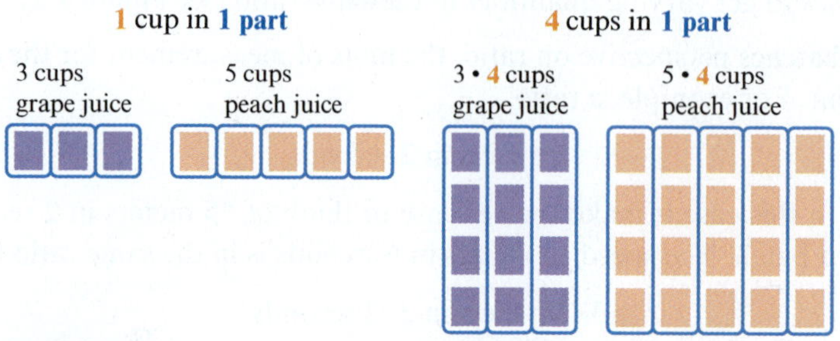

Figure 7.3 A mixture of 3 parts grape juice and 5 parts peach juice has the same color and flavor whether the size of each part is 1 cup or 4 cups.

3 parts grape juice and 5 parts peach juice will have the same flavor and color as the original as long as all parts are the same size as each other. In other words, all mixtures made from

$$3 \cdot N \text{ cups grape juice}, \quad 5 \cdot N \text{ cups peach juice}$$

where N is any positive number of cups, will have the same flavor and color.

Notice that the cups of juice are identical in **Figures 7.1** and **7.3**; they are just grouped differently. In **Figure 7.1**, the cups are grouped into batches of grape and peach juice. In Figure 7.3, the cups are grouped into 3 parts grape juice, 5 parts peach juice.

ratio (variable-parts)

In general, from the variable-parts perspective, if A and B are positive numbers, we say that two quantities are in the **ratio** A to B if there are

$$A \cdot N \text{ units of the first quantity}, \quad B \cdot N \text{ units of the second quantity}$$

for some positive number of units N (where N is the same in $A \cdot N$ and $B \cdot N$). Here we are using our definition of multiplication, so that A and B are numbers of groups (parts) and N is the number of units in 1 group.

With the variable-parts perspective, we think of the number of parts for both quantities as fixed and we *vary the size of the parts* to get varying quantities in the same ratio (see **Figure 7.4**). When we use a strip diagram to represent a ratio, as in **Figure 7.4**, each part can represent any positive number of units, and so is like a variable, such as x, but is visual.

Number of cups in 1 part	Cups grape	Cups peach
1 cup	$3 \cdot 1$	$5 \cdot 1$
2 cups	$3 \cdot 2$	$5 \cdot 2$
3 cups	$3 \cdot 3$	$5 \cdot 3$
4 cups	$3 \cdot 4$	$5 \cdot 4$
N cups	$3 \cdot N$	$5 \cdot N$

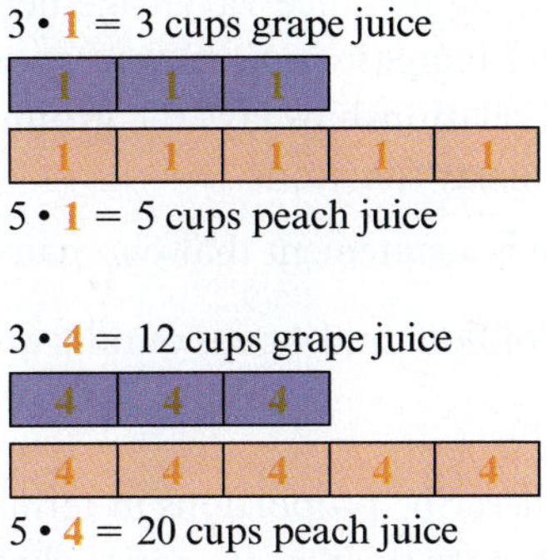

Figure 7.4 A table and strip diagrams showing varying quantities in the same ratio.

Notice that we get the same quantities in a given ratio whether we take a multiple-batches or a variable-parts perspective. The difference is in how we describe the quantities as products and how we describe the variation within a fixed ratio. The different viewpoints will lead to different approaches to solving ratio problems.

Are Part-to-Part Ratios Different from Part-to-Whole?

In the peach-and-grape-juice examples so far we only considered quantities of grape and peach juice, but we could also consider the total quantities of juice in these mixtures. To do so, we could add a column to the tables in **Figures 7.2** and **7.4**. In **Figure 7.2** that third column would have entries

$$1 \cdot 8, \quad 2 \cdot 8, \quad 3 \cdot 8, \quad 4 \cdot 8, \quad N \cdot 8$$

because there are 8 total cups of juice mixture in 1 batch (3 cups grape juice and 5 cups peach juice). We could also add a third number line to the double number line in **Figure 7.2** to show the total cups of juice mixture.

In Figure 7.4 a third column showing the total quantity of juice would have entries

$$8 \cdot 1, \quad 8 \cdot 2, \quad 8 \cdot 3, \quad 8 \cdot 4, \quad 8 \cdot N$$

because there are 8 total parts of juice (3 parts grape juice and 5 parts peach juice). Even though some of those parts represent grape juice and some represent peach juice, each part represents the same number of cups. So we can treat the parts equally in terms of size. If we want to, we can change the strip diagram by putting the grape and peach juice portions of the diagram directly next to each other, side by side, to highlight the total quantity of juice.

The same definitions apply whether we are considering part-to-part or part-to-whole ratios.

What Are Proportional Relationships and Proportions?

When we use a ratio table or a double number line to display several pairs of quantities that are in a given ratio, and when we understand that the parts in a strip diagram stand for various amounts, we begin to get a sense of how the quantities are related and how they change together in a coordinated way. A relationship between two quantities is called a **proportional relationship** if all coordinated pairs of values of the quantities are in the same ratio. So the coordinated pairs of quantities in a ratio table or on a double number line or represented by a strip diagram are in a proportional relationship. We can think of the pairs of quantities in a proportional relationship as *varying together*. For example, if a person walks 5 meters every 2 seconds, then the person's walking times and distances are in a proportional relationship. The person's walking time and walking distance vary together: As the time changes, so does the distance—namely, every 2 second increase in time corresponds with a 5 meter increase in distance. Working with proportional relationships lays the groundwork for understanding the concepts of slope, rate of change, and linear functions.

proportional relationship

A **proportion** is a statement that two pairs of amounts are in the same ratio. For example,

proportion

"14 cups of flour and 4 cups of milk are in the same ratio as 7 cups of flour and 2 cups of milk"

is a proportion.

We can also describe proportions in terms of *equivalent ratios.* If two pairs of amounts are in the same ratio, then we say that the ratios that each pair specifies are **equivalent**. For example, a mixture of 7 cups flour with 2 cups milk is in the same ratio as a mixture of 14 cups flour with 4 cups milk, so the two ratios 7 to 2 and 14 to 4 are equivalent. Proportions are often described as a statement that two ratios are equivalent, such as

equivalent ratios

$$7:2 = 21:X$$

or as a statement that two fractions are equivalent, such as

$$\frac{7}{2} = \frac{21}{X}$$

Section 7.3 will discuss how ratios and fractions are related, and we will explain why it makes sense to specify a proportion in terms of equivalent fractions.

Solving a proportion is the process of finding an unknown value in a proportion in which the other three values are known. If you are asked what the value of X is such that the ratio of 21 cups of flour to X cups of milk is 7 to 2, then you are being asked to solve a proportion. When you find that the value for X is 6, then you have solved the proportion.

solving a proportion

The next section will focus on ways of reasoning about multiplication and division with quantities to solve proportions.

SECTION SUMMARY AND STUDY ITEMS

Section 7.1 Motivating and Defining Ratio and Proportional Relationships

Ratios provide us with a mathematical way to describe qualities of quantities that are combined or related. There are two perspectives on ratio: a multiple-batches perspective and a variable-parts perspective. We can use tables, double number lines, and strip diagrams to represent and compare ratios and proportional relationships.

Key Skills and Understandings

1. Using a ratio table and double number-line as a support, explain what it means for quantities to be in a certain ratio from the multiple-batches perspective. Using a ratio table and strip diagram as a support, explain what it means for quantities to be in a certain ratio from the variable-parts perspective.
2. Given a ratio, find several amounts in that ratio and record them in a table, on a double number line, or on a graph. Recognize that the collection of amounts in the same ratio form a proportional relationship.
3. Compare the qualities of quantities in different ratios by reasoning about tables. For example, compare two walking speeds or two different mixtures of paints or juices.
4. Recognize the common error of making an additive comparison in a ratio situation.

Practice Exercises for Section 7.1

1. Describe mixtures of liquid soap and rose water in the ratio of 5 to 2 from the multiple-batches perspective and from the variable-parts perspective.
2. Lavender oil and bergamot can be mixed in a ratio of 2 to 3. Give three different examples of amounts of lavender oil and bergamot that are mixed in the same ratio. Explain why your examples really are in the same ratio from a variable-parts perspective. Use a strip diagram to illustrate.
3. Give an example of how to use a double number line to show a proportional relationship.
4. Which of the following two mixtures will be more lemony?
   - 2 tablespoons of lemon juice mixed in 3 cups of water
   - 5 tablespoons of lemon juice mixed in 7 cups of water

   Use ratio tables to solve this problem. Explain your reasoning.

Answers to Practice Exercises for Section 7.1

1. Variable parts perspective: The mixtures are made of 5 parts liquid soap and 2 parts rose water, where all the parts are the same size as each other, but can be any size. Visually, we can show this description with a strip diagram, as in Figure 7.5. We can think of each part in the strip diagram as "filled with" any (common) amount such as 5 drops, or $\frac{1}{2}$ cup, or 3 gallons, or any other amount.

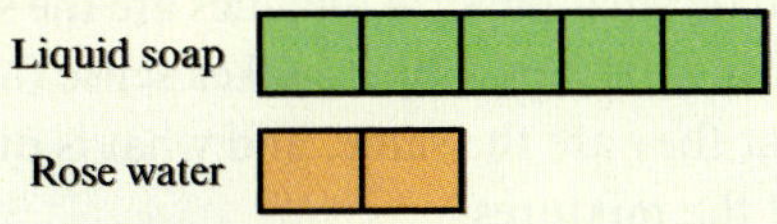

Figure 7.5 The ratio 5 to 2.

Multiple-batches perspective: With this point of view, we think of the mixture as made of a certain number of batches, where 1 batch is made of 5 teaspoons liquid soap and 2 teaspoons rose water. Also, for every 5 teaspoons (say) of liquid soap that are in the mixture, there are 2 teaspoons of rose water in the mixture.

2. The strip diagram in Figure 7.6 can help depict this ratio. If each part is filled with 5 drops, then there will be 10 drops lavender oil and 15 drops bergamot. If each part is filled with $\frac{1}{2}$ cup, then there will be 1 cup lavender oil and $1\frac{1}{2}$ cups bergamot. If each part is filled with $\frac{1}{3}$ of a cup, then there will be $\frac{2}{3}$ of a cup of lavender oil and 1 cup bergamot. All of these pairs are in the same ratio because all apply to the same strip diagram that defines the ratio of 2 to 3. In each example there are 2 parts lavender oil and 3 parts bergamot, where all parts in that example are the same size.

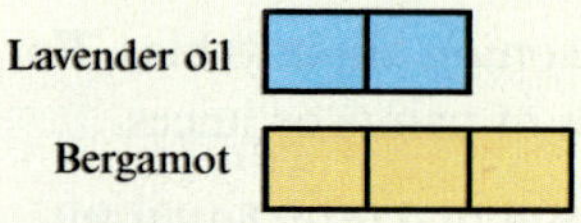

Figure 7.6 The ratio 2 to 3.

3. Suppose you can buy a liquid soap for \$2.25 for every 24 fluid ounces. Figure 7.7 shows a double number line with dollar amounts and numbers of fluid ounces in the ratio 2.25 to 24.

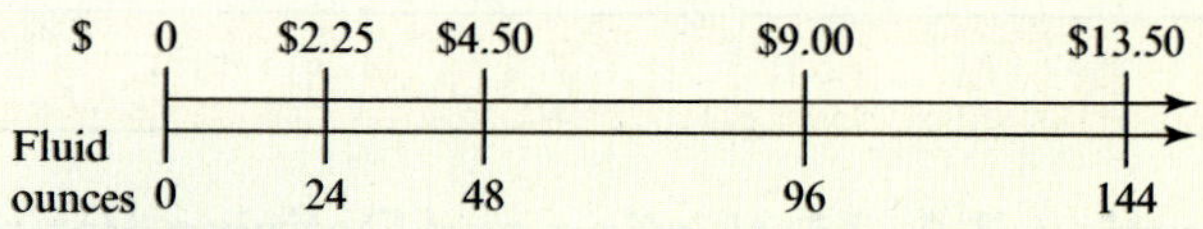

Figure 7.7 A double number line showing a proportional relationship.

4. The mixtures within a ratio table all taste the same because they are just multiple batches of the first mixture. So we can compare any pair in the mixture 1 table with any pair in the mixture 2 table. In the first ratio table, 10 tablespoons of lemon juice requires 15 cups of water, whereas in the second table, the same number of tablespoons of lemon juice requires only 14 cups of water. Therefore, the second mixture is more lemony because it has less water for the same amount of lemon than the first mixture does. You can also compare the two mixtures when they have the same amount of water.

Table for Mixture 1

Tbs lemon juice	2	4	6	8	10	12	14	16	18
Cups water	3	6	9	12	15	18	21	24	27

Table for Mixture 2

Tbs lemon juice	5	10	15	20	25	30	35	40	45
Cups water	7	14	21	28	35	42	49	56	63

PROBLEMS FOR SECTION 7.1

1. Use the multiple-batches perspective to discuss and describe in your own words several different mixtures of chocolate syrup and milk that are in a 2 to 7 ratio. Use a double number line and a table to support your discussion. Include a discussion of what physical qualities are the same for all those mixtures, why it makes sense (intuitively) that they are the same, and what is different about the mixtures.

2. Use the variable-parts perspective to discuss and describe in your own words several different mixtures of chocolate syrup and milk that are in a 2 to 7 ratio. Use a strip diagram and a table to support your discussion. Include a discussion of what physical qualities are the same for all those mixtures, why it makes sense (intuitively) that they are the same, and what is different about the mixtures.

3. John is driving at a constant speed of 10 miles every 12 minutes. On a double number line and an accompanying table show at least three other distances and times in the ratio 10 miles to 12 minutes. Include at least one example that involves one or more decimals, fractions, or mixed numbers. Use the multiple-batches perspective to explain why your entries are all in the 10 to 12 ratio.

4. You can make grape juice by mixing 1 can of frozen grape juice concentrate with 3 cans of water.

a. Make a ratio table that shows at least four other mixtures of grape juice concentrate and water in the same ratio. At least two of your mixtures should involve numbers that are not whole numbers.

b. Explain how to interpret your table in part (a) from the multiple-batches perspective using a double (or triple) number line.

c. Explain how to interpret your table from the variable-parts perspective using a strip diagram.

5. Use ratio tables to determine in *two ways* which of the next two snails is moving faster without using fractions. Explain your reasoning clearly in your own words.

- Snail A moves 3 inches every 2 minutes.
- Snail B moves 4 inches every 3 minutes.

6. Explain how to reason about ratio tables to determine which of the following two laundry detergents is a better buy:

- a box of laundry detergent that washes 80 loads and costs $12.75
- a box of laundry detergent that washes 36 loads and costs $6.75

7. Allie, Benton, and Cathy are planning to mix red and yellow paint. They are considering which of the two following paint mixtures will make a more yellow paint:

- a mixture of 3 cups red and 5 cups yellow
- a mixture of 4 cups red and 6 cups yellow

Allie says that both paints will look the same because to make the second mixture you just add 1 cup of each color to the first mixture. Benton says that the second mixture should be more yellow than the first because it uses more yellow than the first mixture. Cathy says that both paints should look the same because each uses 2 cups more yellow than red.

a. Discuss the students' ideas. Is their reasoning valid or not?

b. Which paint will be more yellow, and why? Use a ratio table to solve this problem in two different ways, explaining in detail why you can solve the problem the way you do.

8. Company A charges $6 for every 100 silly bands. Company B charges $8 for every 125 silly bands.

a. Make two ratio tables to show amounts of dollars and silly bands in the same ratios as offered by Company A and Company B.

b. From which company do you get a better buy on silly bands? Describe several ways you can tell from the tables.

9. You can make a pink paint by mixing $2\frac{1}{2}$ cups white paint with $1\frac{3}{4}$ cups red paint.

a. Make a ratio table that shows at least four other mixtures of white and red paint in the same ratio that will make the same shade of pink paint.

b. Explain how to interpret your table in part (a) from the multiple-batches perspective using a double (or triple) number line.

c. Explain how to interpret your table from the variable-parts perspective using a strip diagram.

You may want to use an equivalent ratio to draw your strip diagram.

7.2 Solving Proportion Problems by Reasoning with Multiplication and Division

CCSS Common Core State Standards Grades 6, 7, and 8

CCSS SMP2

When you think of solving proportions, you may think of the method in which you set two fractions equal to each other, cross-multiply these fractions, and then solve the resulting equation. However, we can also solve proportions by reasoning about multiplication and division with quantities. When students solve proportion problems this way, they have an opportunity to think more deeply about the operations of multiplication and division and when to apply these operations. They also have an opportunity to work with proportional relationships and to develop a feel for how quantities can vary together, which is a foundation for learning about linear relationships and functions.

In this section we will study several methods of reasoning to find quantities in the same ratio and thus to solve proportions. An initial method is to apply repeated addition. More advanced ways of reasoning use multiplication and division. We'll see how strip diagrams and double number lines can support several methods of reasoning about multiplication and division with quantities.

What Are Initial Ways of Reasoning with Tables, Number Lines, and Strip Diagrams to Solve Proportion Problems?

Let's think about ways we can reason to solve a problem:

> *Gas Mileage Problem:* If Joe's car goes 100 miles for every 4 gallons of gas, then how far can Joe drive on 20 gallons of gas?

Using a table or a double number line, as in Figure 7.8, we can repeatedly add "batches" of 100 miles and 4 gallons until we get to 20 gallons.

+ 100 + 100 + 100 + 100

Miles	100	200	300	400	500
Gallons	4	8	12	16	20

+ 4 + 4 + 4 + 4

Every additional 100 miles takes an additional 4 gallons.

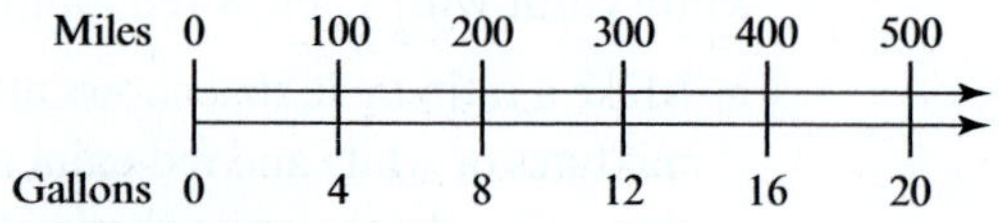

Figure 7.8 Using repeated addition to find amounts in the same ratio.

What if we now ask how many gallons of gas Joe would need for 800 miles? Instead of continuing to add increments of 100 miles and 4 gallons to the table or double number line, we could reason that 800 miles is 2 groups of 400 miles, so Joe will need 2 groups of 16 gallons or 32 gallons of gas, as in Figure 7.9.

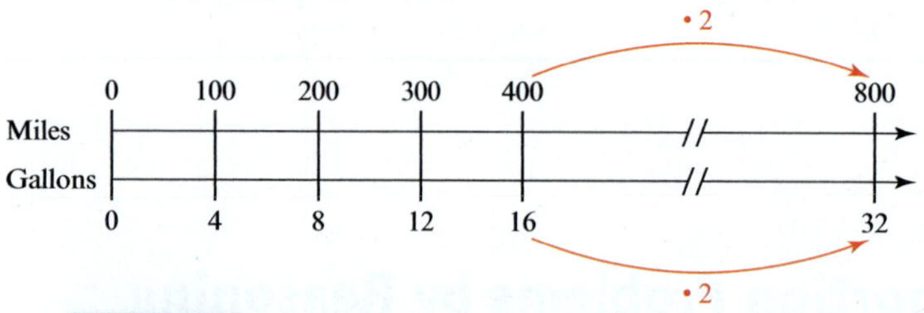

Double the miles takes double the gallons:

• 2

Miles	100	200	300	400	800
Gallons	4	8	12	16	32

• 2

Figure 7.9 Doubling quantities to find new quantities in the same ratio.

What if the Gas Mileage Problem asked about 18 gallons of gas or 15 gallons of gas? We could add more tick marks to the double number line, as in Figure 7.10. Or we could reason that because Joe can drive 400 miles on 16 gallons, and 15 gallons is 1 gallon less than 16 gallons, we should first find out how far Joe can drive on 1 gallon of gas, as in Figure 7.11. By distributing the 100 miles equally across 4 gallons, we see that Joe can drive 25 miles on 1 gallon. So on 15 gallons, Joe will drive 25 miles less than 400 miles, so only 375 miles.

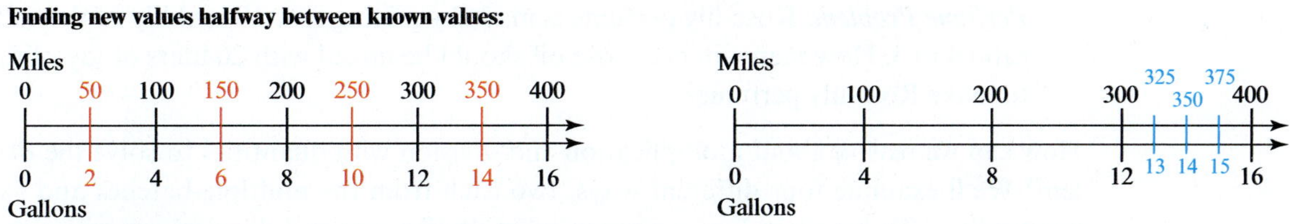

Figure 7.10 Finding new values between known values.

Miles	100	200	300	400	25	375
Gallons	4	8	12	16	1	15

Joe has 1 gallon less than 16 gallons,
so he can go 25 miles less than 400 miles.

Figure 7.11 Using known values to find new values.

We can also use strip diagrams to reason about quantities. For example, consider this problem:

> Rose-lily perfume is made by mixing rose oil and lily of the valley oil in the ratio 4 to 3. How many liters of rose oil should be mixed with 24 liters of lily of the valley oil to make Rose-lily perfume?

To represent the 4 to 3 ratio, we can draw a strip diagram with 4 parts rose oil and 3 parts lily oil. At first it might seem like each part should represent 1 liter of oil, but if we take a variable-parts perspective, a part can represent any amount. In Figure 7.12, the 3 parts of lily of the valley oil represent a total of 24 liters. Because all the parts are the same size, we can divide the 24 liters equally among the 3 parts, so each part contains $24 \div 3$ liters or 8 liters. The rose oil consists of 4 parts, so is $4 \cdot 8$ liters or 32 liters.

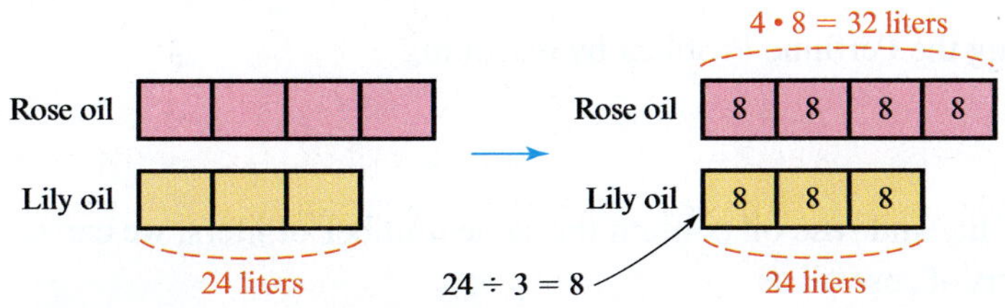

Figure 7.12 Using a strip diagram to solve a proportion problem.

CLASS ACTIVITY

7B Using Double Number Lines to Solve Proportion Problems, p. CA-120

7C Using Strip Diagrams to Solve Proportion Problems, p. CA-121

How Can We Reason about Multiplication and Division with Quantities to Solve Proportion Problems?

Although students may initially use repeated addition to solve simple proportion problems, it's important to progress into more powerful ways of reasoning with multiplication and division. For example, we can't solve the following problem using only repeated addition.

Perfume Problem: Rose-lily perfume is made by mixing rose oil and lily of the valley oil in the ratio 4 to 3. How many liters of rose oil should be mixed with 20 liters of lily of the valley oil to make Rose-lily perfume?

How can we reason about multiplication and division with quantities to solve the Perfume Problem? We'll examine four different ways, two each from the multiple-batches and variable-parts perspectives. To highlight how we are using multiplication and division, we will express the solutions as products of numbers derived from the numbers 4, 3, and 20 in the problem statement, attending closely to our definition of multiplication (and division). This will help us focus on the structure of the solution method rather than on the final numerical answer. A focus on structure becomes important later, when we want to derive and explain equations that use variables. Before you read on, see how many methods you can find!

CCSS
SMP7
6.RP.3

Variable-parts: Multiply 1 Part Taking a variable-parts perspective, we can use a strip diagram like the one in Figure 7.12. This time the 3 parts lily oil represent 20 liters (see Figure 7.13). As before, we use how-many-units-in-1-group division to divide the 20 liters equally among the 3 parts. The number of liters in each part is therefore

$$20 \div 3 \quad \text{or} \quad \frac{20}{3} \quad \text{or} \quad 6\frac{2}{3}$$

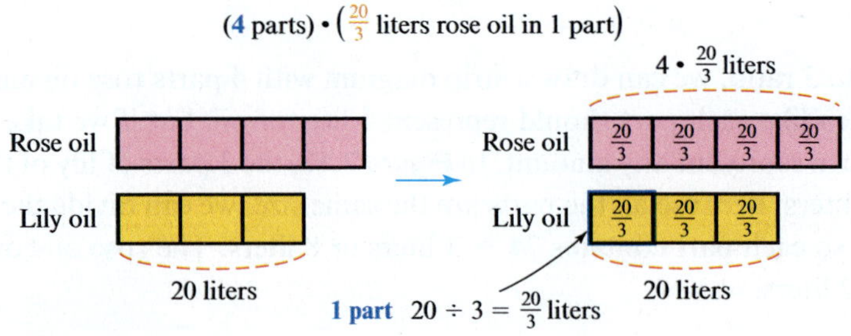

Figure 7.13 Solving the Perfume Problem by reasoning with 1 part.

Because all parts of lily and rose oil contain the same number of liters, we can express the number of liters in the 4 parts of rose oil as

$$4 \cdot \frac{20}{3} \quad \text{or} \quad 4 \cdot 6\frac{2}{3}$$

Multiple-batches: Multiply 1 Batch Taking a multiple-batches perspective, we can use a double number line to show that 1 batch of Rose-lily perfume consists of 4 liters rose oil and 3 liters of lily oil as in Figure 7.14. Because 1 batch of perfume has 3 liters of lily oil and we want 20 liters of lily oil, dividing 20 liters by 3 liters using how-many-groups division tells us how many batches of perfume we need. That number of batches is therefore

$$20 \div 3 \quad \text{or} \quad \frac{20}{3} \quad \text{or} \quad 6\frac{2}{3}$$

Because 1 batch of Rose-lily perfume contains 4 liters of rose oil, we can express the number of liters of rose oil in all those batches as

$$\frac{20}{3} \cdot 4 \quad \text{or} \quad 6\frac{2}{3} \cdot 4$$

Figure 7.14 Solving the Perfume Problem by reasoning with 1 batch.

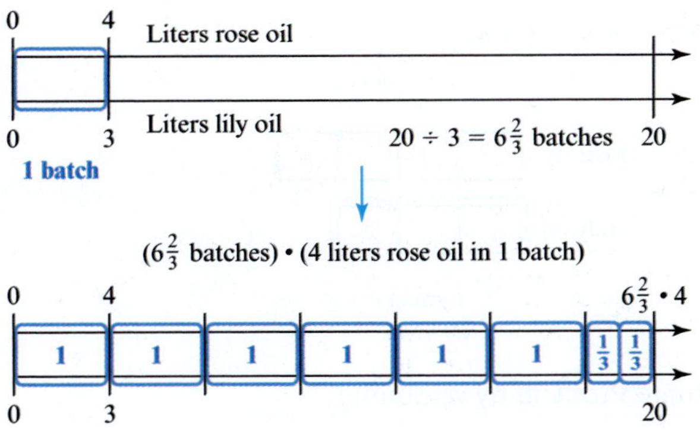

Notice that the roles of the numbers 4 and $\frac{20}{3}$ are the reverse of their roles in the previous method, where we reasoned about 1 part. In the previous method, 4 was the number of groups and $\frac{20}{3}$ was the size of 1 group. In this method, $\frac{20}{3}$ is a number of groups, and 4 is the size of 1 group.

Multiple-batches: Multiply 1 Unit-rate-batch Taking a multiple-batches perspective, we can first find how much rose oil we need for 1 liter of lily oil as in Figure 7.15. We can view the 4 liters of rose oil in 1 batch as distributed equally across the 3 liters of lily oil. How-many-units-in-1-group division therefore tells us that for 1 liter of lily oil the number of liters of rose oil is

$$4 \div 3 \quad \text{or} \quad \frac{4}{3} \quad \text{or} \quad 1\frac{1}{3}$$

Figure 7.15 Solving the Perfume Problem by reasoning with 1 unit-rate-batch.

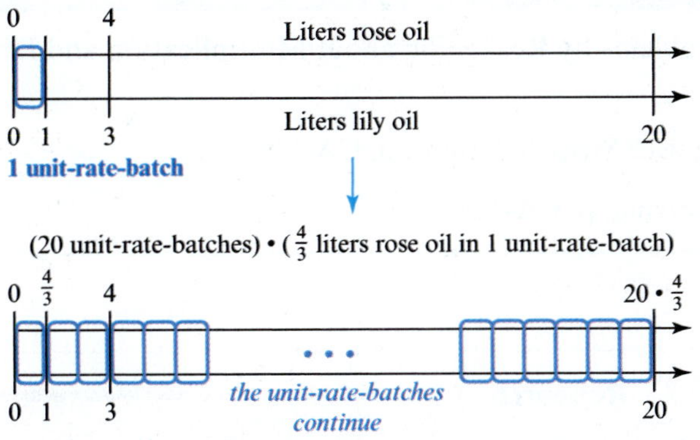

We can think of the combination of $\frac{4}{3}$ liter rose oil and 1 liter lily oil as a "unit-rate-batch." Because we need to use 20 liters of lily oil we will need 20 of these unit-rate-batches, so we can express the number of liters of rose oil as

$$20 \cdot \frac{4}{3} \quad \text{or} \quad 20 \cdot 1\frac{1}{3}$$

Why did we find the amount of rose oil for 1 liter of lily oil? Why not find the amount of lily oil for 1 liter of rose oil? It's because we can easily relate 1 liter of lily oil to any other number of liters of lily oil that we know, such as 20 liters.

Variable-parts: Multiply 1 Total Amount Taking a variable-parts perspective, we can view the total amount of lily oil as 1 group of size 20 liters as in Figure 7.16. Because 1 group of lily oil is 3 parts and we want 4 parts of rose oil, dividing 4 parts by 3 parts using how-many-groups division tells us how many groups of lily oil make the amount of rose oil we need. That number of groups is

$$4 \div 3 \quad \text{or} \quad \frac{4}{3} \quad \text{or} \quad 1\frac{1}{3}$$

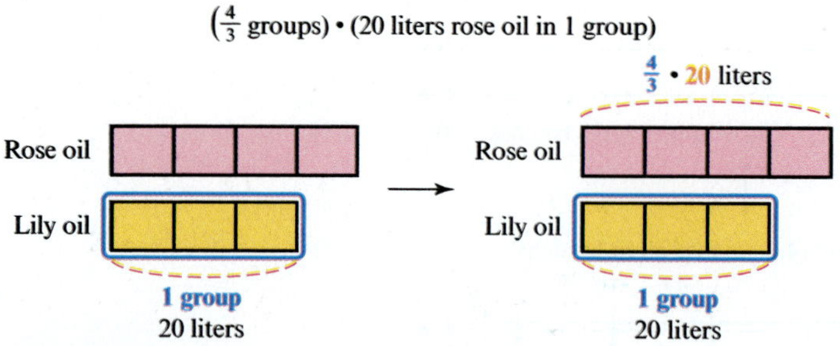

Figure 7.16 Solving the Perfume Problem by reasoning with 1 total amount.

We can also just see the 4-part rose oil strip as 4 parts, each of size $\frac{1}{3}$ of the lily oil strip. Because we need $\frac{4}{3}$ groups of rose oil and 1 group is 20 liters, we can express the number of liters of rose oil we need as

$$\frac{4}{3}\cdot 20 \quad \text{or} \quad 1\frac{1}{3}\cdot 20$$

Notice that the roles of the numbers 20 and $\frac{4}{3}$ are the reverse of their roles in the previous method, where we reasoned about a unit-rate-batch. In the previous method, 20 was the number of groups and $\frac{4}{3}$ was the size of 1 group. In this method, $\frac{4}{3}$ is a number of groups and 20 is the size of 1 group.

CLASS ACTIVITY

7D Solving Proportion Problems by Reasoning about Multiplication and Division with Quantities, p. CA-122.

7E Ratio Problem Solving with Strip Diagrams, p. CA-123

7F More Ratio Problem Solving, p. CA-124

FROM THE FIELD Research

Riehl, S. M., & Steinthorsdottir, O. B. (2014). Revisiting Mr. Tall and Mr. Short. *Mathematics Teaching in the Middle School, 20*(4), p. 220–228.

This article discusses and categorizes the solution strategies of 412 middle-grade students on the classic "Mr. Tall and Mr. Short" problem. The problem asks for the height of Mr. Tall in terms of paperclips given that Mr. Short is both 4 matchsticks and 6 paperclips tall and Mr. Tall is 6 matchsticks tall. The authors identify 6 categories of solution strategies: illogical, additive, build-up, multiplicative, ambiguous, and multiplicative-ambigous, of which the illogical and additive strategies are incorrect. For example, some students used an additive strategy when they reasoned that you have to add 2 to the number of matchsticks to get the number of paperclips. These students concluded incorrectly that Mr. Tall is 8 paperclips tall. Most students from grades 5 through 8 used incorrect illogical or additive strategies, although the percentage of students using correct multiplicative strategies increased across the grades. The authors provide a number of variations on the Mr. Tall and Mr. Short problem and recommend that teachers try some of these variations depending on how they think their students might be reasoning. The variations are designed to help teachers refine their assessments of how students are thinking about proportions. By knowing how their students are reasoning, teachers will be able to guide their students to explore the multiplicative structure of proportional relationships.

SECTION SUMMARY AND STUDY ITEMS

Section 7.2 Solving Proportion Problems by Reasoning with Multiplication and Division

Proportion problems can be solved by reasoning about quantities in several ways. Initial strategies can use repeated addition; more advanced strategies rely on reasoning about multiplication and division. Such strategies provide opportunities for quantitative reasoning, including making sense of quantities and their relationships, not just how to compute them.

Key Skills and Understandings

1. Know a variety of ways to solve proportion problems by reasoning about quantities, including with repeated addition and with strategies that involve multiplication and division. Explain in detail how and why multiplication and division apply and be able to express solutions as products of numbers derived from the problem statement. Use strip diagrams, ratio tables, and double (or triple) number lines as reasoning aids.

Practice Exercises for Section 7.2

1. Traveling at a constant speed, a race car goes 15 miles every 6 minutes. Explain how to reason about a double number line to answer the next questions.

a. How far does the race car go in 15 minutes?

b. How long does it take the race car to go 20 miles?

2. A soda mixture can be made by mixing cola and lemon-lime soda in a ratio of 4 to 3. Explain how to solve each of the following problems in two ways by reasoning about multiplication and division with quantities.

a. How much lemon-lime soda should you use to make the soda mixture with 48 cups of cola? How much soda mixture will this make?

b. How much cola and how much lemon-lime soda should you use to make 105 cups of the soda mixture?

3. Cali mixed $3\frac{1}{2}$ cups of red paint with $4\frac{1}{2}$ cups of yellow paint to make an orange paint. How many cups of red paint and how many cups of yellow paint will Cali need to make 12 cups of the same shade of orange paint? Explain how to reason about multiplication and division with quantities to solve this problem.

4. To make a punch you mixed $\frac{1}{4}$ cup grape juice concentrate with $1\frac{1}{2}$ cups bubbly water. If you want to make the same punch with 2 cups of bubbly water, then how many cups of grape juice concentrate should you use? Explain how to reason about multiplication and division with quantities to solve this problem.

5. In order to reconstitute a medicine properly, a pharmacist must mix 10 milliliters (mL) of a liquid medicine for every 12 mL of water. If one dose must contain 2.5 mL of medicine, then how many milliliters of medicine/water mixture provides one dose of the medicine? Explain how to reason about multiplication and division with quantities to solve this problem.

6. The ratio of Quint's CDs to Chris's CDs was 7 to 3. After Quint gave 6 CDs to Chris, they had an equal number of CDs. How many CDs did Chris have at first? Explain how to reason about a strip diagram to solve this problem.

Answers to Practice Exercises for Section 7.2

1. a. Because the race car goes 15 miles every 6 minutes, it goes 30 miles in 12 minutes and 45 miles in 18 minutes, as shown in the double number line in **Figure 7.17(a)**. Fifteen minutes is halfway between 12 and 18 minutes, so the race car must go halfway between 30 and 45 miles in 15 minutes. Since $45 - 30 = 15$ and $15 \div 2 = 7.5$, the race car goes $30 + 7.5 = 37.5$ miles in 15 minutes.

b. See **Figure 7.17(b)**.

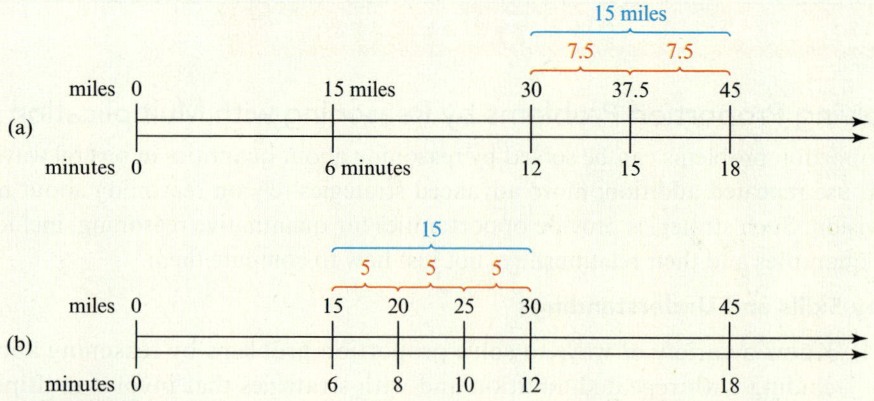

Figure 7.17 Using a double number line to solve a proportion problem.

2. a. The ratio of cola to lemon-lime soda is 4 to 3, so we can think of 4 cups of cola and 3 cups of lemon-lime soda as 1 batch. To make the mixture with 48 cups cola, you'll need 12 batches because $48 \div 4 = 12$. In those 12 batches the number of cups of lemon-lime soda is $12 \cdot 3 = 36$. All together, this will make $12 \cdot (4 + 3) = 84$ cups of soda mixture. **Table 7.1a** summarizes this reasoning.

TABLE 7.1a Solving a proportion by reasoning about quantities in a ratio table

Cups cola	4	·12 batches →	48	4	·12 batches →	48
Cups lemon-lime	3		?	3	·12 batches →	36
Total cups	7		?	7	·12 batches →	84

To solve the problem with a strip diagram, view the soda mixture as made of 4 parts cola and 3 parts lemon-lime soda, as shown in **Figure 7.18**. Because the 4 parts of cola are 48 cups, each part must be $48 \div 4 = 12$ cups. Therefore, the 3 parts lemon-lime soda are $3 \cdot 12 = 36$ cups. The full mixture consists of 7 parts, which is therefore $7 \cdot 12 = 84$ cups.

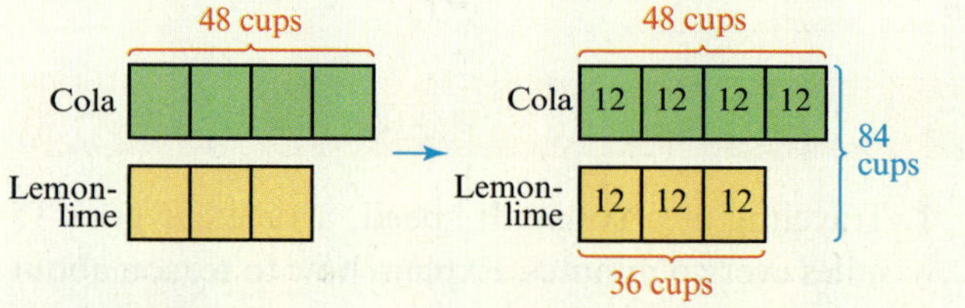

Figure 7.18 Solving a proportion by reasoning about quantities in a strip diagram.

b. To make 105 cups of soda mixture, you must make 15 batches because 1 batch is 7 cups and $105 \div 7 = 15$. Therefore, you must use $15 \cdot 4 = 60$ cups of cola, and $15 \cdot 3 = 45$ cups of lemon-lime soda. **Table 7.1b** summarizes this reasoning.

TABLE 7.1b Solving a proportion by reasoning about quantities in a ratio table

Cups cola	4		?	4	·15 batches →	60
Cups lemon-lime	3		?	3	·15 batches →	45
Total cups	7	·15 batches →	105	7	·15 batches →	105

To solve the problem with a strip diagram, see **Figure 7.19**. Because the 7 parts of soda mixture are 105 cups, each part must be $105 \div 7 = 15$ cups. Therefore, the 4 parts of cola are $4 \cdot 15 = 60$ cups, and the 3 parts of lemon-lime soda are $3 \cdot 15 = 45$ cups.

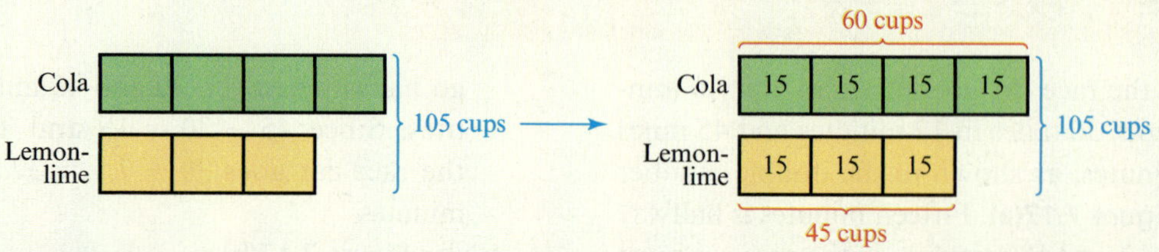

Figure 7.19 Solving a proportion by reasoning about quantities in a strip diagram.

3. The $3\frac{1}{2}$ cups of red paint and $4\frac{1}{2}$ cups of yellow paint combine to make 1 batch of $3\frac{1}{2} + 4\frac{1}{2} = 8$ cups orange paint. Because $12 \div 8 = 1\frac{1}{2}$, Cali will need $1\frac{1}{2}$ batches. Therefore, Cali will need

$$1\frac{1}{2} \cdot 3\frac{1}{2} = \frac{3}{2} \cdot \frac{7}{2} = \frac{21}{4} = 5\frac{1}{4}$$

cups of red paint and

$$1\frac{1}{2} \cdot 4\frac{1}{2} = \frac{3}{2} \cdot \frac{9}{2} = \frac{27}{4} = 6\frac{3}{4}$$

cups of yellow paint. This reasoning is summarized in Table 7.2.

TABLE 7.2 Solving a proportion by reasoning about quantities in a ratio table

Cups red	$3\frac{1}{2}$	?	$3\frac{1}{2} \xrightarrow{\cdot 1\frac{1}{2}\text{ batches}} 5\frac{1}{4}$	
Cups yellow	$4\frac{1}{2}$	?	$4\frac{1}{2} \xrightarrow{\cdot 1\frac{1}{2}\text{ batches}} 6\frac{3}{4}$	
Total cups orange	$8 \xrightarrow{\cdot 1\frac{1}{2}\text{ batches}}$	12	$8 \xrightarrow{\cdot 1\frac{1}{2}\text{ batches}} 12$	

4. You can reason that because $1\frac{1}{2} = \frac{6}{4}$, the $1\frac{1}{2}$ cups bubbly water are 6 parts, each of size $\frac{1}{4}$ cup. The grape juice concentrate is 1 part of size $\frac{1}{4}$ cup. If we take a variable-parts perspective, we can vary the size of the 6 parts bubbly water and 1 part grape juice (keeping all the parts the same size as each other). The 1 part grape juice concentrate will always be $\frac{1}{6}$ the amount of the 6 parts bubbly water. So for 2 cups bubbly water you need $\frac{1}{6} \cdot 2 = \frac{1}{3}$ cup grape juice concentrate.

5. If we distribute the 12 mL of water equally across the 10 mL of medicine, we see that for 1 mL of medicine there will be $12 \div 10 = 1.2$ mL of water. Taking 1 mL medicine and 1.2 mL water as a batch, there are 2.5 of these batches for 2.5 mL of medicine. These 2.5 batches contain $2.5 \cdot 1.2 = 3$ mL of water and therefore a total of $2.5 + 3 = 5.5$ mL of medicine-water mixture.

6. Because the ratio of Quint's CDs to Chris's CDs is 7 to 3, Quint's CDs can be divided into 7 parts and Chris's CDs can be divided into 3 parts, where all parts are the same size, as shown in the strip diagram in Figure 7.20. The boys have the same number of CDs in the end, so two of Quint's parts must go to Chris. Those two parts are 6 CDs, so each part consists of 3 CDs. Therefore, Quint had $7 \cdot 3 = 21$ CDs to start with, and Chris had $3 \cdot 3 = 9$ CDs to start with.

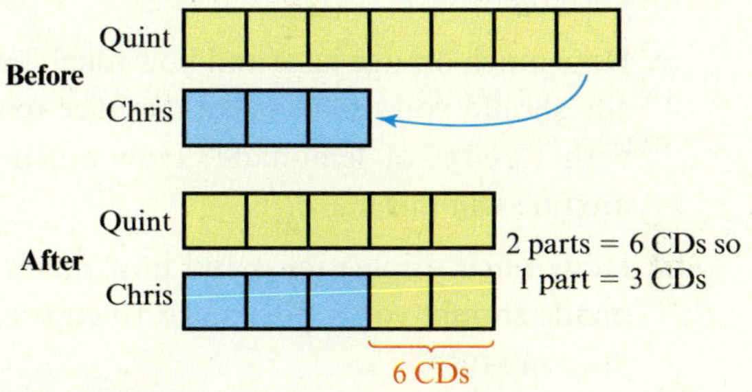

Figure 7.20 Solving a ratio problem with "before" and "after" strip diagrams.

PROBLEMS FOR SECTION 7.2

1. Walking at a constant speed, a person walks $\frac{3}{4}$ of a mile every 12 minutes. Explain how to reason about a double number line to answer the next questions.

a. How far does the person walk in 30 minutes?

b. How long does it take the person to walk $2\frac{1}{2}$ miles?

2. In a terrarium, the ratio of grasshoppers to crickets is 6:5. There are 48 grasshoppers. How many crickets are there? Explain how to solve this problem in two ways, in both cases by reasoning about multiplication and division with quantities. Support your reasoning with math drawings.

3. At a zoo, the ratio of king penguins to emperor penguins is 2:3. In all, there are 45 king and emperor penguins combined. How many emperor penguins are at the zoo? Explain how to solve this problem in two ways, in both cases by reasoning about multiplication and division with quantities. Support your reasoning with math drawings.

4. On a farm, the ratio of grey goats to white goats is 2:5. There are 100 grey goats. How many white goats are there? Explain how to solve this problem in two ways, in both cases by reasoning about multiplication and division with quantities. Support your reasoning with math drawings.

5. An orange-lemon juice mixture can be made by mixing orange juice and lemonade in a ratio of 5 to 2. Explain how to solve each of the following problems about the juice mixture by reasoning about multiplication and division with quantities from the multiple-batches perspective. Use a double or triple number line or a ratio table.

a. How much orange juice and how much lemonade should you use to make the juice mixture with 75 cups of orange juice? How much juice mixture will this make?

b. How much orange juice and how much lemonade should you use to make 140 cups of the juice mixture?

c. How much orange juice and how much lemonade should you use to make the juice mixture with 15 cups of lemonade? How much juice mixture will this make?

d. How much orange juice and how much lemonade should you use to make 10 cups of the juice mixture?

6. Explain how to solve parts a–d of Problem 5 by reasoning about multiplication and division with quantities from the variable-parts perspective. Use a strip diagram as support.

7. *Fertilizer problem:* A type of fertilizer is made by mixing nitrogen and phosphate in an 8 to 3 ratio. For 35 kilograms of nitrogen, how many kilograms of phosphate are needed to make the fertilizer?

Solve the Fertilizer Problem in two ways by reasoning about multiplication and division with quantities, once from the multiple-batches perspective and once from the variable-parts perspective. In each case, describe the number of kilograms of phosphate as a product $A \cdot B$, where A and B are suitable whole numbers, fractions, or mixed numbers that you derive from 8, 3, and 35. Attend carefully to our definition of multiplication when discussing $A \cdot B$. In each case, use a math drawing to support your explanation.

8. Brad made some punch by mixing $\frac{1}{2}$ cup of grape juice with $\frac{1}{4}$ cup of sparkling water. Brad really likes his punch, so he decides to make a larger amount of it by using the same ratio.

a. Brad wants to make 6 cups of his punch. How much grape juice and how much sparkling water should Brad use? Explain how to reason about multiplication and division with quantities to solve this problem in two different ways.

b. Now Brad wants to make 4 cups of his punch. How much grape juice and how much sparkling water should Brad use? Explain how to reason about multiplication and division with quantities to solve this problem.

9. To make grape juice by using frozen juice concentrate, you must mix the frozen juice concentrate with water in a ratio of 1 to 3. How much frozen juice concentrate and how much water should you use to make $1\frac{1}{2}$ cups of grape juice? Solve this problem in two different ways by reasoning about multiplication and division with quantities, explaining your reasoning in each case.

10. You can make a soap bubble mixture by combining 2 tablespoons water with 1 tablespoon liquid dishwashing soap and 4 drops of corn syrup. Using the same ratios, how much liquid dishwashing soap and how many drops of corn syrup should you use if you want to make a soap bubble mixture with 5 tablespoons of water? Solve this problem by reasoning about multiplication and/or division with quantities. Explain your reasoning.

11. You can make concrete by mixing 1 part cement with 2 parts pea gravel and 3 parts sand. Using the same ratios, how much cement and how much pea gravel should you use if you want to make concrete with 8 cubic feet of sand? Solve this problem by reasoning about multiplication and/or division with quantities. Explain your reasoning.

12. Marge made light blue paint by mixing $2\frac{1}{2}$ cups of blue paint with $1\frac{3}{4}$ cups of white paint. Homer poured another cup of white paint into Marge's paint mixture. How many cups of blue paint should Marge add to bring the paint back to its original shade of light blue (mixed in the same ratio as before)? Solve this problem by reasoning about quantities. Explain your reasoning.

13. A batch of lotion was made at a factory by mixing 1.3 liters of ingredient A with 2.7 liters of ingredient B in a mixing vat. By accident, a worker added an extra 0.5 liters of ingredient A to the mixing vat. How many liters of ingredient B should the worker add to the mixing vat so that the ingredients will be in the original ratio? Solve this problem by reasoning about quantities. Explain your reasoning.

14. If a $\frac{3}{4}$ cup serving of snack food gives you 60% of your daily value of calcium, then what percent of your daily value of calcium is in $\frac{1}{2}$ cup of the snack food? Explain how to solve the problem with the aid of a diagram or a table. Explain your reasoning.

15. If \$6000 is 75% of a company's budget for a project, then what percent of the budget is \$10,000? Explain how to solve the problem with the aid of a diagram or a table. Explain your reasoning.

16. Amy mixed 2 tablespoons of chocolate syrup in $\frac{3}{4}$ cup of milk to make chocolate milk. To make chocolate milk that is mixed in the same ratio and therefore tastes the same as Amy's, how much chocolate syrup will you need for 1 gallon of milk? Solve this problem by reasoning about quantities. Express your answer by using appropriate units. Recall that 1 gallon = 4 quarts, 1 quart = 2 pints, 1 pint = 2 cups, 1 cup = 8 fluid ounces, 1 fluid ounce = 2 tablespoons.

17. A 5-gallon bucket filled with water is being pulled from the ground up to a height of 20 feet. The bucket goes up 2 feet every 15 seconds. The bucket has a hole in it, so that 1 quart ($\frac{1}{4}$ gallon) of water leaks out of the bucket every 3 minutes. How much of the water will be left in the bucket by the time the bucket gets to the top? Solve this problem by reasoning about quantities. Explain your reasoning clearly.

18. Explain how to reason about quantities to solve the following problems.

a. John was paid \$250 for $3\frac{3}{4}$ hours of work. At that rate, how much should John make for $2\frac{1}{2}$ hours of work?

b. John was paid \$250 for $3\frac{3}{4}$ hours of work. At that rate, how long should John be willing to work for \$100?

***19.** The ratio of Frank's marbles to Huang's marbles is 3 to 2. After Frank gives $\frac{1}{2}$ of his marbles to another friend, Frank has 30 fewer marbles than Huang. How many marbles does Huang have?

a. Explain how to solve the problem with the aid of a strip diagram.

b. Create an easier problem for your students by changing the ratio, 3 to 2, to a different ratio and by changing the number of marbles, 30, to a different number of marbles. Make sure the problem has a sensible answer. Explain how to solve the problem.

c. Create a problem of about the same level of difficulty as the original problem by changing the ratio, 3 to 2, to a different ratio and by changing the number of marbles, 30, to a different number of marbles. Make sure the problem has a sensible answer. Explain how to solve the problem.

***20.** Asia and Taryn each had the same amount of money. After Asia spent \$14 and Taryn spent \$22, the ratio of Asia's money to Taryn's money was 4 to 3. How much money did each girl have at first?

a. Explain how to solve the problem with the aid of a strip diagram.

b. Create a harder problem for your students by changing the ratio, 4 to 3, to a different ratio and by changing the dollar amounts, \$14 and \$22, to different dollar amounts. Explain how to solve the problem.

c. Create a problem of about the same level of difficulty as the original problem by changing the ratio, 4 to 3, to a different ratio and by changing the dollar amounts, \$14 and \$22, to different dollar amounts. Explain how to solve the problem.

***21.** An aquarium contained an equal number of horseshoe crabs and sea stars. After 15 horseshoe crabs were removed and 27 sea stars were removed, the ratio of horseshoe crabs to sea stars was 5:3. How many horseshoe crabs and sea stars were there at first? Solve this problem and explain your reasoning.

7.3 The Values of a Ratio: Unit Rates and Multipliers

CCSS Common Core State Standards Grades 6, 7, and 8

Ratios are often written as fractions, but we defined ratios and fractions in different ways. In fact, ratios are *pairs* of numbers, but *fractions are numbers*. So what is the connection between ratios and fractions? And why is the common method of solving proportions (by setting two fractions equal to each other, then cross-multiplying and solving the resulting equation) a legitimate way to solve proportions? We will answer these questions in this section.

How Are Ratios Connected to Fractions

In Section 7.2 we reasoned about multiplication and division to solve a proportion problem about fragrant oils mixed in a 4 to 3 ratio. Two of our methods (multiplying 1 unit-rate-batch and multiplying 1 total amount) used the fraction $\frac{4}{3}$, which we obtained by dividing 4 by 3. This connection between ratios and fractions through division works in general and leads to important concepts in mathematics, such as slope, which we will study in the next section.

values of a ratio

constant of proportionality

In general, if two quantities are in the ratio A to B, then as we saw in Section 6.2, the quotient $A \div B$ is equal to the fraction $\frac{A}{B}$ and the quotient $B \div A$ is equal to the fraction $\frac{B}{A}$. These two quotients or fractions, $\frac{A}{B}$ and $\frac{B}{A}$, are called the **values of the ratio** A to B. In situations where we are considering quantities that vary together in a fixed ratio we usually call a value of a ratio a **constant of proportionality**. Because of the two types of division, there are two ways to interpret the value of a ratio or the constant of proportionality: (1) as a *unit rate* and (2) as a *multiplier* with which to compare total amounts.

Notice that the values of equivalent ratios are equivalent fractions, as illustrated in Figure 7.21.

Equivalent ratios	3	6	9	12	15
	2	4	6	8	10

Values of the ratios:

$$\frac{3}{2} = \frac{6}{4} = \frac{9}{6} = \frac{12}{8} = \frac{15}{10}$$

$$\frac{2}{3} = \frac{4}{6} = \frac{6}{9} = \frac{8}{12} = \frac{10}{15}$$

Figure 7.21 The values of equivalent ratios are equivalent fractions.

How Can We Interpret the Values of a Ratio as Unit Rates?

Let's see how to interpret the values of a ratio as unit rates.

Consider a bread recipe in which the ratio of flour to water is 14 to 5, so that if we use 14 cups of flour, we will need 5 cups of water. If we divide each of those amounts into 5 equal parts using how-many-units-in-1-group division then there are

$$14 \div 5 = \frac{14}{5}$$

cups of flour *for 1 cup of water*. In other words, there are $\frac{14}{5}$ cups of flour *per cup of water*. In this way, the ratio 14 to 5 of flour to water is naturally associated with the fraction $\frac{14}{5}$, which tells us the number of cups of flour per cup of water, or the *unit rate* of cups of flour per cup of water.

unit rate

rate

In general, given a ratio A to B relating two (nonzero) quantities, the quotient A/B is the **unit rate** or **rate** that tells us how many units there are of the first quantity for 1 unit of the second quantity.

CCSS
6.RP.2
7.RP.1

Similarly, the quotient B/A is the unit rate that tells us how many units there are of the second quantity for every 1 unit of the first quantity.

In Section 7.2, we found and used a unit rate when we solved a proportion problem by multiplying a unit-rate-batch. Once we find a unit rate, we can reason repeatedly with it to solve many proportion problems. For example, consider this running problem:

> *Running Problem A:* Rich runs 5 meters every 2 seconds (and keeps that steady pace for several minutes). How far does Rich run in 15 seconds? In 35 seconds? In T seconds?

Once we find the unit rate $\frac{5}{2}$ meters for 1 second, we can use it to find Rich's running distance for any number of seconds while he keeps that steady pace (see Figure 7.22(a)). Because each second corresponds to 1 group of $\frac{5}{2}$ meters:

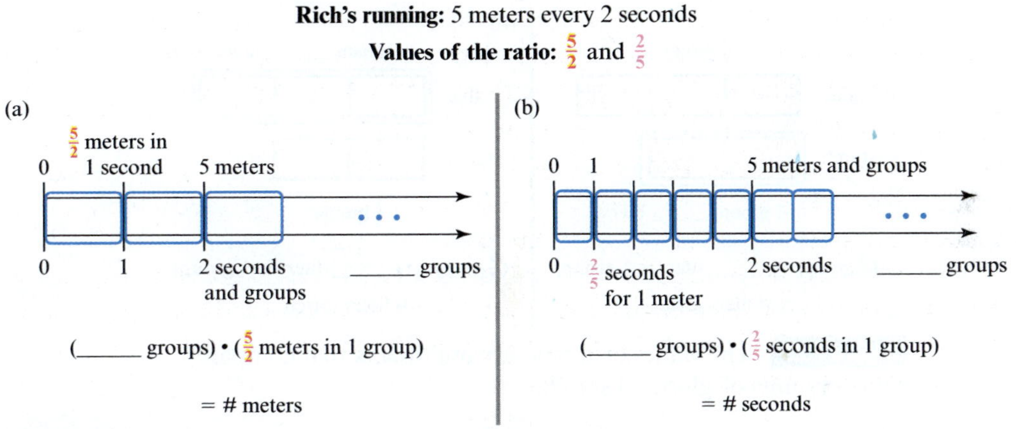

Figure 7.22 The values of a ratio are unit rates that tell how many meters per second $\left(\frac{5}{2}\right)$ and how many seconds per meter $\left(\frac{2}{5}\right)$.

CCSS
SMP8

> In 15 seconds Rich runs 15 groups of $\frac{5}{2}$ meters, so $15 \cdot \frac{5}{2}$ meters;
> in 35 seconds Rich runs 35 groups of $\frac{5}{2}$ meters, so $35 \cdot \frac{5}{2}$ meters;
> in T seconds Rich runs T groups of $\frac{5}{2}$ meters, so $T \cdot \frac{5}{2}$ meters.

Continuing with Rich's running, consider this problem:

> *Running Problem B:* How long does it take Rich to run 24 meters? 32 meters? D meters?

Once we find the unit rate $\frac{2}{5}$ seconds for 1 meter, we can use it to find Rich's running times for any number of meters while he keeps that steady pace (see Figure 7.22(b)). Because each meter corresponds to 1 group of $\frac{2}{5}$ seconds:

> For 24 meters, Rich takes 24 groups of $\frac{2}{5}$ seconds, so $24 \cdot \frac{2}{5}$ seconds;
> for 32 meters, Rich takes 32 groups of $\frac{2}{5}$ seconds, so $32 \cdot \frac{2}{5}$ seconds;
> for D meters, Rich takes D groups of $\frac{2}{5}$ seconds, so $D \cdot \frac{2}{5}$ seconds.

Notice that in solving running problems A and B, the unit rates $\frac{5}{2}$ meters per second and $\frac{2}{5}$ seconds per meter were *multiplicands*, in other words, they were the size of 1 group.

How Can We Interpret the Values of a Ratio as Multipliers that Compare Total Amounts?

When we take a variable-parts perspective on ratio, we view the values of a ratio as *multipliers* that compare total quantities. This extends the method of multiplying 1 total amount that we used in Section 7.2. In that method we used a value of a ratio; now we can reason repeatedly with the value of the ratio to solve many proportion problems. For example, consider this Gloop problem:

Gloop Problem A: A company makes Gloop by mixing glue and starch in a 3 to 2 ratio. How much glue will the company need to mix with 25 liters of starch? With 45 liters of starch? With *S* liters of starch?

Because the starch consists of 2 parts and the glue consists of 3 parts, if we view the starch as 1 group, then the glue is $\frac{3}{2}$ groups, and this is the case no matter what the size of 1 group—whether it is 25 liters or 45 liters or some other number of liters (see Figure 7.23(a)). The fraction $\frac{3}{2}$ answers the question:

How many groups of the total starch amount does it take to make the total glue amount?

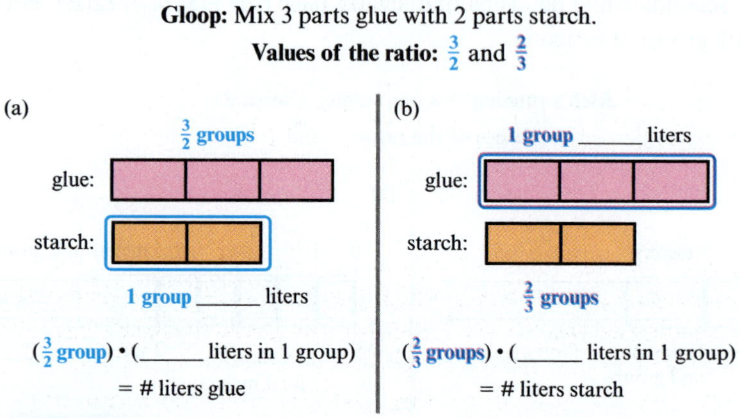

Figure 7.23 The values of a ratio are multipliers that compare total amounts of glue and starch.

CCSS SMP8

We can therefore use the value of the ratio $\frac{3}{2}$ to find the amount of glue for any amount of starch:

For 25 liters of starch, the glue is $\frac{3}{2}$ of 25 liters, so $\frac{3}{2} \cdot 25$ liters;
for 45 liters of starch, the glue is $\frac{3}{2}$ of 45 liters, so $\frac{3}{2} \cdot 45$ liters;
for *S* liters of starch, the glue is $\frac{3}{2}$ of *S* liters, so $\frac{3}{2} \cdot S$ liters.

Notice that we have interpreted the value of the ratio $\frac{3}{2}$ as a *multiplier*, in other words, as a number of groups. The multiplier $\frac{3}{2}$ tells us how many groups of starch are the same as the glue; it is the same number of groups no matter how many liters of starch are used to make Gloop.

Continuing with the company that makes Gloop, consider this problem:

Gloop Problem B: How much starch will the company need to mix with 40 liters of glue? With 50 liters of glue? With *G* liters of glue?

Now if we view the glue as 1 group, then the starch is $\frac{2}{3}$ of a group, and this is the case no matter what the size of 1 group—whether it is 40 liters or 50 liters or some other number of liters (see Figure 7.23(b)). The fraction $\frac{2}{3}$ answers the question:

How many groups of the total glue amount does it take to make the total starch amount?

We can therefore use the value of the ratio $\frac{2}{3}$ to find the amount of starch for any amount of glue:

For 40 liters of glue, the starch is $\frac{2}{3}$ of 40 liters, so $\frac{2}{3} \cdot 40$ liters;
for 50 liters of glue, the starch is $\frac{2}{3}$ of 50 liters, so $\frac{2}{3} \cdot 50$ liters;
for *G* liters of glue, the starch is $\frac{2}{3}$ of *G* liters, so $\frac{2}{3} \cdot G$ liters.

Notice that we have interpreted the value of the ratio $\frac{2}{3}$ as a *multiplier*, in other words, as a number of groups. The multiplier $\frac{2}{3}$ tells us how many groups of glue are the same as the starch; it is the same number of groups no matter how many liters of glue are used to make Gloop.

CLASS ACTIVITY

7G Unit Rates and Multiplicative Comparisons Associated with a Ratio, CA-125

What Is the Logic Behind Solving Proportions by Cross-Multiplying Fractions?

You are probably familiar with the technique of solving proportions by setting fractions equal to each other and cross-multiplying. Why is this a valid technique for solving proportions? We will examine this now.

CLASS ACTIVITY

7H Solving Proportions by Cross-Multiplying Fractions, p. CA-126

Why is the method of solving proportions by setting two fractions equal to each other and cross-multiplying valid? Consider a light-blue paint mixture made with $\frac{1}{4}$ cup blue paint and 4 cups white paint. How much blue paint will we need if we want to use 6 cups white paint and if we are using the same ratio of blue paint to white paint to make the same shade of light-blue paint? A common method for solving such a problem is to set up the following proportion in fraction form:

$$\frac{\frac{1}{4}}{4} = \frac{B}{6}$$

Here, B represents the as-yet-unknown number of cups of blue paint we will need for 6 cups of white paint. We then cross-multiply to get

$$6 \cdot \frac{1}{4} = 4 \cdot B$$

Therefore,

$$B = \left(6 \cdot \frac{1}{4}\right) \div 4 = 1\frac{1}{2} \div 4 = \frac{3}{8}$$

so that we must use $\frac{3}{8}$ cups of blue paint for 6 cups of white paint.

Let's analyze the preceding steps. First, when we set the two fractions

$$\frac{\frac{1}{4}}{4} \quad \text{and} \quad \frac{B}{6}$$

equal to each other, why can we do that, and what does it mean? If we think of the fractions as representing division—that is,

$$\frac{1}{4} \div 4 \quad \text{and} \quad B \div 6$$

then we can interpret each of these expressions as the number of cups of blue paint per 1 cup of white paint. In other words, each fraction stands for the same unit rate. We want to use the same amount of blue paint per cup of white paint either way; therefore, the two fractions should be equal to each other, or

$$\frac{\frac{1}{4}}{4} = \frac{B}{6} \tag{7.1}$$

Next, why do we cross-multiply? We can cross-multiply because two fractions are equal exactly when their "cross-multiples" are equal. Recall that the method of cross-multiplying is really just a shortcut for giving fractions a common denominator. If we give the fractions in Equation 7.1 the common denominator $6 \cdot 4$, which is the product of the two denominators, then we can replace Equation 7.1 with the proportion

$$\frac{\frac{1}{4} \cdot 6}{4 \cdot 6} = \frac{B \cdot 4}{6 \cdot 4} \tag{7.2}$$

In terms of the paint mixture, both sides of this proportion now refer to 24 cups of paint, instead of 4 cups and 6 cups of paint, as in Equation 7.1. But two fractions that have the same denominator are equal exactly when their numerators are equal. Since the denominators of the fractions in Equation 7.2 are equal (because $4 \cdot 6 = 6 \cdot 4$) the proportion will be solved exactly when the numerators are equal—namely, when

$$\frac{1}{4} \cdot 6 = B \cdot 4 \tag{7.3}$$

Therefore, we can solve the proportion in Equation 7.1 by solving Equation 7.3, which was obtained by cross-multiplying.

SECTION SUMMARY AND STUDY ITEMS

Section 7.3 The Values of a Ratio: Unit Rates and Multipliers

Ratios are connected to fractions by division. We can interpret the two values of a ratio either as unit rates or (when the quantities are measured in the same units) as multipliers that relate total amounts. In the standard method of solving proportions, we first set two fractions equal to each other; we can do so because we are equating the unit rates.

Key Skills and Understandings

1. Understand that if two quantities are in a ratio of A to B, then we can interpret $\frac{A}{B}$ as a unit rate; namely, it is the number of units of the first quantity for every 1 unit of the second quantity. Similarly, $\frac{B}{A}$ is a unit rate; it is the number of units of the second quantity for every 1 unit of the first quantity.
2. Identify unit rates and explain what they mean in terms of a context.
3. Use unit rates in solving problems.
4. For ratios viewed as A parts to B parts, use the values $\frac{A}{B}$ and $\frac{B}{A}$ to make multiplicative comparisons between total amounts.
5. Explain why this method of solving proportions is valid: setting two fractions equal to each other, then cross-multiplying, then solving the resulting equation.

Practice Exercises for Section 7.3

1. Jose mixed blue paint and red paint in a ratio of 3 to 4 to make a purple paint. Interpret the fractions $\frac{3}{4}$ and $\frac{4}{3}$ as unit rates and also as multipliers to compare amounts of blue and red paint.
2. Which of the following two mixtures will be more salty?
   - 3 tablespoons of salt mixed in 4 cups of water
   - 4 tablespoons of salt mixed in 5 cups of water

   Solve this problem in two different ways: by comparing unit rates and another way. Explain your reasoning in each case.

Answers to Practice Exercises for Section 7.3

1. From the multiple-batches perspective, if we use cups as a measurement unit, then 1 batch of paint has 3 cups blue paint and 4 cups red paint. Dividing both by 4, we see there are $3 \div 4 = \frac{3}{4}$ cups blue paint for 1 cup red paint. By the same logic, there are $4 \div 3 = \frac{4}{3} = 1\frac{1}{3}$ cups red paint for 1 cup blue paint.

From the variable-parts perspective, viewing the 3 to 4 ratio as 3 parts blue and 4 parts red (a strip diagram would show this nicely), we see that the amount of blue paint is $\frac{3}{4}$ times the amount of red paint no matter how much red paint there is and the amount of red paint is $\frac{4}{3}$ times the amount of blue paint no matter how much blue paint there is.

2. With unit rates: If we think of the 3 tablespoons of salt in the first mixture as being divided equally among the 4 cups of water, then each cup of water contains $3 \div 4 = \frac{3}{4}$ tablespoons of salt. Similarly, each cup of water in the second mixture contains $4 \div 5 = \frac{4}{5}$ tablespoons of salt. Since $\frac{4}{5} = 0.8$ and $\frac{3}{4} = 0.75$, and since $0.8 > 0.75$, the second mixture contains more salt per cup of water. Thus, it is more salty.

Another way: If we make 5 batches of the first mixture and 4 batches of the second mixture, then both will contain 20 cups of water. The first mixture will contain $5 \cdot 3 = 15$ tablespoons of salt, and the second mixture will contain $4 \cdot 4 = 16$ tablespoons of salt. Since both mixtures contain the same amount of water, but the second mixture contains 1 more tablespoon of salt than the first, the second mixture must be more salty.

PROBLEMS FOR SECTION 7.3

1. A company mixes different amounts of blue paint with yellow paint in a ratio of 2 to 5 to make a green paint. For each of the following fractions, interpret the fraction in two ways in terms of the company's paint mixtures (but *not* as ratios). Use double or triple number lines and strip diagrams as supports.

$$\frac{2}{5};\ \frac{5}{2};\ \frac{2}{7};\ \frac{5}{7}$$

2. A company mixes different amounts of grape and peach juice, but always in the ratio 3 to 5.

a. Explain how to reason with a value of the ratio to determine how much peach juice the company should mix with the following amounts of grape juice: 100 liters; 140 liters; *G* liters.

b. Explain how to reason with a value of the ratio in another way to determine how much peach juice the company should mix with the amounts of grape juice in part (a).

c. Explain how to reason with a value of the ratio to determine how much grape juice the company should mix with the following amounts of peach juice: 72 liters; 84 liters; *P* liters.

d. Explain how to reason with a value of the ratio in another way to determine how much grape juice the company should mix with the amounts of peach juice in part (c).

3. a. Which of the following two mixtures will have a stronger lime flavor?

- 2 cups lime juice concentrate mixed in 5 cups water
- 4 cups lime juice concentrate mixed in 7 cups water

Solve this problem in two different ways: with unit rates and another way. Explain your reasoning in each case.

b. A student might say that the second mixture has a stronger flavor than the first mixture because the numbers for the second mixture are greater (in other words, $4 > 2$ and $7 > 5$). Even if the conclusion is correct, is the student's reasoning valid? Explain why or why not.

4. a. Snail A moved 6 feet in 7 hours. Snail B moved 7 feet in 8 hours. Both snails moved at constant speeds. Which snail went faster? Solve this problem in two different ways, explaining in detail why you can solve the problem the way you do. In particular, if you use fractions in your explanation, be sure to explain how the fractions are relevant.

b. A student might say that snail B moved faster than snail A because the numbers for snail B are greater (in other words, $7 > 6$ and $8 > 7$). Even if the conclusion is correct, is the student's reasoning valid? Explain why or why not.

5. A dough recipe calls for 3 cups of flour and $1\frac{1}{4}$ cups of water. You want to use the same ratio of flour to water to make a dough with 10 cups of flour. How much water should you use?

a. Solve this problem by setting up an equation in which you set two fractions equal to each other.

b. Interpret the two fractions that you set equal to each other in part (a) in terms of the recipe. Explain why it makes sense to set these two fractions equal to each other.

c. Why does it make sense to cross-multiply the two fractions in part (a)? What is the logic behind the procedure of cross-multiplying?

d. Now solve the problem of how much water to use for 10 cups of flour in a different way, by using the most elementary reasoning you can. Explain your reasoning clearly.

6. A recipe that serves 6 people calls for $1\frac{1}{2}$ cups of rice. How much rice will you need to serve 8 people (assuming that the ratio of people to cups of rice stays the same)?

a. Solve this problem by setting up an equation in which you set two fractions equal to each other.

b. Interpret the two fractions that you set equal to each other in part (a) in terms of the recipe. Explain why it makes sense to set these two fractions equal to each other.

c. Why does it make sense to cross-multiply the two fractions in part (a)? What is the logic behind the procedure of cross-multiplying?

d. Now solve the same problem in a different way by reasoning about quantities. Explain your reasoning clearly.

7. Explain how to reason with unit rates to solve the following problems:

a. Suppose you drive 4500 miles every half year in your car. At the end of $3\frac{3}{4}$ years, how many miles will you have driven?

b. Mo uses 128 ounces of liquid laundry detergent every $6\frac{1}{2}$ weeks. How much detergent will Mo use in a year?

c. Suppose you have a 32-ounce bottle of weed killer concentrate. The directions say to mix $2\frac{1}{2}$ ounces of weed killer concentrate with enough water to make a gallon. How many gallons of weed killer will you be able to make from this bottle?

8. Buttercup the gerbil drinks $\frac{2}{3}$ of a bottle of water every $1\frac{1}{2}$ days. How many bottles of water will Buttercup drink in 5 days? Explain how to reason with a unit rate to solve this problem.

9. If you used $2\frac{1}{2}$ truck loads of mulch for a garden that covers $\frac{3}{4}$ acre, then how many truck loads of mulch should you order for a garden that covers $3\frac{1}{2}$ acres? Assume that you will spread the mulch at the same rate as before. Explain how to reason with a unit rate to solve this problem.

10. If $2\frac{1}{2}$ pints of jelly filled $3\frac{1}{2}$ jars, then how many jars will you need for 12 pints of jelly? Will the last jar of jelly be completely full? If not, how full will it be? (Assume that all jars are the same size.) Explain how to reason with a unit rate to solve this problem.

*__11.__ A standard bathtub is approximately $4\frac{1}{2}$ feet long, 2 feet wide, and 1 foot deep. If water comes out of a faucet at the rate of $2\frac{1}{2}$ gallons each minute, how long will it take to fill the bathtub $\frac{3}{4}$ full? Use the fact that 1 gallon of water occupies 0.134 cubic feet.

7.4 Proportional Relationships

CCSS Common Core State Standards Grades 7, 8

In many situations, two quantities vary together across many different values. If these varying values are all in the same ratio, then the relationship between them is a proportional relationship. So far we have used double number lines and strip diagrams to get a feel for proportional relationships, and we have reasoned about quantities to find unknown values in proportional relationships. In this section we will describe proportional relationships more generally, with graphs and equations.

In mathematics, we traditionally use graphs and equations to represent relationships succinctly. For proportional relationships, graphs and equations are of a special type and the value of the ratio plays a special role in both. Graphs and equations can represent a wide variety of relationships, not just proportional ones, but the reasoning we use with proportional relationships is fundamental to understanding other relationships in mathematics and science.

How Can We Find and Explain Equations for Proportional Relationships?

CLASS ACTIVITY

7I Representing a Proportional Relationship with Equations, p. CA-127

When we have two quantities that are varying together in a proportional relationship, if we know one quantity, we can find the other quantity using methods we studied in Sections 7.2 and 7.3. Although we can list many pairs of coordinated quantities in a table, or on a double number line, we can also express the relationship between these coordinated quantities succinctly with equations.

CCSS SMP8

Consider biodiesel and petrodiesel (petroleum diesel) mixtures that are 30% biodiesel and 70% petrodiesel, called B30 mixtures. We can imagine making almost any amount of a B30 mixture, but in each mixture the ratio of biodiesel to petrodiesel is 30 to 70, or equivalently, 3 to 7. Figure 7.24 shows how we can reason repeatedly from a variable-parts perspective. Given a number of liters of biodiesel, we can find the corresponding number of liters of petrodiesel in B30 mixtures by using the constant of proportionality (value of the ratio) $\frac{7}{3}$ as a multiplier.

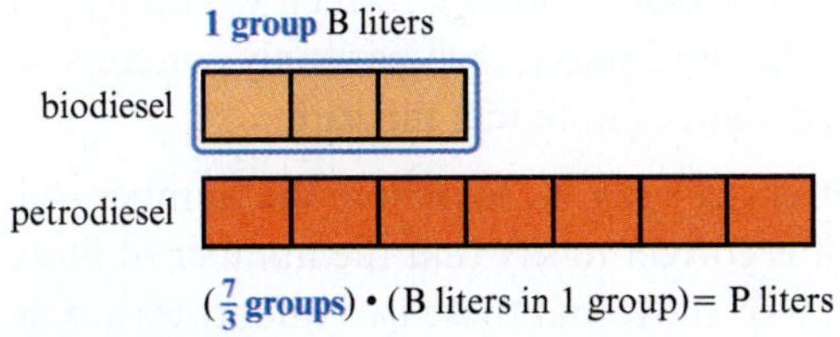

# liters biodiesel	# liters petrodiesel
10	$7/3 \cdot 10 = 23\frac{1}{3}$
15	$7/3 \cdot 15 = 35$
20	$7/3 \cdot 20 = 46\frac{2}{3}$
25	$7/3 \cdot 25 = 58\frac{1}{3}$
B	$7/3 \cdot B = P$

Figure 7.24 Using repeated reasoning to relate quantities of biodiesel and petrodiesel in the ratio 3 to 7.

We can summarize this repeated reasoning by letting B stand for an unspecified number of liters of biodiesel, which can vary, and letting P stand for the corresponding number of liters of petrodiesel. If we let B liters be 1 group, then the petrodiesel consists of $\frac{7}{3}$ groups because it is 7 parts, each $\frac{1}{3}$ of 1 group. The $\frac{7}{3}$ tells us how many groups of 3 parts are the same as 7 parts; it also tells us how many groups of B liters are the same as P liters. Therefore

$$\frac{7}{3} \cdot B = P \quad \text{or} \quad P = \frac{7}{3} \cdot B$$

We can think of this equation as summarizing how to find the number of liters of petrodiesel, P, given a number of liters of biodiesel, B.

CCSS 7.RP.2

Similarly, we could generate the equation

$$\frac{3}{7}\cdot P = B \quad \text{or} \quad B = \frac{3}{7}\cdot P$$

by reasoning repeatedly with $\frac{3}{7}$ as a multiplier. We can think of this equation as summarizing how to find the number of liters of biodiesel, B, given a number of liters of petrodiesel, P.

There are many other equations that relate coordinated quantities of biodiesel, B, and petrodiesel, P, in B30 mixtures. The equations need not be viewed as telling you how to find P if you know B or how to find B if you know P. Equations express that two quantities are equal to each other, so we can view an equation as expressing a *constraint* on which values for P and B go together. For example, can you see how to use a strip diagram to explain the equation

$$\frac{1}{3}\cdot B = \frac{1}{7}\cdot P?$$

This equation tells us that for B and P to go together, $\frac{1}{3}$ of B must be the same amount as $\frac{1}{7}$ of P.

What Are Errors to Watch for in Formulating Equations?

Consider a chocolate company mixing varying quantities of cocoa and milk in the ratio 2 to 5. Let C stand for an unspecified number of grams of cocoa and let M be the corresponding number of liters of milk for such mixtures. What errors do students often make when formulating equations to relate C and M?

Students might be tempted to formulate an equation like this:

Warning, incorrect: $2C = 5M$

Can you see why? They might arrive at this erroneous equation by turning "2 parts cocoa" into $2C$ and "5 parts milk" into $5M$. They might also write the equal sign to denote that 2 parts cocoa are mixed with 5 parts milk.

In mathematics, expressions and equations have very precise meanings. The equal sign means "equals" or "is equal to" or "which is equal to." The equal sign does not mean "goes together with" or "corresponds with." The expression $2C$ means 2 times C, which would be twice the amount of cocoa in a mixture. Because the cocoa takes up 2 parts, each part only contains half of C, not all of C. Similarly $5M$ would be 5 times the amount of milk in a mixture.

Notice that we could use a *different* letter, let's say X, to denote the number of liters in 1 part (of either cocoa or of milk). In that case, it is correct to say that the number of liters of cocoa is $2\cdot X$ and the number of liters of milk is $5\cdot X$. To show that these two quantities are mixed together, we can use a colon

$$2\cdot X \;:\; 5\cdot X$$

It would be incorrect to write $2\cdot X = 5\cdot X$ because the number of liters of cocoa is not equal to the number of liters of milk.

How Do We Use Coordinate Planes?

coordinate plane

axes

origin

A coordinate plane is a plane, together with two perpendicular number lines in the plane that meet at the location of 0 on each number line. The two number lines are called the **axes** of the coordinate plane (singular: axis). Traditionally, one number line is displayed horizontally and the other is displayed vertically. The horizontal axis is often called the *x-axis*, and the vertical axis is often called the *y-axis.* The point where the x- and y-axes meet is called the **origin**.

The main feature of a coordinate plane is that *the location of every point in the plane can be specified by referring to the two axes.* This works in the following way: A pair of numbers, such as $(4.5, 3)$, corresponds to the point in the plane that is located where a vertical line through 4.5 on the horizontal axis meets a horizontal line through 3 on the vertical axis, as shown in Figure 7.25. This point is designated $(4.5, 3)$; or we say that 4.5 and 3 are the **coordinates** of the point. Specifically, 4.5 is the first coordinate, or x-coordinate, of the point and 3 is the second coordinate, or y-coordinate, of the point.

coordinates

Figure 7.25

Locating $(4.5, 3)$ in a coordinate plane.

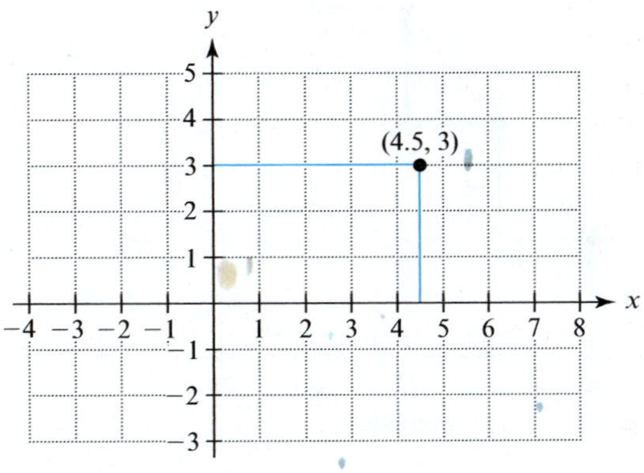

You can use your fingers to locate the point $(4.5, 3)$ by putting your right index finger on 4.5 on the horizontal axis and your left index finger on 3 on the vertical axis, and sliding the right index finger vertically upward and the left index finger horizontally to the right until your two fingers meet. Some people like to use two sheets of paper to locate points in a coordinate plane by holding the edge of one sheet parallel to the y-axis and the edge of the other sheet parallel to the x-axis.

Figure 7.26 shows the coordinates of several different points in a coordinate plane. Notice that in general the coordinates of *a point can include negative numbers.* However, when we work with proportional relationships, we usually use only nonnegative coordinates.

Figure 7.26

Points in a coordinate plane.

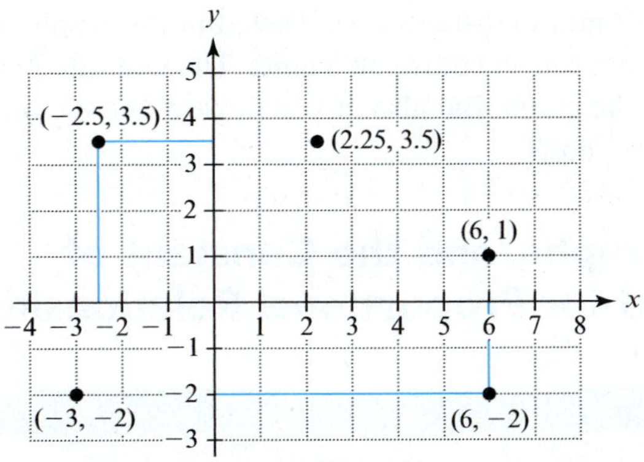

How Are Graphs of Proportional Relationships Special?

How can we represent proportional relationships as graphs in a coordinate plane? For example, suppose Andrew is swimming at a steady pace of 5 meters every 4 seconds. While Andrew keeps that steady pace, the number of seconds and the number of meters he has swum are in the ratio 4 to 5, and so are in a proportional relationship. To represent this relationship as a graph in a coordinate plane, we can choose the horizontal axis to represent either time or

Figure 7.27 Representing a proportional relationship in a table and graph.

Seconds	Meters
4	5
8	10
12	15
16	20
20	25
24	30

(+4 between consecutive Seconds entries; +5 between consecutive Meters entries)

Every 4 seconds, Andrew swims another 5 meters.

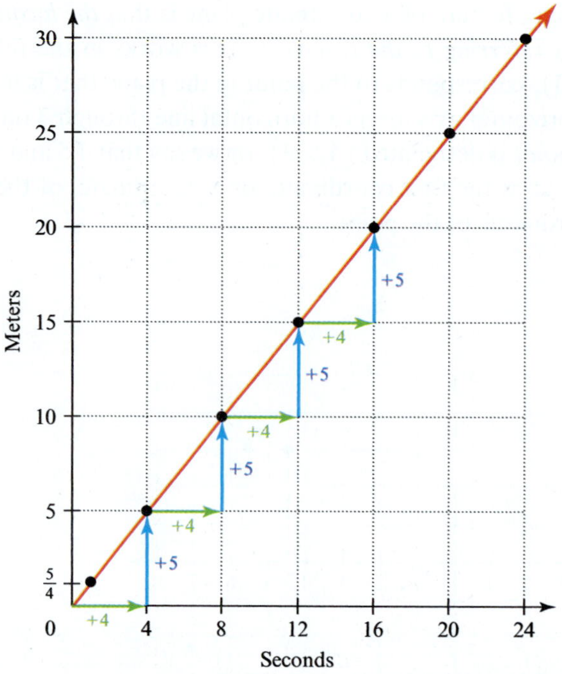

distance; let's say we choose the horizontal axis to represent time. To represent the proportional relationship as a graph we plot the pairs of corresponding times and distances in the coordinate plane as in Figure 7.27.

CCSS 8.EE.5

What do you notice about the graph of the proportional relationship? All the points from the table are on a line that goes through the origin. In Chapter 14 we will use similar triangles to explain why the graph of a proportional relationship always lies on a line. Note that it makes sense to connect the points between the ones plotted from the table. For example, in between 4 seconds and 8 seconds, Andrew swims from 5 meters to 10 meters, so those "in between" points should also be on the graph. Also, note that it makes sense to include the point $(0, 0)$ because at 0 seconds, when Andrew starts swimming, he has swum 0 meters.

Notice also how the 4 to 5 ratio of times to distances is reflected in the graph: moving along the graph, if you move to the right 4 units, you will move up 5 units. The unit rate $\frac{5}{4}$ meter for 1 second corresponds to the point $\left(1, \frac{5}{4}\right)$ on the graph. But also, if you move along the graph, as you move to the right 1 unit, you will move up $\frac{5}{4}$ units.

How Are Equations, Graphs, and the Constant of Proportionality Related for Proportional Relationships?

CLASS ACTIVITY

7J Relating Lengths and Heights of Ramps, p. CA-128.

Consider a collection of ramps, such as wheelchair ramps, all of whose lengths and heights are in the ratio 7 to 2. The lengths and corresponding heights of all these ramps are therefore in a proportional relationship. Figure 7.28 takes a variable-parts perspective on this proportional relationship: the length of each ramp is 7 parts and the height is 2 parts. All the parts are the same number of feet long as each other but that number of feet varies.

Figure 7.28

Ramps whose length and height are in the ratio 7 to 2.

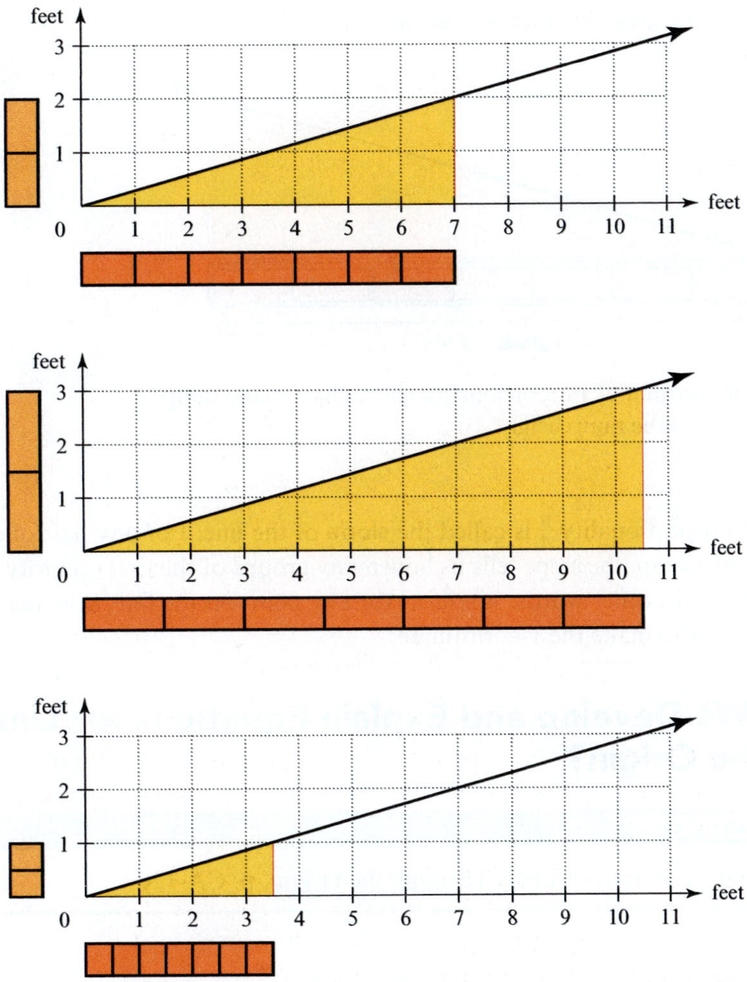

What is the same about all these ramps? The ramps vary in length and height, but they are all equally steep, and their steepness is the same as the steepness of the line that is the graph of the proportional relationship. We usually call the steepness of a line the *slope*. As we'll see, it makes sense to describe this slope with a constant of proportionality (value of a ratio).

We have seen that we can view the constant of proportionality as a multiplier that relates total amounts, which leads to equations for proportional relationships. In this case, let x be an unspecified number of feet, standing for the length of a ramp, and let y be the corresponding height of the ramp in feet. If we view the length of a ramp as 1 group of x feet, as in **Figure 7.29**, then the height of the ramp is $\frac{2}{7}$ of a group. The $\frac{2}{7}$ tells us how many groups of 7 parts are in 2 parts; it also tells us how many groups of x feet are in y feet. Therefore

$$\frac{2}{7} \cdot x = y \quad \text{or} \quad y = \frac{2}{7} \cdot x$$

The constant of proportionality $\frac{2}{7}$ tells us how many groups of the length of a ramp it takes to make the height of a ramp and *this tells us the steepness of the ramp*. If it took fewer groups of the length to make the height then the ramps wouldn't be as steep; if it took more groups, then the ramps would be steeper.

Generalizing the previous reasoning, consider a proportional relationship consisting of quantities in a fixed ratio A to B. The graph of this relationship is a line through the origin and has equation

$$y = \frac{B}{A} \cdot x$$

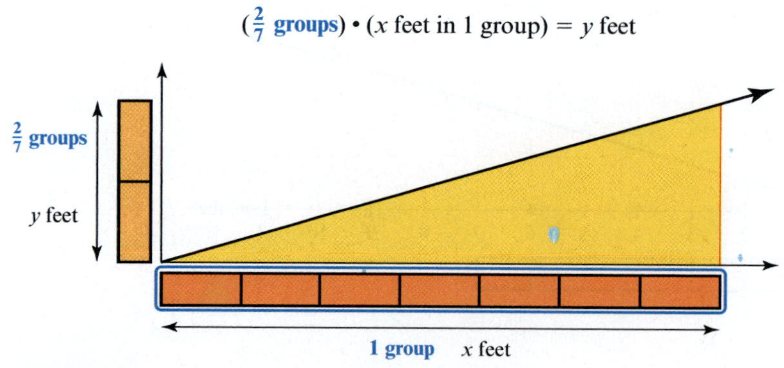

Figure 7.29 The constant of proportionality tells us how many ramp lengths it takes to make the ramp height.

slope The constant of proportionality $\frac{B}{A}$ is called the **slope** of the line. For any pair of quantities in the proportional relationship, the slope tells us how many groups of the first quantity it takes to make the second quantity. In other words, it tells us for any point on the line, how many groups of the x-coordinate it takes to make the y-coordinate.

CCSS 8.EE.6

How Can We Develop and Explain Equations for Lines Through the Origin?

CLASS ACTIVITY

7K Graphs and Equations of Lines Through the Origin, p. CA-129

You may know that equations of (nonvertical) lines have a special form, namely $y = m \cdot x + b$. Why do lines have that kind of equation? Let's consider a special case: lines that go through the origin $(0, 0)$ and have points with positive x and y coordinates. Let's explain why such lines have equations of the form $y = m \cdot x$.

For example, consider the line that goes through the origin and the point $(3, 5)$. In Chapter 14 we will use similar triangles to explain why the x- and y-coordinates of points on such a line are in a proportional relationship. As above, if we take a variable-parts perspective, we can think of the x-coordinate as consisting of 3 parts and the y-coordinate as consisting of 5 parts, as in Figure 7.30. Each part is 1 unit long when the point (x, y) is $(3, 5)$. But if we imagine the point (x, y) moving along the line, away from the origin or toward the origin, the 3 horizontal parts and 5 vertical parts will stretch or shrink, but will always remain the same size as each other.

For a point (x, y) on the line, if we take x to be 1 group, then y is $\frac{5}{3}$ of this group. The $\frac{5}{3}$ tells us how many groups of 3 parts are in 5 parts; it also tells us how many groups of size x units are in y units. Therefore

$$y = \frac{5}{3} \cdot x$$

As before, the constant of proportionality, $\frac{5}{3}$, is the slope of the line.

CLASS ACTIVITY

7L Comparing Tables, Graphs, and Equations, p. CA-130

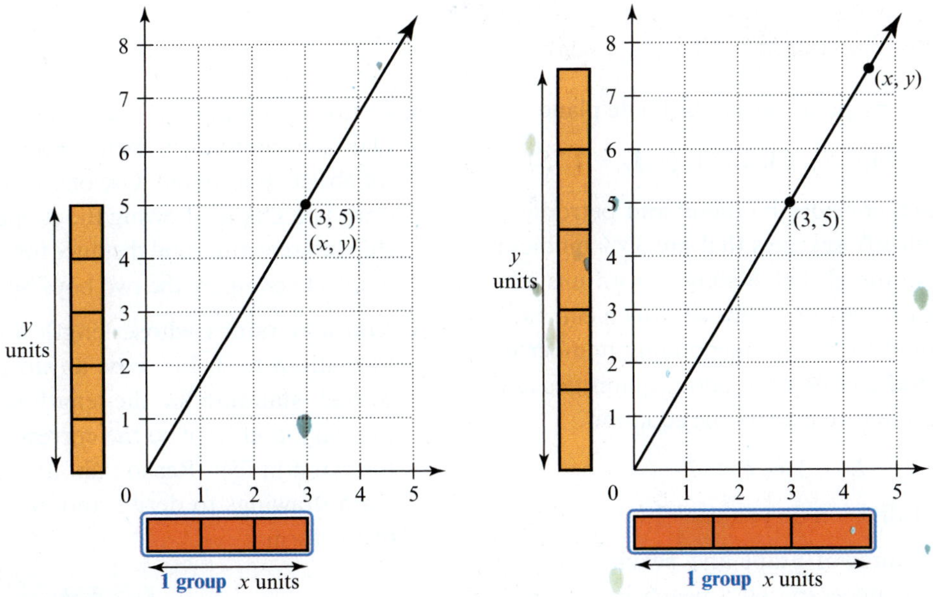

Figure 7.30 The line through the origin and the point $(3, 5)$ viewed from the variable-parts perspective.

SECTION SUMMARY AND STUDY ITEMS

Section 7.4 Proportional Relationships

We can use equations to summarize the relationship between quantities that vary together in a proportional relationship (i.e., they remain in a fixed ratio). We can find equations for proportional relationships by viewing the constant of proportionality (value of the ratio) as a multiplier that relates the quantities.

When we graph a proportional relationship in a plane it turns out that the graph is always a line through the origin $(0, 0)$. If the proportional relationship consists of quantities in the ratio A to B, then the relationship has equation $y = \frac{B}{A} \cdot x$. The constant of proportionality, $\frac{B}{A}$, is also the slope of the line.

It turns out that every line through the origin that has points with positive x- and y-coordinates is the graph of a proportional relationship. By taking a variable-parts perspective and viewing the constant of proportionality as a multiplier that relates quantities, we can explain why such lines have equations of the form $y = m \cdot x$.

Key Skills and Understandings

1. View quantities varying together in a fixed ratio as a proportional relationship, which can be represented with equations and with graphs in the coordinate plane as well as with strip diagrams and double number lines.
2. Find and explain equations to relate quantities that vary together in a proportional relationship.
3. Recognize incorrect equations for proportional relationships and discuss sources of errors.
4. Graph proportional relationships in a coordinate plane and interpret the constant of proportionality as the slope of the line, which tells us how many groups of an x-coordinate it takes to make the y-coordinate of a point (x, y) on the line.
5. Explain why a line through the origin and a given point with positive coordinates has an equation of the form $y = m \cdot x$ for a suitable number m.

Practice Exercises for Section 7.4

1. Plot the following points in a coordinate plane:

$$(4, -2.5), (-4, 2.5), (3, 2.75), (-3, -2.75)$$

2. As in the text, consider biodiesel and petrodiesel (petroleum diesel) mixtures that are 30% biodiesel and 70% petrodiesel. Let B stand for an unspecified number of liters of biodiesel, which can vary, and let P stand for the corresponding number of liters of petrodiesel. Reason about quantities and use math drawings to explain the equation

$$B \div 3 = P \div 7$$

What kind of division do you use?

3. Consider mixtures of grape and peach juice that are 40% grape juice and 60% peach juice. Let G be an unspecified number of liters of grape juice, which can vary, and let P be the corresponding number of liters of peach juice in these mixtures. Reason about quantities and use math drawings to explain an equation of the form

$$c \cdot G = P$$

where c is a suitable constant of proportionality.

4. Carlos is running 25 meters every 3 seconds and Darius is running 30 meters every 4 seconds (each for about 20 seconds). Use one set of coordinates to draw two graphs showing the proportional relationship between time and distance for each boy. Use the graphs to compare the two boys' speeds in two ways.

5. Consider ramps whose length and height are in the ratio 8 to 3. Let x be an unspecified number of feet, standing for the length of a ramp, which can vary, and let y be the corresponding height of the ramp in feet. Reason about quantities and use math drawings to derive and explain an equation of the form

$$y = c \cdot x$$

where c is a suitable constant of proportionality.

6. Consider the line through the origin and the point $(4, 3)$. Derive and explain an equation for the line of the form

$$y = m \cdot x$$

where m is a suitable number.

Answers to Practice Exercises for Section 7.4

1. See Figure 7.31.

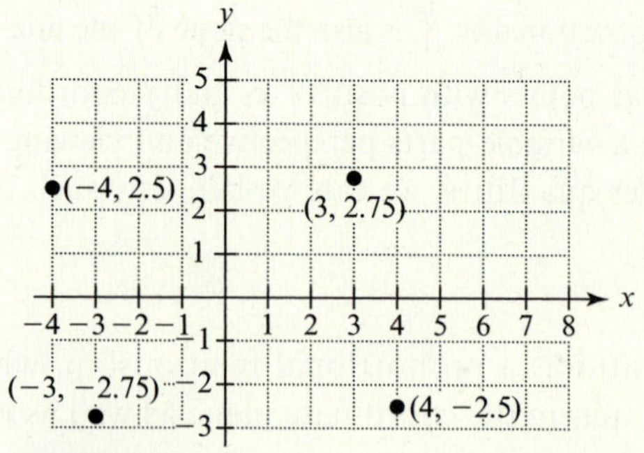

Figure 7.31 Points in a coordinate plane.

2. Taking a variable-parts perspective, we can represent the quantities with a 3-part biodiesel strip and a 7-part petrodiesel strip as in Figure 7.24. Using how-many-units-in-1-group division and focusing on the biodiesel, the number of liters in 1 part is $B \div 3$. Similarly, focusing on the petrodiesel, the number of liters in 1 part is $P \div 7$. But all parts are the same size as each other, so these two ways of describing the size of 1 part must be equal. Therefore

$$B \div 3 = P \div 7$$

3. In all these mixtures, the amounts of grape and peach juice are in the ratio 40 to 60, or equivalently, 2 to 3. Taking a variable-parts perspective, we can draw a 2-part strip for the grape juice and a 3-part strip for the peach juice (not shown). Now reason in the same way as in the biodiesel-petrodiesel example in the text to explain why

$$\frac{3}{2} \cdot G = P$$

4. See Figure 7.32. The vertical dashed line shows that after 12 seconds, Carlos has run 100 meters and Darius has run 90 meters. Since Carlos ran farther in the same amount of time, he is going faster than Darius. The horizontal dashed line shows that it took Carlos 18 seconds to run 150 meters and it took Darius 20 seconds to run 150 meters. Since it took Darius a longer time to run the same distance, he is going slower than Carlos.

Carlos		Darius	
Seconds	Meters	Seconds	Meters
3	25	4	30
6	50	8	60
9	75	12	90
12	100	16	120
15	125	20	150
18	150		

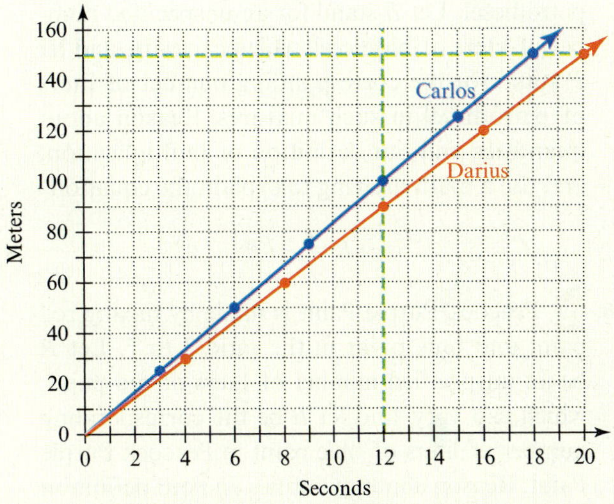

Figure 7.32 Tables and graphs showing Carlos's and Darius's running times and distances.

5. The constant of proportionality c is $\frac{3}{8}$. See the explanation in the text for ramps whose length and height are in the ratio 7 to 2.

6. The number m is $\frac{3}{4}$. See the explanation in the text for the line through the origin and the point $(3, 5)$.

PROBLEMS FOR SECTION 7.4

1. Goblin Green Paint is made by mixing blue and yellow paint in the ratio 4 to 3. Show and discuss at least four different ways to represent varying quantities of blue and yellow paint that will mix to make Goblin Green Paint. (These can include ways discussed in previous sections.) Discuss which aspects of proportional relationships are easier to see in some ways than in others.

2. Aqua Regia is an acid that can dissolve gold. It is made by mixing hydrochloric acid and nitric acid in the ratio 3 to 1 (by volume). Let X be an unspecified number of liters of hydrochloric acid, which can vary, let Y be the corresponding number of liters of nitric acid needed to make Aqua Regia.

a. Sketch a graph in a coordinate plane to show the relationship between X and Y.

b. Reason about quantities and use math drawings to find and explain at least two equations that relate X and Y.

c. Discuss how to interpret the values of the ratio $\frac{3}{1}$ and $\frac{1}{3}$ in terms of your graph in part (a) and your equations in part (b).

3. A type of dark chocolate is made by mixing cocoa and cocoa butter in the ratio 5 to 2. Let C be an unspecified number of grams of cocoa, which can vary, and let B be the corresponding number of grams of cocoa butter needed to make that type of dark chocolate. Reason about quantities and use math drawings to find and explain three different equations that relate C and B (and do not include any other variables).

4. A type of chemical solution is 6% sodium chloride and 21% ammonium hydroxide. The sodium chloride and ammonium hydroxide are therefore in the ratio 6 to 21 or, equivalently, 2 to 7. Chemists made different amounts of the solution. Let S be an unspecified number of grams of sodium chloride, which can vary, and let A be the corresponding number of grams of ammonium hydroxide in this type of chemical solution. Reason about quantities and use math drawings to help you find and explain three different equations that relate S and A (and do not include any other variables).

5. As in the text, consider biodiesel and petrodiesel mixtures that are 30% biodiesel and 70%

petrodiesel. Let B stand for an unspecified number of liters of biodiesel, which can vary, and let P stand for the corresponding number of liters of petrodiesel in such mixtures. Reason about quantities and our definition of multiplication, and use a math drawing to explain the equation

$$\frac{1}{3} \cdot B = \frac{1}{7} \cdot P$$

6. Peacock Purple Paint is made by mixing red paint and blue paint in the ratio 3 to 7. Let R be an unspecified number of liters of red paint, which can vary, and let B be the corresponding number of liters of blue paint in Peacock Purple Paint. Reason about quantities and our definition of multiplication, and use math drawings to find and explain the following types of equations.

a. $c \cdot R = B$, where c is a suitable constant (which you should find).

b. $k \cdot B = R$, where k is a suitable constant (which you should find).

c. $\frac{1}{u} \cdot R = \frac{1}{v} \cdot B$, where $\frac{1}{u}$ and $\frac{1}{v}$ are suitable unit fractions.

7. A company mixes fertilizer and soil in a 4 to 9 ratio by weight, but the amounts of fertilizer and soil the company mixes vary. Let F be an unspecified number of pounds of fertilizer, which can vary, and let S be the corresponding number of pounds of soil the company might use. For each of the following equations and math drawings, discuss whether or not the equation is correct. If it is, explain how to reason about quantities and use a math drawing to obtain the equation. If the equation is not correct, explain why not and discuss how a student might come up with it.

a. $4 \cdot S = 9 \cdot F$. See Figure 7.33.

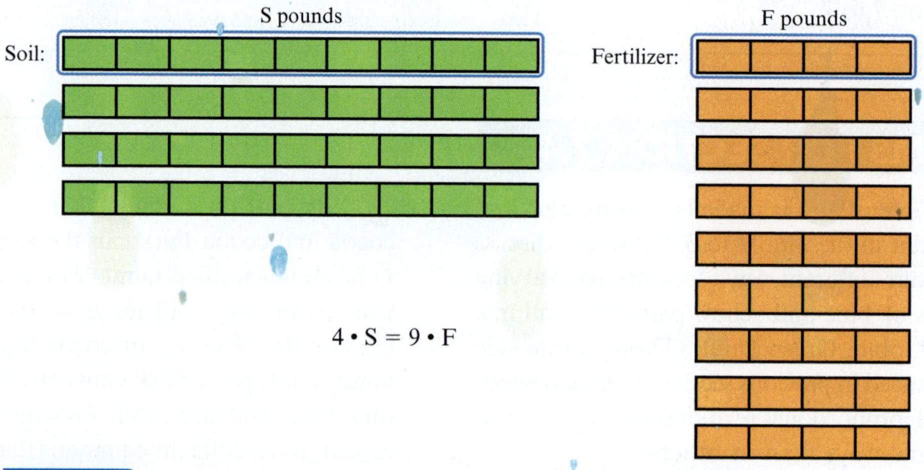

Figure 7.33 Critique or explain the equation and math drawing.

b. $9 \cdot S = 4 \cdot F$. See Figure 7.34.

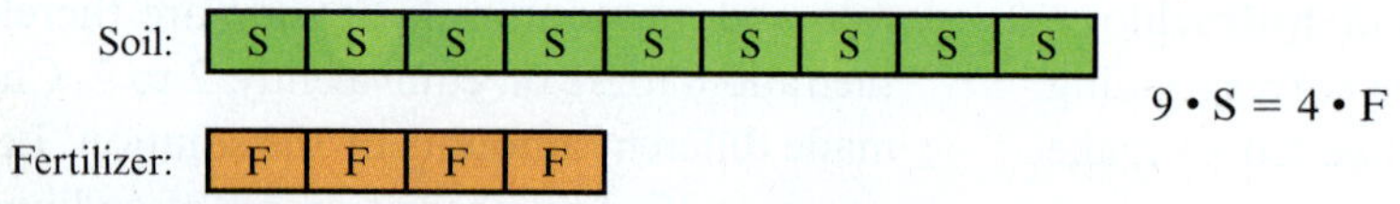

Figure 7.34 Critique or explain the equation and math drawing.

c. $\frac{1}{9} \cdot S = \frac{1}{4} \cdot F$. See Figure 7.35.

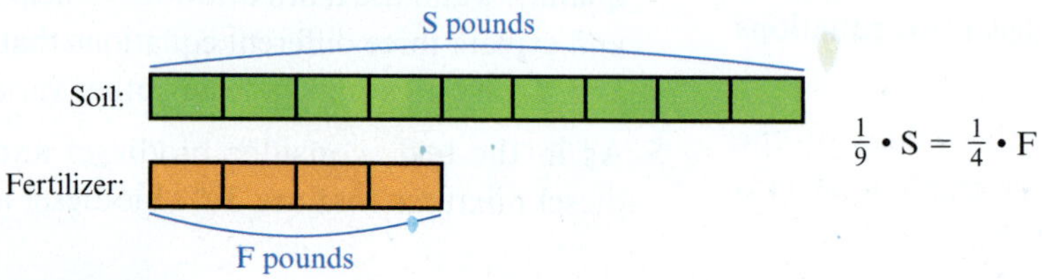

Figure 7.35 Critique or explain the equation and math drawing.

8. Consider ramps whose length and height are in the ratio 11 to 4. Let x be an unspecified number of feet, standing for the length of a ramp, and let y be the corresponding height of the ramp in feet. Think of x and y as varying together to make ramps of different lengths and heights, all in ratio 11 to 4.

a. Sketch a graph in a coordinate plane to show the relationship between the lengths and heights of these ramps.

b. Reason about quantities, and our definition of multiplication, and use math drawings to derive and explain equations of the form

$$y = c \cdot x \quad \text{and} \quad x = k \cdot y$$

where c and k are suitable constants of proportionality (find c and k).

c. Explain how to interpret the constants of proportionality c and k from part (b) in terms of your graph in part (a) in several different ways.

9. Consider the line through the origin and the point $(5, 7)$.

a. Pretend that you do not already know that equations of lines have a particular form. Derive and explain an equation for the line of the form

$$y = m \cdot x$$

where m is a suitable number. In your explanation, attend to our definition of multiplication and explain why the equation holds for all points on the line (that have positive x- and y-coordinates).

b. In your own words, explain how to interpret m as the slope of the line.

10. A company makes mixtures of acetic acid and water such that the acetic acid is 15% of the total mass (weight) of the mixture. Let A be an unspecified number of grams of acetic acid, which can vary, and let W be the corresponding number of grams of water in this type of mixture. Reason about quantities and use math drawings to find and explain at least two different equations that relate A and W and do not include any other variables.

7.5 Proportional Relationships Versus Inversely Proportional Relationships

CCSS Common Core State Standards Grades 7, 8

Some problems may seem as if they should be solved with a proportion, but in fact they cannot be solved that way. How can we tell the difference between problems that can and cannot be solved with a proportion? We will answer this question in this section by contrasting proportional relationships with other relationships, especially *inversely proportional relationships*.

First, recall that proportional relationships are quantities in a fixed ratio. So if we double one quantity, then we also double the other quantity. If we multiply one quantity by 3, then we also multiply the other quantity by 3. If we divide one quantity by 2, then we also divide the other quantity by 2, and so on. But there are other ways in which two quantities can be related. Although we might show the relationship in a table, not every table is a *ratio table* that shows quantities *in the same ratio*.

Before you read on, try Class Activities 7M and 7N.

CLASS ACTIVITY

7M How Are the Quantities Related?, p. CA-131

7N Can You Use a Proportion or Not?, p. CA-133

inversely proportional relationship

Some relationships that may seem at first glance to be proportional relationships are actually *inversely proportional relationships*. A relationship between two quantities is called an **inversely proportional relationship** if whenever one quantity is *multiplied* (respectively, divided) by a positive number N, the other quantity is *divided* (respectively, multiplied) by N.

CLASS ACTIVITY

7O A Proportional Relationship Versus an Inversely Proportional Relationship, p. CA-134

To examine the difference between proportional and inversely proportional relationships, consider an idealized grass-mowing scenario in which people mow grass at the same steady rate and work together at the same time to complete a mowing job. Let's say it takes 2 people 8 hours to mow a 15-acre plot of grass. Now let's consider two kinds of variation: (1) If we vary the number of people, but keep the 8 hours fixed, how does the number of acres they mow vary? (2) If we vary the number of people, but keep the 15 acres fixed, how does the number of hours it takes to complete the job vary?

CCSS 7.RP.2a

What is the relationship between the number of people and the number of acres they mow in 8 hours? The math drawing in Figure 7.36 illustrates how $4 \cdot 2$ people mow $4 \cdot 15$ acres of grass in 8 hours. In general $N \cdot 2$ people will mow $N \cdot 15$ acres of grass in 8 hours, so the number of people and the number of acres they mow is always in the ratio 2 to 15 and is therefore a proportional relationship.

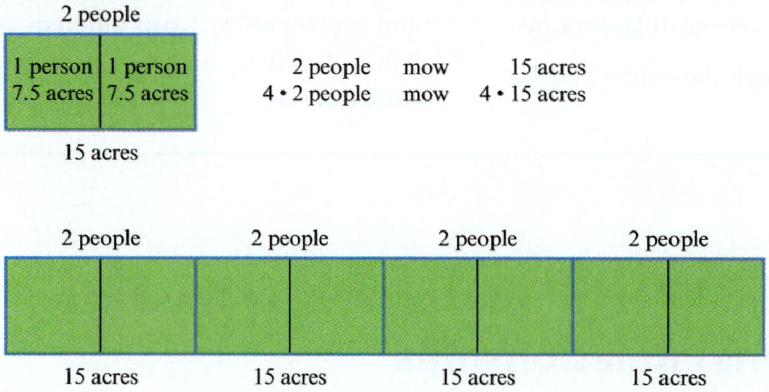

Figure 7.36 The proportional relationship between the number of people mowing and the number of acres they mow in 8 hours.

What is the relationship between the number of people and the number of hours it takes them to mow 15 acres? The math drawing in Figure 7.37 illustrates that if there are 4 times as many people, then the area that would be mowed by 1 person is now divided equally among 4 people. Because they are working together at the same time, the time it takes to finish the mowing is also divided by 4. So if 2 people take 8 hours to mow the plot, then $2 \cdot 4$ people take $8 \div 4$ or $\frac{1}{4} \cdot 8$ hours to mow the plot. In general, N times as many people will divide the work among them by N, so $2 \cdot N$ people will take $\frac{1}{N} \cdot 8$ hours to complete the job. Therefore the relationship between the number of people and the number of hours it takes to mow a 15-acre plot of grass is an inversely proportional relationship.

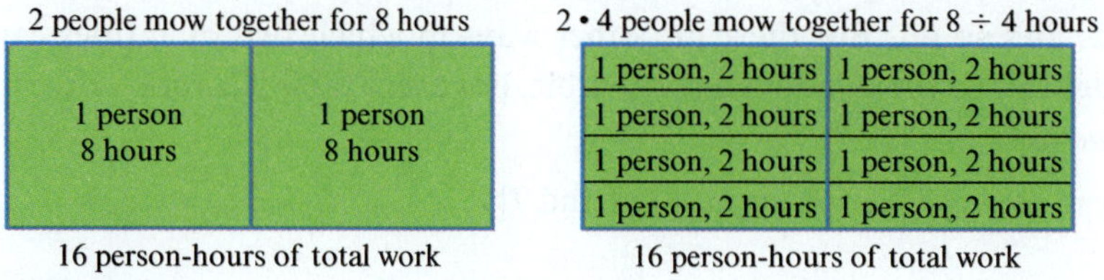

Figure 7.37 The inversely proportional relationship between the number of people mowing and the number of hours it takes them to mow a 15-acre plot together.

You can see where the name "inversely proportional relationship" comes from. "Inversely" fits because as the number of people goes *up*, the number of hours goes *down*. "Proportional" fits because amounts are related through multiplication and division.

person-hours

Notice how we can tell from **Figure 7.37** that if 1 person were to do all the mowing alone, it would take 16 hours. These 16 hours are the number of **person-hours** (or man-hours) it takes to do the work. No matter how many people do the job together, those 16 hours of work must be divided among them. In other words

$$\text{(number of people mowing together)} \cdot \text{(number of hours)} = 16 \text{ person-hours}$$

Person-hours apply more generally, including to situations that involve a fixed amount of work that can be divided among workers. Inversely proportional relationships often arise in such situations.

In general, when quantities vary together in an inversely proportional relationship, they have a *constant product*. In contrast, for proportional relationships, the quantities have a constant quotient, the constant of proportionality, which we can interpret as a unit rate.

CCSS 8.F.3

Figure 7.38 summarizes the two contrasting mowing relationships. Notice that unlike proportional relationships, when we graph an inversely proportional relationship the points *do not* lie on

Figure 7.38 Contrasting a proportional relationship with an inversely proportional relationship.

Suppose that 2 people take 8 hours to mow 15 acres of grass. (Assume all the people work together at the same time and at the same steady pace.)

Proportional Relationship

Relationship: Number of people ⟷ Number of acres when working 8 hours.

2 times as many people mow 2 times as many acres.

$\frac{1}{2}$ as many people mow $\frac{1}{2}$ as many acres.

÷ 2 · 2 · 2

Acres	7.5	15	30	60
People	1	2	4	8

÷ 2 · 2 · 2

· 3

Acres	7.5	15	22.5	30
People	1	2	3	4

· 3

N times as many people mow *N* times as many acres.

Number of acres ÷ Number of people = 7.5

The unit rate of acres per person remains the same (7.5) no matter how many people.

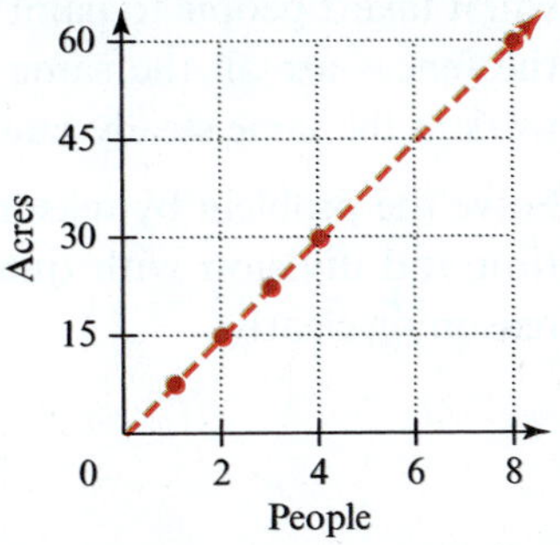

(a)

***Inversely* Proportional Relationship**

Relationship: Number of people ⟷ Number of hours when mowing 15 acres.

2 times as many people take $\frac{1}{2}$ as long.

$\frac{1}{2}$ as many people take 2 times as long.

· 2 ÷ 2 ÷ 2

Hours	16	8	4	2
People	1	2	4	8

÷ 2 · 2 · 2

÷ 3

Hours	16	8	5.33	4
People	1	2	3	4

· 3

N times as many people take $\frac{1}{N}$ as long.

Number of people · Number of hours = 16

The number of person-hours remains the same (16) no matter how many people.

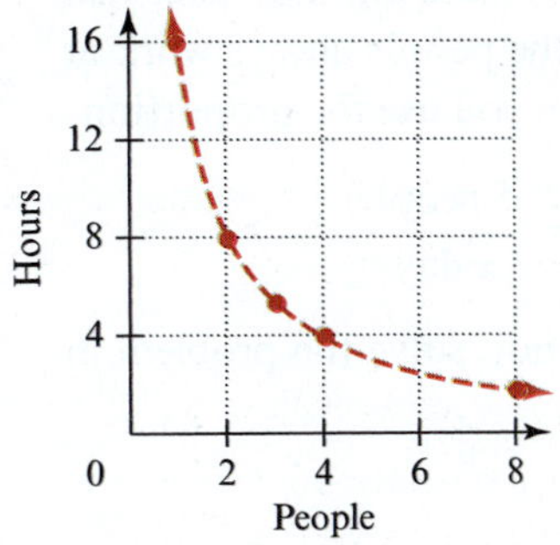

(b)

a line. Notice also that the number of hours it takes 3 people to mow is *not* halfway between the number of hours it takes 2 people and 4 people to mow, in contrast with the way proportional relationships behave.

SECTION SUMMARY AND STUDY ITEMS

Section 7.5 Proportional Relationships Versus Inversely Proportional Relationships

Some relationships between quantities may seem like proportional relationships but are not. Inversely proportional relationships have the property that whenever one quantity is multiplied (respectively, divided) by a positive number N, the other quantity is *divided* (respectively, *multiplied*) by N.

Key Skills and Understandings

1. Recognize problems that should and should not be solved with a proportion.
2. Recognize and give examples of inversely proportional relationships and distinguish them from proportional relationships. Represent inversely proportional and proportional relationships with tables and graphs and reason about quantities to find entries in the tables.
3. Solve problems involving inversely proportional relationships, including multistep problems.

Practice Exercises for Section 7.5

1. What type of relationship is there between the number of house painters and the number of hours it takes to paint a house? (Assume that all house painters work at the same steady rate.) How can you tell?
2. Suppose that 4 house painters take 20 hours to paint a house. (Assume that all house painters work at the same steady rate.) Make a table to show the relationship between the number of house painters and the number of hours it takes to paint a house. Include the case of 3 house painters in your table.
3. If 4 people take 5 days to build a (long) fence, how long will it take 5 people to build another fence just like it? (Assume that all the people always work at the same steady rate.) Can you use the proportion

$$\frac{4 \text{ people}}{5 \text{ days}} = \frac{5 \text{ people}}{x \text{ days}}$$

to solve the problem? If not, solve the problem in another way.

4. Suppose that a logging crew can cut down 5 acres of trees every 2 days. Assume that the crew works at a steady rate. Solve exercises (a) and (b) by reasoning about multiplication and division with quantities. Explain your reasoning clearly.
   a. How many days will it take the crew to cut 8 acres of trees? Give your answer as a mixed number.
   b. Now suppose there are 3 logging crews that all work at the same rate as the original one. How long will it take these 3 crews to cut down 10 acres of trees?
5. If 3 people take 2 days to paint 5 fences, how long will it take 2 people to paint 1 fence? (Assume that the fences are all the same size and the painters work at the same steady rate.)

   Solve the problem by reasoning about multiplication and division with quantities. Explain your reasoning clearly.

Answers to Practice Exercises for Section 7.5

1. The relationship between the number of house painters and the number of hours it takes to paint a house is inversely proportional. This is because 2 times as many painters should take $\frac{1}{2}$ as long to paint a house, and in general N times as many painters should take $\frac{1}{N}$ times as long to paint the house because they can split the work among them.

2. Because half as many house painters take twice as long, we know 2 house painters will take 40 hours and 1 house painter will take 80 hours. Then 3 house painters take $\frac{1}{3}$ as long as 1 house painter, so $80 \div 3 = 26\frac{2}{3}$ hours. Because twice as many house painters take half as long, we can also find that 8 painters take 10 hours and 16 painters take 5 hours.

Hours	80	40	$26\frac{2}{3}$	20	10	5
People	1	2	3	4	8	16

3. The proportion does not apply because the relationship between the number of people and the number of days to build a fence is inversely proportional, not proportional. This is because N times as many people will take $\frac{1}{N}$ times as long to build a fence because they can split the work among them. Instead, we can reason that 1 person will take 4 times as long as 4 people, so it will take them $4 \cdot 5 = 20$ days to build such a fence. Therefore 5 people will take $\frac{1}{5}$ as long—namely, $\frac{1}{5} \cdot 20 = 4$ days to build such a fence.

4. a. Because the crew cuts 5 acres every 2 days, it will cut half as much in 1 day—namely, $2\frac{1}{2}$ acres. To determine how many days it will take to cut 8 acres, we must determine how many groups of $2\frac{1}{2}$ are in 8, which is $8 \div 2\frac{1}{2} = 3\frac{1}{5}$ days.

b. In exercise (a) we saw that 1 crew cuts $2\frac{1}{2}$ acres per day. Therefore, 3 crews will cut 3 times as much per day, which is $3 \cdot 2\frac{1}{2} = 7\frac{1}{2}$ acres per day. To determine how many days it will take to cut 10 acres, we must figure out how many groups of $7\frac{1}{2}$ are in 10. This is solved by the division problem $10 \div 7\frac{1}{2}$. Because $10 \div 7\frac{1}{2} = 10 \div \frac{15}{2} = \frac{10}{1} \cdot \frac{2}{15} = 1\frac{1}{3}$, we conclude that it will take the 3 crews $1\frac{1}{3}$ days to cut 10 acres.

5. If 3 people take 2 days to paint 5 fences, then those 3 people will take $2 \div 5 = \frac{2}{5}$ of a day to paint just 1 fence (dividing the 2 days equally among the 5 fences). If just 1 person were painting, it would take 3 times as long to paint the fence—namely, $3 \cdot \frac{2}{5} = \frac{6}{5}$ days. With 2 people painting, it will take half as much time to paint the fence—namely, $\frac{3}{5}$ of a day.

PROBLEMS FOR SECTION 7.5

1. Suppose that 6 people can stuff flyers into 500 envelopes in 5 minutes. Assume all people work at the same steady rate.

a. The relationship between the number of people stuffing envelopes and the number of minutes it takes to stuff flyers into 500 envelopes is what type of relationship? How can you tell? Make a table and a graph to show the relationship and explain how to find several of the entries. Include an entry for 5 people.

b. The relationship between the number of people stuffing envelopes and the number of envelopes they can stuff in 5 minutes is what type of relationship? How can you tell? Make a table and a graph to show the relationship and explain how to find several of the entries. Include an entry for 5 people.

2. Make up two examples of relationships between two quantities. One example should be a proportional relationship and the other should be an inversely proportional relationship. Make clear which is which. For each example, provide a table to show the relationship and explain how to find several of the entries.

3. *A Sewing Problem:* If 10 workers take 8 hours to sew a store's order of pants, then how long would 15 workers take to sew the store's order of pants? Assume all workers work at the same steady rate and all the pants are the same.

a. Is the proportion

$$\frac{8 \text{ hours}}{10 \text{ workers}} = \frac{X \text{ hours}}{15 \text{ workers}}$$

appropriate for solving the sewing problem? Why or why not?

b. Explain how to reason about multiplication and division with quantities to solve the sewing problem.

4. *A Driving Problem:* Driving at 50 mph, you covered the distance between two markers in 75 seconds. How long would it take you to cover that same distance driving at 60 mph?

a. Is the proportion

$$\frac{50 \text{ mph}}{75 \text{ seconds}} = \frac{60 \text{ mph}}{x \text{ seconds}}$$

appropriate for solving the driving problem? Why or why not?

b. Explain how to reason about multiplication and division with quantities to solve the driving problem.

5. For each of the following relationships, determine if it is proportional, inversely proportional, or neither. Explain your answer in each case.

a. John and David are running around the same track at the same speed. When David started running, John had already run 3 laps. Consider the relationship between the number of laps that David has run and the number of laps that John has run.

b. Kacey is running 100-meter dashes. Each time she runs at a different, but constant speed. Consider the relationship between the speed at which Kacey runs and the time it takes Kacey to run 100 meters.

c. Maleka is running at a constant speed. Consider the relationship between the time Maleka has spent running and the distance she has run.

d. Water is draining out of a tub at a constant rate. Consider the relationship between the time that has elapsed since water started draining out of the tub and the volume of water remaining in the tub.

e. Water is pouring into a tub at a constant rate. Consider the relationship between the time that has elapsed since water started pouring into the tub and the volume of water in the tub.

6. Suppose that you have two square garden plots: One is 10 feet by 10 feet and the other is 15 feet by 15 feet. You want to cover both gardens with a 1-inch layer of mulch. If the 10-by-10 garden took $3\frac{1}{2}$ bags of mulch, could you calculate how many bags of mulch you'd need for the 15-by-15 garden by setting up the following proportion

$$\frac{3\frac{1}{2}}{10} = \frac{x}{15}$$

Explain clearly why or why not. If the answer is no, is there another proportion that you could set up? It may help you to make drawings of the gardens.

7. If a crew of 3 people takes $2\frac{1}{2}$ hours to clean a house, then how long should a crew of 2 take to clean the same house? Assume that all people in the cleaning crew work at the same steady rate. Explain how to reason about multiplication and division with quantities to solve this problem.

8. If you can rent 5 DVDs for 5 nights for $5, then at that rate, how much should you expect to pay to rent 1 DVD for 1 night? Explain how to reason about multiplication and division with quantities to solve this problem.

9. If 6 people take 3 days to dig 8 ditches, then how long would it take 4 people to dig 10 ditches? Assume that all the ditches are the same size and take equally long to dig, and that all the people work at the same steady rate. Explain how to reason about multiplication and division with quantities to solve this problem.

***10.** A candy factory has a large vat into which workers pour chocolate and cream. Each ingredient flows into the vat from its own special hose, and each ingredient comes out of its hose at a constant rate. Workers at the factory know that it takes 20 minutes to fill the vat with chocolate from the chocolate hose, and it takes 15 minutes to fill the vat with cream from the cream hose. If workers pour both chocolate and cream into the vat at the same time (each coming full tilt out of its own hose), how long will it take to fill the vat? Before you find an exact answer to this problem, find an approximate answer, or find a range, such as "between ___ and ___ minutes." Explain your reasoning.

***11.** Jay and Mark run a lawn-mowing service. Mark's mower is twice as big as Jay's, so whenever they both mow, Mark mows twice as much as Jay in a given time period. When Jay and Mark are working together, it takes them 4 hours to cut the lawn of an estate. How long would it take Mark to mow the lawn by himself? How long would it take Jay to mow the lawn by himself? Explain your answers.

***12.** If liquid pouring at a steady rate from hose A takes 15 minutes to fill a vat, and liquid pouring at a steady rate from hose B takes 10 minutes to fill the same vat, then how long will it take for liquid pouring from both hose A and hose B to fill the vat? Explain how to reason about quantities to solve this problem.

***13.** Suppose that there are 400 pounds of freshly picked tomatoes and that 99% of their weight is water. After one day, the same tomatoes only weigh 200 pounds due to evaporation of water. (The tomatoes consist of water and solids. Only the water evaporates; the solids remain.)

a. How many pounds of solids are present in the tomatoes? (Notice that this is the same when they are freshly picked as after one day.)

b. Therefore, when the tomatoes weigh 200 pounds, what percent of the tomatoes is water?

c. Is it valid to use the following proportion to solve for the percent of water, x, in the tomatoes when they weigh 200 pounds?

$$\frac{0.99}{400} = \frac{x}{200}$$

If not, why not?

7.6 Percent Revisited: Percent Increase and Decrease

CCSS **Common Core State Standards** Grade 7

When a quantity increases or decreases, determining the amount of change is a simple matter of subtraction. However, in many situations, the actual value of the increase or decrease is less informative than the *percent* that this increase or decrease represents. For example, suppose that this year, there are 50 more children at Barrow Elementary School than there were last year. If Barrow Elementary had only 100 children last year, then 50 additional children is a very large increase. On the other hand, if Barrow Elementary had 500 children last year, an increase of 50 children is less significant. In this section, we will study increases and decreases in quantities as *percents* rather than as fixed values.

CLASS ACTIVITY

7P How Should We Describe the Change?, p. CA-135

percent increase

If the value of a quantity goes up, then the increase in the quantity, figured as a percent of the original, is the **percent increase** of the quantity. If a school had 400 students and now has 500 students, then that is 100 more students. What is the percent increase in the number of students? Because 100 is 25% of the original 400, the student enrollment increased by 25%.

percent decrease

When a quantity decreases in value, there is the notion of a **percent decrease**. It is the decrease, figured as a percent of the original. If a school had 500 students and now has 400 students, then that is 100 fewer students. What is the percent decrease in the number of students? Because 100 is 20% of the original 500, the student enrollment decreased by 20%.

CCSS 7.RP.3

How Can We Reason to Calculate Percent Increase or Decrease?

To calculate a percent increase or decrease, we can use their definitions, or we can use a method that derives from the distributive property.

CLASS ACTIVITY

7Q Calculating Percent Increase and Decrease, p. CA-135

How Can We Use the Definitions? One way to calculate a percent increase or percent decrease is to use their definitions, as in the previous examples about the increases and decreases in student enrollment at schools. In general, the method works as follows: Suppose a quantity changes from an amount A—the reference amount—to an amount B:

$$\underset{\text{Reference amount}}{A} \quad \rightarrow \quad \underset{\text{Changed amount}}{B}$$

To calculate the percent increase or decrease in the quantity, we use the following procedure:

1. Calculate the *change*, C, in the quantity—namely, either $B - A$ or $A - B$, whichever is positive (or 0).
2. Calculate the percent that the change C is of the reference amount A (using any method from Section 2.5). The result is the percent increase or decrease of the quantity from A to B.

CCSS 7.EE.2

How Can We Reason with the Distributive Property? Another method for calculating percent increase or decrease is often more efficient than using the definitions. Suppose that a quantity increases from an amount A to a larger amount B. If we calculate B as a percent of A, we will find that it is more than 100%. This makes sense because B is more than A, so B must represent more than 100% of A. *The amount by which B, calculated as a percent of A, exceeds 100% is the percent increase from A to B.* We can see why from Figure 7.39, which shows the connection between the percent that B is of A and the percent increase from A to B.

Figure 7.39 Connecting the percent increase from A to B with the percent that B is of A.

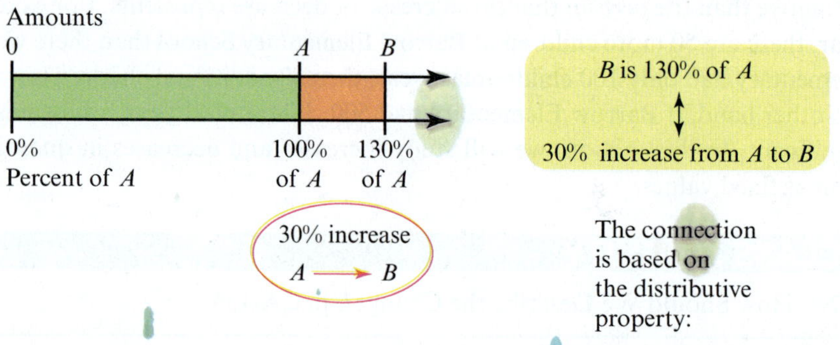

If: $B = 130\% \cdot A$	If: 30% increase from A to B
Then:	Then:
$B - A = 130\% \cdot A - 100\% \cdot A$ $= (130\% - 100\%) \cdot A$ (Distributive property) $= 30\% \cdot A$	$B = A + 30\% \cdot A$ $= 100\% \cdot A + 30\% \cdot A$ $= (100\% + 30\%) \cdot A$ (Distributive property) $= 130\% \cdot A$

Therefore, the following steps describe how to calculate the percent increase from A to a larger amount B:

1. Calculate B as a percent of A.
2. Subtract 100%. The result is the percent increase of the quantity from A to B.

For example, suppose that the population of a city increases from 35,000 to 44,000. What is the percent increase in the population? First, let's calculate what percent 44,000 is of 35,000. Solving

$$\frac{44{,}000}{35{,}000} = \frac{P}{100}$$

We find that 44,000 is about 126% of 35,000. Subtracting 100%, we determine that the population increased by about 26%.

Similarly, if a quantity decreases from an amount A to a smaller amount B, then if we calculate B as a percent of A, we will find that it is less than 100%. This makes sense because B is less than A, so B must represent less than 100% of A. *The amount that B, calculated as a percent of A, is under 100% is the percent decrease from A to B.* We can see why from Figure 7.40, which shows the connection between the percent that B is of A and the percent decrease from A to B.

Figure 7.40

Connecting the percent decrease from A to B with the percent that B is of A.

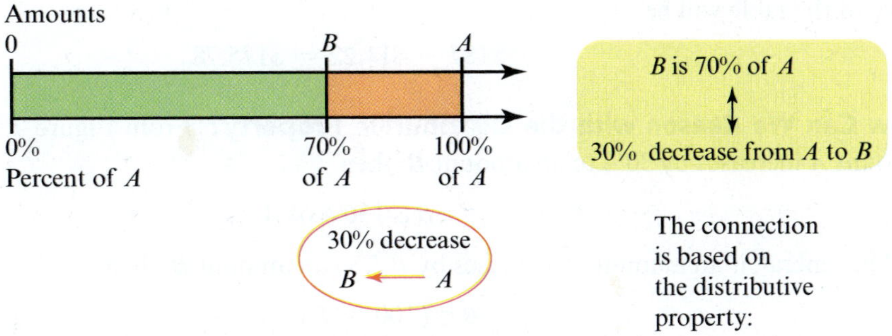

If: $B = 70\% \cdot A$	If: 30% decrease from A to B
Then:	Then:
$A - B = 100\% \cdot A - 70\% \cdot A$ $= (100\% - 70\%) \cdot A$ (Distributive property) $= 30\% \cdot A$	$B = A - 30\% \cdot A$ $= 100\% \cdot A - 30\% \cdot A$ $= (100\% - 30\%) \cdot A$ (Distributive property) $= 70\% \cdot A$

Therefore, the following steps describe how to calculate the percent decrease from A to a smaller amount B:

1. Calculate B as a percent of A.
2. Subtract this percent from 100%. The result is the percent decrease of the quantity from A to B.

For example, suppose that the price of a computer dropped from \$899 to \$825. What is the percent decrease in the price of the computer? First, let's calculate what percent 825 is of 899. Solving

$$\frac{825}{899} = \frac{P}{100}$$

We find that 825 is about 92% of 899. Since $100\% - 92\% = 8\%$, the price of the computer decreased by about 8%.

How Can We Reason to Calculate Amounts When the Percent Increase or Decrease Is Given?

There are also two methods for calculating a new amount when we know the original amount and the percent increase or decrease. The second method is especially important because it allows us to calculate the original amount if we know the final amount and the percent increase or decrease.

CLASS ACTIVITY

7R Calculating Amounts from a Percent Increase or Decrease, p. CA-136

How Can We Use the Definitions of Percent Increase and Decrease? When an amount and its percent increase are given, we can calculate the new amount by calculating the increase and adding this increase to the original amount. Similarly, when an amount and its percent decrease are given, we can calculate the new amount by calculating the decrease and subtracting this decrease from the original amount.

If a table costs \$187 and if the price of the table goes up by 6%, then what will the new price of the table be? We find that 6% of \$187 is

$$0.06 \cdot \$187 = \$11.22$$

Thus, the new price of the table will be

$$\$187 + \$11.22 = \$198.22$$

On the other hand, if a table costs \$187 and the price of the table goes down by 6%, then the new price of the table will be

$$\$187 - \$11.22 = \$175.78$$

How Can We Reason with the Distributive Property? From Figure 7.39 we see that if an amount A increases by 30% to an amount B, then

$$B \text{ is } 130\% \text{ of } A$$

and in general, if an amount A increases by P% to an amount B, then

$$B \text{ is } (100 + P)\% \text{ of } A$$

So, if a table costs \$187 and if the price of the table goes up by 6%, then the new price of the table will be $(100 + 6)\% = 106\%$ of \$187, which is

$$1.06 \cdot \$187 = \$198.22$$

The new price of the table after the 6% increase is \$198.22.

Notice that we can also solve the following type of problem, where the percent increase and the final amount are known, and the initial amount is to be calculated: If the price of gas went up 5% and is now \$3.95 per gallon, then how much did gas cost before this increase? Because the price went up by 5%, the new price is 105% of the initial price. Table 7.3 shows how to use a percent table to solve the problem.

TABLE 7.3 Calculating the initial price of gas before a 5% increase by calculating 1% of the initial price and then 100% of the initial price

			After a 5% increase, the price of gas is \$3.95.
105%	→	\$3.95	Therefore, 105% of the initial price is \$3.95.
1%	→	\$3.95 ÷ 105 = \$0.0376	So, 1% of the initial price is \$0.0376.
100%	→	100 · \$0.0376 = \$3.76	Thus, 100% of the initial price of gas is \$3.76.

The situation is similar when the percent decrease is given. From Figure 7.40, we see that if an amount A decreases by 30% to an amount B, then

$$B \text{ is } 70\% \text{ of } A$$

and in general, if an amount A decreases by P% to an amount B, then

$$B \text{ is } (100 - P)\% \text{ of } A$$

So, if a table costs \$187 and if the price of the table goes down by 6%, then the new price of the table will be $(100 - 6)\% = 94\%$ of \$187, which is

$$0.94 \cdot \$187 = \$175.78$$

Therefore, after the 6% decrease in price, the new price of the table is \$175.78.

Once again, we can also solve the following type of problem, where the percent decrease and the final amount are known, and the initial amount is to be calculated: If the price of gas went down 5% and is now \$3.95 per gallon, then how much did gas cost before this decrease? Table 7.4 shows how to use a percent table to solve the problem.

TABLE 7.4 Calculating the initial price of gas before a 5% decrease by calculating 1% of the initial price and then 100% of the initial price

			After a 5% decrease, the price of gas is \$3.95.
95%	→	\$3.95	Therefore, 95% of the initial price is \$3.95.
1%	→	$\$3.95 \div 95 = \$0.0416$	So, 1% of the initial price is \$0.0416.
100%	→	$100 \cdot \$0.0416 = \$4.16$	Thus, 100% of the initial price of gas is \$4.16.

Why Is It Important to Attend to the Reference Amount?

When you find a percent increase or decrease, make sure you calculate the increase or decrease as a percent *of the reference amount*—namely, the amount for which you want to know the percent increase or decrease. The reference amount is the whole or 100% for the problem. Let's say the price of a Dozey-Chair increases from \$200 to \$300. Then the new price is 50% more than the old price, but the old price is only about 33% less than the new price. In both percent calculations, the change is \$100. The different percentages come about because of the different reference amounts or wholes that are used. In the first case, the reference amount is \$200, and \$100 is 50% of \$200. In the second case, the reference amount is \$300, and \$100 is only about 33% of \$300.

CLASS ACTIVITY

7S Can We Solve It This Way?, p. CA-137

7T Percent Problem Solving, p. CA-138

7U Percent Change and the Commutative Property of Multiplication, p. CA-139

SECTION SUMMARY AND STUDY ITEMS

Section 7.6 Percent Revisited: Percent Increase and Decrease

If the value of a quantity goes up, then to find the increase in the quantity, we subtract. But to find the *percent* increase, we must figure the increase as a percent of the original. If the value of a quantity goes down, then to find the decrease in the quantity, we subtract. But to find the *percent* decrease, we must figure the decrease as a percent of the original. As an application of the distributive property, we can calculate a percent increase by calculating what percent the increased amount is of the original amount and subtracting 100%. Similarly, we can calculate a percent decrease by calculating what percent the decreased amount is of the original amount and subtracting this from 100%. These distributive-property methods allow us to calculate quantities from a given percent increase or decrease.

Key Skills and Understandings

1. Distinguish additive change from percent change.
2. Calculate percent increase and percent decrease in several different ways, and explain why the calculation methods make sense.
3. Calculate quantities from a given percent increase or decrease, and explain why the calculation method makes sense.
4. Distinguish between percent increase/decrease and percent *of*.
5. Solve problems involving percent increase or decrease.

Practice Exercises for Section 7.6

1. Last year, Ken had 2.5 tons of sand in his sand pile. This year, he has 3.5 tons of sand in the pile. By what percent did Ken's sand pile increase from last year to this year? First explain how to solve the problem with a math drawing. Then solve the problem numerically.

2. Last year's profits were \$16 million, but this year's profits are only \$6 million. By what percent did profits decrease from last year to this year? First explain how to solve the problem with a math drawing. Then solve the problem numerically.

3. If sales taxes are 6%, then how much should you charge for an item so that the total cost, including tax, is \$35?

4. A pair of shoes has just been reduced from \$75.95 to \$30.38. Fill in the blanks:

   a. The new price is _____% less than the old price.

   b. The new price is _____ % of the old price.

   c. The old price is _____% higher than the new price.

   d. The old price is _____ % of the new price.

5. John bought a piece of land adjoining the land he owns. Now John has 25% more land than he did originally. John plans to give 20% of his new, larger amount of land to his daughter. Once John does this, how much land will John have in comparison with the amount he had originally? Make a math drawing to help you solve this problem. Then solve the problem numerically, assuming, for example, that John starts with 100 acres of land.

6. The population of a certain city increased by 2% from 2014 to 2015 and then decreased by 2% from 2015 to 2016. By what percent did the population of the city change from 2014 to 2016? Did the population increase, decrease, or stay the same? Make a guess first, then calculate the answer carefully.

7. There are two boxes of chocolate. The chocolate in the second box weighs 15% more than the chocolate in the first box. There are 3 more ounces of chocolate in the second box than in the first. How much does the chocolate in the two boxes together weigh? Explain your reasoning.

8. There are two vats of grape juice. After 20% of the juice in the first vat was poured into the second vat, the first vat had 2 times as much juice as the second vat. By what percent did the amount of juice in the second vat increase when the juice was poured into it? Explain your reasoning.

Answers to Practice Exercises for Section 7.6

1. Using a math drawing, we see that each darker strip in Figure 7.41 represents 20% of last year's sand pile. Since two additional strips have been added since last year, that is a 40% increase.

   To calculate the percent increase numerically, let's first calculate 3.5 as a percent of 2.5.

$$\frac{3.5}{2.5} = 1.4 = 140\%$$

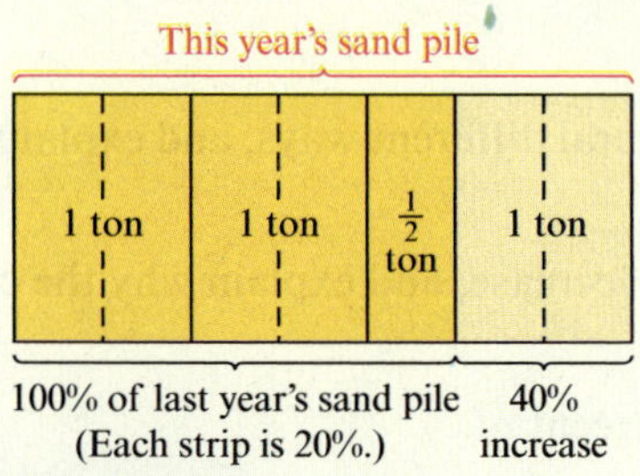

Figure 7.41 Last year's and this year's sand pile.

Subtracting 100%, we see that the percent increase is 40%.

2. Using a math drawing, we see that each strip in Figure 7.42 represents \$2 million. As the drawing shows, the decrease in profits is $\frac{1}{8}$ more than 50%. Since $\frac{1}{8} = 12.5\%$, the profits decreased by 50% + 12.5%—namely, by 62.5%.

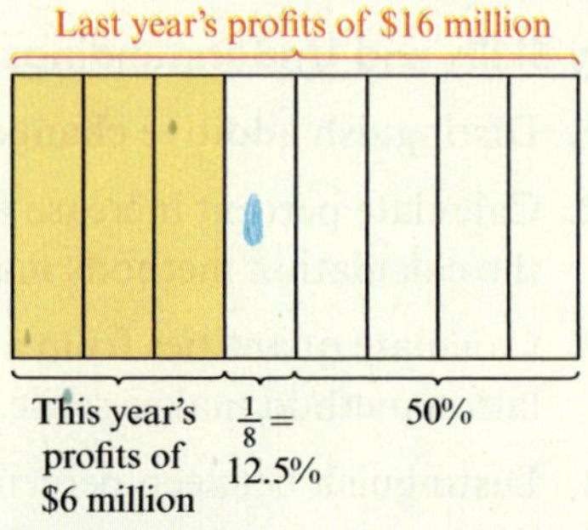

Figure 7.42 Last year's and this year's profits.

To calculate the percent decrease numerically, we can determine what percent the decrease, 10, is of 16.

$$\frac{10}{16} = 0.625 = \frac{62.5}{100} = 62.5\%$$

So, the profits went down by 62.5%.

Another way to calculate the percent decrease is to calculate what percent 6 is of 16 first, and then subtract this from 100%.

$$\frac{6}{16} = 0.375 = \frac{37.5}{100} = 37.5\%$$

Therefore, the percent decrease is 100% − 37.5%, which is 62.5%.

3. The sales tax increases the amount that the customer pays by 6%. Therefore, if P represents the price of the item, 106% of P must equal \$35. Thus,

$$1.06 \cdot P = \$35$$

So,

$$P = \$35 \div 1.06 = \$33.02$$

If the price of the item is \$33.02, then with a 6% sales tax, the total cost to the customer is \$35.

Table 7.5 shows how to solve the problem with a percent table.

TABLE 7.5 A percent table for calculating the amount which becomes \$35 when 6% is added

106%	→	\$35
1%	→	$\$35 \div 106 = \$0.3302$
100%	→	$\$0.3302 \cdot 100 = \$33.02$

4. a. The new price is 60% less than the old price.

b. The new price is 40% of the old price.

c. The old price is 150% higher than the new price.

d. The old price is 250% of the new price.

5. See Figure 7.43. If John starts with 100 acres of land and gets 25% more, then he will have 125 acres. We find that 20% of 125 acres is 25 acres, so if John gives 20% of his new amount of land away, he will have 125 − 25 = 100 acres of land, which is the amount he started with.

6. Although you might have guessed that the population stayed the same, it actually decreased by 0.04%. Try this on a city of 100,000, for example. You can tell that the population will have to decrease because when the population goes back down by 2%, this 2% is *of a larger number* than the original population.

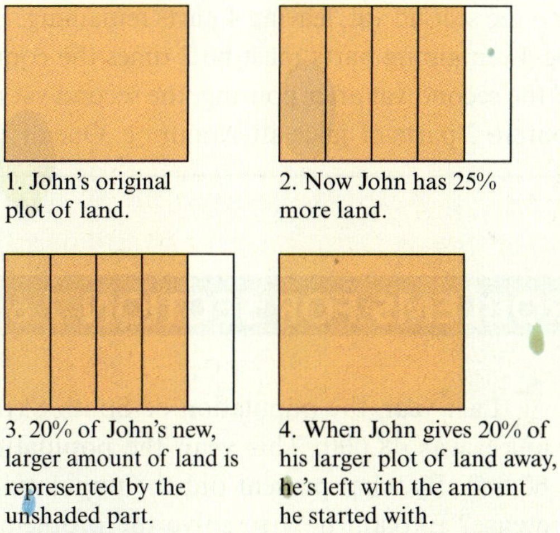

Figure 7.43 Solving the land problem with a drawing.

7. Since 15% of the chocolate in the first box is 3 ounces, the percent table in Table 7.6 shows that the chocolate in the first box weighs 20 ounces. Therefore, the chocolate in the second box weighs 20 + 3 = 23 ounces and the chocolate in the two boxes together weighs 20 + 23 = 43 ounces.

TABLE 7.6 A percent table for the chocolate in box 1

15%	→	3 ounces
5%	→	1 ounce
100%	→	20 ounces

8. See Figure 7.44. The contents of the first vat are represented by a long strip. Since 20% of the juice is poured out, and since $20\% = \frac{1}{5}$, the strip is broken into 5 equal parts. One of those parts will be poured

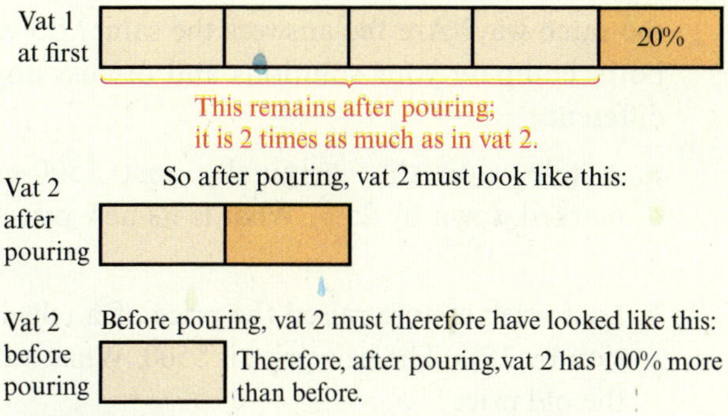

Figure 7.44 Percent change in juice in vats.

into the second vat, leaving 4 parts remaining. Since the 4 remaining parts must be 2 times the contents of the second vat after pouring, the second vat must contain 2 parts of juice after pouring. One of those parts was poured in. Therefore, the other part was there originally. Since an equal amount of juice was poured in as was there originally, the amount of juice in the second vat increased by 100%.

PROBLEMS FOR SECTION 7.6

1. Last year, the population of South Skratchankle was 48,000. This year, the population is 60,000. By what percent did the population increase? Explain how to solve the problem two different ways.

2. Last year's sales were \$7.5 million. This year's sales are only \$6 million. By what percent did sales decrease from last year to this year? Explain how to solve the problem two different ways.

3. Jayna and Lisa are comparing the prices of two boxes of cereal of equal weight. Brand A costs \$3.29 per box, and brand B costs \$2.87 per box. Jayna calculates that

$$\frac{\$3.29}{\$2.87} = 1.15$$

Lisa calculates that

$$\frac{\$2.87}{\$3.29} = 0.87$$

a. Use Jayna's calculation to make *two* correct statements comparing the prices of brand A and brand B with percentages. Explain.

b. Use Lisa's calculation to make *two* correct statements comparing the prices of brand A and brand B with percentages. Explain.

4. Are the two problems that follow solved in the same way? Are the answers the same? Solve both, compare your solutions and discuss any difference.

a. A television that originally cost \$500 is marked down by 25%. What is its new price?

b. Last week a store raised the price of a television by 25%. The new price is \$500. What was the old price?

5. The price of play equipment for the school has just been reduced by 25%. The new, reduced price of the play equipment is \$1500. Bob says he can find the original price (before the reduction) in the following way:

> First I noticed that 25% is $\frac{1}{4}$. Then I found $\frac{1}{4}$ of \$1500, which is \$375. Next I added \$1500 and \$375, so the original price was \$1875.

Is Bob's method correct or not? If it's correct, say so and also explain how to solve the problem in another way. If it's not correct, explain briefly why not and show how to modify Bob's method to solve the problem correctly.

6. How much should Swanko Jewelers charge now for a necklace if they want the necklace to cost \$79.95 when they reduce their prices by 60%? Explain the reasoning behind your method of calculation.

7. If sales taxes are 7%, then how much should you charge for an item if you want the total cost, including tax, to be \$15? Explain the reasoning behind your method of calculation.

8. Connie and Benton bought identical plane tickets, but Benton spent more than Connie.

a. If Connie spent 25% less than Benton, then did Benton spend 25% more than Connie? If not, then what percent more than Connie did Benton spend? Explain.

b. If Benton spent 25% more than Connie, then did Connie spend 25% less than Benton? If not, then what percent less than Benton did Connie spend? Explain.

9. Of the five statements that follow, which have the same meaning? In other words, which of these statements could be used interchangeably (in a news report, for example)? Explain your answers.

a. The price increased by 53%.

b. The price increased by 153%.

c. The new price is 153% of the old price.

d. The new price is higher than the old price by 53%.

e. The old price was 53% less than the new price.

10. Explain the difference between a 150% increase in an amount and 150% of an amount. Give examples to illustrate.

11. Every week, DollarDeals lowers the price of items it has in stock by 10%. Suppose that the price of an item has been lowered twice, each time by 10% of that week's price. Explain why it makes sense that the total discount on the item is *not* 20%, even though the price has been lowered twice by 10% each time. What percent is the total discount?

***12.** The SuperDiscount store is planning a "35%-off sale" in two weeks. This week, a pair of pants costs $59.95.

a. Suppose SuperDiscount raises the price of the pants by 35% this week, and then two weeks from now, lowers the price by 35%. How much will the pants cost two weeks from now? Explain your method of calculation. Explain why it makes sense that the pants won't return to their original price of $59.95 two weeks from now.

b. By what percent does SuperDiscount need to raise the price of the pants this week, so that two weeks from now, when it lowers the price by 35%, the pants will return to the original price of $59.95? Explain your method of calculation.

***13.** One box of cereal contains 12% more cereal than another box. The larger box contains 3 more ounces of cereal than the smaller box. How much does the cereal in each box weigh? Explain your reasoning.

***14.** One box of cereal contains 25% more cereal than another box. Together, the cereal in both boxes weighs 54 ounces. How much does the cereal in each box weigh? Explain your reasoning.

***15.** In a box of chocolate candies, 40% of the candies are dark chocolate; the rest are milk chocolate. There are 6 more milk chocolate candies than dark chocolate candies. In all, how many chocolate candies are in the box? Explain your reasoning.

***16.** In a box of chocolate candies, 30% of the candies are dark chocolate; the rest are milk chocolate. What percent more milk chocolate candies are in the box than dark chocolate candies? Explain why we can't solve this problem by subtracting 70% − 30%. Then solve the problem correctly, explaining your reasoning.

***17.** There are two elementary schools in a county. After 10% of the children at the first school were moved to the second school, both schools had the same number of children. By what percent did the number of children at the second school increase when the children from the first school were added? Explain your reasoning.

***18.** There are two middle schools in a county. The first middle school had 20% more children than the second middle school. Then 10% of the children at the second middle school left the county. What percent more children are now at the first middle school than at the second one? Explain your reasoning.

***19.** One school had 10% more children than another school. After 18 children moved from one school to the other, both schools had the same number of children. How many children are in the two schools together? Explain your reasoning.

***20.** Sue and Tonya started the same job at the same time and earned identical salaries. After one year, Sue got a 5% raise and Tonya got a 6% raise. The following year, the situation was reversed: Sue got a 6% raise and Tonya got a 5% raise. After the first year, Tonya's salary was higher than Sue's, of course, but whose salary was higher after both raises? Explain how the commutative property of multiplication is relevant to comparing the salaries after both raises.

***21.** Suppose that the sales tax is 7%, and suppose that some towels are on sale at a 20% discount. When you buy the towels, you pay 7% tax on the discounted price. What if you were to pay 7% tax on the full price, but you got a 20% discount on the price *including* the tax? Would you pay more, less, or the same amount? Explain how the commutative property of multiplication is relevant to this question.

***22.** A dress is marked down 25%, and then it is marked down 20% from the discounted price.

a. By what percent is the dress marked down after both discounts?

b. Does the dress cost the same, less, or more than if the dress were marked down 45% from the start? Explain how you can determine the answer to this question without doing any calculating.

c. If the dress were marked down 20% first and then 25%, would you get a different answer to part (a)? Explain how the commutative property of multiplication is relevant to this question.

***23.** Frank's Jewelers runs the following advertisement: "Come to our 40%-off sale on Saturday. We're not like the competition, who raise prices by 30% and then have a 70%-off sale!"

a. If two items start off with the same price, which gives you the lower price in the end: taking off 40% or raising the price by 30% and then taking off 70% (of the raised price)?

b. Consider the same problem more generally, with other numbers. For example, if you raise prices by 20% and then take off 50% (of the raised price), how does this compare with taking 30% off of the original price? If you raise a price by 30% and then lower the raised price by 30%, how does that compare with the original price? Try at least two other pairs of percentages by which to raise and then lower a price. Describe what you observe.

Predict what happens in general: If you raise a price by A% and then take B% off of the raised price, does that have the same result as if you had lowered the original price by $(B - A)$%? If not, which produces the lower final price?

c. Use the distributive property or FOIL to explain the pattern you discovered in part (b). Remember that to raise a price by 15%, for example, you multiply the price by $1 + 0.15$, whereas to lower a price by 15%, you multiply the price by $1 - 0.15$.

***24.** The following information about two different snack foods is taken from their packages:

Snack	Serving Size	Calories in 1 Serving	Total Fat in 1 Serving
Small crackers	28 g	140	6 g
Chocolate hearts	40 g	220	13 g

Suppose we want to compare the amount of fat in these foods. One way to compare the fat is to compare the amount of fat in 1 serving of each food; another way is to compare the amount of fat in some fixed number of calories of the foods, such as 1 calorie or 100 calories; yet another way is to compare the amount of fat in a fixed number of grams of the foods, such as 1 gram or 100 grams. Solve the following problems by comparing the small crackers and chocolate hearts in different ways:

a. Explain how to interpret the information in the table so that the chocolate hearts have 117% more fat than the small crackers.

b. Explain how to interpret the information in the table so that the chocolate hearts have 52% more fat than the small crackers.

c. Explain how to interpret the information in the table so that the chocolate hearts have 38% more fat than the small crackers.

***25.** According to the 2000 Census, from 1990 to 2000 the population of Clarke County, Georgia, increased by 15.86% and the population of adjacent Oconee County increased by 48.85%.

a. Can we calculate the percent increase in the total population of the two-county Clarke/Oconee area from 1990 to 2000 by adding 15.86% and 48.85%? Why or why not?

b. Use the census data in the table below to calculate the percent increase in the total population of the two-county Clarke/Oconee area from 1990 to 2000:

County	1990 Population	2000 Population
Clarke	87,594	101,489
Oconee	17,618	26,225

***26.** In 2000, Washington County had a total population—urban and rural populations combined—of 200,000. From 2000 to 2010, the rural population of Washington County went up by 4%, and the urban population of Washington County went up by 8%.

a. Based on the preceding information, make a reasonable guess for the percent increase of the total population of Washington County from 2000 to 2010. Based on your guess, what do you expect the total population of Washington County to have been in 2010?

b. Make up three very different examples for the rural and urban populations of Washington County in 2000 (i.e., pick pairs of numbers that add to 200,000). For each example, calculate the total population in 2010, and calculate the percent increase in the total population of Washington County from 2000 to 2010. Compare these answers with your answers in part (a).

c. If you had only the data given at the beginning of the problem (the 4% and 8% increases and the total population of 200,000 in 2000), would you be able to say exactly what the total population of Washington County was in 2010? Could you give a range for the total population of Washington County in 2010? In other words, could you say that the total population must have been between certain numbers in 2010? If so, what is this range of numbers?

CHAPTER SUMMARY

Section 7.1 Motivating and Defining Ratio and Proportional Relationships — Page 282

- Ratios provide us with a mathematical way to describe qualities of quantities that are combined or related. — Page 282
- There are two perspectives on ratio: a multiple-batches perspective and a variable-parts perspective. — Page 283
- We can use tables, double number lines, and strip diagrams to represent and compare ratios and proportional relationships. — Page 284

Key Skills and Understandings

1. Using a ratio table and double number line as a support, explain what it means for quantities to be in a certain ratio from the multiple-batches perspective. Using a table and strip diagram as a support, explain what it means for quantities to be in a certain ratio from the variable-parts perspective. — Page 283
2. Given a ratio, find several amounts in that ratio and record them in a table, on a double number line, or on a graph. Recognize that the collection of amounts in the same ratio form a proportional relationship. — Page 286
3. Compare the qualities of quantities in different ratios by reasoning about tables. For example, compare two walking speeds or two different mixtures of paints or juices. — Page 287
4. Recognize the common error of making an additive comparison in a ratio situation. — Page 287

Section 7.2 Solving Proportion Problems by Reasoning with Multiplication and Division — Page 289

- Proportion problems can be solved by reasoning about quantities in several ways. Initial strategies can use repeated addition; more advanced strategies rely on reasoning about multiplication and division. Such strategies provide opportunities for quantitative reasoning, including making sense of quantities and their relationships, not just how to compute them. — Page 290

Key Skills and Understandings

1. Know a variety of ways to solve proportion problems by reasoning about quantities, including with repeated addition and with strategies that involve multiplication and division. Explain in detail how and why multiplication and division apply and be able to express solutions as products of numbers derived from the problem statement. Use strip diagrams, ratio tables, and double (or triple) number lines as reasoning aids. — Page 290

Section 7.3 The Values of a Ratio: Unit Rates and Multipliers	Page 300
▪ Ratios are connected to fractions by division.	Page 300
▪ We can interpret the two values of a ratio either as unit rates or (when the quantities are measured in the same units) as multipliers that relate total amounts.	Page 300
▪ In the standard method of solving proportions, we first set two fractions equal to each other; we can do so because we are equating the unit rates.	Page 303
Key Skills and Understandings	
1. Understand that if two quantities are in a ratio of A to B, then we can interpret $\frac{A}{B}$ as a unit rate; namely, it is the number of units of the first quantity for every 1 unit of the second quantity. Similarly, $\frac{B}{A}$ is a unit rate; it is the number of units of the second quantity for every 1 unit of the first quantity.	Page 300
2. Identify unit rates and explain what they mean in terms of a context.	Page 300
3. Use unit rates in solving problems.	Page 300
4. For ratios viewed as A parts to B parts, use the values $\frac{A}{B}$ and $\frac{B}{A}$ to make multiplicative comparisons between total amounts.	Page 301
5. Explain why this method of solving proportions is valid: setting two fractions equal to each other, then cross-multiplying, then solving the resulting equation.	Page 303

Section 7.4 Proportional Relationships	Page 307
▪ We can use equations to summarize the relationship between quantities that vary together in a proportional relationship (i.e., they remain in a fixed ratio). We can find equations for proportional relationships by viewing the constant of proportionality (value of the ratio) as a multiplier that relates the quantities.	Page 307
▪ When we graph a proportional relationship in a plane it turns out that the graph is always a line through the origin $(0, 0)$. If the proportional relationship consists of quantities in the ratio A to B, then the relationship has equation $y = \frac{B}{A} \cdot x$. The constant of proportionality, $\frac{B}{A}$, is also the slope of the line.	Page 309
▪ It turns out that every line through the origin that has points with positive x- and y-coordinates is the graph of a proportional relationship. By taking a variable-parts perspective and viewing the constant of proportionality as a multiplier that relates quantities, we can explain why such lines have equations of the form $y = m \cdot x$.	Page 312
Key Skills and Understandings	
1. View quantities varying together in a fixed ratio as a proportional relationship, which can be represented with equations and with graphs in the coordinate plane as well as with strip diagrams and double number lines.	Page 307
2. Find and explain equations to relate quantities that vary together in a proportional relationship.	Page 307
3. Recognize incorrect equations for proportional relationships and discuss sources of errors.	Page 308
4. Graph proportional relationships in a coordinate plane and interpret the constant of proportionality as the slope of the line, which tells us how many groups of an x-coordinate it takes to make the y-coordinate of a point (x, y) on the line.	Page 309
5. Explain why a line through the origin and a given point with positive coordinates has an equation of the form $y = m \cdot x$ for a suitable number m.	Page 312

Section 7.5 Proportional Relationships Versus Inversely Proportional Relationships Page 317

- Some relationships between quantities may seem like proportional relationships but are not. Page 317
- Inversely proportional relationships have the property that whenever one quantity is multiplied (respectively, divided) by a positive number N, the other quantity is *divided* (respectively, *multiplied*) by N. Page 318

Key Skills and Understandings

1. Recognize problems that should and should not be solved with a proportion. Page 317
2. Recognize and give examples of inversely proportional relationships and distinguish them from proportional relationships. Represent inversely proportional and proportional relationships with tables and graphs and reason about quantities to find entries in the tables. Page 318
3. Solve problems involving inversely proportional relationships, including multistep problems. Page 318

Section 7.6 Percent Revisited: Percent Increase and Decrease Page 323

- If the value of a quantity goes up, then to find the increase in the quantity, we subtract. But to find the *percent* increase, we must figure the increase as a percent of the original. If the value of a quantity goes down, then to find the decrease in the quantity, we subtract. But to find the *percent* decrease, we must figure the decrease as a percent of the original. Page 323
- As an application of the distributive property, we can calculate a percent increase by calculating what percent the increased amount is of the original amount and subtracting 100%. Similarly, we can calculate a percent decrease by calculating what percent the decreased amount is of the original amount and subtracting this from 100%. These distributive-property methods allow us to calculate quantities from a given percent increase or decrease. Page 324

Key Skills and Understandings

1. Distinguish additive change from percent change. Page 323
2. Calculate percent increase and percent decrease in several different ways, and explain why the calculation methods make sense. Page 323
3. Calculate quantities from a given percent increase or decrease, and explain why the calculation method makes sense. Page 325
4. Distinguish between percent increase/decrease and percent *of*. Page 327
5. Solve problems involving percent increase or decrease. Page 327

8

Number Theory

In this chapter, we deepen our study of numbers and number systems. We first look into how the counting numbers decompose through multiplication (or division). This leads us to ideas about factors, multiples, and prime numbers. We then look into decimal representations of fractions. We'll see that decimal representations of fractions are of a special type, so that the numbers that have decimal representations form a larger system of numbers than the fractions do.

We focus on the following topics and practices within the *Common Core State Standards for Mathematics*.

Standards for Mathematical Content in the CCSSM

In the domain of *Operations and Algebraic Thinking* (Kindergarten–Grade 5), students learn about even and odd numbers, about factors and multiples, and about prime numbers. In the domain of *The Number System* (Grades 6–8), students work with common factors and common multiples, including greatest common factors and least common multiples. They also learn that there are irrational numbers, that fractions have decimal representations that eventually repeat (or terminate with repeating zeros), and they learn how to express repeating decimals as fractions.

Standards for Mathematical Practice in the CCSSM

Opportunities to engage in all eight of the Standards for Mathematical Practice described in the *Common Core State Standards* occur throughout the study of number theory, although the following standards may be especially appropriate for emphasis:

- **3 Construct viable arguments and critique the reasoning of others.** Students engage in this practice when they explain how they know they have found all the factors of a number or why a given number must be prime.
- **7 Look for and make use of structure.** Students engage in this practice when they look for how least common multiples and greatest common factors apply to contexts such as gears, spirograph flower designs, and musical rhythms.
- **8 Look for an express regularity in repeated reasoning.** Students engage in this practice when they see that decimal representations of fractions repeat because of the repeating remainders that occur in the standard division algorithm or when they work through multiple explanations for why 0.9999 . . . = 1.

  (From Common Core Standards for Mathematical Practice. Published by Common Core Standards Initiative.)

8.1 Factors and Multiples

CCSS Common Core State Standards Grade 4

What are factors and multiples and why do we have these concepts? Factors and multiples arise when we go beyond solving an individual multiplication or division problem and instead consider the different ways that numbers can be built up or broken down through multiplication and division. Finding the *single* number C such that

$$30 = 5 \times C$$

is a division problem. But finding *all* the counting numbers B and C such that

$$30 = B \times C$$

is the problem of finding all the *factors* of 30. Similarly, finding the *single* number A such that

$$A = 3 \times 8$$

is a multiplication problem. But finding *all* the counting numbers A such that

$$A = B \times 8$$

for some counting number B is the problem of finding all the *multiples* of 8.

We often use factors and multiples when we work with equivalent fractions. But mathematicians have long been intrigued by factors and multiples for what they reveal about the structure of the system of counting numbers.

What Are Factors and Multiples?

If A, B, and C are counting numbers such that

$$A = B \times C$$

multiple
factor
divisor
divide
divisible by

then we say A is a **multiple** of B and of C and that B and C are **factors** or **divisors** of A. We also say that B and C **divide** A, and that A is **divisible by** B and by C. Sometimes we say that A is *evenly divisible* by B and by C.

So the factors of 21 are

$$1, 3, 7, \text{and } 21$$

whereas the multiples of 21 are

$$21, 42, 63, 84, 105, 126, \ldots$$

Note that we cannot explicitly list the full set of multiples of 21 because the list is infinitely long.

We commonly use the concepts of factors and multiples in the context of the counting numbers. However, in some cases, it is useful to include 0 as well. In these cases, it is acceptable to describe 0 as divisible by every counting number, or as a multiple of every whole number.

Notice that the concepts of factors and multiples are closely linked: A counting number A is a multiple of a counting number B exactly when B is a factor of A. For this reason it's easy to get the concepts of factors and multiples confused. Remember that the factors of a number are the numbers you get from writing the original number as a product—i.e., from "breaking down" the number by dividing it. On the other hand, the multiples of a number are the numbers you get by multiplying the number by counting numbers—i.e., by "building up" the number by multiplying it.

To summarize, if A, B, and C are counting numbers, and if

$$A = B \times C$$

then

- A is a multiple of B
- A is a multiple of C
- B is a factor of A
- C is a factor of A

CLASS ACTIVITY

8A Factors and Rectangles, p. CA-140

How Do We Find All Factors?

For a small number, it is usually easy to list all its factors. How do we find all the factors of a big number? What is a systematic way to find all the factors of a counting number?

CLASS ACTIVITY

8B Finding All Factors, p. CA-141

CCSS 4.OA.4

One way to find all the factors of a counting number is to divide the number by all the counting numbers smaller than it, to see which ones divide the number evenly, without a remainder. To make the work efficient, keep track of the quotients, because the whole number quotients will also be factors. For example, to find all the factors of 40, divide 40 by 1, by 2, by 3, by 4, by 5, and so on, recording those numbers that divide 40 and recording the corresponding quotients:

$$\begin{array}{ll} 1, 40 & \text{because } 1 \times 40 = 40 \\ 2, 20 & \text{because } 2 \times 20 = 40 \\ 4, 10 & \text{because } 4 \times 10 = 40 \\ 5, 8 & \text{because } 5 \times 8 = 40 \end{array}$$

After dividing 40 by 1 through 8 and verifying that 1, 2, 4, 5, and 8 are the only factors of 40 up to 8, you don't have to check if 9, 10, 11, 12, and so on divide 40. Why not? Any counting number larger than 8 that divides 40 would have a quotient less than 5 and would already have to appear as one of the factors listed previously. So we can conclude that the full list of factors of 40 is

$$1, 2, 4, 5, 8, 10, 20, 40$$

Notice that every counting number except 1 must have at least two distinct factors—namely, 1 and itself.

So far we've used the word *factor* as a noun. However, the word *factor* can also be used as a verb. If A is a counting number, then **to factor** A means to write A as a product of two or more counting numbers, each of which is less than A. So we can factor 18 as 9 times 2:

to factor

$$18 = 9 \times 2$$

But notice that we can factor 18 even further, because we can factor 9 as 3 times 3, so

$$18 = 3 \times 3 \times 2$$

CLASS ACTIVITY

8C Do Factors Always Come in Pairs?, p. CA-141

SECTION SUMMARY AND STUDY ITEMS

Section 8.1 Factors and Multiples

If A, B, and C are counting numbers and $A = B \times C$, then B and C are factors of A and A is a multiple of B and of C. Also, A is divisible by B and by C, and B and C divide A.

Key Skills and Understandings

1. State the meaning of the terms *factor* and *multiple*.
2. Given a counting number, find all its factors in an efficient way and explain why the method finds all the factors. Given a counting number, list several multiples of the number.
3. Write and solve word problems that can be solved by finding all the factors of a number. Write and solve word problems that can be solved by finding several multiples of a number.

Practice Exercises for Section 8.1

1. Describe an organized, efficient method to find all the factors of a counting number, and explain why this method finds all the factors. Illustrate with the number 72.
2. What are the multiples of the number 72? Explain.
3. Write a word problem such that solving the problem will require finding all the factors of 54. Solve the problem.
4. Write a word problem such that solving the problem will require finding many multiples of 4. Solve the problem.

Answers to Practice Exercises for Section 8.1

1. The factors of the number 72 are the counting numbers that divide 72. To find these numbers, we can simply divide 72 by smaller counting numbers in order, checking to see which ones divide 72. So we can divide 72 by 1, by 2, by 3, by 4, by 5, and so on. To work efficiently, we can record

the factors we find and the corresponding quotients as follows:

1, 72 because $1 \times 72 = 72$
2, 36 because $2 \times 36 = 72$
3, 24 because $3 \times 24 = 72$
4, 18 because $4 \times 18 = 72$
6, 12 because $6 \times 12 = 72$
8, 9 because $8 \times 9 = 72$

Once we have checked the counting numbers up to 9 to see if they divide 72, we won't have to continue dividing 72 by counting numbers greater than 9. This is because any counting number greater than 9 that divides 72 must have a quotient that is less than 9 and therefore must already occur in the column on the right in the previous list. Hence, we can conclude that the list of factors of 72 is

1, 2, 3, 4, 6, 8, 9, 12, 18, 24, 36, 72

2. The multiples of 72 are the numbers that can be expressed as 72 times another counting number. They are

72, 144, 216, 288, . . .

because these numbers are

$72 \times 1, \quad 72 \times 2, \quad 72 \times 3, \quad 72 \times 4, \ldots$

3. There are 54 members of a band. What are the ways of arranging the band members into equal rows? The ways of factoring 54 into a product of two counting numbers are

$1 \times 54, \quad 2 \times 27, \quad 3 \times 18, \quad 6 \times 9$

and the reverse factorizations,

$54 \times 1, \quad 27 \times 2, \quad 18 \times 3, \quad 9 \times 6$

So the band can be arranged in 1 row of 54 people, 2 rows of 27 people, 3 rows of 18 people, 6 rows of 9 people, 9 rows of 6 people, 18 rows of 3 people, 27 rows of 2 people, or 54 rows of 1 person.

4. Tawanda has painted many dry noodles and will put strings through them to make necklaces that are completely filled with noodles, as in Figure 8.1. Each noodle is 4 centimeters long. What are the lengths of the necklaces Tawanda can make?

Whenever Tawanda makes a necklace from these noodles, the length of the necklace will be the number of noodles she used times 4 centimeters. Therefore, this problem is almost the same as the problem of finding the multiples of 4. In Tawanda's case, some of the multiples of 4 would make necklaces that are too short to be practical, and some multiples of 4 would not work because Tawanda doesn't have enough noodles. The multiples of 4 are

4, 8, 12, 16, 20, 24, . . .

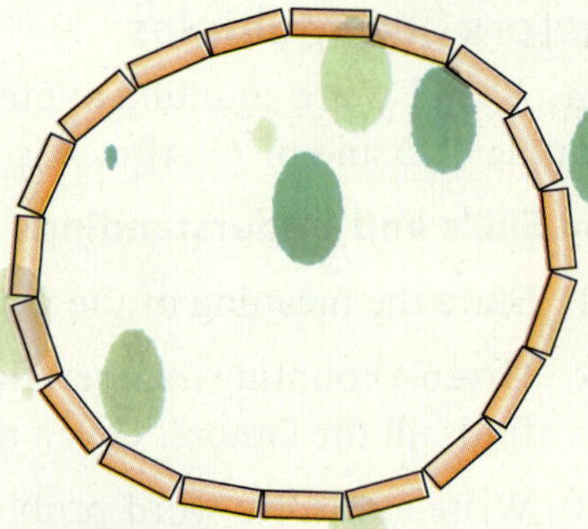

Figure 8.1 A noodle necklace.

PROBLEMS FOR SECTION 8.1

1. Johnny says that 3 is a multiple of 6 because you can arrange 3 cookies into 6 groups by putting $\frac{1}{2}$ of a cookie in each group. Discuss Johnny's idea in detail: In what way does he have the right idea about what the term *multiple* means, and in what way does he need to modify his idea?

2. Manuela is looking for all the factors of 90. So far, she has divided 90 by all the counting numbers from 1 to 10, listing those numbers that divide 90 and listing the corresponding quotients. Here is Manuela's work so far:

1, 90 $1 \times 90 = 90$
2, 45 $2 \times 45 = 90$
3, 30 $3 \times 30 = 90$
5, 18 $5 \times 18 = 90$
6, 15 $6 \times 15 = 90$
9, 10 $9 \times 10 = 90$
10, 9 $10 \times 9 = 90$

Should Manuela keep checking numbers to see if any numbers larger than 10 divide 90, or can

Manuela stop dividing at this point? If so, why? What are all the factors of 90?

3. Show how to find all the factors of the following numbers in an efficient manner. Explain why you can stop checking for factors when you do.

a. 63

b. 75

c. 126

4. a. Write a word problem such that solving your problem will require finding all the factors of 48. Solve your problem.

b. Write a word problem such that solving your problem will require finding several multiples of 15. Solve your problem.

5. a. Write a problem about a realistic situation that involves the concept of factors. Solving your problem should involve finding all the factors of a number. Solve your problem.

b. Write a problem about a realistic situation that involves the concept of multiples. Solving your problem should involve finding multiples of a number. Solve your problem.

6. Solve problems (a) and (b), and determine whether the answers are different or not. Explain why or why not.

a. Josh has 1159 bottle caps in his collection. In how many different ways can Josh arrange his bottle-cap collection into groups so that the same number of bottle caps are in each group and so that there are no bottle caps left over (i.e., not in a group)?

b. How many different rectangles can be made whose side lengths, in centimeters, are counting numbers and whose area is 1159 square centimeters?

7. a. If A and B are counting numbers and B is a factor of A, how are the factors of A and B related? Explain your answer, and give some examples to illustrate.

b. If A and B are counting numbers and A is a multiple of B, how are the multiples of A and B related? Explain your answer, and give some examples to illustrate.

* **8.** If A, B, and C are counting numbers and both A and B are multiples of C, what can you say about $A + B$? Explain why your answer is always true, and give some examples to illustrate. Which property of arithmetic is relevant to this problem?

* **9.** At school there is a long line of closed lockers, numbered 1 to 1000 in order. Outside, 1000 students are lined up, waiting to come in and open or close locker doors. The first student in line opens every locker door. The second student in line closes the doors of lockers that are multiples of 2. The third student in line changes the doors of lockers that are multiples of 3. (The student opens the doors that are closed and closes the doors that are open.) The fourth student in line changes the doors of lockers that are multiples of 4. (The student opens the doors that are closed and closes the doors that are open.) Students keep coming in and changing the locker doors, continuing in the pattern that the Nth student changes the lockers that are multiples of N. Which lockers are open after all 1000 students have gone through changing locker doors? Explain your answer.

8.2 Even and Odd

CCSS Common Core State Standards Grade 2

The study of even and odd numbers is fertile ground for investigating and exploring math. This is an area of math where it's not too hard to find interesting questions to ask about what is true, to investigate these questions, and to explain why the answers are right. Even young children can do this. In this section, in addition to asking and investigating questions about even and odd numbers, we will examine the familiar method for determining whether a number is even or odd.

How would you answer the following questions?

- What does it *mean* for a counting number to be even?
- How can you *tell* if a counting number is even?

Did you give different answers? Then why do these two different characterizations describe *the same numbers*?

even The meaning of *even* is this: A counting number is **even** if it is divisible by 2—in other words, if there is no remainder when you divide the number by 2. There are various equivalent ways to say that a counting number is even:

CCSS 2.OA.3

- A counting number is even if it is divisible by 2—in other words, if there is no remainder when you divide the number by 2.
- A counting number is even if it can be factored into 2 times another counting number (or another counting number times 2).
- A counting number is even if you can divide that number of things into 2 equal groups with none left over.
- A counting number is even if you can divide that number of things into groups of 2 with none left over. (See **Figure 8.2**.)

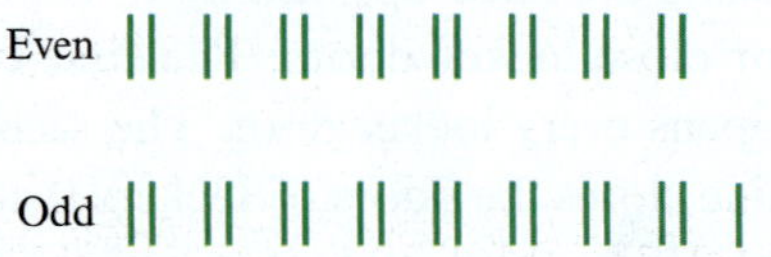

Figure 8.2 Even and odd.

odd If a counting number is not even, then we call it **odd**. Each of the ways previously described of saying what it means for a counting number to be even can be modified to say what it means for a counting number to be odd:

- A counting number is odd if it is not divisible by 2—in other words, if there is a remainder when you divide the number by 2.
- A counting number is odd if it cannot be factored into 2 times another counting number (or another counting number times 2).
- A counting number is odd if there is one thing left over when you divide that number of things into 2 equal groups.
- A counting number is odd if there is one thing left over when you divide that number of things into groups of 2. (See again **Figure 8.2**.)

If you have a particular number in mind, such as 237,921, how do you *tell* if the number is even or odd? Chances are, you don't actually divide 237,921 by 2 to see if there is a remainder or not. Instead, you look at the ones digit. In general, if a counting number has a 0, 2, 4, 6, or 8 in the ones place, then it is even; if it has a 1, 3, 5, 7, or 9 in the ones place, then it is odd. So 237,921 is odd because there is a 1 in the ones place. But *why* is this a valid way to determine that there will be a remainder when you divide 237,921 by 2?

CLASS ACTIVITY

8D Why Can We Check the Ones Digit to Determine Whether a Number Is Even or Odd?, p. CA-142

To explain why we can determine whether a counting number is even or odd by looking at the ones place, think about representing counting numbers with base-ten bundles. For example, we represent 351 with 3 bundles of 100, 5 bundles of 10, and 1 individual item; we represent 4736 with 4 bundles of 1000, 7 bundles of 100, 3 bundles of 10, and 6 individual items. A counting number is even if that number of items can be divided into groups of 2 with none left over; the number is odd if there is an item left over. So consider dividing bundled toothpicks that represent a counting number into groups of 2. To divide these toothpicks into groups of 2, we could divide each of the bundles of 10, of 100, of 1000, and so on, into groups of 2. Each bundle of 10 toothpicks can be divided into 5 groups of 2. Each bundle of 100 toothpicks can be divided into 50 groups of 2. Each bundle of 1000 toothpicks can be divided into 500 groups of 2, and so on, for all the higher places. For each place from the 10s place on up, each bundle of toothpicks can always be divided into groups of 2 with none left over. Therefore, the full number of toothpicks can be divided into groups of 2 with none left over exactly when the number of toothpicks that are in the ones place can be divided into groups of 2 with none left over. This is why we have to check only the ones place of a counting number to see if the number is even or odd.

CLASS ACTIVITY

8E Questions About Even and Odd Numbers, p. CA-143

8F Extending the Definitions of Even and Odd, p. CA-143

SECTION SUMMARY AND STUDY ITEMS

Section 8.2 Even and Odd

A counting number is even if it is divisible by 2, or in other words, if there is no remainder when the number is divided by 2. A counting number is odd if it is not divisible by 2, or in other words, if there is a remainder when it is divided by 2. To check if a counting number is even or odd, we have to look only at the ones digit of the number. We can explain why this way of checking if a number is even or odd is valid by considering base-ten bundles and by recognizing that tens, hundreds, thousands, and so on can all be divided into groups of 2 with none left over.

Key Skills and Understandings

1. State the meaning of even and odd.
2. Explain why it is valid to check if a number is even or odd by examining the ones place of the number.
3. Solve problems about even and odd numbers.

Practice Exercises for Section 8.2

1. Explain why it is valid to determine whether a counting number is divisible by 2 by seeing if its ones digit is 0, 2, 4, 6, or 8.
2. If you add an odd number and an even number, what kind of number do you get? Explain why your answer is always correct.
3. If you multiply an even number and an even number, what kind of number do you get? Explain why your answer is always correct.

Answers to Practice Exercises for Section 8.2

1. Use the idea of representing a counting number with base-ten bundles and think in terms of dividing each bundle into groups of 2. See the text for details.
2. If you add an odd number to an even number, the result is always an odd number. One way to explain why is to think about putting together an odd

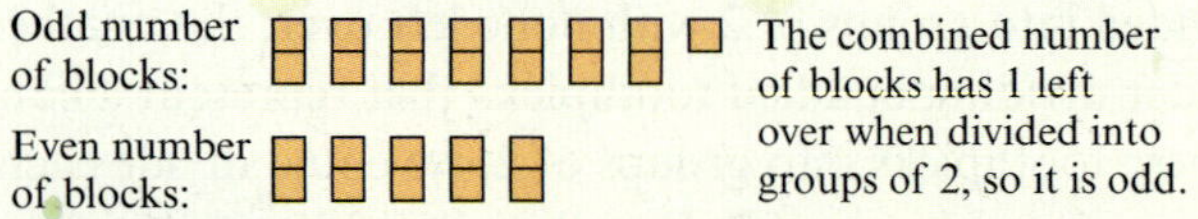

Figure 8.3 Adding an odd number to an even number.

number of blocks and an even number of blocks, as in Figure 8.3. The odd number of blocks can be divided into groups of 2 with 1 block left over. The even number of blocks can be divided into groups of 2 with none left over. When the two block collections are joined together, there will be a bunch of groups of 2, and the 1 block that was left over from the odd number of blocks will still be left over. Therefore, the sum of an odd number and an even number is odd.

3. If you multiply an even number and an even number, the result is always an even number. One way to explain why is to think about creating an even number of groups of blocks, where each group of blocks contains the same even number of blocks, as in Figure 8.4. The total number of blocks is the product of the number of groups with the number of blocks in each group. Because each group of blocks can be divided into groups of 2 with none left over, the whole collection of blocks can be divided into groups of 2 with none left over. Therefore, an even number times an even number is always an even number. (Notice that this argument actually shows that any counting number times an even number is even.)

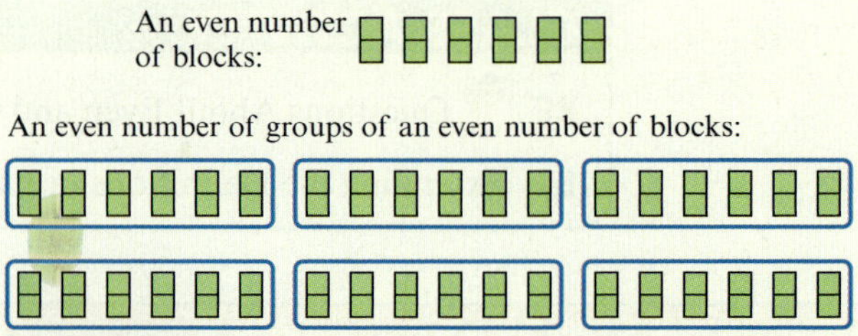

Figure 8.4 Multiplying an even number with an even number.

PROBLEMS FOR SECTION 8.2

1. Describe a way that the children in Mrs. Verner's kindergarten class can tell if there are an even number or an odd number of children present in the class without counting.

2. Explain why an odd counting number can always be written in the form $2N + 1$ for some whole number N.

3. For each of the two designs in Figure 8.5, explain how you can tell whether the number of dots in the design is even or odd without determining the number of dots in the design.

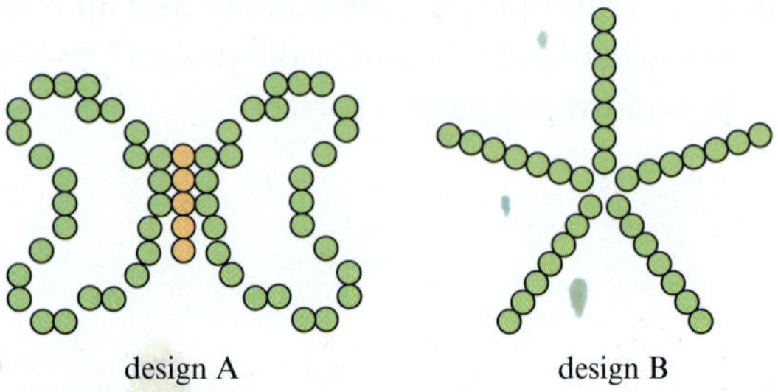

Figure 8.5 Two dot designs.

4. a. Without determining the number of dots in the design in Figure 8.6, determine whether this number is even or odd. Explain.

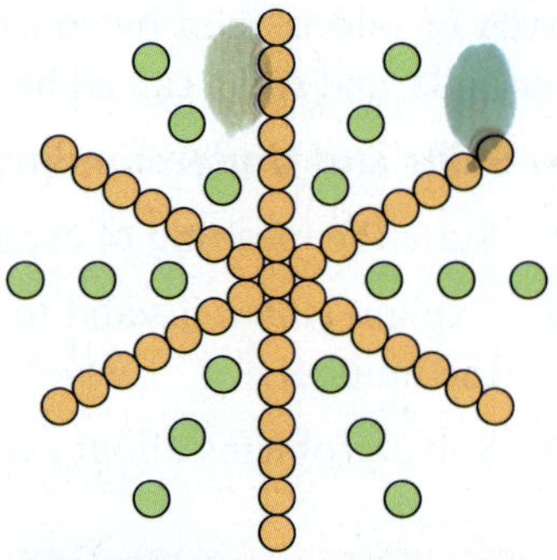

Figure 8.6 A dot design.

b. Give some examples of things where you can tell right away that their number is even or odd without actually counting them.

5. If you add an even number and an even number, what kind of number do you get? Explain why your answer is always correct.

6. If you multiply an odd number and an odd number, what kind of number do you get? Explain why your answer is always correct.

***7.** If you multiply an even number by 3 and then add 1, what kind of number do you get? Explain why your answer is always correct.

*8. If you multiply an odd number by 3 and then add 1, what kind of number do you get? Explain why your answer is always correct.

*9. Suppose that the difference between two counting numbers is odd. What can you say about the sum of the numbers? Explain why your answer is always correct.

*10. Suppose that the difference between two counting numbers is even. What can you say about the sum of the numbers? Explain why your answer is always correct.

*11. If you add a number that has a remainder of 1 when it is divided by 3 to a number that has a remainder of 2 when it is divided by 3, then what is the remainder of the sum when you divide it by 3? Investigate this question by working examples. Then explain why your answer is always correct.

*12. If you multiply a number that has a remainder of 1 when it is divided by 3 with a number that has a remainder of 2 when it is divided by 3, then what is the remainder of the product when you divide it by 3? Investigate this question by working examples. Then explain why your answer is always correct.

8.3 Divisibility Tests

How can you tell if one counting number is divisible by another counting number? In the previous section, we saw a way to explain why we can look at the ones digit to tell whether a counting number is even or odd. Are there easy ways to tell if a counting number is divisible by other numbers, such as 3, 4, or 5? Yes, there are, and we will study some of them in this section.

divisibility test

A method for *checking* or *testing* to see if a counting number is divisible by another counting number, without actually carrying out the division, is called a **divisibility test**.

divisibility test for 2

The **divisibility test for 2** is the familiar test to see if a counting number is even or odd: Check the ones digit of the number. If the ones digit is 0, 2, 4, 6, or 8, then the number is divisible by 2 (it is even); otherwise, the number is not divisible by 2 (it is odd). This divisibility test was discussed in the previous section.

divisibility test for 10

divisibility test for 5

A well-known divisibility test is the **divisibility test for 10**. The test is as follows: Given a counting number, check the ones digit of the number. If the ones digit is 0, then the number is divisible by 10; if the ones digit is not 0, then the number is not divisible by 10. Another familiar divisibility test is the **divisibility test for 5**. The test is as follows: Given a counting number, check the ones digit of the number. If the ones digit is 0 or 5, then the original number is divisible by 5; otherwise, it is not divisible by 5. Problems 2 and 3 at the end of this section focus on why these divisibility tests are valid. The explanations are similar to the explanation developed in the previous section for why we can see if a number is divisible by 2 by checking its ones digit.

CLASS ACTIVITY

8G The Divisibility Test for 3, p. CA-144

divisibility test for 3

The **divisibility test for 3** is as follows: Given a counting number, add the digits of the number. If the sum of the digits is divisible by 3, then the original number is also divisible by 3; otherwise, the original number is not divisible by 3. This divisibility test makes it easy to determine whether a large whole number is divisible by 3. For example, is 127,358 divisible by 3? To answer this, all we have to do is add all the digits of the number:

$$1 + 2 + 7 + 3 + 5 + 8 = 26$$

Because 26 is not divisible by 3, the original number 127,358 is also not divisible by 3. Similarly, is 111,111 divisible by 3? Because the sum of the digits is 6,

$$1 + 1 + 1 + 1 + 1 + 1 = 6$$

and because 6 is divisible by 3, the original number 111,111 must also be divisible by 3.

Why does the divisibility test for 3 work? In other words, why is it a valid way to determine whether a counting number is divisible by 3? As with the divisibility test for 2 (the test for even or odd), let's consider counting numbers as represented by base-ten bundles. A counting number is divisible by 3 exactly when that number of toothpicks can be divided into groups of 3 with none left over. So consider dividing bundled toothpicks that represent a counting number into groups of 3. To divide these toothpicks into groups of 3, we could start by dividing each of the bundles of 10, of 100, of 1000, and so on into groups of 3. Each bundle of 10 can be divided into 3 groups of 3 with 1 left over. Each bundle of 100 can be divided into 33 groups of 3 with 1 left over. Each bundle of 1000 can be divided into 333 groups of 3 with 1 left over. For each place from the 10s place on up, each bundle of toothpicks can always be divided into groups of 3 with 1 left over. Now think of collecting all those leftover toothpicks from each bundle and joining them with the individual toothpicks. How many toothpicks will that be? If the original number was 248, then there will be 2 leftover toothpicks from the 2 bundles of 100, as indicated in Figure 8.7. There will be 4 leftover toothpicks from the 4 bundles of 10. (Even though we could make another group of 3 from these, let's not, because we want to explain why we can add the digits of the number to determine if the number is divisible by 3. So we want to work with the digit 4, not 1.) If we combine the leftover toothpicks with the 8 individual toothpicks, then in all, that is

$$2 + 4 + 8$$

toothpicks. The case is similar for any other counting number: If we combine the individual toothpicks with all the leftover toothpicks that result from dividing each bundle of 10, 100, 1000, and so on into groups of 3, then the number of toothpicks we will have is given by the sum of the digits of the number. All the other toothpicks have already been put into groups of 3, so if we can put these left-over toothpicks into groups of 3, then the original number is divisible by 3. If we can't put these leftover toothpicks into groups of 3 with none left, then the original number is not divisible by 3. In other words, we can determine whether a number is divisible by 3 by adding its digits and seeing if that number is divisible by 3.

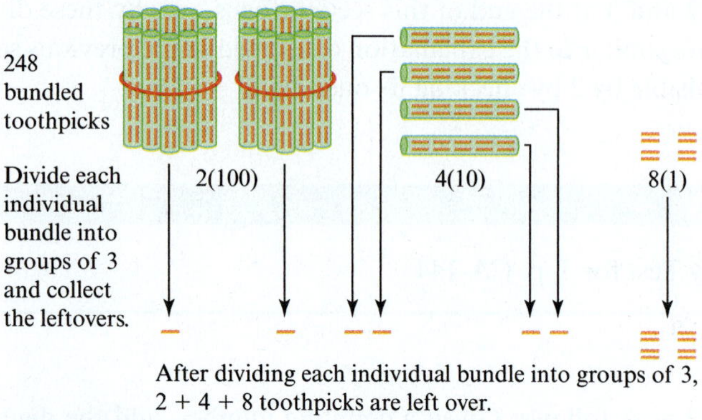

Figure 8.7 Explaining the divisibility test for 3 by dividing individual bundles into groups of 3 and collecting the leftovers.

We can express the previous explanation for why the divisibility test for 3 is valid in a more algebraic way. Let's restrict our explanation to 4-digit counting numbers. Such a number is of the form $ABCD$. Now consider the following equations:

$$\begin{aligned} ABCD &= A \cdot 1000 + B \cdot 100 + C \cdot 10 + D \\ &= (A \cdot 999 + B \cdot 99 + C \cdot 9) + (A + B + C + D) \\ &= (A \cdot 333 + B \cdot 33 + C \cdot 3) \cdot 3 + (A + B + C + D) \end{aligned}$$

The first equation expresses $ABCD$ in its expanded form. After the second = sign, we have separated the leftover toothpicks that result from dividing the bundles of 1000, 100, and 10 into groups of 3, and we have joined these leftovers with the individual toothpicks to make $A + B + C + D$ toothpicks. The expression after the third = sign shows that the toothpicks other than these remaining $A + B + C + D$ can be divided evenly into groups of 3—namely, into $A \cdot 333 + B \cdot 33 + C \cdot 3$ groups of 3. Therefore, the original $ABCD$ toothpicks can be divided into groups of 3 with none left over exactly when the remaining $A + B + C + D$ toothpicks can be divided into groups of 3 with none left over.

divisibility test for 9

The **divisibility test for 9** is similar to the divisibility test for 3: Given a counting number, add the digits of the number. If the sum of the digits is divisible by 9, then the original number is also divisible by 9; otherwise, the original number is not divisible by 9. To explain why this test is a valid way to see if a number is divisible by 9, use the same argument as for the divisibility test for 3, suitably modified for 9 instead of 3. You are asked to explain this divisibility test in Problem 9.

divisibility test for 4

The **divisibility test for 4** is as follows: Given a whole number, check the number formed by its last two digits. For example, given 123,456,789, check the number 89. If the number formed by the last two digits is divisible by 4, then the original number is divisible by 4; otherwise, it is not. So, because 89 is not divisible by 4, according to the divisibility test for 4, the number 123,456,789 is also not divisible by 4. You are asked to explain this divisibility test in Practice Exercise 2.

SECTION SUMMARY AND STUDY ITEMS

Section 8.3 Divisibility Tests

A divisibility test is a quick way to check if a counting number is divisible by a given number without actually dividing. In addition to the divisibility test for 2, there are commonly used divisibility tests for 3, 4, 5, 9, and 10 (plus other divisibility tests). We can explain the divisibility tests by considering base-ten bundles and by considering what happens when tens, hundreds, thousands, and so on are divided by the number in question.

Key Skills and Understandings

1. Describe how to use the divisibility tests for 2, 3, 4, 5, 9, and 10, use the divisibility tests, and explain why these tests work.

Practice Exercises for Section 8.3

1. Use the divisibility test for 3 to determine which of the following numbers are divisible by 3:

a. 125,389,211,464

b. 111,111,111,111,111

c. 123,123,123,123,123

d. 101,101,101,101,101

2. Explain why the divisibility test for 4 described in the text is a valid way to determine if a counting number is divisible by 4.

3. In your own words, explain why the divisibility test for 3 is a valid way to determine if a counting number is divisible by 3.

Answers to Practice Exercises for Section 8.3

1. a. The sum of the digits of 125,389,211,464 is $1 + 2 + 5 + 3 + 8 + 9 + 2 + 1 + 1 + 4 + 6 + 4 = 46$, which is not divisible by 3. Therefore, 125,389,211,464 is not divisible by 3.

   b. The sum of the digits of 111,111,111,111,111 is 5×3, which is divisible by 3. Therefore, 111,111,111,111,111 is divisible by 3.

   c. The sum of the digits of 123,123,123,123,123 is 5×6, which is divisible by 3. Therefore, 123,123,123,123,123 is divisible by 3.

   d. The sum of the digits of 101,101,101,101,101 is 5×2, which is not divisible by 3. Therefore, 101,101,101,101,101 is not divisible by 3.

2. A counting number is divisible by 4 exactly when that number of toothpicks can be divided into groups of 4 with none left over. Consider representing a counting number physically with base-ten bundles, and consider dividing the bundles, from the hundreds place on up, into groups of 4. Each bundle of 100 can be divided into 25 groups of 4. Each bundle of 1000 can be divided into 250 groups of 4. Each bundle of 10,000 can be divided into 2500 groups of 4, and so on, for all the higher places. For each place from the 100s place on up, each bundle can always be divided into groups of 4 with none left over. Therefore, the full number of toothpicks can be divided into groups of 4 with none left over exactly when the number of toothpicks that are together in the tens and ones places can be divided into groups of 4 with none left over.

3. Use the idea of representing a counting number with base-ten bundles. Think of dividing each bundle into groups of 3 and collecting all the leftovers with the individuals in the ones place. See the text for details.

PROBLEMS FOR SECTION 8.3

1. Use the divisibility test for 3 to determine whether the following numbers are divisible by 3:

   a. 7,591,348

   b. 777,777,777

   c. 157,157,157

   d. 241,241,241,241

2. According to the divisibility test for 10, to determine whether a counting number is divisible by 10, you have to check only its ones digit. If the ones digit is 0, then the number is divisible by 10. Otherwise, it is not. Give a clear and complete explanation for why this divisibility test is a valid way to determine whether a number is divisible by 10, using your own words.

3. According to the divisibility test for 5, to determine whether a counting number is divisible by 5, you have to check only its ones digit. If the ones digit is 0 or 5, then the number is divisible by 5. Otherwise, it is not. Give a clear and complete explanation for why this divisibility test is a valid way to determine whether a number is divisible by 5, using your own words.

4. Beth knows the divisibility test for 3. Beth says that she can tell just by looking, and without doing any calculations at all, that the number

   999,888,777,666,555, 444,333, 222,111

   is divisible by 3. How can Beth do that? Explain why it's not just a lucky guess.

5. What are all the different ways to choose the ones digit, A, in the number $572435A$, so that the number will be divisible by 3? Explain your reasoning.

6. Sam used his calculator to calculate

   $$123{,}123{,}123{,}123{,}123 \div 3$$

   Sam's calculator displayed the answer as

   4.1041041041E13

   Sam says that because the calculator's answer is not a whole number, the number 123,123,123,123,123 is not evenly divisible by 3. Is Sam right? Why or why not? How do you reconcile this with Sam's calculator's display? Discuss.

7. Explain how to modify the divisibility test for 3 so that you can determine the remainder of a counting number when it is divided by 3 without

dividing the number by 3. Explain why your method for determining the remainder when a number is divided by 3 is valid. Illustrate your method by determining the remainder of 8,127,534 when it is divided by 3 without actually dividing.

8. For each of the numbers in (a) through (d), verify that the divisibility test for 9 accurately predicts which numbers are divisible by 9.

Example: 52,371 is divisible by 9 because 52,371 ÷ 9 is the whole number 5819, with no remainder. The sum of the digits of 52,371—namely, 5 + 2 + 3 + 7 + 1 = 18—is also divisible by 9.

a. 1,827,364,554,637,281

b. 2,578,109

c. 777,777

d. 777,777,777

9. a. Give a clear and complete explanation for why the divisibility test for 9 is a valid way to determine whether a three-digit counting number ABC is divisible by 9, using your own words.

b. Relate your explanation in part (a) to the following equations:

$$\begin{aligned} ABC &= A \cdot 100 + B \cdot 10 + C \\ &= (A \cdot 99 + B \cdot 9) + (A + B + C) \\ &= (A \cdot 11 + B \cdot 1) \cdot 9 + (A + B + C) \end{aligned}$$

10. a. Find a divisibility test for 25; in other words, find a way to determine if a counting number is divisible by 25 without actually dividing the number by 25.

b. Explain why your divisibility test for 25 is a valid way to determine whether a counting number is divisible by 25.

11. a. What are all the different ways to choose the ones digit, A, in $271854A$ so that the number will be divisible by 9? Explain your reasoning.

b. What are all the different ways to choose the tens digit, A, and the ones digit, B, in the number $631872AB$ so that the number will be divisible by 9? Explain your reasoning.

***12. a.** Find a divisibility test for 8. In other words, find a way to determine if a counting number is divisible by 8 without actually dividing the number by 8.

b. Explain why your divisibility test for 8 is a valid way to determine whether a counting number is divisible by 8.

***13. a.** Is it true that a whole number is divisible by 6 exactly when the sum of its digits is divisible by 6? Investigate this by considering a number of examples. State your conclusion.

b. How could you determine whether the number

111,222,333,444,555,666,777,888,999,000

is divisible by 6 without using a calculator or doing long division? Explain! (*Hint:* $6 = 2 \times 3$.)

***14.** Investigate the questions in the following parts (a) and (b) by considering a number of examples.

a. If a whole number is divisible both by 6 and by 2, is it necessarily divisible by 12?

b. If a whole number is divisible both by 3 and by 4, is it necessarily divisible by 12?

c. Based on your examples, what do you think the answers to the questions in (a) and (b) should be?

d. Based on your answer to part (c), determine whether

3,321,297,402,348,516

is divisible by 12 without using a calculator or doing long division. Explain your reasoning.

***15. a.** If you add 2 consecutive counting numbers (such as 47, 48), will the resulting sum always, sometimes, or never be divisible by 2? Explain why your answer is true.

b. If you add 3 consecutive counting numbers (such as 47, 48, 49), will the resulting sum always, sometimes, or never be divisible by 3? Explain why your answer is true.

c. If you add 4 consecutive counting numbers (such as 47, 48, 49, 50), will the resulting sum always, sometimes, or never be divisible by 4? Explain why your answer is true.

d. If you add 5 consecutive counting numbers (such as 47, 48, 49, 50, 51), will the resulting sum always, sometimes, or never be divisible by 5? Explain why your answer is true.

e. If you add N consecutive counting numbers, will the resulting sum always, sometimes, or never be divisible by N? Explain why your answer is true.

8.4 Prime Numbers

CCSS **Common Core State Standards** Grade 4

CCSS 4.OA.4

In this section, we will discuss the prime numbers, which can be considered the building blocks of the counting numbers. Interestingly, although prime numbers might seem to be only of theoretical interest, they actually have important practical applications to encryption. So, for example, when you use a secure Web site on the Internet, prime numbers are involved.

prime numbers
primes
composite numbers

The **prime numbers**, or simply, the **primes**, are the counting numbers other than 1 that are divisible only by 1 and themselves. The **composite numbers** are the counting numbers other than 1 that are not prime. So the prime numbers are the counting numbers that when factored as $A \times B$, where A and B are counting numbers, then either $A = 1$ or $B = 1$, but not both. For example, 2, 3, 5, and 7 are prime numbers, but 6 is composite because $6 = 2 \times 3$.

The prime numbers are considered to be the *building blocks of the counting numbers.* Why? Because it turns out that every counting number greater than or equal to 2 is either a prime number or can be factored as a product of prime numbers. For example,

$$145 = 29 \times 5$$
$$2009 = 41 \times 7 \times 7$$
$$264{,}264 = 13 \times 11 \times 11 \times 7 \times 3 \times 2 \times 2 \times 2$$

In this way, all counting numbers from 2 onward are "built" from prime numbers. Furthermore, except for rearranging the prime factors, there is only one way to factor a counting number that is not prime into a product of prime numbers. In other words, it is not possible for a product of prime numbers to be equal to a product of different prime numbers. For example, it couldn't happen that $5 \times 7 \times 23 \times 29$ is equal to a product of some different prime numbers, say, involving 11 or 13 or any primes other than 5, 7, 23, and 29. The fact that every counting number greater than 1 can be factored as a product of prime numbers in a unique way as just described is not at all obvious (and beyond the scope of this book to explain). It is a theorem called the **Fundamental Theorem of Arithmetic**.

Fundamental Theorem of Arithmetic

By the way, students sometimes wonder why 1 is not included as a prime number. We don't include 1 as a prime number because if we did, we would have to restate the Fundamental Theorem of Arithmetic in a complicated way.

The prime numbers are to the counting numbers as atoms are to matter—basic and fundamental. And just as many physicists study atoms, so too, many mathematicians study the prime numbers. How do we find prime numbers, and how can we tell if a number is prime? We will consider these questions next.

How Does the Sieve of Eratosthenes Make a List of Prime Numbers?

Sieve of Eratosthenes

How can we find prime numbers? Eratosthenes (275–195 B.C.), a mathematician and astronomer in ancient Greece, discovered a method for finding and listing prime numbers, called the **Sieve of Eratosthenes**. To use the method, list all the whole numbers from 2 up to wherever you want to stop looking for prime numbers. The list in Table 8.1 goes from 2 to 30, so we will look for all the primes up to 30. Next, carry out the following process of circling and crossing out numbers:

1. On the list, circle the number 2, and then cross out every *other* number after 2. (So cross out 4, 6, 8, etc., the multiples of 2.)
2. Then circle the next number that has not been crossed out, 3, and cross out every third number after 3—even if it has already been crossed out. (So cross out 6, 9, 12, etc., the

TABLE 8.1 Using the Sieve of Eratosthenes to find prime numbers

	2	3	~~4~~	5	~~6~~	7	~~8~~	~~9~~	~~10~~
11	~~12~~	13	~~14~~	~~15~~	~~16~~	17	~~18~~	19	~~20~~
~~21~~	~~22~~	23	~~24~~	~~25~~	~~26~~	~~27~~	~~28~~	29	~~30~~

multiples of 3. Notice that you can do this "mechanically," by repeatedly counting, "1, 2, 3," and crossing out on "3.")

3. Circle the next number that has not been crossed out, 5, and cross out every fifth number after 5—even if it has already been crossed out. (So cross out 10, 15, 20, etc., the multiples of 5. Again, notice that you can do this "mechanically," by repeatedly counting, "1, 2, 3, 4, 5," and crossing out on "5.")

4. Continue in this way, going back to the beginning of the list, circling the next number N that has not been crossed out and then crossing out every Nth number after it until every number in the list has either been circled or crossed out.

The circled numbers in the list are the prime numbers.

CLASS ACTIVITY

8H The Sieve of Eratosthenes, p. CA-145

Why are the circled numbers produced by the Sieve of Eratosthenes the prime numbers in the list? If a number is circled, then we did not cross it out, using one of the previously circled numbers. The numbers we cross out are exactly the multiples of a circled number (beyond that circled number). For example, when we circle 3 and cross out every third number after 3, we cross out the multiples of 3 beyond 3, namely, 6, 9, 12, 15, and so on. Therefore, the circled numbers are the numbers that are not multiples of any smaller number (other than 1). Hence, a circled number is divisible only by 1 and itself and so is a prime number.

How Can We Determine If a Number Is Prime?

How can you tell whether a number, such as 239, is a prime number? You could use the Sieve of Eratosthenes to find all prime numbers up to 239, but that would be slow. A faster way is to divide your number by 2. If it's not divisible by 2, divide it by 3; if it's not divisible by 3, divide it by 5; if it's not divisible by 5, divide it by 7, and so on, dividing by consecutive prime numbers. If you ever find a prime number that divides your number, then your number is not prime; otherwise, it is prime.

trial division

This method for determining whether a number is prime is called the method of **trial division**.

CLASS ACTIVITY

8I The Trial Division Method for Determining Whether a Number Is Prime, p. CA-146

Why is trial division a valid way to determine whether a number is prime? Why do we have to check only *prime* numbers to see if they divide the number in question? Why don't we have to check other numbers, such as 4, 6, 8, and 9, which are not prime numbers, to see if they divide the given number? If a number is divisible by 4, for example, then it must also be divisible by 2; if a number is divisible by 6, then it must also be divisible by 2 and by 3. In general, if a number is divisible by some counting number other than 1 and itself, then, according to the Fundamental Theorem of Arithmetic, that divisor is either a prime number or can be factored into a product of prime numbers, and each of those prime factors must also divide the number in question. So, to determine whether a number is prime, we have to find out only whether any prime numbers divide it.

When we use trial division to determine whether a number is prime, how do we know when to stop dividing by primes? Consider the example of 283. To see if 283 is prime or not, divide it by 2, by 3, by 5, by 7, and so on, dividing by consecutive prime numbers.

We find that none of these prime numbers divide 283, but how do we know when we can stop dividing? Let's look at the quotients that result when we divide 283 by prime numbers in order:

$$\begin{aligned}283 \div 2 &= 141.5\\ 283 \div 3 &= 94.33\ldots\\ 283 \div 5 &= 56.6\\ 283 \div 7 &= 40.42\ldots\\ 283 \div 11 &= 25.72\ldots\\ 283 \div 13 &= 21.76\ldots\\ 283 \div 17 &= 16.64\ldots\end{aligned}$$

Notice that as we divide by larger prime numbers, the quotients get smaller. Up to 13, the quotient is larger than the divisor, but when we divide 283 by 17, the quotient, 16.6, is smaller than the divisor. Now imagine that we were to continue dividing 283 by larger prime numbers: by 19, by 23, by 29, and so on. The quotients would get smaller and would be less than 17. So if a prime number larger than 17 were to divide 283, the corresponding quotient would be a whole number less than 17 and thus would have a prime number divisor less than 17. But we already checked all the prime numbers up to 17 and found that none of them divide 283. Therefore, we can already tell that 283 is prime.

In general, to determine whether a number is prime by the trial division method, divide the number by consecutive prime numbers starting at 2. Suppose that none of the prime numbers divide your number. If you reach a point when the quotient becomes smaller than the prime number by which you are dividing, then the previous reasoning tells us your number must be prime.

Of course, it may happen that the number you are checking is not prime. For example, consider 899. When you divide 899 by 2, by 3, by 5, by 7, and so on, dividing by consecutive prime numbers, you will find that none of these primes divides 899 until you get to 29. Then you will discover that $899 = 29 \times 31$, so 899 is not prime.

How Can We Factor Numbers into Products of Prime Numbers?

The prime numbers are the building blocks of the counting numbers because every counting number greater than 1 that is not already a prime number can be factored into a product of prime numbers. How do we write numbers as products of prime numbers?

factor tree

A convenient way to factor a counting number into a product of prime numbers is to create a factor tree. A **factor tree** is a diagram like the one in Figure 8.8, which shows how to factor a number, how to factor the factors, how to factor the factors of factors, and so on, until prime numbers are reached.

Figure 8.8

A factor tree for 600.

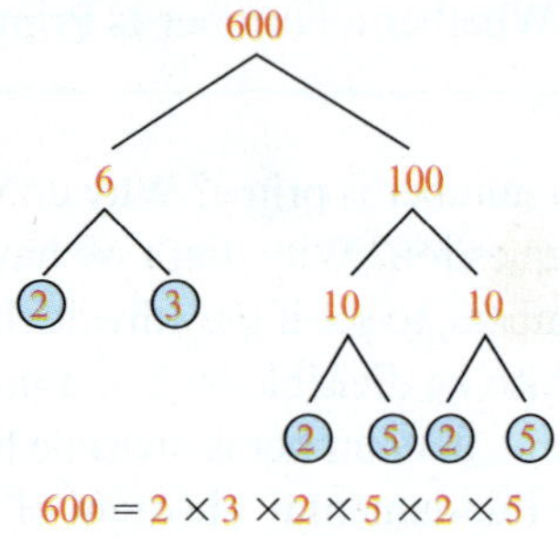

Starting with a given counting number, such as 600, you can create a factor tree as follows: Factor 600 in any way you can think of—for example,

$$600 = 6 \times 100$$

Record this factorization below the number 600, as shown in **Figure 8.8**. Then proceed by factoring each of 6 and 100 separately, and recording these factorizations below each number. Continue until all your factors are prime numbers. Then the numbers at the bottom of the branches of your factor tree are the prime factors of your original number. From **Figure 8.8**, we determine that

$$600 = 2 \times 3 \times 2 \times 5 \times 2 \times 5$$

You might want to rearrange your factorization by placing identical primes next to each other:

$$600 = 2 \times 2 \times 2 \times 3 \times 5 \times 5$$

Using the notation of exponents, you can also write

$$600 = 2^3 \times 3 \times 5^2$$

CLASS ACTIVITY

8J Factoring into Products of Primes, p. CA-147

If two people made two different factor trees, will they still wind up with the same prime factors in the end? Yes, this is what the Fundamental Theorem of Arithmetic tells us.

What if we don't immediately see a way to factor a number? For example, how can we factor 26,741? In this case, we try to divide it by consecutive prime numbers until we find a prime number that divides it evenly. We find that 26,741 is not divisible by 2, by 3, by 5, or by 7, but is divisible by 11:

$$26{,}741 = 11 \times 2431$$

Now we need to factor 2431. It is not divisible by 2, by 3, by 5, or by 7. (If it were, 26,741 would also be divisible by one of these.) But we still must find out whether 2431 is divisible by 11, and sure enough, it is:

$$2431 = 11 \times 221$$

Now we determine whether 221 is divisible by 11. It isn't. So we see if 221 is divisible by the next prime, 13. It is:

$$221 = 13 \times 17$$

Collecting the prime factors we have found, we conclude that

$$26{,}741 = 11 \times 11 \times 13 \times 17$$

The factor tree in **Figure 8.9** records how we factored 26,741.

Figure 8.9

A factor tree for 26,741.

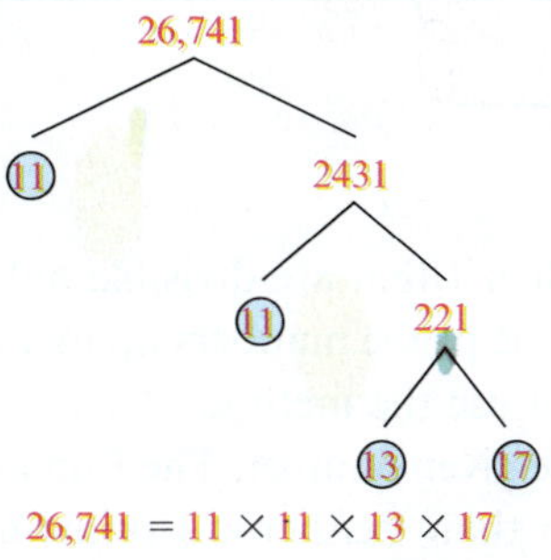

How Many Prime Numbers Are There?

Mathematicians through the ages have been fascinated by prime numbers. There are many interesting facts and questions about them. For instance, does the list of prime numbers go on forever—does it ever come to a stop? Ponder this for a moment before you read on.

Do you have a definite opinion about this or are you unsure? If you do have a definite opinion, can you give convincing evidence for it? This is not so easy.

It turns out that the list of prime numbers does go on forever. How do we know this? It certainly isn't possible to demonstrate this by listing all the prime numbers. It is not at all obvious why there are infinitely many prime numbers, but there is a beautiful line of reasoning that proves this.

In about 300 B.C., Euclid, another mathematician who lived in ancient Greece, wrote the most famous mathematics book of all time, *The Elements*. In it appears an argument that proves there are infinitely many prime numbers (25, Proposition 20 of book IX). The following is a variation of that argument, presented in more modern language:

Suppose that

$$P_1, P_2, P_3, \ldots, P_n$$

is a list of n prime numbers. Consider the number

$$P_1 \cdot P_2 \cdot P_3 \cdot \cdots \cdot P_n + 1$$

which is just the product of the list of prime numbers with the number 1 added on. Then this new number is not divisible by any of the primes $P_1, P_2, P_3, \ldots, P_n$ on our list, because each of these leaves a remainder of 1 when our new number is divided by any of these numbers. Therefore, either $P_1 \cdot P_2 \cdot P_3 \cdot \cdots \cdot P_n + 1$ is a prime number that is not on our list of primes or, when we factor our number $P_1 \cdot P_2 \cdot P_3 \cdot \cdots \cdot P_n + 1$ into a product of prime numbers, each of these prime factors is a new prime number that is not on our list. Hence, no matter how many prime numbers we started with, we can always find new prime numbers, and it follows that there must be infinitely many prime numbers.

FROM THE FIELD Children's Literature

McElligott, M. (2007). *Bean thirteen*. New York, NY: Penguin.

Ralph and Flora gather thirteen beans for dinner, but they struggle to put them into equal groups for dinner. They invite more and more guests, but they cannot find a way to divide the beans equally among themselves. Finally, they place the beans into one large pile and share it.

Schwartz, R. E. (2010). *You can count on monsters*. Natick, MA: A K Peters. The first 100 numbers are represented with drawings of whimsical monsters that correspond to how the numbers factor into products of primes.

SECTION SUMMARY AND STUDY ITEMS

Section 8.4 Prime Numbers

The prime numbers are the counting numbers other than 1 that are divisible only by 1 and themselves. We can use the Sieve of Eratosthenes to find all the prime numbers up to a given number. To determine if a given counting number is prime, we can use the method of trial division, in which we check to see if consecutive prime numbers divide the given number. The Fundamental Theorem of Arithmetic states that every counting number other than 1 either is a prime number or can be factored in a unique way into a product of prime numbers. Factor trees can help find the prime factorization of a counting number. There are infinitely many prime numbers.

Key Skills and Understandings

1. State the meaning of prime number.
2. Use the Sieve of Eratosthenes, and explain why it produces a list of prime numbers.
3. Use trial division to determine if a given counting number is prime. Explain why you have to divide only by prime numbers when using trial division. Explain why you can stop dividing when you do.
4. Given a counting number, determine whether it is prime, or if not, factor it into a product of prime numbers.

Practice Exercises for Section 8.4

1. What is a prime number?
2. What is the Sieve of Eratosthenes, and why does it work?
3. Explain how to use trial division to determine whether 283 is prime. Why do you have to divide only by prime numbers? When can you stop dividing and why?
4. For each of the listed numbers, determine whether it is prime. If it is not prime, factor the number into a product of prime numbers.
   a. 1081
   b. 1087
   c. 269,059
   d. 2081
   e. 1147
5. Given that $550 = 2 \cdot 5 \cdot 5 \cdot 11$, find all the factors of 550, and explain your reasoning.

Answers to Practice Exercises for Section 8.4

1. See text.
2. See text.
3. See text.
4. a. $1081 = 23 \times 47$.
   b. 1087 is prime.
   c. $269{,}059 = 7 \times 7 \times 17 \times 17 \times 19$.
   d. 2081 is prime.
   e. $1147 = 31 \times 37$.
5. 1 and 550 are factors of 550. Given any other factor of 550, when this factor is factored, it must be a product of some of the primes in the list 2, 5, 5, 11. (If not, you would get a different way to factor 550 into a product of prime numbers, which would contradict the Fundamental Theorem of Arithmetic.) Therefore, the factors of 550 are

$$1, 550, 2, 5, 11, 2 \cdot 5 = 10, 2 \cdot 11 = 22,$$
$$5 \cdot 5 = 25, 5 \cdot 11 = 55, 2 \cdot 5 \cdot 5 = 50,$$
$$2 \cdot 5 \cdot 11 = 110, 5 \cdot 5 \cdot 11 = 275$$

PROBLEMS FOR SECTION 8.4

1. For which counting numbers, N, greater than 1, is there only one rectangle whose side lengths, in inches, are counting numbers and whose area, in square inches, is N? Explain.
2. Use trial division to determine whether 251 is prime. In your own words, explain why you have to divide only by prime numbers and why you can stop dividing when you do.

3. For each of the numbers in (a) through (d), determine whether it is a prime number. If it is not a prime number, factor the number into a product of prime numbers.
   a. 8303
   b. 3719
   c. 3721
   d. 80,000
4. Given that $792 = 2^3 \cdot 3^2 \cdot 11$, find all the factors of 792, and explain your reasoning.
5. Without calculating the number $19 \times 23 + 1$, explain why this number is not divisible by 19 or by 23.
6. Following Euclid's proof that there are infinitely many primes, and starting the list with 5 and 7, you form the number $5 \times 7 + 1 = 36$, and you get the new prime numbers 2 and 3 because $36 = 2 \times 2 \times 3 \times 3$. Which new prime numbers would you get if your starting list were the following?
   a. 2, 3, 5
   b. 2, 3, 5, 7
   c. 3, 5

8.5 Greatest Common Factor and Least Common Multiple

CCSS **Common Core State Standards** Grade 6

What do gears, different types of cicadas, and the spirograph drawing toy all have in common? All have aspects that are related to greatest common factors and least common multiples! In Section 8.1, we discussed factors and multiples of individual numbers. If we have two or more counting numbers in mind, we can consider the factors that the two numbers have in common and the multiples that the two numbers have in common. By considering common multiples and common factors of two or more counting numbers, we arrive at the concepts of *greatest common factor* and *least common multiple*, which we will discuss in this section. Greatest common factors and least common multiples can be useful when working with fractions.

Definitions of GCF and LCM

greatest common factor
GCF

If you have two or more counting numbers, then the **greatest common factor**, abbreviated **GCF**, or **greatest common divisor**, abbreviated **GCD**, of these numbers is the greatest counting number that is a factor of all the given counting numbers. For example, what is the GCF of 12 and 18? The factors of 12 are

$$1, 2, 3, 4, 6, 12$$

and the factors of 18 are

$$1, 2, 3, 6, 9, 18$$

CCSS 6.NS.4

Therefore, the common factors of 12 and 18 are the numbers that are common to the two lists, namely,

$$1, 2, 3, 6$$

The greatest of these numbers is 6; therefore, the greatest common factor of 12 and 18 is 6.

least common multiple
LCM

Similarly, if you have two or more counting numbers, then the **least common multiple**, abbreviated **LCM**, of these numbers is the least counting number that is a multiple of all the given numbers. For example, what is the LCM of 6 and 8? The multiples of 6 are

$$6, 12, 18, 24, 30, 36, 42, 48, 54, 60, 66, 72, 78, \ldots$$

and the multiples of 8 are

8, 16, 24, 32, 40, 48, 56, 64, 72, 80, . . .

Therefore, the common multiples of 6 and 8 are the numbers that are common to the two lists, namely,

24, 48, 72, . . .

The least of these numbers is 24; therefore, the least common multiple of 6 and 8 is 24.

Although there are other methods for calculating GCFs and LCMs, we can always use the definition of these concepts to calculate GCFs and LCMs.

What Are Methods for Finding GCFs and LCMs?

To determine the GCF and LCM of counting numbers, we can always use their definitions, as we just saw. Another quicker way to determine GCFs and LCMs is sometimes called the "slide method," which we will study next. Table 8.2 shows the steps for one way to use the slide method to determine the GCF and LCM of 8800 and 10,000. Table 8.3 shows the final result of a more efficient way to use the slide method for determining the GCF and LCM of 8800 and 10,000.

CLASS ACTIVITY

8K The Slide Method, p. CA-148

TABLE 8.2 Using the slide method to calculate the GCF and LCM of 8800 and 10,000

Step 1:

10	8800	10,000
	880	1000

Step 2:

10	8800	10,000
10	880	1000
	88	100

Step 3:

10	8800	10,000
10	880	1000
2	88	100
	44	50

Step 4:

10	8800	10,000
10	880	1000
2	88	100
2	44	50
	22	25

Conclusion:
$GCF = 10 \cdot 10 \cdot 2 \cdot 2 = 400$
$LCM = 10 \cdot 10 \cdot 2 \cdot 2 \cdot 22 \cdot 25 = 220{,}000$

TABLE 8.3 Using the slide method to calculate the GCF and LCM of 8800 and 10,000 in fewer steps

100	8800	10,000
4	88	100
	22	25

$GCF = 100 \cdot 4 = 400$
$LCM = 100 \cdot 4 \cdot 22 \cdot 25 = 220{,}000$

Observe that to use the slide method, you repeatedly find common factors. You write these common factors on the left, and you write the quotients that result from dividing by the common factors on the right. The "slide" stops when the resulting quotients on the right no longer have any common factor except 1. The GCF is then the product of the factors down the left-hand side of the slide, and the LCM is the product of the factors down the left-hand side of the slide *and* the numbers in the last row of the slide.

Notice that we can use slides flexibly. For example, in Table 8.2 we found the common factor 10 of 8800 and 10,000, and at the next step we found another common factor of 10. But in Table 8.3 we found the common factor 100 in one step, rather than taking two steps $(10 \cdot 10 = 100)$.

Why does the slide method give us the GCF? Essentially, it's because we repeatedly take out common factors until there are no more common factors left (except 1). When we multiply all the common factors we collected, that product should form the greatest common factor. By listing successive quotients on the right-hand side of the slide, we ensure that we don't repeat a common factor that has already been accounted for.

Why does the slide method give us the LCM? This is harder to explain, and we will see only roughly why it works. Notice that the product of the numbers on the left in a slide and one of the numbers in the bottom row is equal to the number above it at the top. For example, referring to Table 8.3, we see that $100 \cdot 4 \cdot 22 = 8800$ and $100 \cdot 4 \cdot 25 = 10{,}000$. So the product of the numbers on the left and the numbers in the bottom row of a slide must be a multiple of the initial numbers. Because we don't repeat the factors that the two initial numbers have in common (namely, the factors on the left), we get the smallest possible multiple of the initial numbers.

CLASS ACTIVITY

8L Construct Arguments and Critique Reasoning About GCFs and LCMs, p. CA-149

8M Model With GCFs and LCMs, p. CA-150

8N Spirograph Flower Designs, p. CA-151

FROM THE FIELD **Research**

Bell, C. J., Leisner, H. J., & Shelley, K. (2011). A fruitful activity for finding the greatest common factor. *Mathematics Teaching in the Middle School, 17*(4), pp. 222–229.

This article discusses a Fruit Basket Challenge activity in which seventh graders explore the concept of greatest common factor before this term is introduced. The activity asks students to determine how to make fruit baskets for a food pantry using different numbers and kinds of fruit. The goal is to make as many identical baskets as possible and to have no fruit left over. For example, with 12 apples and 8 oranges, one can make 4 baskets, each containing 3 apples and 2 oranges. Students used colored sticks to represent fruit and found various strategies to group and regroup the fruit to solve the problems. By referencing a prime factorization chart on the classroom wall, students connected their work to prime factors and found a procedure for calculating the greatest common factor. The activity was designed to encourage student inquiry and to allow students to form their own conclusions before the teacher introduces the term *greatest common factor*. The activity also caters to different learning styles and connects the concept of greatest common factor to a real-life example.

How Do We Use GCFs and LCMs with Fractions?

We often use GCFs and LCMs when working with fractions.

To put a fraction in simplest form, we can divide the numerator and denominator of the fraction by the GCF of the numerator and the denominator. For example, to put

$$\frac{24}{36}$$

in simplest form, we can first determine that the greatest common factor of 24 and 36 is 12. Since $24 = 2 \cdot 12$ and $36 = 3 \cdot 12$, we have

$$\frac{24}{36} = \frac{2 \cdot 12}{3 \cdot 12} = \frac{2}{3}$$

So $\frac{2}{3}$ is the simplest form of $\frac{24}{36}$.

It is not necessary to determine the greatest common factor of the numerator and denominator of a fraction in order to put the fraction in simplest form. Instead, we can keep simplifying the fraction until it can no longer be simplified. For example, we can use the following steps to simplify $\frac{24}{36}$:

$$\frac{24}{36} = \frac{12 \cdot 2}{18 \cdot 2} = \frac{12}{18} = \frac{6 \cdot 2}{9 \cdot 2} = \frac{6}{9} = \frac{2 \cdot 3}{3 \cdot 3} = \frac{2}{3}$$

We can use the LCM of the denominators of two fractions to add the fractions. In order to add fractions, we must first find a common denominator. Although any common denominator will do, we may wish to work with the least common denominator. Because common denominators of two fractions must be multiples of both denominators, the least common denominator is the least common multiple of the two denominators. For example, to add

$$\frac{1}{6} + \frac{3}{8}$$

we need a common denominator that is a multiple of 6 and of 8. The least common multiple of 6 and 8 is 24. Using this least common denominator, we calculate

$$\frac{1}{6} + \frac{3}{8} = \frac{1 \cdot 4}{6 \cdot 4} + \frac{3 \cdot 3}{8 \cdot 3} = \frac{4}{24} + \frac{9}{24} = \frac{13}{24}$$

To add $\frac{1}{6} + \frac{3}{8}$ we could also have used the common denominator $6 \cdot 8 = 48$. In this case,

$$\frac{1}{6} + \frac{3}{8} = \frac{1 \cdot 8}{6 \cdot 8} + \frac{3 \cdot 6}{8 \cdot 6} = \frac{8}{48} + \frac{18}{48} = \frac{26}{48}$$

Because we did not use the least common denominator, the resulting sum is not in simplest form. We need the following additional step to put the answer in simplest form:

$$\frac{26}{48} = \frac{13 \cdot 2}{24 \cdot 2} = \frac{13}{24}$$

When we use a common denominator that is not the least common multiple of the denominators, the resulting sum will not be in simplest form. However, even if we use the least common denominator when adding two fractions, the resulting fraction may not be in simplest form. For example,

$$\frac{1}{2} + \frac{1}{6} = \frac{3}{6} + \frac{1}{6} = \frac{4}{6}$$

but $\frac{4}{6}$ is not in simplest form even though we used the least common denominator when adding.

SECTION SUMMARY AND STUDY ITEMS

Section 8.5 Greatest Common Factor and Least Common Multiple

The GCF of two (or more) counting numbers is the greatest counting number that is a common factor of the numbers. The LCM of two (or more) counting numbers is the least counting number that is a common multiple of the numbers. In addition to using the definitions to find the GCF and LCM of counting numbers, we can also use the slide method in which we repeatedly divide by common factors. GCFs and LCMs are often used in fraction arithmetic, although their use is not necessary.

Key Skills and Understandings

1. State the meaning of GCF and LCM.
2. Use the definitions to determine GCFs and LCMs. Use the slide method to determine GCFs and LCMs and give a rough idea of why the method works.
3. Write and solve word problems that can be solved by finding a GCF. Write and solve word problems that can be solved by finding a LCM.
4. Apply GCFs and LCMs to fraction arithmetic, but understand that the arithmetic can also be done without explicitly using GCFs and LCMs.

Practice Exercises for Section 8.5

1. Use the definition of LCM to calculate the least common multiple of 9 and 12.
2. Use the definition of GCF to calculate the greatest common factor of 36 and 63.
3. Use the definition of GCF to calculate the greatest common factor of 16 and 27.
4. Write a word problem such that solving the problem requires finding the GCF of 100 and 75. Solve the problem.
5. Write a word problem such that solving the problem requires finding the LCM of 12 and 10. Solve the problem.
6. Show how to use the slide method to find the GCF and LCM of 224 and 392.
7. Show how to use the slide method to find the GCF and LCM of 12,375 and 16,875.
8. Find the GCF and LCM of $5^{15} \cdot 7^{12} \cdot 11$ and $5^{21} \cdot 7^{10} \cdot 13^2$.

Answers to Practice Exercises for Section 8.5

1. The multiples of 9 are

$$9, 18, 27, 36, 45, 54, 63, 72, 81, \ldots$$

and the multiples of 12 are

$$12, 24, 36, 48, 60, 72, 84, \ldots$$

Therefore, the common multiples of 9 and 12 are

$$36, 72, \ldots$$

The least of these is 36, which is thus the least common multiple of 9 and 12.

2. The factors of 36 are

$$1, 2, 3, 4, 6, 9, 12, 18, 36$$

and the factors of 63 are

$$1, 3, 7, 9, 21, 63$$

Therefore, the common factors of 36 and 63 are

$$1, 3, 9$$

The greatest of these numbers is 9; hence, the greatest common factor of 36 and 63 is 9.

3. The factors of 16 are

$$1, 2, 4, 8, 16$$

and the factors of 27 are

$$1, 3, 9, 27$$

The only common factor of 16 and 27 is 1, so this must be the greatest common factor of 16 and 27.

4. If you have 100 pencils and 75 small notebooks, what is the largest group of children that you can give all the pencils and all the notebooks to so that each child gets the same number of pencils and the same number of notebooks, and so that no pencils or notebooks are left over?

If each child gets the same number of pencils and no pencils are left over, then the number of children you can give the pencils to must be a factor of 100. Similarly, the number of children that you can give the notebooks to must be a factor of 75. Since you are looking for the largest number of children to give the pencils and notebooks to, this will be the greatest common factor of 100 and 75. The factors of 100 are

$$1, 2, 4, 5, 10, 20, 25, 50, 100$$

The factors of 75 are

$$1, 3, 5, 15, 25, 75$$

The common factors of 100 and 75 are

$$1, 5, 25$$

Therefore, the GCF of 100 and 75 is 25, so the largest group of children you can give the pencils and notebooks to is 25.

5. If hot dog buns come in packages of 12 and hot dogs come in packages of 10, what is the smallest number of hot dogs and buns you can buy so that you can pair each hot dog with a bun and so that no hot dogs or buns will be left over? Explain why you can solve the problem the way you do.

The total number of buns you buy will be a multiple of 12, and the total number of hot dogs you buy will be a multiple of 10. You want both multiples to be equal and to be as small as possible. Therefore, the number of hot dogs and buns you will buy will be the least common multiple of 12 and 10. The multiples of 12 are

$$12, 24, 36, 48, 60, 72, 84, 96, 108, \ldots$$

The multiples of 10 are

$$10, 20, 30, 40, 50, 60, 70, 80, 90, \ldots$$

The only common multiple that we see so far in these lists is 60, so this is the LCM of 12 and 10. Therefore, you should buy 5 packs of buns and 6 packs of hot dogs for a total of 60 hot dogs and 60 buns.

6.

2	224	392
2	112	196
2	56	98
7	28	49
	4	7

$\text{GCF} = 2 \cdot 2 \cdot 2 \cdot 7 = 56$

$\text{LCM} = 2 \cdot 2 \cdot 2 \cdot 7 \cdot 4 \cdot 7 = 1568$

7.

5	12,375	16,875
5	2475	3375
5	495	675
3	99	135
3	33	45
	11	15

$\text{GCF} = 5 \cdot 5 \cdot 5 \cdot 3 \cdot 3 = 1125$

$\text{LCM} = 5 \cdot 5 \cdot 5 \cdot 3 \cdot 3 \cdot 11 \cdot 15 = 185{,}625$

8. $\text{GCF} = 5^{15} \cdot 7^{10}$ $\text{LCM} = 5^{21} \cdot 7^{12} \cdot 11 \cdot 13^2$.

PROBLEMS FOR SECTION 8.5

1. Why do we not talk about a *greatest* common multiple and a *least* common factor?

2. Show how to use the definition of GCF to determine the greatest common factor of 27 and 36.

3. Show how to use the definition of GCF to determine the greatest common factor of 30 and 77.

4. Show how to use the definition of LCM to determine the least common multiple of 44 and 55.

5. Show how to use the definition of LCM to determine the least common multiple of 7 and 8.

6. Show how to use the slide method to determine the GCF and LCM of 2880 and 2400.

7. Show how to use the slide method to determine the GCF and LCM of 144 and 2240.

8. Show how to use the slide method to determine the GCF and LCM of 360 and 1344.

9. Find the GCF and LCM of $2^5 \cdot 3^2 \cdot 5$ and $2^3 \cdot 3^4 \cdot 7$ *without* calculating the products. Explain your reasoning.

10. Find the GCF and LCM of $3^4 \cdot 5^2 \cdot 7^5 \cdot 11$ and $3^7 \cdot 5^3 \cdot 7^3$ *without* calculating the products. Explain your reasoning.

11. Describe in general how to find the GCF and LCM of two counting numbers from the prime factorizations of the two numbers, and explain why the method works.

12. Show all the details in the following calculations:

a. Put $\frac{90}{126}$ in simplest form by first determining the greatest common factor of 90 and 126. Show the details of determining this greatest common factor.

b. Put $\frac{90}{126}$ in simplest form without first determining the greatest common factor of 90 and 126.

13. Show all the details in the following calculations:

a. Add $\frac{5}{12} + \frac{1}{8}$ by using the least common denominator. Show the details of determining the least common denominator. Give the answer in simplest form.

b. Add $\frac{5}{12} + \frac{1}{8}$ by using the common denominator $12 \cdot 8$. Give the answer in simplest form.

14. Write a word problem that requires calculating the least common multiple of 45 and 40 to solve. Solve your problem, explaining how the least common multiple is involved.

15. Write a word problem that requires calculating the greatest common factor of 45 and 40 to solve. Solve your problem, explaining how the greatest common factor is involved.

16. Suppose you are teaching students about least common multiples and you use only the following examples to illustrate the concept:

4 and 7
3 and 10
5 and 12
8 and 9

What misconception about the LCM might the students develop? Explain why. What other kinds of examples should the students see? Describe some examples that you think are good, and explain why you think they are good.

17. Kwan and Clevere are playing drums together, making a steady beat. Kwan beats the cymbals on beats that are multiples of 8. Clevere beats the cymbals on beats that are multiples of 12. Find the first 4 beats on which both Kwan and Clevere will beat the cymbals. Use mathematical terms to describe the first beat, and all the beats, on which both Kwan and Clevere beat the cymbals. Explain.

18. At the zoo, the birds must be fed 12 cups of sunflower seeds and 20 cups of millet seed every day. The zookeepers would like to find a container for scooping both the sunflower seeds and the millet seeds with. The container should allow the zookeepers to scoop out the proper amount of sunflower seeds and millet seeds by filling the container completely a certain number of times. What is the largest possible container the zookeepers could use? Use mathematical terms to describe the size of this container. Explain.

19. There are periodical cicadas with 13-year life cycles and other periodical cicadas with 17-year life cycles. The ones with 13-year life cycles become active adults every 13 years, and the ones with 17-year life cycles become active adults every 17 years. Suppose that one year, both the 13-year cicadas and the 17-year cicadas are active adults simultaneously. Find the next four times that both kinds of cicadas will be active adults simultaneously. Use mathematical terms to describe the next time, and all the times, when both kinds of cicadas will be active adults simultaneously. Explain.

20. In a clothing factory, a worker can sew 18 Garment A seams in a minute and 30 Garment B seams per minute. If the factory manager wants to complete equal numbers of Garments A and B every minute, how many workers should she hire for each type of garment? Give three different possibilities, and find the smallest number of workers the manager could hire. Explain your answers.

21. Keiko has a rectangular piece of fabric that is 48 inches wide and 72 inches long. She wants to cut her fabric into identical square pieces, leaving no fabric remaining. She also wants the side lengths of the squares to be whole numbers of inches.

a. Draw rough sketches indicating three different ways that Keiko could cut her fabric into squares.

b. Keiko decides that she wants her squares to be as large as possible. How big should she make her squares? Explain your reasoning. Draw a sketch showing how Keiko should cut the squares from her fabric.

***22.** A large gear will be used to turn a smaller gear. The large gear will make 300 revolutions per minute. The smaller gear must make 1536 revolutions per minute. How many teeth could each gear have? Give three different possibilities, and find the smallest number of teeth each gear could have. Explain your reasoning.

***23.** A large gear is used to turn a smaller gear. The large gear has 60 teeth and makes 12 revolutions per minute; the smaller gear has 50 teeth. How fast does the smaller gear turn? Explain your reasoning.

8.6 Rational and Irrational Numbers

CCSS **Common Core State Standards** Grade 8

How are fractions and decimals related? We saw in Section 6.3 that every fraction can be written as a decimal by dividing the denominator into the numerator. What about the other way around? If we have a decimal, is it the decimal representation of a fraction? In this section, we will analyze the surprising fact that not every decimal comes from a fraction. In fact, the decimals that come from fractions eventually repeat. We will then see that every decimal that eventually repeats can be written as a fraction, and we will see how to write such decimals as fractions. In particular, we will note that the number 0.999999 . . . , where the 9s repeat forever, is equal to 1. Finally, in this section, we will consider why numbers such as $\sqrt{2}$ and $\sqrt{3}$ cannot be written as a fraction, and we'll look at a surprising consequence of this fact reflected in the way pattern tiles—which even the very youngest children play with—behave.

Before we continue, let's review the standard terminology that mathematicians use when discussing fractions versus decimals.

rational numbers The **rational numbers** are those numbers that can be expressed as a fraction or the negative of a fraction. Note that we can write the fraction $\frac{1}{3}$ as the decimal 0.33333 . . . ; these are two different ways to write the same number, and this number is rational no matter which way we write it.

real numbers The **real numbers** are those numbers that can be expressed in a decimal representation or the negative of such a number. Included among the real numbers are those decimals that have infinitely many nonzero entries to the right of the decimal point.

irrational number A real number is called an **irrational number** if it is not rational—in other words, if it cannot be written as a fraction or the negative of a fraction.

Every fraction can be expressed as a decimal, so it is natural to ask: Are there any irrational numbers or are all real numbers rational? As we will see in this section, there are indeed irrational numbers. The discovery of irrational numbers was shocking and disconcerting to people in ancient times. We will also see that there is an exact characterization of the decimals that represent rational numbers.

CLASS ACTIVITY

8O Decimal Representations of Fractions, p. CA-152

Why Do Decimal Representations of Fractions Eventually Repeat?

CCSS 7.NS.2d 8.NS.1

Recall from Chapter 6 that we can use division to write a fraction as a decimal. When we write a fraction as a decimal, the decimal representation is one of two types. For example, consider the decimal representations of the fractions in Table 8.4. The fractions on the left have decimal representations that are **repeating** because it turns out that each of these eventually has a single digit or a fixed string of digits that repeats forever. Notice that the repeating portion of the decimal need not begin right after the decimal point, as in the decimal representations of $\frac{1}{12}$ and $\frac{1}{888}$. The fractions on the right in Table 8.4 have decimal representations that are **terminating** because these decimals only have finitely many nonzero digits. We can also describe terminating decimals as repeating because they have repeating zeros. For example,

repeating / **terminating**

$$\frac{5}{8} = 0.625 = 0.62500000\ldots$$

TABLE 8.4 Decimal representations of fractions are of two types: repeating or terminating

$\frac{5}{9} = 0.555555555555555\ldots$	$\frac{1}{4} = 0.25$
$\frac{1}{7} = 0.142857142857142\ldots$	$\frac{5}{8} = 0.625$
$\frac{41}{333} = 0.123123123123123\ldots$	$\frac{1}{80} = 0.0125$
$\frac{1}{12} = 0.083333333333333\ldots$	$\frac{41}{100} = 0.41$
$\frac{1}{888} = 0.001126126126126\ldots$	$\frac{12,345}{100,000} = 0.12345$

To indicate that a digit or a string of digits repeats forever, we can write a bar above the repeating portion. For example,

$0.\overline{3}$ means $0.333333\ldots$, where the 3s repeat forever

$0.00\overline{126}$ means $0.00126126126\ldots$, where the 126s repeat forever

We will now reason about the standard division algorithm to explain why the decimal representation of every fraction must either terminate or repeat. Consider a proper fraction $\frac{A}{B}$, where A and B are whole numbers and A is less than B. When we use the standard division algorithm to calculate the decimal representation of $\frac{A}{B}$, we will get various remainders along the way. For example, when we divide $1 \div 7$, we get the remainders

$$1, 3, 2, 6, 4, 5, 1, 3, \ldots$$

and when we divide $1 \div 8$, we get the remainders

$$1, 2, 4, 0$$

as shown in Table 8.5. There are three key points to observe about the division process:

1. If we get a remainder of 0, then the decimal representation terminates at that point (i.e., all subsequent digits in the decimal representation of the fraction are 0).
2. If we get a remainder that we got before, the decimal representation will repeat from there on.
3. We can get only remainders that are less than the denominator of the fraction. For example, in finding the decimal representation of $\frac{1}{7}$ we could get only the remainders 0, 1, 2, 3, 4, 5, or 6. (In fact, in that case we get all those remainders except for 0.)

Now imagine carrying out the division process to find the decimal representation of a proper fraction $\frac{A}{B}$. Suppose that we have found $B - 1$ digits to the right of the decimal point. If any of the remainders were 0, then the decimal representation terminates. Now suppose that none of the remainders we found were 0. Each of the $B - 1$ digits we found in the decimal representation was obtained from a remainder by putting a 0 behind the remainder and dividing B into the resulting number. Because of item 3, and because we are assuming that we didn't get 0 as a remainder, the remainders must be among the numbers

$$1, 2, \ldots, B - 1$$

When we calculate the Bth digit to the right of the decimal point, we use the Bth remainder, which either is 0 or must also be among the numbers

$$1, 2, \ldots, B - 1$$

If the Bth remainder is 0, then the decimal representation terminates at that point. But if the Bth remainder is not 0, then since there are only $B - 1$ distinct nonzero remainders that we can possibly get, there must be 2 equal remainders among the first B remainders. When we get a remainder that has appeared before, the decimal representation repeats. So, in this case, the decimal representation of our fraction is repeating. Therefore, for any proper fraction, $\frac{A}{B}$, where A and B are whole

TABLE 8.5 Division process shows that decimal representations of fractions either repeat or terminate

Repeats		Terminates	
↓		↓	
0.1428571		0.125	
7)1.0000000	remainder 1	8)1.000	remainder 1
−7		−8	
30	remainder 3	20	remainder 2
−28		−16	
20	remainder 2	40	remainder 4
−14		−40	
60	remainder 6	0	remainder 0
−56			terminates
40	remainder 4		
−35			
50	remainder 5		
−49			
10	remainder 1		
−7	repeats		
3	remainder 3		

numbers, the decimal representation of $\frac{A}{B}$ either terminates or repeats after at most $B - 1$ places to the right of the decimal point.

We can use the conclusion we just derived to show that there are irrational numbers. For example, consider the number

$$0.28228222822228222228\ldots$$

where the pattern of putting more and more 2s in between 8s continues forever. Notice that even though there is a pattern to this decimal, it does not have a *fixed string* of digits that repeats forever. Therefore, this decimal representation is neither terminating nor repeating, so it cannot be the decimal representation of a fraction, and is therefore irrational. Thus, the set of real numbers, which consists of all decimal numbers, is strictly larger than the set of rational numbers.

How Can We Use Math Drawings to See Decimal Representations of Fractions?

Students sometimes work with large squares that have been subdivided into 100 smaller squares (or even into 1000 tiny rectangles) in order to make sense of decimals. These subdivided squares can also show decimal representations of fractions. If the entire large square in Figure 8.10 represents 1,

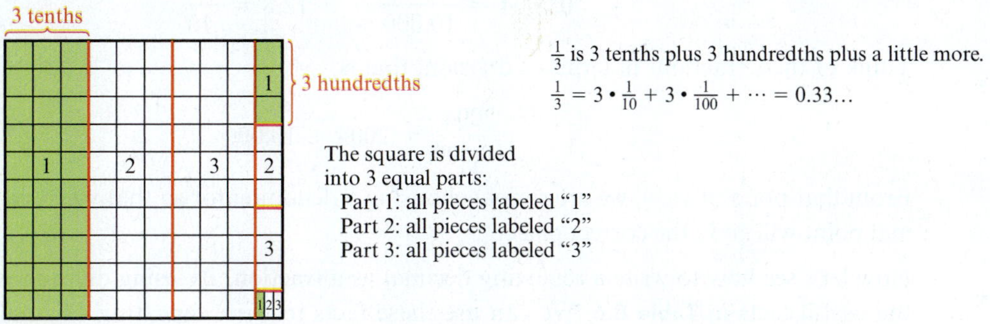

Figure 8.10 Using a subdivided square to determine the decimal representation of $\frac{1}{3}$.

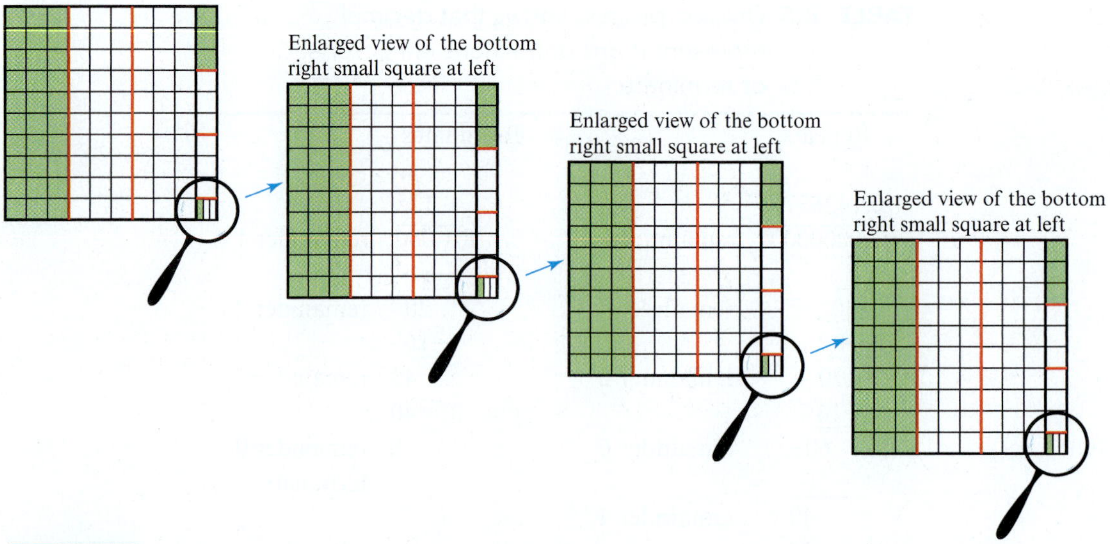

Figure 8.11 Showing the repeating nature of the decimal representation of $\frac{1}{3} = 0.333333\ldots.$

then each of the small squares represents one hundredth, or 0.01, and a strip of 10 small squares represents one tenth, or 0.1. Figure 8.10 shows how to subdivide the square into 3 equal parts to represent $\frac{1}{3}$ as a decimal to the hundredths place.

Interestingly, and perhaps surprisingly, we can use the subdivided square to *see* the repeating nature of the decimal representation of $\frac{1}{3}$, as indicated in Figure 8.11.

Can We Express Repeating and Terminating Decimals as Fractions?

As we have seen, if we start with a fraction, its decimal representation is necessarily either terminating or repeating. But is every terminating or repeating decimal necessarily the decimal representation of a fraction? The answer is yes, as we will see.

CCSS 8.NS.1

CLASS ACTIVITY

8P Writing Terminating and Repeating Decimals as Fractions, p. CA-154

Why can every terminating or repeating decimal be written as a fraction? First, we can write a terminating (i.e., finite) decimal as a fraction by using a denominator that is an appropriate power of 10. For example,

$$0.12345 = \frac{12{,}345}{100{,}000} \qquad 1.2345 = \frac{12{,}345}{10{,}000}$$

$$0.2004 = \frac{2004}{10{,}000} \qquad 12.8 = \frac{128}{10}$$

Think of these fractions in terms of division; that is,

$$\frac{2004}{10{,}000} = 2004 \div 10{,}000$$

From that point of view, we are simply choosing a denominator so that when we divide, the decimal point will go in the correct place.

Now let's see how to write a repeating decimal as a fraction. By using division we can determine the useful facts in Table 8.6. We can use these facts to write repeating decimals as fractions as

TABLE 8.6 Facts that can be used to write repeating decimals as fractions

$\frac{1}{9} = 0.\overline{1} = 0.111111\ldots$	$\frac{1}{9999} = 0.\overline{0001} = 0.00010001\ldots$
$\frac{1}{99} = 0.\overline{01} = 0.010101\ldots$	$\frac{1}{99{,}999} = 0.\overline{00001} = 0.0000100001\ldots$
$\frac{1}{999} = 0.\overline{001} = 0.001001\ldots$	$\vdots$

shown in **Table 8.7**. If we have a repeating decimal whose repeating pattern doesn't start right after the decimal point, then we first shift the decimal point by dividing by a suitable power of 10. We then add on the nonrepeating part, as demonstrated in **Table 8.8**.

TABLE 8.7 Writing repeating decimals as fractions, using facts from Table 8.6

$$0.\overline{12} = 12 \cdot 0.010101\ldots = 12 \cdot \frac{1}{99} = \frac{12}{99}$$

$$0.\overline{123} = 123 \cdot 0.001001001\ldots = 123 \cdot \frac{1}{999} = \frac{123}{999}$$

$$0.\overline{1234} = 1234 \cdot 0.000100010001\ldots = 1234 \cdot \frac{1}{9999} = \frac{1234}{9999}$$

TABLE 8.8 Writing repeating decimals as fractions when the repeating part doesn't start right after the decimal point

How do we write $0.987\overline{12}$ as a fraction?

Step 1: $0.000\overline{12} = \frac{1}{1000} \cdot 0.121212\ldots = \frac{1}{1000} \cdot \frac{12}{99} = \frac{12}{99{,}000}$

Step 2: $0.987\overline{12} = 0.987 + 0.000\overline{12} = \frac{987}{1000} + \frac{12}{99{,}000} = \frac{97{,}725}{99{,}000}$

What Is Another Method for Writing Repeating Decimals as Fractions?

Another method for writing a repeating decimal as a fraction involves shifting the decimal point by multiplying by a suitable power of 10 and then subtracting so that the resulting quantity is a whole number. For example, let N stand for the repeating decimal $0.123123123\ldots = 0.\overline{123}$, so that

$$N = 0.\overline{123} = 0.123123123\ldots$$

Then

$$1000N = 123.123123123\ldots$$

We can subtract N from $1000N$ in two different ways:

$$\begin{array}{rl} 1000N & 123.123123123\ldots \\ -N & -0.123123123\ldots \\ \hline 999N & 123 \end{array}$$

Therefore,

$$999N = 123$$

and

$$N = \frac{123}{999}$$

Thus,

$$0.\overline{123} = \frac{123}{999}$$

Notice that by multiplying N by 1000 we were able to arrange for the decimal portions to cancel when we subtracted. If we had multiplied by 10 or 100, this canceling would not have happened.

If the repeating part of the decimal does not start right after the decimal point, then we can modify the previous method by shifting with additional powers of 10. For example, let

$$N = 0.98\overline{123} = 0.98123123123\ldots$$

Then

$$100{,}000N = 98{,}123.123123123\ldots$$

and

$$100N = 98.123123123\ldots$$

Now subtract $100N$ from $100{,}000N$ in the following two different ways:

$$\begin{array}{rr} 100{,}000N & 98{,}123.123123123\ldots \\ -100N & -98.123123123\ldots \\ \hline 99{,}900N & 98{,}025 \end{array}$$

Therefore,

$$99{,}900N = 98{,}025$$

and

$$N = \frac{98{,}025}{99{,}900}$$

Thus,

$$0.98\overline{123} = \frac{98{,}025}{99{,}900}$$

Notice that by multiplying by 100,000 and by 100, we arranged for the decimal portions to cancel when we subtracted.

CLASS ACTIVITY

8Q What Is 0.9999...?, p. CA-155

Why Is 0.99999 . . . = 1?

We can apply either of the two methods for writing a repeating decimal as a fraction to the repeating decimal

$$0.\overline{9} = 0.99999999\ldots$$

When we do this, we reach the following surprising conclusion:

$$0.\overline{9} = 9 \cdot 0.111111111\ldots = 9 \cdot \frac{1}{9} = \frac{9}{9} = 1$$

Similarly, let

$$N = 0.\overline{9} = 0.999999\ldots$$

Then

$$10N = 9.999999\ldots$$

Now subtract N from $10N$ in the following two ways:

$$\begin{array}{rr} 10N & 9.999999\ldots \\ \underline{-N} & \underline{-0.999999\ldots} \\ 9N & 9 \end{array}$$

Therefore,

$$9N = 9$$

and

$$N = 1$$

Thus,

$$0.\overline{9} = 1$$

Although it may seem like 0.999999 . . . should be just a little bit less than 1, it is in fact *equal to* 1.

Why Is the Square Root of 2 Irrational?

square root Given a positive number N, the **square root** of N, denoted $\sqrt{N}$, is the positive number S such that

$$S^2 = N$$

For example, $\sqrt{2}$ is the positive number such that

$$(\sqrt{2})^2 = 2$$

Square roots of numbers provide many examples of irrational numbers—that is, numbers that cannot be written as fractions. For example, we will see why $\sqrt{3}$ is irrational. In Class Activity 8R you will show that $\sqrt{2}$ is irrational. Note, however, that square roots of some numbers are rational. For example, $\sqrt{4}$ is rational because

$$\sqrt{4} = 2 = \frac{2}{1}$$

CLASS ACTIVITY

8R The Square Root of 2, p. CA-156

If you use a calculator to find $\sqrt{3}$, your calculator's display might read

$$1.73205080757$$

Of course, a calculator can display only a certain number of digits, so the calculator is not telling you that $\sqrt{3}$ is exactly equal to 1.73205080757. Rather, it is telling you the first few digits in the decimal representation of $\sqrt{3}$ (although the last digit may be rounded). When we look at the calculator's display, it appears that the decimal representation of $\sqrt{3}$ does not repeat or terminate, but we can *never tell for sure* by looking at a finite portion of a decimal representation whether or not the decimal repeats or terminates. The decimal might not start to repeat until after 100 places, or the decimal might appear to repeat at first, but in fact not repeat. So to determine whether a number such as $\sqrt{3}$ is rational or irrational, we must do something other than look at some digits in the decimal representation.

We will now show that $\sqrt{3}$ is irrational by supposing that we could write $\sqrt{3}$ as a fraction and then deducing that this could not be the case. So suppose that

$$\sqrt{3} = \frac{A}{B}$$

where A and B are counting numbers. Then

$$3 = \left(\frac{A}{B}\right)^2$$

so

$$3 = \frac{A^2}{B^2}$$

By multiplying both sides of this equation by B^2, we have

$$3 \cdot B^2 = A^2$$

Now imagine factoring A and B into products of prime numbers. Then A^2 has an even number of prime factors because it has twice as many prime factors as A does. For example, if A were 42, then

$$A = 2 \cdot 3 \cdot 7$$

and

$$A^2 = 2 \cdot 3 \cdot 7 \cdot 2 \cdot 3 \cdot 7$$

which has 6 prime factors, twice as many as A. Similarly, $3 \cdot B^2$ has an odd number of prime factors because it has one more than twice as many prime factors as B. For example, if B were 35, then

$$B = 5 \cdot 7$$

and

$$3 \cdot B^2 = 3 \cdot 5 \cdot 7 \cdot 5 \cdot 7$$

which has 5 prime factors, one more than twice as many as A.

But if $3 \cdot B^2$ has an odd number of prime factors and A^2 has an even number of prime factors, then it cannot be the case that

$$3 \cdot B^2 = A^2$$

because of the uniqueness of factorization into products of prime numbers (the Fundamental Theorem of Arithmetic).

Therefore, it cannot be the case that

$$\sqrt{3} = \frac{A}{B}$$

where A and B are counting numbers. Thus, $\sqrt{3}$ is irrational.

As abstract an idea as the irrationality of $\sqrt{3}$ may seem to be, it is actually reflected in pattern tiles, which even children in PreK play with!

How Does Proof by Contradiction Work?

The method of proof that we just used to show that $\sqrt{3}$ is irrational is called **proof by contradiction**. Proof by contradiction works as follows: Assume that the statement you want to prove is *false*. Argue logically until you arrive at a contradiction. Therefore, conclude that your assumption (that the statement you want to prove is false) cannot be true. So the statement you want to prove must in fact be true.

Here is how we used proof by contradiction to prove that $\sqrt{3}$ is irrational. We first assumed that $\sqrt{3}$ is *not* irrational; in other words, we assumed that we could write $\sqrt{3}$ as a fraction, $\frac{A}{B}$, where A and B are whole numbers. We reasoned logically and showed that in that case,

$$3 \cdot B^2 = A^2$$

would be true. We also used logical reasoning to show that the prime factorization of $3 \cdot B^2$ would have odd number of factors and the prime factorization of A^2 would have an even number of factors. But this contradicts the equation $3 \cdot B^2 = A^2$. Therefore, it must be false that $\sqrt{3}$ is *not* irrational, and so $\sqrt{3}$ must be irrational.

Proof by contradiction is an interesting method of proof because it relies on believing that if we can prove that the statement "Statement S is false" is itself false, then statement S must be true. In the past, this method of proof was disputed by some mathematicians.

SECTION SUMMARY AND STUDY ITEMS

Section 8.6 Rational and Irrational Numbers

If $\frac{A}{B}$ is a fraction, then the decimal representation of $\frac{A}{B}$ either terminates or repeats. Conversely, every terminating or repeating decimal can be written as a fraction. Therefore, the rational numbers are exactly those numbers whose decimal representation either terminates or repeats. Surprisingly, the rational number $0.\overline{9}$ is equal to 1. The square root of 2 and the square root of 3 (along with many other square roots) are irrational numbers.

Key Skills and Understandings

1. Understand that the rational numbers (which were defined as numbers that can be expressed as fractions or the negative of a fraction) are the same as those numbers that have a terminating or repeating decimal representation.
2. Use the standard division algorithm to explain why fractions have decimal representations that either terminate or repeat.
3. Given a terminating or repeating decimal, write it as a fraction.
4. Explain why $0.\overline{9} = 1$.
5. Prove that various square roots such as $\sqrt{2}$ and $\sqrt{3}$ are irrational.

Practice Exercises for Section 8.6

1. In your own words, explain why the decimal representation of a fraction must either terminate or repeat.
2. Write the following as fractions. Explain your reasoning.
   - a. $0.00\overline{577}$
   - b. $1.24\overline{3}$
   - c. $13.2\overline{83}$
3. What is the 103rd digit to the right of the decimal point in the decimal representation of $\frac{13}{101}$? Explain your answer.
4. What is another way to write $3.45\overline{9}$ as a decimal? Explain your answer.
5. We have seen that $0.\overline{9} = 1$. Does this mean that there are other decimal representations for the numbers $0.\overline{8}$, $0.\overline{7}$, and $0.\overline{6}$?
6. Find at least two fractions whose decimal representation begins 0.349205986.
7. Are the following numbers rational or irrational?
   - a. 0.252252225252252225252252225 . . . where the pattern of a 2 followed by a 5, two 2s followed by a 5, and three 2s followed by a 5 continues to repeat.

b. 0.2522522252222522222522222225 . . . where the pattern of placing more and more 2s between 5s continues.

8. Suppose you have two decimals and each one is either repeating or terminating. Is it possible that the product of these two decimals could be neither repeating nor terminating?

9. Use your calculator to find the decimal representation of $\frac{1}{29}$. When you look at the calculator's display, does the number look as if its decimal representation either repeats or terminates? Does the number in fact have a repeating or terminating decimal representation?

10. a. Find the square root of 5 on a calculator. Make a guess: Does it look like it is rational, or does it look like it is irrational? Why?

b. Can you tell for sure whether or not $\sqrt{5}$ is rational just by looking at your calculator's display? Explain your answer.

11. Is $\sqrt{1.96}$ rational or irrational?

12. Is the square root of $1.\overline{7}$ rational or irrational?

13. In your own words, explain why $\sqrt{3}$ is irrational.

Answers to Practice Exercise for Section 8.6

1. Be sure your explanation includes the following elements: We find the decimal representation of a fraction by dividing the denominator into the numerator. During the division process, we eventually stop getting new remainders because the remainders are whole numbers between 0 and 1 less than the denominator. Therefore, after several steps, we either get the remainder 0, in which case the decimal terminates at that point, or we get a remainder that we got before, at which point the decimal representation repeats from then on (assuming the division is at the point of only "bringing down zeros").

2. a. Since

$$\frac{1}{999} = 0.\overline{001}$$

it follows that

$$0.\overline{577} = \frac{577}{999}$$

Dividing by 100, we have

$$0.00\overline{577} = \frac{577}{99{,}900}$$

b. Since

$$0.\overline{3} = \frac{3}{9} = \frac{1}{3}$$

dividing by 100, we have

$$0.00\overline{3} = \frac{1}{300}$$

Since

$$1.24 = \frac{124}{100}$$

it follows that

$$1.24\overline{3} = 1.24 + 0.00\overline{3} = \frac{124}{100} + \frac{1}{300} = \frac{373}{300}$$

c. Since

$$0.\overline{83} = \frac{83}{99}$$

dividing by 10, we have

$$0.0\overline{83} = \frac{83}{990}$$

Since

$$13.2 = \frac{132}{10}$$

it follows that

$$13.2\overline{83} = 13.2 + 0.0\overline{83} = \frac{132}{10} + \frac{83}{990} = \frac{13{,}151}{990}$$

3. Since

$$\frac{13}{101} = 0.\overline{1287} = 0.128712871287\ldots$$

has a string of 4 digits that repeats, every 4th digit is a 7. So the 4th, the 8th, the 12th, . . . digit is a 7. Since 100 is divisible by 4, the 100th digit is a 7. The 101st digit is then a 1, the 102nd is a 2, and the 103rd digit is an 8.

4. We know that $0.\overline{9} = 1$. Dividing by 100, we see that $0.00\overline{9} = 0.01$. Therefore,

$$\begin{aligned} 3.45\overline{9} &= 3.45 + 0.00\overline{9} \\ &= 3.45 + 0.01 \\ &= 3.46 \end{aligned}$$

5. No, it turns out that there is no other way to write decimal representations for these numbers. We can, however, write these numbers as the fractions $\frac{8}{9}$, $\frac{7}{9}$, and $\frac{6}{9} = \frac{2}{3}$.

6. There are many examples. The fraction

$$\frac{349{,}205{,}986}{1{,}000{,}000{,}000}$$

is one example. So is

$$\frac{349{,}205{,}986}{999{,}999{,}999}$$

(This one is repeating.) So is

$$\frac{3{,}492{,}059{,}861{,}544}{10{,}000{,}000{,}000{,}000}$$

(This one is terminating.)

7. a. $0.25225222525225222525225 \ldots = 0.\overline{252252225}$ is a repeating decimal, so it is a rational number.

b. $0.252252222522222522222225 \ldots$ is irrational because it is neither repeating nor terminating.

8. No. To see why not, notice that your two decimals can both be written as fractions because they are repeating or terminating. When you multiply fractions, the result is again a fraction. Therefore, the product of the two decimals must have either a repeating or a terminating decimal representation.

9. On a calculator, the decimal representation looks like it is neither terminating nor repeating. But this is just because a calculator can display only so many digits. In fact, since $\frac{1}{29}$ is a fraction, its decimal representation must be either terminating or repeating. (In fact, it is repeating.)

10. a. The calculator display of the decimal representation of the square root of 5 does not appear to be either terminating or repeating, so it looks like the square root of 5 is irrational. But this is not a proof.

b. No, you definitely can't tell for sure whether the square root of 5 is irrational just from looking at the calculator's display. Conceivably, its decimal representation might end after 200 digits. Or it might start to repeat after 63 digits. (It turns out that the square root of 5 is irrational. The proof is the same as for the square root of 3 and the square root of 2.)

11. $\sqrt{1.96} = \sqrt{\frac{196}{100}} = \frac{14}{10}$

So the square root of 1.96 is rational.

12. $1.\overline{7} = 1 + \frac{7}{9} = \frac{16}{9}$

So the square root of $1.\overline{7}$ is $\frac{4}{3}$, which is rational.

13. Be sure your explanation includes the following elements: Suppose that we can express $\sqrt{3}$ as $\frac{A}{B}$, where A and B are some whole numbers. Then show that $3 \cdot B^2 = A^2$. Explain why A^2 has an even number of prime factors whereas $3 \cdot B^2$ has an odd number of prime factors. Explain that this contradicts the Fundamental Theorem of Arithmetic because it would mean one number can be expressed with two different prime factorizations. Therefore conclude that $\sqrt{3}$ can't be expressed as a fraction.

PROBLEMS FOR SECTION 8.6

1. Use the standard division algorithm to determine the decimal representation of $\frac{1}{37}$. Determine whether the decimal representation repeats or terminates; if it repeats, describe the string of repeating digits.

2. Use the standard division algorithm to determine the decimal representation of $\frac{1}{13}$. Determine whether the decimal representation repeats or terminates; if it repeats, describe the string of repeating digits.

3. Use the standard division algorithm to determine the decimal representation of $\frac{1}{101}$. Determine whether the decimal representation repeats or terminates; if it repeats, describe the string of repeating digits.

4. Write the following decimals as fractions. Explain your reasoning (you need not write the fractions in simplest form):

a. $0.\overline{56}$, $0.00\overline{56}$, $1.11\overline{56}$

b. $0.\overline{987}$, $0.00\overline{987}$, $0.40\overline{987}$

c. $1.234\overline{567}$

5. What is the 100th digit to the right of the decimal point in the decimal representation of $\frac{2}{37}$? Explain your reasoning.

6. What is another way to write 1.824 as a decimal? Explain your reasoning.

7. Without actually determining the decimal representation of $\frac{1}{47}$, explain in your own words why its decimal representation must either terminate or repeat after at most 46 decimal places to the right of the decimal point.

8. Give an example of an irrational number, and explain in detail why your number is irrational.

9. In your own words, prove that the square root of 5 is irrational, using the same ideas we used to prove that the square root of 3 is irrational.

10. Carl's calculator displays only 10 digits. Carl punches some numbers and some operations on his calculator and shows you his calculator's display: 0.034482758. Carl has to figure out whether the number whose first 9 digits behind the decimal point are displayed on his calculator is rational or irrational.

a. If you had to make a guess, what would you guess about Carl's number: Do you think it is rational or irrational? Why?

b. Find a fraction of whole numbers whose first nine digits behind the decimal point in its decimal representation agree with Carl's calculator's display. (You do not have to put the fraction in simplest form.)

c. What if Carl's number was actually

0.034483758003448375800034483758000034483758...,

where the pattern of putting in more and more zeros before a block of 34483758 continues forever? In that case, is Carl's number rational or irrational? Why?

d. Without any knowledge of how Carl got his number to display on his calculator, is it possible to say for sure whether or not it is rational? Explain.

11. Fran has a calculator that shows at most 10 digits. Fran punches some numbers and operations on her calculator and shows you the display: 0.232323232.

a. If you had to make a guess, do you think Fran's number is rational or irrational? Why?

b. Find two different rational numbers whose decimal representation begins 0.232323232.

c. Find an irrational number whose decimal representation begins 0.232323232.

12. Tyrone used a calculator to solve a problem. The calculator gave Tyrone the answer 0.217391304. Without any further information, is it possible to tell if the answer to Tyrone's problem can be written as a fraction? Explain.

***13.** Show how to find the exact decimal representation of

$$0.\overline{027} \times 0.\overline{18} = 0.027027027\ldots \times 0.18181818\ldots$$

without using a calculator. Is the product a repeating decimal? If so, what are the repeating digits? *Hint:* Write the numbers in a different form.

***14.** It is a fact that

$$513{,}239 \times 194{,}841 = 99{,}999{,}999{,}999.$$

Use the preceding multiplication fact to find the exact decimal representations of

$$\frac{1}{513{,}239}$$

and

$$\frac{1}{194{,}841}$$

(Your answer should not just show the first bunch of digits, as on a calculator, but should make clear what *all* the digits in the decimal representation are.) If the decimal representation is repeating, describe the string of digits that repeat. Explain your reasoning. *Hint:* Think about rewriting the fractions, using the denominator 99,999,999,999.

***15.** Suppose you have a fraction of the form $\frac{1}{A}$, where A is a counting number. Also, suppose that the decimal representation of this fraction is repeating, that the string of repeating digits starts right after the decimal point, and that this string consists of 5 digits.

a. Explain why $\frac{1}{A}$ is equivalent to a fraction with denominator 99,999.

b. Factor 99,999 as a product of prime numbers.

c. Use parts (a) and (b) to help you find all the prime numbers A such that $\frac{1}{A}$ has a repeating decimal representation with a 5-digit string of repeating digits.

*__16.__ Suppose you have a counting number N that divides 9,999,999 (in other words $9{,}999{,}999 \div N$ is a counting number), but N is not 1, 3, or 9. What can you say about the decimal representation of $\frac{1}{N}$? Explain your reasoning.

*__17.__ **a.** For each of the fractions in the first column of **Table 8.9**, factor the denominator of the fraction as a product of prime numbers.

TABLE 8.9 Decimal representations of various fractions

$\frac{1}{12} = 0.0833333333333333\ldots$	$\frac{1}{4} = 0.25$
$\frac{1}{11} = 0.0909090909090909\ldots$	$\frac{2}{5} = 0.4$
$\frac{113}{33} = 3.424242424242424\ldots$	$\frac{37}{8} = 4.625$
$\frac{491}{550} = 0.8927272727272727\ldots$	$\frac{17}{50} = 0.34$
$\frac{14}{37} = 0.3783783783783783\ldots$	$\frac{1}{125} = 0.008$
$\frac{35}{101} = 0.3465346534653465\ldots$	$\frac{9}{20} = 0.45$
$\frac{1}{41} = 0.0243902439024390\ldots$	$\frac{19}{32} = 0.59375$
$\frac{5}{7} = 0.7142857142857142\ldots$	$\frac{1}{3200} = 0.0003125$
$\frac{1}{14} = 0.0714285714285714\ldots$	$\frac{1}{64{,}000} = 0.000015625$
$\frac{1}{21} = 0.0476190476190476\ldots$	$\frac{1}{625} = 0.0016$

b. For each of the fractions in the second column of **Table 8.9**, factor the denominator of the fraction as a product of prime numbers.

c. Contrast the prime factors that you found in part (a) with the prime factors that you found in part (b). Contrast the decimal representations of the fractions in the first column of **Table 8.9** with the decimal representations of the fractions in the second column of that table. Based on your findings, develop a conjecture (an educated guess) about decimal representations of fractions.

*__18.__ **a.** Use a calculator to calculate the decimal representation of

$$\frac{1}{5^{100}}$$

b. Based on your calculator's display, is it possible to tell whether or not $\frac{1}{5^{100}}$ has a repeating decimal representation or a terminating decimal representation? Why or why not?

c. Find a fraction that is equal to $\frac{1}{5^{100}}$ and that has a denominator that is a power of 10.

d. Based on part (c), determine whether $\frac{1}{5^{100}}$ has a repeating or a terminating decimal representation. Explain your reasoning.

*__19.__ **a.** Use a calculator to calculate the decimal representation of

$$\frac{1}{7^{100}}$$

b. Based on your calculator's display, is it possible to tell whether or not $\frac{1}{7^{100}}$ has a repeating decimal representation or a terminating decimal representation? Why or why not?

c. Can there be a fraction of whole numbers that is equal to $\frac{1}{7^{100}}$ and that has a denominator that is a power of 10? Why or why not?

d. Based on part (c), determine whether $\frac{1}{7^{100}}$ has a repeating or a terminating decimal representation. Explain your reasoning.

*__20.__ **a.** Suppose that a fraction $\frac{A}{B}$ where A and B are counting numbers has a terminating decimal representation and that $\frac{A}{B}$ is in simplest form. Explain why B must divide a power of 10.

b. Suppose that a fraction $\frac{A}{B}$ where A and B are counting numbers has a terminating decimal representation and that $\frac{A}{B}$ is in simplest form. Show that when B is factored into a product of prime numbers, the only prime numbers that can appear are 2 and 5.

c. Given a counting number B, suppose that when B is factored into a product of prime numbers, the only prime numbers that appear are 2, or 5, or both 2 and 5. Let A be another counting number. Explain why $\frac{A}{B}$ must have a terminating decimal representation.

d. Does

$$\frac{1}{2^{100}}$$

have a repeating or a terminating decimal representation? Explain your answer.

e. Does

$$\frac{1}{3^{100}}$$

have a repeating or a terminating decimal representation? Explain your answer.

CHAPTER SUMMARY

Section 8.1 Factors and Multiples	Page 337
▪ If A, B, and C are counting numbers and $A = B \times C$, then B and C are factors of A and A is a multiple of B and of C. Also, A is divisible by B and by C, and B and C divide A.	Page 337
Key Skills and Understandings	
1. State the meaning of the terms *factor* and *multiple*.	Page 337
2. Given a counting number, find all its factors in an efficient way and explain why the method finds all the factors. Given a counting number, list several multiples of the number.	Page 338
3. Write and solve word problems that can be solved by finding all the factors of a number. Write and solve word problems that can be solved by finding several multiples of a number.	Page 339
Section 8.2 Even and Odd	Page 341
▪ A counting number is even if it is divisible by 2, or in other words, if there is no remainder when the number is divided by 2. A counting number is odd if it is not divisible by 2, or in other words, if there is a remainder when it is divided by 2.	Page 341
▪ To check if a counting number is even or odd, we have to look only at the ones digit of the number. We can explain why this way of checking if a number is even or odd is valid by considering base-ten bundles and by recognizing that tens, hundreds, thousands, and so on can all be divided into groups of 2 with none left over.	Page 342
Key Skills and Understandings	
1. State the meaning of even and odd.	Page 341
2. Explain why it is valid to check if a number is even or odd by examining the ones place of the number.	Page 342
3. Solve problems about even and odd numbers.	Page 343
Section 8.3 Divisibility Tests	Page 345
▪ A divisibility test is a quick way to check if a counting number is divisible by a given number without actually dividing. In addition to the divisibility test for 2, there are commonly used divisibility tests for 3, 4, 5, 9, and 10 (plus other divisibility tests).	Page 345
▪ We can explain the divisibility tests by considering base-ten bundles and by considering what happens when tens, hundreds, thousands, and so on are divided by the number in question.	Page 346
Key Skills and Understandings	
1. Describe how to use the divisibility tests for 2, 3, 4, 5, 9, and 10, use the divisibility tests, and explain why these tests work.	Page 345
Section 8.4 Prime Numbers	Page 350
▪ The prime numbers are the counting numbers other than 1 that are divisible only by 1 and themselves.	Page 350
▪ We can use the Sieve of Eratosthenes to find all the prime numbers up to a given number.	Page 350
▪ To determine if a given counting number is prime, we can use the method of trial division, in which we check to see if consecutive prime numbers divide the given number.	Page 351
▪ The Fundamental Theorem of Arithmetic states that every counting number other than 1 either is a prime number or can be factored in a unique way into a product of prime numbers.	Page 350

▪ Factor trees can help find the prime factorization of a counting number.	Page 352
▪ There are infinitely many prime numbers.	Page 354
Key Skills and Understandings	
1. State the meaning of prime number.	Page 350
2. Use the Sieve of Eratosthenes, and explain why it produces a list of prime numbers.	Page 350
3. Use trial division to determine if a given counting number is prime. Explain why you have to divide only by prime numbers when using trial division. Explain why you can stop dividing when you do.	Page 351
4. Given a counting number, determine whether it is prime, or if not, factor it into a product of prime numbers.	Page 351
Section 8.5 Greatest Common Factor and Least Common Multiple	Page 356
▪ The GCF of two (or more) counting numbers is the greatest counting number that is a common factor of the numbers. The LCM of two (or more) counting numbers is the least counting number that is a common multiple of the numbers.	Page 356
▪ In addition to using the definitions to find the GCF and LCM of counting numbers, we can also use the slide method in which we repeatedly divide by common factors.	Page 357
▪ GCFs and LCMs are often used in fraction arithmetic, although their use is not necessary.	Page 358
Key Skills and Understandings	
1. State the meaning of GCF and LCM.	Page 356
2. Use the definitions to determine GCFs and LCMs. Use the slide method to determine GCFs and LCMs and give a rough idea of why the method works.	Page 356
3. Write and solve word problems that can be solved by finding a GCF. Write and solve word problems that can be solved by finding a LCM.	Page 360
4. Apply GCFs and LCMs to fraction arithmetic, but understand that the arithmetic can also be done without explicitly using GCFs and LCMs.	Page 358
Section 8.6 Rational and Irrational Numbers	Page 363
▪ If $\frac{A}{B}$ is a fraction, then the decimal representation of $\frac{A}{B}$ either terminates or repeats. Conversely, every terminating or repeating decimal can be written as a fraction. Therefore, the rational numbers are exactly those numbers whose decimal representation either terminates or repeats.	Page 363
▪ Surprisingly, the rational number $0.\overline{9}$ is equal to 1.	Page 368
▪ The square root of 2 and the square root of 3 (along with many other square roots) are irrational numbers.	Page 369
Key Skills and Understandings	
1. Understand that the rational numbers (which were defined as numbers that can be expressed as fractions or the negative of a fraction) are the same as those numbers that have a terminating or repeating decimal representation.	Page 363
2. Use the standard division algorithm to explain why fractions have decimal representations that either terminate or repeat.	Page 363
3. Given a terminating or repeating decimal, write it as a fraction.	Page 366
4. Explain why $0.\overline{9} = 1$.	Page 368
5. Prove that various square roots such as $\sqrt{2}$ and $\sqrt{3}$ are irrational.	Page 369

9

Algebra

Algebra is the language of mathematics and science, a gateway to mathematical and scientific thinking. Algebra generalizes and builds on arithmetic and its properties. In algebra we consider not just individual calculations but whole collections of calculations all at once. We can reason about collections of calculations to determine which specific ones are especially interesting or useful. Using algebra, we can work with quantities that change and we can relate quantities that change together.

In this chapter, we focus on the following topics and practices within the *Common Core State Standards for Mathematics* (*CCSSM*).

Standards for Mathematical Content in the CCSSM

In the domain of *Operations and Algebraic Thinking* (Kindergarten through Grade 5), students formulate and solve equations to solve problems. They write and interpret expressions that record calculations. They observe, describe, and analyze patterns.

In the domain of *Expressions and Equations* (Grades 6 through 8), students use variables in expressions and equations. They begin to treat expressions as entities in their own right and they consider how expressions are broken into component parts. They represent and analyze quantitative relationships, leading to the study of functions. Students learn to solve equations by viewing equation-solving as a process of determining which values make the equation true. They solve equations to solve real-world problems.

Standards for Mathematical Practice in the CCSSM

Opportunities to engage in all eight of the Standards for Mathematical Practice described in the CCSSM occur throughout the study of algebra, although the following standards may be especially appropriate for emphasis:

- **2 Reason abstractly and quantitatively.** Students engage in this practice when they seek to understand variables, expressions, equations, and functions in terms of a context and when they use contexts to help make sense of how we work with variables, expressions, equations, and functions.
- **4 Model with mathematics.** Students engage in this practice when they define variables and formulate expressions and equations for quantities of interest and when they define, represent, and reason about functions.
- **7 Look for and make use of structure.** Students engage in this practice when they look for the structure of expressions to determine which values are possible or to guide strategic choices in solving equations.
- **8 Look for an express regularity in repeated reasoning.** Students engage in this practice when they recognize that repeated calculations could be described by a "recipe" and when they recognize and use expressions as "calculation recipes."

  (From Common Core Standards for Mathematical Practice. Published by Common Core Standards Initiative.)

9.1 Numerical Expressions

CCSS **Common Core State Standards** Grades 5 and 6

What are mathematical expressions and why do we have them? Expressions are "calculation recipes" that show us how calculations can be carried out. An expression can summarize a multistep calculation compactly and can help us see and analyze the structure of a calculation. Some expressions involve variables such as x or y, whereas others involve only numbers and are called numerical expressions. In this section we study numerical expressions, which are stepping stones to expressions with variables. We'll see that we can visually depict expressions in interesting patterns and that we can write different expressions for the same total amount. An essential skill in algebra is interpreting the meaning of expressions without first evaluating them and viewing expressions as entities in their own right.

How Can We Interpret and Evaluate Numerical Expressions?

numerical expression

A **numerical expression** is a meaningful string of numbers and operation symbols ($+$, $-$, $\times$ or $\cdot$, $\div$ or $/$, or raising to a power, but not $=$) and possibly also grouping symbols such as parentheses or brackets. Students begin working with simple expressions such as

$$2 + 3 \quad \text{and} \quad 3 \times 5 \quad \text{and} \quad 35 \div 7$$

in the early elementary grades. Later, they use more complex expressions such as

$$3 + 7 \cdot 2 \quad \text{and} \quad 8 \cdot (20 + 4) \quad \text{and} \quad 35 - 5^2$$

to record calculations.

CCSS
5.OA.1
5.OA.2
6.EE.1

To interpret the meaning of an expression, we must use the conventions on the order of operations; these conventions were described on page 167. For example,

$$3 + 7 \cdot 2$$

means the sum of 3 with the product $7 \cdot 2$, whereas

$$(3 + 7) \cdot 2$$

means the product of $3 + 7$ with 2.

Because we look for addition and subtraction last when we use the conventions on the order of operations, an expression can be viewed as broken into components that are combined by addition, subtraction, or a combination of addition and subtraction (if there is more than one component).

term

We often call these components the **terms** of the expression. For example,

$$1000 \cdot 5 + 20 \div 4$$

is a sum of two terms, $1000 \cdot 5$ and $120 \div 4$.

evaluate a numerical expression

value of an expression

To **evaluate a numerical expression** means to determine the number to which the numerical expression is equal; we call this number the **value of an expression**. We usually show the process of evaluating an expression with one or more equations. For example, to evaluate $3 + 7 \cdot 2$ we calculate thus:

$$\begin{aligned} 3 + 7 \cdot 2 &= 3 + 14 \\ &= 17 \end{aligned}$$

whereas to evaluate $(3 + 7) \cdot 2$ we can calculate thus:

$$\begin{aligned} (3 + 7) \cdot 2 &= 10 \cdot 2 \\ &= 20 \end{aligned}$$

For success in algebra, students must learn to see the structure in an expression *without first evaluating it* and they must learn to use the structure to reason about the value of the expression. For example, can you tell whether $1 + 2 \cdot 3927$ is even or odd without first evaluating the expression? It must be an odd number because it is 1 more than a multiple of 2.

To help reveal the structure in an expression you can successively circle components according to the order of operations, as shown in Figure 9.1. We see that the expression

$$3 \cdot (7 - 1) + 84 \div 2 - 4 \cdot 2^3$$

has three terms, which are combined by addition and subtraction. The first term is a product of 3 with $7 - 1$; the second term is the quotient, 84 divided by 2; the third term is the product of 4 with 2^3.

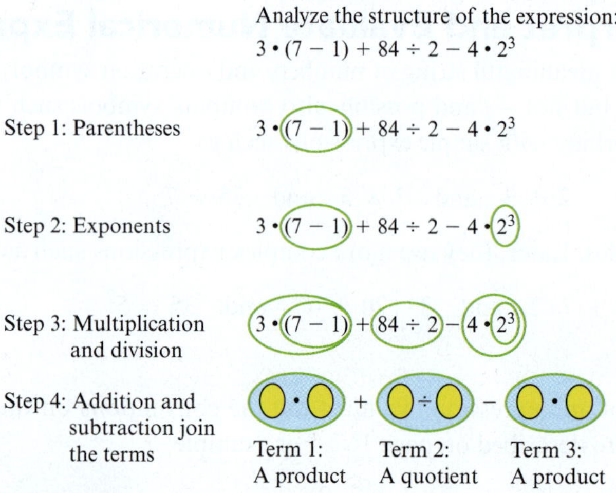

Figure 9.1 Analyzing an expression to reveal its structure.

CLASS ACTIVITY

9A Writing Expressions for Dot, Star, and Stick Designs, p. CA-158

9B How Many High-Fives?, p. CA-161

9C Sums of Odd Numbers, p. CA-162

9D Expressions with Fractions and Percent, p. CA-164

How Can We Evaluate Expressions with Fractions?

CLASS ACTIVITY

9E Explain and Critique Evaluating Expressions with Fractions, p. CA-165

When an expression involves fractions, we can sometimes find ways to make evaluating the expression easy. For example, to evaluate

$$\frac{8}{57}\cdot\frac{57}{61}$$

it would be inefficient to do this:

$$\frac{8}{57}\cdot\frac{57}{61}=\frac{8\cdot 57}{57\cdot 61}=\frac{456}{3477}$$

We might then put $\frac{456}{3477}$ in its simplest form, $\frac{8}{61}$. A much more efficient way to evaluate $\frac{8}{57}\cdot\frac{57}{61}$ is to "cancel" the 57s—namely, to write

$$\frac{8}{57}\cdot\frac{57}{61}=\frac{8}{\cancel{57}}\cdot\frac{\cancel{57}}{61}=\frac{8}{61}$$

Why is this canceling valid? Canceling is just shorthand for several steps, such as the following, that use fraction multiplication, the commutative property of multiplication, and equivalent fractions:

$$\frac{8}{57}\cdot\frac{57}{61}=\frac{8\cdot 57}{57\cdot 61}=\frac{8\cdot 57}{61\cdot 57}=\frac{8}{61}\cdot\overset{=1}{\frac{57}{57}}=\frac{8}{61}$$

Rather than write all these steps, we usually simply cancel the 57s. In effect, the 57s are just "pulled out" to make $\frac{57}{57}$, which is 1.

How can we make

$$\frac{21}{50}\cdot\frac{1}{35}$$

easy to evaluate? In this case, a factor 7 in the 35 and a factor 7 in the 21 cancel:

$$\frac{\overset{3}{\cancel{21}}}{50}\cdot\frac{1}{\underset{5}{\cancel{35}}}=\frac{3\cdot 1}{50\cdot 5}=\frac{3}{250}$$

Again, we can write equations to explain why this canceling is valid:

$$\frac{21}{50}\cdot\frac{1}{35}=\frac{21\cdot 1}{50\cdot 35}=\frac{3\cdot 7}{50\cdot 5\cdot 7}=\frac{3}{50\cdot 5}\cdot\overset{=1}{\frac{7}{7}}=\frac{3}{250}$$

In effect, the 7s are just "pulled out" to make $\frac{7}{7}$, which is 1.

SECTION SUMMARY AND STUDY ITEMS

Section 9.1 Numerical Expressions

A numerical expression is a meaningful string of numbers, operations, and possibly parentheses. Expressions arise in a variety of scenarios and provide a way to formulate a quantity in a scenario mathematically.

Key Skills and Understandings

1. Formulate numerical expressions arising from scenarios.
2. Interpret the meaning of an expression without evaluating it—for example, describe an expression as a sum or product.
3. Evaluate expressions efficiently. Use cancellation correctly when evaluating expressions involving fractions.

Practice Exercises for Section 9.1

1. Write an expression that uses addition and multiplication to describe the total number of dots in Figure 9.2.

Figure 9.2 Write an expression for the total number of dots.

2. Write two different expressions for the number of darkly colored small squares in Figure 9.3. Each expression should use either multiplication or addition.

3. If there are 25 people at a party and everyone "clinks" glasses with everyone else, how many "clinks" will there be? Write two different expressions for the total number of clinks and use one to solve the problem.

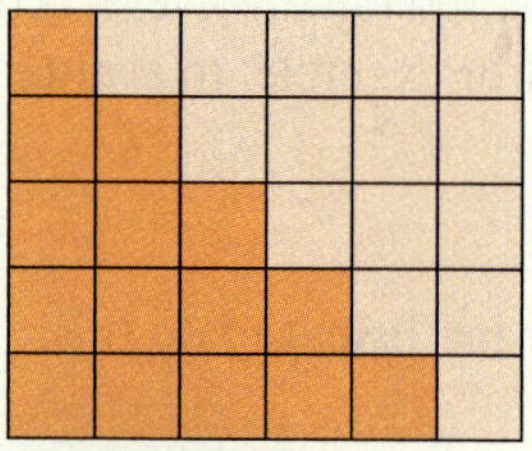

Figure 9.3 Write two expressions for the number of darkly colored squares.

4. Calculate $1 + 2 + 3 + 4 + \cdots + 500$ by finding and evaluating another expression that equals this sum. Explain your method.

5. a. Using the four square patterns in Figure 9.4, find a way to express the following sums in terms of multiplication and subtraction:

$$2 + 4$$
$$2 + 4 + 6$$
$$2 + 4 + 6 + 8$$
$$2 + 4 + 6 + 8 + 10$$

b. Based on your results in part (a), predict the sum

$$2 + 4 + 6 + 8 + \cdots + 98$$

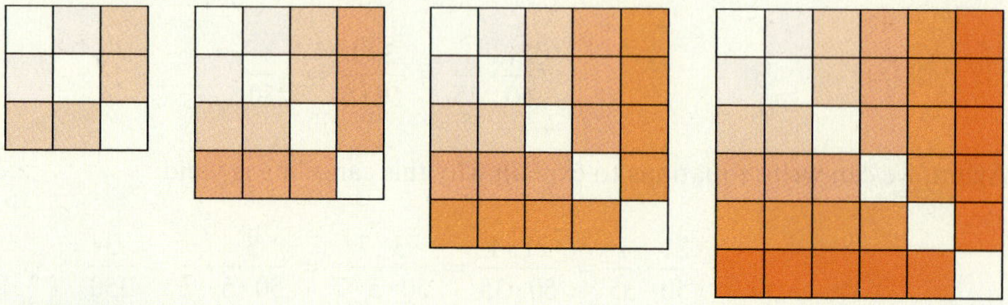

Figure 9.4 Sums of even numbers.

6. Write an expression using multiplication and addition (or subtraction) for the fraction of the area of the rectangle in **Figure 9.5** that is shaded. Explain. You may assume that all parts that appear to be the same size really are the same size.

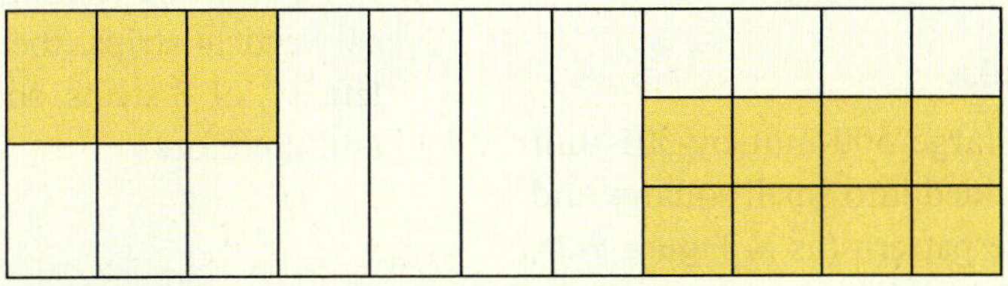

Figure 9.5 What fraction is shaded?

7. Show how to make the expression

$$\frac{37}{80} \cdot \frac{40}{37} + \frac{17}{30} \cdot \frac{15}{34}$$

easy to evaluate without a calculator.

8. Write equations that use rules about operating with fractions, as well as properties of arithmetic, to show why it is valid to cancel to evaluate the expression

$$\frac{35 + 21}{63}$$

as follows:

$$\frac{\overset{5}{\cancel{35}} + \overset{3}{\cancel{21}}}{\underset{9}{\cancel{63}}} = \frac{5 + 3}{9} = \frac{8}{9}$$

9. Explain why it is *not valid* to evaluate

$$\frac{35 \cdot 21}{63}$$

by canceling as follows:

warning, incorrect: $\frac{\overset{5}{\cancel{35}} \cdot \overset{3}{\cancel{21}}}{\underset{9}{\cancel{63}}} = \frac{5 \cdot 3}{9} = \frac{15}{9} = 1\frac{2}{3}$

Answers to Practice Exercises for Section 9.1

1. The 4 triangles arranged vertically in the center that each contain 5 dots contribute $4 \cdot 5$ dots. The 4 elongated triangles that each contain 4 dots contribute $4 \cdot 4$ dots. The 2 quadrilaterals at the left and right that each contain 2 dots contribute $2 \cdot 2$ dots. The square in the center contributes 1 more dot. All together, the number of dots in the figure is

$$4 \cdot 5 + 4 \cdot 4 + 2 \cdot 2 + 1$$

2. On the one hand, the darkly shaded portion of the figure consists of

$$1 + 2 + 3 + 4 + 5$$

small squares, which we see by adding the number of squares in the 1st, 2nd, 3rd, 4th, and 5th rows. On the other hand, the darkly shaded squares are half of all the squares in the 5 by 6 rectangle, so there are

$$\frac{1}{2}(5 \cdot 6)$$

dark squares.

3. One expression for the number of clinks is

$$24 + 23 + 22 + \cdots + 4 + 3 + 2 + 1$$

To see why, imagine the 25 people lined up in a row. The 25th person in the row could go down the row and clink glasses with all other 24 people (and then stand in the corner). The 24th person in the row could go down the row and clink glasses with the other 23 people left in the row (and then stand in the corner also). The 23rd person could clink with the remaining 22, and so on, until the 2nd person clinks with the first. Counting up all the clinks, you get the sum $24 + 23 + \cdots + 2 + 1$.

Another expression for the number of clinks is

$$(25 \cdot 24) \div 2$$

This expression is valid because each of the 25 people clinks with 24 other people. This makes for $25 \cdot 24$ clinks—except that each clink has been counted *twice* this way. Why? Because when Anna clinks with Beatrice, it's the same as Beatrice clinking with Anna, but the $25 \cdot 24$ figure counts those as 2 separate clinks. So, to get the correct count, divide $25 \cdot 24$ by 2. The second expression is easiest to evaluate.

$$(25 \cdot 24) \div 2 = 300$$

so there will be 300 clinks.

4. If you think about a large 500-unit-by-501-unit rectangle that is subdivided into small squares and is shaded in a staircase pattern (as in Figure 9.3), then the number of darkly shaded squares can be expressed in two ways. On the one hand, adding up the darkly shaded squares row by row, there are

$$1 + 2 + 3 + \cdots + 500$$

darkly shaded, small squares. On the other hand, the darkly shaded squares take up half the rectangle, and there are $500 \cdot 501$ small squares in the rectangle, so there are

$$\frac{1}{2}(500 \cdot 501) = 125{,}250$$

darkly shaded, small squares. It's the same number either way you count, so the sum of the first 500 counting numbers must also be 125,250.

5. a. There are $2 + 4$ small, shaded squares in the first square pattern. The shaded squares fill up all but 3 of the $3 \cdot 3$ squares in the pattern. Therefore,

$$2 + 4 = 3^2 - 3$$

Similarly, there are $2 + 4 + 6$ small, shaded squares in the second square pattern. The shaded squares fill up all but 4 of the $4 \cdot 4$ squares in the pattern. Therefore,

$$2 + 4 + 6 = 4^2 - 4$$

Using the same reasoning for the 3rd and 4th square patterns, we see that

$$2 + 4 + 6 + 8 = 5^2 - 5$$
$$2 + 4 + 6 + 8 + 10 = 6^2 - 6$$

b. Assuming that the pattern we found in part (a) continues, it should be the case that

$$2 + 4 + 6 + 8 + \cdots + 98 = 50^2 - 50 = 2450$$

Why $50^2 - 50$? We can find the expressions on the right-hand side of the equations in part (a) by taking half of the last term in the sum (on the left-hand side of the equation) and adding 1. Half of 98 is 49; adding 1, we get 50.

6. If we view the rectangle as consisting of 11 identical vertical strips, then the shaded portion on the left is $\frac{1}{2}$ of 3 strips, so it is $\frac{1}{2}$ of $\frac{3}{11}$ of the rectangle and therefore

$$\frac{1}{2} \cdot \frac{3}{11}$$

of the rectangle. The shaded portion on the right is $\frac{2}{3}$ of 4 strips, so it is $\frac{2}{3}$ of $\frac{4}{11}$ of the rectangle and therefore

$$\frac{2}{3} \cdot \frac{4}{11}$$

of the rectangle. All together, the fraction of the area of the rectangle that is shaded is $\frac{1}{2} \cdot \frac{3}{11} + \frac{2}{3} \cdot \frac{4}{11}$.

7. The 37s cancel each other. The 40 in the numerator cancels with 40 in the denominator, leaving 2 in the denominator. The 17 in the numerator cancels with a 17 in the denominator, leaving 2 in the denominator. The 15 in the numerator cancels with a 15 in the denominator, leaving 2 in the denominator. We can thus write

$$\frac{37}{80} \cdot \frac{40}{37} + \frac{17}{30} \cdot \frac{15}{34} = \frac{1}{2} + \frac{1}{4} = \frac{3}{4}$$

8. We can use the distributive property to "pull out" a 7 in the numerator. We can also "pull out" a 7 in the denominator to make a $\frac{7}{7}$, which is 1. Therefore,

$$\frac{35 + 21}{63} = \frac{(5 + 3) \cdot 7}{9 \cdot 7} = \frac{5 + 3}{9} \cdot \frac{7}{7} = \frac{8}{9} \cdot 1 = \frac{8}{9}$$

9. In the expression

$$\frac{35 \cdot 21}{63} = \frac{(5 \cdot 7) \cdot (3 \cdot 7)}{9 \cdot 7}$$

the 7 in the denominator can cancel only one of the 7s in the numerator, not both of them. So it is correct to write

$$\frac{35 \cdot 21}{63} = \frac{35 \cdot (3 \cdot 7)}{9 \cdot 7} = \frac{35 \cdot 3}{9} \cdot \frac{7}{7} = \frac{35}{3} = 11\frac{2}{3}$$

PROBLEMS FOR SECTION 9.1

1. Write an expression that uses addition and multiplication to describe the total number of stars in Figure 9.6.

Figure 9.6 Write an expression for this picture.

2. Write at least three different expressions for the total number of small squares that make up Design 1 in Figure 9.7. Each expression should fit with the way the squares are arranged in the design and should involve multiplication and either addition or subtraction. In each case, explain briefly why the expression describes the total number of small squares in the design (without counting the squares or evaluating the expression).

3. Write at least three different expressions for the total number of small squares that make up Design 2 in Figure 9.7. Each expression should fit with the way the squares are arranged in the design and should involve multiplication and either addition or subtraction. In each case, explain briefly why the expression describes the total number of small squares in the design (without counting the squares or evaluating the expression).

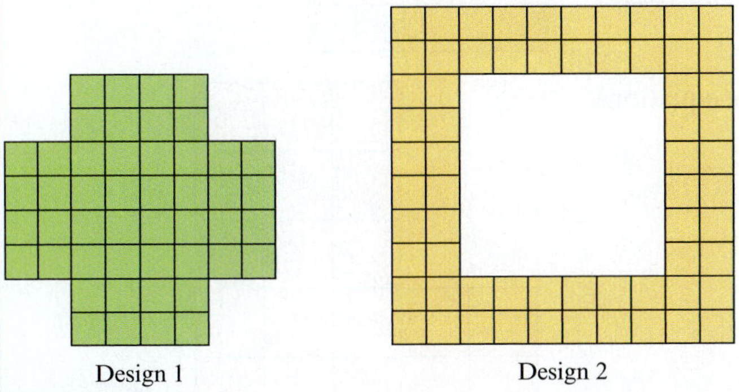

Figure 9.7 Write expressions for the total number of small squares.

4. a. Draw a design made of small circles (or another shape) so that the total number of circles in the design is

$$3 \cdot 2 + 3 \cdot 5$$

and so that you can tell this expression stands for the correct number of circles in the design without counting the circles or evaluating the expression. Explain briefly.

b. Write another expression for the total number of circles in your design in part (a). Explain briefly. (Your expression should not consist of a single number.) Discuss connections to material you have studied previously (see Chapter 4).

5. a. Draw a design made of small circles (or another shape) so that the total number of circles in the design is

$$7 \cdot 8 - 2 \cdot 3$$

and so that you can tell this expression stands for the correct number of circles in the design without counting the circles or evaluating the expression. Explain briefly.

b. Write another expression for the total number of circles in your design in part (a). Explain briefly. (Your expression should not consist of a single number.)

6. Write two different expressions for the total number of small squares in design (a) of Figure 9.8. Each expression should use either multiplication or addition, or both. Explain briefly.

7. Write two different expressions for the total number of small squares in design (b) of Figure 9.8. Each expression should use either multiplication or addition, or both. Explain briefly.

8. Write two different expressions for the total number of small squares in Figure 9.9. Each expression should use either multiplication or addition or both. Explain briefly.

(a) (b)

Figure 9.8 Write equations for these square patterns.

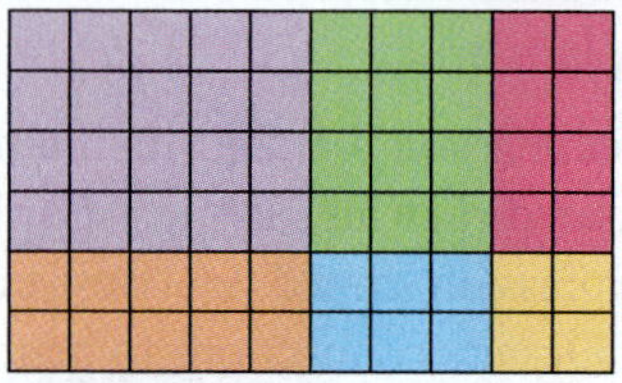

Figure 9.9 A rectangular design.

9. For each of the 4 rectangular patterns in Figure 9.10, find two different ways to express the total number of small squares that the pattern is made of.

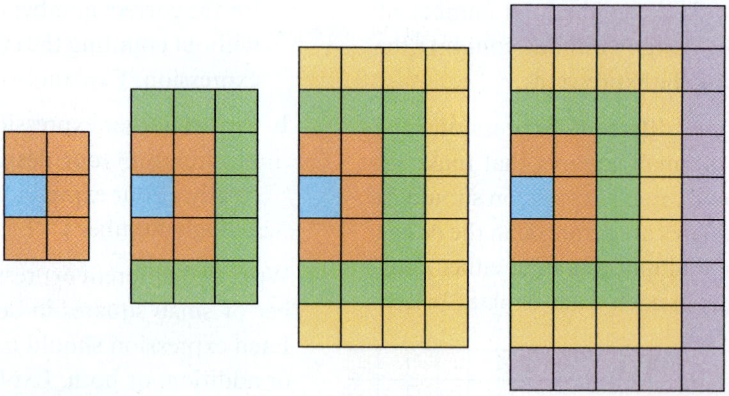

Figure 9.10 Patterns for some equations.

10. An auditorium has 50 rows of seats. The first row has 20 seats, the second row has 21 seats, the third row has 22 seats, and so on, each row having one more seat than the previous row. How many seats are there all together?

   a. Write two different expressions for the total number of seats in the auditorium. One expression should use addition; the other should involve multiplication.

   b. Solve the auditorium problem by evaluating one of the expressions in part (a).

11. Write at least two expressions for the total number of 1-cm-by-1-cm-by-1-cm cubes it would take to build a 6-cm-tall prism (tower) over the base in Figure 9.11. Each expression should use both multiplication and addition (or subtraction). Explain briefly how to obtain the expressions.

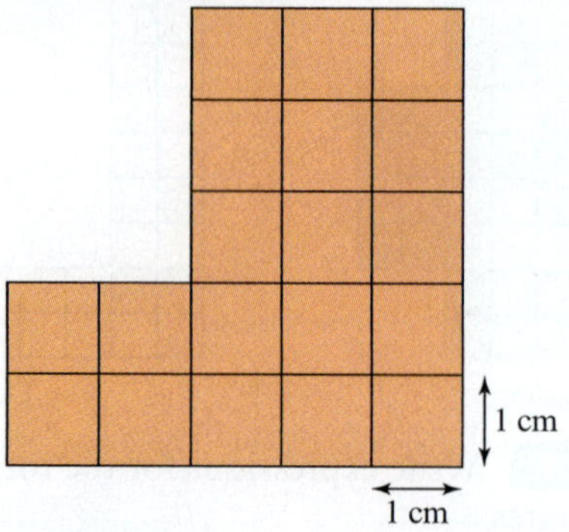

Figure 9.11 A base for a prism.

12. Write an expression using multiplication and addition (or subtraction) for the fraction of the area of Rectangle 1 in **Figure 9.12** that is shaded. Explain briefly. You may assume that all parts that appear to be the same size really are the same size.

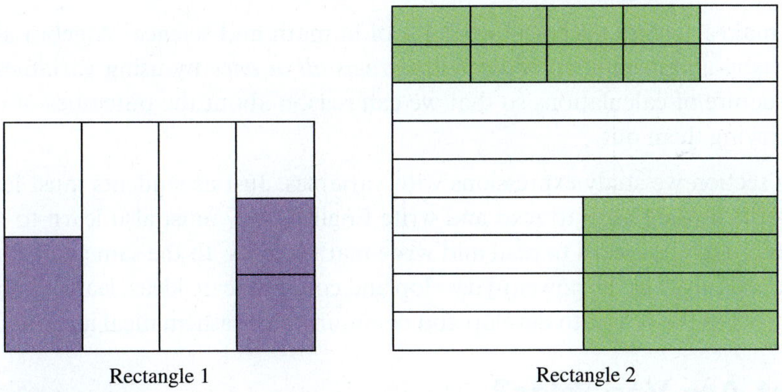

Figure 9.12 Write expressions for the fraction of the area that is shaded.

13. Write an expression using multiplication and addition (or subtraction) for the fraction of the area of Rectangle 2 in **Figure 9.12** that is shaded. Explain briefly. You may assume that all parts that appear to be the same size really are the same size.

14. Shade

$$\frac{2}{3} \cdot \frac{2}{5} + \frac{1}{2} \cdot \frac{1}{5}$$

of a rectangle in such a way that you can tell the correct amount is shaded without evaluating the expression. Explain briefly.

15. Shade

$$\frac{3}{4} - \frac{1}{3} \cdot \frac{1}{2}$$

of a rectangle in such a way that you can tell the correct amount is shaded without evaluating the expression. Explain briefly.

16. Show how to make the expression

$$\frac{21}{13} \cdot \frac{13}{56} + \frac{17}{36} \cdot \frac{18}{34}$$

easy to evaluate without using a calculator. Explain briefly.

17. Show how to make the expression

$$\left(3 - \frac{1}{11}\right) \div \left(\frac{1}{5} + \frac{1}{11}\right)$$

easy to evaluate without using a calculator. Explain briefly.

18. Write equations that use rules about operating with fractions and possibly also properties of arithmetic to show why it is valid to cancel to evaluate the expression

$$\frac{22 \cdot 3}{77}$$

as follows:

$$\frac{\overset{2}{\cancel{22}} \cdot 3}{\underset{7}{\cancel{77}}} = \frac{2 \cdot 3}{7} = \frac{6}{7}$$

19. Explain why it is *not valid* to evaluate

$$\frac{22 + 3}{77}$$

using the canceling shown in the following step:

warning, incorrect: $\frac{\overset{2}{\cancel{22}} + 3}{\underset{7}{\cancel{77}}} = \frac{2 + 3}{7} = \frac{5}{7}$

Show how to evaluate $\frac{22 + 3}{77}$ correctly. (Is canceling possible or not?)

20. Write equations that use rules about operating with fractions as well as properties of arithmetic to show why it is valid to cancel to evaluate the expression

$$\frac{15}{25 + 10}$$

as follows:

$$\frac{\overset{3}{\cancel{15}}}{\underset{5}{\cancel{25}} + \underset{2}{\cancel{10}}} = \frac{3}{5 + 2} = \frac{3}{7}$$

9.2 Expressions with Variables

CCSS Common Core State Standards Grades 6, 7, 8

What makes algebra such a powerful tool in math and science? Algebra allows us to summarize and reason about *infinitely many calculations all at once* by using variables. Algebra helps us see the structure of calculations so that we can reason about the outcomes of calculations even without carrying them out.

In this section we study expressions with variables. Just as students must learn to "decode" letters and words as they learn to read and write English, they must also learn to "decode" variables and expressions as they learn to read and write mathematics. In the same way that learning to read and write words gives us the power to develop and communicate ideas, learning to read and write expressions gives us the power to develop and communicate mathematical ideas about quantities.

What Are Variables?

variable

A **variable** is a letter or other symbol that stands for any number within a specified (or understood) set of numbers. For example, if we say "a circle of radius r has area πr^2," then r is a variable that stands for any positive real number. In this context, it wouldn't make sense for r to be negative, so it's simply understood that r is restricted to positive numbers. We could also write $r > 0$ to make this restriction clear.

In word problems or other real-world contexts, it's important to define the variables that you use. For example, in a problem about water in a tank, you might want to use the variable W to stand for the amount of water. To do so, you first need a unit of measurement, such as gallons or liters. Once you have chosen a unit of measurement, you can define the variable so that it stands for a number, like this:

Let W be the number of liters of water in the tank.

Sometimes students treat a variable as a label instead of as a number. For example, a student might write

Warning, not fully correct: W = water

This use of a variable is not fully correct because it does not indicate how W stands for a number.

How Do We Work with Expressions with Variables?

An expression with variables is a "calculation recipe" that applies to many different numbers. An

expression

expression or **algebraic expression** is a meaningful string of numbers or variables, or both numbers and variables, and operation symbols ($+$, $-$, $\times$ or $\cdot$, $\div$ or $/$, or raising to a power, but not $=$), and possibly also grouping symbols such as parentheses or brackets. For example,

$$x^2 - 13 \quad \text{and} \quad 2 \cdot x + 3 \cdot y + 7 \quad \text{and} \quad 6 \cdot 5 + 6 \cdot 2$$

are expressions. Numerical expressions, such as $6 \cdot 5 + 6 \cdot 2$, were discussed in the previous section; they are special kinds of expressions.

CCSS 6.EE.6

As with numerical expressions, to interpret the meaning of an expression, we must use the conventions on the order of operations. For example, the expression

$$1000 - 10 \cdot (P - 15)^2$$

means 1000 minus the product of 10 with $P - 15$ times $P - 15$.

There are some additional conventions we use when we work with variables. We often drop multiplication symbols between numbers and variables or variables and variables. For example,

$$2x \text{ means } 2 \cdot x$$

and

$$3xy \text{ means } 3 \cdot x \cdot y$$

CCSS 6.EE.2

Also, we usually write numbers before variables when numbers and variables are multiplied in an expression. So instead of writing $x \cdot 2$ or $x2$, we prefer to write $2x$. Since multiplication is commutative, we can always write expressions so that the numbers precede the variables. However, notice that this algebra convention might not agree with the convention we have used in interpreting multiplication. For example, if you buy x pens and each pen costs \$2, then according to our definition of multiplication, the total cost of all the pens is $x \cdot 2$. When we write $2x$ for the cost of all the pens, the 2 still stands for the cost per pen (in dollars) and the x still stands for the number of pens.

Students sometimes have difficulty formulating expressions with variables because this requires working with a letter that stands for an unknown number. For example, a student might feel comfortable with the numerical expression $3 \cdot 8$ yet not feel comfortable with $3 \cdot x$ because x is not a specific number. Sometimes it helps to formulate numerical expressions first before defining and using a variable.

Because we look for addition and subtraction last when we use the conventions on the order of operations, an expression can be viewed as broken into components that are combined by addition, subtraction, or a combination of addition and subtraction (if there is more than one component).

term

As with numerical expressions, we call these components the **terms** of the expression. For example,

$$1000 - 10(P - 15)^2$$

has two terms, 1000 and $10(P - 15)^2$.

coefficient

In a term that involves variables, the number that multiplies the rest of the term is often called a **coefficient**. For example, the coefficient of $3x^2$ is 3 and the coefficient of $10(P - 15)^2$ is 10. As we will see later, it can be especially useful to recognize a "hidden" coefficient of 1. For example, the term x has a coefficient of 1 because $x = 1 \cdot x$.

To help reveal the structure of an expression, you can successively circle components according to the order of operations, as shown in Figure 9.13.

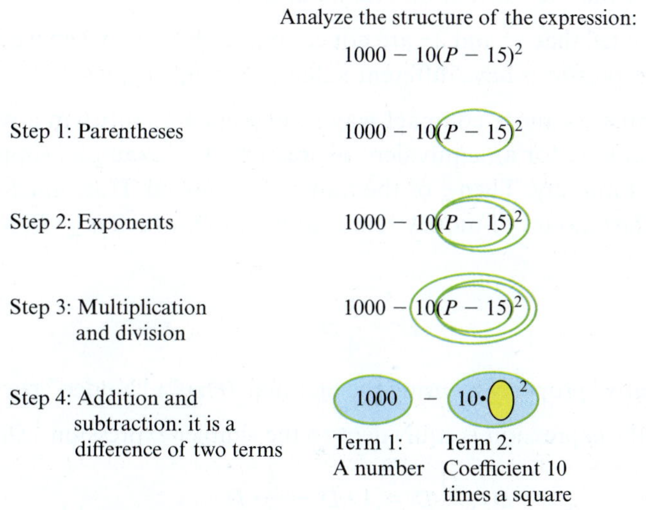

Figure 9.13 Analyzing an expression.

What Does It Mean to Evaluate Expressions with Variables?

evaluate an expression

To **evaluate an expression** that involves variables means to replace the variables with numbers and to determine the number to which the resulting expression is equal. For example, to evaluate

$$3x + 4y$$

at $x = 6$ and $y = 7$, we replace x with 6 and y with 7. We find that

$$\begin{aligned} 3 \cdot 6 + 4 \cdot 7 &= 18 + 28 \\ &= 46 \end{aligned}$$

Note that when a variable occurs several times in an expression, we must replace *all* occurrences of the variable with the same number when evaluating the expression. For example, to evaluate

$$3x^2 + 5x + 1$$

at $x = 2$, we replace each occurrence of x with 2 to obtain

$$\begin{aligned} 3 \cdot 2^2 + 5 \cdot 2 + 1 &= 3 \cdot 4 + 10 + 1 \\ &= 12 + 10 + 1 \\ &= 23 \end{aligned}$$

When we evaluate an expression that has two or more different variables, the values we use for the different variables can be different or they can be the same. Previously we evaluated $3x + 4y$ at $x = 6$ and $y = 7$, but we can also evaluate this same expression at $x = 6$ and $y = 6$:

$$\begin{aligned} 3 \cdot 6 + 4 \cdot 6 &= 18 + 24 \\ &= 42 \end{aligned}$$

Even though the variables x and y are different, we can still use the same numbers for them when we evaluate the expression.

What Are Equivalent Expressions?

equivalent expressions

In most cases, there is more than one way to formulate an expression for a given quantity. Two expressions are called **equivalent expressions** if their values are always equal whenever they are evaluated. For example, $x + x$ and $2x$ are equivalent because $x + x$ and $2x$ give the same number no matter what number is substituted for x.

We can use the commutative, associative, and distributive properties to produce and recognize equivalent expressions. For example, $x \cdot 5$ and $5x$ are equivalent according to the commutative property of multiplication and the convention that $5x$ means $5 \cdot x$.

CCSS
6.EE.3
6.EE.4
7.EE.1

On the other hand, we can tell that x^2 and $2x$ are not equivalent because when we evaluate them at $x = 3$, for example, the expressions have different values: $3^2 = 9$, whereas $2 \cdot 3 = 6$.

Often, we can find a simpler or more compact way to express a quantity in a word problem or real-world scenario by looking for an equivalent expression. For example, suppose there are D dollars in a bank account initially. Then $\frac{3}{4}$ of the money is removed. How much money is in the account at that point? The amount of money removed is $\frac{3}{4}$ of D, which is $\frac{3}{4} \cdot D$. So the remaining amount of money is

$$D - \frac{3}{4}D$$

We can apply the distributive property by using the fact that D has a "hidden" coefficient of 1, i.e., $D = 1 \cdot D$, to show that the expression is equivalent to the simpler expression $\frac{1}{4}D$:

$$\begin{aligned} D - \frac{3}{4}D &= 1 \cdot D - \frac{3}{4} \cdot D \\ &= \left(1 - \frac{3}{4}\right) \cdot D \\ &= \frac{1}{4}D \end{aligned}$$

See also **Figure 9.14**.

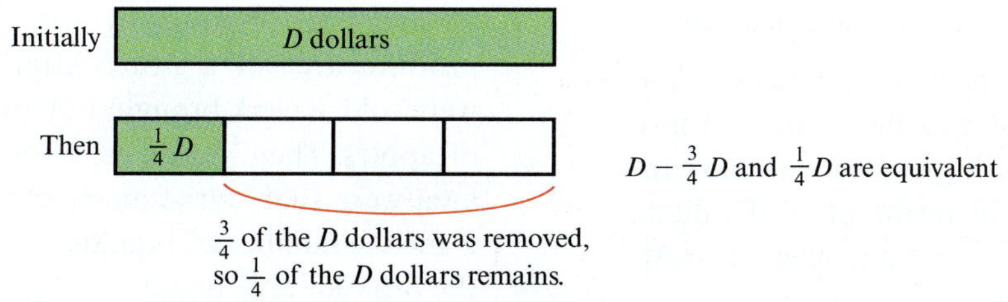

Figure 9.14 Starting with D dollars, when $\frac{3}{4}$ of the money is removed, $\frac{1}{4}D$ dollars remain.

CLASS ACTIVITY

9F Equivalent Expressions, p. CA-166

9G Expressions for Quantities, p. CA-167

SECTION SUMMARY AND STUDY ITEMS

Section 9.2 Expressions with Variables

Variables are letters that stand for numbers. Expressions with variables are "calculation recipes" that apply to many different numbers.

Key Skills and Understandings

1. Interpret the structure of an expression and use mathematical terminology to identify its parts. For example, identify the terms of an expression and describe them—perhaps as products, quotients, or powers. Describe coefficients of terms.
2. Write expressions for quantities and explain why they are formulated the way they are. Evaluate expressions at values of the variables.
3. Generate and recognize equivalent expressions by applying properties of arithmetic. Determine when two expressions are not equivalent.

Practice Exercises for Section 9.2

1. Describe the structure of the expression
$$3xy + 5(x - 4)^2 - 7y^2.$$
2. Are $6 \cdot (A \cdot B)$ and $(6 \cdot A) \cdot (6 \cdot B)$ equivalent or not? How can you tell?
3. Are $6(A + B)$ and $6A + 6B$ equivalent or not? How can you tell?
4. Write two expressions for the area of the floor plan in Figure 9.15. (Assume all the angles are right angles.) Explain. Then explain in two different ways why your expressions are equivalent.
5. Draw, label, and shade a rectangle so that it gives rise to the equivalent expressions $(x + 3) \cdot (y + 4)$ and $xy + 4x + 3y + 12$. Explain your answers.

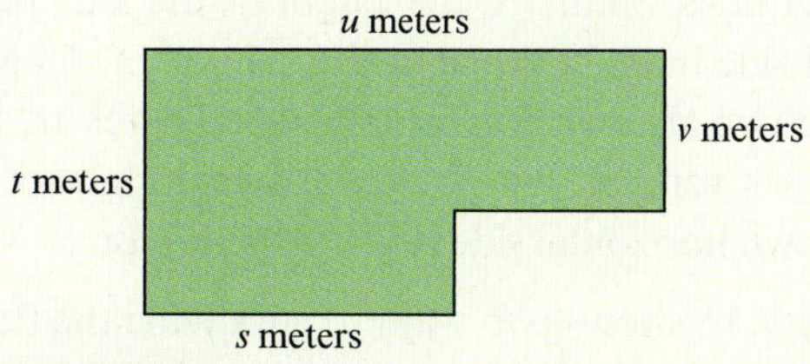

Figure 9.15 Write expressions for the area.

6. Goo-Young has lollipops to distribute among goodie bags. When Goo-Young tries to put L lollipops in each of 6 goodie bags, she is 1 lollipop short. Write an expression for the total number of lollipops Goo-Young has in terms of L. Explain briefly.

7. Initially, there were X liters of gasoline in a can.

a. If $\frac{1}{3}$ of the gasoline in the can is poured out and then another $\frac{1}{2}$ liter of gasoline is poured into the can, then how much gasoline is in the can? Write an expression in terms of X. Evaluate your expression when $X = 3$ and when $X = 3\frac{1}{2}$.

b. If $\frac{1}{2}$ liter of gasoline is poured into the can and then $\frac{1}{3}$ of the gasoline in the can is poured out, how much gasoline is in the can? Write an expression in terms of X. Is your expression the same as in part (a)? Evaluate your expression when $X = 3$ and when $X = 3\frac{1}{2}$.

8. There were P pounds of apples for sale in the produce area of a store. After $\frac{2}{5}$ of the apples were sold, a clerk brought out another 30 pounds of apples. Then $\frac{1}{3}$ of all the apples in the produce area were sold. Write an expression in terms of P for the number of pounds of apples that are in the produce area now.

9. Let P be the initial population of a town. After 10% of the town leaves, another 1400 people move in to the town. Write an expression in terms of P for the new number of people in the town.

Answers to Practice Exercises for Section 9.2

1. The expression $3xy + 5(x - 4)^2 - 7y^2$ is a sum of three terms. The first term, $3xy$, is the product of the coefficient 3 with x and y. The second term, $5(x - 4)^2$, is the product of the coefficient 5 with the square $(x - 4)$ times $(x - 4)$. The third term, $7y^2$, is the product of the coefficient 7 with the square y times y.

2. No, $6 \cdot (A \cdot B)$ and $(6 \cdot A) \cdot (6 \cdot B)$ are not equivalent. We can tell by evaluating these two expressions at $A = 1$ and $B = 1$. The first has value $6 \cdot (1 \cdot 1) = 6$, whereas the second has value $(6 \cdot 1) \cdot (6 \cdot 1) = 6 \cdot 6 = 36$.

3. Yes, $6(A + B)$ and $6A + 6B$ are equivalent. We can tell by applying the distributive property.

4. Let's first determine the unspecified side lengths. The length of the long vertical side of the floor plan must be equal to the sum of the lengths of the two shorter vertical sides. Similarly, the length of the long horizontal side must be equal to the sum of the lengths of the two shorter horizontal sides. Therefore, the unknown vertical side is $t - v$ meters long and the unknown horizontal side is $u - s$ meters long.

Figure 9.16 shows two ways to subdivide the floor plan into two rectangles. Using the subdivision on the left, the area is

$$ts + v(u - s)$$

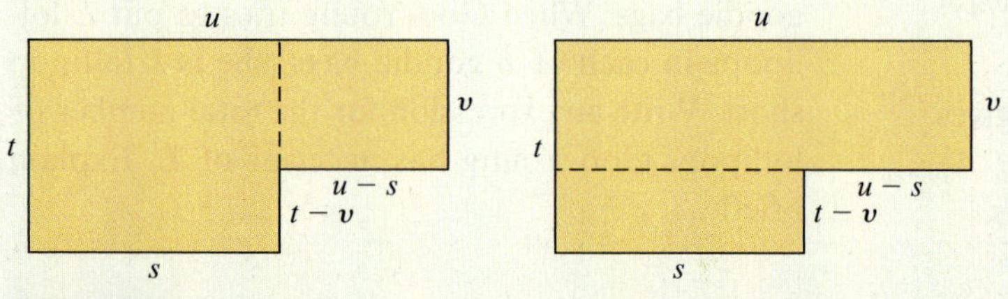

Figure 9.16 Two ways to subdivide the floor plan.

square meters. Using the subdivision on the right, the area is

$$(t - v)s + vu$$

square meters. (We could also find an expression for the area by viewing the floor plan as a large t-by-u rectangle with a $(t - v)$-by-$(u - s)$ rectangular piece missing.) Since both expressions represent the area of the floor plan, they must be equal. Another way to see why the expressions are equal is to apply properties of arithmetic.

5. See Figure 9.17. Overall, the rectangle has dimensions $x + 3$ by $y + 4$ and therefore has area $(x + 3) \cdot (y + 4)$ square units. The 4 portions

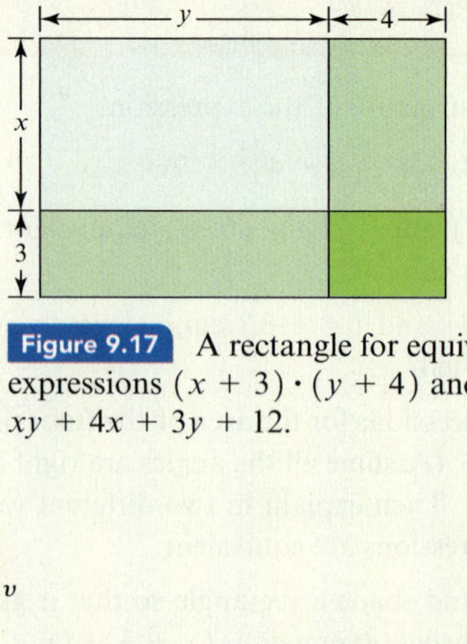

Figure 9.17 A rectangle for equivalent expressions $(x + 3) \cdot (y + 4)$ and $xy + 4x + 3y + 12$.

making up the rectangle have area xy (top left), $4x$ (top right), $3y$ (bottom left), and 12 (bottom right) square units. Therefore, all together, the area of the rectangle is $xy + 4x + 3y + 12$ square units. Since it's the same area either way we calculate it, we have illustrated the expressions $(x + 3) \cdot (y + 4)$ and $xy + 4x + 3y + 12$ are equivalent.

6. If Goo-Young were to put L lollipops in each bag, then she'd have $6L$ lollipops in all. But since she is 1 lollipop short, she only has

$$6L - 1$$

lollipops in all. We could also write the expression

$$5L + (L - 1)$$

for the number of lollipops. This expression is valid because there are L lollipops in each of 5 bags, but the 6th bag contains $L - 1$ lollipops.

7. a. When $\frac{1}{3}$ of the gasoline in the can is poured out, $\frac{2}{3}$ of the gasoline in the can remains. Therefore, after pouring out the gasoline, there are $\frac{2}{3}X$ liters remaining in the can. When another $\frac{1}{2}$ liter is poured in, there are

$$\frac{2}{3}X + \frac{1}{2}$$

liters of gasoline in the can. When $X = 3$, the number of liters of gasoline remaining in the can is

$$\frac{2}{3} \cdot 3 + \frac{1}{2} = 2\frac{1}{2}$$

When $X = 3\frac{1}{2}$, the number of liters of gasoline remaining in the can is

$$\frac{2}{3} \cdot 3\frac{1}{2} + \frac{1}{2} = \frac{2}{3} \cdot \frac{7}{2} + \frac{1}{2} = \frac{17}{6} = 2\frac{5}{6}$$

b. When $\frac{1}{2}$ liter of gasoline is poured into the can, there are $X + \frac{1}{2}$ liters in the can. When $\frac{1}{3}$ of this amount is poured out, $\frac{2}{3}$ of the amount remains in the can. Therefore, after pouring out $\frac{1}{3}$ of the gasoline in the can, there are

$$\frac{2}{3}\left(X + \frac{1}{2}\right)$$

liters left in the can. We can use the distributive property to rewrite this expression for the number of liters of gasoline in the can as

$$\frac{2}{3}X + \frac{2}{3} \cdot \frac{1}{2} = \frac{2}{3}X + \frac{1}{3}$$

Notice that this expression, $\frac{2}{3}X + \frac{1}{3}$, is different from the expression $\frac{2}{3}X + \frac{1}{2}$ in part (a). When $X = 3$, the number of liters of gasoline remaining in the can is

$$\frac{2}{3} \cdot 3 + \frac{1}{3} = 2\frac{1}{3}$$

When $X = 3\frac{1}{2}$, the number of liters of gasoline remaining in the can is

$$\frac{2}{3} \cdot 3\frac{1}{2} + \frac{1}{3} = \frac{2}{3} \cdot \frac{7}{2} + \frac{1}{3} = \frac{7}{3} + \frac{1}{3} = 2\frac{2}{3}$$

8. When $\frac{2}{5}$ of the P pounds of apples were sold, $\frac{3}{5}$ of the apples remain. So the number of pounds of apples remaining at this point is $\frac{3}{5}P$. When 30 more pounds are brought in, there are $\frac{3}{5}P + 30$ pounds of apples. When $\frac{1}{3}$ of these apples are sold, $\frac{2}{3}$ of the apples remain. Notice that this is $\frac{2}{3}$ of the $\frac{3}{5}P + 30$ pounds of apples. Therefore, there are now

$$\frac{2}{3} \cdot \left(\frac{3}{5}P + 30\right)$$

pounds of apples set out. Using the distributive property, we can also write this expression as

$$\frac{2}{5}P + 20$$

9. After 10% of the town leaves, 90% of the town remains, which is 90% of P people, or $0.90P$ people. When another 1400 people move in, the population rises to $0.9P + 1400$.

PROBLEMS FOR SECTION 9.2

1. Describe the structure of the expression $3S^3 + 2S^3$.

2. Describe the structure of the expression $4(T - 8)(T + 1) - 5T^2 + 9$.

3. a. Students sometimes mistakenly think that the expression $4x - x$ is equivalent to 4. Why might a student make such a mistake? What is another expression that is equivalent to $4x - x$?

b. Students sometimes mistakenly think that the expression $8x - 8$ is equivalent to x. Why might a student make such a mistake? What is another expression that is equivalent to $8x - 8$?

4. a. Write at least two expressions for the area of the floor plan in Figure 9.18. Explain briefly how to obtain the expressions. (Assume that all the angles are right angles.)

b. Write two expressions for the perimeter of the floor plan in Figure 9.18. Explain briefly how to obtain the expressions.

c. Explain in two different ways why your expressions in part (a) are equivalent. One way should use properties of arithmetic.

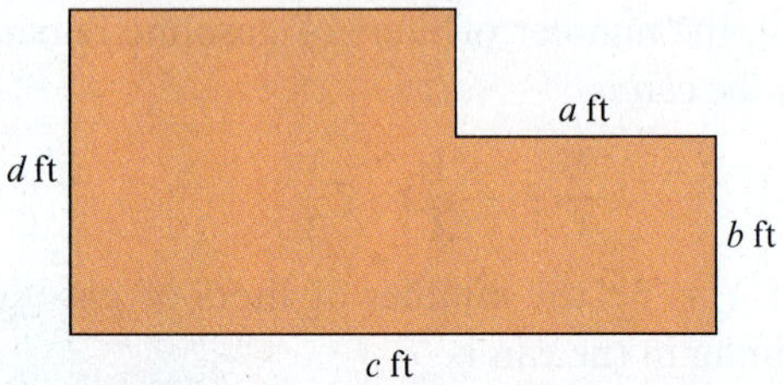

Figure 9.18 Write expressions for the area and perimeter of the floor plan.

5. a. Write at least two expressions for the area of the floor plan in Figure 9.19. Explain briefly how to obtain the expressions. (Assume that all the angles are right angles.)

b. Write two expressions for the perimeter of the floor plan in Figure 9.19. Explain briefly how to obtain the expressions.

c. Explain in two different ways why your expressions in part (a) are equivalent. One way should use properties of arithmetic.

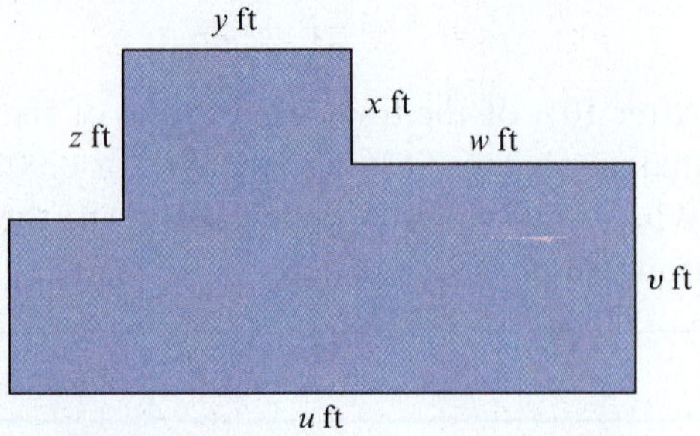

Figure 9.19 Write expressions for the area and perimeter of the floor plan.

6. a. Write at least two expressions for the total number of 1-cm-by-1-cm-by-1-cm cubes it would take to build a v-cm-tall prism (tower) over the base in Figure 9.20. (Assume that all the angles are right angles.) Each expression should be in terms of v, w, x, y, and z. Explain briefly how to obtain the expressions.

b. Explain in two different ways why your expressions in part (a) are equal. One way should use properties of arithmetic.

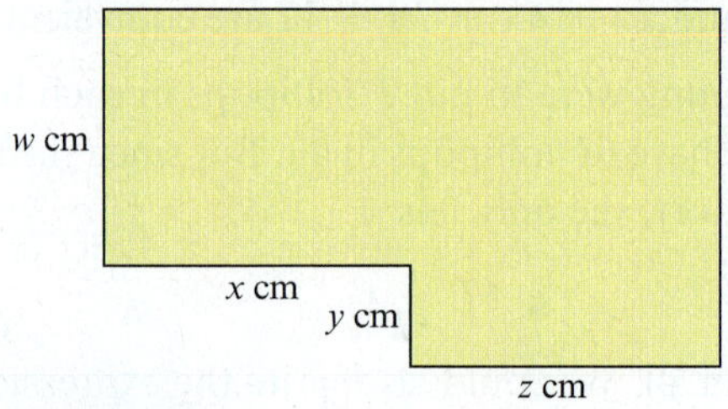

Figure 9.20 A base for a prism.

7. Write and explain two equivalent expressions that the design in Figure 9.21 gives rise to.

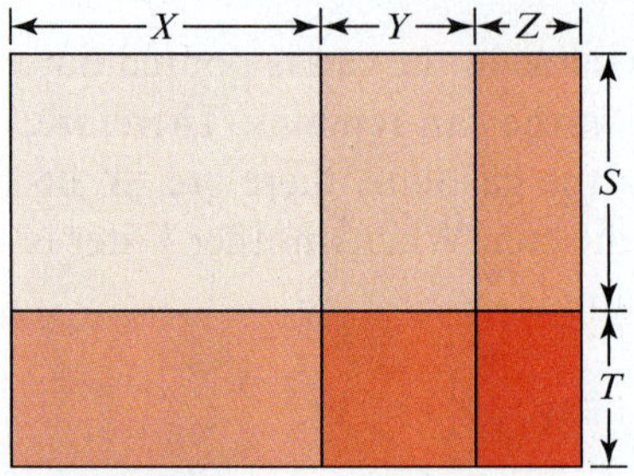

Figure 9.21 A rectangular design.

8. Draw, label, and shade a rectangle so that it gives rise to the equivalent expressions

$$(A + B + C) \cdot (X + Y + Z)$$

and

$$\begin{aligned} &AX + AY + AZ \\ &+ BX + BY + BZ \\ &+ CX + CY + CZ \end{aligned}$$

Explain briefly.

9. Draw, label, and shade a square so that it gives rise to the equivalent expressions

$$(A + B)^2 \text{ and } A^2 + 2AB + B^2$$

Explain briefly.

10. a. Write two equivalent expressions, each using only the numbers 9, 5, and 3, for the unknown length (?) on the left of Figure 9.22. One of your expressions should involve parentheses. Explain briefly.

b. Write two equivalent expressions for the unknown length (?) on the right of Figure 9.22. One of your expressions should involve parentheses. Explain briefly.

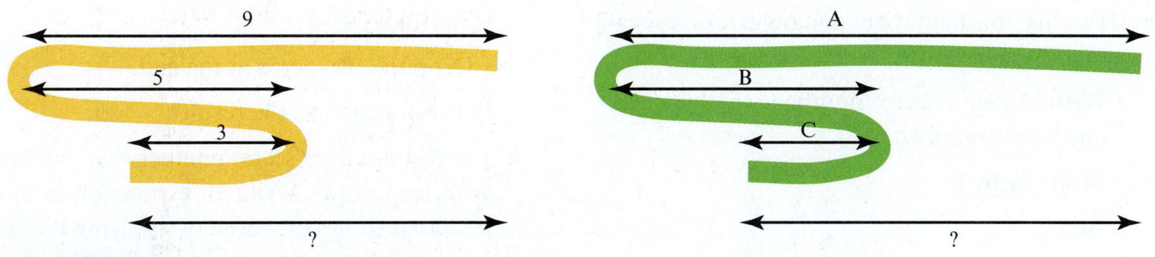

Figure 9.22 Write equivalent expressions arising from the unknown length.

11. Tyrone has a marble collection that he wants to divide among some bags. When Tyrone puts M marbles into each of 8 bags, he has 3 marbles left over. Write an expression for the total number of marbles Tyrone has in terms of M. Evaluate your expression when $M = 3$ and when $M = 4$.

12. Kenny has stickers to distribute among some goodie bags. When Kenny tries to put 5 stickers into each of G goodie bags, he is 2 stickers short of filling the last bag. Write an expression for the total number of stickers Kenny has in terms of G. Explain.

13. Ted is saving for a toy. He has no money now, but he figures that if he saves \$$D$ per week for the next 4 weeks he will have \$1 more than he needs for the toy. Write an expression for the cost of the toy Ted wants to buy in terms of D. Explain.

14. If you pay \$7 for a night light that costs \$0.50 per year in electricity to operate, then how much will you have spent on the night light if you operate it continuously for Y years? Write an expression in terms of Y. Evaluate your expression for the cost of the night light when $Y = 4$ and when $Y = 5\frac{1}{2}$. Explain your work.

15. Bonnie had B stickers. First, Bonnie gave away $\frac{1}{6}$ of her stickers. Then Bonnie got 12 more stickers. Write and explain two equivalent expressions for the number of stickers Bonnie has now.

16. There were X quarts of liquid in a container. First, $\frac{3}{4}$ of the liquid in the container was removed. Then another $\frac{1}{2}$ quart was poured into the container. Write an expression in terms of X for the number of quarts of liquid in the container at the end. Then write another equivalent expression.Explain.

17. Martin had M dollars initially. First, Martin spent $\frac{1}{3}$ of his money. Then Martin spent $\frac{3}{4}$ of what was left. Finally, Martin spent another \$20. Write an expression in terms of M for the final amount of money that Martin has. Then write another equivalent expression. Explain.

18. There are F cups of flour in a bag. First, $\frac{3}{8}$ of the flour in the bag was used. Then $\frac{1}{5}$ of the remaining flour was used. Write and explain two equivalent expressions for the number of cups of flour left in the bag.

19. Write a scenario for the expression

$$(x + 15) - \frac{1}{3}(x + 15)$$

Explain briefly why your scenario fits with this expression.

20. **a.** Write a scenario for the expression

$$\frac{4}{5}x - 25$$

Explain briefly why your scenario fits with this expression.

b. Write a scenario for the expression

$$\frac{4}{5}(x - 25)$$

Explain briefly why your scenario fits with this expression.

21. At a store, the price of an item is \$$P$. Consider three different scenarios:

First scenario: Starting at the price \$$P$, the price of the item was raised by A%. Then the new price was raised by B%.

Second scenario: Starting at the price \$$P$, the price of the item was raised by B%. Then the new price was raised by A%.

Third scenario: Starting at the price \$$P$, the price of the item was raised by $(A + B)$%.

For each scenario, write an expression for the final price of the item. Are any of the final prices the same? Explain.

22. a. Try this mathematical magic trick several times.

- Write down a secret number, which can be any whole number.
- Multiply by 5.
- Add 2.
- Multiply by 4.
- Add 2.
- Multiply by 5.
- Subtract 50.
- Drop the two zeros at the end.

Do you recognize the result?

b. Let S stand for a secret number that you begin with in part (a). Write an expression in terms of S that shows the result of applying the steps in part (a). Apply properties of arithmetic to this expression to explain why the end result of the steps in part (a) is what you noticed.

9.3 Equations

CCSS Common Core State Standards All Grades

Equations are a major idea in mathematics. They are used throughout mathematics—in all topics and at every level. Equations arise whenever we have two different ways to express the same amount. In this section we will first summarize the different ways that we use equations. We then turn our attention to solving equations, which is a major part of algebra. We will focus on the fundamental ideas we need to understand why equation-solving methods work.

equation

An **equation** is a statement that an expression or number is equal to another expression or number. For example, the following are equations:

$$3 + 2 = 5$$
$$x \cdot y = y \cdot x$$
$$y = 4x + 1$$
$$5x + 8 = 2x + 3$$

Unlike expressions, equations involve an equals (=) sign, which always means "equals" or "is equal to" or "which is equal to."

What Are Different Ways We Use Equations?

We use equations in several different ways for different purposes. Depending on the context and purpose of an equation, we give students different tasks about equations.

Some Equations Show Calculations Most commonly, we use equations in the process of calculating the answer to an arithmetic problem—in other words, in evaluating an expression, as in

$$2 + 3 \cdot 4 = 2 + 12 = 14$$

Students are often prompted to calculate with problems like this:

$$5 + 7 = \_\_\_$$

It is very important, however, that students do not develop a view of the equals sign as a "calculate the answer" symbol. Students who view the equals sign this way will not be able to make sense of solving algebraic equations later on. Instead, students must understand that the equals sign tells us that the left and right side of the equation stand for the same number.

Some Equations Are Identities: They Are True for All Values of the Variables We also use equations to describe general relationships, such as the properties of arithmetic. For example, the distributive property states that

$$A \cdot (B + C) = A \cdot B + A \cdot C$$

for all values of the variables A, B, and C.

identity An equation that is true for all values of the variables is often called an **identity**. Therefore the equation for the distributive property is an identity; so are the equations for the commutative and associative properties of addition and of multiplication.

We saw how to explain and use the identities arising from the properties of arithmetic in Chapters 3 and 4. These identities allow us to replace one expression with an equivalent expression. For example, we can replace

$$(100 - 1) \cdot 23$$

with the equivalent expression

$$100 \cdot 23 - 1 \cdot 23$$

which is easier to evaluate.

Some Equations Relate Quantities That Vary Together Another way we use equations is to describe a relationship between two or more quantities that vary together. An equation shows how the relationship between the quantities stays the same even as the quantities vary.

In Sections 7.4 and 7.5 we developed equations for proportional and inversely proportional relationships. We found that a proportional relationship between variables x and y has an equation of the form

$$y = m \cdot x$$

where m is a constant of proportionality. We can think of such an equation as telling us how to get a value for y if we know a value for x. In Section 9.7 we will develop equations for some other kinds of relationships.

formula An equation that relates two or more variables is often called a **formula**. For example, there is a formula for the area, A, (in square units) of a rectangle of length L units and width W units,

$$A = L \times W$$

Some Equations Must Be Solved to Solve a Problem A common way of using equations in algebra is for solving word problems, such as this:

> Tanya had some money. After she spent \$5, she gave half of the remaining money to her mother. At that point, Tanya had \$18 left. How much money did Tanya have at first?

If T stands for the number of dollars that Tanya had at first, then after spending \$5, Tanya has $T - 5$ dollars. After giving half of that to her mother, she has the other half left, which is $\frac{1}{2}(T - 5)$ dollars. That amount of money is also \$18. Therefore

$$\frac{1}{2}(T - 5) = 18$$

To solve the word problem, we'll want to solve the equation; in other words, we'll want to find out what number for T makes the equation true.

Next we'll discuss techniques for solving equations. Then in Section 9.4, we'll discuss how to formulate and solve word problems with strip diagrams and how to connect the strip diagrams to standard algebraic methods.

What Are Solutions of Equations?

What does it mean to solve an equation and what are the solutions of an equation? Some equations are true; for example, the equations

$$3 + 2 = 5 \text{ and } 10 = 2 \times 5$$

are true. Some equations are false; for example, the equations

$$1 = 0 \text{ and } 3 + 2 = 6$$

are false. Most equations with variables are true for some values of the variable and false for other values of the variable. For example, the equation

$$3 + x = 5$$

is true when $x = 2$, but is false for all other values of x.

solve an equation

solutions

To **solve an equation** involving variables means to determine those values for the variables that make the equation true. The values for the variables that make the equation true are called the **solutions** to the equation. To solve the equation

$$3x = 12$$

means to find all those values for x for which $3x$ is equal to 12. Only $3 \cdot 4$ is equal to 12; 3 times any number other than 4 is not equal to 12. So, there is only one solution to $3x = 12$, namely, $x = 4$.

Even young children in elementary school learn to solve simple equations. For example, first- or second-graders could be asked to fill in the box to make the following equation true:

$$5 + \square = 7$$

One source of difficulty in solving equations is understanding that the equals sign does not mean "calculate the answer." For example, when children are asked to fill in the box to make the equation

$$5 + 3 = \square + 2$$

true, many will fill in the number 8 because $5 + 3 = 8$. In order to understand how to solve equations, we must understand that we want to make the expressions to the left and right of the equal sign equal to each other. One helpful piece of imagery for understanding equations is a pan balance, as shown in **Figure 9.23**. If we view the equation $5 + 3 = \square + 2$ in terms of a pan balance, each side of the equation corresponds to a side of the pan balance. To solve the equation means to make the pans balance rather than tilt to one side or the other.

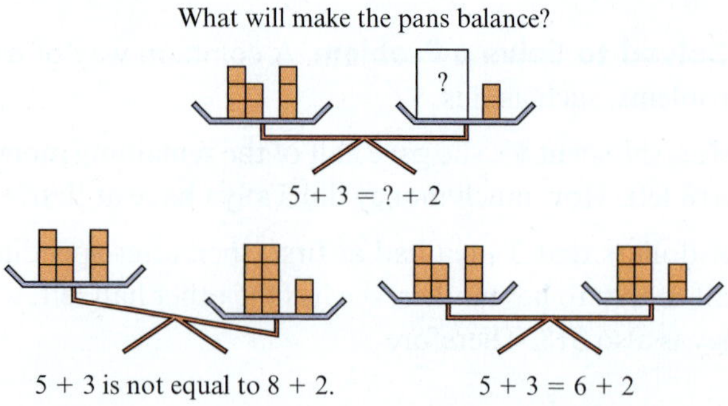

Figure 9.23 Viewing the equation $5 + 3 = ? + 2$ in terms of a pan balance.

CCSS 6.EE.5

How Can We Solve Equations by Reasoning About Relationships?

Although there are general techniques for solving equations, sometimes we can solve an equation just by thinking about what the equation means and by thinking about the relationship between the expressions on both sides of the equal sign. For example, to solve

$$2x = 6$$

we can think: "2 times what number is 6?" The solution is 3, which we could also find by dividing:

$$x = 6 \div 2 = 3$$

To solve

$$x + 3 = 7$$

we can think: "what number plus 3 is 7?" The solution is 4, which we could also find by subtracting:

$$x = 7 - 3 = 4$$

Class Activity 9H asks you to solve equations by using your understanding of expressions and equations.

CLASS ACTIVITY

9H Solving Equations by Reasoning About Expressions, p. CA-169

CCSS 6.EE.5

What Is the Reasoning Behind the Methods We Use to Solve Equations in Algebra?

How can we solve equations? We just saw in Class Activity 9H that in some cases we can reason about relationships to determine the solutions to equations.

But how do we solve an equation like

$$4x + 2 + x = 5 + 3x + 3?$$

In algebra, we use these key ways of thinking to develop general equation-solving strategies:

- We take a "pan-balance view" of equations to see ways to change an equation to a new equation that *has the same solutions*.
- The easiest equations to solve are equations like $x = 3$ and $x = \frac{2}{5}$ because they tell you what their solutions are.
- So, starting from a "messy" equation, we try to change it step by step into an equation that is very easy to solve.

Look at Figure 9.24 to see how these ways of thinking are used to solve the equation $4x + 2 + x = 5 + 3x + 3$. The figure shows a step-by-step process of changing the equation to new equations. Each equation has the same solution. Why? Because a solution is a number that makes both sides equal and the equality doesn't change throughout the process. The very last equation, $x = 3$, tells us that all the equations, including the original one, have the solution 3.

CLASS ACTIVITY

9I Solving Equations Algebraically and with a Pan Balance, p. CA-170

A pan balance is a wonderful piece of imagery, but it doesn't work well for some equations, such as those involving negative numbers. However, the strategy we used in Figure 9.24 by taking a "pan balance view" of equations works in general, for all equations.

In general, to solve an equation, we change the original equation to new equations that have the same solution until we arrive at an equation that is very easy to solve, such as $x = \frac{2}{3}$ or $x = -7$.

Figure 9.24 Solving an equation with equations and with a pan balance.

With Equations

$4x + 2 + x = 5 + 3x + 3$

Change to equivalent expressions

$5x + 2 = 3x + 8$
$\quad -2 \qquad -2$

Take 2 away from both sides.

$5x = 3x + 6$
$-3x \qquad -3x$

Take $3x$ away from both sides.

$2x = 6$

Divide both sides by 2.

$x = 3$

With a Pan Balance

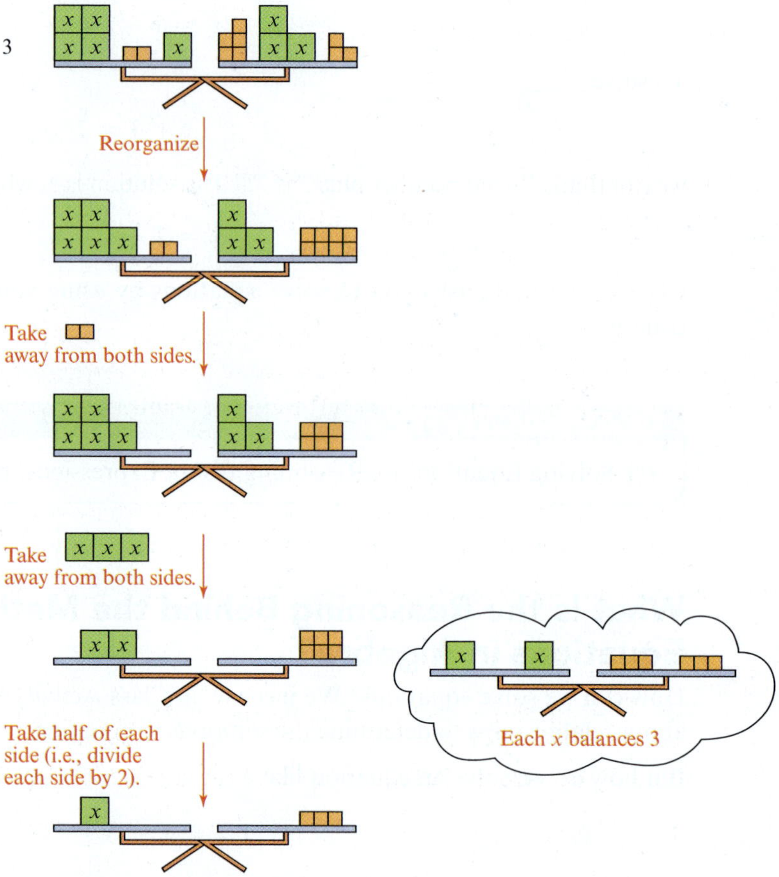

We can change an equation into a new equation that has the same solutions by doing any of the following:

1. Use properties of arithmetic or valid ways of operating with numbers (including fractions) to change an expression on either side of the equals sign into an equivalent expression. For example, we can change the equation
$$3x - 4x + 2 = 1$$
to the equation
$$-x + 2 = 1$$
because, according to the distributive property,
$$3x - 4x = (3 - 4)x = (-1)x = -x$$

2. Add the same quantity to both sides of the equation or subtract the same quantity from both sides of the equation. For example, we can change the equation
$$6x - 2 = 3$$
to the equation
$$6x - 2 + 2 = 3 + 2$$
which becomes
$$6x = 5$$
by using item (1).

3. Multiply both sides of the equation by the same nonzero number, or divide both sides of the equation by the same nonzero number. For example, we can change the equation

$$5x = 3$$

to the equation

$$x = \frac{3}{5}$$

by dividing both sides of the equation by 5. Remember that $5x$ stands for $5 \cdot x$, so when we divide this by 5, we get x.

Why do items (1), (2), and (3) change an equation into a new equation that has the same solutions as the original one? When we use item (1), we don't really change the equation at all; we just write the expression on one side of the equal sign in a different way. To see why items (2) and (3) should produce a new equation that has the same solutions as the original one, remember that the solutions of an equation make the two sides equal. If the two sides are equal and you add the same amount to both sides or take the same amount away from both sides, then the two sides remain equal, and vice versa. If the two sides are equal and you multiply or divide both sides by the same nonzero amount, then the two sides remain equal, and vice versa.

All the equations we have seen in this section have exactly one solution. Does that always happen? Try Class Activity 9J!

CLASS ACTIVITY

9J What Are the Solutions of These Equations?, p. CA-171

SECTION SUMMARY AND STUDY ITEMS

Section 9.3 Equations

An equation is a statement that an expression or number is equal to another expression or number. Equations involve an equal (=) sign. Equations can serve different purposes. Some equations show the result of a calculation. Some equations, called identities, are true for all values of the variables. Some equations describe how quantities that vary together are related. And some equations must be solved to solve a problem.

To solve an equation means to determine values for the variables that make the equation true. Sometimes we can solve an equation just by applying our understanding of expressions and equations. However, a general strategy for solving equations is to change the equation into a new equation that has the same solution and is easy to solve.

Key Skills and Understandings

1. Describe some of the different ways we use equations.
2. Solve appropriate equations by reasoning about numbers, operations, and expressions.
3. Explain different ways to reason to solve "one-step" equations.
4. Solve equations algebraically, and describe how the method can be viewed in terms of a pan balance if appropriate.
5. Give examples of equations in one variable that have no solutions and equations in one variable that have infinitely many solutions.

Practice Exercises for Section 9.3

1. Explain how to solve the next equation by reasoning about numbers, operations, and expressions rather than by using standard algebraic equation-solving techniques.

$$\frac{18}{7} = \frac{18 \cdot 2y}{12 \cdot 7}$$

2. Explain how to solve each of the next "one-step" equations by reasoning in several different ways.

a. $x - \frac{2}{7} = \frac{3}{5}$

b. $\frac{2}{3}x = \frac{9}{16}$

3. Solve $5x + 2 = 2x + 17$ in two ways: with equations and with pictures of a pan balance. Relate the two methods.

4. Solve the next two equations and explain your answers.

a. $2x - 5 + 8x + 3 = 12x - 1 - 2x + 7$

b. $x - 3 - 4x = 1 - 3x - 4$

Answers to Practice Exercises for Section 9.3

1. To make the left side of the equation equal to the right side we must find a y that makes

$$\frac{2y}{12} = 1$$

which is the case when

$$\frac{2y}{12} = \frac{12}{12},$$

so when $y = 6$.

2. **a.** *Undoing the expression:* The equation says that when we subtract $\frac{2}{7}$ from x we get $\frac{3}{5}$. So to undo that and get from $\frac{3}{5}$ back to x, add $\frac{2}{7}$. Therefore $x = \frac{3}{5} + \frac{2}{7} = \frac{31}{35}$.

$$x \xrightarrow{-\frac{2}{7}} \frac{3}{5}$$

$$x \xleftarrow{+\frac{2}{7}} \frac{3}{5}$$

Doing the same operation to both sides of the equation: The equation says that the two sides are equal. So if we add $\frac{2}{7}$ to both sides of the equation, they will still be equal, but now the left side will just be x.

$$x - \frac{2}{7} = \frac{3}{5}$$

$$x = \frac{3}{5} + \frac{2}{7} = \frac{31}{35}$$

b. *Undoing the expression:* The equation says that when you multiply x by $\frac{2}{3}$ you get $\frac{9}{16}$. So to undo that and get from $\frac{9}{16}$ back to x you need to divide by $\frac{2}{3}$. Therefore $x = \frac{9}{16} \div \frac{2}{3} = \frac{27}{32}$.

$$x \xrightarrow{\cdot\frac{2}{3}} \frac{9}{16}$$

$$x \xleftarrow{\div\frac{2}{3}} \frac{9}{16}$$

Doing the same operation to both sides of the equation: The equation says that the two sides are equal. So if we divide both sides of the equation by $\frac{2}{3}$, they will still be equal, but now the left side will just be x.

$$\frac{2}{3}x = \frac{9}{16}$$

$$x = \frac{9}{16} \div \frac{2}{3} = \frac{27}{32}$$

Or we could multiply both sides of the equation by $\frac{3}{2}$ and both sides will still be equal. The left side will just be x because $\frac{3}{2} \cdot \frac{2}{3} = 1$, in other words, because $\frac{3}{2}$ and $\frac{2}{3}$ are multiplicative inverses.

$$\frac{2}{3}x = \frac{9}{16}$$

$$\frac{3}{2} \cdot \frac{2}{3}x = \frac{3}{2} \cdot \frac{9}{16}$$

$$x = \frac{3}{2} \cdot \frac{9}{16} = \frac{27}{32}$$

3. See Figure 9.25. On the pan balance, each small square represents 1 and each larger square, filled

with an x, represents the unknown amount x. At the end, we find that x must stand for 5 small squares in order to make the pans balance. Therefore, the solution of the equation is 5.

4. a. Combining terms on the left and right of the equal sign, the equation becomes $10x - 2 = 10x + 6$. Adding 2 to both sides, we have $10x = 10x + 8$. Subtracting $10x$ from both sides, we have $0 = 8$, which is false. Since the process of adding and subtracting the same amounts from both sides gives us equations that have the same solutions as the original equation, and since the final equation has no solutions, the original equation must have no solutions as well. In other words, the original equation is always false, no matter what the value of x is.

b. Combining terms on the left and right side of the equal sign, the equation becomes $-3x - 3 = -3x - 3$. We can already see that the left and right sides of the equation are the same, and therefore the equation is true for all values of x. So the equation is an identity and has infinitely many solutions.

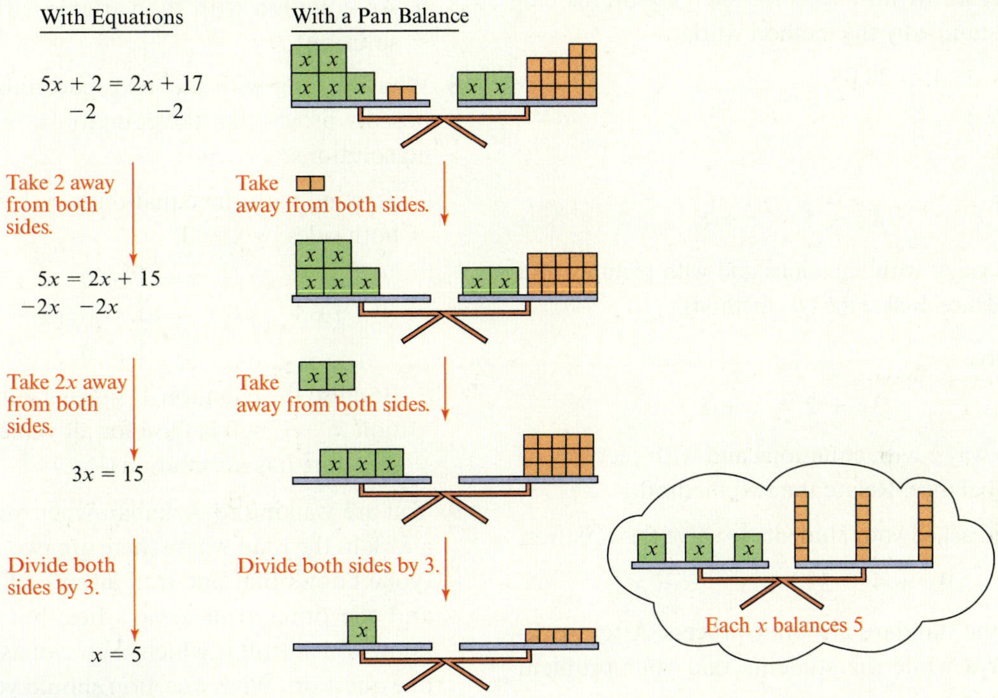

Figure 9.25 Solving $5x + 2 = 2x + 17$ with equations and with a pan balance.

PROBLEMS FOR SECTION 9.3

1. Explain how to solve the next equations by reasoning about numbers, operations, and expressions rather than by using standard algebraic equation-solving techniques.

a. $x + 57 + 94 = 98 + 57$

b. $57 \cdot (94 + x) = 98 \cdot 57$

c. $57 \cdot 94 + x = 58 \cdot 94 + 3$

d. $487 + 176 = x + 490$

e. $333 \cdot 213 = 111A$

f. $C + 34 \cdot 8 = 28 \cdot 34$

g. $\frac{1}{4}x + \frac{7}{16} = 11\frac{7}{16}$

h. $42 = x - 42$

i. $9x + 4x = 4x + 1$

j. $187(x - 419)^2 = 0$

k. $7.9x + 3.2 = 793.2$

2. Write an equation that can be solved by reasoning about numbers, operations, and expressions. Explain how to solve your equation that way rather than by using standard algebraic equation-solving techniques.

3. In your own words, explain how to solve each of the next "one-step" equations by reasoning in at least two different ways. Explain the reasoning in enough detail so that someone who doesn't already know how to solve such equations can understand why the method works.

a. $x + 23.4 = 28.05$

b. $x \div 7 = \frac{3}{5}$

4. Solve

$$4x + 5 = x + 8$$

in two ways: with equations and with pictures of a pan balance. Relate the two methods.

5. Solve

$$3x + 2 = x + 8$$

in two ways: with equations and with pictures of a pan balance. Relate the two methods.

6. You asked your students to solve the equation

$$9x + 4 - 3x = 5x + 3 + x$$

using the standard algebraic process. After working for a while the students said your problem was wrong.

You also asked your students to solve the equation

$$7x + 8 + 5x + 3 = 2(1 + 6x) + 9$$

using the standard algebraic process. After working for a while they were confused about how to answer.

Discuss the reasoning and logic that underlies the standard algebraic equation-solving process. Explain how that helps us interpret the outcome of the process in cases such as the above.

7. Give your own examples of the following equations. Explain why your answers work.

a. An equation with the variable x that has infinitely many solutions.

b. An equation with the variable x that has no solutions.

***8.** What's wrong with the next reasoning that supposedly proves that the equation $x - 1 = 0$ has no solutions?

Starting with the equation $x - 1 = 0$, divide both sides by $x - 1$:

$$x - 1 = 0$$
$$(x - 1)/(x - 1) = 0/(x - 1)$$
$$1 = 0$$

Because the equation $1 = 0$ is false, the equation $x - 1 = 0$ is false for all values of x and therefore has no solutions.

***9.** You are walking to Azkaban when you come to a fork in the road where there are two trolls. Everyone knows that one troll always tells the truth and the other troll always lies, but you don't know which troll is which. You can ask one troll one question. What question should you ask and what should you do in response?

9.4 Solving Algebra Word Problems with Strip Diagrams and with Algebra

CCSS Common Core State Standards Grades 4, 5, 6, 7

In this section, we will see how drawing strip diagrams can make some algebra word problems easy to solve without using variables. But we will also connect the strip-diagram approach and the standard algebraic approach to formulating and solving word problems. Often, a strip diagram can even help us formulate a word problem algebraically, with an equation. And sometimes the steps we use in solving a problem with a strip diagram correspond to the steps we use when we solve the problem algebraically by manipulating equations.

Let's see how a strip diagram can help us solve this word problem:

> **Payment Plan Problem:** The Browns are buying a TV that costs $1300. After they make 4 equal payments, they still owe $400 on the TV. How much was each payment?

CCSS
7.EE.4

Examine Figure 9.26 to see how to solve the word problem with a strip diagram and with algebraic equations, using a variable. Notice that different parts of the strip diagram correspond to different quantities in the problem: the entire strip stands for the full $1300 cost of the TV, the 4 identical small pieces stand for the 4 identical payments, and the remaining piece stands for the remaining $400 that is left to pay after the 4 payments. Observe how the problem is solved by reasoning about the relationships among the quantities represented by the pieces. Observe that the strip diagram is a math drawing for the equation

$$x + x + x + x + 400 = 1300$$

Figure 9.26

Solving a word problem with a strip diagram and with algebraic equations.

Payment Plan Problem: The Browns are buying a TV that costs $1300. After they make 4 equal payments, they still owe $400 on the TV. How much was each payment?

With a strip diagram:

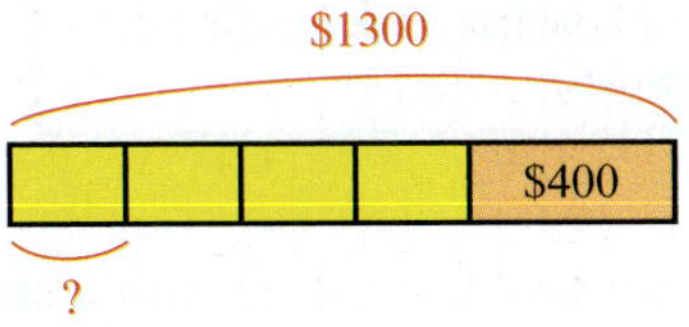

4 parts ⟶ $1300 − $400 = $900

1 part ⟶ $900 ÷ 4 = $225

Each payment was $225.

With algebraic equations, supported by a strip diagram:

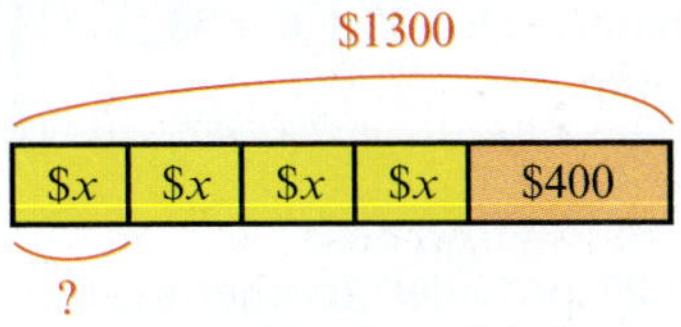

Let x = Number of dollars in one payment.

$4x + 400 = 1300$

$4x = 1300 - 400$

$4x = 900$

$x = 900 \div 4 = 225$

Each payment was $225.

With algebraic equations alone:

Let x = Number of dollars in one payment.

Then 4 payments are $4x$ dollars.

After 4 payments, the Browns still owe $1300 - 4x$ dollars, which is $400.

So: $1300 - 4x = 400$

Add 4x to both sides:

$1300 = 400 + 4x$

Subtract 400 from both sides:

$1300 - 400 = 4x$

$900 = 4x$

Divide both sides by 4:

$900 \div 4 = x$

$x = 225$

Each payment was $225.

or

$$4x + 400 = 1300$$

Compare the different approaches in the figure and notice that all of them involve the same arithmetic: subtract 400 from 1300, then divide the resulting 900 by 4 to obtain 225.

Figure 9.27 Solving a word problem with a strip diagram and with algebraic equations.

Ragworts Problem: All together, Ron and Hermione have 70 ragworts. Hermione has 6 more ragworts than Ron. How many ragworts do they each have?

With a strip diagram:

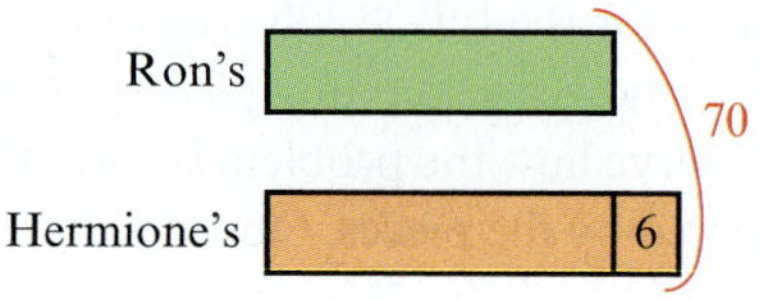

2 parts ⟶ 70 − 6 = 64

1 part ⟶ 64 ÷ 2 = 32

Ron has 32 ragworts.
Hermione has 32 + 6 = 38 ragworts.

With algebraic equations supported by a strip diagram:

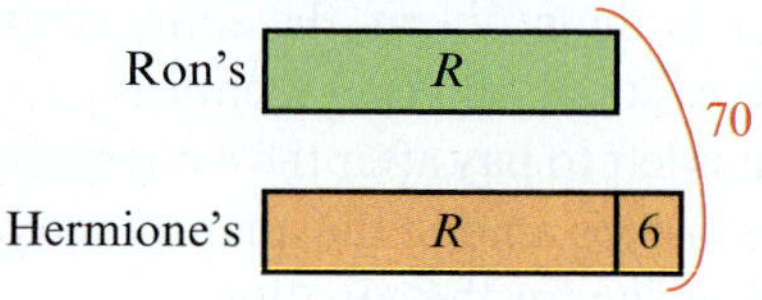

Let R = Number of ragworts Ron has

$$2R + 6 = 70$$
$$2R = 70 - 6$$
$$2R = 64$$
$$R = 64 \div 2$$
$$R = 32$$

Ron has 32 ragworts.
Hermione has 32 + 6 = 38 ragworts.

Figure 9.27 provides another example of a word problem solved with a strip diagram and with algebraic equations. Observe how the strip diagram helps us formulate an equation with a variable. Notice also how solving the algebraic equation parallels the arithmetic we perform when we use the strip diagram. In both cases, we subtract 6 from 70 and then divide the resulting 64 by 2 to obtain 32.

Now try using strip diagrams to solve word problems yourself in Class Activity 9K. Figure 9.28 has some tips for using strip diagrams effectively.

CLASS ACTIVITY

9K Solving Word Problems with Strip Diagrams and with Equations, p. CA-172

Figure 9.28 Strip diagram tips.

Tip 1: Strips represent quantities, just like variables and segments on a number line do.

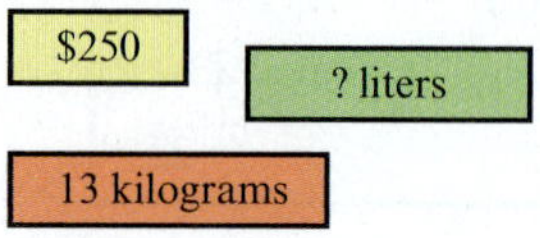

Tip 2: Strips use length, like a number line:
Same length ⟷ same amount
Longer length ⟷ greater amount

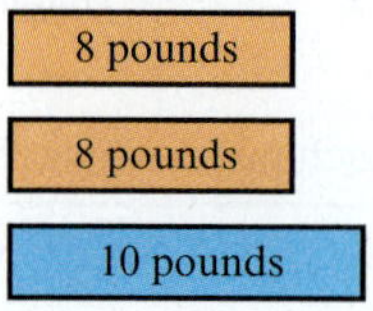

Tip 3: Don't worry about getting size comparisons exactly right.

Tip 4: Sometimes it helps to make several drawings to show how a quantity changes over time:

Markus had some money.

He spent $\frac{1}{4}$ of it.

He gave $\frac{1}{2}$ of what was left to Lila.

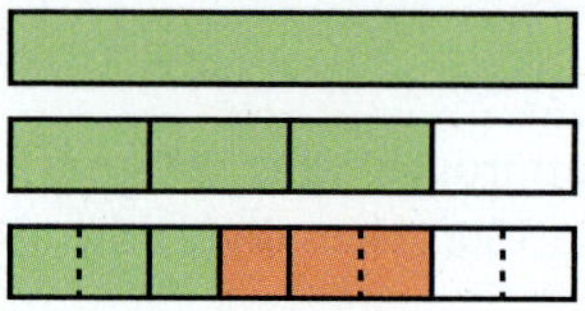

One final thought about strip diagrams. Strip diagrams are a powerful representation that can help students reason about challenging multistep problems even before they learn to use algebraic equations with variables. When students reason with strip diagrams, they often use the very same steps they would use in solving a problem algebraically. Strip diagrams are therefore an excellent springboard toward algebra. But they are not a *substitute* for algebraic equations with variables, which are essential for mathematics and science in high school and beyond.

CLASS ACTIVITY

9L Solving Word Problems in Multiple Ways and Modifying Problems, p. CA-175

FROM THE FIELD Research

Englard, L. (2010). Raise the bar on problem solving. *Teaching Children Mathematics, 17*(3), pp. 156-163.

To improve students' problem-solving abilities the author recommends the Singapore model method (i.e., drawing strip diagrams) because the method's drawings put the focus on quantities, relationships, and actions presented in a problem. For example, if students are asked how many books Karen has if Skylar has 4 times as many books as Karen and Skylar has 36 books, a traditional focus on "key" words may mislead students into multiplying 36 by 4 rather than dividing.

The author conducted a small experimental study with one experimental group of third-graders and control groups of third-, fourth-, and fifth-grade students. While the experimental group received instruction in solving problems with strip diagrams for nineteen days, the control groups received traditional instruction. Based on the pre- and post-test results, third-graders in the experimental group solved more problems correctly than the students in the control groups, including more than the fourth- and fifth-grade students. For example, on the problem about Karen's and Skylar's books, none of the students in the third-grade experimental group used multiplication instead of division but 29% of third-, 26% of fourth-, and 15% of fifth-graders in the control groups mistakenly used multiplication instead of division. Strip diagrams also helped students solve problems involving fractions and multi-step problems. For example, over 40% of the third-graders in the experimental group solved a multi-step word problem correctly whereas only 15% of the control groups of third-, fourth-, and fifth-graders did. Rather than solving a few simple one-step problems after many pages of skills practice, drawing strip diagrams allows students to work on challenging multi-step problems daily and to explore the meaning of operations, fractions, percent, and ratio.

SECTION SUMMARY AND STUDY ITEMS

Section 9.4 Solving Algebra Word Problems with Strip Diagrams and with Algebra

Many algebra word problems can be solved with the aid of strip diagrams. Often, when we use strip diagrams to solve a word problem, our method parallels the algebraic method of setting up and solving an equation with variables.

Key Skills and Understandings

1. Set up and solve word problems with the aid of strip diagrams as well as algebraically, and relate the two.

Practice Exercises for Section 9.4

Some of these Practice Exercises were inspired by problems in the 4th-, 5th-, and 6th-grade mathematics texts used in Singapore (see [78] and [79], volumes 4A–6B).

1. Dante has twice as many jelly beans as Carlos. Natalia has 4 more jelly beans than Carlos and Dante together. All together, Dante, Carlos, and Natalia have 58 jelly beans. How many jelly beans does Carlos have? Solve this problem in two ways: with the aid of a diagram and with algebraic equations. Explain both solution methods, and discuss how they are related.
2. A school has two hip-hop dance teams, the Jelani team and the Zuri team. There are $1\frac{1}{2}$ times as many students on the Jelani team as on the Zuri team. All together, there are 65 students on the two teams combined. How many students are on the Jelani team, and how many are on the Zuri team? Solve this problem in two ways: with the aid of a diagram and with algebraic equations. Explain both solution methods, and discuss how they are related.
3. At first, only $\frac{1}{3}$ of the students who were going on the band trip were on the bus. After another 21 students got on the bus, $\frac{4}{5}$ of the students who were going on the band trip were on the bus. How many students were going on the band trip? Solve this problem in two ways: with the aid of a diagram and with algebraic equations. Explain both solution methods, and discuss how they are related.

Answers to Practice Exercises for Section 9.4

1. Figure 9.29 shows that the total number of jelly beans consists of 6 equal groups of jelly beans and 4 more jelly beans. If we take the 4 away from 58, we have 54 jelly beans that must be distributed equally among the 6 groups. Therefore, each group has $54 \div 6 = 9$ jelly beans. Since Carlos has one group of jelly beans, he has 9 jelly beans. (Dante has $2 \cdot 9 = 18$ jelly beans, and Natalia has $9 + 18 + 4 = 31$ jelly beans.)

   To solve the problem with algebraic equations, let x be the number of jelly beans that Carlos has. Since Dante has twice as many jelly beans as Carlos, Dante has $2x$ jelly beans. Since Natalia has 4 more than Carlos and Dante combined, Natalia has $x + 2x + 4$ jelly beans, which is $3x + 4$ jelly beans. We can also see these expressions in Figure 9.29. Carlos's strip represents x jelly beans. Since Dante has 2 such strips, he has $2x$ jelly beans. Natalia has 3 such strips and 4 more jelly beans, so she has $3x + 4$ jelly beans.

   The total number of jelly beans is Carlos's, Dante's, and Natalia's combined, which is

   $$x + 2x + 3x + 4$$

   which is equivalent to

   $$6x + 4$$

   The total number of jelly beans is 58, so

   $$6x + 4 = 58$$

   We can also deduce this equation from Figure 9.29 because there are 6 strips that each represent x jelly beans and another small strip representing 4 more jelly beans; the combined amount must be 58, so $6x + 4 = 58$.

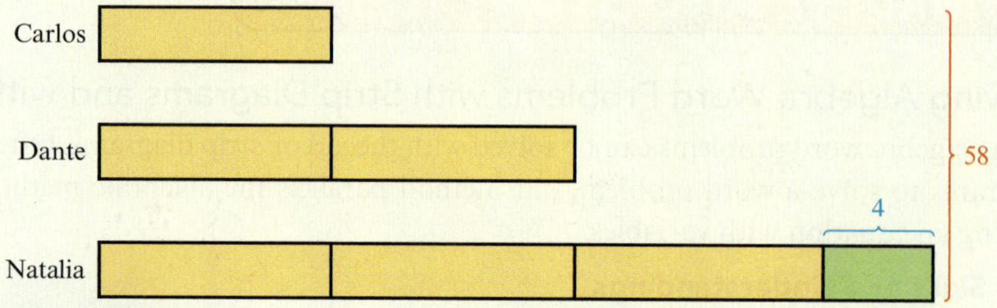

Figure 9.29 A strip diagram for determining how many jelly beans Carlos has.

To solve $6x + 4 = 58$, first subtract 4 from both sides to obtain

$$6x = 58 - 4 \quad \text{after simplifying: } 6x = 54$$

Then divide both sides by 6 to obtain

$$x = 54 \div 6 \quad \text{after simplifying: } x = 9$$

Therefore, Carlos has 9 jelly beans. Notice that the arithmetic we performed to solve the equation $6x + 4 = 58$ is the same as the arithmetic we performed to solve the problem with the diagram: We first subtracted 4 from 58, and then we divided the resulting amount, 54, by 6.

2. **Figure 9.30** shows that there are $1\frac{1}{2}$ times as many students on the Jelani team as on the Zuri team and that there are 65 students in all. Notice that the total number of students is represented by 5 half-strips, each of which is the same size. The 65 students must be divided equally among these 5 portions, so each portion contains $65 \div 5 = 13$ students. The Jelani team consists of 3 of those portions, so there are $3 \times 13 = 39$ students on the Jelani team. Since the Zuri team consist of 2 portions, there are $2 \times 13 = 26$ students on the Zuri team.

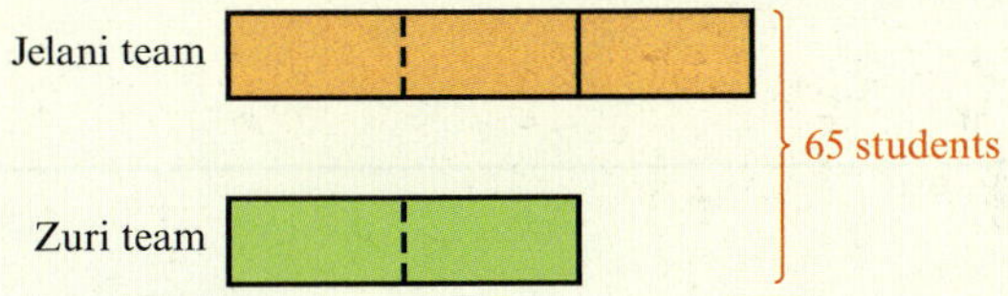

Figure 9.30 A strip diagram for determining how many students are on each dance team.

To solve the problem with equations, let Z stand for the number of students on the Zuri team. Then the number of students on the Jelani team is $1\frac{1}{2} \cdot Z$, or $\frac{3}{2}Z$. Therefore, the two teams combined have $Z + \frac{3}{2}Z$ students. Since this number of students is 65, we obtain the equation

$$Z + \frac{3}{2}Z = 65$$

Applying the distributive property, we have

$$\left(1 + \frac{3}{2}\right)Z = 65$$

$$\text{after simplifying: } \frac{5}{2}Z = 65$$

Notice that we could also formulate this last equation from **Figure 9.30**, as the total number of students is represented by the 5 half-segments. Therefore, the total number of students is $\frac{5}{2}$ of the number of students on the Zuri team, so $\frac{5}{2}Z = 65$. By multiplying both sides of the equation $\frac{5}{2}Z = 65$ by $\frac{2}{5}$ (or by dividing both sides by $\frac{5}{2}$), we obtain

$$Z = 65 \cdot \frac{2}{5} = \overset{13}{\cancel{65}} \cdot \frac{2}{\cancel{5}} = 13 \cdot 2 = 26$$

So there are 26 students on the Zuri team. Therefore, there must be

$$1\frac{1}{2} \cdot 26 = \frac{3}{\cancel{2}} \cdot \overset{13}{\cancel{26}} = 3 \cdot 13 = 39$$

students on the Jelani team.

With both solution methods, we divided the total number of students, 65, by 5 and multiplied the result, 13, by 2 to obtain the number of students on the Zuri team. We also multiplied the 13 by 3 to obtain the number of students on the Jelani team.

3. The full strip in **Figure 9.31** represents all the students going on the band trip. The dark portion represents the students who were on the bus initially, and the lighter shaded portion represents the additional 21 students who got on the bus. This lighter shaded portion must be the difference between $\frac{4}{5}$ and $\frac{1}{3}$ of the students going on the trip. Since the fractions $\frac{4}{5}$ and $\frac{1}{3}$ have 15 as a common denominator, it makes sense to break the total number of students into 15 equal parts. Because

$$\frac{4}{5} - \frac{1}{3} = \frac{12}{15} - \frac{5}{15} = \frac{7}{15}$$

the 21 additional students take up 7 parts. So each part consists of $21 \div 7 = 3$ students. Since there

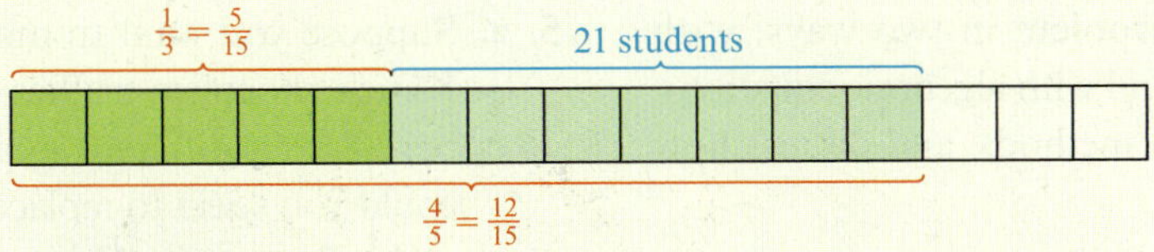

Figure 9.31 A strip diagram for determining how many students are going on the band trip.

are 15 parts making the whole group of students going on the band trip, there are $15 \times 3 = 45$ students going on the band trip.

To solve the problem with equations, let x be the number of students going on the band trip. In terms of the diagram in **Figure 9.31**, x stands for the whole long strip. Then $\frac{1}{3}x$ is the number of students on the bus at first. After 21 students get on the bus, there are

$$\frac{1}{3}x + 21$$

students on the bus. But this number of students is also equal to $\frac{4}{5}x$ because now $\frac{4}{5}$ of the students going on the trip are on the bus. Therefore, we have the equation

$$\frac{1}{3}x + 21 = \frac{4}{5}x$$

Notice that we can also see this equation from the relationship between the shaded portions in **Figure 9.31**. Subtracting $\frac{1}{3}x$ from both sides of the previous equation, we have

$$21 = \frac{4}{5}x - \frac{1}{3}x$$

So

$$21 = \left(\frac{4}{5} - \frac{1}{3}\right)x \quad \text{after simplifying: } 21 = \frac{7}{15}x$$

After multiplying both sides by $\frac{15}{7}$ (or dividing both sides by $\frac{7}{15}$), we have

$$\frac{15}{7}\cdot 21 = \frac{15}{7}\cdot\frac{7}{15}x \quad \text{after simplifying: } 45 = x$$

Therefore, $x = 45$, and there were 45 students going on the band trip.

We perform the same arithmetic by using both solution methods. With each method, we calculated $\frac{4}{5} - \frac{1}{3}$. When we solved the problem by using a diagram, we divided $21 \div 7$ to find the number of students in each $\frac{1}{15}$ part of the total students. Then we multiplied that result, 3, by the total number of parts making up all the students—namely, 15. Similarly, using the algebraic solution method, we had to calculate

$$\frac{15}{7}\cdot 21$$

When we used canceling, we also divided 21 by 7 and multiplied the result, 3 by 15.

PROBLEMS FOR SECTION 9.4

Some of these problems were inspired by problems in the 4th-, 5th-, and 6th-grade mathematics texts used in Singapore (see [78] and [79], volumes 4A–6B).

1. A candy bowl contains 723 candies. Some of the candies are red, and the rest are green. There are twice as many green candies as red candies. How many red candies are in the candy bowl? Solve this problem in two ways: with the aid of a diagram and with algebraic equations. Explain both solution methods, and discuss how they are related.

2. Sarah spends $\frac{2}{5}$ of her monthly income on rent. Sarah's rent is \$750. What is Sarah's monthly income? Solve this problem in two ways: with the aid of a diagram and with algebraic equations. Explain both solution methods, and discuss how they are related.

3. After Joey spent $\frac{3}{8}$ of his money on an electronic game player, he had \$360 left. How much money did Joey have before he bought the electronic game player? How much did the electronic game player cost? Solve this problem in two ways: with the aid of a diagram and with algebraic equations. Explain both solution methods, and discuss how they are related.

4. Katie made some chocolate truffles. She gave $\frac{1}{4}$ of her truffles to Leanne, and then gave $\frac{1}{2}$ of the remaining truffles to Jeff. At that point, Katie had 18 truffles left. How many chocolate truffles did she have at first? Solve this problem in two ways: with the aid of a diagram and with algebraic equations. Explain both solution methods, and discuss how they are related.

5. **a.** Suppose you want to modify Problem 4 about Katie's chocolate truffles by changing the number 18 to a different number. Which numbers could you select to replace the 18 in the problem and still have a sensible problem (without changing anything else in the problem)? Explain.

b. Change the fraction $\frac{1}{2}$ in Problem 4 about Katie's chocolate truffles so that the problem will become easier to solve. You may change the number 18 as well. Show how to solve the new problem, and explain why it is easier to solve than the original problem.

6. Write something similar to Problem 4 in which a fraction of a collection of objects is removed and then a fraction of the remaining objects is removed and a certain number of objects remains. Solve your problem in two ways: with the aid of a diagram and with algebraic equations. Explain both solution methods, and discuss how they are related.

7. By washing cars, Beatrice and Hannah raised \$2000 total. If Beatrice raised \$600 more than Hannah, how much money did Hannah raise? Solve this problem in two ways: with the aid of a diagram and with algebraic equations. Explain both solution methods, and discuss how they are related.

8. There are 1700 bottles of water arranged into 4 groups. The second group has 20 more bottles than the first group. The third group has twice as many bottles as the second group. The fourth group has 10 more bottles than the third group. How many bottles are in each group? Solve this problem in two ways: with the aid of a diagram and with algebraic equations. Explain both solution methods, and discuss how they are related.

9. Write something similar to Problem 8 in which a certain number of objects are divided into a number of unequal groups. Solve your problem in two ways: with the aid of a diagram and with algebraic equations. Explain both solution methods, and discuss how they are related.

10. One day a coffee shop sold 360 cups of coffee, some regular coffee and the rest decaffeinated. The shop sold 3 times as many cups of regular coffee as cups of decaffeinated coffee. The shop made 100 cups of regular coffee with foamed milk. How many cups of regular coffee did the shop make without foamed milk? Solve this problem in two ways: with the aid of a diagram and with algebraic equations. Explain both solution methods, and discuss how they are related.

11. A 1-pound-5-ounce piece of cheese is divided into 2 pieces. The larger piece weighs twice as much as the smaller piece. How much does each piece weigh? Solve this problem in two ways: with the aid of a diagram and with algebraic equations. Explain both solution methods, and discuss how they are related.

12. A hot dog stand sold $\frac{3}{8}$ of its hot dogs before noon and $\frac{1}{2}$ of its hot dogs after noon. The stand sold 280 hot dogs in all. How many hot dogs were left? How many hot dogs were there at first? Solve this problem in two ways: with the aid of a diagram and with algebraic equations. Explain both solution methods, and discuss how they are related.

13. Write something similar to Problem 12 in which the combined fractional amounts of a quantity are equal to a given quantity and the fractions in the problem have different denominators. Solve your problem in two ways: with the aid of a diagram and with algebraic equations. Explain both solution methods, and discuss how they are related.

14. Julie bought donuts for a party—$\frac{1}{6}$ of the donuts were jelly donuts, $\frac{1}{3}$ of the donuts were cinnamon donuts, and $\frac{5}{6}$ of the remainder were glazed donuts. Julie bought 20 glazed donuts. How many donuts did Julie buy in all? Solve this problem in two ways: with the aid of a diagram and with algebraic equations. Explain both solution methods, and discuss how they are related.

15. Shauntay caught twice as many fireflies as Robert. Jessica caught 10 more fireflies than Robert. All together, Shauntay, Robert, and Jessica caught 150 fireflies. How many fireflies did Jessica catch? Solve this problem in two ways: with the aid of a diagram and with algebraic equations. Explain both solution methods, and discuss how they are related.

16. At a frog exhibit, $\frac{3}{5}$ of the frogs are bullfrogs, $\frac{2}{3}$ of the remainder are tree frogs, and the rest are river frogs. There are 36 bullfrogs in the exhibit. How many river frogs are there? Solve this problem in two ways: with the aid of a diagram and with algebraic equations. Explain both solution methods, and discuss how they are related.

17. (See Problem 15 in Section 9.2.) Bonnie had some stickers. First, Bonnie gave away $\frac{1}{6}$ of her stickers. When Bonnie gets 12 more stickers, she has 47 stickers. Determine how many stickers Bonnie had at the start. Explain your solution.

18. (See Problem 18 in Section 9.2.) There was some flour in a bag. First, $\frac{3}{8}$ of the flour in the bag was used. Then $\frac{1}{5}$ of the remaining flour was used. At that point, there were 10 cups of flour in the bag. Determine how much flour was in the bag at the start. Explain your solution.

19. Not every algebra word problem is naturally modeled with a strip diagram. Solve the following problem in any way that makes sense, and explain your solution:

> Kenny has stickers to distribute among some goodie bags. When he tries to put 5 stickers in each goodie bag, he is 2 stickers short. But when he puts 4 stickers in each bag, he has 3 stickers left over. How many goodie bags does Kenny have? How many stickers does he have?

20. Not every algebra word problem is naturally modeled with a strip diagram. Solve the following problem in any way that makes sense, and explain your solution:

> Frannie has no money now, but she plans to save $\$D$ every week so that she can buy a toy she would like to have. Frannie figures that if she saves her money for 3 weeks, she will have \$1 less than she needs to buy the toy, but if she saves her money for 4 weeks, she will have \$3 more than she needs to buy the toy. How much is Frannie saving every week, and how much does the toy Frannie plans to buy cost?

21. In a group of ladybugs there were 10% more male ladybugs than female ladybugs. There were 8 more male ladybugs than female ladybugs. How many ladybugs were there in all? Solve this problem in two ways: with the aid of a diagram and with algebraic equations. Explain both solution methods, and discuss how they are related.

22. A farmer planted 65% of his farmland with onions, 20% of his land with cotton, and the rest with soybeans. The soybeans are on 90 acres of land. How many acres of farmland does the farmer have? Solve this problem in two different ways, one of which involves solving an equation. Explain both solution methods.

23. Some exercise equipment is on sale for 20% off. The sale price is \$340. What was the price of the exercise equipment before the discount? Solve this problem in two different ways, one of which involves solving an equation. Explain both solution methods.

24. Clevere makes 30% more money than Shane. If Clevere makes \$8450 per month, how much does Shane make per month? Solve this problem in two different ways, one of which involves solving an equation. Explain both solution methods.

25. Two stores had a combined total of 210 teddy bears. After the first store sold $\frac{1}{2}$ of its teddy bears and the second store sold $\frac{1}{3}$ of its teddy bears, each store had the same number of teddy bears left. How many teddy bears were sold from the two stores combined? Solve this problem in two different ways, one of which involves solving an equation. Explain both solution methods.

26. If $\frac{2}{3}$ of a cup of juice gives you 120% of your daily value of vitamin C, what percent of your daily value of vitamin C will you get in $\frac{1}{2}$ cup of the juice? Solve this problem in two different ways, one of which involves solving an equation. Explain both solution methods.

27. Erika is weaving a wool blanket. So far, she has used 10 balls of wool and has completed $\frac{5}{8}$ of the blanket. How many more balls of wool will Erika need to finish the blanket? Solve this problem in two different ways, one of which involves solving an equation. Explain both solution methods.

28. Jane had a bottle filled with juice. At first, she drank $\frac{1}{5}$ of the juice in the bottle. After 1 hour, she drank $\frac{1}{4}$ of the juice remaining in the bottle. After another 2 hours, she drank $\frac{1}{3}$ of the remaining juice in the bottle. At that point, Jane checked how much juice was left in the bottle: There was $\frac{2}{3}$ of a cup left. No other juice was added to or removed from the bottle. How much juice was in the bottle originally? Solve this problem, and explain your solution.

29. A flock of geese on a pond was being observed continuously. At 1:00 P.M., $\frac{1}{5}$ of the geese flew away. At 2:00 P.M., $\frac{1}{8}$ of the geese that remained flew away. At 3:00 P.M., 3 times as many geese flew away as had flown away at 1:00 P.M., leaving 28 geese on the pond. At no other time did any geese arrive or fly away. How many geese were in the original flock? Solve this problem, and explain your solution.

9.5 Sequences

CCSS **PreK, Common Core State Standards** Grades K, 1, 4, 5, 8

Most people, including young children, enjoy discovering and experimenting with growing and repeating patterns. But patterns aren't just fun. By thinking about how to describe them generally, we begin to develop algebraic reasoning. In this section, we will study growing and repeating patterns from a mathematical point of view by studying sequences.

sequence

A **sequence** is a list of entries occurring in a specified order. These entries can be numbers, letters, figures, objects, or movements.

CCSS K.CC.1

Students work with different types of sequences in the elementary and middle grades, including repeating patterns, such as those in Figure 9.32, and different kinds of growing patterns, such as those in Figure 9.33. The first, and most important, growing pattern that children learn about is the sequence of counting numbers,

$$1, 2, 3, 4, 5, \ldots$$

which can be represented not only with written numbers but also with physical objects, such as with a staircase pattern of block towers, as in Figure 9.33.

Figure 9.32 Repeating patterns.

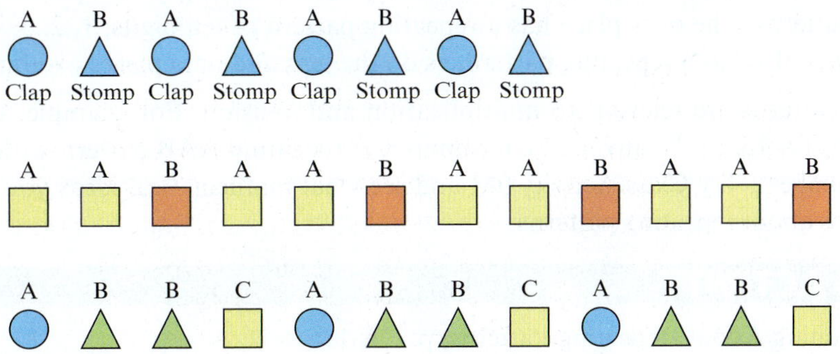

Figure 9.33 Growing sequences.

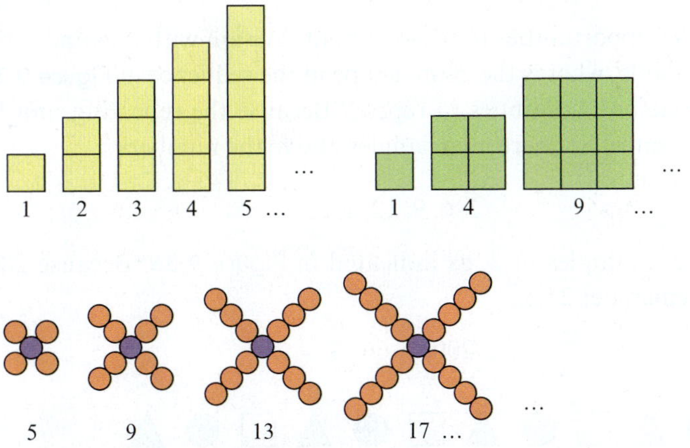

Different kinds of sequences are connected with different mathematical ideas and therefore provide different learning opportunities. Next, we'll examine some of the common types of sequences and the mathematical ideas that can be highlighted while studying them.

CCSS 1.NBT.1 1.NBT.2

How Can We Reason About Repeating Patterns?

Young children in Prekindergarten and Kindergarten often work with repeating patterns of shapes, colors, sounds, or movements. They learn to copy, extend, and fix errors in repeating patterns.

They usually start with simple repeating "AB patterns," then progress to more complex repeating patterns, such as "AAB patterns," and "ABBC patterns," as in Figure 9.32. When young children first grasp the idea of a repeating pattern, they may focus only on the individual shapes, colors, or sounds that are repeated in a certain order. However, repeating patterns offer the opportunity to highlight an important mathematical idea—namely, the idea of a unit.

A repeating pattern is made by repeating a single unit, as indicated in Figure 9.34. The idea of a unit is especially important for young children as they begin to learn about units of 10 in the base-ten system. The idea of a unit is essential for understanding numbers, and repeating patterns offer the opportunity to highlight a unit that is repeated.

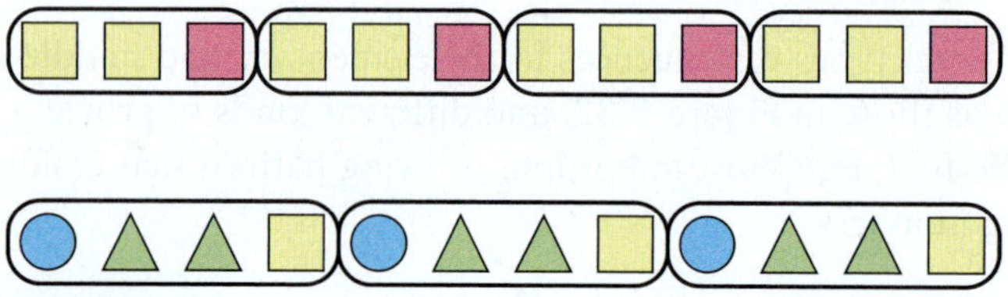

Figure 9.34 Repeating patterns are made by repeating a unit.

Notice that even though the sequence of counting numbers is a growing pattern, it also involves repeating patterns. The ones place has a repeating pattern of ten digits, 1, 2, 3, 4, 5, 6, 7, 8, 9, 0. The tens place also has a repeating pattern, as do the tens and ones places combined, and so on.

CCSS
4.OA.5

Repeating patterns are relevant to multiplication and division. For example, skip counting by 3s—"3, 6, 9, 12, 15, . . ."—involves coordinating a repeating AAB pattern with the sequence of counting numbers. Try Class Activity 9M and see what mathematical ideas you use as you solve the problems about repeating patterns.

CLASS ACTIVITY

9M Reasoning About Repeating Patterns, p. CA-176

CCSS
5.OA.3

Repeating patterns offer opportunities to reason about division with remainder by focusing on the repeating unit. For example, what is the 200th shape in the sequence in Figure 9.35, assuming that the square-circle-triangle unit continues to repeat? Because the repeating unit has three shapes, every third shape is the same. So there are triangles above the numbers

$$3, 6, 9, 12, 15, \ldots$$

and above all the other multiples of 3, as indicated in Figure 9.36. Because $200 \div 3$ has whole number quotient 66, remainder 2 i.e.,

$$200 = 66 \cdot 3 + 2$$

Figure 9.35 A sequence of shapes.

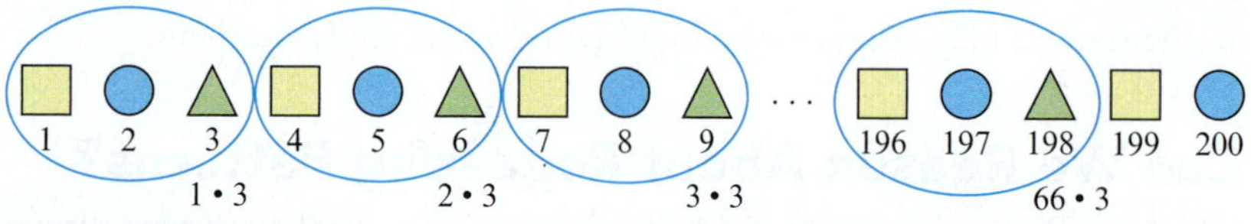

Figure 9.36 Sequence repeating in a pattern of 3.

the unit repeats 66 times among the first 200 shapes in the sequence. The 66th repeat ends on a triangle. The remainder 2 tells us there are 2 more shapes to get to the 200th shape, which must therefore be the second shape in the unit, namely a circle.

CLASS ACTIVITY

9N Solving Problems Using Repeating Patterns, p. CA-177

Next, we'll study some common kinds of growing patterns.

How Can We Reason About Arithmetic Sequences?

Look at the growing sequence of "snap cube trains" in Figure 9.37. The first train is made from a 4-cube engine and one 5-cube train car. Each subsequent train is made by adding on one more 5-cube train car.

Figure 9.37
A sequence of snap cube "trains."

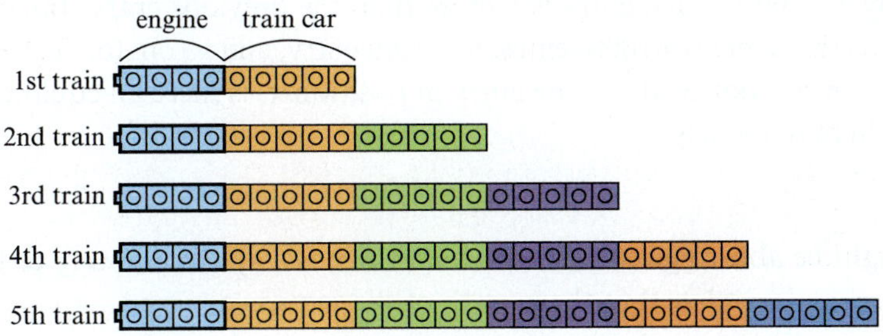

The numbers of cubes in these snap cube trains produce this sequence:

$$9, 14, 19, 24, 29, \ldots$$

which is an example of an arithmetic sequence. Arithmetic sequences are the simplest kind of numerical growing sequence.

arithmetic sequence

We create an **arithmetic sequence** by starting with any number in the first position. The entry in each subsequent position is obtained from the entry in the previous position by adding or subtracting the same fixed number, as shown in Figure 9.38.

Figure 9.38
Arithmetic sequences are produced by repeatedly adding or subtracting the same amount.

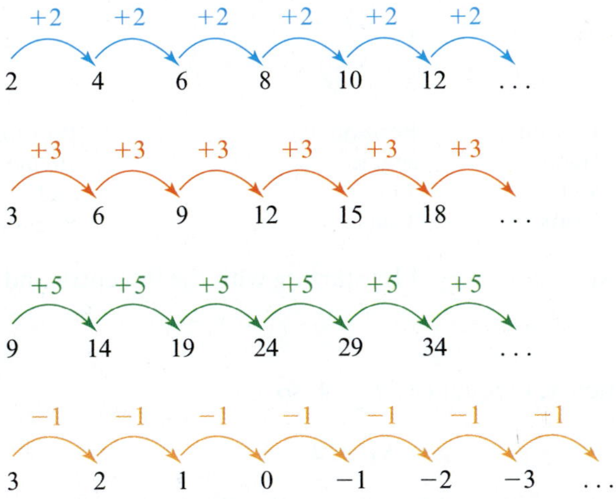

Try Class Activities 9O, 9P, and 9Q to see how arithmetic sequences provide opportunities to develop expressions and equations for entries.

CLASS ACTIVITY

9O Arithmetic Sequences of Numbers Corresponding to Sequences of Figures, p. CA-178

9P How Are Expressions for Arithmetic Sequences Related to the Way Sequences Start and Grow?, p. CA-180

9Q Explaining Equations for Arithmetic Sequences, p. CA-181

Every arithmetic sequence is described by a simple equation. This equation is related to the sequence in a special way. Let's consider the example of the arithmetic sequence

$$9, 14, 19, 24, 29, \ldots$$

of the total number of snap cubes in the snap cube trains of Figure 9.37. This sequence starts with 9 and increases by 5. That is, each entry is 5 more than the previous entry. Imagine this sequence going on forever, so that there is a 100th entry, a 1000th entry, and so on; for each counting number, x, there is an entry in position x. If y is the entry in position x, is there an equation expressing y in terms of x? Yes, there is, namely,

$$y = 5x + 4$$

CCSS 6.EE.6

Although you might be able to guess this equation, there is a systematic way to find it, to explain why it makes sense, and to relate it to the growth of the sequence.

To explain why the equation $y = 5x + 4$ for the foregoing sequence makes sense, think about how to obtain each entry. Notice that the first entry, 9, was obtained by adding one 5 to 4, because a single train car made of 5 snap cubes was added to an engine made of 4 snap cubes. The second entry, 14, was obtained by adding two 5s to 4 because 2 train cars were added to an engine. The third entry, 19, was obtained by adding three 5s to 4. In general, the entry in position x is obtained by adding x 5s to 4. Therefore, the entry in position x is $5x + 4$. So, it's like starting with the 0th entry (just the engine, no train cars), which is 4, and adding x 5s to get the entry in position x, as indicated in Figure 9.39.

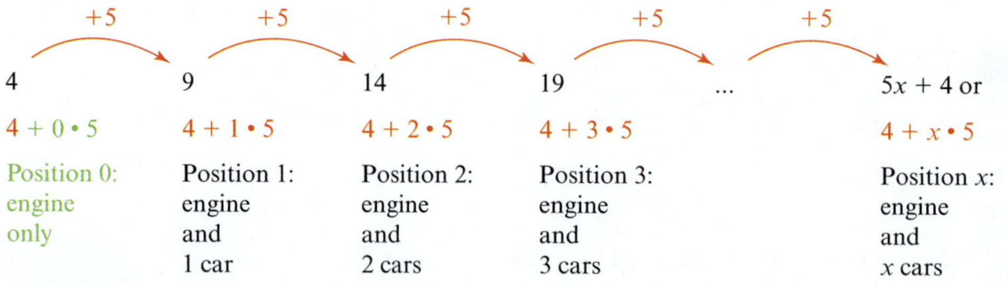

Figure 9.39 We obtain the expression $5x + 4$ by starting with the 0th entry and repeatedly adding 5.

Therefore y, the entry in position x, is equal to $5x + 4$, so

$$y = 5x + 4$$

Figure 9.40 represents the sequence with a table and a graph. Observe how the amount by which the sequence increases appears in the table: whenever the position increases by 1, the entry increases by 5. On the graph, we go to the right 1 unit and up 5 units as we go from point to point in the sequence. Notice that the entry in position 0 corresponds to the point $(0, 4)$, which lies on the y-axis.

Figure 9.40

A table and graph for the arithmetic sequence 9, 14, 19, 24, 29, . . .

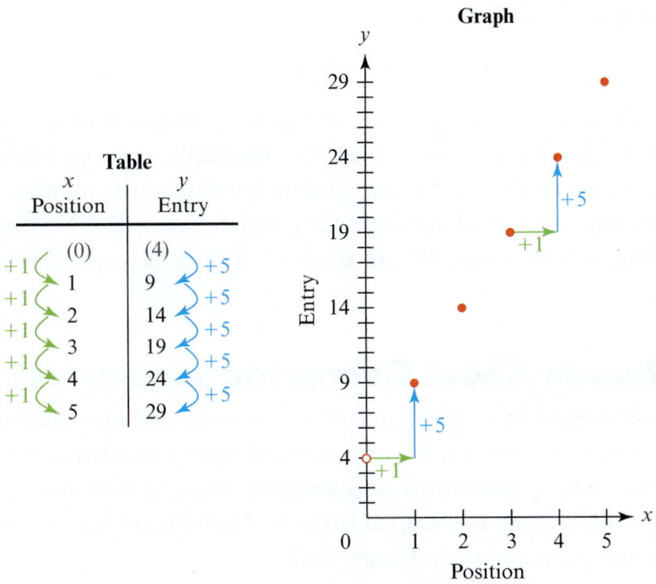

rate of change (sequences)

In general, given an arithmetic sequence, the fixed amount by which the sequence increases (or decreases) from one entry to the next is the **rate of change** of the sequence. For example, the rate of change of the arithmetic sequence

$$4, 7, 10, 13, 16, \ldots$$

is 3. If we let y stand for the entry in position x of an arithmetic sequence, then the equation

$$y = (\text{rate of change})x + (0\text{th entry})$$

CCSS 8.F.4

relates x and y. **Figure 9.41** indicates how we can explain this equation using a table and a graph for the arithmetic sequence

$$4, 7, 10, 13, 16, \ldots$$

We get the 0th entry by subtracting the rate of change, 3, from the first entry, 4. Starting from the 0th entry, to get to the entry in position x, we go x units to the right on the graph, and therefore we go up x groups of 3 units. This gives us the equation

$$y = 1 + x \cdot 3$$

Figure 9.41

Developing the equation for the arithmetic sequence 4, 7, 10, ...

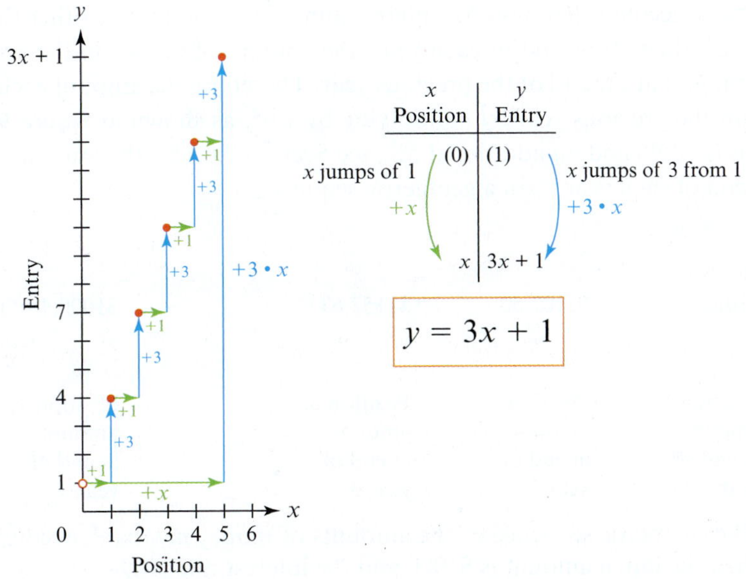

which we more commonly write in the form

$$y = 3x + 1$$

Notice that the graphs of the arithmetic sequences in Figures 9.40 and 9.41 are isolated points, but lie on straight lines, as will all arithmetic sequences. On the graph, as we go from point to point in an arithmetic sequence, we go to the right 1 unit and up (or down) the number of units given by the rate of change. This consistency is what makes the graph fall on a line. Because the graph of an arithmetic sequence always lies on a line, we can think of arithmetic sequences as special kinds of **linear relationships**.

linear relationships

How Can We Reason About Geometric Sequences?

We just saw that arithmetic sequences are obtained by repeatedly adding (or subtracting) the same amount. Geometric sequences are similar, but are obtained by repeatedly *multiplying* (or dividing) by the same amount. We create a **geometric sequence** by starting with any nonzero number as the first entry. Each subsequent entry is obtained from the previous entry by multiplying (or dividing) by the same fixed number, as shown in Figure 9.42.

geometric sequence

Figure 9.42 Geometric sequences are produced by repeatedly multiplying or dividing by the same amount.

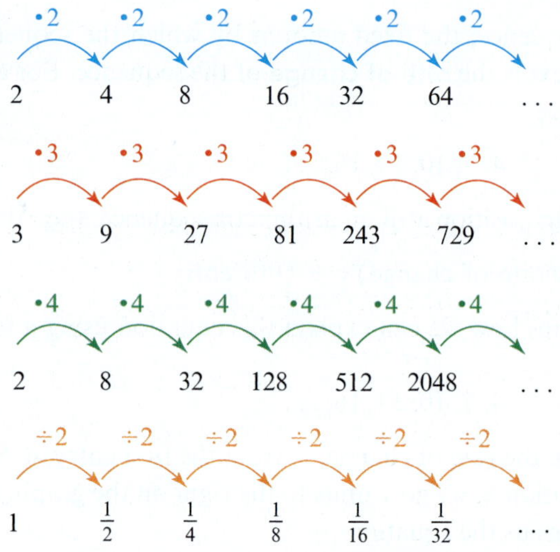

Geometric sequences have important practical applications. For example, suppose that \$1000 is deposited in a bank account that pays 5% interest annually. If no money other than the interest is added or removed, then at the end of each year, the amount of money in the account will be 5% more than the amount at the end of the previous year. Therefore, the amount each year is obtained from the amount the previous year by multiplying by 1.05, as shown in Figure 9.43. (The 1.05 is 1 + 0.05, which is 100% and an additional 5%; see Section 7.6.) So the amounts of money in the account at the end of each year form a geometric sequence.

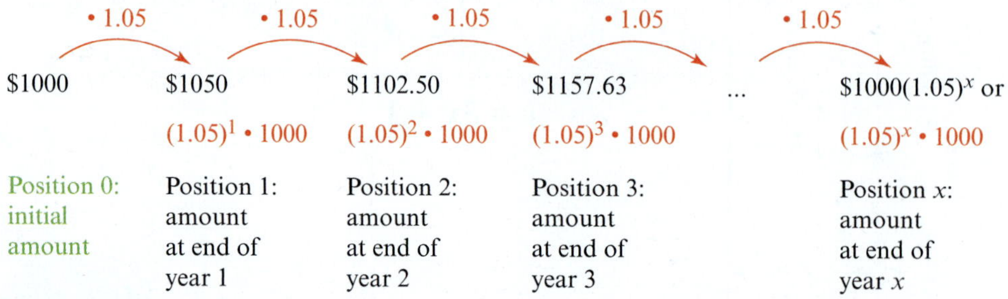

Figure 9.43 The geometric sequence of the amounts of money in a bank account after x years when the initial amount is \$1000 and the interest rate is 5%.

As with arithmetic sequences, geometric sequences can also be described by equation. Consider the sequence

$$1050,\ 1102.50,\ 1157.63,\ 1215.51,\ \ldots$$

which is the dollar amount in the bank account described earlier after 1 year, 2 years, 3 years, and so on. If y is the number of dollars in the account after x years, what is an equation relating y and x? To find this equation, think about how to obtain each entry in the sequence by starting with the initial \$1000 in the account. After 1 year, there is

$$1.05 \cdot 1000$$

dollars in the account. After another year, this amount is multiplied by 1.05, so after 2 years there is

$$1.05 \cdot 1.05 \cdot 1000 = (1.05)^2 \cdot 1000$$

dollars in the account. After another year, this amount is multiplied by 1.05 again, so after 3 years there is

$$1.05 \cdot 1.05 \cdot 1.05 \cdot 1000 = (1.05)^3 \cdot 1000$$

dollars in the account. Continuing in this way, after x years there is

$$\underbrace{1.05 \cdot 1.05 \cdot \cdots \cdot 1.05}_{x \text{ times}} \cdot 1000 = (1.05)^x \cdot 1000$$

dollars in the account.

The reasoning to find the entry in position x of any other geometric sequence is similar. Let's call the factor by which each entry in a geometric sequence is multiplied in order to obtain the next entry the "ratio" of the geometric sequence. Consider a "0th entry" that you obtain from the first entry by dividing by the ratio of the geometric sequence. Then the entry in position x will be obtained by multiplying the 0th entry by the ratio x times. Therefore, the entry in position x is

$$\underbrace{(\text{ratio}) \cdot (\text{ratio}) \cdot \cdots \cdot (\text{ratio})}_{x \text{ times}} \cdot (\text{0th entry}) \quad \text{or} \quad (\text{ratio})^x \cdot (\text{0th entry})$$

Therefore, if we let y be the entry in position x, then

$$y = (\text{ratio})^x \cdot (\text{0th entry})$$

Just as arithmetic sequences are special kinds of linear functions, geometric sequences are special kinds of exponential functions. Exponential functions are studied in high school and beyond.

We have studied repeating patterns, arithmetic sequences, and geometric sequences, but there are also many other types of sequences.

Is a Sequence Determined by Its First Few Entries?

A common mathematics problem for children in elementary school is to determine the next several entries in a sequence when the first few entries are given. For example, what are the next three entries in the sequence

$$3, 5, 7, \ldots ?$$

Be aware that if a rule or expression for the sequence has not been given, and if the sequence has not been specified as of a certain type (such as arithmetic, geometric, or repeating), there are always several different ways to continue the sequence. For example, we could continue the previous sequence like this:

$$3, 5, 7, 9, 11, 13, \ldots$$

viewing the sequence as an arithmetic sequence that increases by 2. Or we could continue the sequence like this:

$$3, 5, 7, 3, 5, 7, \ldots$$

and we could say that a rule for the sequence is to repeat the numbers 3, 5, 7 in order. Yet another possibility is to continue the sequence like this:

$$3, 5, 7, 11, 13, 17, \ldots$$

and we could say that the sequence consists of the odd prime numbers in order. Technically, the next three entries could be *any* three numbers, although the most interesting sequences are ones that have definite rules and patterns.

SECTION SUMMARY AND STUDY ITEMS

Section 9.5 Sequences

A sequence is a list of items (usually numbers or shapes) occurring in a specified order. Repeating sequences offer opportunities to highlight a unit and to reason about division with remainder. Arithmetic sequences are produced by repeatedly adding (or subtracting) the same fixed number. Geometric sequences are produced by repeatedly multiplying (or dividing) by the same fixed number. Arithmetic sequences and geometric sequences both have equations expressing an entry, y, in terms of its position, x. Repeating sequences are frequently studied in elementary school.

Key Skills and Understandings

1. Solve problems about repeating sequences by attending to the repeating unit and reasoning about division and multiplication.
2. Solve problems about growing sequences using algebra and by reasoning about division.
3. Given an arithmetic sequence, find an equation relating entries to their positions, and explain why this equation is valid in terms of the way the sequence grows.
4. Given a geometric sequence, find an equation relating entries to their positions, and explain why this equation is valid in terms of the way the sequence grows.
5. Recognize that specifying a few terms of a sequence doesn't specify the sequence.

Practice Exercises for Section 9.5

1. Assume that the repeating pattern of a circle, two squares, and a triangle in **Figure 9.44** continues. What shape will be above the number 3333? How can you tell?

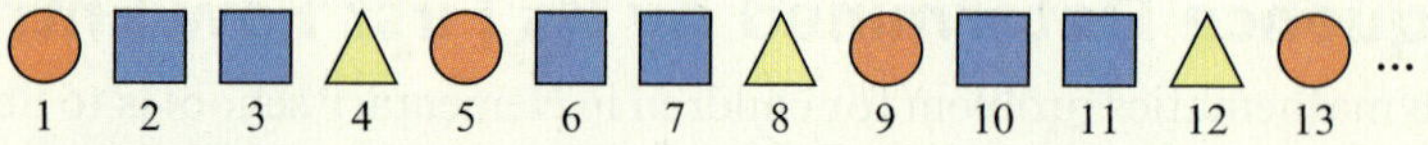

Figure 9.44 A repeating pattern of shapes.

2. **Figure 9.45** shows a sequence of figures made of small squares. Assume that the sequence continues by adding 2 darkly shaded squares to the top of a figure in order to get the next figure in the sequence.

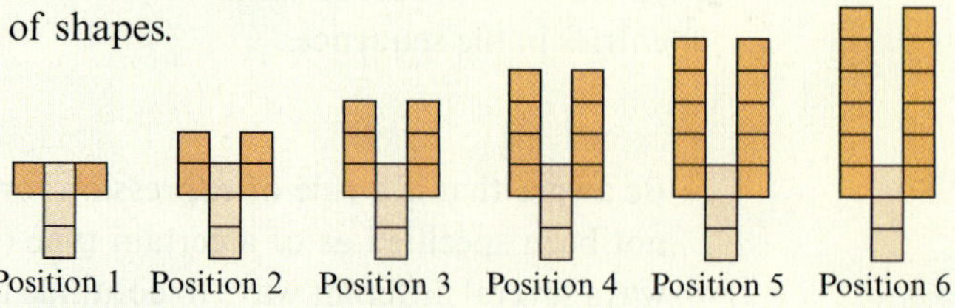

Figure 9.45 A sequence of figures.

a. Write the first 6 entries in the sequence whose entries are the number of small squares making up the figures.

b. Find an expression for the number of small squares making up the Nth figure in the sequence. Explain why your expression makes sense by relating it to the structure of the figures.

c. Will there be a figure in the sequence that is made of 100 small squares? If yes, which one? If no, why not? Determine the answer to these questions in two ways: with algebra and in a way that a student in elementary school might be able to.

d. Will there be a figure in the sequence that is made of 199 small squares? If yes, which one; if no, why not? Determine the answer to these questions in two ways: with algebra and in a way that a student in elementary school might be able to.

3. Consider the arithmetic sequence whose first few entries are

$$1, 4, 7, 10, 13, 16, \ldots$$

Find an expression for the Nth entry in this sequence, and explain in detail why your expression is valid.

4. Consider this arithmetic sequence:

$$2, 5, 8, 11, \ldots$$

a. Let y be the entry in position x of this sequence. Explain how to derive an equation relating x and y. Sketch the graph of the sequence.

b. Discuss how the components of the equation in part (a) are related to the arithmetic sequence and to the graph in part (a).

5. Draw the first 5 figures in a sequence of figures whose entry in position N is made of $3N + 2$ small circles. Describe how subsequent figures in your sequence would be formed.

6. Consider the geometric sequence whose first few entries are

$$6, 12, 24, 48, 96, 192, \ldots$$

Find an equation relating an entry y to its position x and explain in detail why your equation is valid.

7. Describe four different rules for determining the next entries in the following sequence. Give the next three entries in the sequence for each rule.

$$3, 7, 15, \ldots$$

Answers to Practice Exercises for Section 9.5

1. The pattern repeats a unit of 4 shapes, so there is a triangle above 4, 8, 12, 16, and above all the other multiples of 4. Dividing 3333 by 4, we get 833, with remainder 1. This means there are 833 sets of 4 in 3333, and 1 remaining. The shape above $833 \cdot 4$ is a triangle, and 1 shape after that is a circle. So the shape above 3333 is a circle.

2. a. The number of squares in the figures are

$$5, 7, 9, 11, 13, 15, \ldots$$

b. The Nth figure in the sequence is made of $2N + 3$ small squares. We can explain why this expression is valid by relating it to the structure of the figures. Each figure is made of 3 small, lightly shaded squares and 2 "prongs" of darkly shaded squares. In the Nth figure, each prong is made of N squares. So in all, there are $2N + 3$ small squares in the Nth figure.

c. A student in elementary school might realize that the number of squares making up the figures is always odd. Since 100 is even, it could not be the number of small squares that a figure in the sequence is made of. Or the student might try to find how many squares are in each prong if there were a figure made of 100 squares. There would have to be $100 - 3 = 97$ squares in the two prongs, so each prong would have to have $97 \div 2$ squares. Since 97 is not evenly divisible by 2, there can't be such a figure.

To answer the questions with algebra, notice that we want to know if there is a counting number, N, such that

$$2N + 3 = 100$$

Subtracting 3 from both sides, we have the new equation

$$2N = 97$$

which has the same solution as the original equation. Dividing both sides by 2, we have

$$N = \frac{97}{2}$$

But $\frac{97}{2}$ is not a counting number, so there is no such figure.

d. A student in elementary school might try to find the number of squares in each prong. There must be $199 - 3 = 196$ squares in the two prongs combined. Therefore, each prong must have $196 \div 2 = 98$ squares. So the 98th figure in the sequence will be made of 199 small squares.

To answer the questions with algebra, notice that we want to solve the equation

$$2N + 3 = 199$$

Subtracting 3 from both sides, we have the new equation

$$2N = 196$$

which has the same solutions as the original equation. Dividing both sides by 2, we have

$$N = 98$$

So the 98th figure is made of 199 small squares.

3. The entries in the sequence increase by 3 each time. If we imagine an entry in position 0 preceding the first entry, this entry would have to be $1 - 3 = -2$. Starting at the entry in position 0, the 1st entry is obtained by adding one 3, the second entry is obtained by adding two 3s, the third entry is obtained by adding three 3s, and so on, as indicated in Figure 9.46. The entry in position N is obtained by adding N 3s to -2, so that entry is

$$3N + (-2)$$

which we can also write as

$$3N - 2$$

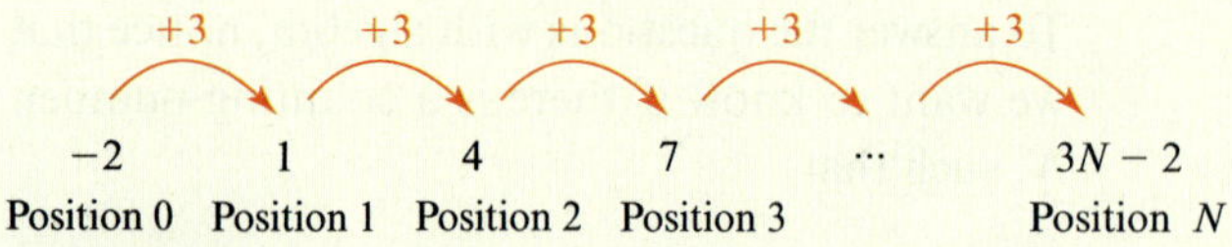

Figure 9.46 We obtain the expression $3N - 2$ by starting with the entry in position 0 and repeatedly adding 3.

4. a. To find an equation, use the reasoning from the text. Because the sequence starts at 2 and increases by 3, the 0th entry of the sequence would be $2 - 3 = -1$. The entry in position 0 is "x jumps of 3 from -1," which is $-1 + 3x$. Therefore, x and y are related by the equation $y = -1 + 3x$. The graph (not shown) consists of isolated points on a line.

b. The 0th entry of the sequence is the y-value where the line that the graph lies on hits the y-axis and is also the term -1 in the equation. The rate of change of the sequence 3, tells us that on the graph, whenever we go to the right 1 unit, we go up 3 units. The rate of change is also the coefficient of x in the equation for the sequence.

5. See Figure 9.47. Of course, many other sequences of figures are also possible. Each figure is formed by adding one circle to each of the three "prongs" of the previous figure.

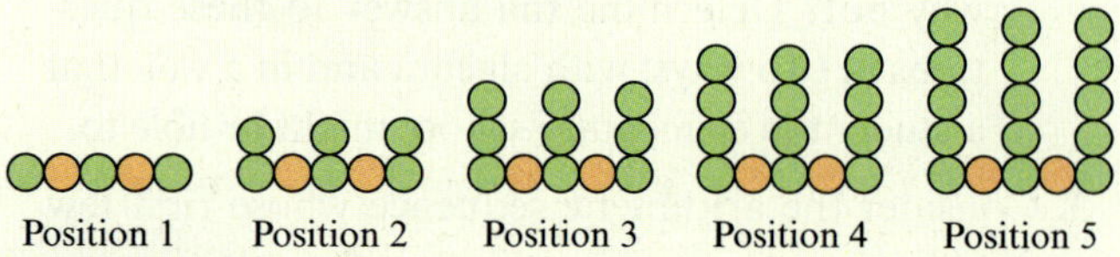

Figure 9.47 A sequence of figures whose entry in position N is made of $3N + 2$ small circles.

6. We multiply each entry in the sequence by 2 to obtain the next entry. If we imagine an entry in position 0 preceding the 1st entry, this entry would have to be 6 divided by 2, namely, 3. Starting at that entry, the 1st entry is obtained by multiplying by 2 once, the second entry is obtained by multiplying by 2 twice, the third entry is obtained by multiplying by 2 three times, and so on, as indicated in Figure 9.48. The entry in position x is obtained by multiplying x 2s with 3, so the entry y in position x is

$$y = 3 \cdot 2^x$$

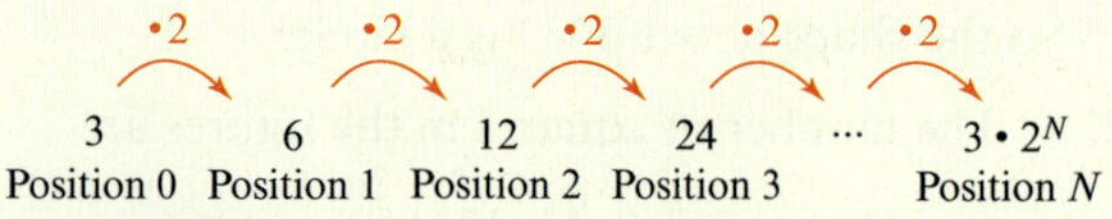

Figure 9.48 We obtain the equation $y = 3 \cdot 2^x$ by starting with the entry in position 0 and repeatedly multiplying by 2.

7. One rule is simply to continue to repeat the sequence 3, 7, 15 in order. In this case, the first 6 entries in the sequence are

3, 7, 15, 3, 7, 15, . . .

Another rule is that each entry is obtained from the previous entry by doubling and adding 1. In this case, the first 6 entries are

3, 7, 15, 31, 63, 127, . . .

A third rule is that the next entries are obtained by adding 4, then adding 8, then adding 16, then adding 32, and so on, so that each time, twice as much is added as the time before. In this case, the first 6 entries are

$$3, 7, 15, 31, 63, 127, \ldots$$

which are the same entries as in the previous rule. A fourth rule is that the next entries are obtained by adding 4, then adding 8, then adding 12, then adding 16, and so on, so that each time, 4 more are added than the time before. In this case, the first 6 entries are

$$3, 7, 15, 27, 43, 63, \ldots$$

PROBLEMS FOR SECTION 9.5

1. Assume that the pattern of a square followed by 2 circles and a triangle continues to repeat in the sequence of shapes in **Figure 9.49** and that the numbers below the shapes indicate the position of each shape in the sequence.

a. What shape will be above the number 150? Explain how you can tell.

b. How many circles will there be above the numbers 1 through 150? Explain your answer.

c. Consider the numbers that are below the squares. What kind of sequence do these numbers form: an arithmetic sequence, a geometric sequence, or neither? Why? If it is either an arithmetic sequence or a geometric sequence, give an expression for the Nth entry.

d. Consider the numbers that are below the circles. What kind of sequence do these numbers form: an arithmetic sequence, a geometric sequence, or neither? Why? If it is either an arithmetic sequence or a geometric sequence, give an expression for the Nth entry.

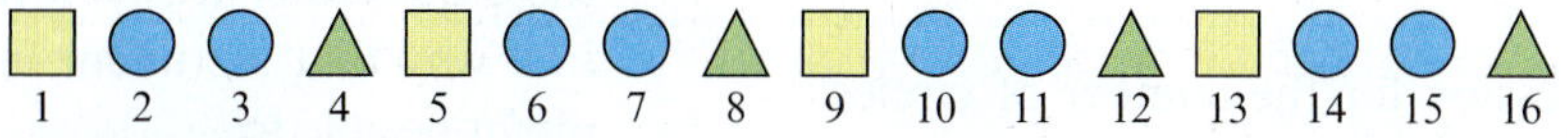

Figure 9.49 A repeating pattern of shapes.

2. Assume that the pattern of a circle followed by 3 squares and a triangle continues to repeat in the sequence of shapes in **Figure 9.50** and that the numbers below the shapes indicate the position of each shape in the sequence.

a. What shape will be above the number 999? Explain how you can tell.

b. How many squares will there be above the numbers 1 through 999? Explain your answer.

c. Consider the numbers that are below the circles. What kind of sequence do these numbers form: an arithmetic sequence, a geometric sequence, or neither? Why? If it is either an arithmetic sequence or a geometric sequence, give an expression for the Nth entry.

d. Consider the numbers that are below the squares. What kind of sequence do these numbers form: an arithmetic sequence, a geometric sequence, or neither? Why? If it is either an arithmetic sequence or a geometric sequence, give an expression for the Nth entry.

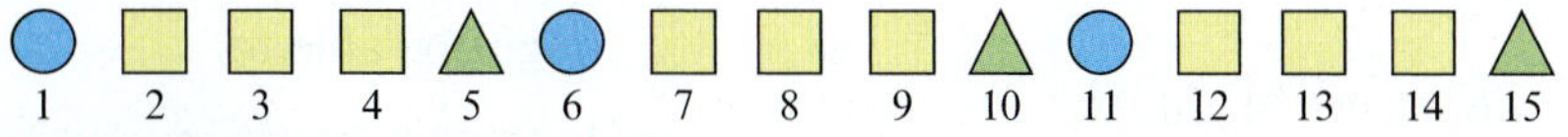

Figure 9.50 A repeating pattern of shapes.

3. Assume that the repeating pattern of 4 squares followed by a circle and a triangle shown in **Figure 9.51** continues to the right. Discuss whether the following reasoning is valid:

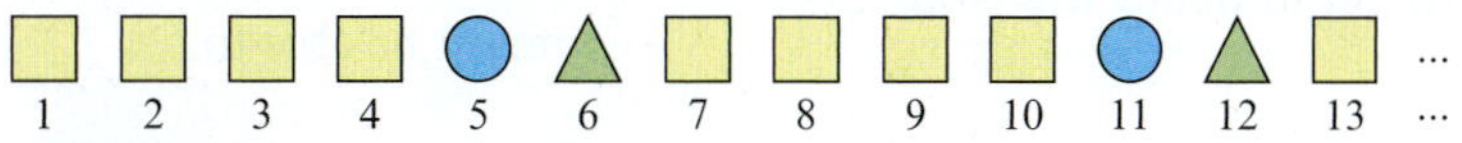

Figure 9.51 A repeating pattern of shapes.

Since there are 8 squares above the numbers 1–10, there will be 24 times as many squares above the numbers 1–240. So there will be $24 \times 8 = 192$ squares above 1–240.

If the reasoning is not valid, explain why not and find a different way to determine the number of squares above the numbers 1–240.

4. What day of the week will it be 1000 days from today? Use math to solve this problem. Explain your answer.

5. **Figure 9.52** shows a sequence of figures made of small circles. Assume that the sequence continues by the addition of one circle to each of the 5 "arms" of a figure in order to get the next figure in the sequence.

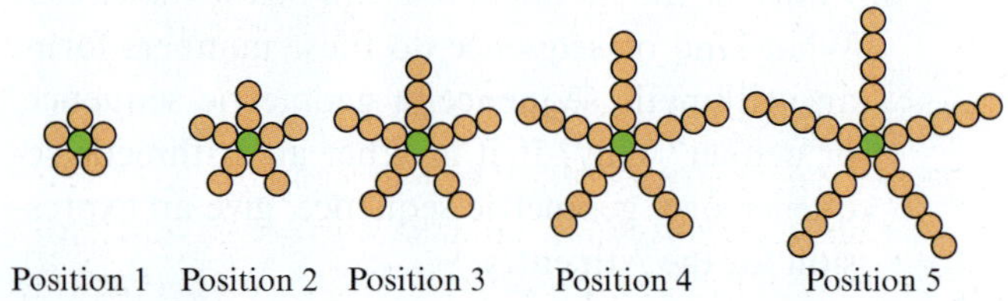

Figure 9.52 A sequence of figures.

a. Find an expression for the number of circles making up the Nth figure in the sequence. Explain why your expression makes sense by relating it to the structure of the figures.

b. Will there be a figure in the sequence that is made of 100 circles? If yes, which one? If no, why not? Determine the answer to these questions in two ways: with algebra and in a way that a student in elementary school might be able to.

c. Will there be a figure in the sequence that is made of 206 circles? If yes, which one? If no, why not? Determine the answer to these questions in two ways: with algebra and in a way that a student in elementary school might be able to.

6. Consider the arithmetic sequence whose first few entries are

$$6, 11, 16, 21, 26, 31, \ldots$$

a. Determine the 100th entry in the sequence, and explain why your answer is correct.

b. Find an expression for the Nth entry in this sequence, and explain in detail why your expression is valid.

c. Is 1000 an entry in the sequence? If yes, which entry? If no, why not? Determine the answer to these questions in two ways: with algebra and in a way that a student in elementary school might be able to.

d. Is 201 an entry in the sequence? If yes, which entry? If no, why not? Determine the answer to these questions in two ways: with algebra and in a way that a student in elementary school might be able to.

7. Consider the arithmetic sequence whose first few entries are

$$3, 7, 11, 15, 19, 23, \ldots$$

a. Determine the 50th entry in the sequence, and explain why your answer is correct.

b. Find an expression for the Nth entry in this sequence, and explain in detail why your expression is valid.

c. Is 207 an entry in the sequence? If yes, which entry? If no, why not? Determine the answer to these questions in two ways: with algebra and in a way that a student in elementary school might be able to.

d. Is 100 an entry in the sequence? If yes, which entry? If no, why not? Determine the answer to these questions in two ways: with algebra and in a way that a student in elementary school might be able to.

8. a. Draw the first 5 figures in a sequence of figures whose Nth entry is made of $6N + 3$ small circles. Describe how subsequent figures in your sequence would be formed.

b. Is there a figure in your sequence in part (a) that is made of 102 small circles? If yes, which one? If no, why not? Determine the answer to these questions in two ways: with algebra and in a way that a student in elementary school might be able to.

c. Is there a figure in your sequence in part (a) that is made of 333 small circles? If yes, which one? If no, why not? Determine the answer to these questions in two ways: with algebra and in a way that a student in elementary school might be able to.

9. a. Consider the arithmetic sequence

$$5, 7, 9, 11, 13, \ldots$$

Let y be the entry in position x. Explain in detail how to reason about the way the sequence grows to derive an equation of the form

$$y = m \cdot x + b$$

where m and b are specific numbers related to the sequence.

b. Sketch a graph for the arithmetic sequence in part (a). Discuss how features of the graph are related to the components of your equation in part (a) and to the way the sequence grows.

10. Consider an arithmetic sequence whose third entry is 10 and whose fifth entry is 16. Use the most elementary reasoning you can to find the first and second entries of the sequence. Explain your reasoning.

11. Consider an arithmetic sequence whose fourth entry is 2 and whose seventh entry is 4. Use the most elementary reasoning you can to find the first and second entries of the sequence. Explain your reasoning.

12. Consider an arithmetic sequence whose fifth entry is 1 and whose tenth entry is 7. Use the most elementary reasoning you can to find the first and second entries of the sequence. Explain your reasoning.

13. Consider the geometric sequence whose first few entries are

$$2, 10, 50, 250, 1250, 6250, \ldots$$

Let y be the entry in position x. Find an equation expressing y in terms of x, and explain in detail why your equation is valid.

14. Consider the geometric sequence whose first few entries are

$$\frac{1}{2}, \frac{1}{4}, \frac{1}{8}, \frac{1}{16}, \frac{1}{32}, \frac{1}{64}, \ldots$$

Let y be the entry in position x. Find an equation expressing y in terms of x, and explain in detail why your equation is valid.

15. Consider the geometric sequence whose first few entries are

$$\frac{1}{6}, \frac{1}{12}, \frac{1}{24}, \frac{1}{48}, \frac{1}{96}, \frac{1}{192}, \ldots$$

Let y be the entry in position x. Find an equation expressing y in terms of x, and explain in detail why your equation is valid.

16. Describe four different rules for determining the next entries in the sequence that follows. Give the next 3 entries in the sequence for each rule.

$$5, 7, 11, \ldots$$

17. Describe four different rules for determining the next entries in the sequence that follows. Give the next 3 entries in the sequence for each rule.

$$1, 2, 4, \ldots$$

18. Describe four different rules for determining the next entries in the sequence that follows. Give the next 3 entries in the sequence for each rule.

$$1, 4, 9, \ldots$$

19. Suppose you owe \$500 on a credit card that charges you 1.6% interest on the amount you owe at the end of every month. Assume that you do not pay off any of your debt and that you do not add any more debt other than the interest you are charged.

a. Explain why you will owe

$$(1.016)^N \cdot 500$$

dollars at the end of N months.

b. Use a calculator to determine how much you will owe after 2 years.

20. Suppose you put \$1000 in an account whose value will increase by 6% every year. Assume that you do not take any money out of this account and that you do not put any money into this account other than the interest it earns. Find an expression for the amount of money that will be in the account after N years. Explain in detail why your expression is valid.

***21.** The Widget Company sells boxes of widgets by mail order. The company charges a fixed amount for shipping, no matter how many boxes of widgets are ordered. All boxes of widgets are the same size and cost the same amount. The company is out of state, so there are no taxes charged. You find out that the total cost (including shipping) for 6 boxes of widgets is \$27, and the total cost (including shipping) for 10 boxes of widgets is \$41. Find the cost of shipping and the price of one box of widgets. Also, find an expression for the total cost (including shipping) of N boxes of widgets. Explain your reasoning.

***22.** Suppose a scientist puts a colony of bacteria weighing 1 gram in a large container. Assume that these bacteria reproduce in such a way that their number doubles every 20 minutes.

a. How much will the colony of bacteria weigh after 1 hour? after 8 hours? after 1 day? after 1 week? In each case, explain your reasoning.

b. Is it plausible that bacteria in a container could double every 20 minutes for more than a few hours? Why or why not?

***23. a.** Use a calculator to compute 2^{161}. Are you able to tell what the ones digit of this number is from the calculator's display? Why or why not?

b. Determine the ones digit of 2^{161}. Explain your reasoning clearly.

9.6 Functions

CCSS Common Core State Standards Grade 8

In this section, we extend our study of proportional relationships in Section 7.4, expressions with variables in Section 9.2, and sequences in Section 9.5 to the study of mathematical functions. We will define the concept of function and consider the four main ways that functions can be represented: with words, with tables, with graphs, and with expressions or equations. Functions are one of the cornerstones of mathematics and science. Functions allow us to describe and study how one quantity determines another. By studying and using functions, mathematicians and scientists have been able to describe and predict a variety of phenomena, such as motions of planets, the distance a rocket will travel, or the dose of medicine that a patient needs. The study of functions goes far beyond what we discuss in this section. But even at the most elementary level, functions provide a way to organize, represent, and study information.

function

A **function** is a rule that assigns one output to each allowable input. Often, these "inputs" and "outputs" are numbers (although they don't have to be). The set of allowable inputs is called the **domain** of the function. The set of allowable outputs is called the **range** of the function.

CCSS 8.F.1

Functions are sometimes portrayed as machines, as in Figure 9.53. With this imagery, it's easy to imagine putting an input into a machine; the machine somehow transforms the input into an output. Figure 9.53 indicates how the *doubling function* works: to each input number, the assigned output is twice the input. Figure 9.54 indicates how to view a sequence as a function. In general, we can consider every sequence a function by assigning to each counting number, N, the output that is in position N in the sequence.

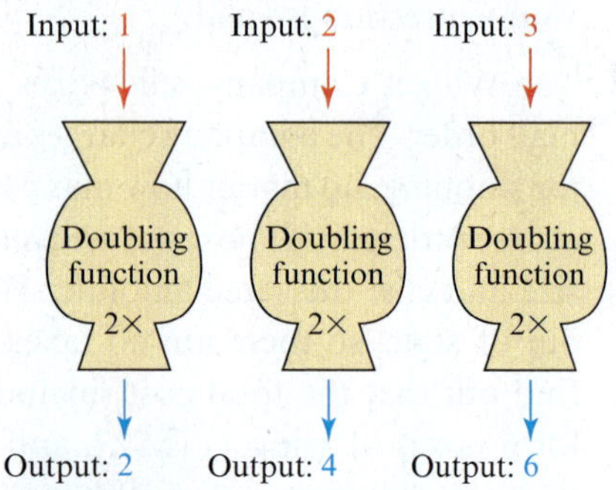

Figure 9.53 Viewing functions as machines.

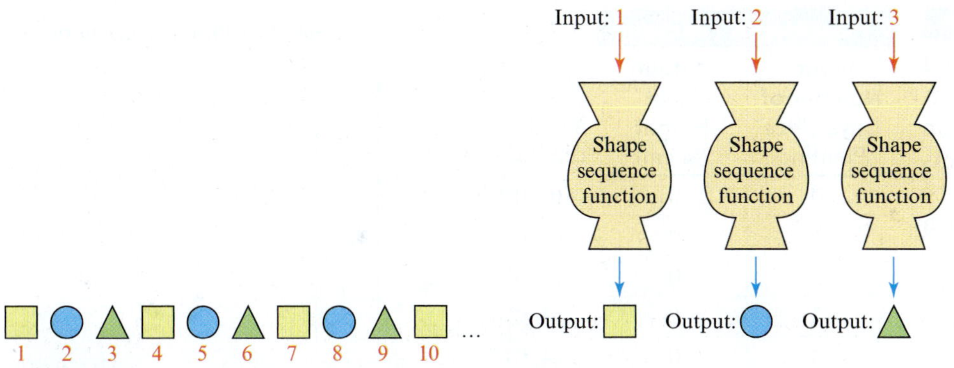

Figure 9.54 A sequence can be viewed as a function.

How Can We Describe Functions with Words?

We use functions to describe how two quantities are related; one way to describe such a relationship is with words. A function tells us how an input determines an output—in other words, how the output depends on the input. To describe a function in words, we say what the allowable inputs are and we give the rule that states what output is associated with each allowable input. Even if our goal is to describe a function with an equation, we can clarify our thinking about how the inputs and outputs are related by first describing the function in words.

Although we often seek to describe a function with an expression or an equation, some functions can't be described that way. For example, there isn't a formula for the function that assigns to each person in the world, their height in centimeters. Or, consider a *blue whale function*, which assigns to a year since 1950 (say), the number of blue whales alive at the end of that year. There is unlikely to be a formula that describes this function exactly. However, biologists might seek a mathematical formula that closely approximates this function to help them make predictions about the blue whale population. A process of mathematical modeling, such as modeling the blue whale population with a function, starts by describing a function in words.

How Do Tables and Graphs Represent Functions?

We have already used tables and graphs to represent proportional relationships in Section 7.4 and sequences in Section 9.5. Tables and graphs for functions follow the same idea. A table helps us record a function's inputs and outputs. A graph displays a function visually, which can help us discern trends and patterns.

To represent a function in a table, we put inputs in the left column and their corresponding outputs in the right column. If a function has inputs and outputs that are numbers, then we can represent it in a graph. The **graph of a function** consists of all those points in a coordinate plane whose first coordinate is an allowable input and whose second coordinate is the output of the first coordinate.

graph of a function

For example, suppose that a class plants a seed, which sprouts and grows into a plant. This situation gives rise to a Plant Height Function, whose inputs are numbers of days since planting and whose outputs are the plant's height. Figure 9.55 shows a table and a graph for this (hypothetical) function.

Notice that the graph is a curve that connects the points obtained from the table. Why does it make sense to connect the points? In between the times that the plant's height was measured, the plant has a height, and that yields a point on the graph. All those points together make a curve that connects the points from the table.

CCSS 8.F.5

Graphs can help us see the story behind a function. For example, we can see from the graph of the Plant Height Function in Figure 9.55 that the plant didn't sprout until after 4 days. From day 7 to

Figure 9.55 A table for a Plant Height Function yields points that can be plotted in a coordinate plane to make a graph of the function.

Plant Height Function		
Input: Number of Days Since Planting	Output: Plant Height (in cm)	Yields the Point
0	0	(0, 0)
1	0	(1, 0)
2	0	(2, 0)
3	0	(3, 0)
4	0	(4, 0)
5	0.5	(5, 0.5)
6	1.5	(6, 1.5)
7	3	(7, 3)
8	5	(8, 5)
9	5.5	(9, 5.5)
10	6	(10, 6)
11	6	(11, 6)
12	6	(12, 6)

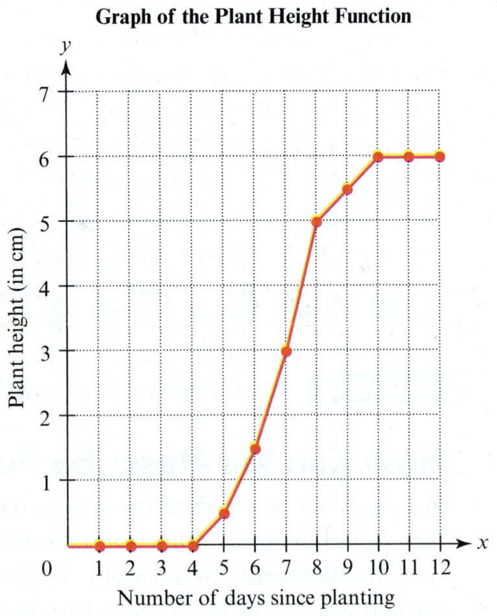

day 8, the graph is steepest. During that time, the plant grew more than in any other 1-day period. After 10 days, the plant stopped growing taller—it reached its maximum height.

In Class Activities 9R and 9S you will see how graphs tell stories and how different graph shapes are associated with different patterns of change.

CLASS ACTIVITY

9R What Does the Shape of a Graph Tell Us About a Function?, p. CA-183

9S Graphs and Stories, p. CA-185

Tables and graphs are helpful, but they usually represent a function only partially because they often do not show all the inputs and outputs. In contrast, expressions and equations often represent a function completely.

How Do Expressions and Equations Represent Functions?

Expressions and equations help us describe functions succinctly. When we use an expression or an equation to describe a function, we rely on some conventions. We also often use the terminology of independent and dependent variables.

When the inputs and outputs of a function are numbers, we can use variables to stand for them. Traditionally, we use x as a variable for the inputs of a function and y as a variable for the outputs, although different variables can be used. We often say that the output variable is a function of the input variable, e.g., "y is a function of x."

independent variable
dependent variable

A variable that is used for the inputs of a function is often called an **independent variable** and a variable that is used for the outputs is then called a **dependent variable**. We can think of the independent variable as ranging freely over the allowable inputs of the function; the value of the dependent variable (the output) depends on the value of the independent variable (the input), hence the names.

CCSS
6.EE.9

In situations in which two quantities vary together, there is often a natural choice for the independent and dependent variables. For example, for an object in motion, we may wish to describe the relationship between the distance traveled and elapsed time. In this case it is natural to think of the distance as depending on the time, so we usually choose time to be the independent variable and distance to be the dependent variable. However, in many cases we are free to choose which variable to take as the independent variable and which to take as the dependent variable.

To represent a function with an equation, we pick an independent variable for the inputs, often x, and a dependent variable for the outputs, often y. We can then reason about quantities and their relationships to find an equation relating the two variables, as we did for proportional relationships in Section 7.4 and for sequences in Section 9.5. Equations for functions often express the dependent variable (y) in terms of the independent variable (x), namely

$$y = (\text{expression in } x)$$

An equation in this form allows us to plug in an input value for x and evaluate the expression in x to determine the corresponding output value for y. We can think of this kind of equation as telling us directly how inputs go to outputs. However, equations for functions can also take other forms. For example, consider the equation

$$x \cdot y = 24$$

This equation tells us how x and y are related, but it does not express y directly in terms of x. Even so, the equation does specify y as a function of x (as long as x is not zero) because given any value for x we can find the corresponding value for y. For example, if $x = 4$, we can find the corresponding value for y by solving $4 \cdot y = 24$. And in general, we can express y in terms of x with the equation

$$y = \frac{24}{x}$$

Notice that when we use an equation to describe a function, an input and its corresponding output are a pair of values for x and y that make the equation for the function true. In other words, an input and its corresponding output are a solution to the function's equation. The graph of the function is therefore a visual display of all the solutions to the function's equation.

CLASS ACTIVITY

9T How Does Braking Distance Depend on Speed?, p. CA-187

9U Is It a Function?, p. CA-188

SECTION SUMMARY AND STUDY ITEMS

Section 9.6 Functions

A function is a rule that assigns one output to each allowable input. We can describe or represent functions with words, with tables and graphs (which usually don't convey complete information), and with expressions or equations (although not every function has an expression or equation).

Key Skills and Understandings

1. Use words to describe a function for a given context.
2. Given a description of a function in words, draw a graph that could be the graph of the function. Explain how the graph fits with the description of the function.
3. Given the graph of a function, describe aspects or features of the function. Explain how the graph informs you about those aspects or features.
4. Determine if a proposed function is or is not a function.

Practice Exercises for Section 9.6

1. For each of the following descriptions, draw the graph of an associated pollen count function for which the input is time elapsed since the beginning of the week and the output is the pollen count at that time. In each case, explain why you draw your graph in the shape that you do.
   a. At the beginning of the week, the pollen count rose sharply. Later in the week, the pollen count continued to rise, but more slowly.
   b. The pollen count fell steadily during the week.
   c. At the beginning of the week, the pollen count fell slowly. Later in the week, the pollen count fell more rapidly.

2. A company that sells juice is interested in a profit function whose inputs are the possible prices that 1 bottle of juice could sell for and whose outputs are the profits the company will make when juice is sold at that price. The graph of this profit function is shown in Figure 9.56.
   a. Based on the graph in Figure 9.56, if juice is sold for \$2.00 per bottle, approximately what will the company's profit be?
   b. Based on the graph in Figure 9.56, which price for one bottle of juice will result in maximal profit?

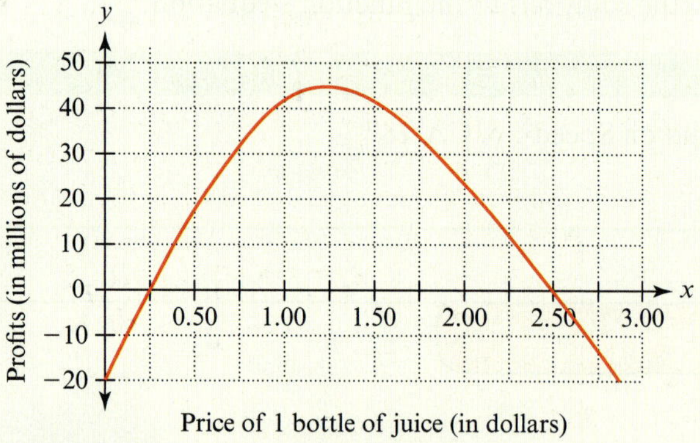

Figure 9.56 A profit function for juice sales.

   c. Where does the graph cross the x-axis, and what is the significance of those points?
   d. What is the significance of the points on the graph that lie above the x-axis? Below the x-axis?

3. Geneva went for a bike ride. She started off riding at a constant speed on level ground. After a while she went down a hill, going faster and faster. But as soon as she got to the bottom of the hill, there was another hill for Geneva to ride up. Geneva had built up speed coming down the other hill, so she started up the hill going fast. But soon Geneva was going more and more slowly. When she got to the top of the hill, Geneva stopped for a long rest.

   Describe three different functions that fit with this story about Geneva's bike ride. In each case, sketch a graph that could be the graph of the function. Indicate how these graphs fit with the story.

4. Fill in the blanks so that the next points lie on the graph of the function which has equation $y = -2x + 1$. Explain.

$$(3, \_), (\_, -13), (a, \_), (\_, b)$$

5. Explain why the next proposed function is *not* a function. Then modify the allowable inputs and/or the rule to describe a function that is relevant to the given context.

   Context: A rocket is launched, goes up to a height of 500 meters above the ground, and then falls back down. Allowable inputs for the proposed function: All numbers between 0 and 500. Rule for the proposed function: To an input number, H, associate the number of seconds, t after the rocket was fired when the rocket is H meters above ground.

Answers to Practice Exercises for Section 9.6

1. See Figure 9.57. Moving from left to right along the graph corresponds to time passing during the week.
   a. For each of the first few days of the week, the pollen count is significantly higher than it was the day before. Thus, the graph slopes steeply upward at first. But later in the week, the pollen count rises only a little from one day to the next. So, the graph must go up less steeply later in the week.

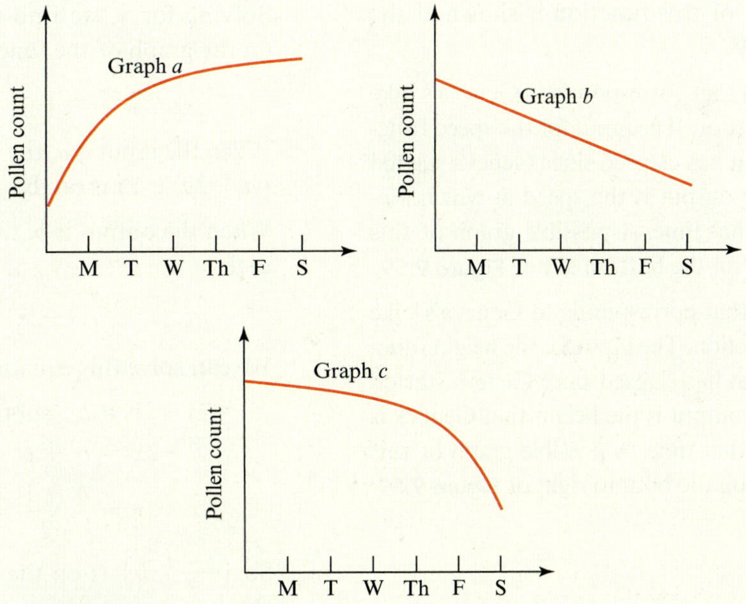

Figure 9.57 Pollen counts.

b. Graph *b* slopes downward (reading from left to right) and shows that for each day that goes by, the pollen count drops by the same amount.

c. For each of the first few days of the week, the pollen count is only a little less than it was the day before. Thus, the graph slopes gently downward at first. But later in the week, the pollen count drops significantly from one day to the next. Thus, toward the right, the graph slopes steeply downward.

2. a. If juice sells for \$2.00 per bottle, the company's profit will be about \$24 million. We can see this from the graph because when the input (x-coordinate) is 2.00, the output (y-coordinate) appears to be about 24, since the point (2.00, 24) seems to be on (or nearly on) the graph, as shown in **Figure 9.58**.

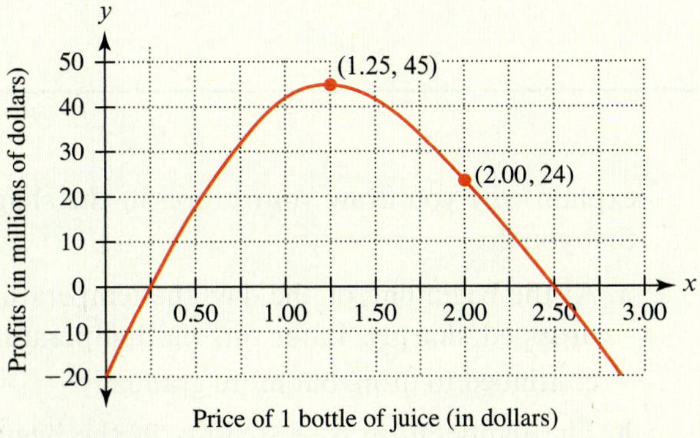

Figure 9.58 Selected points on a profit function for juice sales.

b. The maximal profit will occur when the graph reaches its highest point, which appears to be approximately at the point (1.25, 45), as we see in **Figure 9.58**. So when a bottle of juice sells for \$1.25, the company will make a maximal profit of \$45 million.

c. The graph crosses the x-axis at the points $(0.25, 0)$ and $(2.50, 0)$. At these points the profit is 0. So if juice sells for \$0.25 or for \$2.50 per bottle, the company breaks even but does not make a profit.

d. The points on the graph that lie above the x-axis correspond to prices for a bottle of juice for which the company makes a profit. This is because at these points, the profits (i.e., outputs) are greater than 0. If the company sells 1 bottle of juice for anywhere between \$0.25 and \$2.50, the company will make a profit.

The points on the graph that lie below the x-axis correspond to prices for 1 bottle of juice for which the company loses money. This is because at these points, the profits (i.e., outputs) are less than 0. So, if the company sets the price of 1 bottle of juice either less than \$0.25 or greater than \$2.50, then the company will lose money.

3. One function that corresponds to Geneva's bike ride is a distance function. The input for the distance function is the time that has elapsed since Geneva started her bike ride. The output is the total distance Geneva has traveled at that time.

A possible graph of this function is shown at the top of Figure 9.59.

A second function that corresponds to Geneva's bike ride is a speed function. The input for the speed function is the time that has elapsed since Geneva started her bike ride. The output is the speed at which Geneva is riding at that time. A possible graph of this function is shown on the bottom left of Figure 9.59.

A third function that corresponds to Geneva's bike ride is a height function. The input for the height function is the time that has elapsed since Geneva started her bike ride. The output is the height that Geneva is above sea level at that time. A possible graph of this function is shown on the bottom right of Figure 9.59.

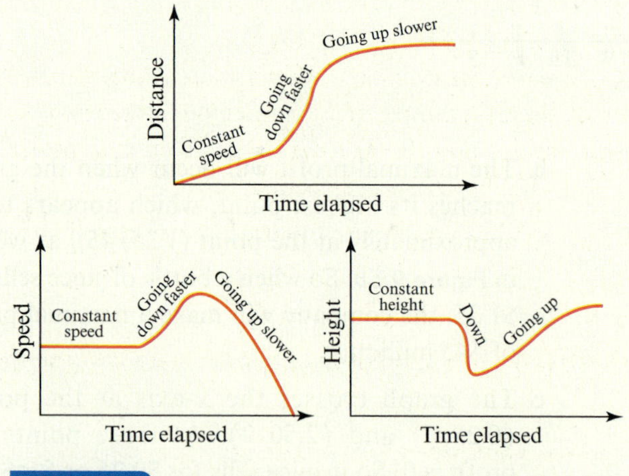

Figure 9.59 Three functions for Geneva's bike ride.

4. When $x = 3$, the output is $y = -2 \cdot 3 + 1 = -6 + 1 = -5$. So $(3, -5)$ is shown on the graph of the function.

When the output y is -13, the input is a number x such that

$$-2x + 1 = -13$$

Solving for x, we find that $x = 7$. So $(7, -13)$ is on the graph of the function. To check,

$$-2 \cdot 7 + 1 = -14 + 1 = -13.$$

When the input is a, the output is $y = -2a + 1$. So $(a, -2a + 1)$ is on the graph of the function.

When the output is b, the input is a number x such that

$$-2x + 1 = b$$

We can solve this equation for x:

$$-2x + 1 = b \quad \text{subtract 1 from both sides}$$

$$-2x = b - 1 \quad \text{divide both sides by } -2$$

$$x = \frac{b-1}{-2} = \frac{1-b}{2}$$

So $\left(\frac{1-b}{2}, b\right)$ is on the graph of the function. To check,

$$-2\left(\frac{1-b}{2}\right) + 1 = -(1-b) + 1$$
$$= -1 + b + 1 = b$$

5. The proposed function is not a function because for most of the heights, there will be two times when the rocket is at that height. So there are 2 outputs for most of the inputs instead of just 1 output for each input. Here is one way to modify the rule so that it becomes a function: To an input number, H, associate the *first* number of seconds after the rocket was fired that the rocket is H meters above the ground. Another option is to reverse the inputs and outputs. Change the allowable inputs to be the numbers of seconds and change the rule as follows: To the input t seconds, associate the height in meters of the rocket t seconds after launch.

PROBLEMS FOR SECTION 9.6

1. Describe your own real or realistic context in which two quantities vary together and use words to describe a function for this context.

2. For each of the following descriptions, draw a possible graph of an associated temperature function, for which the input is time elapsed since the beginning of the day, and the output is the temperature at that time. In each case, explain why you draw your graph in the shape that you do.

   a. At the beginning of the day, the temperature dropped sharply. Later on, the temperature continued to drop, but more gradually.

   b. The temperature rose steadily at the beginning of the day, remained stable in the middle of the day, and fell steadily later in the day.

c. The temperature rose quickly at the beginning of the day until it reached its peak. Then the temperature fell gradually throughout the rest of the day.

3. Consider the following scenario: Over a period of 5 years, the level of water in a lake drops slowly at first, then much more rapidly. During the next 5 years, the water level in the lake remains stable. Over the next 3 years, the water level rises slowly, and after another 2 years of more rapid increase in water level, the lake is back to its original level from 15 years before.

Describe a function that this scenario gives rise to. Sketch a graph that could be the graph of this function. Explain how features of your graph correspond to events described in the previous scenario.

4. A patient receives a drug. This situation gives rise to a drug function for which the input is the time elapsed since the drug was administered and the output is the amount of the drug in the patient's blood (in milligrams per milliliter of blood). The graph of this drug function is shown in **Figure 9.60**.

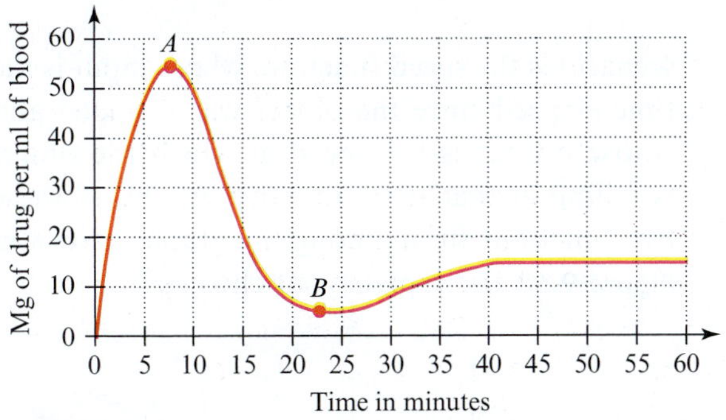

Figure 9.60 The graph of a drug function.

a. Give the approximate coordinates of point A, and discuss the significance of this point in terms of the scenario about the patient and the drug.

b. Give the approximate coordinates of point B, and discuss the significance of this point in terms of the scenario about the patient and the drug.

c. Overall, discuss what the graph tells you about the patient and the drug.

5. Sketch three different graphs that could fit with these three hypothetical descriptions of how the federal debt might change over a 5-year period. For each graph, explain why its shape fits with the description. Your graph need not show specific values for the debt, but be sure to indicate what each axis represents.

a. The debt shrinks, but it will shrink at a decreasing rate.

b. The debt shrinks at a steady rate.

c. The debt shrinks, and it will shrink at an increasing rate.

6. Sketch three different graphs that could fit with these three hypothetical descriptions of how the federal debt might change over a 5-year period. For each graph, explain why its shape fits with the description. Your graph need not show specific values for the debt, but be sure to indicate what each axis represents.

a. The debt grows, and it will grow at an increasing rate.

b. The debt grows at steady rate.

c. The debt grows, but it will grow at a decreasing rate.

7. Solar panels produce electricity by collecting energy from the sun. Over the course of a day, a solar panel might produce a total of 1 kilowatt-hour of electricity. Let's consider a function to describe the electricity production of such a solar panel over the course of a day. This *solar panel function* has independent variable t, the number of hours since sunrise, and dependent variable, E, the total number of kilowatt-hours of electricity produced by the panel over those t hours.

a. Examine the three partial graphs in **Figure 9.61**. Discuss each graph's characteristics in terms of the electricity production it describes. Which graph do you think is most likely to be the graph of the solar panel function between sunrise and about noon on a cloudless day? Explain your reasoning.

b. Complete the graph you selected in part (a) to show the graph of the solar panel function over the course of a full cloudless day, between sunrise and sunset. Explain why your graph has the characteristics and shape that it does.

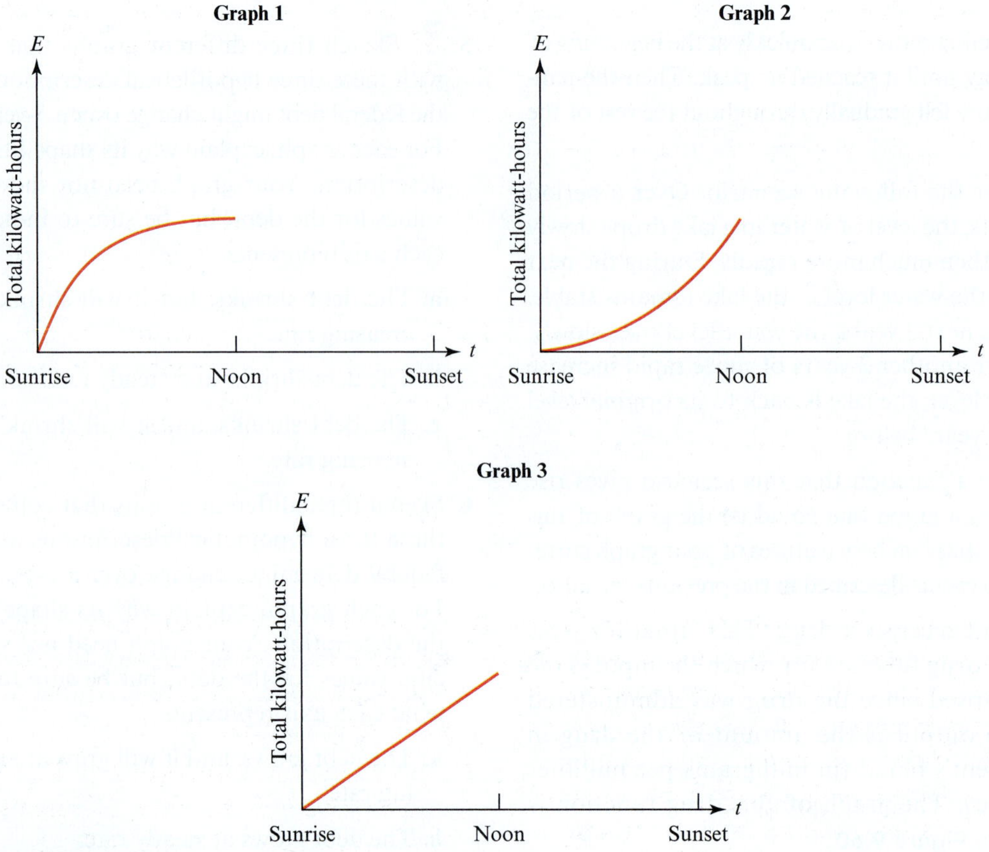

Figure 9.61 Possible graphs of a solar panel function.

8. For each of the three scenarios below, sketch graphs of two different functions, a *distance function* and a *speed function*. For each graph, explain why its shape fits with the description. Your graphs need not show specific values, but be sure to indicate what each axis represents. Make clear which graph represents which function.

a. The car drove faster and faster.

b. The car drove at a steady speed.

c. The car drove slower and slower.

9. An object is dropped from the top of a building. At first the object falls slowly, but as it continues to fall, it falls faster and faster. The speed at which the object falls increases at a steady rate.

One function that arises from this scenario is the height function, whose input is the time elapsed since the object was dropped and whose output is the height of the object above the ground at that time. Another function that arises from this scenario is the speed function, whose input is the time elapsed since the object was dropped and whose output is the speed at which the object is falling at that time. Identify graphs of these two functions from among the graphs shown in Figure 9.62 and explain your choices.

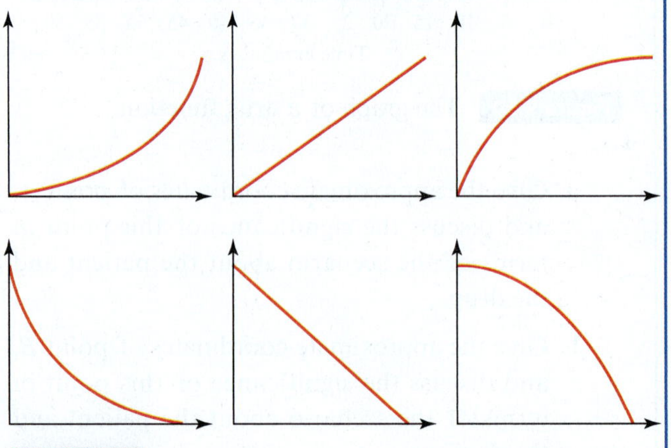

Figure 9.62 Which are graphs of the height and speed functions?

10. A skydiver jumps out of a plane. At first, the skydiver falls faster and faster, but then the skydiver reaches a point where he falls with a constant speed. After falling at a constant speed for a while, the skydiver's parachute opens and the skydiver falls at a slower constant speed until he lands safely on the ground.

Describe two different functions that fit with this story about a skydiver. In each case, sketch a graph that could be the graph of the function. Indicate how these graphs fit with the story.

11. Water is being poured at a steady rate into a round vase whose vertical cross-section is shown in Figure 9.63. One function that arises from this situation is the volume function, whose input is the time elapsed since water began pouring into the vase and whose output is the volume of water in the vase at that time. Another function that arises from this situation is the height function, whose input is the time elapsed since water began pouring into the vase and whose output is the height of the water in the vase at that time. Sketch graphs that could be the graphs of these two functions, and indicate how the graphs fit with the scenario of water pouring into the vase at a steady rate.

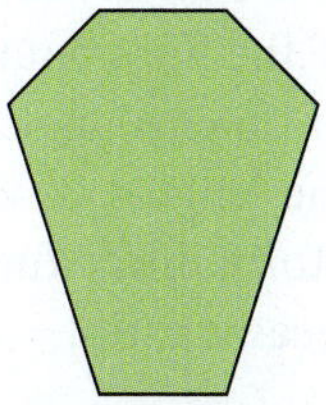

Figure 9.63 The cross-section of a vase.

***12.** Write a story or describe a scenario that gives rise to a function. Describe the function (taking care to describe the inputs and outputs clearly), sketch a possible graph of the function, and indicate how your graph fits with your story or scenario.

***13.** Draw *two different* graphs of two different functions that have the following three properties:

- When the input is 1, the output is 3.
- When the input is 2, the output is 6.
- When the input is 3, the output is 9.

14. Consider the *doubling plus one* function: The output is two times the input, plus one. For example, if the input is 3, the output is 7. Determine which of the following points are on the graph of the doubling plus one function:

$$(2, 5),\quad (2, 8),\quad (-1, -1),\quad (3, 4),$$
$$(-2, -3),\quad (1.5, 4),\quad (0.5, 1.5)$$

Explain your reasoning.

15. Consider the *squaring* function: The output is the input times itself. For example, if the input is 3, the output is 9. Determine which of the following points are on the graph of the squaring function:

$$(6, 36),\quad (4, 8),\quad (-2, 4),\quad (-3, -6),$$
$$(5, 25),\quad (6, 12),\quad (-6, 36)$$

Explain your reasoning.

16. Explain why the next proposed function is *not* a function. Then modify the allowable inputs and/or the rule to describe a function that is relevant to the given context.

Context: Purchasing paint at a paint store. Allowable inputs for the proposed function: Numbers between 0 and 1000. Rule for the proposed function: To an input number, D, associate the number of cans of paint C you will get if you spend $\$D$ at the paint store.

17. The levels of a certain toxin in a lake have been found to go up and down over time. Biologists are interested in studying the number of freshwater mussels in the lake, the level of the toxin in the lake, and any relationship there may be between the two.

a. Explain why the proposed function with this rule might not be a function: Assign to each amount of the toxin found in the lake, the number of mussels present when there is that amount of toxin in the lake. (Note that in Chapter 15 we will study scatterplots, which are often used to look for associations such as those between toxins and populations, and which can sometimes be *approximated* by the graph of a function.)

b. Use words to describe two functions that are relevant to the context of mussels and toxins in a lake.

9.7 Linear and Other Relationships

CCSS Common Core State Standards Grade 8

Many important real-world situations are modeled by linear functions. In this section, we study the characteristics of linear functions and contrast then with other functions. We'll see why straight-line graphs have a certain characteristic type of equation and how these equations relate to the graph and to real-world situations.

linear function

linear relationship

A **linear function** is a function whose graph lies on a line in a coordinate plane. We sometimes refer to linear functions as **linear relationships**.

In Sections 7.4 and 9.5 we saw that graphs of proportional relationships and arithmetic sequences lie on lines and are therefore examples of linear relationships. We developed and explained equations for proportional relationships using the value of the ratio, which is the constant of proportionality and also the slope of the line. We developed and explained equations for arithmetic sequences using their constant rates of change. The same ideas work more generally for linear functions and lead to equations of the form $y = mx + b$.

Why Are Linear Relationships Characterized by Constant Rate of Change?

A key characteristic of linear functions is that *a change in inputs and the corresponding change in outputs are always in the same ratio, so that the rate of change of outputs per change in inputs is constant.* In other words, linear functions have a constant rate of change. Why do linear functions have this characteristic? Let's sketch the reasoning. By definition, the graph of a linear function lies on a line. We can use this line to form many different "slope triangles," which are right triangles that have a horizontal and a vertical side, and a side along the graph as shown in Figure 9.64. Because the horizontal lines are all parallel, the angles they form with the graph of the function are all the same. As we'll see in Chapter 14, all such triangles are similar. Therefore, the lengths of the horizontal sides and the corresponding lengths of the vertical sides are in the same ratio for all these triangles. The lengths of horizontal sides of these triangles represent "changes in inputs"; the lengths of vertical sides of these triangles represent the corresponding "changes in outputs". So, for a linear function, the change in inputs and the corresponding change in outputs are always in the same ratio.

Therefore, the value of this ratio,

$$(\text{change in outputs}) \div (\text{change in inputs})$$

rate of change

slope

is constant. We call this constant value the **rate of change** of the linear function. We also call this constant value the **slope** of the line.

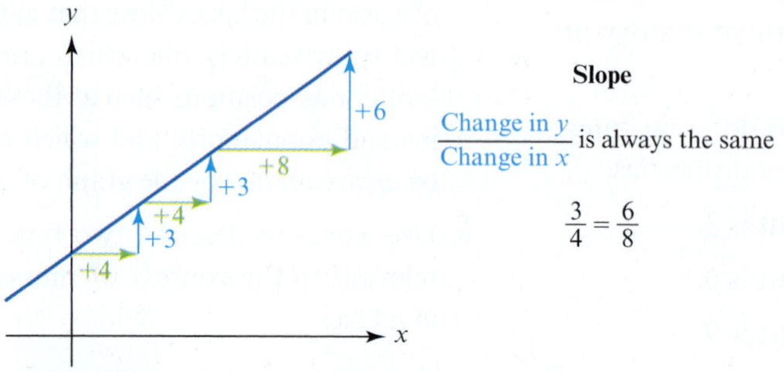

Figure 9.64 For linear functions, the change in input and the corresponding change in output is always in the same ratio due to similar triangles.

Conversely, if a function has the property that the change in inputs and the corresponding change in outputs are always in the same ratio, then it is a linear function. Again, this can be explained by reasoning about similar triangles. In summary, a relationship is linear if and only if it has a constant rate of change.

How Can We Develop and Explain Equations for Linear Relationships Using Constant Rate of Change?

Why do linear functions have equations of the form $y = mx + b$? To explain, we can generalize the reasoning we used for proportional relationships in Section 7.4.

Given a linear function that has independent variable x and dependent variable y, let b be the output that corresponds to the input 0 (which we'll assume is an allowable input). The number b is therefore also the y-coordinate where the line crosses the y-axis and is called the **y-intercept** of the line.

y-intercept

Now let's consider changes in x and y for our linear function. The change in x and the corresponding change in y are always in the same ratio, say A to B, so let's let m be the value of that ratio $\frac{B}{A}$. As we discussed in Section 7.5, that value m tells us how many groups of "change in x" it takes to make the same amount as the corresponding "change in y." Therefore

$$m \cdot (\text{change in } x) = (\text{change in } y) \text{ or } (\text{change in } y) = m \cdot (\text{change in } x)$$

CCSS 8.EE.6 8.F.3

In particular, if we apply this equation to the change from $(0, b)$ to (x, y), we get the equation

$$(y - b) = m \cdot (x - 0)$$

Which we can write in the equivalent form

$$y = mx + b$$

Figure 9.65 summarizes this reasoning.

Figure 9.65 Explaining the $y = mx + b$ equation for linear functions.

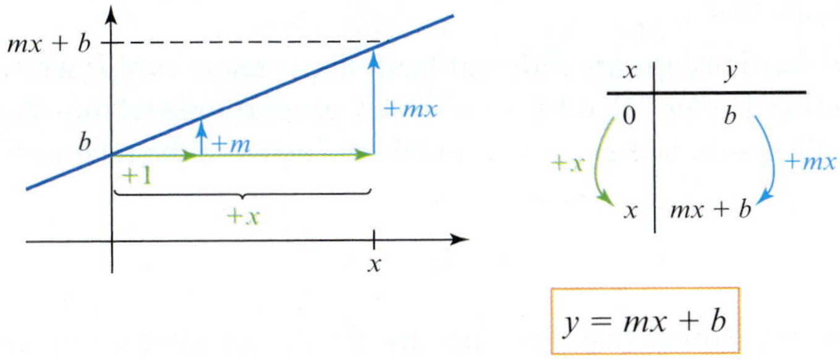

What Kinds of Real-World Situations Do Linear Relationships Model?

Given a real-world situation, how can we tell if it will be modeled by a linear relationship? The answer lies in the key characteristic of linear relationships. If there are two quantities that vary together in the real-world situation, then we can check if the changes in one quantity and corresponding changes in the other quantity are always in the same ratio. If so, the relationship is linear; if not, the relationship is not linear.

For example, suppose the Tropical Orchid Company sells orchids by mail order and charges \$25 for each orchid plus \$9 shipping no matter how many orchids are ordered. Consider the associated function whose independent variable, x, is the number of orchids in an order, and whose dependent variable, y, is the total cost to the customer for the order. This function is linear because for each additional orchid that is ordered, there is an additional cost of \$25, in other words, because the cost increases at a steady rate of \$25 per orchid ordered.

CLASS ACTIVITY

9V Modeling Linear Relationships with Variables and Equations, p. CA-190.

9W Interpreting Equations for Linear Relationships, p. CA-192

How Do Linear Relationships Contrast with Other Relationships?

Linear relationships are characterized by a constant rate of change, so that the change in inputs and the change in outputs are always in the same ratio, and any change in output divided by the corresponding change in inputs gives the same constant rate. Other kinds of relationships have different characteristic patterns of change and these patterns of change determine the kinds of situations the functions model. We will take a very brief look at patterns of change that are characteristic of quadratic functions and those characteristic of exponential functions (which include geometric sequences). We will also contrast inversely proportional relationships with linear relationships whose graphs have a negative slope.

CLASS ACTIVITY

9X Inversely Proportional Relationships Versus Decreasing Linear Relationships, p. CA-193

9Y Is It Linear or Not?, p. CA-194

Inversely Proportional Relationships Versus Linear Relationships In Chapter 7 we saw that if x and y are in an inversely proportional relationship, then as x increases, y decreases. In contrast, when x and y are in a proportional relationship, then as x increases, y must also increase. Do all linear relationships behave that way? No, for linear relationships whose graphs have a negative slope, as x increases, y decreases.

Inversely proportional relationships are different from linear relationships whose graphs have a negative slope. We saw in Chapter 7 that for an inversely proportional relationship between x and y, the *product,* $x \cdot y$ is always the same constant number. Therefore the relationship has an equation of the form

$$x \cdot y = c \quad \text{or} \quad y = \frac{c}{x}$$

for some constant number c. Notice that this is not an equation of a linear function.

Quadratic Functions Quadratic functions are functions that have an equation of the form

$$y = ax^2 + bx + c$$

where a, b, and c are numbers and a is not zero. Although quadratic functions don't have a constant rate of change the way that linear functions do, they nevertheless exhibit an interesting pattern of change, as illustrated in Figure 9.66. It turns out that quadratic functions are characterized by a constant "change in the change."

Quadratic functions model the height of an object that is falling (ignoring air resistance). According to physics, the acceleration due to gravity is (approximately) constant. Velocity is change in distance and acceleration is change in velocity, so situations of constant acceleration, such as falling objects, produce quadratic functions.

Exponential Functions Exponential functions are functions that have an equation of the form

$$y = a \cdot b^x$$

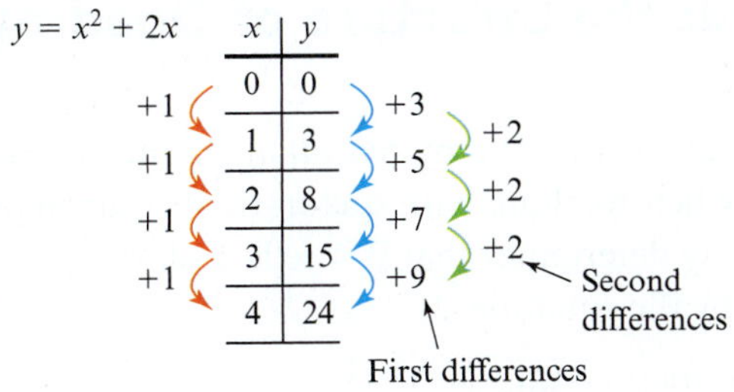

Figure 9.66 Quadratic functions have constant second differences.

where a and b are positive numbers and b is not 1. Exponential functions generalize geometric sequences in the same way that linear functions generalize arithmetic sequences.

Exponential functions grow (and decay) in a characteristic way. On the one hand, like geometric sequences, whenever x increases by 1 unit, y is *multiplied* by a fixed number. But there is another pattern of change as well, which is illustrated in Figure 9.67. The change in an exponential function is again an exponential function, and it is proportional to the original function. Because of this characteristic, exponential functions model many situations in which the growth of a quantity depends on how much of the quantity is present, such as amounts of money in a bank account or borrowed on credit, populations of people or animals, and the decay of radioactive substances.

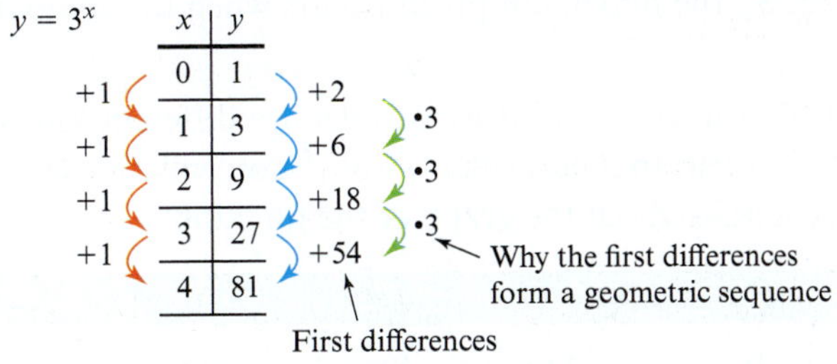

Figure 9.67 Exponential functions have first differences that are also exponential.

Functions Coordinate Values Remember that functions are relationships between values that vary together. Therefore to determine what kind of function a table exhibits we must consider how the independent and dependent variables change together. For example, in Figure 9.68 if we only look at the y-coordinates and ignore the x-coordinates, we might think that the table exhibits a linear function. This conclusion is not valid because the x-coordinates in the table do not have a constant increase.

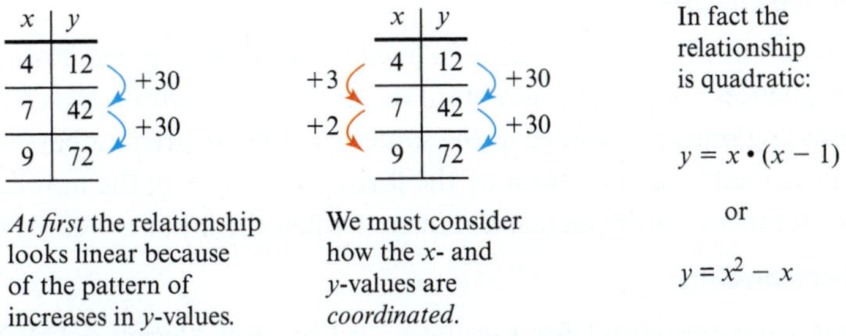

Figure 9.68 Determining the type of a function requires looking at how x- and y-values change together.

CLASS ACTIVITY

9Z What Kind of Relationship Is It?, p. CA-196

CCSS SMP7

How Can We Reason about the Structure of Quadratic Equations?

An important skill in algebra is to deduce information by reasoning about the structure of an expression. This skill is especially useful when working with quadratic relationships. For example, consider this scenario: A T-shirt company determines that if it sells T-shirts for $\$x$ each, then the company's daily profit $\$y$ will be given by the equation

$$y = 1000 - 10(x - 15)^2$$

What can we determine by examining the structure of the expression on the right hand side? The expression is a difference of two terms—namely, it is 1000 minus $10(x - 15)^2$. When you subtract an amount from 1000, the result is less than 1000, unless the amount you subtracted is 0 or negative. Can the subtracted term be negative? No matter what the value of x is, the square $(x - 15)^2$ must be greater than or equal to 0 (because even when you multiply a negative number with itself the result is positive). When that square is multiplied by 10, it is still greater than or equal to 0. So 1000 minus $10(x - 15)^2$ must always be less than or equal to 1000, no matter what the value of x is. This tells us that the maximum possible daily profit for the company is \$1000. How can this profit of \$1000 be achieved? It can be achieved only when the subtracted term $10(x - 15)^2$ is 0. Since $10(x - 15)^2$ is 10 multiplied with $x - 15$ and $x - 15$, it can only be 0 when $x - 15$ is 0 because the only way to multiply numbers together and get 0 is when at least one of the numbers is 0. To make $x - 15$ be 0, x must be 15. So the maximum profit occurs when the company sets the price of a T-shirt at \$15.

In Class Activities 9AA and 9BB you will see that we can deduce different kinds of information when a quantity is expressed in different (but equivalent) ways. You will also see how details in an equation for a function relate to details about the graph of the function.

CLASS ACTIVITY

9AA Doing Rocket Science by Reasoning About the Structure of Quadratic Equations, p. CA-197.

9BB Reasoning About the Structure of Quadratic Equations, p. CA-199.

SECTION SUMMARY AND STUDY ITEMS

Section 9.7 Linear and Other Relationships

A linear function is a function whose graph lies on a line in a coordinate plane. Proportional relationships and arithmetic sequences are linear functions. For any linear function, the change in input and the change in corresponding output is always in the same ratio. There are numbers m and b such that the linear function is described by an equation of the form $y = mx + b$; the number b is the output for the input 0 and the number m is the increase in output when the input increases by 1.

Key Skills and Understandings

1. Understand that variables stand for numbers and are not "labels." Define variables for quantities and specify a unit of measurement (if needed). For example, "let G be the number of gallons of juice in a vat" or "let d be the distance traveled in kilometers."
2. Be aware of common errors in formulating equations.
3. Given a scenario that gives rise to a linear function, describe the function in words, with a table, with a graph, and with an equation, and explain why the equation is valid. Relate details of the situation to details of the table, graph, and equation.

4. Given information about a function, determine if the function is linear or not.
5. Contrast patterns of change in linear and other types of functions.
6. In selected cases, use the structure of an expression to reason about its values, such as its maximum or minimum value or the values of the variable for which the expression is 0.

Practice Exercises for Section 9.7

1. At a yogurt shop, frozen yogurt is 45 cents for each ounce; a waffle cone to hold the yogurt is \$1. Define two variables and write an equation to relate two quantities in this context.
2. A nightlight costs \$7.00 to buy and uses \$0.75 of electricity per year to operate. Describe a function that this situation gives rise to. Make a table for this function, sketch a graph of this function, and find an equation for this function.
3. A juice mixture can be made by mixing juice and lemon-lime soda in a ratio of 3 to 2.
   a. Describe a function that is associated with this situation. Write an equation for the function and explain why the function has this equation. Sketch the graph of the function.
   b. Discuss how the equation for the function in part (a), the graph of the function, and the 3-to-2 ratio (or a different, related ratio) are related to each other.
4. To make orange paint, you need to mix yellow and red paint together. For a certain shade of orange paint, you need to use $2\frac{1}{2}$ times as much yellow paint as red paint. Write an equation that relates quantities in this situation, taking care to define your variables clearly and correctly.
5. At the cafeteria you can get a lunch special that offers a drink and salad. The drink costs \$0.85 and the salad costs \$0.35 for every ounce. Make a table, define two variables, and write an equation to relate two quantities in this situation that vary together.
6. Assuming that the following tables are for linear functions, fill in the blank outputs and find equations for these functions:

x	y
0	
1	
2	
3	4
4	
5	8
6	

x	y
0	
1	
2	3
3	
4	
5	
6	6

x	y
−3	7
−2	
−1	
0	
1	−1
2	
3	

7. An object is dropped from the top of a building. After t seconds, the height h of the object in feet is

$$h = 16(13 + t)(13 - t)$$

Reason about the structure of the equation to determine how long it takes until the object hits the ground.

8. A company that sells snake oil has determined that its profit function is given by the equation

$$y = 32000 - 5(x - 75)^2$$

where x is the number of dollars it sells each bottle of snake oil for and y is the company's annual profit in thousands of dollars. Explain how to reason about the structure of the equation to determine how the company can maximize its profit.

Answers to Practice Exercises for Section 9.7

1. Let Z be the number of ounces of yogurt that a customer puts in a waffle cone. The cost of the yogurt (without the cone) is $0.45Z$ dollars because each of the Z ounces cost 0.45 dollars. Adding in the cost of the waffle cone, the total cost is $0.45Z + 1$ dollars. If C is this total cost (in dollars), then $C = 0.45Z + 1$.

2. The situation gives rise to a nightlight function for which the input, x is the number of years since purchase and the output, y is the total cost (in dollars) of operating the nightlight for that amount of time. The table below shows the nightlight function and a graph appears in Figure 9.69.

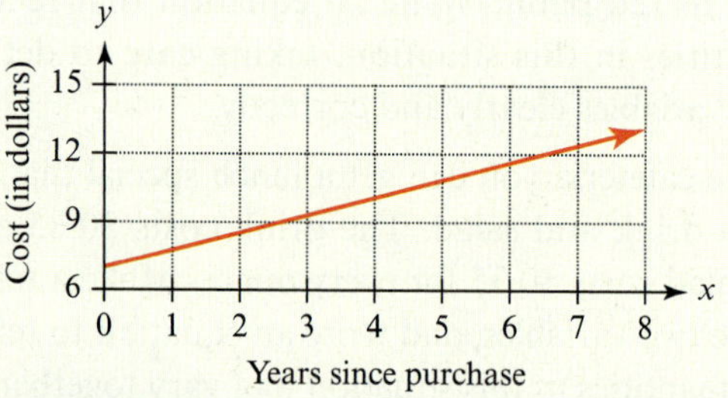

Figure 9.69 The cost of using a nightlight.

Nightlight Function

x Years Since Purchase	y Total Cost of Operating
0	\$7.00
1	\$7.75
2	\$8.50
3	\$9.25
4	\$10.00
5	\$10.75

Because the nightlight costs \$0.75 for each year of operation, it costs $0.75x$ dollars to operate for x years. Adding this amount to the purchase price of \$7.00, we see it costs a total of $7 + 0.75x$ dollars to operate the nightlight for x years. Therefore

$$y = 7 + 0.75x$$

is an equation for the function.

3. a. One function is the function whose independent variable x is a number of liters of lemon-lime soda and whose dependent variable y is the corresponding number of liters of juice that will be needed to mix with that amount of lemon-lime soda. This function has equation

$$y = \frac{3}{2}x$$

This equation is valid because for every 2 cups of lemon-lime soda, we need 3 cups of juice. So for 1 cup of soda we need half as much juice, namely, $\frac{3}{2}$ cups of juice. So for x cups of soda, we need x times as much juice, namely, $\frac{3}{2}x$ cups of juice. The graph of this function (not shown) is a line through the origin with slope $\frac{3}{2}$. (Another possible function has the same inputs but the outputs are the total amount of juice mixture that is made. Two more functions have amounts of juice as inputs and either the corresponding amount of lemon-lime soda as outputs or the total amount of juice mix created as outputs.)

b. The 3-to-2 ratio shows up in the graph this way: Every time we go over 2 units to the right, we go up 3 units. In the equation for the function, the 3-to-2 ratio shows up as the unit rate, $\frac{3}{2}$, of cups of juice per cup of lemon-lime soda. This unit rate is also the slope of the graph: Whenever we go to the right 1 unit on the graph, we go up $\frac{3}{2}$ units.

4. Let Y be the number of liters of yellow paint that you will use and let R be the number of liters of red paint you will use. Then

$$Y = 2.5R$$

as we can see with the aid of the strip diagram in Figure 9.70.

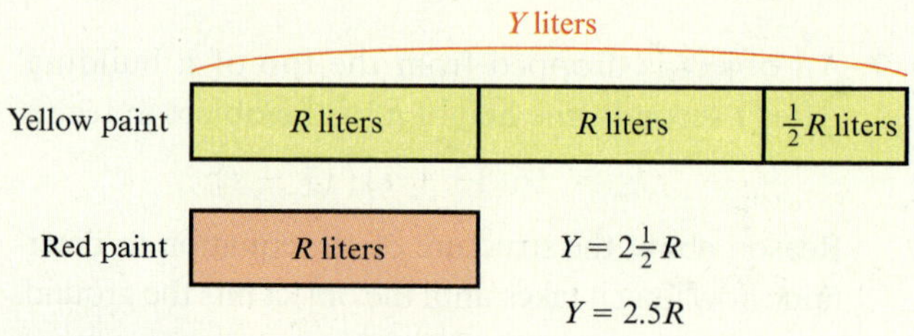

Figure 9.70 A strip diagram to relate yellow and red paint.

5. Let S be the number of ounces of salad one gets and let C be the total cost of the lunch special. Then from the table below we can see that $C = 0.85 + S \cdot 0.35$, which we can rewrite as $C = 0.85 + 0.35S$.

Number of Ounces of Salad	Expression for the Total Cost (in Dollars)
4	$0.85 + 4 \cdot 0.35$
5	$0.85 + 5 \cdot 0.35$
6	$0.85 + 6 \cdot 0.35$
7	$0.85 + 7 \cdot 0.35$
8	$0.85 + 8 \cdot 0.35$

6. See below.

$y = 2x - 2$	
x	y
0	−2
1	0
2	2
3	4
4	6
5	8
6	10

$y = 0.75x + 1.5$	
x	y
0	1.5
1	2.25
2	3
3	3.75
4	4.5
5	5.25
6	6

$y = -2x + 1$	
x	y
−3	7
−2	5
−1	3
0	1
1	−1
2	−3
3	−5

7. The object will hit the ground when the height is 0, so we want to find when the expression $16(13 + t)(13 - t)$ has the value 0. The expression is the product of 16 with $13 + t$ and $13 - t$. The only way for a product to be 0 is for at least one of the factors to be 0, so either $13 + t$ or $13 - t$ must be 0 for $16(13 + t)(13 - t)$ to be 0. For $13 + t$ to be 0, t must be −13. But −13 seconds doesn't make sense in this context. For $13 - t$ to be 0, t must be 13. When $t = 13$, the value of $16(13 + t)(13 - t)$ is 0, which tells us that 13 seconds after the object is dropped, it will hit the ground.

8. The expression $32000 - 5(x - 75)^2$ is a difference of two terms. The first term is just the number 32000. The second term is a product of the coefficient 5 with the square, $(x - 75)^2$. The square must always be greater than or equal to 0 because even when $x - 75$ is negative, a negative times a negative is positive. Therefore the term $5(x - 75)^2$ must always be greater than or equal to 0. Therefore, when this term is subtracted from 32000, the resulting difference must be less than or equal to 32000 no matter what the value of x is. Because the profit, y is equal to $32000 - 5(x - 75)^2$, the greatest that the profit can therefore be is 32000 thousand dollars ($32 million), and this occurs when $5(x - 75)^2$ is 0. But because $5(x - 75)^2$ is a product, it can only be 0 when at least one of its factors is 0, which occurs when $x - 75$ is 0—namely, when $x = 75$. Therefore the snake oil company will maximize its annual profit when it sells snake oil for $75 per bottle, and the profit it will achieve is $32 million.

PROBLEMS FOR SECTION 9.7

1. A small-order coffee company charges $12 for each bag of coffee plus $3 shipping no matter how many bags are ordered. Describe a function that this situation gives rise to. Make a table for this function, sketch a graph of this function, and find an equation for this function. Is the function linear or not? How can you tell?

2. A company prints T-shirts with original designs. To create a T-shirt design for a customer, the company charges $75. For each T-shirt that is made with that design, the company charges $12.

Make a table to show how two quantities in this situation vary together. Then define two variables and write and explain an equation to show how the variables are related.

3. A company sells mail order T-shirts. The company charges \$11 for each T-shirt and \$8 for shipping, no matter how many T-shirts a customer orders.

Examine the two different ways that students made a table:

Work of Student 1	
Number of T-shirts Ordered	Total Cost (in Dollars), with Shipping
1	$11 + 8 = 19$
2	$19 + 11 = 30$
3	$30 + 11 = 41$
4	$41 + 11 = 52$
5	$52 + 11 = 63$

Work of Student 2	
Number of T-shirts Ordered	Total Cost (in Dollars), with Shipping
1	$11 + 8$
2	$2 \cdot 11 + 8$
3	$3 \cdot 11 + 8$
4	$4 \cdot 11 + 8$
5	$5 \cdot 11 + 8$

Discuss the two students' work. Which will be more useful for formulating an equation that relates two variables? Why?

4. Today a T-shirt company has 500 T-shirts in stock. Starting tomorrow, they expect to sell 15 of those T-shirts every day. Define two variables for this situation and write and explain an equation to relate them.

5. In a problem about the weight of animal feed in a bin, a student defined the variable A by writing "A = animal feed." What is wrong with this? What is a correct way to define a variable for this context?

6. Some students are washing cars for a fund-raiser. They will raise \$10 for each car they wash. The expenses for setting up the car wash (for soap, etc.) are \$20. Define two variables and write and explain an equation relating quantities in this context.

7. Initially, a vat is filled with 500 liters of liquid, but the liquid soon starts to drain out of the vat. Every second, 2 liters of liquid drain out. Define two variables and write and explain an equation relating quantities in this context.

8. Max was asked to write an equation corresponding to the following situation:

> There are 2 times as many pencils in Antrice's pencil box as in Ben's pencil box.

Max responded thus:

A = Antrice $\quad B$ = Ben

$2A = B$

Discuss Max's work, describing and correcting any errors he has made. Explain why your corrections are correct.

9. To cook rice, you need to use $1\frac{1}{2}$ times as much water as uncooked rice no matter how much rice you make. Write an equation that describes this relationship. Explain why your equation is correct. Be sure to define your variables with care.

10. A company sells $3\frac{1}{4}$ times as many tissues as napkins. Write an equation that corresponds to this situation. Explain why your equation is correct. Be sure to define your variables with care.

11. Show how to use strip diagrams to help you formulate the equations in parts (a) and (b).

a. Garden Green paint is made by mixing blue paint with yellow paint, using 75% as much blue paint as yellow paint no matter how much Garden Green paint will be made. Write and explain an equation that relates the quantities in this situation.

b. Petunia Pink paint is made by mixing red paint with white paint, using 75% more red paint than white paint no matter how much Petunia Pink paint will be made. Write and explain an equation that relates the quantities in this situation.

12. After a vine that was 18 inches long was planted, the vine grew another $\frac{1}{2}$ inch every day. Make a table to relate the number of days since planting and the length of the vine. Then define two variables and write an equation to show this relationship. Be sure to define your variables with care.

13. A mail order coffee company charges \$12 for each pound of coffee beans. The company charges a \$3 handling fee for any order of coffee beans, no matter how many pounds are ordered. Make a table and then define two variables and write an equation to show a relationship between quantities in this situation. Be sure to define your variables with care.

14. Assuming that the following tables are for linear functions, fill in the blank outputs and find equations for these functions:

Input	Output
0	
1	
2	7
3	
4	10
5	
6	

Input	Output
0	
1	−2
2	
3	
4	
5	
6	2

Input	Output
−3	13
−2	
−1	
0	
1	
2	−2
3	

15. Consider a function that has the following properties:

- When the input is 0, the output is 2.
- Whenever the input increases by 1, the output increases by 4.

a. Make a table, draw the graph, and find an equation for a function that has the given properties.

b. Explain how the properties in the two bulleted items are reflected in the graph of the function.

16. Consider a function that has the following properties:

- When the input is 2, the output is 4.
- Whenever the input increases by 3, the output increases by 5.

a. Make a table, draw the graph, and find an equation for a function that has the given properties.

b. Explain how the properties in the two bulleted items are reflected in the graph of the function.

17. At one stage in a factory's candy production, cocoa is mixed with sugar. For every 20 pounds of cocoa, 13 pounds of sugar are needed.

a. Describe a function that arises from this situation.

b. Make a table and draw the graph of your function.

c. Find an equation for your function, and explain why your equation is valid.

18. A school bought 100 rolls of wrapping paper for a total of \$150. As a fund-raiser, the school will sell the rolls of wrapping paper for \$6 each. This situation gives rise to a profit function whose independent variable, x, is the number of rolls sold and whose dependent variable, y, is the total profit the school makes when that many rolls are sold.

a. Make a table and draw a graph for the profit function.

b. Find an equation for the profit function, and explain why your equation is valid.

c. Where does the graph of the profit function cross the x-axis? What is the significance of this point in terms of the school's fund-raiser?

19. For the spring fling, a school bought 200 slices of pizza for a total of \$100. At the spring fling, pizza will be sold for \$2.50 per slice.

a. Describe a function that arises from this spring fling scenario.

b. Make a table and draw a graph of your function.

c. Find an equation for your function, and explain why your equation is valid.

d. Discuss useful information concerning the spring fling that your graph shows.

20. The Gizmo store sells gizmos. You pay \$3 for the first 3 Gizmos and \$0.75 for each additional Gizmo.

a. Define an independent and a dependent variable and describe a function that arises from this scenario. Give any restrictions on the values of the independent variable.

b. Find an equation for the function and explain why the equation is valid.

21. At a store that sells fences, if you buy 15 feet of fencing or less, the total cost, including delivery, is \$200. Each additional foot of fencing costs an additional \$10. Let F be the number of feet of fencing in an order and let C be the cost (in dollars) of the order.

a. What restriction should be made on F so that the relationship between C and F is linear? Explain.

b. Without writing an equivalent equation, explain how to interpret each side of the equation below and explain why the equation describes the relationship between F and C (with the restrictions of part [a]).

$$C - 200 = 10(F - 15)$$

22. Describe a real-world scenario in which two quantities vary together in a linear relationship. Define an independent and dependent variable for the quantities, write an equation to show how they are related, and explain why the equation is valid.

23. Consider these two relationships:

Relationship 1: A woman paints a fence at a steady rate. Initially, there are 200 feet of fence to paint. After 5 hours, the woman has finished painting the fence. Let x be the number of hours the woman has painted and let y be the number of feet of fence left to paint.

Relationship 2: Some men are painting a long fence. It takes 8 men 4 hours to paint the fence. Assume that all men work at the same steady rate. Let x be the number of men painting the fence and let y be the number of hours it takes to paint the fence with that many men.

Make a table, draw a graph, and write an equation for relationship 1 and for relationship 2, explaining why you can write the equations the way you do. What kinds of relationships are they and how can you tell?

24. For each of the following, either define variables x and y for quantities that vary together in a realistic situation or say why there can't be such an example.

a. As x increases, y decreases and the relationship between x and y *is not* linear.

b. As x increases, y decreases and the relationship between x and y *is* linear.

25. Compare and contrast inversely proportional relationships with linear relationships whose graphs have negative slopes. Illustrate with an example of each. Include a table and a (real-world) context as part of each example.

26. For each of the following tables, determine what kind of relationship the table exhibits and explain how you can tell. You do not need to find equations for the relationships.

Table A

x	y
1	−4
2	−4
3	−4
4	−4
5	−4

Table B

x	y
1	0
2	3
3	8
4	15
5	24

Table C

x	y
1	0
4	6
9	16
16	30
25	48

27. For each of the following tables, determine what kind of relationship the table exhibits and explain how you can tell. You do not need to find equations for the relationships.

Table A

x	y
1	96
2	48
3	32
4	24

Table B

x	y
1	5
2	10
3	20
4	40
5	80

Table C

x	y
1	15
3	30
7	60
15	120
31	240

28. Find *two different* equations for two different functions that have the following two properties:

- When the input is 0, the output is 1.
- When the input is 1, the output is 2.

29. Consider the function whose equation is $y = x^2 - 2x$.

a. Make a table in which you show the outputs for the following inputs:

$$-2, -1, 0, 1, 2, 3, 4$$

b. Plot the points that are associated with the inputs and outputs you found in part (a).

c. Based on the points you found in part (b), draw the graph of the function as best you can.

30. An object is thrown down from the top of a building. A height function for the object is given by the equation

$$h = 16(8 + t)(5 - t)$$

where t is the number of seconds elapsed since the object was thrown and h is the height of the object above the ground (in feet). Explain how to reason about the structure of the equation to determine when the object will hit the ground.

31. An oil company has determined that its profit function is given by the equation

$$y = 4800 - 300(x - 4.50)^2$$

where x is the number of dollars it sells each gallon of oil for and y is the company's annual profit in millions of dollars. Explain how to reason about the structure of the equation to determine how the company can maximize its profit.

32. During a 6-hour rainstorm, rain fell at a rate of 1 inch every hour for the first 3 hours. During the next hour, $\frac{1}{2}$ inch of rain fell. During the last 2 hours, rain fell at a rate of $\frac{1}{4}$ inch every hour.

This scenario gives rise to a rainfall function in which the input is the time elapsed since the rainstorm began and the output is the amount of rain that fell during the storm up to that time.

a. Make a table for the rainfall function.

b. Draw a graph of the rainfall function.

c. Is there a single equation for the rainfall function?

33. In the country of Taxo, each household pays taxes on its annual household income as determined by the following:

- A household with income between \$0 and \$20,000 pays no tax.
- For each \$1 of income beyond \$20,000, and up to \$50,000, the household must pay \$0.20 in tax.
- For each \$1 of income beyond \$50,000, and up to \$100,000, the household must pay \$0.40 in tax.
- For each \$1 of income beyond \$100,000 the household must pay \$0.60 in tax.

This situation gives rise to a tax function in which the input is annual household income and the output is the taxes paid.

a. Make a table for the tax function.

b. Draw a graph of the tax function.

c. Is there a single equation for the tax function?

34. A company that sells coffee has determined that if it sells coffee for C dollars a cup, its daily profit P will be given by

$$P = 1800 - 200(C - 3.55)^2$$

Describe the structure of the equation and reason about this structure to determine how the company can maximize its daily profit.

35. A projectile is launched. After t seconds, the height h of the projectile in feet is given by

$$h = 16(5 + t)(17 - t)$$

Describe the structure of the equation and reason about this structure to determine when the projectile will hit the ground.

CHAPTER SUMMARY

Section 9.1 Numerical Expressions	Page 379
▪ A numerical expression is a meaningful string of numbers, operations, and possibly parentheses. Expressions arise in a variety of scenarios and provide a way to formulate a quantity in a scenario mathematically.	Page 379
Key Skills and Understandings	
1. Formulate numerical expressions arising from scenarios.	Page 379
2. Interpret the meaning of an expression without evaluating it—for example, describe an expression as a sum or product.	Page 380
3. Evaluate expressions efficiently. Use cancellation correctly when evaluating expressions involving fractions.	Page 381
Section 9.2 Expressions with Variables	Page 388
▪ Variables are letters that stand for numbers. Expressions with variables are "calculation recipes" that apply to many different numbers.	Page 388
Key Skills and Understandings	
1. Interpret the structure of an expression and use mathematical terminology to identify its parts. For example, identify the terms of an expression and describe them—perhaps as products, quotients, or powers. Describe coefficients of terms.	Page 388
2. Write expressions for quantities and explain why they are formulated the way they are. Evaluate expressions at values of the variables.	
3. Generate and recognize equivalent expressions by applying properties of arithmetic. Determine when two expressions are not equivalent.	Page 388
Section 9.3 Equations	Page 396
▪ An equation is a statement that an expression or number is equal to another expression or number. Equations involve an equal (=) sign. Equations can serve different purposes. Some equations show the result of a calculation. Some equations, called identities, are true for all values of the variables. Some equations describe how quantities that vary together are related. And some equations must be solved to solve a problem.	Page 396
▪ To solve an equation means to determine values for the variables that make the equation true. Sometimes we can solve an equation just by applying our understanding of expressions and equations. However, a general strategy for solving equations is to change the equation into a new equation that has the same solution and is easy to solve.	Page 398
Key Skills and Understandings	
1. Describe some of the different ways we use equations.	Page 396
2. Solve appropriate equations by reasoning about numbers, operations, and expressions.	Page 398
3. Explain different ways to reason to solve "one-step" equations.	Page 402
4. Solve equations algebraically, and describe how the method can be viewed in terms of a pan balance if appropriate.	Page 399
5. Give examples of equations in one variable that have no solutions and equations in one variable that have infinitely many solutions.	Page 401

Section 9.4 Solving Algebra Word Problems with Strip Diagrams and with Algebra — Page 404

- Many algebra word problems can be solved with the aid of strip diagrams. Often, when we use strip diagrams to solve a word problem, our method parallels the algebraic method of setting up and solving an equation with variables. — Page 404

Key Skills and Understandings

1. Set up and solve word problems with the aid of strip diagrams as well as algebraically, and relate the two. — Page 404

Section 9.5 Sequences — Page 413

- A sequence is a list of items (usually numbers or shapes) occurring in a specified order. Repeating sequences offer opportunities to highlight a unit and to reason about division with remainder. Arithmetic sequences are produced by repeatedly adding (or subtracting) the same fixed number. Geometric sequences are produced by repeatedly multiplying (or dividing) by the same fixed number. Arithmetic sequences and geometric sequences both have expressions or equations that describe their Nth entry. Repeating sequences are frequently studied in elementary school. — Page 413

Key Skills and Understandings

1. Solve problems about repeating sequences by attending to the repeating unit and reasoning about division and multiplication. — Page 413
2. Solve problems about growing sequences using algebra and by reasoning about division. — Page 414
3. Given an arithmetic sequence, find an equation relating entries to their positions and explain why this equation is valid in terms of the way the sequence grows. — Page 415
4. Given a geometric sequence, find an equation relating entries to their positions and explain why this equation is valid in terms of the way the sequence grows. — Page 418
5. Recognize that specifying a few terms of a sequence doesn't specify the sequence. — Page 419

Section 9.6 Functions — Page 426

- A function is a rule that assigns one output to each allowable input. We can describe or represent functions with words, with tables (which usually don't convey complete information), with graphs (which usually don't convey complete information), and with expressions or equations (although not every function has an expression or equation). — Page 426

Key Skills and Understandings

1. Determine if a proposed function is or is not a function. — Page 426
2. Use words to describe a function for a given context. — Page 426
3. Given a description of a function in words, draw a graph that could be the graph of the function. Explain how the graph fits with the description of the function. — Page 426
4. Given the graph of a function, describe aspects or features of the function. Explain how the graph informs you about those aspects or features. — Page 426

Section 9.7 Linear Functions	Page 436
▪ A linear function is a function whose graph lies on a line in a coordinate plane. Proportional relationships and arithmetic sequences are linear functions. For any linear function, the change in input and the change in corresponding output is always in the same ratio. There are numbers m and b such that the linear function is described by an equation of the form $y = mx + b$; the number b is the output for the input 0 and the number m is the increase in output when the input increases by 1.	Page 436
Key Skills and Understandings	
1. Understand that variables stand for numbers and are not "labels." Define variables for quantities and specify a unit of measurement (if needed). For example, "let G be the number of gallons of juice in a vat" or "let d be the distance traveled in kilometers."	Page 438
2. Be aware of common errors in formulating equations.	Page 440
3. Given a scenario that gives rise to a linear function, describe the function in words, with a table, with a graph, and with an equation, and explain why the equation is valid. Relate details of the situation to details of the table, graph, and equation.	Page 437
4. Given information about a function, determine if the function is linear or not.	Page 437
5. Contrast patterns of change in linear and other types of functions.	Page 438
6. In selected cases, use the structure of an expression to reason about its values, such as its maximum or minimum value or the values of the variable for which the expression is 0.	Page 440

10

Geometry

Geometry is the study of space and shapes in space. The word geometry comes from the Greek and means measurement of the earth (geo = earth, metry = measurement). Mathematicians of ancient Greece developed fundamental concepts of geometry to answer basic questions about the earth and its relationship to the sun, the moon, and the planets: How big is the earth? How far away is the moon? How far away is the sun? The geometers of ancient Greece had to be able to visualize the earth and the heavenly bodies in space. They had to extract the relevant relationships from their mental pictures, and they had to analyze these pictures mathematically. The methods the ancient Greeks developed are still useful today for solving modern problems in construction, road building, medicine, and other disciplines.

In this chapter, we focus on the following topics and practices within the *Common Core State Standards for Mathematics* (*CCSSM*).

Standards for Mathematical Content in the CCSSM

In the domain of *Geometry* (Kindergarten through Grade 8), students learn to identify and describe shapes by their geometric properties and they learn about angles and about the concepts of parallel and perpendicular. By attending to shapes' properties, students reason about relationships among categories of shapes, and they organize these categories into hierarchies. In later grades students reason about relationships among angles. They solve problems about angles and they give (informal) explanations for angle relationships, such as that the sum of the angles in a triangle is always 180°.

Standards for Mathematical Practice in the CCSSM

Opportunities to engage in all eight of the Standards for Mathematical Practice described in the *Common Core State Standards* occur throughout the study of shapes and their geometric attributes, although the following standards may be especially appropriate for emphasis:

- **4 Model with mathematics.** Students engage in this practice when they view objects in the real world as approximated by ideal geometric shapes and when they make two-dimensional drawings to represent an aspect of a real-world situation and use geometric aspects of the drawing to analyze the situation.
- **5 Use appropriate tools strategically.** Students engage in this practice when they use tools such as compasses, rulers, protractors, geometry software, or paper and scissors to make shapes and explore geometric properties. For example, students might make a rhombus by folding and cutting paper or by drawing it with a compass. In each case, they could argue that the resulting shape is a rhombus because of the way it was constructed.
- **7 Look for and make use of structure.** Students engage in this practice when they correctly identify a shape based on its attributes rather than on its overall appearance, when they look for parallel lines to determine angle relationships, or when they draw an extra line into a figure so as to make use of known angle relationships.

(From Common Core Standards for Mathematical Practice. Published by Common Core Standards Initiative.)

10.1 Lines and Angles

CCSS Common Core State Standards Grades 2, 4, 7, 8

Geometry gives us tools to mathematize and model aspects of the world around us. For example, think about traveling from one location to another. We might think of the two locations as two points, and we might think of a direct route from one location to the other as a line segment connecting the two points. When we use geometry to model a situation, we often make a math drawing that captures key elements of the situation and shows how those elements are related. We might represent the situation with lines and points and we might analyze the situation by reasoning about lengths of line segments and sizes of angles. In this section we will study some fundamental concepts in geometry, which are especially valuable for modeling navigation, and which will also help us describe and analyze geometric shapes.

What Are Points, Lines, Line Segments, Rays, and Planes?

The terms *point*, *line*, and *plane* are usually considered primitive, undefined terms. Even so, we can describe how to think about and visualize points, lines, and planes (see Figure 10.1):

- To visualize a **point**, think of a tiny dot, such as the period at the end of a sentence. A point is an idealized version of a dot, having no size or shape.
- To visualize a **line**, think of an infinitely long, stretched string that has no beginning or end. A line is an idealized version of such a string, having no thickness.
- To visualize a **plane**, think of an infinite flat piece of paper that has no beginning or end. A plane is an idealized version of such a piece of paper, having no thickness.

line segment

Related to lines are line segments and rays, as pictured in Figure 10.1. A **line segment** is the part of a line lying between two points on a line. These two points are called the **endpoints** of the line segment. Think of a line segment as having both a beginning and an end, even though both points are called endpoints. A **ray** is the part of a line lying on one side of a point on the line. Think of a ray as having a beginning, but no end.

ray

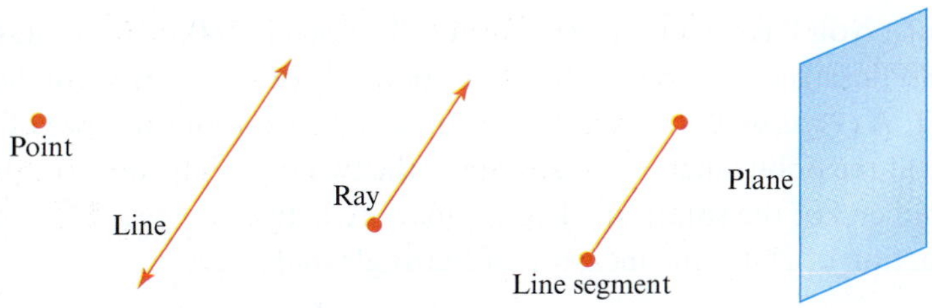

Figure 10.1 Points, lines, planes. Arrow heads indicate that the line or ray extends indefinitely in that direction.

How Can We Define Angles?

In this section we will study some key facts about angles that are produced by configurations of lines in a plane, including the fact that the sum of the angles in a triangle is 180°. We use angles in two ways: to represent an amount of rotation (turning) about a fixed point, and to describe how two rays (or lines, line segments, or even planes) meet.

The two ways of thinking about angles, as amounts of rotation and as rays meeting, are closely related. It is equally valid to define angles from either point of view. Since even very young children have experience spinning around, the "rotation" point of view is perhaps more primitive. So, our first definition is that an **angle** is an amount of rotation about a fixed point. An angle at a point P is said to be **congruent** to an angle at a point Q if both represent the same amount of rotation—even though this rotation takes place around different points. (See Figure 10.2.) Informally, we often simply say that congruent angles are "equal," or "the same size."

angle (rotation)

congruent

Now we turn to the other point of view about angles. Suppose there are two rays in a plane and these rays have a common endpoint P, as illustrated in Figure 10.3(a). The two rays and a region between them is an **angle** at P. The point P is called the **vertex** of the angle.

angle (between rays)

vertex (of an angle)

To relate this definition of angle to the rotation definition, think of rotating one ray around the vertex P until it reaches the other ray, so that the rotating ray "sweeps out" the angle, as indicated in Figure 10.3(b).

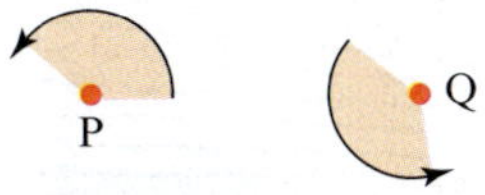

Figure 10.2 The same amount of rotation around different points.

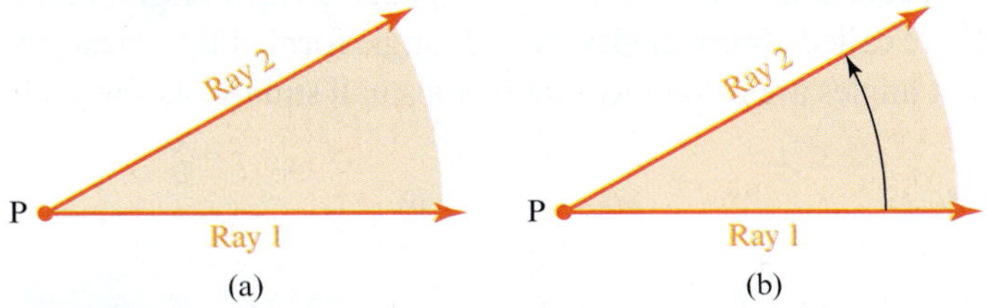

Figure 10.3 An angle formed by two rays meeting at point P.

How Do We Measure Angles and What Are Some Special Angles?

degree

We commonly measure angles in **degrees**, which we indicate with a degree symbol: °. For example, we write 30 degrees as 30°. We define degrees in reference to one full rotation, defined to be 360°. So if you stand at a point and rotate in a full circle, returning to your starting position, you have rotated 360°. We determine other angles by what fraction of a full rotation the angle is. So if you

CCSS 4.MD.5

rotate $\frac{1}{2}$ of a full turn, you have rotated 180° because $\frac{1}{2}$ of 360 is 180. If you rotate $\frac{1}{4}$ of a full turn, you have rotated 90° because $\frac{1}{4}$ of 360 is 90. If you rotate $\frac{1}{360}$ of a full turn, you have rotated 1° because $\frac{1}{360}$ of 360 is 1. As Figure 10.4 indicates, we can also think about the size of an angle in terms of a circle centered at the point where two rays meet (the two rays that show the directions you face at the beginning and end of the rotation). The circular arc between the rays is some fraction of the full circle. That fraction of 360 is the measure of the angle in degrees.

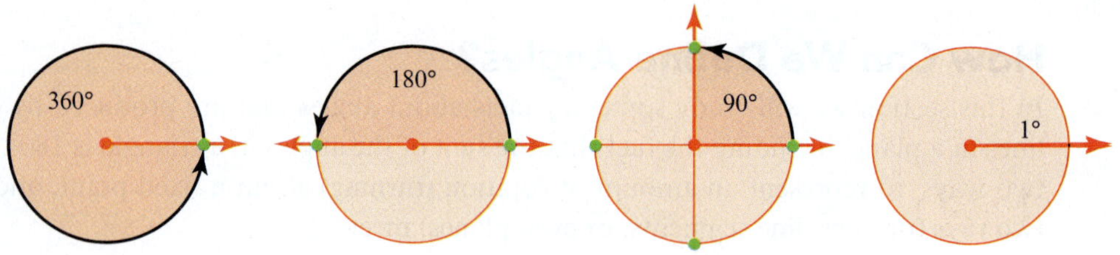

Figure 10.4 We determine the size of an angle by what fraction of a full circle the angle makes.

Now that we have defined what a 1° angle is, for a whole number, N, we can view an $N°$ angle as made from N rotations in the same direction, each of 1°. For example, we can think of 23° as made from 23 rotations, each of 1°, as in Figure 10.5. Angles of less than 1° are also possible. For example, if we view a 1° angle as made of two consecutive, equal rotations, then each of those rotations is $\frac{1}{2}°$.

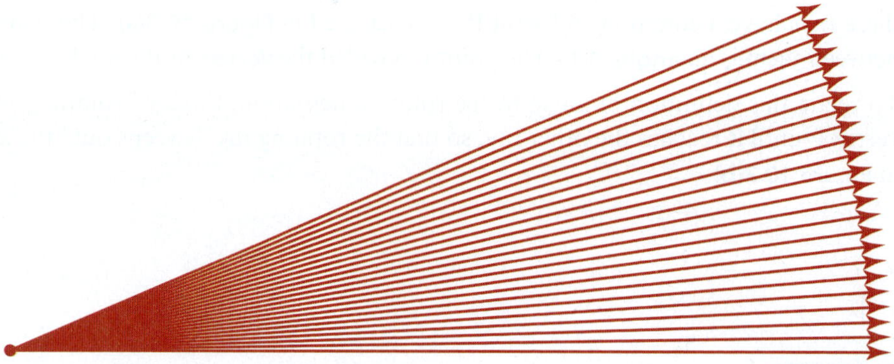

Figure 10.5 Viewing a 23° angle as 23 rotations, each of 1°.

Notice that small angles have a "pointier" look than larger angles, as seen in Figure 10.4. Angles less than 90° are called **acute angles**, an angle of 90° is called a **right angle**, and angles greater than 90° but less than 180° are called **obtuse angles**. A 180° angle formed by a straight line is often called a **straight angle**. Right angles are often indicated by a small square, as shown in Figure 10.6.

right angle

straight angle

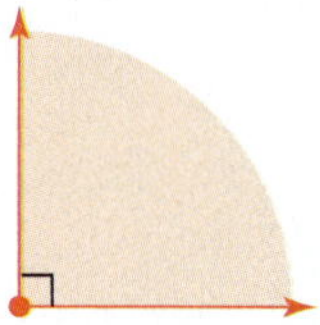

Figure 10.6 Indicating a right angle.

protractor

CCSS 4.MD.6

We can measure angles with a simple device called a **protractor**. Figure 10.7 shows a protractor measuring an angle. Protractors usually have a small hole that should be placed directly over the vertex where the two rays meet. This hole lies on a horizontal line through the protractor. This horizontal line should be lined up with one of the rays that form the angle to be measured. The protractor in Figure 10.7 shows that the angle it is measuring is 65°.

Figure 10.7

A protractor measuring an angle.

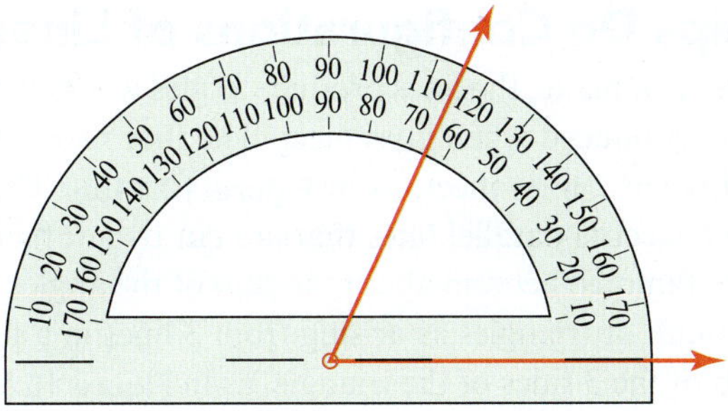

Video: Using a Protractor

Some teachers help their students learn about angles by making and using "angle explorers." An angle explorer is made by attaching 2 strips of cardboard with a brass fastener, as shown in Figure 10.8. By rotating the cardboard strips around, students get a feel for angles.

Figure 10.8

An "angle explorer."

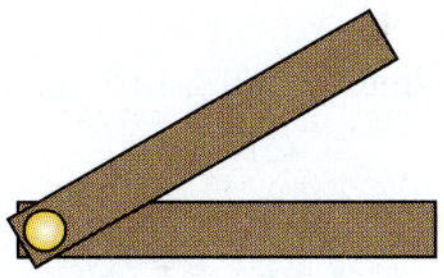

CCSS 4.G.1

perpendicular

When two lines in a plane intersect, they form four angles. Figure 10.9 shows some examples. When all four of the angles are 90°, we say that the two lines are **perpendicular**.

Figure 10.9

Two lines meeting at a point form 4 angles.

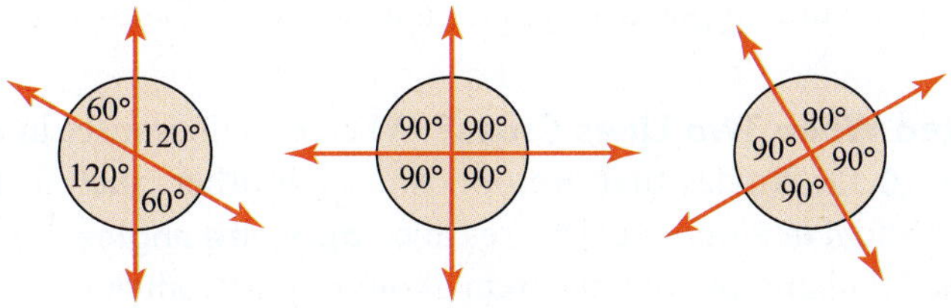

parallel

Two lines in a plane that *never* intersect (even somewhere far off the page they are drawn on) are called **parallel**. Figure 10.10 shows examples of lines that are parallel and lines that are not parallel, even though you can't see where they meet. We often label parallel lines with arrows, as in Figure 10.10(a) and (b).

Figure 10.10

Parallel lines and lines that are not parallel.

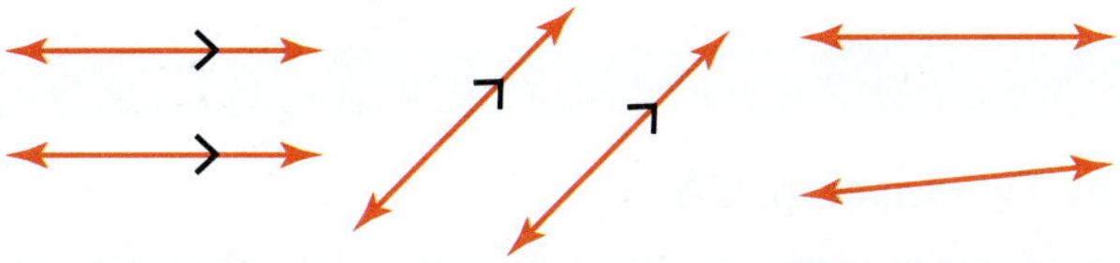

CLASS ACTIVITY

10A Folding Angles, p. CA-200

supplementary angles
complementary angles

We can add the measures of angles that are next to each other and don't overlap. **Supplementary angles** are angles that add to 180°. **Complementary angles** are angles that add to 90°.

What Angle Relationships Do Configurations of Lines Produce?

Whenever there is a collection of lines in a plane, the lines produce angles where they meet. There are three key angle relationships that are produced when lines meet. The first relationship is a theorem about the angles that are produced when 2 lines meet, as in Figure 10.11(a). The second relationship is the Parallel Postulate, which concerns parallel lines that are cut by another line, as in Figure 10.11(b). The third relationship is the famous theorem about the sum of the angles in a triangle. Why does this theorem fit here? We can think of triangles as arising from 3 lines in a plane that don't all meet at a single point, 1 line for each of the 3 sides of the triangle, as in Figure 10.11(c).

Figure 10.11 Three configurations of lines in a plane.

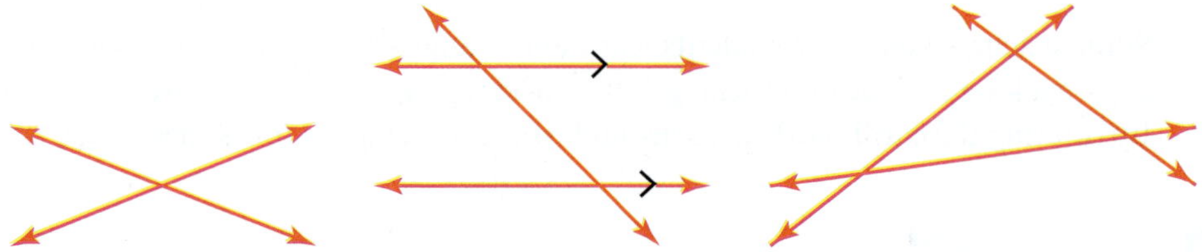

Two lines meeting in a plane (a) | Parallel lines cut by a third line (b) | Three lines forming a triangle (c)

postulate **theorem** **proof**

Before we study these three key relationships, what is a theorem and what is a postulate? A **postulate** (or axiom) is a mathematical statement that is considered foundational and is simply assumed to be true. In mathematics, we want to explain as much as possible why statements are true. But it is necessary to have some foundational assumptions, and these are called axioms or postulates. A **theorem** is a mathematical statement that has been proven to be true by a logical line of reasoning that is based on previously proven theorems and on postulates. Such a logical line of reasoning is called a **proof**.

CCSS 7.G.5

opposite angles **vertical angles**

How Are Angles Related When Two Lines Cross? When two lines meet in a plane they form four angles, as in Figure 10.12. Angles that are opposite each other, such as angles a and c in Figure 10.12 and angles b and d in Figure 10.12, are called **opposite angles** or **vertical angles**. In Class Activity 10B you will find and prove a theorem about opposite angles.

Figure 10.12 Opposite angles produced when two lines cross.

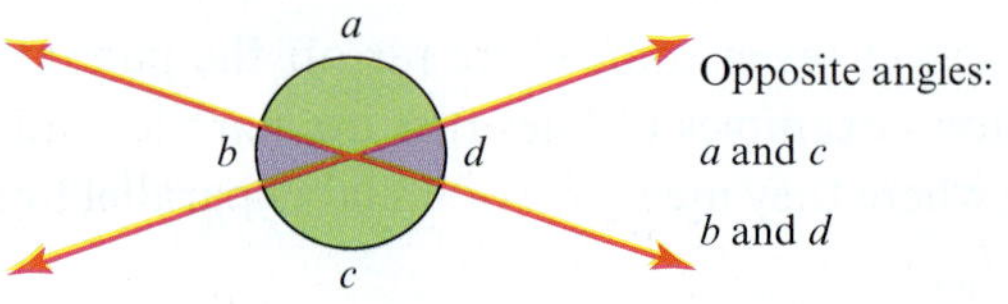

CLASS ACTIVITY

10B Angles Formed by Two Lines, p. CA-201

In Class Activity 10B you saw that opposite angles are equal. You will review a proof that opposite angles are equal in the Practice Exercises at the end of the section.

CCSS 8.G.5

transversal **corresponding angles**

How Are Angles Related When a Line Crosses Two Parallel Lines? If there are two lines in a plane and a third line crosses those two lines, then the third line is called a **transversal** of the other two lines. A transversal line produces 4 pairs of **corresponding angles**, as shown in Figure 10.13. As you can see in Figure 10.13, the two lines m and n are not parallel, and in this case the corresponding angles produced by the transversal line ℓ are not equal.

Figure 10.13

Corresponding angles produced when a transversal line crosses two lines.

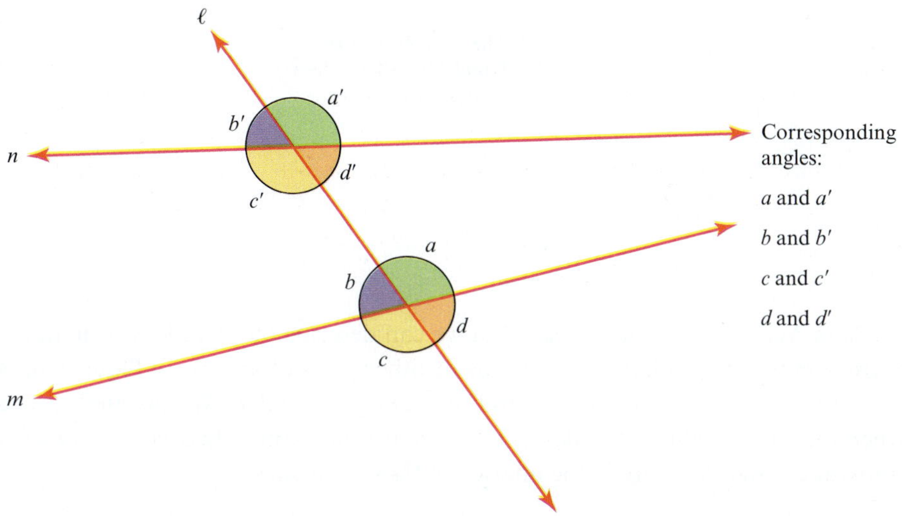

Parallel Postulate

When the two lines are parallel, then the angles produced by a transversal have a special relationship described by the Parallel Postulate. There are different, but equivalent versions of the Parallel Postulate. One version of the **Parallel Postulate** is this:

> If parallel lines are cut by a transversal line, then the corresponding angles produced are equal. In other words, in **Figure 10.14**,
>
> $$a = a', \quad b = b', \quad c = c', \quad d = d'$$

Figure 10.14

The Parallel Postulate.

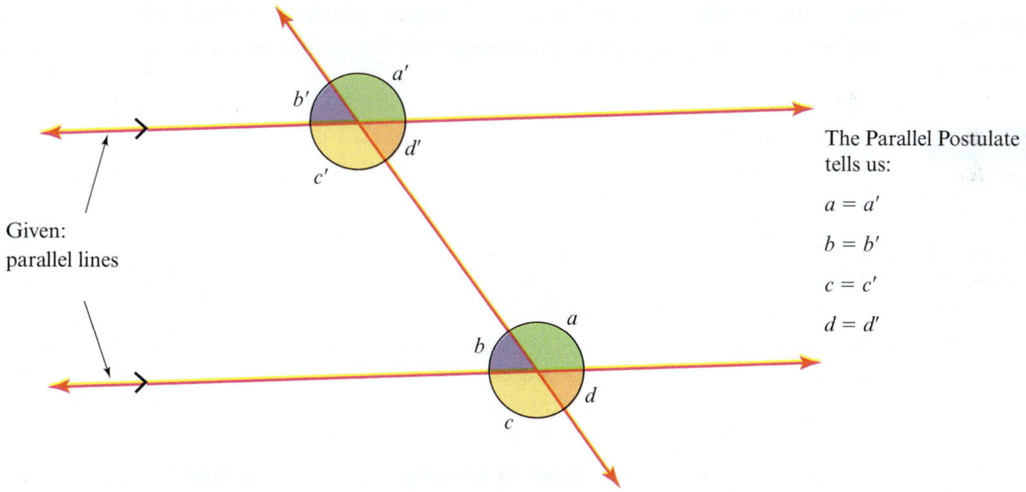

Although it is possible to prove this version of the Parallel Postulate from other equivalent versions of it, we have to assume some version of the postulate because there are versions of geometry in which the Parallel Postulate is actually not true. This was a shocking discovery for mathematicians. In fact, according to current theories in physics, the Parallel Postulate is not completely true in our universe. But in mathematics we often make simplifying assumptions and use models for the world around us that approximate the actual situation. So from here on, we will take the Parallel Postulate to be true in the geometry we are working in.

We will also take the *converse* of the Parallel Postulate to be true. This states that if two lines m and n are crossed by a line ℓ in such a way that two corresponding angles are equal (see **Figure 10.15**), then the lines m and n are parallel. The converse of the Parallel Postulate is useful for constructing parallel lines.

Figure 10.15

The converse of the Parallel Postulate.

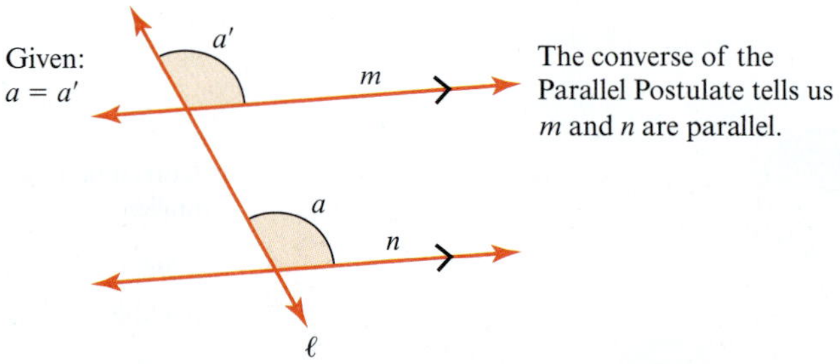

alternate interior angles

In Class Activity 10C you will see that we can describe the Parallel Postulate not just in terms of corresponding angles but also in terms of other pairs of angles. In Figure 10.14, angles a and c' are called *alternate interior angles*, and so are angles b and d'. We say angles are **alternate interior** when they are on alternate sides of a transversal line and in between the two lines crossed by the transversal, and therefore in the interior of those two lines.

CLASS ACTIVITY

10C Angles Formed When a Line Crosses Two Parallel Lines, p. CA-202

CCSS 8.G.5

interior angles

exterior angles

How Are Angles Related When Three Lines Form a Triangle? Three lines in a plane generally form a triangle, as in Figure 10.16(a). Notice that there are many different angles we can focus on. We can consider the angles *in* the triangle, as indicated in Figure 10.16(b). The angles in a triangle are sometimes called the **interior angles** of the triangle. We can also consider the **exterior angles** of the triangle, which are angles d, e, f in Figure 10.16(c), or the angles opposite them. As we'll see, there are interesting relationships among interior angles and among exterior angles of triangles.

Figure 10.16

The interior and exterior angles of a triangle.

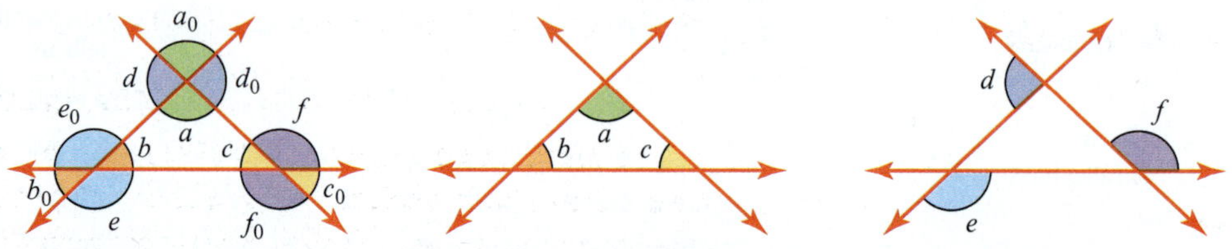

(a) Three lines forming a triangle and many angles.

(b) Angles a, b, c are *interior* angles of the triangle.

(c) Angles d, e, f (or their opposites) are *exterior* angles of the triangle.

What can we say about the angles in a triangle? Before you read on, try Class Activity 10D.

CLASS ACTIVITY

10D How Are the Angles in a Triangle Related, p. CA-203

A beautiful theorem that tells us that for every triangle, the sum of the angles in the triangle is 180° (see Figure 10.17). We can guess this theorem by putting corners of triangles together as in Class Activity 10D, but that doesn't explain why the theorem is true. Stop for a moment to think about the theorem; it is really quite remarkable. No matter what the triangle, no matter how oddly shaped it may be—as long as it's a triangle—the angles *always* add to 180°. This theorem applies to *all* triangles. Surprisingly, we can explain why this theorem is true *without checking every single triangle individually*.

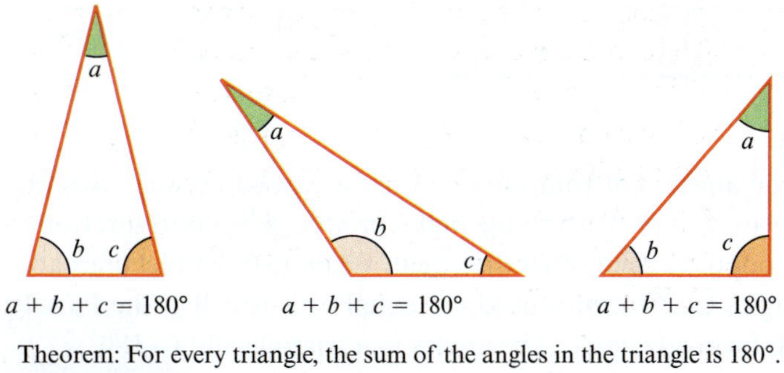

Figure 10.17 The theorem about the sum of the angles in a triangle.

Class Activities 10E and 10F will guide you through two different proofs of the theorem that the sum of the angles in a triangle is always 180°. Class Activity 10F will also help you discover a relationship among the exterior angles of a triangle.

CLASS ACTIVITY

10E Drawing a Parallel Line to Prove That the Angles in a Triangle Add to 180°, p. CA-204

10F Walking and Turning to Explain Relationships Among Exterior and Interior Angles of Triangles, p. CA-205

Notice that the proof in Class Activity 10E involved adding an extra line. Lines in geometry are like catalysts in chemistry—adding them often helps make a process work. When solving geometry problems, see if adding a line can help you solve the problem.

Notice that in Class Activity 10F, you had to think about angles both in terms of navigating a route and in terms of a map for that route. This required you to connect the two ways of thinking about angles: as amounts of turning and as lines meeting.

You will review proofs of the theorem that the angles in a triangle add to 180° in the Practice Exercises.

A common way that we apply the theorem about the sum of the angles in a triangle is this: If we know two of the angles in a triangle, we can figure out what the third angle must be. For example, in Figure 10.18 the triangle was constructed to have an angle of 40° and an angle of 70°. These two angles add to 110°; thus, the third angle must be 70° so that all three angles will add to 180°.

Figure 10.18 What is the third angle?

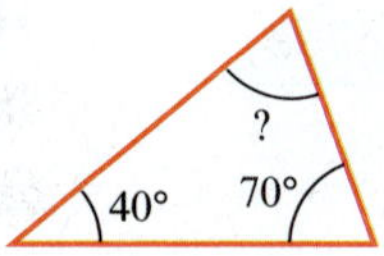

CLASS ACTIVITY

10G Angle Problems, p. CA-207

10H Students' Ideas and Questions About Angles, p. CA-208

SECTION SUMMARY AND STUDY ITEMS

Section 10.1 Lines and Angles

Angles can represent an amount of rotation or a region formed between two rays that meet. There are three important relationships concerning angles produced by configurations of lines. When two lines meet, the angles opposite each other are equal. When two parallel lines are cut by a third line, the corresponding angles that are produced are equal; this is called the Parallel Postulate. Three lines can meet so as to form a triangle. The angles in a triangle add to 180°.

Key Skills and Understandings

1. Be able to discuss the concept of angle.
2. Use the fact that a straight line forms a 180° angle to explain why the angles opposite each other, which are formed when two lines meet, are equal.
3. Know how to show informally that the sum of the angles in a triangle is 180°.
4. Use the Parallel Postulate to prove that the sum of the angles in every triangle is 180°.
5. Use the idea of walking around a triangle to explain why the sum of the angles in every triangle is 180°.
6. Apply facts about angles produced by configurations of lines to find angles.

Practice Exercises for Section 10.1

1. My son once told me that some skateboarders can do "ten-eighties." I said he must mean 180s, not 1080s. My son was right, some skateboarders can do 1080s! What is a 1080, and why is it called that? By contrast, what would a 180 be?

2. Use a protractor to measure the angles formed by the shape in Figure 10.19.

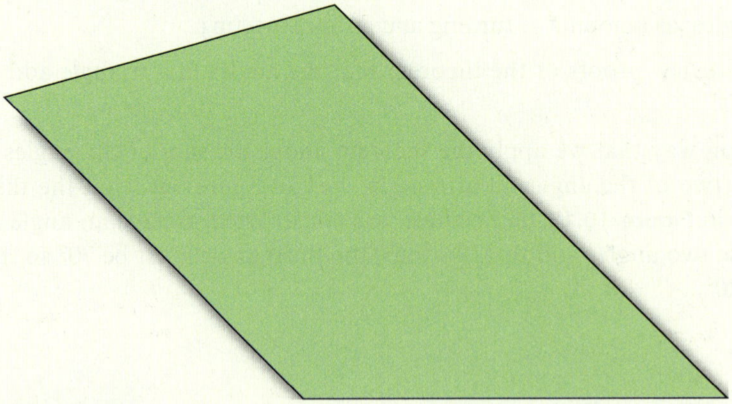

Figure 10.19 Measure the angles.

3. Suppose that two lines in a plane meet at a point, as in Figure 10.20. Use the fact that the angle formed by a straight line is 180° to explain why $a = c$ and $b = d$. In other words, prove that opposite angles are equal.

4. Given that the indicated lines in Figure 10.21 are parallel, determine the unknown angles without actually measuring them. Explain your reasoning briefly.

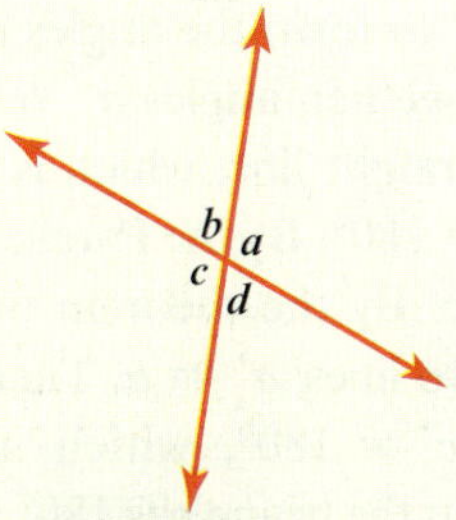

Figure 10.20 Lines meeting at a point.

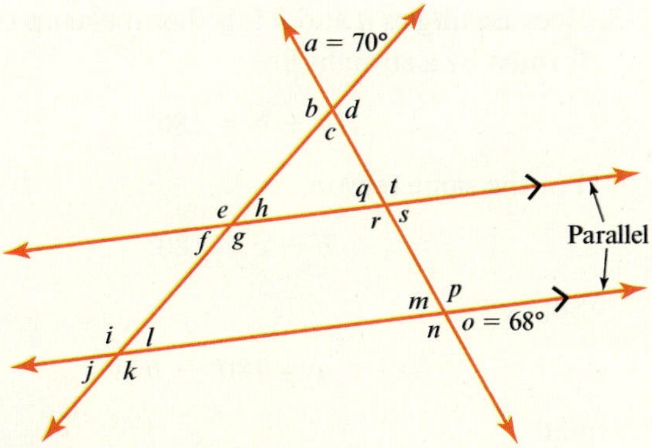

Figure 10.21 Find all angles.

5. Use the Parallel Postulate to prove that for all triangles, the sum of the angles in the triangle is 180°.

6. Dorothy walks from point A to point F along the route indicated on the map in Figure 10.22.

a. Show Dorothy's angles of turning along her route. Use a protractor to measure these angles.

b. Use your answers to part (a) to determine Dorothy's total amount of turning along her route.

c. Now determine Dorothy's total amount of turning along her route in a different way than in part (b).

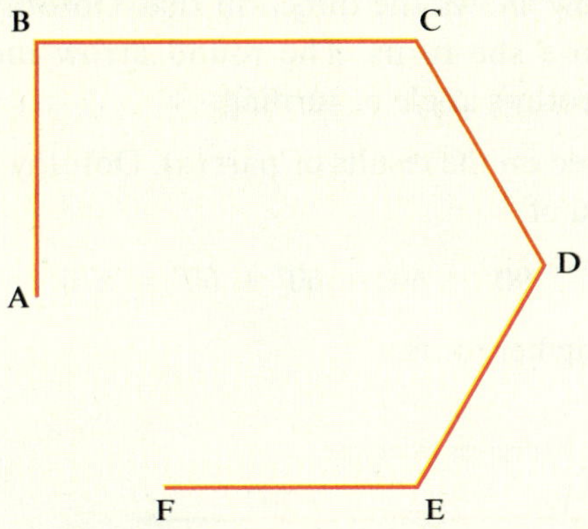

Figure 10.22 Dorothy's route.

7. Use the "walking and turning" method of Class Activity 10F to explain why the angles in a triangle add to 180°.

8. Figure 10.23 shows a square inscribed in a triangle. Since the square is *inside* the triangle, does that mean that the angles in the square add up to fewer degrees than the angles in the triangle?

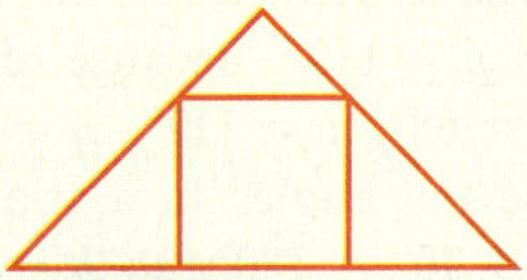

Figure 10.23 A square in a triangle.

Answers to Practice Exercises for Section 10.1

1. A "ten-eighty" is 3 full rotations. This makes sense because a full rotation is 360°, and

$$3 \times 360° = 1080°$$

A "180" would be half of a full rotation, which is not very impressive by comparison (although *I* certainly couldn't do it on a skateboard).

2. See Figure 10.24.

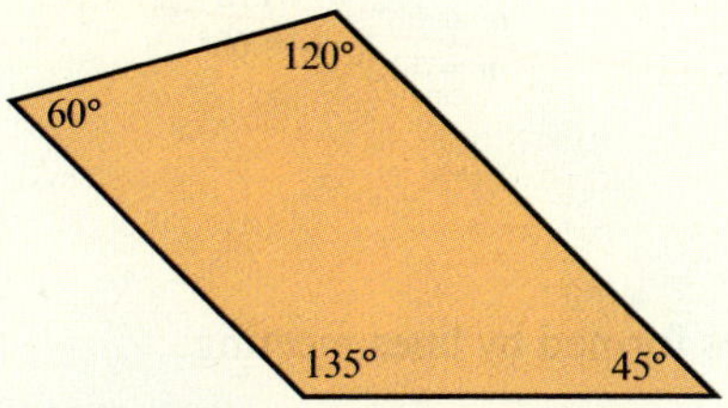

Figure 10.24 Angles in a shape.

3. Because angles a and b together make up the angle formed by a straight line,

$$a + b = 180°$$

For the same reason,

$$b + c = 180°$$

So,

$$a = 180° - b$$

and

$$c = 180° - b$$

Because a and c are both equal to $180° - b$, they are equal to each other (i.e., $a = c$). The same argument (with the letters changed) explains why $b = d$.

4. Solving this problem is a bit like working out a puzzle! See Figure 10.25. Because $a = 70°$ and angle c is opposite a, angle c must also be 70°. Therefore, $b = 110°$, because $b + c = 180°$. Because d is opposite b, it follows that $d = 110°$ too. The very same reasoning allows us to deduce that $m = 68°$ and that $n = 112°$ and $p = 112°$. Because of the Parallel Postulate, $s = 68°$, $t = 112°$, $q = 68°$, and $r = 112°$. Because the angles in a triangle add to 180°, $h + 68° + 70° = 180°$, so $h = 42°$. As before, we deduce that $f = 42°$ and $e = 138°$, because $42° + 138° = 180°$. So $g = 138°$ as well. The Parallel Postulate now allows us to deduce that $i = 138°$, $j = 42°$, $k = 138°$, and $l = 42°$.

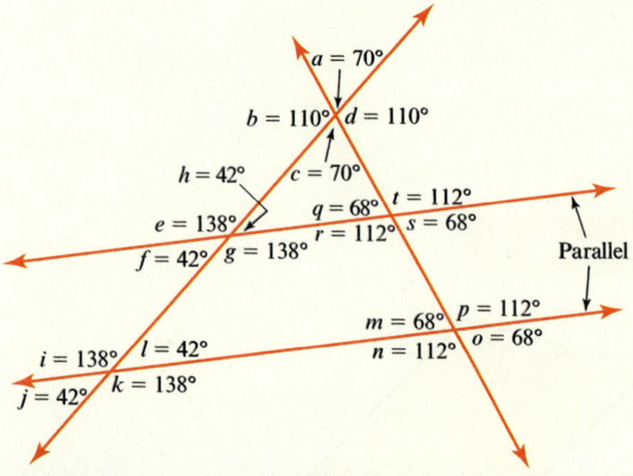

Figure 10.25 Angles formed by lines meeting.

5. Given a triangle with points (vertices) A, B, and C and angles a, b, and c at those points respectively, consider a line that is parallel to the side BC and goes through A, as in Figure 10.26, and extend the line segments AB and AC to form the angles a', b' and c' as in the figure. Together, angles a', b', and c' make the angle of a straight line, which is 180°. Therefore $a' + b' + c' = 180°$. By the Parallel Postulate, $b' = b$ and $c' = c$. By the theorem on opposite angles formed by two lines, $a' = a$. Therefore $a + b + c = a' + b' + c' = 180°$, which shows that the sum of the angles in the triangle is 180°. Even though the figure shows one particular triangle, the argument is general and holds for all triangles.

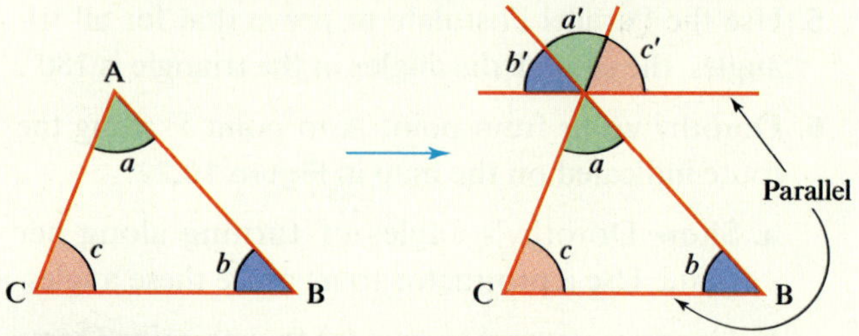

Figure 10.26 Proving that the angles in a triangle add to 180°.

6. a. Figure 10.27 shows Dorothy's angles of turning. At each point where the path turns, the straight arrow shows the direction that Dorothy faces before she turns. The round arrow indicates Dorothy's angle of turning.

b. Based on the results of part (a), Dorothy turns a total of

$$90° + 60° + 60° + 60° = 270°$$

along her route.

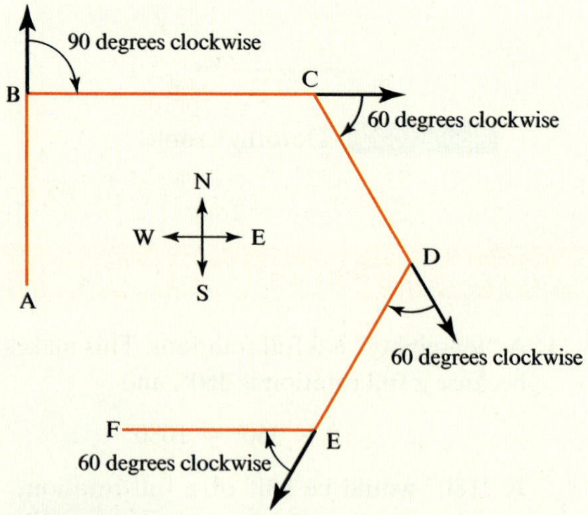

Figure 10.27 Dorothy's angles of turning along her route.

c. Another way to determine Dorothy's total amount of turning is by considering the directions she faces as she walks along her route.

If we think of Dorothy as starting out facing north, then she ends up facing west, and in between she faces all the directions clockwise from north to west. That means Dorothy makes $\frac{3}{4}$ of a full 360° turn during her walk. Since $\frac{3}{4}$ of 360° is 270°, Dorothy turns a total of 270°.

7. As seen in Class Activity 10F, and as indicated in Figure 10.28, if a person were to walk around a triangle, returning to her original position, she would have turned a total of 360°. When the person walks around the triangle, she turns the angles d, e, f, where d, e, and f are the exterior angles, as shown in Figure 10.28. This means that $d + e + f = 360°$. But because

$$a + f = 180°$$
$$b + d = 180°$$
$$c + e = 180°$$

it follows that, on the one hand,

$$(a+f) + (b+d) + (c+e) = 180° + 180° + 180° = 540°$$

while, on the other hand,

$$(a+f)+(b+d)+(c+e) = (a+b+c)+(d+e+f) = (a+b+c) + 360°$$

Because $(a + f) + (b + d) + (c + e)$ is equal to both 540° and $(a + b + c) + 360°$, these last two expressions must be equal to each other. That is,

$$(a + b + c) + 360° = 540°$$

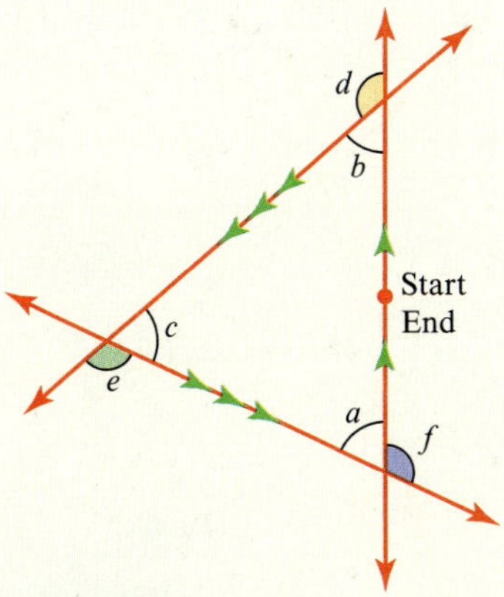

Figure 10.28 Exterior angles of a triangle add to 360°.

Therefore,

$$a + b + c = 540° - 360° = 180°$$

8. Unlike area, the angles in the square and the triangle don't depend on the sizes of either of these shapes. The fact that the square is inside the triangle has nothing to do with the sum of the angles in the square. We could even enlarge the square, so that the triangle fits inside the square, without changing the sum of the angles in the square.

PROBLEMS FOR SECTION 10.1

1. Tiffany says that the angle at A in Figure 10.29 is bigger than the angle at B. Why might she think this? How might you discuss angles with Tiffany?

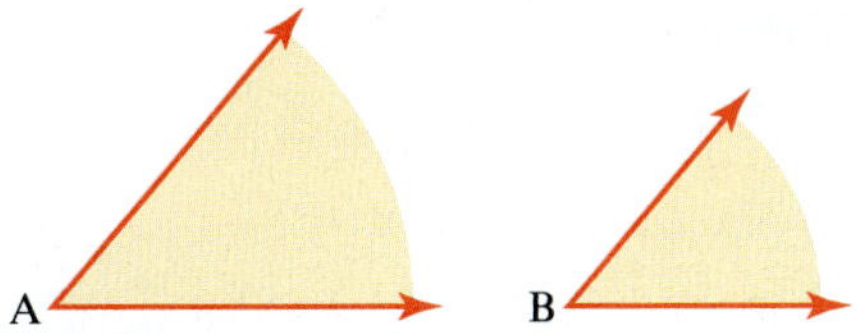

Figure 10.29 Is the angle on the left larger?

2. Given that the indicated lines in Figure 10.30(a) are parallel, determine the unknown angles without actually measuring them. Explain your reasoning briefly.

3. Given that the indicated lines in Figure 10.30(b) are parallel, determine the unknown angles without actually measuring them. Explain your reasoning briefly.

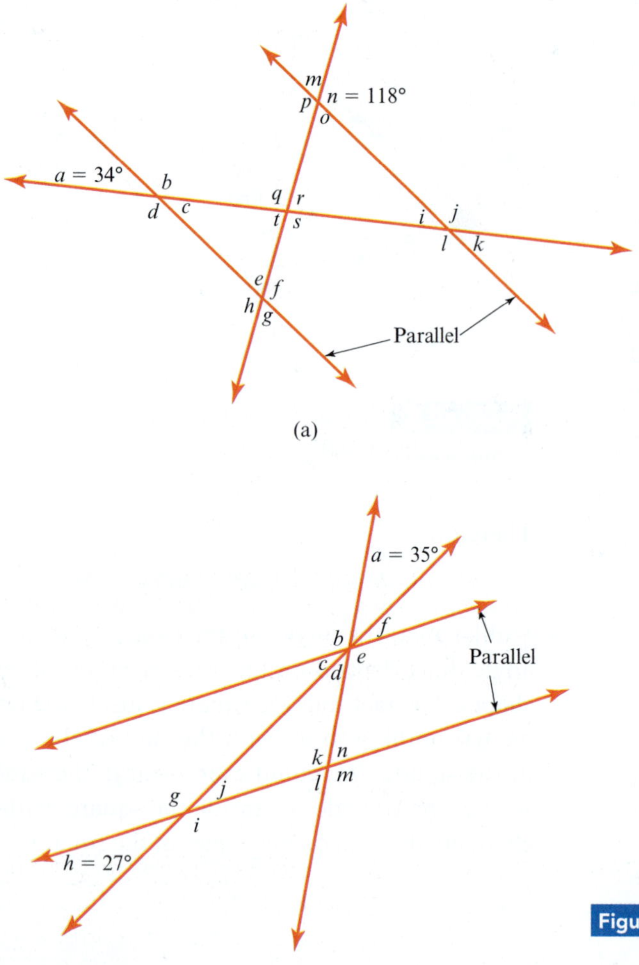

Figure 10.30 Determining the unknown angles in (a) and (b).

4. Amanda got in her car at point A and drove to point F along the route shown on the map in **Figure 10.31**.

a. Trace or redraw Amanda's route shown in **Figure 10.31**; show all of Amanda's angles of turning along her route. Use a protractor to measure these angles.

b. Determine Amanda's total amount of turning along her route by adding the angles you measured in part (a).

c. Now describe a way to determine Amanda's total amount of turning along her route *without* measuring the individual angles and adding them up. *Hint*: Consider the directions that Amanda faces as she travels along her route.

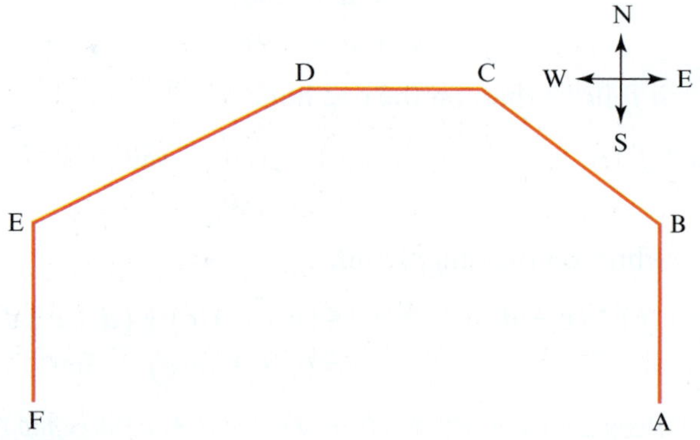

Figure 10.31 Amanda's route.

5. Ed the robot is standing at point A and will walk to point D along the route shown on the map in **Figure 10.32**. On the map, each segment between two dots represents 10 of Ed's paces.

a. Trace or redraw the map in **Figure 10.32**. At the points where Ed will turn, indicate his angle of turning. Use a protractor to measure these angles, and mark them on your map.

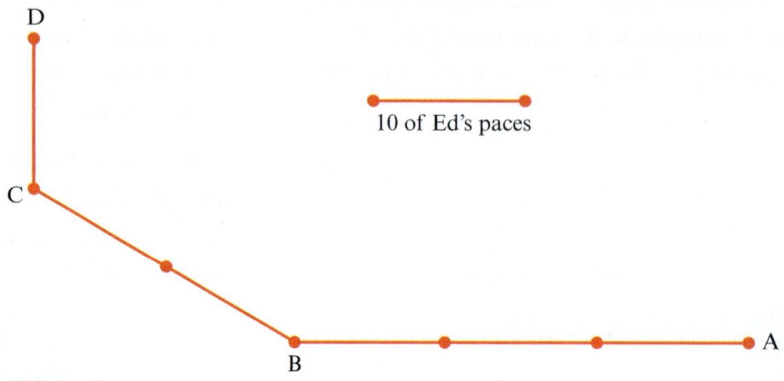

Figure 10.32 Ed's route.

b. Ed is a robot, so you must tell Ed exactly how to walk to point D. At points where Ed must turn, tell him how many degrees to turn, and which way.

c. Determine Ed's total amount of turning along his route by adding the angles you measured in part (a).

d. Now determine Ed's total amount of turning along his route *without* measuring the individual angles and adding them. *Hint*: Consider the directions that Ed faces as he travels along his route.

6. Draw a map that shows the following route leading to buried treasure:

Starting at the old tree, walk 10 paces heading straight for the tallest mountain in the distance. Turn clockwise, 90°. Walk 20 paces. Turn counterclockwise, 120°. Walk 40 paces. Turn clockwise, 60°. Walk 20 paces. This is the spot where the treasure is buried.

7. In your own words, give a clear, logical explanation for why the angles in a triangle must always add to 180°. Do not use the "putting angles together" method of Class Activity 10D.

8. Figure 10.33 shows a rectangle subdivided into four triangles. Jonathan says that because the angles in each of the four triangles add to 180°, the angles in the rectangle add to

$$180° + 180° + 180° + 180° = 720°$$

but Ben says that the angles in a rectangle add to

$$90° + 90° + 90° + 90° = 360°$$

Discuss the boys' ideas.

9. Given that the lines in Figure 10.34 marked with arrows are parallel, determine the sum of the angles $a + b + c$ without measuring.

10. Determine the sum $a + b + c + d + e + f$ of the angles in the 6-sided shape (hexagon) in Figure 10.35 without measuring. Explain your reasoning.

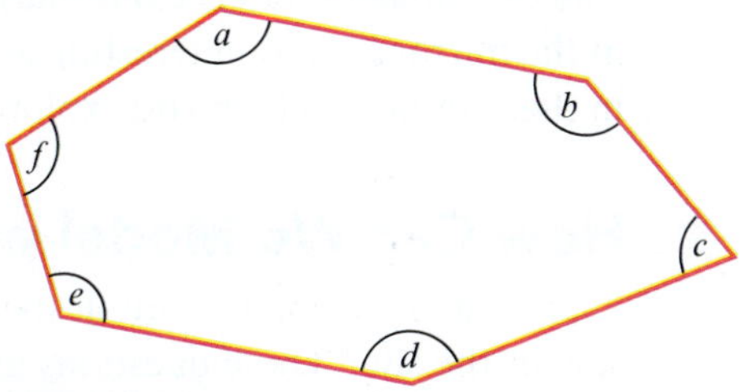

Figure 10.35 What is the sum of the angles in a 6-sided figure?

11. Given that the lines marked with arrows in Figure 10.36 are parallel, determine the sum of the angles $a + b + c + d + e$ without measuring the angles. Explain your reasoning.

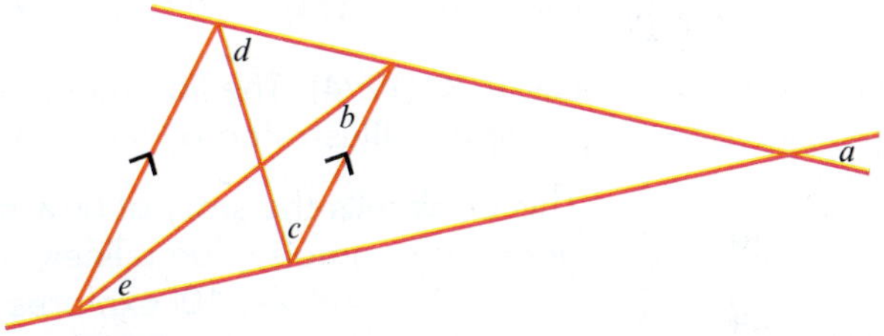

Figure 10.36 Determine the sum of the labeled angles.

12. Determine the sum of the angles at the star points, $a + b + c + d + e$, in Figure 10.37, without measuring. Explain your reasoning.

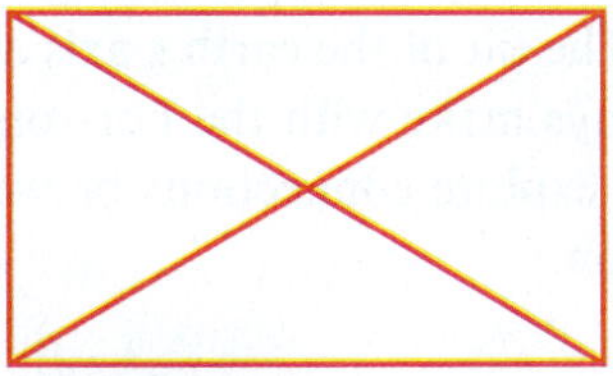

Figure 10.33 A rectangle subdivided into four triangles.

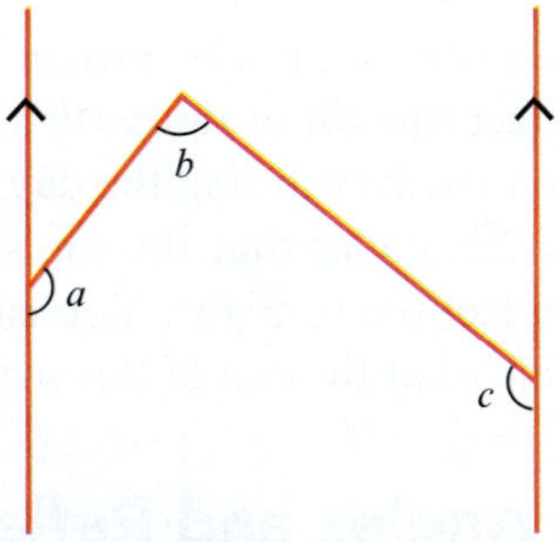

Figure 10.34 What is the sum of the marked angles?

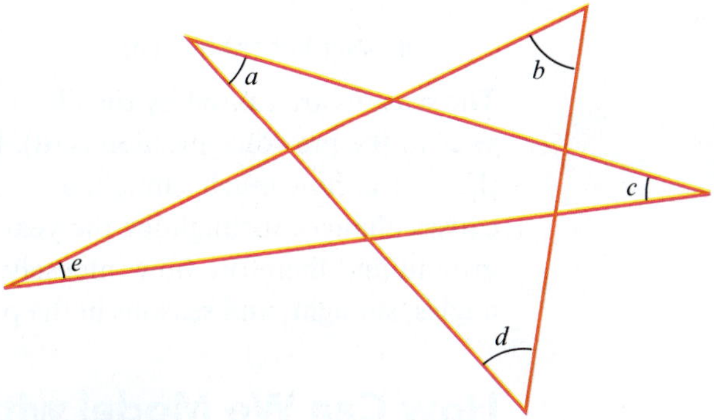

Figure 10.37 What is the sum of the angles in the star points?

10.2 Angles and Phenomena in the World

How are angles relevant to the world around us? An ancient method for determining how big the earth is uses reasoning about angles. The seasons are not caused by variation in the distance from the earth to the sun, but rather by the angles that the sun's rays make with the ground. Another way that angles influence our daily experiences involves reflected light. The way you see your face in the mirror is more interesting and surprising than you might guess. The activities and problems in this section will help you explore how angles help us model the world around us.

How Can We Model with Angles and Sun Rays?

How high or low is the sun in the sky? How long or short is the shadow of a pole standing outside in the sun? These questions are related to the angles that sun rays make with the horizontal ground or the tip of the pole. We can even use these ideas about angles to determine the earth's circumference (i.e., the distance around the earth), as Eratosthenes (275–195 B.C.) of ancient Greece discovered. Class Activity 10I is based on Eratosthenes's method.

CLASS ACTIVITY

10I Eratosthenes's Method for Determining the Circumference of the Earth, p. CA-209

FROM THE FIELD Children's Literature

Lasky, K. (1994). *The librarian who measured the earth.* Boston, MA: Little, Brown and Company. Illustrated by Kevin Hawkes.

This book tells the story of how Eratosthenes, who lived over 2000 years ago, used geometry, camels, plumb lines, and shadows to discover the circumference of the earth. Class Activity 10I explores the geometry behind Eratosthenes's method.

CCSS
SMP4

How would you answer the following question?

Question: What causes the seasons?

a. The seasons are caused by the earth moving closer to and farther from the sun.

b. The seasons are caused by the tilt of the earth's axis. The tilt brings one hemisphere closer to the sun, which makes that hemisphere hotter.

c. Both (a) and (b).

d. Neither (a) nor (b).

The seasons *are* caused by the tilt of the earth, but not by proximity to the sun, so the correct answer to the previous question is (d). In fact, the tilt of the earth's axis causes changes throughout the year in how much sunlight a location receives during the day. The tilt of the earth's axis also causes changes throughout the year in the angles that the sun's rays make with the horizontal ground and therefore how intensely we feel the sunlight. You can explore connections between angles, sunlight, and seasons in the problems at the end of this section.

How Can We Model with Angles and Reflected Light?

When a light ray strikes a smooth reflective surface, such as a mirror, the light ray reflects in a specific way that we can describe with angles.

normal line To describe how light reflection works, we need the concept of a **normal line** to a surface. The normal line at a point on a surface is the line that passes through that point and is perpendicular to the surface at that point. Think of "perpendicular to the surface at that point" as meaning "sticking straight out away from the surface at that point." **Figure 10.38** shows a cross-section of a "wiggly" surface and its normal lines at various points.

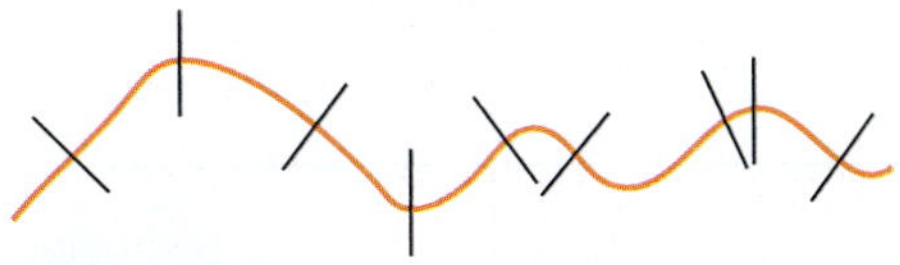

Figure 10.38 Some normal lines to a surface (cross-section shown).

Two fundamental physical laws govern the reflection of light from a surface. These laws also apply to the reflection of similar radiation, such as microwaves and radio waves:

1. The incoming and reflected light rays make the *same angle with the normal line* at the point where the incoming light ray hits the surface.
2. The reflected light ray lies in the same plane as the normal line and the incoming light ray. The reflected light ray is not in the same location as the incoming light ray unless the incoming light ray is in the same location as the normal line.

Figure 10.39 shows examples of how light rays reflect from surfaces. In each case, the figure shows a cross-section of the surface.

Figure 10.39 Light rays reflecting off surfaces.

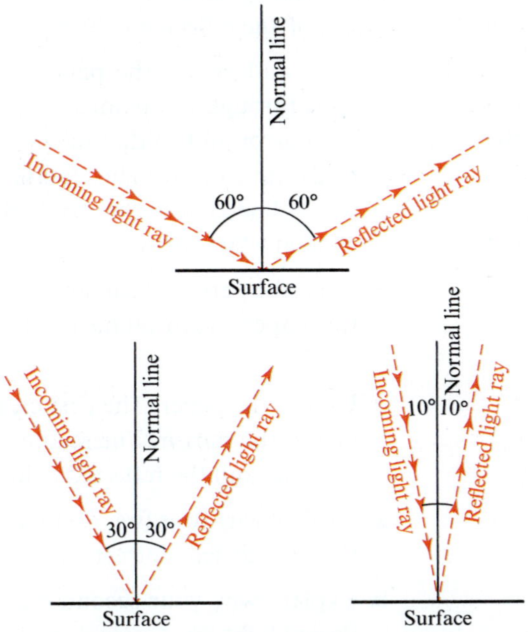

You can demonstrate the reflection laws by putting a mirror on a table, shining a penlight at the mirror, and seeing where the reflected light hits the wall. (Be careful to keep the light beam away from all observers' eyes.) If you raise and lower the light, while still pointing it at the mirror, the light beam traveling toward the mirror will form different angles with the mirror. By observing the reflected light beam on the wall, you can tell that the light beams going toward and from the mirror make the same angle with the normal line to the mirror. (If the mirror is on a horizontal table, then the normal lines to the mirror are vertical.)

Class Activities 10J and 10K show some interesting consequences of the laws of reflection.

CLASS ACTIVITY

10J Why Do Spoons Reflect Upside Down?, p. CA-210

10K How Big Is the Reflection of Your Face in a Mirror?, p. CA-210

SECTION SUMMARY AND STUDY ITEMS

Section 10.2 Angles and Phenomena in the World

Angles arise in real-world situations. For example, angles are formed by the sun's rays and angles describe reflected light rays.

Key Skills and Understandings

1. Apply the laws of reflection to explain what a person looking into a mirror will see.

Practice Exercises for Section 10.2

1. Make math drawings to show the relationship between the angle that the sun's rays make with horizontal ground and the length of the shadow of a telephone pole.

2. Figure 10.40 shows several math drawings (from the point of view of a fly looking down from the ceiling) of a person standing in a room, looking into a mirror on the wall. The direction of the person's gaze is indicated with a dashed line. What place in the room will the person see in the mirror?

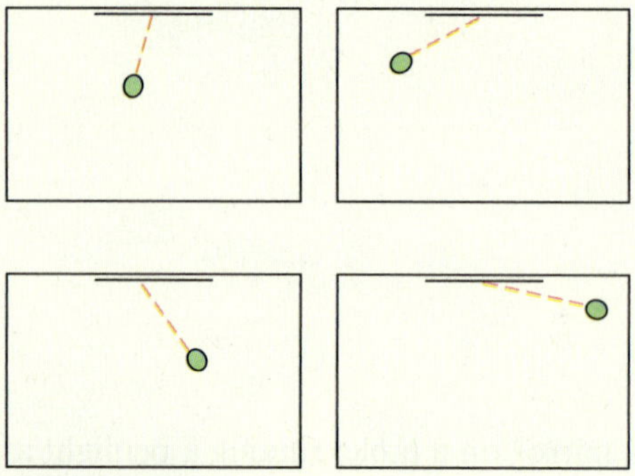

Figure 10.40 Looking into a mirror.

3. Figure 10.41 shows a mirror seen from the top, and a light ray hitting the mirror. Draw a copy of this picture on a blank piece of paper. Use the following paper-folding method to show the location of the reflected light ray:

- Fold and crease the paper so that the crease goes through the point where the light ray hits the mirror and so that the line labeled "mirror" folds onto itself. (This crease is perpendicular to the line labeled "mirror." You'll be asked to explain why shortly.)
- Keep the paper folded, and now fold and crease the paper again along the line labeled "light ray."
- Unfold the paper. The first crease is the normal line to the mirror. The second crease shows the light ray and the reflected light ray.

a. Explain why your first crease is perpendicular to the line labeled "mirror."

b. Explain why your second crease shows the reflected light ray.

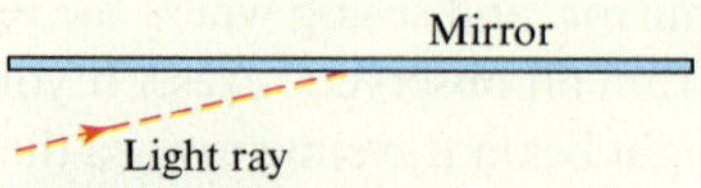

Figure 10.41 Using paper folding to find the reflected ray.

Answers to Practice Exercises for Section 10.2

1. Figure 10.42 shows that when the sun's rays make a smaller angle with horizontal ground, a telephone pole makes a longer shadow than when the sun's rays make a larger angle with the ground.

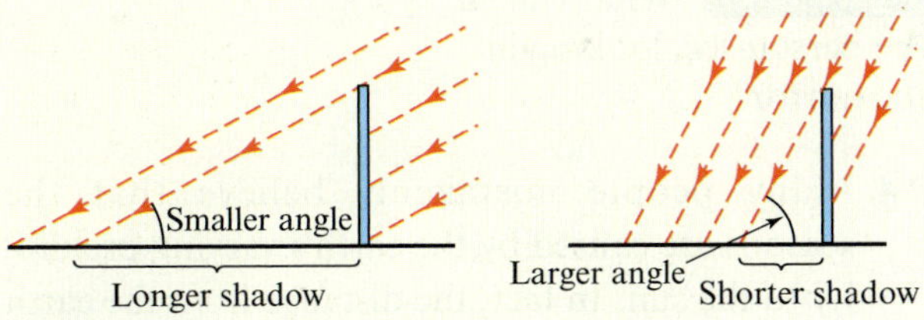

Figure 10.42 Sun rays hitting a telephone pole.

2. Figure 10.43 shows that the person will see locations A, B, C, and D, respectively, which are on various walls. Location C is in a corner. As the drawings show, the incoming light rays and their reflections in the mirror make the same angle with the normal line to the mirror at the point of reflection.

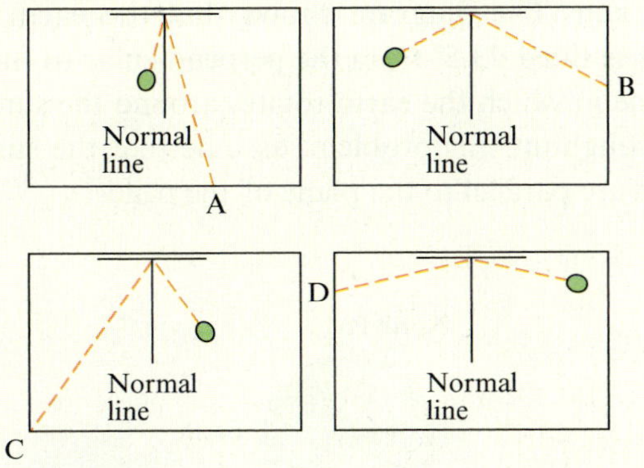

Figure 10.43 What a person sees, looking in a mirror.

3. a. The first crease is made so that angles a and b shown in Figure 10.44 are folded on top of each other and completely aligned. Therefore, angles a and b are equal. But these angles must add up to 180° because the angle formed by a straight line is 180°. Hence, angles a and b must both be half of 180°, which is 90°. Therefore, this first crease is the normal line to the mirror at the point where the light ray hits the mirror.

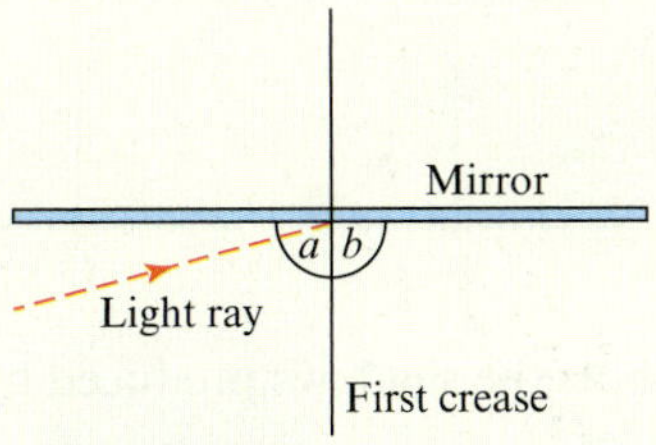

Figure 10.44 First crease.

b. The second crease is made so that angles c and d, shown in Figure 10.45 are folded on top of each other and completely aligned. Therefore, angle c is equal to angle d. Since the first crease is a normal line, by the reflection laws, the second crease shows the reflected light ray.

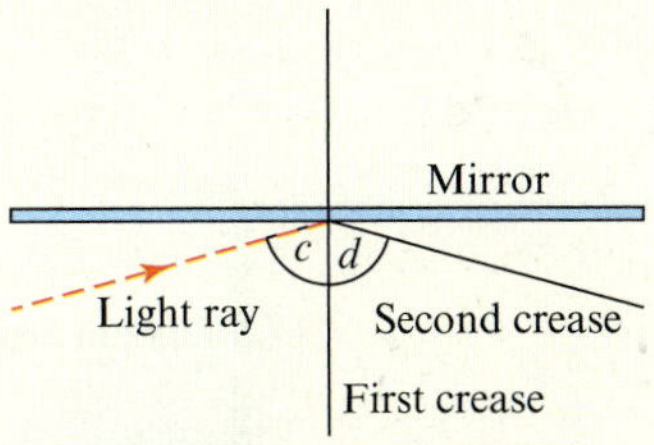

Figure 10.45 Second crease.

PROBLEMS FOR SECTION 10.2

1. Figure 10.46 depicts sun rays traveling toward 3 points on the earth's surface, labeled A, B, and C. Assume that these sun rays are parallel to the plane of the page.

   a. Use a protractor to determine the angle that the sun's rays make with horizontal ground at points A, B, and C.

   b. Suppose there are telephone poles of the same height at locations A, B, and C. Make drawings like the ones in Figure 10.47 on which the vertical lines represent telephone poles, and show on your drawings the different lengths of the shadows of these telephone poles, using the information from part (a).

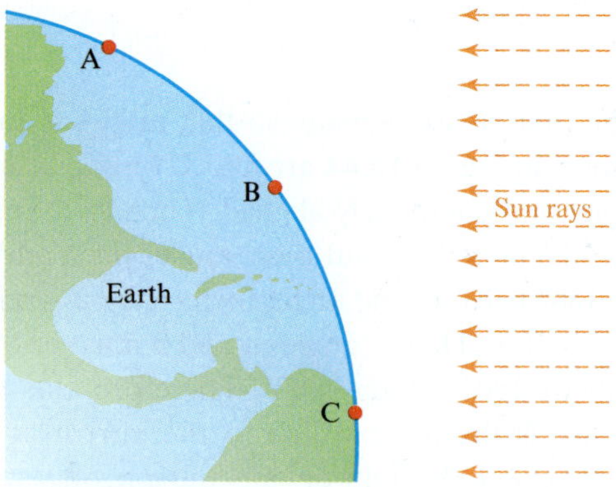

Figure 10.46 Sun rays traveling to the earth.

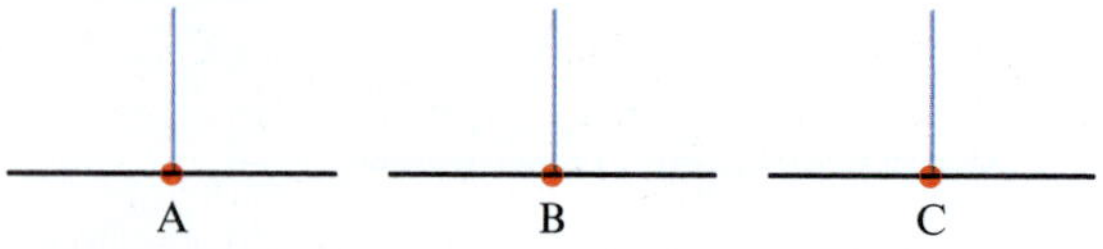

Figure 10.47 Show the shadows produced by the telephone poles.

2. **Figure 10.48** shows a cross-section of Joey's toy periscope. What will Joey see when he looks in the periscope? Explain, using the laws of reflection (trace the periscope). What would be a better way to position the mirror in the telescope?

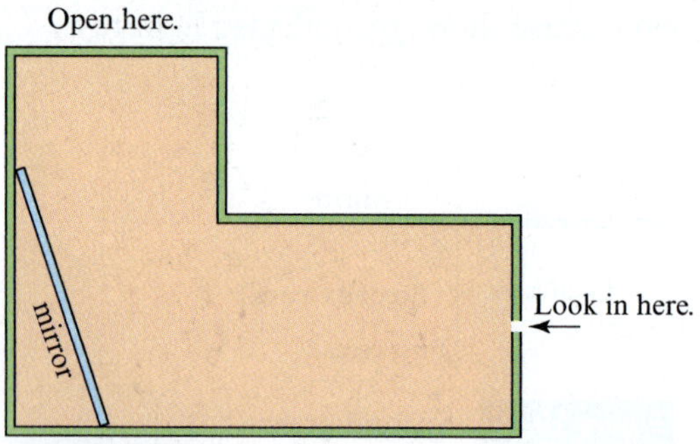

Figure 10.48 Joey's periscope.

3. **Figure 10.49** shows a sketch of a room that has a pair of perpendicular mirrors, drawn from the point of view of a fly on the ceiling. A person is standing in the room, looking into one of the mirrors at the indicated spot. On a separate piece of paper, draw a large sketch that looks (approximately) like **Figure 10.49**. Using your sketch, determine what the person will see in the mirror. (You may want to use paper folding, which is described in Practice Exercise 2.) Explain your answer.

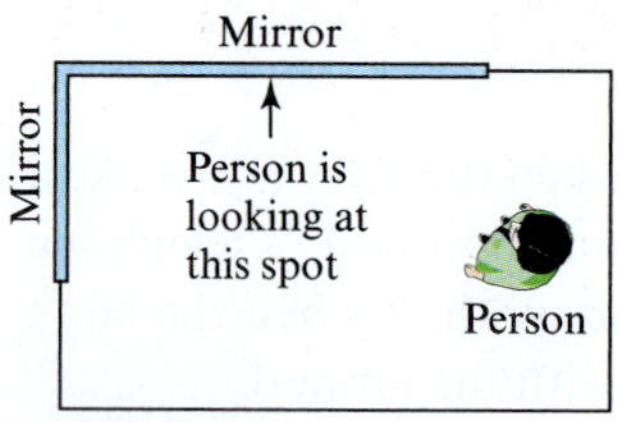

Figure 10.49 What will the person see, looking in the mirror?

*__4.__ Many people mistakenly believe that the seasons are caused by the earth's varying proximity to the sun. In fact, the distance from the earth to the sun varies only slightly during the year, and the seasons are caused by the tilt of the earth's axis. As the earth travels around the sun during the year, the tilt of the earth's axis causes the northern hemisphere to vary between being tilted toward the sun to being tilted away from the sun.

Figure 10.50 shows the earth as seen at a certain time of year from a point in outer space located in the plane in which the earth rotates about the sun. The diagram shows that the earth's axis is tilted 23.5° from the perpendicular to the plane in which the earth rotates around the sun. Throughout this problem, assume that the sun rays are parallel to the plane of the page.

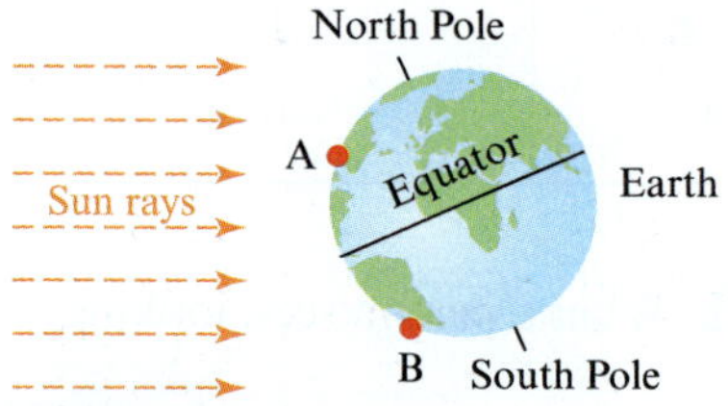

Figure 10.50 Earth and sun.

a. Locations A and B in **Figure 10.50** are shown at noon. Is the sun higher in the sky at noon at location A or at location B? Explain how you can tell.

b. During the day, locations A and B will rotate around the axis through the North and South Poles. Compare the amount of sunlight that locations A and B will receive throughout the day. Which location will receive more sunlight during the day?

c. Based on your answers to parts (a) and (b), what season is it in the northern hemisphere, and what season is it in the southern hemisphere in Figure 10.50? Explain.

d. At another time of year, the earth and sun are positioned as shown in Figure 10.51. At that time, what season is it in the northern and southern hemispheres? Why?

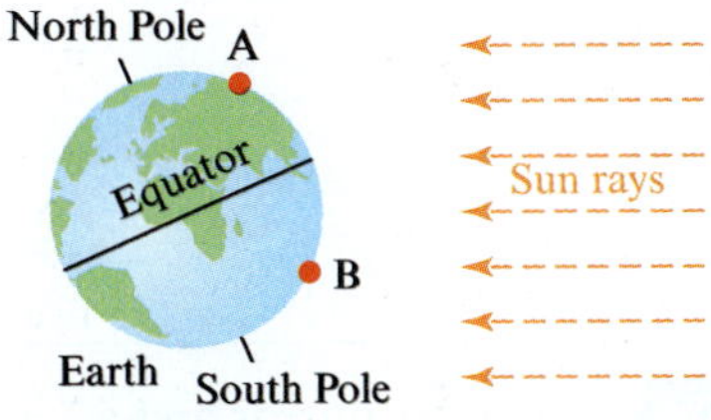

Figure 10.51 Earth and sun at another time of year.

e. At yet another time of year, the earth and sun are positioned as shown in Figure 10.52. At that time, what season is it in the northern and southern hemispheres? Why? Give an either-or-answer. (Notice that Figure 10.51 still shows the tilt of the earth's axis.)

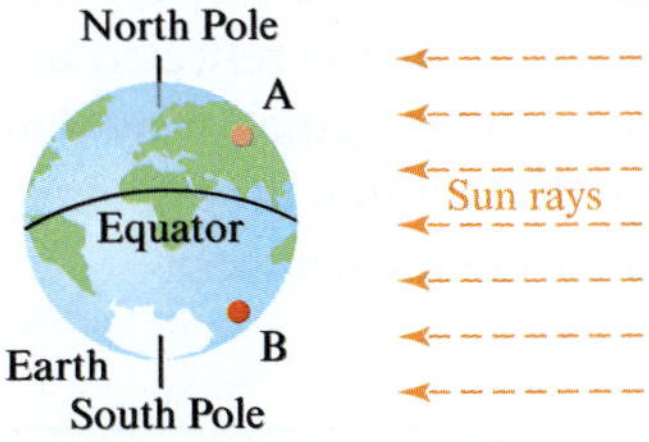

Figure 10.52 Earth and sun at yet another time of year.

f. Refer to Figures 10.50, 10.51, and 10.52 and the results of the previous parts of this problem to answer the following: During which seasons are the sun's rays most intense at the equator? Look carefully before you answer—the answer may surprise you.

*__5.__ Refer to Figures 10.50, 10.51, 10.52, and 10.53 to help you answer the following: There are only certain locations on the earth where the sun can ever be seen *directly* overhead. Where are these locations? How are these locations related to the Tropic of Cancer and the Tropic of Capricorn? Explain.

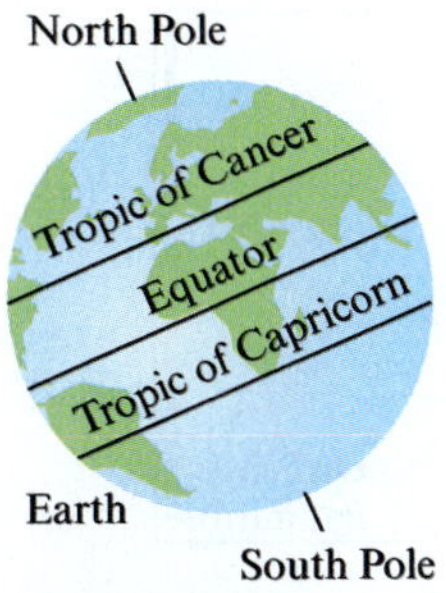

Figure 10.53 The Tropic of Cancer and the Tropic of Capricorn.

*__6.__ Department store dressing rooms often have large mirrors that actually consist of three adjacent mirrors, put together as shown in Figure 10.54 (as seen looking down from the ceiling). Use the laws of reflection to show how you can stand in such a way as to see the reflection of your back. Make a careful drawing, using an enlarged version of Figure 10.54, that shows clearly how light reflected off your back can enter your eyes. Your drawing should show where you are standing, the location of your back, and the direction of the gaze of your eyes. (You may wish to experiment with paper folding before you attempt a final drawing. See Practice Exercise 2.)

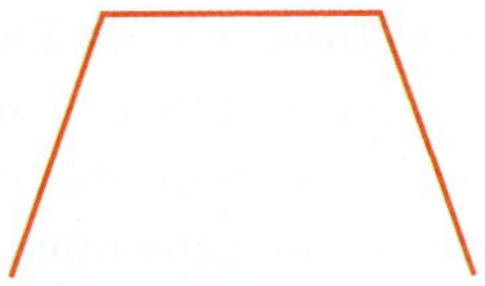

Figure 10.54 Department store mirror.

*__7.__ A **concave** mirror is a mirror that curves in, like a bowl, so that the normal lines on the reflective side of the mirror point toward each other. Makeup mirrors are often concave. The left side of Figure 10.55 shows an eye looking into a concave makeup mirror. The right side of Figure 10.55 shows an eye looking into an ordinary flat mirror. Trace the two diagrams in Figure 10.55 and, in each case, show where a woman applying eye makeup sees her eye in the mirror. Use the laws of reflection to show approximately where the woman sees the top of her reflected eye and where she sees the bottom of her reflected eye. (Assume that the woman sees light that enters the center of her eye.) Notice that the figure shows the normal lines to the concave mirror. Based

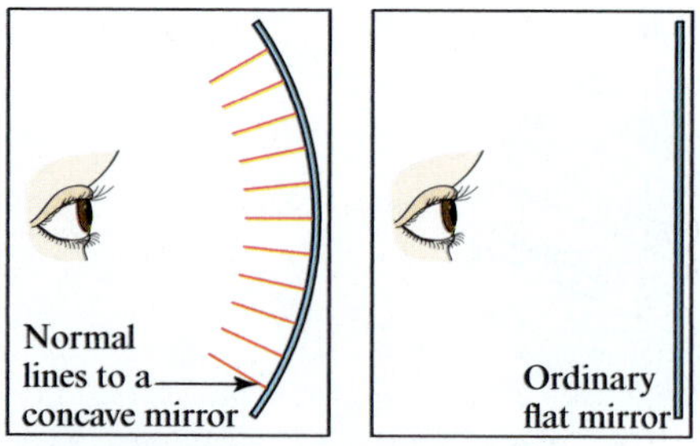

Figure 10.55 An eye looking in a concave mirror and a flat mirror.

on your drawings, explain why a concave mirror makes a good mirror for applying makeup. (By the way, although a concave mirror is curved in the same direction as the bowl of a spoon, the smaller amount of curvature in a concave mirror prevents it from reflecting your image upside down, as a spoon does.)

*8. A **convex** mirror is a mirror that curves out, so that the normal lines on the reflective side of the mirror point away from each other, as shown at the top of Figure 10.56. Convex mirrors are often used as side-view mirrors on cars and trucks. This problem will help you see why convex mirrors are useful for this purpose.

Figure 10.56 shows a bird's-eye view of a cross-section of a convex mirror, a cross-section of a flat mirror, and eyes looking into these mirrors. Draw a copy of these mirrors and the eyes looking into them. Use the laws of reflection to help you compare how much of the surrounding environment each eye can see, looking into its mirror.

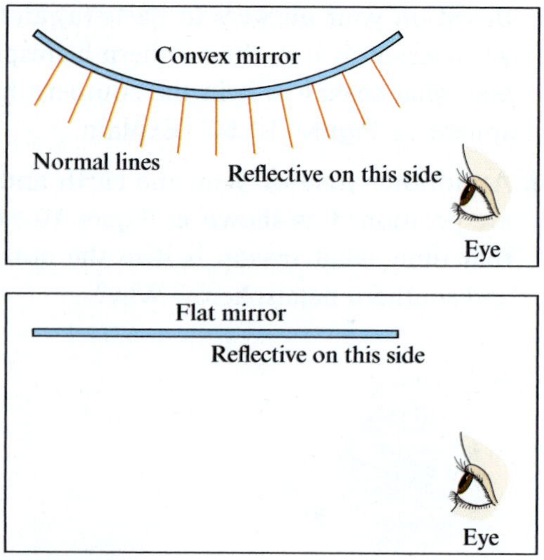

Figure 10.56 A convex mirror and a flat mirror.

Now explain why convex mirrors are often used as side-view mirrors on cars and trucks.

*9. A convex mirror is a mirror that curves out, so that the normal lines on the reflective side of the mirror point away from each other, as shown at the top of Figure 10.56. Convex mirrors are often used as side-view mirrors on cars and trucks, but these mirrors usually carry the warning sign that reads, "Objects are closer than they appear." Explain why objects reflected in a convex mirror appear to be farther away than they actually are. Use the fact that the eye interprets a smaller image as being farther away.

10.3 Circles and Spheres

CCSS Common Core State Standards Grades K, 7

What are circles and spheres? To the eye, circles and spheres are distinguished by their perfect roundness. Informally, we might say that a sphere is the surface of a ball. This describes circles and spheres from an informal or artistic point of view, but there is also a mathematical point of view. As we'll see, the mathematical definitions of circles and spheres yield practical applications that cannot be anticipated by considering only the look of these shapes.

CLASS ACTIVITY

10L Points That Are a Fixed Distance from a Given Point, p. CA-212

What Are Mathematical Definitions of Circle and Sphere?

circle
center
radius
diameter

A **circle** is the collection of all the points in a plane that are a certain fixed distance away from a certain fixed point in the plane. This fixed point is called the **center** of the circle, and the fixed distance is called the **radius** of the circle (plural: **radii**). So the radius is the distance from the center of the circle to any point on the circle. The **diameter** of a circle is two times its radius. Informally, the diameter is the distance "all the way across" the circle, going through the center.

For example, let's fix a point in a plane, and let's call this point P, as in Figure 10.57. All the points in the plane that are 1 unit away from the point P form a circle of radius 1 unit, centered at the point P, as shown in Figure 10.57. The diameter of this circle is 2 units.

Figure 10.57

A circle and a noncircle.

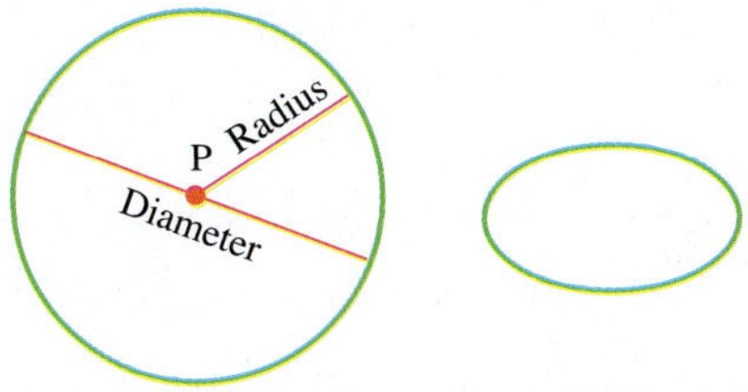

A circle centered at P This is NOT a circle.

sphere
center
radius
diameter

A sphere is defined in almost the same way as a circle, but in space rather than in a plane. A **sphere** is the collection of all the points in space that are a certain fixed distance away from a certain fixed point in space. This fixed point is called the **center** of the sphere, and this distance is called the **radius** of the sphere. So the radius is the distance from the center of the sphere to any point on the sphere. The **diameter** of a sphere is two times its radius. Informally, the diameter is the distance "all the way across" the sphere, through the center.

For example, fix a point P in space. You might want to think of this point P as located in front of you, a few feet away. Then all the points in space that are 1 foot away from the point P form a sphere of radius 1 foot, centered at P. Try to visualize this. What does it look like? It looks like the surface of a very large ball, as illustrated in Figure 10.58. The diameter of this sphere is 2 feet.

Figure 10.58

A sphere and a nonsphere.

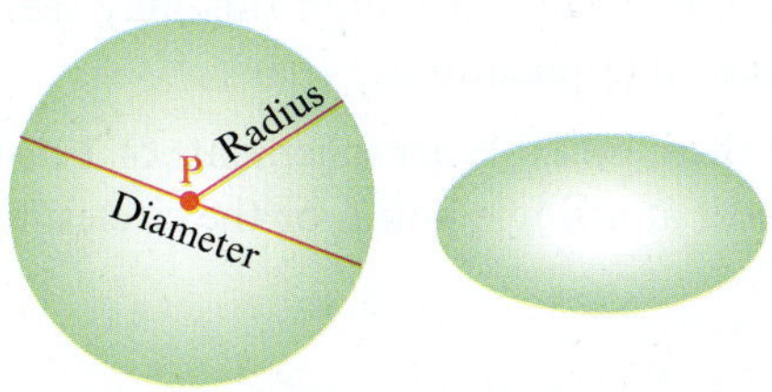

A sphere centered at P This is NOT a sphere.

compass

It's easy to draw almost perfect circles with the use of a common drawing tool called a **compass**, pictured in Figure 10.59. One side of a compass has a sharp point that you can stick into a point

Figure 10.59

Drawing circles and spheres.

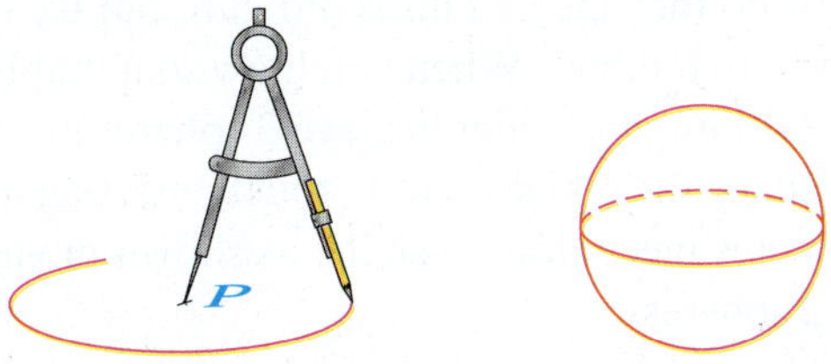

Drawing a circle with a compass A simple drawing of a sphere

on a piece of paper. The other side of a compass has a pencil attached. To draw a circle centered at a point P and having a given radius, open the compass to the desired radius, stick the point of the compass in the point P, and spin the pencil side of the compass in a full revolution around P, all the while keeping the point of the compass at P.

It is also possible to draw good circles with simple homemade tools. For example, you can use 2 pencils and a paper clip to draw a circle, as shown in Figure 10.60. Put a pencil inside one end of a paper clip and keep this pencil fixed at one point on a piece of paper. Put another pencil at the other end of the paper clip and use that pencil to draw a circle.

Figure 10.60 Drawing a circle with a paper clip.

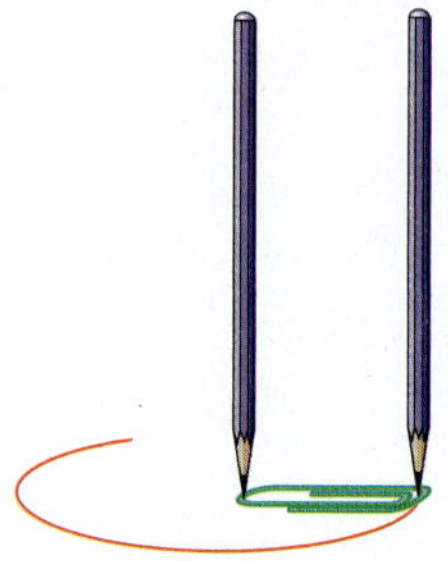

Drawing a circle with a paper clip

Because a sphere is an object in space, and not in a plane, it's harder to draw a picture of a sphere. Unless you are an exceptional artist, something like the simple drawing of a sphere in Figure 10.59 will do.

In Class Activity 10M, you will apply the mathematical definition of circles.

CLASS ACTIVITY

10M Using Circles, p. CA-212

How Can Circles or Spheres Meet?

What happens when 2 circles meet, or 2 spheres meet? Although this may seem like a topic of purely theoretical interest, it actually has practical applications.

As illustrated in Figure 10.61, there are only three possible arrangements of 2 distinct circles: The circles may not meet at all, the circles may meet at a single point, or the circles may meet at 2 distinct points.

Figure 10.61 Three ways that two circles can meet.

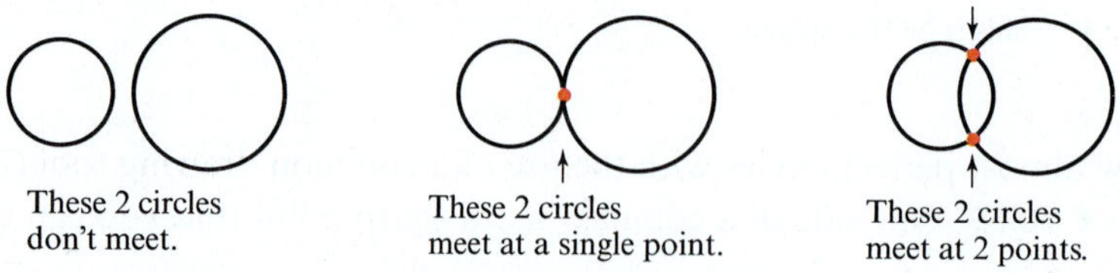

Now what if you have 2 spheres—how do they meet? This is difficult, but try to visualize 2 spheres meeting. It might help to think about soap bubbles. When you blow soap bubbles you occasionally see a "double bubble" that is similar (although not identical) to 2 spheres meeting. As with circles, 2 distinct spheres might not meet at all, or they might barely touch, meeting at a single point. The only other possibility is that the 2 spheres meet along a circle, as shown in Figure 10.62. This is a circle that is common to each of the 2 spheres.

In Class Activity 10N, you will see a practical application of spheres meeting.

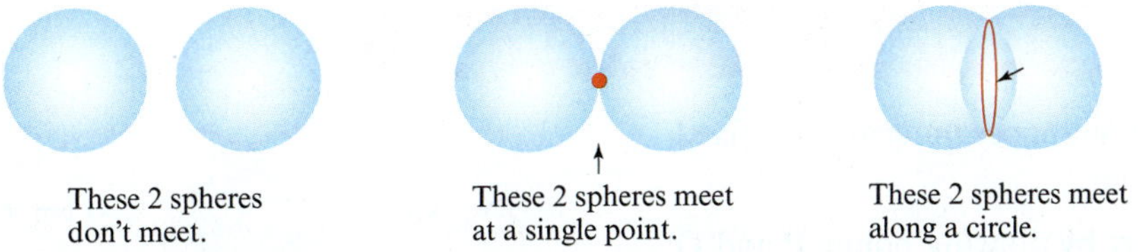

Figure 10.62 Three ways that two spheres can meet.

CLASS ACTIVITY

10N The Global Positioning System (GPS), p. CA-214

10O Circle Designs, p. CA-215

If you did Class Activity 10M, you saw some applications of circles. We will use circles again in the next section to construct special triangles and 4-sided figures. The mathematical definition of circle will allow us to explain why our constructions work.

SECTION SUMMARY AND STUDY ITEMS

Section 10.3 Circles and Spheres

A circle is the collection of all the points in a plane that are a fixed distance away from a fixed point in the plane. A sphere is the collection of all the points in space that are a fixed distance away from a fixed point in space. Two circles in a plane can either not meet, meet at a single point, or meet at two points. Two spheres in space can either not meet, meet at a single point, or meet along a circle.

Key Skills and Understandings

1. Give the definitions of circles and spheres.
2. Given a situation that involves distances from one or more points, use circles or spheres to describe the relevant locations.

Practice Exercises for Section 10.3

1. Give the (mathematical) definitions of the terms *circle* and *sphere*.
2. Points P and Q are 4 centimeters apart. Point R is 2 cm from P and 3 cm from Q. Use a ruler and a compass to draw a precise picture of how P, Q, and R are located relative to each other.
3. A radio beacon indicates that a certain whale is less than 1 kilometer away from boat A and less than 1 kilometer away from boat B. Boat A and boat B are 1 kilometer apart. Assuming that the whale is swimming near the surface of the water, draw a map showing all the places where the whale could be located.
4. Suppose that during a flight an airplane pilot realizes that another airplane is 1000 feet away. Describe the shape formed by all possible locations of the other airplane at that moment.
5. An airplane is in radio contact with two control towers. The airplane is 20 miles from one control tower and 30 miles from another control tower. The control towers are 40 miles apart. Is this information enough to pinpoint the exact location of the airplane? Why or why not? What if you know the altitude of the airplane—do you have enough information to pinpoint the location of the airplane?

Answers to Practice Exercises for Section 10.3

1. Be sure to describe these shapes in terms of a fixed distance to a fixed point.

2. See **Figure 10.63**. Start by drawing points P and Q on a piece of paper, 4 centimeters apart. Now open a compass to 2 cm, stick its point at P, and draw a circle. Since point R is 2 cm from P, it must be located somewhere on that circle. Open a compass to 3 cm, stick its point at Q, and draw a circle. Since point R is 3 cm from Q, it must also be located somewhere on this circle. Thus, point R must be located at one of the two places where your two circles meet. Plot point R at either one of these two locations.

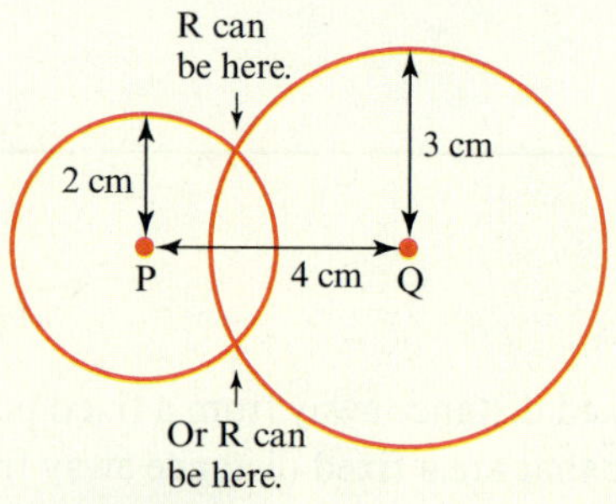

Figure 10.63 Locating points P, Q, and R.

3. Since the whale is less than 1 km from boat A, it must be *inside* a circle of radius 1 km, centered at boat A. Similarly, the whale is *inside* a circle of radius 1 km, centered at boat B. The places where the insides of these two circles overlap are all the possible locations of the whale, as shown in **Figure 10.64**.

4. The other airplane could be at any of the points that are 1000 feet away from the airplane. These points form a sphere of radius 1000 feet.

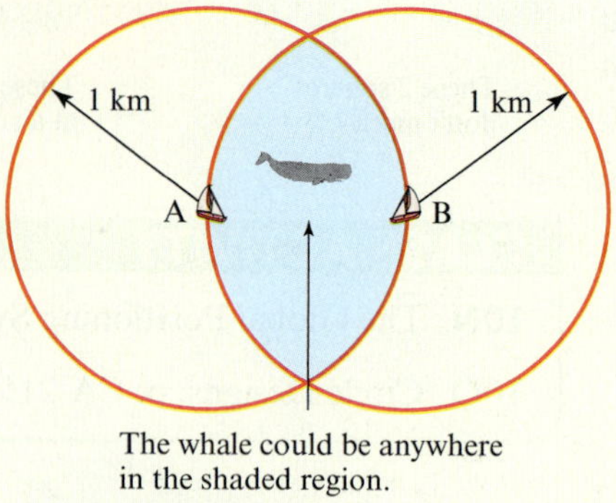

Figure 10.64 Two boats and a whale.

5. The locations that are 20 miles from the first control tower form a sphere of radius 20 miles, centered at the control tower. (Some of these locations are underground, and therefore are not plausible locations for the airplane.) Similarly, the locations that are 30 miles from the second control tower form a sphere of radius 30 miles. The airplane must be located in a place that lies on *both* spheres (i.e., at some place where the two spheres meet). The spheres must meet either at a single point (which is unlikely) or in a circle. So, most likely, the airplane could be anywhere on a circle. Therefore, we can't pinpoint the location of the airplane without more information.

Now suppose that the altitude of the airplane is known. Let's say it's 20,000 feet. The locations in the sky that are 20,000 feet from the ground form a very large sphere around the whole earth. This large sphere and the circle of locations where the airplane might be located either meet in a single point (unlikely) or in two points. So, even with this information, we still can't pinpoint the location of the airplane, but we can narrow it down to two locations.

PROBLEMS FOR SECTION 10.3

1. Smallville is 7 miles south of Gotham. Will is 8 miles from Gotham and 6 miles from Smallville. Draw a map showing where Will could be. Be sure to show a scale for your map. Explain why you draw your map the way you do. Can you pinpoint Will to one location, or not?

2. A new mall is to be built to serve the towns of Sunnyvale and Gloomington, whose centers are 6 miles apart. The developers want to locate the mall not more than 3 miles from the center of Sunnyvale and also not more than 5 miles from the center of Gloomington. Draw a simple map showing Sunnyvale, Gloomington, and *all* potential locations for the new mall, based on the given information. Be sure to show the scale of your map. Explain how you determined the possible locations for the mall.

3. A radio beacon indicates that a certain dolphin is less than 1 mile from boat A and at least $1\frac{1}{2}$ miles from boat B. Boats A and B are 2 miles apart. Draw a simple map showing the locations of the boats and *all* the places where the dolphin might be located. Be sure to show the scale of your map. Explain how you determined all possible locations for the dolphin.

4. A new Giant Superstore is being planned somewhere in the vicinity of Kneebend and Anklescratch, towns that are 10 miles apart. The developers will say only that all the locations they are considering are more than 7 miles from Kneebend and more than 5 miles from Anklescratch. Draw a map showing Kneebend, Anklescratch, and *all* possible locations for the Giant Superstore. Be sure to show the scale of your map. Explain how you determined all possible locations for the Giant Superstore.

5. Part 1 of Class Activity 10O on circle designs shows a design made from circles. In part 2 of the same Class Activity, there is a beautiful circle design that was made by drawing a portion of the design of part 1. Make another interesting circle design by using a compass (or a paper clip) to draw a portion of the design of part 1 of Class Activity 10O. Leave construction marks to show how you made your drawing. You may wish to outline your final drawing in a different color to make it visible and distinct from the construction marks. (Many such designs can be found in church windows and in other art throughout the world.)

10.4 Triangles, Quadrilaterals, and Other Polygons

CCSS PreK, Common Core State Standards Grades K, 1, 2, 3, 4, 5

In this section we study some of the basic two-dimensional geometric shapes that students learn about in elementary school: triangles, squares, rectangles, and other shapes. We will describe attributes of these shapes based on the shapes' sides and angles. By attending to attributes of shapes, we can classify shapes into categories, and we can examine how categories are related. We will see that a given category of shapes can often be described with different lists of attributes. We therefore choose succinct lists of attributes to define shapes. At the end of the section we examine a variety of ways to make shapes.

plane shape

two-dimensional shape

First some terminology: A **plane shape** or **two-dimensional shape** (or **2D shape**, or just **shape** when two dimensions are understood) is a flat shape that lies in a plane and is connected and closed, meaning it has no "loose, dangling ends." For shapes made of line segments, every endpoint of a segment must meet exactly one endpoint of another segment and segments are not allowed to cross.

CCSS K.G.2, K.G.4

CLASS ACTIVITY

10P What Shape Is It?, p. CA-216

How do we know if a shape is a triangle, a square, a rectangle, or some other shape? Very young children decide if a shape is of a specific type just by seeing what it looks like. If it "looks like a door" then they say it's a rectangle; if it doesn't look like a door because it's too long and skinny or because it's tilted, then they say it isn't a rectangle, even if it is. Attending *only* to the overall look of a shape is not a reliable way to identify it.

The *mathematical* study of shapes begins when children learn to consider a shape's parts, the properties of those parts, and the relationships among the parts. It is by considering the specific parts and properties of shapes that we can reliably identify shapes, classify shapes into categories, and determine how categories of shapes are related.

To categorize shapes, we usually look first for how many straight sides the shapes have. Next, we study the especially interesting category of 4-sided shapes.

What Are Quadrilaterals?

quadrilateral A **quadrilateral** is a closed shape in a plane consisting of 4 line segments that do not cross each other. Figure 10.65 shows examples of quadrilaterals and Figure 10.66 shows figures that are not quadrilaterals. The name *quadrilateral* makes sense because it means *four sided* (quad = four, lateral = side).

Figure 10.65

Quadrilaterals.

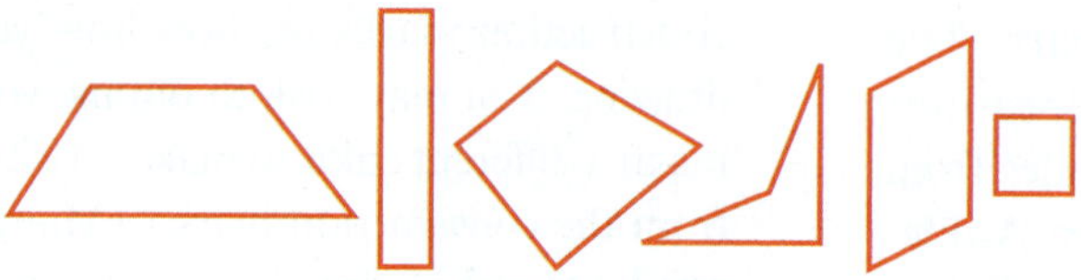

Figure 10.66

Figures that are not quadrilaterals.

Within the category of quadrilaterals, there are several subcategories, for example, squares and rectangles. To describe a category of shapes, we will need to focus on properties. In Class Activity 10Q, think about which properties of shapes are relevant for identifying and categorizing shapes.

CCSS
1.G.1
2.G.1

CLASS ACTIVITY

10Q What Properties Do These Shapes Have?, p. CA-217

When we identify and classify shapes mathematically, we consider the numbers of sides and angles, the lengths of the sides, the sizes of the angles, and the relationships between sides (such as whether they are parallel or not). We may also consider symmetry, which we will study in Chapter 14. But we don't consider some attributes that can be important in artwork or daily life and may stand out to children, such as the color, the orientation, or the overall size of the shape.

In Class Activity 10R, see how considering the parts of shapes and their properties allows us to classify shapes into categories.

CLASS ACTIVITY

10R How Can We Classify Shapes into Categories Based on Their Properties? p. CA-218

When you classified shapes into categories in Class Activity 10R, you probably noticed that some categories contain exactly the same shapes even though they were described by different properties. This leads to our next topic: defining categories of shapes.

How Can We Use Short Lists of Properties to Define Special Quadrilaterals?

What's the difference between the following three categories?

- Quadrilaterals that have 4 right angles
- Quadrilaterals that have 4 right angles and opposite sides parallel
- Quadrilaterals that have 4 right angles and opposite sides of the same length

The surprising answer is that there is no difference. It turns out that each of the three descriptions specifies the exact same collection of shapes—namely, rectangles. If a 4-sided shape has 4 right angles, then *automatically* its opposite sides are parallel and have the same length. How can that be, and why is that so? For now, we leave this as a mystery, to be explored in Chapter 14.

Because we can specify the same category of shapes by listing different collections of properties, we have some choice in how to define shapes. It's worthwhile to pick a short list of properties to define a category of shapes. Why? To decide if a shape is or isn't a rectangle, or some other type of shape, students will check to confirm that all the defining properties hold for the shape. So a short list of properties means less checking.

Look again at the categories in Class Activity 10R, some of which consist of the same shapes. Based on what you found, what *short* lists of properties would you use to define the sets of squares, rectangles, rhombuses, and parallelograms? Compare your definitions with these standard definitions of some special categories of quadrilaterals, which we will use from now on:

square — **square**—quadrilateral with 4 right angles whose sides all have the same length

rectangle — **rectangle**—quadrilateral with 4 right angles

rhombus — **rhombus**—quadrilateral whose sides all have the same length. The name **diamond** is sometimes used instead of rhombus

parallelogram — **parallelogram**—quadrilateral for which opposite sides are parallel

trapezoid — **trapezoid**—quadrilateral for which at least one pair of opposite sides are parallel. (Some books define a trapezoid as a quadrilateral for which *exactly one* pair of opposite sides are parallel.)

Figure 10.67 illustrates some examples.

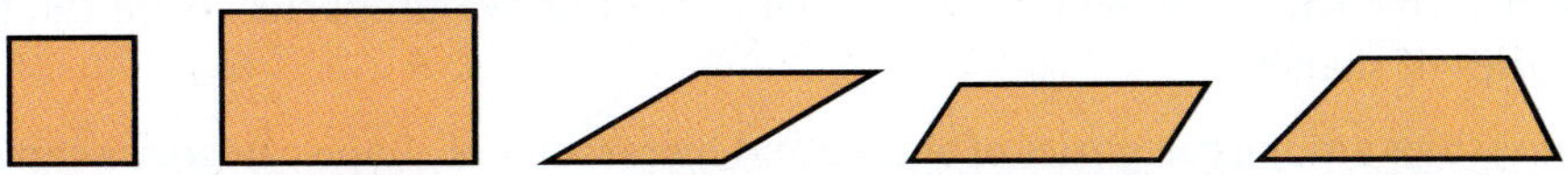

Figure 10.67 Some special quadrilaterals.

How Can We Classify Special Quadrilaterals in a Hierarchy?

CCSS
3.G.1
4.G.2
5.G.3
5.G.4

In Class Activity 10R, you probably noticed that some categories of quadrilaterals are subcategories of other categories. An important consequence of this is that *a given shape can belong to more than one category*. For example, the shape on the left in Figure 10.68 is a square because it has four sides of the same length and four right angles. But it is *also* a rectangle because it has four sides and four right angles. To help young students understand this idea, teachers can point out that squares are special kinds of rectangles, and they can call squares "square rectangles" to make that point.

How do hierarchies of shape categories come about? Think about this question in terms of lists of properties. If the list has only one property, such as "4 straight sides," then many will satisfy just that one property. If we add another property to that list, so it becomes "4 straight sides and 4 right angles," then only some of the shapes in the initial category will satisfy both of those properties. Usually, the more properties we add to the list, the fewer shapes satisfy all those

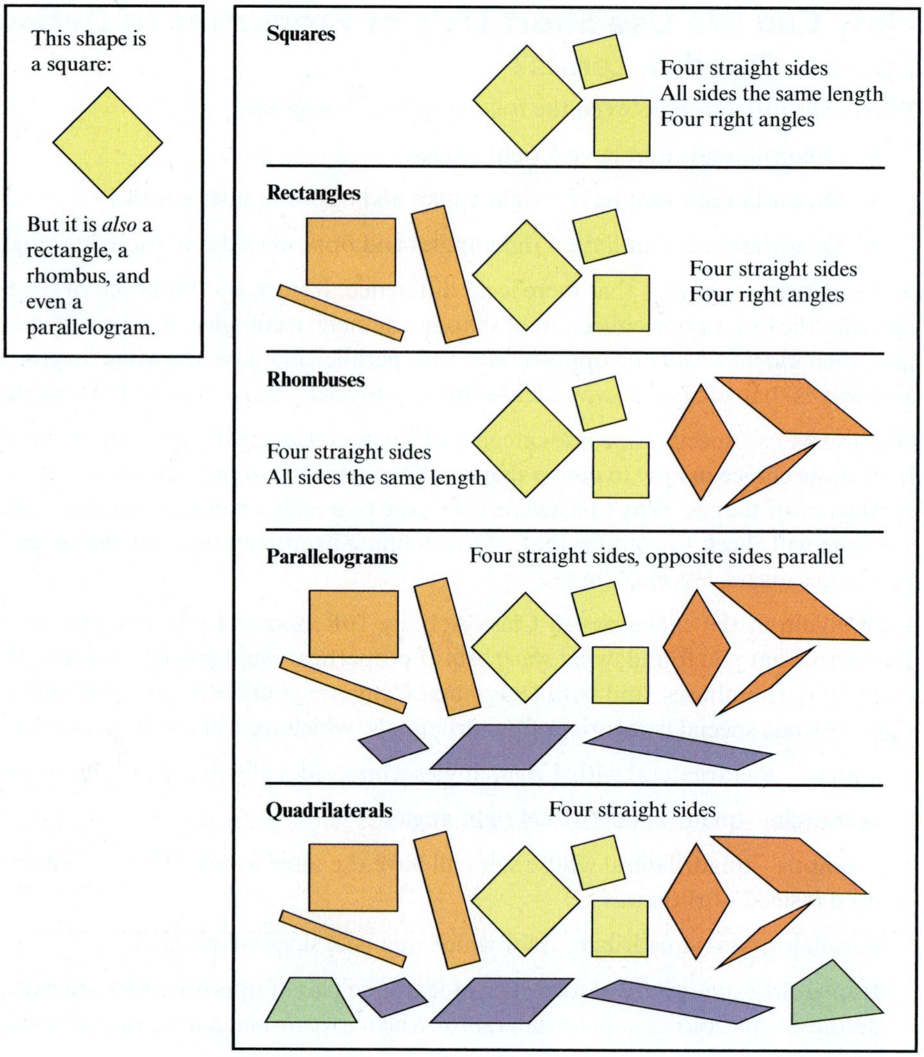

Figure 10.68 Classifying some special quadrilaterals in a hierarchy.

properties. Conversely, if we start with a long list of properties, such as "4 straight sides, all sides the same length, 4 right angles," then as we remove properties from the list we get larger categories of shapes.

Relationships among categories of shapes and hierarchies of shape categories can be described and depicted in various ways—for example, with the diagram in Figure 10.68, with words, and with Venn diagrams as discussed later in this section and in the Practice Exercises.

What Are Triangles?

How can children decide which figures in Figure 10.69 are triangles and which are not? They must see if the figure has all the characteristics of a triangle. A **triangle** is a closed two-dimensional shape made of 3 line segments.

triangle

CCSS 2.G.1

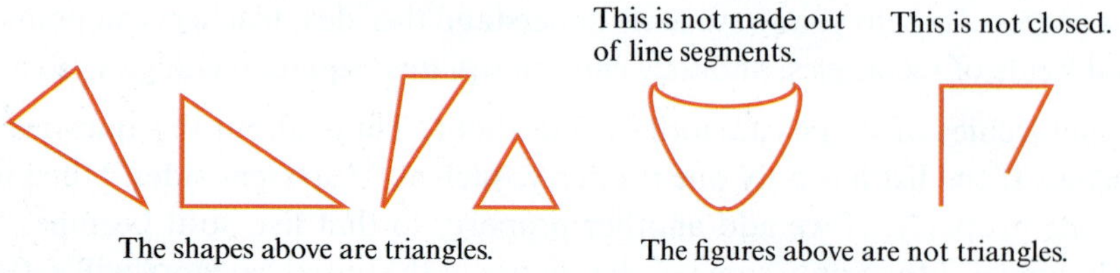

Figure 10.69 Triangles and figures that are not triangles.

Figure 10.70

Special kinds of triangles.

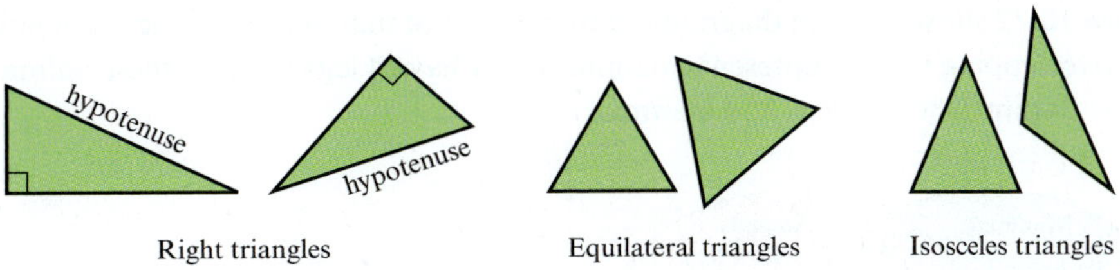

Some kinds of triangles have special names. Figure 10.70 shows some examples of special kinds of triangles. A **right triangle** is a triangle that has a right angle $(90°)$. In a right triangle, the side opposite the right angle is called the **hypotenuse**. A triangle whose sides are the same length is called an **equilateral triangle**. A triangle that has as least 2 sides of the same length is called an **isosceles triangle**. A triangle all of whose angles are smaller than a right angle is called an **acute triangle**, and a triangle that has an angle greater than a right angle is called an **obtuse triangle**.

right triangle
hypotenuse
equilateral triangle
isosceles triangle

CLASS ACTIVITY

10S How Can We Classify Triangles Based on Their Properties? p. CA-220

What Are Polygons?

Triangles and quadrilaterals are kinds of polygons. A **polygon** is a plane shape consisting of a finite number of line segments. The line segments making the polygon are called its **sides** and the points where line segments meet are called the **vertices** (singular: **vertex**) of the polygon. Triangles are polygons with 3 sides, quadrilaterals are polygons with 4 sides, **pentagons** are polygons with 5 sides, **hexagons** are polygons with 6 sides, **octagons** are polygons with 8 sides. Figure 10.71 shows some examples. A polygon with *n* sides can be called an "*n*-gon." For example, a 13-sided polygon is a 13-gon.

polygon
side
vertex
pentagons
hexagons
octagons

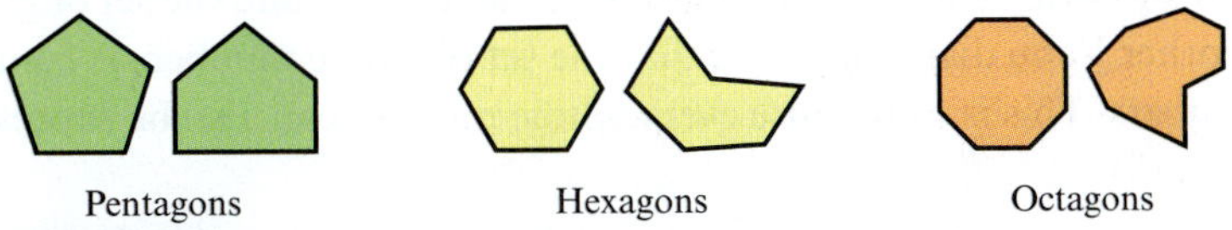

Figure 10.71 Pentagons, hexagons, and octagons.

The name *polygon* makes sense because it means *many angled* (poly = many, gon = angle). Similarly, for the names *pentagon*, *hexagon*, and so on, "penta" means 5, "hexa" means 6, and so on. It would be perfectly reasonable to call triangles *trigons* and quadrilaterals *quadrigons*, although these terms are not used conventionally.

A polygon is called **regular** if all sides have the same length and all angles are equal. In Figure 10.71 the leftmost pentagon, hexagon, and octagon are all regular polygons, whereas the pentagon, hexagon, and octagon on the right are not regular polygons.

regular

How Can We Show Relationships with Venn Diagrams?

A **Venn diagram** is a diagram that uses ovals to show how certain sets or categories are related. A **set** is a collection of things.

Venn diagram
set

Figure 10.72 shows a Venn diagram relating the set of mammals and the set of animals with 4 legs. The overlapping region represents mammals that have 4 legs because these animals fit in both categories: animals with 4 legs and mammals.

Figure 10.72

A Venn diagram showing the relationship between two sets.

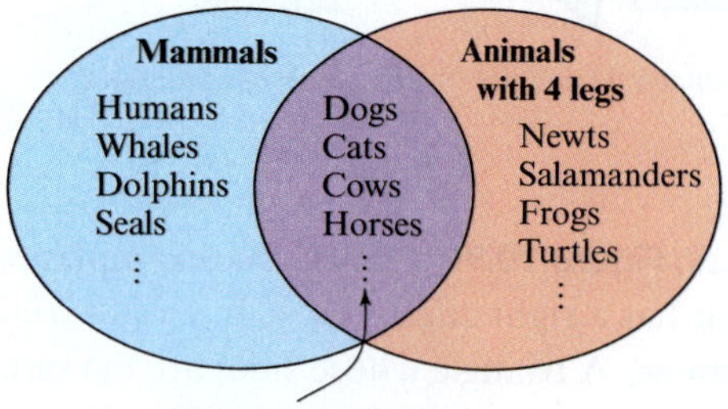

Venn diagrams are not just used in mathematics. Some teachers use Venn diagrams in language arts, for example, to compare the attributes of a child with the attributes of a story character, as in Figure 10.73.

Figure 10.73

A language arts Venn diagram comparing Matilda's and Joey's attributes.

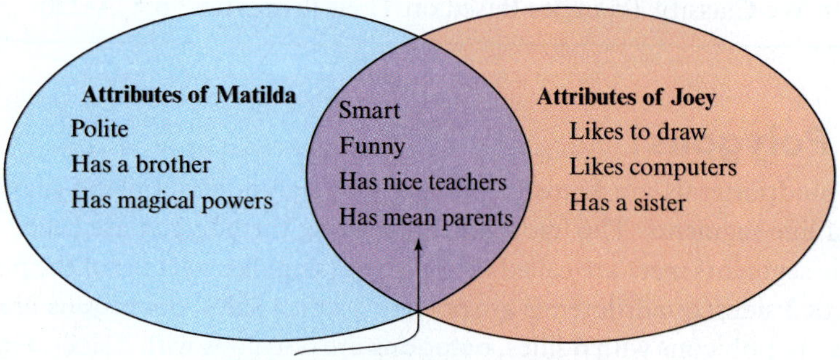

The sets in a Venn diagram do not have to overlap. Figure 10.74 shows two other kinds of Venn diagrams. The first shows a Venn diagram relating the set of boys and the set of girls. These sets do not overlap. The other Venn diagram shows that the set of whole numbers is contained within the set of rational numbers. This is so because every whole number can also be expressed as a fraction (by "putting it over 1").

Figure 10.74

Other Venn diagrams showing the relationships between two sets.

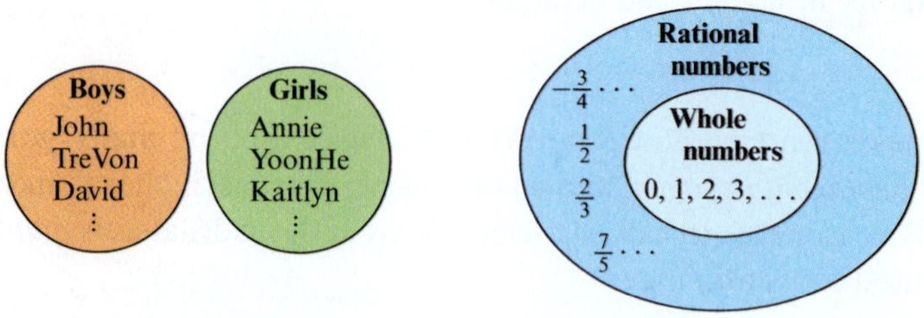

Venn diagram to relate boys and girls

Venn diagram to relate whole numbers and rational numbers

When representing three or more sets, Venn diagrams can become quite complex. In general, there can be double overlaps, triple overlaps, or more (if more than three sets are involved). For example, Figure 10.75 shows a Venn diagram relating the set of warm-blooded animals, the set of animals that lay eggs, and the set of carnivorous animals. There are three double overlaps and one triple overlap.

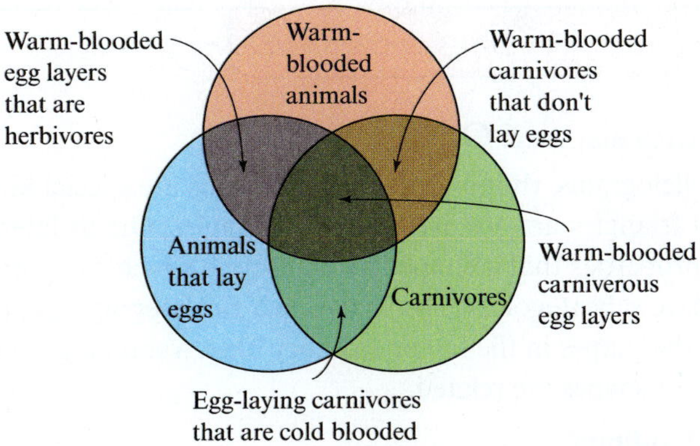

Figure 10.75

A Venn diagram showing the relationships among three sets.

We can use Venn diagrams to show the relationships among the various kinds of special quadrilaterals.

CLASS ACTIVITY

10T Using Venn Diagrams to Relate Categories of Quadrilaterals, p. CA-222

FROM THE FIELD Research

Kimmins, D. L. & Winters, J. J. (2015). Caution: Venn diagrams ahead! *Teaching Children Mathematics, 21*(8), pp. 484–493.

Kimmins and Winters discuss three types of Venn diagrams: overlapping circles, disjoint circles, and concentric circles. They point out that there are different ways to use Venn diagrams, and that the use of Venn diagrams in language arts is often different than in mathematics. The authors argue that students can learn the similarities and differences between mathematical concepts by comparing and contrasting the concepts with Venn diagrams, and that students should have multiple opportunities to use all types of Venn diagrams across subject areas.

How Can We Construct Triangles and Quadrilaterals?

CCSS SMP5

Did you know that circles can help us make triangles? This may seem strange; after all, circles are perfectly round, but triangles have straight sides. In Class Activities U, V, and W, and in the Practice Exercises and Problems, you'll see how to make triangles and quadrilaterals in various ways, such as by folding and cutting paper, by "walking and turning" to trace a route, and by drawing circles. In the process, notice how making shapes provides us with an opportunity to attend closely to their properties.

CLASS ACTIVITY

10U Using a Compass to Construct Triangles and Quadrilaterals, p. CA-223

10V Making Shapes by Folding Paper, p. CA-225

10W Making Shapes by Walking and Turning Along Routes, p. CA-227

SECTION SUMMARY AND STUDY ITEMS

Section 10.4 Triangles, Quadrilaterals, and Other Polygons

Squares, rectangles, parallelograms, rhombuses, and trapezoids are special kinds of quadrilaterals. Isosceles and equilateral triangles are special kinds of triangles. Quadrilaterals and triangles are kinds of polygons. The properties that a shape has determine which categories it belongs to. Categories of shapes often have subcategories. The shapes in a subcategory have properties in addition to all the properties that the shapes in the larger category have. Venn diagrams and other diagrams can show how categories of shapes are related.

Key Skills and Understandings

1. Give our (short) definitions of special quadrilaterals and triangles.
2. Describe how categories of quadrilaterals are related to each other, and show the relationships with the aid of a Venn diagram or other clear diagram. When possible, explain why the relationships hold.
3. Use a compass to construct triangles of specified side lengths (including equilateral and isosceles triangles), and use the definition of circle to explain why the construction must produce the required triangle.
4. Use a compass to construct rhombuses, and use the definition of circle to explain why the construction must produce a rhombus.
5. Fold and cut paper to produce various triangles and quadrilaterals.
6. Describe how to make shapes by walking and turning along routes.

Practice Exercises for Section 10.4

1. Draw a Venn diagram (or other clear diagram) showing the relationship between the categories of rectangles and rhombuses. Which shapes are in both categories? Explain.
2. Draw a Venn diagram (or other clear diagram) showing the relationship between the categories of parallelograms and trapezoids. Explain.
3. Some books define trapezoids as quadrilaterals that have *exactly one* pair of parallel sides. Draw a Venn diagram (or other clear diagram) showing how the categories of parallelograms and trapezoids are related when this alternate definition of trapezoid is used. Explain.
4. Draw a Venn diagram (or other clear diagram) showing the relationship between the categories of rhombuses and parallelograms.
5. **Figure 10.76** shows a method for constructing an equilateral triangle. Explain why this method must always produce an equilateral triangle.

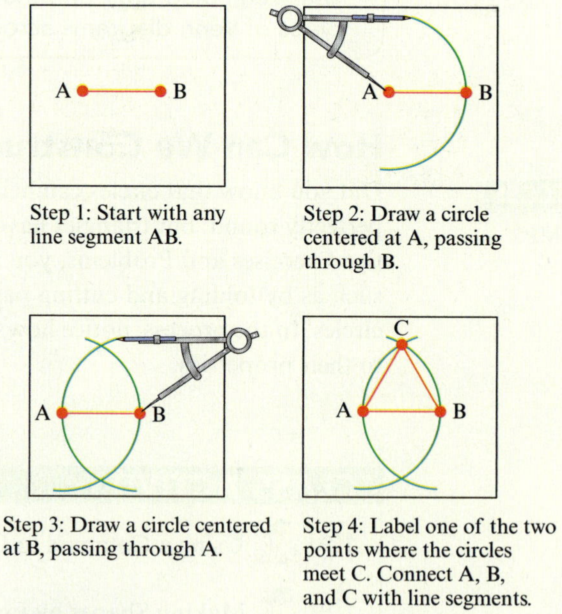

Figure 10.76 Constructing an equilateral triangle.

6. Use a ruler and a compass to construct a triangle that has one side of length 3 inches, one side of length 2 inches, and one side of length 1.5 inches. Describe your method, and explain why it must produce the desired triangle.
7. Use the definition of circles and rhombuses to explain why the quadrilateral ABDC produced by the method of **Figure 10.77** must necessarily be a rhombus.
8. Give instructions telling Robot Rob how to move and turn so that his path is a regular pentagon that has sides of length 2 meters. Explain how to determine the instructions.
9. Give instructions telling Automaton Audrey how to move and turn so that her path is a parallelogram that has a 10-foot side, a 20-foot side, and a 70° angle. Explain how to determine the instructions.

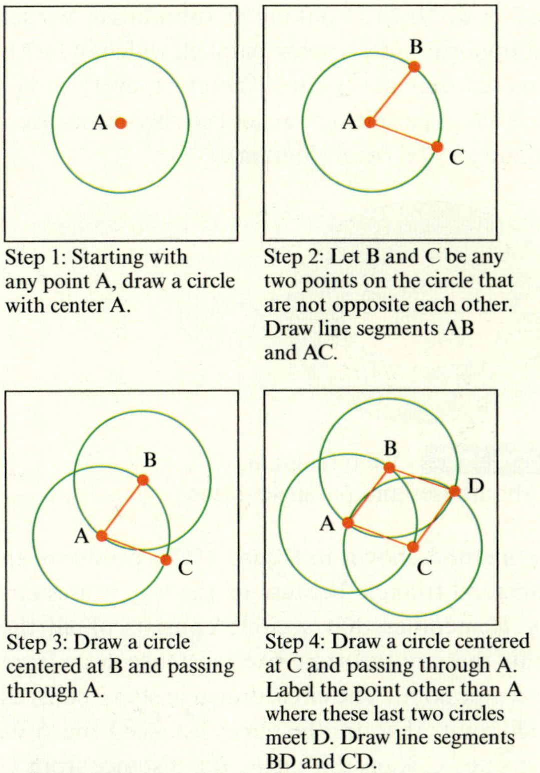

Figure 10.77 Method for constructing rhombuses.

Answers to Practice Exercises for Section 10.4

1. See **Figure 10.78**, which shows that rectangles and rhombuses have squares as a common subcategory. The shapes in both categories are those shapes that have 4 right angles (as rectangles do) and have 4 sides of the same length (as rhombuses do). According to the definition, those are exactly the shapes that are squares.

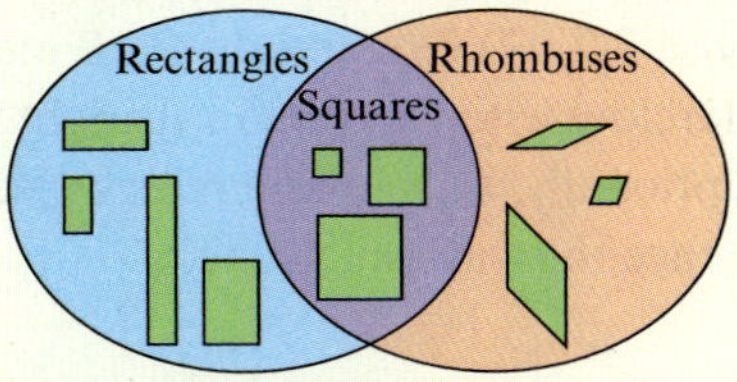

Figure 10.78 Venn diagram of rectangles and rhombuses.

2. See **Figure 10.79**, which shows that parallelograms are a subcategory of trapezoids. According to the definition, every parallelogram is also a trapezoid because parallelograms have two pairs of parallel sides, so they can be said to have at least one pair of parallel sides. Therefore parallelograms are a subcategory of trapezoids.

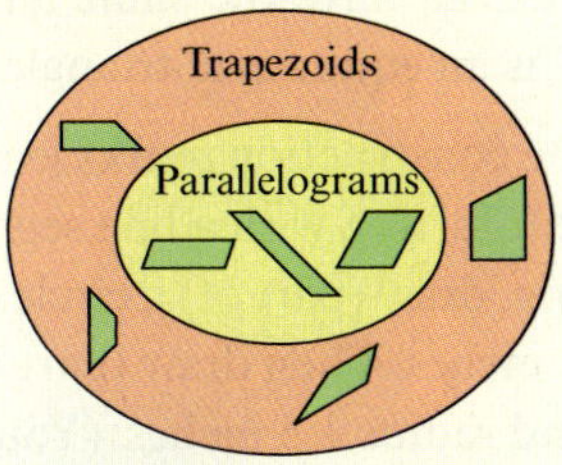

Figure 10.79 Venn diagram of parallelograms and trapezoids.

3. See **Figure 10.80**. According to the alternative definition, no parallelogram is a trapezoid because parallelograms have two pairs of parallel sides, so they don't have exactly one pair of parallel sides. Therefore, with this alternative definition, the set of parallelograms and the set of trapezoids do not have any overlap.

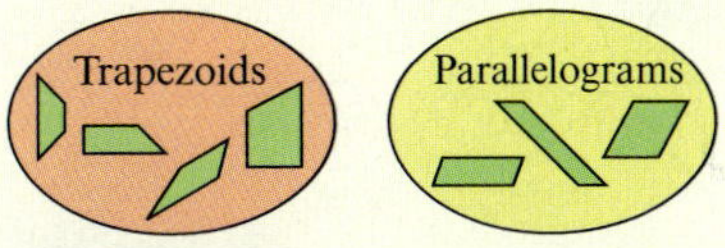

Figure 10.80 Venn diagram of parallelograms and trapezoids, according to the alternative definition.

4. See Figure 10.81. Looking at rhombuses, we see that opposite sides appear parallel, which in fact it turns out they always are. Therefore, every rhombus is also a parallelogram, and so rhombuses are a subcategory of parallelograms.

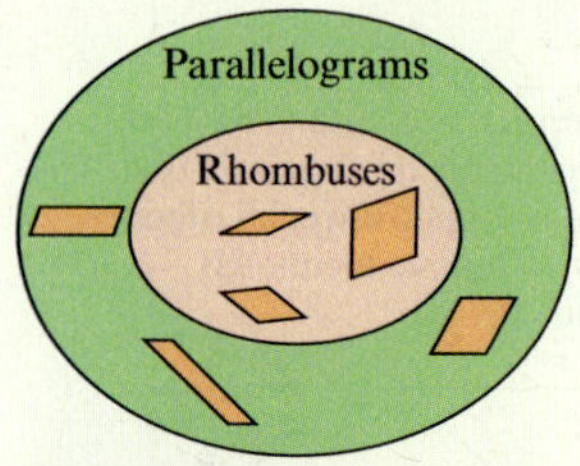

Figure 10.81 Venn diagram of rhombuses and parallelograms.

5. The method shown in Figure 10.76 produces an equilateral triangle because of the way it uses circles. Remember that a circle consists of all the points that are the same fixed distance away from the center point. The circle drawn in step 2 consists of all points that are the same distance from A as B is; since C is on this circle, the distance from C to A is the same as the distance from B to A. The circle drawn in step 3 consists of all points that are the same distance from B as A is; since C is also on this circle, the distance from C to B is the same as the distance from A to B. Therefore, the 3 line segments AB, AC, and BC all have the same length, and the triangle ABC is an equilateral triangle.

6. Figure 10.82 shows the construction marks leading to the desired triangle. Starting with a line segment AB that is 3 inches long, draw (part of) a circle with center A and radius 2 inches. Then draw (part of) a circle with center B and radius 1.5 inches. (You can just as well draw the circle of radius 1.5 inches centered at A and the circle of radius 2 inches centered at B if you want to.) Let C be one of the two locations where the two circles meet. Connect A to C, and connect B to C to create the desired triangle.

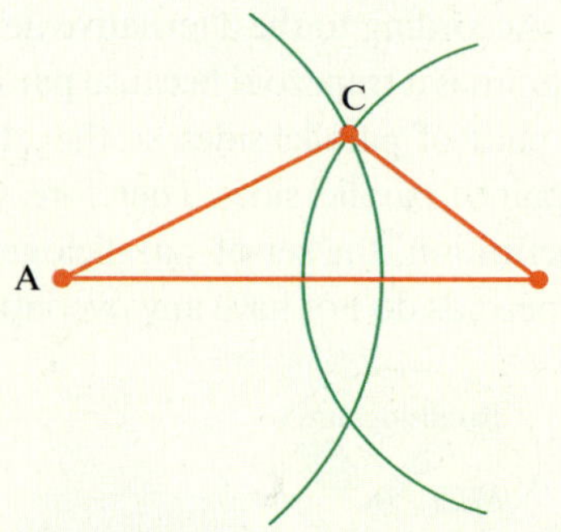

Figure 10.82 Constructing a triangle with sides of lengths 3 inches, 2 inches, and 1.5 inches.

The construction creates the desired triangle because it uses the defining property of circles. When you draw the circle with center A and radius 2 inches, the points on that circle are *all* 2 inches away from A. Similarly, when you draw the circle with center B and radius 1.5 inches, the points on that circle are *all* 1.5 inches away from B. So a point C where the two circles meet is *both* 2 inches from A and 1.5 inches from B. Hence, the side CA of the triangle ABC is 2 inches long, and the side CB is 1.5 inches long. The original side AB was 3 inches long, so the construction creates a triangle with sides of lengths 3 inches, 2 inches, and 1.5 inches.

7. A circle is the set of all points that are the same fixed distance away from the center point. Therefore, because points B and C are on a circle centered at A, the distance from B to A is equal to the distance from C to A. Because the point D is on a circle centered at B and passing through A, the distance from D to B is equal to the distance from B to A. Similarly, because D is on a circle centered at C and passing through A, the distance from D to C is equal to the distance from C to A. Therefore the 4 line segments AB, AC, BD, and CD have the same length. Rhombuses are quadrilaterals that have 4 sides of the same length. So the quadrilateral ABDC formed by the line segments AB, AC, BD, and CD is a rhombus.

8. See Figure 10.83 for Robot Rob's route. Instructions for Robot Rob: (1) Start at the red dot and go straight for 2 meters. (2) Turn counterclockwise 72°. (3) Go straight for 2 meters. (4) Turn counterclockwise 72°. (5) Go straight for 2 meters. (6) Turn counterclockwise 72°. (7) Go straight for 2 meters. (8) Turn counterclockwise 72°. (9) Go straight for 2 meters. (10) Optionally, turn counterclockwise 72° if you want to face the way you started.

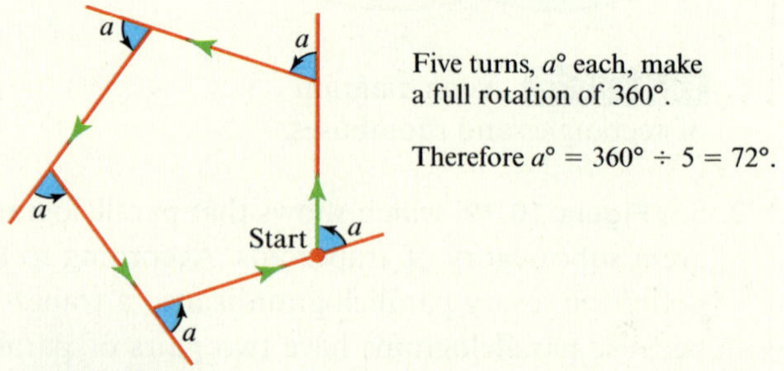

Figure 10.83 A route around a regular pentagon.

We can determine that Robot Rob should turn 72° at each vertex of the pentagon as follows. Each time Rob turns, he turns through an *exterior angle* of the pentagon, labeled *a*. If Rob goes all the way around the pentagon and returns to his original position and orientation, then he will have faced every compass direction once and only once, and so he will have turned a full 360° in the process. Therefore 5 turns of angle *a* make 360°, and so the angle *a* must be $360° \div 5 = 72°$.

9. See Figure 10.84(a) for a drawing that helps us determine Automaton Audrey's route and see Figure 10.84(b) for her route. Instructions for Audrey: (1) Start at the red dot and go straight for 10 feet. (2) Turn counterclockwise 110°. (3) Go straight for 20 feet. (4) Turn counterclockwise 70°. (5) Go straight for 10 feet. (6) Turn counterclockwise 110°. (7) Go straight for 20 feet. (8) Optionally, turn counterclockwise 70° if you want to face the way you started.

Notice that Audrey turns *exterior angles* of the parallelogram as she goes around the route. We can determine Audrey's angles of turning as follows. Because opposite sides of a parallelogram are parallel, the Parallel Postulate tells us that all the angles labeled *a* in Figure 10.84(a) are the same and all of the angles labeled *b* in the figure are the same. So all of the angles labeled *a* are 70°. Because *a* and *b* together make a straight angle, $a + b = 180°$, so all of the angles labeled *b* are 110°.

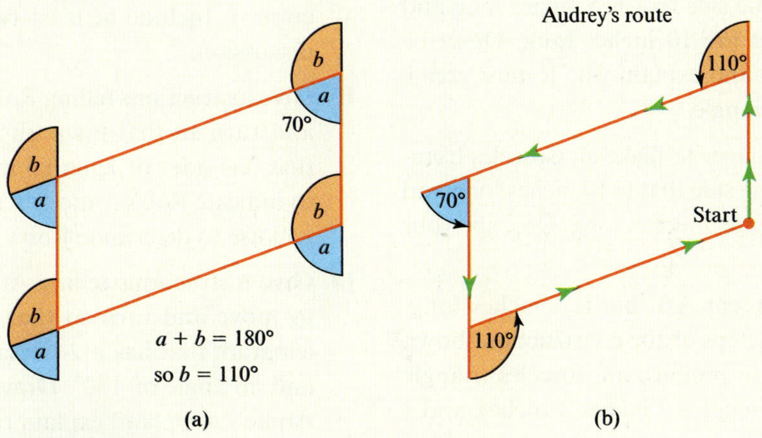

Figure 10.84 A route around a parallelogram.

PROBLEMS FOR SECTION 10.4

1. Students sometimes get confused about the relationship between squares and rectangles. Explain this relationship in your own words, using our (short) definitions of these shapes.

2. Draw a Venn diagram or other clear diagram showing the relationships among the categories of squares, rectangles, parallelograms, and trapezoids. Explain how you use properties of shapes to determine relationships among the categories.

3. Draw a Venn diagram or other clear diagram that shows the relationships among the categories of quadrilaterals, squares, rectangles, parallelograms, rhombuses, and trapezoids. Explain briefly.

4. a. Describe how four children could use 4 pieces of string to show a variety of different rhombuses.

b. Describe how you could show a variety of different rhombuses by threading straws onto string or by fastening sturdy strips of cardboard with brass fasteners.

5. Figure 10.85 shows a method for constructing isosceles triangles.

a. Use the method of Figure 10.85 to draw two different isosceles triangles.

b. Use the definition of circles to explain why this method will always produce an isosceles triangle.

c. Use this method to draw an isosceles triangle that has two sides of length 6 inches and one side of length 4 inches.

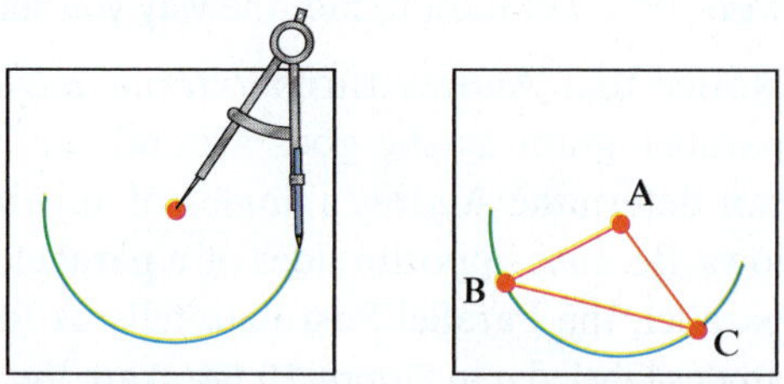

Step 1: Draw part of a circle.

Step 2: Connect the center of the circle with two points on the circle.

Figure 10.85 A method for constructing isosceles triangles.

6. a. Fold and cut paper to make an isosceles triangle that has one side that is 5 inches long and two sides that are 10 inches long. Describe your method, and explain why it must create an isosceles triangle.

b. Fold and cut paper to make an isosceles triangle that has one side that is 10 inches long and two sides that are 7 inches long. Describe your method.

7. Draw a line segment AB that is 4 inches long. Now modify the steps of the construction shown in Figure 10.76 to produce an isosceles triangle that has sides of lengths 7 inches, 7 inches, and 4 inches. Explain why you carry out your construction as you do.

8. a. Use a ruler and compass to help you draw a triangle that has one side of length 5 inches, one side of length 3.5 inches, and one side of length 4 inches.

b. Explain why your method of construction must produce a triangle with the required side lengths.

9. Is there a triangle that has one side of length 4 inches, one side of length 2 inches, and one side of length 1 inch? Explain.

10. Create two different rhombuses, both of which have 4 sides of length 4 inches. You may create your rhombuses by drawing, by folding and cutting paper, or by fastening objects together. In each case, explain how you know your shape really is a rhombus.

11. The line segments AB and AC in Figure 10.86 have been constructed so that they could be two sides of a rhombus.

a. Trace the angle BAC of Figure 10.86 on a piece of paper. Use a compass and straightedge to finish constructing a rhombus that has AB and AC as two of its sides.

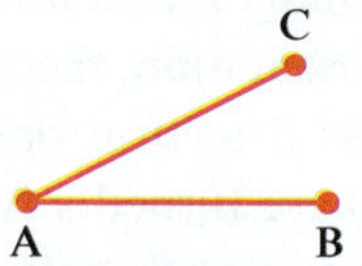

Figure 10.86 Two sides of a rhombus.

b. By referring to the definition of rhombus, explain why your construction in part (a) must produce a rhombus.

12. In a paragraph, discuss how the definition of circle is used in constructing shapes (other than circles). Include at least two examples in your discussion.

13. Give instructions telling Robot Rob how to move and turn so that his path is a regular octagon that has sides of length 3 meters. Draw a sketch to indicate Rob's route and explain the reasoning you use to determine Rob's angles of turning.

14. Give instructions telling Automaton Annie how to move and turn so that her path is a parallelogram that has a 2-meter side, a 3-meter side and an angle of 130°. Draw a sketch to indicate Annie's route and explain the reasoning you use to determine her angles of turning.

15. a. Figure 10.87 shows 3 pairs of parallel line segments of the same length. In each case, describe how to get from point A to point B and from point C to point D by traveling along grid lines. Compare the instructions.

b. Figure 10.88 shows 3 pairs of perpendicular line segments of the same length. In each case, describe how to get from point A to point B and from point C to point D by traveling along grid lines. Compare the instructions and describe how they are related.

c. Copy the 3 line segments in Figure 10.89 onto graph paper (feel free to spread them out across the page) and use what you discovered in parts (a) and (b) to help you draw 3 squares that have those line segments as one side. Say briefly how to use the grid lines to draw the squares.

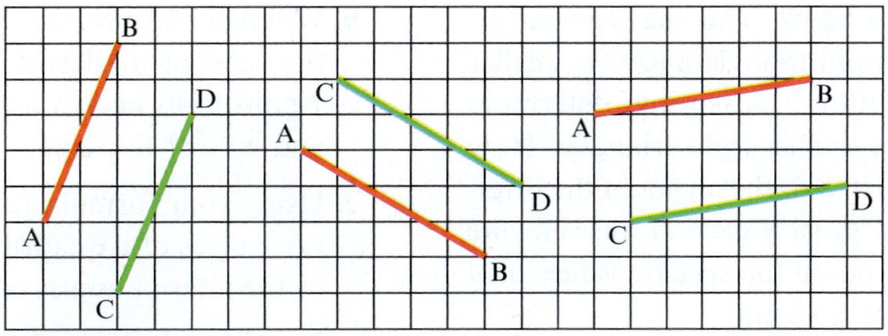

Figure 10.87 Parallel lines on graph paper.

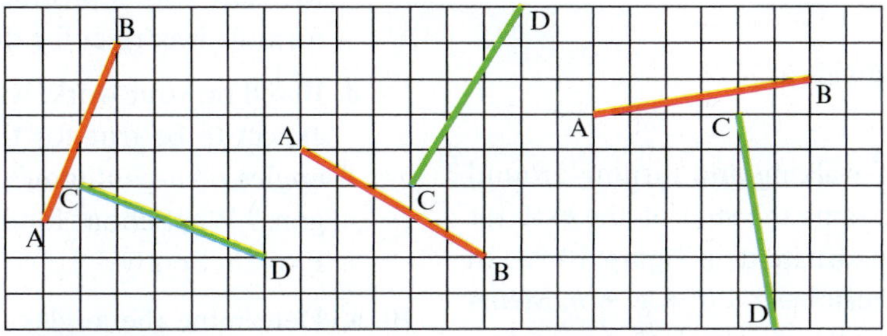

Figure 10.88 Perpendicular lines on graph paper.

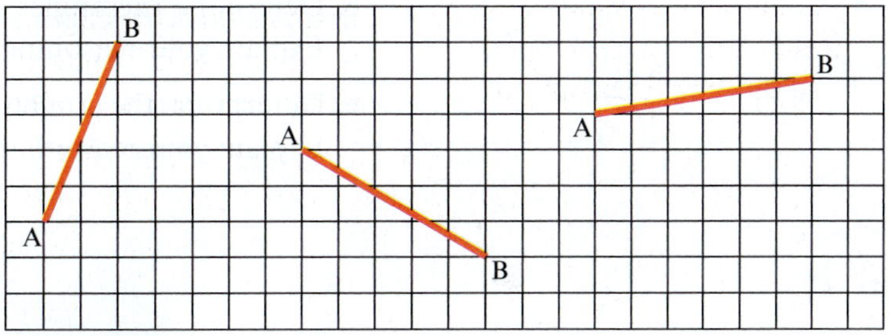

Figure 10.89 Draw squares on graph paper.

16. a. Given a parallelogram with angles *a*, *b*, *c*, and *d*, as in Figure 10.90, describe how angle *a* is related to the other 3 angles. Your answer should be general, in that it would hold for every parallelogram, not just for the particular parallelogram shown in the figure.

b. Use the definition of parallelogram and the Parallel Postulate to explain why the relationships you found in part (a) hold.

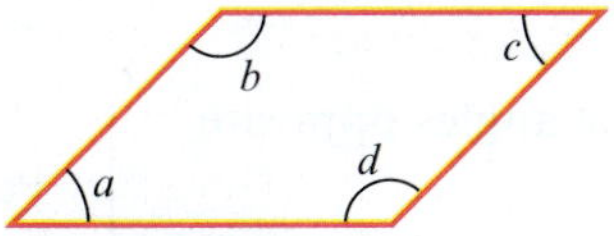

Figure 10.90 How are the angles related?

17. What is wrong with the diagram for the problem in Figure 10.91? Rewrite and redraw the problem to make it correct. Explain briefly.

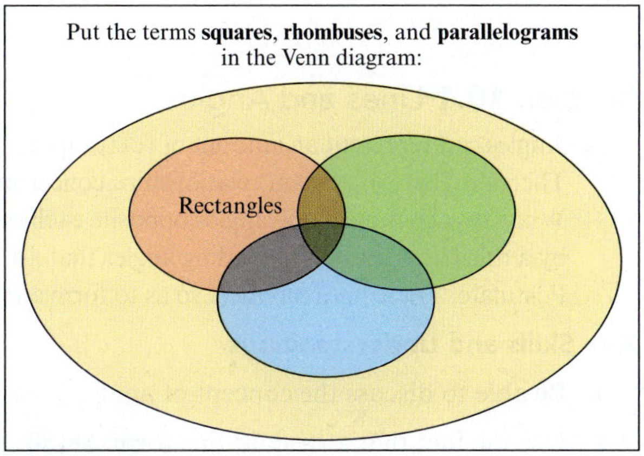

Figure 10.91 A problem with an incorrect Venn diagram.

18. Given a right triangle that has angles *a*, *b*, and 90°, draw a line segment from the corner where the right angle is to the hypotenuse so that this line segment is perpendicular to the hypotenuse, as shown in Figure 10.92. The line segment divides the original right triangle into two smaller right triangles. Without measuring, determine the angles in these smaller right triangles. How are these angles related to the angles in the original large triangle? Give a general answer, one that does not depend on the specific values of a and b. Explain your reasoning.

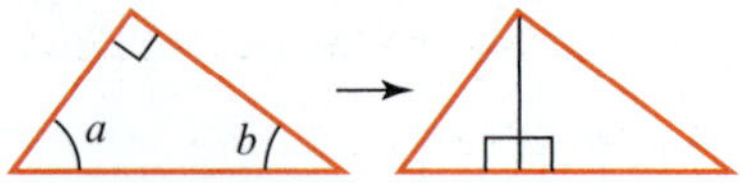

Figure 10.92 Subdivide a right triangle.

19. a. Use the idea of walking and turning around a shape to determine the sum of the exterior angles of the quadrilateral in Figure 10.93. In other words, determine $e + f + g + h$. Measure with a protractor to check that your formula is correct for this quadrilateral.

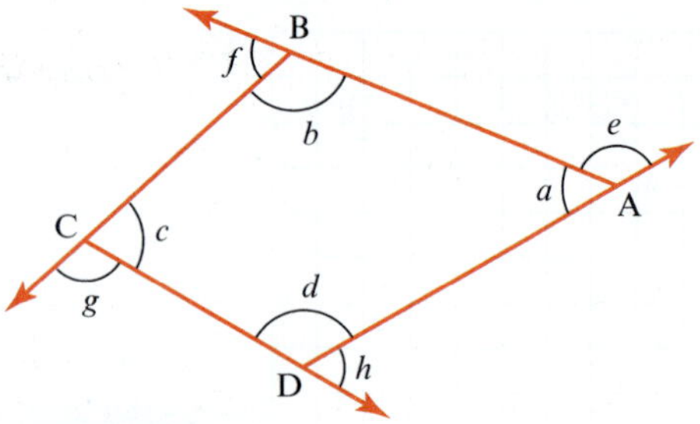

Figure 10.93 A quadrilateral.

b. Will there be a similar formula for the sum of the exterior angles of pentagons, hexagons, 7-gons (also known as heptagons), octagons, and so on? Explain.

c. Using your formula for the sum of the exterior angles of a quadrilateral, deduce the sum of the interior angles of the quadrilateral.

In other words, find $a + b + c + d$, as pictured in Figure 10.93. Explain your reasoning. Measure with a protractor to verify that your formula is correct for this quadrilateral.

d. Based on your work, what formula would you expect to be true for the sum of the interior angles of a pentagon? What about for hexagons? What about for a polygon with n sides? Explain briefly.

20. a. Determine the angles in a regular pentagon. Explain your reasoning.

b. Determine the angles in a regular hexagon. Explain your reasoning.

c. Determine the angles in a regular n-gon. Explain your reasoning.

CHAPTER SUMMARY

Section 10.1 Lines and Angles Page 452

- Angles can represent an amount of rotation or a region formed between two rays that meet. There are three important relationships concerning angles produced by configurations of lines. When two lines meet, the angles opposite each other are equal. When two parallel lines are cut by a third line, the corresponding angles that are produced are equal; this is called the Parallel Postulate. Three lines can meet so as to form a triangle. The angles in a triangle add to 180°. Page 452

Key Skills and Understandings

1. Be able to discuss the concept of angle. Page 452
2. Use the fact that a straight line forms a 180° angle to explain why the angles opposite each other, which are formed when two lines meet, are equal. Page 458
3. Know how to show informally that the sum of the angles in a triangle is 180°. Page 458
4. Use the Parallel Postulate to prove that the sum of the angles in every triangle is 180°. Page 459
5. Use the idea of walking around a triangle to explain why the sum of the angles in every triangle is 180°. Page 459
6. Apply facts about angles produced by configurations of lines to find angles. Page 459

Section 10.2 Angles and Phenomena in the World	Page 466
■ Angles arise in real-world situations. For example, angles are formed by the sun's rays and angles describe reflected light rays.	Page 466
Key Skills and Understandings	
1. Apply the laws of reflection to explain what a person looking into a mirror will see.	Page 467
Section 10.3 Circles and Spheres	Page 472
■ A circle is the collection of all the points in a plane that are a fixed distance away from a fixed point in the plane. A sphere is the collection of all the points in space that are a fixed distance away from a fixed point in space. Two circles in a plane can either not meet, meet at a single point, or meet at two points. Two spheres in space can either not meet, meet at a single point, or meet along a circle.	Page 472
Key Skills and Understandings	
1. Give the definitions of circles and spheres.	Page 472
2. Given a situation that involves distances from one or more points, use circles or spheres to describe the relevant locations.	Page 472
Section 10.4 Triangles, Quadrilaterals, and Other Polygons	Page 477
■ Squares, rectangles, parallelograms, rhombuses, and trapezoids are special kinds of quadrilaterals. Isosceles and equilateral triangles are special kinds of triangles. Quadrilaterals and triangles are kinds of polygons. The properties that a shape has determine which categories it belongs to. Categories of shapes often have subcategories. The shapes in a subcategory have properties in addition to all the properties that the shapes in the larger category have. Venn diagrams and other diagrams can show how categories of shapes are related.	Page 477
Key Skills and Understandings	
1. Give our (short) definitions of special quadrilaterals and triangles.	Page 479
2. Describe how categories of quadrilaterals are related to each other, and show the relationships with the aid of a Venn diagram or other clear diagram. When possible, explain why the relationships hold.	Page 479
3. Use a compass to construct triangles of specified side lengths (including equilateral and isosceles triangles), and use the definition of circle to explain why the construction must produce the required triangle.	Page 480
4. Use a compass to construct rhombuses, and use the definition of circle to explain why the construction must produce a rhombus.	Page 483
5. Fold and cut paper to produce various triangles and quadrilaterals.	Page 483
6. Describe how to make shapes by walking and turning along routes.	Page 483

11

Measurement

Measurement is about the methods and ideas we use to determine the size of things. Because measurement is so familiar, we may take the underlying ideas for granted. Every object has different attributes that can be measured, such as length, area, volume, or weight. In this chapter we study ideas that are involved in all kinds of measurement. We also see how length, area, and volume are different even though they use interrelated units. This lays the groundwork for our additional study of area and volume in Chapters 12 and 13.

In this chapter, we focus on the following topics and practices within the *Common Core State Standards for Mathematics* (*CCSSM*).

Standards for Mathematical Content in the CCSSM

In the domain of *Measurement and Data* (Kindergarten through Grade 5), students describe measurable attributes, and they measure attributes such as length, area, (liquid) volume, and mass. They use different units to measure the same quantity and they describe how the size of a unit affects the numerical value of the measurement. Later, they learn to convert measurements from one unit to another. In the domain of *Ratios and Proportional Relationships* (Grades 6, 7) students use ratio reasoning to convert measurements and they manipulate units appropriately.

Standards for Mathematical Practice in the CCSSM

Opportunities to engage in all eight of the Standards for Mathematical Practice described in the *Common Core State Standards* occur throughout the study of measurement, although the following standards may be especially appropriate for emphasis:

- **4 Model with mathematics.** Students engage in this practice when they make or estimate measurements and then use them in calculations to determine other quantities of interest.
- **5 Use appropriate tools strategically.** Students engage in this practice when they recognize how the markings on rulers determine a count of unit-length segments and when they apply this knowledge in determining lengths, even when one end of the ruler is not at the 0 mark.
- **6 Attend to precision.** Students engage in this practice when they make and report measurements with an appropriate degree of precision depending on the purpose and the tools used.

(From Common Core Standards for Mathematical Practice. Published by Common Core Standards Initiative.)

11.1 Concepts of Measurement

CCSS PreK, Common Core State Standards Grades K–5

Measurement is about determining the size of things. In this section, we focus on the ideas underlying measurement: the need to specify measurable attributes, the order and additive structure of measures, and units of measurement. We will see how these ideas are essential to understanding the process of measurement. Finally, we examine systems of measurement.

Why Does Measurement Require Selecting a Measurable Attribute?

Measurement allows us to describe or determine the sizes of objects and to compare objects. But can we always say unambiguously that one object is larger than another? As we'll see, in some situations there is ambiguity about relative size because it is possible to measure an object with respect to several different kinds of attributes. Before you read on, please do Class Activity 11A.

CLASS ACTIVITY

11A The Biggest Tree in the World, p. CA-228

CCSS K.MD.1

If we compare the trees in Class Activity 11A by volume, then *General Sherman* is the largest tree. But if we compare the trees by height, then the Mendocino tree is largest. We could also compare the trees by the land area they cover, in which case the Banyan tree in Kolkata is probably largest. There is no single measurable attribute that we are required to use to determine which tree is largest. Before we can compare the trees, we must *select* a measurable attribute with which to compare them. Volume and height are natural choices, but we can also compare the circumference of the trees or the area of land they cover. To compare sizes of objects, we must specify, either implicitly or explicitly, a measurable attribute as a basis for the comparison.

A common activity for young children is to sort a collection of objects in different ways, such as by color or shape. In this way children become aware that one object has different attributes and that objects can be compared in different ways.

What Are the Order and Additive Structures of Measures?

CCSS K.MD.2

All measurement involves comparison, and children learn to compare before they learn to measure. For example, before children describe measurements numerically (such as "This desk is 3 feet wide" or "This bottle holds 2 liters of juice"), they learn to compare (such as "This block tower is taller than that block tower" or "This bottle holds less juice than that bottle"). Thus, *order*—namely, the concepts of *greater than, less than,* and *equal to*—fundamentally underlie the concept of measurement.

CCSS 1. MD.2

Another fundamental concept underlying most measurement is *additivity.* For example, when we say that a rock weighs 2 pounds, we mean that it weighs the same as two 1-pound objects together. When we say that a string is 3 meters long, we mean that it is as long as three 1-meter sticks placed end-to-end. Length, area, volume, weight, and time are familiar measures that are additive in the sense that the measure of a combined quantity is equal to the sum of the measures of the individual components. However, some measures, such as temperature, are not additive in that sense. When substances of different temperatures are combined, the temperature of the combined substance is generally not equal to the sum of the temperatures of the component parts.

We will use the additivity of area and volume extensively in the next chapters when we develop area and volume formulas and find areas and volumes of objects.

Why Do We Need Units and How Do We Interpret the Meaning of Measurements?

What does it mean to measure a quantity and what does it mean when we describe the size of an object with a specific measure such as 10 feet or 4 liters? We have discussed the need for specifying an attribute we wish to measure. Measurement involves comparison, but not just any kind of comparison; it is comparison with a fixed *reference amount* of a quantity. This reference amount is called a **unit**. For example, a mile is a unit of length, an acre is a unit of area, a gallon is a unit of capacity (or volume), and a kilowatt-hour is a unit of energy.

unit

measure

To **measure** a given quantity means to compare that quantity with a unit of the quantity. Usually, we must determine how many specific units of the quantity make up the given quantity without any gaps or overlaps. For example, the length of a rug in meters is the number of 1-meter sticks we would have to place end-to-end to traverse the entire length of the rug. The area of a rug in square feet is the number of 1-foot-by-1-foot squares we would need to cover the rug completely without gaps or overlaps. The weight of a piece of bread in grams is the number of 1-gram weights we would have to put on one side of a pan balance to balance the piece of bread on the other side. The volume of a compost pile in cubic yards is the number of 1-yard-by-1-yard-by-1-yard cubes we would need to fill the compost pile.

Once we choose a unit, measurements can be described with numbers. What amounts can be used as units? *Any fixed* (positive) *amount* of a measurable quantity can be a unit, but we usually use commonly accepted standard units in order to communicate clearly.

How Do Measurement Concepts Underlie the Process of Measurement?

How do we measure quantities? The most direct way to measure a quantity is to count how many of a specific unit are in the quantity to be measured. For example, to measure the volume of rice in a small bag, a child could scoop the rice out, cup by cup, and count how many cups of rice she scooped.

Before you read on, do Class Activity 11B, which will help you think about key concepts that students need to understand the process of length measurement.

CLASS ACTIVITY

11B What Concepts Underlie the Process of Length Measurement?, p. CA-229

CCSS 1.MD.2

To measure lengths, young children can place 1-inch long segments end to end, as in **Figure 11.1**, and count the number of segments used. To understand the process and result of measurement, children need to understand and use measurement concepts. Initially, they may not grasp some of the key ideas of measurement. They may not understand the role of a unit, and may therefore use objects of different sizes, as in **Figure 11.2(a).** They may leave gaps (or overlaps) when they are measuring, as in **Figure 11.2(b)**. Or they may not recognize length as the attribute to be measured and therefore not place units end-to-end along a path as in **Figure 11.2(c)**.

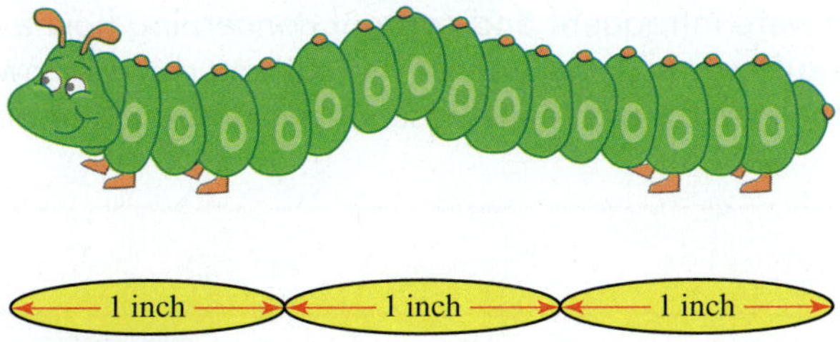

Figure 11.1 Using 1-inch segments to measure a length.

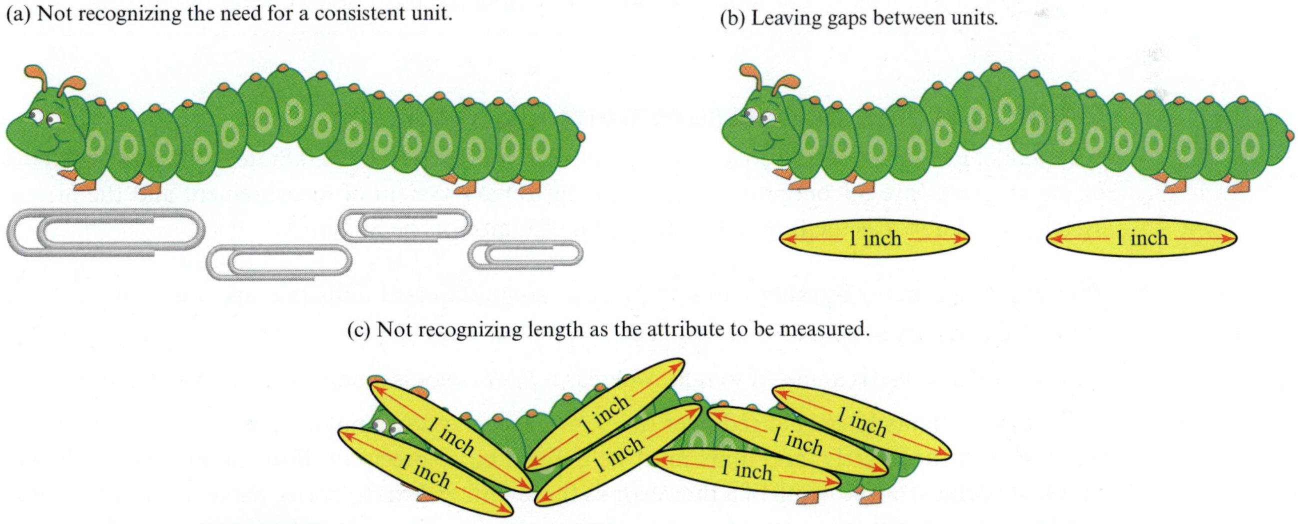

Figure 11.2 Errors in measuring length.

The most familiar measuring device is a standard ruler, for measuring length. A ruler whose unit is an inch displays a number of inch-long lengths. In order to use the ruler to measure how long an object is in terms of inches, we read off the number of inches on the ruler, rather than repeatedly laying down an inch-long object and counting how many it took to cover the length of the object. Similarly, other measuring devices such as scales, calipers, and speedometers allow us to measure quantities indirectly, rather than making direct counts of a number of units.

CLASS ACTIVITY

11C Making and Using a Ruler, p. CA-230

FROM THE FIELD Research

Dietiker, L. C., Gonulates, F., & Smith III, J. P. (2011). Understanding linear measure. *Teaching Children Mathematics, 18*(4), pp. 252–259.

The authors analyzed three representative U.S. elementary textbooks with respect to measurement content and found attention was given to the procedures of measurement and estimation, but there was not much focus on the conceptual principles of measurement that underlie these procedures. Moreover, the authors found that even when conceptual principles were expressed in the textbooks, references to the important idea of unit iteration were infrequent and content concerning how a ruler represents iterated units of length was especially rare. The authors discuss how teachers can modify existing tasks in their textbooks to increase students' understanding of linear measurement.

FROM THE FIELD Children's Literature

Lionni, L. (1960). *Inch by inch*. New York, NY: Astor-Honor.

A clever inchworm uses his measuring ability to avoid being eaten by hungry birds. Launching a measurement lesson with this story can help students make connections to other instances where measurement was useful to them or someone they know.

What Systems of Measurement Do We Use?

A system of measurement is a collection of standard units. In the United States today, two systems of measurement are in common use: the U.S. customary system of measurement and the metric system (also known as the International System of Units, or the SI system).

The U.S. Customary System Table 11.1 shows some standard units that are commonly used in the U.S. customary system of measurement.

Notice that an *ounce* is a unit of weight, whereas a *fluid ounce* is a unit of capacity or volume.

Definitions of units have changed over time. As science and technology progress, we are able to measure more accurately, and units must be defined more precisely. For example, the inch was originally defined as the width of a thumb or as the length of 3 barleycorns. Since widths of thumbs and lengths of barleycorns can vary, these definitions are obviously not very precise. The inch is currently defined to be *exactly* 2.54 centimeters. (See [38].)

The Metric System The metric system was first used in France around the time of the French Revolution (1790). At that time, which is often called the Age of Enlightenment, there was a focus on rational thought, and scientists sought to organize their subjects in a rational way. The metric system is a natural outcome of this desire, as it is an efficient, organized system designed to be compatible with the decimal system for writing numbers.

The metric system gives names to units in a uniform way. For each kind of quantity to be measured, there is a base unit (such as meter, gram, liter). Each related unit is labeled with a prefix that

TABLE 11.1 Common units of measurement in the U.S. customary system

Units in the U.S. Customary System		
Unit	**Abbreviation**	**Some Relationships between Units**
Units of Length		
inch	in.	
foot	ft	1 ft = 12 in.
yard	yd	1 yd = 3 ft
mile	mi	1 mi = 1760 yd = 5280 ft
Units of Area		
square inch	in^2	
square foot	ft^2	$1\ ft^2 = 12^2\ in^2 = 144\ in^2$
square yard	yd^2	$1\ yd^2 = 3^2\ ft^2 = 9\ ft^2$
square mile	mi^2	
acre		$1\ acre = 43{,}560\ ft^2$
Units of Volume		
cubic inch	in^3	
cubic foot	ft^3	$1\ ft^3 = 12^3\ in^3 = 1728\ in^3$
cubic yard	yd^3	$1\ yd^3 = 3^3\ ft^3 = 27\ ft^3$
Units of Capacity		
teaspoon	tsp	
tablespoon	T or Tbs or Tbsp	1 T = 3 tsp
fluid ounce (or liquid ounce)	fl oz	1 fl oz = 2 T
cup	c	1 c = 8 fl oz
pint	pt	1 pt = 2 c = 16 fl oz
quart	qt	1 qt = 2 pt = 32 fl oz
gallon	gal	1 gal = 4 qt = 128 fl oz
Units of Weight (Avoirdupois)		
ounce	oz	
pound	lb	1 pound = 16 ounces
ton	t	1 ton = 2000 pounds
Unit of Temperature		
degree Fahrenheit	°F	water freezes at 32°F water boils at 212°F

indicates the unit's relationship to the base unit. For example, the prefix *kilo-* means *thousand*, so a *kilometer* is 1000 meters and a *kilogram* is 1000 grams. Many of the metric system prefixes are used only in scientific contexts, not in everyday situations. Some of the metric system prefixes are listed in Table 11.2.

TABLE 11.2 The metric system uses prefixes to create larger and smaller units from base units

Some Metric System Prefixes

Prefix	Meaning	
nano-	$10^{-9} = \frac{1}{1{,}000{,}000{,}000}$	billionth
micro-	$10^{-6} = \frac{1}{1{,}000{,}000}$	millionth
milli-	$10^{-3} = \frac{1}{1000}$	thousandth
centi-	$10^{-2} = \frac{1}{100}$	hundredth
deci-	$10^{-1} = \frac{1}{10}$	tenth
deka-	10	ten
hecto-	$10^2 = 100$	hundred
kilo-	$10^3 = 1000$	thousand
mega-	$10^6 = 1{,}000{,}000$	million
giga-	$10^9 = 1{,}000{,}000{,}000$	billion

Table 11.3 shows some of the most commonly used units in the metric system.

cc (cubic centimeter)

Soft drinks often come in 2-liter bottles. One liter is a little more than 1 quart. Doses of liquid medicines are often measured in milliliters. In this case, a milliliter is also often called a **cc**, for cubic centimeter.

metric ton (tonne)

Nutrition labels on food packages often refer to grams. Serving sizes are often given both in grams and in ounces. One ounce is about 28 grams. One kilogram is the weight of 1 liter of water, which is about 2.2 pounds. The weight of vitamins in foods is often given in terms of milligrams. Also, although 1000 kilograms should be called a megagram, it is usually called a **metric ton**, or **tonne**.

In the metric system, the units of length, capacity, and mass (weight) are related in a simple and logical way, as shown in Table 11.4.

Table 11.5 summarizes the basics of the metric system.

Even though the metric system was designed to be more logical and organized than the old customary systems, the definitions of units in the metric system have changed over time, too. For example, the meter was originally defined as one ten-millionth of the distance from the equator to the North Pole, whereas it is currently defined as the length of a path traveled by light in a vacuum in

$$\frac{1}{299{,}792{,}458}$$

of a second (see [38]).

TABLE 11.3 Common units of measurement in the metric system

Units in the Metric System		
Unit	**Abbreviation**	**Some Relationships between Units**
Units of Length		
millimeter	mm	1 mm $= \frac{1}{1000}$ m
centimeter	cm	1 cm $= \frac{1}{100}$ m $=$ 10 mm
meter	m	1 m $=$ 100 cm
kilometer	km	1 km $=$ 1000 m
Units of Area		
square millimeter	mm^2	
square centimeter	cm^2	$1\ cm^2 = 10^2\ mm^2 = 100\ mm^2$
square meter	m^2	$1\ m^2 = 100^2\ cm^2 = 10{,}000\ cm^2$
square kilometer	km^2	$1\ km^2 = 1000^2\ m^2 = 1{,}000{,}000\ m^2$
Units of Volume		
cubic millimeter	mm^3	
cubic centimeter	cm^3 or cc	$1\ cm^3 = 10^3\ mm^3 = 1000\ mm^3$
cubic meter	m^3	$1\ m^3 = 100^3\ cm^3 = 1{,}000{,}000\ cm^3$
cubic kilometer	km^3	$1\ km^3 = 1000^3\ m^3 = 1{,}000{,}000{,}000\ m^3$
Units of Capacity		
milliliter	mL or ml	1 mL $= \frac{1}{1000}$ L
liter	L or l	1 L $=$ 1000 mL
Units of Mass (Weight)		
milligram	mg	1 mg $= \frac{1}{1000}$ g
gram	g	
kilogram	kg	1 kg $=$ 1000 g
Unit of Temperature		
degree Celsius	°C	water freezes at 0°C water boils at 100°C

TABLE 11.4 Relationships among units of capacity, mass, and length in the metric system

Relationships of Liter, Gram, and Meter
1 milliliter of water weighs 1 gram
1 milliliter of water (at 4°C) has volume 1 cm^3 so fills a cube that is 1 cm wide, 1 cm long, and 1 cm high

TABLE 11.5 Summary of the metric system

Metric System					
		Prefix	Length	Capacity	Weight
0.001	$\frac{1}{1000}$	milli-	millimeter	milliliter	milligram
0.01	$\frac{1}{100}$	centi-	centimeter		centigram
0.1	$\frac{1}{10}$	deci-			
1		base unit	meter	liter	gram
10		deka-			
100		hecto-			
1000		kilo-	kilometer		kilogram
Fundamental Relationships					
1 milliliter of water weighs 1 gram and has a volume of 1 cubic centimeter.					

How Are the Metric and U.S. Customary Systems Related?

Figure 11.3 shows lines of length 1 millimeter, 1 centimeter, and 1 inch. The inch and the centimeter are related by 1 inch = 2.54 centimeters. A meter is about 1 yard and 3 inches. A kilometer is about 0.6 miles, so a bit more than half a mile.

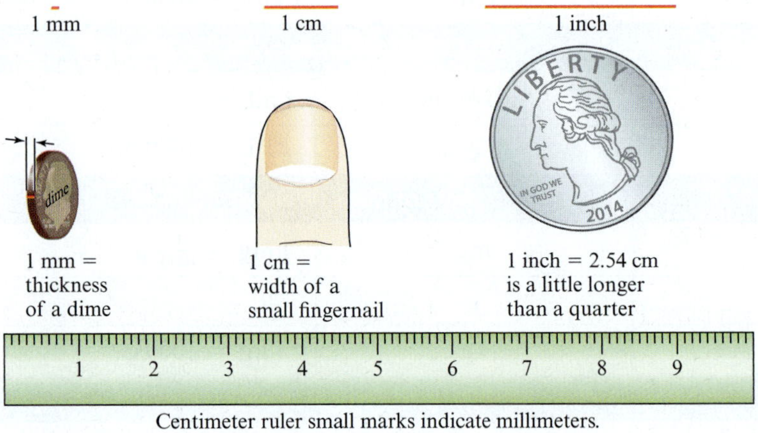

Figure 11.3 Comparing millimeters, centimeters, and inches.

Table 11.6 shows some basic relationships between the U.S. customary and metric systems of measurement.

TABLE 11.6 Relationships between U.S. customary and metric systems

U.S. Customary and Metric	
Length	1 in. = 2.54 cm (exact)
Capacity	1 gal = 3.79 L (good approximation)
Weight (mass)	1 kg = 2.2 lb (good approximation)

How Are Units of Length, Area, Volume, and Capacity Related?

For any unit of length, there are corresponding units of area and volume: a square unit and a cubic unit, respectively. A square unit is the area of a square that is 1 unit wide and 1 unit long. A cubic unit is the volume of a cube that is 1 unit wide, 1 unit long, and 1 unit high. For example, a square inch is the area of a square that is 1 inch wide and 1 inch long and a cubic centimeter is the volume of a cube that is 1 centimeter wide, 1 centimeter long, and 1 centimeter high, as shown in Figures 11.4 and 11.5.

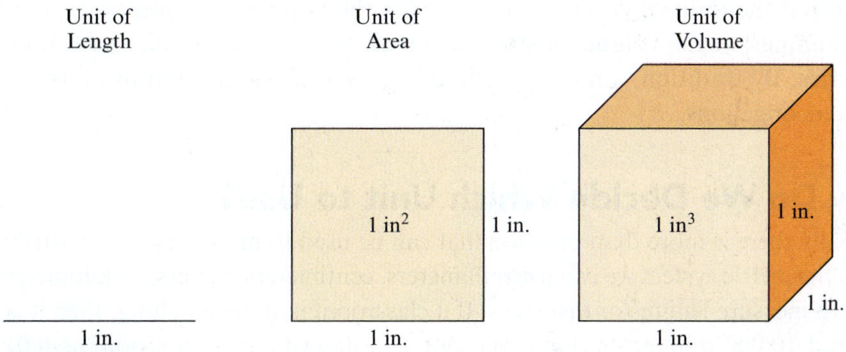

Figure 11.4 Examples of units of length, area, and volume in the U.S. customary system.

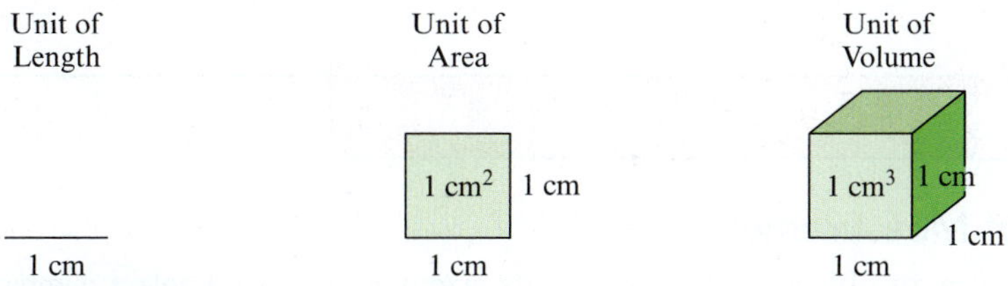

Figure 11.5 Examples of units of length, area, and volume in the metric system.

We often use an exponent of 2 to write the name of a unit of area by writing unit^2 for "square unit." For example, we can write square inches as in^2 and square centimeters as cm^2.

We often use an exponent of 3 to write the name of a unit of volume by writing unit^3 for "cubic unit." For example, we can write cubic inches as in^3 and cubic centimeters as cm^3.

It's easy to misunderstand measurements of areas and volumes. For example, consider this question:

> **Question:** Which of the following describe the same volume?
>
> **a.** A 2-inch-by-2-inch-by-2-inch cube
>
> **b.** 2 cubic inches
>
> **c.** 2 in^3
>
> **d.** 2 in. $\times$ 2 in. $\times$ 2 in.

At first glance, it might seem that they all do. However, this is not the case. Both (a) and (d) describe a volume of 8 cubic inches. On the other hand, (b) and (c) describe a volume of 2 cubic inches, which is the amount of space taken up by two 1-inch-by-1-inch-by-1-inch cubes.

Notice that to avoid confusion, it's best to read 2 in^3 as "2 cubic inches" because reading it as "2 inches cubed" makes it sound like a 2-inch-by-2-inch-by-2-inch cube, which has a volume of 8 cubic inches, not 2 cubic inches. Similar considerations apply to area.

CLASS ACTIVITY

11D What Does "6 Square Inches" Mean?, p. CA-231

What is the difference between volume and capacity? Capacity is essentially the same as volume, except that the term *capacity* can be used for the volume of a container, the volume of liquid in a container, or the volume of space taken up by a substance filling a container, such as flour or berries. By tradition, amounts of liquids are usually measured in units of capacity, such as milliliters or gallons.

How Do We Decide Which Unit to Use?

Generally there is more than one unit that can be used to measure a given attribute. For example, within the metric system we can use millimeters, centimeters, meters, or kilometers (as well as other units) to measure lengths or distances. If a classroom is 10 meters long, then it is 1000 centimeters long and 10,000 millimeters long. We can also describe the classroom as 0.01 or $\frac{1}{100}$ kilometer long. Clearly, reporting the classroom as "10 meters long" is easier to grasp than any of the other options. When measuring or reporting measurements, use a unit that will best help others to comprehend the answer.

SECTION SUMMARY AND STUDY ITEMS

Section 11.1 Concepts of Measurement

In order to measure an object, or to describe its size, we must first select a measurable attribute. To measure a quantity is to compare the quantity with a unit amount. A unit is a fixed, reference amount. The most direct way to measure a quantity is to count how many of the unit amount make the quantity.

The U.S. customary system and the metric system are the two collections of units we use in the United States today. The metric system uses prefixes to name units in a uniform way. A fundamental relationship in the metric system is that 1 milliliter of water weighs 1 gram and has a volume of 1 cubic centimeter.

Key Skills and Understandings

1. Explain that, to measure a quantity, a measurable attribute must first be chosen.
2. Explain what it means for an object to have an area of 8 square centimeters, or a volume of 12 cubic inches, or a weight of 13 grams, or other similar measurements.
3. Discuss units and know that measurement is comparison with a unit.
4. Know and use proper notation for units of area and volume (e.g., square centimeters, cm^2).
5. Explain how the metric system uses prefixes, and state the meaning of common metric prefixes: kilo-, hecto-, deka-, deci-, centi-, and milli-.
6. Describe how capacity, weight (mass), and volume are linked in the metric system.

Practice Exercises for Section 11.1

1. How many feet are in 1 mile? How many ounces are in 1 pound? How many pounds are in 1 ton?
2. What is the difference between 1 ounce and 1 fluid ounce?
3. What is special about the way units in the metric system are named?
4. How is 1 milliliter related to 1 gram and to 1 centimeter?
5. Use square centimeter tiles or centimeter graph paper to show or to draw three different shapes that have an area of 8 cm^2.
6. Use cubic-inch blocks (or cubic-centimeter blocks) to make a variety of solid shapes that have a volume of 24 in^3 (or 24 cm^3).
7. Which of the following have the same area or mean the same as 3 cm^2?
   - a 3-cm-by-3-cm square
   - 3 square centimeters
   - 3 cm $\times$ 3 cm
8. A construction site requires 40 cubic yards of concrete. What does "40 cubic yards of concrete" mean?
9. Explain how it could happen that each of two boxes of cereal could be described as larger than the other.

Answers to Practice Exercises for Section 11.1

1. There are 5280 feet in a mile, 16 ounces in a pound, and 2000 pounds in a ton.
2. An ounce is a unit of weight, whereas a fluid ounce is a unit of capacity (or volume).
3. The metric system uses *base units* such as meter, liter, and gram, and it uses *prefixes* such as milli-, centi-, deci-, deka-, hecto-, and kilo- to create other units such as milliliter and kilogram.
4. One milliliter of water weighs 1 gram and occupies a volume of 1 cubic centimeter.
5. Arrange 8 square-centimeter tiles in any way to form a shape, or draw a shape that encloses 8 squares, as in the scale drawing in Figure 11.6.

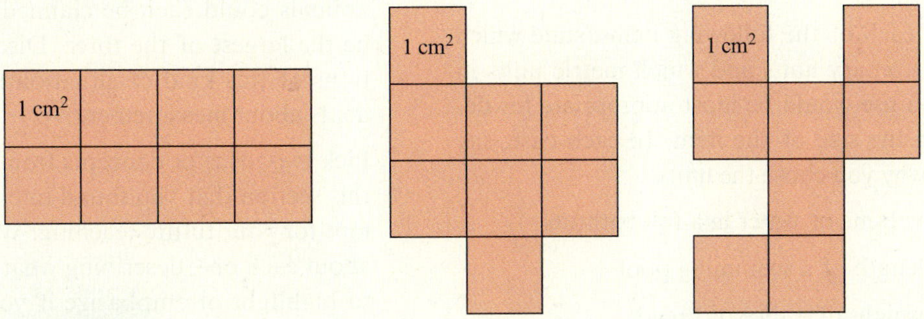

Figure 11.6 Three ways to show 8 cm^2.

6. Arrange 24 cubic-inch blocks (or cubic-centimeter blocks) in any way to make a solid shape.
7. Only "3 square centimeters" means the same thing as 3 cm^2. A 3-cm-by-3-cm square has area 9 cm^2, that is, 9 square centimeters.
8. If you had a 1-yard-by-1-yard-by-1-yard cube that could be filled with concrete, then if you filled that cube up 40 times and poured the contents out into a pile, that big pile of concrete would have a volume of 40 cubic yards.
9. Boxes of cereal could be compared by weight, by volume, by number of servings, by calories, or in other ways. As Table 11.7 shows, if we compare the cereal either by weight or by the number of calories

TABLE 11.7 Comparing boxes of cereal

	Cereal Box 1	Cereal Box 2
Weight of cereal in box	16 oz	11 oz
Volume of cereal in box	4 cups	10 cups
Servings per box	8	10
Calories per box	1600	1100

per box, then cereal box 1 is larger (16 ounces versus 11 ounces and 1600 calories versus 1100 calories). On the other hand, if we compare the cereal by volume or by the number of servings per box, then cereal box 2 is larger (4 cups versus 10 cups and 8 servings per box versus 10 servings per box).

PROBLEMS FOR SECTION 11.1

1. For each of the following metric units, give examples of two objects that you encounter in daily life whose sizes could appropriately be described using that unit.

a. meter

b. gram

c. liter

d. milliliter

e. millimeter

f. kilogram

g. kilometer

2. For each of the following items, state which U.S. customary units and which metric units in common use would be most appropriate for describing the size of the item. In each case, say briefly why you chose the units.

a. The volume of water in a full bathtub

b. The length of a swimming pool

c. The weight of a slice of bread

d. The volume of a slice of bread

e. The weight of a ship

f. The length of an ant

3. What does it mean to say that a shape has an area of 8 square inches? Discuss your answer in as clear and direct a fashion as you can.

4. Discuss why it is easy to give an incorrect solution to the following area problem and what you must understand about measurement to solve the problem correctly:

An area problem: Draw a shape that has an area of 3 square inches.

5. Discuss: Why is it not completely correct to describe volume as "length times width times height"? What is a better way to describe what volume means? Give some examples as part of your discussion, using different units of volume.

6. Describe how it could happen that three different animals could each be claimed—rightfully—to be the largest of the three. Discuss the implications of this kind of situation for teaching students about measurement.

7. Pick two ideas or concepts from your reading of this section that you found interesting or important for your future teaching. Write a paragraph about each one, describing what you would want to highlight or emphasize if you were teaching students these ideas or concepts.

8. Visit a store and write down at least 10 different items and their metric measurements that you find on their package labels. For each of the following units, find at least one item that is described using that unit: grams (g), kilograms (kg), milligrams (mg), liters (L), milliliters (mL).

11.2 Length, Area, Volume, and Dimension

All physical objects have several different measurable attributes. For example, what are some measurable attributes of a spool of wire at a hardware store? There is the length of wire that is wound onto the spool. There is the thickness of the wire. There is the weight of the wire. The most common measurable attributes are length, area, volume, and weight. In this section, we will study length, area, volume, and the related concept of dimension. Although these concepts may seem advanced, the foundations for understanding them can be formed early in elementary school. When children make shapes by placing sticks end-to-end, or when they lay tiles or build with blocks, think of their work as exploring length, area, volume, and dimension.

The units for length, area, and volume are related, so it is easy to get confused about which unit to use when. Most real objects have one or more aspects or parts that should be measured by a unit of length, another aspect or part that should be measured by a unit of area, and yet another aspect or part that should be measured by a unit of volume.

We use a unit of *length*, such as inches, centimeters, miles, or kilometers, when we want to answer one of the following kinds of questions:

- How far?
- How long?
- How wide?

For example, "How much moulding (i.e., *how long* a piece of moulding) will I need to go around this ceiling?" The answer, in feet, is the number of foot-long segments that can be put end-to-end to go all around the ceiling. (See Figure 11.7).

Figure 11.7
A length problem.

How many of these 1-foot lengths — does it take to go around the ceiling of this room (i.e., around the outside)?

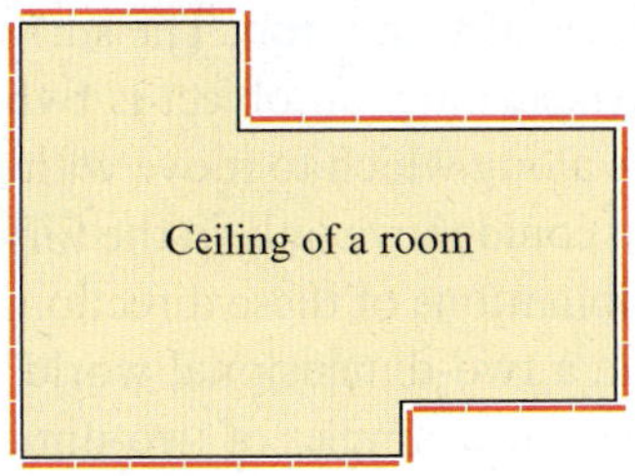

length A **length** describes the size of an object (or a part of an object) that is one-dimensional—the length is how many of a chosen unit of length (such as inches, centimeters, etc.) it takes to span the object from one end to the opposite end without gaps or overlaps. It is understood that we may use parts of a unit, too.

one-dimensional Roughly speaking, an object is **one-dimensional** if at each location, there is only one independent direction along which to move within the object. An imaginary creature living in a one-dimensional world could move only forward and backward in that world. Here are some examples of one-dimensional objects:

A line segment

A circle (only the outer part, not the inside)

The four line segments making a square (not the inside of the square)

A curved line

The equator of the earth

The edge where two walls in a room meet

An imaginary line drawn from one end of a student's desk to the other end

perimeter In general, the "outer edge" around a (flat) shape is one-dimensional. The **perimeter** of a shape is the *distance around a shape*; that is, it is the total length of the outer edge around the shape.

We use a unit of *area*, such as square feet or square kilometers, when we want to answer questions like the following:

- How much flat material (such as paper, fabric, or sheet metal) does it take to make this?
- How much flat material does it take to cover this region?

For example, "How much carpet do I need to cover the floor of this room?" The answer, in square feet, is the number of 1-foot-by-1-foot squares it would take to cover the floor without gaps or overlaps. (See Figure 11.8.)

Figure 11.8

An area problem.

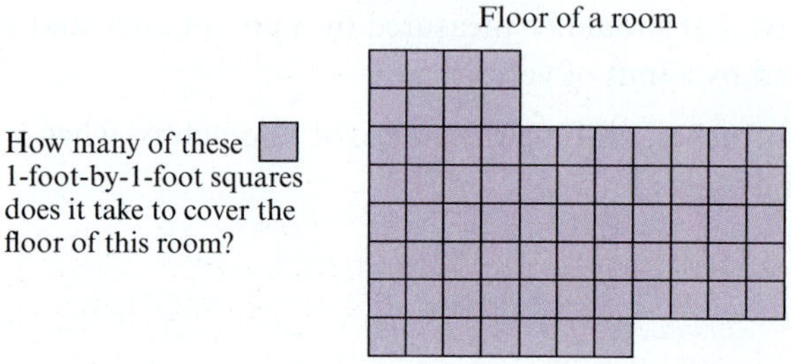

area **surface area** An **area** or a **surface area** describes the size of an object (or a part of an object) that is two-dimensional—the area of that two-dimensional object is how many of a chosen unit of area (such as square inches, square centimeters, etc.) it takes to cover the object without gaps or overlaps. It is understood that we may use parts of a unit, too. The surface area of a solid shape is the total area of its outside surface. Roughly speaking, an object is **two-dimensional** if, at each location, there are two independent directions along which to move within the object. An imaginary creature living in a two-dimensional world could move only in the forward/backward and right/left directions, as well as in "in between" combinations of these directions (such as diagonally), but *not* up/down. What would it be like to live in a two-dimensional world? The book *Flatland* by A. Abbott [1] is just such an account. Here are some examples of two-dimensional objects:

two-dimensional

A coordinate plane

The *inside* of a circle (this is also called a disk)

The *inside* of a square

A piece of paper (only if you think of it as having no thickness)

The *surface* of a balloon (not counting the inside)

The *surface* of a box

Some farmland (only counting the surface, not the soil below)

The surface of the earth

Notice that most of the units that are used to measure area—such as cm^2, m^2, km^2, in^2, ft^2, mi^2—have a superscript "2" in the abbreviation. This 2 reminds us that we are measuring the size of a two-dimensional object.

We use a unit of *volume*, such as cubic yards or cubic centimeters, or even gallons or liters, when we want to answer questions like the following:

- How much of a substance (such as air, water, or wood) is in this object?
- How much of a substance does it take to fill this object?

For example, "How much air is in this room?" The answer, in cubic feet, is the number of 1-foot-by-1-foot-by-1-foot cubes that are needed to fill the room. (See **Figure 11.9**.)

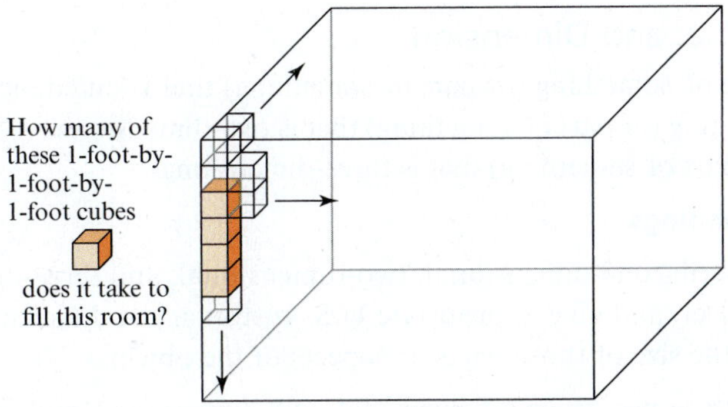

Figure 11.9 A volume problem.

volume A **volume** describes the size of an object (or a part of an object) that is three-dimensional—the volume of that three-dimensional object is how many of a chosen unit of volume (such as cubic inches, cubic centimeters, etc.) it takes to fill the object without gaps or overlaps. It is understood that we may use parts of a unit, too. Roughly speaking, an object is **three-dimensional** if at each location there are three independent directions along which to move within the object. Our world is three-dimensional because we can move in the three independent directions: forward/backward, right/left, and up/down, as well as all "in between" combinations of these directions. Here are some examples of three-dimensional objects:

three-dimensional

The air around us

The inside of a balloon

The inside of a box

The water in a cup

The feathers filling a pillow

The compost in a wheelbarrow

The inside of the earth

Notice that many of the units that are used to measure volume—such as cm^3, m^3, in^3, ft^3—have a superscript "3" in their abbreviation. This 3 reminds us that we are measuring the size of a three-dimensional object.

One of the difficulties with understanding whether to measure a length, an area, or a volume is that many objects have parts of different dimensions. For example, consider a balloon. Its outside *surface* is two-dimensional, whereas the air filling it is three-dimensional. Then there is the circumference of the balloon, which is the length of an imaginary one-dimensional circle drawn around the surface of the balloon. The size of each of these parts should be described in a different way: The air in a balloon is described by its volume, the surface of the balloon is described by its area, and the circumference of the balloon is described by its length.

Another example in which the same object has parts of different dimensions is the earth. The *inside* of the earth is three-dimensional, its *surface* is two-dimensional, and the equator is a one-dimensional imaginary circle drawn around the surface of the earth. Once again, the sizes of these different parts of the earth should be described in different ways: by volume, area, and length, respectively.

CLASS ACTIVITY

11E Dimension and Size, p. CA-232

SECTION SUMMARY AND STUDY ITEMS

Section 11.2 Length, Area, Volume, and Dimension

A length describes the size of something (or part of something) that is one-dimensional. An area describes the size of something (or part of something) that is two-dimensional. A volume describes the size of something (or part of something) that is three-dimensional.

Key Skills and Understandings

1. Given an object, describe one-dimensional, two-dimensional, and three-dimensional parts or aspects of the object, and give appropriate U.S. customary and metric units for measuring or describing the size of those parts or aspects of the object.
2. Compare objects with respect to one-dimensional, two-dimensional, and three-dimensional attributes.
3. Recognize that in some cases, one object can be larger than another object with respect to one attribute, but smaller with respect to a different attribute.

Practice Exercises for Section 11.2

1. a. Use centimeter or inch graph paper to make a pattern for a closed box (rectangular prism). The box should have 6 sides, and when you fold the pattern, there should be no overlapping pieces of paper.

b. How much paper is your box made of? Be sure to use an appropriate unit in your answer.

c. Describe one-dimensional, two-dimensional, and three-dimensional parts or aspects of your box. In each case, give the size of the part or aspect of the box, using an appropriate unit.

2. Describe one-dimensional, two-dimensional, and three-dimensional parts or aspects of a water tower. In each case, name an appropriate U.S. customary unit and an appropriate metric unit for measuring or describing the size of that part or aspect of the water tower. What are practical reasons for wanting to know the sizes of these parts or aspects of the water tower?

3. Describe one-dimensional, two-dimensional, and three-dimensional parts or aspects of a store. In each case, name an appropriate U.S. customary unit and an appropriate metric unit for measuring or describing the size of that part or aspect of the store. What are practical reasons for wanting to know the sizes of these parts or aspects of the store?

Answers to Practice Exercises for Section 11.2

1. a. Figure 11.10 shows a scaled picture of a pattern for one possible box. Each small square represents a 1-inch-by-1-inch square.

b. The box formed by the pattern in Figure 11.10 (in its actual size) is made of 52 square inches of paper because the pattern that was used to create the box consists of 52 1-inch-by-1-inch squares.

c. The edges of the box are one-dimensional parts of the box. Depending on how the box is oriented, the lengths of these edges are the height, width, and length of the box. These measure 2 inches, 3 inches, and 4 inches. Another one-dimensional aspect of the box is its *girth* (i.e., the distance around the box). Depending on where you measure, the girth is either 10 inches, 12 inches, or 14 inches.

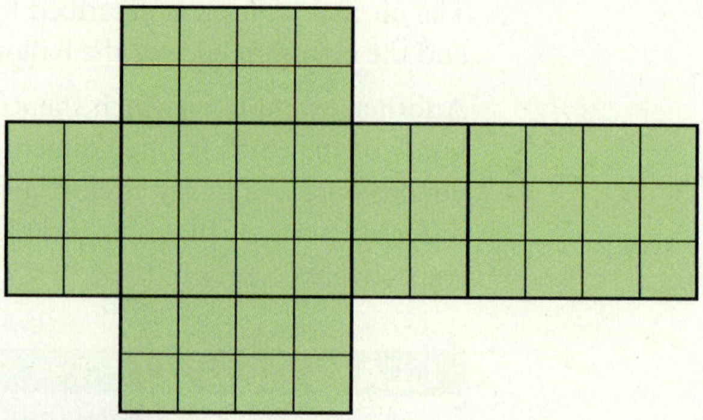

Figure 11.10 A pattern for a box.

The full outer surface of the box is a two-dimensional part of the box. Its area is 52 square inches, as discussed in part (b).

The air inside the box is a three-dimensional aspect of the box. If you filled the box with 1-inch-by-1-inch-by-1-inch cubes, there would be 2 layers with 3×4 cubes in each layer. Therefore, according to the meaning of multiplication, there would be $2 \times 3 \times 4 = 24$ cubes in the box. Thus, the volume of the box is 24 cubic inches.

2. The height of the water tower (or an imaginary line running straight up through the middle of the water tower) is a one-dimensional aspect of the water tower. Feet and meters are appropriate units for describing the height of the water tower. Some towns do not want high structures; therefore, the town might want to know the height of a proposed water tower before it decides whether to build it.

The surface of the water tower is a two-dimensional part of the water tower. Square feet and square meters are appropriate units for describing the size of the surface of the water tower. We would need to know the approximate size of the surface of the water tower to know approximately how much paint it would take to cover it.

The water inside a water tower is a three-dimensional part of a water tower. Cubic feet, gallons, liters, and cubic meters are appropriate units for measuring the volume of water in the water tower. The owner of a water tower will certainly want to know how much water the tower can hold.

3. The edge where the store meets the sidewalk is a one-dimensional aspect of the store. Its length is appropriately measured in feet, yards, or meters. This length is the width of the store front, which is one way to measure how much exposure the store has to the public eye.

The surface of the store facing the sidewalk is a two-dimensional part of the store. Its area is appropriately described in square feet, square yards, or square meters. The area of the store front is another way to measure how much exposure the store has to the public eye. (A store front that is narrow but tall might attract as much attention as a wider, lower store front.)

Another important two-dimensional part of a store is its floor. The area of the floor is appropriately measured in square feet, square yards, or square meters, or possibly even in acres if the store is very large. The area of the floor indicates how much space there is on which to place items to sell to customers.

The air inside a store is a three-dimensional aspect of the store. Its volume is appropriately measured in cubic feet, cubic yards, cubic meters, or possibly even in liters or gallons. The builder of a store might want to know the volume of air in the store in order to figure what size air-conditioning units will be needed.

PROBLEMS FOR SECTION 11.2

1. Describe one-dimensional, two-dimensional, and three-dimensional parts or aspects of a soft drink bottle. In each case, specify an appropriate U.S. customary unit and an appropriate metric unit for measuring or describing the size of that part or aspect of the bottle. What are practical reasons for wanting to know the sizes of these parts or aspects of the soft drink bottle?

2. Describe one-dimensional, two-dimensional, and three-dimensional parts or aspects of a car. In each case, specify an appropriate U.S. customary unit and an appropriate metric unit for measuring or describing the size of that part or aspect of the car. What are practical reasons for wanting to know the sizes of these parts or aspects of the car?

3. Drawing on your reading from this section, describe how it could happen that 3 different cathedrals could each claim—rightfully—to be the largest cathedral. Discuss the implications of this kind of situation for teaching students about measurement.

4. Describe one-dimensional, two-dimensional, and three-dimensional parts or aspects of the blocks in Figure 11.11. In each case, compare the sizes of the 3 blocks, using an appropriate unit. Use this unit to show that each block can be considered biggest of all 3.

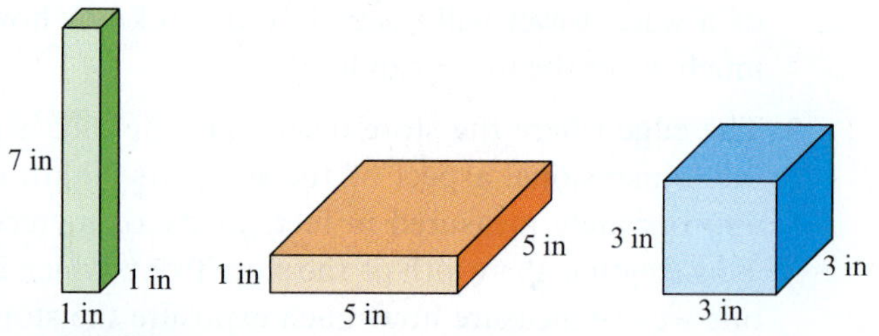

Figure 11.11 Which block is biggest?

5. The Lazy Daze Pool Club and the Slumber-N-Sunshine Pool Club have a friendly rivalry going. Each club claims to have the bigger swimming pool. Make up realistic sizes for the clubs' swimming pools so that each club has a legitimate basis for saying that it has the larger pool. Explain clearly why each club can say its pool is biggest. Be sure to use appropriate units. Considering the function of a pool, what is a good way to compare sizes of pools? Explain.

6. Suppose there are 2 rectangular pools: One is 30 feet wide, 60 feet long, and 5 feet deep throughout, and the other is 40 feet wide, 50 feet long, and 4 feet deep throughout. Show that each pool can be considered "biggest" by comparing the sizes of the pools in two meaningful ways other than by comparing one-dimensional aspects of the pools.

7. Minh says that the rectangle on the left in Figure 11.12 is larger than the one on the right, Sequoia says the rectangle on the right is larger than the one on the left. Explain why Minh and Sequoia can both be right.

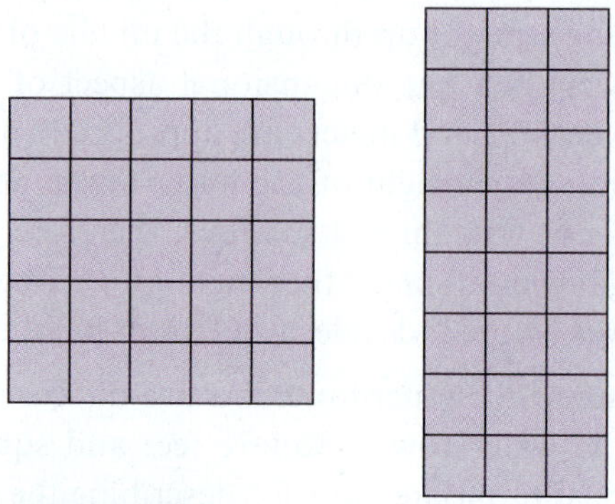

Figure 11.12 Two rectangles.

11.3 Error and Precision in Measurements

CCSS SMP6

Error and precision are natural parts of all actual measurements of physical quantities. It is important to distinguish between *theoretical* and *actual* measurements. When we say that a room that is 4 meters wide and 5 meters long has an area of 20 square meters, we are considering a theoretical room. If we are considering an actual room, we can never say that it is exactly 4 meters wide and 5 meters long. When measuring actual physical quantities, it is impossible to avoid a certain amount of error and uncertainty. In this section, we will see that the way we report a measurement indicates the accuracy of that measurement. Similarly, our interpretation of a measurement depends on the digits reported in the measurement.

Any reported measurement of an actual physical quantity is necessarily only approximate—you can never say that a measurement of a real object is exact. For example, if you use a ruler to measure the width of a piece of paper, you may report that the paper is $8\frac{1}{2}$ inches wide. But the paper isn't *exactly* $8\frac{1}{2}$ inches wide, it is just that the mark on your ruler that is closest to the edge of the paper is at $8\frac{1}{2}$ inches.

How Do We Interpret Reported Measurements?

Because measurements are never exact, we have conventions for reporting measurements. The way that a measurement is reported reflects its precision. For example, suppose that the distance between two cities is reported as 1200 miles. Since there are zeros in the tens and ones places, we assume that this measurement is *rounded to the nearest hundred* (unless we are told otherwise). In other words, the actual distance between the two cities is between 1100 miles and 1300 miles, but closer to 1200 miles than to either 1100 miles or 1300 miles. Thus, the actual distance between the two cities could be anywhere between 1150 and 1250 miles, as indicated in Figure 11.13.

Figure 11.13

Interpreting a reported distance of 1200 miles.

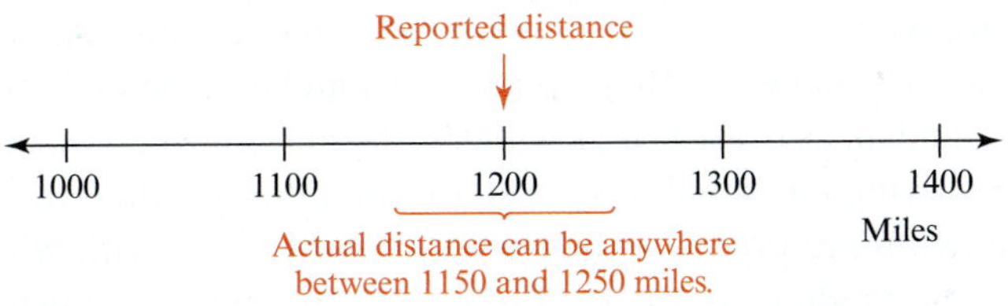

Now suppose that the distance between two cities is reported as 1230 miles, then we assume (unless we are told otherwise) that this measurement is *rounded to the nearest ten.* In other words, the actual distance between the cities is between 1220 miles and 1240 miles, but closer to 1230 miles than to either 1220 miles or 1240 miles. In this case, the actual distance between the two cities could be anywhere between 1225 and 1235 miles, as indicated in **Figure 11.14**. Notice that in this case, we are getting a much smaller range of numbers that the actual distance could be than in the previous example. Here it's a range of 10 miles (1225 miles to 1235 miles), whereas in the previous case it's a range of 100 miles (1150 miles to 1250 miles).

Figure 11.14

Interpreting a reported distance of 1230 miles.

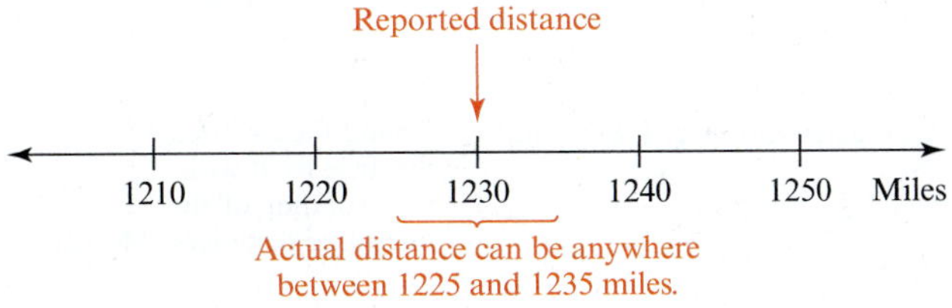

Similarly, if your weight on a digital scale is reported as 130.4 pounds, then (assuming the scale's report is accurate) this is your actual weight, *rounded to the nearest tenth.* In other words, your actual weight is between 130.3 and 130.5 pounds, but closer to 130.4 pounds than to either 130.3 or 130.5 pounds. In this case, your actual weight is between 130.35 and 130.45 pounds.

Notice that there is a difference in reporting that the weight of an object is 130.0 pounds and reporting the weight as 130 pounds. When the weight is reported as 130.0 pounds, we assume that this weight is *rounded to the nearest tenth* of a pound. In this case, the actual weight is between 129.95 and 130.05 pounds. On the other hand, when the weight of an object is reported as 130 pounds, we assume this weight is *rounded to the nearest ten* pounds. In this case, we know only that the actual weight is between 125 and 135 pounds, which allows for a much wider range of actual weights than the range of 129.95 to 130.05 pounds (a 10-pound range versus a 0.1-pound range). So, when a weight is reported as 130.0 pounds, the weight is known with greater accuracy than when the weight is reported as 130 pounds.

What do we do when we know a distance more accurately than the way we write it would suggest? For example, suppose we know that a certain distance is 130,000 km, but that this is rounded to the nearest *hundred* kilometers, not to the nearest *ten-thousand* kilometers, as the way the number is written would suggest. In this case, we can report the distance as "130,000 km to 4 **significant digits**."

How Do We Calculate with Measurements?

In the next section and the chapters that follow we will calculate various lengths, areas, and volumes of objects by using lengths associated with these objects. In some cases, the lengths are taken to be *given* as opposed to actually *measured,* and the area and volume calculations are then purely theoretical. For example, we will be able to calculate the area of a circle that has radius 6.25 inches. In this case, we are imagining a circle that has radius 6.25 inches, and we are calculating the exact area that such an imagined circle would have.

In other cases, we will work with actual or hypothetical measured distances to calculate other distances, or to calculate areas or volumes. In these cases, we should assume that the distances we are working with have been rounded, as described previously. Therefore, when we determine our final answer to a calculation involving measured distances, we should also round that answer, because the answer cannot be known more precisely than the initial distances with which we began the calculations. For example, we might calculate the distance from town A to town C, given that the distance from town A to town B is 250 km, and the distance from town B to town C is 360 km, as shown in Figure 11.15. By calculating, we determine that the distance from town A to town C is 438.29 km. But we should not leave our answer like that, because if we do, we are communicating that we know the distance from town A to town C rounded to the nearest *hundredth* of a kilometer. Because we know only the distances from town A to town B and town B to town C rounded to the nearest *ten* kilometers, we cannot possibly know the resulting distance from town A to town C with greater precision. Instead, we should report the distance from town A to town C as 440 km, which fits with the rounding of the distances we were given.

Figure 11.15

Calculating with measured distances.

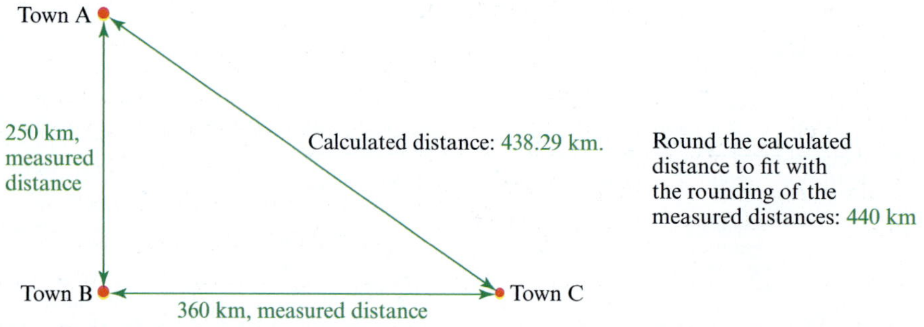

Keep the following point in mind when calculating with actual or hypothetical measurements: Even though you may be rounding your final answer, wait until you are done with all your calculations before you round. Waiting to the end to round will give you the most precise answer that fits with the measurements you started with.

CLASS ACTIVITY

11F Reporting and Interpreting Measurements, p. CA-233

SECTION SUMMARY AND STUDY ITEMS

Section 11.3 Error and Precision in Measurements

When it comes to measuring actual physical quantities, all measurements are only approximate. The way we write a measurement conveys how precisely the measurement is known. When calculating with actual measurements, the final result should be rounded to convey how precisely the result is known.

Key Skills and Understandings

1. Explain how a reported measurement conveys how precisely the measurement is known. For example, if the weight of an object is reported as 5 pounds, the actual weight of the object has been rounded to the nearest whole number and is therefore between 4.5 and 5.5 pounds. But if the weight of an object is reported as 5.0 pounds, the actual weight of the object has been rounded to the nearest tenth and is therefore between 4.95 and 5.05 pounds.

Practice Exercises for Section 11.3

1. What is the difference between reporting that an object weighs 2 pounds and reporting that it weighs 2.0 pounds?

2. If the distance between two cities is reported as 2500 miles, does that mean that the distance is exactly 2500 miles? If not, what can you say about the exact distance?

Answers to Practice Exercises for Section 11.3

1. If an object is reported as weighing 2 pounds, then we assume that the weight has been rounded to the nearest whole pound. Therefore, the actual weight is between 1 and 3 pounds, but closer to 2 pounds than to either 1 pound or 3 pounds. The actual weight must be between 1.5 pounds and 2.5 pounds. On the other hand, if an object is reported as weighing 2.0 pounds, then we assume that the weight has been rounded to the nearest *tenth* of a pound. Therefore, the actual weight is between 1.9 pounds and 2.1 pounds, but closer to 2.0 pounds than to either 1.9 pounds or 2.1 pounds. The actual weight must be between 1.95 pounds and 2.05 pounds. So, when the weight is reported as 2.0 pounds, the weight is known with greater accuracy than when it is reported as 2 pounds.

2. If the distance between two cities is reported as 2500 miles, then we assume that this number is the actual distance, rounded to the nearest hundred. Therefore, the actual distance is between 2400 miles and 2600 miles, but closer to 2500 miles than to either 2400 miles or 2600 miles. The actual distance between the two cities could be anywhere between 2450 miles and 2550 miles.

PROBLEMS FOR SECTION 11.3

1. One source says that the average distance from the earth to the moon is 384,467 kilometers. Another source says that the average distance from the earth to the moon is 384,000 kilometers. Can both of these descriptions be correct, or must at least one of them be wrong? Explain.

2. If an object is described as weighing 6.20 grams, then is this the exact weight of the object? If not, what can you say about the weight of the object? Explain your answer in detail.

3. Tyra is calculating the distance from town A to town C. She is given that the distance from town A to town B is 120 miles, that the distance from town B to town C is 230 miles, that town B is due south of town A, and that town C is due east of town B. She calculates that the distance from town A to town C is 259.4224 miles. Should Tyra leave her answer like that? Why or why not? If not, what answer should she give? Explain. (You may assume that Tyra has calculated correctly.)

*4. John has a paper square that he believes is 100 cm wide and 100 cm long, but in reality, the square is actually 1% longer and 1% wider than he believes. In other words, John's square is actually 101 cm wide and 101 cm long.

   a. Since John believes the square is 100 cm by 100 cm, he calculates that the area of his square is 10,000 cm^2. What percent greater is the actual area of the square than John's calculated area of the square? Is the actual area of John's square 1% greater than John's calculated area, or is it larger by a different percent?

   b. Draw a picture (which need not be to scale) to show why your answer in part (a) is not surprising.

   c. Answer the following questions based on your answers to parts (a) and (b): In general, if you want to know the area of a square to within 1% of its actual area, will it be good enough to know the lengths of the sides of the square to within 1% of their actual lengths? If not, will you need to know the lengths more accurately or less accurately?

*5. Sally has a Plexiglas cube that she believes is 100 cm wide, 100 cm long, and 100 cm tall, but in reality, the cube is actually 1% wider, 1% longer, and 1% taller than she believes. In other words, Sally's cube is actually 101 cm wide, 101 cm deep, and 101 cm tall.

a. Since Sally believes the cube is 100 cm by 100 cm by 100 cm, she calculates that the volume of her cube is 1,000,000 cm^3. What percent greater is the actual volume of the cube than Sally's calculation of the volume of the cube? Is the actual volume of Sally's cube 1% greater than her calculated volume, or is it larger by some different percent?

b. Answer the following questions based on your answer to part (a): In general, if you want to know the volume of a cube to within 1% of its actual volume, will it be good enough to know the lengths of the sides of the cube to within 1% of their actual lengths? If not, will you need to know the lengths more accurately or less accurately?

11.4 Converting from One Unit of Measurement to Another

CCSS Common Core State Standards Grades 2, 4, 5, 6

As we've seen, we can use different units to measure the same quantity. For example, we can measure a length in miles, yards, feet, inches, kilometers, meters, centimeters, or millimeters. If we know a length in terms of one unit, how can we describe it in terms of another unit? This kind of problem is a conversion problem, which is the topic of this section.

For young children it is a big milestone to recognize how a measurement of an object relates (qualitatively) to the size of the unit. For example, suppose a child measures that her desk is 12 sticks long. Next, she chooses a block that is shorter than the stick and plans to use the block to measure how long her desk is. Before she measures, the teacher asks her to predict how long the desk will be in blocks. The child might respond with a number smaller than 12, thinking that a shorter unit will produce a smaller measurement. In fact it is just the opposite: it will take *more* of the shorter unit. Why? Both units must span the same length, so it takes fewer of a long unit and more of a shorter unit to span that length.

CCSS 2.MD.2

Older students learn to convert a given measurement to another unit. To convert a measurement in one unit to another unit, we first need to know the relationship between the two units. For example, if we want to convert the weight of a bag of oranges given in kilograms into pounds, then we need to know the relationship between kilograms and pounds. Referring back to Table 11.6 in Section 11.1, we see that 1 kg = 2.2 lb.

Once we know how the units are related, we either multiply or divide to convert from one unit to another. How do we know which operation to use? Before you read on, please do Class Activity 11G.

CLASS ACTIVITY

11G Conversions: When Do We Multiply? When Do We Divide?, p. CA-234

CCSS 4.MD.1 5.MD.1 6.RP.3d

How Can We Reason About Multiplication and Division to Convert Measurements?

How do we decide whether to multiply or divide to convert from one unit to another? To help us decide we can think about what multiplication and division mean. For example, let's say a bag of oranges weighs 2.5 kg. What is the weight of the bag of oranges in pounds? If 1 kilogram is

2.2 pounds, then we can think of 2.5 kilograms as 2.5 groups of 2.2 pounds each, as illustrated in Figure 11.16. Thus, according to what multiplication means, 2.5 kilograms is 2.5 × 2.2 pounds, which calculates to 5.5 pounds.

Figure 11.16
Converting 2.5 kilograms to pounds.

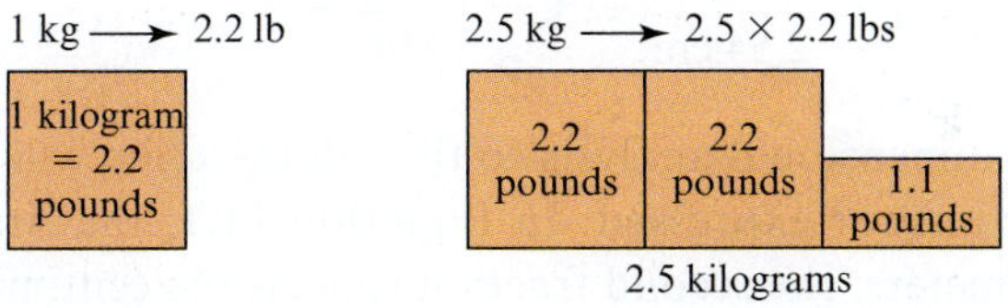

Here's another example: How many gallons are there in a 25-liter container? Referring back to Table 11.6, we see that 1 gal = 3.79 L. That is, every group of 3.79 liters is equal to 1 gallon. Therefore, to find how many gallons are in 25 liters, we need to find out how many groups of 3.79 liters are in 25 liters. This is a division problem, using the how-many-groups interpretation of division. So there are 25 ÷ 3.79 = 6.6 gallons in 25 liters.

CLASS ACTIVITY

11H Conversion Problems, p. CA-235

Why Does Dimensional Analysis Work?

In some cases, the tables in Section 11.1 do not tell us directly how two units are related. For example, Table 11.6 does not tell us how meters relate to feet. To convert a measurement in meters to feet, we must work with the length relationships that we do know from the tables in Section 11.1. In this case, we use the relationship 1 in = 2.54 cm, the relationship between inches and feet, and the relationship between centimeters and meters to convert meters to feet. We go through the following chain of conversions:

meters → centimeters → inches → feet

To organize the process of such a chain of conversions, we can use *dimensional analysis*, as in Class Activity 11I.

CLASS ACTIVITY

11I Using Dimensional Analysis to Convert Measurements, p. CA-236

Dimensional analysis is a process for converting a measurement from one unit to another in which we repeatedly multiply by the number 1 expressed as a fraction that relates two different units. The resulting calculations are identical to those done to convert units by reasoning about multiplication and division, as shown previously.

For example, suppose that a German car is 5.1 meters long. How long is the car in feet? To solve this problem, we will use the information provided in the tables of Section 11.1 to carry out the following chain of conversions:

meters → centimeters → inches → feet

The following equation uses dimensional analysis to convert 5.1 meters to feet:

$$5.1\text{ m} = 5.1\text{ m} \times \frac{100\text{ cm}}{1\text{ m}} \times \frac{1\text{ in.}}{2.54\text{ cm}} \times \frac{1\text{ ft}}{12\text{ in.}} = 16.7\text{ ft} \qquad (11.1)$$

So, the car is 16.7 feet long.

How do we choose the fractions to use in dimensional analysis? Each fraction that we multiply by must be equal to 1. That way, when we multiply by the fraction, we do not change the original amount; we just express the amount in different units. We used the following fractions in Equation 11.1:

$$\frac{100 \text{ cm}}{1 \text{ m}} = 1, \quad \frac{1 \text{ in.}}{2.54 \text{ cm}} = 1, \quad \frac{1 \text{ ft}}{12 \text{ in.}} = 1$$

We also choose the fractions in dimensional analysis so that all the units will "cancel" except the unit in which we want our answer expressed. In Equation 11.1, the first fraction cancels the meters, but leaves centimeters; the second fraction cancels the centimeters, but leaves inches; and the third fraction cancels the inches, leaving feet, which is how we want our answer expressed. Thus,

$$5.1 \text{ m} = 5.1 \cancel{\text{m}} \times \frac{100 \cancel{\text{cm}}}{1 \cancel{\text{m}}} \times \frac{1 \cancel{\text{in.}}}{2.54 \cancel{\text{cm}}} \times \frac{1 \text{ ft}}{12 \cancel{\text{in.}}} = 16.7 \text{ ft}$$

When using a calculator to carry out such a calculation, work from left to right, keeping the result of each step in your calculator to use in the next step. That way you don't introduce an error by punching a number back in. Also, your answer will be more accurate if you wait until the end to round it to the number of places you need. Table 11.8 shows a flowchart for converting 5.1 meters to feet with a calculator.

TABLE 11.8 A calculator flowchart for converting 5.1 meters to feet

Enter 5.1		
Times 100	→	510
Divided by 2.54	→	200.787401. . .
Divided by 12	→	16.732283. . .

How Can We Reason to Convert Areas and Volumes?

CLASS ACTIVITY

11J Area and Volume Conversions, p. CA-237

Once you know how to convert from one unit of length to another, you can then convert between the related units of area and volume. However, to calculate area and volume conversions correctly, it's important to remember the meanings of the units for area and volume (such as square feet, cubic inches, square meters, and so on). Otherwise, it is easy to make the following common kind of mistake:

"Since 1 yard is 3 feet, 1 square yard is 3 square feet."

Figure 11.17 will help you see why this is a mistake. One square yard is the area of a square that is 1 yard wide and 1 yard long. Such a square is 3 feet wide and 3 feet long. Figure 11.17 shows that such a square has the area 3 × 3 = 9 square feet, *not* 3 square feet.

Figure 11.17
One square yard is 9 square feet, not 3 square feet.

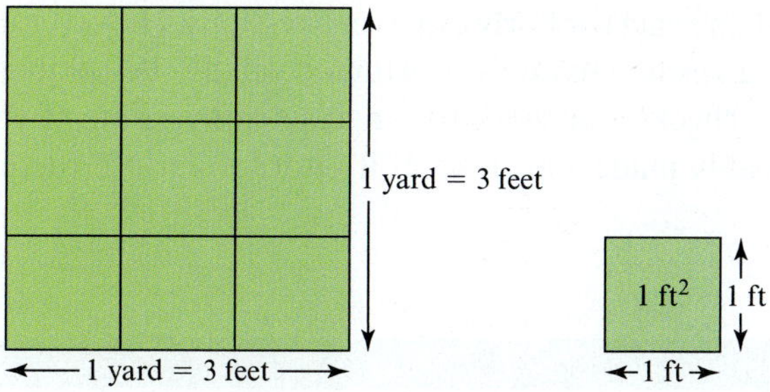

Let's say you are living in an apartment that has a floor area of 800 square feet, and you want to tell your Italian pen pal how big that is in square meters. There are several ways you might approach such an area conversion problem. One good way is this:

> First determine what 1 foot (*linear*) is in terms of meters. Then use that information to determine what 1 *square* foot is in terms of *square* meters.

Here's how to find 1 foot in terms of meters:

$$1 \text{ ft} = 12 \text{ in.}$$
$$12 \text{ in.} = 12 \times 2.54 \text{ cm} = 30.48 \text{ cm}$$

This is because each inch is 2.54 cm and there are 12 inches.

$$30.48 \text{ cm} = (30.48 \div 100) \text{ m} = 0.3048 \text{ m}$$

This is because 1 m = 100 cm, so we must find how many 100s are in 30.48 to find how many meters are in 30.48 cm.

So,

$$1 \text{ ft} = 0.3048 \text{ m}$$

Remember that 1 square foot is the area of a square that is 1 foot wide and 1 foot long. According to the previous calculation, such a square is also 0.3048 meters wide and 0.3048 meters long. Therefore, by the "length times width" formula for areas of rectangles,

$$1 \text{ ft}^2 = (0.3048 \times 0.3048) \text{ m}^2 = 0.0929 \text{ m}^2$$

This means that 800 square feet is $800 \times 0.0929 = 74.32$ square meters, so the apartment is about 74 square meters.

Volume conversions work similarly. Suppose we want to find 13 cubic feet in terms of cubic meters, for example. We already calculated that

$$1 \text{ ft} = 0.3048 \text{ m}$$

Now 1 cubic foot is the volume of a cube that is 1 foot high, 1 foot deep, and 1 foot wide. Such a cube is also 0.3048 meters high, 0.3048 meters long, and 0.3048 meters wide. So, by the "height times width times length" formula for volumes of rectangular prisms (boxes),

$$1 \text{ ft}^3 = 0.3048 \times 0.3048 \times 0.3048 \text{ m}^3 = 0.0283 \text{ m}^3$$

Therefore, 13 cubic feet is 13 times as much as 0.0283 m^3:

$$13 \text{ ft}^3 = 13 \times 0.0283 \text{ m}^3 = 0.37 \text{ m}^3$$

CLASS ACTIVITY

11K Area and Volume Conversions: Which Are Correct and Which Are Not?, p. CA-238

So far this discussion has focused on exact (or fairly exact) conversions. However, in many practical situations, you really don't need an exact answer—an estimate will do. Being able to make a quick estimate also gives you a way to check your work for an exact conversion: If your answer is far from your estimate, you've probably made a mistake. If it's not too far off, there's a good chance your work is correct.

CLASS ACTIVITY

11L Model with Conversions, p. CA-239

SECTION SUMMARY AND STUDY ITEMS

Section 11.4 Converting from One Unit of Measurement to Another

We can use multiplication and division to convert from one unit to another by considering how the units are related and by considering the meanings of multiplication and division. We can also use dimensional analysis to convert from one unit to another. Dimensional analysis relies on repeated multiplication by 1. Length measurements in the U.S. customary and metric systems are related by the fact that 1 inch = 2.54 cm.

Key Skills and Understandings

1. Use multiplication or division or both to convert a measurement from one unit to another. Explain why multiplication or division is the correct operation to use (without using dimensional analysis).
2. Use dimensional analysis to convert from one measurement to another, explaining the method.
3. Explain how to convert areas and volumes properly.

Practice Exercises for Section 11.4

1. A class needs a 175-inch-long piece of rope for a project. How long is the rope in yards?
   a. Use multiplication or division or both to solve the rope problem. Explain your solution.
   b. Describe a number of different correct ways to write the answer to the rope problem. Explain briefly why these different ways of writing the answer mean the same thing.
2. The children in Mrs. Watson's class made chains of small paper dolls, as pictured in Figure 11.18. A chain of 5 dolls is 1 foot long. How long would the following chains of dolls be? In each case, give your answer in either feet or miles, depending on which answer is easiest to understand.
   a. 100 dolls
   b. 1000 dolls
   c. 10,000 dolls
   d. 100,000 dolls
   e. 1 million dolls
   f. 1 billion dolls

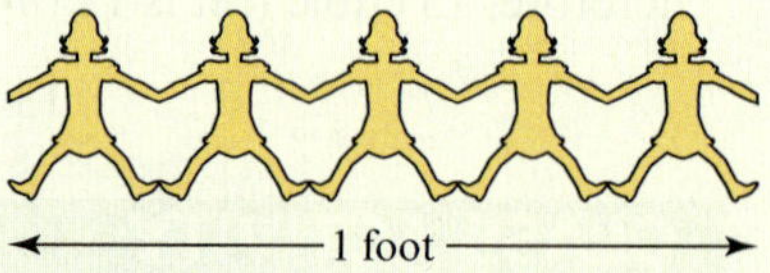

Figure 11.18 A chain of paper dolls.

3. Solve the following conversion problems, using the basic fact 1 inch = 2.54 cm in each case:

a. A track is 100 meters long. How long is it in feet?

b. If the speed limit is 70 miles per hour, what is it in kilometers per hour?

c. Some farmland covers 2.4 square kilometers. How many square miles is it?

d. Convert the volume of a compost pile, 1 cubic yard, to cubic meters.

e. A man is 1.88 meters tall. How tall is he in feet?

f. How many miles is a 10-kilometer race?

4. One mile is 1760 yards. Does this mean that 1 square mile is 1760 square yards? If not, how many square yards are in a square mile? Explain carefully.

5. How many cubic feet of mulch will you need to cover a garden that is 10 feet wide and 4 yards long with 2 inches of mulch? Can you find the amount of mulch you need by multiplying $2 \times (10 \times 4)$?

6. The capacity of a cube that is 10 cm wide, 10 cm deep, and 10 cm tall is 1 liter. One gallon is 0.134 cubic feet. One quart is one-quarter of a gallon. Which is more: 1 quart or 1 liter? Use the basic fact 1 inch = 2.54 cm to figure this out.

7. One acre is 43,560 square feet. The area of land is often measured in acres.

a. What is a square mile in acres?

b. What is a square kilometer in acres?

8. Suppose you want to cover a football field with artificial turf. You'll need to cover an area that's bigger than the actual field—let's say you'll cover a rectangle that's 60 yards wide and 130 yards long. How many square feet (not square yards) of artificial turf will it take? The artificial turf comes in a roll that is 12 feet wide. Let's say the turf costs \$8 per linear foot cut from the roll. (So, if you cut off 10 feet from the roll, it would cost \$80 and you'd have a rectangular piece of artificial turf that's 10 feet long and 12 feet wide.) How much will the turf you need cost? (It has been said that preparing the field costs even more than the artificial turf itself.)

Answers to Practice Exercises for Section 11.4

1. a. One method for determining the length of the rope in yards is to first figure out how many inches are in a yard. One foot is 12 inches. One yard is 3 feet. Since each foot is 12 inches, 3 feet is 3 groups of 12 inches, which is $3 \times 12 = 36$ inches. To find the length of the 175-inch-long piece of rope in yards, we must figure out how many groups of 36 inches are in 175 inches. This is a how-many-groups division problem:

$175 \div 36$ has whole number quotient 4, remainder 31

so there are 4 groups of 36 inches in 175 inches with 31 inches left over. Therefore, 175 inches is 4 yards and 31 inches.

b. We found that we can describe the length of the rope as 4 yards, 31 inches. The 31 inches can also be described as 2 feet, 7 inches, because there are 12 inches in each foot and

$31 \div 12$ has whole number quotient 2, remainder 7

So we can also say that the rope is 4 yards, 2 feet, and 7 inches long. Instead of using a remainder in solving the division problem in part (a), we can give a mixed number or decimal answer to the division problem. So we can also say that the length of the rope is 4.86 yards or $4\frac{31}{36}$ yards, although these answers would probably not be very useful in practice. Similarly, we could give the answer to the division problem $31 \div 12$ as a mixed number or as a decimal and say that the rope is 4 yards and $2\frac{7}{12}$ feet long or 4 yards and 2.6 feet long. Although correct, these answers would probably not be useful in practice.

2. a. Because 5 dolls are 1 foot long, 10 dolls are 2 feet long. Thus, 100 dolls are 10 times as long, namely, 20 feet long.

b. 1000 dolls are 10 times as long as 100 dolls. Because 100 dolls are 20 feet long, 1000 dolls are 200 feet long.

c. 10,000 dolls are 10 times as long as 1000 dolls. Because 1000 dolls are 200 feet long, 10,000 dolls are 2000 feet long.

d. 100,000 dolls are 10 times as long as 10,000 dolls. Because 10,000 dolls are 2000 feet long, 100,000 dolls are 20,000 feet long. One mile is

5280 feet, so 20,000 feet is 20,000 ÷ 5280, about 3.8 miles. So 100,000 dolls are nearly 4 miles long.

e. One million dolls are 10 times as long as 100,000 dolls. Because 100,000 dolls are 3.8 miles long, 1 million dolls are 38 miles long.

f. One billion dolls are 1000 times as long as 1 million dolls. Because 1 million dolls are 38 miles long, 1 billion dolls are 38,000 miles long. The circumference of the earth is about 24,000 miles, so 1 billion dolls would wrap around the earth more than $1\frac{1}{2}$ times!

3. a. 1 m = 100 cm = (100 ÷ 2.54) in. = 39.37 in. = (39.37 ÷ 12) ft = 3.28 ft. So 100 m = 328 ft.

b. 1 mile = 5280 ft = 63,360 in. = 160934.4 cm = 1609.344 m = 1.609 km. So 70 miles per hour is 70 × 1.609 km per hour, which is 113 km per hour.

c. 1 km = 1000 m = 100,000 cm = 39370.08 in. = 3280.84 ft = 0.621 miles. Therefore, 1 square kilometer is 0.621 × 0.621 = 0.39 square miles, and 2.4 square kilometers are 2.4 × 0.39 square miles, which is 0.9 square miles.

d. 1 yd = 3 ft = 36 in. = 91.44 cm = 0.9144 m. So, 1 cubic yard is 0.9144 × 0.9144 × 0.9144 cubic meters = 0.76 cubic meters.

e. 1 m = 100 cm = 39.37 in. = 3.28 ft. So 1.88 m = 6.17 ft, which is 6 ft, 2 in.

Why is 0.17 ft equal to 2 in.? This is *not* explained by rounding the 0.17 to 2. Instead, since 1 ft = 12 in., 0.17 ft = 0.17 × 12 in. = 2.04 in., which rounds to 2 in.

f. 1 km = 1000 m = 100,000 cm = (100,000 ÷ 2.54) in. = 39,370 in. = (39,370 ÷ 12) ft = 3280.8 ft = 3280.8 ÷ 5280 miles = 0.62 miles.

4. One square mile is the area of a square that is 1 mile long and 1 mile wide. Such a square is 1760 yards long and 1760 yards wide, so we can mentally decompose this square into 1760 rows, each having 1760 squares that are 1 yard by 1 yard. Therefore, according to the meaning of multiplication, there are

$$1760 \times 1760 = 3{,}097{,}600$$

1-yard-by-1-yard squares in a square mile. This means that 1 square mile is 3,097,600 square yards and not 1760 square yards.

5. We know that 1 yard = 3 feet and 1 foot = 12 inches. Thus, 4 yards = 12 feet, and 2 inches = $\frac{1}{6}$ feet. So, you can think of the mulch as forming a box shape that is $\frac{1}{6}$ feet high, 10 feet deep (or wide), and 12 feet wide (or deep). This box has volume

$$\frac{1}{6} \times (10 \times 12) \text{ cubic feet} = 20 \text{ cubic feet}$$

So, you'll need 20 cubic feet of mulch.

You can't find the amount of mulch you need by multiplying 2 × 10 × 4 because each number refers to a different unit of length.

6. 1 liter = 1000 cubic centimeters. Convert 1 cm to feet: 1 cm = 0.3937 in. = 0.0328 ft. So 1 cubic centimeter is 0.000035315 cubic feet, and 1 liter = 0.035315 cubic feet. Since each gallon is 0.135 cubic feet, 1 liter = (0.035315 ÷ 0.135) gallons = 0.2635 gallons, which is a little more than $\frac{1}{4}$ gallon. So a liter is a little more than a quart.

7. a. 1 mile = 5280 feet, so 1 square mile = 5280 × 5280 square feet = 27,878,400 square feet. Since each acre is 43,560 square feet, we want to know how many 43,560 are in 27,878,400. This is a division problem. 27,878,400 ÷ 43,560 = 640. So there are 640 acres in a square mile.

b. 1 km = 1000 m = 100,000 cm = (100,000 ÷ 2.54) in. = 39,370 in. = (39,370 ÷ 12) ft = 3280.8399 ft. So 1 km^2 = 3280.8399^2 ft^2 = 10,763,910.42 ft^2 = (10,763,910.42 ÷ 43,560) acres = 247 acres.

8. You'll need 60 × 130 = 7,800 square yards of turf. Each square yard is 3 × 3 = 9 square feet, so you'll need 7,800 × 9 = 70,200 square feet of turf. (Or, notice that 60 yards = 180 feet and 130 yards = 390 feet, so you'll need 180 × 390 = 70,200 square feet.)

$8 buys you 12 square feet of turf, so it will cost at least

$$\$8 \times (70{,}200 \div 12) = \$46{,}800$$

for the turf. But if you lay the turf in strips across the width of the field, then you will need 33 strips that are 60 yards = 180 feet long, which will cost $47,520. (The 33 strips come from the fact that each strip is 4 yards wide and 130 ÷ 4 = 32.5.)

PROBLEMS FOR SECTION 11.4

1. A recipe calls for 4 ounces of chocolate. If you make 15 batches of the recipe (for a large party), then how many pounds of chocolate will you need?

a. Use multiplication or division or both to solve the chocolate problem. Explain why your method of solution makes sense in a way that children could understand. Do not use dimensional analysis.

b. Describe a number of different correct ways to write the answer to the chocolate problem. Explain briefly why these different ways of writing the answer mean the same thing.

2. A class needs 27 pieces of ribbon, each piece 2 feet, 3 inches long. How many yards of ribbon does the class need?

a. Use multiplication or division or both to solve the ribbon problem. Explain why your method of solution makes sense in a way that children could understand. Do not use dimensional analysis.

b. Describe a number of different correct ways to write the answer to the ribbon problem. Explain briefly why these different ways of writing the answer mean the same thing.

3. To convert 24 yards to feet, should you multiply by 3 or divide by 3? Explain your answer in two different ways that a fifth-grader could understand. (Do not use dimensional analysis.)

4. To convert 2000 kilometers to meters, should you multiply by 1000 or divide by 1000? Explain your answer in two different ways that a fifth-grader could understand. (Do not use dimensional analysis.)

5. Shauntay used identical plastic bears to measure the length of a rope and found that the rope was 36 bears long. Next, Shauntay will measure the length of the rope using identical plastic giraffes. Shauntay found that 4 bears are as long as 3 giraffes. How many giraffes long will the rope be? Explain your reasoning clearly and in detail.

6. a. A car is 16 feet, 3 inches long. How long is it in meters? Use the fact that 1 inch = 2.54 centimeters to determine your answer. Explain your reasoning briefly.

b. A car is 415 centimeters long. How long is it in feet and inches? Use the fact that 1 inch = 2.54 centimeters to determine your answer. Explain your reasoning briefly.

7. The distance between two cities is described as 260 kilometers. What is this distance in miles? Make an estimate first, explaining your reasoning briefly. Then calculate your answer using the fact that 1 in. = 2.54 cm. Remember to round your answer appropriately: The way you write your answer should reflect its accuracy. Your answer should not be reported more accurately than your starting data.

8. In Germany, people often drive 130 kilometers per hour on the Autobahn (highway). How fast are they going in miles per hour? Make an estimate first, explaining your reasoning briefly. Then calculate your answer using the fact that 1 in. = 2.54 cm.

9. One foot is 12 inches. Does this mean that 1 square foot is 12 square inches? Draw a picture showing how many square inches are in a square foot. Use the meaning of multiplication to explain why you can calculate the area of 1 square foot in terms of square inches by multiplying.

10. A room has a floor area of 48 square yards. What is the area of the room in square feet? Solve this problem in two different ways, each time referring to the meaning of multiplication.

11. One kilometer is 1000 meters. Does this mean that 1 square kilometer is 1000 square meters? If not, what is 1 square kilometer in terms of square meters? Explain your answer in detail, referring to the meaning of multiplication.

12. One foot is 12 inches. Does this mean that 1 cubic foot is 12 cubic inches? Describe how to use the meaning of multiplication to determine what 1 cubic foot is in terms of cubic inches.

13. How much mulch will you need to cover a rectangular garden that is 20 feet by 30 feet with a 3-inch layer of mulch? Explain.

14. A classroom has a floor area of 600 square feet. What is the floor area of the classroom in square meters? Describe three different correct ways to solve this area problem. Each way of solving should use the fact that 1 in. = 2.54 cm. Explain your reasoning.

15. A house has a floor area of 800 square meters. What is the floor area of this house in square feet? Describe three different correct ways to solve this area problem. Each way of solving should use the fact that 1 in. = 2.54 cm. Explain your reasoning.

16. A house has a floor area of 250 square meters. Convert the house's floor area to square feet. Describe three different correct ways to solve this conversion problem. Each way of solving should use the fact that 1 in. = 2.54 cm. Explain your reasoning.

17. The Smiths will be carpeting a room in their house. In one store, they see a carpet they like that costs \$35 per square yard. Another store has a similar carpet for \$3.95 per square foot. Is this more or less expensive than the carpet at the first store? Explain your reasoning.

18. One acre is 43,560 square feet. If a square piece of land is 3 acres, then what are the length and width of this piece of land in feet? (Remember that the length and width of a square are the same.) Explain your reasoning.

19. A construction company has dump trucks that hold 10 cubic yards. If the company's workers dig a hole that is 8 feet deep, 20 feet wide, and 30 feet long, then how many dump truck loads will they need to haul away the dirt they dug out from the hole? Explain your reasoning.

20. The following question explores why it doesn't really make sense to compare area and perimeter.

A rectangle is 9 feet long and 6 feet wide. Which is greater, the perimeter of the rectangle or the area of the rectangle?

a. Calculate the perimeter and the area of the rectangle in feet and square feet, respectively. Explain briefly.

b. Calculate the perimeter and the area of the rectangle in yards and square yards, respectively. Explain briefly.

c. Use parts (a) and (b) to explain why it doesn't really make sense to compare area and perimeter.

***21.** A penny is $\frac{1}{16}$ of an inch thick.

a. Suppose you have 1000 pennies. If you make a stack of these pennies, how tall will it be? Give your answer in feet and inches (e.g., 7 feet, 3 inches). Explain your reasoning.

b. Suppose you have 1 million pennies. If you make a stack of these pennies, how tall will it be? Give your answer in feet and inches (e.g., 7 feet, 3 inches). Would the stack be more than a mile tall or not? Explain your reasoning.

c. Suppose you have 1 billion pennies. If you make a stack of these pennies, how tall will it be? Give your answer in miles. Explain your reasoning.

d. Suppose you have 1 trillion pennies. If you make a stack of these pennies, how tall will it be? Give your answer in miles. Would the stack reach to the moon, which is about 240,000 miles away? Would the stack reach to the sun, which is about 93 million miles away? Explain your reasoning.

***22. a.** Write 100 zeros on a piece of paper and time how long it takes you. Based on the time it took you to write 100 zeros, approximately how long would it take you to write 1000 zeros? 10,000 zeros? 100,000 zeros? 1 million zeros? 1 billion zeros? 1 trillion zeros? In each case, give your answer either in minutes, hours, days, or years, depending on which answer is easiest to understand. Explain your answers.

b. Recall that a googol is the number whose decimal representation is a 1 followed by 100 zeros. Based on your answers in part (a), explain why it would not be possible for anyone to write a googol zeros.

c. Recall that a googolplex is the number whose decimal representation is a 1 followed by a googol zeros. Based on your answer in part (b), explain why it would not be possible for anyone to write a googolplex in its decimal representation. (Note, however, that it is possible to write a googolplex in scientific notation.)

***23.** For a certain type of rice, about 50 grains fill a 1-cm-by-1-cm-by-1-cm cube. Refer to this type of rice in all parts of this problem.

a. Describe the dimensions of a container that you could use to show your students 1 million grains of rice. Explain. Give a sense of approximately how big this container is by describing it as being about as big as a familiar object.

b. Describe the dimensions of a container that you could use to show your students 1 billion grains of rice. Explain. Give a sense of approximately how big this container is by describing it as being about as big as a familiar object.

c. Describe the dimensions of a container that you could use to show your students 1 trillion grains of rice. Explain. Give a sense of approximately how big this container is by describing it as being about as big as a familiar object.

***24.** Assuming that 1 gram of gold is worth $30, how much would 1 million dollars worth of gold weigh in pounds? Explain your reasoning.

***25.** Imagine that all the people on earth could stand side by side together. How much area would be needed? Would all the people on earth fit in Rhode Island? Explain your reasoning. Rhode Island, the smallest state in terms of area, has a land area of 1000 square miles. Assume that there are 6 billion people on earth, and assume that each person needs 4 square feet to stand in.

***26.** Joey has a toy car that is a 1:64 scale model of an actual car. (In other words, the length of the actual car is 64 times as long as the length of the toy car.) Joey's toy car can go 5 feet per second. What is the equivalent speed in miles per hour for the actual car? Explain.

CHAPTER SUMMARY

Section 11.1 Concepts of Measurement — Page 493

- In order to measure an object, or to describe its size, we must first select a measurable attribute. To measure a quantity is to compare the quantity with a unit amount. A unit is a fixed, reference amount. The most direct way to measure a quantity is to count how many of the unit amount make the quantity. — Page 493
- The U.S. customary system and the metric system are the two collections of units we use in the United States today. The metric system uses prefixes to name units in a uniform way. A fundamental relationship in the metric system is that 1 milliliter of water weighs 1 gram and has a volume of 1 cubic centimeter. — Page 496

Key Skills and Understandings

1. Explain that, to measure a quantity, a measurable attribute must first be chosen. — Page 493
2. Discuss units and know that measurement is comparison with a unit. — Page 494
3. Explain what it means for an object to have an area of 8 square centimeters, or a volume of 12 cubic inches, or a weight of 13 grams, or other similar measurements. — Page 501
4. Know and use proper notation for units of area and volume (e.g., square centimeters, cm^2). — Page 501
5. Explain how the metric system uses prefixes, and state the meaning of common metric prefixes: kilo-, hecto-, deka-, deci-, centi-, and milli-. — Page 498
6. Describe how capacity, weight (mass), and volume are linked in the metric system. — Page 499

Section 11.2 Length, Area, Volume, and Dimension — Page 505

- A length describes the size of something (or part of something) that is one-dimensional. An area describes the size of something (or part of something) that is two-dimensional. A volume describes the size of something (or part of something) that is three-dimensional. — Page 505

Key Skills and Understandings

1. Given an object, describe one-dimensional, two-dimensional, and three-dimensional parts or aspects of the object, and give appropriate U.S. customary and metric units for measuring or describing the size of those parts or aspects of the object. — Page 505
2. Compare objects with respect to one-dimensional, two-dimensional, and three-dimensional attributes. — Page 507
3. Recognize that in some cases, one object can be larger than another object with respect to one attribute, but smaller with respect to a different attribute. — Page 507

Section 11.3 Error and Precision in Measurements	Page 510
■ When it comes to measuring actual physical quantities, all measurements are only approximate. The way we write a measurement conveys how precisely the measurement is known. When calculating with actual measurements, the final result should be rounded to convey how precisely the result is known.	Page 510
Key Skills and Understandings	
■ Explain how a reported measurement conveys how precisely the measurement is known. For example, if the weight of an object is reported as 5 pounds, the actual weight of the object has been rounded to the nearest whole number and is therefore between 4.5 and 5.5 pounds. But if the weight of an object is reported as 5.0 pounds, the actual weight of the object has been rounded to the nearest tenth and is therefore between 4.95 and 5.05 pounds.	Page 510
Section 11.4 Converting from One Unit of Measurement to Another	Page 514
■ We can use multiplication and division to convert from one unit to another by considering how the units are related and by considering the meanings of multiplication and division.	Page 514
■ We can also use dimensional analysis to convert from one unit to another. Dimensional analysis relies on repeated multiplication by 1. Length measurements in the U.S. customary and metric systems are related by the fact that 1 inch = 2.54 cm.	Page 515
Key Skills and Understandings	
1. Use multiplication or division or both to convert a measurement from one unit to another. Explain why multiplication or division is the correct operation to use (without using dimensional analysis).	Page 514
2. Use dimensional analysis to convert from one measurement to another, explaining the method.	Page 515
3. Explain how to convert areas and volumes properly.	Page 516

12

Area of Shapes

In this chapter, we continue our study of measurement by focusing on area of shapes. We examine how students in elementary school determine areas of shapes: progressing from primitive methods to more sophisticated methods and culminating in the use of area formulas. All methods for determining areas of shapes rely on what area means, and all use basic principles about area to decompose and recompose shapes to work with the pieces. These methods function in much the same way as calculation methods in arithmetic rely on what the operations mean and use the basic properties of arithmetic to decompose and recompose calculations. We review some familiar area formulas and discuss how these formulas are derived from the meaning of area and basic principles about area. We also examine the distinction between area and perimeter, which is a common source of confusion for students in elementary school. Finally, we consider how the Pythagorean theorem can be viewed as a fact about areas, and we see how the theorem can be derived from decomposing squares in different ways and equating areas.

We focus in this chapter on the following topics and practices within the *Common Core State Standards for Mathematics* (*CCSSM*).

Standards for Mathematical Content in the CCSSM

In the domain of *Measurement and Data* (Kindergarten–Grade 5), students learn the concept of area. They measure areas by counting squares, and they connect area to multiplication. Students recognize area as additive, and they use additivity to determine areas of shapes. They distinguish linear and area measures, and they consider rectangles with the same perimeter and different areas (and vice versa). In the domain of *Number and Operations—Fractions* (Grades 3–5), students find areas of rectangles of fractional side lengths. In the domain of *Geometry* (Kindergarten–Grade 8), students find areas of triangles and quadrilaterals by composing and decomposing shapes, and they understand and apply the Pythagorean theorem.

Standards for Mathematical Practice in the CCSSM

Opportunities to engage in all eight of the Standards for Mathematical Practice described in the CCSSM occur throughout the study of area, although the following standards may be especially appropriate for emphasis:

- **1 Make sense of problems and persevere in solving them.** Students engage in this practice when they look for multiple ways to determine areas by composing and decomposing shapes.
- **2 Reason abstractly and quantitatively.** Students engage in this practice when they connect the reasoning for deriving an area of a shape with an area formula for that shape, thus developing both a geometric and an algebraic perspective on area.
- **4 Model with mathematics.** Students engage in this practice when they use area and perimeter to find quantities of interest. For example, animals might cluster in a circular pack to keep warm. The circle's circumference and area might give information about the number of animals in the cluster and how many are cold on the boundary.

(From Common Core Standards for Mathematical Practice. Published by Common Core Standards Initiative.)

12.1 Areas of Rectangles Revisited

CCSS Common Core State Standards Grades 3, 4, 5

CCSS 3.MD.5 3.MD.6

Area is a measure of how much two-dimensional space a shape takes up. We have discussed area and the area of rectangles in previous sections (Section 4.3 and Chapter 11). Here, we briefly revisit this important concept, but we examine progressively sophisticated thinking about areas of rectangles and attend more closely to the units of measurement involved.

CLASS ACTIVITY

12A Units of Length and Area in the Area Formula for Rectangles, p. CA-240

Before students in elementary school learn the area formula for rectangles, they must learn what area means. For example, what does it mean to say that the area of a shape is 12 square centimeters? It means that the shape can be covered, without gaps or overlaps, with a total of twelve 1-cm-by-1-cm squares, allowing for squares to be cut apart and pieces to be moved if necessary. Given a 4-cm-by-3-cm rectangle, such as the one in Figure 12.1(a), a primitive way for students to determine its area is to cover the rectangle snugly with 1-cm-by-1-cm square tiles and count that it takes 12 tiles. Although this is a primitive method, it is important to understand it, because the method relies directly on the meaning of area and therefore emphasizes this meaning.

CCSS 3.MD.7a

Covering rectangles with square tiles—though important—is a slow way to determine the area of rectangles. Fortunately, there is a quicker, more advanced way, which is based on the multiplicative structure that rectangles exhibit when they are broken into squares. These squares can be viewed as organized into equal rows (or columns). Because there are equal groups of squares, we can multiply to find the total number of squares. When the 4-cm-by-3-cm rectangle in Figure 12.1 is covered with 1-cm-by-1-cm squares, there are 4 rows of squares with 3 squares in each row. Therefore, there are $4 \cdot 3$ squares covering the rectangle (without overlaps). Each square has area 1 cm^2; therefore, the total area of the rectangle is $4 \cdot 3$ square centimeters, which is 12 square centimeters.

Figure 12.1

Why multiplying the side lengths of a rectangle gives the area of the rectangle.

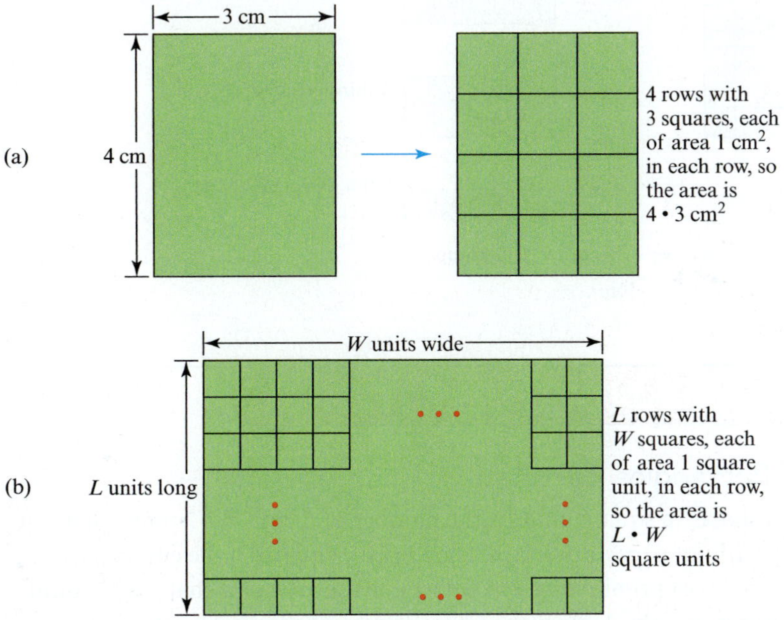

The line of reasoning for the 4-cm-by-3-cm rectangle works generally to explain why a rectangle that is L units long and W units wide has an area of $L \cdot W$ square units. As Figure 12.1(b) indicates, if we cover such a rectangle with 1-unit-by-1-unit squares, then there will be L rows with W squares in each row. Therefore, there are $L \cdot W$ squares covering the rectangle. Each square has area 1 $unit^2$; therefore, the total area of the rectangle is $L \cdot W$ square units.

Notice that in the $L \cdot W$ formula for the area of an L-unit-by-W-unit rectangle, the L and W refer to how many units long the sides of the rectangle are. In other words, in the area formula, the L and W refer to *one-dimensional* attributes of the rectangle. On the other hand, when we explain why we can multiply to find the area of the rectangle, we make equal groups of squares. We find that we can make L groups of squares and that there are W squares in each group, each of area 1 square unit—a *two-dimensional* attribute.

The *length times width* formula,

$$\text{area} = L \cdot W$$

for areas of rectangles is valid not only when L and W are whole numbers, but also when L or W are fractions, mixed numbers, or decimals. For example, consider a rectangle that is $3\frac{1}{2}$ cm by $2\frac{1}{2}$ cm, as shown in Figure 12.2. If we cover such a rectangle with 1-cm-by-1-cm squares and parts of 1-cm-by-1-cm squares, then there will be $3\frac{1}{2}$ groups of squares with $2\frac{1}{2}$ squares in each group. So according to the meaning of multiplication, the area of the rectangle is

$$3\frac{1}{2}\text{ cm} \cdot 2\frac{1}{2}\text{ cm} = \left(3\frac{1}{2} \cdot 2\frac{1}{2}\right)\text{ cm}^2$$

CCSS 5.NF.4b

Numerically, we can calculate that

$$3\frac{1}{2} \cdot 2\frac{1}{2} = \frac{7}{2} \cdot \frac{5}{2} = \frac{35}{4} = 8\frac{3}{4}$$

Hence, the area of the rectangle is $8\frac{3}{4}$ cm^2. If we count up whole squares and collect together the parts of squares in Figure 12.2, we also see that the rectangle's area is $8\frac{3}{4}$ square centimeters.

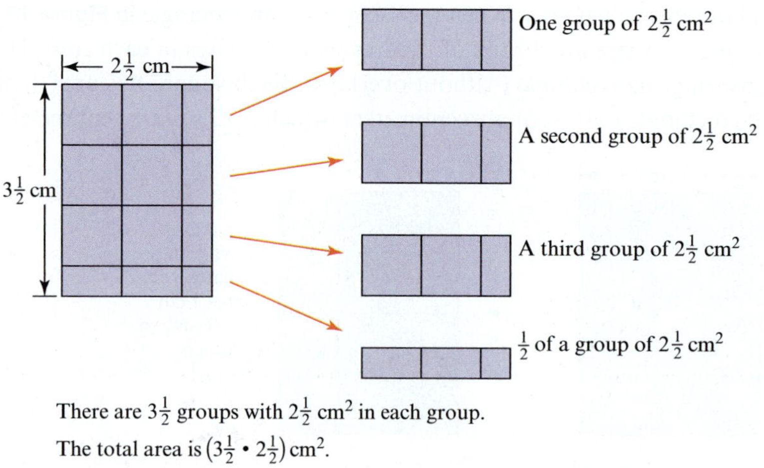

Figure 12.2 Why a $3\frac{1}{2}$-cm-by-$2\frac{1}{2}$-cm rectangle has area $(3\frac{1}{2} \cdot 2\frac{1}{2})$ cm^2.

SECTION SUMMARY AND STUDY ITEMS

Section 12.1 Areas of Rectangles Revisited

The area of a shape, in square units, is the number of 1-unit-by-1-unit squares it takes to cover the shape exactly without gaps or overlaps, allowing for squares to be cut apart and pieces to be moved if necessary. The most primitive way to determine the area of a shape is to count how many squares it takes to cover it exactly. A quicker and more advanced way to determine the area of a rectangle is to multiply its width and length. We can explain this formula by viewing the rectangle as decomposed into rows of squares.

Key Skills and Understandings

1. Know and be able to describe what area is and know the most primitive way to determine the area of a shape.
2. Explain why it is valid to multiply the length and width of a rectangle to determine its area.
3. Attend carefully to units of length and area when discussing and explaining the area formula for rectangles.
4. Explain and use the area formula for rectangles in the case of fractions, mixed numbers, and decimals.

Practice Exercises for Section 12.1

1. Given a 4-cm-by-3-cm rectangle:
   - **a.** What is a primitive way to determine the area of the rectangle?
   - **b.** What is a more advanced way to determine the area of the rectangle and why does it work?
2. Explain why it would be *confusing* to say that the darkly shaded squares in Figure 12.3 represent the 0.6-unit and 0.8-unit *lengths* you use in the length · width formula for the area of the shaded rectangle.

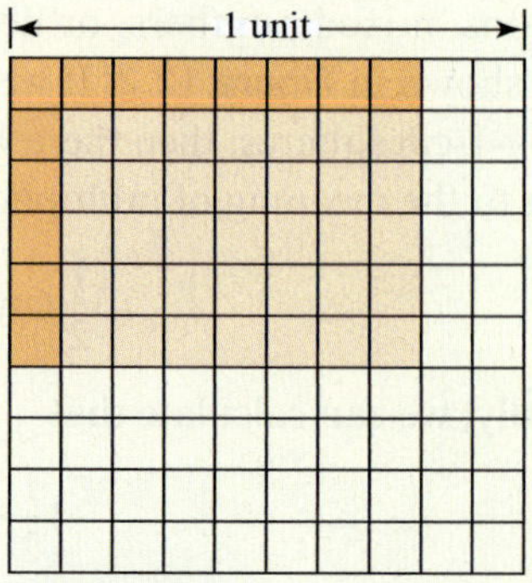

Figure 12.3 Why the darkly shaded squares show *area* better than length and width.

Answers to Practice Exercises for Section 12.1

1. **a.** Cover it (without gaps or overlaps) with 1-cm-by-1-cm squares and count how many it took.
   b. You can multiply $4 \cdot 3$ to find the area of the shape in square centimeters. See the text for details on explaining why.
2. By counting the darkly shaded squares, you set up a potential confusion between the side length of a rectangle, which is a one-dimensional attribute, and the areas of the strips of squares, which are two-dimensional attributes. Furthermore, notice that the *side length* of each small square is 0.1 unit, whereas the *area* of each small square is 0.01 square unit.

PROBLEMS FOR SECTION 12.1

1. You have a 5-foot-by-7-foot rectangular rug in your classroom. You also have a bunch of square-foot tiles and some tape measures.
   a. What is the most primitive way to determine the area of the rug?
   b. What is a less primitive way to determine the area of the rug, and why does this method work?
2. Draw a 3-cm-by-7-cm rectangle. Then discuss the difference in the units we attach to the 3 and the 7 in these two situations:
   - When applying the rectangle area formula to determine the area of the rectangle.
   - When viewing the rectangle as decomposed into equal groups of 1-cm-by-1-cm squares in order to apply the definition of multiplication.

   Use drawings to support your discussion.
3. **a.** Explain how to decompose the large rectangle in Figure 12.4 into $2\frac{1}{2}$ groups with $3\frac{1}{2}$ squares in each group, so as to describe the area of the rectangle as

$$2\frac{1}{2} \cdot 3\frac{1}{2}\ \text{cm}^2$$

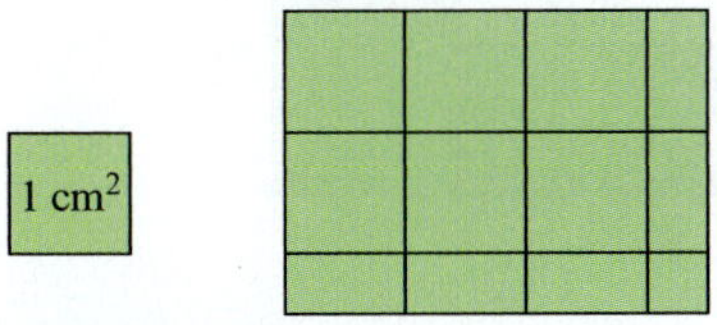

Figure 12.4 Explain why the area is $\left(2\frac{1}{2} \cdot 3\frac{1}{2}\right)\ \text{cm}^2$.

   b. Calculate $2\frac{1}{2} \cdot 3\frac{1}{2}$ without a calculator, showing your calculations. Then verify that this calculation has the same answer as when you determine the area of the rectangle in Figure 12.4 by counting squares.
4. **a.** Explain how to decompose the large rectangle in Figure 12.5 into $4\frac{1}{2}$ groups with $5\frac{3}{4}$ squares in each group, so as to describe the area of the rectangles as

$$\left(4\frac{1}{2} \cdot 5\frac{3}{4}\right)\ \text{cm}^2$$

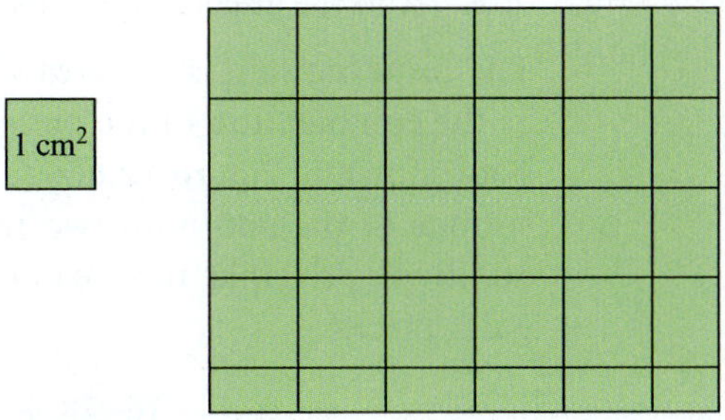

Figure 12.5 Explain why the area is $\left(4\frac{1}{2} \cdot 5\frac{3}{4}\right)\text{cm}^2$.

   b. Calculate $4\frac{1}{2} \cdot 5\frac{3}{4}$ without a calculator, showing your calculations. Then verify that this calculation has the same answer as when you determine the area of the rectangle in Figure 12.5 by counting squares.
5. **a.** Draw a (fairly long) line segment and designate it as being 1 unit long. Then draw a 0.6-unit-by-0.9-unit rectangle.

b. Apply the length · width formula for the area of the rectangle and verify that the formula gives you the correct area for your rectangle in part (a). Attend carefully to units of area.

c. When you applied the length · width formula to find the area of the rectangle in part (b), you used lengths of 0.6 and 0.9 units. Describe these lengths and show them in your drawing.

12.2 Moving and Additivity Principles About Area

CCSS Common Core State Standards Grade 3

How do you figure out the area of a shape? Your first thought might be to look for a formula to apply. But area formulas, as well as other ways of determining area by reasoning, are based on principles about area. In this section, we will study the two most fundamental principles that are used in determining the area of a shape. These principles agree entirely with common sense. In fact, you have probably used them without being consciously aware of it. Students in elementary school use these principles about area, if only informally or implicitly.

- **Moving Principle:** If you move a shape rigidly without stretching it, then its area does not change.
- **Additivity Principle:** If you combine (a finite number of) shapes *without overlapping* them, then the area of the resulting shape is the sum of the areas of the individual shapes.

CCSS 3.MD.7d

Just as the properties of arithmetic allow us to take numbers apart and calculate with the pieces, so too the moving and additivity properties allow us to take shapes apart and calculate areas of the pieces.

A common way to use the additivity principle to find the area of some shape is as follows:

1. Subdivide the shape into pieces whose areas are easy to determine.
2. Add the areas of these pieces.

The resulting sum is the area of the original shape, because we can think of the original shape as the combination of the pieces. For example, what is the area of the surface of the swimming pool shown in Figure 12.6(a) (a bird's-eye view)? To answer this question, imagine dividing the surface of the pool into two rectangular parts, as shown in Figure 12.6(b). According to the additivity principle, the area of the surface of the pool is the sum of the areas of the two rectangular pieces:

$$16 \cdot 28 + 30 \cdot 20 = 448 + 600 = 1048 \text{ square feet}$$

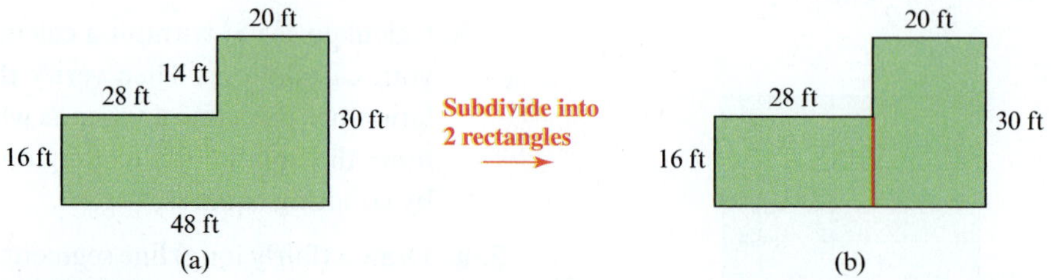

Figure 12.6 Finding the area of the surface of a swimming pool, by subdividing it into two rectangles.

Another way to use the moving and additivity principles to determine the area of a shape is to subdivide the shape into pieces, then move and recombine those pieces, without overlapping, to make a new shape whose area is easy to determine. By the moving and additivity principles, the area of the original shape is equal to the area of the new shape.

For example, what is the area of the patio shown in Figure 12.7(a)? To answer this, imagine slicing off the "bump" and sliding it down to fill in the corresponding "dent," as indicated in Figure 12.7(b). Note that the bump was moved rigidly without stretching, and the bump fills the dent perfectly, without any overlaps. Therefore, by the moving and additivity principles, the area of the resulting rectangle is the same as the area of the original patio. Since the rectangle has area

$$24 \cdot 60 = 1440 \text{ square feet}$$

the patio also has an area of 1440 square feet.

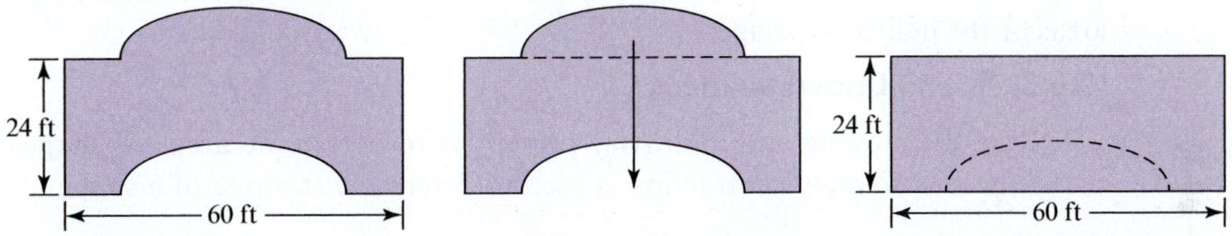

(a) What is the area of this patio?

(b) Subdivide, move, and recombine the pieces to form a rectangle. The area of the rectangle is the same as the area of the original patio.

Figure 12.7 Finding the area of a patio by subdividing, moving, and recombining pieces.

By reversing the reasoning of the patio example and starting with a shape whose area we know, we can subdivide the shape and recombine the pieces without overlapping to make a new shape. This demonstrates that many different shapes can have the same area.

We can use the additivity principle to "take away" area, as in Figure 12.8. Because the shaded region and the "missing hole" combine to make the 3-cm-by-5-cm rectangle, their areas add to 15 cm^2. Because the "missing hole" has area 2 cm^2, the area of the shaded region is $15 \text{ cm}^2 - 2 \text{ cm}^2 = 13 \text{ cm}^2$.

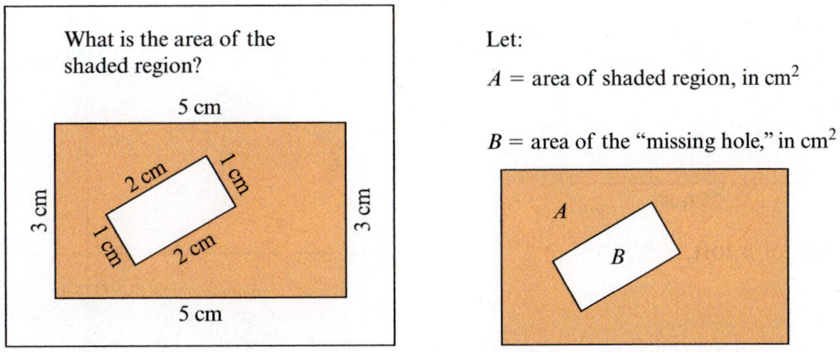

By the additivity principle: $A + B = 15 \text{ cm}^2$

Therefore: $A = 15 \text{ cm}^2 - B$
$= 15 \text{ cm}^2 - 2 \text{ cm}^2$
$= 13 \text{ cm}^2$

Figure 12.8 Using the additivity principle to take area away.

CLASS ACTIVITY

12B Using the Moving and Additivity Principles, p. CA-242

SECTION SUMMARY AND STUDY ITEMS

Section 12.2 Moving and Additivity Principles About Area

Two powerful principles are often used—almost subconsciously—to determine areas. These are the moving and additivity principles. The moving principle states that if you move a shape rigidly without stretching it, then its area does not change. The additivity principle states that if you combine shapes without overlapping them, then the area of the resulting combined shape is the sum of the areas of the individual shapes.

Key Skills and Understandings

1. Use the moving and additivity principles to determine areas of shapes, including cases where ultimately subtraction is used to determine the area of a shape of interest.

Practice Exercises for Section 12.2

1. Figure 12.9 shows the floor plan of a loft that will be getting a new wood floor. How many square feet of flooring will be needed?

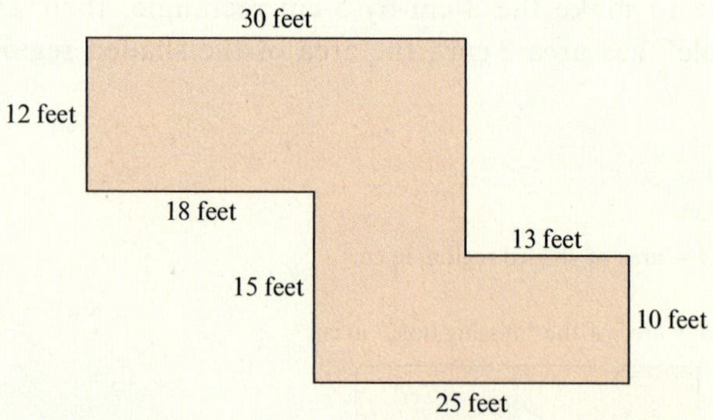

Figure 12.9 Floor plan of a loft.

2. What is the area of the shape in Figure 12.10? (This is not a perspective drawing; it is a flat shape.)

3. Determine the area of the octagon in Figure 12.11 and explain your reasoning.

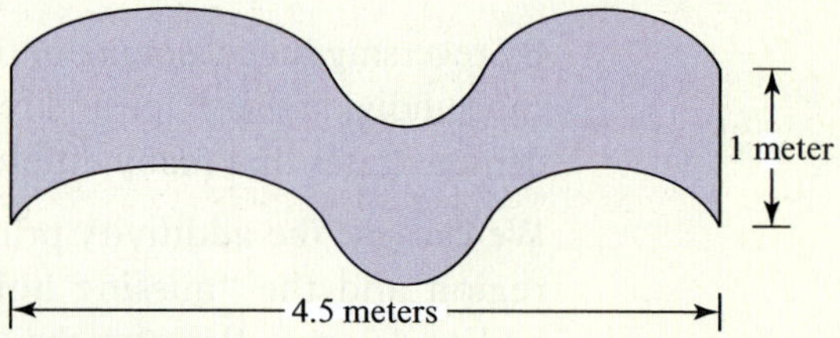

Figure 12.10 A wiggly shape.

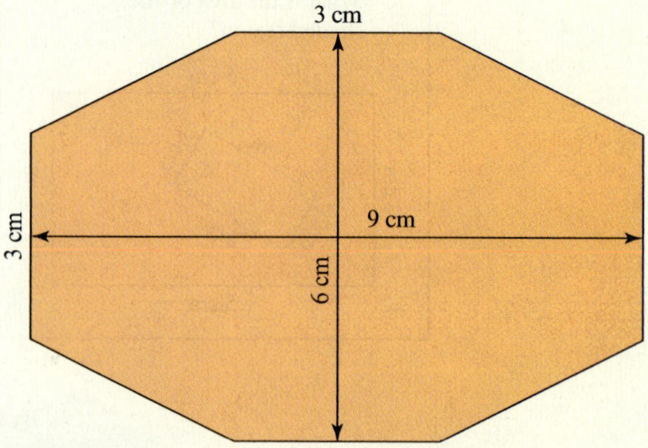

Figure 12.11 An octagon.

Answers to Practice Exercises for Section 12.2

1. First, to find the length of the unlabeled side, compare the "vertical" distance across the floor on the left side of the room with the right side of the room. On the left side, the vertical distance is 12 feet + 15 feet = 27 feet. On the right side, the vertical distance is *unlabeled length* + 10 feet. The two vertical distances across the floor are equal; therefore,

$$\text{Unlabeled length} + 10 = 27$$

So the unlabeled length must be 17 feet.

We can subdivide the floor into three rectangular pieces: one 30 feet by 12 feet, one 12 feet by 5 feet, and one 25 feet by 10 feet. Therefore, according to the additivity principle, the number of square feet of flooring needed is $30 \cdot 12 + 12 \cdot 5 + 25 \cdot 10 = 670$.

2. We can subdivide the wiggly shape of Figure 12.10 into pieces and recombine the pieces as shown in Figure 12.12 to make a rectangle that is 1 meter by $4\frac{1}{2}$ meters. (Slice off the two "bumps" on top and the bump on the bottom, and move them to fill the corresponding "dents.") The rectangle has area $4\frac{1}{2}$ square meters, so by the moving and additivity principles about area, the wiggly shape must also have area $4\frac{1}{2}$ square meters.

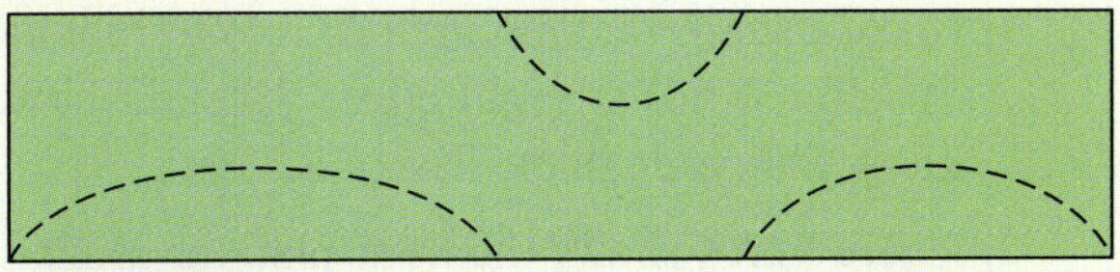

Figure 12.12 A wiggly shape that becomes a rectangle after subdividing and recombining.

3. The area of the octagon is 45 cm^2. Think of the octagon as a 6-cm-by-9-cm rectangle with 4 triangles cut off, as indicated in Figure 12.13(a). Then, according to the additivity principle,

$$\text{area of octagon} + \text{area of 4 triangles} = \text{area of rectangle}$$

But the 4 triangles can be combined, without overlapping, to form 2 small 1.5-cm-by-3-cm rectangles. So, by the previous equation, we have

$$\text{area of octagon} + (2 \cdot 1.5 \cdot 3)\ \text{cm}^2 = (6 \cdot 9)\ \text{cm}^2$$

Therefore,

$$\text{area of octagon} = 54\ \text{cm}^2 - 9\ \text{cm}^2 = 45\ \text{cm}^2$$

Another way to determine the area of the octagon is to subdivide it as shown in Figure 12.13(b) and to combine the 4 triangles to make 2 rectangles.

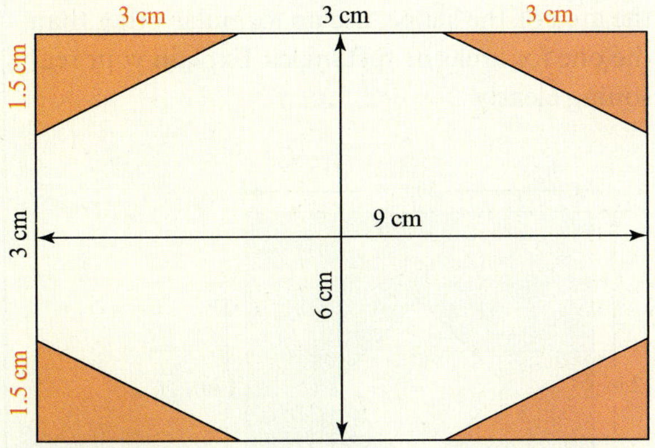

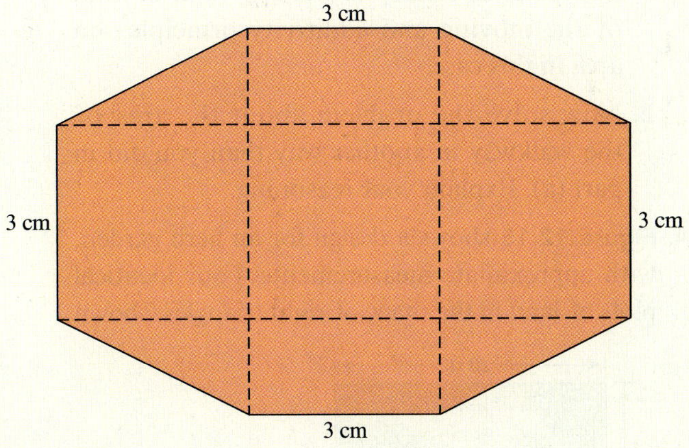

Figure 12.13 (a) The octagon shown as a rectangle with 4 triangles removed. (b) An octagon.

PROBLEMS FOR SECTION 12.2

1. Make a shape that has area 25 in^2 but that has no (or almost no) straight edges. Explain how you know that your shape has area 25 in^2.

2. Figure 12.14 shows the floor plan for a one-story house. Calculate the area of the floor of the house, explaining your reasoning.

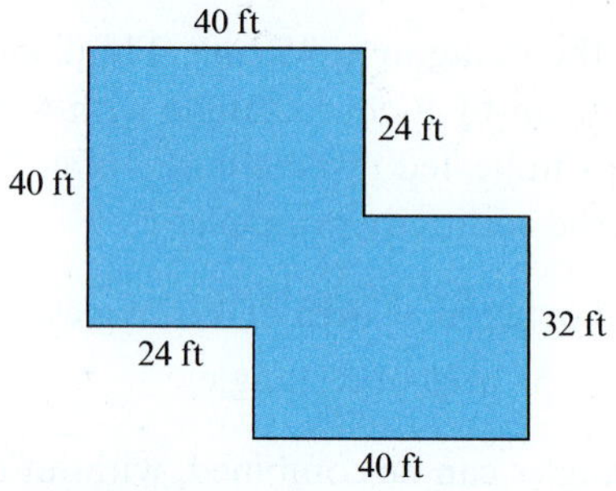

Figure 12.14 Floor plan of a house.

3. An area problem: The Johnsons are planning to build a 5-foot-wide brick walkway around their rectangular garden, which is 20 feet wide and 30 feet long. What will the area of this walkway be? Before you solve the problem yourself, use Kaitlyn's idea:

a. Kaitlyn's idea is to "take away the area of the garden." Explain how to solve the problem about the area of the walkway by using this idea. Explain clearly how to apply one or both of the moving and additivity principles on area in this case.

b. Now solve the problem about the area of the walkway in another way than you did in part (a). Explain your reasoning.

4. Figure 12.15 shows a design for an herb garden, with approximate measurements. Four identical plots of land in the shape of right triangles (shown lightly shaded) are surrounded by paths (shown darkly shaded). Use the moving and additivity principles to determine the area of the paths.

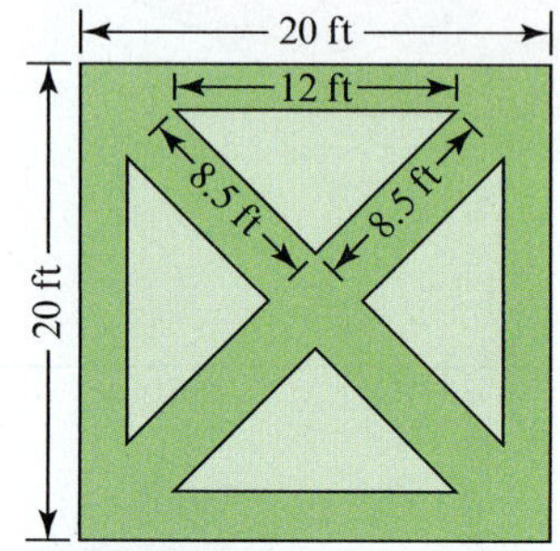

Figure 12.15 An herb garden.

5. Figure 12.16 shows the floor plan for a modern, one-story house. Bob calculates the area of the floor of the house this way:

$$36 \cdot 72 - 18 \cdot 18 = 2268 \text{ ft}^2$$

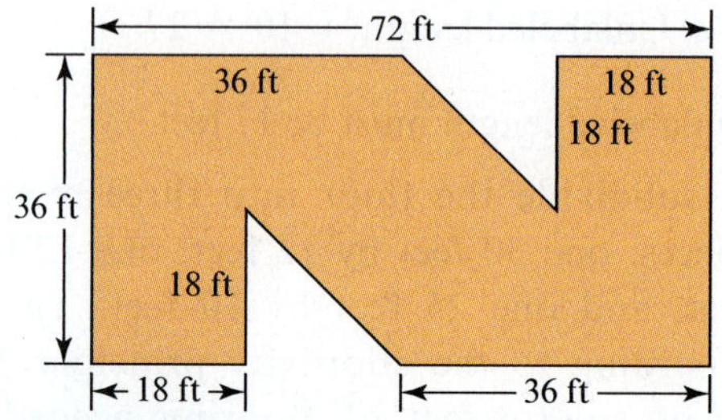

Figure 12.16 Floor plan for a modern house.

What might Bob have in mind? Explain why Bob's method is a legitimate way to calculate the floor area of the house, and explain clearly how one or both of the moving and additivity principles on area apply in this case.

6. Use the moving and additivity principles to determine the area, in square inches, of the shaded region in Figure 12.17. The shape is a 2-inch-by-2-inch square, with a square, placed diagonally inside, removed from the middle. In determining the area of the shape, use no formulas other than the one for areas of rectangles. Explain your reasoning clearly.

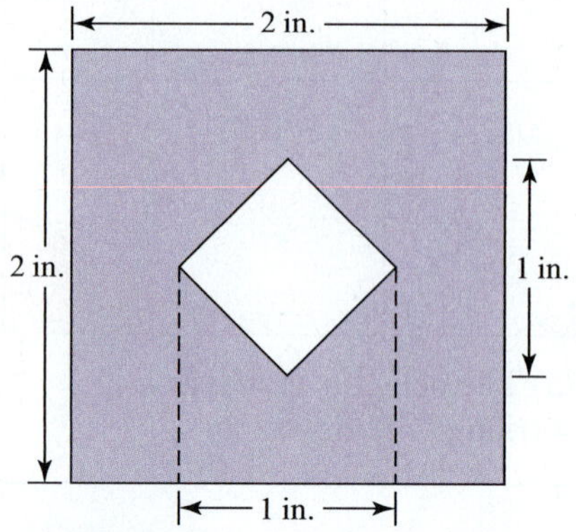

Figure 12.17 A shape with a hole.

7. Use the moving and additivity principles to determine the area, in square inches, of the shaded flower design in Figure 12.18. In determining the area of the shape, use no formulas other than the one for areas of rectangles. Explain your reasoning clearly.

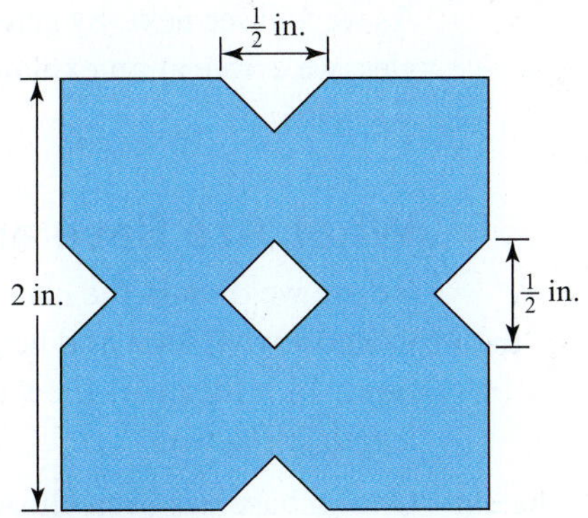

Figure 12.18 A flower design.

12.3 Areas of Triangles

CCSS Common Core State Standards Grade 6

In this section we focus on methods for determining areas of triangles, starting with primitive methods that work only in limited cases, progressing to more sophisticated and powerful methods, and culminating in the area formula for triangles. The reasoning that justifies the formula is a powerful problem-solving tool in its own right.

How Can We Use Moving and Additivity Principles to Determine Areas of Triangles?

Before you read on, do Class Activity 12C about determining areas of triangles using a variety of methods, from primitive to more sophisticated ones.

CLASS ACTIVITY

12C Determining Areas of Triangles in Progressively Sophisticated Ways, p. CA-243

CCSS 6.G.1

Given a triangle, the most primitive method for determining its area is to count how many 1-unit-by-1-unit squares it takes to cover the shape without gaps or overlaps. This method requires squares to be cut apart and pieces to be moved and recombined with other pieces to make as many full squares as possible, as shown in Figure 12.19(a). Although the method is primitive, it is important for students to understand this method because it highlights the meaning of area. More sophisticated methods for determining areas of triangles rely on relating the triangle to a rectangle and applying the rectangle area formula, as shown in Figure 12.19(b).

Figure 12.19 A progression of methods for determining the area of a right triangle.

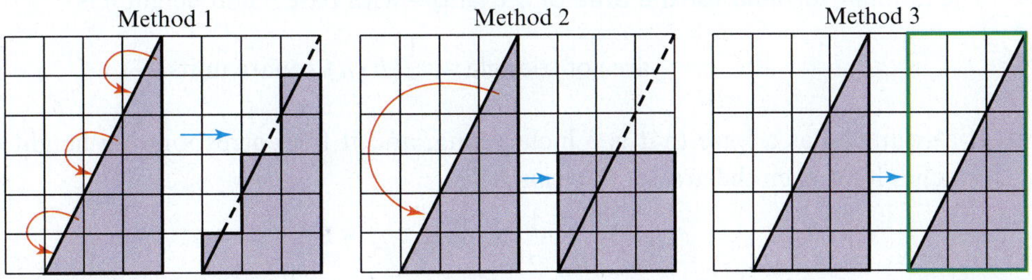

(a) A primitive way to determine the area is to move small pieces and count the total number of squares.

(b) More advanced methods relate the triangle to a rectangle, either by moving a big chunk or by embedding the triangle in a rectangle. These methods lead to the triangle area formula.

As we will see next, by generalizing the more sophisticated methods for determining areas of triangles, we arrive at an explanation for the formula for the area of a triangle—*one-half the base times the height*.

What Are Base and Height for Triangles?

Before we discuss the *one-half the base times the height* formula for the area of a triangle, we need to know what base and height of a triangle mean. The **base** of a triangle can be any one of its three sides. In a formula, the word *base* or a letter, such as *b*, which represents the base, really means *length of the base*.

base

height

Once a base has been chosen, the **height** is the line segment that

- is perpendicular to the base, and
- connects the base, or an extension of the base, to the vertex of the triangle that is not on the base.

In a formula, the word *height* or a letter, such as *h*, which represents the height, actually means *length of the height*.

For example, Figure 12.20 shows two copies of a triangle ABC, and two of the three ways to choose the base *b* and the height *h*. (Notice that the base and height have different lengths for the different choices.) In the second case, the height is the (dashed) line segment CE. In this case, even though the height *h* doesn't meet *b* itself, it meets an *extension* of *b*.

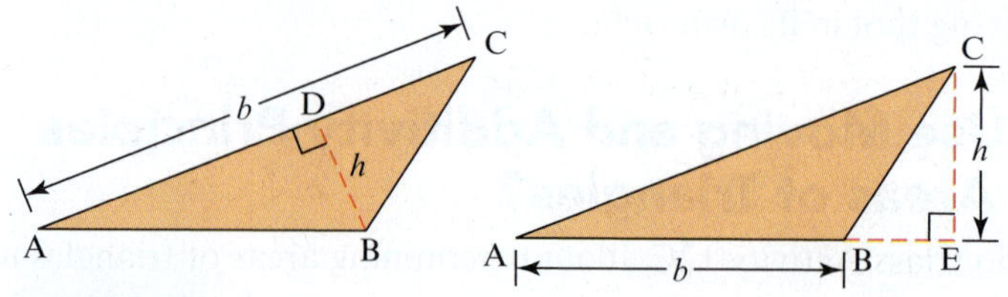

Figure 12.20 Two ways to select the base (*b*) and height (*h*) of triangle ABC.

CLASS ACTIVITY

12D Choosing the Base and Height of Triangles, p. CA-244

What Is a Formula for the Area of a Triangle?

Amazingly, *no matter which side is chosen to be the base of a triangle,* there is a single formula that produces the triangle's area.

formula for the area of a triangle

The familiar **formula for the area of a triangle** with base *b* and height *h* is

$$\text{area of triangle} = \frac{1}{2}(b \cdot h) \text{ square units}$$

If a triangle has a base that is 5 inches long, and if the corresponding height of the triangle is 3 inches long, then the area of the triangle is

$$\frac{1}{2}(5 \cdot 3) \text{ in}^2 = 7.5 \text{ in}^2$$

In the formula

$$\text{area of triangle} = \frac{1}{2}(b \cdot h)$$

the base b and the height h must be described with the *same unit.* For example, both the base and the height could be in feet, or both lengths could be in centimeters. But if one length is in feet and the other is in inches, for example, then you must convert both lengths to a common unit before calculating the area. The area of the triangle is then in *square units* of whatever common unit you used for the base and height. So if the base and height are both in centimeters, then the area resulting from the formula is in square centimeters (cm^2).

Why Is the Area Formula for Triangles Valid?

Try Class Activity 12E before you read on.

CLASS ACTIVITY

12E Explaining Why the Area Formula for Triangles Is Valid, p. CA-245

Suppose we have a triangle, and suppose we have chosen a base b and a height h for the triangle. Why is the area of the triangle equal to one half the base times the height? In some cases, such as those shown in Figure 12.21, two copies of the triangle can be subdivided (if necessary) and recombined, without overlapping, to form a b-by-h rectangle. In such cases, because of the moving and additivity principles about area,

$$2 \cdot \text{area of triangle} = \text{area of rectangle}$$

But since the rectangle has area $b \cdot h$,

$$2 \cdot \text{area of triangle} = b \cdot h$$

Therefore,

$$\text{area of triangle} = \frac{1}{2}(b \cdot h)$$

Figure 12.21
Triangles forming half of a b-by-h rectangle.

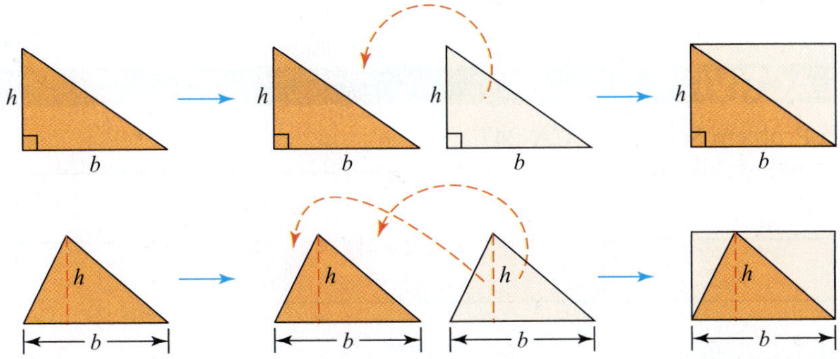

CCSS 6.G.1

But what about the triangle in Figure 12.22? For this triangle, it's not so clear how to turn it into half of a b-by-h rectangle. However, the formula for the area of the triangle is still valid. To explain why the $\frac{1}{2}(b \cdot h)$ formula is still valid for the triangle in Figure 12.22, enclose the triangle in a rectangle, as shown in Figure 12.23(a). The rectangle consists of two copies of the original triangle (darkly shaded) and two copies of another triangle, which are lightly shaded. The rectangle has area $(b + a) \cdot h$, which is equal to $b \cdot h + a \cdot h$ by the distributive property. If we put the two lightly shaded triangles together, as in Figure 12.23(b), they form a rectangle of area $a \cdot h$. If we take this area away from the area of the large rectangle, the remaining area is the area of the two copies of the original triangle combined (by the moving and additivity principles). Therefore, the area of the two copies of the original triangle is

$$(b \cdot h + a \cdot h) - a \cdot h = b \cdot h$$

and so the original triangle has half this area, namely, area

$$\frac{1}{2}(b \cdot h)$$

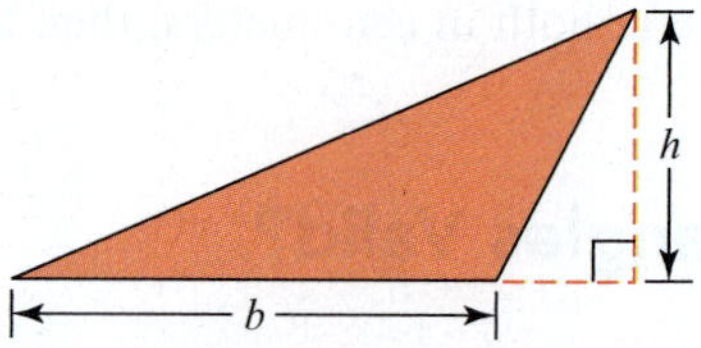

Figure 12.22 Why is the area $\frac{1}{2}(b \cdot h)$?

So, given any triangle, and given any choice of base b and height h for the triangle, one of the arguments we have just given explains why the area of the triangle is $\frac{1}{2}(b \cdot h)$.

Figure 12.23 Enclosing the original triangle in a rectangle of area $(b + a) \cdot h$ to explain why the area of the triangle is $\frac{1}{2}(b \cdot h)$.

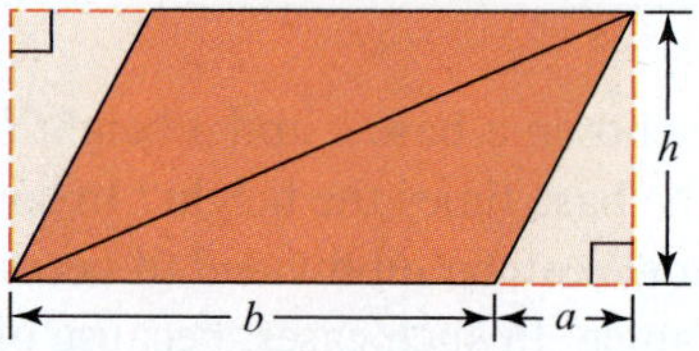

(a) Two copies of the original triangle together with two copies of another triangle form a rectangle of area $(b + a) \cdot h = b \cdot h + a \cdot h$.

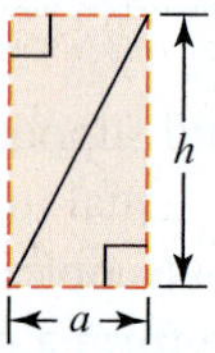

(b) Removing the two additional triangles will remove an area of $a \cdot h$.

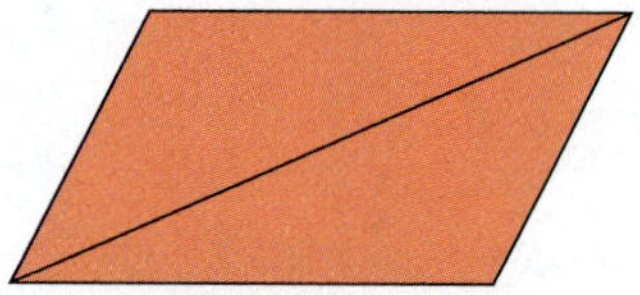

(c) The remaining two copies of the original triangle must therefore have area $b \cdot h$.

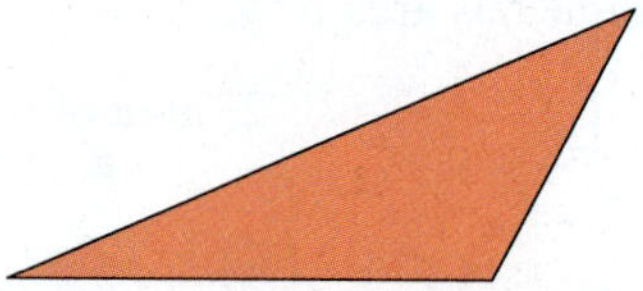

(d) The original triangle therefore has area $\frac{1}{2}(b \cdot h)$.

CLASS ACTIVITY

12F Area Problem Solving, p. CA-247

SECTION SUMMARY AND STUDY ITEMS

Section 12.3 Areas of Triangles

Any of the three sides of a triangle can be the base of the triangle. The height corresponding to a given base is perpendicular to the base and connects the base (or an extension of the base) to the vertex of the triangle that is not on the base. The area of the triangle is $\frac{1}{2}(\text{base} \cdot \text{height})$. The moving and additivity principles explain why this area formula is valid.

Key Skills and Understandings

1. Determine the area of a triangle in various ways, including using less sophisticated and more sophisticated methods.
2. Use the moving and additivity principles to explain why the area formula for triangles is valid for all triangles.
3. Use the area formula for triangles to determine areas and to solve problems.

Practice Exercises for Section 12.3

1. Use the moving and additivity principles about area to determine the area of the triangle in Figure 12.24 in *two different ways*. In both cases, do not use a formula for areas of triangles.

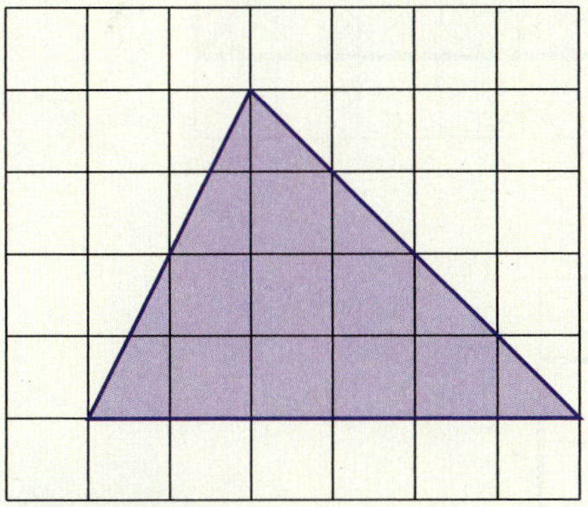

Figure 12.24 Determining the area of a triangle.

2. Show the heights of the triangles in Figure 12.25 that correspond to the bases that are labeled b. Then determine the areas of the triangles.

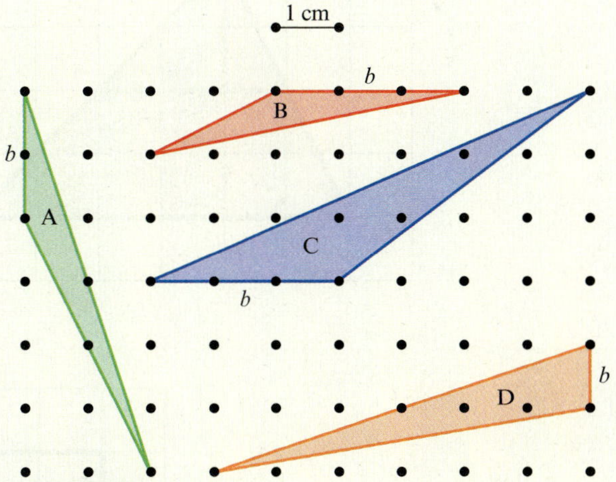

Figure 12.25 Find the heights for the indicated bases.

3. Determine the areas, in square units, of the shaded regions in Figure 12.26.

4. In your own words, explain why the triangles in Figures 12.21 and 12.22 have area $\frac{1}{2}(b \cdot h)$ for the given choices of base b and height h.

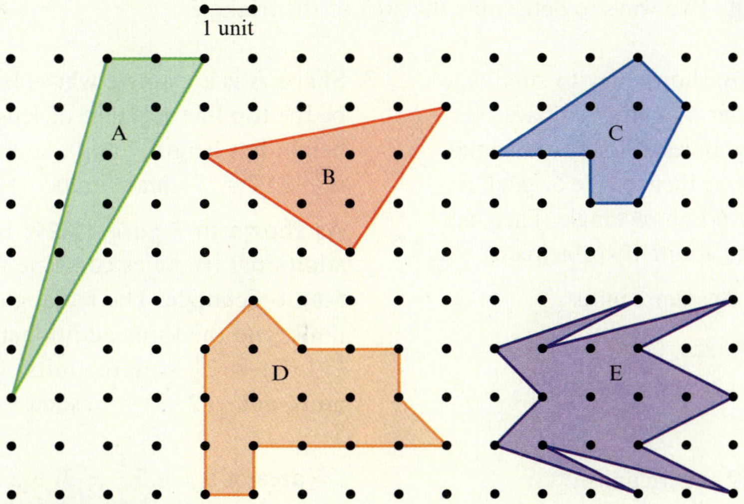

Figure 12.26 Find the areas.

Answers to Practice Exercises for Section 12.3

1. Method 1: Figure 12.27(a) shows how to subdivide the triangle into three pieces. Imagine rotating two of those pieces down to form a 2-unit-by-6-unit rectangle as on the right. Since the rectangle was formed by moving portions of the triangle and recombining them without overlapping, by the moving and additivity principles, the original triangle has the same area as the 2-unit-by-6-unit rectangle—namely, 12 square units.

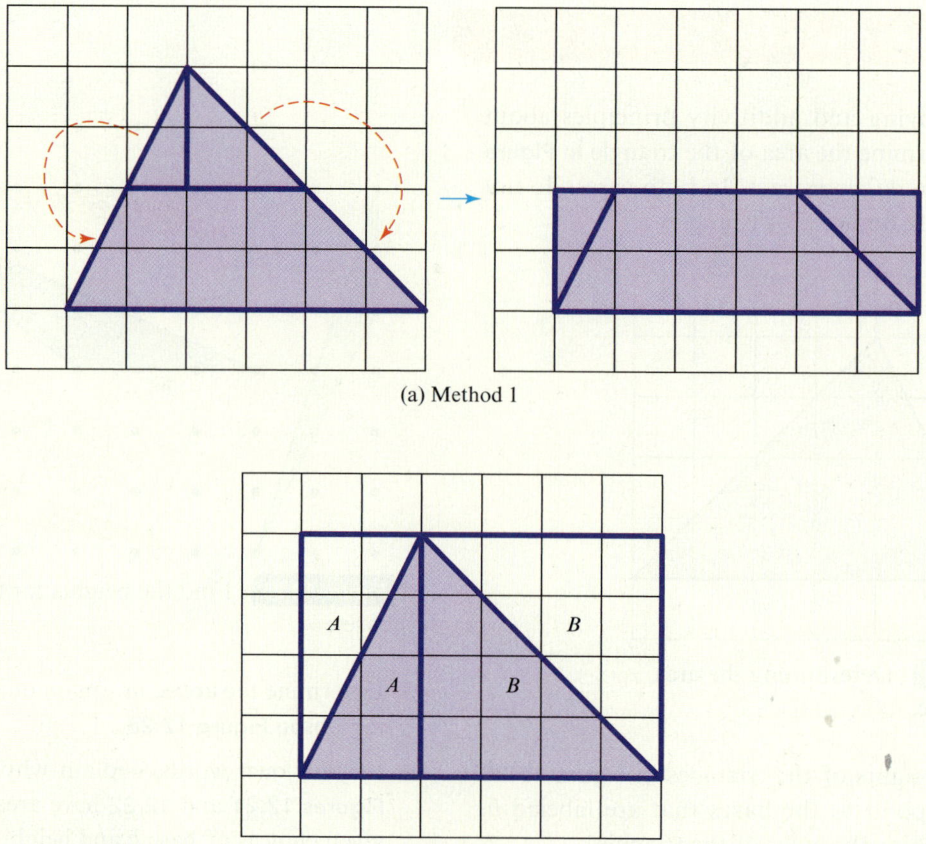

(a) Method 1

(b) Method 2

Figure 12.27 Two ways to determine the area of the triangle.

Method 2: Figure 12.27(b) shows how to subdivide the triangle into two smaller triangles that have areas A and B. Imagine making copies of the two triangles, rotating them, and attaching them to the original triangle to form a 4-unit-by-6-unit rectangle. Then, according to the moving and additivity principles,

$$2A + 2B = 24 \text{ square units}$$

By the distributive property,

$$2 \cdot (A + B) = 24 \text{ square units}$$

Therefore,

$$A + B = 12 \text{ square units}$$

Since $A + B$ is the area of the original triangle, the area of the original triangle is 12 square units.

2. See Figure 12.28. Notice that, for each of these triangles, we must extend the base in order to show the height meeting it at a right angle. Triangle A has height 2 cm and area $(\frac{1}{2} \cdot 2 \cdot 2)\ \text{cm}^2 = 2\ \text{cm}^2$. Triangle B has height 1 cm and area $(\frac{1}{2} \cdot 3 \cdot 1)\ \text{cm}^2 = 1\frac{1}{2}\ \text{cm}^2$. Triangle C has height 3 cm and area $(\frac{1}{2} \cdot 3 \cdot 3)\ \text{cm}^2 = 4\frac{1}{2}\ \text{cm}^2$. Triangle D has height 6 cm and area $(\frac{1}{2} \cdot 1 \cdot 6)\ \text{cm}^2 = 3\ \text{cm}^2$.

3. Shape A is a triangle whose base can be chosen to be the top line segment of length 2 units. Then the height has length 7 units, so the area of triangle A is $\frac{1}{2} \cdot 2 \cdot 7 = 7$ square units.

As shown in Figure 12.29, triangle B and three additional triangles combine to make a 3-unit-by-5-unit rectangle. The rectangle has area 15 square units, and the three additional triangles have areas $\frac{1}{2}(1 \cdot 5) = 2\frac{1}{2}$ square units, $\frac{1}{2}(3 \cdot 2) = 3$ square units, and $\frac{1}{2}(2 \cdot 3) = 3$ square units. Therefore,

$$(\text{area of B}) + 2\frac{1}{2} + 3 + 3 = 15 \text{ square units}$$

So triangle B has area $15 - 8\frac{1}{2} = 6\frac{1}{2}$ square units.

The areas of the remaining shaded shapes can be calculated by subdividing them into rectangles and triangles as indicated in Figure 12.29. Shape C has area $5\frac{1}{2}$ square units. Shape D has area $10\frac{1}{2}$ square units. Shape E has area 10 square units.

4. See text.

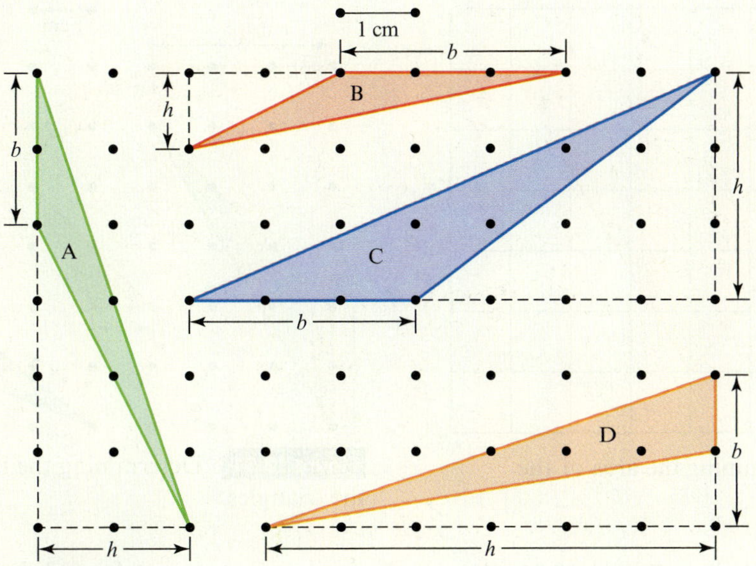

Figure 12.28 Bases and heights.

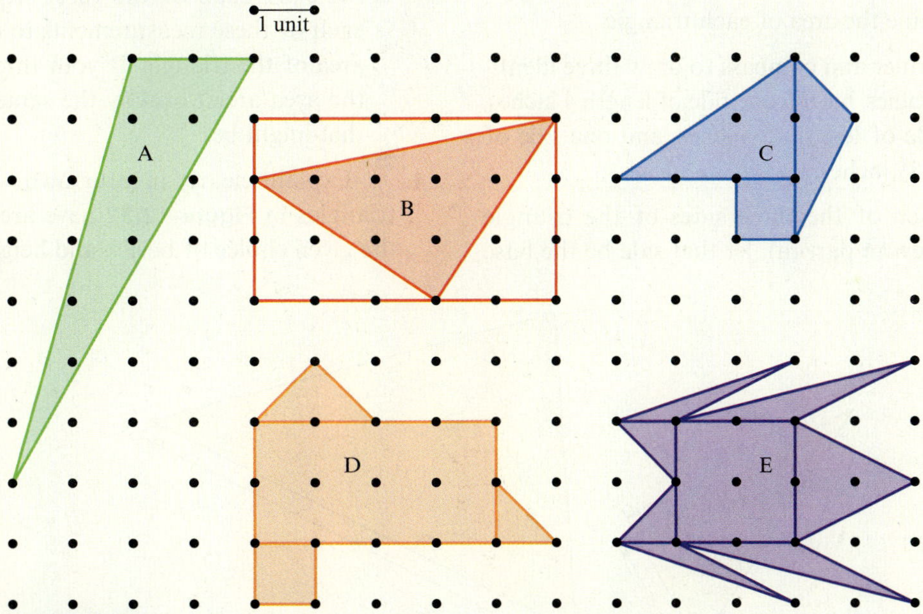

Figure 12.29 Subdividing shapes into triangles and rectangles to calculate their areas.

PROBLEMS FOR SECTION 12.3

1. Use the moving and additivity principles to determine the area (in square units) of the triangle in **Figure 12.30** in *two different ways.* Do not use a formula for areas of triangles. The grid lines in **Figure 12.30** are 1 unit apart. Explain your reasoning.

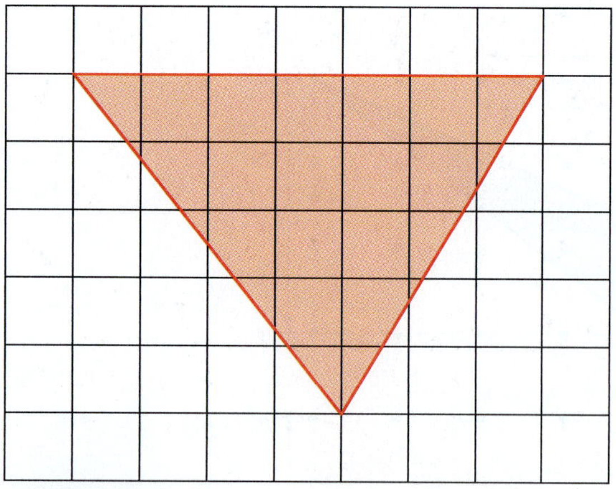

Figure 12.30 Determining the area of the triangle.

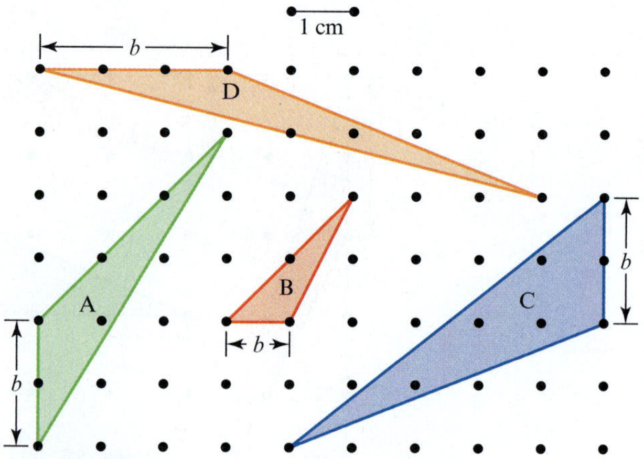

Figure 12.31 Determining the heights and areas of the triangles.

2. For each triangle in Figure 12.31, show the height of the triangle that corresponds to the indicated base b. Then use these bases and heights to determine the area of each triangle.

3. a. Use a ruler and compass to draw three identical triangles, having one side of length 4 inches, one side of length $2\frac{3}{4}$ inches, and one side of length $1\frac{3}{4}$ inches.

b. For each of the three sides of the triangle you drew in part (a), let that side be the base of one triangle, and draw the corresponding height of the triangle.

c. Measure each of the three heights and use each of these measurements to determine the area of the triangle. If your three answers for the area aren't exactly the same, discuss why that might be.

4. Explain clearly in your own words why the triangles in Figure 12.32 have area $\frac{1}{2}(b \cdot h)$ for the given choice of base b and height h.

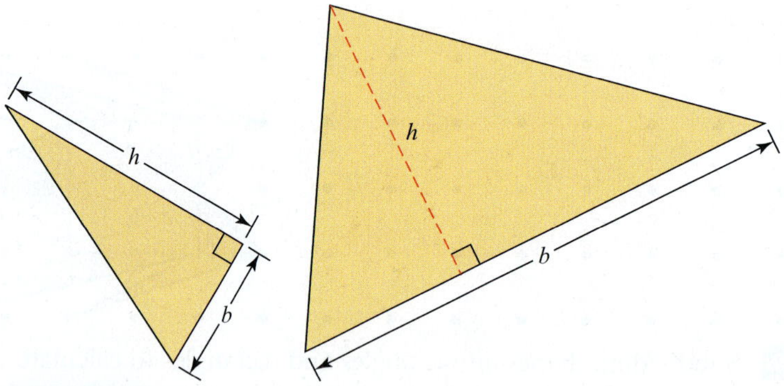

Figure 12.32 Explaining why the area is $\frac{1}{2}(b \cdot h)$.

5. Explain clearly in your own words why the shaded triangle in Figure 12.33 has area $\frac{1}{2}(b \cdot h)$ for the given choice of base b and height h.

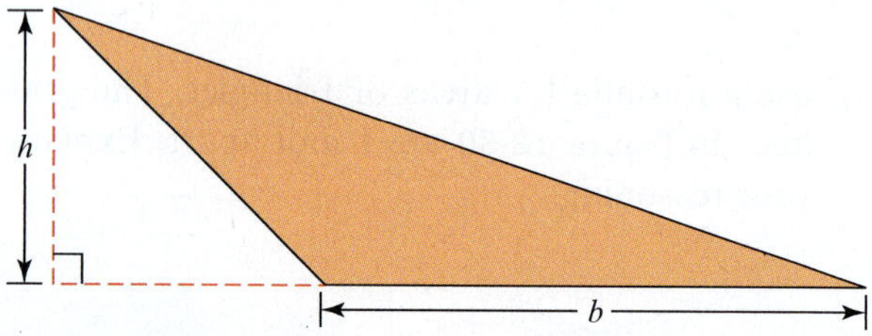

Figure 12.33 Explaining why the area is $\frac{1}{2}(b \cdot h)$.

6. Becky was asked to divide a rectangle into 4 equal pieces and to shade one of those pieces. Figure 12.34 shows her solution. Is Becky right or not? Explain your answer.

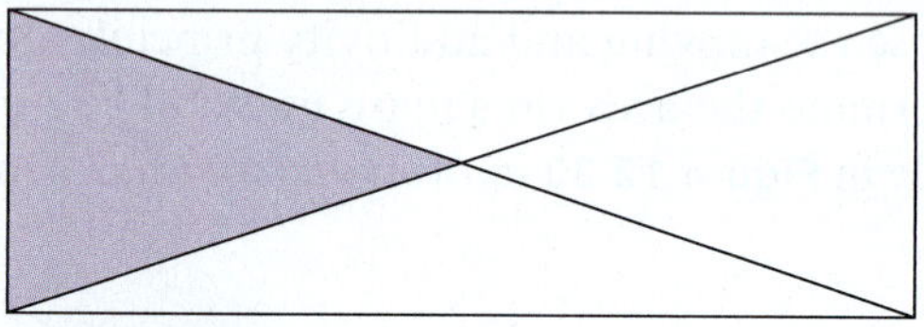

Figure 12.34 Four equal parts or not?

7. Explain how to use the additivity principle to determine the area of the darkly shaded triangle that is inside a rectangle in Figure 12.35.

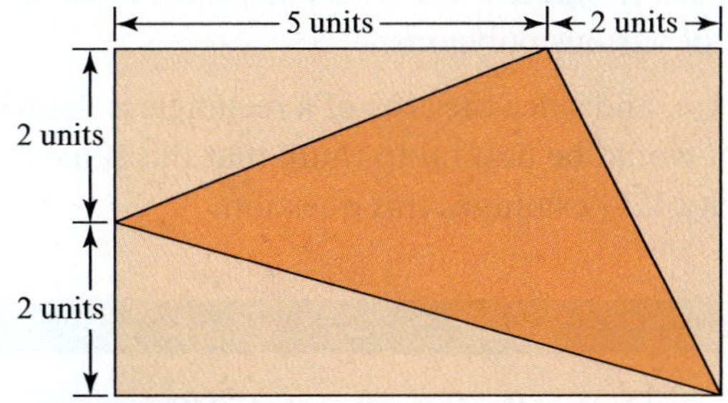

Figure 12.35 Four triangles forming a rectangle.

8. Determine the area of the shaded triangle that is inside a rectangle in Figure 12.36. Explain your reasoning.

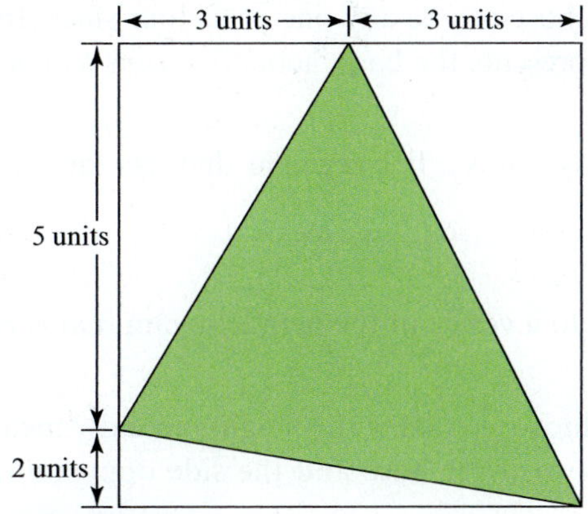

Figure 12.36 Determining the area of the triangle.

9. Determine the area of the shaded shape in Figure 12.37 in *two different ways.* The entire figure consists of two 8-unit-by-8-unit squares. Explain your reasoning in each case.

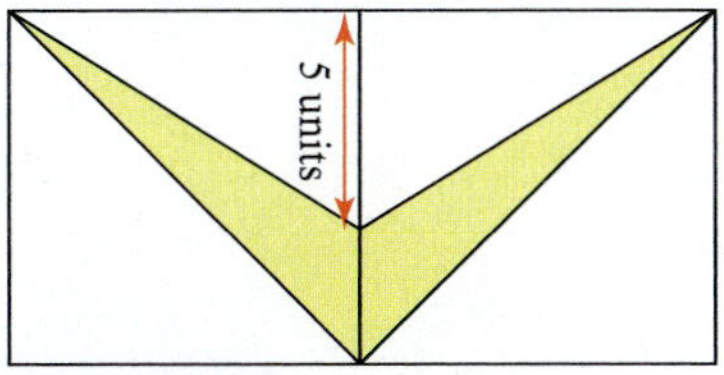

Figure 12.37 Determining the area of the shaded shape.

10. Determine the area of the shaded triangle in Figure 12.38 in *two different ways.* Explain your reasoning in each case.

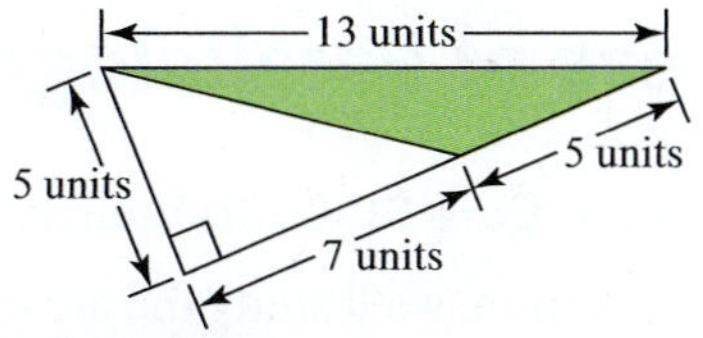

Figure 12.38 Determining the area of the shaded triangle.

11. Determine the area of the shaded shape in Figure 12.39. The entire figure consists of a 3-unit-by-3-unit square and a 5-unit-by-5-unit square. Explain your reasoning.

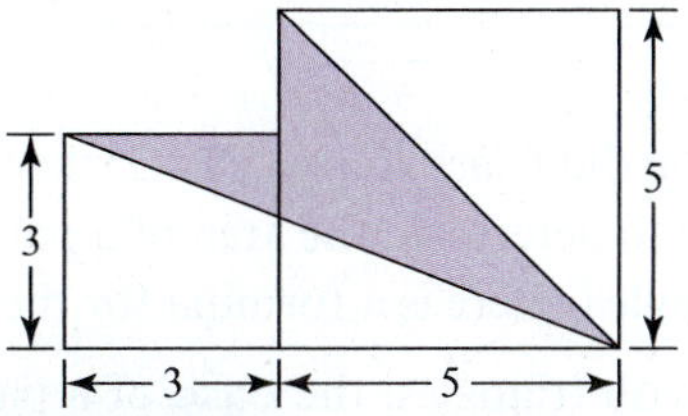

Figure 12.39 Determining the area of the shaded shape.

***12.** Determine the area of the shaded shape in Figure 12.40. The entire figure consists of two 6-unit-by-6-unit squares with a 4-unit-by-4-unit square between them. Explain your reasoning.

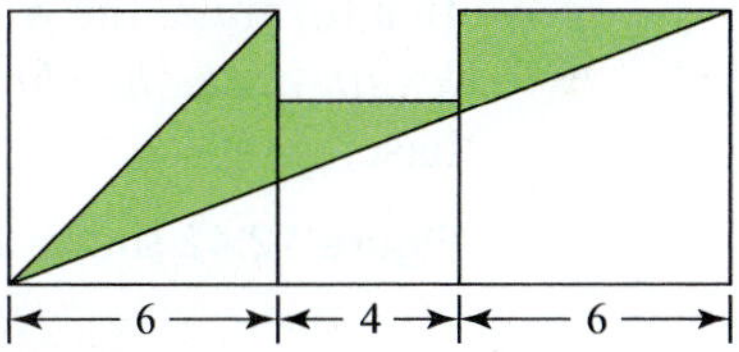

Figure 12.40 Determining the area of the shaded shape.

***13.** Given that the rectangle ABCD in Figure 12.41 has area 108 square units, determine the area of the shaded triangle. Explain your reasoning.

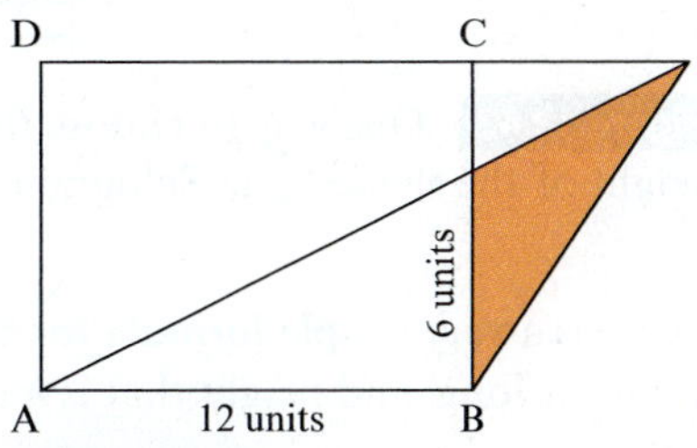

Figure 12.41 Determining the area of the shaded triangle.

12.4 Areas of Parallelograms and Other Polygons

CCSS Common Core State Standards Grades 6, 7

In this section, we'll study the area formula for parallelograms. We'll also continue to use the moving and additivity principles to determine areas of various polygons.

Since a rectangle is a special kind of parallelogram, and since the area of a rectangle is the width of the rectangle times the length of the rectangle, it would be natural to think that this same formula is true for parallelograms. But is it? Class Activity 12G examines this question.

CLASS ACTIVITY

12G Do Side Lengths Determine the Area of a Parallelogram?, p. CA-248

If you did Class Activity 12G, then you saw that, unlike the situation with rectangles, it is not possible to determine the area of a parallelogram from the lengths of its sides alone. However, as with triangles, there is a formula for the area of a parallelogram in terms of a base and a height.

base As with triangles, the **base** of a parallelogram can be chosen to be any one of its four sides. In a formula, the word *base* or a letter, such as b, which represents the base, actually means *length of the base.*

height Once a base has been chosen, the **height** of a parallelogram is a line segment that has these two properties:

- It is perpendicular to the base.
- It connects the base, or an extension of the base, to a vertex of the parallelogram that is not on the base.

In a formula, the word *height* or a letter, such as h, which represents the height, actually means *length of the height.* So the height is the distance between the base and the side opposite the base.

Figure 12.42 shows a parallelogram and one way to choose a base b and a height h.

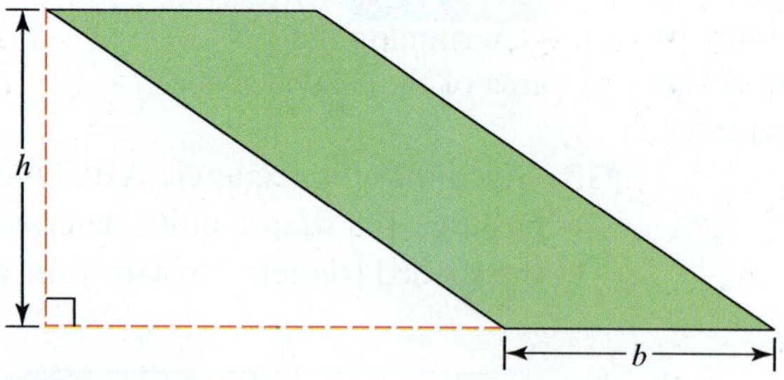

Figure 12.42 One way to choose the base and height of the shaded parallelogram.

formula for the area of a parallelogram There is a very simple **formula for the area of a parallelogram**. If a parallelogram has a base that is b units long, and height that is h units long, then

$$\text{area of parallelogram} = b \cdot h \text{ square units}$$

In this formula, we assume that b and h are described with the same unit (e.g., both in centimeters, or both in feet). If b and h are in different units, then you must convert them to a common unit before using the formula. For example, if a parallelogram has a base that is 2 meters long and a height that is 5 centimeters long, then the area of the parallelogram is

$$200 \cdot 5 \text{ cm}^2 = 1000 \text{ cm}^2$$

because 2 meters = 200 cm.

CLASS ACTIVITY

12H Explaining Why the Area Formula for Parallelograms Is Valid, p. CA-249

12I Finding and Explaining a Trapezoid Area Formula, p. CA-249

Why is the $b \cdot h$ formula for areas of parallelograms valid? In some cases, such as the parallelogram in Figure 12.43(a), we can explain why the area formula is valid by subdividing the parallelogram and recombining it to form a b by h rectangle, as shown in Figure 12.43(b). According to the moving and additivity principles about area, the area of the original parallelogram and the area of the newly formed rectangle are equal. Because the newly formed rectangle has area $b \cdot h$ square units, the original parallelogram also has area $b \cdot h$ square units.

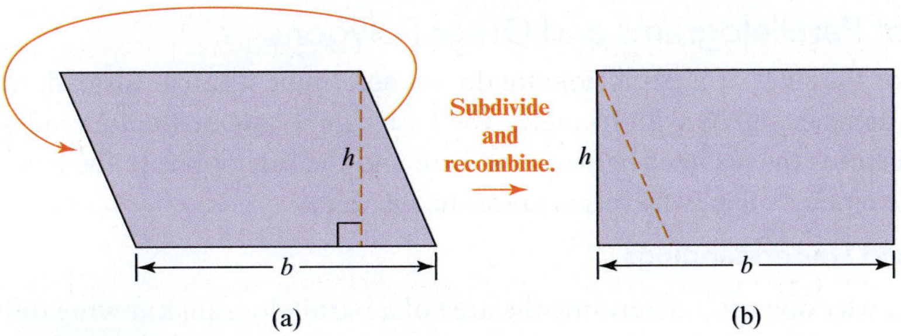

Figure 12.43 Subdividing and recombining a parallelogram (a) to make a rectangle (b).

CCSS 6.G.1

Why is the $b \cdot h$ formula valid for the area of the parallelogram in Figure 12.44, for the given choices of b and h? In this case, we can enclose the parallelogram in a rectangle as shown in Figure 12.45(a). The rectangle consists of the parallelogram (darkly shaded) and two copies of a right triangle. The rectangle has area $(a + b) \cdot h$ which is equal to $a \cdot h + b \cdot h$ by the distributive property. If we put the two triangles together, as in Figure 12.45(b), they form a rectangle of area $a \cdot h$. If we take this area away from the area of the large rectangle, the remaining area is the area of the parallelogram (by the moving and additivity principles). Therefore, the area of the parallelogram is

$$(b \cdot h + a \cdot h) - a \cdot h = b \cdot h$$

and so the parallelogram has area $b \cdot h$, which is what we wanted to show.

So, given any parallelogram, and given any choice of base b and height h for the parallelogram, one of the arguments we have just given explains why the area of the parallelogram is $b \cdot h$.

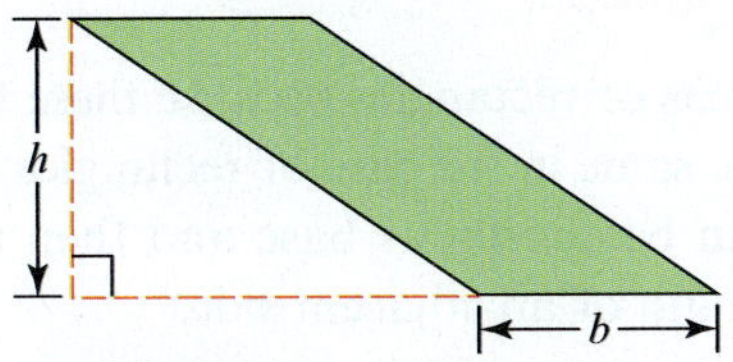

Figure 12.44 Why is the area $b \cdot h$?

Figure 12.45

Enclosing the parallelogram in a rectangle to explain why the area of the parallelogram is $b \cdot h$.

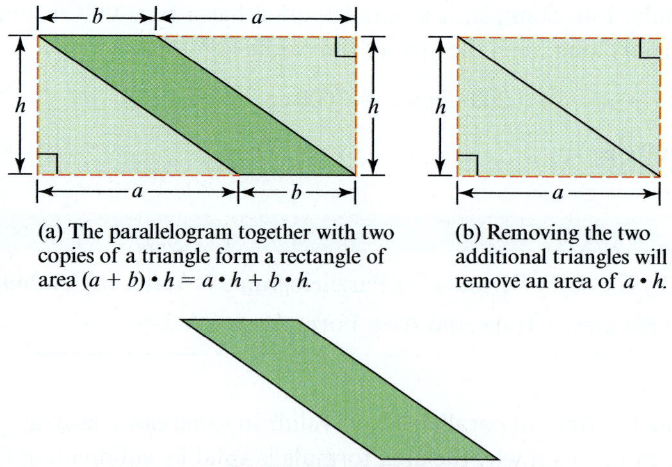

(a) The parallelogram together with two copies of a triangle form a rectangle of area $(a + b) \cdot h = a \cdot h + b \cdot h$.

(b) Removing the two additional triangles will remove an area of $a \cdot h$.

(c) The remaining parallelogram must therefore have area $b \cdot h$.

SECTION SUMMARY AND STUDY ITEMS

Section 12.4 Areas of Parallelograms and Other Polygons

The lengths of the sides of a parallelogram do not determine its area. Instead, the area of a parallelogram is base · height. As with triangles, the base can be any of the four sides of the parallelogram. The height is the distance between the base and the side opposite the base. The moving and additivity principles explain why this area formula is valid.

Key Skills and Understandings

1. Explain why one can't determine the area of a parallelogram knowing only its side lengths.
2. Use the moving and additivity principles to explain why the area formula for parallelograms is valid.
3. Use the area formula for parallelograms to determine areas and to solve problems.
4. Determine areas of various polygons.

Practice Exercises for Section 12.4

1. Every rectangle is also a parallelogram. Viewing a rectangle as a parallelogram, we can choose a base and height for it, as for any parallelogram. In the case of a rectangle, what are other names for *base* and *height*?

2. How are the *base · height* formula for areas of parallelograms and the *length · width* formula for areas of rectangles related?

3. Why can there not be a parallelogram area formula that is expressed only in terms of the lengths of the sides of the parallelogram?

Answers to Practice Exercises for Section 12.4

1. In the case of a rectangle, the *base* and *height* are the *length* and *width* of the rectangle. The base can be either the length or the width.

2. The *base · height* formula for areas of parallelograms generalizes the *length · width* formula for areas of rectangles because these two formulas are the same in the case of rectangles because any side can be chosen as base and then the height is the length of an adjacent side.

3. The three parallelograms in Figure 12.46 all have sides of the same length. If there were a parallelogram area formula that was expressed only in terms of the lengths of the sides of the parallelogram, then all three of these parallelograms would have to have the same area. But it is clear that the three parallelograms have different areas. Therefore, there can be no parallelogram area formula that is expressed only in terms of the lengths of the sides of the parallelogram.

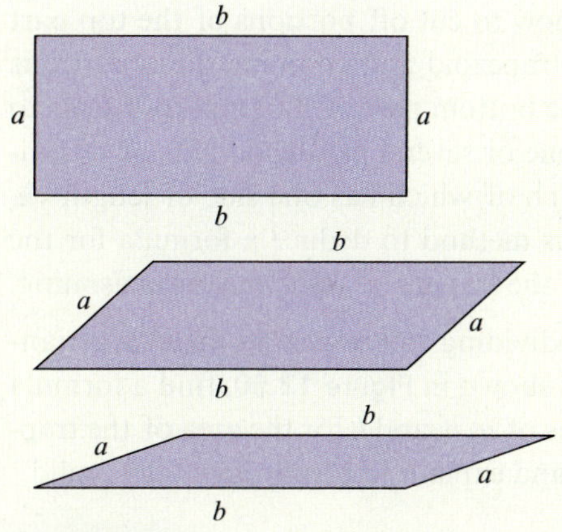

Figure 12.46 Three parallelograms.

PROBLEMS FOR SECTION 12.4

1. Josie has two wooden beams that are 15 feet long and two wooden beams that are 10 feet long. She plans to use these four beams to form the entire border around a closed garden. She likes unusual designs. Without any other information, what is the most you can say about the area of Josie's garden? Explain.

2. Figure 12.47 shows a shaded parallelogram inside a rectangle. Use the moving and additivity principles to find an expression for the area of the shaded parallelogram in terms of all of the lengths x, y, and z. Explain your reasoning. Then simplify the expression you found.

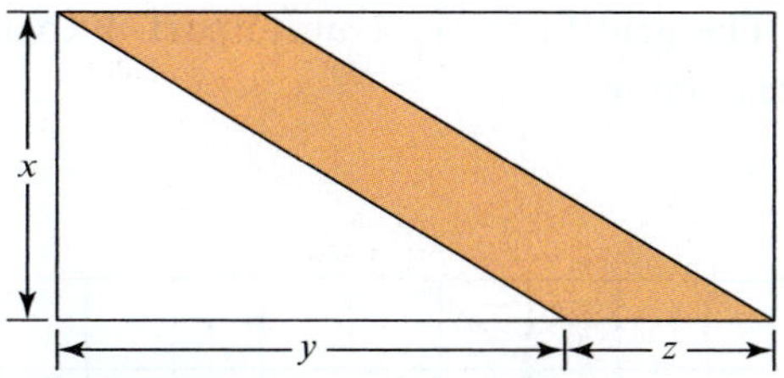

Figure 12.47 Determining an expression for the area of the shaded parallelogram.

3. In the text, we saw a way to explain why the area of the parallelogram in Figure 12.48 is $b \cdot h$. Another way to explain the area formula for parallelograms is to use the area formula for triangles. Show how to subdivide the parallelogram in Figure 12.48 into two triangles. Then use the area formula for triangles to explain why the area of the parallelogram is $b \cdot h$.

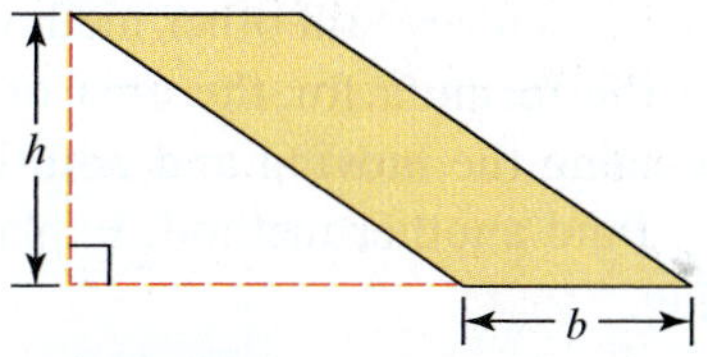

Figure 12.48 Why is the area $b \cdot h$?

4. Figure 12.49 shows a trapezoid. This problem will help you find a formula for the area of the trapezoid and explain in several different ways why this formula is valid.

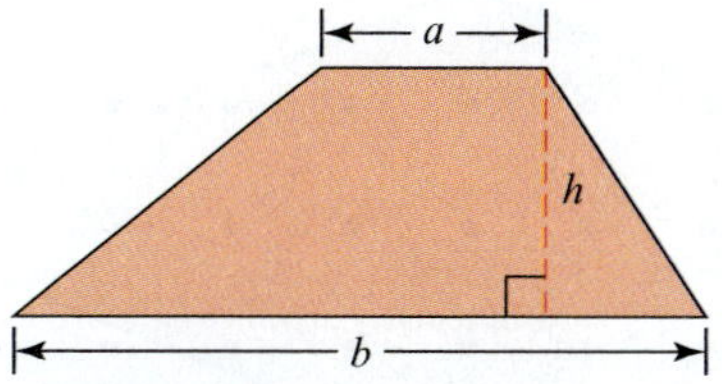

Figure 12.49 A trapezoid.

a. Show how to combine two copies of the trapezoid in Figure 12.49 to make a parallelogram. Then use the formula for the area of a parallelogram to deduce a formula for the area of the trapezoid. Explain your reasoning.

b. Show how to cut off portions of the top part of the trapezoid and combine these portions with the bottom part of the trapezoid so as to make one or several parallelograms or rectangles, each of which has one side of length $\frac{1}{2}h$. Use this method to deduce a formula for the area of the trapezoid. Explain your reasoning.

c. By subdividing the trapezoid into two triangles, as shown in Figure 12.50, find a formula in terms of a, b, and h for the area of the trapezoid, and explain why your formula is valid.

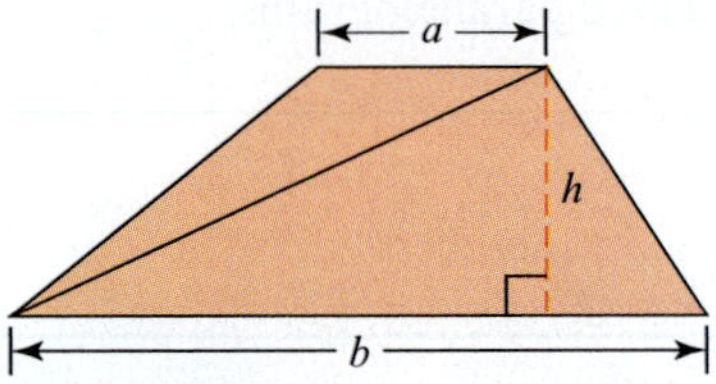

Figure 12.50 A subdivided trapezoid.

d. Besides the methods of parts (a), (b), and (c) of this problem, there are still other methods for deducing the formula for the area of a trapezoid by using the moving and additivity principles. Find another method. Explain your reasoning.

5. Use the moving and additivity principles to determine the area (in square units) of the rhombus in Figure 12.51 *without* using the triangle or parallelogram area formulas. Adjacent dots are 1 unit apart. Explain your reasoning.

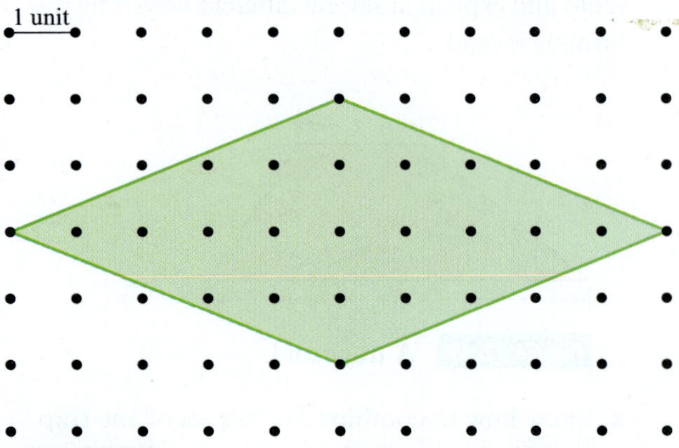

Figure 12.51 Determining the area of the rhombus.

6. Find a formula for the area of a rhombus (see Figure 12.52) in terms of the distances between opposite vertices. Explain why your formula is valid.

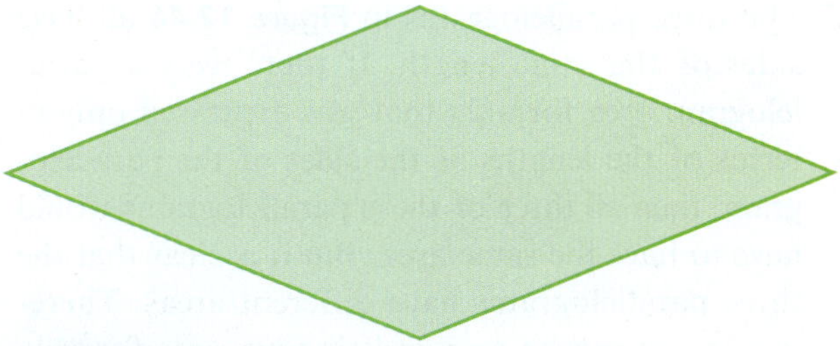

Figure 12.52 Find a formula for the area of a rhombus.

7. a. Determine the areas (in square units) of the 4 lightly shaded triangles in Figure 12.53. The grid lines are 1 unit apart. Explain your reasoning.

b. Use the moving and additivity principles and your results from part (a) to determine the area of the dark shaded quadrilateral in Figure 12.53. Explain your reasoning.

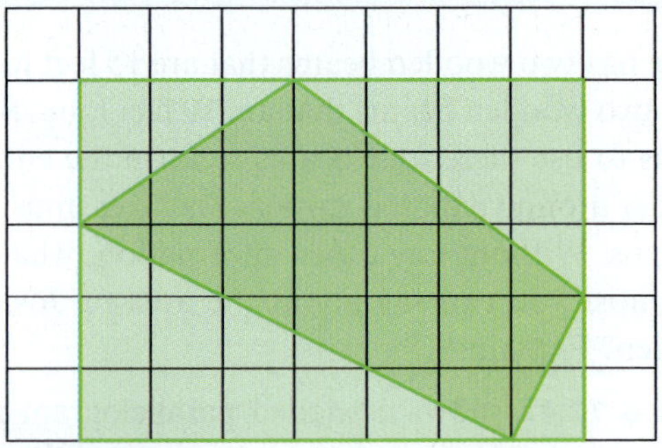

Figure 12.53 Four triangles and a quadrilateral forming a rectangle.

8. Determine the area (in square units) of the quadrilateral in Figure 12.54 in *two different ways*. The grid lines are 1 unit apart. Explain your reasoning.

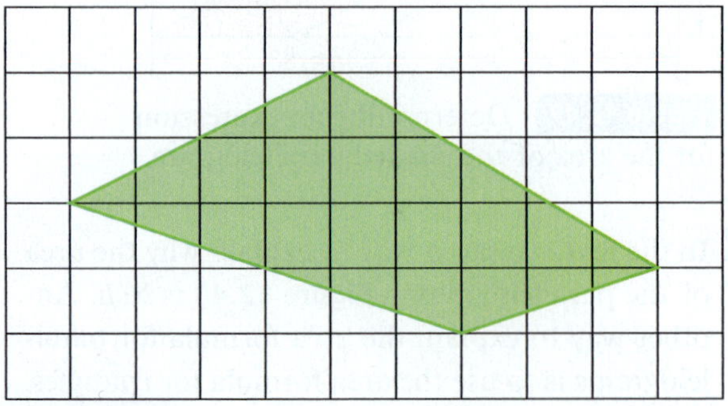

Figure 12.54 Determining the area of the quadrilateral.

9. Determine the area of the shaded shapes in Figure 12.55. Explain your reasoning.

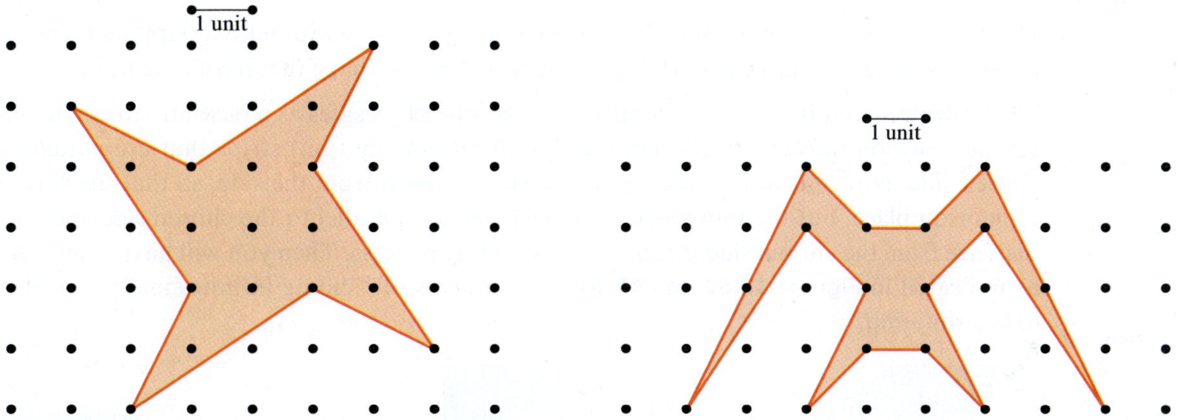

Figure 12.55 Determining the areas of shapes.

10. A rug company weaves rugs that are made by repeating the design in Figure 12.56. Lengths of portions of the design are indicated in the figure. The yarn for the shaded portion of the design costs $5 per square unit, and the yarn for the unshaded portion of the design costs $3 per square unit. How much will the yarn for a 60-unit-by-84-unit rug cost? Explain your reasoning.

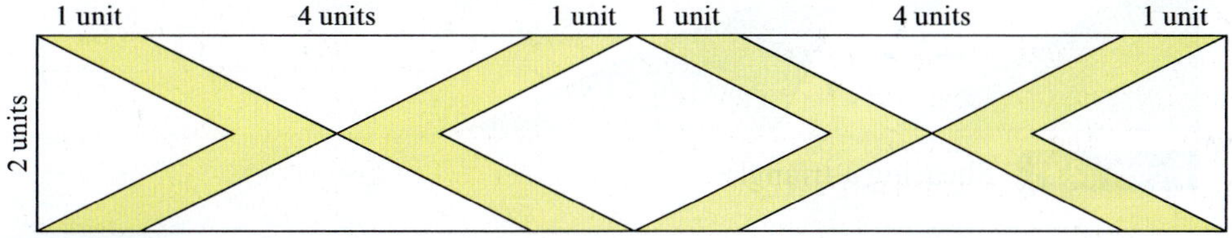

Figure 12.56 A design for rugs.

11. Determine the area of the shaded region in Figure 12.57. Explain your reasoning.

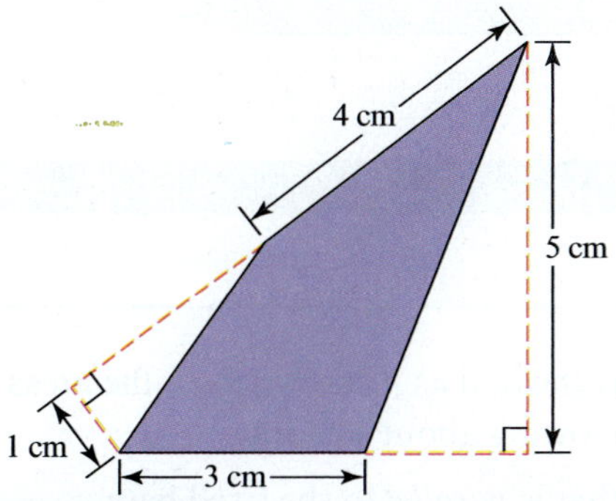

Figure 12.57 Determine the area of the shaded region.

*__12.__ Given that the shaded shape in Figure 12.58 is a parallelogram, find an equation relating the lengths *a*, *b*, *c*, and *d*. Explain why your equation must be true.

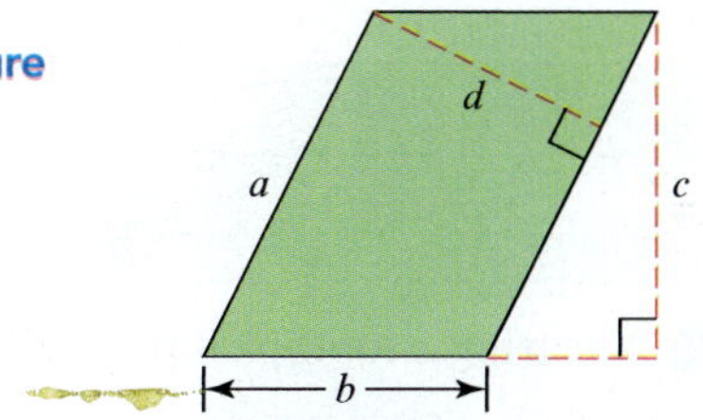

Figure 12.58 Finding an equation relating the given lengths.

*__13.__ Figure 12.59 shows a map of some land. Determine the size of this land in acres. Recall that 1 acre is 43,560 square feet.

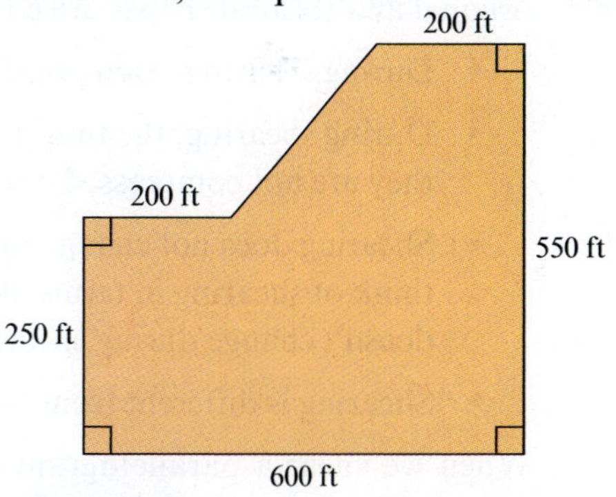

Figure 12.59 How many acres of land?

12.5 Shearing: Changing Shapes Without Changing Area

In addition to subdividing a shape and recombining its parts without overlapping them, there is another way, called *shearing*, to change a shape into a new shape that has the same area.

To illustrate shearing, start with a polygon, pick one of its sides as a base and then imagine slicing the polygon into extremely thin (really, infinitesimally thin) strips that are parallel to the chosen side. Now imagine giving those thin strips a push from the side, so that the chosen side remains in place, but the thin strips slide over, remain parallel to the chosen side and the same distance from the chosen side throughout the sliding process. Then you will have a new polygon, as indicated in Figures 12.60 and 12.61. This process of "sliding infinitesimally thin strips" is called **shearing**.

shearing

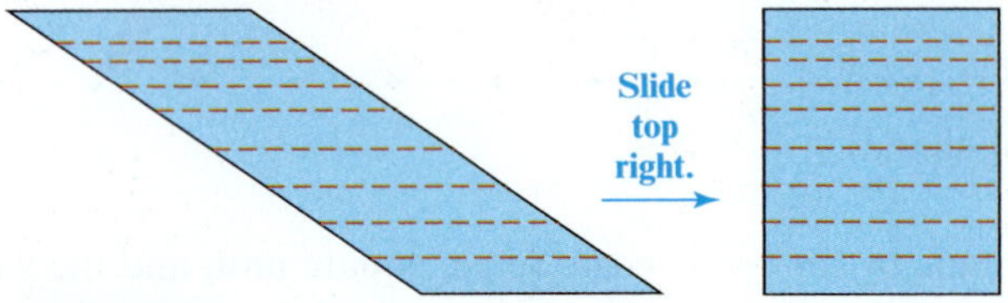

Figure 12.60 Shearing a parallelogram.

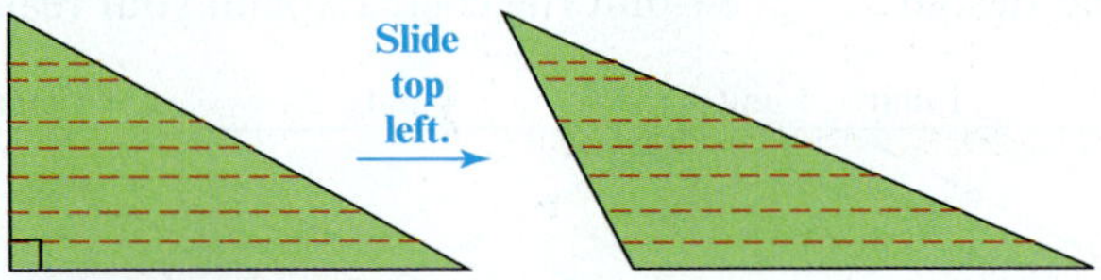

Figure 12.61 Shearing a triangle.

To simulate shearing, we replace the infinitesimally thin strips with toothpicks. If we then give the stack of toothpicks a push from the side, they will slide over, as in shearing. (See Figure 12.62.)

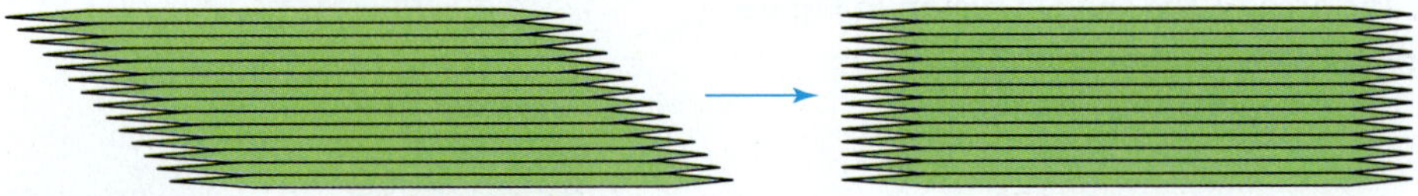

Figure 12.62 Shearing a toothpick parallelogram.

CLASS ACTIVITY

12J Is This Shearing?, p. CA-250

Cavalieri's principle

Cavalieri's principle for areas says that when a shape is sheared as just described, the areas of the original and sheared shapes are equal. Also note the following about shearing:

- During shearing, each point moves along a line that is *parallel* to the fixed base.
- During shearing, the thin strips *remain the same width and length.* The strips just slide over; they are not compressed either in width or in length.
- Shearing does not change the height of the "stack" of thin strips. In other words, if you think of shearing in terms of sliding toothpicks, the height of the stack of toothpicks doesn't change during shearing.
- Shearing is different from "squashing" (see Class Activity 12J).

When we shear a parallelogram or a triangle with respect to a fixed base, some side lengths can change dramatically, but the base, the height, and the area do not change, as indicated in

Figures 12.63 and 12.64. Thus shearing can help to give us a "dynamic" perspective on the parallelogram and triangle area formulas.

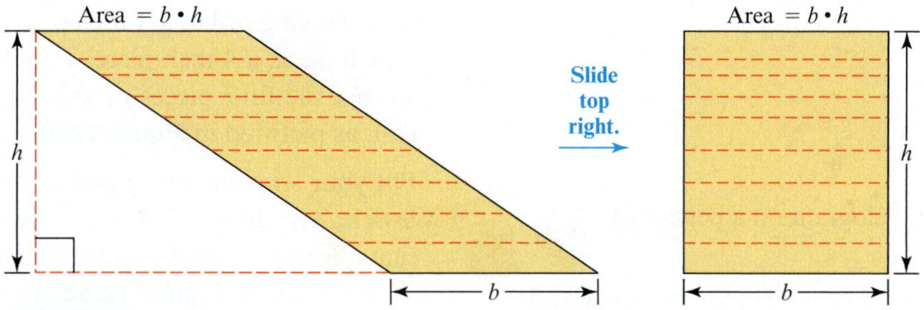

Figure 12.63 Shearing a parallelogram does not change the base, height, or area.

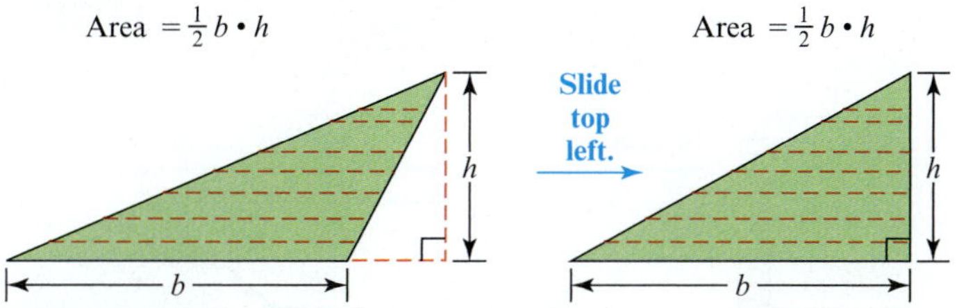

Figure 12.64 Shearing a triangle does not change the base, height, or area.

CLASS ACTIVITY

12K Solving Problems by Shearing, p. CA-251

SECTION SUMMARY AND STUDY ITEMS

Section 12.5 Shearing: Changing Shapes Without Changing Area

Shearing is a process of changing a shape by sliding infinitesimally thin strips of the shape. Cavalieri's principle says shearing does not change areas. When we shear triangles and parallelograms parallel to a base, the base does not change, the height does not change, and the area does not change.

Key Skills and Understandings

1. Show how to shear a shape.
2. Know that when you shear a shape, the original shape and the sheared shape have the same area (Cavalieri's principle).
3. Distinguish shearing from ways of changing a shape that don't preserve area.

Practice Exercises for Section 12.5

1. Figure 12.65 shows a parallelogram on a pegboard. (Think of the parallelogram as made out of a rubber band, which is hooked around four pegs.) Show two ways to move points C and D of the

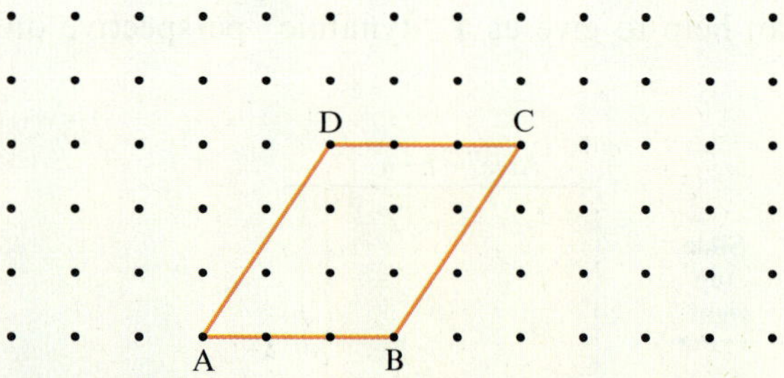

Figure 12.65 Parallelogram on a pegboard.

parallelogram to other pegs, keeping points A and B fixed, in such a way that the new shape is a parallelogram that has the same area as the original parallelogram.

2. Using two ordinary plastic drinking straws, cut two 4-inch pieces of straw and two 3-inch pieces of straw. Lace these pieces of straw onto a string in the following order: a 3-inch piece, a 4-inch piece, a 3-inch piece, a 4-inch piece. Tie a knot in the string so that the four pieces of straw form a quadrilateral, as pictured in Figure 12.66.

Put your straw quadrilateral in the shape of a rectangle. Gradually "squash" the quadrilateral so that it forms a parallelogram that is not a rectangle, as indicated in Figure 12.66. Is this "squashing" process for changing the rectangle into a parallelogram the same as the shearing process? Why or why not?

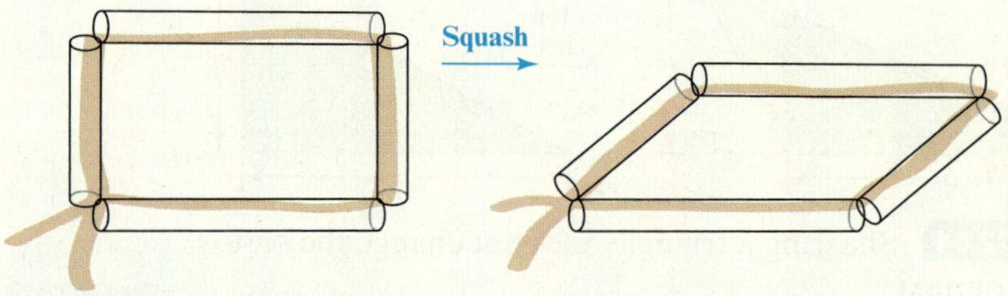

Figure 12.66 "Squashing" a quadrilateral made of straws.

Answers to Practice Exercises for Section 12.5

1. See Figure 12.67. The parallelograms ABEF and ABGH are two examples of parallelograms that have the same area as ABCD. These parallelograms have the same area as ABCD by Cavalieri's principle, because they are just sheared versions of ABCD. In fact, if we move C and D anywhere along the line of pegs that goes through C and D, keeping the same distance between them, the resulting parallelogram will be a sheared version of ABCD and hence will have the same area as ABCD.

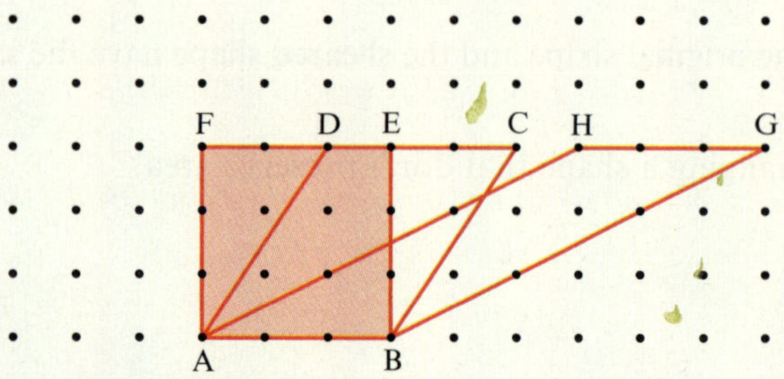

Figure 12.67 Three parallelograms with the same area.

2. No, this "squashing" process is not the same as shearing. You can tell that it's not shearing because if you made the rectangle out of very thin strips and you slid them over to make a parallelogram, they would have to become thinner to make the parallelogram that is formed from the straws (because the parallelogram is not as tall as the rectangle). But in the shearing process, the size of the strips does not change (either in length or in width). Therefore, "squashing" is not the same as shearing.

PROBLEMS FOR SECTION 12.5

1. Figure 12.68 shows a triangle on a pegboard. (Think of the triangle as made out of a rubber band, which is stretched around three pegs.) Describe or draw at least two ways to move point C of the triangle to another peg (keeping points A and B fixed) in such a way that the area of the new triangle is the same as the area of the original triangle. Explain your reasoning.

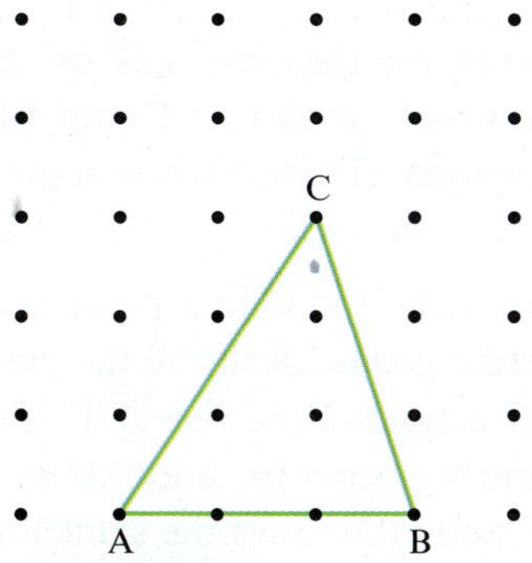

Figure 12.68 A triangle on a pegboard.

2. a. Make a drawing to show the result of shearing the parallelogram in Figure 12.69 into a rectangle. Explain how you know you have sheared the parallelogram correctly.

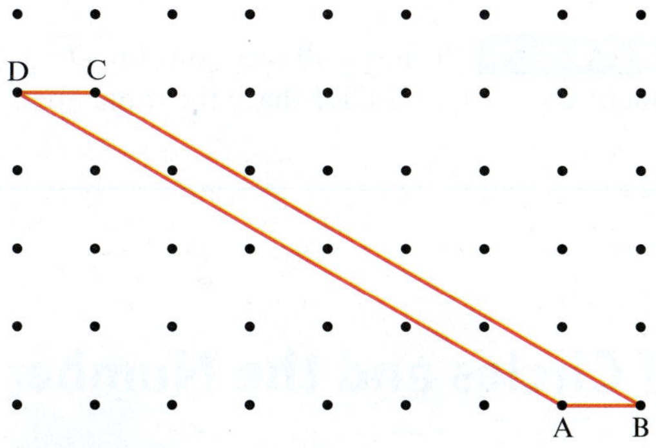

Figure 12.69 Parallelogram to shear.

b. During shearing, what changed and what remained the same?

3. a. Make a drawing to show the result of shearing the parallelogram in Figure 12.70 into a rectangle. Explain how you know you have sheared the parallelogram correctly. Note: Shearing does not have to be horizontal.

b. During shearing, what changed and what remained the same?

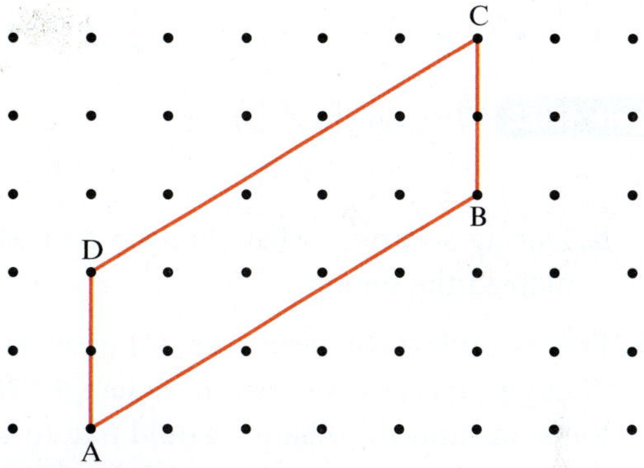

Figure 12.70 Parallelogram to shear.

4. a. Make a drawing to show the result of shearing the triangle in Figure 12.71 into a right triangle. Explain how you know you have sheared the triangle correctly

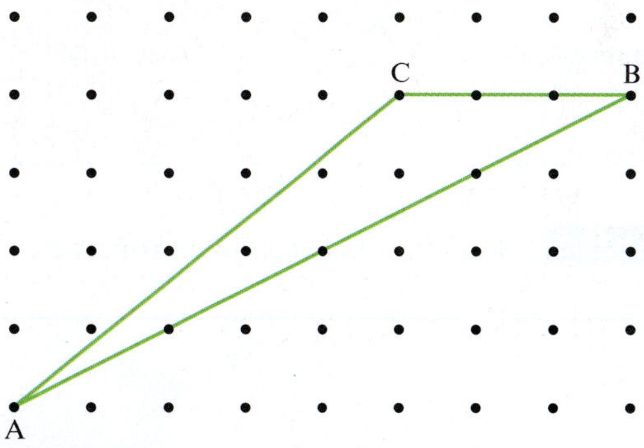

Figure 12.71 Triangle to shear.

b. During shearing, what changed and what remained the same?

5. a. Make a drawing to show the result of shearing the triangle in Figure 12.72 into a right triangle. Explain how you know you have sheared the triangle correctly. *Hint:* Shearing does not have to be horizontal.

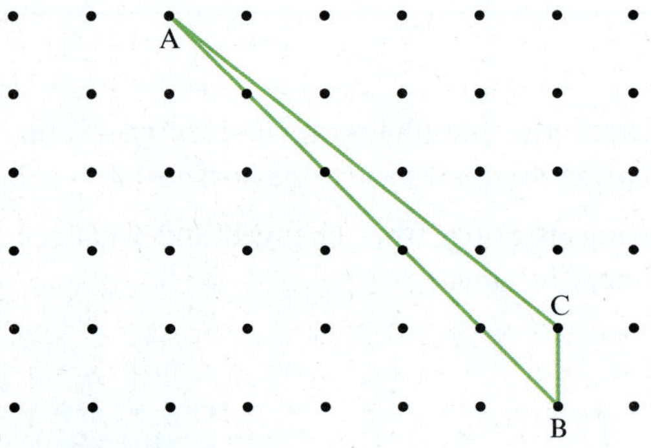

Figure 12.72 Triangle to shear.

b. During shearing, what changed and what remained the same?

6. The boundary between the Johnson and the Zhang properties is shown in Figure 12.73. The Johnsons and the Zhangs would like to change this boundary so that the new boundary is one straight line segment and so that each family still has the same amount of land area. Describe a precise way to redraw the boundary between the two properties. Explain your reasoning. *Hint:* Consider shearing the triangle ABC.

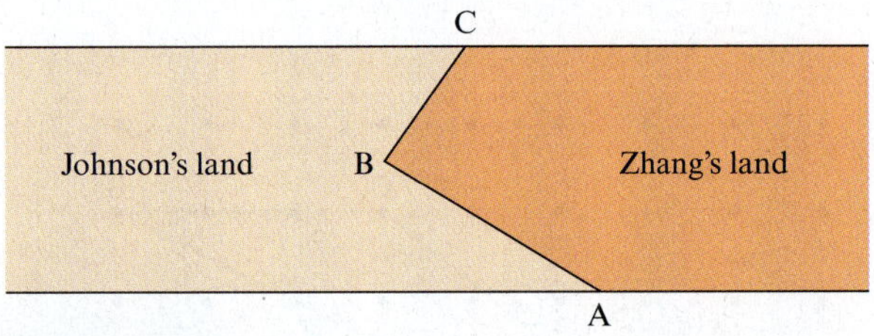

Figure 12.73 Boundary between two properties.

7. Suppose that in a trapezoid ABCD, as in Figure 12.74, AB and CD are parallel.

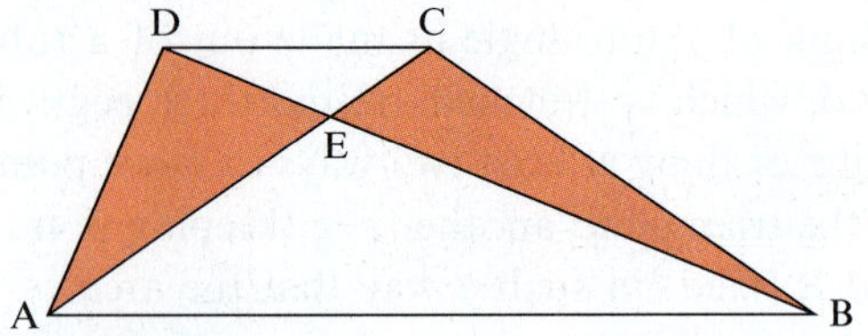

Figure 12.74 Trapezoid and the triangles formed by its diagonals.

Let E be the point where the diagonals AC and BD meet. Explain why triangles AED and BEC must have the same area. (Do this without measuring actual lengths or areas.)

*8. Given three points A, B, and C in a plane, as in Figure 12.75, describe geometrically all the places where you can put a point D so that if E is the point where the line segments BC and AD meet, the triangles ABE and CDE have the same area. Explain your reasoning.

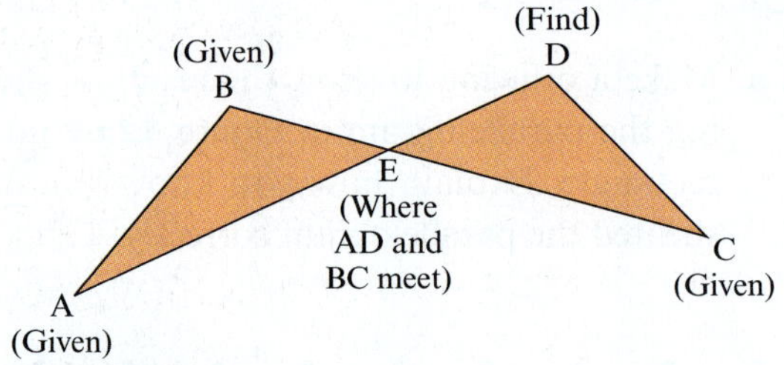

Figure 12.75 Finding all locations for D such that ABE and CDE have the same area.

12.6 Area and Circumference of Circles and the Number Pi

CCSS Common Core State Standards Grade 7

What is the area of a circle? You probably know the familiar formula

$$A = \pi r^2$$

for the area A (in square units) of a circle of radius r units. The Greek letter π, pronounced "pie," stands for a mysterious number that is approximately equal to 3.14159. Since at least the time of the ancient Babylonians and Egyptians, nearly 4000 years ago, people have known about, and been fascinated by, the remarkable number π. In this section, we discuss the number π, observe how it is related to the distance around a circle, and see why the area of a circle of radius r units is πr^2 square units.

circumference First, we need some terminology. The **circumference** of a circle is the distance around the circle. (See **Figure 12.76**.) Recall that the radius of a circle is the distance from the center of the circle to any point on the circle. Recall also that the diameter of a circle is the distance across the circle, going through the center; it is twice the radius.

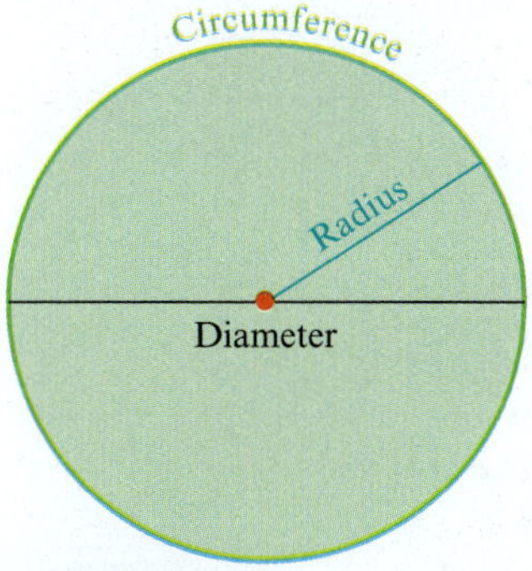

Figure 12.76 The circumference of a circle is the distance around the circle; the diameter is the distance across the circle through the center; the radius is the distance from the center to any point on the circle.

How Are the Circumference, Diameter, and Radius of Circles Related?

Before you read on, try the next two class activities to see how the circumference of a circle is related to its diameter.

CLASS ACTIVITY

12L How Are the Circumference and Diameter of a Circle Related, Approximately?, p. CA-252

12M How Many Diameters Does it Take to Make the Circumference of a Circle?, p. CA-252

In Class Activity 12M you used the diameter of a circle to measure its circumference. You probably noticed that it takes 3 whole diameters plus a little more to make the circumference of the circle, as indicated in **Figure 12.77**. You might have been able to tell that the "little more" is at least $\frac{1}{10}$ of the diameter. Remarkably, for any circle whatsoever—no matter the size—when the circumference of a circle is measured by its diameter the result is always the same number, which

pi, π is approximately 3.14. We call this number **pi** and we denote it with the Greek letter π. There are many formulas for determining the digits in the decimal representation of pi but these are beyond the scope of this book.

Circles can be small or large, so why is the circumference of every circle always the same size when measured by its diameter? In Chapter 14 we will see that all circles are just scaled versions of each other, either scaled smaller or scaled larger. We will see that when we scale shapes, the ratio between the lengths of different parts of the shape remains the same.

CCSS 7.G.4

How can we develop a formula to relate the diameter and circumference of circles? Let's say we have chosen a unit of length measurement, for example, centimeters or feet. Suppose that the diameter of a circle is D units and the circumference of the circle is C units. Is there an equation relating D and C? Taking a *variable-parts* perspective, as in Chapter 7, if we think of the diameter of a circle as 1 part that is D units long, then the circumference is π of those parts, as indicated in **Figure 12.77**. Therefore the number π answers the question, "how many groups of D units are in C units?" or

$$? \cdot D = C$$

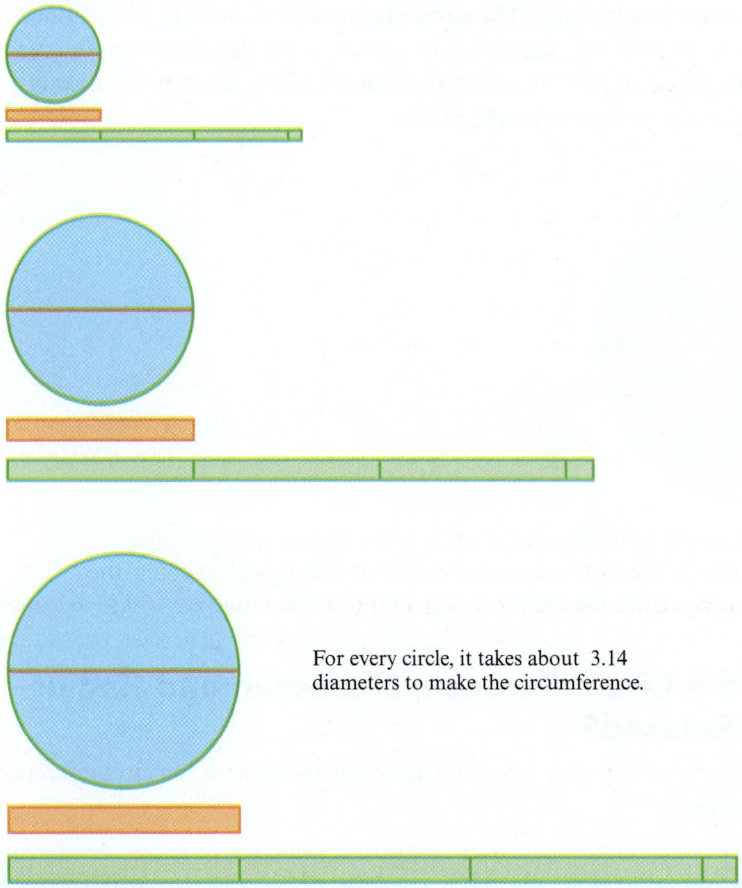

Figure 12.77 For every circle, the circumference is π diameters long.

Therefore D and C are related by the equation

circumference formula

$$\pi \cdot D = C$$

Explained another way, because 1 part is D units long, π parts are $\pi \cdot D$ units long. Because the circumference is π parts, it is $\pi \cdot D$ units long, so $C = \pi \cdot D$. Using the conventions of algebra, we can drop the multiplication symbol and write the formula as

circumference formula

$$C = \pi D$$

Notice that the relationship $C = \pi \cdot D$ between a circle's circumference, C, and diameter, D, holds for all circles, no matter what size. So we can think of C and D as varying together for all circles, small and large. We can therefore view the equation $C = \pi \cdot D$ as expressing a proportional relationship whose **constant of proportionality** is the number π.

constant of proportionality

Sometimes we want to relate the circumference, C, of a circle to its radius r instead of to its diameter D. How can we do that? The diameter of a circle is 2 times its radius, so

$$D = 2r$$

Therefore, $C = \pi \cdot 2r$. We usually rearrange the factors in this formula to express it as

circumference formula

$$C = 2\pi r$$

How can we use a circumference formula? If you plan to make a circular garden with a radius of 10 feet, and you want to enclose the garden with a fence, then how long a fence will you need? You will need C feet of fence, where C is the circumference of a circle of radius 10 feet. Therefore you will need $C = 2\pi \cdot 10 = 20\pi$ feet, or about 63 feet of fence.

How Are the Area and Radius of Circles Related?

Before you read on, try Class Activity 12N, which will help you explain a relationship between the area and radius of a circle.

CLASS ACTIVITY

12N Where Does the Area Formula for Circles Come From?, p. CA-253

CCSS 7.G.4

In Class Activity 12N, which you will review in the practice exercises, you saw that you could cut a circle apart and rearrange it into a shape that is approximately a rectangle. According to the moving and additivity principles about area, that approximate rectangle has the same area as the circle. By finding the area of the rectangle, you could determine that if a circle has radius r units and area A square units, then A and r are related by the equation

$$A = \pi r^2$$

circle area formula

That is just one of several ways to explain the **circle area formula**.

How can we use the circle area formula? If you plan to make a circular patio of diameter 30 feet, then what will the area of this patio be? Because the diameter is 30 feet, the radius is 15 feet. Therefore, if the area of the patio is A square feet, then $A = \pi \cdot 15^2 = 225\pi$, which is about 707 square feet.

When you use the πr^2 area formula, be sure that you square only the value of r, and not the value of πr. For example, if you multiply π times r first, and then square that result, your answer will be π times too large.

CLASS ACTIVITY

12O Area Problems, p. CA-254

SECTION SUMMARY AND STUDY ITEMS

Section 12.6 Area and Circumference of Circles and the Number Pi

For any circle, its circumference measured by its diameter is always the same number; this number is π (pi) and it is approximately 3.14. So the circumference of a circle of diameter D is $\pi \cdot D$ and the circumference of a circle of radius r is $2\pi r$. The area of a circle of radius r is πr^2. We can see why this area formula is plausible by subdividing a circle into "pie pieces" and rearranging them to form an approximate rectangle (or parallelogram) of dimensions r by πr.

Key Skills and Understandings

1. Know how the number π is defined, and use the definition to explain circumference formulas. Explain that π is a constant of proportionality in the relationship between the diameters and circumferences of circles.
2. Explain why the $A = \pi r^2$ formula for the area A square units of a circle of radius r units is plausible by subdividing a circle and rearranging pieces.
3. Use the formulas for circumference and area of a circle to determine lengths and areas and to solve problems.

Practice Exercises for Section 12.6

1. Suppose students don't yet know the formula for the area of a circle, but they do know about areas of squares. What can they deduce about the area of a circle of radius r units from Figure 12.78?

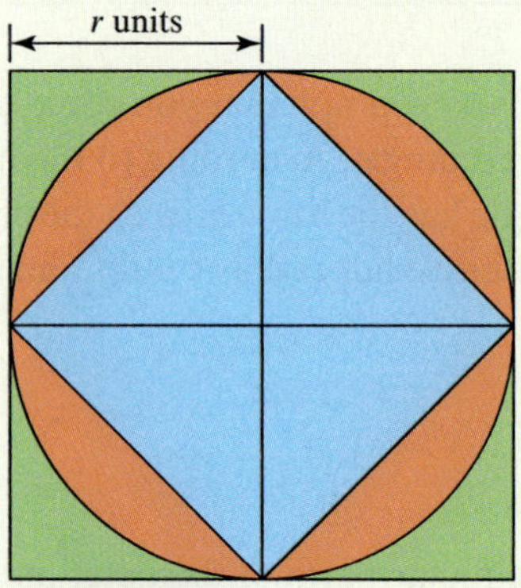

Figure 12.78 Estimating the area of the circle.

2. Using Figure 12.79, explain why it makes sense that a circle of radius r units has area πr^2 square units, assuming we already know that a circle of radius r has circumference $2\pi r$.

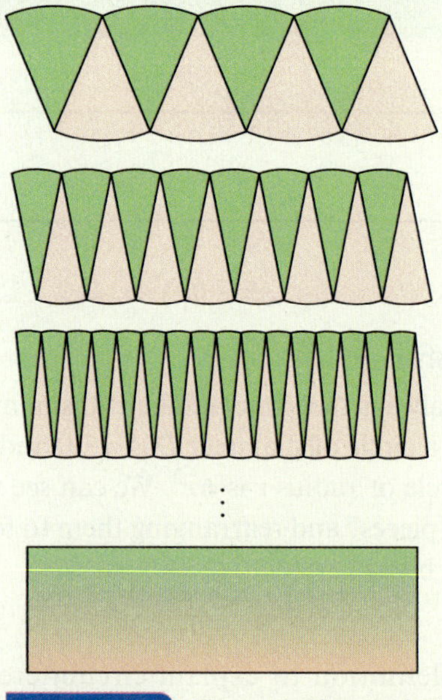

Figure 12.79 Rearranging a circle.

3. Some trees in an orchard need to have their trunks wrapped with a special tape to prevent an attack of pests. Each tree's trunk is about 1 foot in diameter and must be covered with tape from ground level up to a height of 4 feet. The tape is 3 inches wide. Approximately how long a piece of tape will be needed for each tree?

4. The Browns plan to build a 5-foot-wide garden path around a circular garden of diameter 25 feet, as shown in Figure 12.80.

   What is the area of the garden path? Explain your answer.

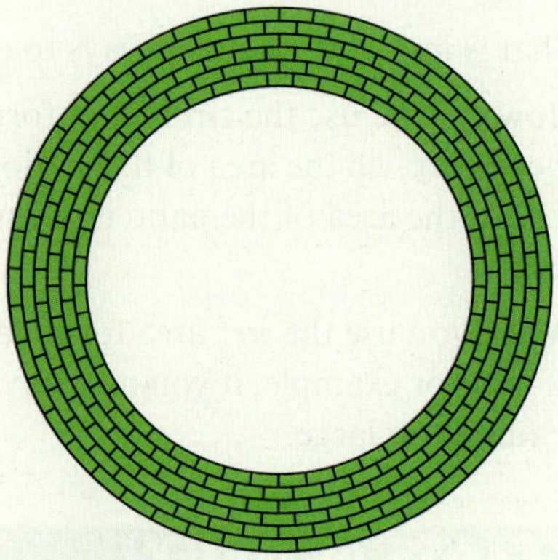

Figure 12.80 A garden path.

5. What is the area of the 4-petal flower in Figure 12.81? The square is 6 cm by 6 cm, and the curves are from half-circles.

Figure 12.81 A flower design.

Answers to Practice Exercises for Section 12.6

1. We can conclude that the area of the circle is between $2r^2$ square units and $4r^2$ square units. The area of the circle is less than $4r^2$ because we can think of the outer square in Figure 12.78 as made up of four r-by-r squares. Since the circle lies completely inside this outer square, the area of

the circle must be less than the area of the outer square. Therefore, the area of the circle is less than $4r^2$ square units. We can subdivide the inner square (diamond) into four triangles, which we can recombine to make two r-by-r squares. Since the circle completely contains the inner square, the area of the circle must be greater than $2r^2$ square units. Just by eyeballing, the area of the circle looks to be roughly halfway between these underestimates and overestimates, so the area of the circle looks to be roughly $3r^2$ (which is pretty close to the actual πr^2).

2. If you cut a circle into 8 "pie pieces" and rearrange them as at the top of Figure 12.79, you get a shape that looks something like a rectangle or a parallelogram. If you cut the circle into 16 or 32 pie pieces and rearrange them as in the middle of Figure 12.79, you get shapes that look even more like rectangles. If you could keep cutting the circle into more and more "pie pieces," and keep rearranging them as before, you would get shapes that look more and more like the rectangle shown at the bottom of Figure 12.79.

   The height of the rectangle in Figure 12.79 is the radius r of the circle. To determine the width of the rectangle (in the horizontal direction), notice that in the rearranged circles at the top and in the middle of Figure 12.79, half of the pie pieces point up and half point down. The circumference of the circle is divided equally between the top and bottom sides of the rectangle. The circumference of the circle is $2\pi r$; therefore, the width (in the horizontal direction) of the rectangle is half as much, which is πr. So the rectangle is r units by πr units, and therefore has area $\pi r \cdot r = \pi r^2$ square units. Since the rectangle is basically a cut-up and rearranged circle of radius r, the area of the rectangle ought to be equal to the area of the circle. Therefore, it makes sense that the area of the circle is also πr^2.

3. One "wind" of tape all the way around a tree trunk makes an approximate circle. Because the tree trunk has diameter 1 foot, each wind around the trunk uses about π feet of tape. The tape is 3 inches wide, so it will take 4 winds for each foot of trunk height to be covered. Therefore, it will take 16 winds to cover the desired amount of trunk. This will use about $16 \cdot \pi$ ft, or about 50 ft of tape.

4. The diameter of the circular garden is 25 feet, so its radius is half as much, which is 12.5 feet. The garden together with the path form a larger circle of radius $(12.5 + 5)$ feet $= 17.5$ feet. By the additivity principle about areas,

$$\begin{aligned}\text{area of garden} + \text{area of path}\\ = \text{area of larger circle}\end{aligned}$$

   Therefore,

$$\begin{aligned}\text{area of path} &= \text{area of larger circle}\\ &\quad - \text{area of garden}\\ &= \pi 17.5^2\ \text{ft}^2 - \pi 12.5^2\ \text{ft}^2\end{aligned}$$

   So the area of the path is about 471 ft^2.

5. You can make the design by covering the square with four half-circles of tissue paper; then the flower petals are exactly the places where *two* pieces of tissue paper overlap. The remaining parts of the square are covered with only *one* layer of tissue paper. Therefore,

$$\begin{aligned}\text{area of four half-circles} &= \text{area of flower}\\ &\quad + \text{area of square}\end{aligned}$$

   So that

$$\begin{aligned}\text{area of flower} &= \text{area of two circles}\\ &\quad - \text{area of square}\\ &= (18\pi - 36)\ \text{cm}^2\end{aligned}$$

   So the area of the flower is about 20.5 cm^2.

PROBLEMS FOR SECTION 12.6

1. In your own words, discuss how the diameter and circumference of all circles—small or large—are related. What is a hands-on way to see this relationship? If a circle has diameter D centimeters and circumference C centimeters, then what is an equation that relates D and C? Explain in detail why your equation holds for all circles, no matter what size.

2. Tim works on the following exercise:

   For each radius r, find the area of a circle of that radius:

$$r = 2\ \text{in.}, \quad r = 5\ \text{ft}, \quad r = 8.4\ \text{m}$$

   Tim gives the following answers:

   39.48, 246.74, 696.399

Identify the errors that Tim has made. How did Tim likely calculate his answers? Discuss how to correct the errors; include a discussion on the proper use of a calculator in solving Tim's exercise. Be sure to discuss the appropriate way to write the answers to the exercise.

3. A large running track is constructed to have straight sections and two semicircular sections with dimensions given in Figure 12.82. Assume that runners always run on the inside line of their lane. A race consists of one full counterclockwise revolution around the track plus an extra portion of straight segment, to end up at the finish line shown. What should the distance x between the two starting blocks be in order to make a fair race? Explain your reasoning.

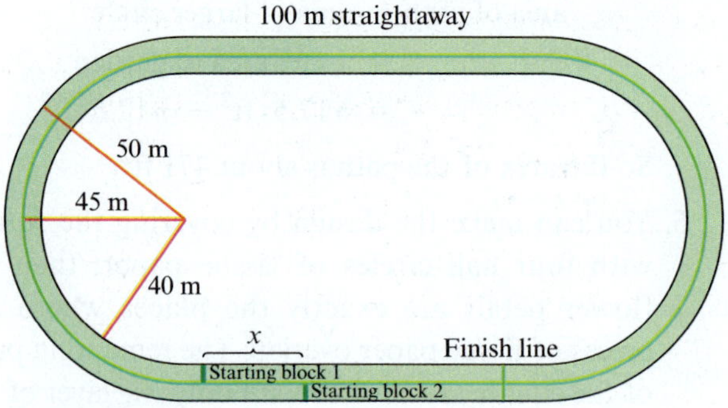

Figure 12.82 A running track.

4. Suppose you have a large spool used for winding rope (just like a spool of thread), such as the one shown in Figure 12.83.

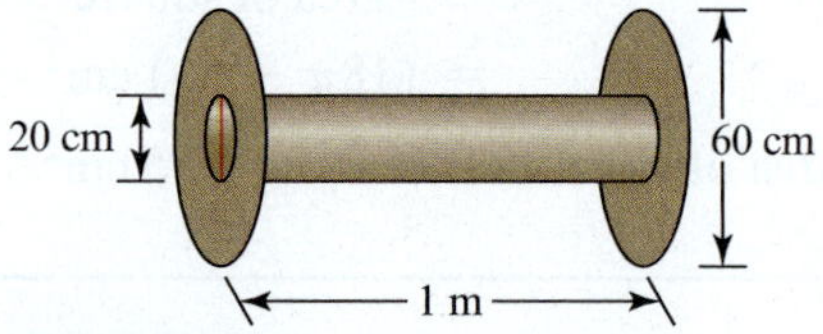

Figure 12.83 A large spool.

Suppose that the spool is 1 m long and has an inner diameter of 20 cm and an outer diameter of 60 cm. Approximately how long a piece of 5-cm-thick rope can be wound onto this spool? (Assume that the rope is wound on neatly, in layers. Each layer will consist of a row of "winds," and each "wind" will be approximately a circle.) Explain your reasoning.

5. Suppose that when pizza dough is rolled out it costs 25 cents per square foot, and that sauce and cheese, when spread out on a pizza, have a combined cost of 60 cents per square foot. Let's say sauce and cheese are always spread out to within 1 inch of the edge of the pizza. Compare the sizes and costs of a circular pizza of diameter 16 inches and a 10-inch-by-20-inch rectangular pizza.

6. Lauriann and Kinsey are in charge of the annual pizza party. In the past, they've always ordered 12-inch-diameter round pizzas, and each 12-inch pizza has always served 6 people. This year, the jumbo 16-inch-diameter round pizzas are on special, so Lauriann and Kinsey decide to get 16-inch pizzas instead. They think that since a 12-inch pizza serves 6 (which is half of 12), a 16-inch pizza should serve 8 (which is half of 16). But when Lauriann and Kinsey see a 16-inch pizza, they think it ought to serve even more than 8 people. Suddenly, Kinsey realizes the flaw in their reasoning that a 16-inch pizza should serve 8. Kinsey has an idea for determining how many people a 16-inch pizza will serve. What mathematical reasoning might Kinsey be thinking of, and how many people should a 16-inch-diameter pizza serve if a 12-inch-diameter pizza serves 6? (See Figure 12.84.) Explain your answers.

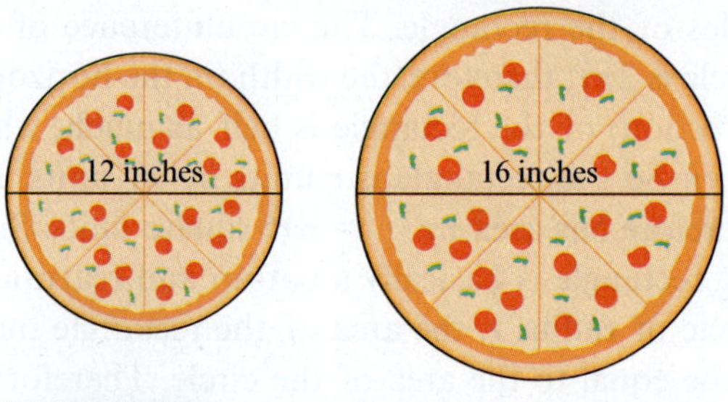

Figure 12.84 A 12-inch-diameter pizza and a 16-inch-diameter pizza.

7. Penguins huddle together to stay warm in very cold weather. (See Figure 12.85.) Suppose that a certain type of penguin has a circular cross-section approximately 14 inches in diameter (so that if you looked down on the penguin from above, the shape you would see would be a circle, 14 inches in diameter). Suppose that a group of this type of penguin is huddling in a large circular cluster, about 20 feet in diameter. (All the penguins are still standing upright on the ground; they are not piled on top of each other.)

Figure 12.85 Penguins in a huge cluster.

a. Assuming that the penguins are packed together tightly, estimate how many penguins are in this cluster. (You might use areas to do this.) Is this an overestimate or an underestimate? Explain.

b. The coldest penguins in the cluster are the ones around the circumference. Approximately how many of these cold penguins are there at any given time? Explain.

c. So that no penguin gets too cold, the penguins take turns being at the circumference. How many minutes per hour does each penguin spend at the circumference if each penguin spends the same amount of time at the circumference? Explain.

***8.** Jack has a truck that requires tires that are 26 inches in diameter. (Looking at a tire from the side of a car, a tire looks like a circle. The diameter of the tire is the diameter of this circle.) Jack puts tires on his truck that are 30 inches in diameter.

a. A car's speedometer works by detecting how fast the car's tires are rotating. Speedometers do not detect how big a car's tires are. When Jack's speedometer reads 60 miles per hour is that accurate, or is Jack actually going slower or faster? Explain your reasoning. An exact determination of Jack's speed is not needed.

b. Determine Jack's speed when his speedometer reads 60 mph. Explain why you can solve the problem the way you do.

***9.** Let r units denote the radius of each circle in Figure 12.86. For each shaded circle portion in this figure, find a formula for its area in terms of r. Use the moving and additivity principles about areas to explain why your formulas are valid.

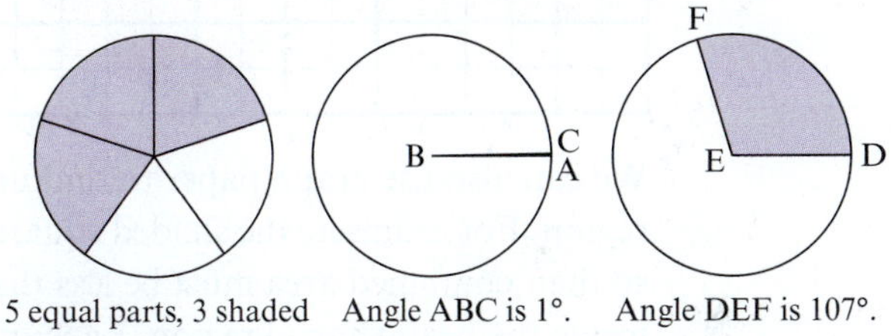

Figure 12.86 Parts of circles.

12.7 Approximating Areas of Irregular Shapes

Moving and additivity principles can help us to find the areas of some shapes and to explain why area formulas are valid, but the principles can't always help us find a precise area for irregular shapes. For example, how could we determine the area of the bean-shaped region in Figure 12.87? There isn't any way to subdivide and recombine the bean-shaped region into regions whose areas are easy to determine. Usually, we cannot determine the area of an irregular region exactly; instead, we must be satisfied with determining the approximate area.

Figure 12.87 A bean-shaped region.

CLASS ACTIVITY

12P Determining the Area of an Irregular Shape, p. CA-255

CCSS SMP5

One hands-on way to approximate the area of an irregular region, such as the bean-shaped region in Figure 12.87, is to cut it out, weigh it, compare its weight with the weight of a full piece of paper, and reason proportionally to find the area of the bean-shaped region. For example, if the irregular region weighs $\frac{3}{8}$ as much as a whole piece of paper, then its area is also $\frac{3}{8}$ as much as the area of the whole piece of paper.

We can also determine the area of an irregular region approximately by using graph paper, as indicated in Figure 12.88. The lines on this graph paper are spaced $\frac{1}{2}$ cm apart, with heavier lines spaced 1 cm apart (so that 4 small squares make 1 square centimeter).

By counting the number of 1-cm-by-1-cm squares (each consisting of 4 small squares) inside the bean-shaped region, and by mentally combining the remaining portions of the region that are near the boundary, we can determine that the bean-shaped region has an area of about 19 square centimeters.

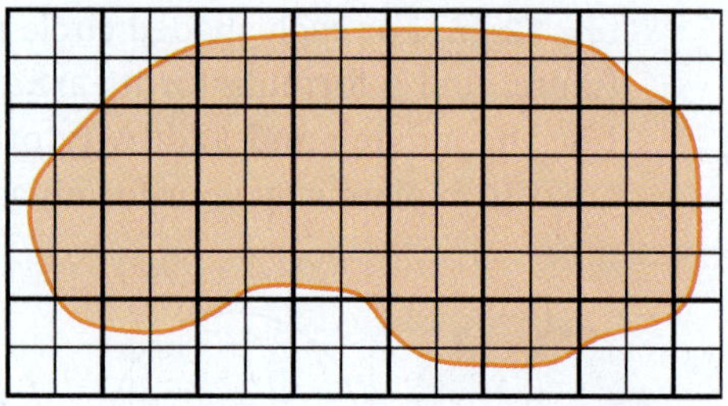

Figure 12.88 Determining the area of a bean-shaped region.

We can also use graph paper to find underestimates and overestimates for the area of an irregular region. For example, the shaded squares in Figure 12.89 lie entirely within the bean-shaped region, so their combined area must be less than the area of the region. There are 58 small squares that lie inside the bean-shaped region. Each of these small squares is $\frac{1}{2}$ cm by $\frac{1}{2}$ cm and thus has an area of

$$\left(\frac{1}{2}\cdot\frac{1}{2}\right)\text{cm}^2 = \frac{1}{4}\text{cm}^2$$

Therefore, the 58 small squares have an area of

$$58\cdot\frac{1}{4}\text{cm}^2 = 14\frac{1}{2}\text{cm}^2$$

So the area of the shaded region must be greater than $14\frac{1}{2}$ cm^2.

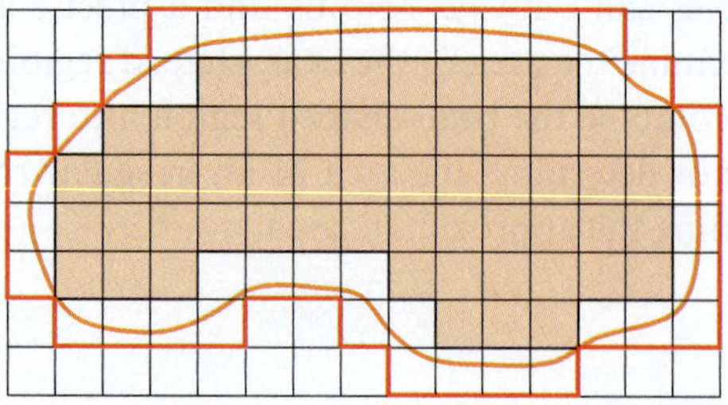

Figure 12.89 Determining underestimates and overestimates for the area of the bean-shaped region.

On the other hand, the colored outline in Figure 12.89 surrounds the squares that contain some portion of the bean-shaped region. The combined area of these squares must therefore be greater than the area of the region. There are 97 such small squares, each of which has area $\frac{1}{4}$ cm^2; thus, these 97 squares have a combined area of

$$97\cdot\frac{1}{4}\text{cm}^2 = 24\frac{1}{4}\text{cm}^2$$

The area of the bean-shaped region must be less than $24\frac{1}{4}$ cm^2. So the area of the bean-shaped region must be between $14\frac{1}{2}$ cm^2 and $24\frac{1}{4}$ cm^2. If we used finer and finer graph paper, we would get narrower and narrower ranges between our underestimates and overestimates for the area of the region.

SECTION SUMMARY AND STUDY ITEMS

Section 12.7 Approximating Areas of Irregular Shapes

When it comes to irregular shapes, we usually have to make do with estimating their areas, because we often cannot determine their areas exactly. We can determine the approximate area of a shape by overlaying it with a grid and determining approximately how many grid squares the shape takes up.

Key Skills and Understandings

1. Use grids to determine approximate areas of shapes, including shapes shown on maps with a scale.
2. Use other informal methods, such as working with modeling dough and weighing card stock, to determine approximate areas of shapes.

Practice Exercises for Section 12.7

1. a. In Figure 12.90, the grid lines are $\frac{1}{2}$ cm apart. Find an underestimate and an overestimate for the area of the shaded region by considering the squares that lie entirely within the region, and the squares that contain a portion of the region.

b. Determine approximately the area of the shaded region in Figure 12.90.

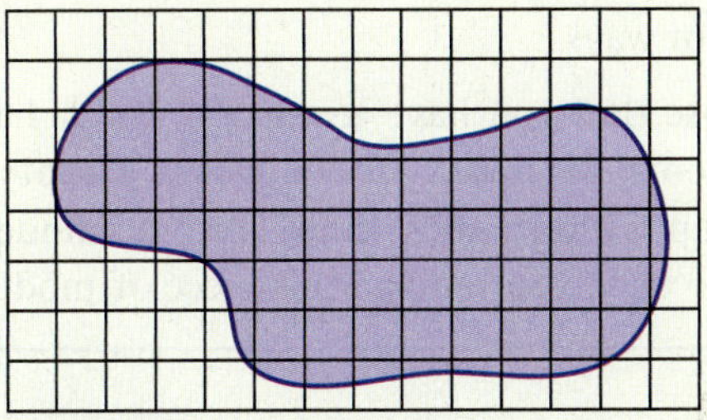

Figure 12.90 Finding an underestimate and overestimate of the area.

2. Suppose that you have a map on which 1 inch represents 50 miles. You trace a state on the map onto $\frac{1}{2}$-inch graph paper. (The grid lines are spaced $\frac{1}{2}$ inch apart.) You count that the state takes up about 91 squares of graph paper. Approximately what is the area of the state? Explain.

Answers to Practice Exercises for Section 12.7

1. a. There are 41 small squares that are contained completely within the shaded region. Each small square is $\frac{1}{2}$ cm by $\frac{1}{2}$ cm; therefore, the area of each small square is $\frac{1}{2}\cdot\frac{1}{2}$ square centimeters, which is $\frac{1}{4}$ cm^2. Consequently, an underestimate for the area of the region is

$$41\cdot\frac{1}{4}\text{cm}^2 = 10\frac{1}{4}\text{cm}^2$$

There are 71 small squares that contain some portion of the region. Therefore, an overestimate for the area of the region is

$$71\cdot\frac{1}{4}\text{cm}^2 = 17\frac{3}{4}\text{cm}^2$$

b. Counting whole squares and combining partial squares, we find that the area is approximately the area of 55 squares. Since each square has an area of $\frac{1}{4}$cm^2, the area of the region is approximately

$$55\cdot\frac{1}{4}\text{cm}^2 = 13\frac{3}{4}\text{cm}^2$$

or about 14 square centimeters.

2. Each square on the graph paper is $\frac{1}{2}$ inch by $\frac{1}{2}$ inch; therefore, the area of each square of graph paper is $\frac{1}{2}\cdot\frac{1}{2}$ square inches, which is $\frac{1}{4}$ square inch.

So, 91 squares of graph paper have a combined area of

$$91 \cdot \frac{1}{4}\text{ in}^2 = 22\frac{3}{4}\text{ in}^2$$

Since 1 inch on the map represents 50 miles, 1 square inch on the map represents

$$50 \cdot 50 \text{ mi}^2 = 2500 \text{ mi}^2$$

of actual land. Thus, $22\frac{3}{4}$ square inches on the map represents

$$22\frac{3}{4} \cdot 2500 \text{ mi}^2 = 56{,}875 \text{ mi}^2$$

or about 57,000 square miles.

PROBLEMS FOR SECTION 12.7

1. Suppose that you have a map on which 1 inch represents 100 miles. You trace a state on the map onto $\frac{1}{4}$-inch graph paper. (The grid lines are spaced $\frac{1}{4}$ inch apart.) You count that the state takes up about 80 squares of graph paper. Approximately what is the area of the state? Explain how to determine this area in two distinctly different ways.

2. Suppose that you have a map on which 1 inch represents 25 miles. You cover a county on the map with a $\frac{1}{8}$-inch-thick layer of modeling dough. Then you re-form this piece of modeling dough into a $\frac{1}{8}$-inch-thick rectangle. The rectangle is $1\frac{3}{4}$ inches by $2\frac{1}{4}$ inches. Approximately what is the area of the county? Explain.

3. Suppose that you have a map on which 1 inch represents 30 miles. You trace a state on the map, cut out your tracing, and draw this tracing onto card stock. Using a scale, you determine that a full $8\frac{1}{2}$-inch-by-11-inch sheet of card stock weighs 10 grams. Then you cut out the tracing of the state that is on card stock and weigh this card stock tracing. It weighs 5 grams. Approximately what is the area of the state? Explain.

12.8 Contrasting and Relating the Perimeter and Area of Shapes

CCSS Common Core State Standards Grades 3, 4

If you know the distance around a shape, can you determine its area? If you know the area of a shape, can you determine the distance around the shape? What is the difference between perimeter and area? Children learn about area and perimeter and the distinction between them in elementary school. As we will see, a deeper investigation into the relationship between area and perimeter is more advanced.

Students sometimes get confused when deciding how to calculate perimeter and area. We *add* to calculate perimeter but we *multiply* to calculate areas of rectangles. When we calculate the perimeter of a rectangle, we add the lengths of all four sides (or we add the lengths of two adjacent sides and multiply by 2). On the other hand, when we calculate the area of a rectangle, we multiply only two of the sides' lengths. It's hard to keep all these different methods of calculation straight unless we have a clear idea of what perimeter and area mean and how they are different. But although they are different, there *is* a relationship between perimeter and area, as we will see in this section.

How Can We Determine Perimeters of Polygons?

CCSS 3.MD.8 4.MD.3

Why do we add the lengths of the sides of a polygon to calculate its perimeter? The perimeter of a polygon (or other shape in the plane) is the total distance around the polygon. A hands-on way to think about perimeter is as the length of a string that wraps snugly once around the shape. When we calculate the perimeter of a polygon by adding the lengths of the sides around the polygon, it is as if we had cut the string into pieces and are determining the total length of the string by adding the lengths of the pieces. We are calculating the total distance around the shape by adding up the lengths of pieces that together encircle the shape.

CLASS ACTIVITY

12Q Critique Reasoning about Perimeter, p. CA-257

12R Find and Explain Perimeter Formulas for Rectangles, p. CA-258

How Is Perimeter Different from Area?

Note the difference between perimeter and area. The shape in Figure 12.91 has perimeter 26 cm, but it has area 21 cm^2. The perimeter of a shape is described by a unit of length, such as centimeters, whereas the area of a shape is described by a unit of area, such as square centimeters. The perimeter of the shape in Figure 12.91 is the number of 1-cm *segments* it takes to go all the way *around the shape*, whereas the area of the shape in Figure 12.91 is the number of 1-cm-by-1-cm *squares* it takes to *cover the shape* (without gaps or overlaps).

Perimeter and area are different, but they *are* related in a complex way.

Figure 12.91

A shape of perimeter 26 cm and area 21 cm^2.

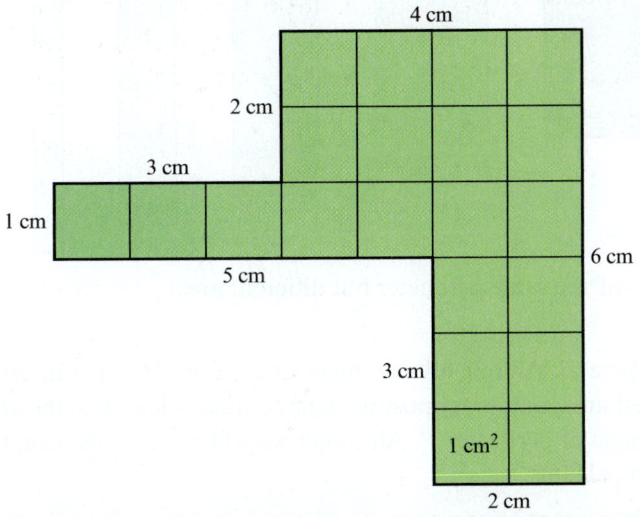

Perimeter = 4 cm + 6 cm + 2 cm + 3 cm + 5 cm + 1 cm + 3 cm + 2 cm = 26 cm

Area = 21 • 1 cm^2 = 21 cm^2

CLASS ACTIVITY

12S How Are Perimeter and Area Related for Rectangles?, p. CA-258

12T How Are Perimeter and Area Related for All Shapes?, p. CA-260

FROM THE FIELD **Research**

Wickstrom, M. H., Nelson, J., & Chumbley, J. (2015). Area conceptions sprout on Earth day. *Teaching Children Mathematics, 21*(8), pp. 466-474.

The authors discuss a measurement and plant growth lesson designed to help second and third graders think about area in a way other than with a formula. Students predicted areas of different garden plots made with the same amount of fencing and used several different strategies to find the areas of the plots.

If We Know the Perimeter What Can We Say About the Area?

CCSS
3.MD.8
4.MD.3

If you did Class Activities 12S and 12T, you discovered that for a given, fixed perimeter, there are many shapes that can have that perimeter, and these shapes can have different areas. Therefore, perimeter does not determine area. But, for a given, fixed perimeter, which areas can occur? The answer depends on which shapes we are considering.

If we consider only *rectangles* of a given, fixed perimeter, what can we say about their areas? If you did Class Activity 12S, then you probably discovered that, of all rectangles of a given, fixed perimeter, the one with the largest area is a square and the greater the difference between the side lengths, the smaller the area becomes. (See Figure 12.92.) For example, among all rectangles of perimeter 24 inches, a square that has four sides of length 6 inches has the largest area, and this area is $6 \cdot 6 \text{ in}^2 = 36 \text{ in}^2$ By moving a 24-inch loop of string to form various rectangles, you can probably tell that *every* positive number less than 36 is the area (in square inches) of *some* rectangle whose perimeter is 24 inches. So, for example, there is a rectangle that has perimeter 24 in. and area 35.723 in^2, and there is a rectangle that has perimeter 24 in. and area 3.72 in^2, even though it would take some work to find the exact lengths and widths of such rectangles.

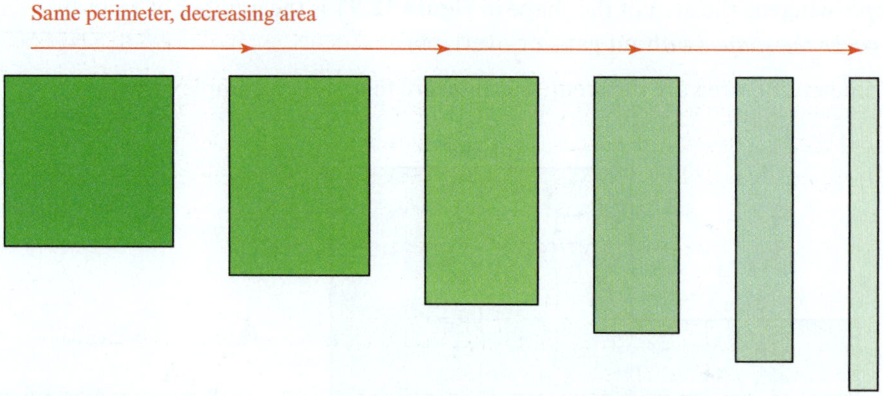

Figure 12.92 Rectangles of the same perimeter but different areas.

The following is true in general: Among all rectangles of a given, fixed perimeter *P*, the square of perimeter *P* has the largest area, and every positive number that is less than the area of that square is the area of some rectangle of perimeter *P*. Although we will not do this here, these facts can be explained with algebra or calculus.

What if we consider *all* shapes that have a given, fixed perimeter? Your intuition probably tells you that of all shapes having a given, fixed perimeter, the circle is the one with the largest area. For example, of all shapes with perimeter 15 inches, a circle of circumference 15 inches is the shape that has the largest area. Since the circumference of this circle is 15 inches, its radius is

$$15 \div 2\pi \text{ in.} = \frac{15}{2\pi} \text{ in.}$$

which is about 2.4 in. Its area is

$$\pi \cdot \left(\frac{15}{2\pi}\right)^2 \text{in}^2$$

which is about 18 in^2. By moving a 15-inch loop of string into various positions on a flat surface, you will probably find it plausible that *every* positive number less than 18 is the area, in square inches, of *some* shape of perimeter 15 in. So, for example, there is a shape that has perimeter 15 in. and area 17.35982 in^2, and there is a shape that has perimeter 15 in. and area 2.7156 in^2. Notice that we can say this *without actually finding shapes* that have those areas and perimeters, which could be quite a challenge.

The following is true in general: Among all shapes of a given, fixed perimeter P, the circle of circumference P has the largest area, and every positive number that is less than the area of that circle is the area of some shape of perimeter P. So, although perimeter does not *determine* area, it does *constrain* which areas can occur. Explanations for why these facts are true are beyond the scope of this book.

SECTION SUMMARY AND STUDY ITEMS

Section 12.8 Contrasting and Relating the Perimeter and Area of Shapes

The perimeter of a polygon (or other shape) is the total distance around the polygon. When we calculate the perimeter of a polygon by adding the lengths of the sides, it is as if we had wrapped a string snugly around the polygon and cut it into pieces corresponding to each side.

For a given, fixed perimeter, there are many shapes that have the same perimeter but have different areas. Thus, perimeter does not determine area. Among all rectangles of a given, fixed perimeter, the square with that perimeter has the largest area. Given a fixed perimeter, every area less than or equal to the area of the square of that perimeter is the area of some rectangle of that perimeter. Among all closed shapes of a given, fixed perimeter, the circle of that circumference has the largest area. Given a fixed perimeter, every area less than or equal to the area of the circle of that circumference is the area of some shape of that perimeter.

Key Skills and Understandings

1. Explain why we calculate perimeters of polygons the way we do. Discuss misconceptions with perimeter calculations.
2. Recognize that perimeter does not determine area.
3. Given a fixed perimeter, determine the areas of all rectangles of that perimeter and determine the areas of all shapes of that perimeter.

Practice Exercises for Section 12.8

1. A rectangle has adjacent sides of lengths x cm and y cm, as in Figure 12.93. Find and explain a formula for the perimeter P of the rectangle in centimeters.

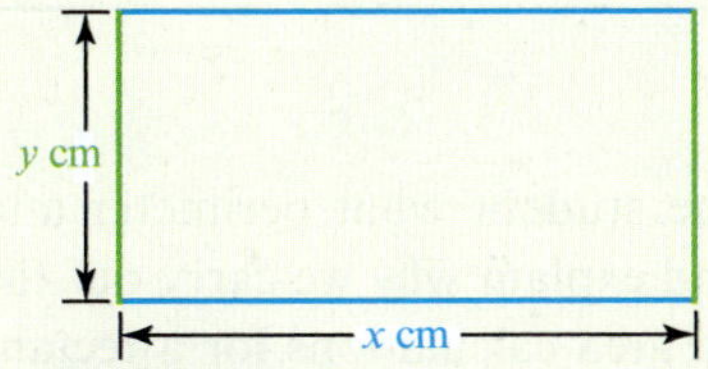

Figure 12.93 What is the perimeter of a rectangle whose adjacent sides are x cm and y cm long?

2. The students in a class have been learning about the perimeter and area of rectangles. The teacher notices that one student consistently gives the wrong answer to problems on the area of rectangles: The student's answer is almost always 2 times the correct answer. For example, if the correct answer is 12 square feet, the student gives the answer 24 square feet. What might be the source of this error?

3. Describe a common error that students make when finding the perimeter of a shape such as the shaded shape in Figure 12.94. Describe the misunderstanding that is at the root of this error.

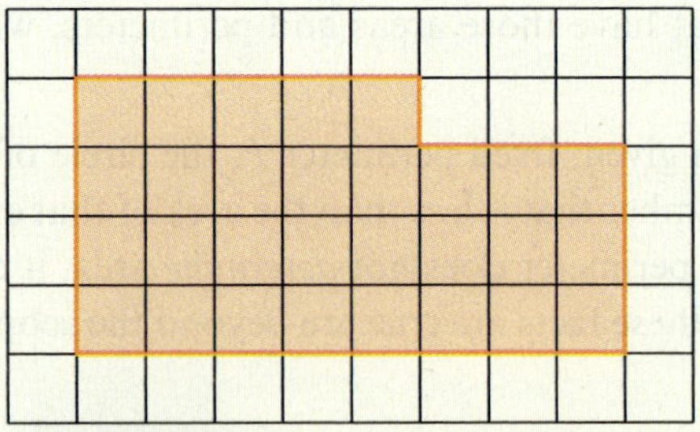

Figure 12.94 What error do students often make concerning perimeter?

4. A piece of property is described as having a perimeter of 4.7 miles. Without any additional information about the property, what is the most informative answer you can give about the area of the property? If you assume that the property is shaped like a rectangle, then what is the most informative answer you can give about its area?

Answers to Practice Exercises for Section 12.8

1. The perimeter of the rectangle is the distance around the rectangle, which is the sum of the lengths of the sides of the rectangle. The two horizontal sides are x cm long and the two vertical sides are y cm long. Therefore the total length of all the sides is $2x + 2y$ and so

$$P = 2x + 2y$$

2. If the student had been calculating perimeters of rectangles by adding the lengths of 2 adjacent sides of the rectangle and multiplying this result by 2 (to account for the other 2 sides), then the student might be attempting to calculate area in an analogous way. Instead of just multiplying the length times width of the rectangle, the student may think that the result must be multiplied by 2, as it was for the perimeter.

3. Students sometimes count the number of squares along the outer border of the shape, as in Class Activity 12R. When they do so, they are not treating perimeter as a one-dimensional attribute. When students count squares, they are determining the area of the region formed by the border squares, which is a two-dimensional attribute.

4. Among all shapes that have perimeter 4.7 miles, the circle with circumference 4.7 miles has the largest area. This circle has radius $4.7 \div 2\pi$ mi^2, which is approximately 0.75 mi^2. Therefore the circle has area approximately $(\pi \cdot 0.75^2)$ mi^2 or about 1.8 mi^2.

Any area that is less than the area of this circle is a possible area for the property. So, without any additional information, the best we can say about the property is that its area is at most 1.8 square miles and the actual area could be anywhere between 0 and 1.8 square miles.

Now suppose that the property is shaped like a rectangle. Among all rectangles of perimeter 4.7 miles, the square of perimeter 4.7 miles has the largest area. This square has 4 sides of length $4.7 \div 4$ miles $= 1.175$ miles, and therefore this square has area $1.175 \cdot 1.175$ mi^2, or about 1.4 mi^2. Any area that is less than the area of this square is a possible area for the property. So, if we know that the property is in the shape of a rectangle, then the most informative answer we can give about the area of the property is that its area is at most 1.4 square miles, and the actual area could be anywhere between 0 and 1.4 square miles.

PROBLEMS FOR SECTION 12.8

1. Suppose that a student in your class wants to know why we multiply only 2 of the lengths of the sides of a rectangle to determine the rectangle's area. After all, when we calculate the perimeter of a rectangle, we add the lengths of the *4 sides* of the rectangle, so why don't we multiply the lengths of the *4 sides* to find the area?

a. Explain to the student what perimeter and area mean, and explain why we carry out the perimeter and area calculations for a rectangle the way we do.

b. Describe some problems or activities that might help the student understand the calculations.

2. Sarah is confused about the difference between the perimeter and the area of a polygon. Explain the two concepts and the distinction between them.

3. a. Describe a concrete way to demonstrate that many different shapes can have the same perimeter.

b. Describe a concrete way to demonstrate that many different shapes can have the same area.

4. Anya wants to draw many different rectangles that have a perimeter of 16 cm. Anya draws a few rectangles, but then she stops drawing and starts looking for pairs of numbers that add to 8.

a. Why does it make sense for Anya to look for pairs of numbers that add to 8? How is she likely to use these pairs of numbers in drawing additional rectangles?

b. How could you adapt Anya's idea if you were going to draw many different rectangles that have a perimeter of 14 cm?

c. Use Anya's idea to help you draw 3 different rectangles that have a perimeter of 5 in. Label your rectangles with their lengths and widths.

5. a. On graph paper, draw 4 different rectangles that have perimeter $6\frac{1}{2}$ in.

b. Without using a calculator, determine the areas of the rectangles you drew in part (a). Show your calculations, or explain briefly how you determined the areas of the rectangles.

6. Which of the lengths that follow could be the length of 1 side of a rectangle that has perimeter 7 in.? In each case, if the length is a possible side length of a rectangle of perimeter 7 in., then determine the lengths of the other 3 sides without using a calculator. If the length is not a possible side length of such a rectangle, then explain why not. Show your calculations.

a. $2\frac{3}{4}$ in.

b. $5\frac{1}{2}$ in.

c. $1\frac{7}{8}$ in.

d. $3\frac{3}{8}$ in.

e. $3\frac{5}{8}$ in.

7. a. Without using a calculator, find the lengths and widths of 5 different rectangles that have perimeter $4\frac{1}{2}$ in. Show your calculations and explain them briefly.

b. Without using a calculator, find the areas of the 5 rectangles you found in part (a). Show your calculations.

8. a. Draw 4 different rectangles, all of which have a perimeter of 8 in. At least 2 of your rectangles should have side lengths that are not whole numbers (in inches). Label your rectangles with their lengths and widths.

b. Determine the areas of each of your 4 rectangles in part (a) without using a calculator. Show your calculations, or explain briefly how you determined the areas. Then label your rectangles A, B, C, and D in decreasing order of their areas, so that A has the largest area and D has the smallest area among your rectangles.

c. Qualitatively, how do the larger-area rectangles you drew in part (a) look different from the smaller-area rectangles? Describe how the shapes of the rectangles change as you go from the rectangle of largest area to the rectangle of smallest area.

9. a. Draw 4 different rectangles, all of which have area 4 square inches. Label your rectangles with their lengths and widths.

b. Determine the perimeters of each of your rectangles in part (a). Then label your rectangles A, B, C, and D in increasing order of their perimeters, so that A has the smallest perimeter and D has the largest perimeter among your rectangles.

c. Qualitatively, how do the smaller-perimeter rectangles you drew in part (a) look different from the larger-perimeter rectangles? Describe how the shapes of the rectangles change as you go from the rectangle of smallest perimeter to the rectangle of largest perimeter.

10. A forest has a perimeter of 210 mi, but no information is given about the shape of the forest. Justify your answers to the following (in all parts of this problem, the perimeter is still 210 mi):

a. Is it possible that the area of the forest is 3000 mi^2? Explain.

b. Is it possible that the area of the forest is 3600 mi^2? Explain.

c. If the forest is shaped like a rectangle, then is it possible that the area of the forest is 3000 mi^2? Explain.

d. If the forest is shaped like a rectangle, then is it possible that the area of the forest is 2500 mi? Explain.

11. Bob wants to find the area of an irregular shape. He cuts a piece of string to the length of the perimeter of the shape. He measures to see that the string is about 60 cm long. Bob then forms his string into a square on top of centimeter graph paper. Using the graph paper, he determines that the area of his string square is about 225 cm^2. Bob says that the area of the irregular shape is therefore also 225 cm^2. Is Bob's method for determining the area of the irregular shape valid or not? Explain. If the method is not valid, what can you determine about the area of the irregular shape from the information that Bob has? Explain.

***12.** Consider all rectangles whose *area* is 4 in^2, including rectangles that have sides whose lengths are not whole numbers. What are the possible perimeters of these rectangles? Is there a smallest perimeter? Is there a largest?

Answer this question either by pure thought or by actual examination of rectangles. Then write a paragraph describing your answer (including drawings, if relevant) and stating your conclusions clearly.

12.9 Using the Moving and Additivity Principles to Prove the Pythagorean Theorem

CCSS Common Core State Standards Grade 8

In this section, we will apply the moving and additivity principles about area to prove the Pythagorean theorem. So the very same ideas that students use to find areas and explain area formulas also apply to the Pythagorean theorem. The Pythagorean theorem is named for the Greek mathematician Pythagoras, who lived around 500 B.C. There is evidence, though, that this theorem was known long before that time, perhaps even by the ancient Babylonians around 2000 B.C.

What Does the Pythagorean Theorem Tell Us?

The Pythagorean theorem (or Pythagoras's theorem) is a theorem about *all* right triangles. Recall that, in a right triangle, the side opposite the right angle is called the hypotenuse. The **Pythagorean theorem** says,

Pythagorean theorem

> In a right triangle, the square of the length of the hypotenuse is equal to the sum of the squares of the lengths of the other two sides. In other words, if c is the length of the hypotenuse of a right triangle, and if a and b are the lengths of the other two sides, then

$$a^2 + b^2 = c^2 \qquad (12.1)$$

For example, in Figure 12.95, the right triangle on the left has sides of lengths a, b, and c, and the hypotenuse has length c, so, according to the Pythagorean theorem,

$$a^2 + b^2 = c^2$$

The right triangle on the right of Figure 12.95 has sides of lengths d, e, and f, and the hypotenuse has length f, so, in this case, according to the Pythagorean theorem,

$$d^2 + e^2 = f^2$$

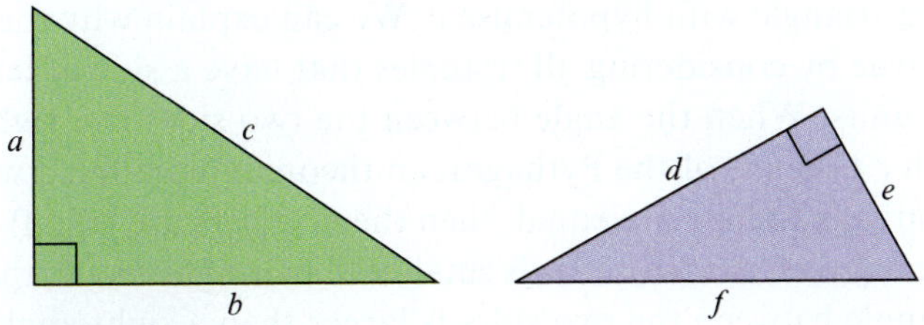

Figure 12.95 Right triangles.

In Equation 12.1 it is understood that *all sides of the triangle are expressed in the same units.* For example, all sides can be given in centimeters or all sides can be given in feet. If you have a right triangle where one side is measured in feet and the other in inches, for example, then you need to convert both sides to inches or to feet (or to some other unit) in order to use Equation 12.1.

Why Is the Pythagorean Theorem True?

If you have a particular right triangle in front of you, such as those in Figure 12.95, you can measure the lengths of its sides and check that the Pythagorean theorem really is true in that case. For example, in the triangle on the left of Figure 12.95, measured in centimeters,

$$a = 3, \quad b = 4, \quad \text{and} \quad c = 5.$$

So

$$a^2 + b^2 = 3^2 + 4^2 = 9 + 16 = 25, \quad \text{and} \quad c^2 = 5^2 = 25.$$

We see that $a^2 + b^2$ really is equal to c^2.

We could continue to check many right triangles to see if the Pythagorean theorem really is true in those cases. If we checked many triangles, it would be compelling evidence that the Pythagorean theorem is always true, but we could never check all the right triangles individually because they are infinite in number.

CCSS 8.G.6

One of the cornerstones of mathematics, an idea that dates back to the time of the mathematicians of ancient Greece, is that a lot of evidence for a statement is not enough to know that the statement is true. *Proof* is required to know for sure that a statement really is true. A **proof** is a thorough, precise, logical explanation for why a statement is true, based on assumptions or facts that we already know or assume to be true. To prove the Pythagorean theorem, we need to develop a general argument that will explain why it is true for *all* triangles.

proof

Professional and amateur mathematicians have found literally hundreds of proofs of the Pythagorean theorem. If you do Class Activity 12V, you will work through and discover one of these proofs. Class Activity 14Y in Chapter 14 and Problem 14 on page 664 in Section 14.6 guide you through two more proofs.

CLASS ACTIVITY

12U Side Lengths of Squares Inside Squares, p. CA-261

12V A Proof of the Pythagorean Theorem, p. CA-262

You will review a proof of the Pythagorean theorem in Practice Exercises 6 and 7.

What Is the Converse of the Pythagorean Theorem?

converse of the Pythagorean theorem

The **converse of the Pythagorean theorem** states that if a triangle has sides of length a, b, and c units, and if

$$a^2 + b^2 = c^2$$

then the triangle is a right triangle with hypotenuse c. We can explain why the converse of the Pythagorean theorem is true by considering all triangles that have a side of length a units and another side of length b units. When the angle between the two sides is a right angle, the hypotenuse will have length c because of the Pythagorean theorem (combined with the fact that if the squares of two positive numbers are equal, then the numbers are equal). As indicated in Figure 12.96, if the angle between the two sides is smaller than a right angle, then the third side is shorter than c; if the angle between the two sides is larger than a right angle, then the third side is longer than c.

CCSS 8.G.6

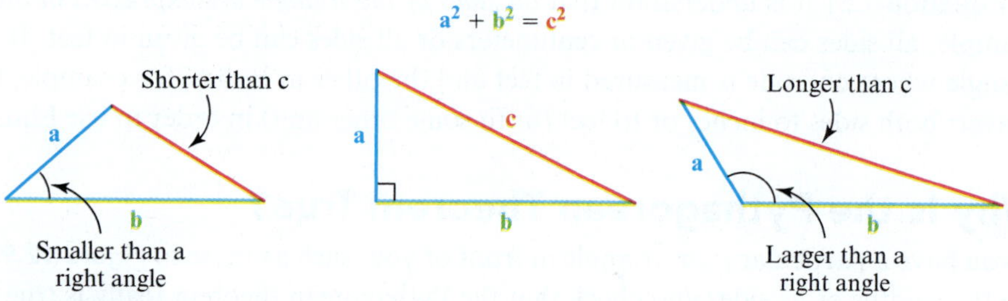

Figure 12.96 Triangles with sides of length a units and b units.

The converse of the Pythagorean theorem is useful for making right angles. For example, if you construct a triangle whose three sides are 5 feet, 12 feet, and 13 feet long, then because $5^2 + 12^2 = 25 + 144 = 169 = 13^2$, the triangle must have a right angle between the 5 foot side and the 12 foot side.

FROM THE FIELD Children's Literature

Ellis, J. (2004). *What's your angle, Pythagoras?* Watertown, MA: Charlesbridge.

Pythagoras is intrigued by right triangles and their usefulness to builders and sailors. He discovers a pattern that helps him build ladders that are tall enough to reach the tops of buildings and find the quickest path when sailing on a boat.

SECTION SUMMARY AND STUDY ITEMS

Section 12.9 Using Moving and Additivity Principles to Prove the Pythagorean Theorem

The Pythagorean theorem states that in any right triangle, the square of the length of the hypotenuse is equal to the sum of the squares of the lengths of the other two sides. We can prove the Pythagorean theorem by applying the moving and additivity principles to two identical squares subdivided into four copies of the right triangle and additional squares.

Key Skills and Understandings

1. State and prove the Pythagorean theorem.
2. Use the Pythagorean theorem to determine lengths and distances.

Practice Exercises for Section 12.9

1. A garden gate that is 3 feet wide and 4 feet tall needs a diagonal brace to make it stable. How long a piece of wood will be needed for this diagonal brace? See Figure 12.97.

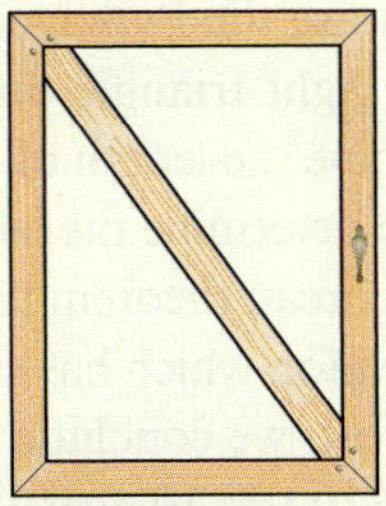

Figure 12.97 Garden gate with diagonal brace.

2. A boat's anchor is on a line that is 75 feet long. If the anchor is dropped in water that is 50 feet deep, then how far away will the boat be able to drift from the spot on the water's surface that is directly above the anchor? Explain.

3. An elevator car is 8 feet long, 6 feet wide, and 9 feet tall. What is the longest pole you could fit in the elevator? Explain.

4. Imagine a right pyramid with a square base (like an Egyptian pyramid). Suppose that each side of the square base is 200 yards and that the distance from one corner of the base to the very top of the pyramid (along an outer edge) is 245 yards. How tall is the pyramid? Explain.

5. The front and back of a greenhouse have the shape and dimensions shown in Figure 12.98. The greenhouse is 40 feet long from front to back, and the angle at the top of the roof is 90°. The entire roof of the greenhouse will be covered with screening to block some of the light entering the greenhouse. How many square feet of screening will be needed?

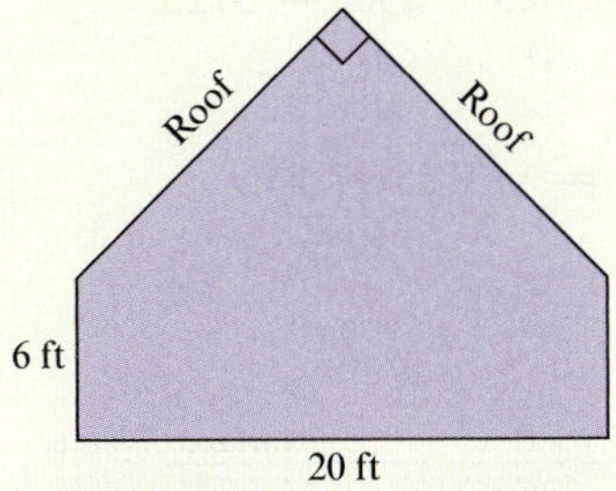

Figure 12.98 A greenhouse.

6. Given a right triangle with short sides of lengths a and b and hypotenuse of length c, use Figure 12.99 to explain why

$$a^2 + b^2 = c^2$$

(You may assume that all shapes that look like squares really are squares and that, in each picture, all four triangles with side lengths a, b, c are identical right triangles.)

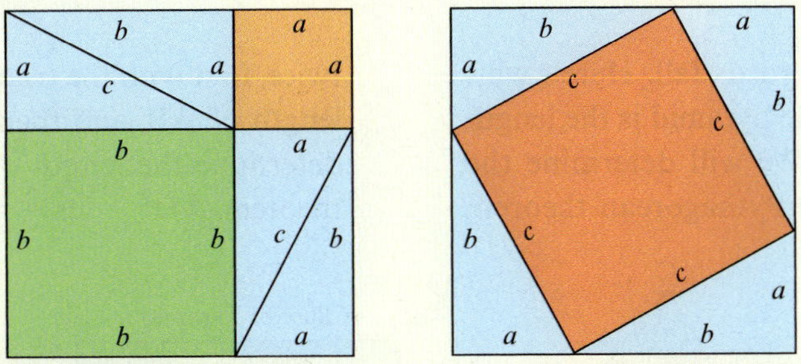

Figure 12.99 Shapes forming identical large squares.

7. In the previous practice exercise, you used Figure 12.99 to provide a proof of the Pythagorean theorem. Even though it is somewhat hidden, discuss how this proof actually relies on the fact that the angles in a triangle add to 180°.

8. Why is the Pythagorean theorem true for *all* right triangles?

9. When Kyle was asked to state the Pythagorean theorem, he responded by writing the equation

$$a^2 + b^2 = c^2$$

Is this a correct statement of the Pythagorean theorem? Why or why not?

Answers to Practice Exercises for Section 12.9

1. If we let x be the length in feet of the diagonal brace, then, according to the Pythagorean theorem,

$$3^2 + 4^2 = x^2$$

Therefore, $x^2 = 25$, so $x = 5$, which means that the diagonal brace is 5 feet long.

2. If we let x be the distance that the boat can drift away from the spot on the water's surface that is directly over the anchor, then, according to Figure 12.100 and the Pythagorean theorem,

$$x^2 + 50^2 = 75^2$$

Therefore,

$$x^2 = 5625 - 2500 = 3125$$

so

$$x = \sqrt{3125} \approx 55.9$$

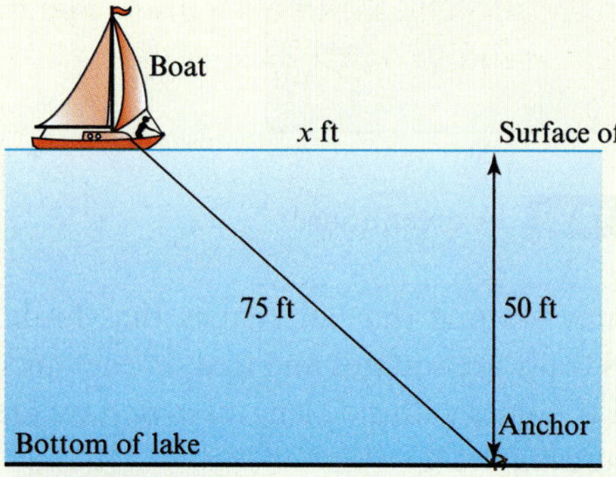

Figure 12.100 How far can an anchored boat drift?

This tells us that the boat can drift 56 feet horizontally.

3. The length of the longest pole that will fit is the distance between point A and point C in Figure 12.101, which is about 13 feet. To determine the distance from A to C, we will use the Pythagorean theorem twice: first with the right triangle on the floor of the elevator, to determine the length of AB, and then with triangle ABC to determine the length of AC. By applying the Pythagorean theorem to the triangle on the floor of the elevator, which has short sides of lengths 8 feet and 6 feet, we conclude that $8^2 + 6^2 = AB^2$. So $AB = \sqrt{100} = 10$ feet. Notice that ABC is also a right triangle, with the right angle at B. Therefore, by the Pythagorean theorem, $AB^2 + BC^2 = AC^2$, so $AC = \sqrt{181}$, about 13.5 feet. Thus, the longest pole that can fit in the elevator is about 13 feet.

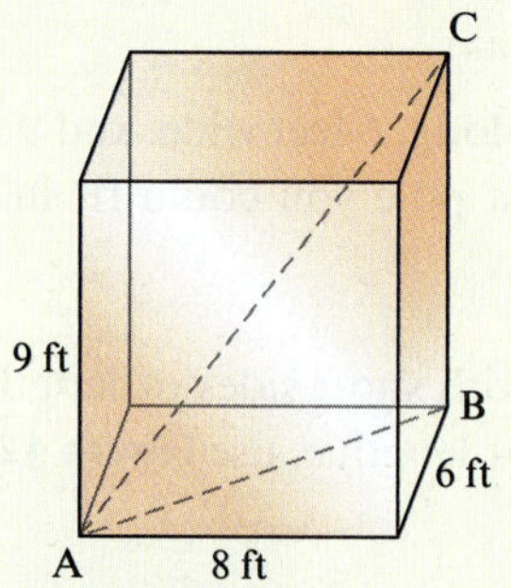

Figure 12.101 A pole in an elevator.

4. The pyramid is about 200 yards tall. Here's why. Notice that the height of the pyramid is the length of BC in Figure 12.102. We will determine the length of BC by using the Pythagorean theorem twice, first with the triangle ABD, to determine the length of AB, and then with the triangle ABC, to determine the length of BC. By the Pythagorean theorem, $AD^2 + BD^2 = AB^2$, so since AD and BD

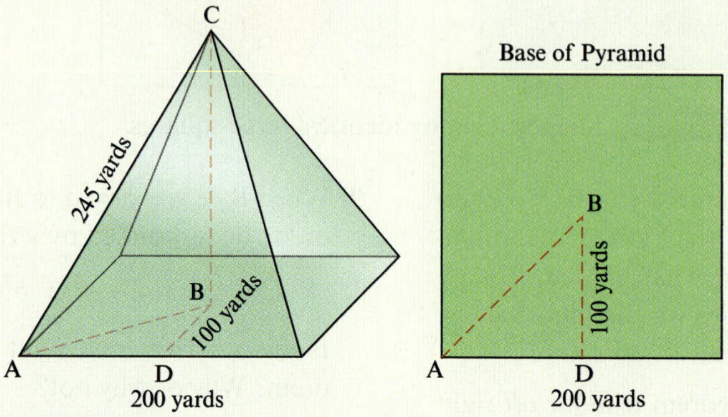

Figure 12.102 A pyramid.

are each 100 yards, $AB = \sqrt{20000}$, about 141.4 yards. (Actually, as you'll see in the next step, we really need only AB^2, not AB, so we don't even have to calculate the square root.) The triangle ABC is a right triangle, with right angle at B. So, by the Pythagorean theorem, $(\sqrt{20000})^2 + BC^2 = 245^2$, and it follows that $BC = \sqrt{40025}$, about 200 yards. So the pyramid is about 200 yards tall.

5. We need 1131 square feet of screening. The roof is made out of two rectangular pieces, each of which is 40 feet long and A feet wide, where A is shown in Figure 12.103. Because the angle at the top of the roof is 90°, we can use the Pythagorean theorem to determine A:

$$A^2 + A^2 = 20^2$$

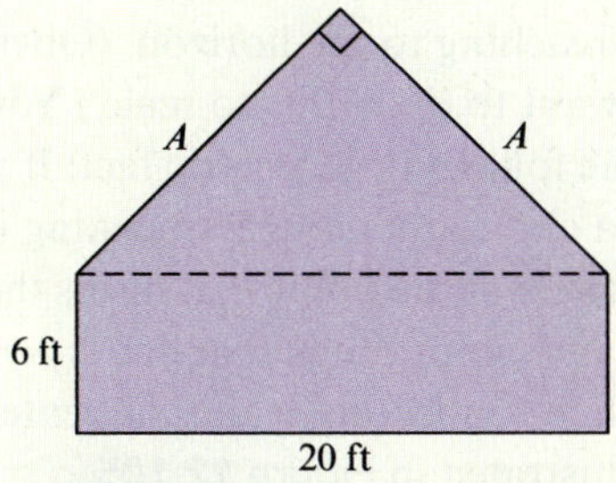

Figure 12.103 A greenhouse.

Therefore, $2A^2 = 400$, so $A^2 = 200$, which means that A is about 14.14 ft. Hence, each rectangular piece of roof needs $40 \cdot 14.14 \text{ ft}^2 = 565.6 \text{ ft}^2$ of screening. The two pieces of roof require twice as much. Rounding our answer, we see that we need about 1131 ft^2 of screening.

6. One way to prove that $a^2 + b^2 = c^2$ is to imagine taking away the 4 triangles in each large square of Figure 12.99. Both of the large squares in this figure have sides of length $a + b$, so both large squares have the same area. Hence, according to the moving and additivity principles, if we remove the 4 copies of the right triangle from each large square, the remaining areas will still be equal. From the square on the left, two smaller squares remain, one with sides of length a and one with sides of length b. Thus, the remaining area on the left is $a^2 + b^2$. From the square on the right, a single square with sides of length c remains. Therefore, the remaining area on the right is c^2. The remaining area on the left is equal to the remaining area on the right, therefore $a^2 + b^2 = c^2$.

7. When we used Figure 12.99 to prove the Pythagorean theorem, we assumed that the 2 large squares (each of which is made of 4 copies of the right triangle and either 1 or 2 more squares) really are squares of side length $a + b$ in order to know that they have the same area. Now refer to Figure 12.104. Focusing on point P on the square on the right, we see that the edge there should be a straight line in order for the square really to be a square. If we let A be the angle in the triangle opposite the side of length a and we let B be the angle opposite the side of length b (as shown on the triangle on the left in Figure 12.104), then

$$A + B + 90 = 180$$

because the angles in the right triangle add to 180°. Therefore at point P on the right in Figure 12.104, the angles add to 180°, so we can conclude that, in fact, we do have a straight line there. The situation will be similar at the other points along the edge of this square, as well as the points inside and on the edges of the square on the left in Figure 12.99.

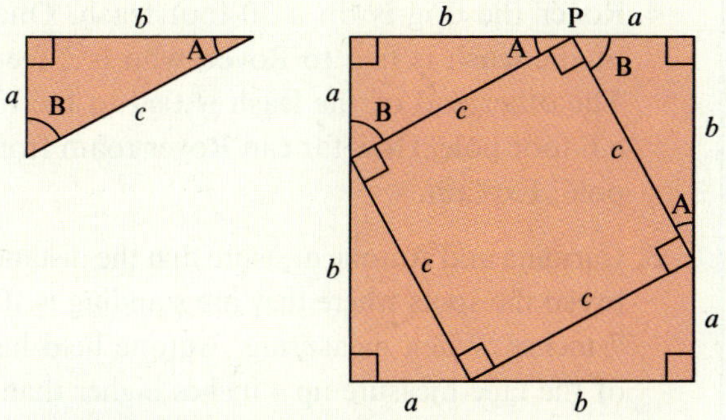

Figure 12.104 Explaining why the edges of the square really are straight.

8. Although the explanations of Practice Exercises 6 and 7 refer to one specific right triangle, they are general, in that they would work in the very same way for *any* right triangle. They do not involve any specific numbers or calculations.

9. Kyle's statement of the Pythagorean theorem is not a correct statement because it is not complete. Kyle should say that the Pythagorean theorem is about the lengths of the sides of right triangles. If a right triangle has short sides of lengths a and b and hypotenuse of length c, then $a^2 + b^2 = c^2$.

PROBLEMS FOR SECTION 12.9

1. Jessica says she doesn't understand the Pythagorean theorem. She put numbers in for a, b, and c in the equation $a^2 + b^2 = c^2$, but the equation isn't always true. For example, when she put in $a = 2$, $b = 3$, and $c = 4$, she got $4 + 9 = 16$, which isn't correct. Jessica wants to know why the Pythagorean theorem isn't working. Write an informative paragraph discussing Jessica's conundrum. Why does the Pythagorean theorem appear not to be correct? What misunderstanding does Jessica have?

2. Town B is 380 km due south of town A. Town C is 460 km due east of town B. What is the distance from town A to town C? Explain your reasoning. Be sure to round your answer appropriately.

3. What length ribbon will you need to stretch from the top of a 25-foot pole to a spot on the ground that is 10 feet from the bottom of the pole? Explain.

4. Rover the dog is on a 30-foot leash. One end of the leash is tied to Rover, who is 2 feet tall. The other end of the leash is tied to the top of a 6-foot pole. How far can Rover roam from the pole? Explain.

5. Carmina and Antone measure that the distance between the spots where they are standing is 10 feet, 7 inches. When measuring, Antone held his end of the tape measure up 4 inches higher than Carmina's end. If Carmina and Antone had measured the distance between them along the floor, would the distance be appreciably different than what they measured? Explain.

6. Use the Pythagorean theorem to help you determine the area of an equilateral triangle with sides of length 1 unit. Explain your reasoning.

* 7. Assuming that the earth is a perfectly round, smooth ball of radius 4000 miles and that 1 mile = 5280 feet, how far away does the horizon appear to be to a 5-foot-tall person on a clear day? Explain your reasoning. To solve this problem, start by making a math drawing that shows the cross-section of the earth, a person standing on the surface of the earth, and the straight line of the person's gaze reaching to the horizon. (Obviously, you won't want to draw this to scale.) You will need to use the following geometric fact: If a line is *tangent* to a circle at a point P (meaning it just "grazes" the circle at the point P; it meets the circle only at that one point), then that line is *perpendicular* to the line connecting P and the center of the circle, as illustrated in Figure 12.105.

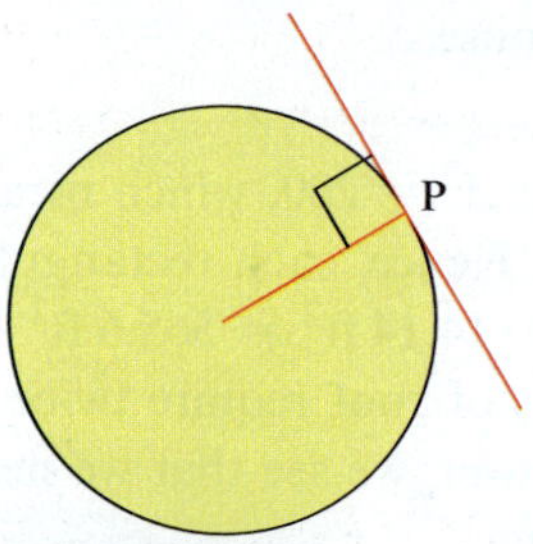

Figure 12.105 A line tangent to a circle.

CHAPTER SUMMARY

Section 12.1 Areas of Rectangles Revisited

Page 526

- The area of a shape, in square units, is the number of 1-unit-by-1-unit squares it takes to cover the shape exactly without gaps or overlaps, allowing for squares to be cut apart and pieces to be moved if necessary. The most primitive way to determine the area of a shape is to count how many squares it takes to cover it exactly. A quicker and more advanced way to determine the area of a rectangle is to multiply its width and length. We can explain this formula by viewing the rectangle as decomposed into rows of squares. Page 526

Key Skills and Understandings

1. Know and be able to describe what area is and know the most primitive way to determine the area of a shape. Page 526
2. Explain why it is valid to multiply the length and width of a rectangle to determine its area. Page 527

3. Attend carefully to units of length and area when discussing and explaining the area formula for rectangles.	Page 528
4. Explain and use the area formula for rectangles in the case of fractions, mixed numbers, and decimals.	Page 527

Section 12.2 Moving and Additivity Principles About Area

Page 530

- Two powerful principles are often used—almost subconsciously—to determine areas. These are the moving and additivity principles. The moving principle states that if you move a shape rigidly without stretching it, then its area does not change. The additivity principle states that if you combine shapes without overlapping them, then the area of the resulting combined shape is the sum of the areas of the individual shapes. Page 530

Key Skills and Understandings

1. Use the moving and additivity principles to determine areas of shapes, including cases where ultimately subtraction is used to determine the area of a shape of interest. Page 530

Section 12.3 Areas of Triangles

Page 535

- Any of the three sides of a triangle can be the base of the triangle. The height corresponding to a given base is perpendicular to the base and connects the base (or an extension of the base) to the vertex of the triangle that is not on the base. The area of the triangle is $\frac{1}{2}(\text{base} \cdot \text{height})$. The moving and additivity principles explain why this area formula is valid. Page 535

Key Skills and Understandings

1. Determine the area of a triangle in various ways, including using less sophisticated and more sophisticated methods. Page 535
2. Use the moving and additivity principles to explain why the area formula for triangles is valid for all triangles. Page 536
3. Use the area formula for triangles to determine areas and to solve problems. Page 538

Section 12.4 Areas of Parallelograms and Other Polygons

Page 544

- The lengths of the sides of a parallelogram do not determine its area. Instead, the area of a parallelogram is base · height. As with triangles, the base can be any of the four sides of the parallelogram. The height is the distance between the base and the side opposite the base. The moving and additivity principles explain why this area formula is valid. Page 544

Key Skills and Understandings

1. Explain why one can't determine the area of a parallelogram knowing only the side lengths. Page 544
2. Use the moving and additivity principles to explain why the area formula for parallelograms is valid. Page 545
3. Use the area formula for parallelograms to determine areas and to solve problems. Page 546
4. Determine areas of various polygons. Page 546

Section 12.5 Shearing: Changing Shapes Without Changing Area

Page 550

- Shearing is a process of changing a shape by sliding infinitesimally thin strips of the shape. Cavalieri's principle says shearing does not change areas. When we shear triangles and parallelograms parallel to a base, the base does not change, the height does not change, and the area does not change. Page 550

Key Skills and Understandings

1. Show how to shear a shape. Page 550
2. Know that when you shear a shape, the original shape and the sheared shape have the same area (Cavalieri's principle). Page 551
3. Distinguish shearing from ways of changing a shape that don't preserve area. Page 550

Section 12.6 Areas and Circumference of Circles and the Number Pi — Page 554

- For any circle, its circumference measured by its diameter is always the same number; this number is π (pi) and it is approximately 3.14. So the circumference of a circle of diameter D is $\pi \cdot D$ and the circumference of a circle of radius r is $2\pi r$. The area of a circle of radius r is πr^2. We can see why this area formula is plausible by subdividing a circle into "pie pieces" and rearranging them to form an approximate rectangle (or parallelogram) of dimensions r by πr. Page 554

Key Skills and Understandings

1. Know how the number π is defined, and use the definition to explain circumference formulas. Explain that π is a constant of proportionality in the relationship between diameters and circumferences of circles. Page 555
2. Explain why the $A = \pi r^2$ formula for the area A square units of a circle of radius r units is plausible by subdividing a circle and rearranging pieces. Page 557
3. Use the formulas for circumference and area of a circle to determine lengths and areas and to solve problems. Page 557

Section 12.7 Approximating Areas of Irregular Shapes — Page 561

- When it comes to irregular shapes, we usually have to make do with estimating their areas, because we often cannot determine their areas exactly. We can determine the approximate area of a shape by overlaying it with a grid and determining approximately how many grid squares the shape takes up. Page 561

Key Skills and Understandings

1. Use grids to determine approximate areas of shapes, including shapes shown on maps with a scale. Page 561
2. Use other informal methods, such as working with modeling dough and weighing card stock, to determine approximate areas of shapes. Page 561

Section 12.8 Contrasting and Relating the Perimeter and Area of Shapes — Page 564

- The perimeter of a polygon (or other shape) is the total distance around the polygon. When we calculate the perimeter of a polygon by adding the lengths of the sides, it is as if we had wrapped a string snugly around the polygon and cut it into pieces corresponding to each side. Page 564
- For a given, fixed perimeter, there are many shapes that have the same perimeter but have different areas. Thus, perimeter does not determine area. Among all rectangles of a given, fixed perimeter, the square with that perimeter has the largest area. Given a fixed perimeter, every area less than or equal to the area of the square of that perimeter is the area of some rectangle of that perimeter. Among all closed shapes of a given, fixed perimeter, the circle of that circumference has the largest area. Given a fixed perimeter, every area less than or equal to the area of the circle of that circumference is the area of some shape of that perimeter. Page 566

Key Skills and Understandings	
1. Explain why we calculate perimeters of polygons the way we do. Discuss misconceptions with perimeter calculations.	Page 564
2. Recognize that perimeter does not determine area.	Page 565
3. Given a fixed perimeter, determine the areas of all rectangles of that perimeter and determine the areas of all shapes of that perimeter.	Page 566
Section 12.9 Using Moving and Additivity Principles to Prove the Pythagorean Theorem	Page 570
▪ The Pythagorean theorem states that in any right triangle, the square of the length of the hypotenuse is equal to the sum of the squares of the lengths of the other two sides. We can prove the Pythagorean theorem by applying the moving and additivity principles to two identical squares subdivided into four copies of the right triangle and additional squares.	Page 570
Key Skills and Understandings	
1. State and prove the Pythagorean theorem.	Page 570
2. Use the Pythagorean theorem to determine lengths and distances.	Page 572

13

Solid Shapes and Their Volume and Surface Area

Even very young children learn about the common solid shapes of geometry when they play with a set of building blocks and observe the different characteristics and qualities of the shapes. Later in elementary school, students learn how to make patterns for shapes as they distinguish the outer surface of a shape and its surface area from the interior of a shape and its volume. And later still, students learn formulas for volumes and surface areas of shapes. Why do we study the common solid shapes of geometry? Most objects in the world around us can be thought of as approximated by a combination of the perfect shapes of geometry. For example, a tree trunk is approximately a cylinder, a pile of sand could be roughly in the shape of a cone, and a soft drink bottle is roughly like the combination of a cylinder and a cone. When children create buildings or other structures from a set of blocks, they are discovering how to compose complex new shapes from simple ones.

In this chapter, we focus on the following topics and practices within the *Common Core State Standards for Mathematics* (*CCSSM*).

Standards for Mathematical Content in the CCSSM

In the domain of *Measurement and Data* (Kindergarten–Grade 5), students learn the concept of volume, and they use multiplication and addition to find volumes. In the domain of *Geometry* (Kindergarten–Grade 8), students identify three-dimensional (solid) shapes and they build and compose three-dimensional shapes. They find volumes and apply volume formulas, they represent three-dimensional figures with nets, and they use nets to find surface areas. They describe two-dimensional figures that result from slicing three-dimensional figures.

Standards for Mathematical Practice in the CCSSM

Opportunities to engage in all eight of the Standards for Mathematical Practice described in the CCSSM occur throughout the study of solid shapes and their volume and surface area, although the following standards may be especially appropriate for emphasis:

- **1 Make sense of problems and persevere in solving them.** Students engage in this practice when they design and construct patterns for shapes and persevere in visualizing solid shapes and distinguishing the outer surface of a shape from its interior.
- **2 Reason abstractly and quantitatively.** Students engage in this practice when they investigate the properties of shapes, reason that certain shapes will stack and others will not, or explain why a solid shape having more than a certain number and type of polygons coming together at a vertex will have to have indentations or protrusions.
- **4 Model with mathematics.** Students engage in this practice when they use solid shapes and their volume or surface area to model quantities. For example, toothpaste extruded from a tube forms a very long cylinder. The volume of toothpaste in a tube is relevant for how long a tube will last, as is the length of toothpaste that is extruded each time one brushes.

  (From Common Core Standards for Mathematical Practice. Published by Common Core Standards Initiative.)

13.1 Polyhedra and Other Solid Shapes

CCSS PreK Common Core State Standards Grades K, 1, 2

In this section we study some of the most familiar solid shapes of geometry, most of which are common in sets of building blocks but some of which are less common yet surprisingly beautiful.

Initially, young children may only view solid shapes holistically. As teachers draw attention to component parts of solid shapes, such as points, edges, and flat or curved outer surfaces, children become more aware of these different aspects of solid shapes and how they affect whether the shapes can be stacked on top of or under other shapes, or whether the shapes roll or not. In this Section we study how to describe some common solid shapes and their component parts.

Some solid shapes have curved outer surfaces, whereas others have only flat surfaces consisting of triangles, squares, rectangles, pentagons, or other polygons. These latter kinds of shapes are called *polyhedra.*

What Are Polyhedra?

CCSS K.G.4, K.G.5, 1.G.2, 2.G.1

polyhedron (polyhedra) **faces** **edge** **vertex (vertices)**

A closed, connected shape in space whose outer surfaces consist of polygons (such as triangles, squares, or pentagons) is called a **polyhedron**. The plural of polyhedron is polyhedra. **Figure 13.1** shows two polyhedra and a shape that is not a polyhedron. The polygons that make up the outer surface of the polyhedron are called the **faces** of the polyhedron. The place where two faces come together is called an **edge** of the polyhedron. A corner point where several faces come together is called a **vertex** or corner of the polyhedron. The plural of vertex is vertices. (See **Figure 13.2**.)

The name *polyhedron* comes from the Greek: *poly* means *many* and *hedron* means *base,* so *polyhedron* means *many bases.*

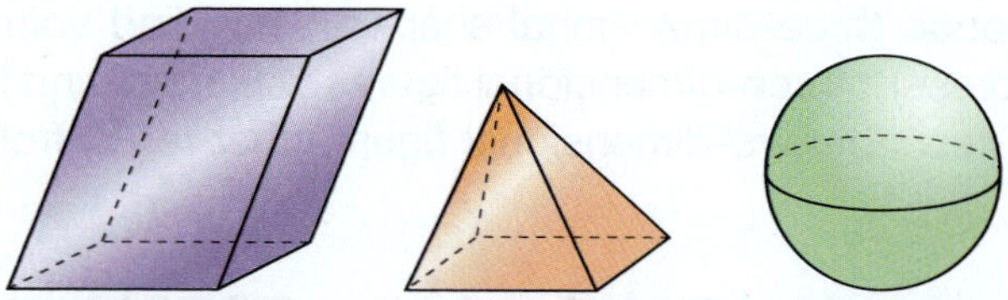

Figure 13.1 Polyhedra and a shape that is not a polyhedron.

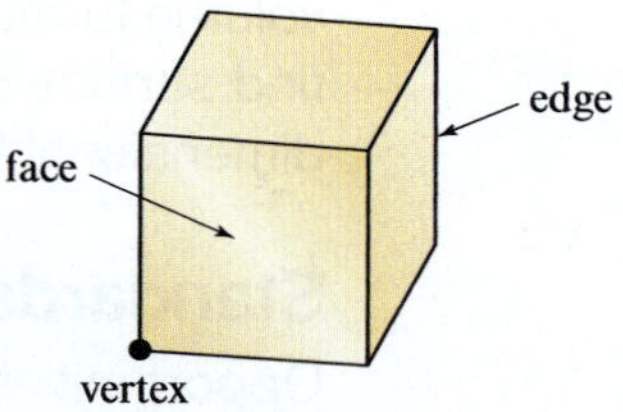

Figure 13.2 A face, edge, and vertex of a polyhedron.

Given a polyhedron, students can study its characteristics and qualities by observing its faces, edges, and vertices. What kinds of shapes are the faces? How many faces are there? How many vertices and how many edges does the shape have?

Two common types of polyhedra are prisms and pyramids. Related to these shapes are cylinders and cones, but since they have curved surfaces, they are *not* polyhedra.

What Are Prisms, Cylinders, Pyramids, and Cones?

One special type of polyhedron is a prism. Some people hang glass or crystal prisms in their windows to bend the incoming light and cast rainbow patterns around the room.

right prisms

From a mathematical point of view, a *right prism* is a polyhedron that can be thought of as "going straight up over a polygon," as shown in Figure 13.3. Think of **right prisms** as formed in the following way: Take two paper copies of any polygon and lay both flat on a table, one on top of the other so that they match up. Move the top polygon *straight up* above the bottom one. If vertical rectangular faces are now placed so as to connect corresponding sides of the two polygons, then the shape formed this way is a right prism. The two polygons that you started with are called the **bases** of the right prism.

bases

Figure 13.3 Three right prisms.

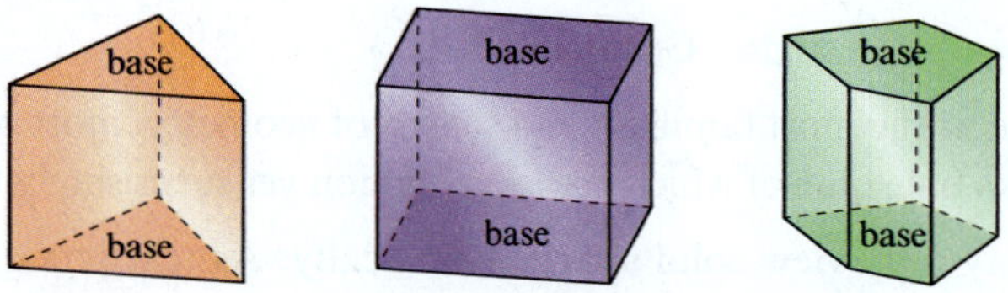

By modifying the previous description, we get a description of *all* prisms, not just right prisms. As before, start with two paper copies of any polygon and lay both flat on a table, one on top of the other so that they match up. Move the top polygon up without twisting, away from the bottom polygon, keeping the two polygons parallel. This time, the top polygon does not need to go straight up over the bottom polygon. Instead, it can be positioned to one side, as long as it is not twisted and it remains parallel to the bottom polygon. Once again, if faces are now placed so as to connect corresponding sides of the two polygons, then the shape formed this way is a **prism**. (This time the faces will be parallelograms.) Every right prism is a prism, but Figure 13.4 shows prisms that are not right prisms. A prism that is not a right prism can be called an **oblique prism**. As before, the two polygons that you started with are called the bases of the prism.

prism

oblique prism

Figure 13.4 Two oblique prisms.

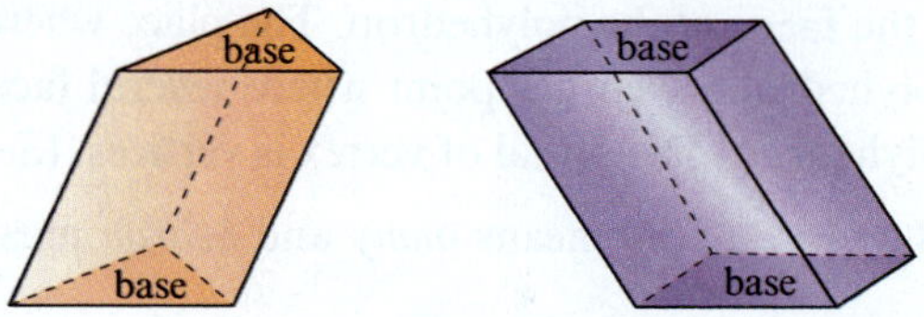

If a prism is moved to a different orientation in space, it is still a prism. So, for example, the polyhedra shown in Figure 13.5 are prisms, although you may need to rotate them mentally to be convinced that they really are prisms.

Figure 13.5
When a prism is rotated sideways, it is still a prism.

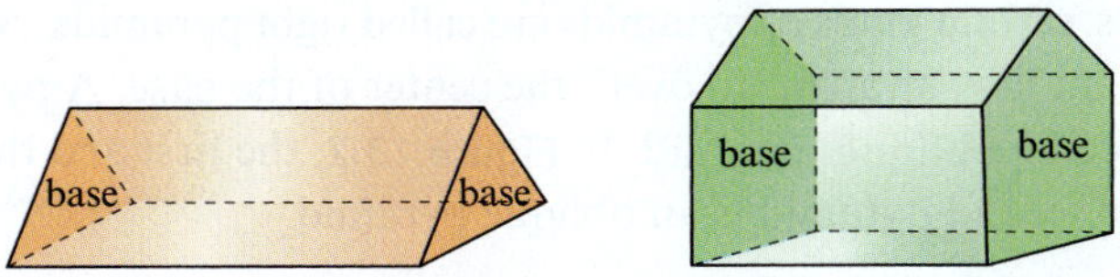

triangular, rectangular prism

Prisms are often named according to the kind of polygons that make the bases of the prism. So, for example, a prism with a triangle base can be called a **triangular prism**, and a prism with a rectangle base can be called a **rectangular prism**. Take care to use proper terminology when discussing solid shapes. For example, a common error is to call a rectangular prism a rectangle. A rectangle is a two-dimensional shape; the *faces* of a right rectangular prism are rectangles.

cylinders

A cylinder is a kind of shape that is related to prisms. Roughly speaking, a cylinder is a tube-shaped object, as shown in Figure 13.6. The tube inside a roll of paper towels is an example of a cylinder. **Cylinders** can be described in the following way: Draw a closed curve on paper (such as a circle or an oval), cut it out, make a copy of it, and lay the two copies on a table, one on top of the other, so that they match. Next, take the top copy and move it up without twisting, away from the bottom copy, keeping the two copies parallel. Now imagine paper or some other kind of material connecting the two curves in such a way that every line between corresponding points on the curves lies on this paper or material. The shape formed by this paper or other material is a cylinder. The regions formed by the two starting curves can again be called the bases of the cylinder. In some cases, you want to consider the two bases as part of the cylinder; in other cases, you do not. Both kinds of shapes can be called cylinders.

right cylinder
oblique cylinder

As with prisms, a cylinder can be a **right cylinder** or an **oblique cylinder**, according to whether one base is or is not "straight up over" the other. The first and third cylinders in Figure 13.6 are right cylinders, whereas the second cylinder is an oblique cylinder.

Figure 13.6
Cylinders.

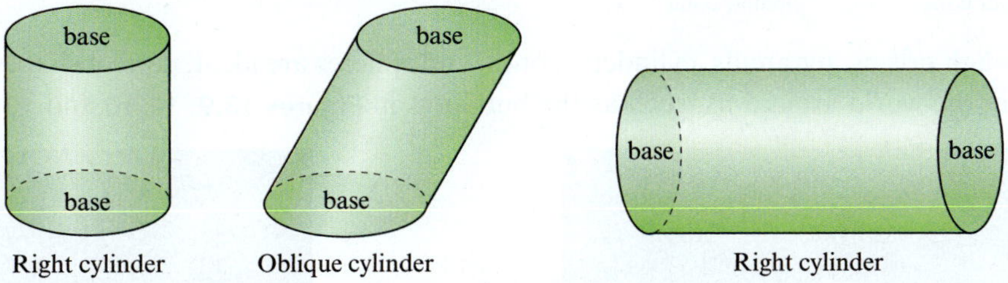

pyramid
apex

In addition to prisms, another special type of polyhedron is a *pyramid.* You have probably seen pictures of the famous pyramids in Egypt. Mathematical pyramids include shapes like the Egyptian pyramids as well as variations on this kind of shape, as shown in Figure 13.7. **Pyramids** can be described as follows: Start with any polygon and a separate single point, called an **apex**, that does not lie in the plane of the polygon. Now use additional polygons to connect the apex to the original

Figure 13.7
Pyramids.

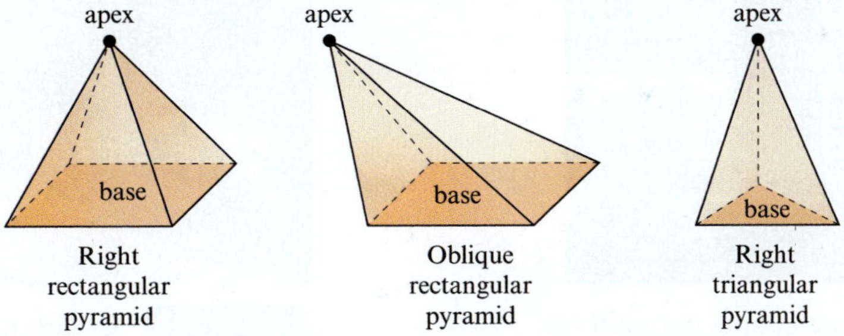

polygon. This should be done in such a way that all the lines that connect the apex to the polygon lie on the sides of these additional polygons. These new polygons, together with the original polygon, form a pyramid. As usual, the original polygon is called the base of the pyramid.

right pyramid As with prisms and cylinders, certain kinds of pyramids are called right pyramids. A **right pyramid** is a pyramid for which the point lies "straight up over" the center of the base. A pyramid that is not a right pyramid can be called an **oblique pyramid**. In Figure 13.7, the first and third pyramids are right pyramids, whereas the second pyramid is an oblique pyramid.

oblique pyramid

cones Just as cylinders are shapes that are related to prisms, *cones* are shapes that are related to pyramids. Roughly speaking, cones are objects like ice cream cones or cone-shaped paper cups, as well as related objects, as shown in Figure 13.8. As with pyramids, **cones** can be described by starting with a closed curve in a plane and a separate point that does not lie in that plane. Now imagine paper or some other kind of material that joins the point to the curve in such a way that all the lines that connect the point to the curve lie on that paper (or other material). That paper (or other material), together with original curve, form a cone. As usual, the starting curve and the region inside it is called the base of the cone. Sometimes the base of a cone is considered a part of the cone, and sometimes it isn't. Either shape (with or without the base) can be called a cone.

right cone As with prisms, cylinders, and pyramids, certain kinds of cones are called right cones. A **right cone** is a cone whose point lies directly over the center of its base. A cone that is not a right cone can be called an **oblique cone**. In Figure 13.8, the first and third cones are right cones, whereas the second cone is an oblique cone.

oblique cone

Figure 13.8 Cones.

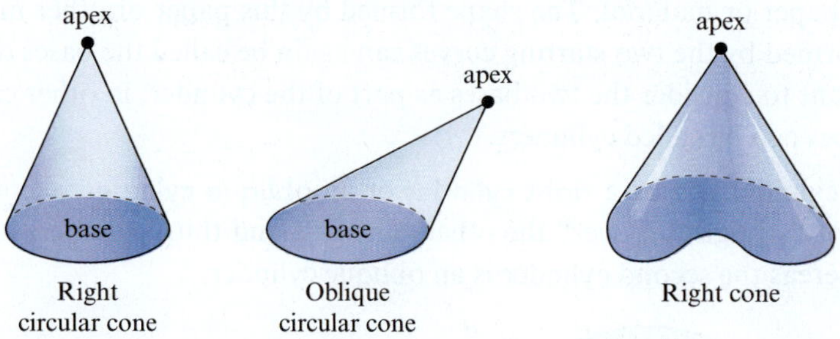

Although prisms, pyramids, cylinders, cones, and spheres are ideal, perfect shapes, objects that we find in the world around us, such as the buildings in Figures 13.9, 13.10 and 13.11, can often be

Figure 13.9 An octagonal pyramid on top of an octagonal prism.

Figure 13.10 Cones on top of cylinders.

Figure 13.11 Half of a sphere on top of a cylinder on top of an octagonal prism.

viewed as approximated by combinations of portions of those shapes. We can view a tree trunk as approximated by a cylinder or the bottom part of a very tall cone. The legs of a chair might be square prisms, approximately, or a bottle of juice might be roughly like a cylinder with a cone on top of it. Discovering how objects are composed of approximate versions of ideal shapes provides us with a way to analyze objects and to think about how to make them.

CLASS ACTIVITY

13A Making Prisms and Pyramids, p. CA-264

13B Analyzing Prisms and Pyramids, p. CA-265

13C What's Inside the Magic 8 Ball?, p. CA-266

What Is Special About the Platonic Solids?

Five special polyhedra that are completely regular and uniform all around are called the **Platonic solids**, in honor of Plato, who thought of these solid shapes as associated with earth, fire, water, air, and the whole universe. Figure 13.12 illustrates the five Platonic solids, which are described as follows:

- **Tetrahedron** has 4 equilateral triangle faces, with 3 triangles coming together at each vertex.
- **Cube** has 6 square faces, with 3 squares coming together at each vertex.
- **Octahedron** has 8 equilateral triangle faces, with 4 triangles coming together at each vertex.
- **Dodecahedron** has 12 regular pentagon faces, with 3 pentagons coming together at each vertex.
- **Icosahedron** has 20 equilateral triangle faces, with 5 triangles coming together at each vertex.

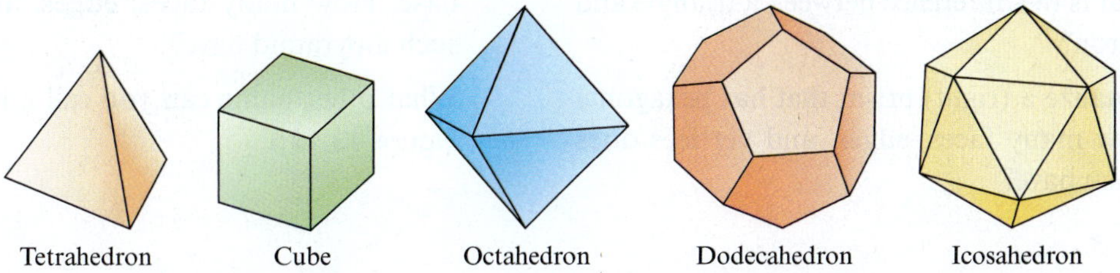

Figure 13.12 The Platonic solids.

The names of the Platonic solids make sense because *hedron* comes from the Greek for "base" or "seat," while *tetra, octa, dodeca*, and *icosa* mean 4, 8, 12, and 20, respectively. So, for example, *icosahedron* means *20 bases*, which makes sense because an icosahedron is made of 20 triangular "bases."

The Platonic solids are special because each one is made of only one kind of regular polygon, and the same number of polygons come together at each vertex. In fact, the Platonic solids are the only convex polyhedra having these two properties. A shape in the plane or in space is **convex** if any line segment connecting two points on the shape lies entirely within the shape. So, convex polyhedra do not have any protrusions or indentations.

convex

If you make the Platonic solids with plastic or paper polygons, you may begin to see why there are only five of these special shapes. When you put polygons together to make polyhedrons, you may notice that when the angles at a vertex add to more than 360°, the shape you make will be forced to have indentations and protrusions, and therefore will not be convex. So for a polyhedron to be convex, the angles coming together at a vertex must add to at most 360°, and this puts limits on the convex polyhedra that are possible.

Although cubes are commonly found in daily life (e.g., boxes and ice cubes), the other Platonic solids are not commonly seen. However, these shapes do occur in nature occasionally. For example, the mineral pyrite can form a crystal in the shape of a dodecahedron. (This crystal is often called a pyritohedron.) The mineral fluorite can form a crystal in the shape of an octahedron, and although rare, gold can, too. Some viruses are shaped like an icosahedron.

CLASS ACTIVITY

13D Making Platonic Solids, p. CA-267

SECTION SUMMARY AND STUDY ITEMS

Section 13.1 Polyhedra and Other Solid Shapes

Some of the basic solid shapes are prisms, cylinders, pyramids, and cones. There are five Platonic solids: the tetrahedron, the cube, the octahedron, the dodecahedron, and the icosahedron.

Key Skills and Understandings

1. Describe prisms, cylinders, pyramids, and cones and distinguish them from two-dimensional shapes.
2. Determine the number of vertices, edges, and faces of a given type of prism or pyramid, and explain why the numbers are correct.
3. Explain, based on angles, why putting too many faces together at a vertex results in a nonconvex shape.

Practice Exercises for Section 13.1

1. What is the difference between a square and a cube? What is the difference between a triangle and a tetrahedron?
2. Try to visualize a (right) prism that has hexagonal bases. How many faces, edges, and vertices does such a prism have?
3. Try to visualize a pyramid that has an octagonal base. How many faces, edges, and vertices does such a pyramid have?
4. What other name can you call a tetrahedron? (See **Figure 13.13**.)

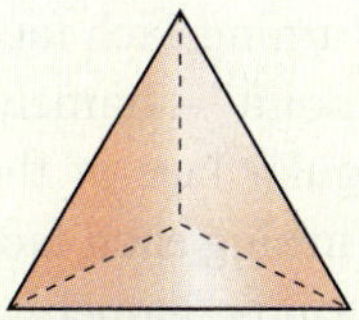

Figure 13.13 A tetrahedron.

5. What happens if you try to make a convex polyhedron whose faces are all equilateral triangles and for which 6 triangles come together at every vertex?
6. Is it possible to make a convex polyhedron so that 7 or more equilateral triangles come together at every point? Explain.

Answers to Practice Exercises for Section 13.1

1. A cube is a solid three-dimensional shape that has square faces. A square is a flat two-dimensional shape. A tetrahedron is a solid three-dimensional shape that has triangles as faces. A triangle is a flat, two-dimensional shape.
2. The prism has 8 faces: 6 rectangular faces and 2 hexagonal faces. The prism has 18 edges and 12 vertices.
3. The pyramid has 9 faces: 8 triangular faces and 1 octagonal face. The pyramid has 16 edges and 9 vertices.
4. A tetrahedron is also a pyramid with a triangle base.
5. If you put 6 equilateral triangles together so that they all meet at one point, you will find that they make a flat hexagon, as seen in Figure 13.14. It makes sense that they will make a flat shape because the angle at a vertex of an equilateral triangle is 60°, so 6 of these angles side by side will make a full 360°. If you tried to make a convex polyhedron in such a way that 6 equilateral triangles came together at every vertex, you could never get the polyhedron to "close up."

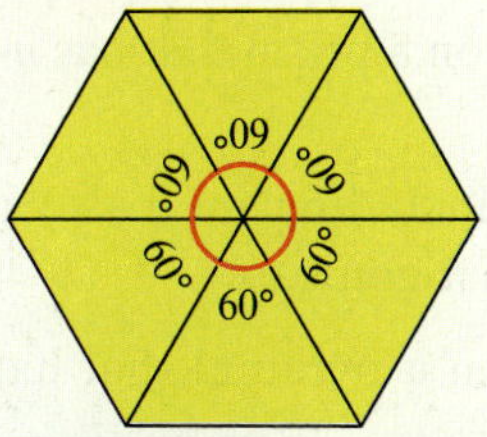

Figure 13.14 Six equilateral triangles put together at one point.

6. If you put 7 or more equilateral triangles together so that they all meet at one point, you will find that you have to create indentations or protrusions. Therefore, the polyhedron will not be convex. It makes sense that 7 triangles meeting at a point will do this because the angle at a vertex of an equilateral triangle is 60°, so 7 or more of these angles side by side will make more than 360°. This forces indentations and protrusions to occur in order to put so many triangles together at one point.

PROBLEMS FOR SECTION 13.1

1. For each of the following solid shapes, find at least two examples of a real-world object that has that shape or has a part that is that shape:
   a. Prism
   b. Cylinder
   c. Pyramid
   d. Cone
2. Answer the following questions without using a model:
   a. How many faces (including the bases) does a prism with a parallelogram base have? What shapes are the faces? Explain briefly.
   b. How many edges does a prism with a parallelogram base have? Explain.
   c. How many vertices does a prism with a parallelogram base have? Explain.
3. Answer the following questions without using a model:
   a. How many faces (including the base) does a pyramid with a rhombus base have? What shapes are the faces? Explain briefly.
   b. How many edges does a pyramid with a rhombus base have? Explain.
   c. How many vertices does a pyramid with a rhombus base have? Explain.

4. Recall that an n-gon is a polygon with n sides. For example, a polygon with 17 sides can be called a 17-gon. A triangle could be called a 3-gon. Find formulas (in terms of n) for the following, and explain why your formulas are valid:

   a. Number of faces (including the bases) on a prism that has n-gon bases

   b. Number of edges on a prism that has n-gon bases

   c. Number of vertices on a prism that has n-gon bases

   d. Number of faces (including the base) on a pyramid that has an n-gon base

   e. Number of edges on a pyramid that has an n-gon base

   f. Number of vertices on a pyramid that has an n-gon base

5. This problem goes with Class Activity 13C on the Magic 8 Ball. Make a closed, three-dimensional shape out of some or all of the triangles in Download 1. Make any shape you like—be creative. Answer the following questions:

   a. Do you think your shape could be the one that is actually inside the Magic 8 Ball? Why or why not?

   b. If you were going to make your own advice ball, but with a (possibly) different shape inside, what would be the advantages or disadvantages of using your shape?

6. Referring to the descriptions of the Platonic solids on page 585, construct these solids by cutting out the triangles, squares, and rectangles in Downloads 1, 2, and 3 and taping these shapes together. You may wish to print the pages onto card stock to make sturdier models that are easier to work with.

7. A cube is a polyhedron that has 3 square faces meeting at each vertex. What would happen if you tried to make a polyhedron that has 4 square faces meeting at each vertex? Explain.

8. The Platonic solids are convex polyhedra with faces that are equilateral triangles, squares, or regular pentagons. If you try to make a convex polyhedron whose faces are all regular hexagons, what will happen? Explain.

***9.** Two gorgeous polyhedra can be created by *stellating* an icosahedron and a dodecahedron. *Stellating* means *making starlike.* Imagine turning each face of an icosahedron into a "star point"—namely, a pyramid whose base is a triangular face of the icosahedron. Likewise, imagine turning each face of a dodecahedron into a star point—namely, a pyramid whose base is a pentagonal face of the dodecahedron. A stellated icosahedron will then have 20 star points, whereas a stellated dodecahedron will have 12 star points.

   a. Make 20 copies of Figure 13.15 on card stock. Cut, fold, and tape them to make 20 triangle-based pyramid star points. (The pattern makes star-point pyramids that don't have bases.) Tape the star points together as though you were making an icosahedron out of their (open) bases.

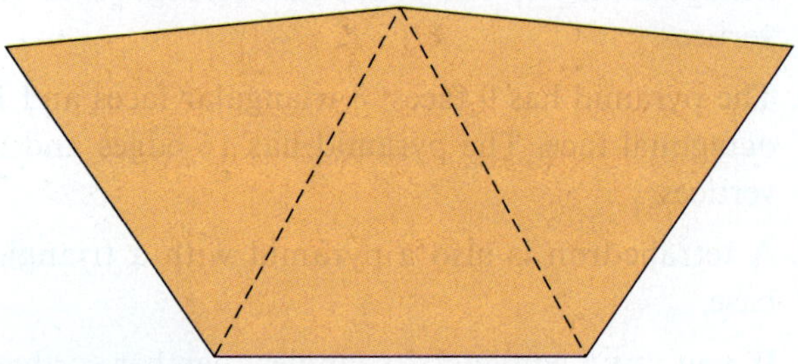

Figure 13.15 Pattern for a star point of an icosahedron.

   b. Make 12 copies of Figure 13.16 on card stock. Cut, fold, and tape them to make 12 pentagon-based pyramid star points. (The pattern makes star point pyramids that don't have bases.) Tape the star points together as though you were making a dodecahedron out of their (open) bases.

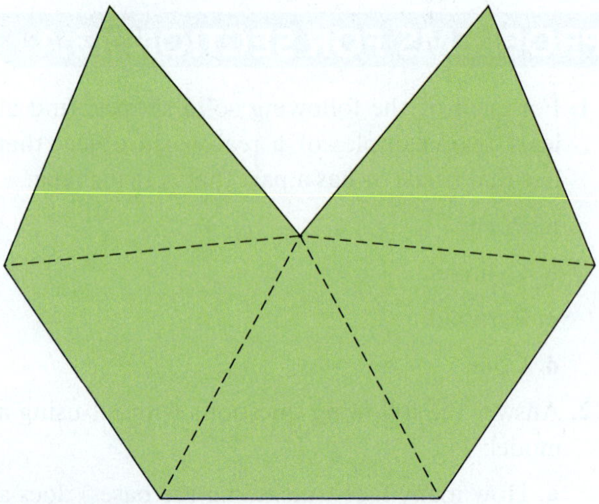

Figure 13.16 Pattern for a star point of a dodecahedron.

13.2 Patterns and Surface Area

CCSS **Common Core State Standards** Grades 6 and 7

In the previous section we saw that important characteristics of polyhedra are their faces, edges, and vertices. So three-dimensional shapes have essential characteristics that are two-dimensional, one-dimensional, and even zero-dimensional (points). In this brief section, we focus on a two-dimensional aspect of solid shapes: their outer surface. We also consider other two-dimensional aspects of solid shapes: cross-sections and shadows.

CCSS 6.G.4

One way to make a solid shape is to make a pattern on paper (or light cardboard) for its outer surface, cut out the pattern, fold it as needed, and join the cut edges. Making a shape in this way helps to concentrate on the characteristics of the outer surface of the shape, and the process provides an opportunity to practice visualization. A pattern for a polyhedron is sometimes called a **net** if it has the property that whenever polygons are joined in the pattern, they are joined along a full edge, not just at a vertex.

net

Patterns are useful not only for making models of shapes but also for determining and finding formulas for surface areas of solid shapes. Recall that the surface area of a solid shape is the total area of the outer surface of the shape. To determine the surface area of a shape, we can make or think about how to make a pattern for the shape. We can then apply the additivity principle for area and add the areas of the component parts of the pattern to determine the surface area of the shape.

CLASS ACTIVITY

13E What Shapes Do These Patterns Make?, p. CA-268

13F Patterns and Surface Area for Prisms and Pyramids, p. CA-269

13G Patterns and Surface Area for Cylinders, p. CA-269

13H Patterns and Surface Area for Cones, p. CA-270

Given a solid shape, a cross-section of it is formed by slicing the shape with a plane. The places where the plane meets the solid shape form a shape in the plane. This shape in the plane is a **cross-section** of the solid shape. Both patterns and cross-sections provide two-dimensional information about a solid shape, but unlike patterns, cross-sections give information about the *inside* of a shape. Thinking about cross-sections can help us recognize that solid shapes have an interior in addition to an outer surface. Cross-sections are especially important in calculus, where they are used to determine the volume of a shape. In the next section we will determine volumes of prisms by thinking about them as decomposed into layers. These layers are much like thickened cross-sections.

cross-section

CCSS 7.G.3

CLASS ACTIVITY

13I Cross-Sections of a Pyramid, p. CA-271

13J Cross-Sections of a Long Rectangular Prism, p. CA-272

The outer surface of a solid shape and its cross-sections are two-dimensional aspects of the shape. Additional two-dimensional aspects of a solid shape are the shadows the shape casts when light shines on it. Students who are learning about light and shadows are often asked to predict the shadow that a solid shape will make. Depending on how a shape is oriented with respect to the light source, a single shape can have several very different shadows. Experiment by holding solid shapes under a light and seeing what shadows you can get!

SECTION SUMMARY AND STUDY ITEMS

Section 13.2 Patterns and Surface Area

An essential two-dimensional aspect of a solid shape is its outer surface. Patterns can be made to form the outer surface of a solid shape. The surface area of a solid shape is determined by adding the areas of all the component parts of the outer surface of the shape. Additional two-dimensional aspects of solid shapes are cross-sections and shadows.

Key Skills and Understandings

1. Visualize what shape a pattern will make.
2. Make patterns for prisms, cylinders, pyramids, and cones of specified dimensions.
3. Determine the surface area of prisms, cylinders, pyramids, and cones.

Practice Exercises for Section 13.2

1. Figure 13.17 shows small versions of several patterns for shapes. Download 4 has larger versions. Before you cut, fold, and tape the larger patterns, visualize the folding process to help you visualize the final shapes. Then cut out the patterns along the solid lines, fold down along the dotted lines, and tape sides with matching labels together. Does the shape match your visualized shape?

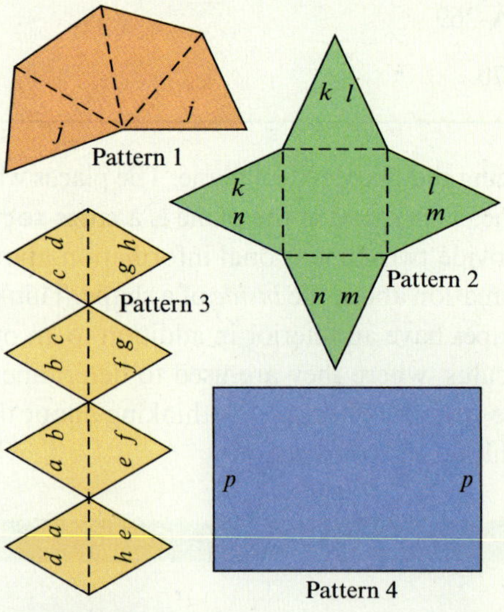

Figure 13.17 Small version of patterns for shapes.

2. Figure 13.18 shows small versions of 6 patterns for shapes. Download 5 has larger versions. Before you cut and tape the larger patterns, try to visualize the final shapes. Predict how the shapes will be alike and how they will be different. Then cut out the patterns along the solid lines, and tape sides with matching labels together. (Don't do any folding!)

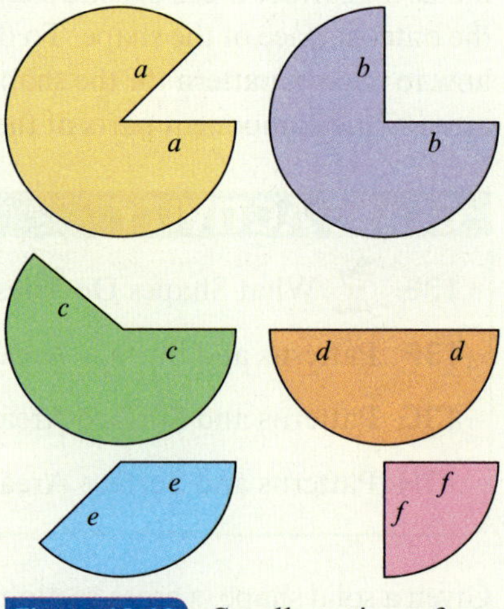

Figure 13.18 Small version of patterns for shapes.

3. What familiar parts of clothing are made from the patterns in Figure 13.19 when sides with matching labels are sewn together? Visualize!

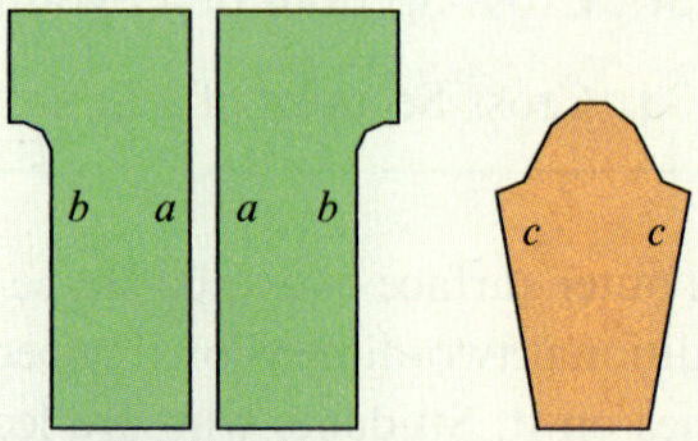

Figure 13.19 What parts of clothes do these patterns make?

4. Examine the small patterns in Figure 13.20. Try to visualize the polyhedra that would result if you were to cut these patterns out on the heavy lines, fold them on the dotted lines, and tape various sides together.

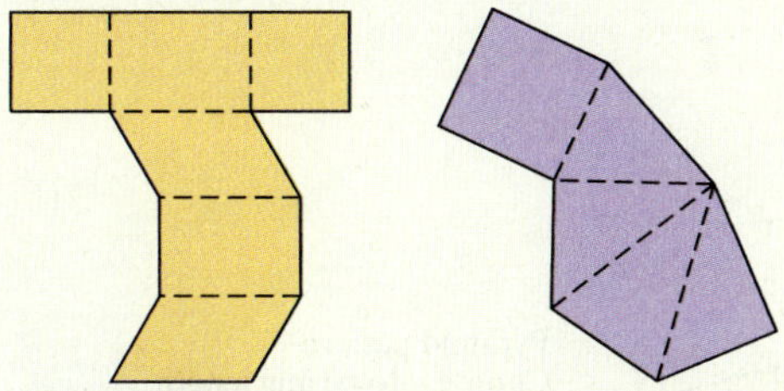

Figure 13.20 Small patterns for two polyhedra.

Download 6 has larger versions you can use to make the polyhedra. Were your predictions correct? If not, undo your polyhedra and try to visualize again how they turn into their final shapes.

5. Make a pattern for a prism whose two bases are identical to the triangle in Figure 13.21. Include the bases in your pattern. (Use a ruler and compass to make a copy of the triangle in Figure 13.21.) Label all sides that have length a, b, and c on your pattern.

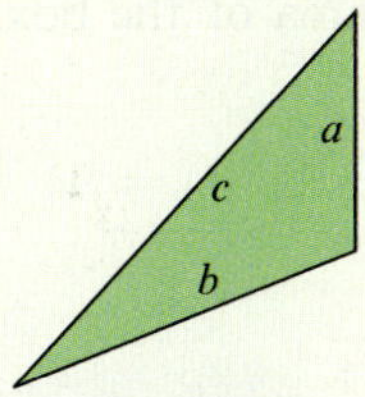

Figure 13.21 A base for a prism and a pyramid.

6. Make a pattern for a pyramid whose base is identical to the triangle in Figure 13.21. Include the base in your pattern. (Use a ruler and compass to make a copy of the triangle in Figure 13.21.) Label all sides that have length a, b, and c on your pattern. You may use graph paper.

7. What is the surface area of a closed box (rectangular prism) that is 4 ft wide, 3 ft deep, and 5 ft tall? Draw a small version of a pattern for the box to help you determine its surface area.

8. A right pyramid has a square base with sides 60 m long. The distance from one vertex on the base to the apex of the pyramid (along an edge) is 50 m. Determine the surface area of the pyramid (not including the base).

9. Suppose you take a rectangular piece of paper, roll it up, and tape two ends together, without overlapping them, to make a cylinder. If the cylinder is 12 inches long and has a diameter of $2\frac{1}{2}$ in. then what were the length and width of the original rectangular piece of paper?

10. Find a formula for the surface area of a cylinder of radius r units and height h units by reasoning about a pattern for a cylinder. (Include the bases of the cylinder.)

11. A cone is to be made from a circle of radius 3 cm (for the base) and a quarter-circle (for the lateral portion). Determine the radius of the quarter-circle.

***12.** Make a paper model of a cone. Now visualize a plane slicing through the cone making a cross-section. Describe all possible shapes in the plane that can be made this way, as cross-sections of the cone. Use your model to help you, but also visualize each case without the use of your model.

***13.** Make a paper model of a cube. Now visualize a plane slicing through the cube making a cross-section. Describe how to choose a plane so that the cross-section is the following shape:

- Square
- Rectangle that is not a square
- Triangle
- Rhombus that is not a square
- Hexagon

Answers to Practice Exercises for Section 13.2

1. Patterns 1 and 2 make pyramids, without and with a base, respectively. Pattern 3 makes an octahedron. Pattern 4 makes a cylinder without bases.

2. All 6 patterns make cones. Some are shorter and wider; some are taller and narrower.

3. The 2 patterns on the left form a pant leg when they are sewn together. The curved portion of side b makes the crotch. The pattern on the right makes a sleeve. The curved portion at the top makes the arm hole. The seam at edge c runs straight down the arm, from the armpit to the wrist.

4. The pattern on the left of Figure 13.20 makes an oblique square prism. The pattern on the right makes an oblique pyramid with a square base.

5. & 6. See Figure 13.22.

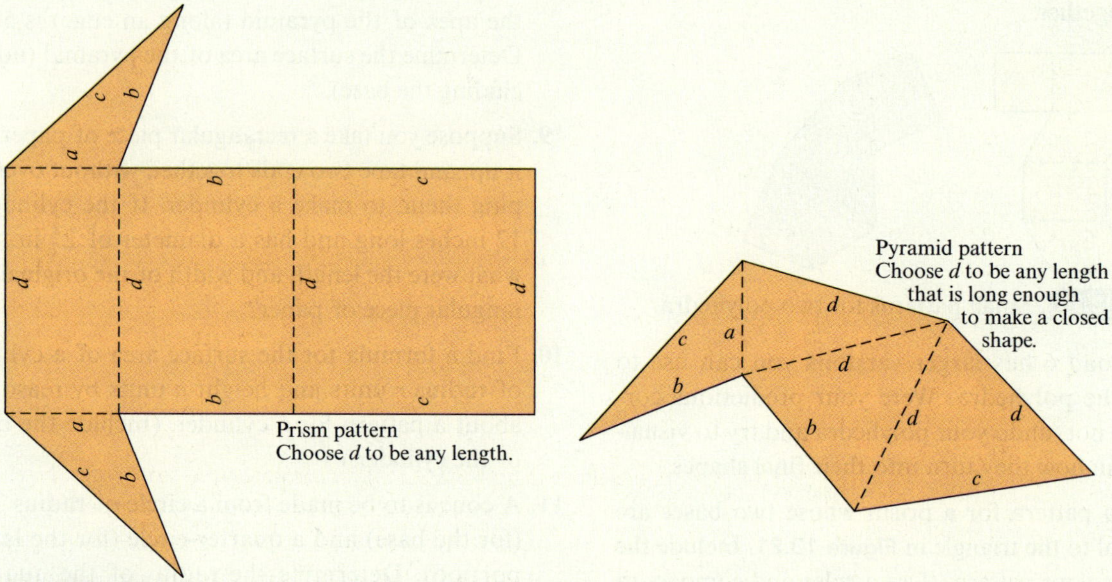

Figure 13.22 Patterns for a prism and a pyramid with triangle bases.

7. Figure 13.23 shows a scaled-down pattern for the desired box. By subdividing the pattern into rectangles and using the additivity principle about areas, we see that the surface area of the box, in square feet, is

$$2 \cdot (3 \cdot 4) + 2 \cdot (5 \cdot 4) + 2 \cdot (5 \cdot 3) = 94$$

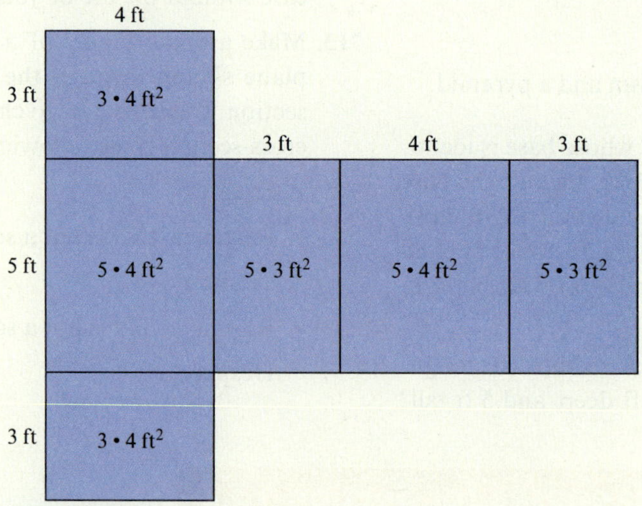

Figure 13.23 A pattern for a rectangular prism.

8. The surface of the pyramid (not including the base) consists of 4 isosceles triangles, each of which has a base of 60 m and 2 sides of length 50 m, as shown in Figure 13.24. To find the area of one of these triangles, we must determine the triangle's height. If h stands for the height of the triangle, then, by the Pythagorean theorem,

$$30^2 + h^2 = 50^2$$

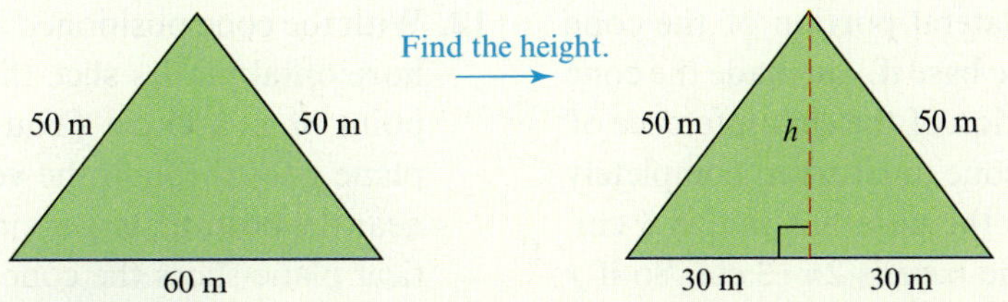

Figure 13.24 Determining the height of a triangular face of a pyramid in order to determine the surface area of the pyramid.

Therefore,

$$h^2 = 50^2 - 30^2 = 2500 - 900 = 1600$$

and $h = 40$, so the height of the triangle is 40 m. The area of each of the 4 triangles making the surface of the pyramid is therefore

$$\frac{1}{2}(60 \cdot 40) = 1200$$

square meters. So the surface area of the pyramid (not including the base) is

$$4 \cdot 1200 = 4800$$

square meters.

9. The two edges of paper that are rolled up make circles of diameter 2.5 in. Therefore, the lengths of these edges are $\pi \cdot 2.5$ in. (the circumference of the circles), which is about 8 in. The other two edges of the paper run along the length of the cylinder, so they are 12 in. long. Thus, the original piece of paper was about 8 in. by 12 in.

10. The surface of the cylinder consists of two circles of radius r units (the bases) and a tube. Imagine slitting the tube open along its length and unrolling it, as indicated in Figure 13.25. The tube then becomes a rectangle. The height, h, of the tube becomes the length of two sides of the rectangle. The circumference of the tube, $2\pi r$, becomes the length of two other sides of the rectangle. Therefore, the rectangle has area $2\pi rh$. According to the moving and additivity principles about area, the surface area of the cylinder is equal to the sum of the areas of the two circles (from the top and bottom) and the area of the rectangle (from the tube), which is

$$(2\pi r^2 + 2\pi rh) \text{ units}^2$$

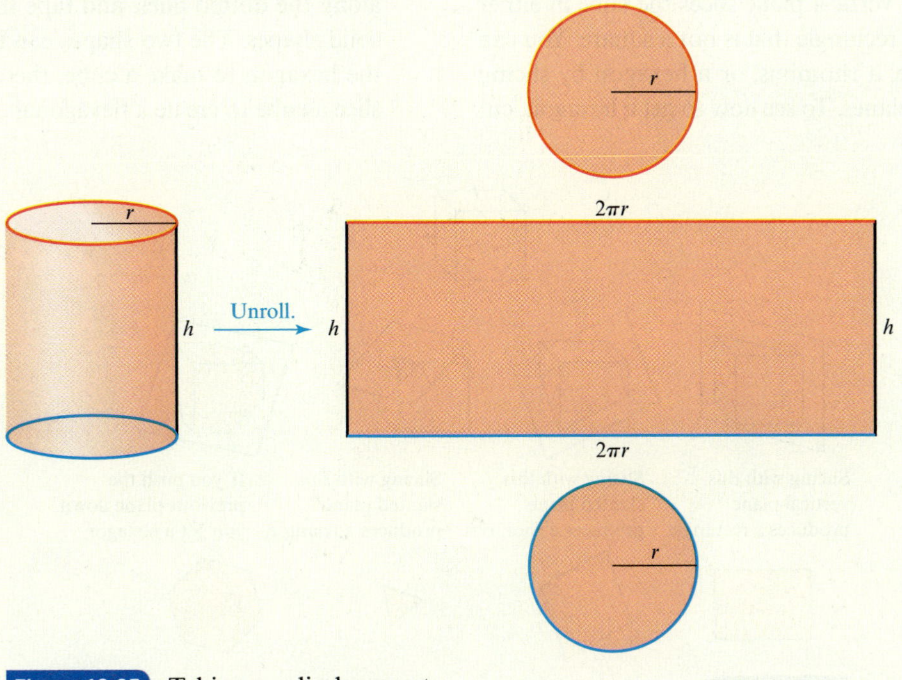

Figure 13.25 Taking a cylinder apart.

11. Think about how the lateral portion of the cone would be attached to the base if you made the cone out of paper. The $\frac{1}{4}$ portion of the circumference of the circle making the cone must wrap completely around the base. Since the base has radius 3 cm, the circumference of the base is $2\pi \cdot 3$ cm. So if r is the radius of the larger circle, $\frac{1}{4}$ of $2\pi r$ must be equal to $2\pi \cdot 3$. In other words,

$$\frac{1}{4} \cdot 2\pi r = 2\pi \cdot 3$$

Therefore, $r = 12$, and so the radius of the quarter-circle making the cone is 12 cm.

12. With the cone positioned as shown in Figure 13.26, horizontal planes slice the cone either in a single point or in a circle. (You get a single point if the plane goes through the very bottom, small circles near the bottom, larger circles as you go up.) A vertical plane slices the cone in either a V-shape or a curve called a **hyperbola**. A slanted plane slices the cone in either an **ellipse** (oval shape) or in a curve called a **parabola**. This is why circles, ellipses, hyperbolas, and parabolas are collectively referred to as **conic sections**, because they come from slicing a cone.

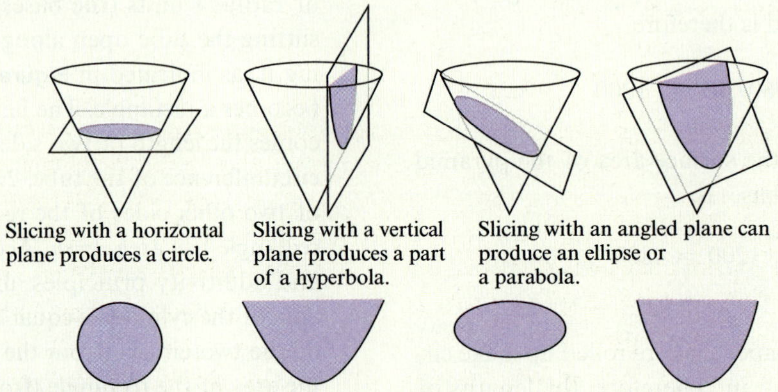

Figure 13.26 Slicing a cone with a plane.

13. See Figure 13.27. Horizontal planes slice the cube in squares. A vertical plane slices the cube in either a square or a rectangle that is not a square. You can get a triangle, a rhombus, or a hexagon by slicing with angled planes. To see how to get a hexagon, cut out the two patterns in Download 7, fold them down along the dotted lines, and tape them to create two solid shapes. The two shapes can be put together at the hexagon to make a cube; therefore, a plane can slice a cube to create a hexagonal cross-section.

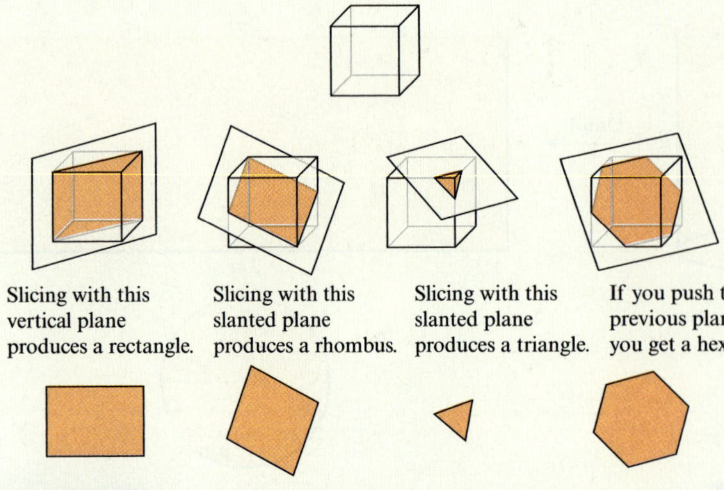

Figure 13.27 Slicing a cube with a plane.

PROBLEMS FOR SECTION 13.2

1. Find all the different nets for a tetrahedron made from 4 equilateral triangles. Remember that for a pattern to be a net, each triangle must be joined to another triangle along a whole edge, not just at a vertex. For two patterns to be considered different, you should not be able to match the patterns up when they are cut out.

2. Make three different nets that could be cut out, folded, and taped to make an open-top 1-in.-by-1-in.-by-1-in. cube. To be nets, each square should be joined to another square along a whole edge of the square, not just at a vertex. For two patterns to be considered different, you should not be able to match up the patterns when they are cut out. Download 8 has 1-in. graph paper you can use.

3. **a.** Describe or show how to make a cylinder without bases out of a 4-in.-by-5-in. rectangular piece of paper in such a way that all the paper is used and the paper does not overlap. Then describe or show how to make a different cylinder without bases out of another 4-in.-by-5-in. rectangular piece of paper. As before, all the paper should be used and there should be no overlaps. Determine the surface area (without the bases) of each cylinder and explain your reasoning.

   b. Describe or show how to make a pyramid without a base out of a 4-in.-by-5-in. rectangular piece of paper in such a way that all the paper is used and the paper does not overlap. (You may form faces of the pyramid by joining pieces of paper.) Determine the surface area (without the base) of your pyramid.

4. If a cardboard box (rectangular prism) with a top, a bottom, and 4 sides is W ft wide, L ft long, and H ft tall, then what is the surface area of this box? Find a formula for the surface area in terms of W, L, and H. Explain clearly why your formula is valid.

5. Use a ruler and compass to help you make a pattern for a prism whose two bases are identical to the triangle in Figure 13.21 on page 591. You may wish to draw your pattern on graph paper such as Downloads 8 or 9. Make your pattern significantly different from the one for a prism shown in Figure 13.21 on page 591. Include the bases in your pattern. (Use a ruler and compass to make a copy of the triangle in Figure 13.21.) Label all sides that have length a, b, and c on your pattern.

6. Use a ruler and compass to help you make a pattern for a pyramid whose base is identical to the triangle in Figure 13.21. Make your pattern significantly different from the one for a pyramid shown in Figure 13.22. Include the base in your pattern. (Use a ruler and compass to make a copy of the triangle in Figure 13.21.) Label all sides that have length a, b, and c on your pattern.

7. Use a ruler and compass to help you make a pattern for a prism whose two bases are triangles that have one side of length 2.5 in., one side of length 2 in., and one side of length 1.5 in. It may help you to draw your pattern on $\frac{1}{4}$-in. graph paper such as Download 9. Indicate which sides of your pattern would be joined to make the prism.

8. Use a ruler and compass to help you make a pattern for a pyramid with a triangular base that has one side of length 2 in., one side of length 3 in., and one side of length 4 in. Indicate which sides of your pattern would be joined to make the pyramid.

9. Make a pattern for the "bottom portion" **(frustum)** of a right cone, as pictured in Figure 13.28. What article of clothing is often shaped like this?

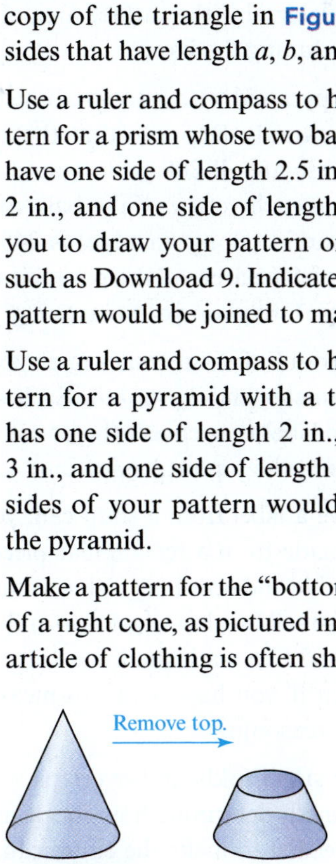

Figure 13.28 Bottom portion of a cone.

10. A company will manufacture a tent that will have a square, 15-ft by 15-ft base, 4 vertical walls, each 10 ft high, and a pyramid-shaped top made out of 4 triangular pieces of cloth, as shown in Figure 13.29. At its tallest point in the center of the tent, the tent will reach a height of 15 ft.

    Draw a small version of a pattern for one of the 4 triangular pieces of cloth that will make the pyramid-shaped top of the tent. Label your pattern with enough information so that someone making the tent would know how to measure and cut the material for the top of the tent.

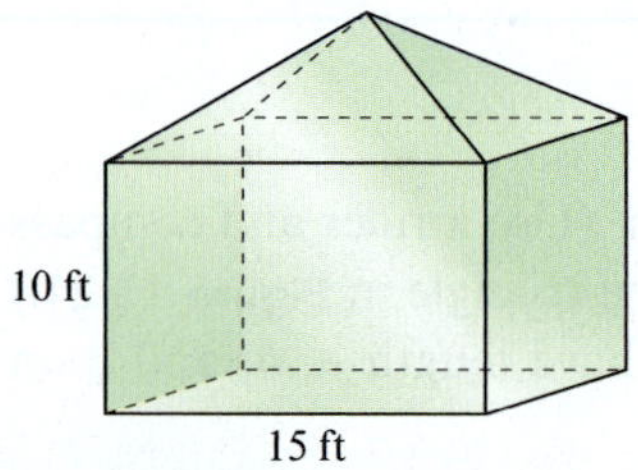

Figure 13.29 A tent.

11. Imagine an Egyptian right pyramid with a square base. Suppose the sides of the square base are 80 m long and the apex of the pyramid is 80 m above ground. Determine the surface area of the pyramid (not including the base).

12. Imagine a right pyramid with a square base (like an Egyptian pyramid). Suppose that the sides of the square base are all 200 yd long and that the distance from each vertex of the base to the apex of the pyramid (along an edge) is 245 yd. Determine the surface area of the pyramid (not including the base). Explain your reasoning.

13. A popular brand of soup comes in cans that are $2\frac{5}{8}$ in. in diameter and $3\frac{3}{4}$ in. tall. Each such can has a paper label that covers the entire side of the can (but not the top or the bottom).

a. If you remove a label from a soup can, you'll see that it's made from a rectangular piece of paper. How wide and how long is this rectangle (ignoring the small overlap where two ends are glued together)? Use mathematics to solve this problem, even if you have a can to measure. Explain your reasoning.

b. Ignoring the small folds and overlaps where the can is joined, determine how much metal sheeting is needed to make the entire can, including the top and bottom. Use the moving and additivity principles about area to explain why your answer is correct. Be sure to use an appropriate unit to describe the amount of metal sheeting.

14. The lateral portion of a cone (the part other than the base) is made from $\frac{1}{4}$ of a circle of radius 8 inches (by joining radii). By calculating, determine the radius of the circle that will make a base for the cone, and determine the total surface area of the cone (with the base). Explain your reasoning. You may wish to make the cone and check that the base you propose really does work.

15. a. Make a pattern for a cone such that the lateral portion of the cone (the part other than the base) is made from a portion of a circle of radius 4 in. (by joining radii) and such that the base of the cone is a circle of radius 3 in. Show any relevant calculations, explaining your reasoning.

b. Determine the total surface area (including the base) of your cone in part (a). Explain your reasoning.

16. a. Make a pattern for a cone such that the lateral part of the cone (the part other than the base) is made from $\frac{2}{3}$ of a circle (by joining radii) and such that the base of the cone is a circle of radius 6 cm. Show any relevant calculations, explaining your reasoning.

b. Determine the total surface area (including the base) of your cone in part (a). Explain your reasoning.

17. A cone with a circular base of radius 6 cm is to be made so that the distance from the apex of the cone to the center of its base is 8 cm.

a. Make a rough sketch of a pattern for the cone and determine the total surface area of the cone (including the base). Explain your reasoning.

b. Carefully draw a precise pattern for the lateral portion of the cone (the part other than the base), labeling the pattern with relevant lengths and angles. Explain all relevant calculations.

***18.** Tim needs a sturdy cardboard box that is 3 ft tall by 2 ft long by 1 ft wide. He wants to make the box out of a large piece of cardboard that he will cut, fold, and tape. The box must close up completely, so it needs a top and a bottom. Show Tim how to make such a box out of one rectangular piece of cardboard: Tell Tim what size cardboard he'll need to get (how wide, how long) and explain or show how he should cut, fold, and tape the cardboard to make the box. Be sure to specify exact lengths of any cuts Tim will need to make. Include pictures where appropriate. Your instructions and box should be practical; Tim should be able to actually make and use the box. You might want to make a scale model for your box out of paper. *Note:* Many boxes have flaps around all 4 sides that one can fold down and interlock to make the top and bottom of the box. You can make this kind of top and bottom, or something else if you prefer, but make sure it will make a sturdy box that closes completely.

*__19.__ Carla wants to make the triangular faces for a right pyramid with a $2\frac{3}{4}$-in.-by-$2\frac{3}{4}$-in. square base. The tallest point in the center of Carla's pyramid must be $2\frac{1}{2}$ in. above the base. Using a ruler that has tick marks every $\frac{1}{16}$ of an inch, carefully measure and draw one of the triangular faces for Carla's pyramid, labeling the lengths you measured. Show all calculations, explaining your reasoning. Be sure to explain how to measure decimal lengths, using a ruler that has tick marks every $\frac{1}{16}$ of an inch.

*__20.__ Describe in detail or draw precise patterns for two different cones that each has a surface area of approximately 30 cm^2, not including the base. Explain why your patterns produce cones of the desired surface area.

*__21.__ **a.** Make a pattern for an oblique cylinder with a circular base. You may leave the bases off your pattern. (*Advice:* Be willing to experiment first. You might start by making a pattern for a right cylinder and modifying this pattern.)

b. The sleeves of most shirts and blouses are more or less in the shape of an oblique cylinder. If your pattern in part (a) were to make a sleeve, what part of your pattern would be at the shoulder? What part of your pattern would be at the armpit?

*__22.__ Make a pattern for an oblique cone with a circular base. You may leave the base off your pattern. (*Advice:* Be willing to experiment first. You might start by making a pattern for a right cone and modifying this pattern.)

*__23.__ How many different nets for an open-top cube as described in Problem 2 are there? Find all such patterns.

*__24.__ **a.** Make three different nets that could be cut out, folded, and taped to make a closed 1-in.-by-1-in.-by-1-in. cube. To be nets, each square should be joined to another square along a whole edge of the square, not just at a vertex. For two patterns to be considered different, you should not be able to match the patterns up when they are cut out. Download 8 has 1-in. graph paper you can use.

b. How many different nets for a closed cube as described in part (a) are there? Find all such patterns.

13.3 Volumes of Solid Shapes

CCSS Common Core State Standards Grades 5, 6, 7, 8

The volume of a solid shape is a measure of how much three-dimensional space the shape takes up. As with area, there is a progression of increasingly sophisticated methods for determining volumes of solid shapes, culminating in the development of volume formulas for various shapes. Basic principles underlie all these methods.

What Is Volume?

CCSS
5.MD.3
5.MD.4

To understand how to determine volumes and to understand volume formulas, students must first know what volume means. For example, what does it mean to say that the volume of a solid shape is 30 cubic centimeters? It means that the solid shape could be made (without leaving any gaps) with a total of 30 1-cm-by-1-cm-by-1-cm cubes, allowing cubes to be cut apart and pieces to be moved if necessary. So the most primitive, basic way to determine the volume of a solid shape is to make the shape out of unit cubes (filling the inside completely) and to count how many cubes it took. Although primitive, this method is important, because it relies directly on the definition of volume and therefore emphasizes the meaning of volume. Even though students should know this approach to determining volume, they need to move beyond this primitive method to more efficient ways of determining volumes that can be used to solve problems. The first step in understanding how to determine volumes in other ways is to use the moving and additivity principles for volume, if only implicitly and subconsciously.

How Do the Moving and Additivity Principles Apply to Volumes?

As with shapes in a plane, fundamental principles determine how volumes behave when solid shapes are moved or combined:

1. If we move a solid shape rigidly without stretching or shrinking it, then its volume does not change.
2. If we combine (a finite number of) solid shapes *without overlapping* them, then the volume of the resulting solid shape is the sum of the volumes of the individual solid shapes.

We have already used these principles implicitly, in explaining why we can determine the volume of a box by multiplying its height times its width times its length (see Chapter 4). We thought of the box as subdivided into layers, and each layer as made up of 1-unit-by-1-unit-by-1-unit cubes. Each small cube has volume 1 cubic unit, and the volume of the whole box (in cubic units) is the sum of the volumes of the cubes, which is just the number of cubes.

CCSS 5.MD.5c

If you have a solid lump of clay, you can mold it into various different shapes. Each of these different shapes is made of the same volume of clay. From the point of view of the moving and additivity principles, it is as if the clay had been subdivided into many tiny pieces, and then these tiny pieces were recombined in a different way to form a new shape. Therefore, the new shape is made of the same volume of clay as the old shape. Similarly, if you have water in a container and you pour the water into another container, the volume of water stays the same, even though its shape changes.

How Does Cavalieri's Principle About Shearing Apply to Volumes?

In Section 12.5 we saw how we could shear plane shapes to obtain new shapes with the same area. Similarly, we can shear a solid shape and obtain a new solid shape that has the same volume.

Let's start with a polyhedron, pick one of its faces, and then imagine slicing the polyhedron into extremely thin (really, infinitesimally thin) slices that are parallel to the chosen face—this is rather like slicing a salami with a meat slicer. Now imagine giving those thin slices a push from the side, so that the chosen slice remains in place but so that the other thin slices slide over, remaining parallel to the chosen face and remaining the same distance from the chosen slice throughout the sliding process. The result is a new solid shape, as indicated in Figure 13.30. This process of sliding infinitesimally thin slices is called **shearing**.

shearing

You can show shearing nicely with a stack of paper. Give the stack of paper a push from the side, so that the sheets of paper slide over as shown in Figure 13.30. To understand shearing of solid shapes, think of the thin slices as made of paper. Note that in the shearing process, each thin slice remains unchanged: each slice is just slid over and is not compressed or stretched.

Figure 13.30 Shearing a stack of paper.

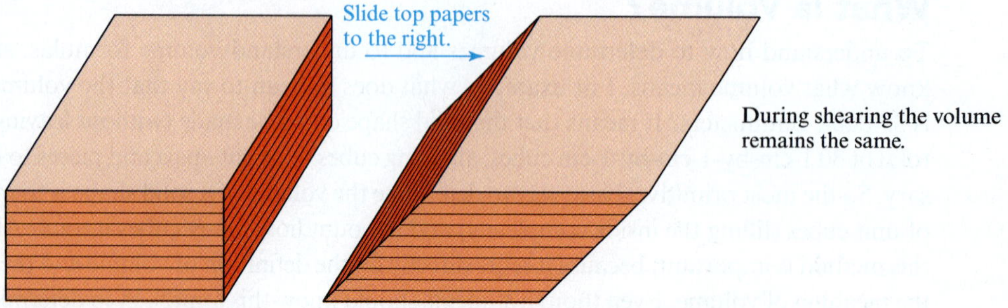

Cavalieri's principle

Cavalieri's principle for volume says that when you shear a solid shape as described, the volume of the original and sheared solid shapes are equal. This makes sense because during shearing volume is not added or taken away just; it is just "shifted over."

How Does the Volume Formula for Prisms and Cylinders Develop?

height (of prism or cylinder)

Before introducing the simple volume formula for prisms and cylinders, recall that prisms and cylinders can be thought of as formed by joining two parallel, identical bases. The **height of a prism or cylinder** is the distance between the planes containing the two bases of the prism or cylinder, measured in the direction perpendicular to the bases, as indicated in **Figure 13.31**. *The height is measured in the direction perpendicular to the bases, not along the slant.*

Figure 13.31

The base and height of a prism or cylinder.

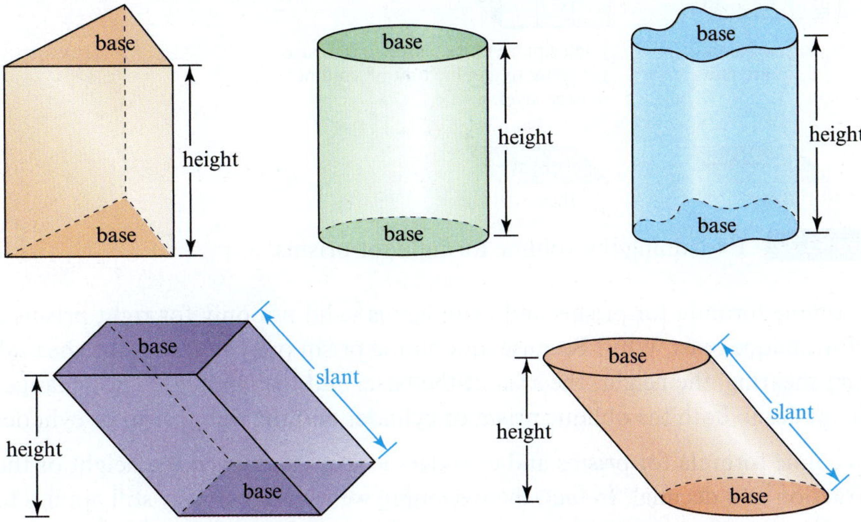

Class Activity 13K will help you explain why the height times area of the base volume formula for prisms and cylinders makes sense.

CLASS ACTIVITY

13K Why the Volume Formula for Prisms and Cylinders Makes Sense, p. CA-273

prism and cylinder volume formula

The **formula for volumes of prisms and cylinders** can be expressed as follows:

$$\text{volume} = (\text{height}) \cdot (\text{area of base})$$

In the volume formula it is understood that if the height is measured in some unit, then the area of the base is measured in square units of the same unit. The volume of the prism or cylinder resulting from the formula is then in cubic units of the same basic unit. For example, what is the volume of a 4-inch-tall can that has a circular base of radius 1.5 inches? The area of the base is $\pi(1.5)^2$ in^2, about 7.07 in^2; therefore, the volume of the can is $4 \cdot 7.07$ in^3 $= 28$ in^3 (approximately).

CCSS 5.MD.5a

Why do we multiply the height by the area of the base to calculate the volume of a prism or cylinder? Consider a rectangular prism that is 4 units high and has a base of area 6 square units, as pictured in **Figure 13.32**. If we fill this prism with 1-unit-by-1-unit-by-1-unit cubes, then the prism will be made of 4 layers. Each layer has 6 cubes in it—1 cube for each square unit of area in the base. So the whole prism is made of 4 groups (layers), with 6 cubes in each group, and therefore there are

$$4 \cdot 6$$

cubes in the prism. Each cube has a volume of 1 cubic unit; therefore, the volume of the prism is $4 \cdot 6$ cubic units. Notice that the height tells us how many layers the prism can be made of and the area of the base tells us how many unit cubes are in each layer. The same reasoning explains why the volume formula for any right prism or cylinder is

$$\text{volume} = (\text{height}) \cdot (\text{area of base})$$

cubic units.

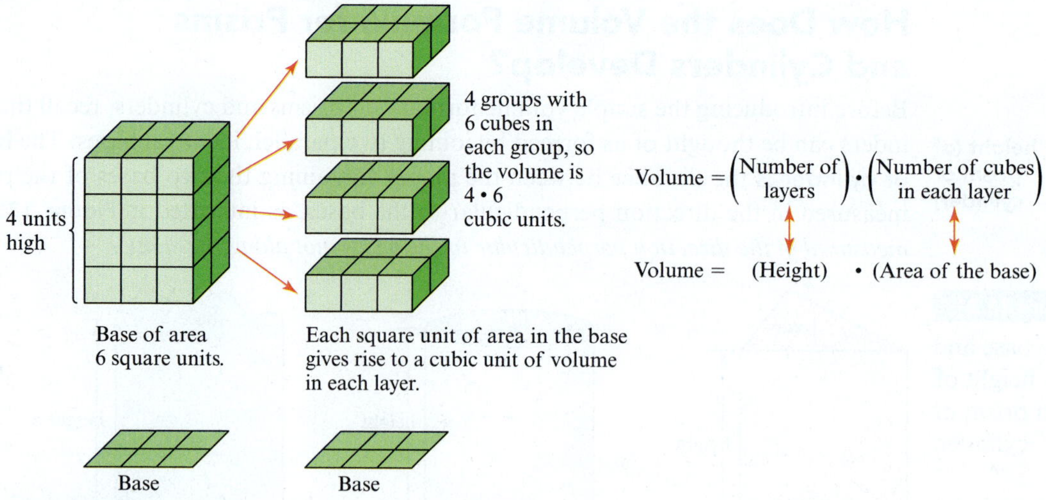

Figure 13.32 Explaining the volume formula for prisms.

CCSS
6.G.2

The volume formula for prisms and cylinders is valid not only for right prisms and cylinders but also for oblique ones. Why? Because an oblique prism or cylinder can be sheared into a right one. During shearing, the height, the area of the base, and the volume do not change, so the same formula applies to both the oblique prism or cylinder and the right prism or cylinder.

The volume formula for prisms and cylinders is also valid when the height or the area of the base is a fraction or a decimal. In fact, the reasoning we used previously still applies to the case of fractions and decimals, except that we must consider layers that contain partial cubes, such as layers containing $7\frac{1}{2}$ cubes. And we must consider partial layers. For example, if a prism is $4\frac{1}{2}$ units high instead of 4 units high, then we would have $4\frac{1}{2}$ groups of cubes instead of 4 groups of cubes. So we would have 4 full groups of cubes and another half-group of cubes, meaning another group of cubes that has only $\frac{1}{2}$ as much volume as the other groups.

How Does the Volume Formula for Pyramids and Cones Develop?

height (of pyramid or cone)

Recall that pyramids and cones can be thought of as formed by joining a base with an apex (a point). The **height of a pyramid or cone** is the perpendicular distance between the apex of the pyramid or cone and the plane containing the base. As indicated in Figure 13.33, *the height is measured in the direction perpendicular to the base, not along the slant.*

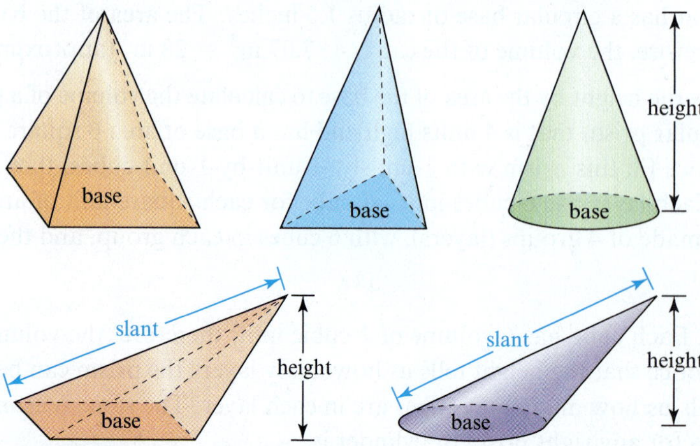

Figure 13.33 The base and height of a pyramid or cone.

CLASS ACTIVITY

13L Comparing the Volume of a Pyramid with the Volume of a Rectangular Prism, p. CA-274

13M The $\frac{1}{3}$ in the Volume Formula for Pyramids and Cones, p. CA-274

13N Volume Problem Solving, p. CA-275

pyramid and cone volume formula

The **formula for volumes of pyramids and cones** is

$$\text{volume} = \frac{1}{3} \cdot (\text{height}) \cdot (\text{area of base})$$

CCSS 8.G.9

In the volume formula it is understood that if the height is measured in some unit, then the area of the base is measured in square units of the same unit. The volume of the pyramid or cone resulting from the formula is then in cubic units of the same basic unit. For example, what is the volume of sand in a cone-shaped pile that is 15 feet high and has a radius at the base of 7 feet? According to the volume formula, the volume of sand is

$$\frac{1}{3} \cdot 15 \cdot \pi(7)^2 \text{ ft}^3$$

which is about 770 cubic feet of sand.

Where does the $\frac{1}{3}$ in the volume formula for pyramids and cones come from? Class Activities 13L and 13M help you see why the $\frac{1}{3}$ is plausible.

How Does the Volume Formula for a Sphere Develop?

The volume in cubic units of a sphere of radius r units is given by the formula

$$\text{volume} = \frac{4}{3}\pi r^3$$

sphere volume formula

How do we know this formula? Because volume is a three-dimensional attribute, it is not surprising that the formula involves r^3, which has an exponent of 3. Also, because spheres are related to circles, it is not surprising that the formula involves π. But where does the fraction $\frac{4}{3}$ come from?

One way to derive the volume formula for a sphere is with calculus. But there is another ingenious way that works by showing how the volume of a half-sphere is the same as the volume of a cylinder with a cone removed from it. To explain why those two volumes are the same, we can use Cavalieri's principle, which applies not only to shearing but actually to any case where two solid shapes have the same cross-section at every height. So to derive the volume of a sphere, we show that a half-sphere and a "cylinder minus a cone" have the same cross-section at every height. Because they have the same cross-section at every height, they have the same volume. But we can find the volume of a "cylinder minus a cone" because we know the volume formulas for cylinders and cones. Class Activity 13O will help you examine the details of this explanation.

CLASS ACTIVITY

13O Deriving the Volume of a Sphere, p. CA-276

How Are Volume and Surface Area Different?

A common source of confusion is the distinction between volume and surface area of a solid shape. Informally, we can think of surface area as a measure of how much paper, cloth, or other thin substance it would take to cover the outer surface of the shape. On the other hand, the volume of a shape is a measure of how much stuff it takes to fill the shape. You will think more about the distinction between surface area and volume in Class Activity 13P, as well as in the exercises and problems that follow.

CLASS ACTIVITY

13P Volume Versus Surface Area and Height, p. CA-277

FROM THE FIELD Research

Georgeson, J. (2011). Fold in origami and unfold math. *Mathematics Teaching in the Middle School, 16*(6), pp. 354–361.

The author discusses an engaging activity in which middle-school students use origami to fold three-dimensional shapes and explore various mathematical relationships along the way, including how the volume and surface area of a folded shape change with the size of paper used to make the shape.

SECTION SUMMARY AND STUDY ITEMS

Section 13.3 Volumes of Solid Shapes

The volume of a solid shape, in cubic units, is the number of 1-unit-by-1-unit-by-1-unit cubes it takes to make the shape (without leaving any gaps), allowing for cubes to be cut apart and pieces to be moved if necessary. The most primitive way to determine the volume of a shape is to count how many cubes it takes to make it.

The volume of a prism or a cylinder is (height) · (area of base). The volume of a pyramid or a cone is $\frac{1}{3}\cdot$ (height · area of base). In all cases, the height is measured perpendicular to the base, not on the slant. We can explain the volume formula for prisms and cylinders by viewing a prism or cylinder as cut into layers parallel to the base. We can explain the $\frac{1}{3}$ in the volume formula for pyramids and cones by putting 3 oblique pyramids together to form a cube and then shearing the oblique pyramids. We can also see that this $\frac{1}{3}$ is plausible because the contents of a prism will fill a pyramid of the same base and height 3 times.

The volume of a sphere of radius r is $\frac{4}{3}\pi r^3$.

The volume of a solid shape is distinct from its surface area and its height, although these distinctions are a source of confusion for some students.

Key Skills and Understandings

1. Know what volume is and know the most primitive way to determine the volume of a solid shape.
2. Know and use the moving and additivity principles for volume.
3. Explain why the volume formula for prisms and cylinders is valid.
4. Explain why the $\frac{1}{3}$ in the volume formula for pyramids and cones is plausible.
5. Use the volume formulas for prisms, cylinders, pyramids, cones, and spheres to determine volumes and to solve problems.
6. Discuss the distinction between the volume, the surface area, and the height of a solid shape.

Practice Exercises for Section 13.3

1. Your students have an open-top box that has a 2-in.-by-4-in. rectangular base and is 3 in. high. They also have a bunch of cubic-inch blocks and some rulers.
   a. What is the most primitive way for your students to determine the volume of the box?
   b. What is a more advanced way for your students to determine the volume of the box, and why does this method work?
2. Explain why the volume = (height) · (area of the base) formula is valid for right prisms using the example of a prism that is 4 units high and that has a base of area 6 square units.
3. Discuss why the $\frac{1}{3}$ in the volume formula for pyramids (and cones) is plausible.
4. What is the difference between the surface area of a solid object and its volume?
5. The water in a full bathtub is roughly in the shape of a rectangular prism that is 54 in. long, 22 in. wide, and 9 in. high. Use the fact that 1 gallon = 0.134 ft^3 to determine how many gallons of water are in the bathtub.
6. A concrete patio will be made in the shape of a 15-ft-by-15-ft square with half-circles attached at two opposite ends, as shown in Figure 13.34. If the concrete will be 3 in. thick, how many cubic feet of concrete will be needed?

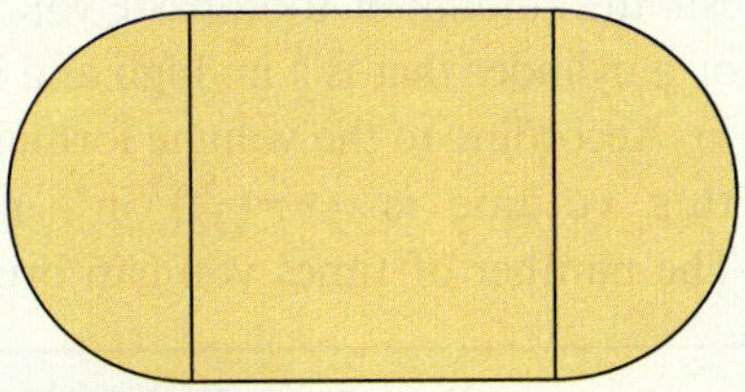

Figure 13.34 A patio.

7. Suppose that a tube of toothpaste contains 15 in^3 of toothpaste and that the circular opening where the toothpaste comes out has a diameter of $\frac{5}{16}$ inch. If every time you brush your teeth you squeeze out a $\frac{1}{2}$-inch-long piece of toothpaste, how many times can you brush your teeth with this tube of toothpaste?
8. A typical ice cream cone is $4\frac{1}{2}$ in. tall and has a diameter of 2 in. How many cubic inches does it hold (just up to the top)? How many fluid ounces is this? (One fluid ounce is 1.8 cubic inches.)

Answers to Practice Exercises for Section 13.3

1. The most primitive way for your students to determine the volume of the box is to count how many cubes it takes to fill the box. That number of cubes is the volume of the box in cubic inches. A more advanced way to determine the volume of the box is to multiply the height (3 in.) by the area of the base (8 in^2) to obtain the volume, 24 in^3. See text for how to explain why this method works.
2. Relate the formula to the number of layers and number of cubes in each layer. See text for details.
3. As the experiment in Class Activity 13L showed, you must pour the contents of a pyramid into a prism of the same height and base three times to fill the prism. Class Activity 13M showed that 3 oblique pyramids that have square bases fit together to make a cube.
4. See the text in this section as well as the previous section (and Section 11.2) for a discussion of surface area. See this section for a discussion of volume.
5. There are about 46 gallons of water in the bathtub. Because the water in the tub is $(54 \div 12)$ ft $= 4.5$ ft long, $(22 \div 12)$ ft, about 1.83 ft wide, and $(9 \div 12)$ ft $= 0.75$ ft tall, the volume of water in the tub is $4.5 \cdot 1.83 \cdot 0.75$ $ft^3 = 6.1875$ ft^3. Since each gallon of water is about 0.134 ft^3, the number of gallons of water in the tub is the number of 0.134 ft^3 in 6.1875 ft^3, which is $6.1875 \div 0.134$ gallons. This is about 46 gallons.
6. We need about 100 ft^3 of concrete for the patio. To explain why, notice that the patio is a kind of cylinder, so its volume can be calculated with the (height) · (area of base) formula. The base, which is the surface of the patio, consists of a square and two half-circles. Therefore, by the moving and additivity principles for areas, the area of the base is the area of the square plus the area of the circle that is created by putting the two half-circles together. All together, the area of the patio (the base) is $15^2 + \pi(7.5)^2$, about

401.7 ft^2. The height of the concrete patio is $\frac{1}{4}$ of a foot—notice that we must convert 3 in. to feet to have consistent units. Therefore, according to the (height) · (area of base) formula, the volume of concrete needed for the patio is $\frac{1}{4} \cdot 401.7$ ft^3, which is about 100 ft^3 of concrete.

7. You will be able to brush your teeth 391 times. Each time you brush, the amount of toothpaste you use is the volume of a cylinder that is $\frac{1}{2}$ in. high and has a radius of $\frac{5}{32}$ in. According to the volume formula for cylinders, this volume is $\frac{1}{2} \cdot \pi (\frac{5}{32})^2$ in^3, about 0.0383 in^3. The number of times you can brush is the number of 0.0383 in^3 in 15 in^3, which is 15 ÷ 0.0383, so about 391 times.

8. The cone holds about 4.7 in^3, which is about 2.6 fluid ounces. The diameter of an ice cream cone is 2 in., therefore its radius is 1 in., and so the area of the base of an ice cream cone (the circular hole that holds the ice cream) is $\pi \cdot 1^2$ in^2, about 3.14 in^2. According to the volume formula for cones, the volume of an ice cream cone is $\frac{1}{3} \cdot 4.5 \cdot 3.14$ in^3, about 4.7 in^3. Since each fluid ounce is 1.8 in^3, the number of fluid ounces the cone holds is the number of 1.8 in^3 in 4.7 in^3, which is 4.7 ÷ 1.8, about 2.6 fluid ounces.

PROBLEMS FOR SECTION 13.3

1. Suppose that a student in your class wants to know why we multiply only three of the lengths of the edges of a box in order to calculate the volume of the box. Why don't we have to multiply *all* the lengths of the edges?

a. Explain to this student why it makes sense to calculate the volume of a box as we do.

b. Describe some problems or activities that might help the student understand the calculation.

2. **a.** Students are sometimes confused about the difference between the *surface area* and the *volume* of a box. Explain the two concepts in a way that could help students learn to distinguish between them.

b. Determine the surface area and the volume of a closed box that is 5 in. wide, 4 in. deep, and 6 in. tall. Explain in detail why you calculate as you do.

3. Young children sometimes think that tall containers necessarily hold more than shorter containers. Use graph paper to make two patterns for open-top boxes so that one box is taller than the other, but so that the shorter box has the greater volume. Explain briefly why your boxes meet the required conditions.

4. Students often confuse the surface area and the volume of a solid shape.

a. Describe what surface area and volume are and discuss how they are different.

b. Describe a hands-on activity that could help students understand the distinction between surface area and volume.

5. In your own words, explain why the volume = (height) · (area of the base) formula is valid for right prisms using the example of a prism that is 5 cm high and that has a base of area 8 cm^2.

6. Discuss how to use blocks to explain why the volume = (height) · (area of base) formula for the volume of a prism is true (for the case of right prisms where the height and area of the base are whole numbers). In your explanation, be sure to attend to the meaning of multiplication, the meaning of volume, and the units you are using.

7. A cylindrical container has a base that is a circle of diameter 8 cm. When the container is filled with 250 mL of liquid, the container becomes $\frac{2}{5}$ full. How tall is the container? Explain your reasoning.

8. One liter of water is in a cylindrical container. The water is poured into a second cylindrical container whose radius is $\frac{3}{4}$ the radius of the first container. Describe how the water level in the second container compares to the water level in the first container. Be specific and explain your reasoning.

9. **a.** Measure how fast water comes out of some faucet of your choice. Give your answer in gallons per minute, and explain how you arrived at your answer. (You do not have to fill up a gallon container to do this.)

b. Figure 13.35 shows a bird's-eye view of a swimming pool in the shape of a cross (four 10-ft-by-10-ft squares surrounding a 10-ft-by-10-ft square). The pool is 4 ft deep but doesn't have any water in it. There is a small faucet on one side with which to fill the pool. How long would it take to fill up this pool, assuming that the water runs out of this faucet at the same rate as water from your faucet? Use the fact that 1 gallon is about 0.134 ft^3. Explain your answer.

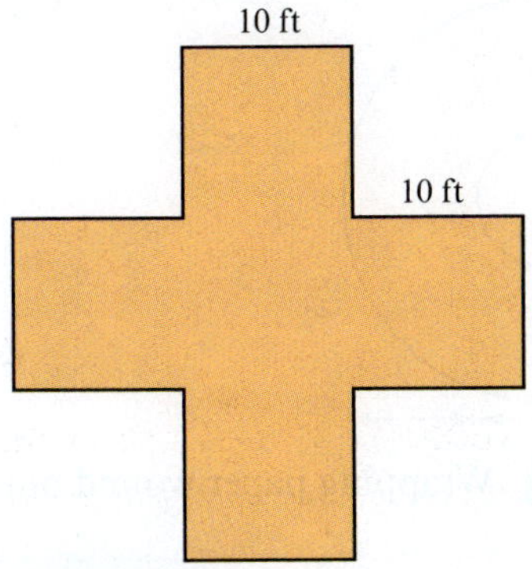

Figure 13.35 Swimming pool.

10. Find a gallon container, a half-gallon container, a quart container, a pint container, or a one-cup measure. Measure the lengths of various parts of your chosen container and use these measurements to determine the volume of your container in cubic inches. Use your answer to estimate how many gallons are in a cubic foot. Explain your answer. (There are about $7\frac{1}{2}$ gallons in a cubic foot.)

11. One gallon is 3.79 L, and 1 cm^3 holds 1 mL of liquid. Use these facts to determine the number of gallons in a cubic foot. Explain.

12. A cake recipe will make a round cake that is 6 in. in diameter and 2 in. high.

a. If you use the same recipe but pour the batter into a round cake pan that is 8 in. in diameter, how tall will the cake be? Explain.

b. Suppose you want to use the same cake recipe to make a rectangular cake. If you use a rectangular pan that is 8 in. wide and 10 in. long, and if you want the cake to be about 2 in. tall, then how much of the recipe should you make? (For example, should you make twice as much of the recipe, half as much, three-quarters as much, or some other amount?) Give an approximate, but *practical*, answer. Explain.

13. A recipe for gingerbread makes a 9-in.-by-9-in.-by-2-in. pan full of gingerbread. Suppose that you want to use this recipe to make a gingerbread house. You decide that you can either use a 10-in.-by-15-in. pan or a 11-in.-by-17-in. pan, and that you can either make a whole recipe or half of the recipe of gingerbread.

a. Which pan should you use, and should you make the whole recipe or just half a recipe, if you want the gingerbread to be between $\frac{1}{4}$ in. and $\frac{1}{2}$ in. thick? Explain.

b. Draw a careful diagram showing how you would cut the gingerbread into parts to assemble into a gingerbread house. Indicate the different parts of the house (front, back, roof, etc.). When assembled, the gingerbread house should look like a real three-dimensional house—but be as creative as you like in how you design it. You do not have to use every bit of the gingerbread to make the house, but try to use as much as possible.

c. If you want to put a solid, 2-in.-tall fence around your gingerbread house so that the fence is 6 in. away from the house, what will be the perimeter of this fence? Explain.

d. To make the fence in part (c), how many batches of gingerbread recipe will you need, and what pan (or pans) will you use? Indicate how you will cut the fence out of the pan (or pans). Explain.

14. The front (and back) of a greenhouse have the shape and dimensions shown in Figure 13.36. The greenhouse is 40 ft long, and the angle at the top of the roof is 90°. A fungus has begun to grow in the greenhouse, so a fungicide will need to be sprayed. The fungicide is simply sprayed into the air. To be effective, 1 tablespoon of fungicide is needed for every cubic yard of volume in the greenhouse. How much fungicide should be used? Give your answer in terms of units that are practical. (For example, it would not be practical to have to measure 100 tablespoons, nor would it be practical to have to measure 3.4 quarts. But it would be practical to measure 1 quart and 3 fluid ounces.) Explain.

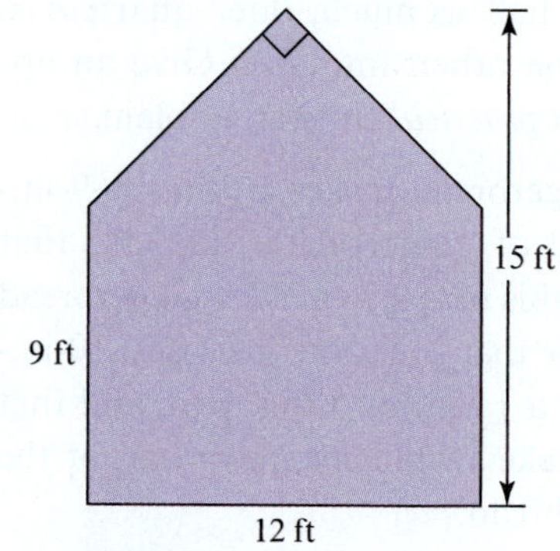

Figure 13.36 A greenhouse.

15. A volume problem:

The front and back of a storage shed are shaped like isosceles right triangles with two sides of length 10 ft, as shown on the left of Figure 13.37. The storage shed is 40 ft long. Determine the volume of the storage shed.

Qing solves the volume problem as follows: First, he uses the Pythagorean theorem to determine that the length of the unknown side of the triangle shown at the left is $\sqrt{200}$ ft long. Then, Qing uses the Pythagorean theorem again to calculate the height of the triangle h if the base is the side of length $\sqrt{200}$. Qing carries this out as follows:

$$\left(\frac{\sqrt{200}}{2}\right)^2 + h^2 = 10^2$$

$$h^2 = 100 - \frac{200}{4} = 100 - 50 = 50$$

$$h = \sqrt{50}$$

Next, Qing determines that the floor area of the storage shed is

$$40 \cdot \sqrt{200}$$

square feet. Finally, Qing calculates that the volume of the storage shed is given by the floor area of the shed times the height of the triangle:

$$40 \cdot \sqrt{200} \cdot \sqrt{50}$$

Figure 13.37 Determining volume.

a. Is Qing's method of calculation correct? Discuss Qing's work. Which parts (if any) are correct; which parts (if any) are incorrect?

b. Solve the volume problem in a different way than Qing did, explaining your reasoning.

16. Fifty pounds of wrapping paper are wound onto a roll. Looking at the roll from the side, we see that the outer diameter of the roll is 3 times the inner diameter, as indicated in Figure 13.38. The inner (unshaded) part of the roll is not filled with paper. How much would the same kind of wrapping paper weigh if the outer diameter of the roll were 2 times the inner diameter? Explain your reasoning.

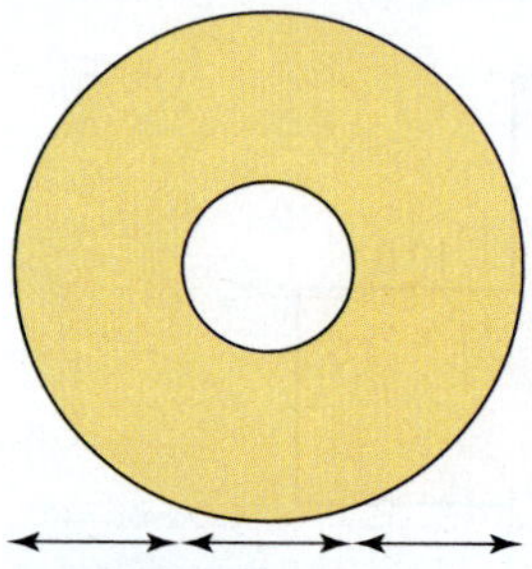

Figure 13.38 Wrapping paper wound onto a roll.

17. A conveyor belt dumps 2500 yd^3 of gravel to form a cone-shaped pile. How high could this pile of gravel be, and what could the circumference of the pile of gravel be at ground level? Give *two different realistically possible pairs of answers* for the height and circumference of the pile of gravel (both piles of volume 2500 yd^3), and compare how the piles of gravel would look in the two cases. Explain.

18. A construction company wants to know how much sand is in a cone-shaped pile. The company measures that the (slanted) distance from the edge of the pile at ground level to the very top of the pile is 55 ft. The company also measures that the distance around the pile at ground level is 220 ft.

a. How much sand is in the pile? (Be sure to state the units in which you are measuring this.) Explain.

b. The construction company has trucks that carry 10 cubic yards in each load. How many loads will it take to move the pile of sand? Explain.

19. One of the Hawaiian volcanoes is 30,000 ft high (measured from the bottom of the ocean) and has volume 10,000 mi^3. Assuming that the volcano is shaped like a cone with a circular base, find the distance around the base of the volcano (at the bottom of the ocean). In other words, if a submarine were to travel all the way around the base of the volcano, at the bottom of the ocean, how far would the submarine go? Explain.

20. Make a pattern for a cone-shaped cup that will hold about 120 mL when filled to the rim. Indicate measurements on your cup's pattern, and explain why the pattern will produce a cup that holds 120 mL. (A centimeter ruler may be helpful.)

21. Cut out the pattern on Download 10 and tape the two straight edges together to make a small cone-shaped cup. You might want to leave a small "tab" of paper on one of the edges, especially if you use glue instead of tape. In any case, make sure that the two straight edges are joined.

a. Determine approximately how many fluid ounces the cone-shaped cup holds by filling it with a dry, pourable substance (such as rice, sugar, flour, or even sand), and pouring the substance into a measuring cup.

b. The pattern for the cone-shaped cup indicates that the two straight edges you taped together are 10 cm long, and the curved edge that makes the rim of the cup is 25 cm long. Use these measurements to determine the volume of the cone-cup in milliliters. Explain your method.

c. If you read the label on a can of soda, you'll see that 12 fluid ounces is 355 mL. Use this relationship, and your answer to part (b), to determine how many fluid ounces the cone-cup holds. Compare your answer to your estimate in part (a).

22. A cone without a base is made from a quarter-circle. The base of the cone is a circle of radius 3 cm. What is the volume of the cone? Explain your reasoning.

23. a. Determine the volume of a cone that has a circular base of radius 6 cm and height 8 cm.

b. Make a pattern for the cone (without its base) in part (a). Show all relevant calculations, explaining your reasoning.

24. A cone without a base is made from $\frac{3}{4}$ of a circle of radius 20 cm. Determine the volume of the cone. Explain your reasoning.

25. a. A cone-shaped cup has a circular opening at the top of diameter 10 cm. When the cup is filled with 240 mL of liquid, it is $\frac{4}{5}$ full in terms of volume (so 240 mL is $\frac{4}{5}$ of the volume of the cup). What is the height of the cup? Explain your reasoning.

b. A cone-shaped cup has a circular opening at the top of diameter 10 cm. When the cup is filled with 240 mL of liquid, it is filled to $\frac{4}{5}$ of its height and the diameter of the surface of the liquid is 8 cm. What is the height of the cup? Explain your reasoning.

26. a. Eight identical spherical raindrops join together to form a larger spherical raindrop. How do the diameters compare? Explain.

b. A large spherical raindrop breaks apart into 64 identical spherical raindrops. How do the diameters compare? Explain.

13.4 Volume of Submersed Objects Versus Weight of Floating Objects

In this brief section we consider a hands-on way to determine volumes: submersing an object in water. When we do this, we can obtain information about the volume of the object. On the other hand, when an object *floats* in water, we can get information about the *weight* of the object.

CLASS ACTIVITY

13Q Underwater Volume Problems, p. CA-278

13R Floating Versus Sinking: Archimedes's Principle, p. CA-279

A hands-on way to determine the volume of an object is to submerse the object in a known volume of water and measure how much the water level goes up. For example, if we have a measuring cup filled with 300 mL of water and if the water level goes up to 380 mL when we put a plastic toy that sinks in the water, then the volume of the toy is 80 mL. Recall that 1 cubic centimeter holds 1 mL of liquid, so we can also say that the volume of the toy is 80 cm^3. Why does it make sense that the volume of the toy is 80 mL or 80 cm^3? When we place the toy in the water, the toy takes up space where water had been. The water that had been where the toy is now, is the "extra" water at the top of the measuring cup. So the volume of this "extra" water at the top of the measuring cup is the volume of the toy.

But what if an object is not *submersed* in water but *floats* on the water instead? For example, suppose we have a measuring cup filled with 300 mL of water and suppose that when we put a toy in the water, the toy floats, and the water level goes up to 380 mL. In this case, the amount of water that the toy displaced does not have the same *volume* as the toy; instead, it has the same *weight* as the toy. This physical fact, that *an object that floats displaces an amount of water that weighs as much as the object*, is called **Archimedes's principle**, in honor of Archimedes, the great mathematician and physicist who lived in ancient Greece, 287–212 B.C. So in the case of this floating toy, the toy weighs 80 grams because 80 mL of water weighs 80 grams.

Archimedes's principle

Archimedes's principle indicates why some objects float. Think about gradually lowering an object into water. Will the object float or not? As you lower the object into the water, it displaces more and more water. At some point, the object may have displaced a volume of water that weighs as much as the object. If so, then the object will float at that point. But if you lower an object into water, and if the amount of water it displaces never weighs as much as the object, then the object will sink. Notice that floating is not just a matter of how *light* the object is: It has to do with the *shape* of the submersed part of the object, and whether this shape displaces enough water. Otherwise, heavy ships made of steel would never be able to float.

SECTION SUMMARY AND STUDY ITEMS

Section 13.4 Volume of Submersed Objects Versus Weight of Floating Objects

A hands-on way to determine the volume of an object is to *submerse* the object in water and to determine how much water the object displaced. If an object *floats*, it does not displace its volume, but rather the amount of water that *weighs* as much as it does (Archimedes's principle).

Key Skills and Understandings

1. Determine the volume of an object that sinks in liquid from how much liquid the object displaces. Solve problems using this idea.
2. Determine the weight of an object that floats from how much liquid the floating object displaces. Solve problems using this idea.

Practice Exercises for Section 13.4

1. A fish tank in the shape of a rectangular prism is 40 cm tall, 60 cm long, and 25 cm wide. A rock is placed on the bottom of the tank. Then 25 liters of water are poured into the tank. At that point, the tank is $\frac{2}{3}$ full. What is the volume of the rock in cubic meters?

2. A measuring cup contains 300 mL of water. When a ball is put into the water in the measuring cup, the ball sinks to the bottom and the water level rises to 400 mL.

   a. What, if anything, can you deduce about the volume of the ball?

b. What, if anything, can you deduce about the weight of the ball?

3. Suppose you have a paper cup floating in a measuring cup that contains water. When the paper cup is empty, the water level in the measuring cup is at 250 mL. When you put some flour into the measuring cup, the cup is still floating and the water level in the measuring cup goes up to 350 mL. What information about the flour in the measuring cup can you deduce from this experiment? Explain.

Answers to Practice Exercises for Section 13.4

1. The volume of the rock is 0.015 cubic meters, which we can see as follows: The total volume of the tank is $40 \cdot 60 \cdot 25 = 60{,}000\text{ cm}^3$. Since 1 liter fills a 10-cm-by-10-cm-by-10-cm cube, 1 liter is 1000 cm^3. Therefore, $60{,}000\text{ cm}^3$ is 60 liters. So when the tank is $\frac{2}{3}$ full, it is filled with $\frac{2}{3} \cdot 60 = 40$ liters. Since 25 liters of water were poured in, the rock must take up $40 - 25 = 15$ liters, which is $15{,}000\text{ cm}^3$. But we want the volume in cubic meters. Since $1\text{ m} = 100\text{ cm}$,

$$1\text{ m}^3 = 100 \cdot 100 \cdot 100\text{ cm}^3 = 1{,}000{,}000\text{ cm}^3$$

Therefore, the volume of the rock is

$$\frac{15{,}000}{1{,}000{,}000} = \frac{15}{1000} = 0.015$$

cubic meters.

2. a. Since the water level rose 100 mL, the ball displaced 100 mL of water. One hundred milliliters has a volume of 100 cubic centimeters because 1 mL has a volume of 1 cm^3. So the ball has a volume of 100 cm^3.

b. If the ball were *floating*, then we would be able to say that the ball weighs 100 grams, because according to Archimedes's principle, a floating body displaces an amount of water that weighs as much as the body. But because the ball is not floating, we cannot determine the exact weight of the ball from this experiment. However, we can say that the ball must weigh more than 100 grams. Here's why. Think of gradually lowering the ball into the water. When we lower the ball into the water, there can never be a time when the amount of water that the ball displaces weighs as much as the ball—otherwise, the ball would float, according to Archimedes's principle. So the amount of water that the ball displaces as it is lowered into the water must always weigh less than the ball. Since the ball displaces 100 mL of water, and since 100 mL of water weighs 100 g, the ball must weigh more than 100 g.

3. Since the water level rose from 250 mL to 350 mL, the flour floating in the cup displaced 100 mL of water. Because the cup with the flour in it is floating, the weight of the displaced water is equal to the weight of the flour, according to Archimedes's principle. Because 1 mL of water weighs 1 gram, 100 mL of water weighs 100 g, so the flour weighs 100 g.

PROBLEMS FOR SECTION 13.4

1. A tank in the shape of a rectangular prism has a base that is 20 cm wide and 30 cm long. The tank is partly filled with water. When a rock is put in the tank and sinks to the bottom, the water level in the tank goes up 2 cm. What is the volume of the rock? Explain your reasoning.

2. A container holds 5 liters. Initially, the container is $\frac{1}{2}$ full of water. When an object is placed in the container, the object sinks to the bottom and the container becomes $\frac{7}{8}$ full. What is the volume of the object in cubic centimeters? Explain your reasoning.

3. A fish tank in the shape of a rectangular prism is 40 cm wide and 60 cm tall. At first, the tank is empty. Then, some stones with a total volume of $15{,}000\text{ cm}^3$ are put in the tank. Finally, 120 liters of water are poured into the tank. At that point, the tank is $\frac{3}{4}$ full. How long is the tank? Explain your reasoning.

4. Suppose that you have a recipe that calls for 200 grams of flour, but you don't have a scale. Explain in detail how to apply Archimedes's principle to measure 200 grams of flour by using a measuring cup that has metric markings.

5. Suppose that you have a recipe that calls for $\frac{1}{2}$ pound of flour, but you don't have a scale. Explain in detail how to apply Archimedes's principle and the fact that 1 kilogram = 2.2 pounds to measure $\frac{1}{2}$ pound of flour using a measuring cup that has metric markings. As a point of interest, we often say that "a pint is a pound"; therefore, you might think that 1 cup of flour (which is $\frac{1}{2}$ of a pint) weighs $\frac{1}{2}$ pound. However, the "pint is a pound" rule applies to water and to similar liquids like milk and clear juices. But ordinary flour is less dense than water, so 1 cup of flour actually weighs less than $\frac{1}{2}$ pound.

CHAPTER SUMMARY

Section 13.1 Polyhedra and Other Solid Shapes

Page 581

- Some of the basic solid shapes are prisms, cylinders, pyramids, and cones. There are five Platonic solids: the tetrahedron, the cube, the octahedron, the dodecahedron, and the icosahedron. Page 581

Key Skills and Understandings

1. Describe prisms, cylinders, pyramids, and cones and distinguish them from two-dimensional shapes. Page 581
2. Determine the number of vertices, edges, and faces of a given type of prism or pyramid, and explain why the numbers are correct. Page 585
3. Explain, based on angles, why putting too many faces together at a vertex results in a nonconvex shape. Page 586

Section 13.2 Patterns and Surface Area

Page 589

- An essential two-dimensional aspect of a solid shape is its outer surface. Patterns can be made to form the outer surface of a solid shape. The surface area of a solid shape is determined by adding the areas of all the component parts of the outer surface of the shape. Additional two-dimensional aspects of solid shapes are cross-sections and shadows. Page 589

Key Skills and Understandings

1. Visualize what shape a pattern will make. Page 589
2. Make patterns for prisms, cylinders, pyramids, and cones of specified dimensions. Page 589
3. Determine the surface area of prisms, cylinders, pyramids, and cones. Page 589

Section 13.3 Volumes of Solid Shapes

Page 597

- The volume of a solid shape, in cubic units, is the number of 1-unit-by-1-unit-by-1-unit cubes it takes to make the shape (without leaving any gaps), allowing for cubes to be cut apart and pieces to be moved if necessary. The most primitive way to determine the volume of a shape is to count how many cubes it takes to make it. Page 597
- The volume of a prism or a cylinder is (height) · (area of base). The volume of a pyramid or a cone is $\frac{1}{3}$ · (height · area of base). In all cases, the height is measured perpendicular to the base, not on the slant. We can explain the volume formula for prisms and cylinders by viewing a prism or cylinder as cut into layers parallel to the base. We can explain the $\frac{1}{3}$ in the volume formula for pyramids and cones by putting 3 oblique pyramids together to form a cube and then shearing the oblique pyramids. We can also see that this $\frac{1}{3}$ is plausible because the contents of a prism will fill a pyramid of the same base and height 3 times. Page 599

▪ The volume of a sphere of radius r is $\frac{4}{3}\pi r^3$.	Page 601
▪ The volume of a solid shape is distinct from its surface area and its height, although these distinctions are a source of confusion for some students.	Page 601
Key Skills and Understandings	
1. Know what volume is and know the most primitive way to determine the volume of a solid shape.	Page 597
2. Know and use the moving and additivity principles for volume.	Page 597
3. Explain why the volume formula for prisms and cylinders is valid.	Page 599
4. Explain why the $\frac{1}{3}$ in the volume formula for pyramids and cones is plausible.	Page 600
5. Use the volume formulas for prisms, cylinders, pyramids, cones, and spheres to determine volumes and to solve problems.	Page 601
6. Discuss the distinction between the volume, the surface area, and the height of a solid shape.	Page 601
Section 13.4 Volume of Submersed Objects Versus Weight of Floating Objects	Page 607
▪ A hands-on way to determine the volume of an object is to *submerse* the object in water and to determine how much water the object displaced. If an object *floats*, it does not displace its volume, but rather the amount of water that *weighs* as much as it does (Archimedes's principle).	Page 607
Key Skills and Understandings	
1. Determine the volume of an object that sinks in liquid from how much liquid the object displaces. Solve problems using this idea.	Page 607
2. Determine the weight of an object that floats from how much liquid the floating object displaces. Solve problems using this idea.	Page 608

14

Geometry of Motion and Change

In this chapter we extend our study of shapes by allowing them to move and to change size. The movements of shapes lead to the subject of symmetry: By copying and moving shapes in different ways, we can create designs with different kinds of symmetry. We will also consider what information we need to make an exact replica of a shape and the structural stability of shapes. Certain shapes are rigid and inflexible, whereas others are "floppy" and movable. Finally, we will ask what happens when objects are scaled so that they change size but otherwise retain the same shape. Scaling has widely used practical applications, such as the measurement of distances in land surveying.

In this chapter, we focus on the following topics and practices within the *Common Core State Standards for Mathematics* (*CCSSM*).

Standards for Mathematical Content in the CCSSM

In the domain of *Geometry* (Kindergarten–Grade 8), students recognize and draw lines of symmetry and line-symmetric figures. They draw geometric shapes with given conditions, and they focus on constructing triangles from three pieces of information about angles or side lengths, observing that some conditions determine a unique triangle, whereas others do not. They experiment with rotations, reflections, and translations and they use these transformations in describing congruence.

Students also solve problems involving scale drawings of geometric figures and they learn about dilations and the role of dilations in similarity.

Standards for Mathematical Practice in the CCSSM

Opportunities to engage in all eight of the Standards for Mathematical Practice described in the CCSSM occur throughout the study of transformations, symmetry, congruence, constructions, and similarity, although the following standards may be especially appropriate for emphasis:

- **2 Reason abstractly and quantitatively.** Students engage in this practice when they reason about scale models or scale drawings and their relationship to full-sized objects—for example, by comparing like parts in the scale model and the full-sized object or by comparing different parts within the scale model and within the full-sized object.
- **5 Use appropriate tools strategically.** Students engage in this practice when they use transparencies or geometry software to explore rotations, reflections, and translations and when they use a compass or protractor and ruler to construct shapes that have given conditions.
- **7 Look for and make use of structure.** Students engage in this practice when they investigate which attributes of a shape will specify the shape uniquely and when they relate the findings to structural aspects of shapes—for example, by seeing that triangles constructed to have fixed side lengths are structurally rigid whereas quadrilaterals are not.

(From Common Core Standards for Mathematical Practice. Published by Common Core Standards Initiative.)

14.1 Reflections, Translations, and Rotations

CCSS Common Core State Standards Grade 8

One especially attractive property of some shapes and designs is *symmetry*. In order to study symmetry, we will first examine certain transformations of planes—namely, reflections, translations, rotations, and glide-reflections. In the next section, we will use reflections, translations, rotations, and glide-reflections to define what we mean by symmetry, as well as to create symmetrical designs. From this point of view, these transformations are the building blocks of symmetry. Later in this chapter we will use reflections, translations, and rotations to define and study *congruence*.

What Are Reflections, Translations, Rotations, and Glide-Reflections?

Roughly speaking, a **transformation** (of a plane) is just what the name implies: an action that changes or transforms the locations of points in a plane. A transformation takes each point in the plane to another point in the plane (which could be the same point). The three kinds of transformations that we study in this section are reflections, translations, and rotations. We also briefly consider a fourth kind of transformation: glide-reflections. Throughout, assume we are working in a plane.

reflection **line of reflection**

A **reflection** (or **flip**) across a chosen line, called the **line of reflection**, is the following kind of transformation: For any point Q in the plane, imagine a line that passes through Q and is perpendicular to the line of reflection, as shown in Figure 14.1. Now imagine moving the point Q to the other point on the line that is the same distance from the line of reflection. (If point Q had been on the line of reflection, then it would stay in its place.) If we move each point in the plane as just described, then the end result is a reflection.

image

One way to think about reflections is that, under a reflection, each point in the plane goes to its "mirror image" on the other side of the line of reflection (hence the name "reflection"). In fact, if a reflection takes a point Q to a point Q', then we call Q' the **image** of Q. We use the term image for other transformations as well.

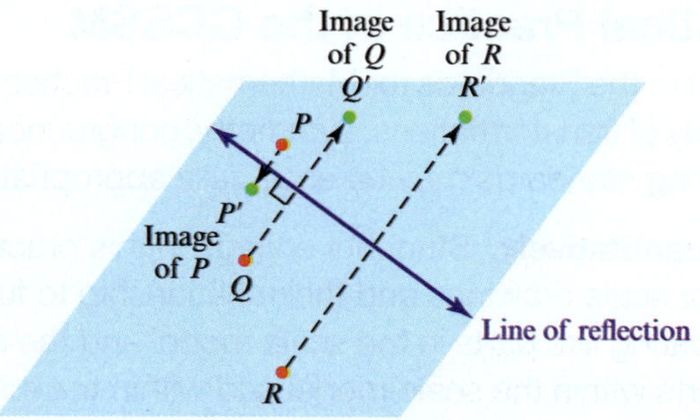

Figure 14.1 Effect of a reflection on points P, Q, and R.

An informal way to see the effect of a reflection is to draw points or a shape on paper in wet ink or paint and quickly fold the paper along the desired line of reflection: The location where the wet ink rubs off onto the paper shows the image (the final position) of the points or shape after reflecting. Another way to see the effect of a reflection is by drawing points or a shape on a transparency and flipping the transparency upside down by twirling it around the desired line of reflection: You will see the location of the reflected points or shapes.

translation A **translation** (or **slide**) by a given distance in a given direction is the end result of moving each point in the plane the given distance in the given direction. To illustrate a translation, put a transparency on a tabletop and slide the transparency in some direction (without rotating it). Now imagine doing this sliding process with all the points on a plane instead of just the points on a transparency: This would produce a translation. Figure 14.2 indicates initial and final positions—the images—of various points under a translation. Notice that the direction and distance of a translation can be specified by an arrow, as in this figure.

rotation A **rotation** (or **turn**) about a point through a given angle is a transformation that is the end result of rotating all points in the plane about a fixed point, through a fixed angle. To illustrate a rotation about a point, put a transparency on a tabletop, hold one point on the transparency fixed by pressing the end of a paperclip down on the point, and rotate the transparency about that point. Now imagine rotating all points on a plane instead of just the points on a transparency: This movement would produce a rotation about the fixed point. Figure 14.3 shows the initial and final positions—the images—of some points in a plane under a rotation about the point A. Notice that points farther from the point A move a greater *distance* than points closer to A, even though all points rotate through the same *angle*.

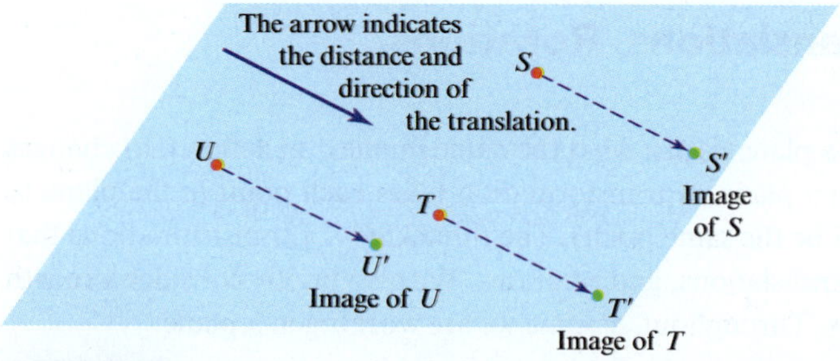

Figure 14.2 Effect of a translation on points S, T, and U. The solid arrow represents the distance and direction of the translation.

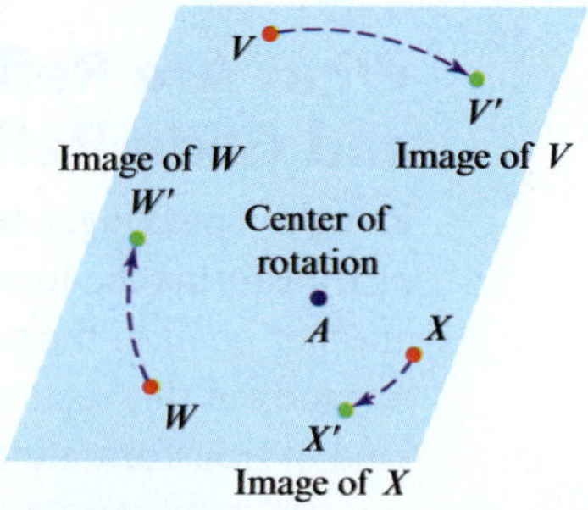

Figure 14.3 Effect of a rotation about point A on points V, W, and X.

glide-reflection A **glide-reflection** is the end result of combining a reflection and then a translation in the direction of the line of reflection. See Figure 14.4.

We defined reflections, translations, rotations, and glide-reflections in terms of what they do to points. What do these transformations do to other geometric objects, such as lines, line segments, shapes, and angles? You will explore this question in Class Activities 14A through 14D. Figure 14.5 shows examples of the initial and final positions—the images—of parallelograms under several transformations.

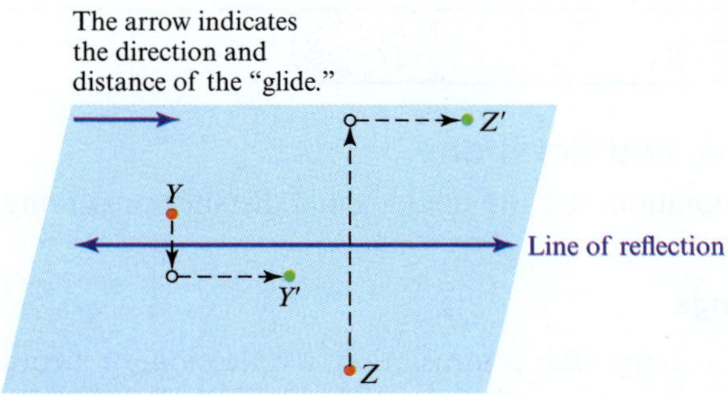

Figure 14.4 Effect of a glide-reflection on points Y and Z. The solid arrow indicates the direction and distance of the "glide" (translation) portion of the transformation and the line indicates the line of reflection.

Figure 14.5 Effect of a reflection, a translation, and a rotation on parallelograms.

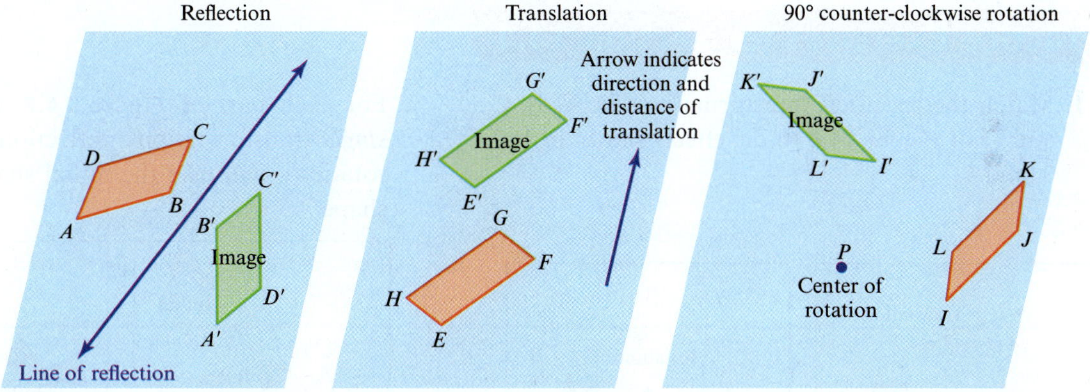

CLASS ACTIVITY

14A Exploring Reflections, Rotations, and Translations with Transparencies, p. CA-280

14B Reflections, Rotations, and Translations in a Coordinate Plane, p. CA-282

14C Rotating and Reflecting with Geometry Tools, p. CA-284

14D Which Transformation Is It?, p. CA-285

What Is Special About Reflections, Translations, Rotations, and Glide-Reflections?

CCSS 8.G.1

Rotations, reflections, translations, and glide-reflections are special because these transformations *do not change distances and do not change the size of angles*. When we model reflections, translations, and rotations with transparencies, as in Class Activity 14A, we see that these transformations take lines to lines, line segments to line segments of the same length, angles to angles of the same size, and parallel lines to parallel lines. For example, in Figure 14.5, when line segment AB is reflected its image is line segment $A'B'$, and both line segments are the same length. When the parallel sides EF and GH are translated their images are $E'F'$ and $G'H'$, which are still parallel. When angle IJK is rotated its image is angle $I'J'K'$, and both angles are the same size.

In Section 14.5 we will study another kind of transformation that *doesn't* preserve distances but does preserve the size of angles. Interestingly, mathematicians have proven that every transformation of the plane that preserves distances between all pairs of points must be either a rotation, a translation, a reflection, or a glide-reflection.

SECTION SUMMARY AND STUDY ITEMS

Section 14.1 Reflections, Translations, and Rotations

Reflections, translations, and rotations are the fundamental distance-preserving transformations of the plane.

Key Skills and Understandings

1. Determine the image of a shape after a translation, a reflection, or a rotation is applied.
2. Determine the effect on the coordinates of a point after a translation, a reflection across the *x*- or *y*-axis, or a rotation about the origin of 180° or 90° clockwise or counterclockwise is applied.
3. Given the location of a shape before and after a transformation, determine the transformation.

Practice Exercises for Section 14.1

1. Match the specified transformations in A, B, C, and D of Figure 14.6 to the effects shown in 1, 2, 3, and 4.

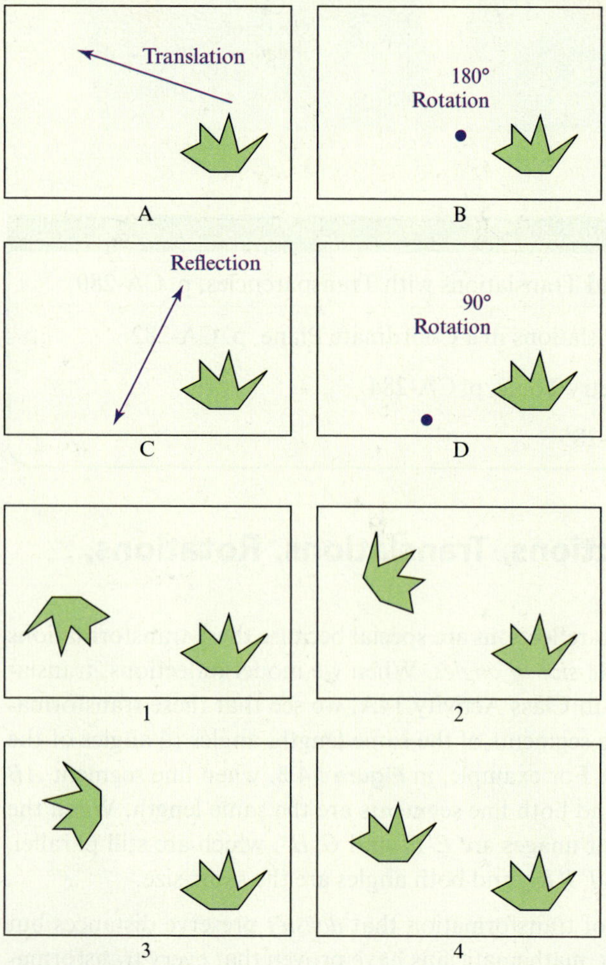

Figure 14.6 Various transformations and their effects.

2. For each part of Figure 14.7, determine what *single* transformation—reflection, translation, or rotation—will take the initial shape to the image shape.

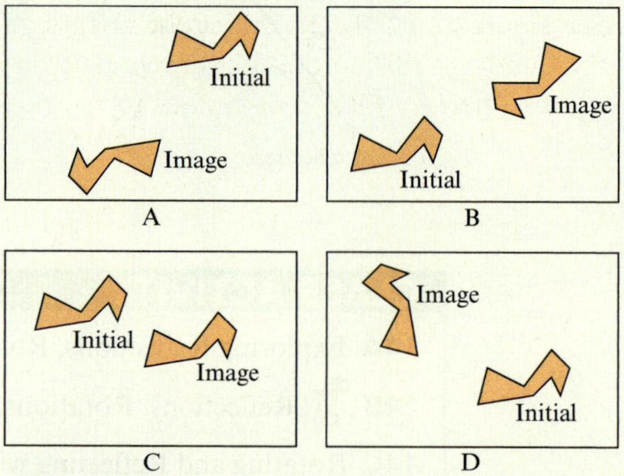

Figure 14.7 Identifying kinds of transformations.

3. Draw the images after rotating the shaded shapes in Figure 14.8 by 90° counterclockwise around the origin. Explain how you know where to draw the images.

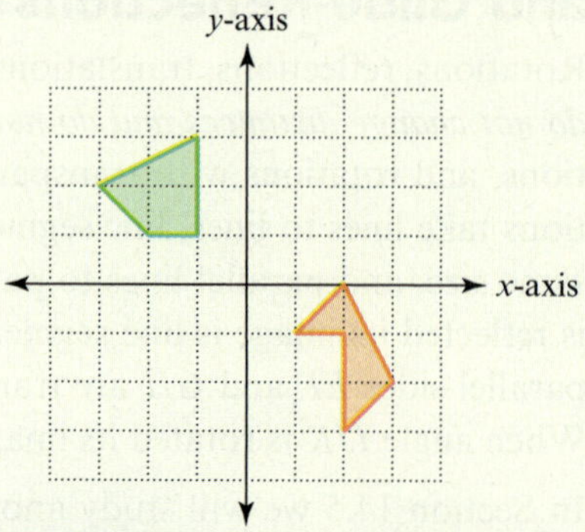

Figure 14.8 Determining the images of shapes after a 90° counterclockwise rotation.

Answers to Practice Exercises for Section 14.1

1. A–4, B–1, C–2, D–3.

2. See Figure 14.9.

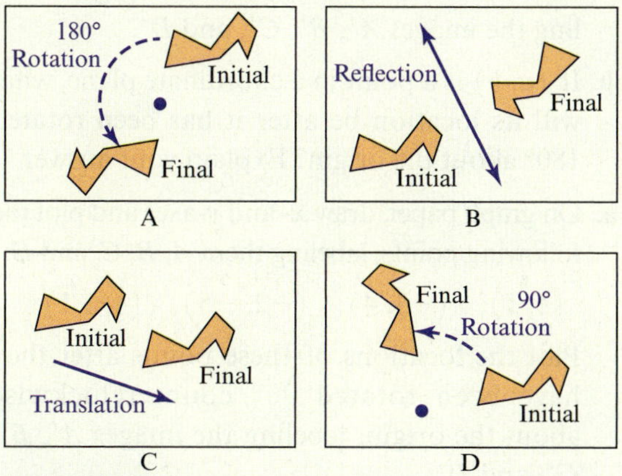

Figure 14.9 Effects of various transformations on a shape.

3. See Figure 14.10. To determine the images, notice that under a 90° rotation counterclockwise the positive part of the x-axis rotates to the positive part of the y-axis and the positive part of the y-axis rotates to the negative part of the x-axis. The rotation takes point A, which is 2 units to the right of the origin, to point A' which is 2 units up from the origin. Point B, which is 3 units below point A, must therefore rotate to point B', which is 3 units to the right of point A'. Point C, which is 3 units up and 1 unit to the left of the origin, will rotate to point C', which is 3 units to the left and 1 unit down from the origin. Similarly, by considering the location of each vertex in the shaded shapes relative to the x- and y-axes, we can determine the images of points when they are rotated, thereby determining the images of the shapes.

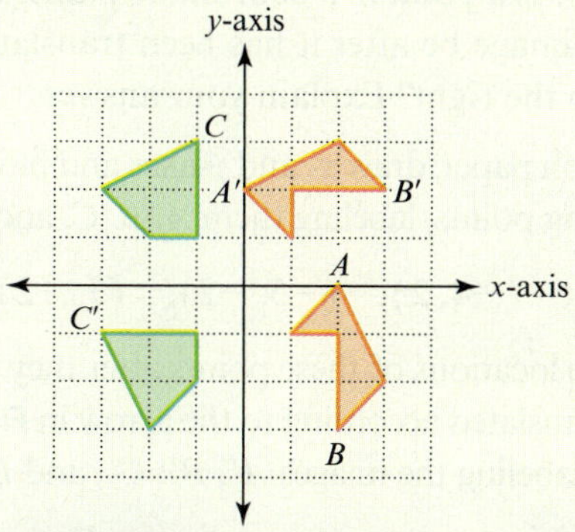

Figure 14.10 Shapes and their images after a 90° counterclockwise rotation.

PROBLEMS FOR SECTION 14.1

1. On graph paper, draw x- and y-axes, and draw two shapes that are not symmetrical. Draw the result of reflecting the shapes across the x-axis. Explain how you know where to draw the images.

2. On graph paper, draw x- and y-axes, and draw two shapes that are not symmetrical. Draw the result of reflecting the shapes across the y-axis. Explain how you know where to draw the images.

3. a. On graph paper, draw x- and y-axes, and draw two shapes that are not symmetrical. Draw the images after translating the shapes 5 units to the right.

b. On graph paper, draw x- and y-axes, and draw two shapes that are not symmetrical. Draw the result of translating the shapes in the direction given by the arrow in Figure 14.11. Explain how you know where to draw the images.

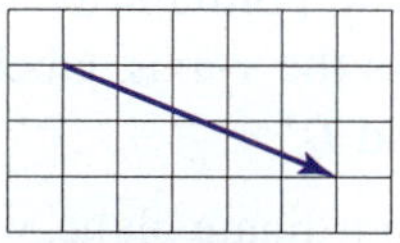

Figure 14.11 An arrow for a translation.

4. On graph paper, draw x- and y-axes, and draw two shapes that are not symmetrical. Draw the result of rotating the shapes 180° around the origin. Explain how you know where to draw the images.

5. On graph paper, draw x- and y-axes, and draw two shapes that are not symmetrical. Draw the result of rotating the shapes 90° counterclockwise around the origin. Explain how you know where to draw the images.

6. On graph paper, draw x- and y-axes, and draw two shapes that are not symmetrical. Draw the result of rotating the shapes 90° clockwise around the origin. Explain how you know where to draw the images.

7. a. On graph paper, draw x- and y-axes, and plot the following points, labeling them A, B, C, and D:

$$(5, 2), \quad (-3, 4), \quad (-4, -2), \quad (4, -5)$$

Plot the locations of these points after they have been translated 3 units to the right, labeling the images A', B', C', and D'.

b. If (a, b) is a point in a coordinate plane, what will its image be after it has been translated 3 units to the right? Explain your answer.

8. a. On graph paper, draw x- and y-axes and plot the following points, labeling them A, B, C, and D:

$$(3, 1), \quad (-4, 2), \quad (-5, -2), \quad (3, -2)$$

Plot the locations of these points after they have been translated according to the arrow in Figure 14.11, labeling the images A', B', C', and D'.

b. If (a, b) is a point in a coordinate plane, what will its location be after it has been translated according to the arrow in Figure 14.11. Explain your answer.

9. a. On graph paper, draw x- and y-axes and plot the following points, labeling them A, B, C, and D:

$$(4, 5), \quad (-4, 3), \quad (-3, -5), \quad (2, -1)$$

Plot the locations of these points after they have been reflected across the x-axis, labeling the images A', B', C', and D'.

b. If (a, b) is a point in a coordinate plane, what will its location be after it has been reflected across the x-axis? Explain your answer.

10. a. On graph paper, draw x- and y-axes and plot the following points, labeling them A, B, C, and D:

$$(3, 4), \quad (-5, 2), \quad (-4, -5), \quad (3, -4)$$

Plot the locations of these points after they have been reflected across the y-axis, labeling the images A', B', C', and D'.

b. If (a, b) is a point in a coordinate plane, what will its location be after it has been reflected across the y-axis? Explain your answer.

11. a. On graph paper, draw x- and y-axes and plot the following points, labeling them A, B, C, and D:

$$(3, 4), \quad (-5, 2), \quad (-4, -5), \quad (3, -4)$$

Plot the locations of these points after they have been rotated 180° about the origin, labeling the images A', B', C', and D'.

b. If (a, b) is a point in a coordinate plane, what will its location be after it has been rotated 180° about the origin? Explain your answer.

12. a. On graph paper, draw x- and y-axes and plot the following points, labeling them A, B, C, and D:

$$(5, 1), \quad (-3, 4), \quad (-2, -5), \quad (4, -1)$$

Plot the locations of these points after they have been rotated 90° counterclockwise about the origin, labeling the images A', B', C', and D'.

b. If (a, b) is a point in a coordinate plane, what will its location be after it has been rotated 90° counterclockwise about the origin? Explain your answer.

13. a. On graph paper, draw x- and y-axes and plot the following points, labeling them A, B, C, and D:

$$(3, 4), \quad (-1, 4), \quad (-3, -5), \quad (2, -3)$$

Plot the locations of these points after they have been rotated 90° clockwise about the origin, labeling the images A', B', C', and D'.

b. If (a, b) is a point in a coordinate plane, what will its location be after it has been rotated 90° clockwise about the origin? Explain your answer.

14. On a piece of paper, draw a point P and a separate irregular quadrilateral $ABCD$. Use a compass and protractor to draw the image $A'B'C'D'$ of the quadrilateral after a 140° rotation clockwise around P. Describe your method.

15. On a piece of paper, draw a point Q and a separate irregular quadrilateral $EFGH$. Use a compass and protractor to draw the image $E'F'G'H'$ of the quadrilateral after a 110° rotation counterclockwise around Q. Describe your method.

16. On a piece of paper, draw a line ℓ and a separate irregular quadrilateral $IJKL$. Use a right angle ruler (or the corner of a piece of paper) and a compass to draw the image $I'J'K'L'$ of the quadrilateral after reflecting across line ℓ. Describe your method.

17. For each of the following transformations, describe what the transformation does to the coordinates of points. How are the coordinates of a point and its image related? Explain in detail how you can tell. Does the answer depend on whether the coordinates are positive or negative or can you give a single rule that applies to all points in the plane?

a. A reflection across the x-axis.

b. A reflection across the y-axis.

18. For each of the following transformations, describe what the transformation does to the coordinates of points. How are the coordinates of a point and its image related? Explain in detail how you can tell. Does the answer depend on whether the coordinates are positive or negative or can you give a single rule that applies to all points in the plane?

a. A 180° rotation about the origin.

b. A 90° counterclockwise rotation about the origin.

c. A 90° clockwise rotation about the origin.

19. Describe what a reflection across the diagonal line $y = x$ does to the coordinates of points. How are the coordinates of a point and its image related? Explain in detail how you can tell.

20. Describe what a 90° counterclockwise rotation about the origin does to a line through the origin. How are the slopes of a line and its image related? Explain in detail how you can tell.

21. Describe what a reflection across the diagonal line $y = x$ does to a line through the origin. How are the slopes of a line and its image related? Explain in detail how you can tell.

***22.** Investigate the following questions, either with geometry software or by making drawings on (graph) paper. Start by drawing a shape (or design) that is not symmetrical. What is the *net effect* if you rotate your shape about some point by 180° and then rotate the resulting shape by 180° *about some other point?* What *single* transformation (reflection, translation, or rotation) will take your initial shape to your final shape?

***23.** Investigate the following questions, either with geometry software or by making drawings on (graph) paper. Start by drawing a shape (or design) that is not symmetrical. What is the *net effect* if you reflect your shape across a line and then reflect the resulting shape across another line that is perpendicular to the first line? What *single* transformation (reflection, translation, or rotation) will take your initial shape to your final shape?

***24.** Investigate the questions that follow, either with geometry software or by making drawings on (graph) paper. Draw a shape (or design) that is not symmetrical, draw a separate line, and draw a point on the line.

a. If you reflect the shape across the line and then rotate the reflected shape 180° about the point, will the final position of the shape be the same as if you had *first* rotated the shape 180° and *then* reflected the rotated shape across the line?

b. If you reflect the shape across the line and then rotate the reflected shape 90° counterclockwise about the point, will the final position of the shape be the same as if you had *first* rotated the shape 90° counterclockwise and *then* reflected the rotated shape across the original, unrotated line?

14.2 Symmetry

CCSS Common Core State Standards Grade 4

Symmetry is an area shared by mathematics, the natural world, and art, so it offers opportunities for cross-disciplinary study. Why is symmetry deeply appealing to most people? Maybe it is because objects with symmetry seem more perfect than objects that don't have symmetry. Or maybe it is because objects with symmetry involve repetition. Is there something about human nature that causes us to enjoy repetition? For example, almost all music involves repetition of themes—different themes are repeated throughout a piece of music in a certain pattern. Young children love to hear the same story over and over again. Maybe we like repetition because it helps us learn and understand, and maybe our enjoyment of repetition makes symmetrical designs appealing.

When we look at natural objects in the world around us, we find a mix of symmetry and asymmetry. Nearly all creatures are mostly symmetrical. Plants typically have symmetrical parts, even if they are not symmetrical over all: Leaves and flowerheads are usually symmetrical. On the other hand, geological features are rarely symmetrical. It would be surprising to see a mountain, a lake, a river, or a rock that we would describe as symmetrical, even though these objects are usually made up of smaller, repeated parts. Other natural objects such as some volcanoes, snowflakes, crystals, beaches, river stones, and waves are symmetrical, or nearly so.

Just as the concept of *circle* has an informal or artistic interpretation as well as a mathematical definition, the notion of *symmetry* has both an informal interpretation and a specific mathematical definition that applies to shapes in a plane or in space. Children in elementary school first learn about symmetry informally, in terms of matching parts. We consider a more precise mathematical definition of symmetry, which requires that we use the transformations just studied: rotations, reflections, and translations. First, we use rotations, reflections, and translations to *define* symmetry. In other words, we use these transformations to say what it means for a shape or a design to have specific kinds of symmetry. Next, we use these transformations to *create* shapes and designs that have symmetry.

What Is Symmetry?

There are four kinds of symmetry that a shape or design in a plane can have: reflection symmetry, translation symmetry, rotation symmetry, and glide-reflection symmetry. We use reflections, translations, and rotations to say what type of symmetry a shape or design has.

reflection (mirror) symmetry

line of symmetry

CCSS 4.G.3

A shape or design in a plane has **reflection symmetry** (or **mirror symmetry**) if there is a line in the plane such that there are matching parts when the shape or design is folded along the line; in other words, the shape or design as a whole occupies the same place in the plane both before and after reflecting across the line. This line is called a **line of symmetry**. For example, **Figure 14.12** shows a design and a line of symmetry of the design. Notice that reflecting across the line of symmetry causes most points on the design to swap locations with another point on the design, but the design *as a whole* occupies the same place in the plane. Shapes or designs can have more than one line of symmetry, as shown in **Figure 14.13**.

Figure 14.12 A shape with reflection symmetry.

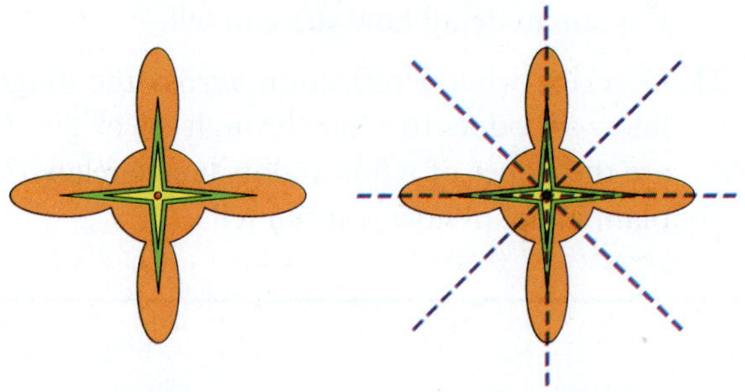

Figure 14.13 A design with four lines of symmetry.

To determine whether a design or shape drawn on a semitransparent piece of paper has reflection symmetry with respect to a certain line on the paper, fold the paper along that line. If you can see that the parts of the design on the two parts of the paper match one another, then the design has reflection symmetry, and the line you folded along is a line of symmetry. Another way to see whether a design or shape has reflection symmetry is to use a mirror that has a straight edge. Place the straight edge of the mirror along the line that you think might be a line of symmetry, and hold the mirror so that it is perpendicular to the design. When you look in the mirror, is the design the same as it was without the mirror in place? If so, then the design has reflection symmetry. These methods may help students get a better feel for reflection symmetry, but it is also important to use *visualization* to determine whether a design has reflection symmetry.

translation symmetry

A design or pattern in a plane has **translation symmetry** if there is a translation of the plane such that the design or pattern *as a whole* occupies the same place in the plane both before and after the translation. True translation symmetry occurs only in designs or patterns that take up an infinite amount of space. So when we draw a picture of a design or pattern that has translation symmetry, we can only show a small portion of it; we must imagine the pattern continuing on indefinitely. Figure 14.14 shows a pattern with translation symmetry. In fact, notice that this pattern has two independent translations that take the pattern as a whole to itself: One is "shift right," another is "shift up." Wallpaper patterns generally have translation symmetry with respect to translations in two independent directions.

The pattern continues forever in all directions.

Figure 14.14 A wallpaper pattern with translation symmetry.

Frieze patterns—often seen on narrow strips of wallpaper positioned around the top of the walls of a room—provide additional examples of designs with translation symmetry. Figure 14.15 shows a frieze pattern that has translation symmetry. Unlike a wallpaper pattern, a frieze pattern will have translation symmetry only in one direction (and its "reverse") instead of in two independent directions.

The pattern continues forever to the right and to the left.

Figure 14.15 A frieze pattern with translation symmetry.

rotation symmetry

A shape or design in a plane has **rotation symmetry** if there is a rotation of the plane of more than 0° but less than 360°, such that the shape or design *as a whole* occupies the same points in the plane both before and after rotation. For example, in Figure 14.16, rotating 72° about the center of the design results in the same design in the plane, even though each individual curlicue in the design moves to the position of another curlicue.

Figure 14.16 A design with 5-fold rotation symmetry.

The design of Figure 14.16 is said to have 5-fold rotation symmetry because, by applying the $72° = 360° \div 5$ rotation about the center of the design 5 times, all points on the design return to their starting positions. Generally, shapes or designs can have 2-fold, 3-fold, 4-fold, and so on, rotation symmetry.

n-fold rotation symmetry

A design has **n-fold rotation symmetry**, provided that a rotation of $360 \div n$ degrees takes the design as a whole to the same location. Applying this rotation n times returns every point on the design to its initial position. Figure 14.17 shows some examples of designs with various rotation symmetries.

Figure 14.17 Designs with rotation symmetry.

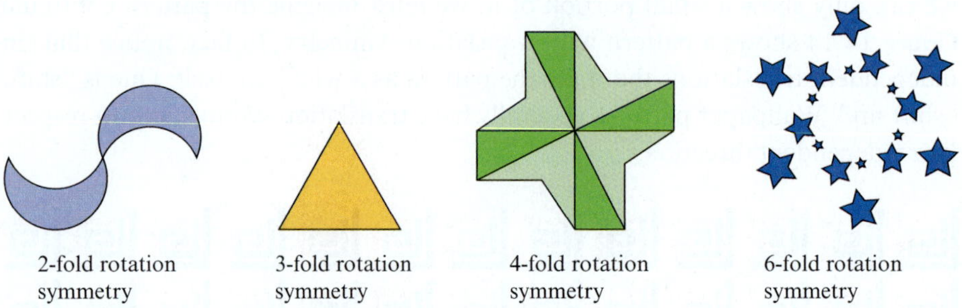

You can determine whether a design or shape drawn on a piece of paper has rotation symmetry: Make a copy of the design on a transparency. Place this transparency over the original design so that the two designs match one another. If you can rotate the transparency less than a full turn so that the two designs match one another again, then the design has rotation symmetry. If you can keep rotating by the same amount for a total of 2, 3, 4, 5, and so on, times until the design on the transparency is back to its initial position, then the design has 2-fold, 3-fold, 4-fold, 5-fold, and so on, rotation symmetry, respectively. When you are first learning about rotation symmetry, it may help you to physically rotate designs; however, you should also use *visualization* to determine if a design has rotation symmetry.

glide-reflection symmetry

A design or pattern in a plane has **glide-reflection symmetry** if there are a reflection and a translation such that, after applying the reflection followed by the translation, the *design as a whole* occupies the same location in the plane. Glide-reflection symmetry is often seen on frieze patterns. For example, Figure 14.18 shows an example of a frieze pattern with glide-reflection symmetry. If this design is reflected across a horizontal line through its middle and then translated to the right (or left), the design as a whole will occupy the same location in the plane as it did originally. (See Figure 14.19.)

Figure 14.18 A frieze pattern with glide-reflection symmetry.

Figure 14.19 Understanding glide-reflection symmetry.

Notice that the frieze pattern in Figure 14.18 also has translation symmetry. Many designs have more than one type of symmetry. For example, the design shown in Figure 14.20 has both 2-fold rotation symmetry as well as reflection symmetry with respect to two lines: one horizontal and one vertical.

Figure 14.20

A design on Egyptian fabric.

CLASS ACTIVITY

14E Checking for Symmetry, p. CA-286

SECTION SUMMARY AND STUDY ITEMS

Section 14.2 Symmetry

A shape or design has reflection symmetry (or mirror symmetry), if there is a line in the plane such that the shape or design as a whole occupies the same place in the plane both before and after reflecting across the line. A shape or design has n-fold rotation symmetry if there is a point such that, after rotating by $360° \div n$ around the point, the shape or design as a whole occupies the same place in the plane both before and after rotation. A shape or design has translation symmetry if there is a translation such that the shape or design as a whole occupies the same place in the plane both before and after translation.

Key Skills and Understandings

1. Given a shape or design, determine its symmetries.

Practice Exercises for Section 14.2

1. Symmetrical designs are found throughout the world in all cultures. Figure 14.23 shows a small sample of such designs. Determine the kinds of symmetry these designs have.

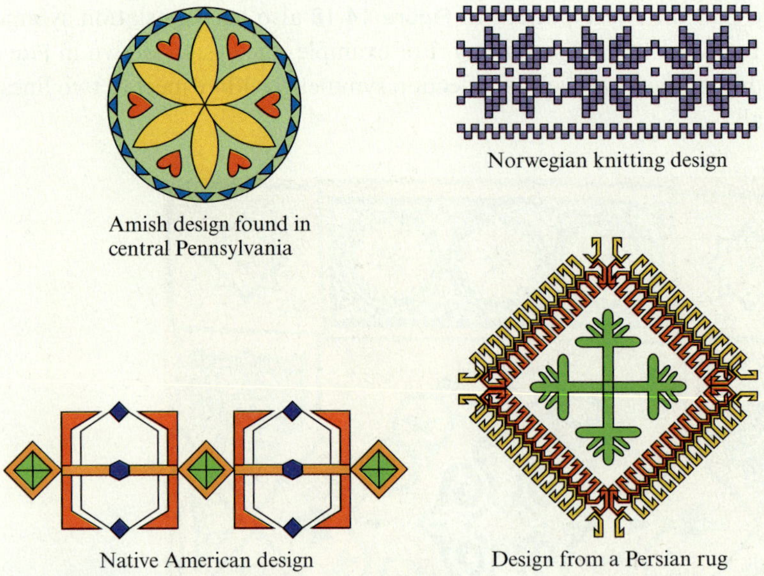

Figure 14.21 Symmetrical designs from around the world.

Figure 14.22 Create a symmetrical design.

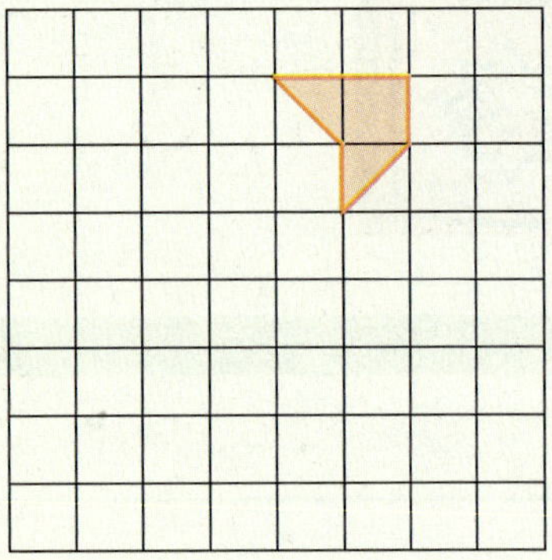

2. Draw a design that is made out of copies of the shaded shape in Figure 14.22 and has 4-fold rotation symmetry but no reflection symmetry.
3. Draw a design that is made out of copies of the shaded shape in Figure 14.22 and has two lines of symmetry.
4. What is the difference between a reflection and reflection symmetry?

Answers to Practice Exercises for Section 14.2

1. The Amish design has 6-fold rotation symmetry in addition to reflection symmetry with respect to 6 different lines: 3 lines that pass through the middle of opposite flower petals and 3 lines that pass through the middle of opposite hearts. (Notice that because the design has 6-fold rotation symmetry, it also has 3-fold and 2-fold rotation symmetry.)

   The Norwegian knitting design has 2-fold rotation symmetry in addition to reflection symmetry with respect to 2 different lines: one horizontal line and one vertical line. (Notice that the individual "snowflake" designs within the design have 4-fold rotation symmetry and have reflection symmetry with respect to horizontal, vertical, and 2 diagonal lines, but the entire Norwegian knitting design has less symmetry.)

   The Native American design has the same symmetries as the Norwegian knitting design—namely, 2-fold rotation symmetry and reflection symmetry with respect to a horizontal and a vertical line.

   The design from a Persian rug has 2-fold rotation symmetry in addition to reflection symmetry with respect to 2 different lines: one vertical and one horizontal. (The design at first appears to have 4-fold rotation symmetry, but it does not because of the way the "teeth" on the border are oriented.)

2. See Figure 14.23. The center of rotation is marked.

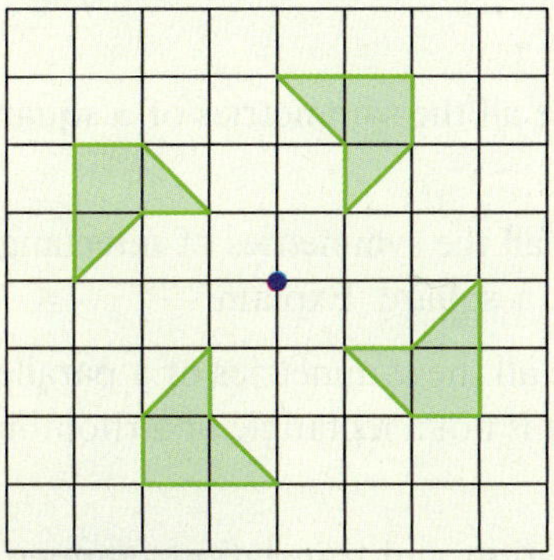

Figure 14.23 A design with 4-fold rotation symmetry.

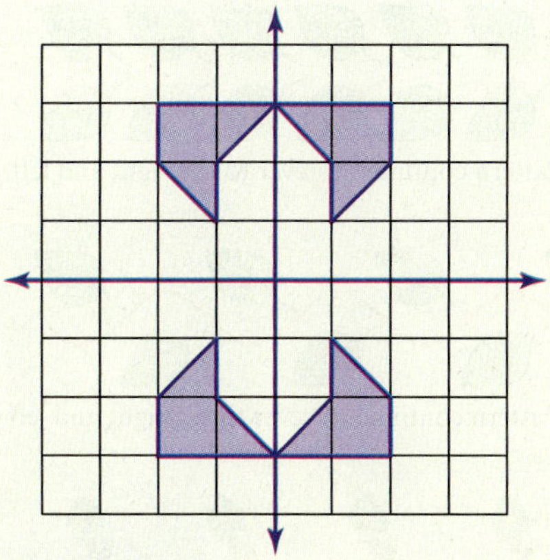

Figure 14.24 A design with 2 lines of symmetry.

3. See Figure 14.24. The 2 lines of symmetry are marked. Notice that the design also has 2-fold rotation symmetry.

4. A reflection is a *transformation* of the plane. Reflection symmetry is a *property* that certain shapes or designs have.

PROBLEMS FOR SECTION 14.2

1. Find examples of symmetrical designs from a modern-day culture or a historical one. Make copies of the designs (either by photocopying or by drawing them) and determine what kinds of symmetry the designs have.

2. Determine all the symmetries of Design 1 and Design 2 in Figure 14.25. Consider each design as a whole. For each design, describe all lines of symmetry (if the design has reflection symmetry), and determine whether the design has 2-fold, 3-fold, 4-fold, or other rotation symmetry. Explain your answers.

3. Determine all the symmetries of Design 3 and Design 4 in Figure 14.25. Consider each design as a whole. For each design, describe all lines of symmetry (if the design has reflection symmetry), and determine whether the design has 2-fold, 3-fold, 4-fold, or other rotation symmetry. Explain your answers.

4. Determine all the symmetries of Design 5 and Design 6 in Figure 14.25. Consider each design as a whole. For each design, describe all lines of symmetry (if the design has reflection symmetry), and determine whether the design has 2-fold, 3-fold, 4-fold, or other rotation symmetry. Explain your answers.

Figure 14.25 Some symmetrical designs.

5. Determine all the symmetries of Design 7 and of Design 8 in Figure 14.26. Consider each design

Design 7: Pattern continues forever to the right and left.

Design 8: Pattern continues forever to the right and left.

Design 9: Pattern continues forever in all directions.

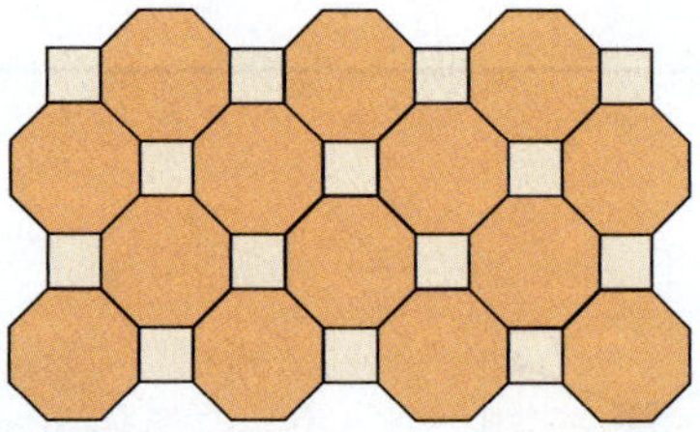

Design 10: Pattern continues forever in all directions.

Figure 14.26 Some symmetrical designs.

as a whole. For each design, describe all lines of symmetry (if the design has reflection symmetry), and determine whether the design has 2-fold, 3-fold, 4-fold, or other rotation symmetry. Don't forget to look for translation symmetry and glide-reflection symmetry, too. Explain your answers.

6. Determine all the symmetries of Design 9 and Design 10 in Figure 14.26. Consider each design as a whole. For each design, describe all lines of symmetry (if the design has reflection symmetry), and determine whether the design has 2-fold, 3-fold, 4-fold, or other rotation symmetry. Don't forget to look for translation symmetry and glide-reflection symmetry, too. Explain your answers.

7. **a.** What types of triangles have reflection symmetry? Explain.

b. What types of triangles have rotation symmetry? Explain.

8. **a.** Determine all the symmetries of a square. Explain.

b. Determine all the symmetries of a rectangle that is not a square. Explain.

c. Determine all the symmetries of a parallelogram that is not a rectangle or a rhombus. Explain.

9. Compare translation and translation symmetry. Explain how these two concepts are related.

10. Compare rotation and rotation symmetry. Explain how these two concepts are related.

11. Compare (mathematical) reflection and reflection symmetry. Explain how these two concepts are related.

***12.** Draw a simple asymmetrical shape or design. Then copy your shape or design so as to create a single new design that has both 2-fold rotation symmetry and translation symmetry *simultaneously*. (You may wish to use graph paper or software.)

***13.** Draw a simple asymmetrical shape or design. Then copy your shape or design so as to create a single new design that has both 4-fold rotation symmetry and translation symmetry *simultaneously*. (You may wish to use graph paper or software.)

***14.** Draw a simple asymmetrical shape or design. Then copy your shape or design so as to create a single new design that has both 2-fold rotation symmetry and reflection symmetry *simultaneously*. (You may wish to use graph paper.)

***15.** Draw a simple asymmetrical shape or design. Then copy your shape or design so as to create a single new design that has both 4-fold rotation symmetry and reflection symmetry *simultaneously*. (You may wish to use graph paper or software.)

***16.** Draw a simple asymmetrical shape or design. Then copy your shape or design so as to create a single new design that has rotation, reflection, and translation symmetry *simultaneously*. (You may wish to use graph paper or software.)

14.3 Congruence

CCSS **Common Core State Standards** Grades 7, 8

In this section we focus on *congruence*, which is the mathematical term for saying two geometrical objects are identical or exact replicas of each other. We can define congruence in terms of rotations, translations, and reflections, which we studied in Section 14.1. What information about two shapes will guarantee that they are congruent? Surprisingly, the answer for triangles differs from the answer for other shapes. Triangles, unlike other polygons, are structurally rigid, and a criterion for the congruence of triangles is related to this property. This distinction between triangles and other polygons explains some standard practices in building construction. So even though the study of congruence may seem purely abstract and theoretical, congruence has many practical applications.

What Is Congruence?

Informally, two shapes (either in a plane or solid shapes in space) that are the same size and shape are called congruent. But if we have two shapes or designs drawn on two semitransparent pieces of paper, how would we determine whether the shapes are congruent? Before you read on, try Class Activity 14F.

CLASS ACTIVITY

14F Motivating a Definition of Congruence, p. CA-287

CCSS 8.G.2

To check if two paper shapes are congruent we would probably slide one shape over the other and rotate the shapes to see if they match up. If not, we might flip one piece of paper over and try again. So to check if two shapes are congruent, we would use translations, rotations, and reflections. Therefore, we define congruence formally in the following way: Two shapes or designs in a plane are **congruent** if there is a rotation, a reflection, a translation, or a combination of these transformations that takes one shape or design to the other shape or design. For example, Figure 14.27 shows how shapes 1 and 3 are congruent. They are congruent because shape 3 is obtained by first translating shape 1 horizontally to the right to make shape 2 and then rotating shape 2 90° about the point shown.

congruent

Figure 14.27 Congruent shapes.

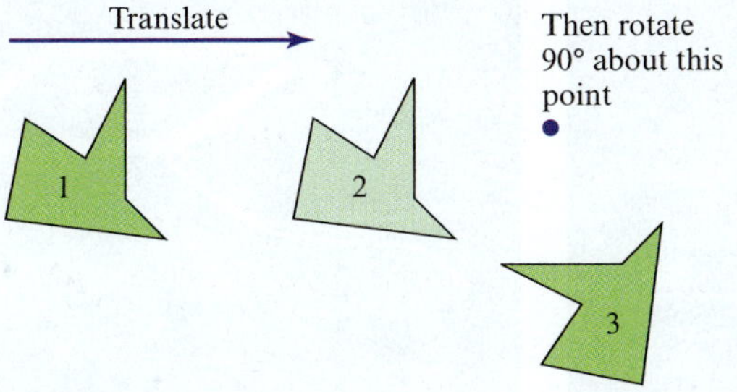

What Are Criteria for Congruence?

In design and construction, people often need to describe how to make an exact replica of an object. What information will be enough to specify how to make an exact replica? We look at this question from a mathematical perspective in the special case of triangles, and contrast the behavior of triangles and quadrilaterals. Before you read on, try Class Activities G and H.

CLASS ACTIVITY

14G Triangles and Quadrilaterals of Specified Side Lengths, p. CA-287

14H What Information Specifies a Triangle?, p. CA-288

CCSS 7.G.2

In Class Activities G and H, you saw that some pieces of information are enough to specify how to make an exact replica of a triangle, but other pieces of information are not enough. Sometimes there is more than one distinct triangle that fits with the given information; other times there is no such triangle. A set of information that specifies a unique triangle is often called a *congruence criterion* because any two triangles that fit that information are exact replicas of each other—in other words, the triangles are congruent. Next we study three congruence criteria for triangles. We then contrast these congruence criteria with information that does not specify a unique triangle.

The Side-Side-Side (SSS) Congruence Criterion If you form a triangle by threading a 3-inch, a 4-inch, and a 5-inch piece of straw together and a quadrilateral by threading a 3-inch, a 4-inch, a 3-inch, and a 4-inch piece of a straw together in that order (see Figure 14.28), you will find that the triangle is rigid, whereas the quadrilateral is "floppy." That is, the triangle's sides cannot be moved independently, whereas the quadrilateral's sides can be moved independently, forming many different quadrilaterals. (See Figure 14.29.) We can interpret this difference about triangles and quadrilaterals in terms of congruence. *All triangles with sides of length 3 inches, 4 inches, and 5 inches are congruent*, whereas there are quadrilaterals with sides of length 3 inches, 4 inches, 3 inches, 4 inches (in that order) that are *not* congruent.

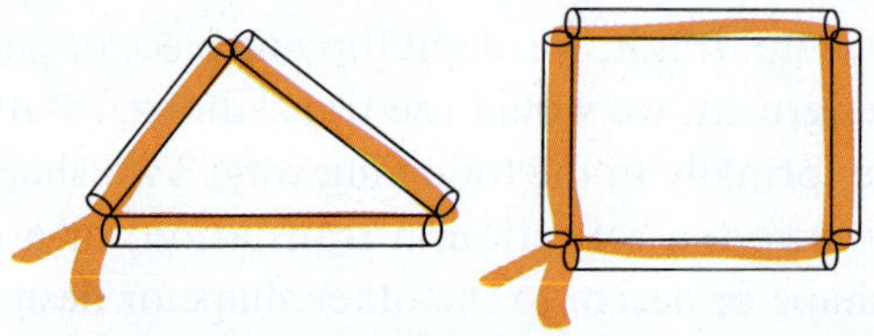

Figure 14.28 A triangle and a quadrilateral made of straws.

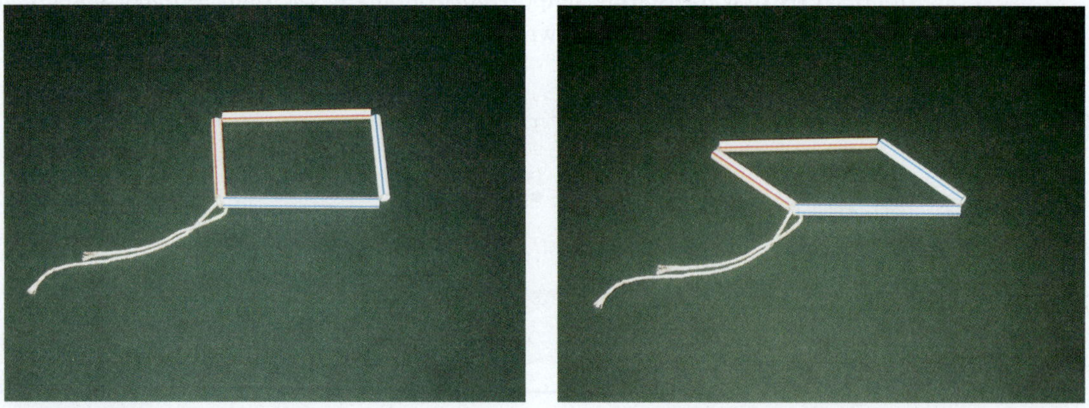

Figure 14.29 Straws forming a rectangle and the same straws forming a parallelogram.

In fact, any triangle, no matter what the lengths of its sides are, is structurally rigid. This fact about triangles can be stated in terms of congruence.

> Given a triangle that has sides of length a, b, and c units, it is congruent to all other triangles that have sides of length a, b, and c units.

SSS congruence

This last statement is the **side-side-side congruence** criterion, which is often simply called **SSS congruence** for triangles; it is the mathematical way to say that triangles are structurally rigid and determined by their side lengths.

The structural rigidity of triangles makes them common in building construction. For example, builders temporarily brace the frame of a house under construction with additional pieces of wood to keep the walls from falling over. (See Figure 14.30.) These additional pieces of wood create triangles; as triangles are rigid, the walls are held securely in place. The triangles in the crane in Figure 14.31 make the crane strong without using solid metal, which would be very heavy.

Figure 14.30 Extra pieces of wood create triangles for stability.

Figure 14.31 A crane is made out of many triangles.

In contrast, the structural flexibility of quadrilaterals makes them useful in objects that must move and change shape. For example, the folding laundry rack in Figure 14.32 is able to collapse because it is made from hinged rhombuses. The triangle at the top keeps the rack from collapsing when it is in use. When the rack is ready to be stowed, the triangle at the top is disconnected and the rhombuses collapse.

Figure 14.32 A folding laundry rack uses hinged rhombuses in its construction. The triangle at the top keeps the rack from collapsing when in use.

We've seen from a physical point of view that SSS congruence for triangles should be true, but we can also see this mathematically. When we are given the three side lengths of a triangle, we can use a compass and ruler to construct such a triangle. We start by drawing one of the sides, and then we draw circles centered at its endpoints. The radii of the circles are the lengths of the other two sides. See **Figure 14.33**. Because the circles meet in two points, there are only two ways to finish drawing the triangle. These two triangles are reflections of each other and so are congruent.

Figure 14.33 Constructing a triangle with given side lengths.

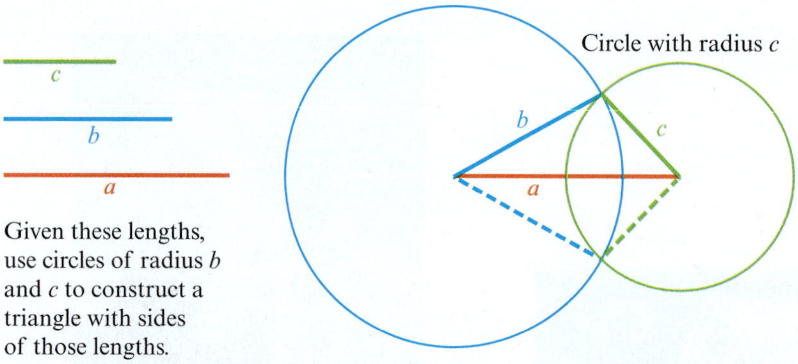

Will any three given lengths create a triangle? From **Figure 14.34** we can see that if one side's length is greater than the sum of the lengths of the other two sides, then when we try to construct such a triangle, it won't close up because the circles don't meet. In general, the **triangle inequality** says that if a, b, and c are the side lengths of a triangle, then

triangle inequality

$$a < b + c$$

Figure 14.34 Some side lengths not forming a triangle.

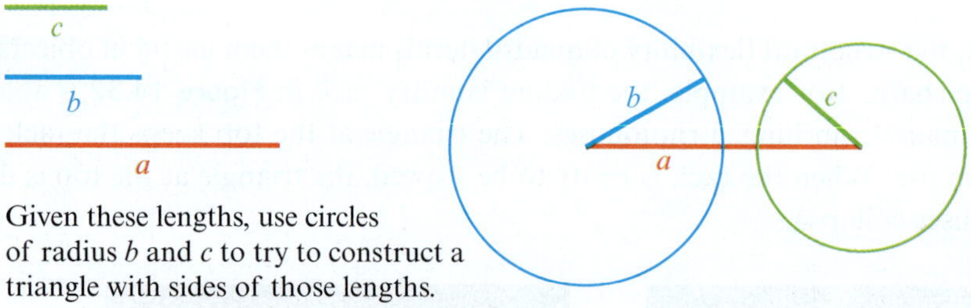

The Angle-Side-Angle (ASA) Congruence Criterion Another way to specify a triangle is to specify the length of one side and the two angles at either end of this side. For example, given a line segment AB, which triangles have a 40° angle at A and a 60° angle at B? **Figure 14.35** shows that there are two such triangles: one triangle is "above segment" AB and the other is "below" it. However, these two triangles are congruent, as you can see by reflecting one triangle across segment AB.

Figure 14.35 The two triangles that have segment AB as one side and have specified angles at A and B.

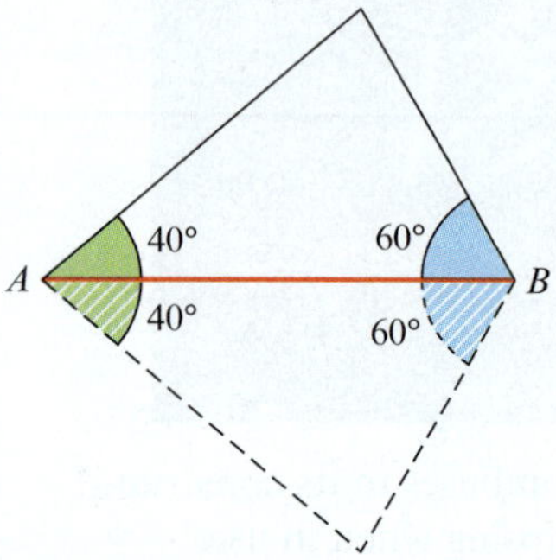

So, all triangles that have segment AB as one side and that have a 40° angle at A and a 60° angle at B are congruent. If a triangle has a different side of the same length as AB—say, CD—and if the triangle has a 40° angle at C and a 60° degree angle at D, then we can translate and rotate the triangle so that sides CD and AB match up (C landing at A and D landing at B). Once again, either the two triangles already match up, or one triangle can be reflected so as to match the other, as in Figure 14.35. The same will be true for other side lengths and other angles as well—as long as the two specified angles add to less than 180°. (Otherwise, a triangle cannot be formed, because all three positive angles in a triangle must add to 180°.) Thus, in general,

> if we specify the length of a side of a triangle and two angles that add to less than 180° at the two ends of a line segment, then all triangles formed with those specifications are congruent.

ASA congruence This last statement is the **angle-side-angle congruence** criterion, often simply called **ASA congruence**, for triangles.

The Side-Angle-Side (SAS) Congruence Criterion There is yet another way to specify a triangle:

> If we specify the lengths of two sides of a triangle and the angle (less than 180°) between these two sides, then all triangles formed with those specifications are congruent.

SAS congruence This last statement is the **side-angle-side congruence** criterion, often simply called **SAS congruence**, for triangles. Why is SAS congruence valid? Think about creating two triangles from two sides of given lengths and with a given angle between them, as at the top of Figure 14.36. By translating and rotating one of the triangles, we can arrange for one side of the same length on each triangle to be matched up, as at the bottom of Figure 14.36. If the triangles aren't already matched up, then after reflecting one of them, they must match.

Figure 14.36 Showing why SAS congruence is valid.

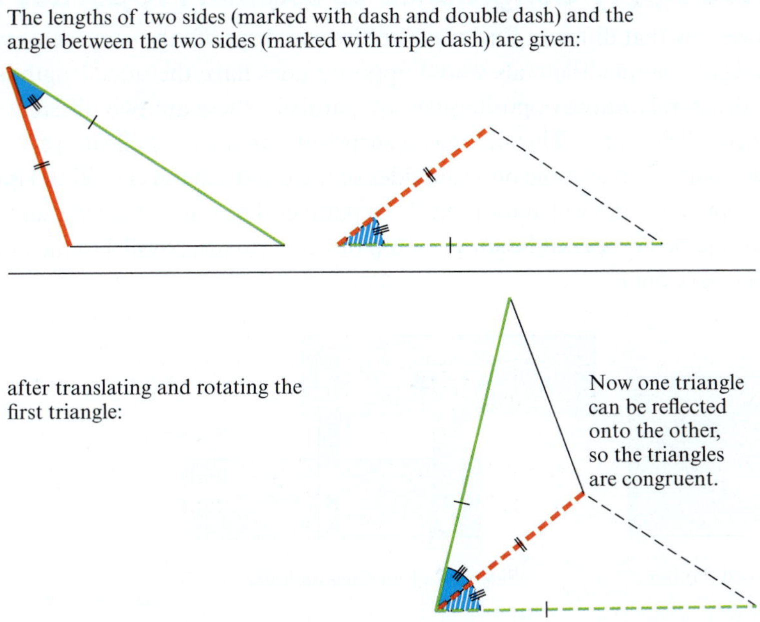

What Conditions on Side Lengths and Angles Do Not Specify a Unique Triangle?

We have just seen how some pieces of information (SSS, ASA, SAS) specify a unique triangle. In those cases all triangles that meet given conditions about side lengths and angles are congruent. When does information about side lengths and angles not specify a unique triangle?

We have already seen some restrictions on the side lengths and angles we can use with the SSS, ASA, and SAS congruence criteria. When we use the SSS criterion, the longest side length must be shorter than the sum of the other two (the triangle inequality). When we use ASA and SAS, the sum of the angles or the single angle must be less than 180°.

What if we specify the three angles of a triangle? In Section 14.5 we will see that three angles do not determine a *unique* triangle, but rather a collection of similar triangles.

In the SAS criterion we specify the angle *between* two sides. What if we specify the lengths of two sides of a triangle and the size of an angle that is *not* between those two sides? In general, such information does not specify a unique triangle, as shown in the example in Figure 14.37.

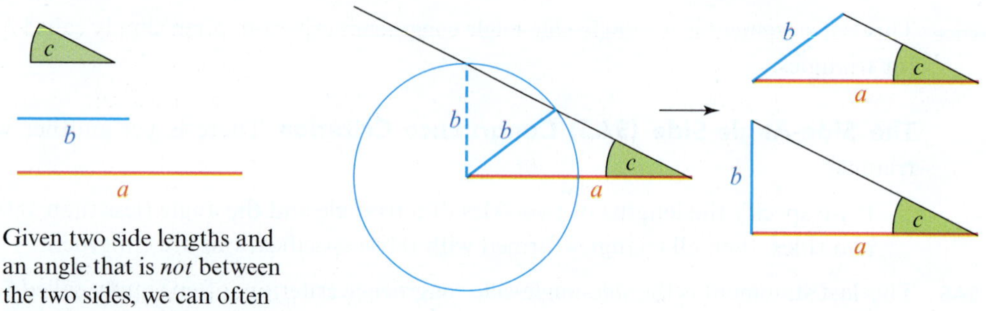

Given two side lengths and an angle that is *not* between the two sides, we can often construct *two* different triangles that fit that information.

Figure 14.37 There is not a side-side-angle congruence criterion because two side lengths and an angle that is not between the two sides often do not specify a unique triangle.

How Can We Apply Congruence to Explain Properties of Shapes?

In Section 10.4 we saw that different lists of properties could specify the same category of shapes. For example, the category of quadrilaterals whose opposite sides have the same length is the same as the category of quadrilaterals whose opposite sides are parallel—these are two different ways to describe the category of parallelograms. This fact has a surprising practical application: When a toolbox or sewing box is constructed so that the opposite sides of the quadrilateral $ABCD$ in Figure 14.38 are the same length, the opposite sides will automatically be parallel. The tools or sewing supplies will therefore not fall out when the box is opened because the top drawer remains parallel to the bottom drawer and therefore remains horizontal.

Figure 14.38 How congruence explains why a toolbox's drawers remain parallel.

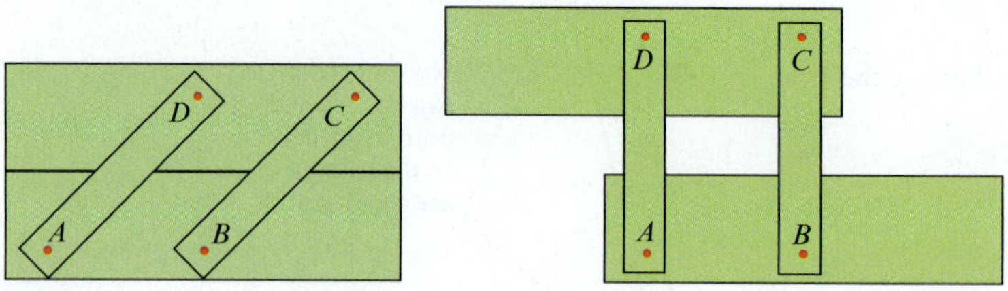

Side view of a closed toolbox. Side view of an open toolbox.

We can use congruence criteria or symmetry to explain why categories of shapes can be described in different ways or why shapes with certain properties will automatically have some other properties.

Class Activity 14I will help you explain why every isosceles triangle can be decomposed into two congruent right triangles and every rhombus can be decomposed into four congruent right triangles, as indicated in Figure 14.39. Because of this property of rhombuses, the *diagonals* of a rhombus are perpendicular and cut the angles of a rhombus in half. In general, the **diagonals** of a quadrilateral are the line segments that connect opposite vertices of the quadrilateral.

diagonals

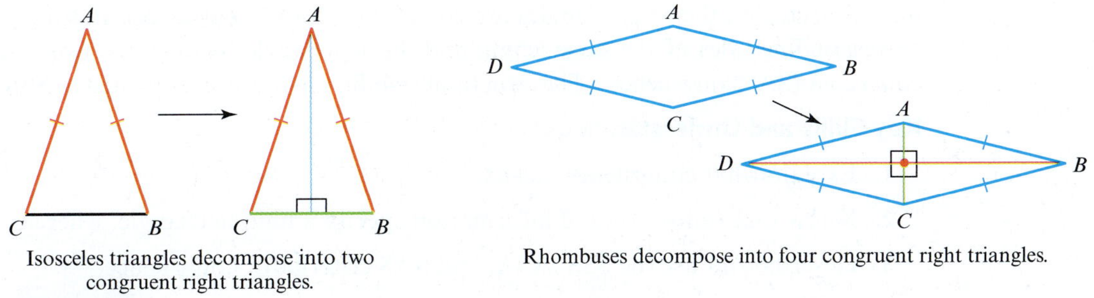

Figure 14.39 How isosceles triangles and rhombuses decompose into congruent right triangles.

CLASS ACTIVITY

14I Why Do Isosceles Triangles and Rhombuses Decompose into Congruent Right Triangles?, p. CA-289

14J Sewing Boxes and Congruence, p. CA-290

14K What Was the Robot's Path?, p. CA-290

We will use the fact that diagonals of rhombuses are perpendicular in the next section to explain why some of the common ruler and compass constructions work. This aspect of rhombuses is also essential to the functioning of the rhombus-shaped car jack in Figure 14.40. Even as the horizontal diagonal of the car jack is shortened or lengthened to raise or lower the car, the vertical diagonal remains perpendicular to the horizontal diagonal and thus remains vertical, so that the car remains stable.

Figure 14.40 Two positions for a rhombus-shaped car jack, a structure that remains stable because its diagonals remain perpendicular.

SECTION SUMMARY AND STUDY ITEMS

Section 14.3 Congruence

Informally, we say two shapes are congruent if they are the same size and shape. More formally, we describe two shapes as congruent if there is a rotation, a reflection, a translation, or a combination of these transformations that takes one shape to the other. There are several standard criteria for two triangles to be congruent. If two triangles have the same side lengths (so that corresponding sides have the same lengths), then they are congruent (SSS congruence). If two triangles have a side of the same length, and the corresponding angles at each end of the sides have the same measure in both triangles, then the triangles are congruent (ASA congruence). If two triangles have two corresponding sides of the same length and the same angle between the two sides, then they are congruent (SAS congruence). The structural stability of triangles is related to SSS congruence.

Key Skills and Understandings

1. Explain what congruence means.
2. Know that some pieces of information specify a unique triangle, whereas others do not.
3. Describe and use the SSS, ASA, and SAS criteria for congruence.
4. Relate the structural stability of triangles to the SSS congruence criterion.
5. Contrast the structural stability of triangles with the lack of stability for quadrilaterals.

Practice Exercises for Section 14.3

1. Suppose someone tells you that he has a garden with 4 sides, 2 of which are 10 feet long and opposite each other, and the other 2 of which are 15 feet long and opposite each other. With only this information, can you determine the exact shape of the garden? Take into account that the person might like gardens in unusual shapes.
2. Explain why isosceles triangles can be decomposed into two congruent right triangles.
3. Use a triangle congruence criterion and a fact about angles that we established in Section 10.1 to prove that every rhombus is also a parallelogram.
4. Some barbecue tongs are made by fastening a pair of identical metal pieces halfway across their lengths, as shown in Figure 14.41. The tongs are fastened at point E, halfway between points A and D and halfway between points B and C. Visually, it appears that the distance between A and B is equal to the distance between C and D, and that this will be the case no matter what angle the tongs are spread apart by. Use a triangle congruence criterion and previous facts we established about angles in Section 10.1, to prove that the distance between A and B must always be equal to the distance between C and D, no matter what angle the tongs are spread apart by. Be sure to state which triangle congruence criterion you use and why it applies.

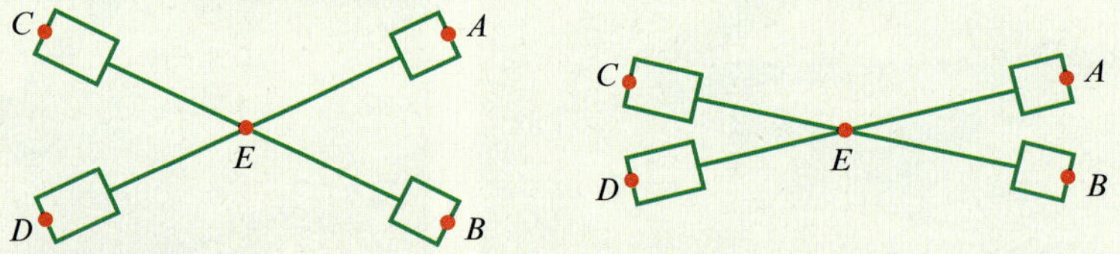

Figure 14.41 Applying triangle congruence to barbecue tongs.

Answers to Practice Exercises for Section 14.3

1. No, this information alone does not allow you to determine the shape of the garden. Although most people would probably make the garden rectangular, someone with an artistic flair might make the garden in the shape of a parallelogram that is not a rectangle. You can simulate this with 4 strung-together straws, as in Class Activity 14G and Figure 14.29.
2. See Figure 14.42. Assume that AB and AC are the same length. Let M be the midpoint between B and C, so that BM and CM are the same length. Therefore, the triangles AMB and AMC have the same side lengths. According to the SSS congruence criterion, the triangles AMB and AMC must therefore be congruent. Since these triangles are congruent, all their angles must match, too. Therefore, angles AMB and AMC are the same size. Since they add to 180°, they must each be 90°. Therefore triangles AMB and AMC are congruent right triangles that compose to make triangle ABC.

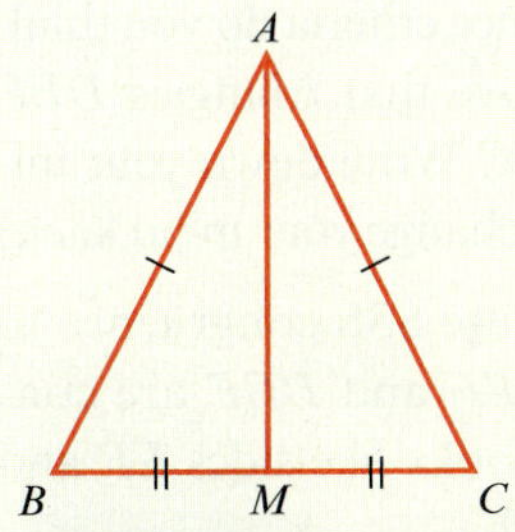

Figure 14.42 Using SSS to explain why isosceles triangles decompose into congruent right triangles.

3. Given a rhombus $ABCD$, as in **Figure 14.43(a)**, we want to prove that its opposite sides are parallel. Draw the diagonal AC. Sides AB, BC, CD, and DA are all the same length by definition of rhombus. By SSS congruence, triangles ABC and CDA are congruent. Therefore angles a and b in **Figure 14.43(b)** are the same size. Viewing the diagonal AC as a transversal to lines AB and DC, angles a and b are alternate interior angles, so the converse of the Parallel Postulate tells us that sides AB and DC are parallel. The same way of reasoning explains why the other two sides of the rhombus are parallel.

4. See **Figure 14.44**. In Chapter 10, we saw that when two lines meet at a point, the opposite angles formed that way are the same size. Therefore, the angles AEB and CED have the same size. Because the metal pieces are identical and are fastened halfway across their lengths, the lengths AE, BE, CE, and DE are equal. Now we can apply the SAS congruence criterion to the triangles AEB and CED. Because AE and CE are the same length, and BE and DE are the same length, and because the angle between segments AE and BE is congruent to the angle between segments CE and DE, the SAS criterion tells us that the triangles AEB and CED are congruent. Because these triangles are congruent, all their side lengths must match, too. Therefore, the distance between A and B must be equal to the distance between C and D.

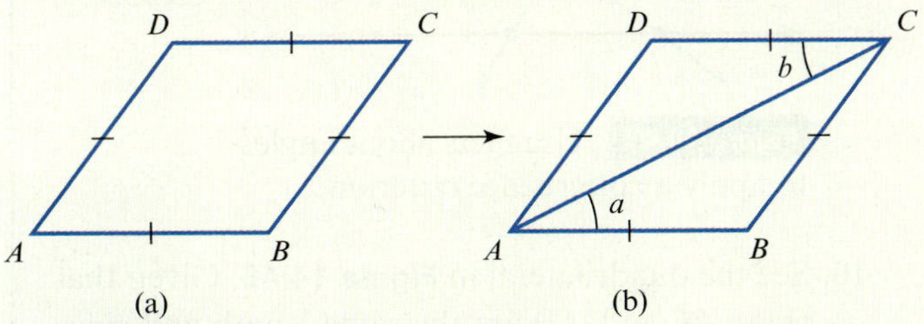

Figure 14.43 Using SSS congruence and the converse of the Parallel Postulate to prove that every rhombus is a parallelogram.

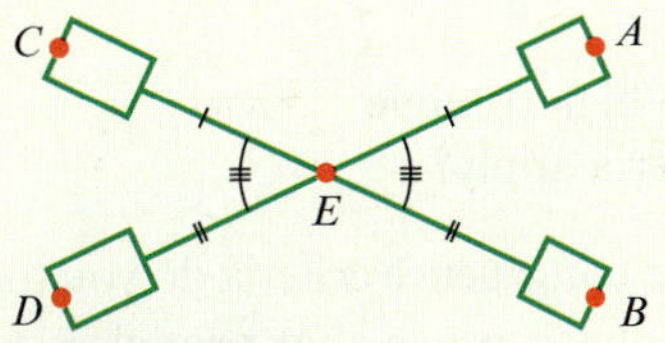

Figure 14.44 Applying SAS to barbecue tongs.

PROBLEMS FOR SECTION 14.3

1. Suppose that Ada, Bada, and Cada are three cities and that Bada is 20 miles from Ada, Cada is 30 miles from Bada, and Ada is 40 miles from Cada. There are straight-line roads between Ada and Bada, Ada and Cada, and Bada and Cada.

   a. Draw a careful and precise map showing Ada, Bada, and Cada and the roads between them, using a scale of 10 miles to 1 inch. Describe how to use a compass to make a precise drawing.

   b. If you were to draw another map, or if you were to compare your map to a classmate's, how would they compare? In what ways might the maps differ? In what ways would they be the same? Which criterion for triangle congruence is most relevant to these questions?

2. Use a ruler, a compass, and a protractor to construct triangles with the following characteristics. In each case, discuss whether or not all triangles with those characteristics are congruent.

   a. Triangle ABC has a side AB of length 6 cm, the angle at A is 50° and the angle at B is 120°.

   b. Triangle DEF has a side DE of length 7 cm, the angle at D is 15°, and side EF is 6 cm.

c. Triangle GHI has side GH of length 5 cm, the angle at H is 50°, and side HI is 8 cm.

3. Write a paragraph in which you discuss, in your own words, what you learned about triangles and quadrilaterals from Class Activity 14I, and how the concept of congruence is related to what you learned.

4. Is there a side-side-side-side congruence criterion? Discuss.

5. Is there an angle-angle-angle congruence criterion? Discuss.

6. Give your own example to explain why there is not a side-side-angle congruence criterion.

7. See Figure 14.45. We are given that sides AD and AB are the same length and sides CD and CB are the same length.

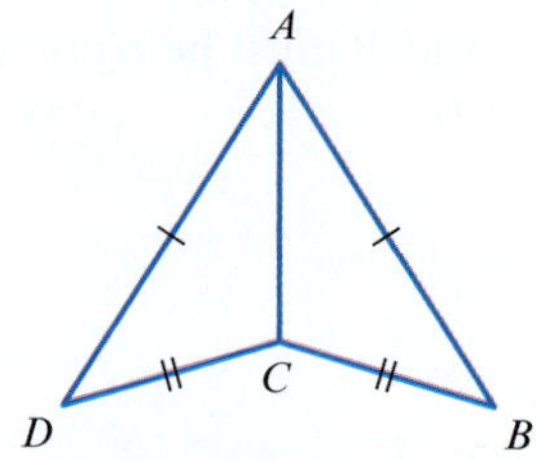

Figure 14.45 Which triangle congruence criteria apply?

a. Which triangle congruence criteria do you think you could apply to prove that triangles ABC and ADC are congruent? Write down your initial thoughts. (You may change your mind later.)

b. Explain why you *can't* use SAS congruence to prove that triangles ABC and ADC are congruent even though it looks like angles BAC and DAC are the same size.

c. Use a congruence criterion to *prove* that angles BAC and DAC are the same size.

8. See Figure 14.46. We are given that sides DG and DE are the same length and angles GDF and EDF are the same size.

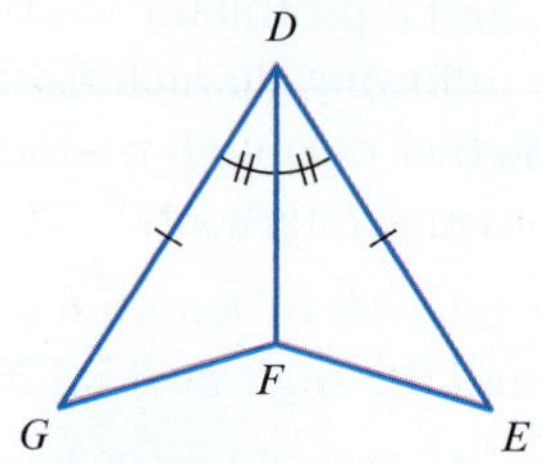

Figure 14.46 Which triangle congruence criteria apply?

a. Which triangle congruence criteria do you think you could apply to prove that triangles DEF and DGF are congruent? Write down your initial thoughts. (You may change your mind later.)

b. Explain why you *can't* use SSS congruence to prove that triangles DEF and DGF are congruent even though it looks like sides EF and GF are the same length.

c. Use a congruence criterion to *prove* that sides EF and GF are the same length.

9. See Figure 14.47. Given that QR and TR are the same length and that lines ℓ and m are parallel, use facts about angles that we established in Section 10.1 so that you can apply a triangle congruence criterion to prove that triangles PQR and STR are congruent.

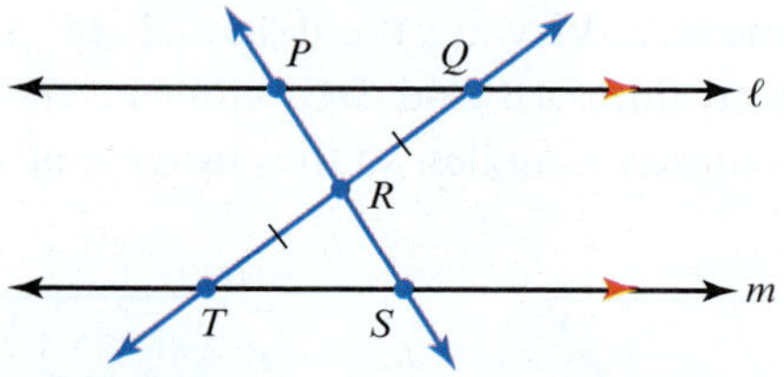

Figure 14.47 Use facts about angles to apply a congruence criterion.

10. See the quadrilateral in Figure 14.48. Given that sides AB and AD are the same length and sides BC and DC are the same length, use a triangle congruence criterion to prove that the diagonal AC divides angles a and c each in half. In other words, prove that angles DAC and BAC are the same size and angles BCA and DCA are the same size.

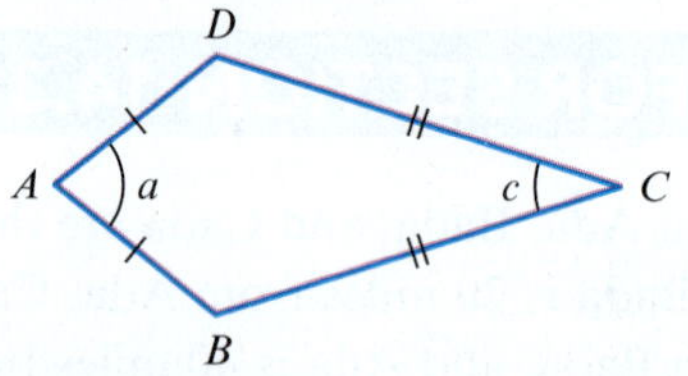

Figure 14.48 Use congruence to prove that a diagonal divides angles a and c in half.

11. This problem continues the investigation of Class Activity 10J in Section 10.2 on the size of your reflected face in a mirror. Figure 14.49 shows a side view of a person looking into a mirror. The mirror is parallel to line AE. Lines BF and DG are normal lines to the mirror.

a. What is the significance of the points F and G in Figure 14.49? Explain why, referring to the laws of reflection.

b. Using the theory of congruent triangles discussed in this section, explain why triangles ABF and CBF are congruent and explain why triangles CDG and EDG are congruent.

c. Use your answer to part (b) to explain why the length BD is half of the length AE. Therefore, explain why the reflection of your face in a mirror is half as long as your face's actual length. (You may assume that BD and FG are the same length.)

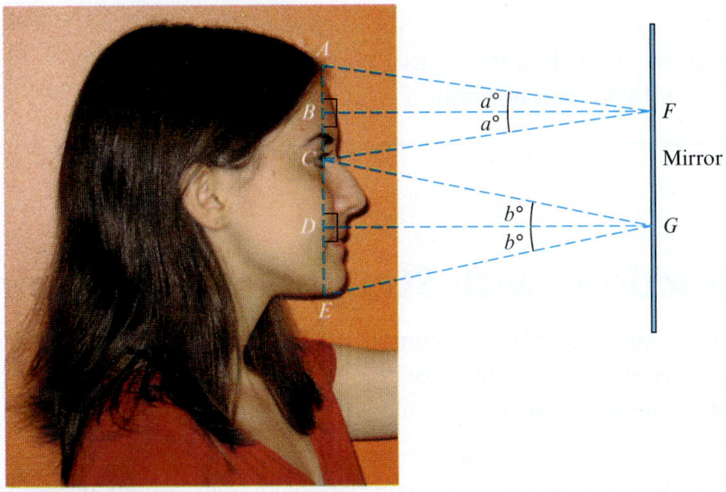

Figure 14.49 A side view of a person looking in a mirror.

12. Suppose you fasten two rods of different lengths at their midpoints, as shown in Figure 14.50. Let the endpoints of the rods be A, B, C, and D. Visually, it appears that the distance between A and B is equal to the distance between C and D and that the distance between A and C is equal to the distance between B and D, no matter what angle the rods are spread apart by. Use a triangle congruence criterion, and previous facts we have studied about angles, to explain why the previous statements about distances must always be true, no matter what angle the rods are spread apart by. Be sure to state which triangle congruence criterion you used and why it applies.

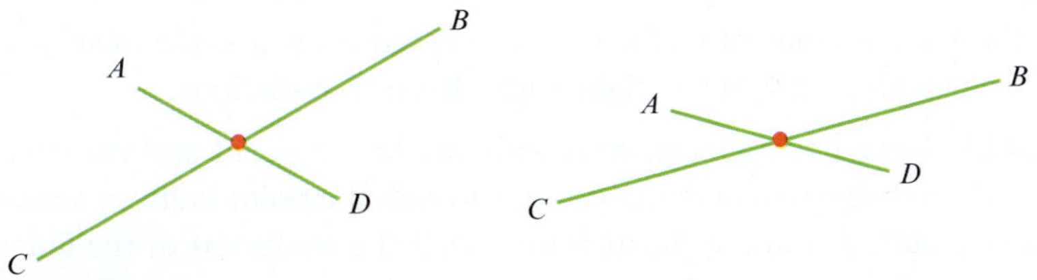

Figure 14.50 Applying triangle congruence to rods fastened at their midpoints.

13. Ann and Kelly are standing on a river bank, wondering how wide the river is. Ann is wearing a baseball cap, so she comes up with the following idea: She lowers her cap until she sees the tip of the visor just at the opposite bank of the river. She then turns around 180°, to face away from the river, being careful not to tilt her head or cap, and has Kelly walk to the spot where she can just see Kelly's shoes. By pacing off the distance between them, Ann and Kelly figure that Kelly was 50 feet away from Ann. If the ground around the river is level, what, if anything, can Ann and Kelly conclude about how wide the river is? Relate this to triangle congruence.

*14. We defined a parallelogram to be a quadrilateral for which opposite sides are parallel. When we look at parallelograms, it appears that opposite sides also have the same length. Use a triangle congruence criterion and facts we studied previously about angles to explain why opposite sides of a parallelogram really must have the same length. In order to do so, consider triangles formed by a diagonal.

*15. Here is an old-fashioned way to make a rectangular foundation for a house. Take a pair of identical pieces of wood for the length of the house and another pair of identical pieces of wood for the width of the house. Place the wood on the ground to show approximately where the foundation will go. The pieces of wood now form a quadrilateral whose opposite sides are the same length. Measure the two diagonals of the quadrilateral, and keep adjusting the quadrilateral until the two diagonals are the same length. Explain why the quadrilateral must now be a rectangle. In other words, explain why the quadrilateral must have 4 right angles.

14.4 Constructions with Straightedge and Compass

straightedge

In this section, we explore some common constructions that can be done with a compass and a straightedge. These constructions work because of special properties of rhombuses. A **straightedge** is just a "straight edge"—namely, a ruler, except that it need not have any markings for making measurements.

Constructions with straightedge and compass have their origins in the desire of those who wish to make precise drawings by using only simple, reliable tools. Think back to the times before there were computers. If you were studying plane shapes, and if you wanted to discover additional properties these shapes had beyond those properties given in the definitions, then you would need some way to make precise drawings. A sloppy drawing could hide interesting features, or worse, could seem to show features that aren't really there. Nowadays computer technology can help us make accurate drawings easily, but we can still benefit from learning some of the old hands-on techniques.

How Can We Divide a Line Segment in Half and Construct a Perpendicular Line?

Stop to think for a moment about how you could draw a perfect (or nearly perfect) right angle without using a previously made right angle. It's not so obvious.

midpoint

bisect (line segment)

perpendicular bisector

Figure 14.51 shows how you can start with any line segment and construct a line that is perpendicular to the line segment *and* that passes through the point halfway across the line segment. The point halfway across a line segment is often called a **midpoint** of the line segment. So the procedure shown in Figure 14.51 actually does two things at once: It constructs a perpendicular line, and it finds a midpoint of a line segment. Instead of saying that the construction in Figure 14.51 divides the line segment *in half,* we can say that the construction **bisects** the line segment or that it constructs the **perpendicular bisector** to the line segment. Notice that the word *bisect* is similar to the word *dissect*. The former is used in geometry, whereas the latter in used in biology.

How Can We Divide an Angle in Half?

bisect (angle)

angle bisector

Figure 14.52 shows how you can start with any angle at a point P and divide the angle in half. In other words, given 2 rays that meet at a point P, the construction in Figure 14.52 produces another ray, also starting at P, that is halfway between the 2 original rays. Instead of saying that the construction of Figure 14.52 divides an angle in half, we can say that the construction **bisects** the angle, or that it constructs an **angle bisector**.

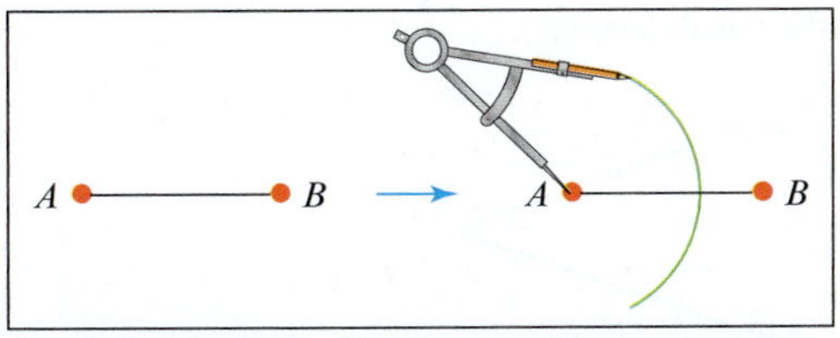

Step 1: Starting with a line segment AB, open a compass to a radius greater than half the length of the line segment, and draw part of a circle centered at A.

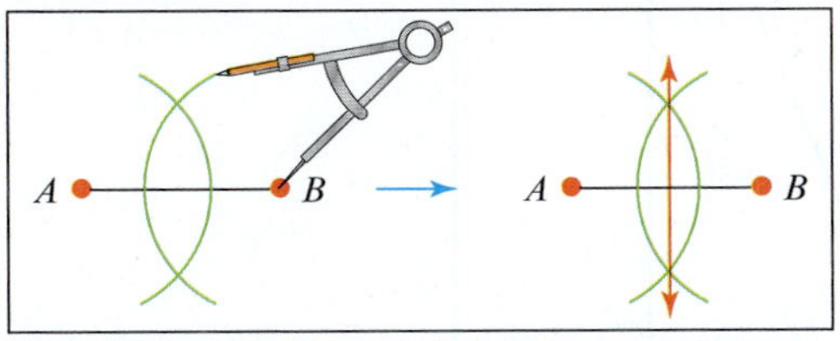

Step 2: Keep the compass opened to the same radius, and draw part of a circle centered at B. Draw a line through the two points where the two circles meet. This line is perpendicular to line segment AB and divides it in half.

Figure 14.51 Constructing the perpendicular bisector of a line segment.

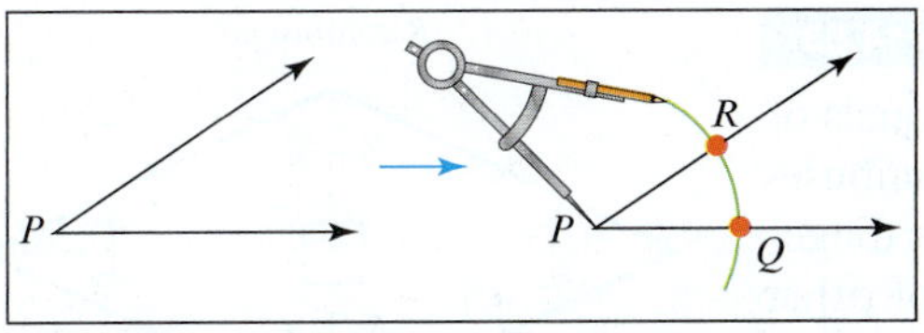

Step 1: Starting with two rays that meet at a point P, draw part of a circle centered at P. Let Q and R be the points where the circle meets the rays.

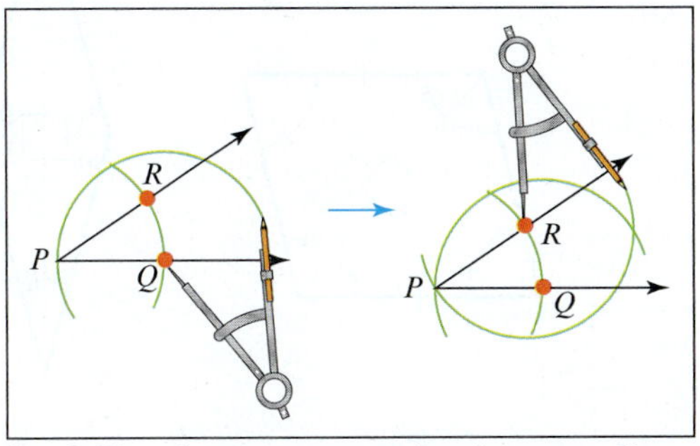

Step 2: Keep the compass opened to the same radius. Draw a circle centered at Q and another circle centered at R.

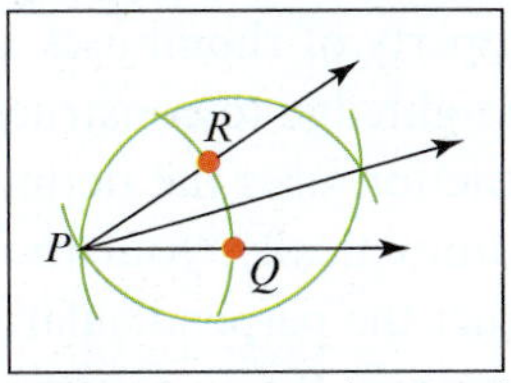

Step 3: Draw a line through point P and the other point where the last two circles drawn meet. This line cuts the angle RPQ in half.

Figure 14.52 Constructing an angle bisector.

How Do Special Properties of Rhombuses Explain Why Constructions Work?

The construction in Figure 14.51 certainly looks like it creates a line that is perpendicular to the line segment AB and cuts it in half. But how do we know it really does? And *why* does it work? We can explain why this and other constructions work in terms of special properties of rhombuses.

Table 14.1 summarizes some special properties of rhombuses that follow from Section 14.3. In that section we used triangle congruence criteria to prove that every rhombus decomposes into 4 congruent right triangles, as indicated in Figure 14.53(a). We can formulate this special property of rhombuses in terms of diagonals; see 2–4 in Table 14.1. Observe in Figures 14.53(a) and (b) how the diagonals of rhombuses are special compared to diagonals of other quadrilaterals. We also proved in Section 14.3 (Practice Exercise 3) that every rhombus is a parallelogram. See (5) in Table 14.1.

TABLE 14.1 Special properties of rhombuses.

1. All 4 sides of a rhombus are the same length (by definition of rhombus).
2. The diagonals of a rhombus are perpendicular.
3. The diagonals of a rhombus bisect each other.
4. The diagonals of a rhombus bisect the rhombus's angles.
5. Opposite sides of a rhombus are parallel.

Figure 14.53

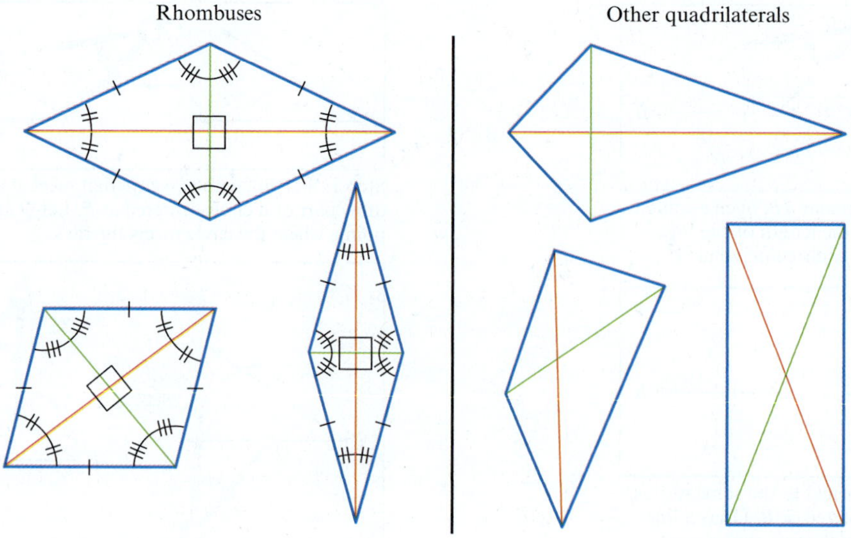

Diagonals of rhombuses versus diagonals of other quadrilaterals.

You can construct perpendicular and parallel lines and angle bisectors by selecting a relevant property of rhombuses and then constructing a suitable rhombus. When you use a compass and straightedge to construct a rhombus, you do so by creating sides of the same length. That construction uses the definition of rhombus—what it means for a quadrilateral to be a rhombus. Automatically, your rhombus will *also* have properties 2–5 in **Table 14.1**. For example, to construct the perpendicular bisector of a line segment, think of the line segment as the diagonal of a rhombus. If you construct a rhombus so that the given line segment is a diagonal, then the other diagonal will necessarily be perpendicular to the given line segment and will bisect it.

In Class Activity 14L you will show that the perpendicular bisector of **Figure 14.51** and the angle bisector of **Figure 14.52** are diagonals of rhombuses, so that these constructions rely on special properties of rhombuses. In Class Activity 14M you will figure out how to do some straightedge and compass constructions by selecting an appropriate property of rhombuses and constructing a suitable rhombus.

CLASS ACTIVITY

14L How Are Constructions Related to Properties of Rhombuses? p. CA-291

14M Construct Parallel and Perpendicular Lines by Constructing Rhombuses, p. CA-292

14N How Can We Construct Shapes with a Straightedge and Compass? p. CA-293

SECTION SUMMARY AND STUDY ITEMS

Section 14.4 Constructions with Straightedge and Compass

Using only a straightedge and a compass, we can construct the perpendicular bisector of a given line segment, and we can bisect a given angle. Both of these constructions produce an associated rhombus. The constructions give the desired results because of special properties of rhombuses: The diagonals in a rhombus are perpendicular and bisect each other, and the diagonals in a rhombus bisect the angles in the rhombus.

Key Skills and Understandings

1. Given a line segment, construct the perpendicular bisector by using a straightedge and compass. Describe the associated rhombus, and relate the construction to a special property of rhombuses: In a rhombus, the diagonals are perpendicular and bisect each other.
2. Use a straightedge and compass to bisect a given angle. Describe the associated rhombus, and relate the construction to a special property of rhombuses: In a rhombus, the diagonals bisect the angles in the rhombus.
3. Use the basic straightedge and compass constructions. For example, construct a square.

Practice Exercises for Section 14.4

1. Draw a line segment. Use a straightedge and compass to construct a line that is perpendicular to your line segment and divides your line segment in half. Repeat with several more line segments.

2. Draw an angle. Use a straightedge and compass to divide your angle in half. Repeat with several more angles.

3. a. Draw a rhombus that is naturally associated with the construction of a perpendicular bisector of a line segment shown in Figure 14.51.

b. Explain why the quadrilateral that you identify as a rhombus in part (a) really must be a rhombus, according to the definition of rhombus and according to the way it was constructed.

c. Which special properties of rhombuses explain why the construction of a perpendicular bisector of a line segment shown in Figure 14.52 works? Explain.

Answers to Practice Exercises for Section 14.4

1. See Figure 14.51.

2. See Figure 14.52.

3. a. Starting with the finished construction of a perpendicular bisector shown in Figure 14.54, let C and D be the points in the construction where the two circles meet. Form the 4 line segments AC, BC, AD, and BD, as shown in Figure 14.55.

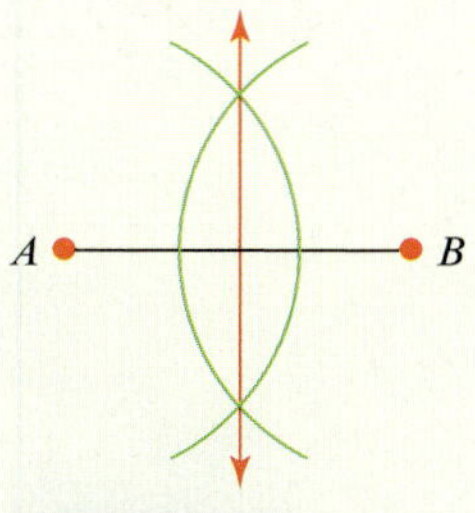

Figure 14.54 The result of using a straightedge and compass to construct a perpendicular bisector.

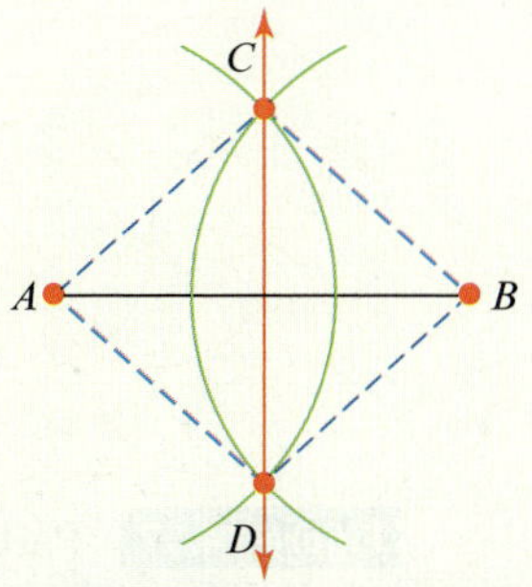

Figure 14.55 The construction for a perpendicular line forms a rhombus.

b. Notice that sides AC and AD in Figure 14.55 are radii of the circle centered at A. Therefore, AC and AD are the same length. Similarly, BC and BD are the same length because they are radii of the circle centered at B. But the circles centered at A and at B have the same radius because they were constructed that way (without changing the width of the compass). Therefore, all 4 line segments AC, BC, AD, and BD have the same length. By definition, this means that the quadrilateral $ACBD$ is a rhombus.

c. The original line segment AB is one of the diagonals of the rhombus $ACBD$ of Figure 14.55. The line segment CD is the other diagonal of the rhombus $ACBD$. Because rhombuses decompose into 4 congruent right triangles, the diagonals of a rhombus meet at a point that is halfway across each diagonal, and the diagonals of a rhombus are perpendicular. These properties tell us that segment CD is perpendicular to segment AB and divides AB in half.

PROBLEMS FOR SECTION 14.4

1. a. Draw a ray with endpoint A. Use a straightedge and compass to carefully construct a 45° angle to your ray at A. Briefly describe your method, and explain why it makes sense.

b. Draw a ray with endpoint B. Use a straightedge and compass to carefully construct a 22.5° angle to your ray at B. Briefly describe your method, and explain why it makes sense. (*Hint:* Notice that 22.5 is half of 45.)

2. a. On a blank piece of paper, draw a ray with endpoint A. Carefully fold your paper to create a fold line that makes a 90° angle with your ray at A. Briefly describe your method, and explain why it makes sense.

b. On a blank piece of paper, draw a ray with endpoint B. Carefully fold your paper to create a fold line that makes a 45° angle with your ray at B. Briefly describe your method, and explain why it makes sense.

c. On a blank piece of paper, draw a ray with endpoint C. Carefully fold your paper to create a fold line that makes a 22.5° angle with your ray at C. Briefly describe your method, and explain why it makes sense.

3. a. Draw a rhombus that is naturally associated with the straightedge and compass construction of a ray dividing an angle in half, as shown in Figure 14.54.

b. Explain why the quadrilateral that you identify as a rhombus in part (a) really must be a rhombus, according to the definition of rhombus and according to the way it was constructed.

c. Which special properties of rhombuses explain why the construction of a ray dividing an angle in half that is shown in Figure 14.52 works? Explain.

4. On a piece of paper, draw a point P and a separate line ℓ. Use a straightedge and compass to construct a line through P parallel to ℓ by constructing a suitable rhombus. Describe your method and explain why it works.

5. On a piece of paper, draw a point Q and a separate line m. Use a straightedge and compass to construct a line through Q perpendicular to m by constructing a suitable rhombus. Describe your method and explain why it works.

6. On a piece of paper, draw a line n and a point R on the line. Use a straightedge and compass to construct a line through R perpendicular to n by constructing a suitable rhombus. Describe your method and explain why it works.

7. Use a straightedge and compass (but not a protractor for measuring angles!) to construct a square. Leave the marks showing your construction. Describe your method.

8. Use a straightedge and compass (but not a protractor for measuring angles!) to construct a 15° angle. Describe your method.

9. Use a straightedge and compass (but not a protractor for measuring angles!) to construct a triangle whose angles are 15°, 75°, 90°. Describe your method.

10. Describe how to use a compass to construct the pattern of 3 circles shown in Figure 14.56 so that the triangle shown inside the circles is an equilateral triangle. Then explain why the construction guarantees that the triangle really is an equilateral triangle.

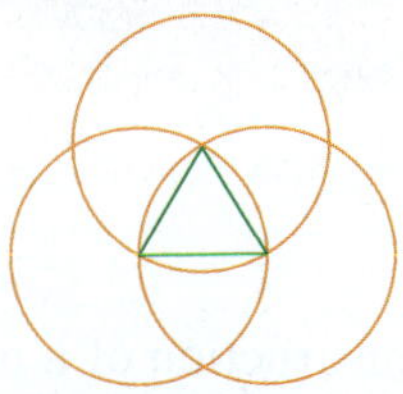

Figure 14.56 Pattern of three circles.

11. a. Use a compass to draw a pattern of circles like the one in Figure 14.57. Briefly describe how to create this pattern.

Figure 14.57 Pattern of circles.

b. Use a straightedge and compass to carefully construct a regular hexagon. Construct your hexagon by first drawing another pattern of circles like the one in Figure 14.57, but this time draw only parts of the circles, so that you don't clutter your drawing.

12. Use a straightedge and compass to carefully construct a regular 12-gon. Briefly describe your method, and explain why it makes sense. It may help you to examine Figure 14.57.

13. Draw a large circle on a piece of paper. This circle will represent a pie. Use a straightedge and compass to carefully subdivide your pie into 6 equal pie pieces. Briefly describe your method, and explain why it makes sense. It may help you to examine Figure 14.57.

14. Draw a large circle on a piece of paper. This circle will represent a pie. Use a straightedge and compass to carefully subdivide your pie into 12 equal pieces. Briefly describe your method, and explain why it makes sense. It may help you to examine Figure 14.57.

14.5 Similarity

CCSS **Common Core State Standards** Grades 7, 8

Model cars, trains, and planes are often just like their real-life counterparts, only smaller. We can imagine scaling down the real cars, trains, and planes to become identical to the models, as in Figure 14.58. In geometry, we use the notion of *congruence* as a precise way to discuss shapes that are identical and the notion of *similarity* as a precise way to discuss shapes that are identical except for their size.

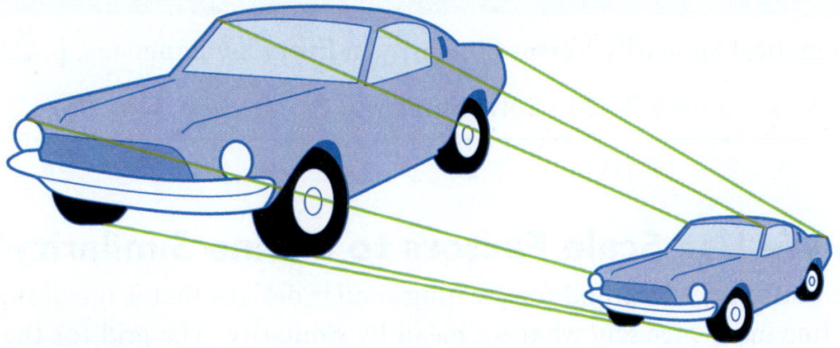

Figure 14.58 If a real car is scaled down, it becomes identical (on the outside) to a model car.

Similarity is a key concept in mathematics. We need it for understanding lines and their equations and for solving a wide range of problems, including problems about surveying, mapmaking, and representational drawings. In this section we define what it means for geometric shapes or objects in the real world to be similar and we contrast this mathematical definition with everyday language. We then study two methods for solving similarity problems. In the next section we continue our study of similarity by providing another geometric definition of similarity for shapes in a plane and developing a criterion for shapes to be similar.

What Is Similarity Informally?

similar (informal)

Informally, we sometimes say that two objects that have the same shape, but not necessarily the same size, are **similar**. However, this language can be misleading, as you will see in Class Activity 14O. Another way to describe similarity informally is as follows: Two objects or shapes are similar if one object represents a scaled version of the other (scaled up or down). For example, a scale model of a train is similar (at least on the outside) to the actual train on which it is modeled. The network of streets on a street map is similar to the network of real streets it represents—at least if the streets are on flat ground.

One way to create a shape or design that is similar to another shape or design is by using grids of lines. If you enjoy doing crafts, then you may be familiar with this technique. Draw a grid of equally spaced parallel and perpendicular lines over the design you want to scale (or use an overlay of such grid lines), as shown in Figure 14.59. If you want the new design to be, say, twice as wide and twice as long, then make a new grid of parallel and perpendicular lines that are spaced twice as wide as the original grid lines. Now copy the design onto the new grid lines, square by square. The new design will be similar to the original design.

Figure 14.59 Using a grid to make similar pictures.

CLASS ACTIVITY

14O Mathematical Similarity Versus Similarity in Everyday Language, p. CA-295

14P What Ways Can We Find to Solve Similarity Problems? p. CA-295

How Can We Use Scale Factors to Define Similarity?

Examining the craft project method for creating similar designs that is illustrated in Figure 14.59 will help us define more precisely what we mean by similarity. The grid for the large bunny was made so that the width and height of the large bunny would be twice the width and height of the small bunny. But notice that *all* lengths on the large bunny are twice as long as the corresponding lengths on the small bunny. This scaling of all lengths is a key aspect of similarity.

In contrast, the two bunnies on the right in Figure 14.60 are *not* similar to the original bunny on the left, because in each case the width has a different relationship to the original than the height does. The middle bunny is just as tall as the original bunny, but it is only half as wide. The bunny on the right is just as wide as the original bunny, but it is only half as tall.

Figure 14.60 Examples of figures that are *not* similar.

The original bunny

These bunnies are *not* similar to the original bunny.

similar

scale factor

Mathematically, we say two shapes or objects (in a plane or in space) are **similar** if every point on one object corresponds to a point on the other object and there is a positive number, k, such that the distance between any two points on the second object is k times as long as the distance between the corresponding points on the first object. This number k is called the **scale factor** from the first object to the second object. Because of the consistent scale factor, a collection of distances within the second object (e.g., length, width, height) are in the same ratio as the corresponding distances within the first object.

For example, the box of a toy model car might indicate that the toy car is a 24:1 scale model of an actual car. This means that the toy car and the actual car are similar and the scale factor from the toy car to the actual car is 24, or equivalently, that the scale factor from the actual car to the toy car is $\frac{1}{24}$. In particular, the length, width, and height of the actual car are 24 times the length, width, and height, respectively, of the toy car. Similarly, for other distances, such as the width of the windshield or the length of the hood, each distance on the actual car is 24 times the corresponding distance on the toy car.

Warning: Scale factors apply only to lengths *not* to areas or volumes. In Section 14.7 we will see how to adjust scale factors to relate areas or volumes.

How Can We Reason to Solve Problems About Similar Objects or Shapes?

In many practical situations, two shapes or objects are given to be similar. If we know various lengths on one object, then we can determine all the corresponding lengths on the other object, as long as we know at least one of the corresponding lengths. We study two ways to do this, and both use the scale factor, even if indirectly. Consider this problem.

Problem: Suppose an artist creates a scale model of a sculpture. The scale model is 10 inches wide and 20 inches tall. If the actual sculpture is to be 60 inches wide, then how tall will the actual sculpture be?

Solutions:

Scale Factor Method

1. **Scale Factor Method** (See **Figure 14.61**.) As the scale model and the actual sculpture are to be similar, there is a scale factor, k, from the model to the actual sculpture such that every length on the actual sculpture is k times as long as the corresponding length on the scale model. Therefore, the width of the actual sculpture is $k \cdot 10$ inches and the height of the actual sculpture is $k \cdot 20$ inches. Because the width of the sculpture is given as 60 inches,

$$k \cdot 10 \text{ inches} = 60 \text{ inches}$$

Figure 14.61

Using the scale factor method.

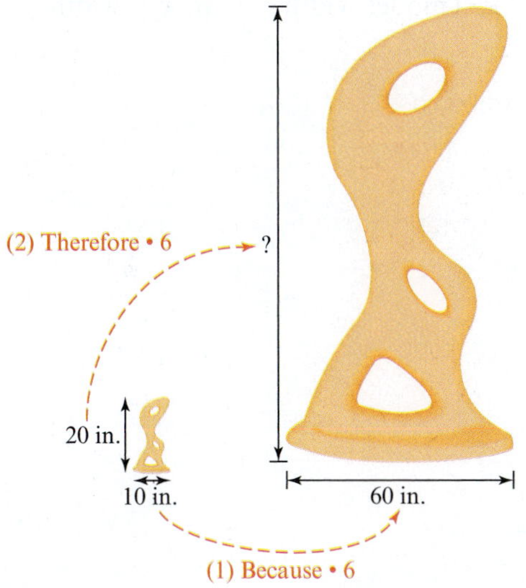

so

$$k = 60 \div 10 = 6$$

Thus, the height of the sculpture is

$$k \cdot 20 \text{ inches} = 6 \cdot 20 \text{ inches} = 120 \text{ inches}$$

We can relate the scale factor method to a proportion as follows. We know that the scale factor, k, relates the width and height of the model and the sculpture in a consistent way:

$$(\text{sculpture width}) = k \cdot (\text{model width})$$
$$(\text{sculpture height}) = k \cdot (\text{model height})$$

Therefore,

$$\frac{\text{sculpture width}}{\text{model width}} = k \qquad \frac{\text{sculpture height}}{\text{model height}} = k$$

and so we obtain the proportion

$$\frac{\text{sculpture width}}{\text{model width}} = \frac{\text{sculpture height}}{\text{model height}} \tag{14.1}$$

In terms of our problem, Equation 14.1 becomes

$$\begin{array}{r} \text{sculpture} \\ \text{model} \end{array} \cdot 6 \curvearrowright \overset{\text{width}}{\frac{60}{10}} = \overset{\text{height}}{\frac{?}{20}} \curvearrowright \cdot 6 \begin{array}{l} \text{sculpture} \\ \text{model} \end{array} \tag{14.2}$$

We can interpret Equations 14.1 and 14.2 as telling us that the scale factor is the same for the width as well as for the height.

Notice that with the scale factor method we *compare like quantities.* We can think of it as an "external" comparison method because it compares like parts across two different shapes.

Internal Factor Method

2. **Internal Factor Method** (See Figure 14.64.) As the scale model is 10 inches wide and 20 inches tall, it is $20 \div 10 = 2$ times as tall as it is wide. The actual sculpture should therefore also be 2 times as tall as it is wide. Because the sculpture is to be 60 inches wide, it should be

$$2 \cdot 60 \text{ inches} = 120 \text{ inches}$$

tall.

How do we know the actual sculpture should be 2 times as tall as it is wide? It is because the sculpture's height and width are in the same ratio as the models height and width. The scale factor, k, scales the height and the width in the same way:

$$\frac{\text{sculpture height}}{\text{sculpture width}} = \frac{k \cdot (\text{model height})}{k \cdot (\text{model width})} = \frac{\text{model height}}{\text{model width}}$$

Figure 14.62

Using the internal factor method.

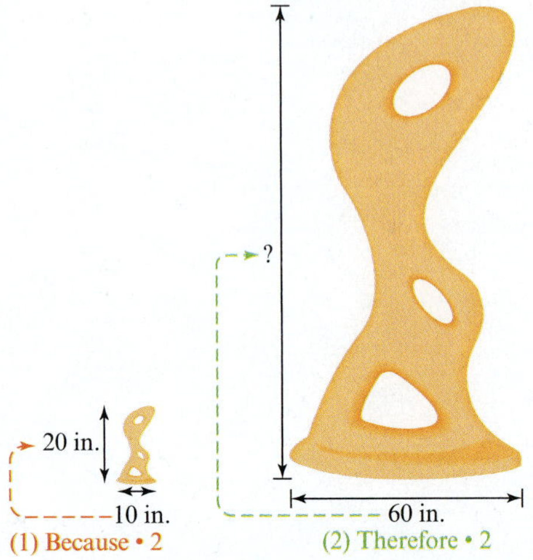

Therefore we obtain the proportion

$$\frac{\text{sculpture height}}{\text{sculpture width}} = \frac{\text{model height}}{\text{model width}} \tag{14.3}$$

In terms of our problem, Equation 14.3 becomes

$$\begin{array}{lcccl} & \text{sculpture} & & \text{model} & \\ \text{height} & ? & & 20 & \text{height} \\ & \cdot 2 \curvearrowleft\ \overline{} & = & \overline{}\ \curvearrowright \cdot 2 & \\ \text{width} & 60 & & 10 & \text{width} \end{array} \tag{14.4}$$

We can interpret Equations 14.3 and 14.4 as telling us that the factor or multiplier relating the width and height of the sculpture is the same as the factor or multiplier relating the width and height of the model. This factor or multiplier is the value of the height-to-width ratio.

We can also relate the internal factor method to the proportion in Equation 14.2. When we multiply the denominator by 2 we must also multiply the numerator by 2 to maintain equality:

$$\begin{array}{lcccl} & \text{width} & \xrightarrow{\cdot 2} & \text{height} & \\ \text{sculpture} & 60 & & ? & \text{sculpture} \\ & \overline{} & = & \overline{} & \\ \text{model} & 10 & \xrightarrow[\cdot 2]{} & 20 & \text{model} \end{array} \tag{14.5}$$

Notice that with the internal factor method we *compare unlike quantities*. We can think of it as an "internal" comparison method because it compares different parts within one shape. Internal factors are studied in trigonometry, where they are viewed as sines, cosines, and values of other trigonometric ratios. Internal factors are also used to describe slopes of lines.

Figure 14.63 summarizes how we can interpret the scale factor method and the internal factor method in terms of a proportion.

Applying the scale factor 6:	**model** / **sculpture**; Therefore • 6 (top arrow: 20 → ?); height/width: $\frac{20}{10} = \frac{?}{60}$; Because • 6 (bottom arrow: 10 → 60) Sculpture height = 6 • 20 in. = 120 in.
Applying the internal factor 2:	**model** / **sculpture**; height/width: Because • 2 (10 → 20) $\frac{20}{10} = \frac{?}{60}$ Therefore • 2 (60 → ?) Sculpture height = 2 • 60 in. = 120 in.

Figure 14.63 Interpreting the scale factor method and the internal factor method in terms of a proportion.

CLASS ACTIVITY

14Q Reasoning with the Scale Factor and Internal Factor Methods, p. CA-286

14R Critique Reasoning About Similarity, p. CA-297

FROM THE FIELD Research

Cox, D. C., & Edwards, M. T. (2012). Sizing up the Grinch's heart. *Mathematics Teaching in the Middle School, 18*(4), pp. 228-235.

In class discussions about what "two sizes too small" means, the authors find middle-school students eventually conjecture that one of the hearts is "exactly like a normal heart, only smaller." They build on this discussion by asking students to draw a heart that is proportional to a given heart drawing. Students come up with visual strategies and measurement strategies, indicating how they are thinking about proportion. The authors give further tasks to help students see that additive strategies distort figures, but multiplicative strategies do not. The authors recommend using complex figures in similarity tasks to challenge rote procedures for scaling. The complex figures help students develop more robust strategies, investigate global rules, and develop creative measurement strategies.

Cox, D. C., & Lo, J. J. (2012). Discuss similarity using visual intuition. *The Mathematics Teacher, 18*(1), pp. 30-37.

Based on their work with urban middle-school students, the authors recommend providing opportunities for proportional reasoning with (1) simple and complex figures, (2) distortion and proportion, and (3) visual reasoning and more analytical strategies. When students tried to apply incorrect additive reasoning to more complex shapes, such as an L-shape, they realized their strategy didn't work. Students gave more reasoned and specific justifications when explaining why distorted shapes were not similar than when explaining why similar shapes were similar. The authors recommend categorization tasks, where students decide whether shapes are similar, and tasks where students make predictions about what a figure will look like when it is scaled.

FROM THE FIELD Children's Literature

Schwartz, D. (1999). *If you hopped like a frog.* New York, NY: Scholastic.

Using ratios and proportions, students can have conversations about scale factors and similarity as they compare themselves to the nifty creatures mentioned in this book.

Sundby, S. (2000). *Cut down to size at high noon.* Watertown, MA: Charlesbridge.

Louie was known as the best barber around town! His unpredictable hair cuts were scaled down models of huge objects. However, a stranger strolls into town and challenges him to a hair showdown at high noon to decide which barber has to pack up his shop and leave town.

SECTION SUMMARY AND STUDY ITEMS

Section 14.5 Similarity

We call two shapes or objects similar if all distances between corresponding parts of the shapes or objects are scaled by the same factor called a scale factor.

We studied two methods for solving problems about similar shapes or objects: the scale factor method and the internal factor method. We related both methods to setting up and solving a proportion. To use the scale factor method, we use corresponding known lengths in the two similar objects in order to determine the scale factor. We then apply the scale factor to find an unknown length. To use the internal factor method, we find the factor by which two known lengths within one of the objects are related. The corresponding lengths in the other, similar object will be related by the same factor. We then apply this factor to find an unknown length in the other object.

Key Skills and Understandings

1. Discuss what it means for shapes or objects to be similar.
2. Use the scale factor and internal factor methods to solve problems about similar shapes or objects, explain the rationale for each method, and relate them to proportions.

Practice Exercises for Section 14.5

1. Amber has drawn a simple design on a rectangular piece of paper that is 4 inches wide and 12 inches long. She wants to make a scaled-up version of her design on a rectangular piece of paper that is 10 inches wide. She must figure out how long to make the larger rectangle. Amber says that since the larger rectangle is 6 inches wider than the smaller rectangle (because $10 - 4 = 6$), she should make her larger rectangle 6 inches longer than the smaller rectangle. She figures that she should make the larger rectangle $12 + 6 = 18$ inches long. Will Amber be able to make a correctly proportioned scaled-up version of her design on a rectangle that is 10 inches wide and 18 inches long? If not, how long should she make her rectangle, and why?

2. *A Postcard Problem:* A 24-inch-by-48-inch rectangular picture will be scaled down to fit on a postcard. On the postcard, the short side of the picture will be 4 inches long. What will the long side of the picture be on the postcard?

Solve the postcard problem in two ways: with the scale factor method and with the internal factor method. In each case:

- Briefly explain the idea and the reasoning of the method.
- Link the method to a proportion.

3. Hannah wants to make a scale drawing of herself standing straight with her arms down. Hannah is 4 feet 6 inches tall. Her arm is 22 inches long. If Hannah wants to make the drawing of herself 10 inches tall, then how long should she draw her arm?

Solve this problem in two different ways: one using the scale factor method and one using the internal factor method, explaining your reasoning both times.

4. Ms. Bullock's class is making a display of the sun, the planets, and Pluto. Each planet and Pluto will be depicted as a circle. The students in Ms. Bullock's class want to show the earth as a circle of diameter 10 cm, and they want to show the correct relative sizes of the planets, Pluto, and the sun. Given the information in Table 14.2, what should the diameters of these scale drawings be? Since the sun is so big, it wouldn't be practical to make a full model of the sun; how could the students make a sliver of the sun of the correct size?

TABLE 14.2 Diameters of heavenly bodies

Heavenly Body	Approximate Diameter in km
Sun	1,392,000
Mercury	4,900
Venus	12,100
Earth	12,700
Mars	6,800
Jupiter	138,000
Saturn	115,000
Uranus	52,000
Neptune	49,500
Pluto	2,300

5. A sculptor makes a scale model for a sculpture she plans to carve out of marble that is 15 inches long, 8 inches wide, and 27 inches high. She finds a block of marble that is 5 feet long, 4 feet wide, and 10 feet high. What will the finished dimensions of the sculpture be if she makes the sculpture as large as possible?

Answers to Practice Exercises for Section 14.5

1. Amber will not be able to make a scaled-up version of her design on paper that is 10 inches wide and 18 inches long. As we see in Figure 14.64, if Amber makes the larger rectangle 10 inches by 18 inches, it will not be proportioned in the same way that the original rectangle is. Adding the same amount to the length and width generally does not preserve the ratio of length to width in a rectangle. Instead, Amber could reason that since the original rectangle is 3 times as long as it is wide, the larger rectangle should also be 3 times as long as it is wide. So if the larger rectangle is 10 inches wide, it should be $3 \cdot 10 = 30$ inches long.

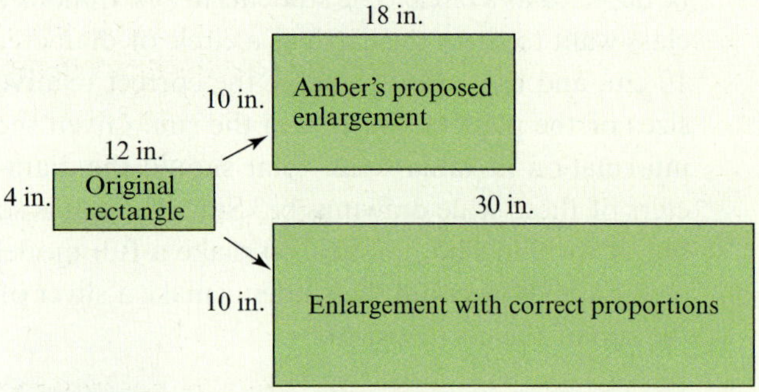

Figure 14.64 How should Amber scale up her 4-inch-by-12-inch rectangle?

2. *Scale Factor Method:* The short side of the postcard is $\frac{1}{6}$ as long as the short side of the picture, so the scale factor from the picture to the postcard is $\frac{1}{6}$. The long side of the postcard must also be $\frac{1}{6}$ as long as the long side of the picture because all side lengths must be scaled down by the same factor. Therefore, the long side of the postcard must be $\frac{1}{6} \cdot 48 = 8$ inches. This solution method is linked to a proportion students could set up to solve the problem, as shown below:

$$\cdot \tfrac{1}{6} \text{ or } \div 6$$
$$\frac{48}{24} = \frac{x}{4}$$
$$\cdot \tfrac{1}{6} \text{ or } \div 6$$

Internal Factor Method: The long side of the picture is 2 times as long as the short side of the picture. The long side of the postcard must therefore also be 2 times as long as the short side of the postcard in order to be proportioned in same way as the picture. Therefore, the long side of the postcard must be $2 \cdot 4 = 8$ inches. This solution method is linked to a proportion students could set up to solve the problem, as shown below:

$$\cdot 2 \left(\frac{48}{24} = \frac{x}{4} \right) \cdot 2$$

3. First, let's determine Hannah's height in inches. Since there are 12 inches in 1 foot, 4 feet is $4 \cdot 12 = 48$ inches. Therefore, 4 feet 6 inches is $48 + 6 = 54$ inches, so Hannah is 54 inches tall.

Internal Factor Method: Hannah's arm is 22 inches long, which is $\frac{22}{54}$ of her height. Therefore, in her drawing, Hannah's arm should also be $\frac{22}{54}$ of her height in the drawing, namely,

$$\frac{22}{54} \cdot 10 \text{ inches} = 4 \text{ inches}$$

approximately. So Hannah should draw her arm 4 inches long.

Scale Factor Method: Because Hannah is 54 inches tall and the scale drawing of herself is to be 10 inches tall, the height of Hannah's drawing is $\frac{10}{54}$ of Hannah's actual height. In other words, the scale factor from Hannah to the drawing of herself is $\frac{10}{54}$. The length of Hannah's arm in her drawing should also be $\frac{10}{54}$ of the length of her actual arm. Therefore, the drawing of Hannah's arm should be

$$\frac{10}{54} \cdot 22 \text{ inches} = 4 \text{ inches}$$

long approximately.

4. From the problem statement, we infer that we want the collection of circles representing the planets, Pluto, and the sun to be similar to cross-sections of the actual planets—all with the same scale factor. If we use the internal factor method, we don't have to worry about converting kilometers to centimeters (or vice versa), and we won't have to work with huge numbers.

Because the earth's diameter is 12,700 km and Mercury's diameter is 4,900 km, Mercury's diameter is $\frac{4{,}900}{12{,}700}$ times as long as the earth's diameter. The same relationship should hold for the models of Mercury and the earth. Since the model earth is to have a diameter of 10 cm, the model Mercury should have a diameter of

$$\frac{4{,}900}{12{,}700} \cdot 10 \text{ cm, about } 3.9 \text{ cm}$$

Table 14.3 shows the calculations for the other heavenly bodies.

TABLE 14.3 Diameters of scale of heavenly bodies

Heavenly Body	Approximate Diameter of Circle Representing It
Sun	$\frac{1{,}392{,}000}{12{,}700} \cdot 10$ cm, about 1096 cm = 10.96 m
Mercury	$\frac{4{,}900}{12{,}700} \cdot 10$ cm, about 3.9 cm
Venus	$\frac{12{,}100}{12{,}700} \cdot 10$ cm, about 9.5 cm
Earth	given as 10 cm
Mars	$\frac{6{,}800}{12{,}700} \cdot 10$ cm, about 5.4 cm
Jupiter	$\frac{138{,}000}{12{,}700} \cdot 10$ cm, about 108.7 cm
Saturn	$\frac{115{,}000}{12{,}700} \cdot 10$ cm, about 90.6 cm
Uranus	$\frac{52{,}000}{12{,}700} \cdot 10$ cm, about 40.9 cm
Neptune	$\frac{49{,}500}{12{,}700} \cdot 10$ cm, about 39 cm
Pluto	$\frac{2{,}300}{12{,}700} \cdot 10$ cm, about 1.8 cm

The circle representing the sun should have a diameter of about 11 meters and therefore a radius of about 5.5 meters. The students could measure a piece of string $5\frac{1}{2}$ meters long. One student could hold one end and stay in a fixed spot, while another student could attach a pencil to the other end and use it to draw a piece of a 5.5-meter-radius circle representing the sun.

5. One way to solve this problem is to think about the scale factors we might use. Since each foot is 12 inches, 5 feet is $5 \cdot 12 = 60$ inches, 4 feet is $4 \cdot 12 = 48$ inches, and 10 feet is $10 \cdot 12 = 120$ inches. Thinking only about the length, we see that the scale factor from the model to the sculpture, k, should be such that

$$k \cdot 15 = 60$$

so

$$k = \frac{60}{15} = 4$$

Similarly, thinking only about the width and the height, we find that the scale factors would be

$$k = \frac{48}{8} = 6$$

and

$$k = \frac{120}{27}, \text{ about } 4.4$$

respectively. The artist will need to use the smallest of these scale factors; otherwise her sculpture would require more marble than she has. So the artist should use $k = 4$, in which case the dimensions of the sculpture are as follows:

$$\text{width} = 4 \cdot 15 \text{ in.} = 60 \text{ in.}$$
$$\text{depth} = 4 \cdot 8 \text{ in.} = 32 \text{ in.}$$
$$\text{height} = 4 \cdot 27 \text{ in.} = 108 \text{ in.}$$

PROBLEMS FOR SECTION 14.5

1. Using your own examples, discuss the mathematical meaning of the term "similar" and discuss how this meaning is different from the everyday meaning.

2. Frank's dog, Fido, is 16 inches tall and 30 inches long. Frank wants to draw Fido 4 inches tall. Frank figures that because Fido's height in the drawing will be 12 inches less than Fido's actual height (because $16 - 12 = 4$), Fido's length in the drawing should also be 12 inches less than Fido's actual length. Therefore, Frank figures that he should draw Fido $30 - 12 = 18$ inches long. Will the drawing of Fido be correctly proportioned if Frank makes it 18 inches long? If not, how long should Frank draw Fido? Explain.

3. *Tyler's Flag Problem:* Tyler has designed his own flag on a rectangle that is 3 inches tall and 6 inches wide. Now he wants to draw a larger version of his flag on a rectangle that is 9 inches tall. How wide should Tyler make his larger flag?

Solve this problem in two ways: with the scale factor method and with the internal factor method. In each case:

- Explain clearly the idea and the reasoning of that method.
- Link the method to a proportion.

4. *Jasmine's Flag Problem:* Jasmine has designed her own flag on a rectangle that is 8 inches tall and 12 inches wide. Now she wants to draw a smaller version of her flag on a rectangle that is 6 inches tall. How wide should Jasmine make her smaller flag?

Solve this problem in two ways: with the scale factor method and with the internal factor method. In each case:

- Explain clearly the idea and the reasoning of that method.
- Link the method to a proportion.

5. Kelsey wants to make a scale drawing of herself standing straight and with her arms down. Kelsey is 4 feet 4 inches tall. Her leg is 24 inches long. If Kelsey wants to make the drawing of herself 10 inches tall, then approximately how long should she draw her leg?

Solve this problem in two ways: with the scale factor method and with the internal factor method. In each case:

- Explain clearly the idea and the reasoning of that method.
- Link the method to a proportion.

6. A painting that is 4 feet 3 inches by 6 feet 4 inches will be reproduced on a card. The side that is 4 feet 3 inches long will become 3 inches long on the card. Determine how long the 6 feet 4 inches side will become on the card. Explain why your method of solution is valid.

7. Write two problems about similar shapes or figures:

a. For the first problem, choose numbers so that the problem is especially easy to solve using the scale factor method.

b. For the second problem, choose numbers so that the problem is especially easy to solve using the internal factor method.

For both problems, show how to solve the problem with that method and explain clearly the logic and reasoning of the method.

8. Mr. Jackson's class wants to make a display of the solar system showing the correct relative distances of the planets and Pluto from the sun (in other words, showing the distances of the planets and Pluto from the sun to scale). Table 14.4 shows the approximate actual distances of the planets and Pluto from the sun.

TABLE 14.4 Distances of planets and Pluto from the sun

Heavenly Body	Approximate Distance from Sun in Miles
Mercury	36,000,000
Venus	67,000,000
Earth	93,000,000
Mars	141,000,000
Jupiter	484,000,000
Saturn	887,000,000
Uranus	1,783,000,000
Neptune	2,794,000,000
Pluto	3,666,000,000

If the distance from the sun to Mercury is to be represented in the display as 1 inch, then how should the distances from the sun to the other planets and Pluto be represented? Explain your reasoning in detail for one of the planets in such a way that fifth graders who understand multiplication and division but do not know about setting up proportions might be able to understand. Calculate the answers for the other planets without explanation.

9. An art museum owns a painting that it would like to reproduce in reduced size onto a 24-inch-by-36-inch poster. The painting is 85 inches by 140 inches. Give your recommendation for the size of the reproduced painting on the poster—how wide and long do you suggest that it be? Draw a scale picture showing how you would position the reproduced painting on the poster. Explain your reasoning.

10. Cameras that use film produce a negative, which is a small picture of the scene that was photographed (with reversed colors). A photograph is produced from a negative by printing onto photographic paper. When a photograph is printed from a negative, one of two things happens: Either the printed picture shows the full picture that was captured in the negative, or the printed picture is cropped, showing only a portion of the full picture on the negative. In either case, the rectangle forming the printed picture is similar to the rectangular portion *of the negative* that it comes from.

An ordinary 35-mm camera produces a negative in the shape of a $1\frac{7}{16}$-inch-by-$\frac{15}{16}$-inch rectangle. Some of the most popular sizes for printed pictures are $3\frac{1}{2}$ in. by 5 in., 4 in. by 6 in. and 8 in. by 10 in. Can any of these size photographs be produced without either cropping the picture or leaving blank space around the picture? Why, or why not? Explain your reasoning clearly.

11. If the map in Figure 14.65 has a scale such that 1 inch represents 15 miles, then what is the scale of the map in Figure 14.66, which shows the same region? Explain your reasoning.

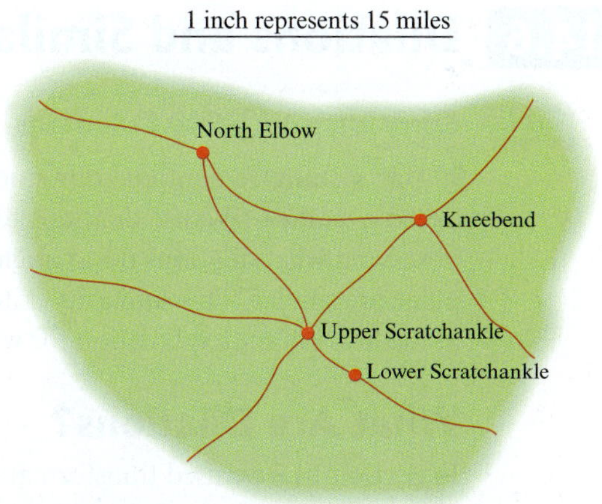

Figure 14.65 Map with a scale of 1 inch $\leftrightarrow$ 15 miles.

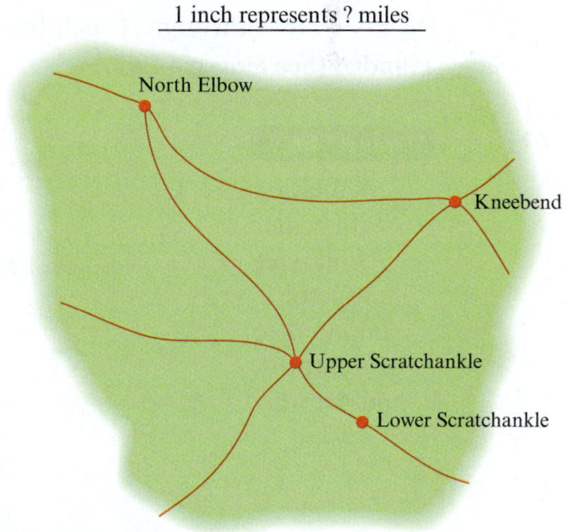

Figure 14.66 What is the scale of this map?

12. Sue has a rectangular garden. If she makes her garden twice as wide and twice as long as it is now, will the area of her garden be twice as big as its current area? Examine this problem *carefully* by working out some examples and drawing pictures. Explain your conclusion. If the garden is not twice as big (in terms of area), how big is it compared with the original?

14.6 Dilations and Similarity

CCSS Common Core State Standards Grade 8

In this section we continue our study of similarity from Section 14.5 by defining a transformation called a *dilation*. We use dilations to describe and discuss similarity for shapes in a plane. Based on observing what dilations do to angles, we develop a criterion for determining when two shapes in a plane are similar. This similarity criterion applies in many practical problem-solving situations and is at the heart of the explanation for why lines in a plane have a certain characteristic kind of equation.

What Are Dilations?

In Section 14.3 we used transformations to define what it means for shapes in a plane to be *congruent*. We can also use transformations to describe or define what it means for shapes in a plane to be *similar*. To do so, we need to introduce another transformation called a *dilation*.

To define a dilation of the plane, fix a point P in the plane to be the center of the dilation, and specify a positive number k to be the scale factor of the dilation. A **dilation** centered at the point P with the scale factor k is a transformation of the plane that takes each point A to the point A' on the ray from P through A such that the distance between P and A' is k times the distance between P and A (See Figure 14.67.)

dilation

Figure 14.67

A dilation with center P and scale factor 2, taking A to A', B to B', and C to C'.

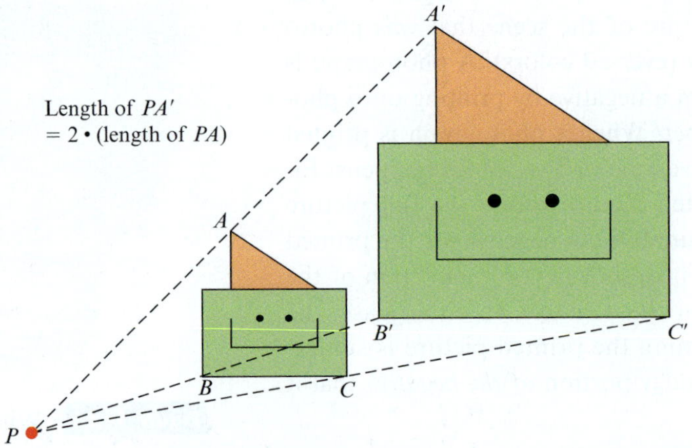

CLASS ACTIVITY

14S Dilations Versus Other Transformations: Lengths, Angles, and Parallel Lines, p. CA-298

14T Reasoning about Proportional Relationships with Dilations, p. CA-300

What Properties Do Dilations Have?

If you did Class Activity 14S then you probably noticed that dilations have special properties, which you can also observe in Figure 14.67. Although we will not do so here, it is possible to prove that dilations have the following properties (among others):

1. Dilations scale *every* distance in a plane by the scale factor of the dilation (not just the distances from the center of the dilation).
2. Dilations preserve the sizes of angles. In other words, when we apply a dilation to an angle, the transformed angle may be in a different location than the original, but it has the same size as the original.

Note that although the term *dilation* sounds like it should apply only when the scale factor is greater than 1, we still use this term for scale factors less than 1, even though these dilations actually shrink figures.

How Can We Use Transformations to Discuss Similarity?

CCSS 8.G.4

similar (transformation definition)

In Section 14.5 we defined similarity in terms of a scale factor that relates all corresponding distances in two shapes or objects. Because dilations have property (1) above, another way to define similarity for shapes in a plane is in terms of transformations: two shapes in a plane are **similar** if there is a sequence of transformations, each of which is a translation, reflection, rotation, or dilation, that take one shape to the other.

With our transformation definition of similarity we can explain why all circles are similar (which was useful to know in Section 12.6). Given two circles in a plane, with the first centered at P with radius r and the second centered at Q with radius s, translate the second one so that Q goes to P, as indicated in **Figure 14.68**. After it is translated, the points on the second circle are s units from P. Then apply the dilation centered at P with scale factor $\frac{r}{s}$ to the translated second circle. After the dilation, the points on the second circle are $\frac{r}{s} \cdot s = r$ units from P. Therefore the dilation takes the translated second circle to the first circle and so the two circles are similar.

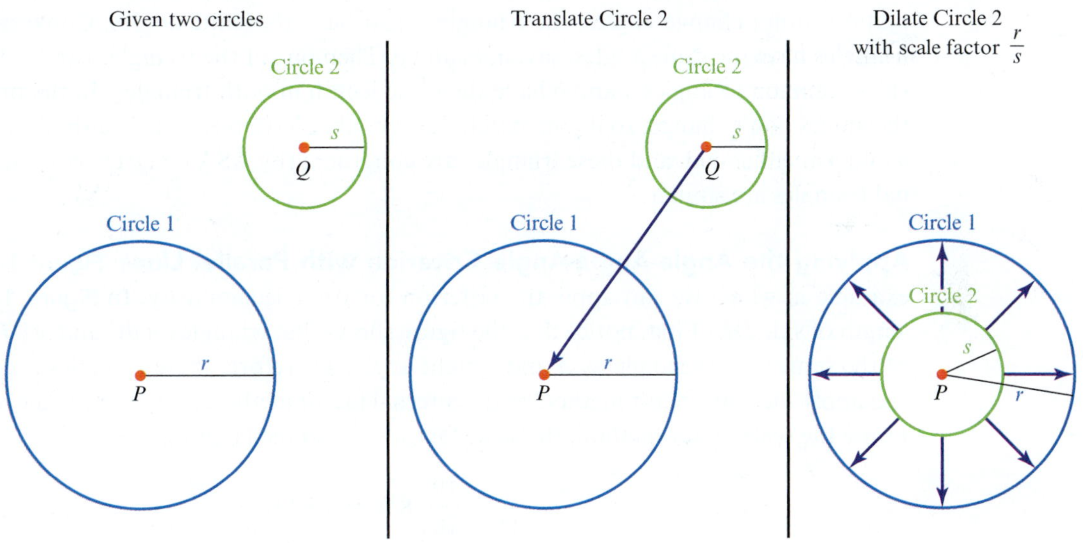

Figure 14.68 Applying transformations to show that two circles are similar.

When Are Two Shapes Similar?

In the situations in Section 14.5, objects were *given* as being similar. In other words, the very nature of the situation told us that the objects were similar. However, there can be cases where it is not entirely obvious or clear that two shapes in question are similar. In those cases, how can we tell whether the shapes are similar? It is tempting to think that we can always tell "by eye," but this method is not reliable. **Figure 14.69** shows two triangles that appear to be similar but are not.

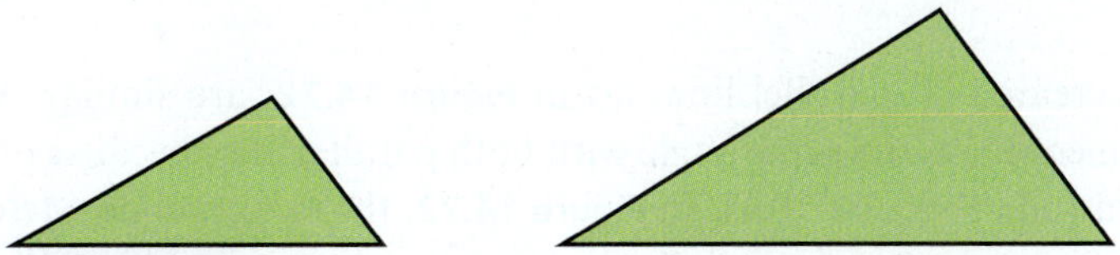

Figure 14.69 Two triangles that may look similar but are not.

triangle similarity criterion

The Angle-Angle-Angle Criterion for Triangle Similarity Here is a **criterion for similarity of triangles**: Two triangles are similar exactly when the two triangles have the same-size angles. To clarify, two triangles are similar exactly when it is possible to match each angle of the first triangle with an angle of the second triangle in such a way that matched angles have equal measures. So, for example, the criterion tells us that the triangles in Figure 14.70 are similar because they have the same-size angles.

Figure 14.70

Similar triangles.

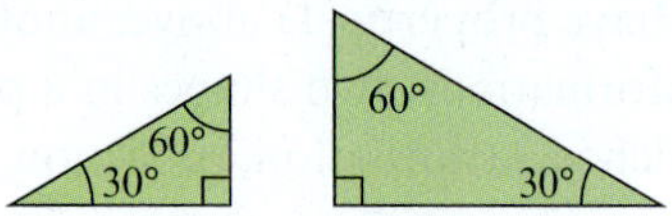

Actually, notice that to determine whether two triangles are similar, we really need to check that only *two* of their angles are the same size, because the sum of the angles in a triangle is always 180°. If two angles are known, the third one is determined as the angle that makes the three add to 180°. Thus, if two angles in one triangle are the same size as two respective angles in another triangle, then the third angles will automatically also be equal in size.

Why does the angle-angle-angle criterion for triangle similarity work? Basically, it is because dilations don't change angles. If two triangles are similar, then one triangle can be transformed into the other by a sequence of translations, reflections, rotations, or dilations. Because these transformations don't change angles, the triangles must have the same angles. Conversely, suppose two triangles have the same angles, say, a, b, and c. Then one of the triangles can be dilated so that the sides common to angles a and b have the same length in both triangles. In the process of dilating, the angles don't change, so if you match the sides that have the same length, they both have angles a and b at either end, and these triangles are congruent (by ASA congruence). Therefore, the original triangles are similar.

Applying the Angle-Angle-Angle Criterion with Parallel Lines Figure 14.71 provides an example in which we can apply the criterion for triangle similarity. In Figure 14.71, what is the length of side DE? First, notice that the figure shows that triangles ABC and ADE have two angles of the same size: the angle at A and a right angle. Therefore, these triangles are similar. Now we can apply the scale factor method or the internal factor method to determine the length of side DE. Using the scale factor method, the scale factor is $\frac{19}{10}$, so the length DE is

$$\frac{19}{10} \cdot 8 \text{ ft} = 15.2 \text{ ft}$$

Figure 14.71

What is the length of DE?

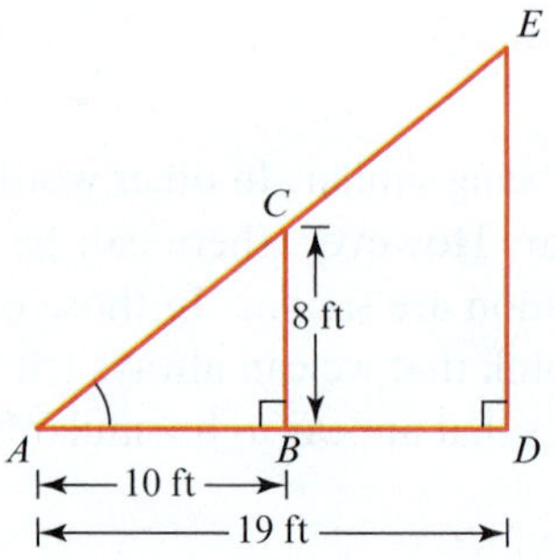

In general, triangles created by parallel lines, as in Figure 14.72, are similar. Why? A line that crosses two parallel lines makes the same angle with both parallel lines because of the Parallel Postulate (which we discussed in Section 10.1). In Figure 14.72, the two triangles ABC and ADE share an angle at A, and sides BC and DE are parallel, so the angles at B and D are the same size, and the triangles ABC and ADE are similar.

Figure 14.72 Parallel lines create similar triangles.

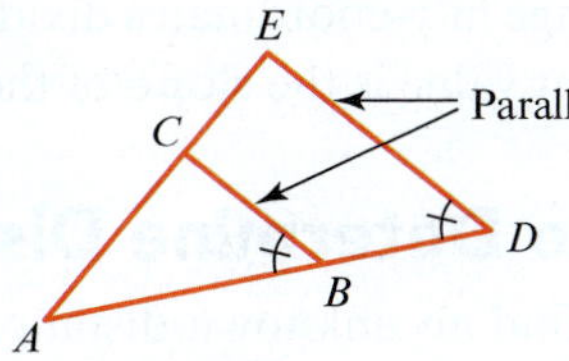

Another common situation in which similar triangles are created is illustrated in Figure 14.73. In this case, two lines cross at a point A. If the lines BC and DE are parallel, then the triangles ABC and ADE are similar. This is because angles BAC and EAD are opposite and therefore the same size, as we discussed in Section 10.1. Also, the angles at B and at D are equal in size because these are where the parallel line segments BC and DE meet the line segment BD. Therefore, two out of three angles in ABC and ADE have the same size, so all three must have the same size, and the triangles are similar.

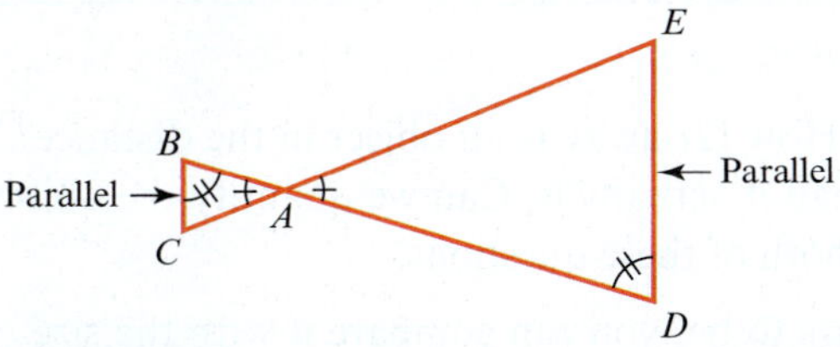

Figure 14.73 Parallel lines on opposite sides of the point where two lines meet create similar triangles.

How Does the Angle-Angle-Angle Criterion Apply to Lines and Slope?

CCSS 8.EE.6

In Section 9.7 we explained why linear relationships have equations of the form $y = mx + b$. At the heart of this explanation is the fact that non-vertical lines in a coordinate plane have a consistent slope, which we can now explain with the angle-angle-angle criterion. Given a non-vertical line in a coordinate plane, consider "slope triangles," which are right triangles that have a horizontal side parallel to the x-axis, a vertical side parallel to the y-axis, and a side along the line, as in Figure 14.74. Because all the horizontal segments are parallel, they make the same angle with the line by the

Figure 14.74 Applying the angle-angle-angle criterion to deduce that lines have a consistent slope.

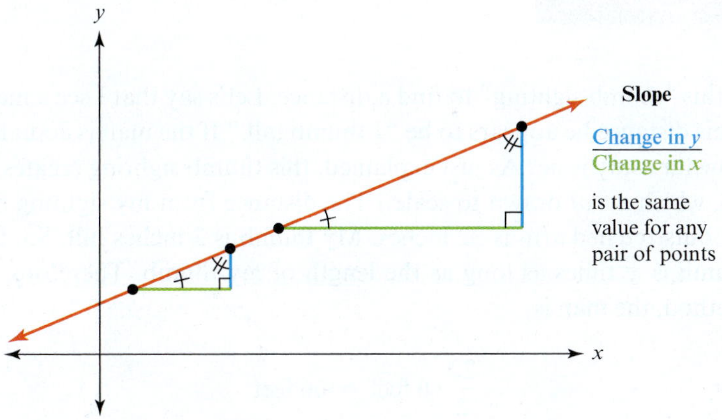

Parallel Postulate. Similarly, because all the vertical segments are parallel, they also make the same angle with the line. By the angle-angle-angle criterion, all these slope triangles are similar. Therefore, for all of these slope triangles, the lengths of a vertical side and a horizontal side are in the same ratio. So for any pair of points on the line, the change in y-coordinates divided by the change in x-coordinates is always the same value. This consistent value is the **slope** of the line.

slope

How Can We Use Similar Triangles to Determine Distances?

Similar triangles can be used in practical situations to find an unknown distance or length when several other related distances and lengths are known. Often, the tricky part in applying the theory of similar triangles is locating the similar triangles and sketching them. You will need to apply your visualization skills to help you sketch the situation, showing the relevant components. In the following examples, think carefully about why the drawings were drawn the way they were. If they had been drawn from a different perspective, would you see the relevant similar triangles? (In some cases, yes, but in many cases, no.)

CLASS ACTIVITY

14U How Can We Measure Distances by "Sighting"? p. CA-301

Finding Distances by "Sighting" How far away is an object in the distance? When we look at a distant object, it appears smaller than it actually is. Can we quantify the object's apparent size? Similar triangles can help us answer both of these questions.

sighting

To describe how big an object appears to be, you can compare it with the size of your thumb by stretching your arm out straight in front of you, closing one eye, and "**sighting**" from your thumb to the object you are considering. This situation creates a pair of similar triangles, as shown in Figure 14.75. The similar triangles are ABC (eye, base of thumb, top of thumb) and ADE (eye, bottom of picture, top of picture). These triangles are similar because the angles at A are equal and the angles at B and D are equal if the thumb is held parallel to the picture. (See Figure 14.72.)

Figure 14.75
"Thumb sighting" a picture on a wall.

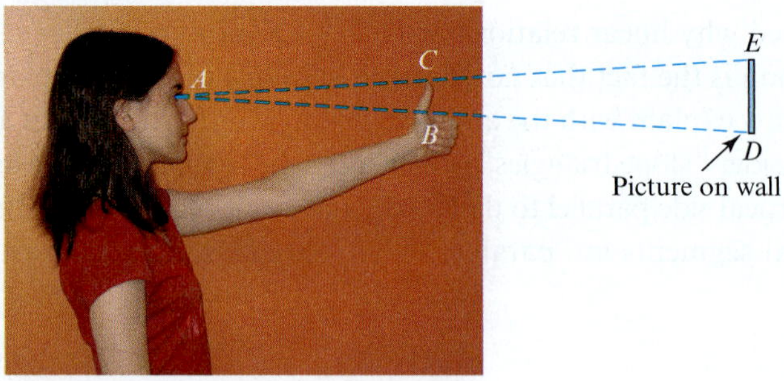

You can also use this "thumb sighting" to find a distance. Let's say that I see a man standing in the distance, and at this distance he appears to be "1 thumb tall." If the man is actually 6 feet tall, then approximately how far away is he? As just explained, this thumb sighting creates similar triangles. (See Figure 14.76, which is *not* drawn to scale.) The distance from my sighting eye to the base of my thumb on my outstretched arm is 22 inches. My thumb is 2 inches tall. So, the distance from my eye to my thumb is $\frac{22}{2}$ times as long as the length of my thumb. Therefore, according to the internal factor method, the man is

$$\frac{22}{2} \cdot 6 \text{ feet} = 66 \text{ feet}$$

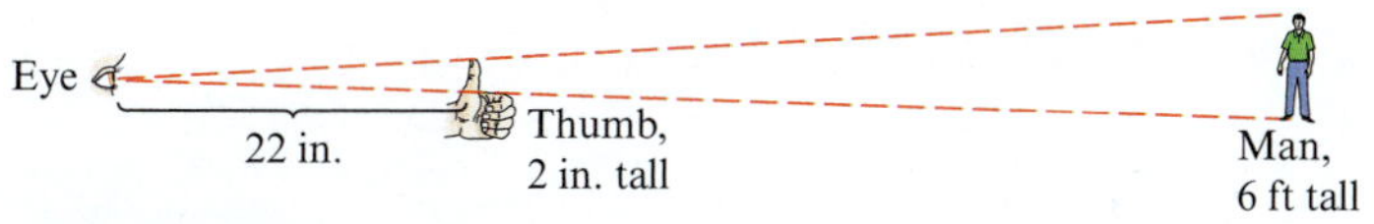

Figure 14.76 "Thumb sighting" a man.

away. You might object that this tells us the distance from my eye to the man's feet, but, because the man is fairly far away, this distance is almost identical to the distance along the ground from my feet to his feet. Plus, we are only determining *approximately* how far away the man is, because all the measurements used are very rough. So this extra bit of inaccuracy is insignificant.

How could you modify the method of thumb sighting to make it more useful and more accurate? You could use objects besides your thumb for sighting—a ruler held vertically would be much more accurate, for example. And what if you didn't just hold the ruler in your hand but had it attached to a fixed length of pole? That would also create greater accuracy. These sorts of improvements have led to some of the surveying equipment used today. People at construction sites who are looking through a device on a tripod at a pole in the distance (as in Figure 14.77) are surveying.

Figure 14.77 Surveying at a construction site.

Some kinds of surveying equipment measure distances based on the theory of similar triangles using a more elaborate method than primitive thumb sighting. For example, to measure the distance between the surveying equipment and a pole of known height in the distance, the person surveying looks through the equipment and "sights" the pole. A kind of ruler inside the surveying equipment replaces the thumb.

Finding Heights by Using Sun Rays and Shadows Is there a way we can find the height of a tree, pole, or other tall object without climbing up it to measure how tall it is? In fact, yes. We can use the shadow that an object casts, together with the theory of similar triangles, to determine how tall the object is.

CLASS ACTIVITY

14V How Can We Use a Shadow or a Mirror to Determine the Height of a Tree? p. CA-302

Suppose Ms. Ovrick takes her class outside on a sunny day to determine the height of the flagpole. The students hold meter sticks perpendicular to ground and measure the length of the shadows that the meter sticks cast. Although the numbers vary, all the shadows are about 120 cm long. Why should all the shadows be about the same length? Since the sun is very far away, when its light rays reach the earth, the rays are virtually parallel. So, as we see in Figure 14.78, the angles that the sun rays make with the meter sticks should all be the same, and so the shadows cast by the meter sticks should all be the same too.

Next, the students measure the length of the flagpole's shadow and find that it is 7 meters and 65 centimeters long. How tall is the flagpole? Let's use the theory of similar triangles to determine the height of the flagpole. Think about how shadows of the meter sticks and the flagpole are

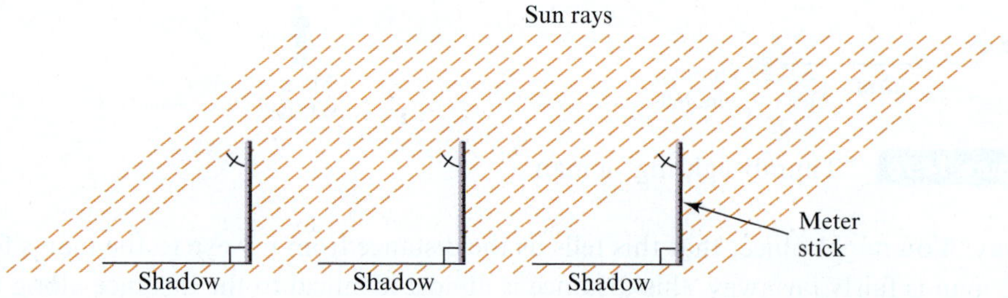

Figure 14.78
Sun rays make the same angles with vertical meter sticks.

created by the sun's rays. Because the sun rays are virtually parallel when they get to the earth, the rays make the same angle with the top of the flagpole as they do with the tops of the meter sticks, as we see in Figure 14.79. The meter sticks and the flagpole are perpendicular to the ground, so the triangle formed by a meter stick and its shadow (ABC) and the triangle formed by the flagpole and its shadow (DEF) have the same angles and are therefore similar. Since the length of the shadow of the meter stick is 1.2 times its height, the length of the shadow of the flagpole must also be 1.2 times its height. Therefore, the flagpole is

$$7.65 \div 1.2 = 6.375$$

meters tall, or about 6.4 meters tall.

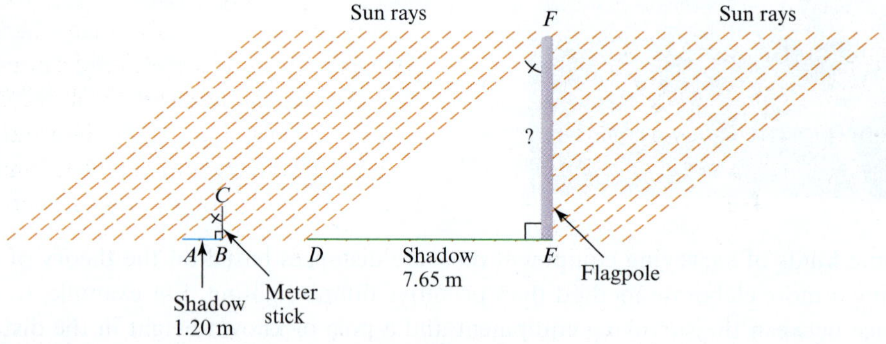

Figure 14.79
Similar triangles are formed by objects and their shadows.

SECTION SUMMARY AND STUDY ITEMS

Section 14.6 Dilations and Similarity

Dilations are transformations of a plane that take shapes to similar shapes. Dilations scale all distances by the same scale factor and preserve sizes of angles.

The angle-angle-angle criterion states that two triangles are similar exactly when the two triangles have corresponding angles of the same size. Similar triangles can arise when parallel lines cross a pair of lines that cross. We can explain why triangles formed this way are similar by explaining why the corresponding angles have the same size (using the Parallel Postulate). We can use similar triangles to determine distances and heights.

Key Skills and Understandings

1. Use transformations to discuss what it means for shapes in a plane to be similar.
2. Explain why triangles are similar by showing that corresponding angles must have the same size (the angle-angle-angle criterion). When the Parallel Postulate applies, use it to show that corresponding angles are produced by a line crossing two parallel lines and are therefore of the same size.
3. Use similar triangles to determine heights and distances.
4. Use similarity to solve problems.

Practice Exercises for Section 14.6

1. Istabrag is 4 feet 9 inches tall, and her shadow is 3 feet 6 inches long. At the same time, the shadow of a tree (measured from the base of the tree to the tip of the shadow) is 24 feet 6 inches long. How tall is the tree?

   a. Make a math drawing of the similar triangles that are relevant to solving this problem and explain why the triangles are similar.

   b. Determine the height of the tree and explain your reasoning.

2. Suppose that you go outside on a clear night when much of the moon is visible and use a ruler to "sight" the moon (as described in Class Activity 14U). Use the information that follows to determine how big the moon will appear to you on your ruler. (You will also need to measure the distance from your eye to a ruler that you are holding with your arm stretched out.) On a night when the moon is visible, try this and verify it.

   The distance from the earth to the moon is approximately 384,000 km. The diameter of the moon is approximately 3,500 km.

3. Go online to learn how to make a **camera obscura** from a Pringles potato chip can. A camera obscura is a primitive camera that projects images onto a screen through a small hole. It illustrates a fundamental idea of photography: projecting an image through a small hole. You can make a camera obscura from any tube by cutting off a piece of the tube (about 2 inches from one end), placing semitransparent paper or plastic over the place where you cut, and taping the tube back together, with the semitransparent paper now in the middle of the tube. Cover one end of the tube with aluminum foil, and use a pin to poke a small hole in the middle of the foil. Cover the side of the tube with aluminum foil to keep light out. Now look through the open end of the tube at a well-lit scene. You should see an upside-down projected image of the scene on the semitransparent paper.

   Suppose you make a camera obscura out of a tube of diameter 3 inches, and suppose you put the semitransparent paper or plastic 2 inches from the hole in the aluminum foil. How far away would you have to stand from a 6-foot-tall man in order to see the entire man on the camera obscura's screen?

Answers to Practice Exercises for Section 14.6

1. As explained with the meter stick and flagpole in Figure 14.79, the triangle formed by Istabrag and her shadow is similar to the triangle formed by the tree and its shadow. Istabrag is 4 feet 9 inches, or 4.75 feet tall. (The 0.75 is because 9 inches is $\frac{9}{12} = \frac{3}{4} = 0.75$ of a foot.) Her shadow is 3 feet 6 inches, or 3.5 feet long. So, Istabrag is

$$\frac{4.75}{3.5} = 1.357\ldots$$

   times as tall as her shadow is long. The same relationship must hold for the tree. So, the tree must be about 1.357 times as tall as its shadow is long. Therefore, the tree is

$$1.357 \cdot 24.5, \text{ about } 33.25$$

   feet tall, or 33 feet 3 inches tall. (The 3 inches is because $0.25 = \frac{1}{4}$ and $\frac{1}{4}$ of a foot is 3 inches.)

2. When you sight the moon with a ruler, holding the ruler parallel to the moon's diameter, you create similar triangles ABC and ADE, as shown in Figure 14.80 (not to scale). These triangles are similar as explained on p. 657 with Figure 14.72. Since the diameter of the moon is 3,500 km and the distance to the moon is 384,000 km, the moon's diameter (DE) is $\frac{3{,}500}{384{,}000}$ times as long as the distance to the moon (AD). The same relationship must hold for the apparent size of the moon on the ruler (BC) and the distance from your eye to the ruler (AB): The apparent size of the moon on the ruler must be $\frac{3{,}500}{384{,}000}$ times as long as the distance from your eye to the ruler. (This is the internal factor method.) Notice that $\frac{3{,}500}{384{,}000}$ is about $\frac{1}{100}$. So, if the distance from your eye to the ruler is about 22 inches, then the moon's diameter will appear to be about 0.2 inches on your ruler, or about $\frac{1}{5}$ of an inch.

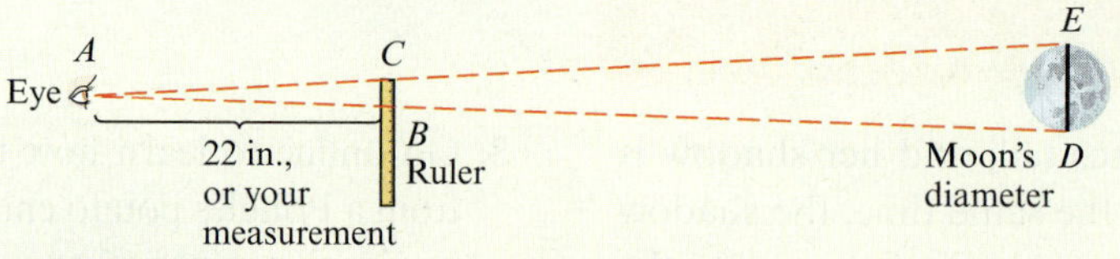

Figure 14.80 Sighting the moon.

3. Figure 14.81 shows how light from an object DE enters the pinhole (at point A) and projects onto the screen inside a camera obscura. If the object being looked at and the screen BC of the camera obscura are both vertical, then BC and DE are both vertical and are thus parallel. Therefore, as described in the text, triangles ABC and ADE are similar. Since these two triangles are similar, all corresponding distances on the triangle scale by the same scale factor. Since the distance from the screen of the camera obscura to the pinhole is $\frac{2}{3}$ times the height of the screen, by the internal factor method, the distance from the pinhole to the man should be $\frac{2}{3}$ times the length of the man, which is $\frac{2}{3} \cdot 6$ feet $= 4$ feet. So you need to stand at least 4 feet from the man to see the full length of him on the camera obscura.

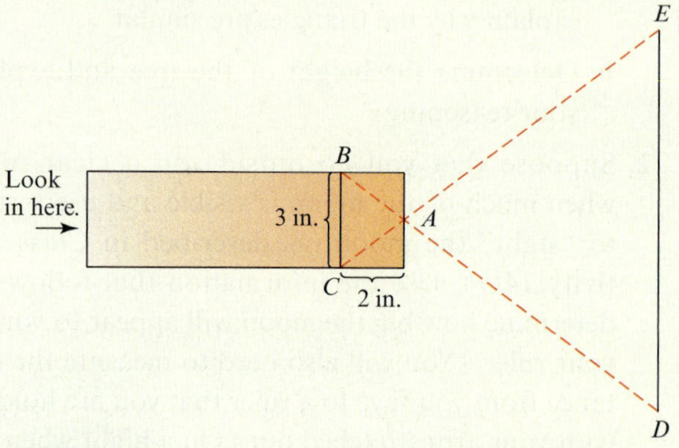

Figure 14.81 Side view of a camera obscura.

PROBLEMS FOR SECTION 14.6

1. **a.** If two shapes are congruent, are they also similar? Explain.

   b. If two shapes are similar, are they necessarily also congruent? Explain.

2. Is there an angle-angle-angle-angle similarity criterion for quadrilaterals? Explain why or why not.

3. After applying a dilation centered at O, the figure in Figure 14.82 becomes 35 cm wide and H cm tall.

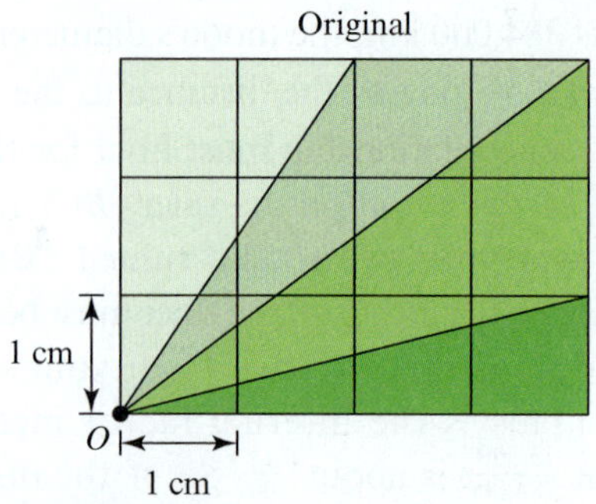

Figure 14.82 After applying a dilation, how are the new width and height related?

   a. What is the new spacing between the grid lines?

   b. How tall does the figure become? Use our definition of multiplication (and division) to write different equations that relate H to other quantities. Interpret the parts of your equations in terms of the figure and the grid lines.

4. After applying a dilation centered at O, the figure in Figure 14.82 becomes X cm wide and Y cm tall.

   a. What is the new spacing between the grid lines?

   b. How are X and Y related? Use our definition of multiplication (and division) to write different equations relating X and Y. Interpret the parts of your equations in terms of the figure and the grid lines.

5. Ms. Winstead's class went outside on a sunny day and measured the lengths of some of their classmates' shadows. The class also measured the length of a shadow of a tree. Inside, the students arranged the numbers as shown in Table 14.5.

TABLE 14.5 Heights of students and tree and lengths of shadows

	Tyler	Jessica	Sunjae	Lameisha	Tree
Shadow length	33 in.	34 in.	32 in.	34 in.	22 feet
Height	53 in.	57 in.	52 in.	58 in.	?

In your own words, explain clearly how the theory of similar triangles applies in this situation. Include a math drawing as part of your explanation. Use the theory of similar triangles to determine the approximate height of the tree.

6. *A Thumb Sighting Problem:* Suppose you are looking down a road and you see a person ahead of you. You hold out your arm and "sight" the person with your thumb, finding that the person appears to be as tall as your thumb is long. Assume that your thumb is 2 inches long and that the distance from your sighting eye to your thumb is 22 inches. If the person is 5 feet 4 inches tall, then how far away are you from the person?
   a. Make a math drawing showing that the thumb sighting problem involves similar triangles. Explain why the triangles are similar.
   b. Solve the thumb sighting problem in two different ways. In both cases, explain the logic behind the method you use.

7. Explain and draw a picture to show how you could use the theory of similar triangles to determine the height of a flagpole by looking into a mirror lying flat on the ground with its reflective side up. Indicate the similar triangles that would be involved and explain why these triangles really must be similar. Say which measurements you will need to make, and show or describe briefly how you will use these measurements to determine the height of the pole. (You may either make up numbers for these measurements or use letters to stand for these numbers.)

8. Suppose you have a TV whose screen is 36 inches wide. You decide that you want to simulate the experience of watching a movie on a 30-foot-wide screen as closely as possible. How far away should you sit from the TV so that the TV screen will appear as wide in your field of vision as a 30-foot-wide screen would from 40 feet away? Explain how the theory of similar triangles applies to help you answer this question. Include a sketch of similar triangles in your explanation, and explain in detail why these triangles really are similar.

9. Which TV screen appears bigger: a 50-inch screen viewed from 8 feet away or a 25-inch screen viewed from 3 feet away? Explain carefully. (TV screens are typically measured on the diagonal, so to say that a TV has a "50-inch screen" means that the diagonal of the screen is 50 inches long.)

10. A city has a large cone-shaped Christmas tree that stands 20 feet tall and has a diameter of 15 feet at the bottom. The lights will be wound around the tree in a spiral, so that each "row" of lights is about 2 feet higher than the previous row of lights. Approximately how long a strand of lights (in feet) will the city need? To answer this, it might be helpful to think of each "row" (one "wind") of lights around the tree as approximated by a circle. Explain your reasoning.

11. Let's say that you are standing on top of a mountain looking down at the valley below, where you can see cars driving on a road. You stretch out your arm, use your thumb to "sight" a car, and find that the car appears to be as long as your thumb is wide. Estimate how far away you are from the car. Explain your method clearly. (You will need to make an assumption to solve this problem. Make a realistic assumption, and make it clear what your assumption is.)

12. **a.** During a total solar eclipse, the moon moves in front of the sun, obscuring the view of the sun from the earth (at some locations on the earth). Surprisingly, the moon seems to be superimposed on the sun during a solar eclipse, appearing to be almost identical in size as seen from the earth. How does this situation give

rise to similar triangles? Draw a sketch. (It does not have to be to scale.)

b. Use the data that follow and either the scale factor method or the internal factor method to determine the approximate distance of the earth to the sun. Explain the reasoning behind the method you use.

- The distance from the earth to the moon is approximately 384,000 km.
- diameter of the moon is approximately 3500 km.
- The diameter of the sun is approximately 1,392,000 km.

13. A pinhole camera is a very simple camera made from a closed box with a small hole on one side. Film is put inside the camera, on the side opposite the small hole. The hole is kept covered until the photographer wants to take a picture, when light is allowed to enter the small hole, producing an image on the film. Unlike an ordinary camera, a pinhole camera does not have a view finder, so with a pinhole camera, you can't see what the picture you are taking will look like.

a. Suppose that you have a pinhole camera for which the distance from the pinhole to the opposite side (where the film is) is 3 inches. Suppose that the piece of film to be exposed (opposite the pinhole) is about 1 inch tall and $1\frac{1}{2}$ inches wide. Let's say you want to take a picture of a bowl of fruit that is about 15 inches wide and piled 6 inches high with fruit and that you want the bowl of fruit to fill up most of the picture. Approximately how far away from the fruit bowl should you locate the camera? Explain, using similar triangles.

b. Explain why the image produced on film by a pinhole camera is upside down.

*__14.__ This problem will help you prove the Pythagorean theorem with the aid of similar triangles instead of with the moving and additivity principles. Given any right triangle with short sides of length a and b and hypotenuse of length c, as on the left in Figure 14.83, subdivide the triangle into two smaller right triangles, as shown on the right in Figure 14.83. Let s and t be the lengths of sides of these smaller triangles, as shown in the figure. To prove the Pythagorean theorem, you must prove that $a^2 + b^2 = c^2$, which you will do by completing parts (a), (b), and (c).

a. Use angles to explain why the two smaller right triangles on the right in Figure 14.83 are similar to the original triangle on the left in this figure. (Do not use any actual measurements of angles because the proof must be general—it must work for *any* initial right triangle.)

b. Use proportions to relate $\frac{s}{a}$ to a and c, and to relate $\frac{t}{b}$ to b and c. Use part (a) to explain why your proportions are true.

c. Use your proportions in part (b) to show that

$$c = \frac{a^2}{c} + \frac{b^2}{c}$$

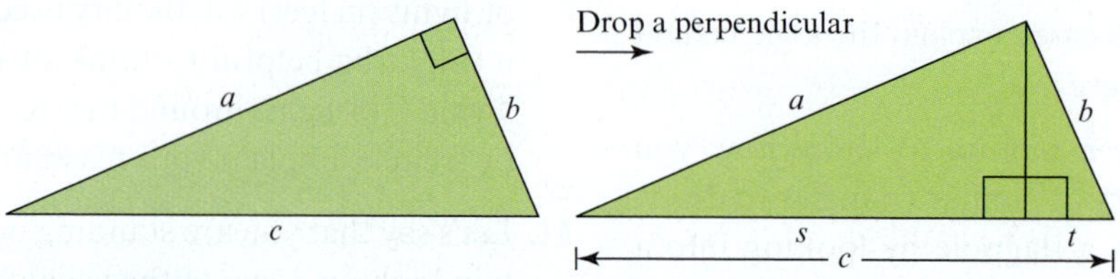

Figure 14.83 Right triangle subdivided into two right triangles.

Then use this equation to prove the Pythagorean theorem.

*__15.__ Complete the following area puzzle.

a. Trace the 8-unit-by-8-unit square in Figure 14.84.

b. Cut out the pieces and reassemble them to form the 5-unit-by-13-unit rectangle in Figure 14.85.

c. Calculate the area of the square and the area of the rectangle. Is there a problem? Can the pieces from your 8-unit-by-8-unit square really fit together perfectly, without overlaps or gaps, to form a 5-unit-by-13-unit rectangle?

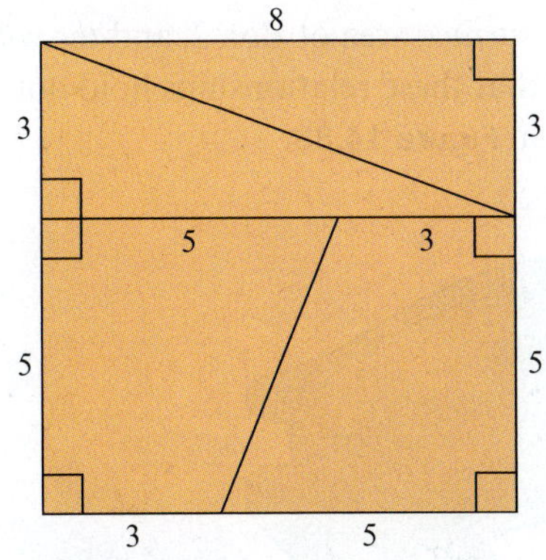

Figure 14.84 A subdivided 8-by-8-unit square.

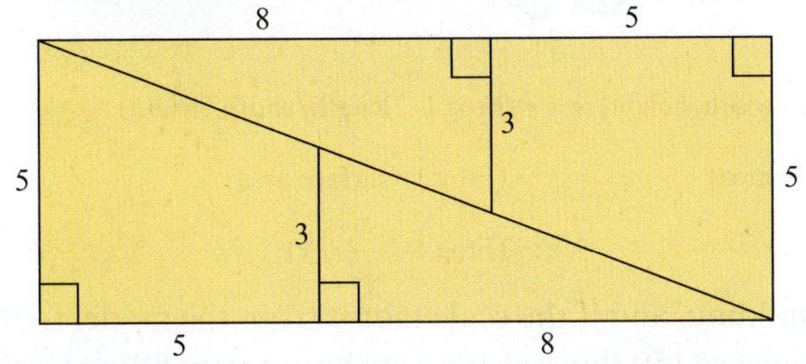

Figure 14.85 A subdivided 5-by-13-unit rectangle.

d. Explain the problem that you discovered in part (c). *Hints:* Can the two pieces shown in Figure 14.86, which you cut out of the square, really fit together as shown in the rectangle to form an actual triangle? If so, wouldn't there be some similar triangles? Look at the lengths of corresponding sides of the supposedly similar triangles. Do these numbers work the way they should for similar triangles?

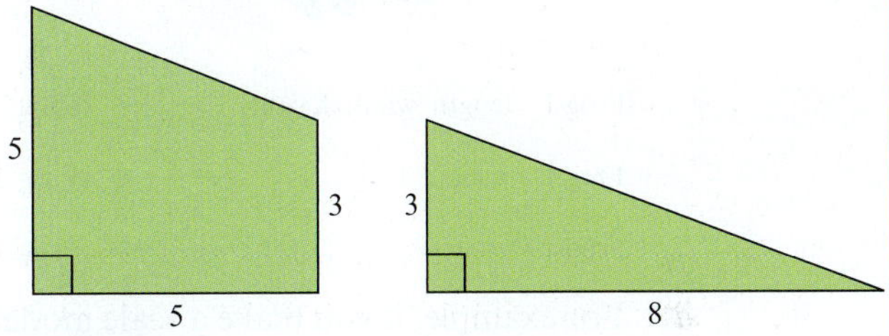

Figure 14.86 Combining two pieces.

14.7 Areas, Volumes, and Similarity

How are the surface areas and volumes of similar objects related?

When two objects are similar, as in Figure 14.87, there is a scale factor such that corresponding *lengths* on the objects are related by multiplying by the scale factor. Does this mean that the *surface areas* of the objects are related by multiplying by the scale factor? Does this mean that the *volumes* of the objects are related by multiplying by the scale factor? We will examine these questions in this section.

CLASS ACTIVITY

14W How Are Surface Areas and Volumes of Similar Boxes Related?, p. CA-303

Figure 14.87 Two similar cups.

If you did Class Activity 14W, you probably discovered that if a box is scaled with a scaled factor k, then even though the *lengths* of various parts of the box scale by the factor k, the *surface area* of the box scales by the factor k^2, and the *volume* of the box scales by the factor k^3. In other words, if Box 1 is similar to Box 2 with scale factor k, so that it is k times as wide, k times as long, and k times as tall as Box 2, then *the surface area of Box 1 is* k^2 *times as big as the surface* area of Box 2, and *the volume of Box 1 is* k^3 *times as big* as the volume of Box 2. It turns out that these relationships hold not only for similar boxes but also for other similar shapes as indicated in **Figure 14.88**.

Figure 14.88

When lengths are multiplied by the scale factor k, surface area is multiplied by k^2 and volume is multiplied by k^3.

Frog 1 length/width/height $\xrightarrow{k\,\cdot}$ (Frog 2 length/width/height) = $k \cdot$ (Frog 1 length/width/height)

Frog 1 surface area $\xrightarrow{k^2\,\cdot}$ (Frog 2 surface area) = $k^2 \cdot$ (Frog 1 surface area)

Frog 1 volume $\xrightarrow{k^3\,\cdot}$ (Frog 2 volume) = $k^3 \cdot$ (Frog 1 volume)

For example, if you make a scale model of a building, and if the scale factor from the model to the building is 150 (so that all lengths on the building are 150 times as long as the corresponding length on the model), then the surface area of the building is $150^2 = 22{,}500$ times as large as the surface area of the model, and the volume of the building is $150^3 = 3{,}375{,}000$ times as large as the volume of the model.

Notice how nicely these scale factors for surface area and volume fit with their dimensions and with the units used to measure surface area and volume: The surface of an object is two-dimensional. It is measured in units of in.^{2}, cm^2, . . . , and when an object is scaled with scale factor k, the surface area scales by k^2. Similarly, the size of a full, three-dimensional object can be measured by volume, which is measured in units of in.^{3}, cm^3, . . . , and when an object is scaled with scale factor k, the volume scales by k^3.

CLASS ACTIVITY

14X How Can We Determine Surface Areas and Volumes of Similar Objects? p. CA-304

14Y How Can We Prove the Pythagorean Theorem with Similarity? p. CA-305

14Z Area and Volume Problem Solving, p. CA-306

SECTION SUMMARY AND STUDY ITEMS

Section 14.7 Areas, Volumes, and Similarity

If the lengths in one object are k times the corresponding lengths in a second similar object, then the surface area of the first object is k^2 times the surface area of the second object and the volume of the first object is k^3 times the volume of the second object. We can see why these facts about how areas and volumes scale are plausible by extrapolating from patterns and models for prisms and cylinders and from formulas for their surface areas and volumes.

Key Skills and Understandings

1. Use patterns, models, and formulas to explain how areas and volumes of similar objects are related.
2. Apply the way that areas and volumes scale in similar objects to solve problems.

Practice Exercises for Section 14.7

1. If there were a new Goodyear blimp that was $2\frac{1}{2}$ times as long, $2\frac{1}{2}$ times as wide, and $2\frac{1}{2}$ times as tall as the current one, then how much paint would be needed to paint the new blimp, compared with the current one? How much gas would be needed to fill the new blimp (at the same pressure), compared with the current one? (See Figure 14.89.)

Figure 14.89 A blimp and the blimp scaled with scale factor $2\frac{1}{2}$.

2. Suppose you make a scale model for a pyramid by using a scale of 100:1 (so that the scale factor from the model to the actual pyramid is 100). If your model pyramid is made of 4 cardboard triangles, using a total of 60 square feet of cardboard, then what is the surface area of the actual pyramid?

3. If a giant were 12 feet tall, but proportioned like a typical 6-foot-tall man, about how much would you expect the giant to weigh? (See Figure 14.90.)

Figure 14.90 A 6-foot-tall man and a 12-foot-tall giant.

4. Suppose that a larger box is 3 times as wide, 3 times as long, and 3 times as high as a smaller box. Use a formula for the surface area of a box to explain why the larger box's surface area is 9 times the smaller box's surface area. Use a formula for the volume of a box to explain why the larger box's volume is 27 times the smaller box's volume.

5. Suppose a large cylinder has twice the radius and twice the height of a small cylinder. Use a formula for the surface area of a cylinder to explain why the surface area of the large cylinder is 4 times the surface area of the small cylinder. Use the formula for the volume of a cylinder to explain why the volume of the large cylinder is 8 times the volume of the small cylinder.

Answers to Practice Exercises for Section 14.7

1. According to the way surface area behaves under scaling, the new blimp's surface area would be

$$\left(2\frac{1}{2}\right)^2 = \frac{25}{4} = 6\frac{1}{4}$$

times as large as the current blimp's surface area. So the new blimp would require $6\frac{1}{4}$ times as much paint as the current blimp. According to the way volume behaves under scaling, the new blimp's volume would be

$$\left(2\frac{1}{2}\right)^3 = \frac{125}{8} = 15\frac{5}{8}$$

times as large as the current blimp's volume. So the new blimp would require $15\frac{5}{8}$ times as much gas to fill it as the current blimp.

2. According to the way surface areas behave under scaling, the actual pyramid's surface area is 100^2, or 10,000 times as large as the surface area of the pyramid. The model is made of 60 square feet of cardboard, so this is its surface area. Therefore, the surface area of the actual pyramid is $10{,}000 \cdot 60$ square feet, which is 600,000 square feet.

3. Since the giant is twice as tall as a typical 6-foot-tall man, and since the giant is proportioned like a 6-foot-tall man, the giant should be a scaled version of a 6-foot-tall man, with scale factor 2. Therefore, the volume of the giant should be $2^3 = 8$ times the volume of the 6-foot-tall man. Assuming that weight is proportional to volume, the giant's weight should be 8 times the weight of a typical 6-foot-tall man. If a typical 6-foot-tall man weighs between 150 and 180 pounds, then the giant should weigh 8 times as much, or between 1200 and 1440 pounds.

4. A box that is w units wide, l units long, and h units high has a volume of wlh cubic units. A box that is 3 times as wide, 3 times as long, and 3 times as high is $3w$ units wide, $3l$ units deep, and $3h$ units high, so it has volume

$$(3w)(3l)(3h) = 27wlh$$

cubic units. Therefore, a box that is 3 times as wide, 3 times as long, and 3 times as high as another smaller box has a volume that is 27 times the volume of the smaller box.

A box that is w units wide, l units long, and h units high has a surface area of

$$2wl + 2wh + 2lh$$

square units. This is because the box has two faces that are w units by l units, two faces that are w units by h units, and two faces that are l units by h units, as you can see by looking at the pattern for a box in Figure 14.91. If another box is 3 times as wide,

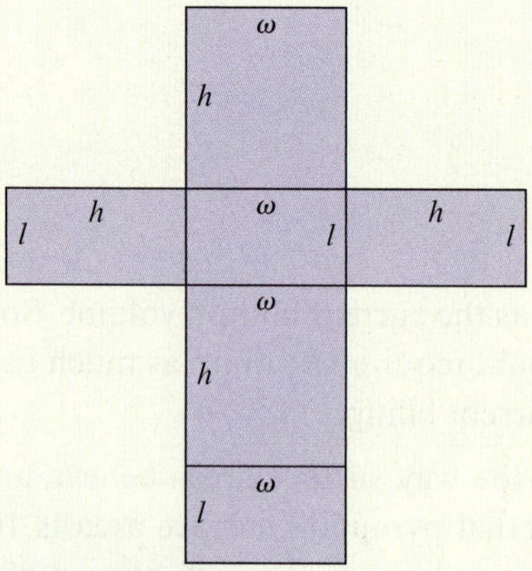

Figure 14.91 Pattern for a box of width w, height h, and length l.

3 times as long, and 3 times as high, then this bigger box will have surface area

$$2(3w)(3l) + 2(3w)(3h) + 2(3l)(3h)$$

square units because its width, length, and height are $3w$, $3l$, and $3h$ units, respectively. But

$$\begin{aligned} 2(3w)(3l) + 2(3w)(3h) + 2(3l)(3h) \\ = 9(2wl + 2wh + 2lh) \end{aligned}$$

Therefore, a box that is $3w$ units wide, $3l$ units long, and $3h$ units high has 9 times the surface area of a box that is w units wide, l units long, and h units high.

5. Let's call the radius of the small cylinder r units and the height of the small cylinder h units. Then the volume of the small cylinder is

$$h\pi r^2$$

cubic units, according to the (*height*) · (*area of base*) volume formula. The larger cylinder has twice the radius and twice the height of the smaller cylinder; therefore, the larger cylinder has radius $2r$ units and height $2h$ units. So the larger cylinder has volume

$$(2h)\pi(2r)^2 = 8(h\pi r^2)$$

cubic units. Because $8h\pi r^2$ is 8 times $h\pi r^2$, the volume of the large cylinder is 8 times the volume of the small cylinder.

The surface area of the small cylinder of radius r and height h is

$$2\pi r^2 + 2\pi rh$$

square units. (See the answer to Practice Exercise 10 of Section 13.2.) Using the same formula again, but now with $2r$ substituted for r and $2h$ substituted for h, we obtain the surface area of the big cylinder as

$$\begin{aligned} 2\pi(2r)^2 + 2\pi(2r)(2h) &= 8\pi r^2 + 8\pi rh \\ &= 4(2\pi r^2 + 2\pi rh) \end{aligned}$$

square units. Because $4(2\pi r^2 + 2\pi rh)$ is 4 times $2\pi r^2 h + 2\pi rh$, the surface area of the large cylinder is 4 times the surface area of the small cylinder.

PROBLEMS FOR SECTION 14.7

1. In triangle ADE shown in Figure 14.92, the point B is halfway from A to D and the point C is halfway from A to E. Compare the areas of triangles ABC and ADE. Explain your reasoning clearly.

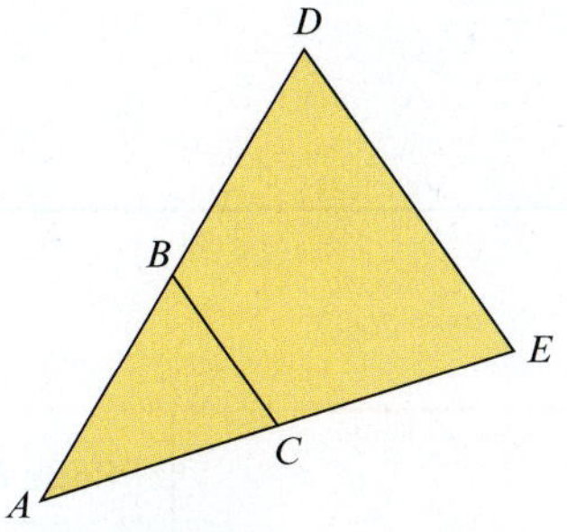

Figure 14.92 Triangles.

2. A scale model is constructed for a domed baseball stadium with a scale of 1 foot to 100 feet. The model's dome is made of 40 square feet of cardboard. The model contains 50 cubic feet of air. How many square feet of material will the actual stadium's dome be made of? How many cubic feet of air will the actual stadium contain?

3. Og, a giant mentioned in the Bible, might have been 13 feet tall. If Og was proportioned like a 6-foot-tall man weighing 200 pounds, then how much would Og have weighed? Explain your reasoning.

4. An artist plans to make a large sculpture of a person out of solid marble. She first makes a small scale model out of clay, using a scale of 1 inch for 2 feet. The scale model weighs 1.3 pounds. Assuming that a cubic foot of clay weighs 150 pounds and a cubic foot of marble weighs 175 pounds, how much will the large marble sculpture weigh? Explain your reasoning.

5. According to one description, King Kong was 19 feet 8 inches tall and weighed 38 tons. Typical male gorillas are about 5 feet 6 inches tall and weigh between 300 and 500 pounds. Assuming that King Kong was proportioned like a typical male gorilla, does his given weight of 38 tons agree with what you would expect? Explain.

6. If you know the volume of an object in cubic inches, can you find its volume in cubic feet by dividing by 12? Discuss.

7. Suppose that a gasoline-powered engine has a gas tank in the shape of an inverted cone, with radius 10 inches and height 20 inches, as shown in Figure 14.93. The gas flows out through the tip of the cone, which points down. The shaded region represents gas in the tank. The height of this "cone of gas" is 10 inches, which is half the height of the gas tank.

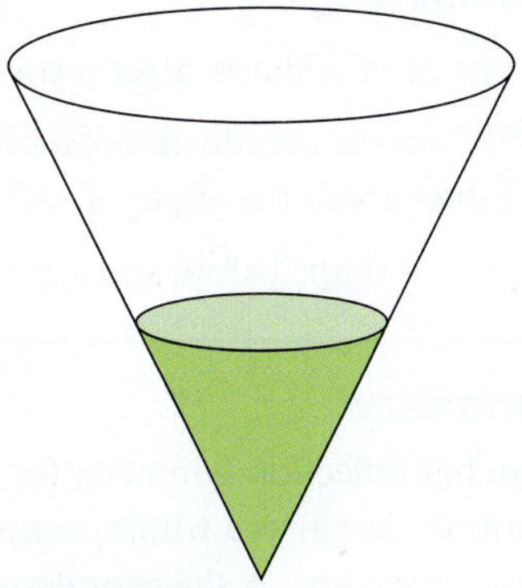

Figure 14.93 A cone-shaped gas tank.

a. In terms of volume, is the tank half full, less than half full, or more than half full? Draw a picture of the gas tank, and mark approximately where you think the gas would be when the tank is $\frac{3}{4}$ full, $\frac{1}{2}$ full, and $\frac{1}{4}$ full in terms of volume.

b. Find the volume of the gas in the tank. Is this half, less than half, or more than half the volume of the tank?

c. Now find the volume of gas if the height of the gas was $\frac{3}{4}$ of the height of the tank. Is this more or less than half the volume of the tank? Is this surprising?

8. A cup has a circular opening and a circular base. A cross-section of the cup and the dimensions of the cup are shown in Figure 14.94.

a. Determine the volume of the cup. Explain your reasoning.

b. If the cup is filled to $\frac{1}{2}$ of its height, what percent of the volume of the cup is filled? Explain your reasoning.

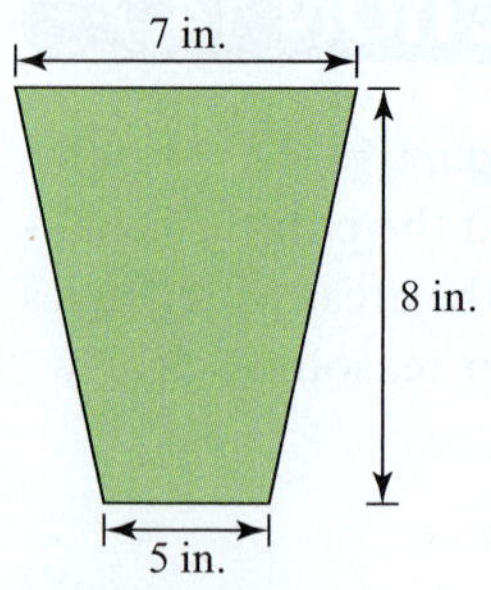

Figure 14.94 A cup.

CHAPTER SUMMARY

Section 14.1 Reflections, Translations, and Rotations — Page 613

- Reflections, translations, and rotations are the fundamental distance-preserving transformations of the plane. — Page 613

Key Skills and Understandings

1. Determine the image of a shape after a translation, a reflection, or a rotation is applied. — Page 613
2. Determine the effect on the coordinates of a point after a translation, a reflection across the x- or y-axis, or a rotation about the origin of 180° or 90° clockwise or counterclockwise is applied. — Page 615
3. Given the location of a shape before and after a transformation, determine the transformation. — Page 614

Section 14.2 Symmetry — Page 619

- A shape or design has reflection symmetry (or mirror symmetry), if there is a line in the plane such that the shape or design as a whole occupies the same place in the plane both before and after reflecting across the line. A shape or design has n-fold rotation symmetry if there is a point such that, after rotating by $360° \div n$ around the point, the shape or design as a whole occupies the same place in the plane both before and after rotation. A shape or design has translation symmetry if there is a translation such that the shape or design as a whole occupies the same place in the plane both before and after translation. — Page 619

Key Skills and Understandings

1. Given a shape or design, determine its symmetries. — Page 619

Section 14.3 Congruence — Page 627

- Informally, we say two shapes are congruent if they are the same size and shape. More formally, we describe two shapes as congruent if there is a rotation, a reflection, a translation, or a combination of these transformations that takes one shape to the other. There are several standard criteria for two triangles to be congruent. If two triangles have the same side lengths (so that corresponding sides have the same lengths), then they are congruent (SSS congruence). If two triangles have a side of the same length, and the corresponding angles at each end of the sides have the same measure in both triangles, then the triangles are congruent (ASA congruence). If two triangles have two corresponding sides of the same length and the same angle between the two sides, then they are congruent (SAS congruence). The structural stability of triangles is related to SSS congruence. — Page 627

Key Skills and Understandings

1. Explain what congruence means. Page 627
2. Know that some pieces of information specify a unique triangle, whereas others do not. Page 627
3. Describe and use the SSS, ASA, and SAS criteria for congruence. Page 627
4. Relate the structural stability of triangles to the SSS congruence criterion. Page 628
5. Contrast the structural stability of triangles with the lack of stability for quadrilaterals. Page 628

Section 14.4 Constructions with Straightedge and Compass Page 638

- Using only a straightedge and a compass, we can construct the perpendicular bisector of a given line segment, and we can bisect a given angle. Both of these constructions produce an associated rhombus. The constructions give the desired results because of special properties of rhombuses: The diagonals in a rhombus are perpendicular and bisect each other, and the diagonals in a rhombus bisect the angles in the rhombus. Page 638

Key Skills and Understandings

1. Given a line segment, construct the perpendicular bisector by using a straightedge and compass. Describe the associated rhombus, and relate the construction to a special property of rhombuses: In a rhombus, the diagonals are perpendicular and bisect each other. Page 638
2. Use a straightedge and compass to bisect a given angle. Describe the associated rhombus, and relate the construction to a special property of rhombuses: In a rhombus, the diagonals bisect the angles in the rhombus. Page 638
3. Use the basic straightedge and compass constructions. For example, construct a square. Page 640

Section 14.5 Similarity Page 643

- We call two shapes or objects similar if all distances between corresponding parts of the shapes or objects are scaled by the same factor called a scale factor. Page 643
- We studied two methods for solving problems about similar shapes or objects: the scale factor method and the internal factor method. We related both methods to setting up and solving a proportion. To use the scale factor method, we use corresponding known lengths in the two similar objects in order to determine the scale factor. We then apply the scale factor to find an unknown length. To use the internal factor method, we find the factor by which two known lengths within one of the objects are related. The corresponding lengths in the other, similar object will be related by the same factor. We then apply this factor to find an unknown length in the other object. Page 645

Key Skills and Understandings

1. Discuss what it means for shapes or objects to be similar. Page 643
2. Use the scale factor and internal factor methods to solve problems about similar shapes or objects, explain the rationale for each method, and relate them to proportions. Page 645

Section 14.6 Dilations and Similarity Page 654

- Dilations are transformations of a plane that take shapes to similar shapes. Dilations scale all distances by the same scale factor and preserve sizes of angles. Page 654

- The angle-angle-angle criterion states that two triangles are similar exactly when the two triangles have corresponding angles of the same size. Similar triangles can arise when parallel lines cross a pair of lines that cross. We can explain why triangles formed this way are similar by explaining why the corresponding angles have the same size (using the Parallel Postulate). We can use similar triangles to determine distances and heights. Page 656

Key Skills and Understandings

1. Use transformations to discuss what it means for shapes in a plane to be similar. Page 654
2. Explain why triangles are similar by showing that corresponding angles must have the same size (the angle-angle-angle criterion). When the Parallel Postulate applies, use it to show that corresponding angles are produced by a line crossing two parallel lines and are therefore of the same size. Page 656
3. Use similar triangles to determine heights and distances. Page 658
4. Use similarity to solve problems. Page 658

Section 14.7 Areas, Volumes, and Similarity Page 665

- If the lengths in one object are k times the corresponding lengths in a second similar object, then the surface area of the first object is k^2 times the surface area of the second object and the volume of the first object is k^3 times the volume of the second object. We can see why these facts about how areas and volumes scale are plausible by extrapolating from patterns and models for prisms and cylinders and from formulas for their surface areas and volumes. Page 665

Key Skills and Understandings

1. Use patterns, models, and formulas to explain how areas and volumes of similar objects are related. Page 665
2. Apply the way that areas and volumes scale in similar objects to solve problems. Page 666

15

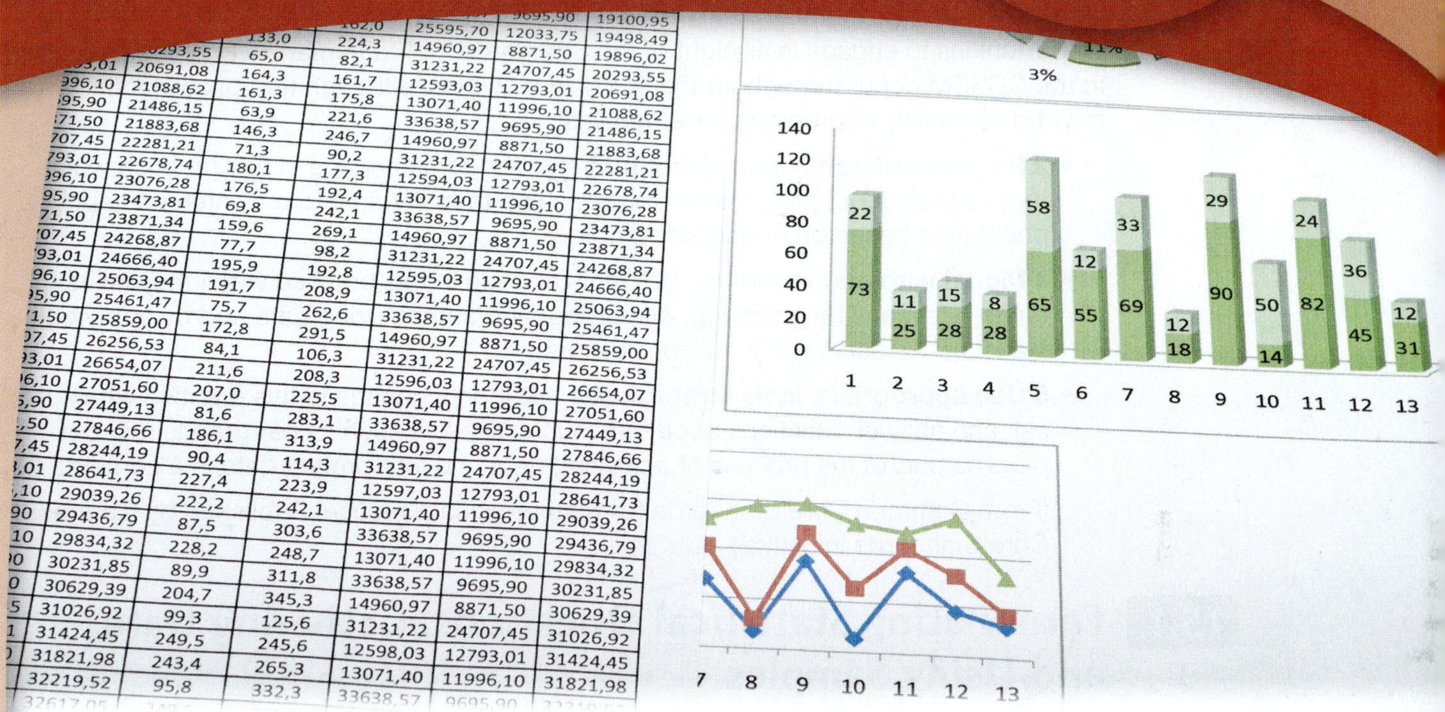

Statistics

The field of statistics provides tools for studying questions that can be answered with data. A seemingly endless variety of data is available for populations, health, financial and business activities, and the environment. It is virtually impossible to think of a social activity or a physical phenomenon for which we cannot collect data. Statistical concepts help to interpret these data and to recognize trends.

In this chapter, we focus on the following topics and practices within the *Common Core State Standards for Mathematics* (*CCSSM*).

Standards for Mathematical Content in the CCSSM

In the domain of *Measurement and Data* (Kindergarten–Grade 5), students organize categorical and numerical measurement data and they represent data in picture graphs, bar graphs, and dot plots (line plots). They use these representations as they ask and answer questions in order to interpret data.

In the domain of *Statistics and Probability* (Grades 6–8), students appreciate that statistical questions anticipate variability. They work with distributions displayed in dot plots or histograms, they use measures of center (median or mean) and variation (interquartile range or mean absolute deviation) to summarize distributions and to compare sets of data, and they interpret the comparison in terms of the context. Students learn that we can use random samples to draw inferences about a population. They use scatterplots and their understanding of relationships, especially linear relationships, to investigate patterns of association between two quantities.

Standards for Mathematical Practice in the CCSSM

Opportunities to engage in all eight of the Standards for Mathematical Practice described in the CCSSM occur throughout the study of statistics, although the following standards may be especially appropriate for emphasis:

- **2 Reason abstractly and quantitatively.** Students engage in this practice when they reason about proportional relationships to make an informal inference about a population based on a random sample.
- **4 Model with mathematics.** Students engage in this practice when they investigate questions by gathering, displaying, and summarizing data and by applying statistical reasoning to draw conclusions.
- **5 Use appropriate tools strategically.** Students engage in this practice when they ask and answer questions about statistical displays such as bar graphs and dot plots as part of the process of analyzing data and interpreting data in a context.

(From Common Core Standards for Mathematical Practice. Published by Common Core Standards Initiative.)

15.1 Formulating Statistical Questions, Gathering Data, and Using Samples

CCSS Common Core State Standards Grades 6, 7

We humans are naturally curious about each other and about the world. We want to know what others around us think about the political questions of the day; the latest movies, TV shows, and music; and similar topics of interest. Local, state, and national governments want to know facts about population, employment, income, and education in order to design appropriate programs. Scientists want to know how to cure and prevent diseases, as well as to know about the world around us. Therefore, news organizations, government bodies, and scientists regularly design investigations and gather data to help answer specific questions.

According to guidelines for assessment and instruction in statistics education endorsed by the American Statistical Association (see [3]), statistical problem solving involves four components:

1. Formulate questions
2. Collect data
3. Analyze data
4. Interpret results

In this section we start by examining the first two of these components, formulating questions and collecting data. Then we turn our attention to analyzing data and interpreting results in the case where a sample is used to predict characteristics of a full population. The rest of this chapter is devoted to tools we can use to analyze data and interpret results.

What Are Statistical Questions?

The first step in a statistical investigation is to formulate a question that can be answered by collecting and analyzing data.

data
numerical data

Data are pieces of information. Some data are **numerical**, such as the weights of apples picked at an orchard, the heights of students in a class, or the volume of juice poured into bottles at a bottling categorical data

plant. But other data are **categorical**, such as the colors of the plastic dinosaurs in a bucket, the yes or no vote of each person in a class on some issue, or the favorite piece of playground equipment of each student at school. Students can ask a variety of statistical questions. For example, if a class is planning a party, the students might like to know what the favorite snacks are among the students in the class. If everyone in a class received a small packet of candies as a treat, the students might be curious whether all the bags contain the same number of candies. If the candies come in different colors, the students might wonder which color generally appears the most. Students might be curious about how heavy their backpacks are. Are girls' names typically longer than boys' names? Do bean seeds sprout better in the dark or in light? How does the length of a pendulum affect the time it takes to swing back and forth? All of these questions can be answered (at least in part) by collecting and analyzing data, which make them statistical questions.

CCSS 6.SP.1

Another distinguishing characteristic of statistical questions is that they *anticipate variability* in the data that will be collected. For example, we don't expect all backpacks to weigh the same or all girls' names to be the same length. So statistical investigations must anticipate and take into account the variability that naturally exists in most data.

CLASS ACTIVITY

15A Statistical Questions Versus Other Questions, p. CA-307

What Data Do We Collect for Observational Studies and Experiments?

There are two basic types of statistical studies: *observational studies* and *experiments*.

Observational Studies In doing observational studies we observe characteristics and quantities, but do not attempt to influence these characteristics or quantities. For example, in an observational study, researchers might count the number of turtle eggs found in a certain area, measure the height of a plant every day, or collect participants' responses to a questionnaire in a survey.

Writing survey questions to learn facts about people or to determine people's opinion may seem like a simple matter, but it is surprisingly difficult. Different people may interpret the same question in different ways. For example, even a simple question such as "How many children are in your family?" can be interpreted in different ways. Does it mean how many people under the age of 18 are in your family? Does it mean how many siblings you have, plus yourself? Should stepchildren be included? A good survey question must be clear.

Experiments In contrast to observational studies, in experiments researchers don't simply observe characteristics or quantities; instead, they try to determine if certain factors will influence the characteristics or quantities. For example, in an experiment to determine if a new drug is safe and effective, a group of volunteers will be divided into two (or more) groups. One group receives the new drug, and the other group receives a placebo—an inert substance that is made to look like a real drug. Data are collected on the volunteers to determine if their symptoms improve and if there are any side effects in the group receiving the new drug.

What Are Populations and Samples?

population

Regardless of the nature of the investigation, a statistical study focuses on a certain population. The **population** of a statistical study is the full set of people or things that the study is designed to investigate. For example, before a presidential election, a study might investigate how registered voters plan to vote. In this case, the population of the study is all registered voters in the country.

Another study might investigate properties of ears of corn in a cornfield. In this case, the population of the study is all ears of corn in the cornfield. In most cases, it is not possible to study every single member of a population; instead, a sample is chosen for study. A **sample** of a population of a study consists of some collection of members of the full population. A pollster will select a sample of voters to survey about how they will vote in an upcoming election instead of trying to ask every voter. A scientist studying ears of corn in a cornfield will select a sample of ears of corn to study rather than try to study every single ear of corn in the cornfield.

sample

CLASS ACTIVITY

15B Choosing a Sample, p. CA-309

CCSS 7.SP.1

How Do We Use Random Samples to Predict Characteristics of a Full Population?

If you did Class Activity 15B, then you probably realize that some samples may not accurately reflect the full population. The best way to pick a sample so that it will be likely to reflect the full population accurately is to pick a *random sample*.

Random Samples For a sample of a population under study to be useful, the characteristics of the sample must reflect the characteristics of the full population. In other words, the sample should be **representative** of the full population. But unless you choose a sample carefully, the sample's characteristics might not be representative of the full population's characteristics. For example, if a group studying child nutrition in a county took their sample of children from a school in an affluent neighborhood, their findings could differ significantly from the actual status of child nutrition in the county. The best way to choose a representative sample of a population is to pick a **random sample**—choosing the sample in such a way that every member of the population has an equal chance of becoming a member of the sample.

representative sample

random sample

Sometimes students may interpret *random sample* incorrectly as a sample that is chosen in an unconventional, haphazard, or out-of-the-blue way. For example, say there are 100 students in the seventh grade and a random sample of 10 students is to be chosen. Although Johnny may have an unusual method in mind for picking the 10 students, the method won't produce a random sample of 10 students unless each student was equally likely to be chosen for the sample. A simple way for students to pick a random sample is to cut a list of the students' names into identical slips of paper, put the slips of paper in a bag, mix them well, and then pick out exactly 10 of the slips without looking. The slips of paper could instead be numbered from 1 to 100, with each number corresponding to a student in a class list. The 10 names or numbers that are picked correspond to the 10 students that are to be in the random sample.

In settings where the population is too large for it to be practical to label a slip of paper for each individual in the population, random samples are often chosen with the aid of a list of random numbers generated by computers or calculators.

CCSS 7.SP.2

Predicting the Characteristics of a Full Population If a random sample of a population is chosen, then the characteristics of the sample are likely to be very close to the characteristics of the full population if the sample is large enough. In other words, sufficiently large random samples are likely to be representative of the full population. For example, if 54% of a (large enough) random sample of voters said they were likely to vote for candidate X, then it is likely that close to 54% of the full population would also vote for candidate X.

CLASS ACTIVITY

15C How Can We Use Random Samples to Draw Inferences about a Population?, p. CA-310

15D Using Random Samples to Estimate Population Size by Marking (Capture–Recapture), p. CA-311

Here is an example of how we can use a random sample to predict the characteristics of a full population. Suppose that at a factory, 450 transistors have just been produced. A random sample of 75 transistors is selected for testing. Of the 75 transistors, 3 are found to be defective. How many defective transistors should we expect to find among the full population of 450?

Because the sample of 75 was chosen randomly, it is likely to be representative of the full population. There are a number of ways we can determine approximately how many of the 450 transistors are likely to be defective. If the sample is perfectly representative of the full batch, then the fraction of defective transistors in the sample and in the full population will be equivalent. Thus, we should solve the proportion

$$\frac{3}{75} = \frac{x}{450} \tag{15.1}$$

for x, where x stands for the number of defective transistors in the full population. We can solve this proportion algebraically, but we can also reason in other ways that will help us think about how samples and populations are related.

One way to reason about the relationship between the sample and the population is to think about repeatedly picking out "batches" of 75 transistors until all 450 transistors have been picked out. If each batch is just like the first one, then each batch of 75 would have 3 defective transistors. The two tables below can help us think about repeatedly picking batches of 75 and finding 3 defective transistors in each batch:

75	→	3	75	→	3
75	→	3	150	→	6
75	→	3	225	→	9
75	→	3	300	→	12
75	→	3	375	→	15
75	→	3	450	→	18

We can reason more efficiently by using division. Since $450 \div 75 = 6$, we can pull out 6 batches of 75. If each of the 6 batches of 75 yields 3 defective transistors, then there would be a total of $6 \cdot 3 = 18$ defective transistors in the full population of 450. The expression

$$\frac{450}{75} \cdot 3$$

summarizes how we estimated the number of defective transistors. Of course, if we were to pick out batches of 75, it is unlikely that each and every batch would contain exactly 3 defective transistors, and most likely the full population wouldn't contain exactly 18 defective transistors either. However, this method gives us a good way to determine the approximate number of defective transistors we should expect.

Another way to reason about the relationship between the sample and the population is to imagine distributing the sample of 75 transistors to 75 bins. Then 3 of those bins are for defective transistors. Now imagine distributing the full population of 450 transistors among the 75 bins so that the defective transistors go in those 3 bins. If the sample is representative of the full population, then we should be able to distribute the transistors equally among the bins. Because $450 \div 75 = 6$, each bin will have 6 transistors in it, and so the 3 bins for defective transistors will have $3 \cdot 6 = 18$ defective transistors in all. The expression

$$3 \cdot \frac{450}{75}$$

summarizes how we estimated the number of defective transistors.

We can also reason about the bins in another way. Because 3 out of 75 bins are for defective transistors and the transistors are distributed equally among the bins, $\frac{3}{75}$ of the all the transistors are defective. Therefore, there should be about

$$\frac{3}{75} \cdot 450$$

defective transistors in all. Of course, this number is again equal to 18.

Why Is Randomness Important?

Consider an experiment to determine if a new drug is safe and effective. A group of volunteers is divided into two (or more) groups. One group receives the new drug; the other group receives a placebo. Such an experiment is useful only if it accurately predicts the outcome of the use of the drug in the population at large. Therefore, it is crucial that the two groups, the one receiving the drug and the one receiving the placebo, are as alike as possible—in all ways except in whether or not they receive the drug. The two groups should also be as much as possible like the general population who would take the drug. Therefore, the two groups are usually chosen randomly (such as by flipping a coin—heads goes to group 1, tails to group 2). In addition, such experiments are usually *double-blind*; that is, neither the volunteer nor the treating physician knows who is in which group.

In controlled drug experiments, the randomness of the assignment of patients to the two groups and the double-blindness ensure that the study will accurately predict the outcome of the use of the new drug in the whole population. If the patients were not assigned randomly to the two groups, then the person assigning the patients to a group might subconsciously pick patients with certain attributes for one or the other group, thereby making the group that is treated by the new drug slightly different from the group receiving the placebo. This difference could result in inaccurate predictions about the safety and effectiveness of the drug in the population at large. Similarly, if patients know whether or not they have received the new drug, their resulting mental states may affect their recovery. If the physician knows whether or not the new drug has been administered, her treatment of the patient may be different, and the patient may again be affected differently.

SECTION SUMMARY AND STUDY ITEMS

Section 15.1 Formulating Statistical Questions, Gathering Data, and Using Samples

A statistical study begins by formulating a question that anticipates variability and that can be answered (at least partially) by collecting and analyzing data. In statistics, there are observational studies and experiments; both involve considering a population. Often it is not practical to study the entire population of interest, so a sample is studied instead. Large enough random samples are generally representative of the full population. Therefore, one can reason about proportional relationships to make reasonable inferences about a full population on the basis of a random sample.

Key Skills and Understandings

1. Ask and recognize statistical questions as distinct from other questions.
2. Recognize that some samples may not be representative of a whole population, but that large enough random samples generally are.
3. Use random samples to make predictions about a full population by reasoning about proportional relationships.

Practice Exercises for Section 15.1

1. What distinguishes statistical questions from other questions?
2. Kaitlyn, a fifth-grader, asked five of her friends in class which book is their favorite. All of them said *Harry Potter and the Sorcerer's Stone*. Can Kaitlyn conclude that most of the children at her school would say *Harry Potter and the Sorcerer's Stone* is their favorite book? Why or why not?
3. A large bin is filled with 200 table-tennis balls. Some of the balls are white and some are orange. Tyler reaches into the bin and randomly pulls out 10 table-tennis balls. Three of the balls are orange and 7 are white. Based on Tyler's sample, what is the best estimate we can give for the number of orange table-tennis balls in the bin? Explain your reasoning.
4. There is a large bin filled with table-tennis balls, but we don't know how many. There are 40 orange table-tennis balls in the bin; the rest are white. Amalia reaches into the bin and randomly picks out 20 table-tennis balls. Of the 20 she picked, 6 are orange. Based on Amalia's sample, what is the best estimate we can give for the number of table-tennis balls in the bin? Explain your reasoning.

Answers to Practice Problems for Section 15.1

1. Statistical questions anticipate variability and can be addressed by collecting data.
2. No, Kaitlyn can't conclude that most of the children at her school would say *Harry Potter and the Sorcerer's Stone* is their favorite book. Kaitlyn's friends may all have a similar taste in books, and this may not be representative of the tastes of the whole school. Also, younger children at lower reading levels might prefer different books because *Harry Potter and the Sorcerer's Stone* is too advanced for them to read.
3. Tyler picked a random sample, so the characteristics of his sample should reflect the characteristics of the full population of the balls in the bin (more or less). Three of the 10 balls that Tyler picked are orange, so 30% of the balls in Tyler's sample are orange. The full population of balls in the bin should also be about 30% orange. Since 30% of 200 is 60, the best estimate is that 60 balls are orange and 140 are white.

   Another way to solve this problem is to think about repeatedly picking out "batches" of 10 table-tennis balls until all 200 balls have been picked out. If each batch of 10 is just like the first one, then each batch will contain 3 orange balls and 7 white ones. Since $200 \div 10 = 20$, we can pick 20 batches of 10 balls. This would yield a total of $20 \cdot 3 = 60$ orange balls and $20 \cdot 7 = 140$ white balls. Of course, since not every set of 10 balls will contain exactly 3 orange balls, this is just an estimate, and it is not likely to be exactly correct.
4. Amalia picked a random sample, so the characteristics of her sample should reflect the characteristics of the full population of the balls in the bin (more or less). Since 6 of 20 balls that Amalia picked are orange, 30% of the balls she picked are orange. Therefore, approximately 30% of the balls in the bin should be orange. Since there are 40 orange table-tennis balls in the bin, and since these 40 balls should be about 30% of all the balls in the bin, we have

   $$30\% \cdot (\text{balls in the bin}) = 40$$

   So

   $$\text{balls in the bin} = 40 \div 0.30 = 133$$

   Therefore, there should be approximately 133 balls in the bin.

Another way to solve this problem is to think about repeatedly picking out batches of 20 table-tennis balls until all the orange balls have been picked out. If each batch of 20 is just like the first one, then each batch will contain 6 orange balls. The next table shows what would happen:

$$20 \rightarrow 6$$
$$40 \rightarrow 12$$
$$60 \rightarrow 18$$
$$80 \rightarrow 24$$
$$100 \rightarrow 30$$
$$120 \rightarrow 36$$

Once we get up to 36 orange balls, there are only 4 orange balls left, which is $\frac{4}{6}$ as much as 6. So it should take $\frac{4}{6}$ or $\frac{2}{3}$ of a batch to get those 4 orange balls. If we pick another $\frac{2}{3}$ of 20 balls, which is about another 13 balls, we should get the remaining orange balls. Thus, in all, we would pick $120 + 13 = 133$ balls in order to get the full 40 orange balls. So there should be approximately 133 table-tennis balls in the bin.

PROBLEMS FOR SECTION 15.1

1. Give two examples of a statistical question and contrast each with a related question that is not statistical.

2. A class has a collection of 100 bottle caps and wants to pick a random sample of 20 of them. Kyle has the idea that he will pick every third one he looks at in order to pick these 20. Discuss whether Kyle's method is appropriate for picking a random sample. If not, how else could the class pick a random sample of 20 bottle caps?

3. Neil, a third-grader, asked 10 of his classmates whether they prefer to wear shoes with laces or without laces. Most of the classmates Neil asked prefer to wear shoes without laces. Discuss whether it would be reasonable to assume that most of Neil's class or most of the students at Neil's school would also prefer to wear shoes without laces.

4. An announcer of a TV program invited viewers to vote in an Internet poll, indicating whether or not they are better off economically this year than last year. Most of the people who participated in the poll indicated they are worse off this year than last year. Based on this information, can we conclude that most people are worse off this year than last year? If so, explain why. If not, explain why not.

5. There is a bowl containing 300 plastic tiles. Some of the tiles are yellow and the rest are blue. You take a random sample of 10 tiles and find that 2 are yellow and 8 are blue. How many yellow tiles are there? Write three different expressions $A \cdot B$ for the number of yellow tiles you predict there are based on the random sample, where A and B are numbers derived from among 300, 10, 2, and 8. Explain each expression in detail in terms of quantities in the situation. Support your explanations with math drawings.

6. There is a bowl containing 80 green tiles and an unknown number of purple tiles. You take a random sample of 10 tiles and find that 4 are green and 6 are purple. How many purple tiles are there? Write at least two different expressions $A \cdot B$ for the number of purple tiles you predict there are based on the random sample, where A and B are numbers derived from among 80, 10, 4, and 6. Explain each expression in detail in terms of quantities in the situation. Support your explanations with math drawings.

7. At a factory that produces doorknobs, 1500 doorknobs have just been produced. To check the quality of the doorknobs, a random sample of 100 doorknobs is selected to test for defects. Of these 100 doorknobs, two were found to be defective. Based on these results, what is the best estimate you can give for the number of defective doorknobs among the 1500? Solve this problem in three different ways, explaining your reasoning in each case.

8. At a factory that produces switches, 3000 switches have just been produced. To check the quality of the switches, a random sample of 75 switches is selected to test for defects. Of these 75 switches, two were found to be defective. Based on these results, what is the best estimate you can give for the number of defective switches among the 3000? Solve this problem in three different ways, explaining your reasoning in each case.

9. At a light bulb factory, 1728 light bulbs are ready to be packaged. A worker randomly pulls 60 light bulbs out and tests each one. Two light bulbs are found to be defective.

a. Based on the information given, what is your best estimate for the number of defective light bulbs among the 1728 light bulbs? Explain your reasoning.

b. If the light bulbs are put in boxes with 144 light bulbs in each box (in 12 packages of 12), then approximately how many defective light bulbs would you expect to find in a box? Explain your reasoning.

10. Carter has a large collection of marbles. He knows that he has exactly 20 blue marbles in his collection, but he does not know how many marbles he has in all. Carter mixes up his marbles and randomly picks out 40 marbles. Of the 40 marbles he picked, four are blue. Based on these results, what is the best estimate you can give for the total number of marbles in Carter's collection? Explain your reasoning. Solve this problem in three different ways, explaining your reasoning in each case.

11. The following problem is an example of the *capture–recapture* method of population estimation. A researcher wants to estimate the number of rabbits in a region. The researcher sets some traps and catches 30 rabbits. After the rabbits are tagged, they are released unharmed. A few days later the researcher sets some traps again and this time catches 35 rabbits. Of the 35 rabbits trapped, 5 are tagged, which indicates that they had been trapped a few days earlier. Based on these results, what is the best estimate you can give for the number of rabbits in the region? Solve this problem in three different ways, explaining your reasoning in each case.

12. A group studying violence wants to determine the attitudes toward violence of all the fifth-graders in a county. The group plans to conduct a survey, but the group does not have the time or resources to survey every fifth-grader. The group can afford to survey 120 fifth-graders. For each of the methods in (a) through (c) of choosing 120 fifth-graders to survey, discuss advantages and disadvantages of that method. Based on your answer, which of the three methods will be best for determining the attitudes toward violence of all the fifth-graders in the county? (There are 6 elementary schools in the county, and each school has 4 fifth grade classes.)

a. Have each elementary school principal in the county select 20 fifth-graders.

b. Have each fifth-grade teacher in the county select 5 fifth-graders.

c. Obtain a list of all fifth-graders in the county. Obtain a list of 120 random numbers from 1 to the number of fifth-graders in the county. Use the list of random numbers to select 120 random fifth-graders in the county.

15.2 Displaying Data and Interpreting Data Displays

CCSS Common Core State Standards Grades 2, 3, 4, 5, 6, 8

Every day we see the results of surveys in newspapers, on television newscasts, or on the Internet. Sometimes results are reported with numbers (such as a count or a percentage), but other times results are shown in a table, a chart, or a graph. Charts and graphs help us get a sense of what data mean because a chart or graph can show us the "big picture" about data in a way that tables and numbers often can't.

In this section we study common data displays as well as different types of questions that we can ask about displays of data. Interpreting data can range from reading information directly off a graph to comparing and combining this information or thinking about implications of this information. We produce data displays to help us analyze data and interpret results, not as an end goal in itself. So learning to read and interpret data displays is essential. Also in this section, we examine some common errors and misleading practices that sometimes occur with data displays.

Different types of displays are appropriate for categorical and numerical data. Next we examine displays suitable for categorical data.

How Can We Display Categorical Data?

Categorical data can be displayed in "real graphs," pictographs, bar graphs, or pie graphs.

Real Graphs, Pictographs, and Bar Graphs The most basic kinds of graphs are real graphs and pictographs. Both of these are suitable for use with young children and can serve as a springboard for understanding other kinds of graphs. A **real graph**, or *object graph*, displays actual objects in a graph form. A **pictograph** is like a real graph, except that it uses icons or pictures of objects instead of actual objects. In some pictographs, a single icon may represent more than one object. We can use real graphs and pictographs to show how a collection of related objects is sorted into different groups. The real graph or pictograph allows us to see at a glance which groups have more or fewer objects in them.

real graph

pictograph

Suppose we have a tub filled with small beads that are alike, except that they have different colors: yellow, pink, purple, and green. We can mix up the beads and scoop some out. How many beads of each color do we have? Which color beads do we have the most of? Which do we have the least of? To answer these questions at a glance, we can place the beads as indicated in Figure 15.1. This creates a real graph.

Figure 15.1

A real graph: beads of different colors.

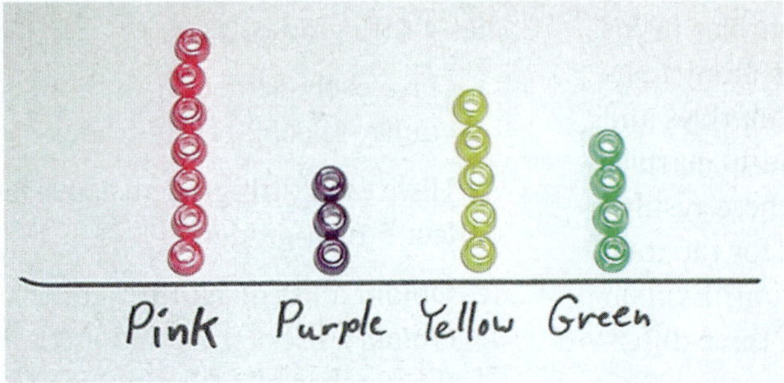

CCSS 2.MD.10 3.MD.3

We can turn a real graph into a pictograph by replacing the objects with icons or simple drawings of the objects, as in Figure 15.2(a).

Figure 15.2

Colors of beads displayed in a pictograph and bar graph.

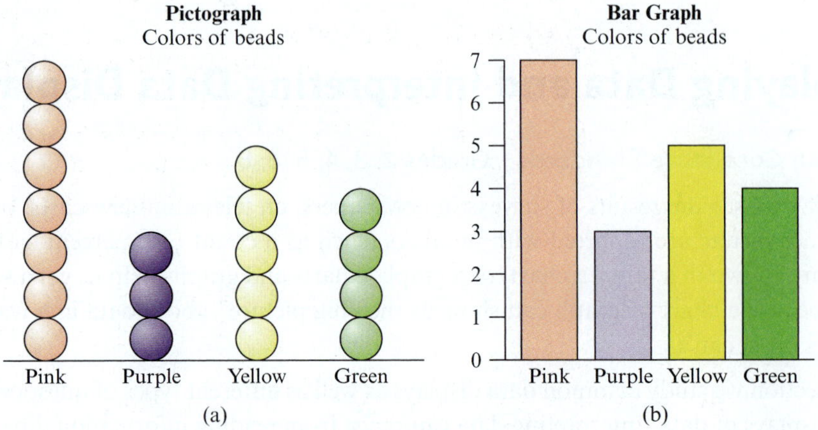

bar graph A **bar graph**, or *bar chart*, is essentially just a streamlined or fused version of a pictograph. A bar graph shows at a glance the relative sizes of different categories. We can turn a pictograph into a bar graph by replacing the icons with a bar that shows how many objects the icons represent, as in **Figure 15.2(b)**.

When we want to represent a large number of objects in a pictograph, it makes sense to allow each icon to stand for more than one object. Each icon could stand for 2, 5, 10, 100, 1000, or some other convenient number of objects. If each icon stands for more than 1 object, then a key near the graph should show clearly how many objects the icon stands for.

To compare populations of some states we could make a list such as this one (Source: U.S. Census Bureau, Population Division):

Wyoming	586,107
Idaho	1,659,930
Oregon	4,028,977
Washington	7,170,351
Georgia	10,214, 860

Although this list shows the (estimated) populations quite precisely, to get a good feel for the relative populations of the states, we can first round the populations to the nearest half-million and then make a pictograph as in **Figure 15.3(a)**. In this pictograph, the picture of a person represents 1 million people, so half a million people can be represented by half a person. In the pictograph we can see at a glance how much bigger in population some of these states are than others. The bar graph in **Figure 15.3(b)** shows the state populations a little more precisely than does the pictograph.

Pictographs and bar graphs can be vertical as in **Figure 15.2** or horizontal as in **Figure 15.3**.

double bar graph A **double bar graph** is a bar graph in which each category has been subdivided into two subcategories. For example, if the bar graph is about people, then the two subcategories might be male and female.

Figure 15.3

Populations of some states.

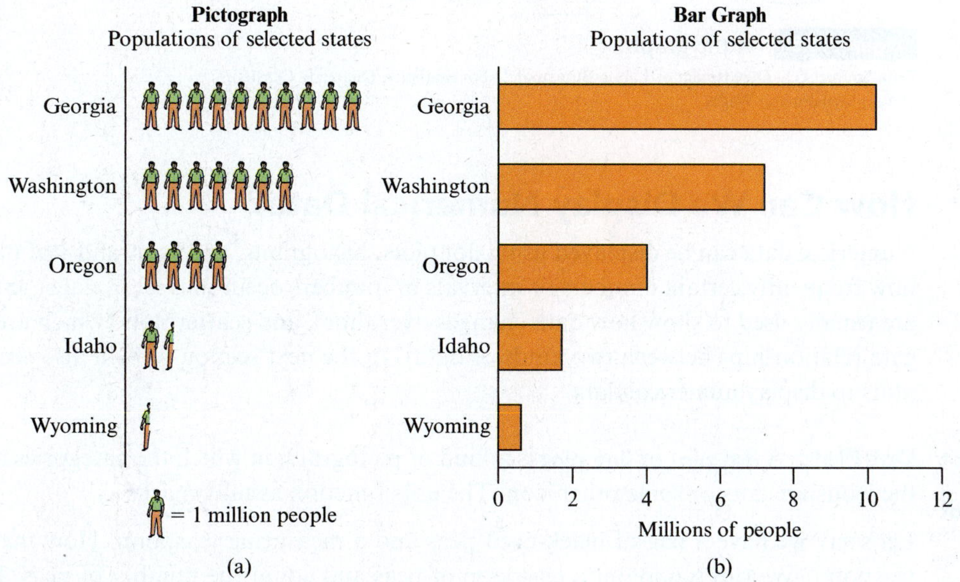

Figure 15.4 shows median weekly earnings in 2014. The different colored bars for men and women show the different earnings of these two groups in two age ranges.

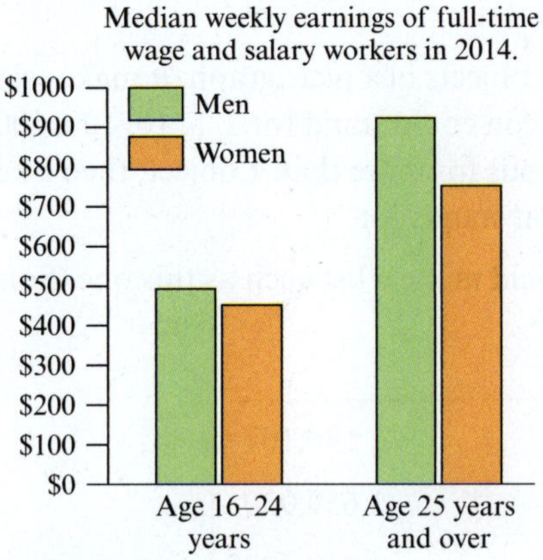

Figure 15.4 A double bar graph.

Data source: U.S. Department of Labor, Bureau of Labor Statistics, Labor Force Statistics from the Current Population Survey.

pie graph **Pie Graphs** A **pie graph**, *pie chart*, or *circle graph* uses a subdivided circle to show how data falls into categories. Figure 15.5 shows in a pie graph how U.S. firms in 2015 were divided according to how many employees they had. We can see at a glance that about half of the firms had fewer than 5 employees and about a quarter of the firms had 10 or more employees.

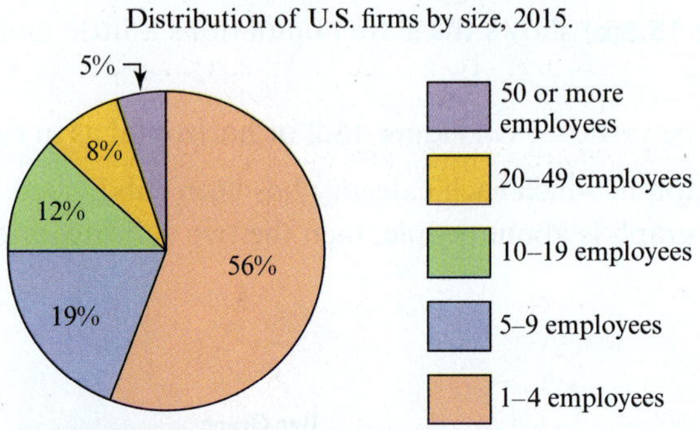

Figure 15.5 A pie graph.

Data Source: U.S. Department of Labor, Bureau of Labor Statistics, Quarterly Census of Employment and Wages.

How Can We Display Numerical Data?

CCSS
2.MD.9
3.MD.4
4.MD.4
5.MD.2
6.SP.4

Numerical data can be displayed using dot plots, histograms, and stem-and-leaf plots (which show how frequently certain numbers or intervals of numbers occur in a set of data), line graphs (which are usually used to show how data changes over time), and scatterplots (which are used to investigate relationships between two kinds of data). In the next section we will also see how to use box plots to display numerical data.

dot plot **line plot** **Dot Plots** A **dot plot** or **line plot** is a kind of pictograph in which the categories are numbers and the icons are dots or some other icon. The dots function as tally marks.

Let's say we have a bag of black-eyed peas and a measuring teaspoon. How many peas are in a teaspoon? We can scoop out a teaspoon of peas and count the number of peas. If we do it again and again, we may get different numbers of peas in a teaspoon. The following numbers could be the number of peas in a teaspoon that we get from many different trials:

18, 19, 19, 17, 19, 18, 20, 18, 19, 19, 20, 20, 19,
20, 18, 19, 18, 17, 18, 20, 18, 18, 16, 18, 19

We can organize these data in a dot plot, as shown in **Figure 15.6**. Each dot in the dot plot represents a number in the previous list. We can see right away from the dot plot that a teaspoon of black-eyed peas usually consists of 18 or 19 peas. By counting that there are 5 dots above 20, we see that there were 5 times when the teaspoon held 20 peas.

Figure 15.6
A dot plot.

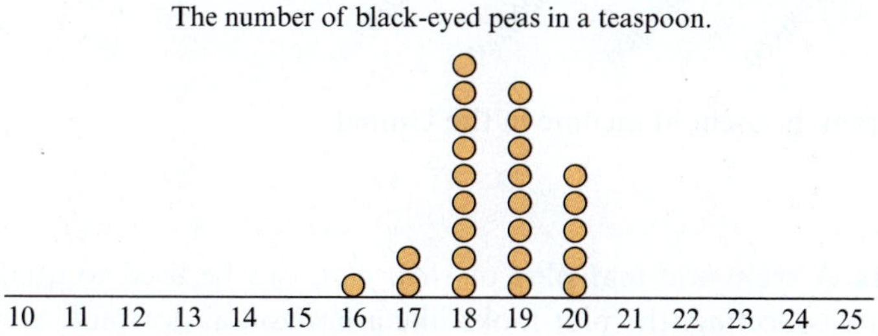

Dot plots are easy to draw, and you can even draw a dot plot as you collect data—you don't necessarily have to write your data down first before you draw a dot plot. For this reason, dot plots can be a quick and handy way to organize numerical data.

We can also use dot plots to record how data falls into intervals. Consider the average January temperatures in degrees Fahrenheit of 25 U.S. cities from this list:

22, 43, 30, 18, 33, 26, 44, 30, 7, 20, 73, 40, 45,
38, 57, 40, 50, 31, 36, 54, 29, 58, 14, 61, 63

We can group these data by the intervals 0–9, 10–19, 20–29, and so on, up to 70–79 and make the dot plot on the left of **Figure 15.7**. By counting that there are 6 dots over the interval 30 to 39, we learn that there were 6 cities (among the cities in the list) that have average January temperatures between 30 and 39 degrees Fahrenheit.

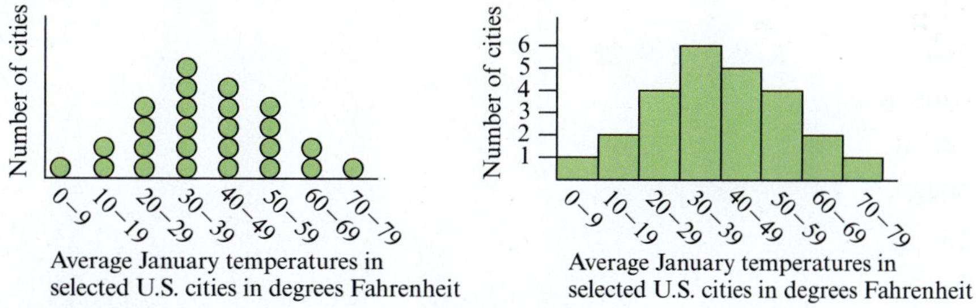

Figure 15.7 A dot plot and histogram with intervals of temperatures as categories.

histogram **Histograms** A **histogram** is a bar graph for which the categories are individual numbers or equal-length intervals of numbers. Just as we can think of fusing the pictures in a pictograph to make a bar graph, we can think of fusing the dots in a dot plot to make a histogram. **Figure 15.7** shows a dot plot and corresponding histogram of average January temperatures in some U.S. cities. But histograms allow for a bit more flexibility than dot plots: In histograms, the height of a bar can refer either to the *number* of pieces of data in that category, as in **Figure 15.7**, or to the *percentage* of the data in that category, as in the histogram on household income in **Figure 15.8**. Note that the implied intervals in the household income histogram are \$0–\$9,999, \$10,000–\$19,999, \$20,000–\$29,999, and so on. From this histogram, we can see that about 7% of U.S. households had income between \$10,000 and \$19,999 and about 4% of households had income between \$100,000 and \$109,999.

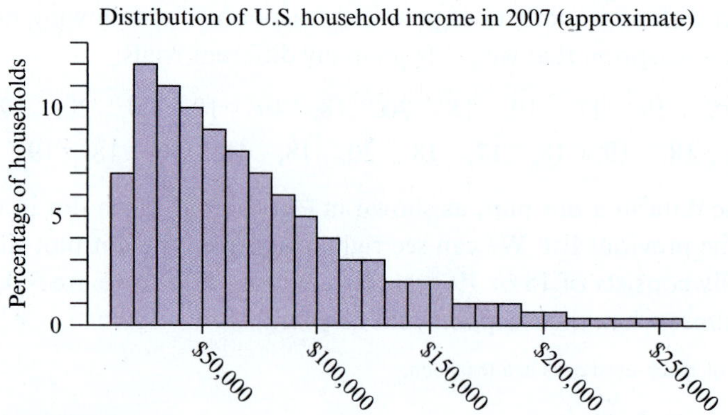

Figure 15.8 Histogram: household income in the United States in 2007.

stem-and-leaf plot

Stem-and-Leaf Plots A **stem-and-leaf plot** or *stem plot*, can be used to quickly organize the numerical data you are collecting; the plot looks like a horizontal dot plot. Stem-and-leaf plots are most useful for organizing two-digit data, such as the average January temperatures discussed previously, which are displayed in Figure 15.7. Table 15.1 shows these temperatures organized into a stem-and-leaf plot. The *stem* in a stem-and-leaf plot consists of the numbers to the left of the vertical line, and the *leaves* are the numbers to the right. For example, the third row in Table 15.1 shows the stem 2 and the leaves 2, 6, 0, and 9, which stand for the data 22, 26, 20, and 29. Although you can put the leaves in order, it is also acceptable simply to put them in as you encounter them. In general, the stem should consist of the digits in the places above the ones place (the tens and hundreds places and perhaps even the thousands place or higher), and the leaves should consist of ones digits in the data.

TABLE 15.1 Stem-and-leaf plot of average January temperatures in 25 U.S. cities in degrees Fahrenheit

0	7	
1	84	
2	2609	Key: 2 | 2 = 22
3	030816	
4	34050	
5	7048	
6	13	
7	3	
8		
9		

line graph

Line Graphs A **line graph** is a graph in which adjacent data points are connected by a line. Often, a line graph is just a graph of a function, such as the graph of a population of a region over a period of time or the graph of the temperature of something over a period of time.

CCSS 5.G.2

Line graphs are appropriate for displaying “continuously varying” data—for example, data that vary over a period of time. Figure 15.9 shows that the average math scores of 9-year-old children have been slowly increasing, as measured by the National Assessment of Educational Progress (NAEP), an organization that is often called “The Nation’s Report Card,” conducting a national assessment of what children know and can do in various subjects. Notice that the points that are plotted in Figure 15.9 indicate that the NAEP math test was given in 1978, 1982, 1986, 1990, 1992,

1994, 1996, 1999, 2004, 2008, and 2012. It makes sense to connect these points because if the test were given at times in between, and if the children's average math scores were plotted as points, these points would probably be close to the lines drawn in the graph of Figure 15.9.

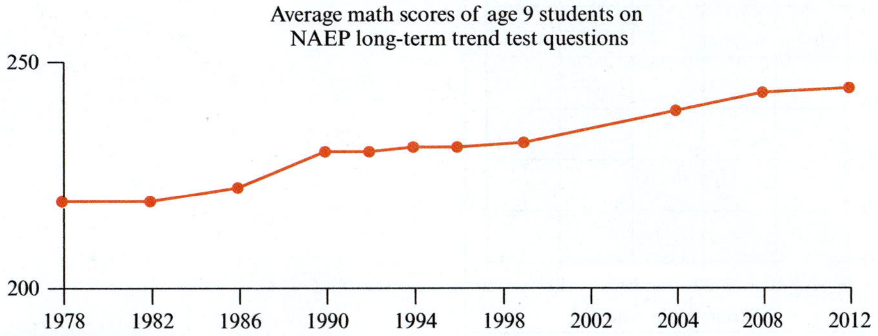

Figure 15.9 A line graph.

Data Source: U.S. Department of Education, Institute of Education Sciences, National Center for Education Statistics, National Assessment of Educational Progress (NAEP), Long Term Trend Mathematics Assessments.

Line graphs are not appropriate for displaying data in categories that don't vary continuously. For example, it would not be appropriate to use a line graph to display the data on state populations that is displayed in the pictograph and bar graph of Figure 15.3. Figure 15.10 shows an *inappropriate* line graph of this sort. It doesn't make sense to use a line graph in this case because there is no logical reason to connect the state populations—there is no continuous variation between the population in Georgia and the population in Washington, for example.

Figure 15.10 An inappropriate use of a line graph: connecting state populations.

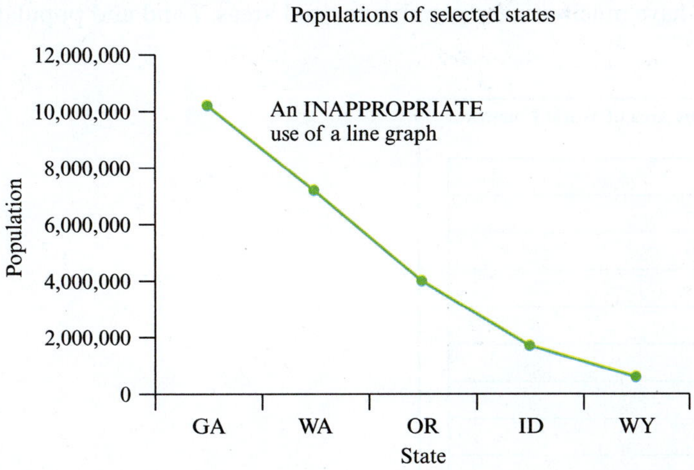

scatterplot

Scatterplots A **scatterplot** consists of a collection of data points that are plotted in a plane. We use scatterplots to see how two kinds of data are related. For example, for each of the 50 states, how is the percentage of students in eighth grade who are proficient in reading on the NAEP exam related to the percent of students who are proficient in mathematics on the NAEP exam? Figure 15.11 shows a scatterplot that consists of 51 points, one for each of the 50 states and one for the District of Columbia. Each point consists of a pair of numbers, namely,

$$(\text{percentage proficient in reading, percentage proficient in math})$$

CCSS 8.SP.1

For example, the District of Columbia contributes the point $(19, 19)$, because 19% of its eighth-grade students are proficient in reading and 19% are proficient in math. Massachusetts contributes the point $(46, 51)$, because 46% of its eighth-grade students are proficient in reading and 51% are proficient in math.

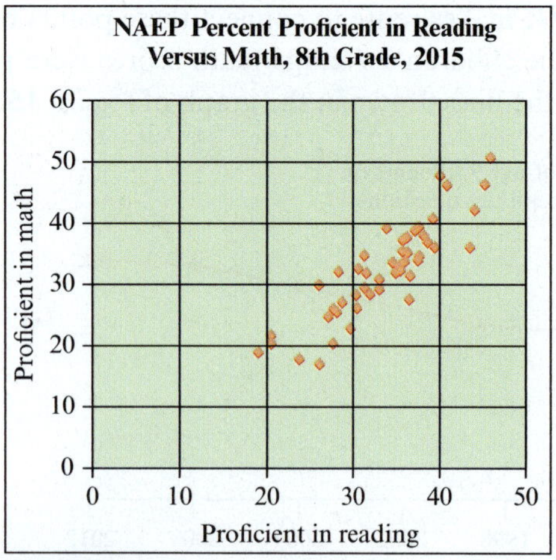

Figure 15.11 A scatterplot.

Data Source: U.S. Department of Education, Institute of Education Sciences, National Center for Education Statistics, National Assessment of Educational Progress (NAEP).

We can see from the scatterplot in **Figure 15.11** that states with a low percentage of students proficient in reading also have a low percentage of students proficient in math, and states with a higher percentage of students proficient in reading also have a higher percentage of students proficient in math. So proficiency in math seems to be correlated with proficiency in reading.

In contrast, the scatterplot in **Figure 15.12** shows a different kind of situation. This scatterplot shows populations and areas of countries in the world. We see that there are a few countries with very large populations and large areas and that there are a few countries with very large areas, but that most countries have much smaller populations and areas. Land and population do not seem to be correlated.

Figure 15.12 Scatterplot of population versus area of world countries, 2005.

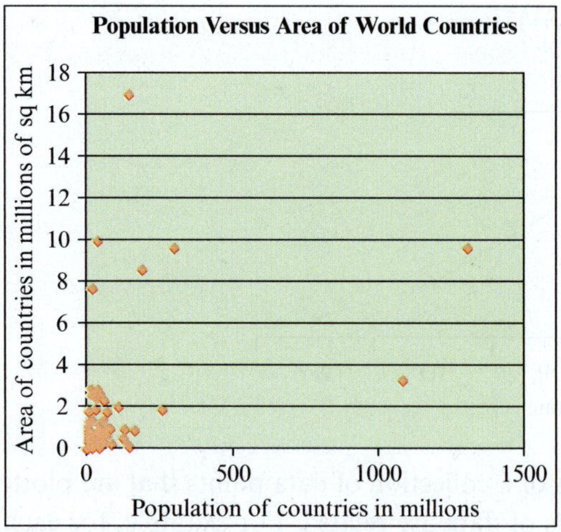

How Can We Create and Interpret Data Displays?

Data displays are created to summarize data and help us analyze and interpret them. Therefore, in creating a data display, choose a display that will communicate the point you want to make most clearly. Give your display a title so that the reader will know what it is about. Label your display clearly.

Read data displays carefully before interpreting them. For example, the vertical axis of the line graph in **Figure 15.9** starts at 200, not at 0, which can make changes appear greater than they are. You might initially think that the math scores in 1999 were almost 50% higher than they were in 1978. In fact, the math scores in 1999 were only about 6% higher than they were in 1978.

Be careful when drawing conclusions about cause and effect from a data display. For example, in Figure 15.4, we see that women earn less money than men in both age ranges. We might be tempted to conclude that this is due to discrimination against women on the basis of gender. Although this could be the case, there could be other reasons for the difference in earnings. For example, many of the women in the age 25 and over category might have taken time off from work to raise children. This break in employment could cause many women to have less work experience and therefore earn less. In any case, further study would be needed to determine the cause of the difference in earnings.

Although we can read specific numerical information from data displays, the main value of displays is in showing qualitative information about the data. So when you read a data display, look for qualitative information that it conveys and think about the implications of this information. The pie chart in Figure 15.5, for example, shows that only a small percentage of U.S. firms have 50 or more employees. Therefore, a law that would apply only to firms with 50 or more employees would not affect most businesses. On the other hand, from the display in Figure 15.5, we do not know what percent of all *employees* work in firms with 50 or more employees. Further data would need to be gathered to determine this.

CLASS ACTIVITY

15E Critique Data Displays or Their Interpretation, p. CA-312

How Can Students Read Graphs at Different Levels?

Data displays are produced to organize data so that we can make sense of the data and analyze it to answer questions. Merely producing data displays is not an important mathematical goal. To make data displays valuable in mathematics instruction, we must apply mathematical ideas to analyze the displays and answer questions.

To help us analyze graphs, we can ask and answer questions at three different levels of graph comprehension: reading the data, reading between the data, and reading beyond the data. The following descriptions of these three levels are adapted from Curcio [19, p. 7]:

1. *Reading the data.* This level of comprehension requires a literal reading of the graph. The reader simply "lifts" the facts explicitly stated in the graph, or the information found in the graph title and axes labels, directly from the graph. No interpretation occurs at this level.
2. *Reading between the data.* This level of comprehension includes the interpretation and integration of the data in the graph. It requires the ability to compare quantities (e.g., greater than, tallest, smallest) and the use of other mathematical concepts and skills (e.g., addition, subtraction, multiplication, division) that allow the reader to combine and integrate data and identify the mathematical relationships expressed in the graph.
3. *Reading beyond the data.* This level of comprehension requires the reader to predict or infer from the data by tapping existing knowledge and knowledge developed from "reading the data" and "reading between the data" for information that is neither explicitly nor implicitly stated in the graph.

CCSS
2.MD.10
3.MD.3
4.MD.4
5.MD.2

For example, here are some questions we could ask about the bar graph in Figure 15.13:

- How many students chose pizza for lunch? (Read the data.)
- How many students chose a school lunch? (Read between the data.)
- How many more students chose pizza than hamburger? (Read between the data.)
- If students had the same choices tomorrow, would the graph come out the same? (Read beyond the data.)

Figure 15.13

Students' school lunch choices: pizza or hamburger.

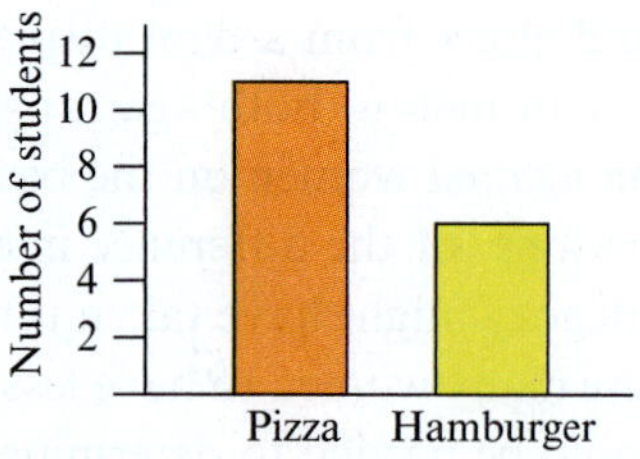

CLASS ACTIVITY

15F Display and Ask Questions About Graphs of Random Samples, p. CA-315

15G Displaying Data About Pets, p. CA-317

15H Investigating Small Bags of Candies, p. CA-318

15I The Length of a Pendulum and the Time It Takes to Swing, p. CA-319

15J Balancing a Mobile, p. CA-320

SECTION SUMMARY AND STUDY ITEMS

Section 15.2 Displaying Data and Interpreting Data Displays

Common displays of categorical data include real graphs, pictographs, bar graphs, and pie graphs. Displays of numerical data include dot plots, histograms, stem-and-leaf plots, line graphs, and scatterplots. All such graphical displays can assist in the interpretation of data. To help students think about data displays, one can pose three levels of graph-reading questions: read the data, read between the data, and read beyond the data.

Key Skills and Understandings

1. Make data displays to help convey information about data.
2. For a given data display, formulate and answer questions at the three levels of graph reading.
3. Recognize erroneous or misleading data displays.

Practice Exercises for Section 15.2

1. A class plays a fishing game in which there is a large tub filled with plastic fish that are identical, except that some are red and the rest are white. A student is blindfolded and pulls 10 fish out of the tub. The student removes the blindfold, writes down how many of each color fish she got, and then puts the fish back in the tub. Each student takes a turn. The results are shown in Table 15.2.

   a. Display these data in a dot plot.

   b. Write and answer (to the extent possible) at least three questions about the data display in part (a); include at least one question at each of the three graph-reading levels discussed in this section.

2. When is it appropriate to use a line graph?

3. If you have data consisting of percentages, is it always possible to display these data in a single pie graph?

TABLE 15.2 Numbers of red and white fish picked out of a tub by students

Name	Fish	Name	Fish	Name	Fish
Michelle	9 red, 1 white	Peter	6 red, 4 white	Sarah	8 red, 2 white
Tyler	7 red, 3 white	Brandon	9 red, 1 white	Adam	7 red, 3 white
Antrice	7 red, 3 white	Brittany	6 red, 4 white	Lauren	6 red, 4 white
Yoon-He	6 red, 4 white	Orlando	4 red, 6 white	Letitia	9 red, 1 white
Anne	6 red, 4 white	Chelsey	7 red, 3 white	Jarvis	7 red, 3 white

Answers to Practice Exercises for Section 15.2

1. a. Each dot in Figure 15.14 represents one child. The dot is plotted over the number of red fish that were picked (of course, the dot plot could also be organized according to the number of white fish that were picked).

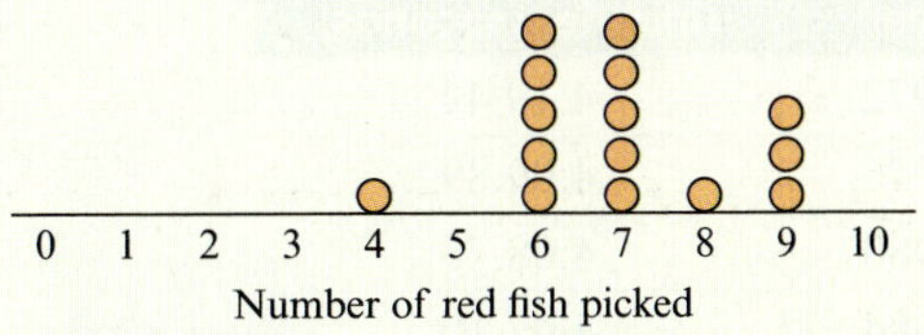

Figure 15.14 Dot plot showing how many students picked each number of red fish.

b. How many people picked 9 red fish? (Level: Read the data.) Answer: Three students picked 9 red fish, which we can see from the 3 dots above 9.

How many students picked either 6 or 7 red fish? (Level: Read between the data.) Answer: Ten students picked either 6 or 7 fish because there are 5 dots over 6 and another 5 dots over 7.

Are there more red fish or white fish in the tub? (Level: Read beyond the data.) Answer: There are probably more red fish than white fish because most students picked more red fish than white fish. These random samples are probably representative of the fish in the tub.

2. See text. Line graphs are appropriate for displaying only data that vary continuously.

3. It is not always possible to display data consisting of percentages in a single pie graph. If the percentages aren't *separate parts of the same whole*, then you can't display them in a pie graph. For example, see the (hypothetical) table on children's eating, shown below.

Food	Percentage of 4- to 6-Year-Olds Meeting the Dietary Recommendation for a Food Group
Grains	28
Vegetables	15
Fruits	28
Saturated fat	29

Even though the percentages add to 100%, they probably do not represent *separate* groups of children. For example, many of the 15% of the children who meet the recommendations on vegetables probably also meet the recommendations on fruits. Therefore, it would not be appropriate to display these data in a pie graph.

PROBLEMS FOR SECTION 15.2

1. Three third-grade classes are having a contest to see which class can read more pages during the month of February. The classes want to create a display, to be posted in the hall, that will show their progress. What do you recommend? Be specific, and make sure your recommendation can be carried out realistically. Show what the display might look like at some point during February. Explain briefly why you think the display might look like that.

2. Find 3 coins of any type (as long as all 3 have a head side and a tail side).

   a. Take 2 of the coins and flip the pair 30 times. While you flip the coins, make a dot plot to show how many times there were 0 heads, 1 head, and 2 heads.

   b. Now take all 3 coins and flip the triple 30 times. While you flip the coins, make a dot plot to show how many times there were 0 heads, 1 head, 2 heads, and 3 heads.

   c. Write at least six questions about your dot plots in parts (a) and (b); include at least two at each of the three levels of graph reading discussed in this section. Label each question with its approximate level. Your questions may be suitable for each dot plot separately or for the two dot plots together. Answer each question (to the extent possible).

3. Table 15.3 shows women's 400-meter freestyle Olympic winning times in recent history.

   a. Make a graphical display of the data in Table 15.3.

   b. Write at least four questions about your graph in part (a); include at least one at each of the three graph-reading levels. Label each question with its level. Answer your questions (to the extent that it is possible) and explain your answers briefly.

TABLE 15.3 Women's 400-meter freestyle Olympic winning times.
Source: Olympic.org, Official Website of the Olympic Movement.

Year	Winning Time (minutes:seconds)	Year	Winning Time (minutes:seconds)
1920	4:34.0	1972	4:19.44
1924	6:02.2	1976	4:09.89
1928	5:42.8	1980	4:08.76
1932	5:28.5	1984	4:07.10
1936	5:26.4	1988	4:03.85
1948	5:17.8	1992	4:07.18
1952	5:12.1	1996	4:07.25
1956	4:54.6	2000	4:05.80
1960	4:50.6	2004	4:05.34
1964	4:43.3	2008	4:03.22
1968	4:31.8	2012	4:01.45

4. Using an Internet browser, go to census.gov, the webpage of the U.S. Census Bureau, click on "Data" and then on "Visualizations" to find a collection of graphs or data displays. Select a data display that interests you. Describe or include a picture of the data display and provide a link to it. Write at least four questions about the data display, including at least one question at each of the three graph-reading levels discussed in this section. Label each question with the level you think fits best. Answer each of your questions to the extent possible and explain your answers. Now repeat the entire process with a different data display.

5. Using an Internet browser, go to the webpage of the National Assessment of Educational Progress (NAEP), nationsreportcard.gov, and find a graph or data display that interests you. For example, you could look under "Dashboards" or "Reports." Describe or include a picture of the data display and provide a link to it. Write at least four questions about the data display, including at least one question at each of the three graph-reading levels discussed in this section. Label each question with the level you think fits best. Answer each of your questions to the extent possible and explain your answers. Now repeat the entire process with a different data display.

6. Using an Internet browser, go to one of the following websites and find a graph or data display that interests you:

   • census.gov, the website of the U.S. Census Bureau, click on "Data" and then on "Visualizations";

- nationsreportcard.gov, the website of the National Assessment of Educational Progress (NAEP), and look under "Dashboards" or "Reports";
- cdc.gov, the website of the Centers for Disease Control and Prevention (CDC), and click on "Data & Statistics";
- pewresearch.org, the website of the Pew Research Center, and click on "Data";
- theharrispoll.org, the website of the Harris Poll;
- YouGov.com, the website of YouGov.

Describe or include a picture of the data display and provide a link to it. Write at least four questions about the data display, including at least one question at each of the three graph-reading levels discussed in this section. Label each question with the level you think fits best. Answer each of your questions to the extent possible and explain your answers. Now repeat the entire process with a different data display.

7. Using an Internet browser, go to one of the following websites and find a table or list of data that interests you and may help answer a question you have:

- census.gov, the website of the U.S. Census Bureau, click on "Data," then on "Data Tools & Apps," and then for example on "American FactFinder" or "Easy Stats";
- nationsreportcard.gov, the website of the National Assessment of Educational Progress (NAEP), click on "Data Tools" and then on "NAEP Data Explorer";
- cdc.gov, the website of the Centers for Disease Control and Prevention, click on "Data & Statistics";
- fbi.gov, the website of the Federal Bureau of Investigation, click on "Stats & Services" and then on "Crime Statistics";
- Olympic.org, the website of the Olympic Movement;
- theharrispoll.org, the website of the Harris Poll.

Provide a link to the data. Write a paragraph about your question and discuss how the data address that question. Include a display of the data to help convey your points.

8. a. Describe *in detail* an activity suitable for use with elementary or middle school students in which the students pose a question, gather data, and create an appropriate display for the data to help answer the question.

b. Write at least four questions that you could ask students about the proposed graph in part (a), once the graph is completed. Include at least one question at each of the three graph-reading levels. Label each question with its level.

15.3 The Center of Data: Mean, Median, and Mode

CCSS Common Core State Standards Grades 5, 6, 7

When we have a collection of numerical data, such as the collection of scores on a test or the heights of a group of 6-year-old children, our first questions about the data are usually, "What is typical or representative of these data?" and "What is the center of these data?" When a data set is large, it is especially helpful to know the answers to these questions. For example, if the data consist of the test scores of all students taking the SAT in a given year, no one person would want to look at the entire data set (even if he or she could); it would be overwhelming and impossible to comprehend. Instead, we want ways of understanding the nature of the data. In particular, it is helpful to have a single number that summarizes a set of data.

To say what is representative, in the center, or typical of a list of numbers, we commonly use the terms *mean, median,* or *mode*. The mean, median, and mode each provide a single-number summary of a set of numerical data and are often called **measures of center**.

measures of center

What Is the Mean and What Does It Tell Us?

mean
average

To calculate the **mean**, also called the *arithmetic mean,* or the **average**, of a list of numbers, add all the numbers and divide this sum by the number of numbers in the list. For example, the mean of

$$7, \quad 10, \quad 11, \quad 8, \quad 10$$

is

$$(7 + 10 + 11 + 8 + 10) \div 5 = 46 \div 5 = 9.2$$

We divide by 5 because there are 5 numbers in the list: 7, 10, 11, 8, 10. This is how we *calculate* the mean, but what does the mean *tell us*? Try Class Activity 15K.

CLASS ACTIVITY

15K The Mean as "Making Even" or "Leveling Out," p. CA-321

CCSS
5.MD.2
6.SP.3

The Mean as "Leveling Out" Although we defined the mean in terms of adding and dividing, we can think of the mean of a list of numbers as obtained by "leveling out" the numbers so that all numbers become the same. For example, consider the list

$$4, \quad 3, \quad 6, \quad 5, \quad 2$$

as represented by towers of blocks: one 4 blocks tall, one 3 blocks tall, one 6 blocks tall, one 5 blocks tall, and one 2 blocks tall, as shown in **Figure 15.15**. If we rearrange the blocks so that all 5 towers become the same height, the common height is the mean. Why? When we calculate the mean numerically, we first add the numbers

$$4 + 3 + 6 + 5 + 2$$

Figure 15.15
The mean as "leveling out."

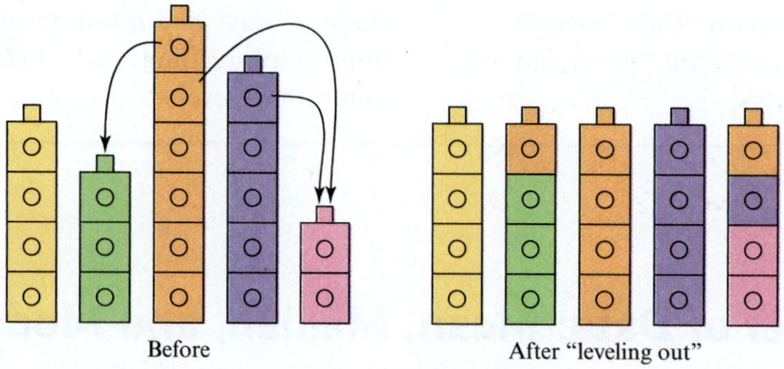

In terms of the block towers, adding the numbers tells us the total number of blocks in all 5 towers. When we then divide the sum of the numbers by 5,

$$(4 + 3 + 6 + 5 + 2) \div 5 = 20 \div 5 = 4$$

we are determining the number of blocks in each tower when the full collection of blocks is divided equally among 5 towers. In other words, we are determining the number of blocks in each tower when the 5 towers are made even. As you will see in Class Activity 15L, some problems about the mean will be easy to solve by thinking of the mean in terms of leveling out.

CLASS ACTIVITY

15L Solving Problems About the Mean, p. CA-322

The Mean as Balance Point Another way to think about the mean is as a "balance point," an approach that is especially appropriate when data are represented in a dot plot or a histogram. Think of a dot plot or histogram as being like a seesaw that will balance on a fulcrum when the fulcrum is placed at an appropriate spot. The location for the fulcrum—the balance point—turns out to be the mean of the set of data. For example, consider the data set

4, 4, 4, 12

which is represented in the dot plot in **Figure 15.16**. You can probably tell that the dot plot will balance if a fulcrum is placed under the number 6. And in fact, the mean of the data is also 6.

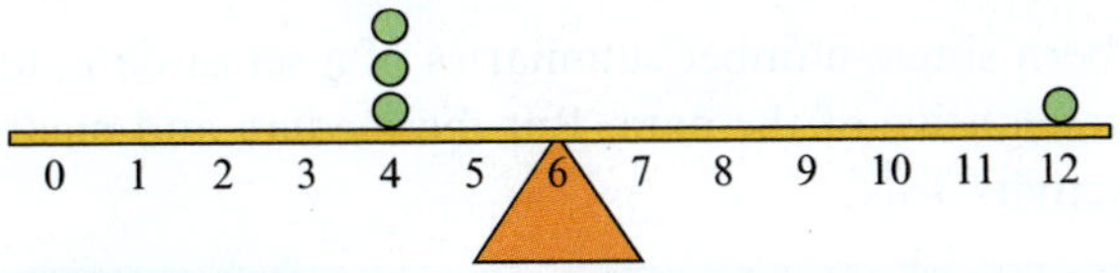

Figure 15.16 The mean as "balance point."

Why is the mean at the location of a fulcrum? We will not discuss this subject in detail here, but two factors are involved: (1) knowing how levers work and (2) knowing that for data to the right of the mean, the sum of all the distances from the mean of these data is equal to the sum of all the distances from the mean of the data that lie to the left of the mean.

By thinking about the mean as a balance point, we can quickly estimate the mean of data that are displayed in a dot plot or a histogram.

CLASS ACTIVITY

15M The Mean as "Balance Point," p. CA-323

FROM THE FIELD Research

Hudson, R. A. (2012/2013). Finding balance at the elusive mean. *Mathematics Teaching in the Middle School, 18*(5), pp. 300-306.

The author introduced a diverse group of fifth- through seventh-grade students attending a summer enrichment program to the mean as a balance point. With this view of the mean, for data in a dot plot, the sum of the distances to the left of the mean equals the sum of the distances to the right of the mean. To develop the students' conceptions of the mean as a mathematical point of balance, the author designed "target practice" tasks in which students (1) estimated the value of the mean, (2) found the actual value of the mean using software and compared it with their estimated value, (3) added a point to the data set so that the mean would change to a given "target" mean, and (4) guessed where they believed the target mean would be located and checked the outcome using software. Students went on to develop and test conjectures about how adding data points could affect the mean, and they posed target practice problems for other groups to solve.

What Is The Median and What Does It Tell Us?

median To calculate the **median** of a list of numbers, organize the list from smallest to largest (or vice versa). The number in the middle of the list is the median. If there is no number in the middle, then the median is halfway between the two middle numbers. For example, to find the median of

7, 10, 11, 8, 10

rearrange the numbers from smallest to largest as follows:

7, 8, 10, 10, 11

The number 10 is exactly in the middle, so it is the median. When we put the six numbers in the list 4, 6, 5, 3, 5, 3 in order, we see there is no single number in the middle of the list:

3, 3, 4, 5, 5, 6

The median is halfway between the two middle numbers 4 and 5; therefore, the median is 4.5.

How Is the Median Different from the Mean?

The mean and the median are both single-number summaries of a set of data, and both give us a sense of what is typical or representative of the data. But the median and mean need not be the same, as you will see in Class Activity 15N.

CLASS ACTIVITY

15N Same Median, Different Mean, p. CA-324

CCSS 6.SP.5d

The median is usually not affected much by a few pieces of data that are very far from the majority of the data. On the other hand, the mean can be significantly affected by data that are far from the majority of the data, as Figure 15.17 shows.

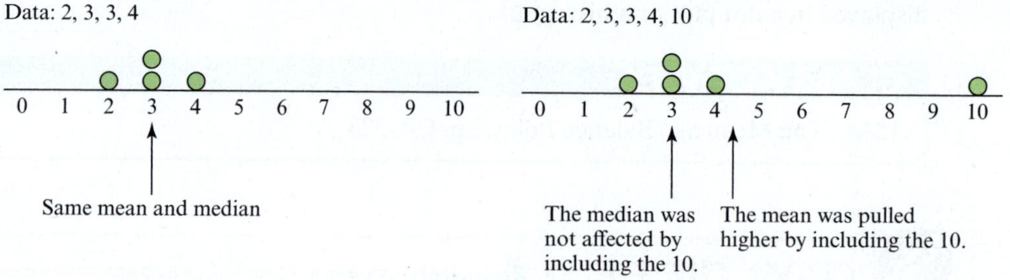

median of 2, 3, 3, 4 is 3
mean = (2 + 3 + 3 + 4) ÷ 4
= 12 ÷ 4
= 3

median of 2, 3, 3, 4, 10 is 3
mean = (2 + 3 + 3 + 4 + 10) ÷ 5
= 22 ÷ 5
= 4.4

Figure 15.17 Data that are far from the majority of the data affect the mean more than the median.

Because of the mean's sensitivity to extreme values, the median sometimes gives a better sense of what is typical or representative for a set of data. For example, when discussing household income of families in the United States, the median is generally preferred over the mean, because a small percentage of very wealthy households pulls the mean up. The vast majority of households have a much more modest income. Thus, the median is more typical, or representative, of the household income of most families in the United States than is the mean.

On the other hand, when a total combined amount is to be calculated from a representative amount, the mean is a better choice for the representative amount than is the median. For example, to calculate the total income of all U.S. households, we should multiply the *mean* household income by the number of households. If we multiplied the median household income by the number of households we would get a value that is lower than the total income of all households. Some problems at the end of this section further explore the mean versus the median.

CLASS ACTIVITY

15O Can More Than Half Be Above Average?, p. CA-325

To help remember which is the median and which is the mean, think of this example: On a road, the median line is the line down the middle of the road. Similarly, the median is the middle of all the data values when they are put in order.

What Are Common Errors with the Mean and the Median?

Students who are beginning to learn statistical concepts sometimes mistakenly try to apply the median in cases where it does not apply (see [47]). The median (and the mean) apply only to numerical data, not to categorical data. For example, if students in a class have each chosen a favorite sport and constructed a bar graph showing how many students chose football, basketball, baseball, soccer, or another sport as favorite, there is no "median favorite sport" because there is no natural or preferred way to order the sports. Similarly, a "mean favorite sport" would not make sense either. Expressed more generally, it would not make sense to consider a balance point on a bar graph of categorical data as the mean of the data.

CLASS ACTIVITY

15P Critique Reasoning About the Mean and the Median, p. CA-326

When data are displayed in a dot plot or a histogram, students sometimes calculate means and medians incorrectly by finding the mean or median of the frequency counts instead of the data. For example, a student might attempt to find the mean of the data shown in the dot plot in Figure 15.18 incorrectly by averaging the frequencies 3, 5, 4 instead of by averaging the data values

6, 6, 6, 7, 7, 7, 7, 7, 8, 8, 8, 8,

to determine the mean correctly.

Figure 15.18

What is the mean of the data in the dot plot?

Incorrect mean:
$(3 + 5 + 4) \div 3 = 4$

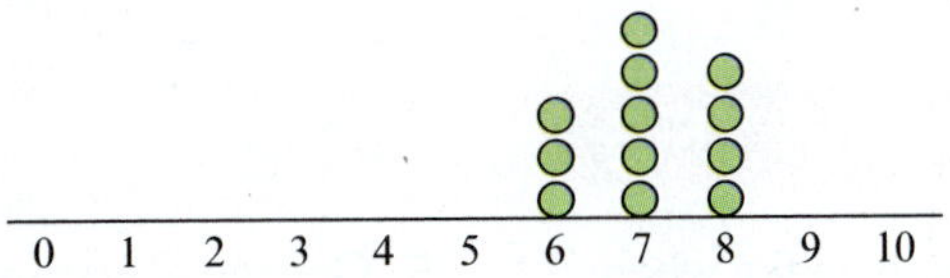

Correct mean:
$(6 + 6 + 6 + 7 + 7 + 7 + 7 + 7 + 8 + 8 + 8 + 8) \div 12 = 7.083\ldots$

What Is the Mode?

mode To calculate the **mode** of a list of numbers, reorganize the list so as to group equal numbers in the list together. The mode is the number or numbers that occur most frequently in the list. For example, to find the mode of

11, 9, 10, 8, 11, 10, 11, 9, 12

first reorganize the list as follows:

8, 9, 9, 10, 10, 11, 11, 11, 12

The number 11 occurs 3 times, which is more than the occurrence of any other number in the list. Therefore 11 is the mode of this list of numbers. Statisticians do not use the mode as much as the mean and the median, because relatively small changes in data can produce large changes in the mode.

modal category Unlike the mean or the median, the mode can be used with categorical data to identify a **modal category**, which is a category (or categories) that contains the largest number of entries. For example, suppose 20 students in a class vote on which of baseball, basketball, football, or soccer is their favorite sport to play. Let's say that 9 students choose basketball and each of the other sports receives fewer than 9 votes. Then basketball is the modal category in that case.

We can use the modal category to predict which category a new piece of data is most likely to fall into. Return to the example of voting for a favorite sport to play. If a new student were to join the class and we had to guess about what sport this student would like to play, our best guess is that it would be the modal category, basketball.

SECTION SUMMARY AND STUDY ITEMS

Section 15.3 The Center of Data: Mean, Median, and Mode

The mean (average) and median are good one-number summaries of a collection of numerical data. We can think of the mean in terms of "leveling out" the data. This way of thinking about the mean agrees with the way we calculate the mean. It also provides a way of solving problems about the mean. Another way to think about the mean is as a "balance point." The mean and median can be the same or different for a given set of data.

Key Skills and Understandings

1. Describe how to view the mean as leveling out, and explain why this way of viewing the mean agrees with the way we calculate the mean (by adding and then dividing).
2. Use the leveling out view of the mean to solve problems about the mean. Also use the standard way of calculating the mean to solve problems about the mean.
3. View the mean as a balance point and understand that this point of view is especially appropriate for data displayed in histograms and dot plots.
4. Create data sets with different means and medians.
5. Discuss errors that students commonly make with the mean and the median.

Practice Exercises for Section 15.3

1. What is a common error that students make when they compute the mean of numerical data that are displayed in a dot plot?
2. Describe some common errors students make with the median.
3. Ten students take a 10-point test. The mean score is 8. Could the median be 10? Could the median be 5? In each case, explain why or why not.
4. Juanita read an average of 3 books a day for 4 days. How many books will Juanita need to read on the fifth day so that she will have read an average of 5 books a day over 5 days? Solve this problem in several ways, and explain your solutions.
5. George's average score on his math tests in the first quarter is 60. His average score on his math tests in the second quarter is 80. George's semester score in math is the average of all the tests he took in the first and second quarters. Can he necessarily calculate his semester score by averaging 60 and 80? If so, explain why. If not, explain why not. Explain what other information you would need to calculate George's semester average, and show how to calculate this average.
6. Suppose that all fourth-graders in a state take a writing competency test that is scored on a 5-point scale. Is it possible (in theory) for 80% of the fourth-graders to score below average? If so, show how that could occur. If not, explain why not.

Answers to Practice Exercises for Section 15.3

1. Students sometimes calculate the mean of the frequencies instead of using the data values. See the text and Class Activities.
2. See the text and the Class Activities. Students sometimes forget to put the data in order before determining the middle value. They sometimes try

to apply the median to categorical data, where it does not apply. As with the mean, students may mistakenly work with frequencies rather than with data values when calculating the median.

3. Since the mean score is 8, the total number of points scored by all 10 students is $10 \cdot 8 = 80$. For the median to be 10, at least 6 students would have to score a 10. These 6 students contribute a total of 60 points. The remaining 20 points could then be distributed in many ways among the remaining 4 students—for example, say all 4 remaining students scored a 5. So yes, it's possible for the median to be 10.

For the median to be 5, at least 6 students would have to score 5 or less. For example, if there are 6 students who each score a 5 or less, they contribute a total of at most 30 points. For the mean to be 8, the total number of points scored by all students is 80, so the remaining 4 students would have to contribute the remaining 50 points, which is not possible since it's a 10-point test. So no, it is not possible for the median to be 5.

4. *Method 1:* Imagine putting the books that Juanita has read into 4 stacks, one for each day. Because she has read an average of 3 books per day, the books can be redistributed so that there are 3 books in each of the 4 stacks. This uses the "average as leveling out" point of view. Now picture a fifth stack with an unknown number of books in it—the number of books Juanita must read to make the average 5 books per day. If the books in this fifth stack are distributed among all 5 stacks so that all 5 stacks have the same number of books, then this common number of books is the average number of books Juanita will have read over 5 days. Therefore, this fifth stack must have 5 books plus another $4 \cdot 2 = 8$ books, in order to distribute 2 books to each of the other 4 stacks. So Juanita must read $5 + 8 = 13$ books on the fifth day.

Method 2: To read an average of 5 books a day over 5 days, Juanita must read a total of 25 books. This is because

$$(\text{total books}) \div 5 = \text{average}$$

Therefore,

$$(\text{total books}) \div 5 = 5$$

so

$$\text{total books} = 5 \cdot 5 = 25$$

Similarly, because Juanita has already read an average of 3 books per day for 4 days, she has read a total of $4 \cdot 3 = 12$ books. Therefore, Juanita must read $25 - 12 = 13$ books on the fifth day.

5. George cannot necessarily calculate his semester score by averaging 60 and 80. We don't know if George took the same number of tests in the first and second quarters. To calculate George's average score over the whole semester, we need to know how many tests he took in the first and second quarters. Suppose George took 3 tests in the first quarter and 5 tests in the second quarter. Then the sum of his first 3 test scores is $3 \cdot 60 = 180$, and the sum of his next 5 test scores is $5 \cdot 80 = 400$. The sum of all 8 of his test scores is thus $180 + 400 = 580$ and his average score on all 8 tests is $580 \div 8 = 72.5$. Notice that this is not the same as the average of 60 and 80, which is only 70.

6. It is theoretically possible for 80% of the fourth-graders to score below average. For example, suppose there are 200,000 fourth-graders and that

10,000 score 1 point,
150,000 score 2 points,
10,000 score 3 points,
25,000 score 4 points, and
5000 score 5 points,

as shown in the histogram of Figure 15.19. You can see from the histogram that the "balance point" is between 2 and 3, and 160,000 children—namely, 80% of the 200,000 children—score below this average.

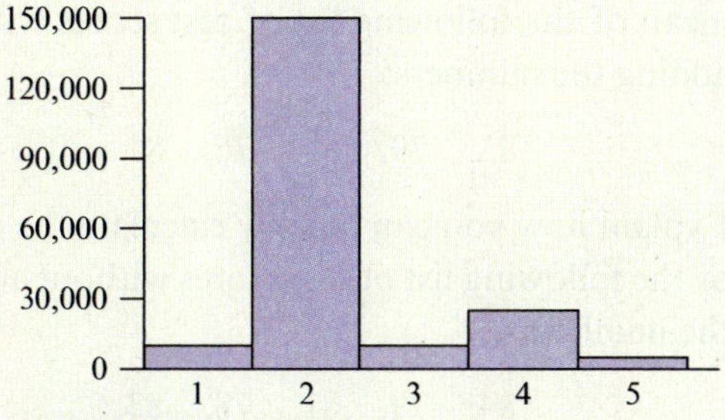

Figure 15.19 Scores on a writing test.

PROBLEMS FOR SECTION 15.3

1. In your own words, explain why the following two ways of finding the mean of a set of numerical data lead to the same result:

a. add the numbers and divide by how many there are

b. "level out" all the data to be the same number (as in Class Activity 15K)

Give an example to illustrate.

2. Explain why the mean of a list of numbers must always be in between the smallest and largest numbers in the list.

3. Explain why the mean of two numbers is exactly halfway between the two numbers.

4. Shante caught 17 ladybugs every day for 4 days. How many ladybugs does Shante need to catch on the fifth day so that she will have caught an average of 20 ladybugs per day over the 5 days? Solve this problem in two different ways, and explain your solutions.

5. John's average annual income over a 4-year period was $25,000. What would John's average annual income have to be for the next 3 years so that his average annual income over the 7-year period would be $50,000? Solve this problem in two different ways, and explain your solutions.

6. Tracy's times swimming 200 yards were as follows:

2:45, 2:47, 2:44

How fast will Tracy have to swim her next 200 yards so that her mean time for the 4 trials is 2:45? Explain how you can solve this problem without adding the times.

7. Explain how you can quickly calculate the mean of the following list of test scores without adding the numbers:

83, 79, 81, 76, 81

8. Explain how you can quickly calculate the mean of the following list of test scores without adding the numbers:

84, 79, 81, 78, 83

9. Julia's average on her first 3 math tests was 80. Her average on her next 2 math tests was 95. What is Julia's average on all 5 math tests? Solve this problem in two different ways, and explain your solutions.

10. A teacher gives a 10-point test to a class of 10 children.

a. Is it possible for 9 of the 10 children to score above average on the test? If so, give an example to show how. If not, explain why not.

b. Is it possible for all 10 of the children to score above average on the test? If so, give an example to show how. If not, explain why not.

11. In your own words, describe how to view the mean of a set of numerical data in two different ways: in terms of leveling out and as a balance point. In each case, give an example to illustrate.

12. Discuss Jessica's reasoning about calculating the mean of the data displayed in the dot plot in Figure 15.20.

I took 2 dots from above the 8 and moved one to 6 and one to 7. Then all the towers were leveled out and all were 4 tall, so the mean is 4.

Is Jessica's reasoning valid? If not, discuss how her idea could be used to make a correct statement about the mean.

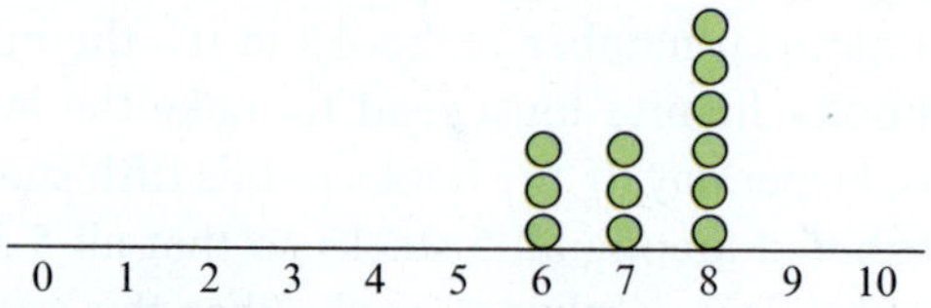

Figure 15.20 What is the mean of the data displayed in this dot plot?

13. The dot plot in Figure 15.21 represents the number of small candies found in several packets.

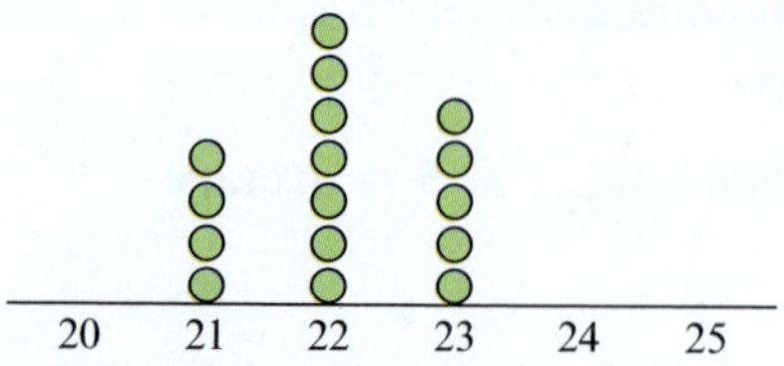

Figure 15.21 Dot plot of the number of candies in packets.

Michael found the mean number of candies in a packet by calculating this way:

$$\frac{21 + 22 + 23}{3} = 22$$

Anne found the mean number of candies in a packet by calculating this way:

$$\frac{4 + 7 + 5}{3} = 5\frac{1}{3}$$

Is either of these methods correct? If not, explain what is wrong, and explain how to calculate the mean number of candies in the packets correctly.

14. For each of the following situations, decide whether the mean or the median would likely give a better sense of what is typical or representative of the data for the purpose that is described. Discuss your reasoning in each case.

a. The starting salaries of a class of college graduates in which one of the graduates is drafted by the National Football League. Prospective students would like a sense of the starting salary they might expect when they get a job after they graduate.

b. The net weights of one type and size of boxes of laundry detergent produced at a factory over the course of a year. Government regulators would like to know if consumers who use the laundry detergent on a regular basis are getting the amount of detergent they have paid for in the long run.

c. The number of turtle eggs in nests along a beach. Researchers estimate that 80% of all the turtle eggs will hatch and they want to estimate how many baby turtles there will be.

15. A teacher gives a 10-point test to a class of 9 children. All scores are whole numbers. If the median score is 8, then what are the highest and lowest possible mean scores? Explain your answers.

16. A teacher gives a 10-point test to a class of 9 children. All scores are whole numbers. If the mean score is 8, then what are the highest and lowest possible median scores? Explain your answers.

17. In Ritzy County, the average annual household income is \$100,000. In neighboring Normal County, the average annual household income is \$30,000. Does it follow that in the two-county area (consisting of Ritzy County and Normal County), the average annual household income is the average of \$100,000 and \$30,000? If so, explain why; if not, explain why not. What other information would you need to calculate the average annual household income in the two-county area? Show how to calculate this average if you had this information.

18. In county A, the average score on the grade 5 Iowa Test of Basic Skills (ITBS) was 50. In neighboring county B, the average score on the grade 5 ITBS was 71. Can you conclude that the average score on the grade 5 ITBS in the two-county region (consisting of counties A and B) can be calculated by averaging 50 and 71? If so, explain why; if not, explain why not. What other information would you need to know to calculate the average grade 5 ITBS score in the two-county region? Show how to calculate this average if you had this information.

19. a. The histogram at the top of Figure 15.22 shows the distribution of annual household incomes in Swank County. Without calculating the median or average annual household income in Swank County, make an educated guess about which one is greater. Describe your thinking.

b. Determine the approximate values of the median and average annual household incomes in Swank County from the histogram at the top of Figure 15.22. Which is greater, the average or the median? Compare with part (a).

c. The histogram at the bottom of Figure 15.22 shows the distribution of annual household incomes in Middle County. Without calculating the median or average annual household income in Middle County, make an educated guess about which one is greater. Describe your thinking.

d. Determine the approximate values of the median and average annual household income in Middle County from the histogram at the bottom of Figure 15.22. Which is greater, the average or the median?

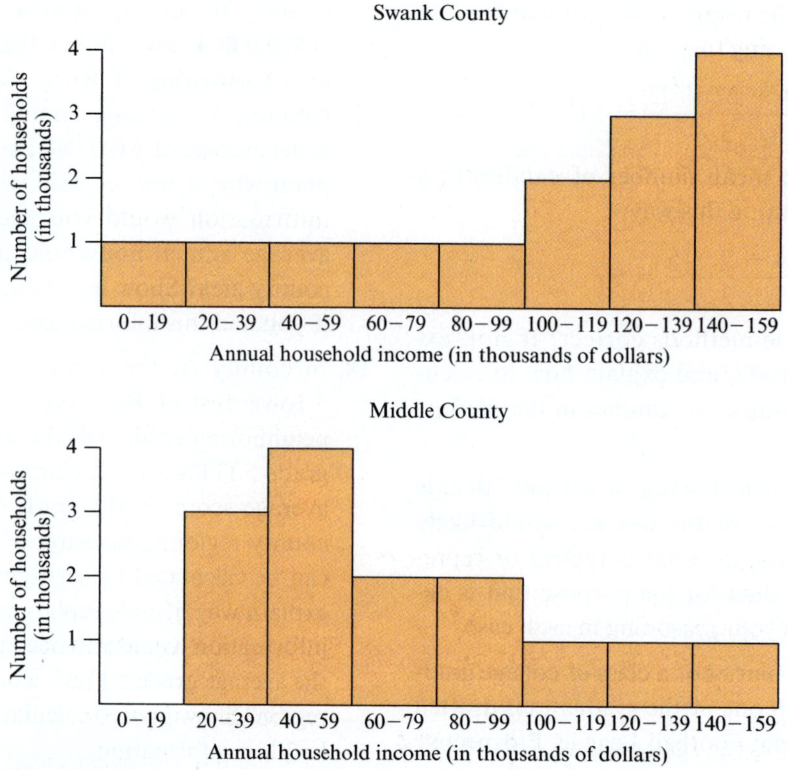

Figure 15.22 Annual household incomes in Swank and Middle Counties.

*__20.__ The students in Mrs. Marshall's class are estimating the number of blades of grass in the school's lawn. Each student places a piece of cardboard with a 1-inch-by-1-inch square cut out of the middle over a patch of grass, cuts off the grass that pokes through the hole, and counts the number of blades of grass that were cut. The students determine that the entire lawn consists of about 21,600 1-inch-by-1-inch squares. The students also determine that the mean number of blades of grass in their samples is 37 and the median number is 34. The students want to use their samples to estimate the number of blades of grass in the lawn, but they can't decide if they should use the mean or the median number of blades of grass in their samples. In other words, should they multiply 37 by 21,600 or should they multiply 34 by 21,600 in order to estimate the number of blades of grass in the lawn? Parts (a), (b), and (c) will help you determine whether the mean or the median will be better for the purpose of estimating the number of blades of grass.

a. Suppose Mrs. Marshall's students use their method to determine the number of blades of grass in a 5-inch-by-5-inch square of grass. In this case, they can subdivide the patch of grass into $5 \cdot 5 = 25$ 1-inch-by-1-inch squares and determine the number of blades of grass in each such square. The students can then determine the mean number of blades of grass in the squares and multiply this number by 25. Explain why this method, which uses the mean, will always produce the correct number of blades of grass in the 5-inch-by-5-inch patch of grass.

b. As in part (a), suppose Mrs. Marshall's students use their method to determine the number of blades of grass in a 5-inch-by-5-inch square of grass. But this time, suppose the students use the *median* number of blades of grass in the 25 1-inch-by-1-inch squares instead of the mean. Give an example to show that multiplying the median number of blades of grass in the 1-inch-by-1-inch squares by 25 may not produce the correct number of blades of grass in the 5-inch-by-5-inch patch. Explain your example.

c. Returning to the original problem about the grass in the school lawn, how should Mrs. Marshall's students estimate the number of blades of grass in the lawn? Explain.

***21.** Ms. Smith needs to figure her students' homework grades. There have been 5 homework assignments: the first with 30 points, the second with 40 points, the third with 30 points, the fourth with 50 points, and the fifth with 25 points. Ms. Smith is debating between two different methods of calculating the homework grades.

Method 1: Add up the points scored on all 5 of the homework assignments, and divide by the total number of points possible. Then multiply by 100 to get the total score out of 100.

Method 2: Find the percentage correct on each homework assignment, and find the average of these five percentages.

a. Jodi's homework grades are $\frac{27}{30}, \frac{25}{40}, \frac{26}{30}, \frac{30}{50}, \frac{24}{25}$, where, for example, $\frac{27}{30}$ stands for 27 points out of 30 possible points in a homework assignment. Compare how Jodi does with each of the two methods. Which method gives her a higher score?

b. Make up scores (on the same assignments) for Robert so that Robert scores higher with the other method than the one that gave Jodi the higher score.

***22.** The *average speed* of a moving object during a period of time is the distance the object traveled divided by the length of the time period. For example, if you left on a trip at 1:00 P.M., arrived at 3:15 P.M., and drove 85 miles, then the average speed for your trip would be

$$85 \div 2.25 \approx 37.8$$

miles per hour.

a. Jane drives 100 miles from Philadelphia to New York with an average speed of 60 miles per hour. When she gets to New York, Jane turns around immediately and heads back to Philadelphia along the same route. Using this information, find several different possible average speeds for Jane's *entire trip* from Philadelphia back to Philadelphia.

b. Ignoring practical issues, such as speed limits and how fast cars can go, would it be theoretically possible for Jane to average 100 miles per hour for the whole trip from Philadelphia to New York and back that is described in part (a)? If so, explain how; if not, explain why not.

c. Theoretically, what are the largest and smallest possible average speeds for Jane's entire trip from Philadelphia back to Philadelphia that is described in part (a)? Explain your answers clearly.

15.4 Summarizing, Describing, and Comparing Data Distributions

CCSS **Common Core State Standards** Grades 6, 7

When we collect data, we don't expect all the data to be the same; rather, we expect to see variability in the data. If a state gives a test to all fourth-graders, the test scores will surely vary from student to student. If a company measures the amount of mercury in cans of tuna fish, the amount will likely vary from can to can. We have seen that we can summarize a set of numerical data with the mean or the median of the data. The mean and the median give us an idea of what is typical or representative for a set of data, but two sets of data can have the same mean or median and yet may be distributed very differently across a dot plot or histogram.

data distribution

We often call numerical data that are displayed in a dot plot or histogram a **data distribution**. In this section we examine some of the common shapes that data distributions take. We also study some statistical tools for summarizing how data are distributed within a data set. Some of these tools involve the use of percentiles. Although you may not teach students about percentiles, you will probably receive reports about your students' performance on standardized tests in terms of percentiles. You may also receive reports about your students in terms of normal distributions; therefore, we study normal distributions briefly at the end of the section.

Why Do Data Distributions Have Different Shapes?

A large set of numerical data often has a particular type of shape when it is plotted in a histogram or a dot plot. The overall shape of the data distribution can give insight into the nature of the data. Before you read on, do Class Activity 15Q to get a feel for what different shapes of data distributions tell about the data.

CLASS ACTIVITY

15Q What Does the Shape of a Data Distribution Tell About the Data?, p. CA-327

CCSS 6.SP.2

Some data distributions have a long tail that extends far to the right or far to the left of the majority of the data, as shown in Figure 15.23. These distributions are called **skewed to the right** and **skewed to the left** respectively. For example, in a mature forest, the majority of the trees may be approximately the same height—the height of the tree canopy. There may be a far smaller number of shorter trees, perhaps because there is not enough sunlight below the canopy for many new trees to start growing or to survive. A histogram of tree heights in this forest would be skewed to the left. On the other hand, if most of the fish in a lake are small and there are far fewer large fish, then a histogram of the weights of fish in this lake would be skewed to the right.

Figure 15.23 Distributions skewed to the right (a) and left (b).

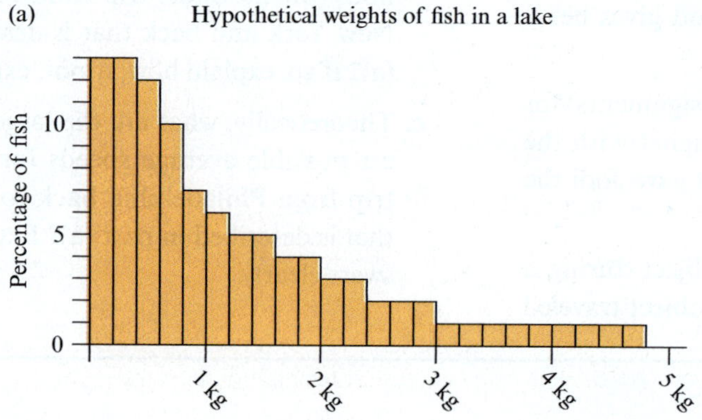

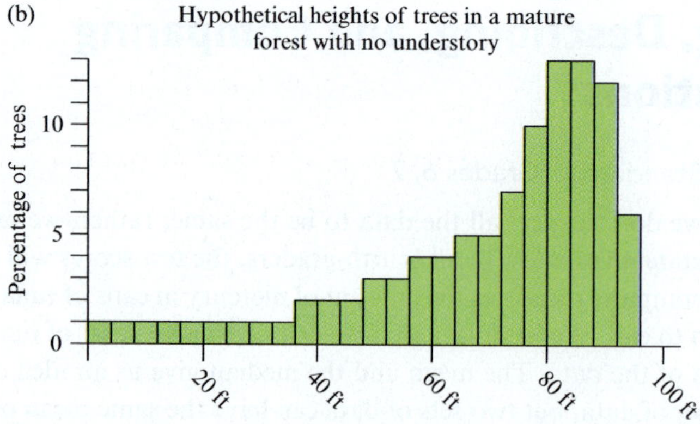

Some data distributions have two (or more) peaks, as shown in Figure 15.24. A distribution that has two peaks is called **bimodal**. Some forests have some tall trees that form the tree canopy and other shorter trees that form an understory. A histogram of tree heights in such a forest is bimodal.

Data distributions that are symmetrical, or nearly so, such as the one in Figure 15.25, are called **symmetric**. Data sets such as the weights of children of a fixed age, the lengths of a certain species of fish in a lake, or the scores of students on a large, standardized test, tend to be symmetric. One especially important type of symmetric distribution arises from random samples of a population.

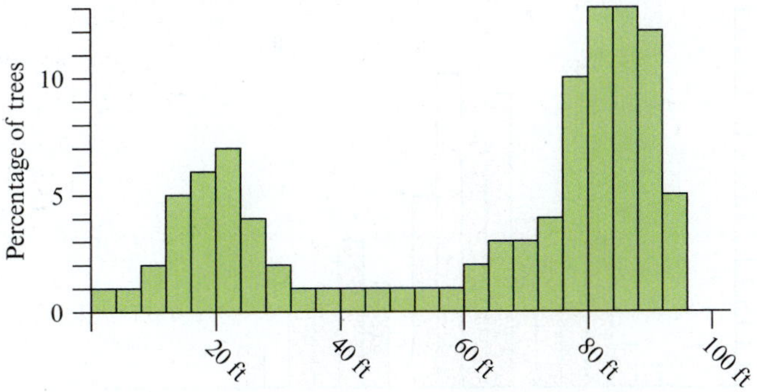

Figure 15.24

A bimodal distribution showing the hypothetical heights of trees in a forest that has an understory.

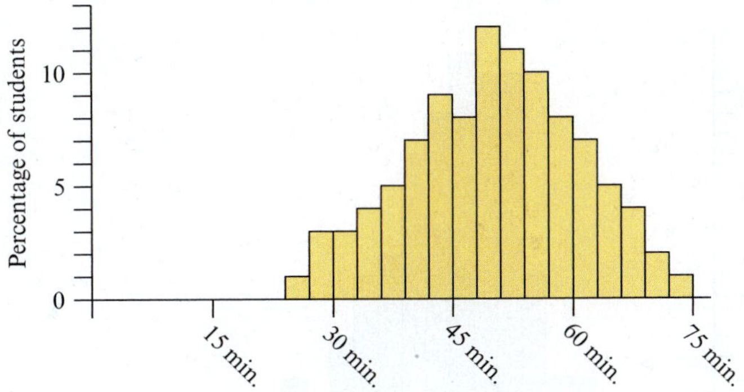

Figure 15.25

A symmetric distribution showing the time taken to complete a test.

What Are Characteristics of Distributions Arising from Random Samples of a Population?

Before you read on, try Class Activity 15R.

CLASS ACTIVITY

15R Distributions of Random Samples, p. CA-329

CCSS 7.SP.2

Suppose a population is divided into two categories. The population could be light bulbs produced at a factory and the categories could be "defective" and "not defective," or the population could be the deer in a region and the categories could be "infected" and "not infected" with a certain disease. For concreteness and simplicity, let's think of the population as voters, and let's say the categories are "yes" and "no," as if the voters will vote for or against a certain referendum. Now suppose we take a bunch of reasonably large random samples of the same size. In each case, we determine the percentage of the sample that voted "yes", and we plot these data in a histogram or a dot plot, as in **Figure 15.26**. An amazing and very useful fact is displayed: These random samples tend to form a distribution that is symmetric about the percentage of "yes" votes in the full population. Furthermore, the larger the sample size, the more tightly clustered the distribution tends to be about its center of symmetry. These facts about random samples allow statisticians to use samples to make predictions with a certain degree of confidence (e.g., when predicting the outcome of an election based on a poll). The details about making predictions based on samples is beyond the scope of this book, but they can be found in any book on statistics in a chapter on statistical inference.

Notice that both of the distributions in **Figure 15.26** are symmetrical, both have approximately the same mean, and both have approximately the same median. So although the shape, mean, and median are useful summaries of each data set, these pieces of information are not enough to

Figure 15.26

Random samples of voters.

Samples of 200 voters from a population of 100,000 voters, of which 45% vote "yes" (100 such samples were taken.)

Samples of 400 voters from a population of 100,000 voters, of which 45% vote "yes" (100 such samples were taken.)

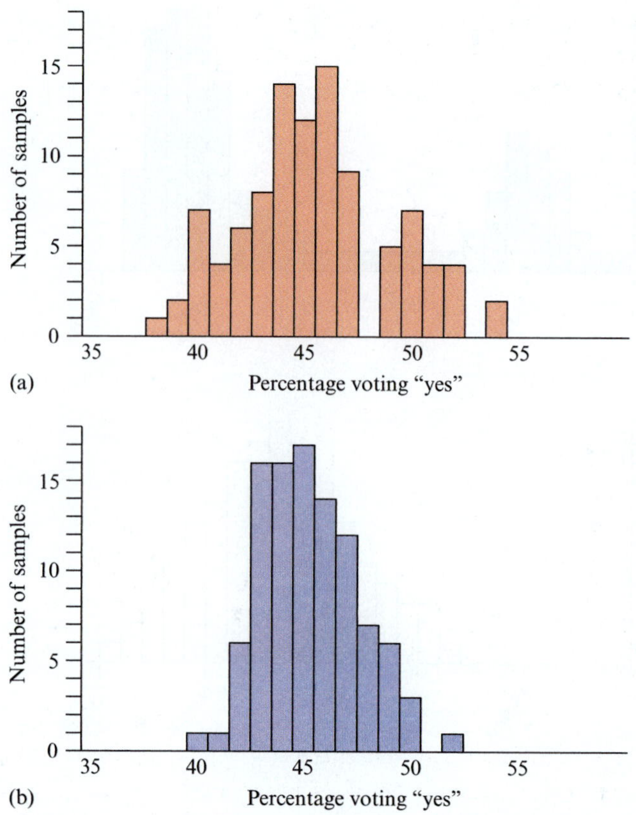

distinguish the two distributions. The two distributions are distinguished by how dispersed they are. The second distribution is less dispersed and more tightly clustered than the data in the first distribution. We can summarize this difference in dispersion with measures of variation, which we examine next.

How Can We Summarize Distributions with the Median, Interquartile Range, and Box Plots?

Try Class Activity 15S before you read on.

CLASS ACTIVITY

15S Comparing Distributions: Mercury in Fish, p. CA-332

In Class Activity 15S, you probably noticed that the means or medians of two data distributions may not provide enough information to compare the data adequately. It can be helpful to have additional information, such as the high and low values of the data, as well as information about how spread out the data are. Although a histogram can show such information, summarizing a histogram is also helpful. We can summarize data sets in several ways with the aid of quartiles.

CCSS
6.SP.3
6.SP.5c

Quartiles and Percentiles Consider the three dot plots in Figure 15.27. The mean and median of the data in all three dot plots is 7, so neither the mean nor the median help distinguish one data set from another. But quartiles are special kinds of percentiles that can help distinguish these data sets.

percentile A **percentile** is a generalization of the median. The 90th percentile of a set of numerical data is the number such that 90% of the data is less than or equal to that number and 10% is greater than or equal to that number. (If there is no single number that satisfies this criterion, then use the midpoint of all numbers that satisfy it, as in computing the median.) Other percentiles are defined similarly. The median of a list of numbers is the 50th percentile, because 50% of the list is at or below the median, and 50% is at or above the median. The **first quartile** is the 25th percentile. The **third quartile** is the 75th percentile.

first quartile
third quartile

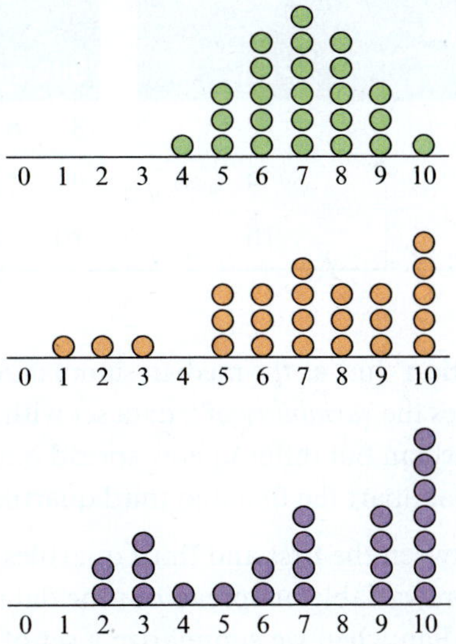

Figure 15.27 Dot plots of three different hypothetical test results, each with median 7 and mean 7.

How can we determine the medians and the first and third quartiles for the three data sets in Figure 15.27? Each data set consists of 24 numbers, and since 25% of 24 is 6, we should break the data after every 6th value to find the 25th, 50th, and 75th percentiles, as indicated in Figure 15.28. (See also Table 15.4.) Notice that in the second data set, the lowest 25% of the data ends at 5 and the rest of the data begin at 6, so the 25th percentile is 5.5, because 5.5 is halfway between 5 and 6. Similarly, in the third data set the 25th percentile is 4.5, which is halfway between 4 (where the lowest 25% of the data ends) and 5 (where the rest of the data begin).

Figure 15.28 Finding the 25th, 50th, and 75th percentiles of three sets of data.

75% of data
50% of data
25% of data

First data set:	4 5 5 5 6 6	6 6 6 7 7 7	7 7 7 8 8 8	8 8 9 9 9 10
Second data set:	1 2 3 5 5 5	6 6 6 7 7 7	7 8 8 8 9 9	9 10 10 10 10 10
Third data set:	2 2 3 3 3 4	5 6 6 7 7 7	7 9 9 9 9 10	10 10 10 10 10 10

↑ 25th percentile (first quartile)
↑ 50th percentile (median)
↑ 75th percentile (third quartile)

Results of standardized tests are often reported as percentiles. So if a student scores at the 60th percentile on a test, then that student has scored the same as or better than 60% of the students taking the test.

Refer again to the three data sets displayed in the dot plots in Figure 15.27. Even though we cannot distinguish the data sets by their medians or means because all of these numbers are 7, we can see a difference among the data sets when we examine their first and third quartiles in Table 15.4. From the table, we can see that the first data set is more tightly clustered around the median than the other two data sets, because the 25th and 75th percentiles are closer to the 50th percentile in the first data set than in the other two. So if we didn't have the dot plots for the three data sets, but we had the values of the 25th, 50th, and 75th percentiles, we could still get a sense of how the data are distributed in these data sets.

TABLE 15.4 The median and first and third quartiles for three data sets

Data Set in Figure 15.27	Percentile 25th	50th	75th	Interquartile Range
First data set	6	7	8	$8 - 6 = 2$
Second data set	5.5	7	9	$9 - 5.5 = 3.5$
Third data set	4.5	7	10	$10 - 4.5 = 5.5$

The Interquartile Range as Measure of Variation Just as the median summarizes a data set with a single number, the *interquartile range* summarizes the *variability* of a data set with a single number. The data sets in Figure 15.28 all have the same median but differ in how spread out or variable they are. We can detect this variability by seeing how far apart the first and third quartiles are.

interquartile range

The **interquartile range (IQR)** is the distance between the first and third quartiles (see again Table 15.4). The bigger the interquartile range, the more variable or spread out the data are. We usually use the interquartile range together with the median when we summarize a set of numerical data. The median tells us what is typical or representative for the data and the interquartile range tells us how dispersed the data are.

CLASS ACTIVITY

15T Using Medians and Interquartile Ranges to Compare Data, p. CA-333

box plot

Box Plots Another way to summarize how data are distributed is with a **box plot**, or *box-and-whiskers* plot. In simple cases, a box plot shows the location of the lowest data value, the 25th percentile (first quartile), the 50th percentile (median), the 75th percentile (third quartile), and the highest data value. A box is drawn from the 25th percentile to the 75th percentile, and "whiskers" are drawn from the lowest data value to the 25th percentile and from the 75th percentile to the highest data value, as in Figure 15.29, which shows box plots for the three data sets in Figure 15.27.

Figure 15.29

Box plots for the three data sets in Figure 15.27.

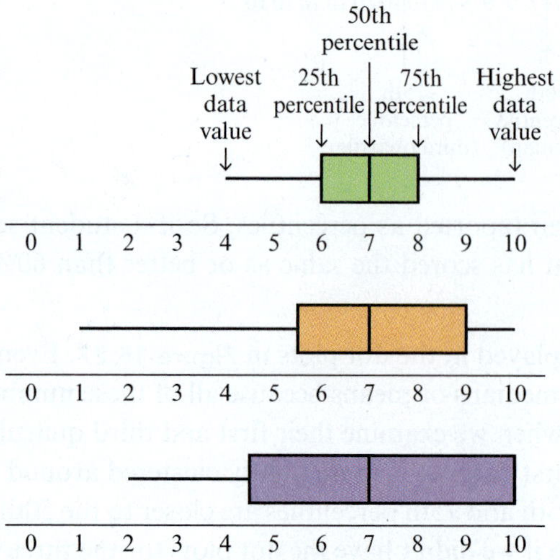

CCSS 6.SP.4

Although we will not work with examples of this type here, in general, the whiskers in a box plot extend to the lowest and highest data values *that are not outliers* instead of to the lowest and highest values. Informally, an **outlier** is a data value that is much higher or lower than most of the data. An outlier is indicated in a box plot by placing a star at the location of its value.

Notice that even if we didn't have the dot plots in Figure 15.27, we could still obtain information about how the data are distributed from the box plots in Figure 15.29. We can see that the first data set is clustered more closely around 7 and is spread over a narrower range than either of the other two data sets.

A box plot shows at a glance where the middle 50% of the data lie. This is so because the data that lie between the 25th and 75th percentiles form the middle 50% of the data $(75\% - 25\% = 50\%)$. Box plots highlight the median and allow us to determine quickly whether the middle 50% of the data are symmetrically distributed about the median.

CLASS ACTIVITY

15U Using Box Plots to Compare Data, p. CA-334

Percentiles Versus Percent Correct A common source of confusion is the distinction between a student's percentile on a test and what percent of the test the student got correct. The percentile and the percent correct report very different things about a student's performance on a test. When a student's performance on a test is reported as a percentile, that information tells us how the student did in comparison with other students who took the test. If a student scored at the 75th percentile, then that means that 75% of the students who took the test had the same or a lower score than this student. But this percentile information only tells us how the student did relative to other students who took the test; it doesn't tell us what percent of the test the student did correctly. For example, it could be that the student got 90% of the test correct and that 75% of the students who took the test scored 90% or less too. Or it could be that the student got 50% of the test correct and that 75% of the students who took the test scored 50% or less too.

In interpreting students' scores on standardized tests, be sure to distinguish between percentiles and percent correct.

CLASS ACTIVITY

15V Percentiles Versus Percent Correct, p. CA-335

How Can We Summarize Distributions with the Mean and the Mean Absolute Deviation (MAD)?

CCSS
6.SP.3
6.SP.5c

We saw that when we use the *median* to summarize a data set with one number it is natural to use the interquartile range to describe how spread out the data are. But when we use the *mean* to summarize a data set, it is more natural to use other measures to describe the variability of a set of numerical data. One such measure that we can use together with the mean to summarize a set of numerical data is the **mean absolute deviation (MAD)**—the mean of the distances of the data from the mean. So to find the MAD, carry out these steps, as illustrated with the data set

mean absolute deviation (MAD)

$$3, \quad 4, \quad 6, \quad 7, \quad 10$$

1. Calculate the mean:

$$(3 + 4 + 6 + 7 + 10) \div 5 = 6$$

2. For each piece of data, find its distance to the mean:

$$\begin{array}{ccccc} 3 & 4 & 6 & 7 & 10 \\ \downarrow & \downarrow & \downarrow & \downarrow & \downarrow \\ 3 & 2 & 0 & 1 & 4 \end{array}$$

3. The MAD is the mean of those distances:

$$(3 + 2 + 0 + 1 + 4) \div 5 = 2$$

So the data set 3, 4, 6, 7, 10 has mean 6 and MAD 2. When data are presented in a dot plot, it is nice to compute the MAD as indicated in Figure 15.30.

When we use the mean and the MAD to summarize a data set, the mean tells us what is typical or representative for the data and the MAD tells us how dispersed the data are. The bigger the MAD, the more spread out the data are.

CLASS ACTIVITY

15W Comparing Paper Airplanes, p. CA-336

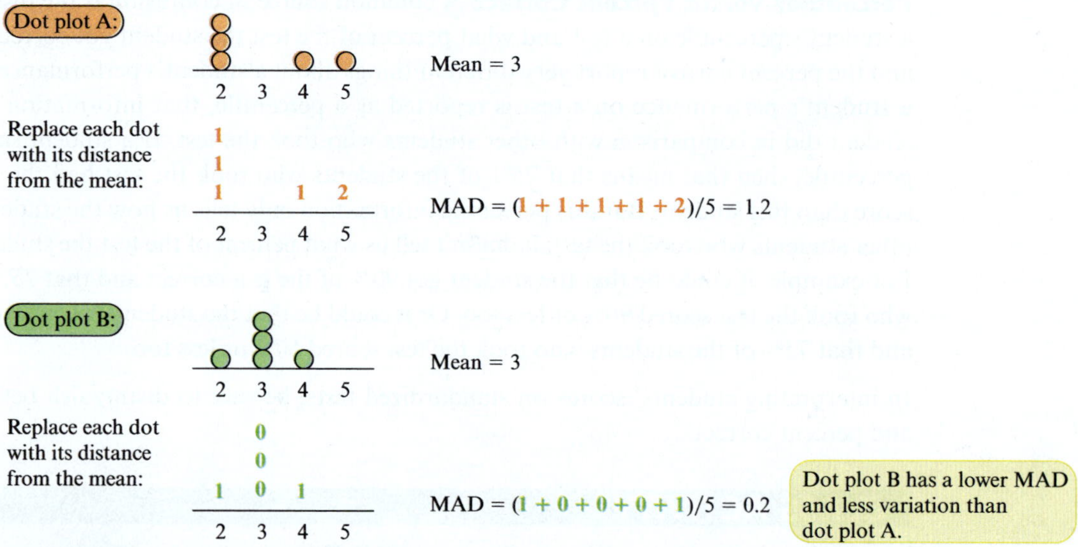

Figure 15.30 Computing the MAD from a dot plot by first replacing dots with their distances from the mean.

How Can We Use Measures of Center and Variation to Make Comparisons?

CCSS 7.SP.3 7.SP.4

Measures of center and variation are especially important for investigating differences between groups, such as a treatment group and a control group in an experiment. For example, suppose researchers want to determine whether giving students grapes and water before taking a test helps the students do better. They randomly assign students to a grapes-and-water group or a control group, give all the students a 10-point quiz, and find that the mean scores are 7 for the grapes-and-water group and 6 for the control group. So on average, the grapes-and-water group scored 1 point higher than the control group. But how much of an effect did the grapes and water really have on the students' performance?

Before you read on, try Class Activity 15X to think about why we should consider variability when making comparisons between groups.

CLASS ACTIVITY

15X Why Does Variability Matter When We Compare Groups?, p. CA-337

Let's return to the experiment in which a grapes-and-water group scored an average of 1 point higher than a control group on a 10-point quiz. **Figure 15.31** shows two hypothetical, idealized scenarios to illustrate why variability matters in comparing two groups.

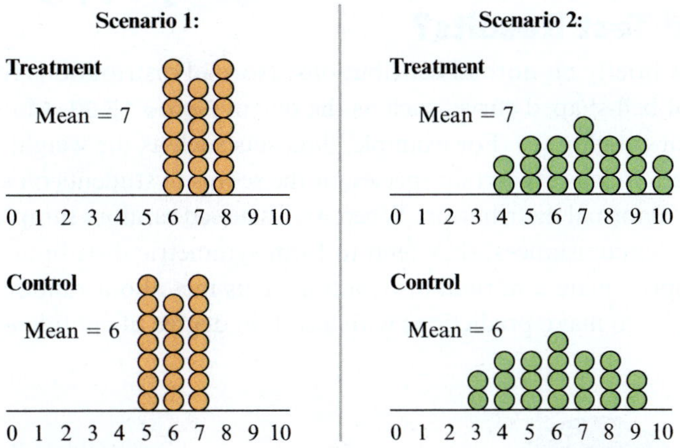

Figure 15.31 Two hypothetical scenarios in which the means of treatment and control groups differ by 1 point.

In both scenarios, the difference in means is 1 point, but that 1 point makes more of a difference in the first scenario than in the second. There is much less overlap in the two distributions in the first scenario than in the second because there is much less variability in the first scenario than in the second. How can we quantify this distinction? Because the MAD is a measure of variation, let's relate the difference in means to the MAD, as summarized in **Figure 15.32**.

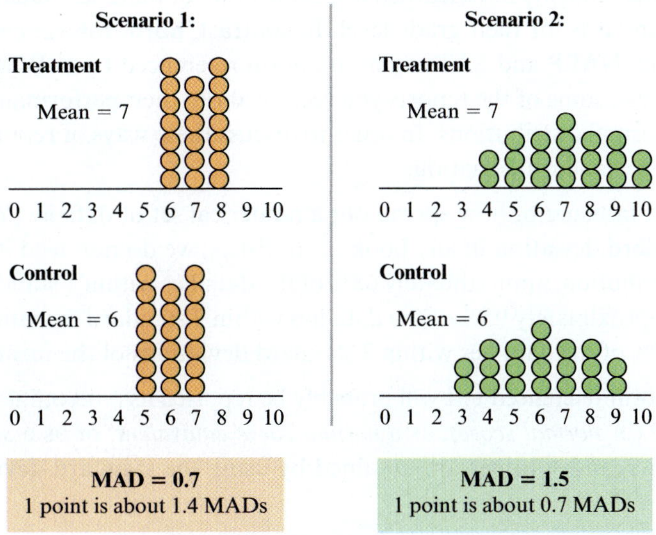

Figure 15.32 Comparing a 1 point difference in means to the MAD.

In Scenario 1, the MAD of each distribution is 0.7, so the 1 point difference in the means is *more than* 1 MAD. It is about 1.4 MADs. In Scenario 2, the MAD of each distribution is 1.5, so the 1 point difference in means is *less than* 1 MAD. It is about 0.7 MADs. Viewing the difference in the means in terms of the MAD gives us a way to describe how important the 1 point difference is in terms of the scenario. In Scenario 1 the grapes and water had an effect of 1.5 MADs, which is more than the effect of 0.7 MADs in Scenario 2.

CLASS ACTIVITY

15Y How Well Does Dot Paper Estimate Area?, p. CA-339

What Are Normal Curves and How Do They Apply to Standardized Test Results?

In this section, we focus briefly on normal distributions. Normal distributions form the shape of the familiar symmetrical bell-shaped curve, such as the one in Figure 15.33. Many data sets have a normal distribution (or roughly so). For example, data sets such as the weights of children of a fixed age, the length of adult fish of a certain species, or the scores of students on a large, standardized test, generally have a normal distribution. When we discussed random samples previously, we noted that under certain circumstances, they tend to form symmetric distributions. In fact, these distributions tend to approximate a normal distribution. This fact about random samples allows statisticians to use samples to make predictions with a certain degree of confidence.

Figure 15.33

A normal distribution.

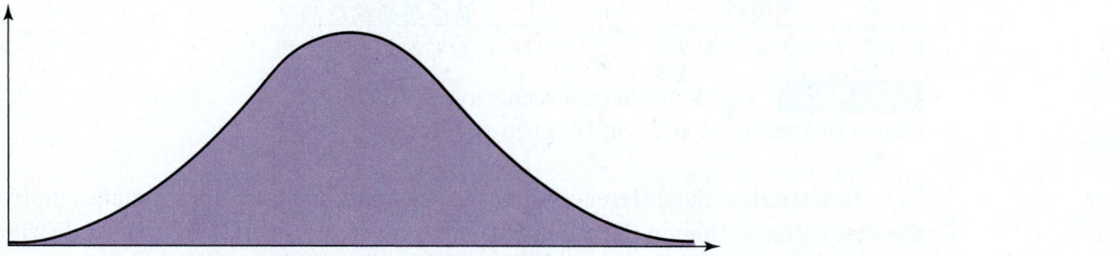

When you teach, your students may take different kinds of large-scale standardized tests. They may take criterion referenced tests as well as norm referenced tests. **Criterion referenced tests** are used to determine whether students have learned a specific body of material, such as subject material specified in state standards for their grade level. In contrast, **norm referenced tests** are used to compare students. The NAEP and SAT exams are norm referenced tests. When your students take a norm referenced test, some of the reports you receive about their performance may be based on information about normal distributions. In order to discuss these ways of reporting test results, we must first consider the standard deviation.

criterion referenced tests

norm referenced tests

A **standard deviation** is a measure of how spread out a numerical set of data is. (You can read the exact definition of standard deviation in any book on statistics; we do not need it here.) For data that are in a normal distribution, approximately 68% of the data lies within 1 standard deviation of the mean of the data, approximately 95% of the data lies within 2 standard deviations of the mean, and approximately 99.7% of the data lies within 3 standard deviations of the mean.

A student's score on a norm referenced test will probably be reported as a percentile, but it may also be reported as a *z-score* (or *normal score*), as a *normal curve equivalent*, or as a *stanine*. Z-scores, normal curve equivalents, and stanines are obtained by using the standard deviation to rescale test scores.

A student's **z-score** tells how many standard deviations away from the mean the student's test score is. So a z-score of 1.5 is 1.5 standard deviations above the mean, and a z-score of −0.6 is 0.6 standard deviations below the mean. If a test had an average of 250 points and a standard deviation of 50 points, then a student with a z-score of 1.5 scored

$$250 + 1.5 \cdot 50 = 325$$

points on the test, and a student with a z-score of −0.6 scored

$$250 - 0.6 \cdot 50 = 220$$

points on the test. See Figure 15.34 for the locations of the z-scores 0, 1, 2, −1, and −2.

Figure 15.34 Z-scores, normal curve equivalents, and stanines for a normal distribution.

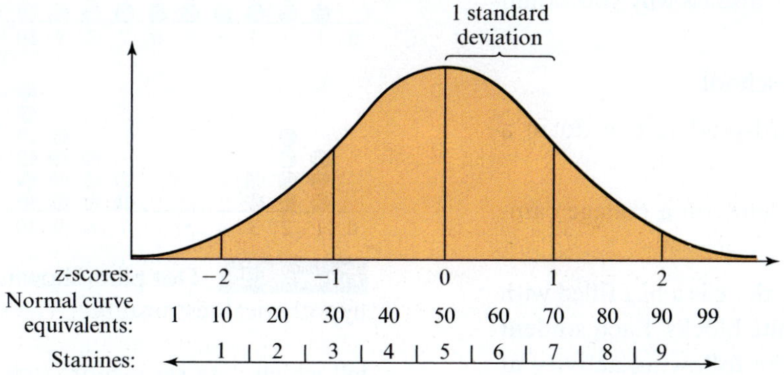

A student's **normal curve equivalent** score on a test is a score between 1 and 99. Normal curve equivalent scores are like z-scores, except that they are scaled so that a student who scores the average score on the test receives a 50 and a student who scores at the 99th percentile receives a 99. To do this, it turns out that 1 standard deviation needs to be 21.06 points. So for a student who has a z-score of 1.5, the normal curve equivalent score is

$$50 + 1.5 \cdot 21.06 = 81.59$$

or 82 points. See Figure 15.34 for locations of some normal curve equivalent scores.

A student's **stanine** score on a test is a whole number score between 1 and 9. Except for the 1st and 9th stanine, each stanine covers $\frac{1}{2}$ of a standard deviation worth of scores.

Students who score within $\frac{1}{4}$ of a standard deviation from the mean receive a stanine score of 5. See Figure 15.34 for the locations of other stanines.

SECTION SUMMARY AND STUDY ITEMS

Section 15.4 Summarizing, Describing, and Comparing Data Distributions

Data can be distributed in different ways, so two sets of data with the same means and medians can be quite different. Measures of variation such as the interquartile range and the mean absolute deviation (MAD) can help to discern differences in such data sets.

Quartiles and percentiles are generalizations of the median. A student's percentile on a test is not the same as the percent the student got correct on the test. Reports about student performance on standardized tests often include percentiles and may also include other measures that are based on rescaling normal distributions.

Key Skills and Understandings

1. Indicate the shape a data distribution is likely to take, especially in the case of random samples.
2. Given a data set, determine the median, the first and third quartiles, and the interquartile range. Use medians and interquartile ranges to discuss and compare data sets.
3. Make box plots and use box plots to discuss and compare data sets.
4. Discuss the difference between percentiles and percent correct.
5. Given a data set, determine the mean and the mean absolute deviation (MAD). Use means and MADs to discuss and compare data sets.

Practice Exercises for Section 15.4

1. For each of the following situations, describe what shape you would expect a histogram or dot plot of the data to take, and briefly discuss why you would expect that shape.
   a. Student arrival times at school
   b. Number of sit-ups third-graders can do in a minute
   c. Amount of money students on a college campus carry with them
2. At a math center in a class, there is a bag filled with 35 black blocks and 15 white blocks. Each student in a class of 30 will do the following activity at the math center: randomly pick 10 blocks out of the bag without looking and write the number of black blocks picked on a sticky note.

   Make a hypothetical dot plot that could arise from this situation. Make your dot plot so that it has characteristics you would expect to see in a dot plot arising from actual data. Briefly describe these characteristics of the dot plot.
3. a. Determine the median, the first and third quartiles, and the interquartile range for the hypothetical test scores shown in the dot plots of Figure 15.35.

   b. Suppose you only have the median and interquartile range from part (a) and you don't have the dot plots in Figure 15.35. Discuss what you can tell and not tell about the data sets. Can you tell which data set is more tightly clustered and which is more dispersed? Can you tell about symmetry or gaps in the data?

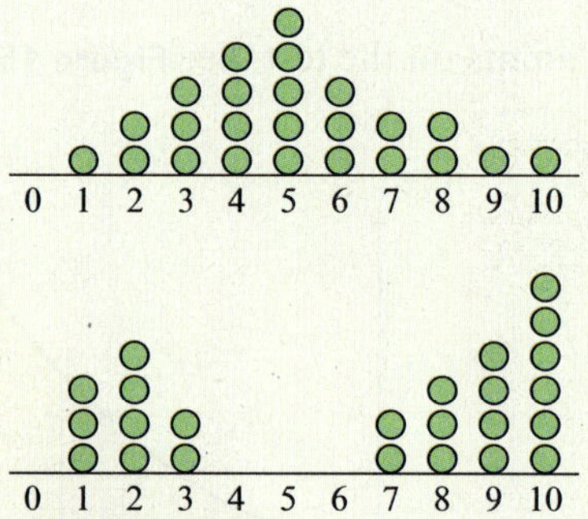

Figure 15.35 Dot plots showing hypothetical test results.

4. Some participants in a focus group watched two proposed new TV shows and rated each show on a 0 to 5 scale. For show A, the mean rating was 3.6 and the MAD was 0.7. For show B, the mean rating was 3.4 and the MAD was 0.9. Discuss what these numbers tell us.
5. A test is given to students across the country. The test has a distribution that is approximately normal. How can we characterize a student's performance on this test relative to other students if the student's score is 1 standard deviation above the mean? What if the score is 1 standard deviation below the mean? What if the score is 2 standard deviations above the mean?

Answers to Practice Exercises for Section 15.4

1. a. A few students arrive early, but most students probably arrive just before school starts. So these data would probably be skewed to the left, with a long tail to the left consisting of those students who arrive early.

   b. These data will probably be symmetrical around the mean number of sit-ups that third-graders can do.

   c. Most students probably don't carry a lot of money with them, so the data will probably be skewed to the right, with a long tail to the right consisting of those students who do have a lot of money on them.
2. See Figure 15.36. Since 70% of the blocks in the bag are black and since the samples are random samples, we expect that in most of the samples, the percent of black blocks will also be close to 70%. The dot plot should be roughly symmetrical around 7.

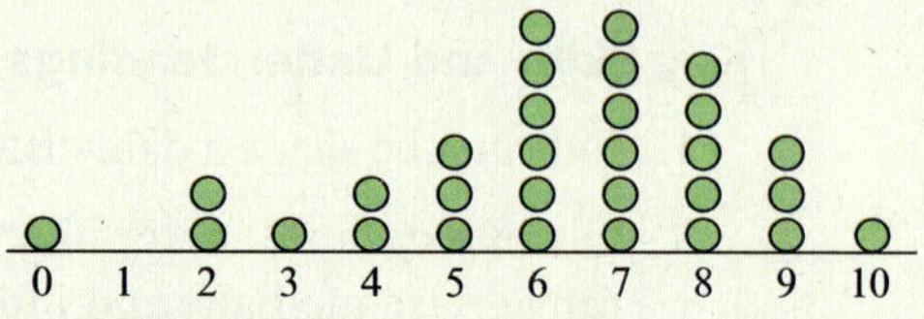

Figure 15.36 Dot plot of the number of black blocks picked in samples of 10 by 30 students from a bag in which 70% of the blocks were black.

3. a. For both of the hypothetical tests, there are 24 test scores. Since 25% of 24 is 6, the 25th percentile (first quartile) is the score between the score of the 6th lowest and 7th lowest test scores, the 50th percentile (median) is the score between the 12th and 13th lowest scores, and the 75th percentile (third quartile) is between the 18th and 19th lowest scores. The following table shows the results:

	First Quartile 25th Percentile	Median 50th Percentile	Third Quartile 75th Percentile	Interquartile Range
Test 1	3.5	5	6.5	6.5 − 3.5 = 3
Test 2	2	8	9.5	9.5 − 2 = 7.5

b. Because the first data set has a smaller interquartile range, we can tell it is more tightly clustered and less dispersed than the second data set. But we can't tell from the median and interquartile range alone whether the test scores on the second test are distributed in a long tail or have a gap and form a second clump (which they do). We also can't tell whether there is symmetry (as in the first distribution).

4. Comparing the means, which are single-number summaries of the data sets, we see that show A was rated more highly than show B. Comparing the MADs, we see there was more variation in the ratings for show B because its ratings had a higher MAD. Because the difference in the means (0.2) is small compared with the variability in the data (as expressed by the MADs, which are 0.7 and 0.9), we can't be very confident that overall, the participants really do prefer show A to show B.

5. In a normal distribution, approximately 68% of the data lie within 1 standard deviation of the mean. Since the data are symmetrical about the mean, half of the 68%, namely 34% of the scores, lie above the mean. Another 50% of the scores lie below the mean. So a student who scored 1 standard deviation above the mean scored higher than about 50% + 34% = 84% of the students who took the test. In comparison with the other students, this student did quite well.

Similarly, a student who scored 1 standard deviation below the mean scored lower than about 84% of the students who took the test. In comparison with the other students, this student did poorly.

In a normal distribution, approximately 95% of the data lie within 2 standard deviations of the mean. Since the data are symmetrical about the mean, half of the 95%—namely about 48% of the scores—lie above the mean. Another 50% of the scores lie below the mean. So a student who scored 2 standard deviations above the mean scored higher than about 50% + 48% = 98% of the students who took the test. In comparison with the other students, this student did exceptionally well.

PROBLEMS FOR SECTION 15.4

1. What is the difference between scoring in the 90th percentile on a test and scoring 90% correct on a test? Discuss this question carefully, giving examples to illustrate.

2. What is the purpose of reporting a student's percentile on a state or national standardized test? How is this purpose different from reporting the student's percent correct on a test?

3. The three histograms in Figure 15.37 show the hypothetical performance of students in three different school districts on the same test. A score below 40 on the test is considered failing. A score of 80 or above is considered excellent.

a. Estimate the mean score on the test for each school district by viewing the mean as a balance point, as discussed in Section 15.3.

b. Discuss what information you can glean from the histograms that wouldn't be apparent just from knowing the mean or median scores on the test.

c. Discuss how each school district could argue that it did better than at least one other school district.

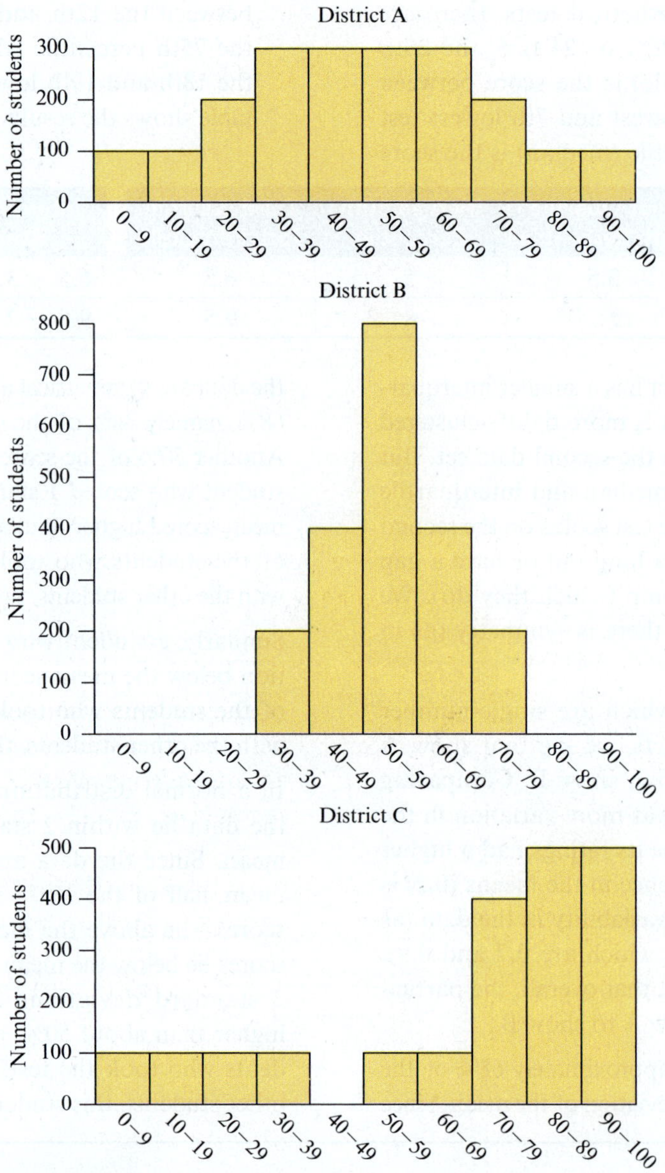

Figure 15.37 Performance of three hypothetical school districts on a test.

4. At a math center in a class, there is a bag filled with 40 red blocks and 10 blue blocks. Each child in the class of 25 will do the following activity at the math center: Randomly pick 10 blocks out of the bag without looking, and write on one sticky note the number of red blocks and the number of blue blocks picked.

   a. Describe a good way to display the data for the whole class. Your proposed display should be a realistic and practical way to show every child's piece of data (in a class of 25).

   b. Sketch a graph that could be the graph you proposed in part (a). Briefly describe the characteristics you expect the graph to have.

   c. Write at least four questions about your hypothetical graph in part (b) that the teacher could ask the children. Include at least one question at each of the three levels discussed in Section 15.2 (Read the data, Read between the data, and Read beyond the data). Label each question with its level. Answer your questions.

5. Refer to Figure 15.26 on page 706.

   a. Refer to the first histogram: samples of 200. What percent of those samples would predict correctly that 45% of the voters would vote "yes"? Explain briefly.

b. Refer to the second histogram: samples of 400. What percent of those samples would predict correctly that 45% of the voters would vote "yes"? Explain briefly.

c. Refer to the first histogram: samples of 200. What percent of those samples have percentages of "yes" votes between 43% and 47%? Explain briefly.

d. Refer to the second histogram: samples of 400. What percent of those samples have percentages of "yes" votes between 43% and 47%? Explain briefly.

6. Refer to **Figure 15.26** on page 706.

a. Refer to the first histogram: samples of 200. What percent of those samples would predict (incorrectly) that more than 50% of the voters would vote "yes"? Explain briefly.

b. Refer to the second histogram: samples of 400. What percent of those samples would predict (incorrectly) that more than 50% of the voters would vote "yes"? Explain briefly.

7. Refer to **Figure 15.26** on page 706.

a. Write at least three questions that can be asked about either the first or the second histogram. Include at least one question for each of the three graph-reading levels discussed in Section 15.2. Answer each question for the first histogram and for the second histogram (as best you can).

b. Write a "read beyond the data" question about the two histograms together. Answer your question as best you can.

8. Use the NAEP long-term trend data about mathematics performance of 9-year-olds in **Table 15.5** to make box plots for the years 1978, 1992, 2004, and 2008. You may assume that in each year, scores ranged from 0 to 400. Use your box plots to compare the mathematics performance of 9-year-olds in these years.

TABLE 15.5 NAEP long-term trend mathematics scores for 9-year-olds at selected percentiles in selected years

Percentile	1978	1992	2004	2008
25th	195	208	220	222
50th	220	231	243	246
75th	244	253	264	266

9. Determine the median and interquartile range for the amounts of vitamin C found in each of the two different hypothetical brands of vitamin pills shown in **Figure 15.38**. For each brand, 40 samples were tested. Then use the medians and interquartile ranges to compare the two brands of vitamins.

10. a. Make a box plot for the data in **Figure 15.38**.

b. Suppose you only have the box plot from part (a) and you don't have the dot plots in **Figure 15.38**. Use the box plots to discuss how the data sets compare.

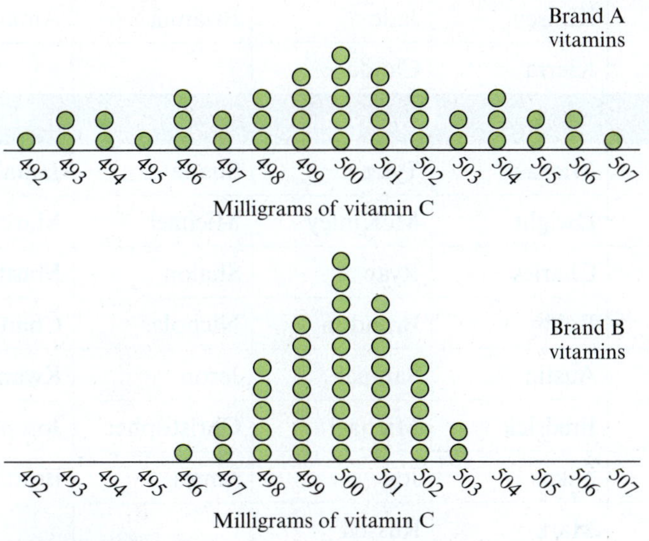

Figure 15.38 Amount of vitamin C found in hypothetical samples of two different brands of vitamin pills.

11. Determine the medians and interquartile ranges for the hypothetical test scores shown in each of the dot plots of Figure 15.39. Each dot plot has 28 pieces of data. Use the medians and interquartile ranges to discuss and compare the results on the two tests.

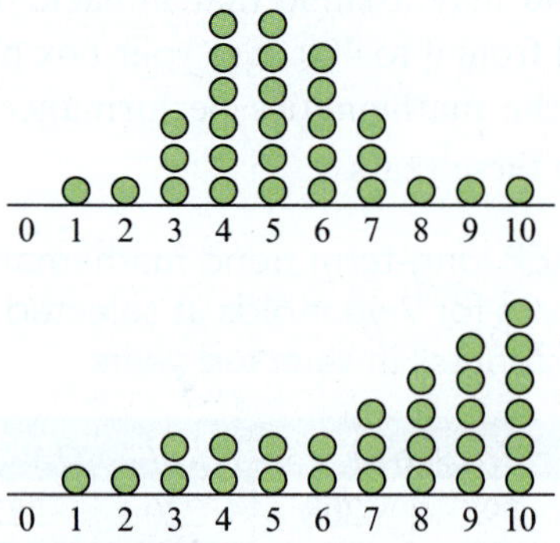

Figure 15.39 Dot plots showing hypothetical test results.

12. a. Make box plots for the dot plots in Figure 15.39.

b. Suppose you only have the box plot from part (a) and you don't have the dot plots in Figure 15.39. Use the box plots to discuss how the data sets compare.

13. A fifth-grade class is wondering if girls' names tend to be longer than boys' names. Table 15.6 shows the names of all the fifth-grade girls and boys at the school.

a. Display the lengths of the fifth-grade girls' and boys' names in dot plots. Compare the dot plots.

b. Write and answer at least four questions about your dot plots from part (a), including at least one question at each of the three graph-reading levels discussed in Section 15.2 (Read the data, Read between the data, and Read beyond the data).

TABLE 15.6 Fifth-grader names

Girls' Names				
Felecia	Laquiera	Kayla	Sandra	Brittany
Shanique	Kelly	Darica	Crystal	Sasha
Shana	Gloria	Autumn	Katlyn	Shalamar
Shelli	Lauren	Alexis	Keshaundra	Brittney
Brandi	Kendra	Olivia	Sierra	Beth
Jessica	Ebonique	Christiana	Talia	Morgan
Ashley	Kimsey	Jade	Brianna	Amanda
Adebola	Kierra	Candace		
Boys' Names				
Kody	Michael	Tyler	Shane	Jamal
Adam	Dwight	McKinley	Michael	Marcus
Troy	Charles	Ryan	Shalon	Matthew
Mario	Bobby	Brandon	Nicholas	Chandler
Ali	Austin	Samuel	Jeron	Kwame
Everett	Fredrick	Michael	Christopher	Joseph
Tyler	Nikos	Jon	James	Bradley
Orlando	Mark	Russell		

c. Display the lengths of the fifth-grade girls' and boys' names in box plots.

d. Compare the box plots from part (c). What do they tell you about the fifth-grade girls' and boys' names?

14. *Target drop problem.* For this problem, tape two pieces of paper together and draw a target like the one in **Figure 15.40**. Get a paperclip and another small object (e.g., a crumpled piece of tissue) that you can drop onto the target.

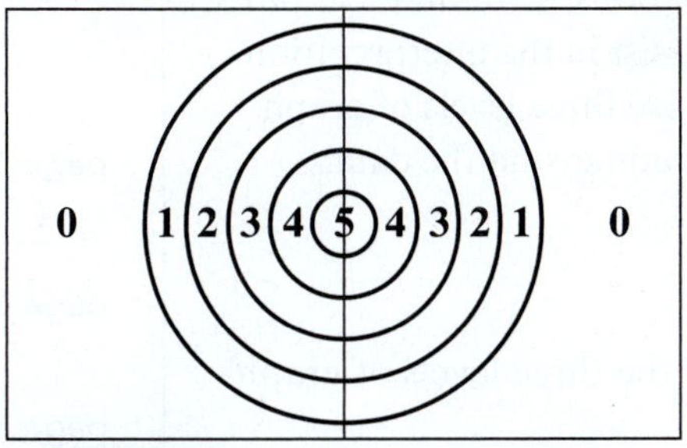

Figure 15.40 A target for dropping paperclips and other small objects onto.

a. Pick a distance above the target and drop your paperclip onto the target 20 times from that distance. Record the number of the ring that your paperclip falls in and make a dot plot of the data.

b. Repeat part (a) with a different distance above the target.

c. Repeat parts (a) and (b) with another small object.

d. Discuss your data using statistical tools from this section.

15. a. Compute the mean and the MAD for each of the dot plots in **Figure 15.41**.

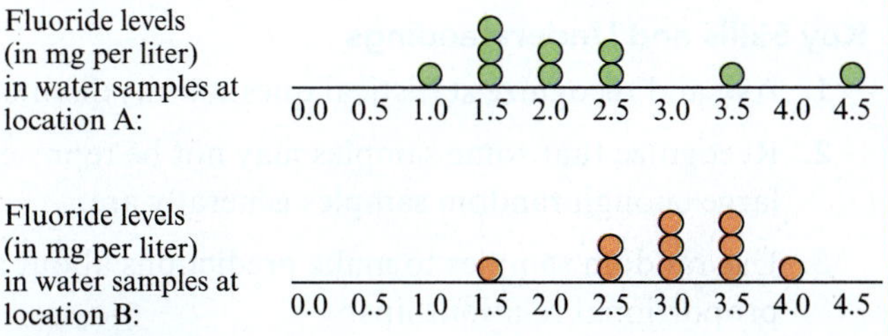

Figure 15.41 Dot plots of fluoride levels in water at two locations.

b. Use the means and MADs from part (a) to discuss how the fluoride levels at the two locations compare. What do the means tell us? What do the MADs tell us? How confident can we be in distinguishing the levels of fluoride in the water at the two locations?

16. On a state-wide test graded on a 10-point scale, the seventh-graders at school A had mean score 7.5 with a MAD of 1.2 and the seventh-graders at school B had a mean score of 8.5 with a MAD of 1.1. Discuss what these numbers tell us.

17. A 400-point test is given to a group of students. The mean score is 240. A student scores a 300 on the test.

a. Is it possible that the student's score falls within the middle 50% of the scores on the test? Explain, discuss, and illustrate your answer using tools and terminology developed in this section.

b. Is it possible that the student scored higher than 90% of the students taking the test? Explain, discuss, and illustrate your answer using tools and terminology developed in this section.

CHAPTER SUMMARY

Section 15.1 Formulating Statistical Questions, Gathering Data, and Using Samples

page 674

- A statistical study begins by formulating a question that anticipates variability and that can be answered (at least partially) by collecting and analyzing data. In statistics, there are observational studies and experiments; both involve considering a population. Often it is not practical to study the entire population of interest, so a sample is studied instead. Large enough random samples are generally representative of the full population. Therefore, one can reason about proportional relationships to make reasonable inferences about a full population on the basis of a random sample.

page 674

Key Skills and Understandings	
1. Ask and recognize statistical questions as distinct from other questions.	page 674
2. Recognize that some samples may not be representative of a whole population, but that large enough random samples generally are.	page 676
3. Use random samples to make predictions about a full population by reasoning about proportional relationships.	page 676
Section 15.2 Displaying Data and Interpreting Data Displays	page 681
▪ Common displays of categorical data include real graphs, pictographs, bar graphs, and pie graphs. Displays of numerical data include dot plots, histograms, stem-and-leaf plots, line graphs, and scatterplots. All such graphical displays can assist in the interpretation of data. To help students think about data displays, one can pose three levels of graph-reading questions: read the data, read between the data, and read beyond the data.	page 681
Key Skills and Understandings	
1. Make data displays to help convey information about data.	page 682
2. For a given data display, formulate and answer questions at the three levels of graph reading.	page 689
3. Recognize erroneous or misleading data displays.	page 689
Section 15.3 The Center of Data: Mean, Median, and Mode	page 693
▪ The mean (average) and median are good one-number summaries of a collection of numerical data. We can think of the mean in terms of "leveling out" the data. This way of thinking about the mean agrees with the way we calculate the mean. It also provides a way of solving problems about the mean. Another way to think about the mean is as a "balance point." The mean and median can be the same or different for a given set of data.	page 693
Key Skills and Understandings	
1. Describe how to view the mean as leveling out, and explain why this way of viewing the mean agrees with the way we calculate the mean (by adding and then dividing).	page 694
2. Use the leveling out view of the mean to solve problems about the mean. Also use the standard way of calculating the mean to solve problems about the mean.	page 694
3. View the mean as a balance point and understand that this point of view is especially appropriate for data displayed in histograms and dot plots.	page 695
4. Create data sets with different means and medians.	page 696
5. Discuss errors that students commonly make with the mean and the median.	page 697
Section 15.4 Summarizing, Describing, and Comparing Data Distributions	page 703
▪ Data can be distributed in different ways, so two sets of data with the same means and medians can be quite different. Measures of variation such as the interquartile range and the mean absolute deviation (MAD) can help to discern differences in such data sets.	page 703
▪ Quartiles and percentiles are generalizations of the median. A student's percentile on a test is not the same as the percent the student got correct on the test. Reports about student performance on standardized tests often include percentiles and may also include other measures that are based on rescaling normal distributions.	page 706

Key Skills and Understandings

1. Indicate the shape a data distribution is likely to take, especially in the case of random samples.	page 704
2. Given a data set, determine the median, the first and third quartiles, and the interquartile range. Use medians and interquartile ranges to discuss and compare data sets.	page 706
3. Make box plots and use box plots to discuss and compare data sets.	page 708
4. Discuss the difference between percentiles and percent correct.	page 709
5. Given a data set, determine the mean and the mean absolute deviation (MAD). Use means and MADs to discuss and compare data sets.	page 709

16

Probability

The study of probability arises naturally in a variety of situations. Games in which we flip coins, roll dice, spin spinners, or pick cards all have an element of chance. We don't know what the spinner will land on, and we don't know how the dice will roll. This element of chance adds excitement to the game and makes it easy to design class activities and problems that present probability in a way that can be fun for students to learn.

If probability were used only in analyzing games of chance, it would not be an important subject. But probability has far wider applications. In business and finance, probability can be used in determining how best to allocate assets. In medicine, probability can be used to determine how likely it is that a person actually has a certain disease, given the outcomes of test results.

In this chapter, we focus on the following topics and practices within the *Common Core State Standards for Mathematics* (*CCSSM*).

Standards for Mathematical Content in the CCSSM

In the domain of *Statistics and Probability* (Grades 6–8), students learn that the probability of a chance event expresses the likelihood that the event will occur. They collect data on chance processes and they recognize that the fraction of times an event is observed to occur approximates the event's probability. They find probabilities, including those for compound events. They use organized lists, tables, and tree diagrams to display and analyze compound events and to determine their probabilities.

Standards for Mathematical Practice in the CCSSM

Opportunities to engage in all eight of the Standards for Mathematical Practice described in the CCSSM occur throughout the study of probability, although the following standards may be especially appropriate for emphasis:

- **1 Make sense of problems and persevere in solving them.** Students engage in this practice when they solve challenging probability problems, both practical and whimsical.
- **2 Reason abstractly and quantitatively.** Students engage in this practice when they use the definition of multiplication in reasoning about compound events and when they capitalize on the relationship between the long term relative frequency of an event and its theoretical probability, applying it in both directions.
- **4 Model with mathematics.** Students engage in this practice when they make a probability model for a chance process and use it to make predictions about likelihoods of events.

(From Common Core Standards for Mathematical Practice. Published by Common Core Standards Initiative.)

16.1 Basic Principles of Probability

CCSS **Common Core State Standards** Grade 7

In this section, we focus on some of the basic principles of probability. We begin with some simple contexts in which to introduce basic ideas of probability. We then discuss how to use basic principles of probability to calculate probabilities in simple situations. Finally, we examine the relationship and distinction between theoretical and experimental probability.

In probability, we study chance processes, which concern experiments or situations where we know which outcomes are possible, but we do not know precisely which outcome will occur at a given time. Examples include flipping a coin and seeing if it lands with a head or a tail facing up, spinning a spinner and seeing which color the arrow lands on, and randomly picking a person out of a group and determining if the person has a certain disease.

sample space

For a chance process, such as flipping a coin or spinning a spinner, the set of different possible individual outcomes is called a **sample space**. For example, the sample space for a board game's spinner is the set of colors red, blue, yellow, green if those are all the possible colors that the spinner can land on. We often consider not only individual outcomes of a chance process but also **events**, which are collections of outcomes. For example, spinning either red or blue is an event.

events

probability (theoretical)

The theoretical **probability** of a given event is a number quantifying how likely that event is; it is the fraction or percentage of times that event should occur ideally. For example, the probability that a coin lands on heads is the fraction of times that the coin should ideally land on a head when tossed many times.

probability model

When we study a chance process, we usually seek to develop a **probability model**, which means that we make assumptions about the situation (e.g., all outcomes are equally likely) and we want to determine a probability for each outcome in the sample space.

CCSS 7.SP.5

Probabilities are always between 0 and 1, or equivalently, when they are given as percentages, between 0% and 100%. A probability of 0% means that event will not occur, and a probability of 100% means that event is certain to occur. The greater the probability, the more likely an event is to occur. A probability of 50% means the event is as likely to occur as not to occur. For example, when we flip a coin, the probability of getting *either* a head *or* a tail is 100% because a head or a tail is certain to occur. The probability of getting a head is 50%, or $\frac{1}{2}$, because heads and tails are equally likely to occur. For example, if you toss a coin 100 times, then ideally, the coin would land on heads 50 times.

If we have a coin in hand, how do we know for sure that a coin is *fair* (i. e., that heads and tails are equally likely to occur when we flip the coin)? In fact, we never know for sure that the coin is fair; however, unless the coin has been weighted lopsidedly, there is no reason to believe that either heads or tails should be more likely to occur than the other. If it were important for the coin to be fair—for example, if the coin were to be used in a state lottery—then the coin would be flipped many times, perhaps thousands of times, to determine if the coin is likely to be fair. Statistical methods exist for determining the likelihood that a coin is fair, but we will not discuss these methods.

What Are Principles That Underlie How We Determine Probabilities?

principles of probability

Several key **principles of probability** can be used to calculate probabilities, and we will discuss and apply these principles in the rest of this section. Here are the principles:

1. If two events are equally likely, then their probabilities are equal.
2. The probability of an event is the sum of the probabilities of the distinct outcomes that compose the event.
3. If an experiment is performed many times, then the fraction of times that an event occurs is likely to be close to the event's probability. The greater the number of times the experiment is performed, the more likely it is that the fraction of times an event occurs is close to the event's probability. In other words, the long-run relative frequency with which an event occurs approximates the event's probability.

How Do We Develop and Use Uniform Probability Models?

CCSS 7.SP.6

uniform probability model

If the distinct possible outcomes of a chance process are all equally likely, then we call a probability model for such a process a **uniform probability model**. For uniform probability models there is a simple way to describe probabilities of events as fractions.

Why Is the Probability of an Outcome a Unit Fraction? In a uniform probability model, if there are N possible outcomes, then for each outcome, the probability of that outcome is $\frac{1}{N}$. This statement follows from the first two principles of probability, as the following simple case shows.

Suppose you have an ordinary number cube (die) that has dots on each of its six faces. Each face has a different number of dots on it, ranging from 1 dot to 6 dots. You can roll the number cube and count the number of dots on the side that lands face up. Let's assume that the number cube is not weighted, so that each side is equally likely to land face up. Then by principle 1, the probability of rolling a 1 is equal to the probability of rolling a 2, which is equal to the probability of rolling a 3, and so on, up to 6.

CCSS 7.SP.7a

So if we let

$$P(\text{roll an } N)$$

stand for the probability of rolling an N, then

$$P(\text{roll a } 1) = P(\text{roll a } 2) = P(\text{roll a } 3) = P(\text{roll a } 4) = P(\text{roll a } 5) = P(\text{roll a } 6)$$

Now when you roll a number cube, the probability of rolling either a 1 or 2 or 3 or 4 or 5 or 6 is 1, because *some* face has to land up. And you can't *simultaneously* roll a 2 and a 3 on one number cube, or any other pair of distinct numbers. Therefore, by principle 2,

$$P(\text{roll a } 1) + P(\text{roll a } 2) + P(\text{roll a } 3) + P(\text{roll a } 4) + P(\text{roll a } 5) + P(\text{roll a } 6) = 1$$

Because all these probabilities are equal and add up to 1, each probability must be $\frac{1}{6}$. That is,

$$P(\text{roll a } 1) = \frac{1}{6} \qquad P(\text{roll a } 4) = \frac{1}{6}$$

$$P(\text{roll a } 2) = \frac{1}{6} \qquad P(\text{roll a } 5) = \frac{1}{6}$$

$$P(\text{roll a } 3) = \frac{1}{6} \qquad P(\text{roll a } 6) = \frac{1}{6}$$

In general, consider a uniform probability model in which there are N possible outcomes. Because the outcomes are equally likely and their probabilities sum to 1, by principles 1 and 2 the probability of each individual outcome is $\frac{1}{N}$ or

$$\frac{1}{\text{number of possible outcomes}}$$

Importantly, *this applies only to uniform probability models*, where all outcomes are equally likely.

Why Is the Probability of an Event the Fraction of Outcomes that Compose the Event? Let's think about how we can find probabilities of events in a uniform probability model. For example, what is the probability of rolling an odd number with a number cube? The only way to roll an odd number with a number cube is to roll a 1, 3, or 5. Since the outcomes 1, 3, 5 are distinct, principle 2 applies and tells us that the probability of rolling a 1, 3, or 5 is the sum of the probabilities of rolling a 1, of rolling a 3, and of rolling a 5. Therefore, the probability of rolling an odd number is

$$\frac{1}{6} + \frac{1}{6} + \frac{1}{6} = \frac{3}{6} = \frac{1}{2}$$

In general, for a uniform probability model in which there are N possible outcomes, consider an event composed of A of those outcomes. Because the probability of each individual outcome is $\frac{1}{N}$ (as discussed above), by principle 2 the probability of the event is the sum of A $\frac{1}{N}$s, which is $\frac{A}{N}$. In other words,

$$\text{Probability of event} = \frac{\text{number of outcomes in the event}}{\text{number of possible outcomes}}$$

Importantly, *this applies only to uniform probability models,* where all outcomes are equally likely.

CLASS ACTIVITY

16A Probabilities with Spinners, p. CA-340

16B Critique Probability Reasoning, p. CA-342

How Is Empirical Probability Related to Theoretical Probability?

Although the theoretical probability of an event is the fraction of times that the event occurs in the ideal, when you perform an experiment a number of times, the *actual* fraction of times that the event occurs will usually *not* be equal to its probability. When you perform an experiment a number

empirical probability
experimental probability

of times, the fraction or percentage of times that a given event occurs is called the **empirical probability** or **experimental probability** of that event. For example, if you flipped a coin 100 times and you got 54 heads, then the empirical probability of getting heads was 54%.

CLASS ACTIVITY

16C Empirical Versus Theoretical Probability: Picking Cubes from a Bag, p. CA-343

16D Using Empirical Probability to Make Predictions, p. CA-344

16E If You Flip 10 Pennies, Should Half Come Up Heads?, p. CA-345

CCSS
7.SP.6

Probability principle 3 says that the empirical probability is likely to be close to the theoretical probability of an event if a chance process occurred many times. In other words, if you perform an experiment many times, the actual fraction that a given event occurs is usually close to the probability of that event occurring. This statement is similar to the statement that a random sample of a population is likely to be representative of the full population if the sample is large enough. (See Sections 15.1 and 15.4.)

Because of principle 3, we can use empirical probabilities to estimate theoretical probabilities. For example, suppose there are 10 blocks in a bag. Some of the blocks are red and some are blue, but we don't know how many of each color block are in the bag. Let's say we reach into the bag 50 times, each time randomly removing one block, recording its color, and putting it back into the bag. If we picked 16 red blocks, then the experimental probability of picking a red block was 32%. This 32% is likely to be close to the theoretical probability of picking a red block, which is

$$\frac{(\text{number of red blocks})}{10}$$

because there are 10 blocks and each block is equally likely to be chosen. If there are 3 red blocks in the bag, the probability of picking red is 30%, which is close to 32%. Our best guess for the number of red blocks in the bag is therefore 3. Even though we based our guess on the available evidence, our guess could turn out to be wrong because of the variability that is inherent in random choices.

When we perform an experiment a number of times to determine an empirical probability, the greater the number of times we perform the experiment, the more likely it is that the empirical probability is close to the theoretical probability. For example, suppose that the probability of winning a certain game is 25%. Imagine that you could play the game 100 times; imagine that you could play the game 1000 times. If you play the game 100 times, which is a fairly large number of times, you should expect to win close to 25% of the games. But if you play the game 1000 times, it's even more likely that the percent of games you win will be close to 25%. We can see these statements reflected in the dot plots in Figure 16.1, which were obtained by repeatedly playing the game 100 times and then repeatedly playing the game 1000 times (in a simulation). Each dot in Figure 16.1(a) represents 100 games; the dot is plotted at the percentage of the 100 games that were won. Each dot in Figure 16.1(b) represents 1000 games; the dot is plotted at the percentage of the 1000 games that were won (rounded to the nearest whole number). Notice that in both dot plots in Figure 16.1, the dots cluster around 25%, but that more of the dots in Figure 16.1(b) are close to 25% than in Figure 16.1(a). Therefore, when 1000 games are played, the percentage of games won is more likely to be close to 25% than when 100 games are played.

Figure 16.1 Percentage of games won in 100 sets of 100 games and in 100 sets of 1000 games, where the probability of winning each game was 25%.

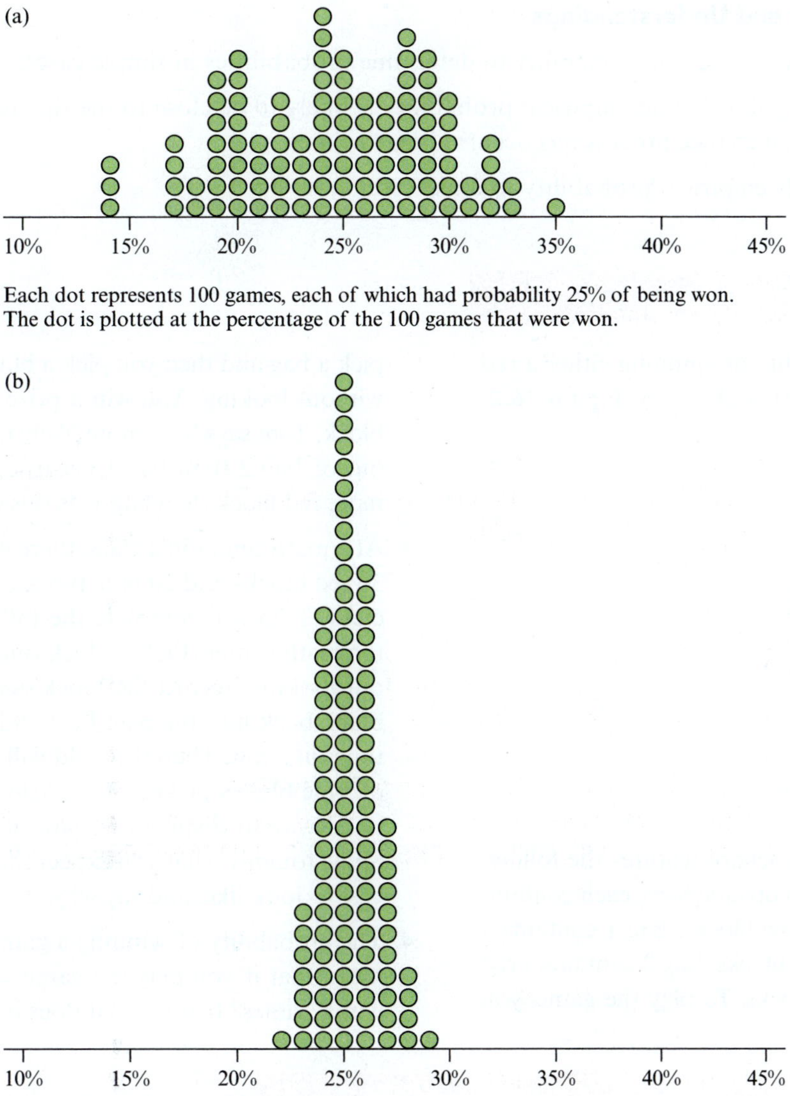

(a) Each dot represents 100 games, each of which had probability 25% of being won. The dot is plotted at the percentage of the 100 games that were won.

(b) Each dot represents 1000 games, each of which had probability 25% of being won. The dot is plotted at the percentage of the 1000 games that were won.

SECTION SUMMARY AND STUDY ITEMS

Section 16.1 Basic Principles of Probability

The probability that a given event of a chance process will occur is the fraction of times that event should occur "in the ideal." Several principles of probability are key: (1) if two events of a chance process are equally likely, then their probabilities are equal; (2) if there are several events of a chance process that cannot occur simultaneously, then the probability that one of those events will occur is the sum of the probabilities of each individual event; and (3) if a chance process occurs many times, the fraction of times that a given event occurs is likely to be close to the probability of that event occurring. Using these principles, we can determine probabilities in many simple cases. When a chance process occurs a number of times, the fraction of times that a given event occurred is the empirical probability of that event occurring.

Key Skills and Understandings

1. Use principles of probability to determine probabilities in simple cases.
2. Recognize that an empirical probability is likely to be close to the theoretical probability when a chance process has occurred many times.
3. Apply empirical probability to make estimates.

Practice Exercises for Section 16.1

1. Determine the probability of spinning either a red or a yellow on the spinner shown in Figure 16.2. Explain briefly.

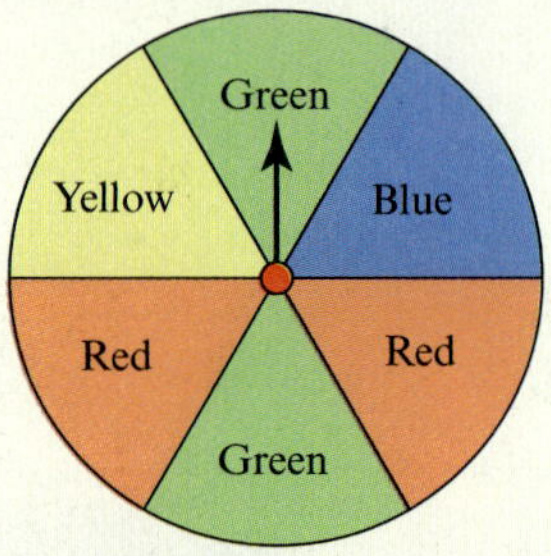

Figure 16.2 A spinner.

2. A family math night at school features the following game. There are two opaque bags, each containing red blocks and yellow blocks. Bag 1 contains 3 red blocks and 5 yellow blocks. Bag 2 contains 4 red blocks and 9 yellow blocks. To play the game, you pick a bag and then you pick a block out of the bag without looking. You win a prize if you pick a red block. Tom says he is more likely to pick a red block out of bag 2 than bag 1 because bag 2 contains 1 more red block than bag 1. Is this correct?

3. At a math center in a class, there is a bag filled with 30 red blocks and 20 blue blocks. Each child in the class of 25 will complete the following activity at the math center: Pick a block out of the bag without looking, record the block's color, and put the block back into the bag. Each child will do this 10 times in a row. Then the child will write the number of blue blocks picked on a sticky note. Describe a good way to display the data of the whole class. Show roughly what you expect the display you suggest to look like, and say why.

4. The probability of winning a game is $\frac{17}{100}$. Does this mean that if you play the game 100 times you will win 17 times? If not, what does it mean?

Answers to Practice Exercises for Section 16.1

1. The spinner is equally likely to land in each of the 6 sections. Therefore, by principle 1, the probability of landing on a particular one of those sections is $\frac{1}{6}$. Since red and yellow together make up 3 of the 6 sections, the probability of spinning red or yellow is $\frac{1}{6} + \frac{1}{6} + \frac{1}{6} = \frac{3}{6} = \frac{1}{2}$ by principle 2.

2. No, the probability of picking a red block out of bag 1 is $\frac{3}{8}$ by principles 1 and 2 because 3 out of 8 blocks are red, whereas the probability of picking a red block out of bag 2 is $\frac{4}{13}$ by principles 1 and 2 because 4 out of 13 blocks are red. Since $\frac{3}{8}$ is greater than $\frac{4}{13}$, the probability of picking a red block is higher for bag 1 than for bag 2.

3. A dot plot like the one in Figure 16.3 would be a good way to display these data. The class could make this dot plot on the chalkboard, using sticky notes instead of dots. Since $\frac{2}{5}$ of the blocks in the bag are blue, the probability of picking a blue block is $\frac{2}{5} = 40\%$. So, in the ideal, there would be 4 blue blocks chosen in 10 picks of a block. However, since the choices are random, most children probably won't get exactly 4 blue blocks, but most will probably get somewhere around 4 blocks. A few children may even get 8, 9, or 10 blue blocks in their

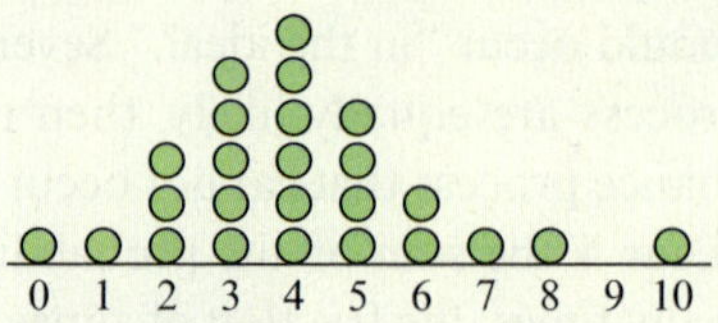

Figure 16.3 Dot plot of the number of blue blocks picked in 10 trials out of a bag containing 20 blue blocks and 30 red blocks.

10 picks. So the dot plot will probably look something like the one in Figure 16.3, where most dots cluster around 4, but some dots are not close to 4.

4. No, if you play the game 100 times you probably won't win exactly 17 times. The fraction of times that you do win is the empirical probability, which is usually not exactly equal to the theoretical probability. The theoretical probability of winning, $\frac{17}{100}$, is the number of times you would win "in the ideal" if you played 100 times. If you play the game many times, the fraction of times you win will usually be close to this theoretical probability and will usually get closer and closer to this theoretical probability the more games you play.

PROBLEMS FOR SECTION 16.1

1. Some games have spinners. When the arrow in a spinner is spun, it can land in any one of several different colored regions. Determine the probability of spinning each of the following on a spinner like the one in Figure 16.4:

   a. Red

   b. Either red or green

   c. Either red or yellow

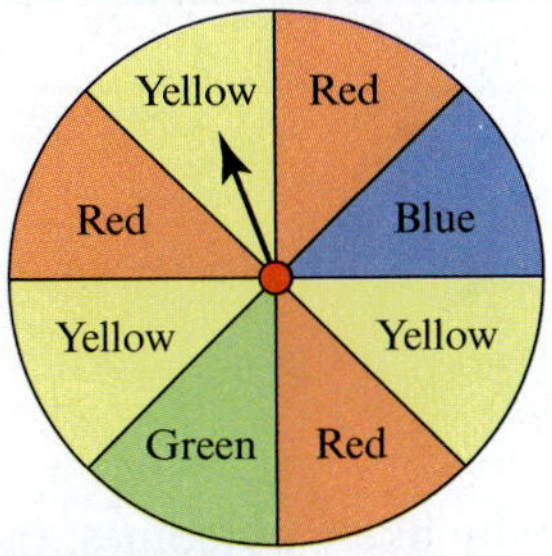

Figure 16.4 A spinner.

   Explain your answers, referring to ideas discussed in this section.

2. a. Draw a spinner such that the probability of landing on red is $\frac{1}{3}$, the probability of landing on green is $\frac{1}{4}$, and the only other color that the spinner could land on is yellow.

   b. For your spinner in part (a), what is the probability of landing on yellow? Explain.

   c. Now draw a different spinner that has the same probabilities as in part (a). Explain briefly.

3. a. Draw a 4-color spinner (red, green, yellow, blue) such that

   - landing on green is two times as likely as landing on red;
   - landing on yellow is two times as likely as landing on green;
   - landing on blue is more likely than landing on red.

   b. Determine the probabilities of landing on each of the colors on your spinner in part (a). Briefly explain your reasoning.

   c. Draw another spinner that satisfies the conditions in part (a) and is different from your spinner in part (a).

4. Write a paragraph discussing the following:

   a. Michael says that if you flip a coin 100 times, it will have to come up heads 50 times out of the 100. Is Michael correct?

   b. If the probability that a certain outcome of an experiment will occur is 30%, does that mean that when you carry out the experiment 100 times, the outcome will occur 30 of those times?

5. A family math night at school features the following game. There are two opaque bags, each containing red blocks and yellow blocks. Bag 1 contains 3 red blocks and 5 yellow blocks. Bag 2 contains 5 red blocks and 15 yellow blocks. To play the game, you pick a bag and then you pick a block out of the bag without looking. You win a prize if you pick a red block. Kate thinks she should pick from bag 2 because it contains more red blocks. Is Kate more likely to pick a red block if she picks from bag 2 rather than bag 1? Explain why or why not.

6. There are 50 small balls in a tub. Some balls are white and some are orange. Without being able to see into the tub, each student in a class of 25 is allowed to pick a ball out of the tub at random.

The color of the ball is recorded and the ball is put back into the tub. At the end, 7 orange balls and 18 white balls were picked. What is the best estimate you can give for the number of orange balls and the number of white balls in the tub? Describe how to calculate this best estimate, and explain why your method of calculation makes sense in a way that a seventh-grader might understand. Is your best estimate necessarily accurate? Why or why not?

7. In a classroom, there are 100 plastic fish in a tub. The tub is hidden from the students' view. Some fish are green and some are yellow. The students know that there are 100 fish, but they don't know how many of each color there are. The students go fishing, each time picking a random fish from the tub, recording its color, and throwing the fish back in the tub. At the end of the day, 65 fish have been chosen, 12 green and 53 yellow. What is the best estimate you can give for the number of green fish and the number of yellow fish in the tub? Describe how to calculate this best estimate, and explain why your method of calculation makes sense in a way that a seventh-grader might understand. Is your best estimate necessarily accurate? Why or why not?

8. There is a bag filled with 4 red blocks and 16 yellow blocks. Each child in a class of 25 will pick a block out of the bag without looking, record the block's color, and put the block back into the bag. Each child will repeat this for a total of 10 times. Then the child will write on a sticky note the number of red blocks picked. Describe a good way to display the data for the whole class. Show roughly what you expect the display you suggest to look like, and say why.

9. Write several paragraphs in which you describe and discuss some of the misconceptions that students often have about probability and that you learned about from the Class Activities and Practice Exercises in this section.

16.2 Counting the Number of Outcomes

CCSS Common Core State Standards Grade 7

In the previous section, we saw that if a chance process has equally likely outcomes, then we can determine the probability that one or several of those outcomes occurs. In simple cases, probabilities are easy to determine—think of a spinner that is equally likely to land in its sections or a number cube for which each face is equally likely to land up. But what if we want to determine the probability that several chance processes in a row each have a certain outcome? For example, what is the probability that a certain spinner lands on red, and then a number cube lands with 6 up? To calculate such a probability, we need to first determine how many different, equally likely outcomes there are.

In this section, we study some methods for calculating numbers of outcomes. Our main tools will be multiplication and the ways of displaying multiplicative structure that we studied in Section 4.1. We can use these ways of thinking about multiplication and showing multiplicative structure to calculate numbers of outcomes in a variety of situations, including situations commonly encountered in the study of probability. In Section 16.3 we apply our techniques for determining numbers of outcomes to calculate probabilities.

Note that we'll be working with two-stage and, more generally, with multistage experiments. **Multistage experiments** consist of performing several experiments in sequence, such as spinning a spinner and then rolling a number cube (a two-stage experiment) or flipping a coin three times in sequence (a three-stage experiment). When we view the outcomes of a multistage experiment as forming a sample space, then we often call the events of this sample space **compound events**.

multistage experiments

compound events

CCSS
7.SP.8b

How Are Two-Stage Experiments Related to Ordered Pair Problems?

Section 4.1 described how ordered pair problems are one type of problem that can be solved by multiplication. For example, if there are 3 pants and 4 shirts, how many outfits consisting of pants and a shirt can be made? Each outfit can be viewed as an ordered pair (pants, shirt). Each pants can be paired with 4 shirts, so each pants contributes a group of 4 outfits. There are 3 pants, so there are 3 groups of 4 outfits that can be made. According to the meaning of multiplication, this is $3 \cdot 4$ outfits, which is 12 outfits.

Probability problems about two-stage experiments usually involve ordered pair problems. For example, suppose we first spin a spinner that is equally likely to land on 1 of 4 colors (red, blue, yellow, or green) and then we roll a number cube for which each of the 6 faces labeled with 1 through 6 dots is equally likely to land face up. The problem of determining how many different possible outcomes there are to this two-stage experiment is an ordered pair problem. Each outcome consists of a pair

(color, number of dots)

For each color that can be spun, there are 6 possible outcomes for the number cube, so each color contributes 6 ordered pairs. Since there are 4 colors, there are 4 groups of 6 ordered pairs altogether, which is $4 \cdot 6$ ordered pairs according to the meaning of multiplication. Therefore, there are 24 possible outcomes for the two-stage experiment of spinning the spinner and rolling the number cube. We can use a tree diagram or an organized list of the ordered pairs to display the outcomes visually and to show their multiplicative structure, as in Figure 16.5.

Figure 16.5
Outcomes of a two-stage experiment.

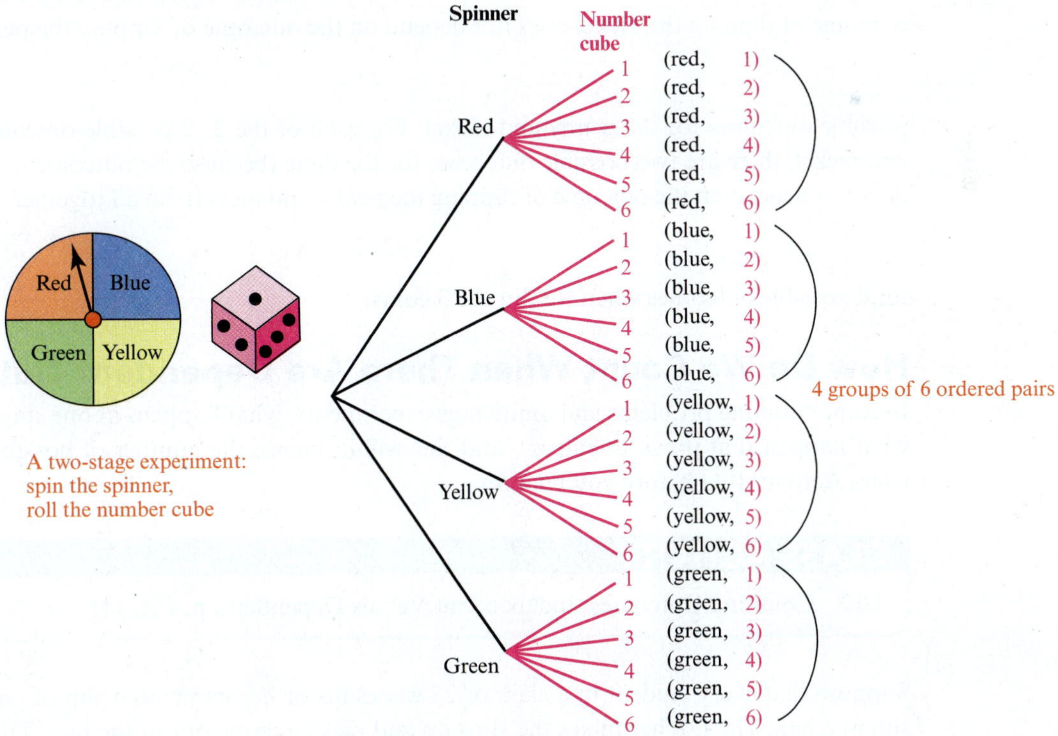

How Can We Count Outcomes of Multistage Experiments?

In some situations we want to determine the number of possible outcomes when multiple stages are involved. Try Class Activity 16F before you read on.

CLASS ACTIVITY

16F How Many Keys Are There?, p. CA-346

In the same way that we can use multiplication to determine a total number of ordered pairs, we can use multiplication to determine a total number of ordered triples or ordered 4-tuples, and so on, which is useful for determining the total number of outcomes in multistage experiments.

For example, suppose you have 3 coins: a penny, a nickel, and a dime. How many possible outcomes are there if you flip all 3 coins? Think of each outcome as an ordered triple, consisting of an outcome on the penny, an outcome on the nickel, and an outcome on the dime. As with ordered pairs, ordered triples can be displayed in organized lists and tree diagrams (as in **Figure 16.6**), and they have a multiplicative structure.

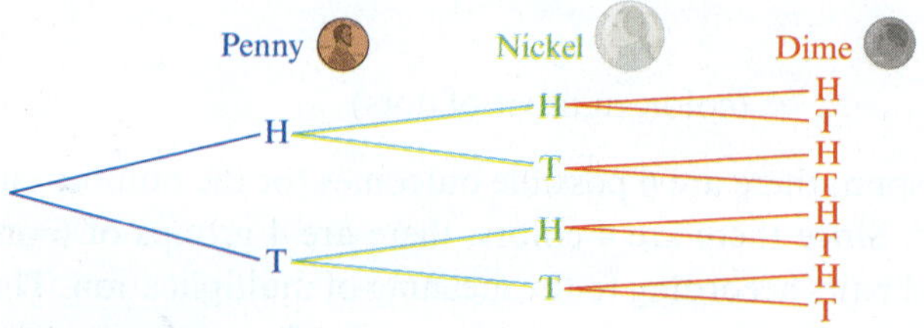

HHH
HHT
HTH
HTT
THH
THT
TTH
TTT

Figure 16.6 A tree diagram and an organized list showing all possible outcomes from flipping 3 coins.

For each of the two outcomes for the penny, the nickel has two possible outcomes (because the outcome of flipping the nickel does not depend on the outcome of flipping the penny), so there are

$$2 \cdot 2$$

possible outcomes for the penny and nickel. For each of the $2 \cdot 2$ possible outcomes for the penny and nickel, there are two possible outcomes for the dime (because the outcome of flipping the dime does not depend on the outcome of flipping the penny and nickel). So all together, there are

$$2 \cdot 2 \cdot 2 = 8$$

total possible outcomes when we flip the 3 coins.

How Do We Count When There Are Dependent Outcomes?

In some counting problems and multistage experiments, what happens at one stage may depend on what happened at previous stages—and this will influence the number of possible outcomes. Try Class Activity 16G before you read on.

CLASS ACTIVITY

16G Counting Outcomes: Independent Versus Dependent, p. CA-347

Suppose that every student in a class of 25 writes his or her name on a slip of paper and puts the slip in a bag. The teacher mixes the slips up and picks a name out of the bag. This student will be class president for the day. The teacher reaches into the bag again and picks another name out of the bag. This student will be class vice president for the day. How many possible outcomes are there? There are 25 possible outcomes for class president. But once the president has been chosen, only 24 additional possibilities remain for the vice-president. So altogether, each of the 25 possible presidents can be paired with 24 possible vice presidents. This activity involves a situation of 25 groups of 24 pairs, each pair consisting of a president and vice president. Therefore, there are $25 \cdot 24 = 600$ possible different outcomes when a class president and vice president are picked as described. Even though there are 25 different students who could become vice-president, we multiply $25 \cdot 24$ rather than $25 \cdot 25$ because the choice of vice-president depends on the choice of president.

FROM THE FIELD **Children's Literature**

Anno, M., & Anno, M. (1983). *Anno's mysterious multiplying jar.* New York, NY: Philomel Books.

A mysterious multiplying jar that contains an island introduces students to the mathematical idea of factorials. The island has 2 countries, the countries each have 3 mountains, the mountains each have 4 walled kingdoms, and so on, until there are 9 boxes that each contain 10 jars. Following the story there is a discussion of the mathematical concept of factorial. For example, 10 factorial, which is written as 10! means $10 \cdot 9 \cdot 8 \cdot 7 \cdot 6 \cdot 5 \cdot 4 \cdot 3 \cdot 2 \cdot 1$.

SECTION SUMMARY AND STUDY ITEMS

Section 16.2 Counting the Number of Outcomes

We can often use multiplication to calculate the number of outcomes in two-stage and multistage experiments and other situations.

Key Skills and Understandings

1. Apply multiplication to count the total number of outcomes in various situations, including those of multistage experiments and cases where the outcome at one stage depends on the outcome at previous stages.

Practice Exercises for Section 16.2

1. How many different 3-digit numbers can you write using only the digits 1, 2, and 3 if you do not repeat any digits (so that 121 and 332 are not counted)? Show how to solve this problem with an organized list and with a tree diagram. Explain why this problem can be solved by multiplying.
2. How many different 3-digit numbers can be made by using only the digits 1, 2, and 3, where repeated digits are allowed (so that 121 and 332 are counted)? Show how to solve this problem with an organized list and with a tree diagram. Explain why this problem can be solved by multiplying.
3. How many different keys can be made if there are 10 places along the key that will be notched and if each notch will be 1 of 8 depths?
4. Annette buys a wardrobe of 3 skirts, 3 pants, 5 shirts, and 3 sweaters, all of which are coordinated so that she can mix and match them any way she likes. How many different outfits can Annette create from this wardrobe? (Every day Annette wears either a skirt or pants, a shirt, and a sweater.)
5. A delicatessen offers 4 types of bread, 20 types of meats, 15 types of cheese, a choice of mustard, mayonnaise, both or neither, and a choice of lettuce, tomato, both or neither for their sandwiches. How many different types of sandwiches can the deli make with 1 meat and 1 cheese? Should the deli's advertisement read "hundreds of sandwiches to choose from," or would it be better to substitute thousands or even millions for hundreds?

6. A pizza parlor offers 10 toppings to choose from. How many different large pizzas are there with exactly 2 different toppings? (For example, you could order pepperoni and mushroom, but not double pepperoni.)

Answers to Practice Exercises for Section 16.2

1. Figure 16.7 shows a tree diagram and an organized list displaying all such 3-digit numbers. Both show a structure of 3 big groups, 1 group for each possibility for the first digit. Once a first digit has been chosen, there are 2 choices for the second digit (since repeats aren't allowed). This means there are 3 groups of 2 possibilities for the first 2 digits. So by the meaning of multiplication, there are $3 \cdot 2 = 6$ possibilities for the first 2 digits. Once the first 2 digits are chosen, there is only 1 choice for the third digit—it must be the remaining unused digit. So there are $3 \cdot 2 \cdot 1 = 6$ three-digit numbers using only the digits 1, 2, 3, with no digit repeated.

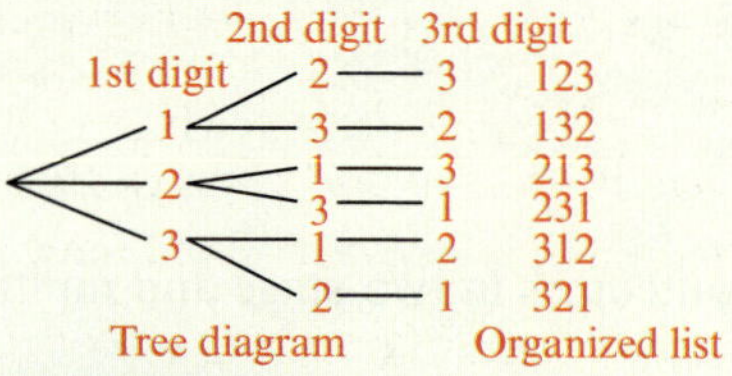

Figure 16.7 Tree diagram and an organized list showing all 3-digit numbers using 1, 2, and 3 with no digit repeated.

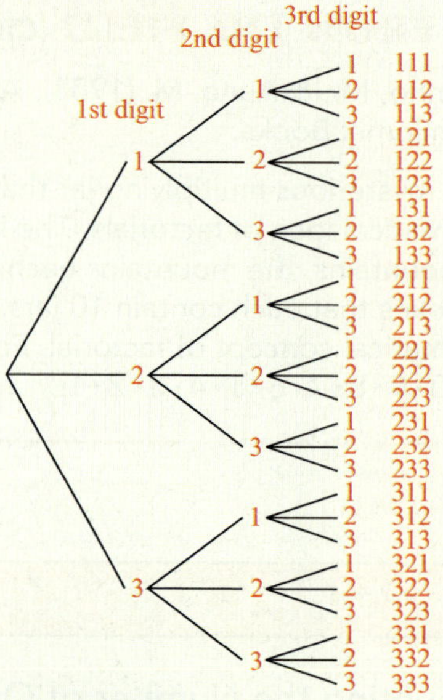

Figure 16.8 Tree diagram and an organized list showing all 3-digit numbers using 1, 2, and 3.

2. The tree diagram and the organized list in Figure 16.8 both show 3 big groups. Each of those 3 big groups has 3 groups of 3. Therefore, according to the meaning of multiplication, there are

$$3 \cdot (3 \cdot 3) = 27$$

3-digit numbers that use only the digits 1, 2, and 3.

3. Figure 16.9 shows part of a tree diagram for all such keys. There are 8 primary branches corresponding to the 8 choices for the first notch. For each of those 8 choices on the first notch, there are 8 choices for the second notch. Therefore, there are 8 groups of 8 choices for the first 2 notches, and so, by the meaning of multiplication, there are $8 \cdot 8$ choices for the first 2 notches. For each of the $8 \cdot 8$ choices for the first 2 notches, there are 8 choices for the third notch. Therefore, there are $8 \cdot 8 \cdot 8$ choices for the first 3 notches. And so on. When all 10 notches are taken into account, there are $8 \cdot 8 \cdot 8 \cdot 8 \cdot 8 \cdot 8 \cdot 8 \cdot 8 \cdot 8 \cdot 8 = 8^{10} = 1{,}073{,}741{,}824$ that can be made, which is more than a billion keys!

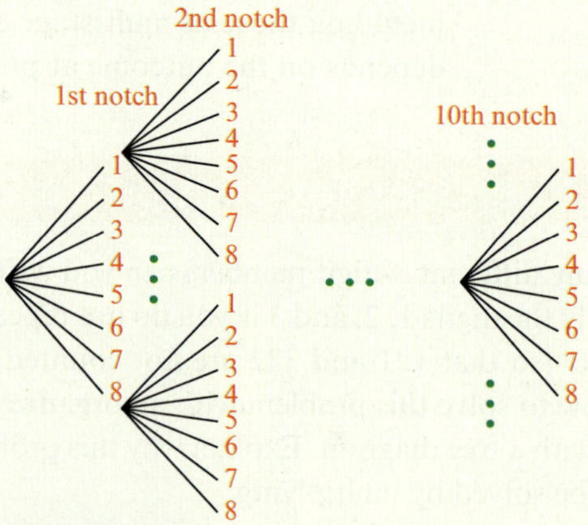

Figure 16.9 Partial tree diagram showing all keys with 10 notches of 8 depths.

4. Annette can make $6 \cdot 5 \cdot 3 = 90$ different outfits.

5. There are $4 \cdot 20 \cdot 15 \cdot 4 \cdot 4 = 19{,}200$ different sandwiches that can be made with a meat and a cheese. The deli should substitute thousands for hundreds.

6. For each of the 10 choices for the first topping, there are 9 choices for the second topping (there are only 9 because repeats are not allowed). This forms 10 groups of 9, so there are $10 \cdot 9$ choices of double toppings. However, when they are counted this way, pepperoni and mushrooms counts as different from mushrooms and pepperoni. Therefore, we must divide by 2 so as not to double-count pizzas. So there are actually only $\frac{10 \cdot 9}{2} = 45$ different large pizzas with 2 different toppings.

PROBLEMS FOR SECTION 16.2

1. A bakery makes 4 different kinds of cake. Each cake can have 3 different kinds of frosting. Each frosted cake can be decorated in 2 different ways. How many ways are there of ordering a decorated, frosted cake?

Show how to solve the problem by using an organized list and a tree diagram. Explain why you can solve the problem by multiplying.

2. Allie and Betty want to know how many 3-letter combinations, such as BMW or DDT, are possible. (Letters are allowed to repeat, as in DDT or BOB.) Allie thinks there can be $26 + 26 + 26$ three-letter combinations, whereas Betty thinks the number is $26 \cdot 26 \cdot 26$. Which girl, if either, is right, and why? Explain your answers clearly and thoroughly, drawing on the meaning of multiplication.

3. Explain your answers to the following:

a. How many 9-digit numbers are there that use only the digits 1, 2, 3, . . . , 8, 9? (Repetitions are allowed, so, for example, 123211114 is allowed.)

b. How many 9-digit numbers are there that use each of the digits 1, 2, 3,..., 8, 9 exactly once?

4. In all 3 parts in this problem, explain your solution clearly.

a. How many whole numbers are there that have exactly 10 digits and that can be written by using only the digits 8 and 9?

b. How many whole numbers are there that have at most 10 digits and that can be written by using only the digits 0 and 1? (Is this situation related to part (a)?)

c. How many whole numbers are there that have exactly 10 digits and that can be written by using only the digits 0 and 1?

5. Most Georgia car license plates currently use the format of 3 numbers followed by 3 letters (such as 123 ABC). How many different license plates can be made this way? Explain your solution clearly.

6. a. A 40-member club will elect a president and then elect a vice-president. How many possible outcomes are there?

b. A 40-member club will elect a pair of co-presidents. How many possible outcomes are there?

c. Are the answers to parts (a) and (b) the same or different? Explain why they are the same or why they are different.

7. A dance club has 10 women and 10 men. In each of the following parts, give a clear explanation.

a. How many ways are there to choose one woman and one man to demonstrate a dance step?

b. How many ways are there to pair each woman with a man (so that all 10 women and all 10 men have a dance partner at the same time)?

***8.** *A pizza parlor problem.* How many different large pizzas with no double toppings can be made? (For example, mushroom, pepperoni, and sausage is one possibility, and so is mushroom and sausage, but not double mushroom and sausage. Your count should also include the case of no toppings.) You can't answer the question yet because you don't know how many toppings the pizza parlor offers.

a. Determine the number of different pizzas when there are exactly 3 toppings to choose from (e.g., mushroom, pepperoni, and sausage).

b. Determine the number of different pizzas when there are exactly 4 toppings to choose from.

c. Determine the number of different pizzas when there are exactly 5 toppings to choose from.

d. Look for a pattern in your answers in parts (a), (b), and (c). Based on the pattern you see, predict the number of different pizzas when there are 10 toppings to choose from.

e. Now find a different way to determine the number of different pizzas when there are 10 toppings to choose from. This time, think about the situation in the following way: Pepperoni can be either on or off, mushrooms can be either on or off, sausage can be either on or off, and so on, for all 10 toppings. Explain clearly how to use this idea to answer the question and why this method is valid.

* **9.** A pizza parlor offers 10 different toppings to choose from. How many different large pizzas can be made if double toppings, but not triple toppings or more, are allowed? For example, double pepperoni and (single) mushroom is one possibility, as is double pepperoni and double mushroom, but triple pepperoni is not allowed. Notice that you can choose each topping in one of three ways: off, as a single topping, or as a double topping.

16.3 Calculating Probabilities of Compound Events

CCSS Common Core State Standards Grade 7

Now we will apply the counting techniques we studied in the previous section to solve probability problems involving multistage experiments. Before we do so though, we continue the discussion from Section 16.2 on the distinction between independent and dependent outcomes in multistage experiments. We conclude this section with a discussion of expected value—a useful application of probability.

How Are Independent and Dependent Outcomes Different?

An important factor in calculating probabilities for multistage experiments is the distinction between independent and dependent outcomes at a stage. If you flipped a coin 10 times in a row and all 10 flips came up heads, would you think that your next flip is more likely to be a tail because a tail is "due"? Would you think that you were on a "hot streak" and so are more likely to get another head on your next coin flip? In fact, on your next flip you are just as likely to get a tail as a head. There is nothing in the flipping history of the coin that can influence the next flip. The outcome of one coin flip is **independent** of the outcome of any other coin flip. In other words, the outcome of one coin flip has no bearing on the outcome of any other coin flip.

independent

Suppose you have a bag filled with 9 blue marbles and 1 green marble. If you randomly pick a marble out of the bag, record its color, and put it back in the bag, then your next random marble pick is independent of your first marble pick. On the other hand, if you leave the first marble out of the bag after selecting it, then you *do* influence the next marble pick. If the first marble you selected was the green one, then your second marble must be blue. If the first marble you selected was blue, then on your second marble pick, the probability of picking the green marble is $\frac{1}{9}$ instead of $\frac{1}{10}$ because now there are only 9 marbles in the bag. When the first marble is left out of the bag, the second marble pick is **dependent** on the first.

dependent

How Can We Calculate Probabilities of Compound Events?

Let's consider the case of flipping 3 coins: a penny, a nickel, and a dime. What is the probability that all 3 will land heads up when flipped? In Section 16.2 we saw that there are

$$2 \cdot 2 \cdot 2 = 8$$

possible outcomes when 3 coins are flipped, and all these outcomes are equally likely. In only 1 of those outcomes do all 3 coins land heads up. Therefore, the probability of all 3 coins landing heads up is $\frac{1}{8}$, or 12.5%.

CCSS 7.SP.8a

We can use the organized list and tree diagram of **Figure 16.6** to calculate the probabilities of other events as well. For example, what is the probability of getting 2 heads and 1 tail when we toss the 3 coins? Two heads and one tail occur in 3 of the 8 possible outcomes—namely, HHT, HTH, and THH. So the probability that either 1 of these 3 outcomes will occur is

$$\frac{1}{8} + \frac{1}{8} + \frac{1}{8} = \frac{3}{8}$$

or 37.5%.

In some games that use spinners, such as the one in **Figure 16.10**, the spinner is equally likely to land on red, yellow, blue, or green. What is the probability that in 2 spins, the spinner will land first on red and then on blue? We could make an organized list or a tree diagram to show all possible outcomes. We can also make an array to show all possible outcomes, as in **Figure 16.11**. Because the outcome of the second spin is independent of the outcome of the first spin, the array shows 4 rows of 4 possible outcomes, so there are $4 \cdot 4 = 16$ possible outcomes for the 2 spins. Each of the 16 spins is equally likely, so the probability of spinning a red and then a blue is $\frac{1}{16} = 6.25\%$.

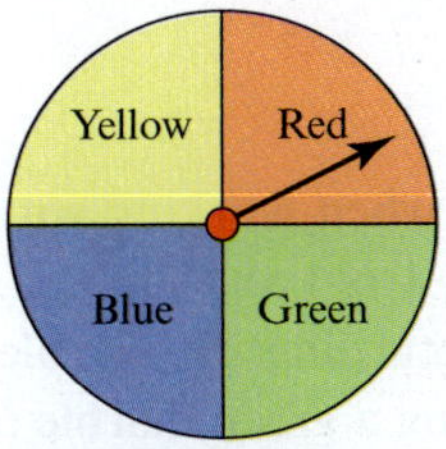

Figure 16.10 A spinner for a children's game.

Spin 2

Spin 1	R	G	B	Y
R	RR	RG	RB	RY
G	GR	GG	GB	GY
B	BR	BG	BB	BY
Y	YR	YG	YB	YY

Figure 16.11 Array showing all possible outcomes on 2 spins of a spinner.

CLASS ACTIVITY

16H Number Cube Rolling Game, p. CA-348

16I Picking Two Marbles from a Bag of 1 Black and 3 Red Marbles, p. CA-349

Be careful about using an organized list, a tree diagram, or an array when working with outcomes that are not equally likely. For example, suppose there are 3 marbles in a bag, 1 red and 2 green. If we randomly pick a marble from the bag, record its color, put it back in the bag, and then randomly pick another marble, can we use the tree diagram in **Figure 16.12** to calculate the probability of picking a green marble followed by the red marble? No, we cannot use this tree diagram, because even though there are only 2 possible outcomes when picking a marble—picking red and

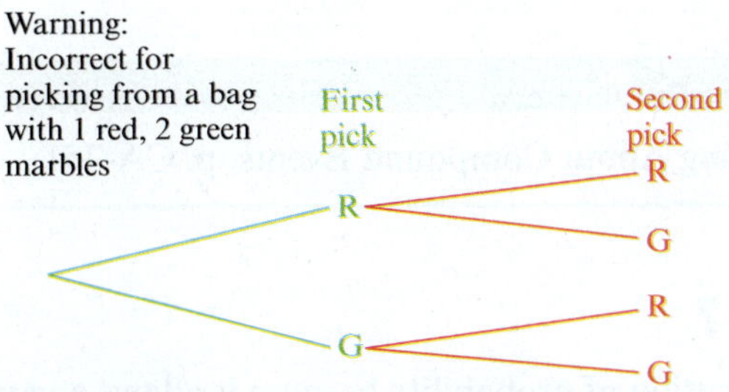

Figure 16.12 An incorrect tree diagram for picking 2 marbles from a bag with 1 red and 2 green marbles (if the first marble is replaced after picking).

picking green—these 2 outcomes are not equally likely. We are more likely to pick a green marble than the red one. If we think of the 2 green marbles as labeled "green 1" and "green 2," then we can use the tree diagram in Figure 16.13. The 9 outcomes in Figure 16.13 are equally likely. Two of these outcomes are a green marble followed by the red marble. Therefore, the probability of picking green followed by red is $\frac{2}{9} = 22\%$.

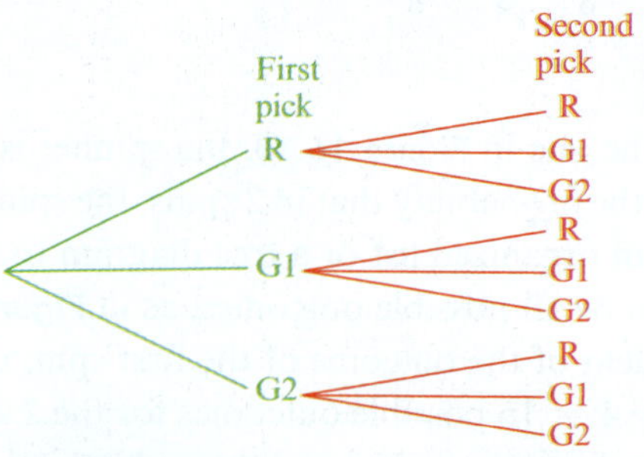

Figure 16.13 A correct tree diagram for picking 2 marbles from a bag with 1 red and 2 green marbles (if the first marble is replaced after picking).

We can use organized lists, tree diagrams, and arrays even when working with dependent outcomes. For example, suppose there are 3 marbles in a bag, 1 red and 2 green. If we randomly pick a marble from the bag, record its color, and *without* returning the marble to the bag, randomly pick another marble, what is the probability of picking a green marble followed by a red marble? There are 6 equally likely outcomes, as shown in Figure 16.14. Two of these outcomes are a green marble followed by the red marble. Therefore, the probability of picking green followed by red is $\frac{2}{6} = \frac{1}{3} = 33\%$.

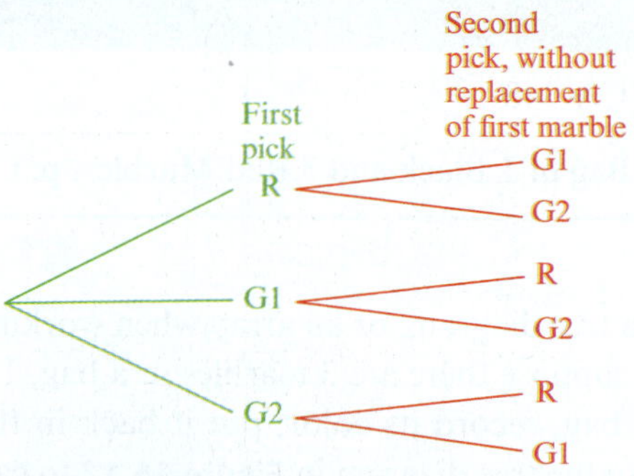

Figure 16.14 A tree diagram for picking 2 marbles from a bag with 1 red and 2 green marbles (if the first marble is *not* replaced after picking).

CLASS ACTIVITY

16J Critique Probability Reasoning About Compound Events, p. CA-351

What Is Expected Value?

expected value

Expected value is an important application of probability because it allows a variety of businesses to make educated decisions about how to allocate money. For an experiment that has numerical outcomes, the **expected value** of the experiment is the average outcome of the experiment in the ideal, over the long term.

CLASS ACTIVITY

16K Expected Earnings from the Fall Festival, p. CA-351

Suppose a farmer is considering planting a crop. Let's say that the probability the crop is successful is 75%. If the crop is successful, the crop will earn the farmer a net amount of \$100,000, but if the crop is not successful, then the farmer will lose the \$20,000 he invested in seed and supplies. The two outcomes of this experiment are \$100,000 and −\$20,000. Although one or the other of these outcomes will occur, if we think about the farmer planting the crop *repeatedly over the long term*, we can ask what amount the farmer should expect to earn on average, in the ideal. This amount is the expected value.

Think about the farmer planting the crop over and over, many times. In the ideal (according to the given probabilities), the crop will be successful 75% of those times and will fail in the remaining 25% of the times. Let's say that the farmer could plant the crop 100 times. Then of those 100 times, in 75 of them, the farmer would earn \$100,000, but in 25 of them the farmer would lose \$20,000. Altogether, over those 100 times, the farmer would earn a total of

$$75 \cdot \$100{,}000 - 25 \cdot \$20{,}000$$

If we divide by 100, we will have the average amount the farmer should expect to earn per crop in the long run, namely

$$\frac{75}{100} \cdot \$100{,}000 - \frac{25}{100} \cdot \$20{,}000 = \$70{,}000$$

So on average, over the long run, the farmer should expect to earn \$70,000 from planting the crop. Notice that we found this average amount by multiplying the probability of each outcome by the outcome and then adding over all outcomes. This is how to calculate expected value in general.

SECTION SUMMARY AND STUDY ITEMS

Section 16.3 Calculating Probabilities of Compound Events

We can apply the counting techniques of Section 16.2 to calculate probabilities for multistage experiments. A useful application of probability is the calculation of expected value.

Key Skills and Understandings

1. Apply the counting techniques of Section 16.2 to calculate probabilities for multistage experiments, including cases where the outcome at one stage depends on the outcome at previous stages.
2. Calculate an expected amount of earnings when probabilities are involved.

Practice Exercises for Section 16.3

1. Consider the experiment of rolling a number cube 2 times.

a. Determine the probability of getting a 1 on both rolls of the number cube. Explain your answer.

b. Determine the probability of getting a 1 on either the first or the second roll of the number cube. Explain your answer.

c. Determine the probability of getting either a 1 or a 2 on the rolls of the number cube.

2. What is the probability of getting 4 heads in a row on 4 tosses of a coin?

3. Suppose you have 3 marbles in a bag, 1 red and 2 green. If you reach into the bag without looking and randomly pick out 2 marbles at once, what is the probability that both of the marbles you pick will be green?

4. A school's fall festival includes the following game. There are 3 enclosed tubs, each containing red, yellow, and blue plastic bears. To play the game, a contestant reaches once into each tub and randomly picks out a bear. The contestant must reach through a sleeve attached to the tub, so that the color of the bear being chosen cannot be seen. If all 3 bears are red, the contestant wins a prize. The first tub contains 5 red bears, 1 yellow bear, and 1 blue bear. The second tub contains 3 red bears, 2 yellow bears, and 1 blue bear. The third tube contains 1 red bear, 2 yellow bears, and 2 blue bears. What is the probability of winning the prize?

5. Continue Practice Exercise 4: The school is expecting 350 people to play this game. Each contestant pays 50 cents to play. Each prize costs the school 75 cents. How much money (net) should the school expect to earn from this game? Explain.

Answers to Practice Exercises for Section 16.3

1. a. Since the 2 rolls of the number cube are independent, the following array shows all possible outcomes:

(1,1)	(1,2)	(1,3)	(1,4)	(1,5)	(1,6)
(2,1)	(2,2)	(2,3)	(2,4)	(2,5)	(2,6)
(3,1)	(3,2)	(3,3)	(3,4)	(3,5)	(3,6)
(4,1)	(4,2)	(4,3)	(4,4)	(4,5)	(4,6)
(5,1)	(5,2)	(5,3)	(5,4)	(5,5)	(5,6)
(6,1)	(6,2)	(6,3)	(6,4)	(6,5)	(6,6)

Here, (4,3) stands for a 4 on the first roll and a 3 on the second roll. There are $6 \cdot 6 = 36$ possible outcomes, all of which are equally likely. So, by principles 1 and 2, the probability of getting a 1 on both rolls is $\frac{1}{36}$, or about 2.8%.

b. Referring to the array in part (a), there are 11 ways to get a 1 on either the first roll or the second roll, or both (the first row together with the first column). Therefore, the probability of getting at least one 1 on 2 rolls of a number cube is $\frac{11}{36}$, or about 30.6% (so you'd expect to get at least one 1 a little less than one-third of the time).

c. Referring to the array in part (a), there are 4 ways to get either a 1 or a 2 on the rolls—namely, (1,1), (1,2), (2,1), and (2,2). Therefore, the probability of getting either a 1 or a 2 each time on 2 rolls of a number cube is $\frac{4}{36} = \frac{1}{9}$, or about 11%.

2. The tree diagram in **Figure 16.15** shows all possible outcomes of the 4 coin tosses, since the coin tosses are independent. There are $2 \cdot 2 \cdot 2 \cdot 2 = 16$ possible outcomes, all of which are equally likely. Only one of those outcomes produces 4 heads in a row. So the probability of getting 4 heads in a row on 4 tosses of a coin is $\frac{1}{16}$, or 6.25%.

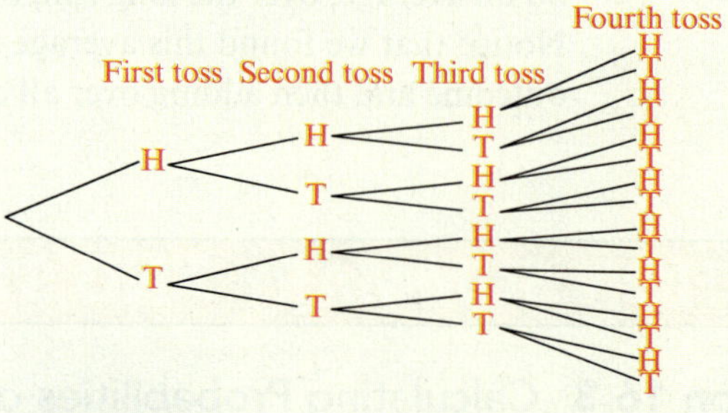

Figure 16.15 Tree diagram for tossing a coin 4 times.

3. We can use the tree diagram in **Figure 16.14** to solve this problem by thinking of one of the marbles as the first one chosen and the other as the second one chosen (where the first marble is not replaced before the second one is chosen). The tree diagram shows that there are 6 equally likely outcomes. Two of those outcomes correspond to both green marbles being chosen. So the probability of picking the 2 green marbles is $\frac{2}{6} = \frac{1}{3}$, about 33%.

Another way to solve the problem is to create the following organized list, which shows all possible ways of picking a pair of marbles from the bag:

$(G1, G2)$ $(G1, R)$

$(G2, R)$

(Notice that $(G2, G1)$ would be the same as $(G1, G2)$. Results are similar for other cases.) These 3 possible ways of picking 2 marbles are all equally likely. Out of the 3 ways of picking 2 marbles, there is 1 way to pick 2 green marbles. Therefore, the probability of picking 2 green marbles is $\frac{1}{3}$.

4. From the (partial) tree diagram in **Figure 16.16**, we see that for each of the 7 bears that could be picked from the first tub, there are 6 bears that could be picked from the second tub, and for each of these $7 \cdot 6$ possibilities, there are 5 bears that could be chosen from the third tub. Therefore, there are $7 \cdot 6 \cdot 5$ equally likely possible outcomes for the game. How many of these outcomes consist of picking a winning combination of 3 red bears? The tree diagram shows that for each of the 5 red bears that could be chosen from the first tub, there are 3 red bears that could be chosen from the second tub, and for each of these $5 \cdot 3$ possibilities, there is only 1 red bear that could be chosen from the third tub. So there are $5 \cdot 3 \cdot 1$ outcomes in which all three bears are red. Since there are $7 \cdot 6 \cdot 5$ equally likely outcomes and since $5 \cdot 3 \cdot 1$ of those outcomes result in winning the game, the probability of winning the game is

$$\frac{5 \cdot 3 \cdot 1}{7 \cdot 6 \cdot 5} = \frac{1}{14}$$

Figure 16.16

Tree diagram for picking bears from 3 tubs.

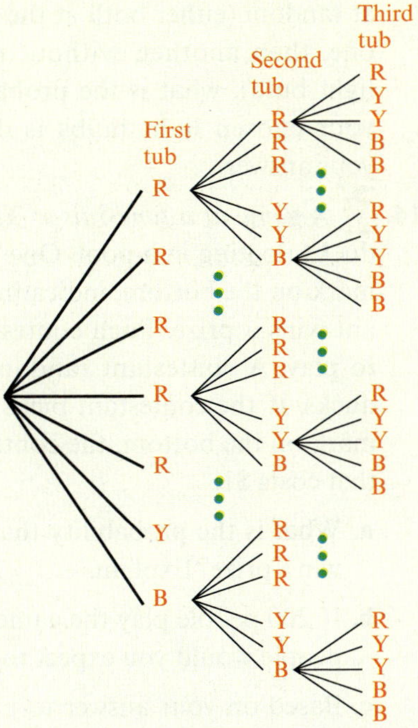

5. If 350 people play the game, then in the ideal, $\frac{1}{14}$ of the people should win the game. So in the ideal,

$$\frac{1}{14} \cdot 350 = 25$$

people should win. The 25 prizes the school would give out to the winners would cost

$$25 \cdot \$0.75 = \$18.75$$

But the school will collect

$$350 \cdot \$0.50 = \$175$$

if 350 people play the game. So net, the school can expect to earn

$$\$175 - \$18.75 = \$156.25$$

from this game in the ideal case. Of course, in actuality, the school might earn more or less than this amount since more people or fewer people might win the game and since more than 350 or fewer than 350 people might play the game.

PROBLEMS FOR SECTION 16.3

1. A children's game has a spinner that is equally likely to land on any 1 of 4 colors: red, blue, yellow, or green. Determine the probability of spinning a red both times on 2 spins. Explain your reasoning.

2. A children's game has a spinner that is equally likely to land on any 1 of 4 colors: red, blue, yellow, or green. Determine the probability of spinning a red at least once on 2 spins. Explain your reasoning.

3. A children's game has a spinner that is equally likely to land on any 1 of 4 colors: red, blue, yellow, or green. Determine the probability of spinning a red 3 times in a row on 3 spins. Explain your reasoning.

4. A children's game has a spinner that is equally likely to land on any 1 of 4 colors: red, blue, yellow, or green. Determine the probability of spinning either a red followed by a yellow or a yellow followed by a red in 2 spins. Explain your reasoning.

5. A children's game has a spinner that is equally likely to land on any 1 of 4 colors: red, blue, yellow, or green. Determine the probability of *not* spinning a red on either of 2 spins. Explain your reasoning.

6. Determine the probability of spinning a blue followed by a green in 2 spins on the spinner in Figure 16.17. Explain your reasoning.

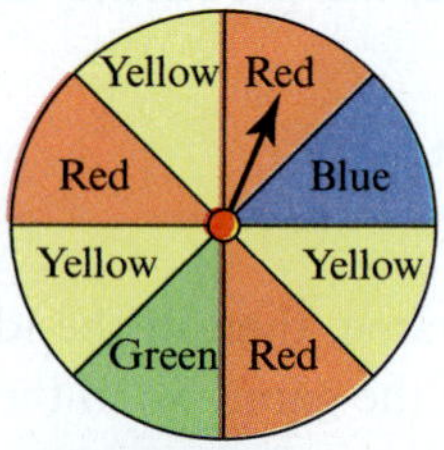

Figure 16.17 A spinner.

7. Determine the probability of spinning a blue followed by a red in 2 spins on the spinner in Figure 16.17. Explain your reasoning.

8. Determine the probability of spinning a red followed by a yellow in 2 spins on the spinner in Figure 16.17. Explain your reasoning.

9. Suppose you have a penny, a nickel, a dime, and a quarter in a bag. You shake the bag and dump out the coins. What is the probability that exactly 2 coins will land heads up and 2 coins will land heads down? Is the probability 50% or not? Solve this problem by drawing a tree diagram that shows all the possible outcomes when you dump the coins out. Explain.

10. You have a bag containing 2 yellow and 3 blue blocks.

a. You reach in and randomly pick a block, record its color, and put the block back in the bag. You then reach in again and randomly pick another block. What is the probability of picking 2 yellow blocks? Explain your answer.

b. You reach in and randomly pick a block, record its color, and without replacing the first block, you reach in again and randomly pick another block. What is the probability of picking 2 yellow blocks? Explain your answer.

11. There are 3 plastic bears in a bag. The teacher tells Bob that there are 3 bears in the bag, but she doesn't tell him what color the bears are. Bob picked a bear out of the bag 3 times and each time he got a red bear. (He puts the bear he picked back in the bag before picking the next bear).

a. Bob says that the 3 bears in the bag must all be red. Is he necessarily correct?

b. Now suppose that of the 3 bears in the bag, 2 are red and 1 is blue. Calculate the probability of picking 3 red bears in 3 picks.

12. There are 4 black marbles and 5 red marbles in a bag. If you reach in and randomly select 2 marbles, what is the probability that both are red? Explain your reasoning clearly.

13. Suppose you have 100 light bulbs and one of them is defective. If you pick out 2 light bulbs at random (either both at the same time, or first one, then another, without replacing the first light bulb), what is the probability that one of your chosen light bulbs is defective? Explain your answer.

14. *A game at a fund-raiser:* There are 20 rubber ducks floating in a pool. One of the ducks has a mark on the bottom, indicating that the contestant wins a prize. Each contestant pays 25 cents to play. A contestant randomly picks 2 of the ducks. If the contestant picks the duck with the mark on the bottom, the contestant wins a prize that costs $1.

a. What is the probability that a contestant will win a prize? Explain.

b. If 200 people play the game, about how many people would you expect to win? Why?

c. Based on your answer to part (b), how much money should the duck game be expected to earn for the fund-raiser (net) if 200 people play it? Explain.

15. You are making up a game for a fund-raiser. You take 5 table-tennis balls, number them from 1 to 5, and put all 5 in a brown bag. Contestants will pick the 5 balls out of the bag one at a time, without looking, and line the balls up in the order they were picked. If the contestant picks 1, 2, 3, 4, 5, in that order, then the contestant wins a prize of $2. Otherwise, the contestant wins nothing. Each contestant pays $1 to play.

a. What is the probability that a contestant will win the prize? Explain your reasoning.

b. If 240 people play your game, then approximately how many people would you expect to win the prize? Why?

c. Based on your answer to part (b), about how much money would you expect your game to earn (net) for the fund-raiser if 240 people play? Explain.

16. a. A waitress is serving 5 people at a table. She has the 5 dishes they ordered (all 5 are different), but she can't remember who gets what. How many different possible ways are there for her to give the 5 dishes to the 5 people? Explain your answer.

b. Based on part (a), if the waitress just hands the dishes out randomly, what is the probability that she will hand the dishes out correctly?

17. Social Security numbers have 9 digits and are presented in the form

$$xxx - xx - xxxx$$

where each x can be any digit from 0 to 9.

a. How many different Social Security numbers can be made this way? Explain.

b. If the population of the United States is 300 million people, and if everybody living in the United States has a Social Security number, then what fraction of the possible Social Security numbers are in current use? Explain your answer.

***18.** Standard dice are shaped like cubes. They have 6 faces, each of which is a square. Because there are 6 faces on a standard die and because each face is equally likely to land face up, the probability that a given face will land up is $\frac{1}{6}$. Could you make other three-dimensional "dice" so that for each face, the probability of that face landing up is $\frac{1}{4}$? $\frac{1}{8}$? $\frac{1}{10}$? $\frac{1}{12}$? Other fractions? Explain your answers. *Hint:* See Section 13.1.

16.4 Using Fraction Arithmetic to Calculate Probabilities

The probability of an outcome of an experiment can be calculated by determining the ideal fraction of times the given outcome should occur if the experiment were performed many times. When we apply this point of view, we will see how to use fraction arithmetic to calculate probabilities.

Why Can We Multiply Fractions to Calculate Probabilities?

We can use basic principles of probability and the meaning of fraction multiplication to calculate certain probabilities. In order to do so, we will rely on the following way of thinking about the outcome of an experiment: Consider what would happen ideally if the experiment were repeated a very large number of times. Calculate directly the ideal fraction of times the outcome you are interested in occurs. This fraction is the probability of that outcome.

For example, if we roll a number cube (die) 2 times, what is the probability that we will get two 1s in a row? Think about doing the experiment of rolling a number cube 2 times in a row many times over. The numbers 1, 2, 3, 4, 5, and 6 are all equally likely to come up on one roll of a number cube; therefore, in the ideal, the first roll will be a 1 in $\frac{1}{6}$ of the rolls. Now consider those $\frac{1}{6}$ of the experiments when the first roll is a 1. Of those double rolls in which the first roll is a 1, in the ideal, the second roll will be a 1 in $\frac{1}{6}$ of the time, too, because the outcomes of the first and second rolls are independent. Therefore, in the ideal, in $\frac{1}{6}$ of $\frac{1}{6}$ of all the experiments, both rolls will be a 1. According to the meaning of multiplication, "$\frac{1}{6}$ of $\frac{1}{6}$ of all the experiments" is

$$\frac{1}{6} \cdot \frac{1}{6} = \frac{1}{36}$$

of all the experiments. Therefore, in the ideal, in $\frac{1}{36}$ of all the experiments, you will roll two 1s, so the probability of rolling two 1s in a row on 2 rolls of a number cube is $\frac{1}{36}$.

We can also see the "$\frac{1}{6}$ of $\frac{1}{6}$ of all the experiments" pictorially in **Figure 16.18**. The large rectangle represents "all the experiments." The lightly shaded region represents $\frac{1}{6}$ of the experiments. The darkly shaded region represents $\frac{1}{6}$ of the lightly shaded region, or $\frac{1}{6}$ of $\frac{1}{6}$ of all the experiments. In this case, this probability can also be calculated with an array or a tree diagram.

Figure 16.18

Showing how $\frac{1}{6}$ of $\frac{1}{6}$ of the experiments is $\frac{1}{6} \cdot \frac{1}{6} = \frac{1}{36}$ of the experiments.

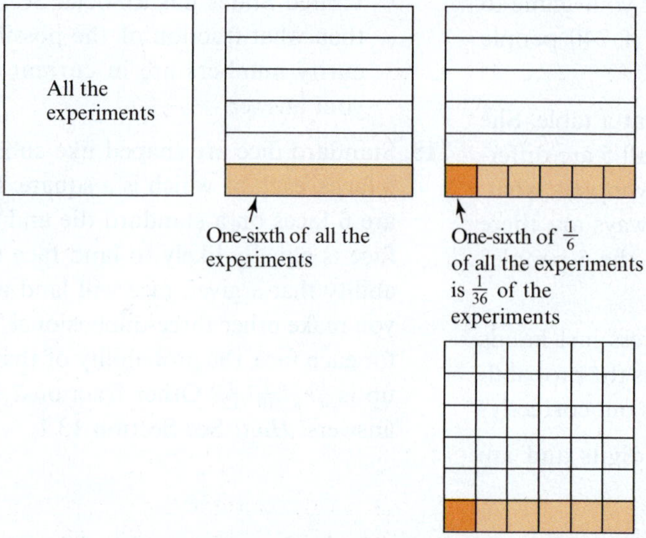

CLASS ACTIVITY

16L Using the Meaning of Fraction Multiplication to Calculate a Probability, p. CA-352

FROM THE FIELD Research

Dowd, D. S. (2013). A circle model for multiplying probabilities. *Mathematics Teaching in the Middle School*, *18*(8), pp. 464-466.

The author describes a model that helps her teach probabilities of compound events. To illustrate, she uses the problem of picking two chocolates out of a box containing 1 milk chocolate and 3 dark chocolates. Her model involves partitioning a circle into parts based on the number of choices at the first stage. The circle is expanded to include a ring around the circle. The circle and ring are partitioned based on the choices for the second stage, thus creating a dart-board–like model with the outer band indicating all possible outcomes. Using this model, students can find the probability of a compound event. The author uses the model to help students understand why calculating two-stage probabilities involves multiplying probabilities and why multiplying by a fraction between one and zero reduces the original number. The model applies to compound events with and without replacements.

Why Can We Use Fraction Multiplication and Addition to Calculate Probabilities?

CLASS ACTIVITY

16M Using Fraction Multiplication and Addition to Calculate a Probability, p. CA-354

Besides using fraction multiplication, we can use a combination of multiplication and addition to calculate probabilities of compound events. Again, think about performing an experiment a large number of times. View the probability of a given event as the fraction of times the event should occur in the ideal.

For example, consider the game that consists of choosing a rubber duck from among 5 rubber ducks floating in a duck pond and then choosing a rubber frog from among 4 frogs in a frog pond. One of the ducks and 3 of the frogs are marked with Xs on the bottom, as indicated in Figure 16.19. The other 4 ducks and the other frog are marked with Os on the bottom. To win the game, both the duck and the frog that are chosen should have the same marking on the bottom (either both X or both O). The markings cannot be seen when the ducks and frogs are chosen; all the ducks look alike and all the frogs look alike. What is the probability of winning this game?

Figure 16.19

A game: picking a duck and a frog.

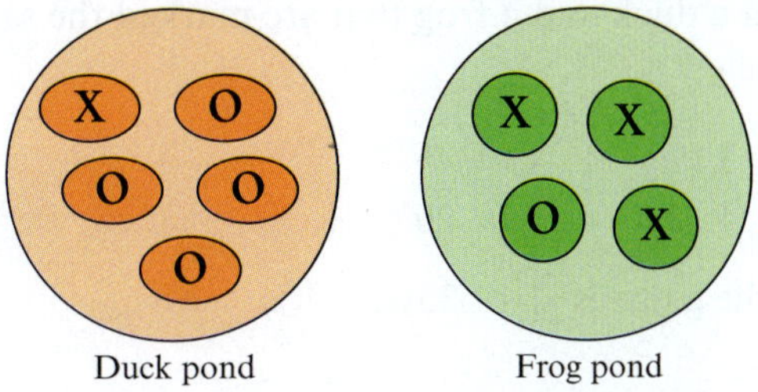

Think about playing this game of picking a duck and a frog many times over. Sometimes both animals chosen will be marked with Xs; sometimes both will be Os; sometimes one will be marked with an X and the other with an O. Each of these events occurs a certain fraction of the time, and none of these different events can occur at the same time. So the fraction of times that the game will be won is the fraction of times that both animals have an X plus the fraction of times both animals have an O.

So let's first find the fraction of times when both the duck and the frog that are chosen should have Xs, in the ideal. Since 1 out of the 5 ducks is marked with an X, the duck with an X will be chosen $\frac{1}{5}$ of the time, in the ideal. Of those times when the duck with an X is chosen, the fraction of times that the frog chosen has an X is $\frac{3}{4}$, in the ideal, since 3 out of the 4 frogs have Xs. So in the ideal, in $\frac{3}{4}$ of $\frac{1}{5}$ of the times, both the duck and the frog that are chosen will have an X. Since "$\frac{3}{4}$ of $\frac{1}{5}$" is

$$\frac{3}{4} \cdot \frac{1}{5}$$

we can find the probability that both the duck and the frog that are chosen have an X by multiplying $\frac{3}{4}$ and $\frac{1}{5}$. We can also see this probability represented pictorially on the left in Figure 16.20.

Figure 16.20

Probability that both the duck and the frog chosen are marked the same.

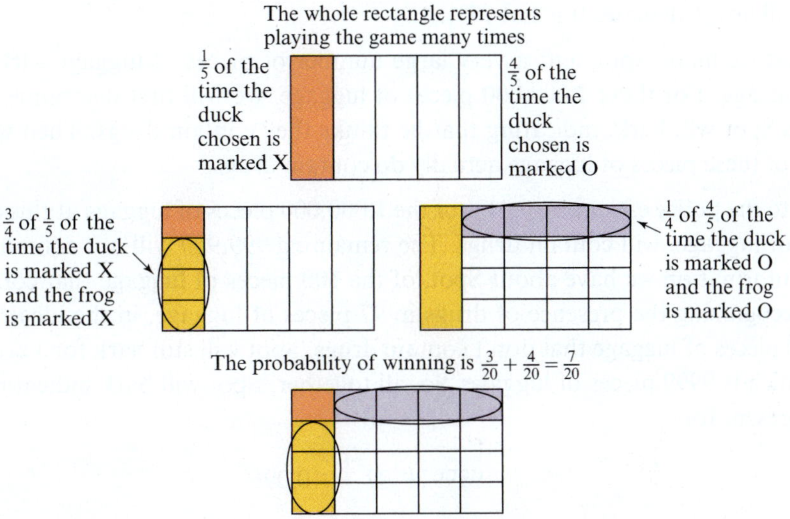

What is the probability that both animals chosen will be marked with an O? In the ideal, the duck that is chosen will be marked with an O $\frac{4}{5}$ of the time, since 4 out of the 5 ducks are marked with an O. Of those times when a duck with an O is chosen, the frog that is chosen should be marked with an O $\frac{1}{4}$ of the time, in the ideal, since 1 out of the 4 frogs has an O. So in the ideal, in $\frac{1}{4}$ of $\frac{4}{5}$ of the times, both the duck and the frog will have an O. Since "$\frac{1}{4}$ of $\frac{4}{5}$" is

$$\frac{1}{4} \cdot \frac{4}{5}$$

this product is the probability that both the duck and the frog that are chosen have an O. We can also see this probability represented pictorially on the right in Figure 16.20.

Finally, the probability of picking both a duck and a frog that are marked the same is the sum of the 2 probabilities just calculated, namely,

$$\frac{3}{4} \cdot \frac{1}{5} + \frac{1}{4} \cdot \frac{4}{5} = \frac{7}{20}$$

Therefore, the probability of winning the game is $\frac{7}{20}$, or 35%.

The Surprising Case of Spot, the Drug-Sniffing Dog

Now let's look at a more advanced probability calculation. As in the previous example, we first consider the outcome of a large number of experiments in the ideal. We then divide the number of times the outcome we are interested in occurs by the total number of experiments to calculate the probability that the outcome we are interested in occurs.

Spot is a dog who has been trained to sniff luggage to detect drugs. But Spot isn't perfect. His trainers conducted careful experiments and found that in luggage that *doesn't* contain any drugs, Spot will nevertheless bark to indicate the presence of drugs for 1% of such luggage. In luggage that *does* contain drugs, Spot will bark to indicate he thinks drugs are present for about 97% of this luggage. (So he fails to recognize the presence of drugs in about 3% of luggage containing drugs.)

Think of the consequences of putting Spot in a busy airport where police estimate that 1 in 10,000 pieces of luggage (or 0.01%) passing through that airport contain drugs. Here's a question to answer before placing Spot in a busy airport:

> If Spot barks to indicate that a piece of luggage contains drugs, what is the probability that the luggage actually does contain drugs? (Remember, Spot is not perfect; sometimes he barks when there are no drugs.)

Before you read on, guess what this probability is. You might be surprised at the actual answer, which we will now calculate. It is not 97%.

Imagine that we made Spot sniff a very large number of pieces of luggage—let's say, 1,000,000 pieces of luggage. For these 1,000,000 pieces of luggage, we will first determine in the ideal how many times Spot will bark, indicating that he thinks they contain drugs. Then we will determine how many of these pieces of luggage actually do contain drugs.

According to the police estimates, 0.01% of the 1,000,000 pieces of luggage at this airport, or about 100 pieces of luggage, will contain drugs. The remaining 999,900 will not contain drugs. According to the information we have about Spot, of the 100 pieces of luggage that contain drugs, Spot will bark recognizing the presence of drugs in 97 pieces of luggage, in the ideal. In the ideal, for the 999,900 pieces of luggage that don't contain drugs, Spot will still bark for 1% of these pieces of luggage, which is 9999 pieces of luggage. So, all together, Spot will bark indicating that he thinks drugs are present for

$$9999 + 97 = 10{,}096$$

pieces of luggage, in the ideal. But of those 10,096 pieces of luggage, only 97 actually contain drugs (because the other 9999 of the 10,096 were ones where Spot barked, but they didn't actually contain any drugs). So, of the 10,096 times that Spot barks, indicating he thinks there are drugs, only 97 of those times are drugs *actually present.* Now,

$$\frac{97}{10{,}096} = 0.0096\ldots,\ \text{about } 0.96\%$$

so that, of the times when Spot barks, indicating he thinks drugs are present, only 0.96% of the time are drugs actually present. So if Spot barks to indicate he *thinks* drugs are present, the probability that drugs actually *are* present is 0.96%, which is *less than 1%.* Even though Spot's statistics sounded pretty good at the start, we might think twice about putting him on the job, since many innocent people could be delayed at the airport.

SECTION SUMMARY AND STUDY ITEMS

Section 16.4 Using Fraction Arithmetic to Calculate Probabilities

One way to determine the probability of an outcome of an experiment is to think about performing the experiment many times. The probability of that outcome occurring is the fraction of times that the outcome should occur in the ideal. By applying this point of view, we can see how to calculate probabilities with fraction arithmetic.

Key Skills and Understandings

1. Explain why certain probabilities can be calculated by the multiplication of fractions.
2. Use fraction arithmetic to determine certain probabilities, and explain why the method of calculation makes sense.

Practice Exercises for Section 16.4

1. What is the probability of spinning a yellow followed by a blue in 2 spins on the spinner in Figure 16.21? Explain how to solve this problem with fraction multiplication, and explain why this method makes sense.

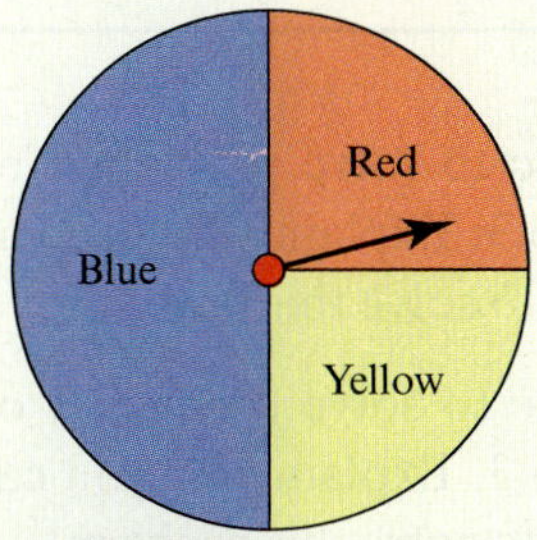

Figure 16.21 A spinner.

2. At a doughnut factory, 3 of the 200 boxes of doughnuts are bad. Each bad box contains only 10 doughnuts (instead of the usual 12), and in each bad box, 3 of the 10 doughnuts are tainted. (No doughnuts in any other boxes are tainted.) If a person buys a random box of these doughnuts and eats a random doughnut from the box, what is the probability that the person will eat a tainted doughnut? Explain how to solve this problem with fraction multiplication, and explain why the method makes sense.

3. You are on a game show where you can pick 1 of 3 doors. A prize is placed randomly behind 1 of the 3 doors; the other doors do not have a prize behind them. You pick a door. The host opens 1 of the other doors and it does not have the prize behind it. The host then offers you the opportunity to switch to the other unopened door before revealing where the prize is.

a. What is the probability that you will win the prize if you stick with your door?

b. What is the probability that you will win the prize if you switch to the other unopened door? Solve this problem by imagining that you could play the game a large number of times; determine what fraction of the time you would win if you switched to the other unopened door each time you played.

c. If you want to win the prize, what strategy is better: sticking with the first door you chose or switching to the other unopened door? Explain.

Answers to Practice Exercises for Section 16.4

1. Imagine spinning the spinner twice in a row many times. In the ideal, in $\frac{1}{4}$ of the times, the first spin will land on yellow. In the ideal, in $\frac{1}{2}$ of those times when the first spin is yellow, the second spin will be blue, since the second spin does not depend on the first spin. Therefore, in the ideal, in $\frac{1}{2}$ of $\frac{1}{4}$ of the times, the spinner will land on yellow followed by blue. Since $\frac{1}{2}$ of $\frac{1}{4}$ is

$$\frac{1}{2} \cdot \frac{1}{4}$$

the probability of spinning yellow followed by blue is

$$\frac{1}{2} \cdot \frac{1}{4} = \frac{1}{8}$$

2. Imagine that the person could buy a random box of doughnuts and eat a random doughnut from the box many times. Then in the ideal, in $\frac{3}{200}$ of the times, the person will get a bad box. Of the $\frac{3}{200}$ of the times the person gets a bad box, the person will pick a tainted doughnut $\frac{3}{10}$ of the time, in the ideal. Therefore, in $\frac{3}{10}$ of $\frac{3}{200}$ of the times, the person will eat a tainted doughnut from a bad box. Since $\frac{3}{10}$ of $\frac{3}{200}$ is

$$\frac{3}{10} \cdot \frac{3}{200}$$

the probability of eating a tainted doughnut is

$$\frac{3}{10} \cdot \frac{3}{200} = \frac{3 \cdot 3}{10 \cdot 200} = \frac{9}{2000} = 0.45\%$$

which is less than half a percent.

3. a. Since the prize is placed randomly behind 1 of the 3 doors, your probability of winning the prize is $\frac{1}{3}$ if you stick with the door you picked originally.

b. If you played the game many times, then in the ideal, in $\frac{1}{3}$ of those times, the prize would be behind the first door you pick. So in these cases, when you switch, you will not get the prize. But in the $\frac{2}{3}$ of the times when the door you picked does not have the prize behind it, the other unopened door must have the prize behind it. So when you switch doors, you will get the prize. Therefore, when you switch doors, you will win the prize $\frac{2}{3}$ of the times, in the ideal.

c. The switching strategy is better because when you switch, your probability of winning is $\frac{2}{3}$, whereas when you don't switch, your probability of winning is only $\frac{1}{3}$.

PROBLEMS FOR SECTION 16.4

1. A children's game has a spinner that is equally likely to land on any 1 of 4 colors: red, blue, yellow, or green. What is the probability of spinning a red followed by a green in 2 spins? Explain how to solve this problem with fraction multiplication, and explain why this method makes sense.

2. Suppose you flip a coin and roll a number cube (die). What is the probability of getting a head and rolling a 6? (The numbers 1 through 6 are all equally likely to occur.) Explain how to solve this problem with fraction multiplication, and explain why this method makes sense.

3. Use fraction arithmetic to solve problem 1 on page 741 of Section 16.3. Explain why you can use fraction arithmetic to solve this problem.

4. Use fraction arithmetic to solve problem 3 on page 741 of Section 16.3. Explain why you can use fraction arithmetic to solve this problem.

5. Use fraction arithmetic to solve problem 6 on page 742 of Section 16.3. Explain why you can use fraction arithmetic to solve this problem.

6. Use fraction arithmetic to solve problem 7 on page 742 of Section 16.3. Explain why you can use fraction arithmetic to solve this problem.

7. Use fraction arithmetic to solve problem 8 on page 742 of Section 16.3. Explain why you can use fraction arithmetic to solve this problem.

8. There are 3 boxes, one of which contains 2 envelopes, and the other 2 boxes contain 3 envelopes each. In the box with the 2 envelopes, 1 of the 2 envelopes contains a prize. No other envelope contains a prize. If you pick a random box and a random envelope from that box, what is the probability that you will win the prize? Explain how to solve this problem with fraction multiplication, and explain why this method makes sense.

9. A game consists of spinning a spinner and then rolling a number cube. The spinner is equally likely to land on any 1 of the 4 colors red, yellow, green, or blue. The number cube is equally likely to land with any of the 6 sides labeled 1 through 6 up. To win the game, a contestant must either spin red and roll any number or spin a color other than red and roll a 6. In other words, the contestant must either spin red or roll a 6 to win. What is the probability of winning this game? Explain why you can use fraction arithmetic to solve this problem.

10. A game consists of spinning a spinner and then rolling a number cube. The spinner is equally likely to land on any 1 of the 4 colors red, yellow, green, or blue. The number cube is equally likely to land with any of the 6 sides labeled 1 through 6 up. To win the game, a contestant must either spin red and roll a 1 or spin green and roll a 6. What is the probability of winning this game? Explain why you can use fraction arithmetic to solve this problem.

11. A game consists of spinning a spinner and then rolling a number cube. The spinner is equally likely to land on any 1 of the 4 colors red, yellow, green, or blue. The number cube is equally likely to land with any of the 6 sides labeled 1 through 6 up. To win the game, a contestant must either spin a red or a yellow and roll a 1 or a 2 or spin a green or blue and roll a 5 or a 6. What is the probability of winning this game? Explain why you can use fraction arithmetic to solve this problem.

12. Use fraction arithmetic to solve problem 2 on page 741 of Section 16.3. Explain why you can use fraction arithmetic to solve this problem.

13. Use fraction arithmetic to solve problem 4 on page 741 of Section 16.3. Explain why you can use fraction arithmetic to solve this problem.

14. Use fraction arithmetic to solve problem 5 on page 742 of Section 16.3. Explain why you can use fraction arithmetic to solve this problem.

15. Suppose you have 2 boxes, 50 black pearls and 50 white pearls. You can mix the pearls up any way you like and put them all back into the 2 boxes. You don't have to put the same number of pearls in each box, but each box must have at least 1 pearl in it. You will then be blindfolded, and you will get to open a random box and pick a random pearl out of it.

a. Suppose you put 25 black pearls in box 1 and 25 black pearls and 50 white pearls in box 2. Calculate the probability of picking a black pearl by imagining that you were going to pick a random box and a random pearl in the box a large number of times.

In the ideal, what fraction of the time would you pick box 1? What fraction of those times that you picked box 1 would you pick a black pearl from box 1, in the ideal?

On the other hand, in the ideal, what fraction of the time would you pick box 2? What fraction of those times that you picked box 2 would you pick a black pearl from box 2, in the ideal?

Overall, what fraction of the time would you pick a black pearl, in the ideal?

b. Describe at least 2 other ways to distribute the pearls than what is described in part (a). Find the probability of picking a black pearl in each case.

c. Try to find a way to arrange the pearls so that the probability of picking a black pearl is as large as possible.

16. Due to its high population, China has a stringent policy on having children. In rural China, couples are allowed to have either 1 or 2 children according to the following rule: If their first child is a boy, they are not allowed to have any more children. If their first child is a girl, then they are allowed to have a second child. Let's assume that all couples follow this policy and have as many children as the policy allows.

a. Do you think this policy will result in more boys, more girls, or about the same number of boys as girls being born? (Answer without performing any calculations—just make a guess.)

b. Now consider a random group of 100,000 rural Chinese couples who will have children. Assume that any time a couple has a child the probability of having a boy is $\frac{1}{2}$.

i. In the ideal, how many couples will have a boy as their first child, and how many will have a girl as their first child?

ii. In the ideal, of those couples who have a girl first, how many will have a boy as their second child and how many will have a girl as their second child?

iii. Therefore, overall, what fraction of the children born in rural China under the given policy should be boys and what fraction should be girls?

17. The Pretty Flower Company starts plants from seed and sells the seedlings to nurseries. They know from experience that about 60% of the calla lily seeds they plant will sprout and become a seedling. Each calla lily seed costs 20 cents, and a pot containing at least one sprouted calla lily seedling can be sold for \$2.00. Pots that don't contain a sprouted seedling must be thrown out. The company figures that costs for a pot, potting soil, water, fertilizer, fungicide and labor are \$0.30 per pot (whether or not a seed in the pot sprouts). The Pretty Flower Company is debating between planting 1 or 2 seeds per pot. Help them figure out which choice will be more profitable by working through the following problems:

a. Suppose the Pretty Flower Company plants 1 calla lily seed in each of 100 pots. Using the previous information, approximately how much profit should the Pretty Flower Company expect to make on these 100 pots? Profit is income minus expenses.

b. Now suppose that the Pretty Flower Company plants 2 calla lily seeds (1 on the left, 1 on the right) in each of 100 pots. Assume that whether or not the left seed sprouts has no influence on whether or not the right seed sprouts. So, the right seed will still sprout in about 60% of the pots in which the left seed does not sprout. Explain why the Pretty Flower Company should expect about 84 of the 100 pots to sprout at least 1 seed.

c. Using part (b), determine how much profit the Pretty Flower Company should expect to make on 100 pots if 2 calla lily seeds are planted per pot. Compare your answer with part (a). Which is expected to be more profitable: 1 seed or 2 seeds per pot?

d. What if calla lily seeds cost 50 cents each instead of 20 cents each (but everything else stays the same)? Now which is expected to be more profitable: 1 seed or 2 seeds per pot?

***18.** Suppose that in a survey of a large, random group of people, 17% were found to be smokers and 83% were not smokers. Suppose further that 0.08% of the smokers and 0.01% of the nonsmokers from the group died of lung cancer. (Be careful: Notice the decimal points in these percentages—they are not 8% and 1%.)

a. What percent of the group died of lung cancer?

b. Of the people in the group who died of lung cancer, what percent were smokers? (It may help you to make up a number of people for the large group and to calculate the number of people who died of lung cancer and the number of people who were smokers and also died of lung cancer.)

***19.** Suppose that 1% of the population has a certain disease. Also suppose that there is a test for the disease, but it is not completely accurate: It has a 2% rate of false positives and a 1% rate of false negatives. This means that the test *reports* that 2% of the people who don't have the disease do have it, and the test *reports* that 1% of the people who do have the disease don't have it. This problem is about the following question:

If a person tests positive for the disease, what is the probability that he or she actually has the disease?

Before you start answering the next set of questions, go back and read the beginning of the problem again. Notice that there is a difference between *actually having* the disease and *testing positive* for the disease, and there is a difference between *not having* the disease and *testing negative* for it.

a. What do you think the answer to the previous question is? (Answer without performing any calculations—just make a guess.)

b. For parts (b) through (f), suppose there is a random group of 10,000 people and all of them are tested for the disease.

Of the 10,000 people, in the ideal, how many would you expect to actually have the disease and how many would you expect not to have the disease?

c. Continuing part (b), of the people from the group of 10,000 who do not have the disease, how many would you expect to test positive, in the ideal? (Give a number, not a percentage.)

d. Continuing part (b), of the people from the group of 10,000 who have the disease, how many would you expect to test positive? (Give a number, not a percentage.)

e. Using your work in parts (c) and (d), of the people from the group of 10,000, how many in total would you expect to test positive, in the ideal?

f. Using your previous work, what percent of the people who test positive for the disease should actually have the disease, in the ideal? How does this compare to your guess in part (a)? Are you surprised at the actual answer?

CHAPTER SUMMARY

Section 16.1 Basic Principles of Probability

Page 723

- The probability that a given event of a chance process will occur is the fraction of times that event should occur "in the ideal." Several principles of probability are key: (1) if two events of a chance process are equally likely, then their probabilities are equal; (2) if there are several events of a chance process that cannot occur simultaneously, then the probability that one of those events will occur is the sum of the probabilities of each individual event; and (3) if a chance process occurs many times, the fraction of times that a given event occurs is likely to be close to the probability of that event occurring. Using these principles, we can determine probabilities in many simple cases. When a chance process occurs a number of times, the fraction of times that a given event occurred is the empirical probability of that event occurring. Page 723

Key Skills and Understandings

1. Use principles of probability to determine probabilities in simple cases. Page 724
2. Recognize that an empirical probability is likely to be close to the theoretical probability when a chance process has occurred many times. Page 725
3. Apply empirical probability to make estimates. Page 725

Section 16.2 Counting the Number of Outcomes

Page 730

- We can often use multiplication to calculate the number of outcomes in two-stage and multistage experiments and other situations. Page 730

Key Skills and Understandings

1. Apply multiplication to count the total number of outcomes in various situations, including those of multistage experiments and cases where the outcome at one stage depends on the outcome at previous stages. Page 731

Section 16.3 Calculating Probabilities of Compound Events	Page 736
▪ We can apply the counting techniques of Section 16.2 to calculate probabilities for multistage experiments. A useful application of probability is the calculation of expected value.	Page 736
Key Skills and Understandings	
1. Apply the counting techniques of Section 16.2 to calculate probabilities for multistage experiments, including cases where the outcome at one stage depends on the outcome at previous stages.	Page 736
2. Calculate an expected amount of earnings when probabilities are involved.	Page 738
Section 16.4 Using Fraction Arithmetic to Calculate Probabilities	Page 743
▪ One way to determine the probability of an outcome of an experiment is to think about performing the experiment many times. The probability of that outcome occurring is the fraction of times that the outcome should occur in the ideal. By applying this point of view, we can see how to calculate probabilities with fraction arithmetic.	Page 743
Key Skills and Understandings	
1. Explain why certain probabilities can be calculated by the multiplication of fractions.	Page 743
2. Use fraction arithmetic to determine certain probabilities, and explain why the method of calculation makes sense.	Page 744

DOWNLOADS

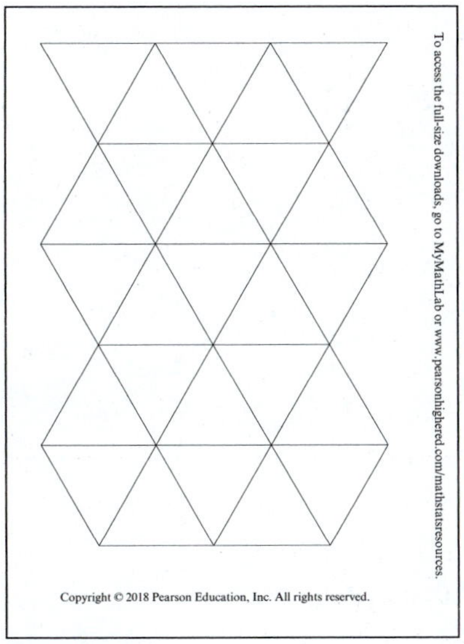

Download 1

Download 2

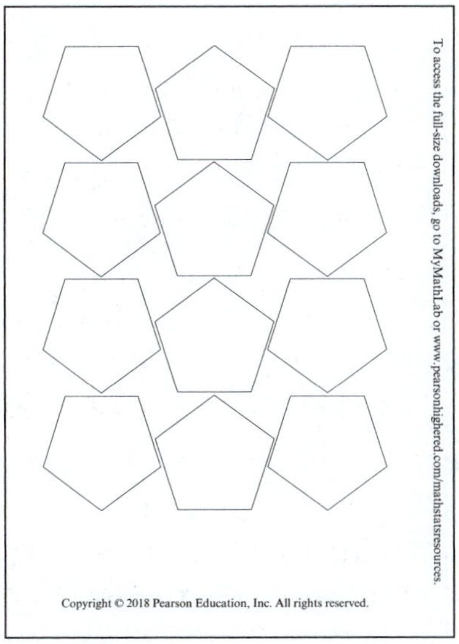

Download 3

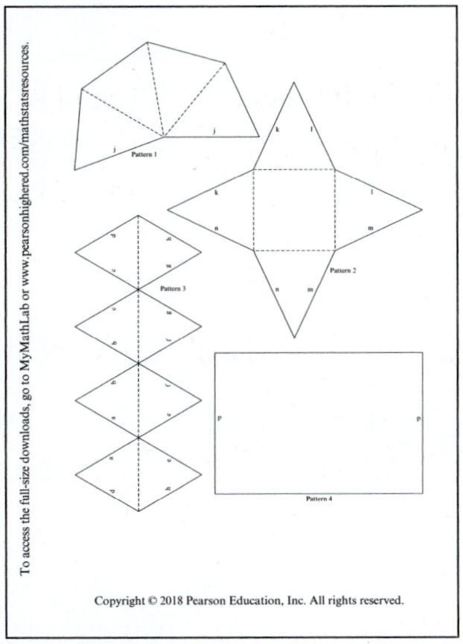

Download 4

Download 5

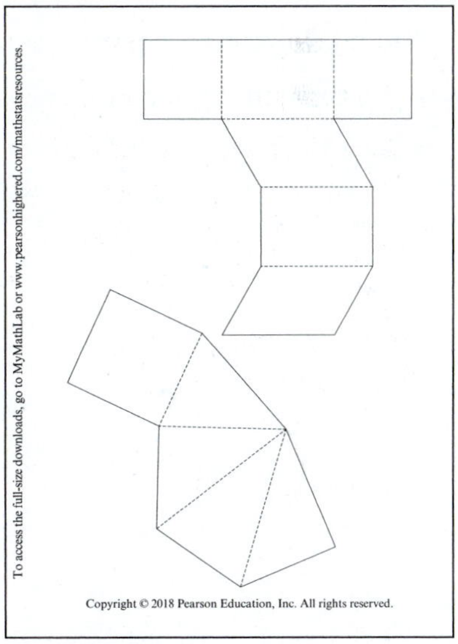

Download 6

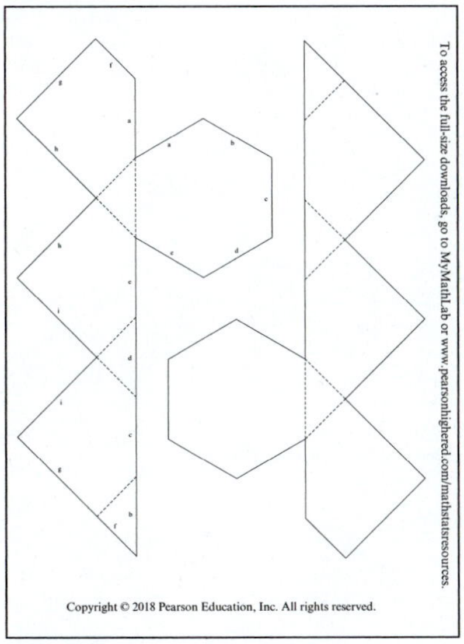

Download 7

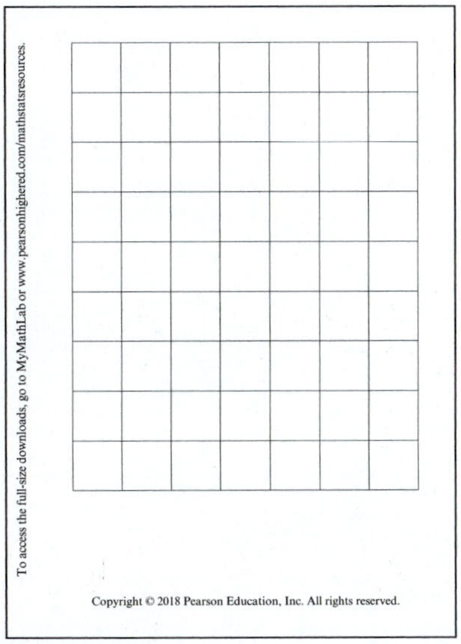

Download 8

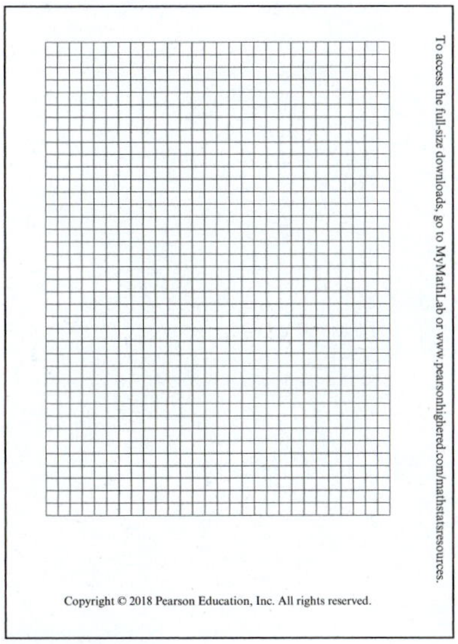

Download 9

To access the full-size downloads, go to MyMathLab or www.pearsonhighered.com/mathstatsresources.

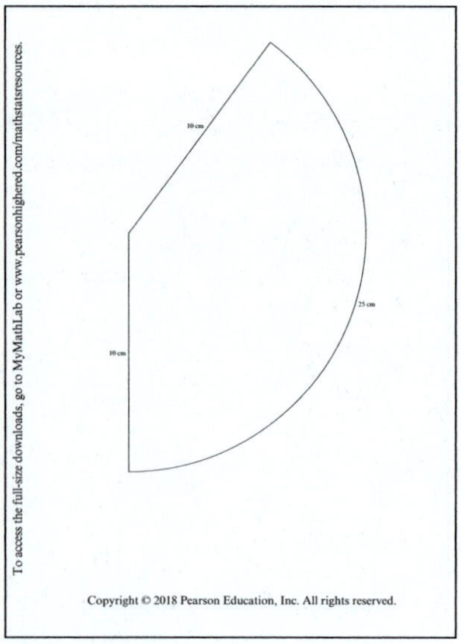
10 cm
25 cm
10 cm
To access the full-size downloads, go to MyMathLab or www.pearsonhighered.com/mathstatsresources.
Copyright © 2018 Pearson Education, Inc. All rights reserved.

Download 10

BIBLIOGRAPHY

[1] Edwin Abbott. *Flatland.* Princeton University Press, 1991.

[2] Alan Agresti and Christine Franklin. *Statistics, the Art and Science of Learning from Data.* Pearson Prentice Hall, 2007.

[3] American Statistical Association. A Curriculum Framework for PreK–12 Statistics Education. Available at *http://www.amstat.org/education/gaise/* and *http://www.amstat.org/education/gaise/GAISEPreK-12.htm*

[4] W. S. Anglin. *Mathematics: A Concise History and Philosophy*. Springer-Verlag, 1994.

[5] Deborah Loewenberg Ball. Prospective elementary and secondary teachers' understanding of division. *Journal for Research in Mathematics Education*, 21(2):132–144, 1990.

[6] P. Barnes-Svarney, *New York Public Library Science Desk Reference*. Stonesong Press, 1995.

[7] Tom Bassarear. *Mathematics for Elementary School Teachers*. Houghton Mifflin, 1997.

[8] Peter Beckmann. *A History of Pi*. St. Martin's Press, 1971.

[9] Sybilla Beckmann, *Focus in Grade 5: Teaching with Curriculum Focal Points.* National Council of Teachers of Mathematics, 2009.

[10] E. T. Bell. *The Development of Mathematics*. Dover, 1992. Originally published in 1945.

[11] George W. Bright. Helping elementary- and middle-grades preservice teachers understand and develop mathematical reasoning. In *Developing Mathematical Reasoning in Grades K–12*, pages 256–269. National Council of Teachers of Mathematics, 1999.

[12] Bureau of Labor Statistics, U.S. Department of Labor. Core subjects and your career. *Occupational Outlook Quarterly*, pages 26–40, Summer 1999. Available online at *http://stats.bls.gov/opub/ooq/ooqhome.htm*

[13] Bureau of Labor Statistics, U.S. Department of Labor. More education: Higher earnings, lower unemployment. *Occupational Outlook Quarterly*, page 40, Fall 1999. Available online at *http://www.bls.gov/opub/ooq/ooqhome.htm*

[14] California State Board of Education. *Mathematics Framework for California Public Schools*, 1999.

[15] The Carnegie Library of Pittsburgh Science and Technology Department. *Science and Technology Desk Reference*, 1993.

[16] T. P. Carpenter, E. Fennema, M. L. Franke, S. B. Empson, and L. W. Levi. *Children's Mathematics: Cognitively Guided Instruction*. Heinemann, Portsmouth, NH, 1999.

[17] Suzanne H. Chapin, Catherine O'Connor, and Nancy Canavan Anderson. *Classroom Discussions, Using Math Talk to Help Students Learn.* Math Solutions Publications, 2003.

[18] Conference Board of the Mathematical Sciences (CBMS). *The Mathematical Education of Teachers*– Volume 11 of *CBMS Issues in Mathematics Education*. The American Mathematical Society and the Mathematical Association of America, 2001. See also *http://www.cbmsweb.org/MET_Document/index.htm*

[19] Frances Curcio. *Developing Data-Graph Comprehension in Grades K–8*. National Council of Teachers of Mathematics, 2nd ed., 2001.

[20] Stanislas Dehaene. *The Number Sense*. Oxford University Press, 1997.

[21] Demi. *One Grain of Rice*. Scholastic, 1997.

[22] Carol S. Dweck. *Mindset, The New Psychology of Success.* Random House, Inc., 2006.

[23] Tatiana Ehrenfest-Afanassjewa. *Uebungensammlung zu einer Geometrischen Propaedeuse*. Martinus Nijhoff, 1931.

[24] Encyclopaedia Britannica *Encyclopaedia Britannica*. Encyclopaedia Britannica, Inc., 1994. Available at *www.britannica.com*

[25] Euclid. *The Thirteen Books of the Elements, Translated with Commentary by Sir Thomas Heath*. Dover, 1956.

[26] David Freedman, Robert Pisani, and Roger Purves. *Statistics*. W. W. Norton and Company, 1978.

[27] K. C. Fuson. Developing mathematical power in whole number operations. In J. Kilpatrick, W. G. Martin, and D. Schifter, eds. *A Research Companion to Principles and Standards for School Mathematics*. National Council of Teachers of Mathematics, 2003.

[28] K. C. Fuson. *Math Expressions.* Houghton Mifflin Company, 2006.

[29] K. C. Fuson, S. T. Smith, and A. M. Lo Cicero. Supporting first graders' ten-structured thinking in urban classrooms. *Journal for Research in Mathematics Education*, vol. 28, Issue 6, pages 738–766, 1997.

[30] Georgia Department of Education. Georgia Performance Standards, 2005. Available at *http://www.georgiastandards.org*

[31] Georgia Department of Education. *Georgia Quality Core Curriculum.*

[32] Gersten, R., Beckmann, S., Clarke, B., Foegen, A., Marsh, L., Star, J. R., & Witzel, B. *Assisting students struggling with mathematics: Response to Intervention (RtI) for elementary and middle schools* (NCEE 2009-4060), 2009.

[33] Alexander Givental. The Pythagorean theorem: What is it about? *The American Mathematical Monthly*, vol. 113, number 3, March 2006.

[34] Sir Thomas Heath. *Aristarchus of Samos*. Oxford University Press, 1959.

[35] N. Herscovics and L. Linchevski. A cognitive gap between arithmetic and algebra. *Educational Studies in Mathematics*, 27:59–78, 1994.

[36] Heisuke Hironaka and Yoshishige Sugiyama, ed., *Mathematics for Elementary School.* Tokyo Shoseki Co., Ltd. 2006 Available at *www.globaledresources.com*, 2006.

[37] Infoplease. *http://www.infoplease.com*

[38] H. G. Jerrard and D. B. McNeill. *Dictionary of Scientific Units*. Chapman and Hall, 1963.

[39] Graham Jones and Carol Thornton. *Data, Chance, and Probability, Grades 1–3 Activity Book*. Learning Resources, 1992.

[40] Constance Kamii, Barbara A. Lewis, and Sally Jones Livingston. Primary arithmetic: Children inventing their own procedures. *Arithmetic Teacher*, 41:200–203, 1993.

[41] Felix Klein. *Elementary Mathematics from an Advanced Standpoint*. Dover, 1945. Originally published in 1908.

[42] Morris Kline. *Mathematics and the Physical World.* Dover, 1981.

[43] Kunihiko Kodaira, ed. *Japanese Grade 7 Mathematics.* UCSMPTextbook Translations. The University of Chicago School Mathematics Project, 1992.

[44] Kunihiko Kodaira, ed. *Japanese Grade 8 Mathematics.* UCSMPTextbook Translations. The University of Chicago School Mathematics Project, 1992.

[45] Susan J. Lamon. *Teaching Fractions and Ratios for Understanding.* Lawrence Erlbaum Associates, 1999.

[46] Glenda Lappan, James T. Fey, William M. Fitzgerald, Susan N. Friel, and Elizabeth Difanis Phillips. *Connected Mathematics 2.* Pearson Prentice Hall, 2006.

[47] Aisling M. Leavy, Susan N. Friel, and James D. Marner. *It's a Fird! Can You Compute a Median of Categorical Data?* Vol. 14, 6, February 2009.

[48] Liora Linchevski and Drora Livneh. Structure sense: The relationship between algebraic and numerical contexts. *Educational Studies in Mathematics*, 40(2):173–196, 1999.

[49] Liping Ma. *Knowing and Teaching Elementary Mathematics*. Lawrence Erlbaum Associates, 1999.

[50] Edward Manfre, James Moser, Joanne Lobato, and Lorna Morrow. *Heath Mathematics Connections*. D. C. Heath and Company, 1994.

[51] Mathematical Association of America (MAA) Committee on the Mathematical Education of Teachers. *A Call For Change: Recommendations for the Mathematical Preparation of Teachers of Mathematics*. The Mathematical Association of America, 1991.

[52] Francis H. Moffitt and John D. Bossler. *Surveying*. Addison-Wesley, 1998.

[53] Joan Moss and Robbie Case. Developing children's understanding of the rational numbers: A new model and an experimental curriculum. *Journal for Research in Mathematics Education*, 30(2): 122–147, 1999.

[54] Gary Musser and William F. Burger. *Mathematics for Elementary Teachers*, 4th ed., Prentice Hall, 1997.

[55] National Center for Education Evaluation and Regional Assistance, Institute of Education Sciences, U.S. Department of Education. Retrieved from *http://ies.ed.gov/ncee/wwc/publications/practiceguides/*, Washington, DC.

[56] National Center for Education Statistics. Findings from Education and the Economy: An indicators Report. Technical Report NCES 97-939, National Center for Education Statistics, 1997.

[57] National Center for Education Statistics. *Highlights from the Third International Mathematics and Science Study*. Technical Report NCES 2001-027, National Center for Education Statistics, 2000. Available online at *http://nces.ed.gov/timss/*

[58] National Council of Teachers of Mathematics. *Curriculum and Evaluation Standards for School Mathematics*. National Council of Teachers of Mathematics, 1989.

[59] National Council of Teachers of Mathematics. *Curriculum Focal Points for PreKindergarten through Grade 8 Mathematics: A Quest for Coherence.* National Council of Teachers of Mathematics, 2006.

[60] National Council of Teachers of Mathematics. *Navigating through Algebra in Grades 3–5*. National Council of Teachers of Mathematics, 2001.

[61] National Council of Teachers of Mathematics. *Navigating through Algebra in Prekindergarten–Grade 2*. National Council of Teachers of Mathematics, 2001.

[62] National Council of Teachers of Mathematics. *Navigating through Geometry in Grades 3–5*. National Council of Teachers of Mathematics, 2001.

[63] National Council of Teachers of Mathematics. *Navigating through Geometry in Prekindergarten–Grade 2*. National Council of Teachers of Mathematics, 2001.

[64] National Council of Teachers of Mathematics. *Principles and Standards for School Mathematics*. National Council of Teachers of Mathematics, 2000. See the Web site at *www.nctm.org*

[65] National Mathematics Advisory Panel. *Foundations for Success.* U.S. Department of Education, 2008.

[66] National Research Council. *Adding It Up: Helping Children Learn Mathematics*. J. Kilpatrick, J. Swafford, and B. Findell, eds. Mathematics Learning Study Committee, Center for Education, Division of Behavioral and Social Sciences and Education. National Academy Press, 2001.

[67] National Research Council. *How People Learn*. National Academy Press, 1999.

[68] National Research Council. *Mathematics Learning in Early Childhood: Paths Toward Excellence and Equity.* Committee on Early Childhood Mathematics, Christopher T. Cross, Taniesha A. Woods, and Heidi Schweingruber, eds. Center for Education, Division of Behavioral and Social Sciences and Education. The National Academies Press, 2009.

[69] Phares G. O'Daffer, Randall Charles, Thomas Cooney, John Dossey, and Jane Schielack. *Mathematics for Elementary School Teachers*. Addison-Wesley, 2004.

[70] A. M. O'Reilley. Understanding teaching/teaching for understanding. In Deborah Schifter, ed., *What's Happening in Math Class?*, vol. 2: Reconstructing Professional Identities, pages 65–73. Teachers College Press, 1996.

[71] Dav Pilkey. *Captain Underpants and the Attack of the Talking Toilets*. Scholastic, 1999.

[72] George Polya. *How To Solve It; A New Aspect of Mathematical Method.* Princeton University Press, 1988. Reissue.

[73] Susan Jo Russell and Karen Economopoulos. *Investigations in Number, Data, and Space.* Pearson Scott Foresman, 2008.

[74] Deborah Schifter. Reasoning about operations, early algebraic thinking in grades K–6. In *Developing Mathematical Reasoning in Grades K–12*, 1999 Yearbook. National Council of Teachers of Mathematics, 1999.

[75] Steven Schwartzman. *The Words of Mathematics*. The Mathematical Association of America, 1994.

[76] The Secretary's Commission on Achieving Necessary Skills. *What Work Requires of Schools, A SCANS Report for America 2000*. U.S. Department of Labor, 1991.

[77] Singapore Ministry of Education. EPB Pan Pacific, Panpac Education Private Limited. *The Singapore Model Method for Learning Mathematics.* Singapore: 2009.

[78] Singapore Curriculum Planning and Development Division, Ministry of Education. *Primary Mathematics*, vol. 1A–6B. Times Media Private Limited, Singapore, 3d ed., 2000. Available at *http://www.singaporemath.com*

[79] Singapore Curriculum Planning and Development Division, Ministry of Education. *Primary Mathematics Workbook*, vol. 1A–6B. Times Media Private Limited, Singapore, 3d ed., 2000. Available at *http://www.singaporemath.com*

[80] K. Stacey. Traveling the road to expertise: a longitudinal study of learning. In H. L. Chick and J. L. Vincent, eds. *Proceedings of the 29th Conference of the International Group for the Psychology of Mathematics Education*, vol. 1, pp. 19–36. PME, 2005.

[81] K. Stacey, S. Helme, S. Archer, and C. Condon. The effect of epistemic fidelity and accessibility on teaching with physical materials: A comparison of two models for teaching decimal numeration. *Educational Studies in Mathematics*, 47, pp. 199–221, 2001.

[82] K. Stacey, S. Helme, V. Steinle, A. Baturo, K. Irwin, and J. Bana. Preservice teachers' knowledge of difficulties in decimal numeration. *Journal of Mathematics Teacher Education*, 4(3), 205–225, 2001.

[83] K. Stacey, S. Helme, and V. Steinle. Confusions between decimals, fractions and negative numbers: A consequence of the mirror as a conceptual metaphor in three different ways. In M. van den Heuvel-Panhuizen, ed., *Proceedings of the 25th Conference of the International Group for the Psychology of Mathematics Education*, vol. 4, 217–224. PME, 2001.

[84] V. Steinle, K. Stacey, and D. Chambers. *Teaching and Learning about Decimals* [CD-ROM]: Department of Science and Mathematics Education, The University of Melbourne, 2002. Online sample at *http://extranet.edfac.unimelb.edu.au/DSME/decimals/*

[85] John Stillwell. *Mathematics and Its History*. Springer-Verlag, 1989.

[86] Dina Tirosh. Enhancing prospective teachers' knowledge of children's conceptions: The case of division of fractions. *Journal for Research in Mathematics Education*, 31(1):5–25, 2000.

[87] Ron Tzur. An integrated study of children's construction of improper fractions and the teacher's role in promoting learning. *Journal for Research in Mathematics Education*, 30(4):390–416, 1999.

[88] The University of Chicago School Mathematics Project. *Everyday Mathematics*. McGraw Hill, 2006.

[89] U.S. Department of Education. *Mathematics Equals Opportunity*. Technical report, U.S. Department of Education, 1997. Available online at *http://www.ed.gov/pubs/math/index.html*

INDEX

O

P

T

CONTENTS

CLASS ACTIVITIES 1

Numbers and the Base-Ten System

1.1 The Counting Numbers

Class Activity 1A The Counting Numbers as a List

CCSS CCSS SMP3, SMP7

One way to think about the counting numbers is as a list. What are the characteristics of this list?

1. Let's first examine errors that very young children commonly make when they are first learning to say the list of counting numbers. Here are some examples of these errors:
   a. Child 1 says: "1, 2, 3, 4, 5, 8, 9, 4, 5, 2, 6, . . ."
   b. Child 2 says: "1, 2, 3, 1, 2, 3, . . ."

   Identify the nature of the errors. What are characteristics of the correct list of counting numbers?

2. Children usually learn the list of counting numbers at around the same time they learn the alphabet. Compare and contrast the alphabet and the counting numbers. In particular, why is the order of the list of counting numbers more important than the order of the letters of the alphabet?

Class Activity 1B Connecting Counting Numbers as a List with Cardinality

CCSS CCSS SMP2, K.CC.4

To determine the number of objects in a set, we generally count the objects one by one. The process of counting the objects in a set connects the list view of the counting numbers with cardinality. Surprisingly, this connection is more subtle and intricate than we might think.

1. Spend a moment thinking about this question: If a child can correctly say the first five counting numbers—"one, two, three, four, five"—will the child necessarily be able to determine how many blocks are in a collection of 5 blocks? Why or why not? Return to this question after completing parts 2 and 3.
2. Let's examine some errors that very young children commonly make when they are first learning to count the number of objects in a set. Examples of errors follow. The picture of a pointing hand indicates a child pointing to the object. A number indicates a child saying the number.

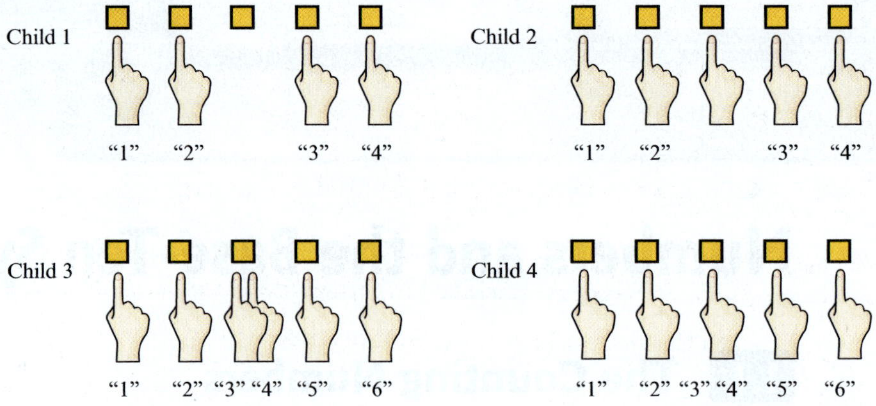

What are characteristics of correctly counting a set of objects and how does this process connect the counting numbers as a list with cardinality?

3. Compare the responses of the following two children to a teacher's request to determine how many blocks there are. Even though both children make a one-to-one correspondence between the 5 blocks and the list 1, 2, 3, 4, 5, do both children appear to understand counting equally well? If not, what is the difference?

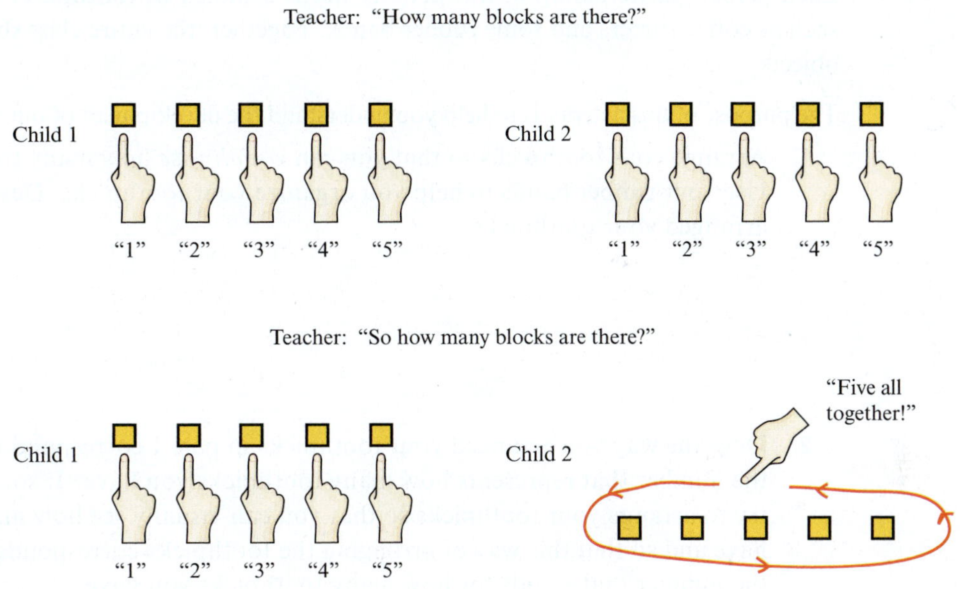

4. Return to the question in part 1 of this activity.

5. What do children need to understand about the connection between the "list" and "cardinality" views of the counting numbers to answer the question posed by the teacher in the next scenario?

The teacher shows a child some toy bears:

The child counts 6 bears. Then the teacher covers the bears and puts one more bear to the side:

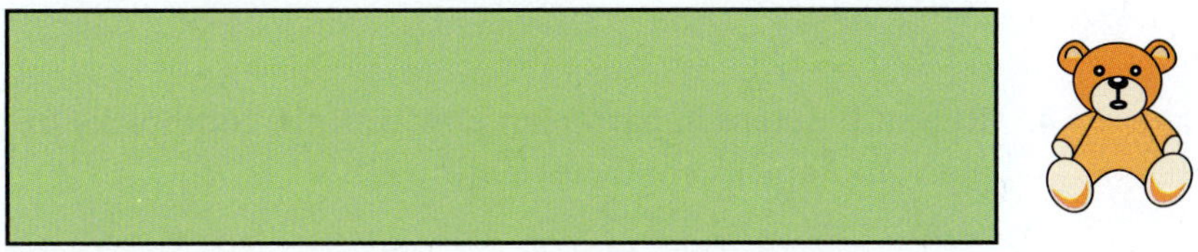

The teacher says: "Now how many bears are there in all?"

a. Child 1 is unable to answer.

b. Child 2 says "1, 2, 3, 4, 5, 6" while pointing to the covered bears, then points at the new bear and says "7," and finally says "there are 7 bears."

c. Child 3 says "6" while pointing to the covered bears, then points at the new bear and says "7," and finally says "there are 7 bears."

Compare the children's different responses.

Class Activity 1C How Many Are There?

CCSS CCSS SMP7, 1.NBT.2, 2.NBT.1

Each person participating in this activity needs a bunch of toothpicks or other small objects, such as coffee stirrers and some rubber bands. Together, the entire class should have at least 1000 objects.

The purpose of this activity is to help you understand the development of our way of writing numbers.

1. Arrange your toothpicks so that you can *visually see* how many toothpicks you have. Use your rubber bands to help you organize your toothpicks. Describe how you arranged your toothpicks.

2. Does the way you arranged your toothpicks in part 1 correspond to the way you write the number that represents how many toothpicks you have? If so, explain how. If not, try to arrange your toothpicks so that you can visually see how many toothpicks you have and so that this way of arranging the toothpicks corresponds to the way we write the number that stands for how many toothpicks you have.

3. Put your toothpicks together with the toothpicks of several other people. Once again, arrange the toothpicks so that you can visually see how many toothpicks there are and so that your way of arranging the toothpicks corresponds to the way we write the number that stands for how many toothpicks there are. Use rubber bands to help organize the toothpicks. Describe how you arranged the toothpicks.

4. Repeat the steps in part 3 but now with the toothpicks from everyone in the class. How many toothpicks are there in all?

5. If you give a child in kindergarten a bunch of counting chips and ask the child to show you what the 2 in 23 stands for, the child might show you 2 of the counting chips. You might be tempted to respond that the 2 really stands for "twenty" and not 2. It's true that the 2 does stand for twenty, but is there a better way you can respond so as to draw attention to the base-ten system?

6. Draw rough pictures showing how to bundle 137 toothpicks so that the way the toothpicks are organized corresponds to the way we write the number 137.

7. Explain why the way the bagged and loose toothpicks pictured here are organized does not correspond to the way we write that number of toothpicks. Show how to reorganize these bagged and loose toothpicks to correspond to the way we write the number that stands for how many toothpicks there are.

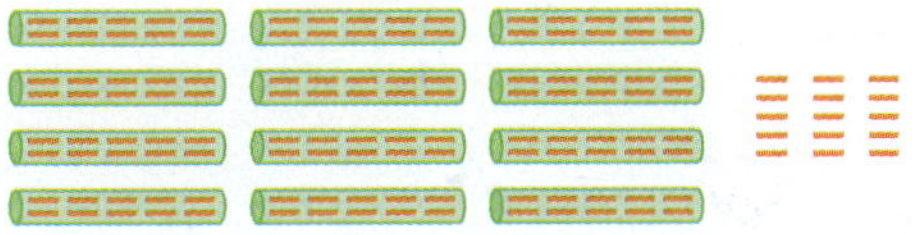

1.2 Decimals and Negative Numbers

Class Activity 1D Representing Decimals with Bundled Objects

CCSS CCSS SMP2, SMP5

1. It is not a strange idea to use 1 object to represent an amount less than 1. After all, 1 penny stands for $0.01, or one hundredth of a dollar.

   Let's make 1 paper clip stand for 0.001, or one thousandth. Show simple drawings of bundled paper clips so that the way of organizing the paper clips corresponds to the way we write the following decimals:

   0.034

   0.134

   0.13

2. Let's let 1 small bead stand for 0.0001, or one ten-thousandth. Show simple drawings of bundled beads so that the way of organizing the beads corresponds to the way we write the following decimals:

   0.0028

   0.012

3. List at least three different decimals that the toothpicks pictured below could represent. In each case, state the value of the single toothpick.

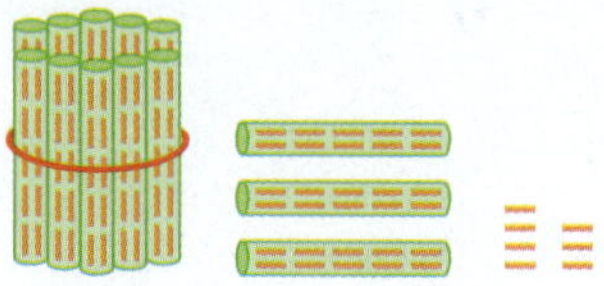

Class Activity 1E Representing Decimals as Lengths

CCSS CCSS SMP2, SMP5

A good way to represent positive decimal numbers is as lengths. Cut out the 5 long strips on Activity Download 1 and tape them end-to-end without overlaps to make one long strip. The length of this long strip is 1 unit. Cut out the ten 0.1-unit-long strips.

1. By placing strips end-to-end without gaps or overlaps, verify the following:
   a. The 1-unit-long strip is as long as 10 of the 0.1-unit-long strips.

   b. A 0.1-unit-long strip is as long as 10 of the 0.01-unit-long strips.

   c. A 0.01-unit-long strip is as long as 10 of the 0.001-unit-long strips.

   d. Now cut apart the 0.01- and 0.001-unit-long strips.

   Represent the following decimals as lengths by placing appropriate strips end-to-end without gaps or overlaps (as best you can). In each case, draw a rough sketch (which need not be to scale) to show how you represented the decimal as a length.

2. 1.234

3. 0.605

4. 1.07

5. 1.007

6. A student was asked to place tenths on this number line:

The student put ten tick marks between 0 and 1:

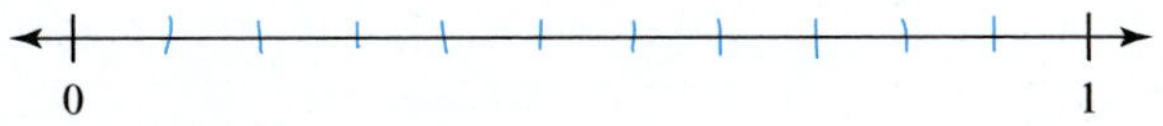

What's wrong with that?

Class Activity 1F Zooming In on Number Lines

CCSS SMP5, SMP7, SMP8

1. Label the tick marks on the number lines that follow with appropriate decimals. The second, third, and fourth number lines should be labeled as if they are "zoomed in" on the indicated portion of the previous number line.

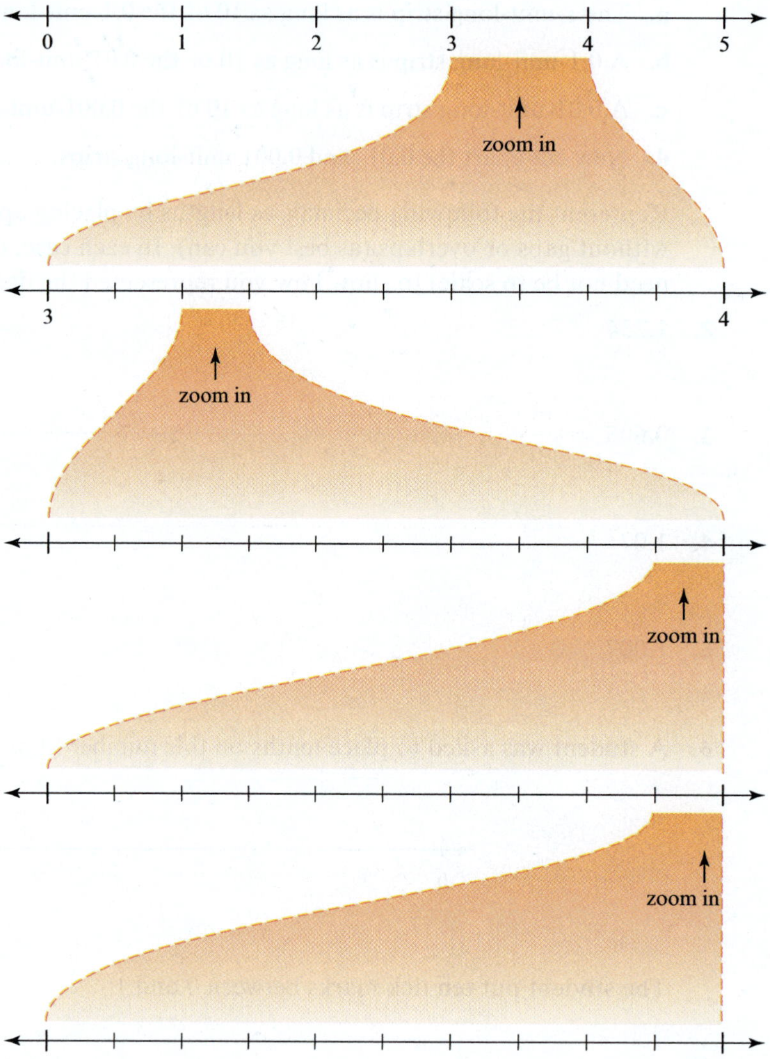

2. Now plot 3.2996 on each of the number lines in part 1 (it's easiest to start at the last number line and work backward). Use the number lines to answer the following questions:
   a. Which whole numbers does 3.2996 lie between?

   b. Which tenths does 3.2996 lie between?

   c. Which hundredths does 3.2996 lie between?

3. Label the tick marks on the next three number lines in three different ways. In each case, your labeling should fit with the structure of the base-ten system and the fact that the tick marks at the ends of the number lines are longer than the other tick marks. (You may further lengthen the tick marks at either end as needed.) It may help you to think about zooming in on the number line.

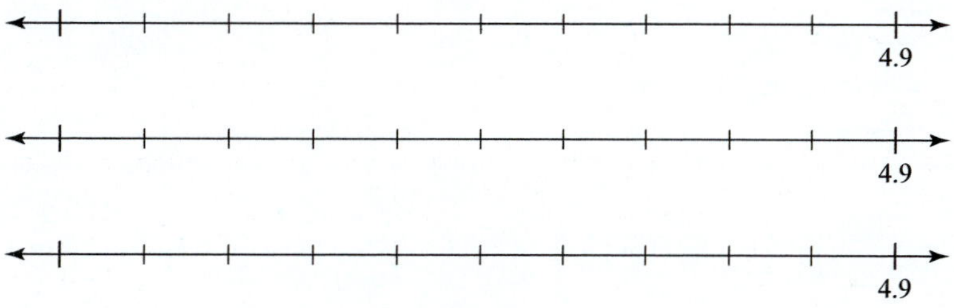

4. Why does the labeling on the next number line not fit with the structure of the base-ten system?

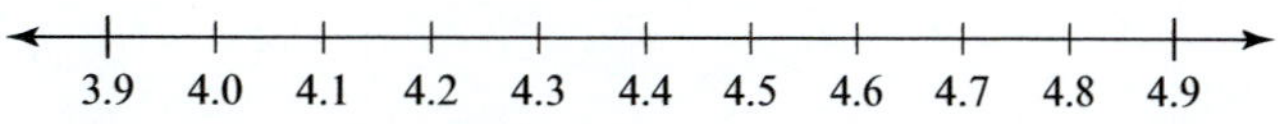

5. Label the tick marks on the next three number lines in three different ways. In each case, your labeling should fit with the structure of the base-ten system and the fact that the tick marks at the ends of the number lines are longer than the other tick marks. (You may further lengthen the tick marks at either end as needed.) It may help you to think about zooming in on the number line.

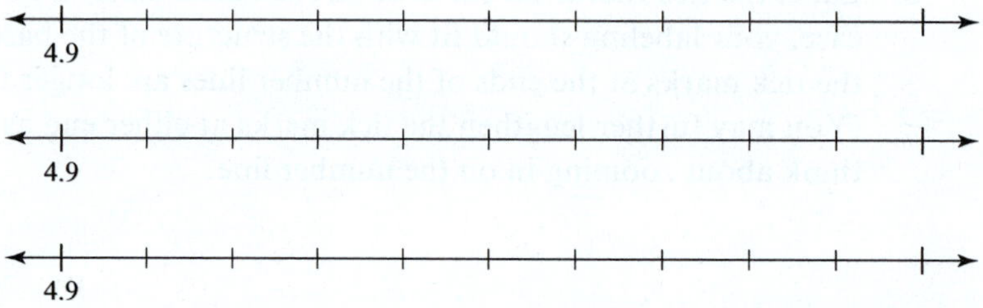

Class Activity 1G Numbers Plotted on Number Lines

CCSS CCSS SMP7

1. What number could the point labeled A on the next number line be? Among the numbers in this list, which ones could A possibly be? Which ones could A definitely not be? Why?

1.14, 1.3915, 1.834, 1.4, 1.4263, 1.43, 1.644

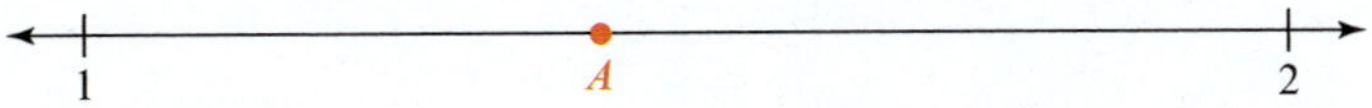

2. Label the tick marks on the following number lines so that the tick marks fit with the structure of the base-ten system and are as specified, and so that the given number can be plotted on the number line. The number need not land on a tick mark.

Plot 23.84

Long ticks: whole numbers
Short ticks: tenths

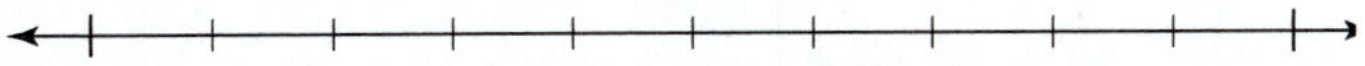

Plot 0.03402

Long ticks: hundredths
Short ticks: thousandths

Plot 0.005

Long ticks: tenths
Short ticks: hundredths

Plot 7.0095

Long ticks: whole numbers
Short ticks: tenths

3. Label the tick marks on the next number lines appropriately (so that the long and short tick marks fit with the structure of the base-ten system) and explain why you labeled them that way.

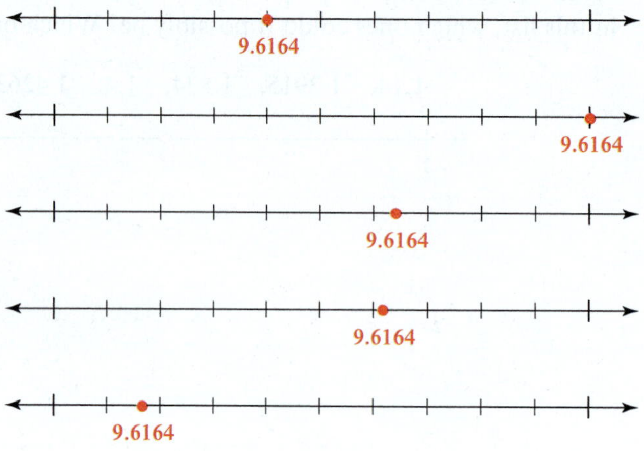

Class Activity 1H Negative Numbers on Number Lines

CCSS CCSS SMP3, 6.NS.6

1. Katie, Matt, and Parna were asked to label the tick marks on a number line on which one tick mark was already labeled as −7. What's wrong with their work? Show how to label the number line appropriately. Is there more than one way to label it appropriately?

Katie's work:

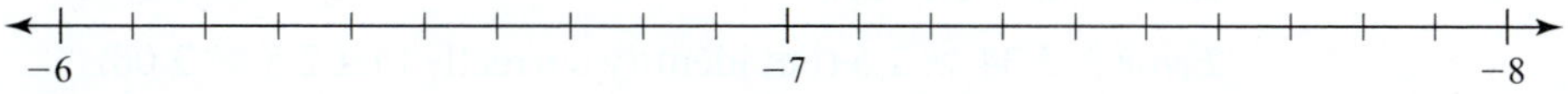

Matt's work:

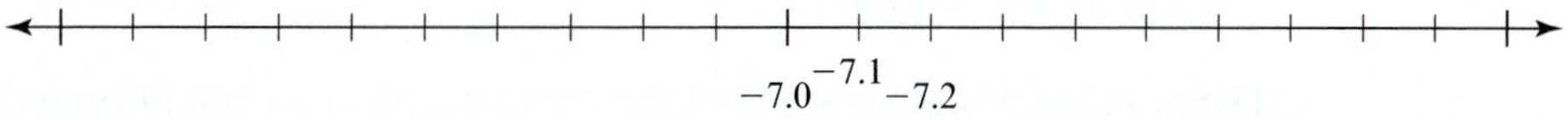

Parna's work:

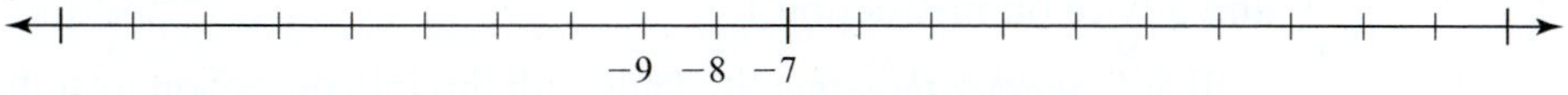

Appropriate labeling:

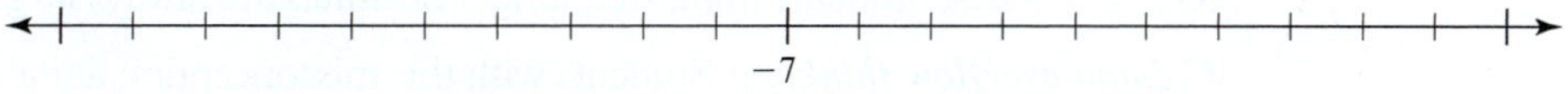

Another appropriate labeling:

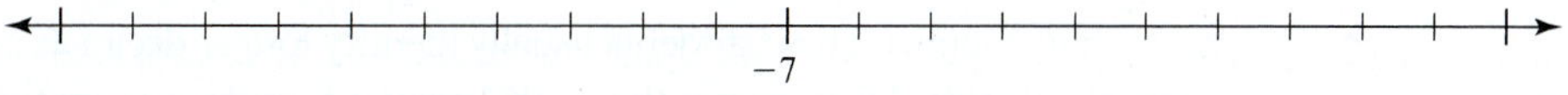

2. Sometimes students get confused about the relative locations of certain decimals, negative numbers, and zero. Plot −1 on the next number line. Then give a few examples of numbers that lie between 0 and 1 including some numbers that lie between 0 and −1. Include examples of numbers that land *between* tick marks.

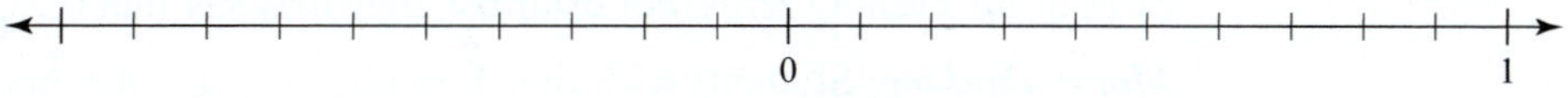

3. Use a number line to explain why $-(-3) = 3$. More generally, explain why $-(-N) = N$ is true for all numbers N.

1.3 Reasoning to Compare Numbers in Base Ten

Class Activity 1I Critique Reasoning About Comparing Decimals

CCSS CCSS SMP3

The list that follows describes some common errors students make when comparing decimals. For each error, discuss why students might make that error.

Error 1: $2.352 > 2.4$

Error 2: $2.34 > 2.5$ (but identify correctly that $2.5 > 2.06$)

Error 3: $5.47 > 5.632$

Error 4: $1.8 = 1.08$

The next list describes some of the misconceptions students can develop about comparing decimals. (See [13] and [14] for further information, including additional misconceptions and advice on instruction.)

Whole number thinking: Students with this misconception treat the portion of the number to the right of the decimal point as a whole number, thus thinking that $2.352 > 2.4$ because $352 > 4$. These students think that longer decimals are always larger than shorter ones.

Column overflow thinking: Students with this misconception name decimals incorrectly by focusing on the first nonzero digit to the right of the decimal point. For example, they say that 2.34 is "two and thirty-four tenths." These students think that $2.34 > 2.5$ because 34 tenths is more than 5 tenths. These students usually identify longer decimals as larger; they will, however, correctly identify 2.5 as greater than 2.06 because 5 tenths is more than 6 hundredths.

Denominator-focused thinking: Students with this misconception think that any number of tenths is greater than any number of hundredths and that any number of hundredths is greater than any number of thousandths, and so on. These students identify 5.47 as greater than 5.632, reasoning that 47 hundredths is greater than 632 thousandths because hundredths are greater than thousandths. Students with this misconception identify shorter decimal numbers as larger.

Reciprocal thinking: Students with this misconception view the portion of a decimal to the right of the decimal point as something like the fraction formed by taking the reciprocal. For example, they view 0.3 as something like $\frac{1}{3}$ and thus identify 2.3 as greater than 2.4 because $\frac{1}{3} > \frac{1}{4}$. These students usually identify shorter decimal numbers as larger, except in cases of intervening zeros. For example, they may say that $0.03 > 0.4$ because $\frac{1}{3} > \frac{1}{4}$.

Money thinking: Students with this difficulty truncate decimals after the hundredths place and view decimals in terms of money. If two decimals agree to the hundredths place, these students simply guess which one is greater—sometimes guessing correctly, sometimes guessing incorrectly. Most of these students recognize that 1.8 is like \$1.80, although some view 1.8 incorrectly as \$1.08.

Class Activity 1J Finding Decimals Between Decimals

CCSS CCSS SMP8

1. Contemplate the questions in the following paragraph for a few minutes before you continue with the rest of the activity. Then return to these questions at the end.

   There aren't any whole numbers between 2 and 3, but there are plenty of decimals in between 2 and 3. If you are given two decimals, will there always be another decimal in between the two? Are there some decimals that don't have any other decimals in between them?

2. Work with a partner and take turns giving your partner a pair of decimals and challenging him or her to find a decimal in between your pair. For example, you could give your partner the pair 1.2, 1.4; your partner could respond with 1.3. Try to stump your partner! Continue taking turns until one of you has stumped the other or you both agree that neither of you will be able to stump the other.

3. Return to the questions in part 1.

Class Activity 1K Decimals Between Decimals on Number Lines

For each of the pairs of numbers that follow, find a number in between the two numbers. Label the longer tick marks on the number line so that all three numbers can be plotted visibly and distinctly. The labeling should fit with the structure of the base-ten system. Plot all three numbers. The numbers need not land on tick marks.

1. The numbers 1.6 and 1.7

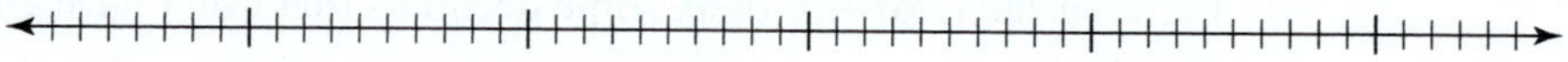

Now, describe how to use money to find a number between 1.6 and 1.7.

2. The numbers 12.54 and 12.541

3. The numbers 2.781 and 2.7342

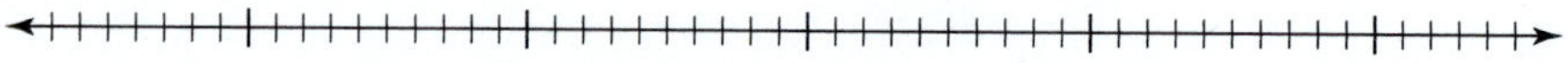

4. The numbers 23.99 and 24

5. The numbers −1.9 and −2.

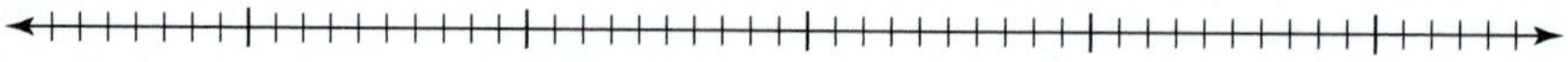

6. The numbers −3.49 and −3.489

Class Activity 1L "Greater Than" and "Less Than" with Negative Numbers

CCSS CCSS SMP2, SMP7

1. Explain in two different ways why a negative number is always less than a positive number.

2. Johnny says that $-5 > -2$. Describe two different ways to explain to Johnny why this is not correct.

3. For each of the following pairs of numbers, write an inequality to show how the numbers compare. Explain your reasoning.

a. -6 and -5.9

b. -4.777 and -4.8

c. -0.555 and -0.1

4. Given the following number line, write inequalities to show how each of the pairs of numbers below compare. Explain your reasoning.

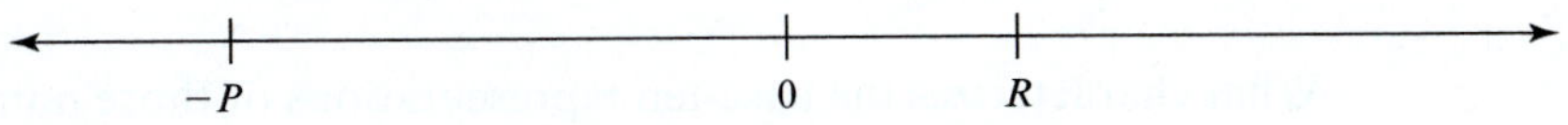

a. $-R$ and $-P$

b. R and P

1.4 Reasoning about Rounding

Class Activity 1M Explaining Rounding

CCSS CCSS 3.NBT.1

1. Round points A and B to the nearest hundred.

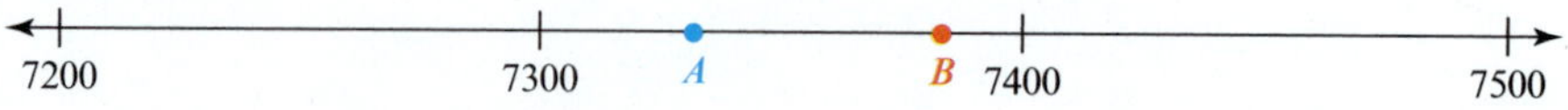

What might the base-ten representations of A and B be? Why?

2. Which numbers in the interval between 7300 and 7400 round to 7300, when rounding to the nearest hundred? Indicate that range of numbers on the number line.

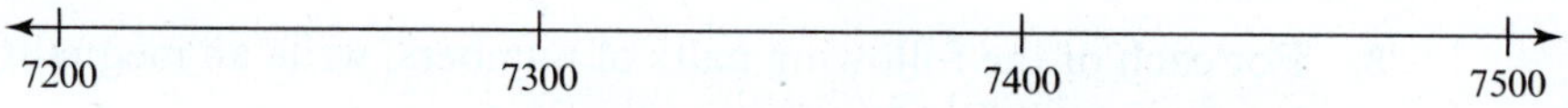

What characterizes the base-ten representations of those numbers?

3. Which numbers in the interval between 7300 and 7400 round to 7400, when rounding to the nearest hundred? Indicate that range of numbers on the number line.

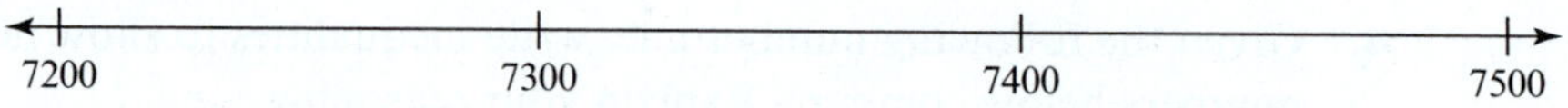

What characterizes the base-ten representations of those numbers?

4. When we round a number to the nearest hundred, why do we look to the tens place? What does the digit in the tens place tell us about the location of the number on a number line labeled with tick marks representing hundreds?

Class Activity 1N Rounding with Number Lines

CCSS CCSS 5.NBT.4

Using number lines to round can help us focus on the base-ten system and understand its structure better.

1. The tick marks on the following number line are labeled with thousands. Plot 38721 in its approximate location on this number line. Then use the number line to explain how to round 38721 to the nearest thousand.

2. Label the unlabeled tick marks on the following number line with appropriate tens so that you can plot 5643 in its approximate location on this number line. Then use the number line to explain how to round 5643 to the nearest ten.

3. The tick marks on the following number line are labeled with hundredths. Plot 2.349 in its approximate location on this number line. Then use the number line to explain how to round 2.349 to the nearest hundredth.

4. Label the unlabeled tick marks on the following number line with appropriate tenths so that you can plot 2.349 in its approximate location on this number line. Then use the number line to explain how to round 2.349 to the nearest tenth.

5. Label the tick marks on the following number line so that you can use the number line to explain how to round 54,831 to the nearest hundred.

6. Label the tick marks on the following number line so that you can use the number line to explain how to round 16.936 to the nearest hundredth.

7. Label the tick marks on the following number line so that you can use the number line to explain how to round 16.936 to the nearest tenth.

Class Activity 10 Can We Round This Way?

CCSS CCSS SMP3

Maureen has made up her own method of rounding. Starting at the right-most place in a decimal number, she keeps rounding to the value of the next place to the left until she reaches the place to which the decimal number was to be rounded.

For example, Maureen would use the following steps to round 3.2716 to the nearest tenth:

$$3.2716 \rightarrow 3.272 \rightarrow 3.27 \rightarrow 3.3$$

Try Maureen's method on several examples. Is her method valid? That is, does it always round decimal numbers correctly? Or are there examples of decimal numbers where Maureen's method does not give the correct rounding?

CLASS ACTIVITIES

2

Fractions and Problem Solving

2.1 Solving Problems and Explaining Solutions

2.2 Defining and Reasoning About Fractions

Class Activity 2A Getting Familiar with Our Definition of Fractions

CCSS CCSS SMP6, 3.NF.1

1. There are different ways to define fractions. One common way of defining $\frac{A}{B}$ is as "A out of B equal parts." Discuss limitations of this definition. For example, consider fractions such as $\frac{4}{3}$ or $\frac{5}{2}$. (This is why we will use the different definition given in the text.)

2. Work with a partner. Use *our definition of fraction given in the text* to describe the three fractions on the left (or the right) to your partner. Listen to your partner's description of the other three fractions.

Unit amount or whole	Unit amount or whole

Show $\frac{1}{3}$ of the unit amount:	Show $\frac{1}{5}$ of the unit amount:
Show $\frac{2}{3}$ of the unit amount:	Show $\frac{3}{5}$ of the unit amount:
Show $\frac{4}{3}$ of the unit amount:	Show $\frac{6}{5}$ of the unit amount:

3. For each of the following, draw the requested unit amount and explain your answer using our definition of fraction.

a. $\frac{1}{5}$ cup butter: 
Draw 1 cup butter:

b. $\frac{3}{5}$ of a liter of juice: 
Draw 1 liter of juice:

c. $\frac{4}{9}$ of a pound of flour: 
Draw 1 pound of flour:

d. $\frac{6}{5}$ of a kilogram of sugar: 
Draw 1 kilogram of sugar:

4. Discuss how a 3 in the numerator of a fraction $\frac{3}{\square}$ is like a digit 3 in a number $3\square\square$ in base ten. Also, how are unit fractions like base-ten units?

Class Activity 2B Using Our Fraction Definition to Solve Problems

CCSS CCSS SMP1, SMP6

You will need pattern tiles for parts 3 and 4.

1. Take a blank piece of paper and imagine that it is $\frac{4}{5}$ of some larger piece of paper. Fold your piece of paper to show $\frac{3}{5}$ of the larger (imagined) piece of paper. Do this as carefully and precisely as possible without using a ruler or doing any measuring. Explain why your answer is correct. Could two people have different-looking solutions that are both correct?

2. Benton used $\frac{3}{4}$ cup of butter to make a batch of cookie dough. He rolled his cookie dough out into a rectangle. Now Benton wants a portion of the dough that contains $\frac{1}{4}$ cup of butter. How could Benton cut the dough? Explain your answer.

3. **a.** The yellow hexagon pattern tile is $\frac{2}{3}$ of the area of a pattern tile design. Use pattern tiles to make what could be the design. Explain your reasoning.

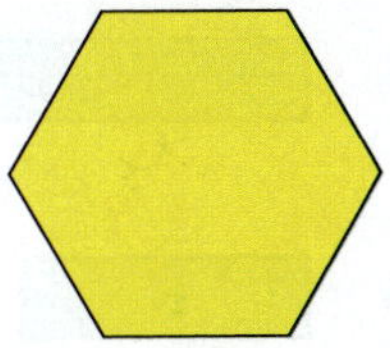

   b. What if the yellow hexagon is $\frac{3}{2}$ of the area of another design?

   c. What if the yellow hexagon is $\frac{6}{10}$ of the area of another design?

4. Suppose 3 yellow hexagons are $\frac{2}{7}$ of the area of a pattern tile design. Use pattern tiles to make what could be the design. Explain your reasoning.

Class Activity 2C Why Are Fractions Numbers? A Measurement Perspective

CCSS CCSS SMP2, SMP8

1. Contemplate and discuss the following question: Why are fractions numbers? For example, why are $\frac{2}{3}$ and $\frac{8}{3}$ numbers just like 2 and 3 are numbers?

2. How many of the unit strip does it takes to make Strip X exactly?

How many of the unit strip does it takes to make Strip Y exactly?

How many of the unit strip does it takes to make Strip Z exactly?

How many of the unit strip does it takes to make Strip W exactly?

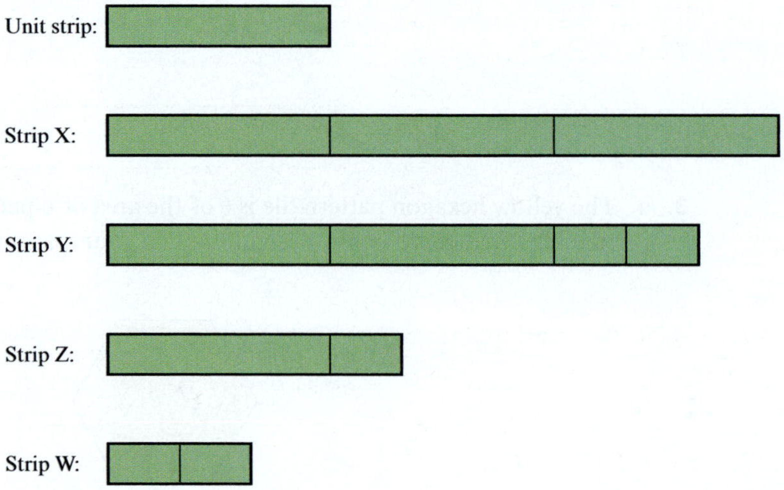

3. Return to part 1 and discuss it some more.

We can think of numbers as the result of measuring by a unit amount or a 1. If you have 4 pounds of flour, the 4 tells you how many "1 pounds" you have. Similarly, if you have $\frac{1}{4}$ of a pound of flour, the $\frac{1}{4}$ tells you how many (i.e., how much) of "1 pound" you have. Use this measurement perspective *and our definition of fraction* as you relate quantities in the following problems.

4. **a.** How many of Strip A does it take to make Strip B?

b. How many of Strip B does it take to make Strip A? Explain briefly.

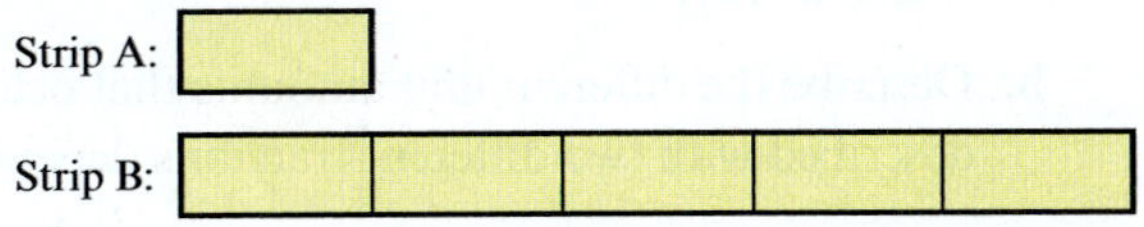

5. **a.** How many of Strip C does it take to make Strip D? Explain.

b. How many of Strip D does it take to make Strip C? Explain.

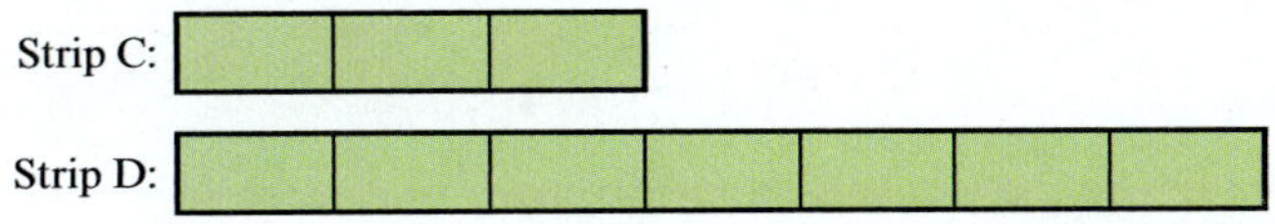

6. **a.** How many of Strip E does it take to make Strip F? Explain.

b. How many of Strip F does it take to make Strip E? Explain.

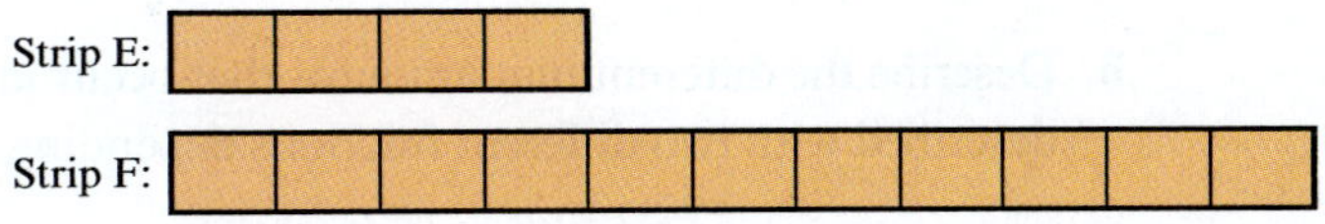

7. For each of the above, what do you notice about the fractions in parts (a) and (b)?

Class Activity 2D Relating Fractions to Wholes

CCSS CCSS SMP1, SMP6

1. At a neighborhood park, $\frac{1}{3}$ of the area of the park is to be used for a new playground. Swings will be placed on $\frac{1}{4}$ of the area of the playground. What fraction of the neighborhood park will the swing area be?
   a. Make a math drawing to help you solve the problem and explain your solution. Use our definition of fraction in your explanation and attend to the unit amount that each fraction is *of*.

   b. Describe the different unit amounts that occur in part (a). Discuss how one amount can be described with two different fractions depending on what the unit amount is taken to be.

2. Ben is making a recipe that calls for $\frac{1}{3}$ cup of oil. Ben has a bottle that contains $\frac{2}{3}$ cup of oil. Ben does not have any measuring cups. What fraction of the oil in the bottle should Ben use for his recipe?
   a. Make a math drawing to help you solve the problem and explain your solution. Use our definition of fraction in your explanation and attend to the unit amount that each fraction is *of*.

   b. Describe the different unit amounts that occur in part (a). Discuss how one amount can be described with two different fractions depending on what the unit amount is taken to be.

Class Activity 2E Critiquing Fraction Arguments

CCSS CCSS SMP3, SMP6

The unit amount that a fraction refers to need not be a single contiguous object. Instead, the unit amount can consist of several pieces that need not even be the same size. Working with a noncontiguous unit amount provides an opportunity to think more deeply about the definition of fraction.

Recall these definitions:

- An amount is $\frac{1}{B}$ of a unit amount if B copies of it joined together are the same size as the unit amount.
- An amount is $\frac{A}{B}$ of a unit amount if it can be formed by A parts, each of which is $\frac{1}{B}$ of the unit amount.

1. *Peter's garden problem:* The drawing below is a map of Peter's garden, which consists of two plots. The two plots have each been divided into 5 pieces of equal area. The shaded parts show where carrots have been planted. What fraction of the area of Peter's garden is planted with carrots?

Spend a few minutes solving this problem yourself. Then move on to the next parts, which show methods that some students used as they attempted to solve this problem.

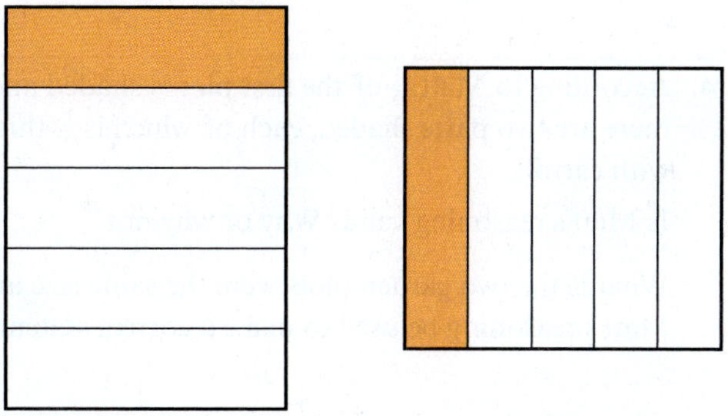

2. Mariah thought about Peter's garden problem this way: "There are 2 parts out of a total of 10 parts, so $\frac{2}{10}$ of the garden is planted with carrots. Since $\frac{2}{10} = \frac{1}{5}$, then $\frac{1}{5}$ of the garden is planted with carrots."

Is Mariah's reasoning valid? Why or why not?

If the two garden plots had been the same size, then would her reasoning be valid?

3. Aysah drew this picture as she was thinking about Peter's garden problem:

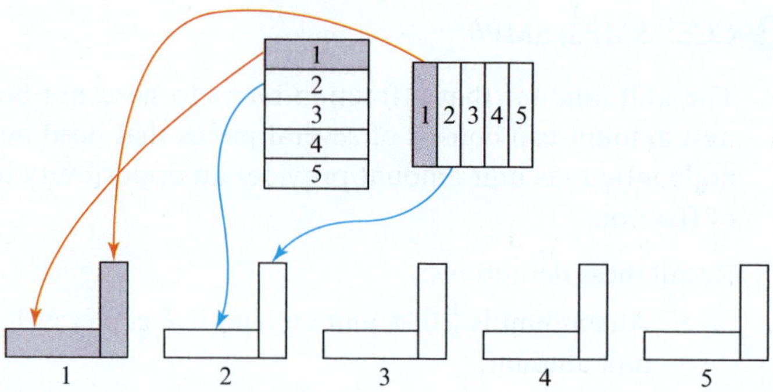

Can Aysah's picture be used to solve the problem? Explain.

4. According to Matt, $\frac{1}{5}$ of the first plot is shaded and $\frac{1}{5}$ of the second plot is shaded. Because there are two parts shaded, each of which is $\frac{1}{5}$, this means $\frac{2}{5}$ of Peter's garden are planted with carrots.

Is Matt's reasoning valid? Why or why not?

What if the two garden plots were the same size and each plot had an area of 1 acre? Could Matt's reasoning be used to make a correct statement? Explain.

Class Activity 2F Critique Fraction Locations on Number Lines

CCSS CCSS SMP6, 3.NF.2

1. Students were asked to plot the fractions $\frac{1}{4}$, $\frac{2}{4}$, and $\frac{3}{4}$ on a number line.

   Eric's work:

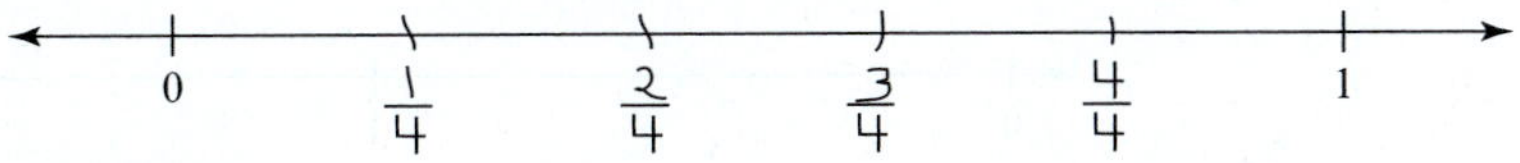

   Kristin's work:

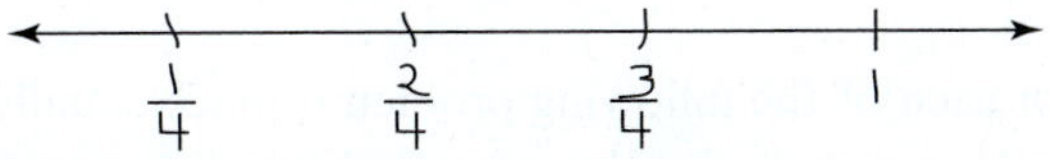

   How might Eric be thinking? Although Kristin's labeling is not incorrect, what might she not be attending to?

2. When Tyler was asked to plot $\frac{3}{4}$ on a number line showing 0 and 2, he plotted it as shown on the next number line. How might Tyler be thinking?

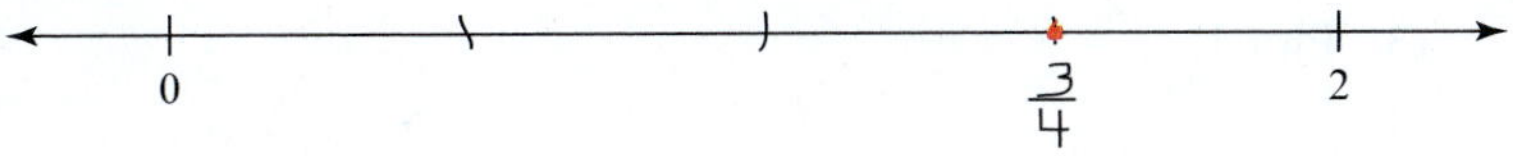

3. When Amy was asked what fraction should go in the box, she wrote $\frac{2}{6}$. Why might she have done so? What idea might she not be attending to?

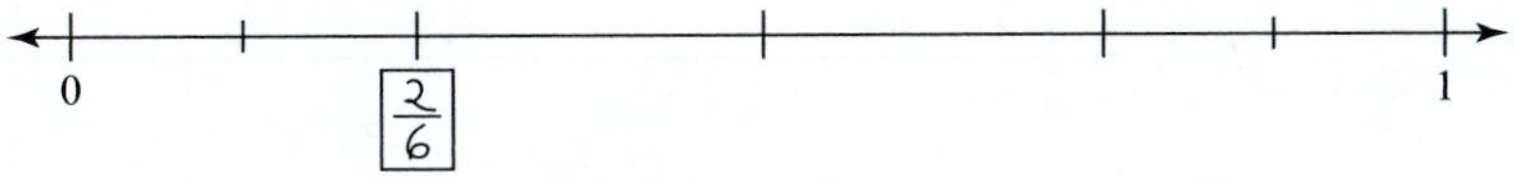

4. Discuss ideas you have for helping students understand number lines. How might you draw their attention to the lengths of intervals and to distance from 0, and away from merely counting tick marks without attending to length and distance?

Class Activity 2G Fractions on Number Lines

CCSS CCSS SMP1, SMP3, 3.NF.2

1. Explain in detail how to determine where to plot $\frac{3}{4}$ and $\frac{5}{4}$ on the number line below, and explain why those locations fit with the definition of fraction.

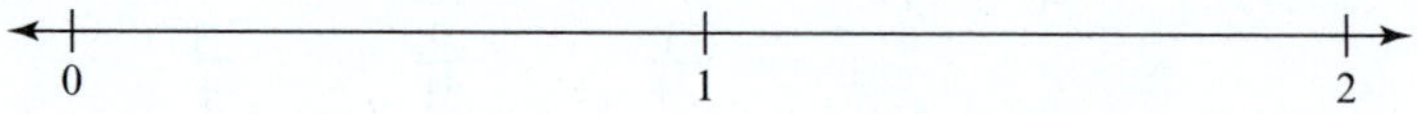

For each of the following problems, place equally spaced tick marks on the number line so that you can plot the requested fraction on a tick mark. You may place the tick marks "by eye"; precision is not needed. Explain your reasoning.

2. Plot $\frac{5}{4}$.

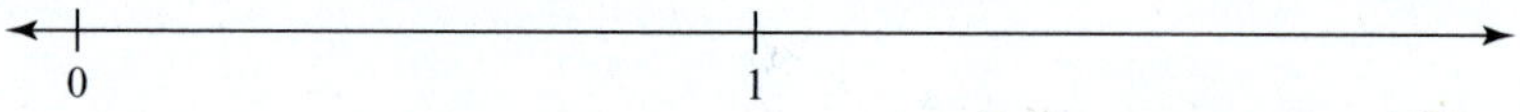

3. Plot 1.

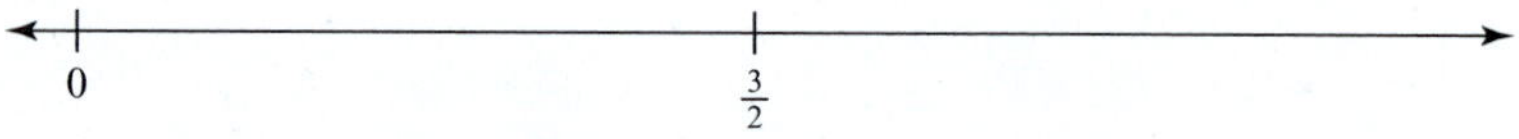

4. Plot 1.

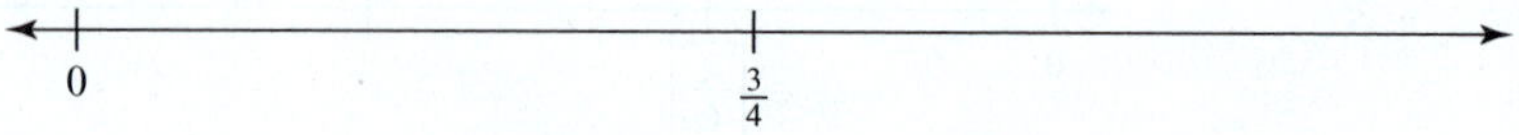

5. Plot $\frac{3}{5}$.

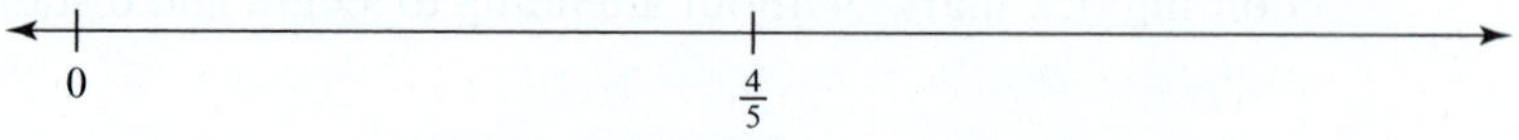

Class Activity 2H Improper Fraction Problem Solving with Pattern Tiles

CCSS CCSS SMP1, SMP3

You need a set of pattern tiles for this activity (yellow hexagons, red trapezoids, blue rhombuses, and green triangles only).

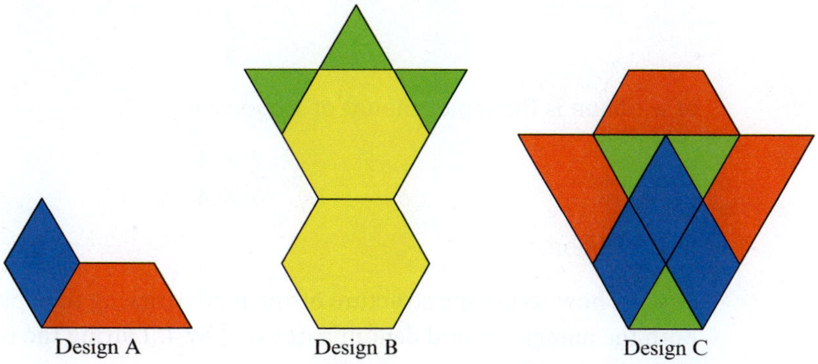

1. The area of Design A is $\frac{5}{4}$ of the area of another design. Make a design that could be the other design. Explain your reasoning.

2. The area of Design B is $\frac{5}{3}$ of the area of another design. Make a design that could be the other design. Explain your reasoning.

3. The area of Design C is $\frac{9}{7}$ of the area of another design. Make a design that could be the other design. Explain your reasoning.

2.3 Reasoning About Equivalent Fractions

Class Activity 2I Explaining Equivalent Fractions

CCSS CCSS SMP2, SMP7, 4.NF.1

1. Use our definition of fraction and the math drawing below to give a detailed conceptual explanation for why

$$\frac{2}{3}$$

of a ribbon is the same amount of ribbon as

$$\frac{2 \times 4}{3 \times 4}$$

of the ribbon.

Discuss how to use the structure of the math drawing to explain the process of multiplying both the numerator and denominator of $\frac{2}{3}$ by 4. During the process, what changes and what stays the same?

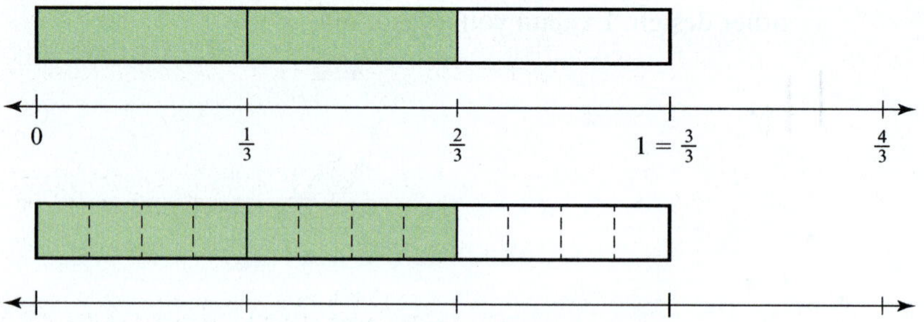

2. Use a carefully structured math drawing to help you explain why $\frac{4}{5} = \frac{4 \times 3}{5 \times 3}$. Take special care to explain why the two fractions are *equal.*

3. When you discuss the ideas of parts 1 and 2 with students, they may find it confusing that in the math drawing, the fraction pieces are *divided*, but in the numerical work, we *multiply*. Address this confusing point by discussing what happens to the *size* of the pieces and what happens to the *number* of pieces when we create equivalent fractions.

Class Activity 2J Critique Fraction Equivalence Reasoning

CCSS CCSS SMP3

1. Anna says that

$$\frac{2}{3} = \frac{6}{7}$$

because, starting with $\frac{2}{3}$, you get $\frac{6}{7}$ by adding 4 to the top and the bottom. If you do the same thing to the top and the bottom, the fractions must be equal.

Is Anna right? If not, why not? What should we be careful about when talking about equivalent fractions?

2. Don says that $\frac{11}{12} = \frac{16}{17}$ because both fractions are one part away from a whole. Is Don correct? If not, what is wrong with Don's reasoning?

3. Peter says that $\frac{6}{6}$ is greater than $\frac{5}{5}$ because $\frac{6}{6}$ has more parts. Is Peter correct? If not, what is wrong with Peter's reasoning?

Class Activity 2K Interpreting and Using Common Denominators

CCSS CCSS SMP2

1. Write $\frac{3}{4}$ and $\frac{5}{6}$ with two different common denominators. In terms of the strips and the number lines, what are we doing when we give the two fractions common denominators? During the process, what changes and what stays the same?

first common denominator:

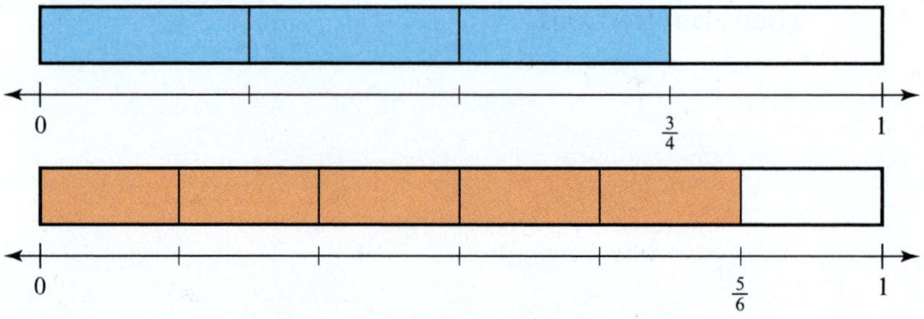

second common denominator:

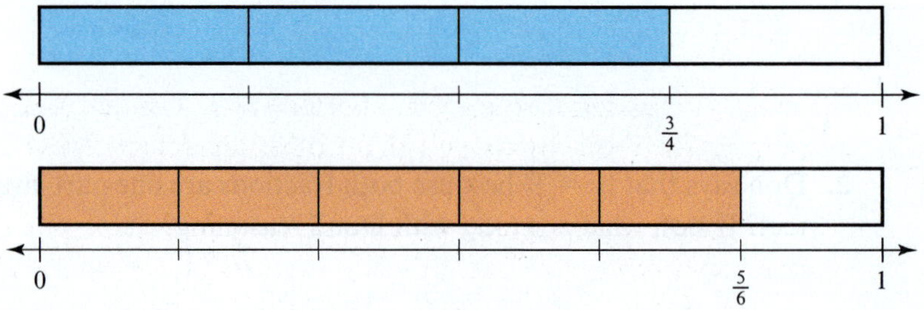

2. Plot 1, $\frac{2}{3}$, and $\frac{5}{2}$ on the number line for this problem in such a way that each number falls on a tick mark. Lengthen the tick marks of whole numbers.

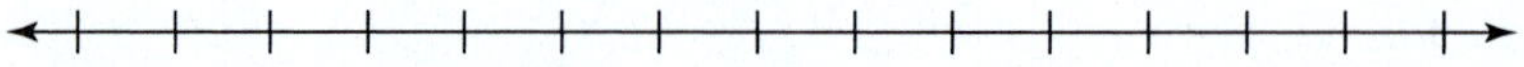

3. Plot 9, $\frac{55}{6}$, and $\frac{33}{4}$ on the number line for this problem in such a way that each number falls on a tick mark. Lengthen the tick marks of whole numbers.

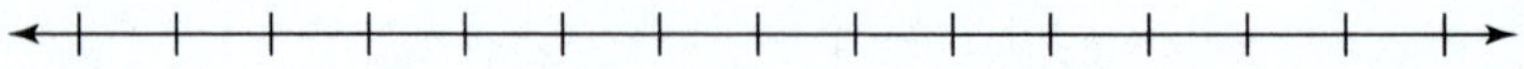

4. Plot 1, 0.7, and $\frac{3}{4}$ on the number line for this problem in such a way that each number falls on a tick mark. Lengthen the tick marks of whole numbers.

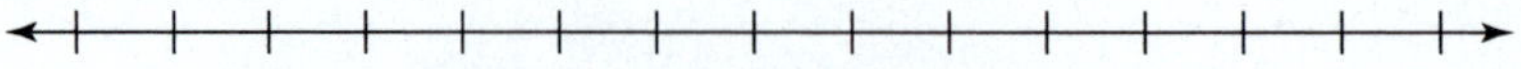

Class Activity 2L Solving Problems by Using Equivalent Fractions

CCSS CCSS SMP1, SMP2

1. Take a blank piece of paper and imagine that it is $\frac{2}{3}$ of some larger piece of paper. Fold your piece of paper to show $\frac{1}{6}$ of the larger (imagined) piece of paper. Do this as carefully and precisely as possible without using a ruler or doing any measuring. Explain why your answer is correct.

   In solving this problem, how does $\frac{2}{3}$ appear as an equivalent fraction? Could two people have different solutions that are both correct?

2. Jean has a casserole recipe that calls for $\frac{1}{2}$ cup of butter. Jean only has $\frac{1}{3}$ cup of butter. Assuming that Jean has enough of the other ingredients, what fraction of the casserole recipe can Jean make? Make math drawings to help you solve this problem. Explain why your answer is correct. In your explanation, attend carefully to the unit amount that each fraction is *of*.

   In solving this problem, how do $\frac{1}{2}$ and $\frac{1}{3}$ each appear as equivalent fractions?

3. One serving of SugarBombs cereal is $\frac{3}{4}$ cup. Joey wants to eat $\frac{1}{2}$ of a serving of SugarBombs cereal. How much of a cup of cereal should Joey eat? Make math drawings to help you solve this problem. Explain why your answer is correct. In your explanation, attend carefully to the unit amount that each fraction is *of*.

   In solving the problem, how does $\frac{3}{4}$ appear as an equivalent fraction?

Class Activity 2M Problem Solving with Fractions on Number Lines

CCSS CCSS SMP1, SMP2

Explain how to solve the following problems by *reasoning with equivalent fractions*. Place equally spaced tick marks on the number line so that the given fraction and the requested fraction both land on tick marks. You may place the tick marks "by eye"; precision is not needed.

1. Plot $\frac{3}{4}$ without first plotting 1.

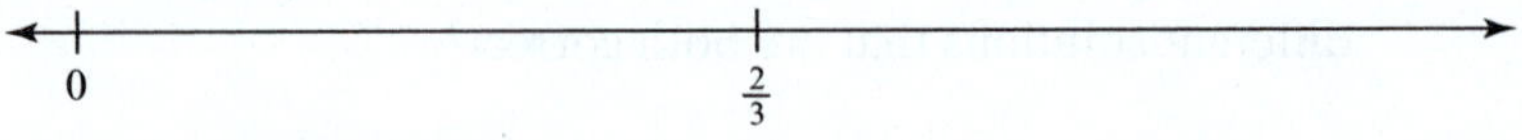

2. Plot $\frac{3}{5}$ without first plotting 1.

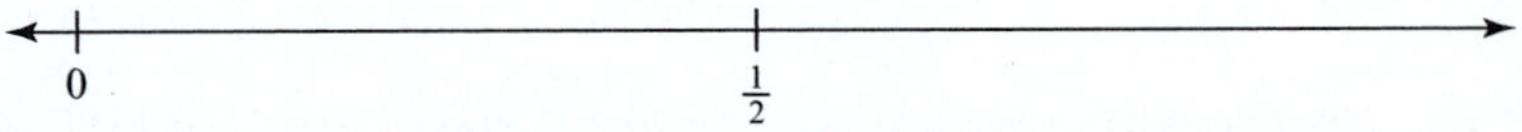

3. Plot $\frac{3}{8}$ without first plotting 1.

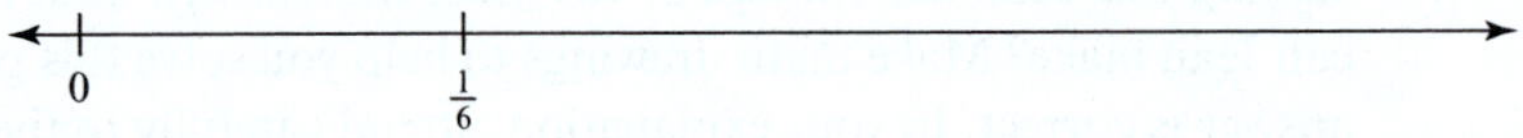

4. So far, Sue has run $\frac{1}{4}$ of a mile, but that is only $\frac{2}{3}$ of the total distance she will run. What is her total running distance?

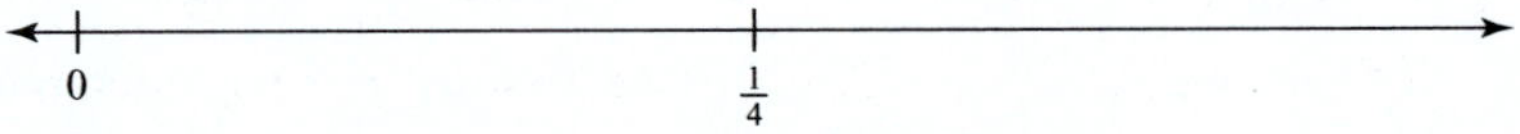

5. So far, Tyler has run $\frac{2}{3}$ of a mile, but that is only $\frac{3}{4}$ of his total running distance. What is his total running distance?

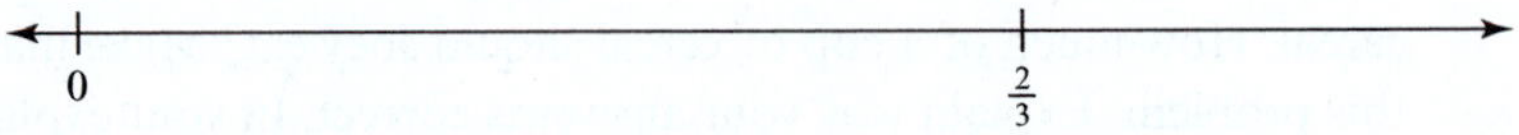

Class Activity 2N Measuring One Quantity with Another

CCSS CCSS SMP2, SMP8

Suppose that a company makes buttons and puts them on cardboard strips.

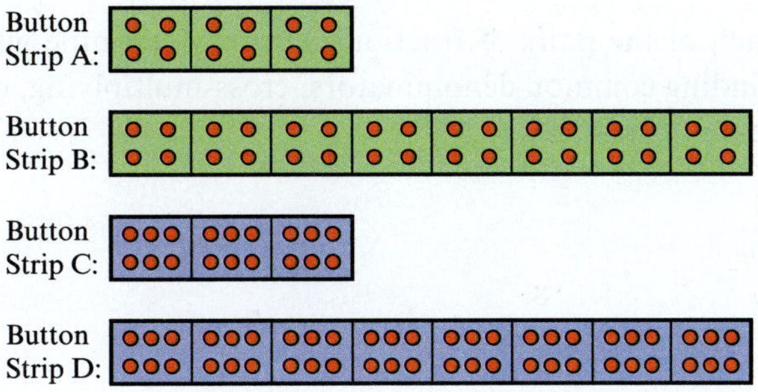

1. Discuss: What are ways to compare or relate Strips A and B? How is the relationship between Strip A and Strip B the same as the relationship between Strip C and Strip D? How is the relationship different?

2. One way to relate two quantities is to choose one quantity as a unit amount and measure the other quantity with this unit amount. Use this measurement perspective *and our definition of fraction* in the following.
   a. How many of Strip A does it take to have the same number of buttons as Strip B?
   b. How many of Strip B does it take to have the same number of buttons as Strip A?
   c. How many of Strip C does it take to have the same number of buttons as Strip D?
   d. How many of Strip D does it take to have the same number of buttons as Strip C?
   e. If you didn't already, explain how to answer part (a) with two equivalent fractions. Likewise, explain how to answer each of parts (b) through (d) with two equivalent fractions.

3. Return to part 1 of this activity and discuss some more!

2.4 Reasoning to Compare Fractions

Class Activity 2O What Is Another Way to Compare These Fractions?

CCSS CCSS SMP2, SMP3, 3.NF.3d

For each of the pairs of fractions shown, determine which fraction is greater in a way other than finding common denominators, cross-multiplying, or converting to decimals. Explain your reasoning.

$\frac{1}{49}$ $\frac{1}{39}$

$\frac{7}{37}$ $\frac{7}{45}$

Class Activity 2P Comparing Fractions by Reasoning

CCSS CCSS SMP2, SMP3, 4.NF.2

Use reasoning other than finding common denominators, cross-multiplying, or converting to decimals to compare the sizes (=, <, or >) of the following pairs of fractions:

$\frac{27}{43}$ $\frac{26}{45}$

$\frac{13}{25}$ $\frac{34}{70}$

$\frac{17}{18}$ $\frac{19}{20}$

$\frac{9}{40}$ $\frac{12}{44}$

$\frac{51}{53}$ $\frac{65}{67}$

$\frac{13}{25}$ $\frac{5}{8}$

$\frac{37}{35}$ $\frac{27}{25}$

Class Activity 2Q Can We Reason This Way?

CCSS CCSS SMP3, 3.NF.3d, 4.NF.2

Claire says that

$$\frac{4}{9} > \frac{3}{8}$$

because

$$4 > 3 \text{ and } 9 > 8$$

Discuss whether Claire's reasoning is correct.

2.5 Reasoning About Percent

Class Activity 2R Math Drawings, Percentages, and Fractions

1. For each of diagrams 1 through 5, determine the percent of the diagram that is shaded, explaining your reasoning. Write each percent as a fraction in simplest form, and explain how to see that this fraction of the diagram is shaded. You may assume that portions of each diagram which appear to be the same size really are the same size.

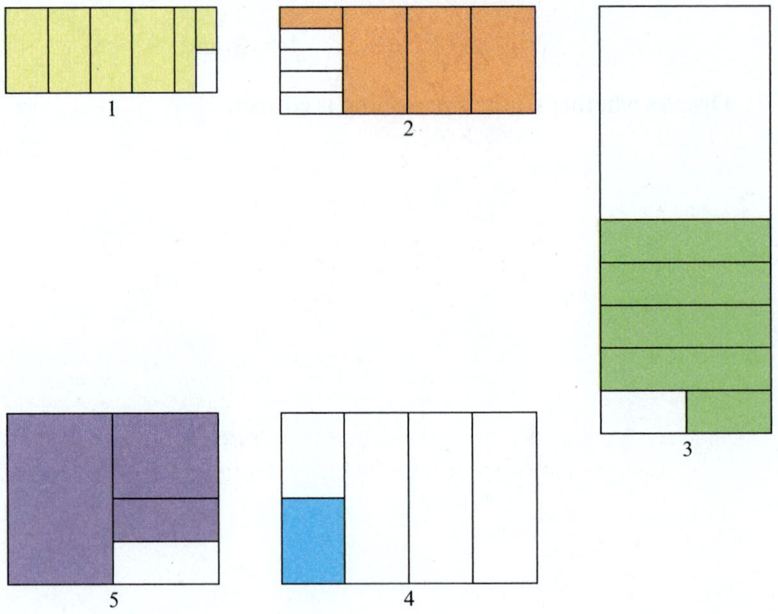

Class Activity 2S Reasoning About Percent Tables to Solve "Portion Unknown" Percent Problems

CCSS CCSS SMP2, SMP5, 6.RP.3c

We can make some percent problems easy to solve mentally by working with benchmark percentages. Use math drawings and percent tables to help you reason and record your thinking.

1. Calculate 95% of 80,000 by calculating $\frac{1}{10}$ of 80,000 and then calculating half of that result. Use the following math drawing and the percent table to help you explain why your method makes sense and to record your thinking. Is there more than one way to find 95%?

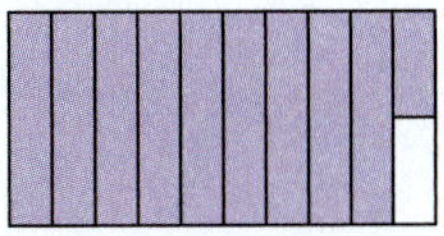

Percent table

100%	→	80,000
10%	→	______
5%	→	______
95%	→	______

2. Calculate 15% of 6500 by first calculating $\frac{1}{10}$ of 6500. Use the math drawing and a percent table to help you explain why your method makes sense and to record your thinking.

Percent table

3. Calculate 7% tax on a \$25 purchase by first finding 1%. Use a percent table to record the reasoning.

Percent table

100%	→	\$25
1%	→	______
7%	→	______

4. Calculate 60% of 810 by first calculating $\frac{1}{2}$ of 810. Use the math drawing and a percent table to help you explain why this method makes sense.

Percent table

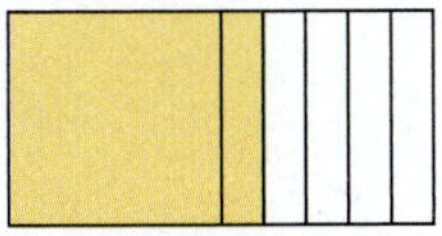

5. Find another way to use a percent table to calculate 60% of 810.

6. Use a percent table to calculate 55% of 180.

7. Make up a percent calculation problem, and explain how to solve it with a percent table.

Class Activity 2T Reasoning About Percent Tables

CCSS CCSS SMP1, SMP2, 6.RP.3c

1. Lenny has received 6 boxes of paper, which is 30% of the paper he ordered. How many boxes of paper did Lenny order? Make a math drawing and use a percent table to help you solve this problem. Explain your reasoning in each case.

   Math drawing: Percent table:

   $30\% \rightarrow 6$

2. Ms. Jones paid \$2.10 in tax on an item she purchased. The tax was 7% of the price of the item. What was the price of the item (not including the tax)? Solve this problem with the aid of a percent table. Explain your reasoning.

3. In Green Valley, the average daily rainfall is typically $\frac{5}{8}$ of an inch. Last year, the average daily rainfall in Green Valley was only $\frac{3}{8}$ of an inch. What percent of the typical average daily rainfall fell last year in Green Valley? Solve this problem with the aid of either a math drawing or a percent table, or both. Explain your reasoning.

4. If a $\frac{1}{3}$ cup serving of cheese provides your full daily value of calcium, then what percentage of your daily value of calcium is provided by $\frac{3}{4}$ cup of the cheese? Solve this problem with the aid of either a math drawing or a percent table, or both. Explain your reasoning.

Class Activity 2U Percent Problem Solving

CCSS CCSS SMP1

1. There are 30 blue marbles in a bag, which is 40% of the marbles in the bag. How many marbles are in the bag? Solve this problem in at least one of three ways: (1) with the aid of a math drawing, (2) with the aid of a percent table, and (3) by making equivalent fractions (without cross-multiplying). Explain your reasoning.

2. Andrew ran 40% as far as Marcie. How far did Marcie run as a percentage of Andrew's running distance? Explain your answer.

3. In a terrarium, there are 10% more female bugs than male bugs. If there are 8 more female bugs than male bugs, then how many bugs are in the terrarium?

   a. Solve the bug problem and explain your solution.

   b. An easy error to make is to say that there are 80 bugs in all in the terrarium. Why is this not correct and why is it an easy error to make?

4. The animal shelter has only dogs and cats. There are 25% more dogs than cats. What percentage of the animals at the animal shelter are cats? Explain your solution.

Addition and Subtraction

3.1 Interpretations of Addition and Subtraction

Class Activity 3A Relating Addition and Subtraction—The Shopkeeper's Method of Making Change

CCSS CCSS SMP7

When a patron of a store gives a shopkeeper \$$A$ for a \$$B$ purchase, we can think of the change owed to a patron as what is left from \$$A$ when \$$B$ are taken away. In contrast, the shopkeeper might make change by starting with \$$B$ and handing the customer money while adding on the amounts until they reach \$$A$.

Explain how the shopkeeper's method links subtraction to addition.

Class Activity 3B Writing Add to and Take from Problems

CCSS CCSS 1.OA.1, 2.OA.1

For each of the following equations, write a problem that is formulated naturally by the equation.

1. $6 + 9 = ?$

2. $6 + ? = 15$

3. $? + 6 = 15$

4. $15 - 6 = ?$

5. $15 - ? = 6$

6. $? - 6 = 9$

Class Activity 3C Writing Put Together/Take Apart and Compare Problems

CCSS CCSS 1.OA1, 2.OA.1, Table 1

For each of the following types, write a problem of that type, draw a strip diagram or number bond (as appropriate), and formulate *two* equations for the problem, one using addition and one using subtraction.

1. A Put Together/Take Apart, Addend Unknown problem.

2. Two versions of a Compare, Difference Unknown problem: one formulated with "more" and one with "fewer."

3. Two versions of a Compare, Bigger Unknown problem: one formulated with "more" and one with "fewer."

4. Two versions of a Compare, Smaller Unknown problem: one formulated with "more" and one with "fewer."

Class Activity 3D Identifying Problem Types and Difficult Language

CCSS CCSS SMP4, SMP7, 1.OA.1, 2.OA.1, Table 1

Do the following for each problem:

- Identify the type and subtype of the problem.
- Formulate an equation that fits with the language of the story problem.
- Identify the language that might cause students to solve the problem incorrectly if they rely only on keywords.
- Make a math drawing for the problem.

1. Clare had 3 bears. After she got some more bears, Clare had 12 bears. How many bears did Clare get?

2. Clare has 12 bears altogether; 3 of the bears are red and the others are blue. How many blue bears does Clare have?

3. Kwon had some bugs. After he got 3 more bugs, Kwon had 12 bugs altogether. How many bugs did Kwon have at first?

4. Kwon has 12 red bugs. He has 3 more red bugs than blue bugs. How many blue bugs does Kwon have?

5. Nemili has 12 red triangles and 3 blue triangles. How many more red triangles does Nemili have than blue triangles?

6. Matt had some dinosaurs. After he gave away 5 dinosaurs, he had 9 dinosaurs left. How many dinosaurs did Matt have at first?

7. Matt has 5 fewer dinosaurs than bears. Matt has 9 dinosaurs. How many bears does Matt have?

3.2 The Commutative and Associative Properties of Addition, Mental Math, and Single-Digit Facts

Class Activity 3E Mental Math

Try to find ways to make the problems that follow easy to do *mentally*. In each case, explain your method.

1. 7999 + 857 + 1

2. 367 + 98 + 2

3. 153 + 19 + 7

4. 7.89 + 6.95 + 0.05

Class Activity 3F Children's Learning Paths for Single-Digit Addition

CCSS CCSS SMP7, K.OA, 1.OA.3, 1.OA.6, 2.OA.2

In order for children to develop fluency with the basic addition facts (from 1 + 1 = 2 up to 9 + 9 = 18), they first need extensive experience solving these basic addition problems in ways that make sense to them and that become increasingly sophisticated. The following levels describe the increasingly sophisticated methods children learn. (See [2] and [4].)

Level 1: Direct modeling, count all To add 5 + 4, a child at this level counts out 5 things (or fingers), counts out another 4 things (or fingers), and then counts the total number of things (or fingers).

Level 2: Count on To add 6 + 3, a child at this level imagines 6 things, says "six" (possibly elongating it, as in "siiiiix," while perhaps thinking about pointing along a collection of 6 things). Then the child says the next 3 number words, "seven, eight, nine," usually keeping track on fingers.

Count on from larger After children can count on from the first addend, they learn to count on from the larger addend. So to add 2 + 7, a child at this level would count on from 7 instead of counting on from 2.

Level 3: Derived fact methods Children at this level use addition facts they already know to find related facts.

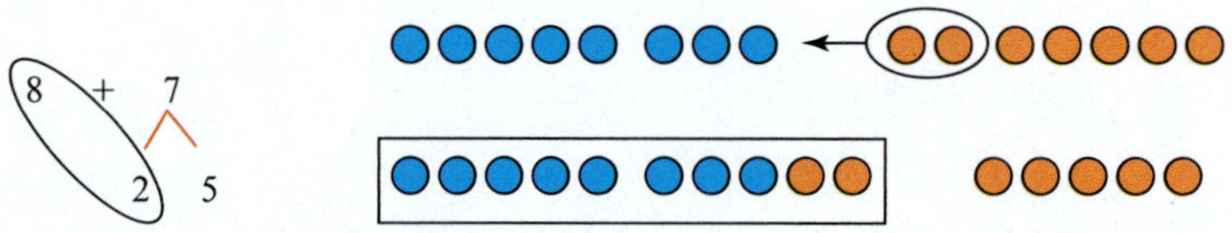

Make-a-ten method To calculate 8 + 7 with this strategy, a child breaks 7 apart into 2 + 5 so that a 10 can be made from the 8 and the 2. So the total is a 10 and 5 ones, which is 15 (as shown above).

Make-a-ten from larger The child makes a 10 with the larger number instead of just with the first addend.

Doubles ± 1 Children who know the doubles facts (1 + 1, 2 + 2, up to 9 + 9) can use these facts to find a related fact in which one of the addends is one more or one less than the addends in the double. For example, a child could determine that 6 + 7 is 1 more than 6 + 6.

1. For each method in the three levels, act out a few examples of how a child could use that method to solve a basic addition problem (adding two 1-digit numbers). For example, try $3 + 8$ at levels 2 and 3 and try $9 + 4$ and $7 + 5$ at level 3.

2. What property of addition does the count on from larger method rely on? Explain, writing equations to demonstrate how the property is used.

3. Why is the property of addition you discussed in the previous part especially important for lightening the load of learning the table of basic addition facts (from $1 + 1$ up to $9 + 9$)?

4. What property of addition does the make-a-ten method rely on? Explain, writing equations to demonstrate how the property is used.

5. Why is the make-a-ten method especially important?

6. What property of addition does the doubles $+1$ method rely on? Explain, writing equations to demonstrate how the property is used.

7. Explain why the following three items are prerequisites for children to be able to understand and use the make-a-ten method fluently:
   a. For each counting number from 1 to 9, know the number to add to it to make 10—the partner to 10.
   b. For each counting number from 11 to 19, know that the number is a 10 and some ones. For example, know that $10 + 3 = 13$ without counting and know that 13 decomposes as $10 + 3$.
   c. For each counting number from 2 to 9, know all the ways to decompose it as a sum of two counting numbers (and know all the basic addition facts with sums up to 9).

Class Activity 3G Children's Learning Paths for Single-Digit Subtraction

CCSS CCSS SMP7, K.OA.1, K.OA.2, 1.OA.4, 1.OA.6, 2.OA.2

For each of the subtraction methods listed below, act out a few examples of how a child could use that method to solve a basic subtraction problem.

Level 1: Direct modeling, take from To subtract 9 − 4, a child at this level counts out 9 things (or fingers), takes 4 things from 9 (or puts down 4 fingers), and then counts the number of things (or fingers) remaining.

Level 2: Count on to find the unknown addend The child views a subtraction problem as an unknown addend problem and counts on from the known addend to the total. So 13 − 9 = ? becomes 9 + ? = 13. The child counts on from 9, using fingers to keep track of how many have been counted on, and stops when the total is said. (This method is easier than counting down, which is difficult for most children. It makes subtraction as easy as addition.)

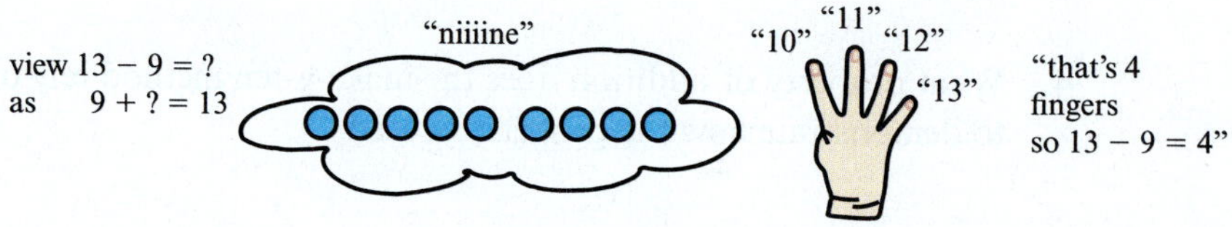

Level 3: Make-a-ten methods

Make-a-ten with the unknown addend 14 − 8 = ? becomes 8 + ? = 14. The child figures that adding 2 to 8 makes 10, then adding another 4 makes 14, so the unknown addend is 2 + 4 = 6.

Subtract from ten This is like the previous method but the child breaks 14 into 10 and 4 and subtracts 8 from 10, leaving 2, then combines this 2 with the remaining 4 from 14 to make 6.

Subtract down to ten first To solve 12 − 3 = ? the child breaks 3 into 2 + 1, takes 2 from 12 to get down to 10, then takes the remaining 1 from 10 to get 9.

Compare and contrast the methods at level 3. When are the first two methods easier than the third? When is the third method easier than the first two?

Class Activity 3H Reasoning to Add and Subtract

CCSS CCSS SMP2

1. John and Anne want to calculate 253 − 99 by first calculating

$$253 - 100 = 153$$

John says that they must now *subtract* 1 from 153, but Anne says that they must *add* 1 to 153.

a. Draw a number line (which need not be perfectly to scale) to help you explain who is right and why. Do not just say which answer is numerically correct; use the number line to help you explain why the answer must be correct.

b. Explain in another way who is right and why.

2. Jamarez says that he can calculate 253 − 99 by adding 1 to both numbers and calculating 254 − 100 instead.

a. Draw a number line (which need not be perfectly to scale) to help you explain why Jamarez's method is valid.

b. Explain in another way why Jamarez's method is valid.

c. Could you adapt Jamarez's method to other subtraction problems, such as to the problem 324 − 298? Explain, and give several other examples.

3. Find ways to solve the addition and subtraction problems that follow *other than* by using the standard addition or subtraction algorithms. In each case, explain your reasoning, and—except for part (g)—write equations that correspond to your line of reasoning.

a. 183 + 99

b. 268 + 52

c. 600 − 199

d. 164 − 70

e. 999 + 9999

f. \$10.00 − \$2.99

g. (No equations are needed.)

$$\begin{array}{r} 2.99 \\ 3.99 \\ 1.99 \\ +4.99 \\ \hline \end{array}$$

3.3 Why the Standard Algorithms for Addition and Subtraction in Base Ten Work

Class Activity 3I Adding and Subtracting with Base-Ten Math Drawings

CCSS CCSS SMP3, SMP5, 1.NBT.4

The hypothetical student work below is similar to actual first graders' work. These urban Latino first-graders performed substantially above first-graders of higher socioeconomic status and older children. (See [5].)

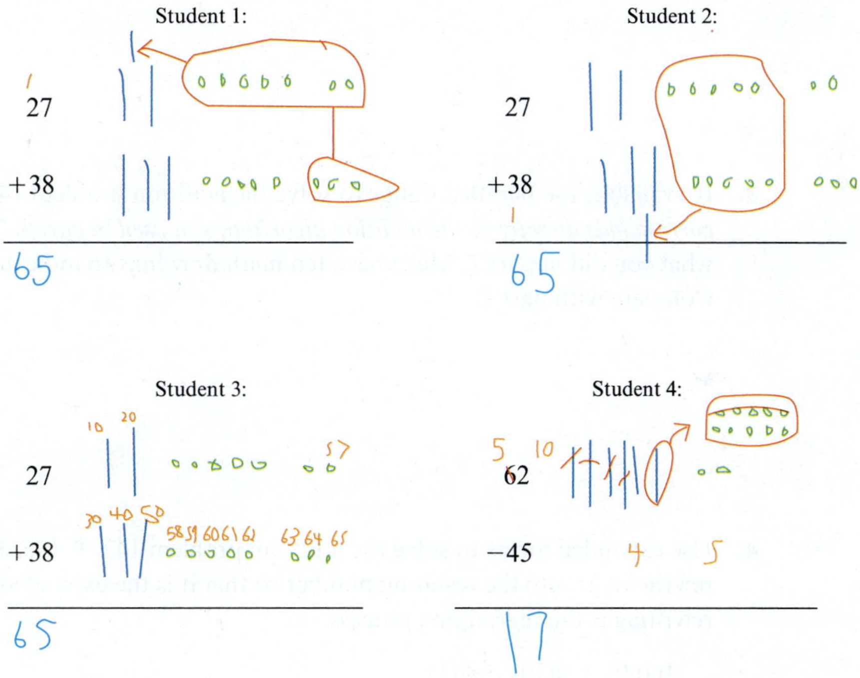

1. Examine and discuss the students' work. Compare the work of students 1, 2, and 3. In particular, compare the methods of student 1 and student 2 for adding 7 + 8. What mental method for subtracting 12 − 5 does the work of student 4 suggest?

2. Show how students 1, 2, and 3 might solve the addition problem 36 + 27 and how student 4 might solve the subtraction problem 43 − 18.

Class Activity 3J Understanding the Standard Addition Algorithm

CCSS CCSS SMP3, SMP5, 2.NBT.7, 2.NBT.9

Bundled toothpicks or base-ten blocks would be useful for this activity.

1. Add the numbers, using the standard paper-and-pencil method. Notice that regrouping is involved.

$$\begin{array}{r} 147 \\ +\ 195 \\ \hline \end{array}$$

2. If available, use bundled things to solve the addition problem 147 + 195. Make base-ten math drawings to indicate the process you used.

3. If available, use bundled things to solve the addition problem 147 + 195 *in a way that corresponds directly to the addition algorithm you used in part 1.* This may be different from what you did in part 2. Make base-ten math drawings to indicate the process you used. Compare with part 2.

4. Use expanded forms to solve the addition problem 147 + 195. First add like terms; then rewrite (regroup) the resulting number so that it is the expanded form of a number. This rewriting is the regrouping process.

$$\begin{array}{r} 1(100) + 4(10) + 7(1) \\ +\ 1(100) + 9(10) + 5(1) \\ \hline \end{array}$$

← First add like terms, remaining in expanded form.

← Then regroup so that you have the expanded form of a decimal number. You might want to take several steps to do so.

5. Compare and contrast your work in parts 1–4.

Class Activity 3K Understanding the Standard Subtraction Algorithm

CCSS CCSS SMP3, SMP5, 2.NBT.7, 2.NBT.9

Bundled toothpicks or base-ten blocks would be useful for this activity.

1. Subtract the following numbers, using the standard paper-and-pencil method. Notice that regrouping is required.

$$\begin{array}{r} 125 \\ -\ 68 \\ \hline \end{array}$$

2. If available, use bundled things to solve the subtraction problem 125 − 68. Make base-ten math drawings to indicate the process.

3. If available, use bundled things to solve the subtraction problem 125 − 68 *in a way that corresponds directly to the subtraction algorithm you used in part 1.* This may be different from what you did in part 2. Make base-ten math drawings to indicate the process. Compare with part 2.

4. Solve the subtraction problem 125 − 68, but now use expanded forms. Start by rewriting the number 125 in expanded form. Rewrite the number in several steps, so that it will be easy to take 68 from 125. This rewriting is the regrouping process.

$$125 = 1(100) + 2(10) + 5(1) =$$

Write your regrouped number here ⟶

Subtract 68: $\underline{-\ [6(10) + 8(1)]}$

5. Compare and contrast your work in parts 1–4.

6. Use the standard algorithm to solve the following subtraction problem. Notice that regrouping across 0 is required.

$$\begin{array}{r} 104 \\ -\ 47 \\ \hline \end{array}$$

Now explain the process in terms of bundled things or base-ten drawings.

How else could you solve the subtraction problem?

Class Activity 3L A Third-Grader's Method of Subtraction

CCSS CCSS SMP3

When asked to compute 423 − 157, Pat (a third-grader) wrote the following:

4−
30−
34−
300
266

"You can't take 7 from 3; it's 4 too many, so that's negative 4. You can't take 50 from 20; it's 30 too many, so that's negative 30; and with the other 4, it's negative 34. 400 minus 100 is 300, and then you take the 34 away from the 300, so it's 266."[1]

1. Discuss Pat's idea for calculating 423 − 157. Is her method legitimate? Analyze Pat's method in terms of expanded forms.

2. Could you use Pat's idea to calculate 317 − 289? If so, write what you think Pat might write, and also use expanded forms.

[1]This is taken from [1, p. 263]

Class Activity 3M Regrouping in Base 12

CCSS CCSS SMP3

We can use the regrouping idea when objects are bundled in groups of a dozen instead of in groups of ten, as in the following problem.

> A store owner buys small, novelty party favors in bags of 1 dozen and boxes of 1-dozen bags (for a total of 144 favors in a box). The store owner has 5 boxes, 4 bags, and 3 individual party favors at the start of the month. At the end of the month, the store owner has 2 boxes, 9 bags, and 7 individual party favors left. How many favors did the store owner sell? Give the answer in terms of boxes, bags, and individual favors.

We can use a sort of *expanded form* for these party favors:

$$5(\text{boxes}) + 4(\text{bags}) + 3(\text{individual})$$
$$2(\text{boxes}) + 9(\text{bags}) + 7(\text{individual})$$

Solve this problem by regrouping among the boxes, bags, and individual party favors.

Class Activity 3N Regrouping in Base 60

CCSS CCSS SMP3

We can use the regrouping idea with time. Just as 1 hundred is 10 tens and 1 ten is 10 ones, 1 hour is 60 minutes and 1 minute is 60 seconds. The following problem asks you to regroup among hours, minutes, and seconds:

> Ruth runs around a lake two times. The first time takes 1 hour, 43 minutes, and 38 seconds. The second time takes 1 hour, 48 minutes, and 29 seconds. What is Ruth's total time for the two laps? Give the answer in hours, minutes, and seconds.

We can use a sort of *expanded form* for time:

$$1(\text{hour}) + 43(\text{minutes}) + 38(\text{seconds})$$
$$1(\text{hour}) + 48(\text{minutes}) + 29(\text{seconds})$$

Solve this problem by regrouping among hours, minutes, and seconds.

3.4 Reasoning About Fraction Addition and Subtraction

Class Activity 30 Why Do We Add and Subtract Fractions the Way We Do?

CCSS CCSS SMP3, 4.NF.3a, 5.NF.1, 5.NF.2

For part 2 of this activity, each person will need at least 5 identical strips of paper (or card stock).

1. When two fractions have the same denominator, we add or subtract them by keeping the same denominator and adding or subtracting the numerators. For example,

$$\frac{1}{5} + \frac{3}{5} = \frac{1+3}{5} = \frac{4}{5}$$

Patti says: "We should add the tops *and* the bottoms." She shows you this picture to explain why:

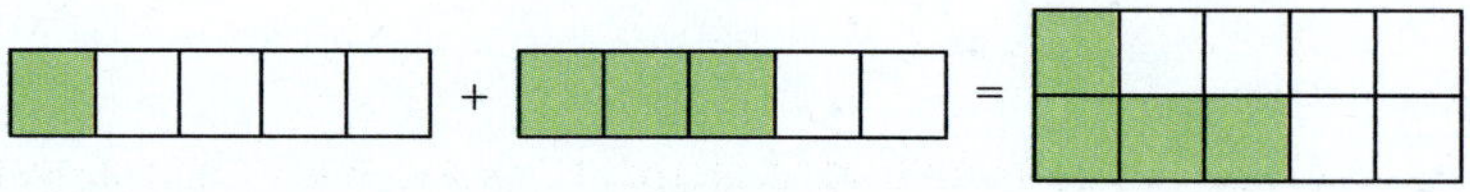

So according to Patti:

$$\frac{1}{5} + \frac{3}{5} = \frac{1+3}{5+5} = \frac{4}{10}$$

a. Why is Patti's method *not* a valid way to add fractions, and why doesn't Patti's picture prove that fractions can be added in her way? Critique Patti's reasoning.

b. How could Patti use estimation and benchmark fractions to see if her answer is reasonable?

c. Explain why the proper way to add $\frac{1}{5} + \frac{3}{5}$ and $\frac{2}{5} + \frac{4}{5}$ makes sense, using our definition of fractions.

2. You need 5 identical strips of paper for this part.
   a. Label one of the strips "1 whole," and fold and label two other strips as indicated:

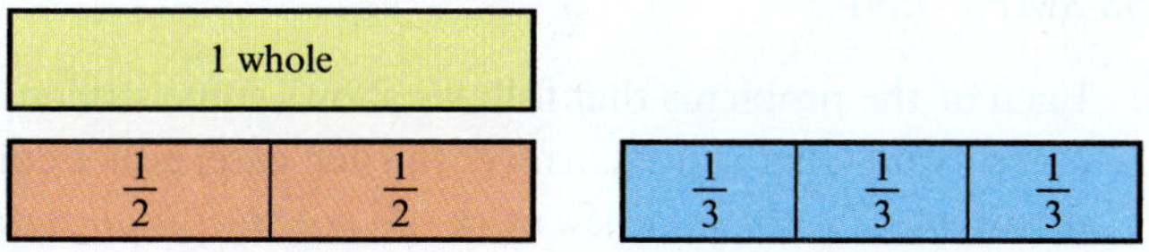

   b. Fold and place your halves and thirds strips to show these lengths:

$$\frac{1}{2} + \frac{1}{3}, \quad \frac{2}{3} - \frac{1}{2}, \quad \frac{2}{3} + \frac{1}{2}$$

   Make drawings to record your work.

   c. Discuss: Your folded strips in part b show the requested lengths, so why are you not done solving the problem?

   d. Fold the remaining two strips, one in halves, the other in thirds. Then fold each of them again into parts that will be better for expressing the lengths in part b. Relate this to the numerical procedure for adding and subtracting fractions.

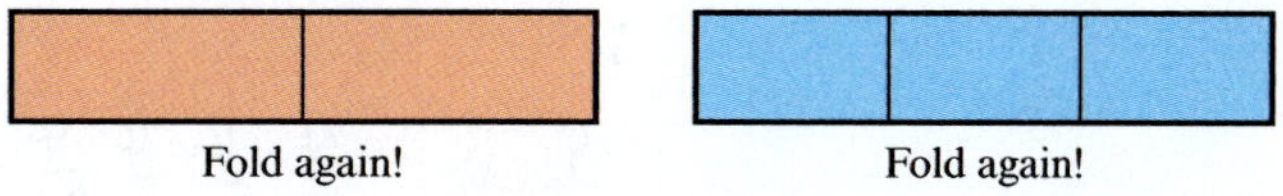

Class Activity 3P Critiquing Mixed Number Addition and Subtraction Methods

CCSS CCSS SMP3, 5.NF.1

1. Each of the problems that follows shows some student work. Discuss the work: What is correct and what is not correct? In each case, either complete the work or modify it to make it correct. *Use the student's work; do not start from scratch.*

a. Subtract: $3\frac{1}{4} - 1\frac{3}{4}$.

student 1:

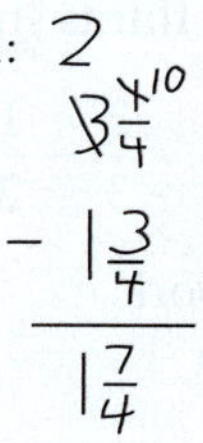

student 2:

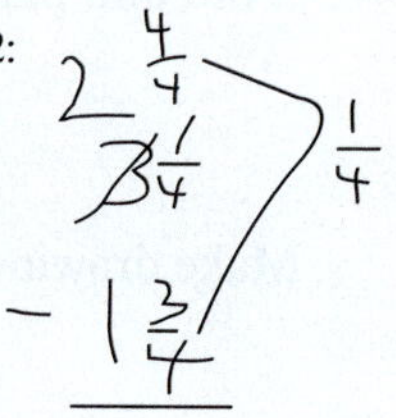

b. There are $2\frac{1}{3}$ cups of milk in a bowl. How much milk must be added to the bowl so that there will be 3 cups of milk in the bowl?

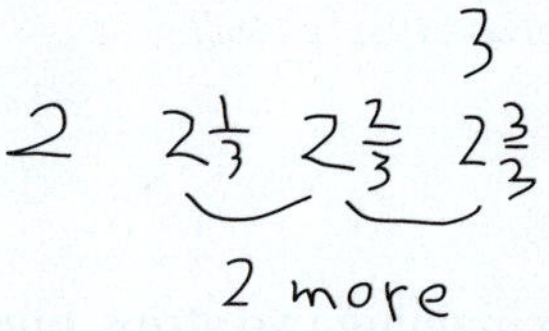

c. There were 5 pounds of apples in a bag. After some of the apples were removed from the bag, there were $3\frac{1}{4}$ pounds of apples left. How many pounds of apples were removed?

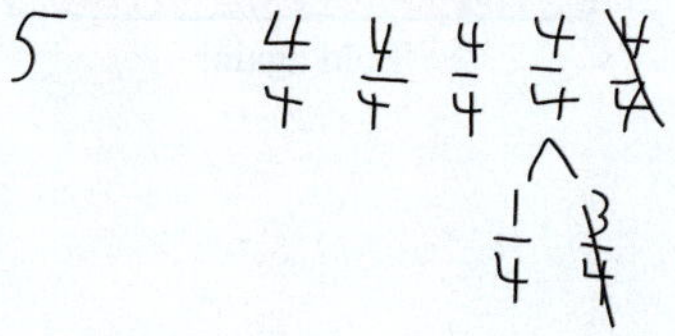

d. Add: $2\frac{2}{3} + 1\frac{2}{3}$.

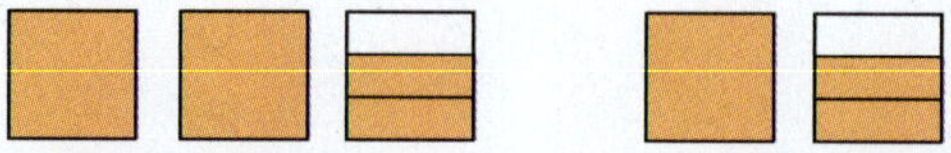

$2\frac{2}{3} + 1\frac{2}{3} = 3\frac{4}{6}$ because the fraction part is 4 out of 6.

2. Find at least two different ways to calculate

$$7\frac{1}{3} - 4\frac{1}{2}$$

and to give the answer as a mixed number. In each case, explain why your method makes sense.

Class Activity 3Q Are These Word Problems for $\frac{1}{2} + \frac{1}{3}$?

CCSS CCSS SMP4, 5.NF.2

For each problem, determine if it is a problem for $\frac{1}{2} + \frac{1}{3}$. If not, explain why not, and explain how to solve the problem if possible. If there is not enough information to solve the problem, explain why not.

1. Tom pours $\frac{1}{2}$ cup of water into an empty bowl and then pours $\frac{1}{3}$ cup of water into the bowl. How many cups of water are in the bowl now?

2. Tom pours $\frac{1}{2}$ cup of water into an empty bowl and then pours in another $\frac{1}{3}$. How many cups of water are in the bowl now?

3. $\frac{1}{2}$ of the land in Heeltoe County is covered with forest, $\frac{1}{3}$ of the land in the adjacent Toejoint County is covered with forest. What fraction of the land in the two-county Heeltoe–Toejoint region is covered with forest?

4. $\frac{1}{2}$ of the land in Heeltoe County is covered with forest, $\frac{1}{3}$ of the land in the adjacent Toejoint County is covered with forest. Heeltoe and Toejoint counties have the same land area. What fraction of the land in the two-county Heeltoe–Toejoint region is covered with forest?

5. Students at Martin Luther King Elementary School could vote for all the lunch choices they like. $\frac{1}{2}$ of the children say they like to have pizza for lunch, $\frac{1}{3}$ of the children say they like to have a hamburger for lunch. What fraction of the children at Martin Luther King Elementary School would like to have either pizza or a hamburger for lunch?

6. $\frac{1}{2}$ of the children at Timothy Elementary School like to have pizza for lunch, and the other half does not like to have pizza for lunch. Of the children who do not like to have pizza for lunch, $\frac{1}{3}$ like to have a hamburger for lunch. What fraction of the children at Timothy Elementary School like to have either pizza or a hamburger for lunch?

Class Activity 3R Are These Word Problems for $\frac{1}{2} - \frac{1}{3}$?

CCSS CCSS SMP4, 5.NF.2

For each of the following story problems, determine whether the problem can be solved by subtracting $\frac{1}{2} - \frac{1}{3}$. If not, explain why not, and explain how the problem should be solved if there is enough information to do so. If there is not enough information to solve the problem, explain why not.

1. Zelha pours $\frac{1}{2}$ cup of water into an empty bowl and then pours out $\frac{1}{3}$. How much water is in the bowl now?

2. Zelha pours $\frac{1}{2}$ cup of water into an empty bowl and then pours out $\frac{1}{3}$ cup of water. How much water is in the bowl now?

3. Zelha pours $\frac{1}{2}$ cup of water into an empty bowl and then pours out $\frac{1}{3}$ of the water that is in the bowl. How much water is in the bowl now?

4. Yesterday James ate $\frac{1}{2}$ of a pizza, and today he ate $\frac{1}{3}$ of a pizza of the same size. How much more pizza did James eat yesterday than today?

5. Yesterday James ate $\frac{1}{2}$ of a pizza, and today he ate $\frac{1}{3}$ of the whole pizza. Nobody else ate any of that pizza. How much pizza is left?

6. Yesterday James ate $\frac{1}{2}$ of a pizza, and today he ate $\frac{1}{3}$ of the pizza that was left over from yesterday. Nobody else ate any of that pizza. How much pizza is left?

Class Activity 3S What Fraction Is Shaded?

CCSS CCSS SMP1, 5.NF.2

For each square shown, determine the fraction of the square that is shaded. Explain your reasoning. You may assume that all lengths that appear to be equal really are equal. Try not to use any area formulas. Apply your knowledge of how to add and subtract fractions!

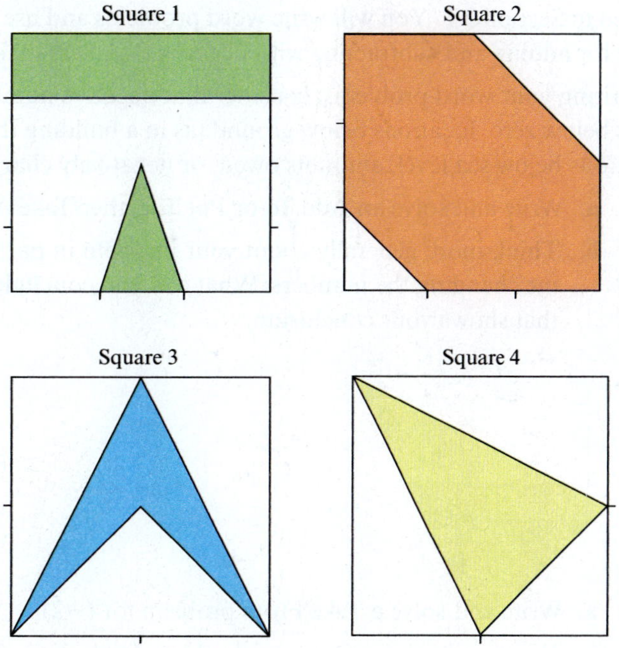

Class Activity 3T Addition with Whole Numbers, Decimals, Fractions, and Mixed Numbers: What Are Common Ideas?

CCSS CCSS SMP8

Think about how we add (and subtract) whole numbers, decimals, fractions, and mixed numbers, and think about the ideas and reasoning that are involved. What is common in the way we add (and subtract) across all these different kinds of numbers?

3.5 Why We Add and Subtract with Negative Numbers the Way We Do

Class Activity 3U Using Word Problems to Find Rules for Adding and Subtracting Negative Numbers

CCSS CCSS SMP2, SMP8, 7.NS.1

This activity relies on some of the different types of addition and subtraction word problems discussed in Section 3.1. You will write word problems and use these problems to see why some of the rules for adding and subtracting with negative numbers make sense.

In writing your word problems, consider that negative numbers are nicely interpreted as temperatures below zero, locations below ground (as in a building that has basement stories, for example), locations below sea level, amounts owed, or negatively charged particles.

1. **a.** Write and solve an Add To or Put Together/Take Apart problem for $(-5) + 5 = ?$.

   b. Think more generally about your problem in part (a) and about its solution, and imagine changing the numbers. What can you conclude about $(-N) + N$? Write an equation that shows your conclusion.

2. **a.** Write and solve a Take From problem for $(-2) - 5 = ?$.

   b. Write and solve a Compare problem for $(-2) - 5 = ?$ in which one quantity is -2 and the other quantity is 5 less and is unknown.

   c. Think more generally about your problems in parts (a) and (b) and about their solutions, and imagine changing the numbers. What can you conclude about how $(-A) - B$ and $A + B$ are related? Write an equation to show this relationship.

3. **a.** Write and solve a Compare problem for $2 - (-5) = ?$ in which one quantity is 2, the other quantity is -5, and the difference between the two quantities is unknown.

   b. Think more generally about your problem in part (a) and about its solution, and imagine changing the numbers. What can you conclude about $A - (-B)$? Write an equation that shows your conclusion.

CLASS ACTIVITIES 4

Multiplication

4.1 Interpretations of Multiplication

Class Activity 4A Showing Multiplicative Structure

CCSS CCSS SMP2, SMP6, 3.OA.3, 4.OA.1, 4.OA.2, 7.RP.8b

Recall our definition of multiplication.

M	•	N	=	P
number of equal **groups**		number of **units** in **1 group**		number of **units** in **M groups**

1. Using the definition of multiplication, explain why you can determine the number of ladybugs in the following picture by multiplying. Write a corresponding equation.

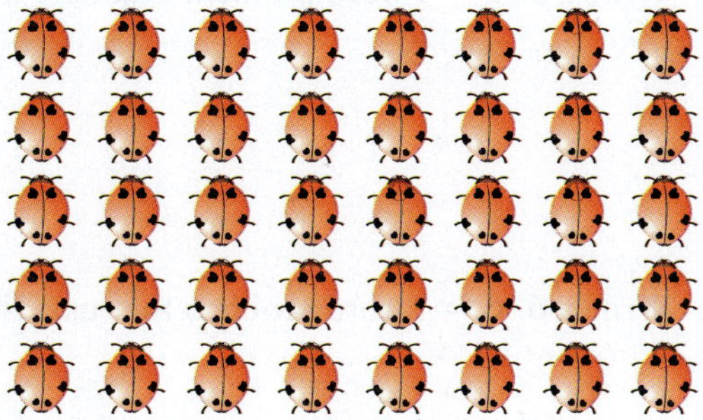

2. Eva's puppy weighs 9 pounds. Micah's puppy weighs 4 times as much as Eva's puppy. How much does Micah's puppy weigh? Draw a strip diagram for this problem and use it to help explain how to solve the problem. Write a corresponding equation.

3. Fran has 3 pairs of pants (pants 1, 2, and 3) that coordinate perfectly with 4 different shirts (shirts A, B, C, and D). How many different outfits consisting of a pair of pants and a shirt can Fran make from these clothes?

   Apply the definition of multiplication to explain why you can solve this problem with multiplication. Write a corresponding equation.

4. If 1 meter of a rope weighs 9 grams, how much do 7 meters of the same kind of rope weigh?

   Apply the definition of multiplication to explain why you can solve this problem with multiplication. Write a corresponding equation.

5. If a piece of rope is 4 feet long and weighs 1 pound, then how long is a piece of rope of the same type that weighs 7 pounds?

   Apply the definition of multiplication to explain why you can solve this problem with multiplication. Write a corresponding equation.

6. You have 6 pairs of gloves, one pair in each of these colors: red, blue, yellow, green, orange, and purple. Each pair consists of a left glove and a right glove.

   a. Write a $2 \cdot 6 = ?$ word problem that uses this context.

   b. Write a $6 \cdot 6 = ?$ word problem that uses this context.

Class Activity 4B Writing Multiplication Word Problems

CCSS CCSS SMP2, SMP6

Recall our definition of multiplication.

M	•	N	=	P
number of equal groups		number of units in 1 group		number of units in M groups

1. Write a simple word problem for $4 \cdot 8 = ?$. Explain why the problem can be solved by multiplying $4 \cdot 8$ (make a math drawing to aid your explanation, if possible).

2. Write an Array problem for $3 \cdot 7 = ?$. Explain why the problem can be solved by multiplying $3 \cdot 7$. How else can the problem be solved? Explain.

3. Write a Multiplicative Comparison problem for $5 \cdot 9 = ?$. Draw a strip diagram for the problem and explain why the problem can be solved by multiplying $5 \cdot 9$.

4. Write an Ordered Pair problem for $8 \cdot 5 = ?$. Explain why the problem can be solved by multiplying $8 \cdot 5$. How else can the problem be solved? Explain.

5. Write a multiplication word problem that concerns the volume (e.g., in cups, milliliters, or liters) and weight (e.g., in pounds, ounces, grams, or kilograms) of some kind of food.

4.2 Why Multiplying by 10 Is Special in Base Ten

Class Activity 4C Multiplying by 10

CCSS CCSS SMP3, 4.NBT.1, 5.NBT.1, 5.NBT.2

1. Which of the following statements are correct and appropriate to use with all numbers in base ten?
   a. To multiply a number by 10, put a 0 at the end of the number.
   b. To multiply a number by 10, move the decimal point one place to the right.
   c. To multiply a number by 10, move all the digits one place to the left.

2. Explain why statement (c) about multiplying by 10 is true. To do so, use the math drawing below, which represents 10×23. What happens to each of the 2 tens and what happens to each of the 3 ones when we multiply 23 by 10?

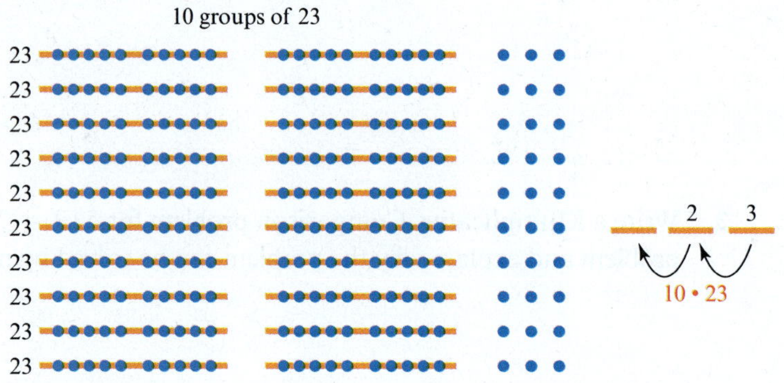

4.3 The Commutative and Associative Properties of Multiplication, Areas of Rectangles, and Volumes of Boxes

Class Activity 4D Explaining the Commutative Property of Multiplication with Arrays and Area

CCSS CCSS SMP7, 3.OA.5, 3.MD.7a

1. Write one word problem for $7 \cdot 4$ and another for $4 \cdot 7$. Before determining the answers to the problems, discuss whether it would necessarily be obvious to a student that the answers are the same.

2. View the following array in two ways to explain why $4 \cdot 7 = 7 \cdot 4$.

3. The large rectangle shown here is 9 centimeters wide and 5 centimeters tall. Use the definition of multiplication to explain why you can find the area of the rectangle by multiplying.

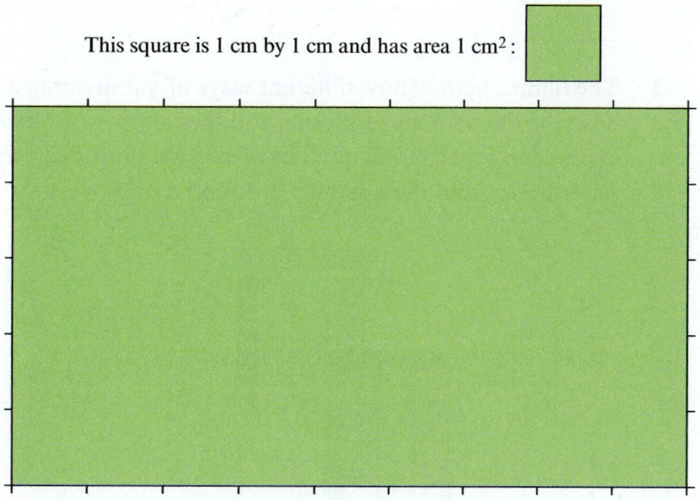

4. Using the rectangle in part 3, explain why

$$5 \cdot 9 = 9 \cdot 5$$

5. Why does the commutative property hold more generally? Why is it true that

$$A \cdot B = B \cdot A$$

no matter what the counting numbers A and B are?

Class Activity 4E Describing the Volume of a Box with Multiplication and Explaining the Associative Property

CCSS CCSS SMP7, 5.MD.5a

You will need a set of blocks for parts 1 and 2 of this activity.

1. If you have cubic-inch blocks available, build a box that is 3 inches wide, 2 inches long, and 4 inches tall. It should look like this:

2. Subdivide your box into natural groups of blocks, and describe how you subdivided the box. How many groups were there and how many blocks were in each group? Using multiplication, write the corresponding expressions for the total number of blocks in the box. Now repeat, this time subdividing your box into natural groups in a different way.

3. The figures below show different ways of subdividing a box into groups. In each case, describe the number of groups and the number of blocks in each group. Then write an expression for the total number of blocks. Your expressions should use multiplication, parentheses, and the numbers 2, 3, and 4 only.

A.

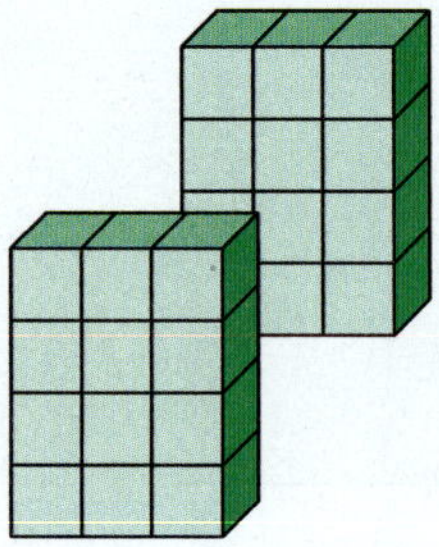

_______ groups of _________ blocks

Write an expression using multiplication, parentheses, and the numbers 2, 3, and 4 for the total number of blocks:

____

B.

_______ groups of _________ blocks

Write an expression using multiplication, parentheses, and the numbers 2, 3, and 4 for the total number of blocks:

____

C.

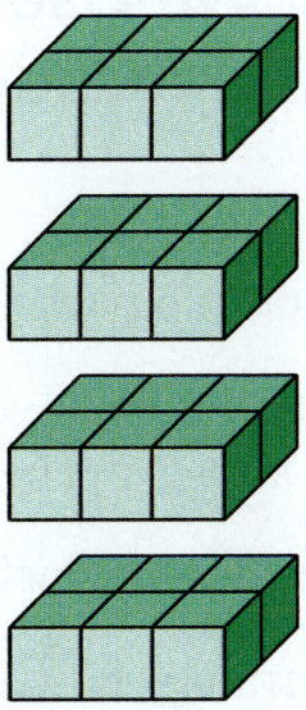

________ groups of ________ blocks

Write an expression using multiplication, parentheses, and the numbers 2, 3, and 4 for the total number of blocks:

____

D.

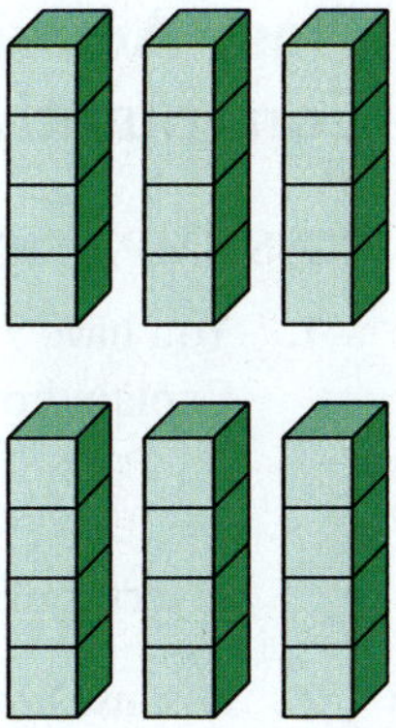

________ groups of ________ blocks

Write an expression using multiplication, parentheses, and the numbers 2, 3, and 4 for the total number of blocks:

____

E.

________ groups of ________ blocks

Write an expression using multiplication, parentheses, and the numbers 2, 3, and 4 for the total number of blocks:

____

F.

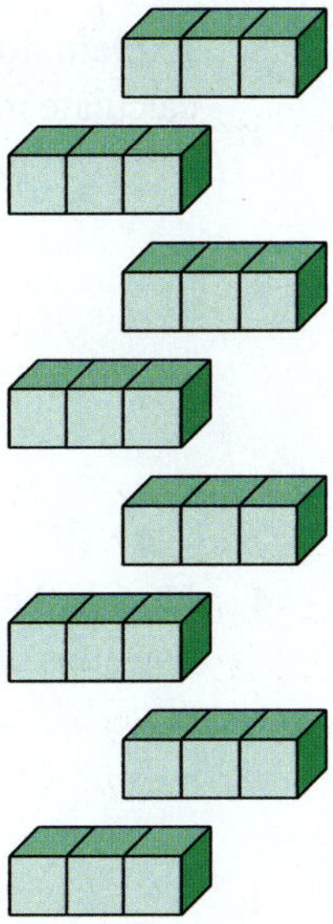

________ groups of ________ blocks

Write an expression using multiplication, parentheses, and the numbers 2, 3, and 4 for the total number of blocks:

____

4. Use some of the figures and the expressions you wrote for part 3 to help you explain why

$$(A \times B) \times C = A \times (B \times C)$$

for counting numbers A, B, C.

Class Activity 4F How Can We Use the Associative and Commutative Properties of Multiplication?

CCSS SMP2, SMP7, 3.NBT.3

1. You have 7 bags of gum. Each bag contains 6 packs. Each pack contains 5 pieces of gum. Explain how to interpret

$$(7 \cdot 6) \cdot 5 \quad \text{and} \quad 7 \cdot (6 \cdot 5)$$

in terms of the gum. Why must they be equal? Which of the two is easier to calculate?

2. Discuss how to multiply $4 \cdot 60$ using a strategy based on place value and properties of operations. Discuss how the equations and math drawing below are relevant. Which property of multiplication is used?

$$\begin{aligned} 4 \cdot 60 &= 4 \cdot (6 \cdot 10) \\ &= (4 \cdot 6) \cdot 10 \\ &= 24 \cdot 10 \\ &= 240 \end{aligned}$$

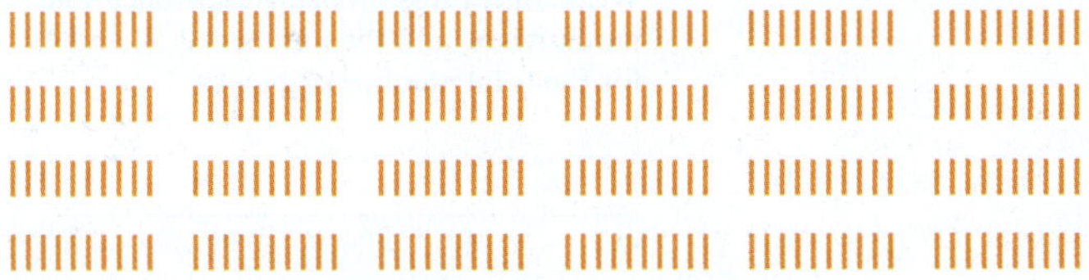

3. Explain how to use the associative property of multiplication to make $28 \cdot 0.25$ easy to calculate mentally. Write equations to show how the associative property is used. How else can you think about the problem?

4. How is the associative property of multiplication involved when viewing 20 groups of 10 as 2 groups of 100? Write equations to help explain.

5. There are 21 bags with 2 marbles in each bag. Ben calculates the number of marbles there are in all by counting by twos 21 times:

$$2, 4, 6, 8, 10, \ldots, 40, 42$$

Kaia calculates $21 + 21 = 42$ instead.

Discuss the two calculation methods. Are both legitimate? How could you relate the two methods? Is this related to a property of multiplication? Explain.

4.4 The Distributive Property

Class Activity 4G Explaining the Distributive Property

CCSS CCSS SMP7, 3.MD.7c

1. There are 6 goodie bags. Each goodie bag contains 3 eraser tops and 4 stickers.

Explain why each of the two expressions

$$6 \cdot 3 + 6 \cdot 4 \quad \text{and} \quad 6 \cdot (3 + 4)$$

describes the total number of items in the goodie bags and therefore why

$$6 \cdot (3 + 4) = 6 \cdot 3 + 6 \cdot 4$$

Your explanation should *not* involve calculating sums or products.

Then discuss how to view your explanation as explaining why the distributive property makes sense for all counting numbers.

2. Use the different shading shown in the rectangle, and use the definition of multiplication, to explain why

$$3 \cdot (2 + 4) = 3 \cdot 2 + 3 \cdot 4$$

Your explanation should be general in the sense that you could use it to explain why

$$A \cdot (B + C) = A \cdot B + A \cdot C$$

for *all* counting numbers A, B, and C.

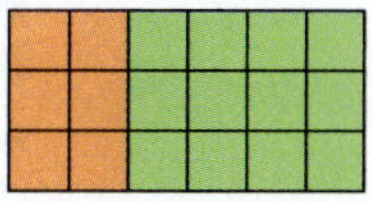

3. Make a rough drawing of an array and shade it to illustrate the equation

$$8 \cdot (10 + 5) = 8 \cdot 10 + 8 \cdot 5$$

Class Activity 4H Applying the Distributive Property to Calculate Flexibly

CCSS CCSS SMP2, SMP7

1. Use the multiplication facts

$$15 \cdot 15 = 225$$
$$2 \cdot 15 = 30$$

to help you mentally calculate

$$17 \cdot 15$$

Explain how your calculation method is related to the array below, which consists of 17 rows of dots with 15 dots in each row. Also write equations showing how your mental strategy for calculating $17 \cdot 15$ involves the distributive property.

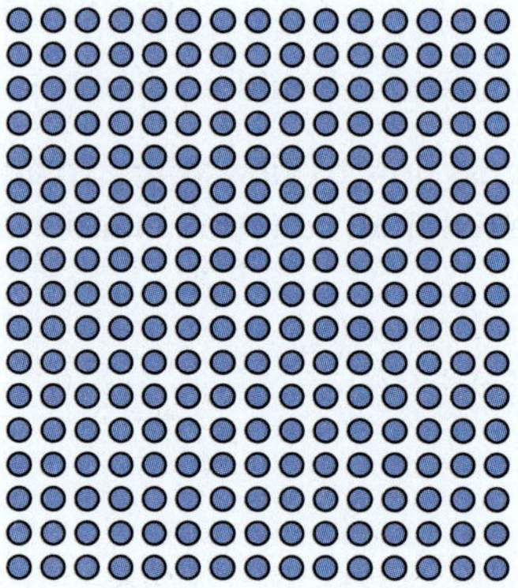

2. Write an equation that uses subtraction and the distributive property and that goes along with the following array:

3. Mentally calculate $20 \cdot 15$, and use your answer to mentally calculate $19 \cdot 15$. Write an equation that uses subtraction and the distributive property and that goes along with your strategy. Without drawing all the detail, draw a rough picture of an array that illustrates this calculation strategy.

Class Activity 4I Critique Multiplication Strategies

CCSS CCSS SMP3

1. Kylie has an idea for how to calculate $23 \cdot 23$. She says,

   Twenty times 20 is 400, and 3 times 3 is 9; so $23 \cdot 23$ should be 400 plus 9, which is 409.

   Is Kylie's method valid? If not, how could you modify her work to make it correct? Don't just start over in a different way; work with Kylie's idea. Use the large square below, which consists of 23 rows with 23 small squares in each row, to help you explain your answer.

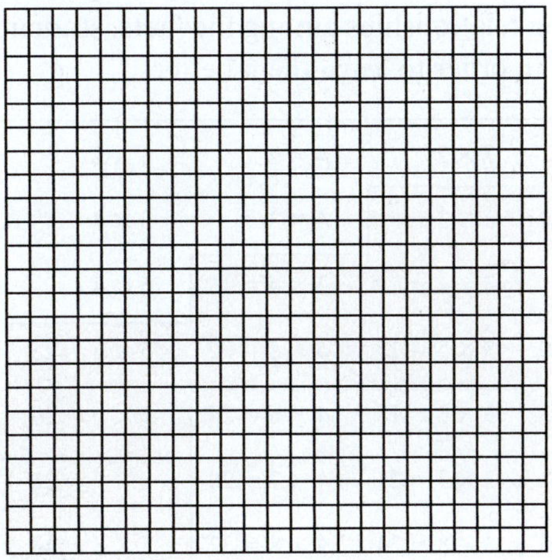

2. Annie is working on the multiplication problem $19 \cdot 21$. She says that $19 \cdot 21$ should equal $20 \cdot 20$ because 19 is one less than 20 and 21 is one more than 20.

   This is a wonderful idea, but is Annie correct? If not, use the diagram below to help you explain to Annie why not. There are 20 rows of dots with 21 dots in each row.

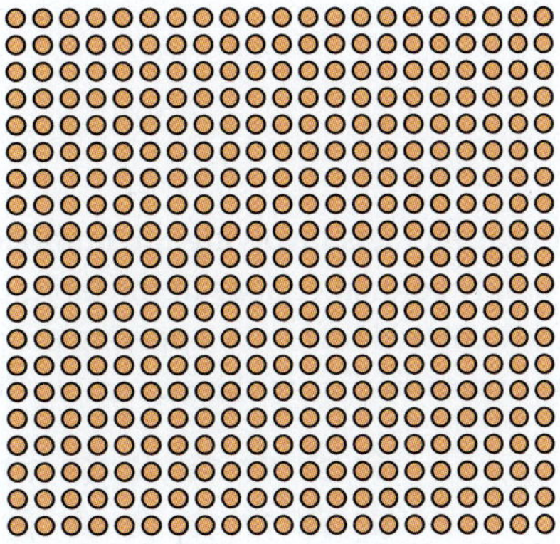

3. Mary is working on the multiplication problem $19 \cdot 21$. She says that $19 \cdot 21$ is 21 less than $20 \cdot 21$, and $20 \cdot 21$ is 20 more than $20 \cdot 20$, which she knows is 400. Mary thinks this ought to help her calculate $19 \cdot 21$, but she can't quite figure it out.

   Discuss Mary's idea in detail. Can you make her idea work?

4.5 Properties of Arithmetic, Mental Math, and Single-Digit Multiplication Facts

Class Activity 4J Using Properties of Arithmetic to Aid the Learning of Basic Multiplication Facts

CCSS CCSS 3.OA.5

In school, students must learn the single-digit multiplication facts from $1 \times 1 = 1$ to $9 \times 9 = 81$. By learning relationships among the facts, students can structure their understanding of the single-digit facts in order to learn them better.

2 × 2 = 4	3 × 2	4 × 2	5 × 2	6 × 2	7 × 2	8 × 2	9 × 2
2 × 3 = 6	3 × 3 = 9	4 × 3	5 × 3	6 × 3	7 × 3	8 × 3	9 × 3
2 × 4 = 8	3 × 4 = 12	4 × 4 = 16	5 × 4	6 × 4	7 × 4	8 × 4	9 × 4
2 × 5 = 10	3 × 5 = 15	4 × 5 = 20	5 × 5 = 25	6 × 5	7 × 5	8 × 5	9 × 5
2 × 6 = 12	3 × 6 = 18	4 × 6 = 24	5 × 6 = 30	6 × 6 = 36	7 × 6	8 × 6	9 × 6
2 × 7 = 14	3 × 7 = 21	4 × 7 = 28	5 × 7 = 35	6 × 7 = 42	7 × 7 = 49	8 × 7	9 × 7
2 × 8 = 16	3 × 8 = 24	4 × 8 = 32	5 × 8 = 40	6 × 8 = 48	7 × 8 = 56	8 × 8 = 64	9 × 8
2 × 9 = 18	3 × 9 = 27	4 × 9 = 36	5 × 9 = 45	6 × 9 = 54	7 × 9 = 63	8 × 9 = 72	9 × 9 = 81

1. Examine the darkly colored, lightly colored, and uncolored regions in the above multiplication table. Explain how to obtain the uncolored facts quickly and easily from the shaded facts. In doing so, what property of arithmetic do you use? How does knowing this property of arithmetic lighten the load for students of learning the single-digit multiplication facts?

2. Multiplication facts involving the numbers 6, 7, and 8 are often hard to learn. For each fact in the lightly colored regions in the table, describe one or more ways to derive it from facts in the darkly colored region by applying properties of arithmetic. Use arrays and equations to show the reasoning, as in these examples for 3×7, which use the distributive property:

$3 \times 7 = 2 \times 7 + 1 \times 7$
$= 14 + 7 = 21$

$3 \times 7 = 3 \times 3 + 3 \times 4$
$= 9 + 12 = 21$

3. The *5× table* is easy to learn because it is "half of the *10× table*."

$$5 \times 1 = 5 \qquad 10 \times 1 = 10$$
$$5 \times 2 = 10 \qquad 10 \times 2 = 20$$
$$5 \times 3 = 15 \qquad 10 \times 3 = 30$$
$$5 \times 4 = 20 \qquad 10 \times 4 = 40$$
$$5 \times 5 = 25 \qquad 10 \times 5 = 50$$
$$5 \times 6 = 30 \qquad 10 \times 6 = 60$$
$$5 \times 7 = 35 \qquad 10 \times 7 = 70$$
$$5 \times 8 = 40 \qquad 10 \times 8 = 80$$
$$5 \times 9 = 45 \qquad 10 \times 9 = 90$$

Write an equation showing the relationship between 10×7 and 5×7 that fits with the statement about the *5× table* being half of the *10× table*. Which property of arithmetic do you use?

4. The *9× table* is easy to learn because you can just subtract a number that is to be multiplied by 9 from that number with a 0 placed behind it, as shown next. Explain why this way of multiplying by 9 is valid.

$$9 \times 1 = 10 - 1 = 9$$
$$9 \times 2 = 20 - 2 = 18$$
$$9 \times 3 = 30 - 3 = 27$$
$$9 \times 4 = 40 - 4 = 36$$
$$9 \times 5 = 50 - 5 = 45$$
$$9 \times 6 = 60 - 6 = 54$$
$$9 \times 7 = 70 - 7 = 63$$
$$9 \times 8 = 80 - 8 = 72$$
$$9 \times 9 = 90 - 9 = 81$$

5. What is another pattern (other than the one described in part 4) in the *9× table*?

Class Activity 4K Solving Arithmetic Problems Mentally

CCSS CCSS SMP7

For each of the following arithmetic problems, describe a way to make the problem easy to solve mentally:

1. 4×99

2. 16×25 (Try to find several ways to solve this problem mentally.)

3. $45\% \times 680$

4. 12×125 (Try to find several ways to solve this problem mentally.)

5. $125\% \times 120$

Class Activity 4L Writing Equations That Correspond to a Method of Calculation

CCSS CCSS SMP7

Each arithmetic problem in this activity has a description of the problem solution. In each case, write a sequence of equations that corresponds to the given description. Which properties of arithmetic were used and where? Write your equations in the following form:

$$\begin{aligned} \text{original} &= \text{some expression} \\ &= \vdots \\ &= \text{some expression} \end{aligned}$$

1. What is 55% of 120?

   Half of 120 is 60. 10% of 120 is 12, so 5% of 120 is half of that 10%, which is 6. So the answer is 60 plus 6, which is 66.

2. What is 35% of 80?

   25% is $\frac{1}{4}$, so 25% of 80 is $\frac{1}{4}$ of 80, which is 20. 10% of 80 is 8. So 35% of 80 is 20 plus 8, which is 28.

3. What is 90% of 350?

   10% of 350 is 35. Taking 35 away from 350 leaves 315. So the answer is 315.

4. What is 12.5% of 1800?

   Half is 900 and half of that is 450, so that's 25%. Then half of that is 225 which is 12.5%.

Class Activity 4M Showing the Algebra in Mental Math

CCSS CCSS SMP7

For each arithmetic problem in this activity, find ways to use properties of arithmetic to make the problem easy to do mentally. Describe your method in words, and write equations that correspond to your method. Write your equations in the following form:

$$\begin{aligned}\text{original} &= \text{some expression}\\ &= \vdots\\ &= \text{some expression}\end{aligned}$$

1. 6×12 (Try to find several different ways to solve this problem mentally.)

2. $24 \cdot 25$ (Try to find several different ways to solve this problem mentally.)

3. $5\% \cdot 48$

4. $15\% \cdot \$44$

5. $26\% \cdot 840$

6. $9 \cdot 99$ (Try to find several different ways to solve this problem mentally.)

4.6 Why the Standard Algorithm for Multiplying Whole Numbers Works

Class Activity 4N How Can We Develop and Understand the Standard Multiplication Algorithm?

CCSS CCSS SMP7, 4.NBT.5

1. Imagine you are a fourth-grade student who is ready to learn about multiplying multi-digit numbers. Let's say you are fluent with the one-digit multiplications 1×1 through 9×9 and you understand multiplication with one-digit multiples of 10 such as 7×90 and 30×80. How could you use what you know to figure out how many dots are in the array below, which consists of 6 rows with 38 dots in each row?

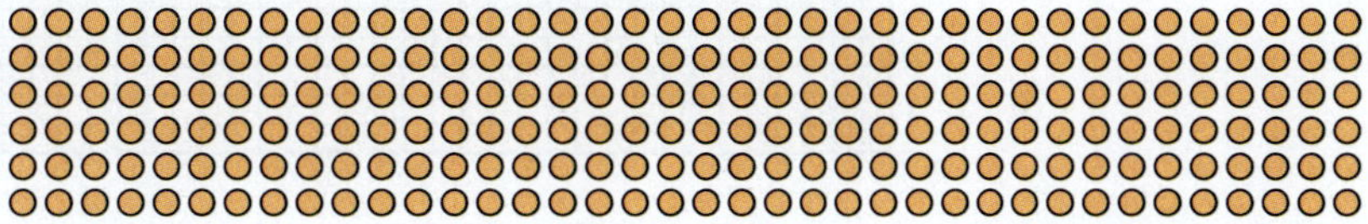

2. Use the distributive property to calculate 6×38 in one or more ways.

3. Use the partial-products method to calculate 6×38. Does this method correspond to any of your methods in parts 1 or 2? If not, go back to parts 1 and 2 and look for corresponding methods.

4. Imagine once again that you are that fourth-grade student from part 1. How could you use what you know to figure out how many small squares are in the array below, which consists of 23 rows with 45 small squares in each row?

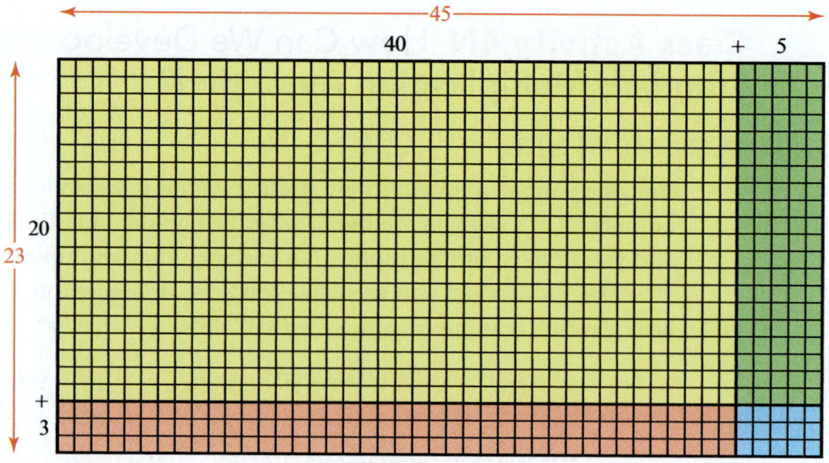

5. Use the distributive property to calculate 23 × 45 in one or more ways.

6. Use the partial-products method to calculate 23 × 45. Does this method correspond to any of your methods in parts 4 or 5? If not, go back to parts 4 and 5 and look for corresponding methods. In part 5, decompose 23 and 45 into their place value parts.

7. Calculate 23 × 45 using the common method for writing the standard algorithm. Relate the two lines in the calculation to the array in part 4 and to a way to apply the distributive property to 23 × 45.

8. Make a rough sketch to indicate an array of small squares for 46 × 53. Your sketch should not show all the small squares and it does not have to be to scale. Use your sketch to explain why the partial-products method calculates the correct answer to 46 × 53. Begin your explanation by using the definition of multiplication to relate the array to the multiplication problem.

CLASS ACTIVITIES

5

Multiplication of Fractions, Decimals, and Negative Numbers

5.1 Making Sense of Fraction Multiplication

Class Activity 5A Extending Multiplication to Fractions, Part I

CCSS CCSS SMP6, 4.NF.4

In this Class Activity, use our definitions of multiplication and of fractions to explain your answers.

M	•	N	=	P
Number of equal **groups**		Number of **units** in **1 group**		Number of **units** in **M groups**

1. If 1 serving of popcorn is 2 cups, then how many cups are in 3 servings of popcorn? Write and annotate an equation for this problem.

2. If 1 serving of cereal is $\frac{2}{3}$ cups, then how many cups are in 5 servings of cereal? Complete the math drawing and write and annotate an equation for this problem.

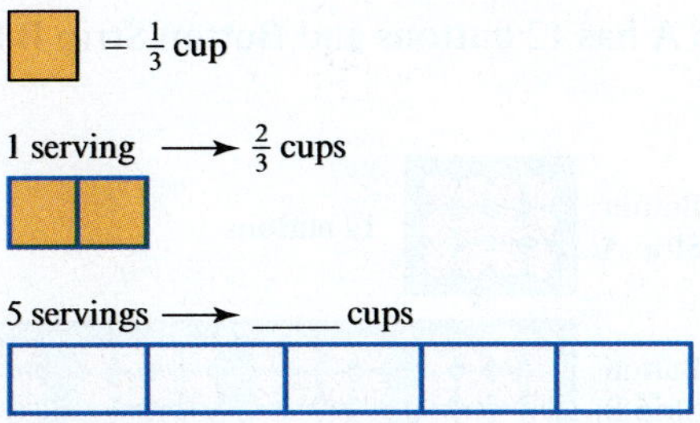

3. Write a word problem and make and explain a math drawing for

$$4 \cdot \frac{3}{5} = ?$$

4. If you had 3 servings of a sports drink and one serving is $\frac{4}{5}$ liters, then how many liters did you have? Make a math drawing and write and annotate an equation for this problem.

Class Activity 5B Making Multiplicative Comparisons

CCSS CCSS SMP2, 4.OA.2

Button Strip A has 12 buttons and Button Strip B has 72 buttons.

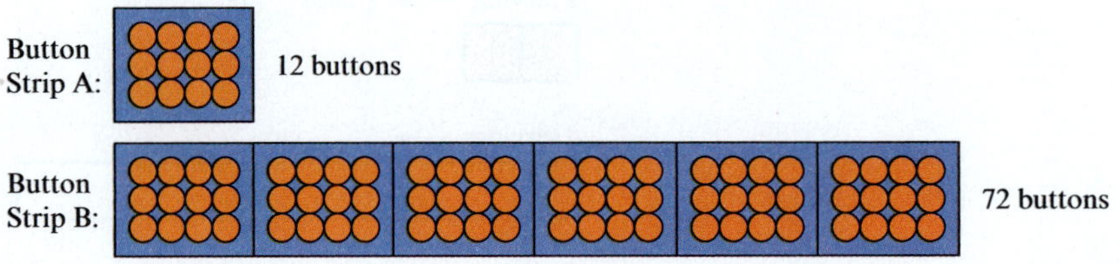

1. How many groups of Button Strip A does it take to make the exact same number of buttons as Button Strip B?

   How many groups of Button Strip B does it take to make the exact same number of buttons as Button Strip A?

2. Use your answers to part 1 to write multiplication equations that relate 12 and 72. Explain your equations with our definition of multiplication.

Button Strip C has 75 buttons and Button Strip D has 120 buttons.

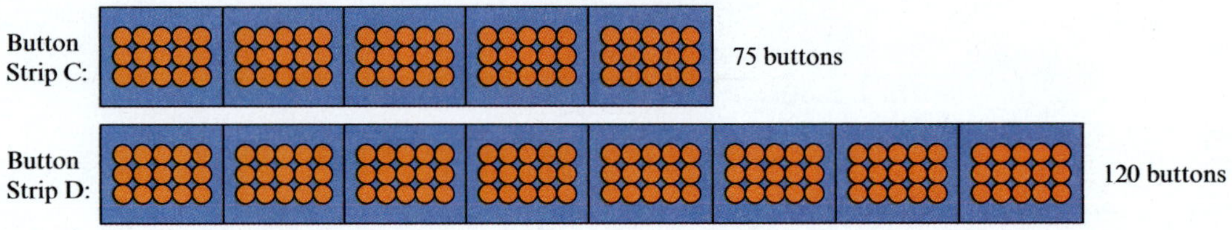

3. How many groups of Button Strip C does it take to make the exact same number of buttons as Button Strip D?

   How many groups of Button Strip D does it take to make the exact same number of buttons as Button Strip C?

4. Use your answers to part 3 to write multiplication equations that relate 75 and 120. Explain your equations with our definition of multiplication.

Class Activity 5C Extending Multiplication to Fractions, Part II

CCSS CCSS SMP6, 5.NF.4a, 5.NF.6

In this Class Activity, *use equivalent fractions wherever they help you.* Use our definitions of multiplication and of fractions to explain your answers. Pretend you don't yet know how to compute products with fractions.

1. If 1 serving of juice has 12 grams of sugar, then how many grams of sugar are in $\frac{1}{4}$ of a serving? Complete the math drawing and write and annotate an equation for this situation.

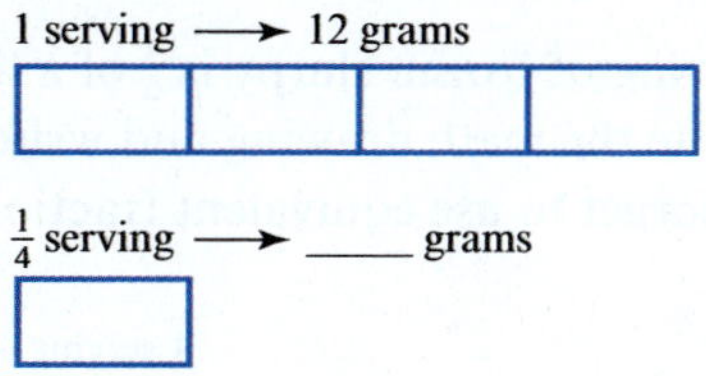

2. Write a word problem and make and explain a math drawing for

$$\frac{1}{3} \cdot 15 = ?$$

3. If 1 serving of cheese is $\frac{8}{3}$ ounces, then how many ounces are in $\frac{1}{4}$ of a serving? Complete the math drawing and write and annotate an equation for this situation.

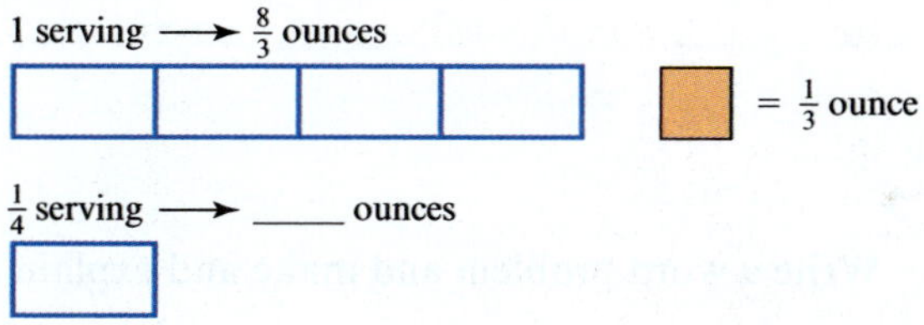

4. If 1 serving of juice is $\frac{1}{5}$ liter, then how many liters are in $\frac{1}{3}$ of a serving? *Turn* $\frac{1}{5}$ *into an equivalent fraction* to help you complete the math drawing. Write and annotate an equation for this situation.

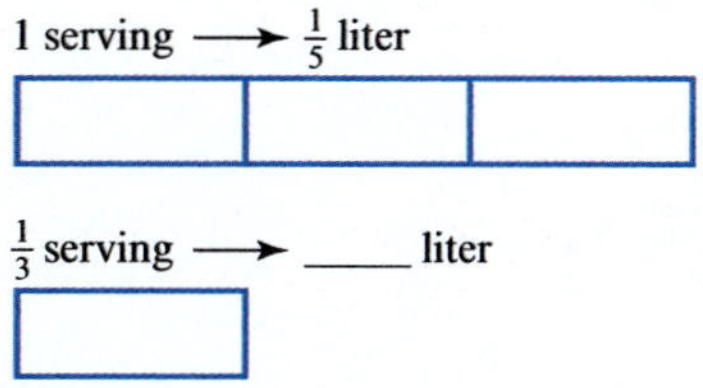

5. Write a word problem and make and explain a math drawing for

$$\frac{1}{5} \cdot \frac{1}{3} = ?$$

Don't forget that you can use equivalent fractions!

6. Compare and contrast the previous two parts. How can you tell the difference between $\frac{1}{3} \cdot \frac{1}{5}$ and $\frac{1}{5} \cdot \frac{1}{3}$?

7. If 1 serving of frozen slurpy is $\frac{4}{5}$ of a liter, then how many liters are in $\frac{2}{3}$ of a serving? Complete the math drawing and write and annotate an equation for this situation. Don't forget to use equivalent fractions.

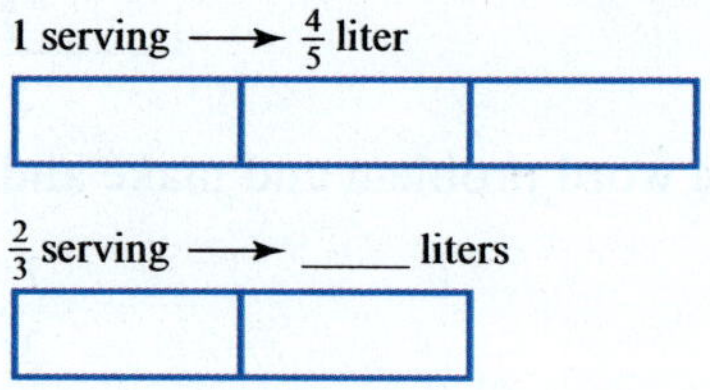

8. Write a word problem and make and explain a math drawing for

$$\frac{4}{5} \cdot \frac{2}{3} = ?$$

9. Compare and contrast the previous two parts.

Class Activity 5D Explaining Why the Procedure for Multiplying Fractions Is Valid

CCSS CCSS SMP3, 5.NF.4a

1. For each of the rectangles below, fill in the blanks so you can use multiplication to describe the fraction of the rectangle that is shaded. You may assume that all lengths that appear to be the same really are the same.

(a)

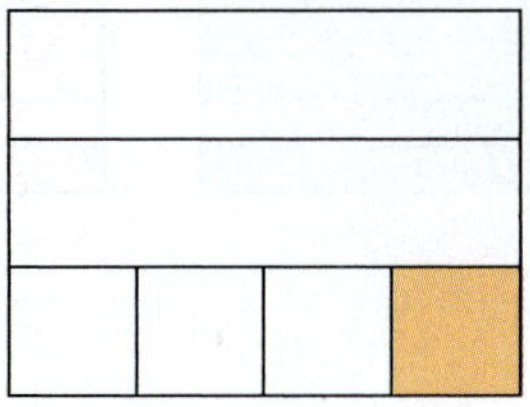

_____ of _____ is shaded.

_____ × _____ is shaded.

(b)

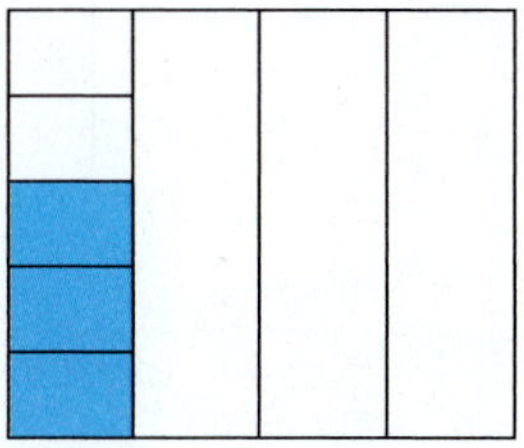

_____ of _____ is shaded.

_____ × _____ is shaded.

(c)

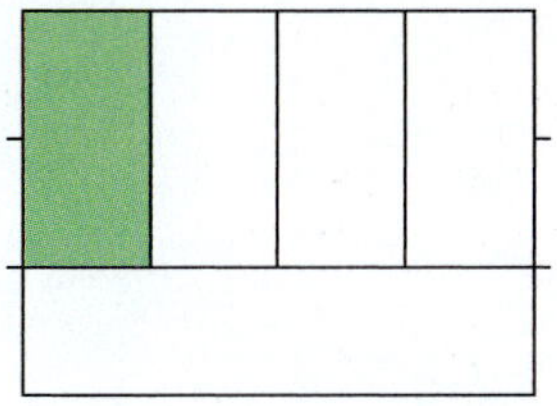

_____ of _____ is shaded.

_____ × _____ is shaded.

(d)

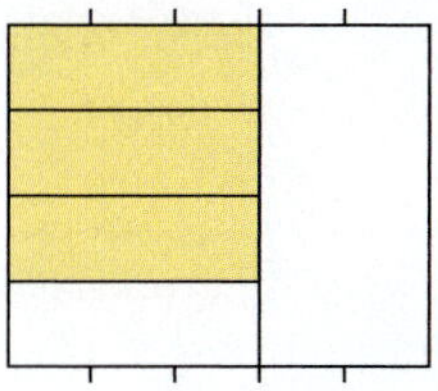

_____ of _____ is shaded.

_____ × _____ is shaded.

2. Use the math drawings to explain why

$$\frac{1}{4} \cdot \frac{1}{3} = \frac{1}{4 \cdot 3}$$

Discuss how you are interpreting $\frac{1}{4} \cdot \frac{1}{3}$ and how that fits with the definition of multiplication.

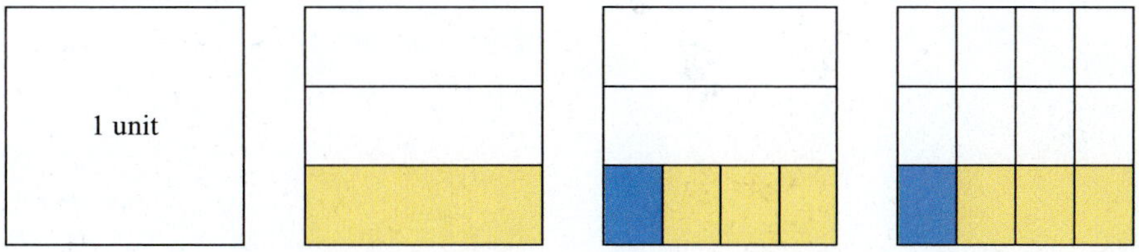

3. Use the math drawings to explain why

$$\frac{2}{3}\cdot\frac{5}{8} = \frac{2\cdot 5}{3\cdot 8}$$

In particular, explain why we multiply the denominators and why we multiply the numerators. Discuss how you are interpreting $\frac{2}{3}\cdot\frac{5}{8}$ and how that fits with the definition of multiplication.

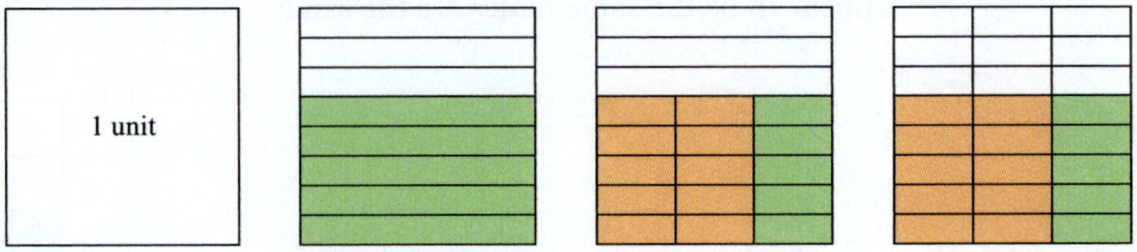

4. Use the definition of multiplication and math drawings to explain why

$$\frac{2}{5}\cdot\frac{3}{7} = \frac{2\cdot 3}{5\cdot 7}$$

In particular, explain why we multiply the denominators and why we multiply the numerators. Discuss how you are interpreting $\frac{2}{5}\cdot\frac{3}{7}$ and how that fits with the definition of multiplication.

Class Activity 5E When Do We Multiply Fractions?

CCSS CCSS SMP2, SMP4

As a teacher, you will probably write word problems for your students. This activity will help you see how slight changes in the wording of a problem can produce big changes in meaning and in the operation that is used to solve the problem.

1. *A mulch pile problem:* Originally, there was $\frac{3}{4}$ of a cubic yard of mulch in a mulch pile. Then $\frac{1}{3}$ of the mulch in the mulch pile was removed. Now how much mulch is left in the mulch pile?

   a. Is the mulch pile problem a problem for $\frac{1}{3} \cdot \frac{3}{4}$, is it a problem for $\frac{3}{4} - \frac{1}{3}$, or is it not a problem for either of these? Explain.

   b. Write a new mulch pile problem for $\frac{1}{3} \cdot \frac{3}{4}$ and write a new mulch pile problem for $\frac{3}{4} - \frac{1}{3}$. Make clear which is which.

2. Which of the following problems are word problems for $\frac{2}{3} \cdot \frac{1}{4}$, and which are not? Why?

   a. Joe is making $\frac{2}{3}$ of a recipe. The full recipe calls for $\frac{1}{4}$ cup of water. How much water should Joe use?

   b. $\frac{1}{4}$ of the students in Mrs. Watson's class are doing a dinosaur project. $\frac{2}{3}$ of the children doing the dinosaur project have completed it. How many children have completed a dinosaur project?

   c. $\frac{1}{4}$ of the students in Mrs. Watson's class are doing a dinosaur project. $\frac{2}{3}$ of the children doing the dinosaur project have completed it. What fraction of the students in Mrs. Watson's class have completed a dinosaur project?

   d. There is $\frac{1}{4}$ of a cake left in Mrs. Watson's class. $\frac{2}{3}$ of the class would like to have some cake. What fraction of the cake does each student who wants cake get?

   e. Carla is making snack bags that each contain $\frac{1}{4}$ package of jelly worms. $\frac{2}{3}$ of her snack bags have been bought. What fraction of her jelly worms have been bought?

Class Activity 5F What Fraction Is Shaded?

CCSS CCSS SMP1, SMP7

1. For each of the figures below, write an expression that uses both multiplication and addition (or subtraction) to describe the total fraction of the figure that is shaded. (For example, $\frac{5}{7} \cdot \frac{2}{9} + \frac{1}{3}$ is an expression that uses both multiplication and addition). Explain your reasoning. Then compute what fraction of the figure is shaded (in simplest form). In each figure, you may assume that lengths appearing to be equal really are equal.

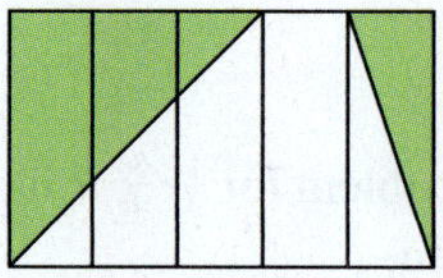
(a)

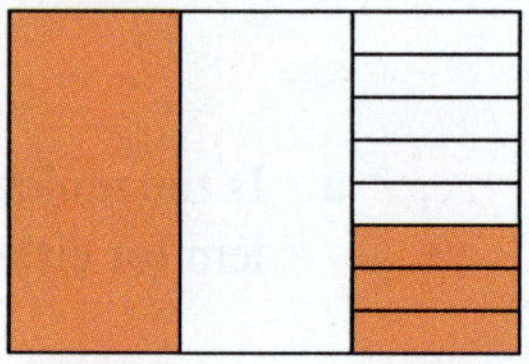
(b)

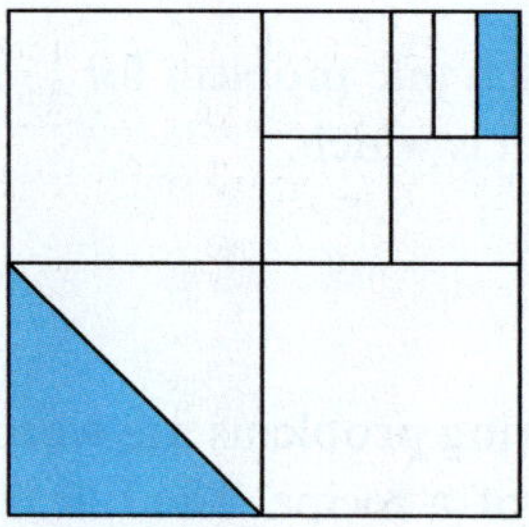
(c)

2. Draw a figure in which you shade $\frac{1}{3} \cdot \frac{2}{7} + \frac{1}{2} \cdot \frac{3}{7}$ of the figure.

5.2 Making Sense of Decimal Multiplication

Class Activity 5G Decimal Multiplication Word Problems and Estimation

CCSS CCSS SMP2

1. Write a word problem for 2.7×1.35.

2. Ben wants to multiply 3.46×1.8. He first multiplies the numbers by ignoring the decimal points:

$$\begin{array}{r} 3.46 \\ \times\ 1.8 \\ \hline 6228 \end{array}$$

   Ben knows that he only needs to figure out where to put the decimal point in his answer, but he can't remember the rule about where to put the decimal point. Explain how Ben can reason about the sizes of the numbers to determine where to put the decimal point in his answer.

3. What if Ben's original problem in part 2 was 3.46×0.18? How can he then reason about the sizes of the numbers to determine where to put the decimal point?

4. Using a calculator, Lameisha finds that

$$1.5 \times 1.2 = 1.8$$

   She wants to know why the rule about adding the number of places behind the decimal point doesn't work in this case. Why aren't there 2 digits to the right of the decimal point in the answer? Is Lameisha right that the rule about adding the number of places behind the decimal points doesn't work in this case? Explain.

Class Activity 5H Explaining Why We Place the Decimal Point Where We Do When We Multiply Decimals

CCSS CCSS SMP7, 5.NBT.7

1. As indicated in the diagram below, to get from 1.36 to 136, we multiply by 10×10. To get from 2.7 to 27, we multiply by 10. In other words,

$$136 = 10 \times 10 \times 1.36 \quad \text{and} \quad 27 = 10 \times 2.7$$

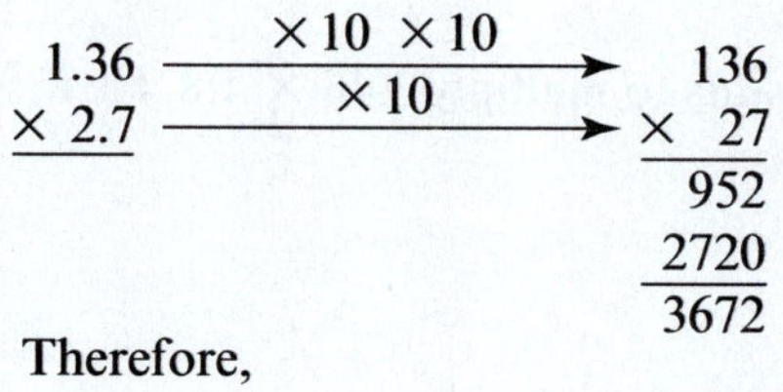

Therefore,

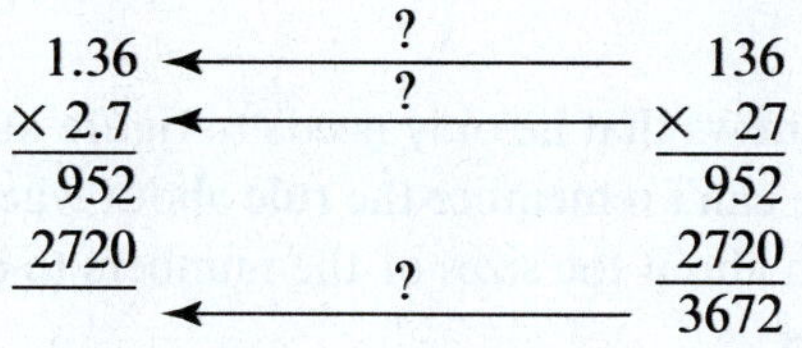

Explain what we should do now to

$$136 \times 27 = 3672$$

to get back to

$$1.36 \times 2.7$$

Use your answer to explain the placement of the decimal point in 1.36×2.7.

2. More generally, explain why the following is valid: If you multiply a number that has 3 digits to the right of its decimal point by a number that has 4 digits to the right of its decimal point, you should place the decimal point $3 + 4 = 7$ places from the end of the product calculated without the decimal points.

Class Activity 5I Decimal Multiplication and Areas of Rectangles

CCSS CCSS SMP2

1. Find the area of the 2.3-unit-by-1.8-unit rectangle *without* multiplying. Explain.

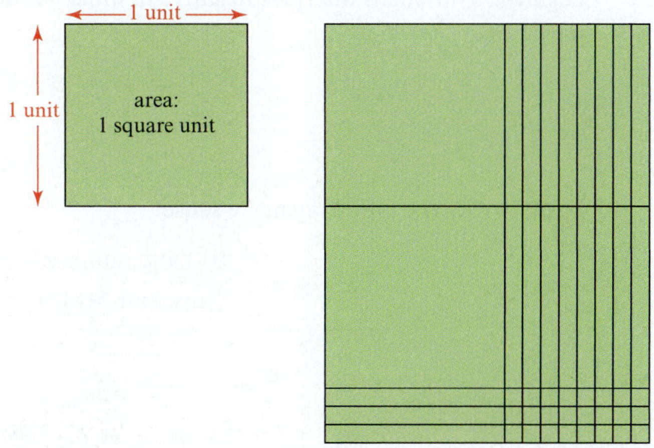

2. Discuss the following questions: How is the 2.3-unit-by-1.8-unit rectangle related to an array or rectangle for 23×18? How is 2.3×1.8 related to 23×18?

   How are decimal, whole number, and mixed number multiplication alike?

Class Activity 5J Using the Distributive Property to Explain Multiplication with Negative Numbers (and 0)

CCSS CCSS SMP3, 7.NS.2a,c

1. Write a word problem for $3 \cdot -5$. Solve the word problem, thereby explaining why $3 \cdot -5$ is negative. You might interpret negative numbers as amounts owed or as negatively charged particles.

2. Explain why the following make sense:

$$0 \cdot (\text{any number}) = 0$$
$$(\text{any number}) \cdot 0 = 0$$

3. Assume that you don't yet know what $(-3) \cdot 5$ is, but you do know that $3 \cdot 5 = 15$. Use the distributive property to show that the expression

$$(-3) \cdot 5 + 3 \cdot 5$$

is equal to 0. Then use that result to determine what $(-3) \cdot 5$ must be equal to.

4. Assume that you don't yet know what $(-3) \cdot (-5)$ is, but you do know that $(-3) \cdot 5 = -15$ from part 3. Use the distributive property to show that the expression

$$(-3) \cdot (-5) + (-3) \cdot 5$$

is equal to 0. Then use that result to determine what $(-3) \cdot (-5)$ must be equal to.

5.4 Powers and Scientific Notation

Class Activity 5K Multiplying Powers of 10

CCSS CCSS SMP3, SMP8, 8.EE.1

1. Use the meaning of powers of 10 to show how to write each of the expressions in (a), (b), and (c) as a single power of 10 (i.e., in the form 10^A for some exponent A). For example, 10^2 means 10×10, and 10^3 means $10 \times 10 \times 10$; therefore,

$$10^2 \times 10^3 = (10 \times 10) \times (10 \times 10 \times 10) = 10^5$$

a. $10^3 \times 10^4$ **b.** $10^2 \times 10^5$ **c.** $10^3 \times 10^3$

2. In (a), (b), and (c) in part 1, relate the exponents in the product with the exponent in the answer. (For the example given at the beginning of part 1, relate 2 and 3 to 5.) In each case, how are the three exponents related?

3. Explain why it is always true that $10^A \times 10^B = 10^{A+B}$ when A and B are counting numbers.

4. Assume now that we want the equation $10^A \times 10^B = 10^{A+B}$ to be true not just when A and B are counting numbers, but even when A or B is 0. With this assumption, explain why it makes sense that 10^0 should be equal to 1.

5. Assume now that we want the equation $10^A \times 10^B = 10^{A+B}$ to be true not just when A and B are whole numbers, but even when A or B is a negative integer. Also assume that $10^0 = 1$.

   a. With these assumptions, explain why the following make sense: 10^{-1} should be equal to $\frac{1}{10}$.

   b. 10^{-2} should be equal to $\frac{1}{100}$.

   c. 10^{-N} should be equal to $\frac{1}{10^N}$.

Division

6.1 Interpretations of Division

Class Activity 6A What Does Division Mean?

CCSS CCSS 3.OA.2

1. Write a simple word problem and make a math drawing that you could use to help children understand what $10 \div 2$ means.

2. Reformulate the division problem $10 \div 2 = ?$ as

$$2 \times ? = 10 \qquad \text{or as} \qquad ? \times 2 = 10$$

whichever fits with your word problem in part 1 and with the way we have described multiplication.

3. Now write another simple division word problem for $10 \div 2$, one that fits with the *other* multiplication equation identified in part 2. Make a math drawing that fits with your new problem.

Class Activity 6B Division Word Problems

CCSS CCSS SMP4, 3.OA.2

For each problem, determine if it is a how-many-groups or a how-many-units-in-1-group division problem and write annotated division and multiplication equations. Discuss how the annotated division equations are similar for the how-many-groups problems and how they are similar for the how-many-units-in-1-group problems.

$15 \div 5 = ?$

How-many-groups division

There are 15 stickers to put in bags. 5 stickers go in each bag. How many bags do we need?

?	•	5	=	15
How many **bags**?		Number of **stickers** in **1 bag**		Number of **stickers** to be put **equally in bags**

How-many-units-in-one-group division

There are 15 stickers to distribute equally among 5 bags. How many stickers go in each bag?

5	•	?	=	15
Number of equal **bags**		How many **stickers** in **1 bag**?		Number of **stickers** in **5 bags**

1. Bill has a muffin recipe that calls for 2 cups of flour. How many batches of muffins can Bill make if he has 8 cups of flour available? (Assume that Bill also has all the other ingredients.)

2. One foot is 12 inches. If a piece of rope is 96 inches long, then how long is it in feet?

3. Francine has 32 yards of rope that she wants to cut into 8 equal pieces. How long will each piece be?

4. If 1 gallon of water weighs 8 pounds, how many gallons will there be in 400 pounds of water?

5. If you drive 220 miles at a constant speed and it takes you 4 hours, then how fast did you go?

6. If 6 limes cost $3, then how much should 1 lime cost (assuming that all limes are priced equally)?

7. If 6 limes cost $3, then how many limes can you buy for $1?

8. Write a situation for $4 \cdot 8 = 32$. Then write two related division word problems, one for each of the two types of division.

Class Activity 6C Why Can't We Divide by Zero?

CCSS CCSS SMP2, 7.NS.2b

1. Use the fact that every division problem can be rewritten as a multiplication problem with a factor unknown to explain why $2 \div 0$ is not defined.

2. Write word problems for the two interpretations of $2 \div 0 = ?$. Use your problems to explain why $2 \div 0$ is not defined. Link your word problems to your multiplication equations from part 1.

3. Write word problems for the two interpretations of $0 \div 2 = ?$. Use your problems to explain why $0 \div 2$ *is* defined. Explain the difference between $2 \div 0$ and $0 \div 2$.

4. Explain why $0 \div 0$ is undefined by viewing division in terms of multiplication. (We can also say that $0 \div 0$ is "indeterminate.") Can you give the same explanation as for why $2 \div 0$ is not defined? If not, how are the explanations different?

6.2 Division and Fractions and Division with Remainder

Class Activity 6D Relating Whole Number Division and Fractions

CCSS CCSS SMP6, SMP8, 5.NF.3

There are 3 pizzas that will be divided equally among 4 people. How much pizza will each person get? Explain. What does your answer tell you about $3 \div 4$?

Class Activity 6E Using Measurement Ideas to Relate Whole Number Division and Fractions

CCSS CCSS SMP8, 5.NF.3

1. Measurement problem: How many of the first strip does it take to make the second strip?

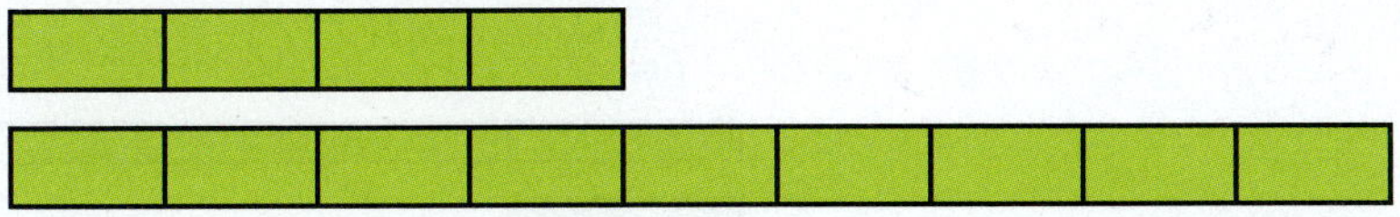

a. Explain how to interpret the measurement problem as asking $? \cdot 4 = 9$ and $9 \div 4 = ?$.

b. Solve the measurement problem by reasoning about the strips with our definition of fraction.

c. Repeat parts (a) and (b) but now with this measurement problem: How many of the second strip does it take to make the first strip? (You will need to modify the equations in part (a).)

2. Use how-many-groups division and the idea of measuring one strip by another strip to explain why $3 \div 7 = \frac{3}{7}$ and why $7 \div 3 = \frac{7}{3}$.

Class Activity 6F Relating Whole Number Division and Fraction Multiplication

CCSS CCSS SMP2, SMP8, 5.NF.4a

1. A company puts beads on strips of different lengths. Each section has the same number of beads. Write and explain several multiplication and division equations to show how the numbers of beads on the two strips are related. Include an equation that involves a fraction.

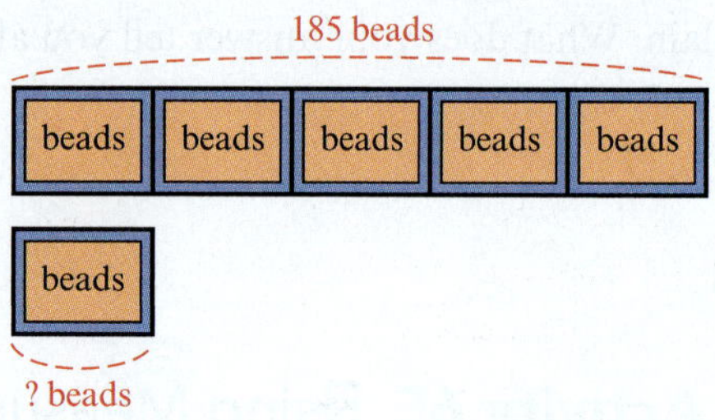

2. A company sells ropes in different lengths. All the sections of rope weigh the same. Write and explain several multiplication and division equations to show how the weights of the two ropes are related. Include an equation that involves a fraction.

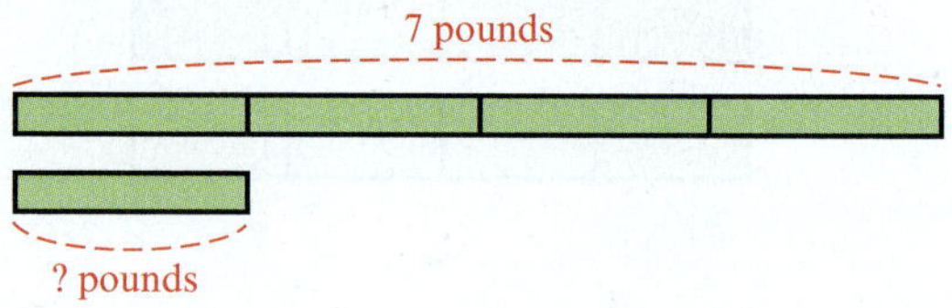

3. How are $185 \div 5$ and $\frac{1}{5} \cdot 185$ related? How are $7 \div 4$ and $\frac{1}{4} \cdot 7$ related? Explain!

4. Explain why $17 \div 3 = \frac{1}{3} \cdot 17$ with the aid of a situation and a math drawing.

5. By reflecting on the previous parts, explain more generally why dividing a number by a natural number N is equivalent to multiplying the number by $\frac{1}{N}$.

Class Activity 6G What to Do with the Remainder?

CCSS CCSS SMP4, SMP6, 4.OA.3

1. Consider these two word problems for 14 ÷ 3:

   A baking problem: A batch of cookies requires 3 cups of flour. How many batches of cookies can you make if you have 14 cups of flour (and all the other ingredients you need)?

   A brownie problem: You have 14 brownies which you will divide equally among 3 bags. How many brownies should you put in each bag?

   a. In the table below, write your interpretation of the whole number quotient 4, remainder 2, and the mixed number quotient $4\frac{2}{3}$ for both problems. In each case, what does the 4 stand for? What does the 2 stand for? What does the $\frac{2}{3}$ stand for? Could the $\frac{2}{3}$ stand for something else (not connected to 4)?

	Baking problem: 1 batch ⟶ 3 cups ? batches ⟶ 14 cups	**Brownie problem:** 3 bags ⟶ 14 brownies 1 bag ⟶ ? brownies
4, R 2		
$4\frac{2}{3}$		

   b. What is different about how the whole number quotient 4 and remainder 2 are interpreted in the two problems?

   c. Discuss how the remainder 2 and the $\frac{2}{3}$ are related for each problem.

2. Write a word problem for which you would calculate $14 \div 3$ in order to solve the problem, but which has the answer 5.

3. *A calendar problem:* What day of the week will it be 31 days from today? Explain how $31 \div 7$ is relevant to solving the calendar problem.

4. Consider these three problems about distance, speed, and time:

i. How long will it take you to drive 180 miles if you drive at the constant speed of 55 mph?

ii. How long will it take you to drive 195 miles if you drive at the constant speed of 60 mph?

iii. How long will it take you to drive 105 miles if you drive at the constant speed of 30 mph?

a. Write numerical division problems to solve problems (i), (ii), and (iii) and give the mixed number answers and whole-number-with-remainder answers to these numerical problems.

b. Interpret the meaning of the mixed number answers and whole-number-with-remainder answers you gave in part (a) in terms of the original word problems.

c. For problem (i) Josh says: "The 3, remainder 15, answer tells you that it will take 3 full hours, and the remainder 15 tells you it will take another 15 minutes." Explain why Josh's comment is approximately correct, but not completely correct.

6.3 Why Division Algorithms Work

Class Activity 6H Discuss Division Reasoning

CCSS CCSS SMP3, 4.NBT.6, 5.NBT.6

1. There are 260 pencils to be put in packages of 12. How many packages of pencils can be made, and how many pencils will be left over? Antrice's solution:

> 10 packages will use up 120 pencils. After another 10 packages, 240 pencils will be used up. After 1 more package, 252 pencils are used. Then there are only 8 pencils left, and that's not enough for another package. So the answer is 21 packages of pencils with 8 pencils left over.

Explain why the equations below correspond to Antrice's work, and explain why the last equation shows that $260 \div 12$ has whole number quotient 21, remainder 8.

$$\begin{aligned}(10 \cdot 12) + (10 \cdot 12) + (1 \cdot 12) + 8 &= 260\\ (10 + 10 + 1) \cdot 12 + 8 &= 260\\ 21 \cdot 12 + 8 &= 260\end{aligned}$$

2. Ashley's work on the division problem $258 \div 6$ is shown below. Explain what Ashley did and why her strategy makes sense. Then write equations that correspond to Ashley's work and demonstrate that $258 \div 6 = 43$.

$$258 \div 6 = ?$$

$$\begin{array}{rcl} 10 & \longrightarrow & 60\\ 20 & \longrightarrow & 120\\ 40 & \longrightarrow & 240 \end{array} \qquad \begin{array}{rcr} 240 & \longleftarrow & 40\\ \underline{+12} & \longleftarrow & 2\\ 252 & \longleftarrow & 42\\ \underline{+6} & & \\ 258 & \longleftarrow & 43 \end{array}$$

3. Zane's work on the division problem $245 \div 15$ is shown below. Explain why Zane's strategy makes sense. Then write equations that correspond to Zane's work and demonstrate that $245 \div 15$ has whole number quotient 16, remainder 5.

$$\begin{array}{r} 15\\ \underline{\times\ 2}\\ 30\\ \underline{\times\ 2}\\ 60\\ \underline{\times\ 4}\\ 240\\ \text{5 left} \end{array} \qquad 2 \times 2 \times 4 = 16\ \text{R}5$$

4. Pretend that you don't have a calculator and have forgotten how to do longhand division. Explain how you can calculate $5170 \div 6$.

Class Activity 6I Why the Scaffold Method of Division Works

CCSS CCSS SMP3, 4.NBT.6

1. Interpret each of the steps in the next scaffold in terms of the following word problem:

   You have 3475 marbles, and you want to put these marbles into bags with 8 marbles in each bag. How many bags of marbles can you make, and how many marbles will be left over?

$$
\begin{array}{r}
4 \\
30 \\
400 \\
8\overline{)3475} \\
-\ 3200 \\
\hline
275 \\
-\ 240 \\
\hline
35 \\
-\ 32 \\
\hline
3
\end{array}
$$

   Then relate these equations to the scaffold and the division problem:

$$
\begin{aligned}
3475 - 400\cdot 8 - 30\cdot 8 - 4\cdot 8 &= 3 \\
3475 - (400 + 30 + 4)\cdot 8 &= 3 \\
3475 - 434\cdot 8 &= 3
\end{aligned}
$$

2. Use the scaffold method to calculate $8321 \div 6$. (You may use the method flexibly or in standard algorithm form.) Interpret each step in your scaffold in terms of the following word problem:

   You have 8321 pickles, and you want to put these pickles in packages with 6 pickles in each package. How many packages can you make, and how many pickles will be left over?

   Then write equations like those in part 1 and relate them to the scaffold and the division problem.

Class Activity 6J Interpreting the Standard Division Algorithm as Dividing Bundled Toothpicks

CCSS CCSS SMP3, SMP5, 4.NBT.6

For each part in this activity, use the *common method* for implementing the standard algorithm. Explain how to interpret each step in terms of dividing bundled toothpicks equally among a number of groups. View the toothpicks as ones, bundles of tens, bundles of hundreds (which are 10 bundles of ten), and bundles of 1000s (which are 10 bundles of one hundred).

Pay special attention to these points:

- Do not say "goes into." Instead of "4 goes into 9," you can say "there are 4 groups and we have 9 bundles of ten to put into the groups."
- Interpret the "bringing down" steps as unbundling and combining with the bundles at the next lower place.

1. $4\overline{)93}$

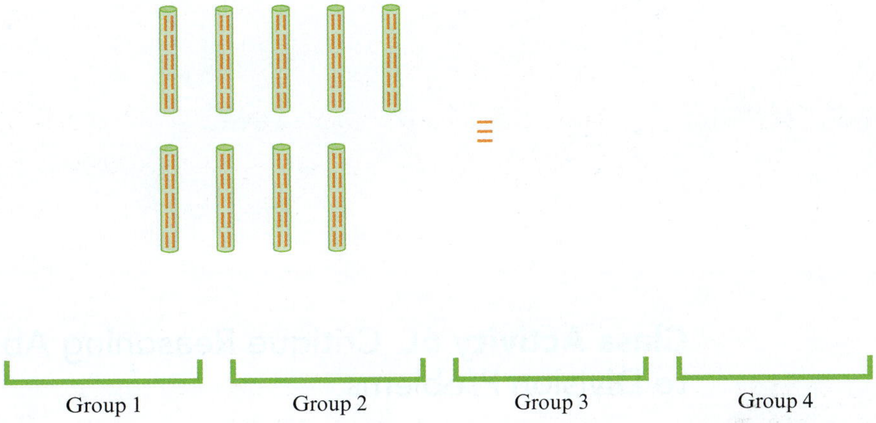

2. $3\overline{)143}$

3. $3\overline{)1372}$

4. $6\overline{)1230}$

Class Activity 6K Interpreting the Calculation of Decimal Answers to Whole Number Division Problems in Terms of Money

CCSS CCSS SMP3, 5.NBT.7

Use the standard division algorithm to determine the decimal answer to 2674 ÷ 3 to the hundredths place. Interpret each step in your calculation in terms of dividing \$2674 equally among 3 people by imagining that you distribute the money in stages: First distribute hundreds, then tens, then ones, then dimes (tenths), then pennies (hundredths).

Class Activity 6L Critique Reasoning About Decimal Answers to Division Problems

CCSS CCSS SMP3

1. Here is how Ben answered some division problems:

$$251 \div 6 = 41.5$$
$$269 \div 7 = 38.3$$
$$951 \div 21 = 45.6$$

   Is Ben right? How might Ben be thinking?

2. If you know that the answer to a whole number division problem is 4, remainder 1, can you tell what the decimal answer to the division problem is without any additional information? If not, what other information would you need to determine the decimal answer?

6.4 Fraction Division from the How-Many-Groups Perspective

Class Activity 6M How-Many-Groups Fraction Division Problems

CCSS CCSS SMP3, SMP6, 5.NF.7, 6.NS.1

Example for $\frac{6}{4} \div \frac{3}{4} = ?$

$$? \cdot \frac{3}{4} = \frac{6}{4}$$

? servings

How many servings of juice do you have if 1 serving is $\frac{3}{4}$ cups and you have $\frac{6}{4}$ cups?

$\frac{3}{4}$ cups is 1 serving of juice

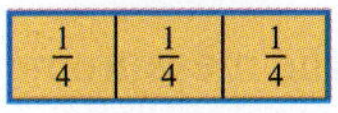

3 parts, each $\frac{1}{4}$ of a cup

$\frac{6}{4}$ cups is ? servings

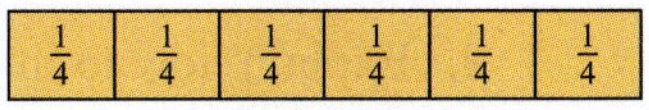

6 parts, each $\frac{1}{4}$ of a cup

Use equivalent fractions wherever they help you.

1. Explain how to use the math drawing to solve the problem in the example.
2. Write a simple how-many-groups word problem for $3 \div \frac{3}{4} = ?$ and solve the problem with the aid of a math drawing, a table, or a double number line.
3. Tonya and Chrissy are trying to understand $1 \div \frac{2}{3} = ?$ by using the following problem:

   One serving of rice is $\frac{2}{3}$ of a cup. I ate 1 cup of rice. How many servings of rice did I eat?

   To solve the problem, Tonya and Chrissy draw a square divided into three equal pieces, and they shade two of those pieces.

   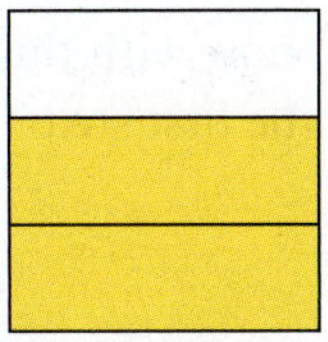

   Tonya says, "There is one $\frac{2}{3}$-cup serving of rice in 1 cup, and there is $\frac{1}{3}$ cup of rice left over, so the answer should be $1\frac{1}{3}$."

   Chrissy says, "The part left over is $\frac{1}{3}$ cup of rice, but the answer is supposed to be $\frac{3}{2} = 1\frac{1}{2}$. Did we do something wrong?"

   Help Tonya and Chrissy.
4. Write a how-many-groups word problem for $1\frac{1}{2} \div \frac{1}{3} = ?$ and solve your problem with the aid of a math drawing, a table, or a double number line. Explain your reasoning.
5. Write a how-many-groups word problem for $\frac{1}{3} \div \frac{3}{4} = ?$ and solve your problem with the aid of a math drawing, a table, or a double number line. Explain your reasoning.

Class Activity 6N Equivalent Division Problems

CCSS CCSS SMP8, 6.NS.1

Measurement problem: How many of the first strip does it take to make the second strip?

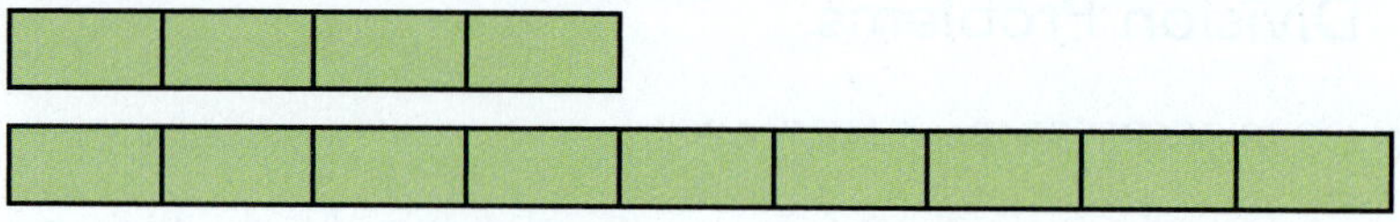

1. Explain how to interpret the measurement problem as asking $9 \div 4 = ?$.

2. Explain how to interpret the measurement problem as asking $\frac{9}{5} \div \frac{4}{5} = ?$ by viewing each part as having size $\frac{1}{5}$.

3. Give at least 3 other examples of fraction division equations for the measurement problem. View each part as having the same size—some size other than 1.

4. Solve the measurement problem by reasoning about the strips. In solving the problem, does it matter what we say the size of each part is (as long as all the parts are the same size)? What does that tell you about the solutions to the division problems in parts 1–3?

5. Repeat parts 1–4 but now with this measurement problem: how many of the second strip does it take to make the first strip? (You will need to adjust the division equations.)

6. Reflecting on parts 1–5, in general, how are the division problems

$$\frac{A}{C} \div \frac{B}{C} = ? \quad \text{and} \quad A \div B = ?$$

related?

Class Activity 6O Dividing Fractions by Dividing the Numerators and Dividing the Denominators

CCSS CCSS SMP3, 6.NS.1

1. Consider the two division problems

$$\frac{6}{5} \div \frac{2}{5} = ? \qquad \text{and} \qquad 6 \div 2 = ?$$

Explain in two ways why these division problems must have the same solution:

- By interpreting the small rectangles below in two ways.

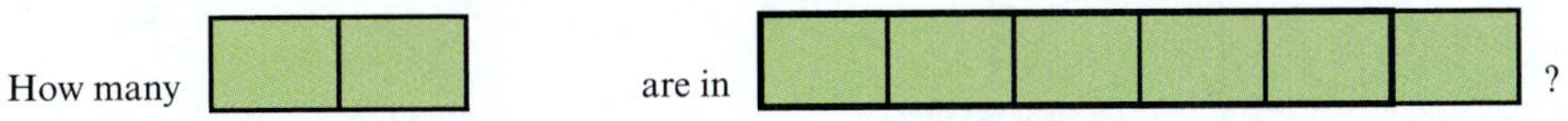

- By rewriting the division problems as multiplication problems with unknown factors.

2. View

$$\frac{6}{20} \div \frac{3}{4} = \frac{?}{?} \qquad \text{as} \qquad \frac{?}{?} \cdot \frac{3}{4} = \frac{6}{20}$$

and explain how to deduce that

$$\frac{6}{20} \div \frac{3}{4} = \frac{6 \div 3}{20 \div 4} = \frac{2}{5}$$

3. Give another example where you can divide fractions by dividing the numerators and dividing the denominators. Use the reasoning of part 2 to explain why this method works.

4. Explain how to use equivalent fractions so that you can apply the method of parts 2 and 3 to other cases, such as

$$\frac{5}{7} \div \frac{3}{4}$$

6.5 Fraction Division from the How-Many-Units-in-1-Group Perspective

Class Activity 6P How-Many-Units-in-1-Group Fraction Division Problems

CCSS CCSS SMP3, SMP6, 5.NF.7, 6.NS.1

Example for $\frac{6}{5} \div \frac{3}{4} = ?$

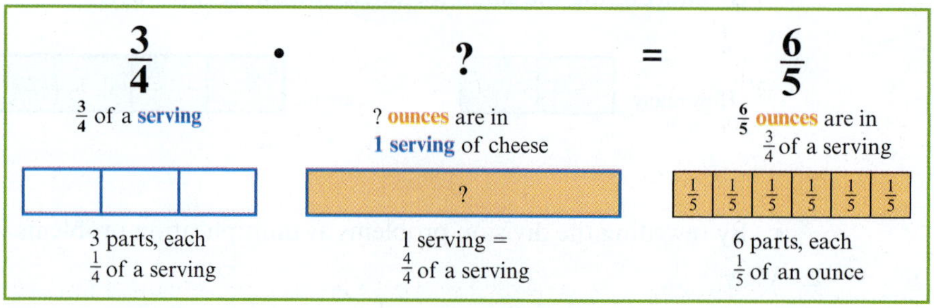

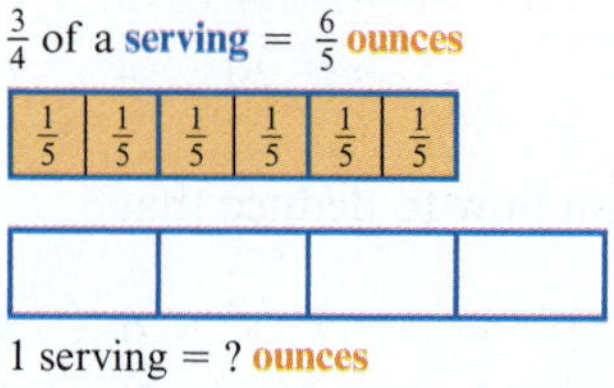

1. Finish solving the problem in the example. Then explain how to see the solution as multiplying $\frac{6}{5}$ by the reciprocal of $\frac{3}{4}$, or in other words as $\frac{4}{3} \cdot \frac{6}{5}$.

2. Write a how-many-units-in-1-group problem for $3 \div \frac{1}{4} = ?$ and explain how to solve the problem with the aid of a math drawing.

   Use your math drawing to explain why $3 \div \frac{1}{4}$ is equivalent to multiplying 3 by the reciprocal of $\frac{1}{4}$, in other words to $\frac{4}{1} \cdot 3$.

3. Write a how-many-units-in-1-group problem for $6 \div \frac{3}{4} = ?$ and explain how to solve the problem with the aid of a math drawing.

   Use your math drawing to explain why $6 \div \frac{3}{4}$ is equivalent to $\frac{4}{3} \cdot 6$.

4. Write a how-many-units-in-1-group problem for $\frac{1}{4} \div \frac{2}{3} = ?$ and explain how to solve the problem with the aid of a math drawing. You may use equivalent fractions.

   Use your math drawing to explain why $\frac{1}{4} \div \frac{2}{3}$ is equivalent to $\frac{3}{2} \cdot \frac{1}{4}$.

5. Write a how-many-units-in-1-group problem for $\frac{3}{5} \div \frac{2}{3} = ?$ and explain how to solve the problem with the aid of a math drawing. You may use equivalent fractions.

   Use your math drawing to explain why $\frac{3}{5} \div \frac{2}{3}$ is equivalent to $\frac{3}{2} \cdot \frac{3}{5}$.

Class Activity 6Q Are These Division Problems?

CCSS CCSS SMP4

Which of the following are word problems for the division problem $\frac{3}{4} \div \frac{1}{2}$? For those that are, which interpretation of division is used? For those that are not, determine how to solve the problem if it can be solved.

1. Beth poured $\frac{3}{4}$ cup of cereal in a bowl. The cereal box says that 1 serving is $\frac{1}{2}$ cup. How many servings are in Beth's bowl?

2. Beth poured $\frac{3}{4}$ cup of cereal in a bowl. Then Beth took $\frac{1}{2}$ of that cereal and put it into another bowl. How many cups of cereal are in the second bowl?

3. A crew is building a road. So far, the road is $\frac{3}{4}$ mile long. This is $\frac{1}{2}$ the length that the road will be when it is finished. How many miles long will the finished road be?

4. A crew is building a road. So far, the crew has completed $\frac{3}{4}$ of the road, and this portion is $\frac{1}{2}$ mile long. How long will the finished road be?

5. If $\frac{3}{4}$ cup of flour makes $\frac{1}{2}$ of a batch of cookies, then how many cups of flour are required for a full batch of cookies?

6. If $\frac{1}{2}$ cup of flour makes 1 batch of cookies, then how many batches of cookies can you make with $\frac{3}{4}$ cup of flour?

7. If $\frac{3}{4}$ cup of flour makes 1 batch of cookies, then how much flour is in $\frac{1}{2}$ of a batch of cookies?

6.6 Dividing Decimals

Class Activity 6R Reasoning and Estimation with Decimal Division

CCSS CCSS SMP2

1. Describe a way to calculate 32.5 ÷ 0.5 mentally. *Hint:* Think in terms of fractions or in terms of money.

2. Describe ways to calculate 1.2 ÷ 0.25 mentally.

3. Describe a way to estimate 7.2 ÷ 0.333 mentally.

4. Fran must calculate 2.45 ÷ 1.5 longhand, but she can't remember what to do about decimal points. Instead, Fran solves the division problem 245 ÷ 15 longhand and gets the answer 16.33. Fran knows that she must shift the decimal point in 16.33 somehow to get the correct answer to 2.45 ÷ 1.5. Explain how Fran could use estimation to determine where to put the decimal point.

Class Activity 6S Decimal Division

CCSS CCSS SMP3, 5.NBT.7

1. Write one how-many-groups word problem and another how-many-units-in-1-group word problem for $23.45 \div 2.7 = ?$.

2. How are the problems

"How many \$0.25s are in \$12.37?" and "How many 25s are in 1237?" related?

What can you conclude about how the two division problems

$$0.25\overline{)12.37} \quad \text{and} \quad 25\overline{)1237}$$

are related?

3. Explain how the figure below can be interpreted as:

$$0.06 \div 0.02 = ?$$

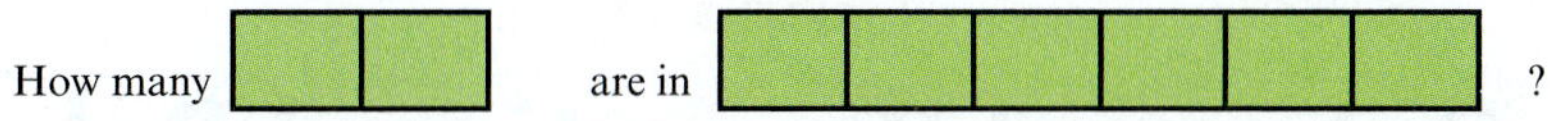

Explain how the same figure can be interpreted as:

$0.6 \div 0.2 = ?$ or as $6 \div 2 = ?$ or as $6{,}000{,}000 \div 2{,}000{,}000 = ?$

What other division problems can the figure illustrate?

What's the moral here?

4. Make rough drawings of bundled objects to represent "How many 0.15 are in 1.2?" Then describe what other questions your drawing could represent and how it is useful to calculating $0.15\overline{)1.2}$.

Ratio and Proportional Relationships

7.1 Motivating and Defining Ratio and Proportional Relationships

Class Activity 7A Mixtures: The Same or Different?

CCSS CCSS SMP2, 6.RP.1, 6.RP.3a

Cups and two juices or collections of small square tiles or beads in two colors would be helpful.

There are two containers, each holding a mixture of 1 cup red punch and 3 cups lemon-lime soda. The first container is left as it is. That is Mixture A. Somebody adds 2 cups red punch and 2 cups lemon-lime soda to the second container. That becomes Mixture B.

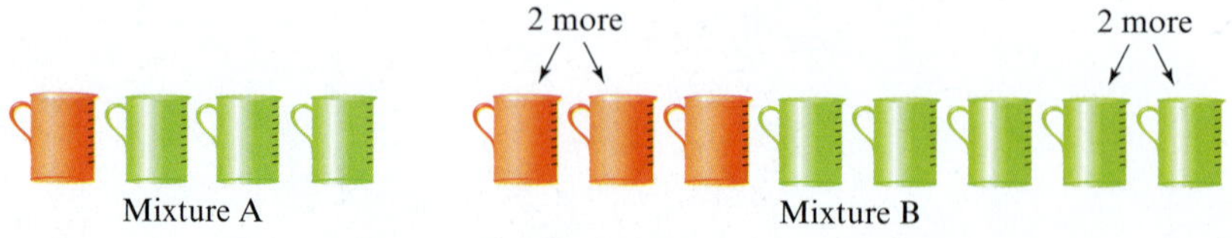

1. Do you think Mixture A and Mixture B will taste the same and have the same color? Why or why not? Try to think about these questions in the way that a student who has not yet studied ratios might. What ideas do you think such a student might have?

If possible, make the mixtures to see if they taste and look the same or not. You can simulate the juice mixtures by mixing cups containing equal numbers of tiles or beads. Use tiles or beads in two different colors.

2. Suppose you mix 4 cups red punch with 12 cups lemon-lime soda. Make a math drawing showing how to *organize* those cups *so that you can tell from the way the cups are organized* that this mixture will have the same flavor and color as Mixture A.

3. Suppose you have 4 identical containers, 1 with red punch and 3 with lemon-lime soda, all poured to the same level. You are about to mix them in a pitcher when you decide you want to make a little bit more. You want this larger amount to have the exact same flavor and color. How can you do that? How will the mixture compare to Mixture A?

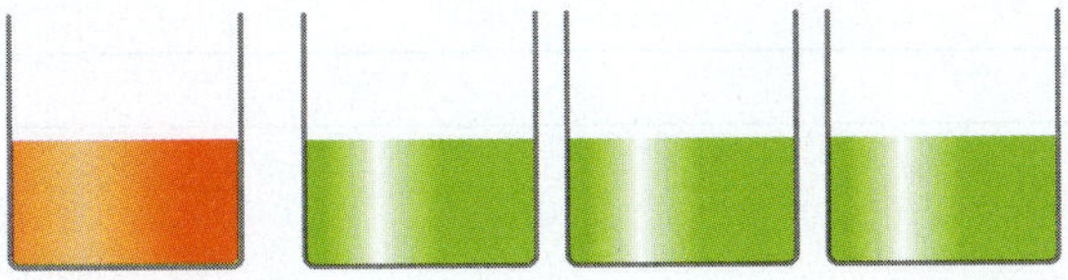

4. Find mixtures of red punch and lemon-lime soda that taste the same as mixture A. Find other mixtures that taste the same as Mixture B. Show these mixtures in the columns of the two tables below.

Explain why the mixtures in one table will taste the same *without using the words "ratio" or "fraction" in your explanation.*

Mixtures that taste the same as Mixture A

Cups of red punch	1					
Cups of lemon-lime	3					
Total number of cups	4					

Mixtures that taste the same as Mixture B

Cups of red punch	3					
Cups of lemon-lime	5					
Total number of cups	8					

5. Explain how to use the tables in part 4 to compare the flavors of mixtures A and B in several ways *without* using the terms "ratio" or "fraction." Which mixture is more red-punchy and which is more lemon-limey? Why is it useful to look for common entries in the two tables? Why is it legitimate to compare different columns from each table?

7.2 Solving Proportion Problems by Reasoning with Multiplication and Division

Class Activity 7B Using Double Number Lines to Solve Proportion Problems

CCSS CCSS SMP2, SMP3, 6.RP.3

Explain how to solve each of the following problems by reasoning about the quantities. Support your reasoning with double number lines.

1. If 3 yards of rope weigh 2 pounds, then how much do the following lengths of the same kind of rope weigh?

   a. 18 yards **b.** 16 yards **c.** 14 yards

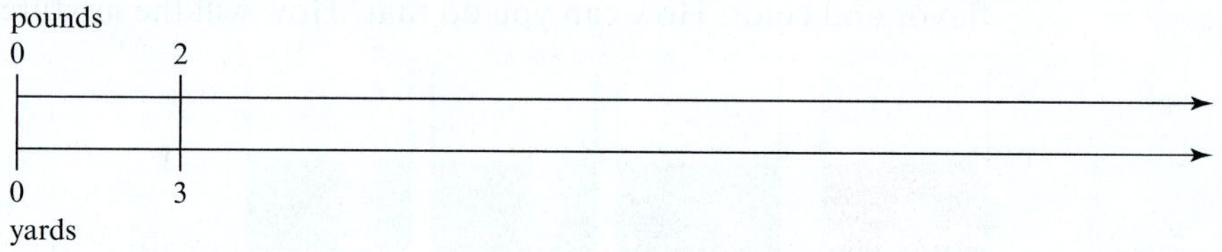

2. If 2 meters of wire weigh 24.8 grams, then how much do 15 meters of that same kind of wire weigh? Try to find several ways of reasoning about the quantities to solve this problem.

3. A scooter is going $\frac{3}{4}$ of a mile every 4 minutes. How far does the scooter go in the following amounts of time?

   a. 12 minutes **b.** 17 minutes

   How long does it take the scooter to go the following distances?

   c. 1 mile **d.** 2 miles

Class Activity 7C Using Strip Diagrams to Solve Proportion Problems

CCSS CCSS SMP2, SMP3, 6.RP.3

1. Suppose a certain shade of green paint is made by mixing blue paint with yellow paint in a ratio of 2 to 3.

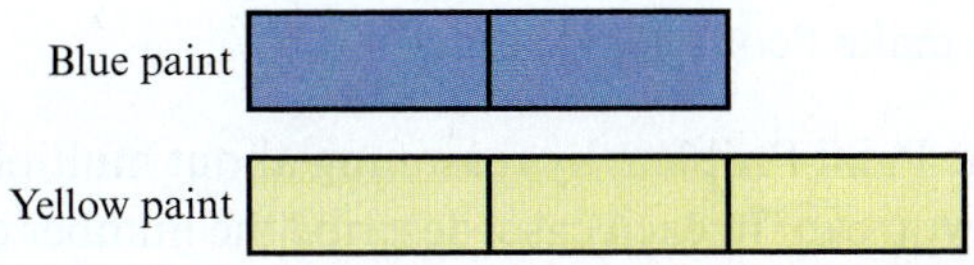

For each of parts (a), (b), and (c), use the same shade of green paint as above, which is made by mixing blue paint with yellow paint in a ratio of 2 to 3. Explain how to solve the problems by using the strip diagram.

a. If you use 26 pails of blue paint, how many pails of yellow paint will you need?

b. If you use 48 pails of yellow paint, how many pails of blue paint will you need?

c. If you want to make 125 pails of green paint, how many pails of blue paint and how many pails of yellow paint will you need?

2. Explain how to solve the following problems by reasoning about a strip diagram. If you mix fruit juice and bubbly water in a ratio of 3 to 5 to make a punch, then how many liters of fruit juice and how many liters of bubbly water will you need to make the following amounts of punch:

a. 32 liters **b.** 10 liters

Class Activity 7D Solving Proportion Problems by Reasoning about Multiplication and Division with Quantities

CCSS CCSS SMP2, SMP3, 6.RP.3

Paint Problem: A paint company makes Peony Pink Paint by mixing red and white paint in the ratio 4 to 7. How many liters of white paint does the company need to mix with 35 liters of red paint to make Peony Pink Paint?

Solve the Paint Problem by reasoning about multiplication and division with quantities in as many ways as you can. In each case, describe the number of liters of white paint as a product $A \cdot B$, where A and B are suitable whole numbers, fractions, or mixed numbers that you derive from 4, 7, and 35. Attend carefully to our definition of multiplication. When you use division, explain what kind it is (how-many-groups or how-many-units-in-1-group). Use math drawings to support your explanations.

Class Activity 7E Ratio Problem Solving with Strip Diagrams

CCSS CCSS SMP1, SMP3, 7.RP.3

1. At lunch, there was a choice of pizza or a hot dog. Three times as many students chose pizza as chose hot dogs. All together, 160 students got lunch. How many students got pizza and how many got a hot dog? Draw a strip diagram to help you solve this problem. Explain your reasoning.

2. The ratio of Shauntay's cards to Jessica's cards is 5 to 3. After Shauntay gives Jessica 15 of her cards, both girls have the same number of cards. How many cards do Shauntay and Jessica each have now? Draw a strip diagram to help you solve this problem. Explain your reasoning.

3. The ratio of Shauntay's cards to Jessica's cards is 5 to 2. After Shauntay gives Jessica 12 of her cards, both girls have the same number of cards. How many cards do Shauntay and Jessica each have now? Draw a strip diagram to help you solve this problem. Explain your reasoning.

4. Make a new problem for your students by modifying part 2 or part 3. Change the ratio and change the number of cards that Shauntay gives to Jessica. When you make these changes, which ratios will make the problem easier, and which ratios will make it harder? Once you have chosen a ratio, can the number of cards that Shauntay gives to Jessica be any number, or do you need to take care in choosing this number? Explain.

Class Activity 7F More Ratio Problem Solving

CCSS CCSS SMP1, SMP2, 7.RP.3

1. Chandra made a milkshake by mixing $\frac{1}{2}$ cup of ice cream with $\frac{3}{4}$ cup of milk. Reason about quantities to determine how many cups of ice cream and milk Chandra should use if she wants to make the same milkshake (i.e., using the same ratio) with the following amounts:
   a. using 3 cups of ice cream

   b. to make 3 cups of milkshake

2. Russell was supposed to mix 3 tablespoons of weed killer concentrate with $1\frac{3}{4}$ cups of water to make a weed killer. By accident, Russell put in an extra tablespoon of weed killer concentrate, mixing 4 tablespoons of weed killer concentrate with $1\frac{3}{4}$ cups of water. How much water should Russell add to his mixture so that the ratio of weed killer concentrate to water will be the same as in the correct mixture? Reason about quantities to solve this problem.

7.3 The Values of a Ratio: Unit Rates and Multipliers

Class Activity 7G Unit Rates and Multiplicative Comparisons Associated with a Ratio

CCSS CCSS SMP2, SMP8, 6.RP.2, 6.RP.3, 7.RP.1

For a certain shade of orange paint, the ratio of red to yellow is 3 to 5.

1. Taking a multiple-batches perspective, fill in the ratio table below. Make a double number-line to illustrate.

Cups of red	3		1
Cups of yellow	5	1	

2. Now take a variable-parts perspective. Draw a strip diagram to illustrate quantities of red and yellow paint in the fixed 3 to 5 ratio.

3. Explain how to Interpret each of the following fractions in two ways:

- As a unit rate
- As a multiplier that compares total amounts of paint.

a. $\frac{3}{5}$

b. $\frac{5}{3}$

4. **a.** Given an amount of red paint, how can you use a value of the ratio to find the corresponding amount of yellow paint? Explain why your method works.

b. Given an amount of yellow paint, how can you use a value of the ratio to find a corresponding amount of red paint? Explain why your method works.

Class Activity 7H Solving Proportions by Cross-Multiplying Fractions

CCSS CCSS SMP3

Recipe Problem: A recipe that serves 6 people calls for $2\frac{1}{2}$ cups of flour. How much flour will you need to serve 10 people, assuming that the ratio of people to cups of flour remains the same?

One familiar way to solve this problem is by letting x be the number of cups of flour we need to serve 10 people and setting two fractions equal to each other:

$$\frac{x}{10} = \frac{2\frac{1}{2}}{6}$$

Next, we cross-multiply to obtain the equation

$$6 \cdot x = 10 \cdot 2\tfrac{1}{2}$$

Finally, we solve for x by dividing both sides of the equation by 6. Therefore,

$$x = \frac{10 \cdot 2\frac{1}{2}}{6} = \frac{10 \cdot \frac{5}{2}}{6} = \frac{25}{6} = 4\tfrac{1}{6}$$

We need $4\frac{1}{6}$ cups of flour to serve 10 people.

Let's investigate the rationale for this method of solving proportions.

1. Interpret the meaning of the fractions

$$\frac{x}{10} \quad \text{and} \quad \frac{2\frac{1}{2}}{6}$$

in terms of the recipe problem. (Remember that we can interpret fractions in terms of division.) Explain why these two fractions should be equal.

2. What is the rationale behind the procedure of cross-multiplying?

3. We could have set the equation up as

$$\frac{10}{6} = \frac{x}{2\frac{1}{2}}$$

Interpret the fractions $\frac{10}{6}$ and $\frac{x}{2\frac{1}{2}}$ in terms of the recipe problem. Why should they be equal?

7.4 Proportional Relationships

Class Activity 7I Representing a Proportional Relationship with Equations

CCSS CCSS SMP2, 7.RP.2c

Suppose that Strip 1 and Strip 2 are stretchy, so they can get longer and shorter, but their lengths always remain in a 3 to 5 ratio. Strip 1 is X centimeters long and Strip 2 is Y centimeters long, so X and Y vary together in a proportional relationship.

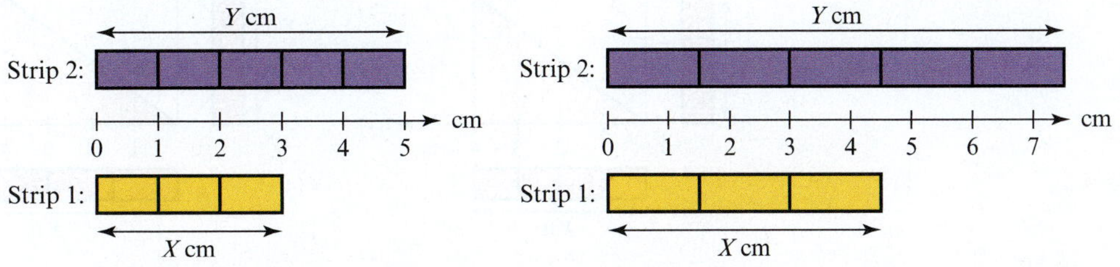

1. How many of Strip 1 does it take to make Strip 2? How many of Strip 2 does it take to make Strip 1? Do the answers depend on how many centimeters long the strips are? Explain!

2. Given a value for X, how can you find the corresponding value of Y? Given a value for Y, how can you find the corresponding value of X? Explain!

3. Write and explain as many equations as you can that relate X and Y. Also write incorrect equations that you think students might write to relate X and Y.

Class Activity 7J Relating Lengths and Heights of Ramps

CCSS CCSS SMP2, 7.RP.2

Suppose a ramp is 4 feet long and 3 feet high. Imagine other ramps that are *X* feet long and *Y* feet high but whose length and height are in that same 4 to 3 ratio. So *X* and *Y* vary together in a proportional relationship.

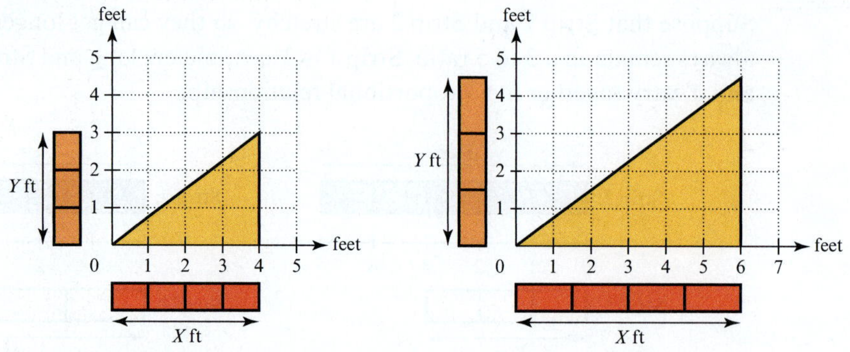

1. For these ramps, how many groups of their length amounts (*X* feet) does it take to make their height amounts (*Y* feet)? How many groups of their height amounts does it take to make their length amounts? Do the answers depend on the specific number of feet in their lengths and heights? Explain!

2. Write and explain as many equations as you can that relate *X* and *Y* for these ramps.

3. Describe in as many ways as you can what is the same for all the ramps.

4. Now pick another fixed ratio. Make sketches of ramps and write and explain equations to relate the lengths and heights of ramps in that ratio.

Class Activity 7K Graphs and Equations of Lines through the Origin

CCSS CCSS SMP7, 8.EE.5, 8.EE.6

Why do lines have a specific type of equation? During this activity, pretend that you don't yet know how to formulate equations of lines.

Consider the line that goes through the origin, $(0, 0)$, and the point $(2, 3)$. For all the points (X, Y) on this line, X and Y are in the ratio 2 to 3 (in Chapter 14 we will use similar triangles to explain why).

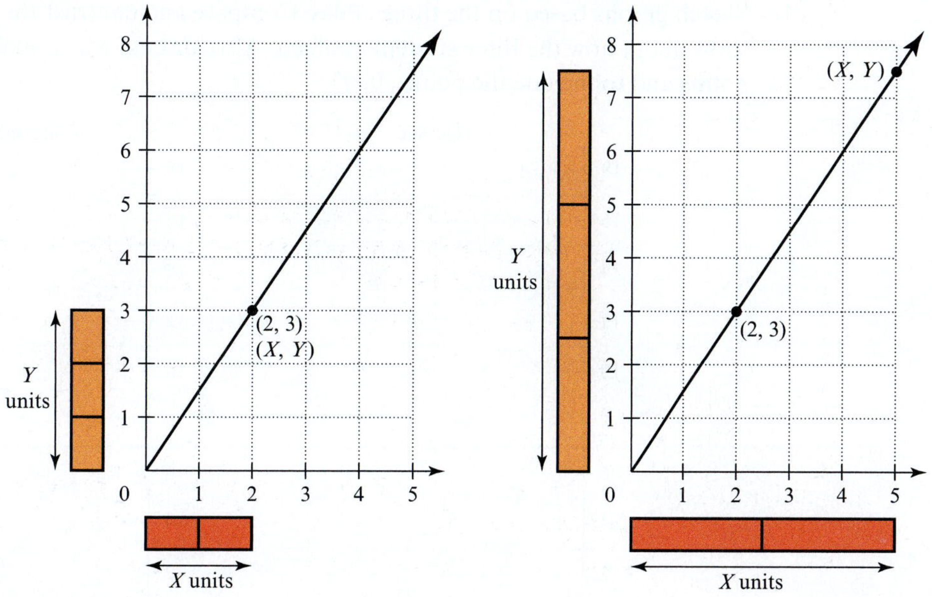

Explain why for all points (X, Y) on the line (for nonnegative values of X and Y only), X and Y are related by the equations

$$Y = \frac{3}{2} \cdot X \quad \text{and} \quad X = \frac{2}{3} \cdot Y$$

In your explanation, take care to use our definition of multiplication and to interpret the meaning of $\frac{3}{2}$ and $\frac{2}{3}$ suitably.

Class Activity 7L Comparing Tables, Graphs, and Equations

CCSS CCSS SMP2, 8.EE.5

The tables below show total distances and elapsed times as Kellie, Devonte, and Heather walked along a running track.

Devonte						
Meters	3	6	9	12	15	18
Seconds	2	4	6	8	10	12

Kellie				
Meters	4	8	12	16
Seconds	3	6	9	12

Heather						
Meters	1	2	4	7	11	16
Seconds	2	4	6	8	10	12

1. Sketch graphs based on the three tables. Compare and contrast the graphs and what they show about how the three students walked. Also, discuss why it makes sense to connect the points and to include the point $(0, 0)$.

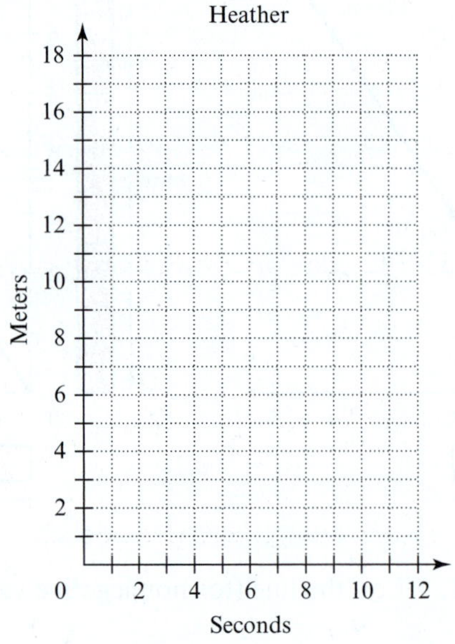

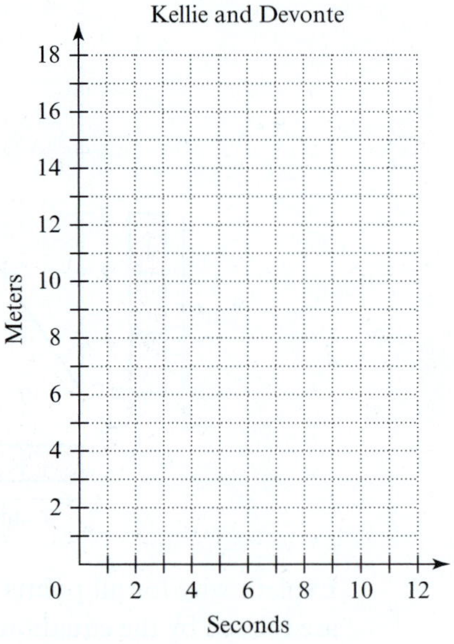

2. Which students are walking at a constant speed? How can you tell?

3. Who walked faster, Kellie or Devonte? Explain in several different ways how you can tell from their graphs.

4. A fourth student, Ashga, also walked along the running track with total distances and elapsed times shown below. Who walked faster, Ashga, Kellie, or Devonte? Explain how you can tell in several different ways.

Ashga			
Meters	11	22	33
Seconds	7	14	28

5. A fifth student, Brittany, walked along the track so that her total distance, *d*, in meters, and elapsed time, *t*, in seconds are given by the equation

$$d = \frac{9}{5} \cdot t$$

How does Brittany's walking compare to the others? Explain!

7.5 Proportional Relationships Versus Inversely Proportional Relationships

Class Activity 7M How Are the Quantities Related?

CCSS CCSS SMP3, 7.RP.2a

Use our definition of multiplication throughout this activity. When you use division, say what kind it is.

1. A painting company will send a crew of painters to paint a large wall. All painters work at the same steady pace. If a crew of 6 painters does the job, they will each paint an area of 300 square meters.

   a. If a crew of 18 painters does the job, how much area will each painter paint? Make a math drawing to help you explain.

   b. If a crew of 10 painters does the job, how much area will each painter paint?

   c. If each painter on the crew will paint 200 square meters, how many painters will be on the crew?

   d. Suppose X painters will paint Y square meters each to complete the job, where X and Y are unspecified numbers, which can vary. Find and explain an equation that relates 6, 300, X, and Y.

   e. Are X and Y in a proportional relationship? What are several different ways you can tell?

2. A restaurant will hire a crew of cooks to stuff dumplings for a large party. All the cooks stuff dumplings at the same steady pace and they all work at the same time. If a crew of 8 cooks does the job, it will take 6 hours to stuff all the dumplings.

   a. If a crew of 2 cooks does the job, how long will it take? Make a math drawing to help you explain.

   b. If a crew of 3 cooks does the job, how long will it take?

   c. If the job needs to be done in 4 hours, how many cooks should be on the crew?

   d. Suppose X cooks will work for Y hours to do the job, where X and Y are unspecified numbers, which can vary. Find and explain an equation that relates 8, 6, X, and Y.

   e. Are X and Y in a proportional relationship? What are several different ways you can tell?

Class Activity 7N Can You Use a Proportion or Not?

CCSS CCSS SMP2, SMP4, 7.RP.2a

1. Ken used 3 loads of stone pavers to make a 10-foot-by-10-foot square patio. Ken wants to make another square patio, this one 20 feet by 20 feet, so he sets up the proportion

$$\frac{3 \text{ loads}}{10 \text{ feet}} = \frac{x \text{ loads}}{20 \text{ feet}}$$

Is this correct? If not, why not? Is there another way that Ken could solve the problem?

2. In a cookie factory, 4 assembly lines make enough boxes of cookies to fill a truck in 10 hours. How long will it take to fill the truck if 8 assembly lines are used? Is the proportion

$$\frac{10 \text{ hours}}{4 \text{ lines}} = \frac{x \text{ hours}}{8 \text{ lines}}$$

appropriate for this situation? Why or why not? If not, can you solve the problem another way? (Assume that all assembly lines work at the same steady rate.)

3. In the cookie factory of part 2, how long will it take to fill a truck if 6 assembly lines are used? (If you get stuck here, move on to the next problem and come back.)

4. Robyn used the following reasoning to solve the previous problem:

> "Four assembly lines fill a truck in 10 hours, so 8 assembly lines should fill a truck in half that time, so, in 5 hours. Since 6 assembly lines is halfway between 4 and 8, it ought to take halfway between 10 hours and 5 hours, or $7\frac{1}{2}$ hours, to fill a truck."

Robyn's reasoning seems quite reasonable, but is it really correct? Let's look carefully.

Fill in the following table by thinking logically about the assembly lines:

Number of hours			10			
Number of lines	1	2	4	8	16	32

Now apply Robyn's reasoning again, but to 1 assembly line versus 32. Sixteen assembly lines is approximately halfway between 1 and 32. But is the corresponding number of hours also approximately halfway between?

What can you conclude about Robyn's reasoning?

Class Activity 7O A Proportional Relationship Versus an Inversely Proportional Relationship

CCSS CCSS SMP2, SMP4, 7.RP.2a, 8.F.3

1. At a bakery, 2 people can frost a total of 50 cupcakes in 12 minutes. Assume that all people work at the same steady rate.

a. Make math drawings to help you explain how to fill in the blanks in part (a). Fill in the tables and answer the questions. Then compare and contrast the two relationships.

Relationship: Number of people ↔ Number of cupcakes when working for 12 minutes.

(a)
2 times as many people frost ________ as many cupcakes.

$\frac{1}{2}$ as many people frost ________ as many cupcakes.

N times as many people frost ________ as many cupcakes.

(b)

Cupcakes		50						
People	1	2	3	4	5	6	7	8

(c) Find a • or ÷ relationship between number of people, number of cupcakes:

(d) What type of relationship is it between number of people, number of cupcakes? How can you tell?

Relationship: Number of people ↔ Number of minutes when frosting 50 cupcakes.

(a)
2 times as many people take ________ as long.

$\frac{1}{2}$ as many people take ________ as long.

N times as many people take ________ as long.

(b)

Minutes		12						
People	1	2	3	4	5	6	7	8

(c) Find a • or ÷ relationship between number of people, number of minutes:

(d) What type of relationship is it between number of people, number of minutes? How can you tell?

b. On separate paper, graph the points in the tables. Use horizontal *x*-axes for the number of people. How are the shapes of the graphs different?

2. For each of the following, determine if the relationship between *X* and *Y* is proportional, inversely proportional, or neither. Explain!

a. There is a large pile of sand that needs to be hauled away. *X* trucks will each take *Y* loads of sand to haul away the sand.

b. There are 20 trucks. Of those trucks, *X* trucks will haul sand and the rest, *Y* trucks, will haul gravel.

c. There are 20 trucks. Each truck will haul *X* loads of sand. All loads are the same size. All together, the trucks will haul *Y* tons of sand.

d. There is a large sand pile. After *X* identical truck-loads of sand have been hauled away, there are *Y* tons of sand left in the pile.

7.6 Percent Revisited: Percent Increase and Decrease

Class Activity 7P How Should We Describe the Change?

CCSS CCSS SMP2, 4.OA.2, 7.RP.3

A store raised some of its prices:

- A carton of milk went from $2 to $3.
- A box of laundry detergent went from $5 to $6.
- A small tube of makeup went from $10 to $15.
- A large tube of makeup went from $20 to $30.

The milk and the laundry detergent each went up by $1. But does that $1 increase seem equally significant in both cases?

The small tube of makeup went up by $5 and the large tube went up by $10. Does that mean that the price of the large tube of makeup went up more? Discuss!

Class Activity 7Q Calculating Percent Increase and Decrease

CCSS CCSS SMP3, 7.RP.3, 7.EE.2

1. Brand A cereal used to be sold in a 20-ounce box. Now Brand A cereal is sold in a 23-ounce box.

   a. Calculate the increase in the weight of cereal in a Brand A box as a percentage of the original weight.

   b. Now calculate the new weight of a Brand A box of cereal as a percentage of the original weight, and subtract 100%.

   c. Why does it make sense that the calculations in (a) and (b) come out the same?

2. There were 20 gallons of gas in a tank. Now there are only 15 gallons left.

   a. Calculate the decrease in the amount of gas in the tank as a percentage of the original.

   b. Now calculate the new amount of gas in the tank as a percentage of the original, and subtract it from 100%.

   c. Why does it make sense that the calculations in (a) and (b) come out the same?

Class Activity 7R Calculating Amounts from a Percent Increase or Decrease

CCSS CCSS SMP3, 7.RP.3, 7.EE.2

1. The price of a Loungy Chair was \$400. The price of this chair has just gone up by 20%. Complete the percent table (fill in steps as needed) and explain why the blank must be the new price of the Loungy Chair.

   100% → \$400

   120% → ____

2. A set of sheets was \$60. The sheets are now on sale for 15% off. Complete the percent table (filling in steps as needed) and explain why the blank must be the new price of the sheets. Where does the 85% in the percent table come from?

   100% → \$60

   85% → ____

3. The price of a suit just went up by 20%. The new price, after the increase, is \$180.
   a. Complete the percent table and explain why the blank will be the price of the suit before the increase.

   120% → \$180

   100% → ____

   Where does the 120% come from, and why is it equated with \$180?

   b. Explain why you *can't* calculate the price of the suit before the increase by decreasing \$180 by 20%.

4. The price of a sofa went down by 20%. The new reduced price is \$400.
   a. Complete the percent table and explain why the blank will be the price of the sofa before the reduction.

   80% → \$400

   100% → ____

   Where does the 80% come from, and why is it equated with \$400?

   b. Explain why you *can't* calculate the price of the sofa before the reduction by increasing \$400 by 20%.

Class Activity 7S Can We Solve It This Way?

CCSS CCSS SMP3

1. The price of a cruise increased by 15%. The new price is \$2300. What was the price before the increase?

   Here is how Matt solved the problem:

   > First I found 10% and that's \$230. Then 5% is half of that, so \$115. So 15% is \$345. So I took \$345 away from \$2300, which leaves \$1955, and that's the answer.

   Discuss Matt's method. Is it correct or not? Show how to solve the problem in another way.

2. Whoopiedoo makeup used to be sold in 4-ounce tubes. Now it's sold in 5-ounce tubes for the same price. Ashlee says the label should read "25% more," whereas Carolyn thinks it should read "20% more." Who is right, who is wrong, and why?

3. The Film Club increased from 15 members to 45 members. Amy says that's a 300% increase. Kaia says it's a 200% increase. Who is right, who is wrong, and why?

Class Activity 7T Percent Problem Solving

CCSS CCSS SMP1

Strip diagrams may help you solve some of these problems.

1. At first, Prarie had 10% more than the cost of a computer game. After Prarie spent \$7.50, she had 15% less than the cost of the computer game. How much did the computer game cost? How much money did Prarie have at first? Explain your reasoning.

2. One mouse weighs 20% more than another mouse. Together, the two mice weigh 66 grams. How much does each mouse weigh? Explain your reasoning.

3. There are two vats of orange juice. After 10% of the orange juice in the first vat is poured into the second vat, the first vat has 3 times as much orange juice as the second vat. By what percent did the amount of juice in the second vat increase when the juice from the first vat was poured into it? Explain your reasoning.

Class Activity 7U Percent Change and the Commutative Property of Multiplication

CCSS CCSS SMP3, 7.EE.2

Which, if either, of the following two options will result in the lower price for a pair of pants?

- The price of the pants is marked up by 10% and then marked down by 20% from the increased price.
- The price of the pants is marked down by 20% and then marked up by 10% from the discounted price.

Both options involve marking up by 10% and marking down by 20%. The difference is the order in which the marking up and marking down occur.

1. Before you do any calculations, make a guess about which of the two options should result in a lower price.

2. Suppose the pants cost $50 to start with. Which of the two options will result in a lower price?

3. How are

$$0.80 \cdot 1.10 \cdot 50 \quad \text{and} \quad 1.10 \cdot 0.80 \cdot 50$$

relevant to the question about the pants? How is the commutative property of multiplication relevant?

Number Theory

8.1 Factors and Multiples

Class Activity 8A Factors and Rectangles

CCSS CCSS SMP8, 4.OA.4

1. Elsie has 24 square tiles that she wants to arrange in the shape of a rectangle in such a way that the rectangle is completely filled with tiles. What are the different rectangles that Elsie can make and what do they tell you about the factors of 24?

2. If Elsie has more than 24 square tiles, will she necessarily be able to make more rectangles than she could in part 1? Try some experiments. What does this tell you about factors?

Class Activity 8B Finding All Factors

CCSS CCSS SMP3, 4.OA.4

1. Tyrese is looking for all the factors of 156. So far, Tyrese has divided 156 by all the counting numbers from 1 to 13, listing those numbers that divide 156 and listing the corresponding quotients. Here is Tyrese's work so far:

1, 156	$1 \times 156 = 156$
2, 78	$2 \times 78 = 156$
3, 52	$3 \times 52 = 156$
4, 39	$4 \times 39 = 156$
6, 26	$6 \times 26 = 156$
12, 13	$12 \times 13 = 156$
13, 12	$13 \times 12 = 156$

Should Tyrese keep checking to see if numbers larger than 13 divide 156, or can Tyrese stop dividing at this point? If so, why? What are all the factors of 156?

2. Find all the factors of 198 in an efficient way.

Class Activity 8C Do Factors Always Come in Pairs?

CCSS CCSS SMP1

Carmina noticed that factors always seem to come in pairs. For example,

$48 = 1 \times 48$, 1 and 48 are a pair of factors of 48.
$48 = 2 \times 24$, 2 and 24 are a pair of factors of 48.
$48 = 3 \times 16$, 3 and 16 are a pair of factors of 48.
$48 = 4 \times 12$, 4 and 12 are a pair of factors of 48.
$48 = 6 \times 8$, 6 and 8 are a pair of factors of 48.

The number 48 has 10 factors that come in 5 pairs. Carmina wants to know if every counting number always has an even number of factors. Investigate Carmina's question carefully. When does a counting number have an even number of factors, and when does it not?

8.2 Even and Odd

Class Activity 8D Why Can We Check the Ones Digit to Determine Whether a Number Is Even or Odd?

CCSS CCSS SMP3

Remember that a counting number is called *even* if that number of objects can be divided into groups of 2 with none left over:

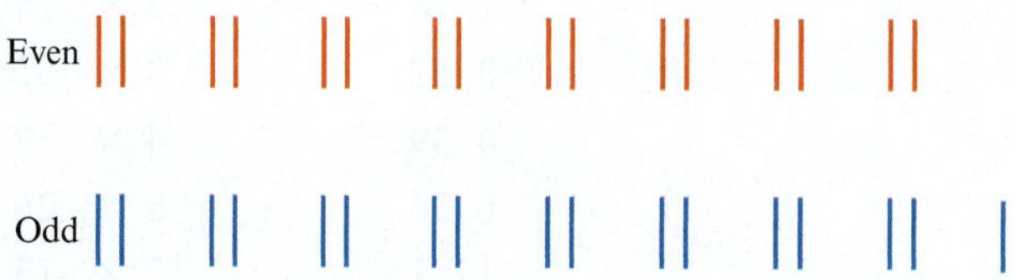

Why is it valid to determine whether a number of objects can be divided into groups of 2 with none left over by checking the ones digit of the number? We will investigate this question in several ways in this Class Activity.

1. What happens to the ones digits when we count by twos?
2. Recall that we can represent a whole number with base-ten bundles and think about putting a number of toothpicks into groups of 2 by working with such bundles. Why does it come down to the ones place to determine if a toothpick will be left over?

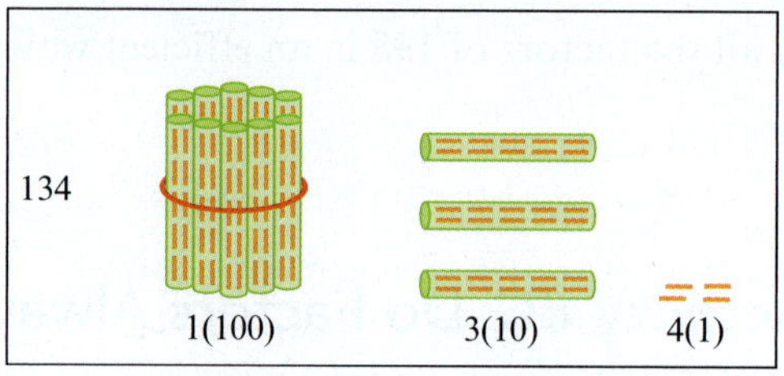

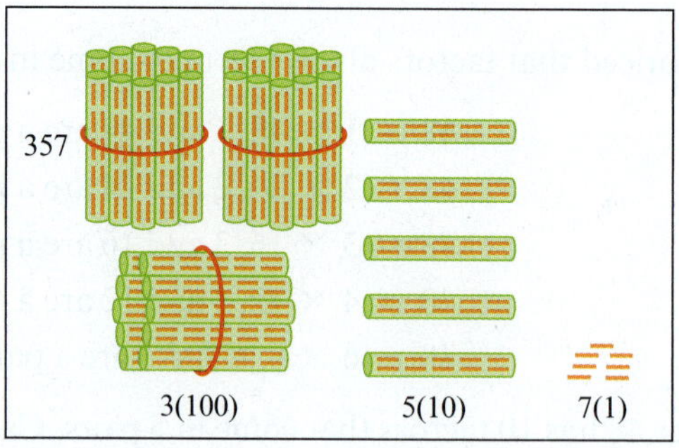

3. Working more generally, let ABC be a 3-digit whole number with A hundreds, B tens, and C ones. Use the idea of representing ABC with base-ten bundles to help explain why ABC is divisible by 2 exactly when C is either 0, 2, 4, 6, or 8. What can we say about each of the B bundles of 10 toothpicks and each of the A bundles of 100 toothpicks when we divide the toothpicks into groups of 2?

Class Activity 8E Questions About Even and Odd Numbers

CCSS CCSS SMP3, SMP8

1. If you add an odd number and an odd number, what kind of number do you get? Investigate this question by working out examples. Then explain why your answer is always correct. Try to find several different explanations by working with the various equivalent ways of saying that a number is even or odd.

2. If you multiply an even number and an odd number, what kind of number do you get? Investigate this question by working out examples. Then explain why your answer is always correct. Try to find several different explanations by working with the various equivalent ways of saying that a number is even or odd.

Class Activity 8F Extending the Definitions of Even and Odd

CCSS CCSS SMP3, SMP6

We have defined even and odd only for counting numbers. What if we wanted to extend the definition of even and odd to other numbers?

1. If we extend the definitions of even and odd to all the integers, what should 0 be, even or odd? What should -5 be, even or odd? Explain.

2. Give definitions of even and odd that apply to all integers, not just to the counting numbers.

3. Would it make sense to extend the definitions of even and odd to fractions? Why or why not?

8.3 Divisibility Tests

Class Activity 8G The Divisibility Test for 3

CCSS CCSS SMP3, SMP8

1. Is it possible to tell if a whole number is divisible by 3 just by checking its last digit? Investigate this question by considering a number of examples. State your conclusion.

2. The divisibility test for 3 is this: Given a counting number, add its digits. If the sum is divisible by 3, then the original number is, too; if the sum is not divisible by 3, then the original number is not either.

   For each of the numbers listed, check that the divisibility test for 3 accurately predicts which numbers are divisible by 3.

   2570 14,928 11,111

3. Explain why the divisibility test for 3 is valid for 3-digit counting numbers. In other words, explain why you can determine whether a 3-digit counting number, *ABC*, is divisible by 3 by adding its digits, $A + B + C$, and determining if this sum is divisible by 3.

   To develop your explanation, consider the following:

   a. A counting number is divisible by 3 exactly when that many objects can be divided into groups of 3 with none left over.

   b. Think about representing 3-digit numbers with base-ten bundles.

   c. Think about dividing bundled toothpicks into groups of 3 by dividing *each individual bundle* of 10 and *each individual bundle* of 100 into groups of 3. How many toothpicks are left over from each bundle?

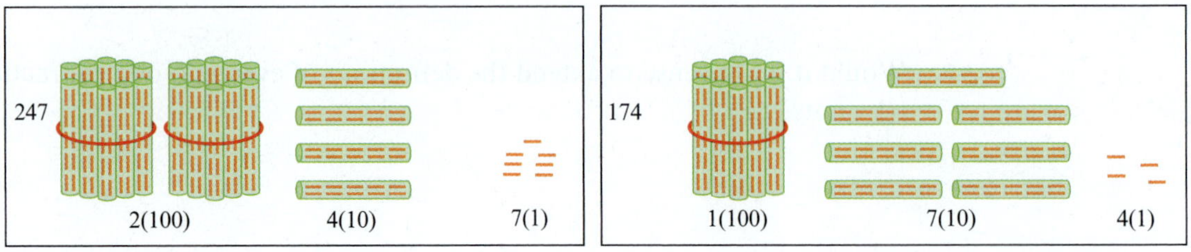

8.4 Prime Numbers

Class Activity 8H The Sieve of Eratosthenes

CCSS CCSS SMP3, 4.OA.4

1. Use the Sieve of Eratosthenes to find all the prime numbers up to 120. Start by circling 2 and crossing off every 2nd number after 2. Then circle 3 and cross off every 3rd number. (Cross off a number even if it already has been crossed off.) Continue in this manner, going back to the beginning of the list, circling the next number N that hasn't been crossed off, and then crossing off every Nth number until every number in the list is either circled or crossed off. The numbers that are circled at the end are the prime numbers from 2 to 120.

	2	3	4	5	6	7	8	9	10
11	12	13	14	15	16	17	18	19	20
21	22	23	24	25	26	27	28	29	30
31	32	33	34	35	36	37	38	39	40
41	42	43	44	45	46	47	48	49	50
51	52	53	54	55	56	57	58	59	60
61	62	63	64	65	66	67	68	69	70
71	72	73	74	75	76	77	78	79	80
81	82	83	84	85	86	87	88	89	90
91	92	93	94	95	96	97	98	99	100
101	102	103	104	105	106	107	108	109	110
111	112	113	114	115	116	117	118	119	120

2. Explain why the circled numbers must be prime numbers and why the numbers that are crossed off are not prime numbers.

Class Activity 8I The Trial Division Method for Determining Whether a Number Is Prime

CCSS CCSS SMP3, 4.OA.4

You will need the list of prime numbers from Class Activity 8H on the Sieve of Eratosthenes for this activity.

1. Using the trial division method and your list of primes from the Sieve of Eratosthenes, determine whether the 3 numbers listed are prime numbers. Record the results of your trial divisions below the number. (The first few are done for you.) You will need these results for part 3.

239	323	4001
$239 \div 2 = 119.5$	$323 \div 2 = 161.5$	$4001 \div 2 = 2000.5$
$239 \div 3 = 79.67\ldots$	$323 \div 3 = 107.67\ldots$	$4001 \div 3 = 1333.67\ldots$

2. How do you know when to stop with the trial division method? To help you answer, look at the list of divisions you did in part 1. As you go down each list, what happens to the divisor and the quotient? If your number *was* divisible by some whole number (other than 1), at what point would that whole number be known?

3. In the trial division method, you determine only whether your number is divisible by *prime* numbers. Why is this legitimate? Why don't you also have to find out if your number is divisible by other numbers such as 4, 6, 8, 9, 10, and so on?

Class Activity 8J Factoring into Products of Primes

CCSS CCSS SMP3

1. Make a factor tree for 240 and use it to write 240 as a product of prime numbers.

2. Compare your factor tree for 240 to a classmate's (or try to make a factor tree for 240 in another way). Are the factor trees identical in all respects? Do they produce the same end result?

   Will it be obvious to students that the end result must be the same, no matter how you make your factor tree? Discuss!

3. Lindsay factored 637 as $637 = 7 \times 7 \times 13$. When Lindsay was then asked to factor 637^2 as a product of prime numbers, she first multiplied $637 \times 637 = 405{,}769$; then she divided 405,769 by 2, then by 3, then by 5, and so on, in order to factor 405,769. Is there an easier way for Lindsay to factor 637^2 into a product of prime numbers? Explain.

4. Given that $527 = 17 \times 31$, is there a way to check if 77 is a factor of 527 without actually dividing 527 by 77? Explain.

8.5 Greatest Common Factor and Least Common Multiple

Class Activity 8K The Slide Method

CCSS CCSS SMP8

1. Examine the initial and final steps of a "slide" that was used to find the GCF and LCM of 900 and 360. Try to determine how it was made. Then make another slide to find the GCF and LCM of 900 and 360.

A Slide

initially:

	900	360

final:

10	900	360
2	90	36
3	45	18
3	15	6
	5	2

$\text{GCF} = 10 \cdot 2 \cdot 3 \cdot 3 = 180$

$\text{LCM} = 10 \cdot 2 \cdot 3 \cdot 3 \cdot 5 \cdot 2 = 1800$

2. Use the slide method to find the GCF and LCM of 1080 and 1200 and to find the GCF and LCM of 675 and 1125.

3. Why does the slide method work?

Class Activity 8L Construct Arguments and Critique Reasoning About GCFs and LCMs

CCSS CCSS SMP3

1. Find the GCF and LCM of $2^6 \cdot 3^4 \cdot 7^3$ and $2^5 \cdot 3^{10} \cdot 7^5$ and explain your reasoning.

2. Some students are finding GCFs and LCMs of numbers written as products of powers of primes.

   - Knowshon says he has a way to find the GCF and LCM: For the GCF, take the smallest power of each prime; for the LCM take the largest power of each prime.
   - Abbey is wondering if Knowshon's method is backward. She asks: Why don't you take the greatest power for the greatest common factor?
   - Matt says that Knowshon's method won't work for numbers like $2^7 \cdot 3^8 \cdot 5^3$ and $2^9 \cdot 3^4 \cdot 7$, because they don't both have 5s and 7s and also the 7 doesn't have an exponent.

   Discuss the students' ideas and objections. (Recall that any nonzero number raised to the 0 power is 1.)

Class Activity 8M Model with GCFs and LCMs

CCSS CCSS SMP4

Solve each problem and explain your solution. Say whether the problem involves the GCF or the LCM.

1. Pencils come in packages of 18; erasers that fit on top of these pencils come in packages of 24. What is the smallest number of pencils and erasers that you can buy so that all the pencils and erasers can be paired? (Assume that you can't buy partial packages.)

2. A class is clapping and snapping to a steady beat. Half of the class uses the pattern

   snap, snap, clap, snap, snap, clap, . . .

   The other half of the class uses the pattern

   snap, clap, snap, clap, snap, clap, . . .

   When will the whole class be clapping together?

3. Mary will make a small 8-inch-by-12-inch rectangular quilt for a doll house out of identical square patches. Each square patch must have side lengths that are a whole number of inches and no partial squares are allowed in the quilt. Other than using 1-inch-by-1-inch squares, what size squares can Mary use to make her quilt? Show Mary's other options below. What are the largest squares that Mary can use?

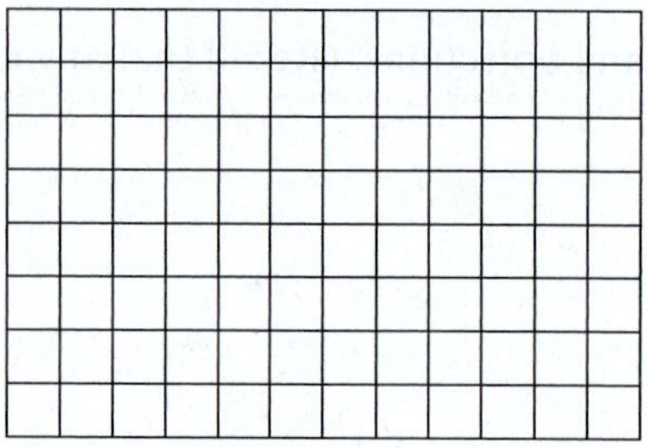

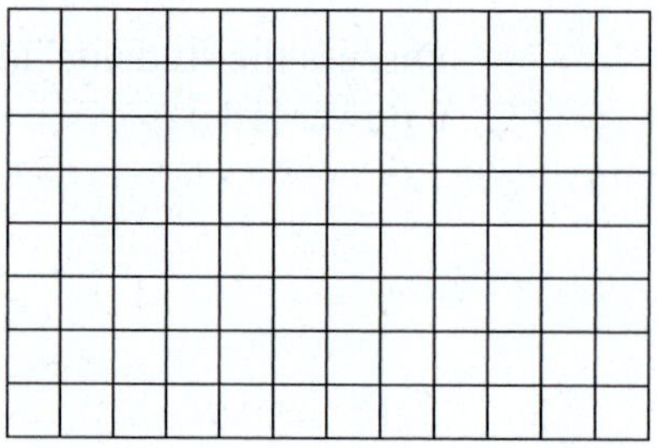

4. Two gears are meshed, as shown in the figure below, with the stars on each gear aligned. The large gear has 36 teeth, and the small gear has 15 teeth. Each gear rotates around a pin through its center. How many revolutions will the large gear have to make and how many revolutions will the small gear make in order for the stars to be aligned again?

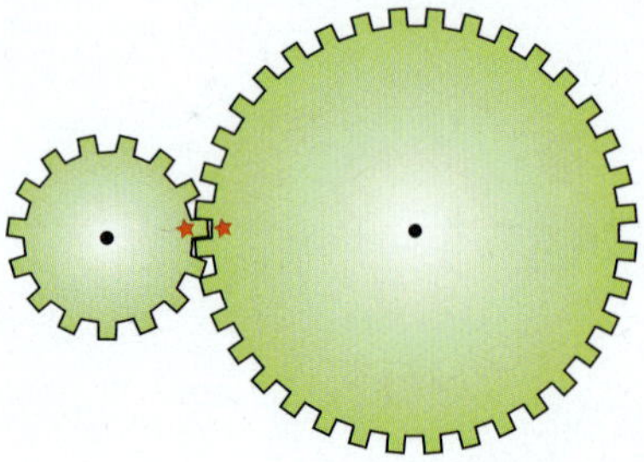

Class Activity 8N Spirograph Flower Designs

CCSS CCSS SMP1

Each flower design below is created by starting at a dot, and connecting each subsequent Nth dot until returning to the starting dot.

Design 1:
36 dots; a petal connects every 8th dot, 9 petals

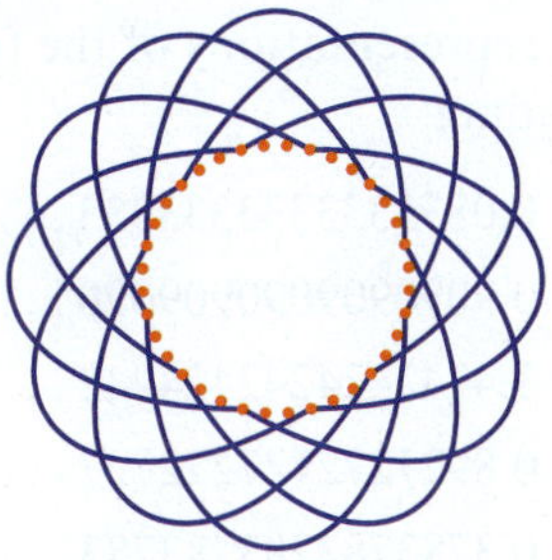

Design 2:
36 dots; a petal connects every 15th dot, 12 petals

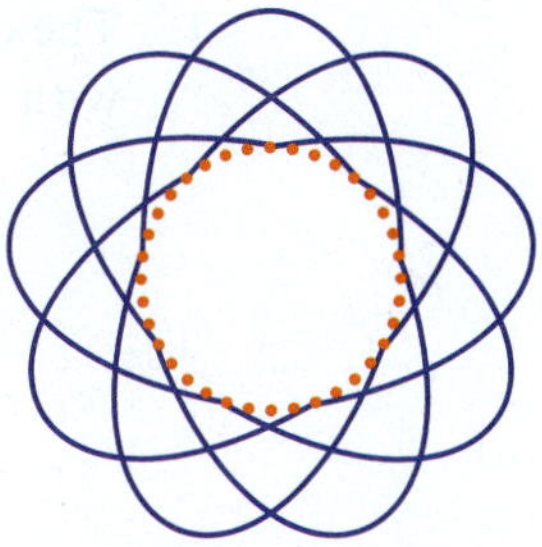

Design 3:
36 dots; a petal connects every 16th dot, 9 petals

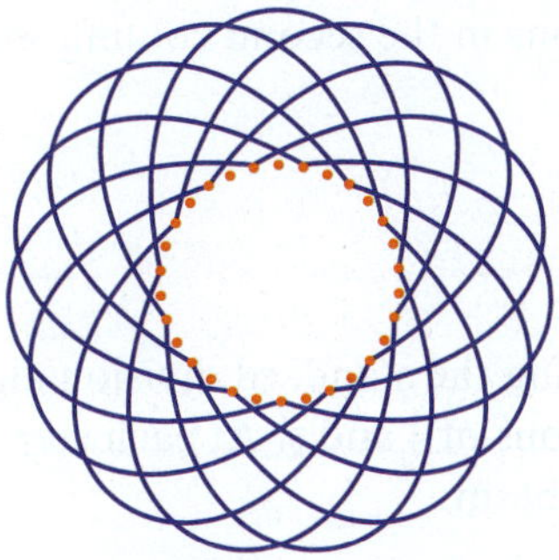

Design 4:
30 dots; a petal connects every 14th dot, 15 petals

Design 5:
30 dots; a petal connects every 4th dot, ____ petals

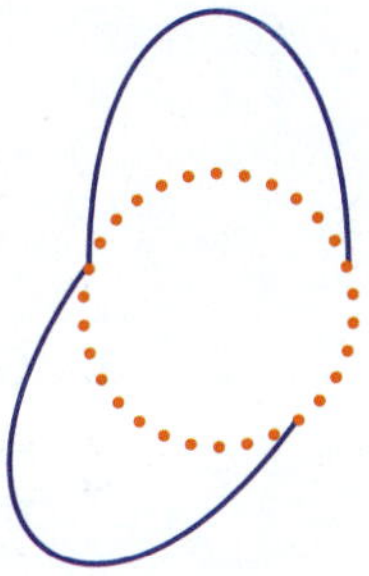

Design 6:
30 dots; a petal connects every 12th dot, ____ petals

1. For the first four flower designs, find a relationship between the number of dots at the center of the flower, the way the petals were made, and the number of petals in the flower design. These numbers are listed in the following table:

Design	Number of Dots	A Petal Connects Every ____ th Dot	Number of Petals
1	36	8	9
2	36	15	12
3	36	16	9
4	30	14	15
5	30	4	
6	30	12	

2. Predict the number of petals that the 5th and 6th flower designs will have. Then complete the designs to see if your prediction was correct. (To complete the designs, you might find it easiest to count dots forward and then draw the petal backward.) See if you can explain why the numbers are related the way they are!

8.6 Rational and Irrational Numbers

Class Activity 8O Decimal Representations of Fractions

CCSS CCSS SMP8, 7.NS.2d, 8.NS.1

1. The decimal representations of the following fractions are shown to 16 decimal places, with no rounding:

$\frac{1}{12} = 0.0833333333333333\ldots$	$\frac{1}{4} = 0.25$
$\frac{1}{11} = 0.9090909090909090\ldots$	$\frac{2}{5} = 0.4$
$\frac{113}{33} = 3.4242424242424242\ldots$	$\frac{37}{8} = 4.625$
$\frac{491}{550} = 0.8927272727272727\ldots$	$\frac{17}{50} = 0.34$
$\frac{14}{37} = 0.3783783783783783\ldots$	$\frac{1}{125} = 0.008$
$\frac{35}{101} = 0.3465346534653465\ldots$	$\frac{9}{20} = 0.45$

In what way are the decimal representations in the first column similar? In what way are the decimal representations of the fractions in the second column similar?

2. Complete the next set of calculations, using the standard division algorithm (not a calculator!) to find the decimal representations of $\frac{4}{7}$ and $\frac{3}{8}$. At each step in the long-division process, write down the remainder you obtain.

$7\overline{)4.0000000}$ with quotient 0.5	remainder 4	$8\overline{)3.0000000}$ with quotient $0.$	remainder 3
-35		___	
5	remainder 5		remainder ___
___	remainder___	___	remainder___
___	remainder___	___	remainder___
___	remainder___	___	remainder___
___	remainder___	___	remainder___
___	remainder___	___	remainder___
___	remainder___	___	remainder___

- What happened to the decimal representation of the fraction when you got a remainder that you had before?
- What happened to the decimal representation of the fraction when you got a remainder of 0?

3. Without actually carrying it out, imagine doing division to find the decimal representation of $\frac{7}{31}$.

a. What remainders could you possibly get in the division process when finding $7 \div 31$? For example, could you possibly get a remainder of 45 or 73 or 32? How many different remainders are theoretically possible?

b. If you were doing division to find $7 \div 31$ and you got a remainder of 0 somewhere along the way, what would that tell you about the decimal representation of $\frac{7}{31}$?

c. If you were doing division to find $7 \div 31$ and you got a remainder you had gotten before, what would then happen in the decimal representation of $\frac{7}{31}$?

d. Now use your answer to parts (a), (b), and (c) to explain why the decimal representation of $\frac{7}{31}$ must either terminate or eventually repeat after at most 30 decimal places.

4. In general, suppose that $\frac{A}{B}$ is a proper fraction, where A and B are whole numbers. Explain why the decimal representation of $\frac{A}{B}$ must either terminate or begin to repeat after at most $B - 1$ decimal places.

5. Could the number

$$0.1010010001000010000010000001\ldots$$

where the decimal representation continues forever with the pattern of more and more 0s in between 1s, be the decimal representation of a fraction? Explain your answer.

Class Activity 8P Writing Terminating and Repeating Decimals as Fractions

CCSS CCSS SMP8, 8.NS.1

1. By using denominators that are suitable powers of 10, show how to write the following terminating decimals as fractions:

   $0.137 =$ $\qquad$ $0.25567 =$

   $13.89 =$ $\qquad$ $329.2 =$

2. Write the following fractions as decimals, and observe the pattern:

   $\frac{1}{9} =$

   $\frac{1}{99} =$

   $\frac{1}{999} =$

   $\frac{1}{9999} =$

   $\frac{1}{99{,}999} =$

3. Using the decimal representations of $\frac{1}{9}, \frac{1}{99}, \ldots$, that you found in part 2, show how to write the following decimals as fractions:

   $0.\overline{2} = 0.222222\ldots =$ $\qquad$ $00.\overline{08} = 0.080808\ldots =$

   $0.\overline{003} = 0.003003\ldots =$ $\qquad$ $0.\overline{52} = 0.525252\ldots =$

   $0.\overline{1234} =$ $\qquad$ $0.\overline{123456} =$

4. Use the fact that $0.\overline{49} = \frac{49}{99}$ to write the next four repeating decimals as fractions. *Hint:* Shift the decimal point by dividing by suitable powers of 10.

   $0.0\overline{49} =$ $\qquad$ $0.00\overline{49} =$

   $0.000\overline{49} =$ $\qquad$ $0.0000\overline{49} =$

5. Use the results of part 4, together with facts such as

   $$0.3\overline{49} = 0.3 + 0.0\overline{49}$$

   to write the following repeating decimals as fractions:

   $7.3\overline{49} =$ $\qquad$ $0.12\overline{49} =$

   $1.2\overline{49} =$ $\qquad$ $0.111\overline{49} =$

Class Activity 8Q What Is 0.9999...?

CCSS CCSS SMP3, 8.NS.1

1. Use the fact that $\frac{1}{9} = 0.\overline{1} = 0.111111111\ldots$ to determine the decimal representations of the following fractions:

$\frac{2}{9} =$ $\frac{3}{9} =$ $\frac{4}{9} =$ $\frac{5}{9} =$

$\frac{6}{9} =$ $\frac{7}{9} =$ $\frac{8}{9} =$ $\frac{9}{9} =$

What can you conclude about $0.\overline{9}$?

2. Add longhand:

$$
\begin{array}{r}
0.9999999999\ldots \\
+\ 0.1111111111\ldots \\
\hline
\end{array}
$$

Note that the nines and ones repeat forever.

Now subtract longhand: $\underline{-.1111111111\ldots}$

Look back at what you just did: Starting with $0.\overline{9}$, you added and then subtracted $0.\overline{1}$. What does this tell you about $0.\overline{9}$?

3. Let N stand for the number $0.\overline{9} = 0.999999\ldots$, so $N = 0.999999999\ldots$

Below, write the decimal representation of $10N$ and then subtract N from $10N$ in two ways—in terms of N and as decimals.

In terms of N:	As decimals:
$10N$	
$\underline{-N}$	$\underline{-0.999999999\ldots}$

What can you conclude about $0.\overline{9}$?

4. Given that the number 1 has two different decimal representations—namely, 1 and $0.\overline{9}$—find different decimal representations of the following numbers:

$17 =$ $23.42 =$ $139.8 =$

Class Activity 8R The Square Root of 2

CCSS CCSS SMP3

1. If the sides of a square are 1 unit long, then how long is the diagonal of the square?

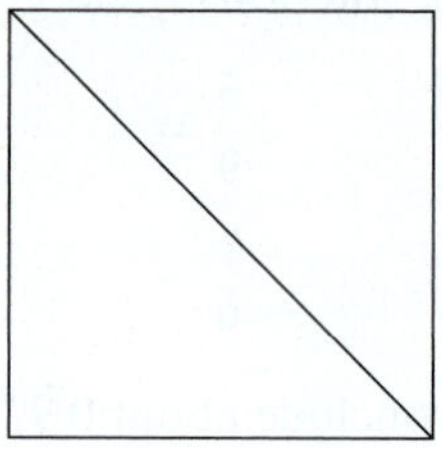

2. Use a calculator to find the decimal representation of $\sqrt{2}$. Based on your calculator's display, does it look like $\sqrt{2}$ is rational or irrational? Why? Can you tell *for sure* just by looking at your calculator's display?

3. Find the decimal representation of

$$\frac{1{,}414{,}213{,}562}{999{,}999{,}999}$$

Is this number rational or irrational? Compare with part 2.

4. Suppose that it were somehow possible to write the square root of 2 as a fraction $\frac{A}{B}$, where A and B are counting numbers:

$$\sqrt{2} = \frac{A}{B}$$

Show that, in this case, we would get the equation

$$A^2 = 2 \cdot B^2$$

5. Suppose A is a counting number, and imagine factoring it into a product of prime numbers. For example, if A is 30, then you factor it as

$$A = 2 \cdot 3 \cdot 5$$

Now think about factoring A^2 as a product of prime numbers. For example, if $A = 30$, then

$$A^2 = 2 \cdot 3 \cdot 5 \cdot 2 \cdot 3 \cdot 5$$

Could A^2 have an odd number of prime factors? Make a general qualitative statement about the number of prime factors that A^2 has.

6. Now suppose that B is a counting number, and imagine factoring the number $2 \cdot B^2$ into a product of prime numbers. For example, if $B = 15$, then

$$2 \cdot B^2 = 2 \cdot 3 \cdot 5 \cdot 3 \cdot 5$$

Could $2B^2$ have an even number of prime factors? Make a general qualitative statement about the number of prime factors that $2 \cdot B^2$ has.

7. Now use your answers in parts 5 and 6 to explain why a number in the form A^2 can never be equal to a number in the form $2 \cdot B^2$, when A and B are counting numbers.

8. What does part 7 lead you to conclude about the assumption in part 4 that it is somehow possible to write the square root of 2 as a fraction, where the numerator and denominator are counting numbers? Now what can you conclude about whether $\sqrt{2}$ is rational or irrational?

Algebra

9.1 Numerical Expressions

Class Activity 9A Writing Expressions for Dot, Star, and Stick Designs

CCSS CCSS SMP7, 5.OA.1, 5.OA.2, 6.EE.1

1. For each flower design, write an expression for the total number of dots in the design. Each expression should involve both multiplication and addition. What do all the expressions have in common?

2. Write two different expressions for the total number of dots in the dot design below. Show or explain why your expressions work without evaluating them.

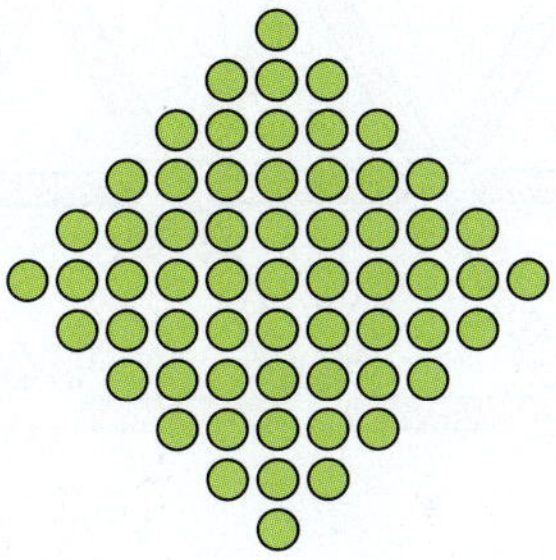

Expression 1: ____________________

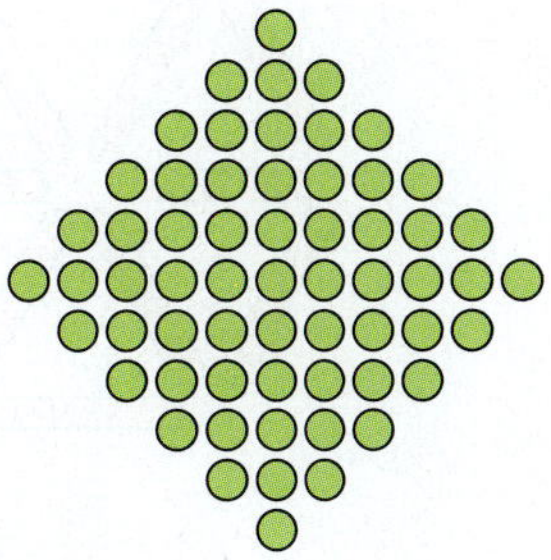

Expression 2: ____________________

3. Write an expression for the total number of stars in the star design below. Show or explain why the other two expressions give the total number of stars without evaluating the expressions.

Your expression: ____________________

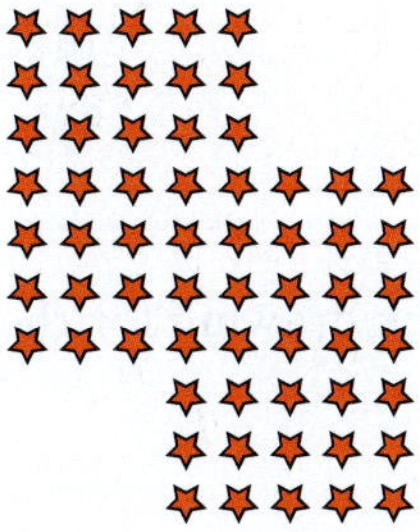

Explain this expression:

$10 \cdot 8 - 2 \cdot 3^2$

Explain this expression:

$2 \cdot (7 \cdot 5) - 4 \cdot 2$

4. Write two different expressions for the total number of dots in the dot design below. Try to find an expression that involves subtraction and a power.

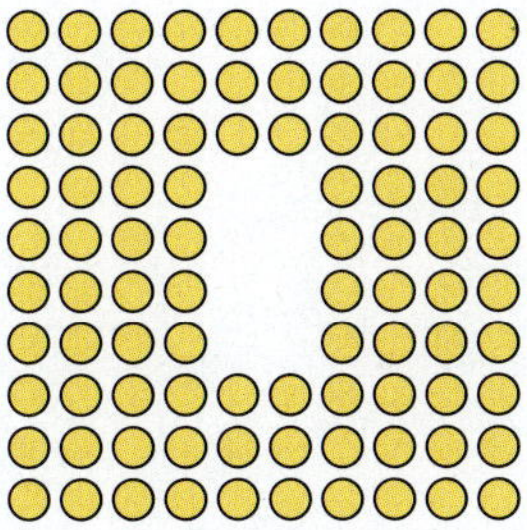

Expression 1: ____________________

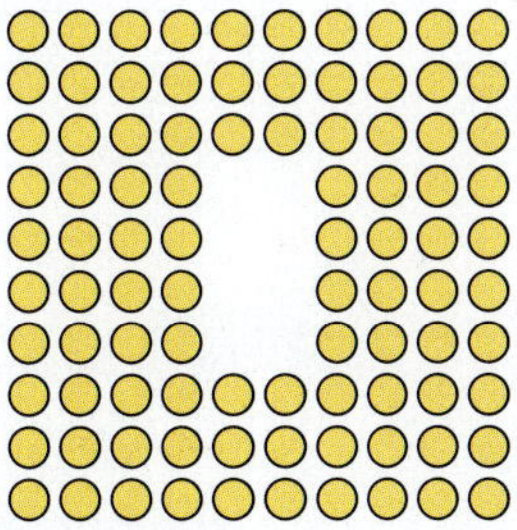

Expression 2: ____________________

5. Write an expression for the total number of sticks in the stick design below.

6. The stick design above is 6 sticks wide. How would you modify your expression in part 5 if the design were 8 sticks wide?

7. Make a dot or stick drawing for the expression

$$4^2 + 4^3$$

8. Make your own dot or stick drawing. Ask a partner to write an expression for the number of dots or sticks in your drawing.

Class Activity 9B How Many High-Fives?

CCSS CCSS SMP1, SMP4

The *high-five problem* (traditionally known as the *handshake problem*): There are 20 students in a class. If every student high-fives with every other student, how many high-fives will there be?

1. Try to find two different expressions for the total number of high-fives among 20 students. One expression should involve addition, and the other expression should involve multiplication. Explain why each expression stands for the total number of high-fives.

2. Solve the high-five problem in part 1 by evaluating one of the expressions.

   Which expression is easiest to evaluate?

3. What if there were 50 students in the class? Write two expressions for the number of high-fives in this case; feel free to use an ellipsis (. . .) in one of your expressions! Explain why the expressions stand for the number of high-fives. Then solve the high-five problem in this case.

   Which expression is easiest to use to solve the high-five problem?

Class Activity 9C Sums of Odd Numbers

CCSS CCSS SMP7, SMP8

Is there a quick way to add a bunch of consecutive odd numbers? This activity will help you find and explain another way to express a sum of odd numbers.

1. Calculate each of the next sums.

$$1 + 3 = \_\_\_\_\_$$
$$1 + 3 + 5 = \_\_\_\_\_$$
$$1 + 3 + 5 + 7 = \_\_\_\_\_$$
$$1 + 3 + 5 + 7 + 9 = \_\_\_\_\_$$
$$1 + 3 + 5 + 7 + 9 + 11 = \_\_\_\_\_$$
$$1 + 3 + 5 + 7 + 9 + 11 + 13 = \_\_\_\_\_$$

2. What is special about the solutions to the sums in part 1?

3. Based on your answer in part 2, predict the sum of the first 100 odd numbers.

4. Based on your answer in part 2, predict the next sum:

$$1 + 3 + 5 + 7 + 9 + \cdots + 91 + 93 + 95 + 97 + 99 = \_\_\_\_\_$$

5. Use the square designs below to explain why there are two ways to express sums of consecutive odd numbers.

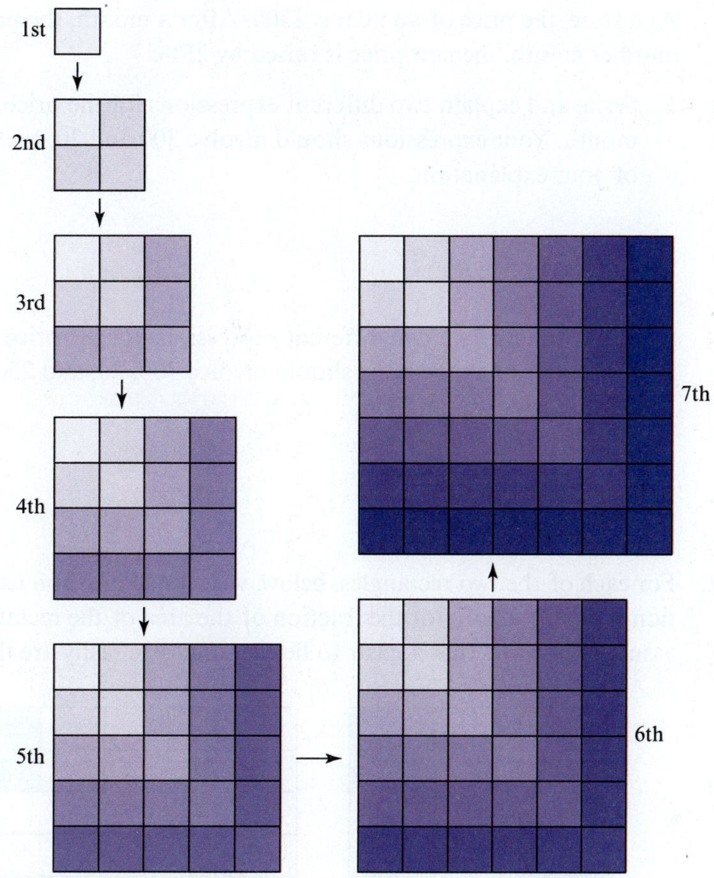

Class Activity 9D Expressions with Fractions and Percent

CCSS CCSS SMP7

1. At a store, the price of an item is \$300. After a month, the price is raised by 20%. After another month, the new price is raised by 25%.

   a. Write and explain two different expressions for the price of the item after the first month. Your expressions should involve 300 and 20. Include a math drawing as part of your explanation.

   b. Write and explain two different expressions for the price of the item after the second month. Your expressions should involve 300, 20, and 25. Include a math drawing as part of your explanation.

2. For each of the two rectangles, below, write an expression using multiplication and addition (or subtraction) for the fraction of the area of the rectangle that is shaded. You may assume that parts that appear to be the same size really are the same size.

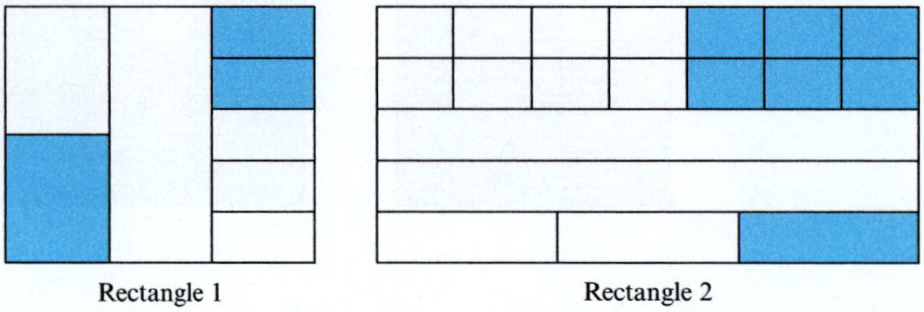

3. Shade

$$\frac{2}{3}\cdot\frac{4}{5}+\frac{1}{4}\cdot\frac{1}{10}$$

of a rectangle in such a way that you can tell the correct amount is shaded without evaluating the expression.

Class Activity 9E Explain and Critique Evaluating Expressions with Fractions

CCSS CCSS SMP3

1. In order to evaluate

$$\frac{8}{35} \cdot \frac{35}{61}$$

we can cancel thus:

$$\frac{8}{\cancel{35}} \cdot \frac{\cancel{35}}{61} = \frac{8}{61}$$

Discuss the following equations and explain why they demonstrate that the canceling shown previously is legitimate:

$$\frac{8}{35} \cdot \frac{35}{61} = \frac{8 \cdot 35}{35 \cdot 61} = \frac{8 \cdot 35}{61 \cdot 35} = \frac{8}{61} \cdot \frac{35}{35} = \frac{8}{61}$$

2. Write equations to demonstrate that the canceling shown in the following equations is legitimate:

$$\frac{\overset{2}{\cancel{18}}}{5} \cdot \frac{7}{\underset{11}{\cancel{99}}} = \frac{2}{5} \cdot \frac{7}{11} = \frac{14}{55}$$

3. Which of the cancellations in parts (a) through (d) are correct, and which are incorrect? Explain your answers.

a. $\dfrac{\overset{6}{\cancel{36}} \cdot \overset{16}{\cancel{96}}}{\underset{1}{\cancel{6}}} = \dfrac{6 \cdot 16}{1} = 96$

b. $\dfrac{\overset{6}{\cancel{36}} \cdot 96}{\underset{1}{\cancel{6}}} = \dfrac{6 \cdot 96}{1} = 576$

c. $\dfrac{\overset{6}{\cancel{36}} + \overset{16}{\cancel{96}}}{\underset{1}{\cancel{6}}} = \dfrac{6 + 16}{1} = 22$

d. $\dfrac{\overset{6}{\cancel{36}} + 96}{\underset{1}{\cancel{6}}} = \dfrac{6 + 96}{1} = 102$

9.2 Expressions with Variables

Class Activity 9F Equivalent Expressions

CCSS CCSS SMP7, 6.EE.3, 6.EE.4, 7.EE.1

1. There are J liters of juice in a vat. Someone pours $\frac{1}{3}$ of the juice out of the vat. Which of the next expressions give the number of liters of juice that remain in the vat? Explain.

   a. $J - \frac{1}{3}$ **b.** $(J - \frac{1}{3}) \cdot J$ **c.** $J - \frac{1}{3} \cdot J$ **d.** $J - \frac{1}{3}J$ **e.** $\frac{2}{3}J$

2. For each pair, determine if the expressions are equivalent or not. If they are equivalent, use properties of arithmetic to show why. If they are not, explain why not.

   a. $P - \frac{2}{5}P$ and $\frac{3}{5}P$ **b.** $P - \frac{2}{5}$ and $\frac{3}{5}P$

   c. $7x - x$ and 7 **d.** $7x - 7$ and x

   e. $3 \cdot (x \cdot y)$ and $(3 \cdot x) \cdot (3 \cdot y)$ **f.** $3 \cdot (x + 2)$ and $3x + 6$

   g. $(x + y)^2$ and $x^2 + y^2$ **h.** $x^4 y^4$ and $(xy)^8$

3. For each expression below, see if you can write an equivalent expression that is a product.

   a. $(x^2 + y^2) + (x^2 + y^2)z^2$

   b. $(x^2 + y^2) + (x^5 + y^5)$

   c. $x^7y^3z^4 + x^4y^5z^7$

Class Activity 9G Expressions for Quantities

CCSS CCSS SMP2, SMP4, 6.EE.6, 7.EE.4

1. a. There are x tons of sand in a pile initially. Then $\frac{1}{4}$ of the sand in the pile is removed from the pile and, after that, another $\frac{2}{3}$ of a ton of sand is dumped onto the pile. Write two equivalent expressions in terms of x for the number of tons of sand that are in the pile now.

   b. There are x tons of sand in a pile initially. Then $\frac{2}{3}$ of a ton of sand is dumped onto the pile and, after that, $\frac{1}{4}$ of the sand in the new, larger pile is removed. Write two equivalent expressions in terms of x for the number of tons of sand that are in the pile now.

   c. Are your expressions in parts (a) and (b) equivalent or not? How can you tell?

2. For each of the following expressions, describe a corresponding situation. Be sure to say what x means in each situation.

   a. $x - \frac{1}{4}x + 30 = 150$

   b. $x - \frac{1}{4} + 30 = 150$

   c. $(x + 30) - \frac{1}{4}(x + 30) = 150$

   d. $\frac{2}{3}(x - 60) + 20 = 80$

3. A T-shirt company has found that if it sells T-shirts for $\$x$ each, then it will sell $280 - 10x$ T-shirts per day. The company has daily fixed operating costs of \$690. Each T-shirt costs the company \$2 to make.

a. Interpret the expression $280 - 10x$ in terms of the situation. What does it tell you?

b. Write and explain an expression for the company's daily profit (income minus expenses) in terms of x.

4. **4.** At a store, the price of an item is $\$P$. Consider three different scenarios:

First scenario: Starting at the price $\$P$, the price of the item was lowered by $A\%$. Then the new price was lowered by $B\%$.

Second scenario: Starting at the price $\$P$, the price of the item was lowered by $B\%$. Then the new price was lowered by $A\%$.

Third scenario: Starting at the price $\$P$, the price of the item was lowered by $(A + B)\%$.

For each scenario, write an expression for the final price of the item. Are any of the expressions for the final prices equivalent?

9.3 Equations

Class Activity 9H Solving Equations by Reasoning About Expressions

CCSS CCSS SMP7, 6.EE.5

Solve each of the equations by thinking about the expressions on both sides of the equal sign and reasoning about which value of the variable will make them equal. Do not use any standard algebraic techniques for solving equations that you may know. Explain your reasoning in each case.

1. $382 + 49 = x + 380$

2. $7 \cdot (x + 5) = 7 \cdot 38$

3. $23 + 36 + x = 24 + 36$

4. $23 \cdot (x - 36) = 46$

5. $23 \cdot 36 + x = 24 \cdot 36$

6. $12 \cdot 84 = 2 \cdot 84 + A$

7. $\frac{5}{16} + \frac{2}{3} = 2x + \frac{5}{16}$

8. $14Z = 7 \cdot 48$

9. $17 = 17(x - 12)$

10. $4 \cdot (x - 19) = 8$

Class Activity 9I Solving Equations Algebraically and with a Pan Balance

CCSS CCSS SMP2, 6.EE.5

1. If this balances:

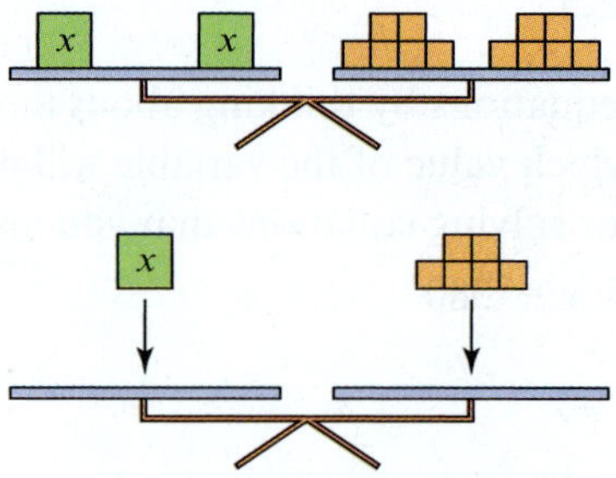

What about this? Will it tilt to the left or to the right, or will it balance? Why?

2. If this balances:

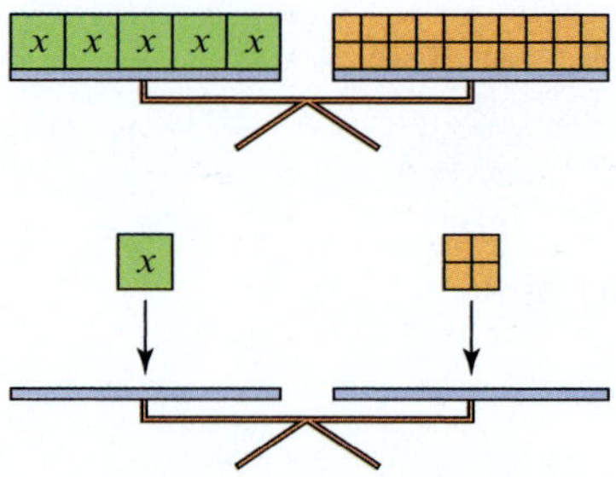

What about this? Will it tilt to the left or to the right, or will it balance? Why?

3. Solve $5x + 1 = 2x + 7$ in two ways, with equations and with pictures of a pan balance. Relate the two methods.

With equations

$5x + 1 = 2x + 7$

With a pan balance

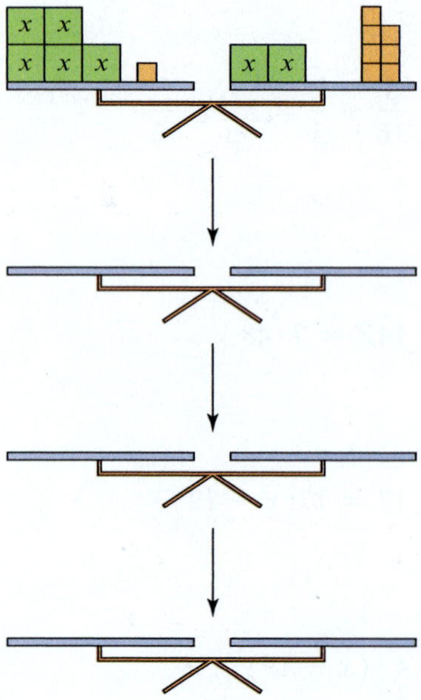

Class Activity 9J What Are the Solutions of These Equations?

CCSS CCSS SMP3, 6.EE.5, 8.EE.7

Solve the next equations using the standard algebraic equation-solving process. Discuss how to interpret the outcome of that process. Think about these issues:

- Does every equation have a solution? If an equation doesn't have a solution, how can we tell that?
- Can an equation have more than one solution? Can an equation have infinitely many solutions? How could that happen?

1. $5x + 7 = 3x + 5 + x$

2. $5x + 7 = 3x + 5 + 2x$

3. $4x - 8 + x = 2x - 3$

4. $4x - 8 + x = 2x - 3 + 3x - 5$

5. Write your own equation in x that has no solutions. Explain.

6. Write your own equation in x that has infinitely many solutions. Explain.

9.4 Solving Algebra Word Problems with Strip Diagrams and with Algebra

Class Activity 9K Solving Word Problems with Strip Diagrams and with Equations

CCSS CCSS SMP1, SMP2, 4.OA.3, 5.NF.2, 5.NF.6, 6.EE.6, 6.EE.7, 7.NS.3, 7.EE.3, 7.EE.4

The problems in this activity were inspired by problems in the mathematics textbooks used in Singapore in grades 4–6 (see [12], volumes 4A–6B).

1. At a store, a hat costs 3 times as much as a T-shirt. Together, the hat and T-shirt cost $35. How much does the T-shirt cost?

   Solve this problem in two ways: by using the strip diagram shown here and with algebraic equations. Explain both solution methods, and discuss how they are related.

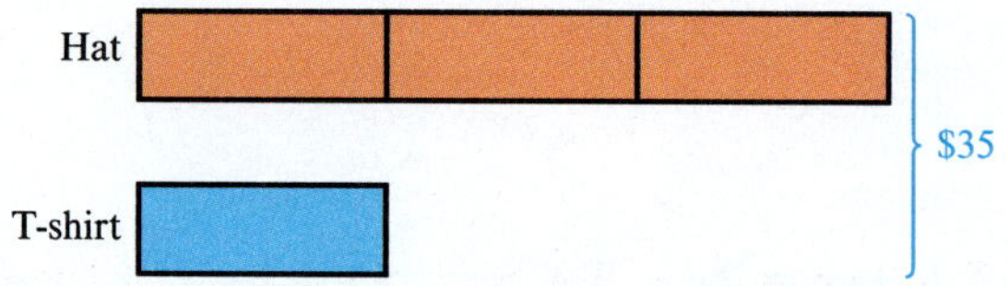

2. There are 180 blankets at a shelter. The blankets are divided into two groups. There are 30 more blankets in the first group than in the second group. How many blankets are in the second group?

   Solve this problem in two ways: by using the strip diagram shown here and with algebraic equations. Explain both solution methods, and discuss how they are related.

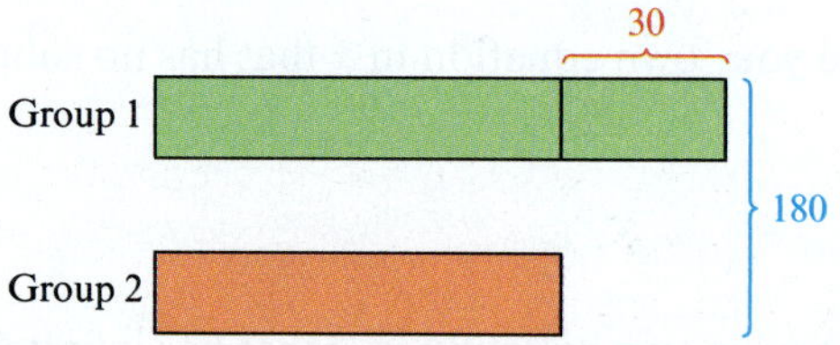

3. On a farm, $\frac{1}{7}$ of the sheep are gray, $\frac{2}{7}$ of the sheep are black, and the rest of the sheep are white. There are 36 white sheep. How many sheep in all are on the farm?

   Solve this problem in two ways: by using the strip diagram shown here and with algebraic equations. Explain both solution methods, and discuss how they are related.

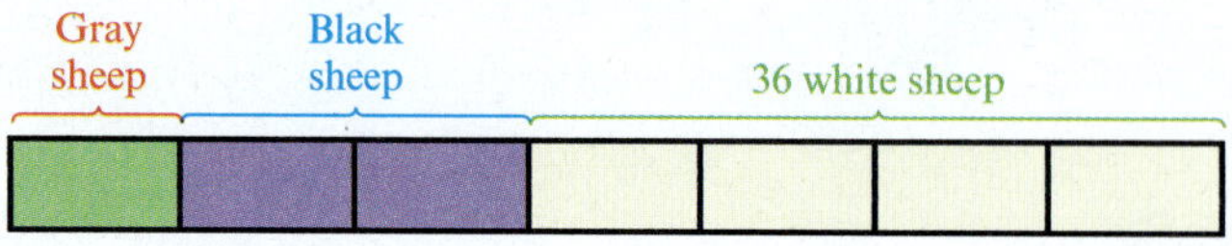

4. Ms. Jones gave $\frac{1}{4}$ of her money to charity and $\frac{1}{2}$ of the remainder to her mother. Then Ms. Jones had \$240 left. How much money did she have at first?

Solve this problem in two ways: by using the strip diagram shown here and with algebraic equations. Explain both solution methods, and discuss how they are related.

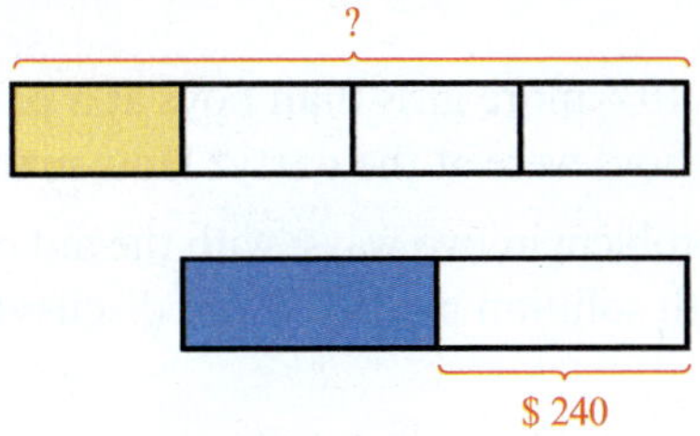

5. After Carmen spent $\frac{1}{6}$ of her money on a CD, she then had \$45 left. How much money did Carmen have at first?

Solve this problem in two ways: with the aid of a strip diagram and with algebraic equations. Explain both solution methods, and discuss how they are related.

6. When a box of chocolates was full, it weighed 1.1 kilograms. After $\frac{1}{2}$ of the chocolates were eaten, the box (with the remaining chocolates) weighed 0.7 kilograms. How much did the box weigh without the chocolates?

Solve this problem in two ways: with the aid of a strip diagram and with algebraic equations. Explain both solution methods, and discuss how they are related.

7. Allie, Barbara, and Carson will divide \$440 among themselves as follows: Barbara gets 2 times as much as Allie. Carson gets $\frac{1}{3}$ as much as Barbara. How much does each person get?

Solve this problem in two ways: with a strip diagram and with algebraic equations. Explain both solution methods and discuss how they are related.

8. There were 25 more girls than boys at a party. All together, 105 children were at the party. How many boys were at the party? How many girls were at the party?

Solve this problem in two ways: with the aid of a strip diagram and with algebraic equations. Explain both solution methods, and discuss how they are related.

9. There were 10% more girls than boys at a party. All together, 168 children were at the party. How many boys were at the party? How many girls were at the party?

Solve this problem in two ways: with the aid of a strip diagram and with algebraic equations. Explain both solution methods, and discuss how they are related.

10. A bakery sold $\frac{3}{5}$ of its muffins. The remaining muffins were divided equally among the 3 employees. Each employee got 16 muffins. How many muffins did the bakery have at first?

Solve this problem in two ways: with the aid of a strip diagram and with algebraic equations. Explain both solution methods, and discuss how they are related.

11. Quint had 4 times as many math problems to do as Agustin. After Quint did 20 problems and Agustin did 2 problems, they each had the same number of math problems left to do. How many math problems did Quint have to do at first?

Solve this problem in two ways: by using the strip diagram shown here and with algebraic equations. Explain both solution methods, and discuss how they are related.

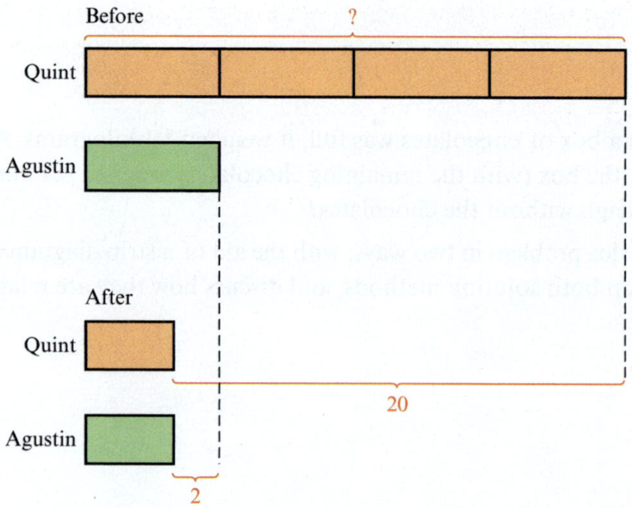

Class Activity 9L Solving Word Problems in Multiple Ways and Modifying Problems

CCSS CCSS SMP1

1. In the morning, Ms. Wilkins put some pencils for her students in a pencil box. After a while, she found that $\frac{1}{2}$ of the pencils were gone. A little later, she found that $\frac{1}{3}$ of the pencils that were left from when she checked before were gone. Still later, Ms. Wilkins found that $\frac{1}{4}$ of the pencils that were left from the last time she checked were gone. At that point there were 15 pencils left. No pencils were ever added to the pencil box. How many pencils did Ms. Wilkins put in the pencil box in the morning?

   Solve this problem in as many different ways as you can think of, and explain each solution. Try to relate your different solution methods to each other.

2. Suppose you want to modify the pencil problem in part 1 for your students by changing the number 15 to a different number. Which numbers could you replace the 15 in the problem with and still have a sensible problem (without changing anything else in the problem)? Explain.

3. Experiment with changing some or all of the fractions—$\frac{1}{2}$, $\frac{1}{3}$, and $\frac{1}{4}$—in the pencil problem in part 1 to some other "easy" fractions. When you make a change, do you also need to change the number 15? Which changes make the problem harder? Which changes make the problem easier?

9.5 Sequences

Class Activity 9M Reasoning About Repeating Patterns

CCSS CCSS SMP3, 5.OA.3

Assume that the following pattern of a square followed by 3 circles and 2 triangles continues to repeat:

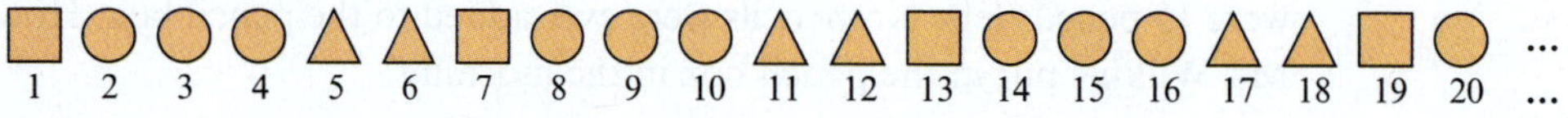

1. What will be the 100th shape in the pattern? Explain how you can tell.

2. How many circles will there be among the first 100 entries of the sequence? Explain your reasoning.

3. Here are three ways that students answered part 2:

    Amanda: There are 6 circles among the first 10 shapes. Because 100 is 10 sets of 10, there will be 10 sets of 6 circles. So there are $10 \times 6 = 60$ circles among the first 100 shapes.

    Robert: My idea was like Amanda's but I got a different answer. I said there were 10 circles among the first 20 shapes. Because 100 is 5 sets of 20, there will be 5 sets of 10 circles. So there are $5 \times 10 = 50$ circles among the first 100 shapes.

    Kayla: I got the same answer as Robert but I thought about it in a different way. The pattern repeats in sets of 6 and 3 of those 6 are circles. So half of the shapes are circles and $\frac{1}{2}$ of 100 is 50, so there are 50 circles.

    Discuss these students' ways of reasoning. Are any of their methods valid? Why or why not?

Class Activity 9N Solving Problems Using Repeating Patterns

CCSS CCSS SMP1, SMP4

1. On a train, the seats are numbered as indicated below. Assume the numbering of the seats continues in this way and that all rows are arranged in the same way. Will seat number 43 be a window seat or an aisle seat? What about seat number 137? What about seat number 294? Describe different ways that you can figure out the answers to these questions.

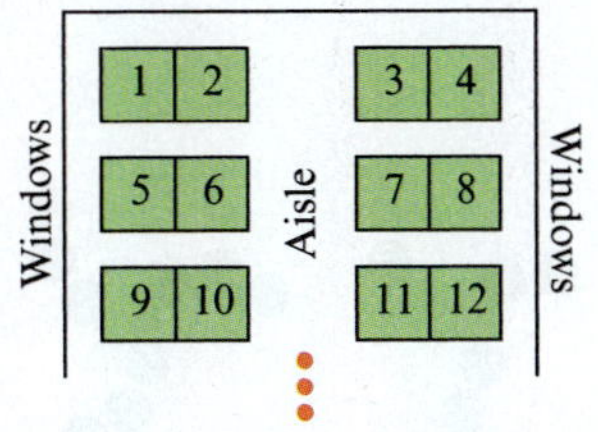

2. What day of the week will it be 100 days from today? Determine the answer with math. Explain your reasoning. How is this problem related to repeating patterns?

3. Five friends are sitting in a circle as shown. Antrice sings a song that has 22 syllables and, starting with Benton, and going clockwise, points to one person for each syllable of the song. The last person that Antrice points to will be "it."

   a. Who will be "it"? Explain how to predict the answer by using math.

   b. If Fran comes and sits between Ellie and Antrice before Antrice sings her song, who will be "it"?

   c. Antrice switches to a song that has 24 syllables. Now who will be "it"? Use math to predict.

4. What is the digit in the ones place of 2^{100}? Explain how you can tell.

Class Activity 90 Arithmetic Sequences of Numbers Corresponding to Sequences of Figures

CCSS SMP7, 5.OA.3, 6.EE.6, 7.EE.4

In the following sequence of figures made of small circles, assume that the sequence continues by adding a green circle to the end of each of the three "arms" of a figure in order to get the next figure in the sequence.

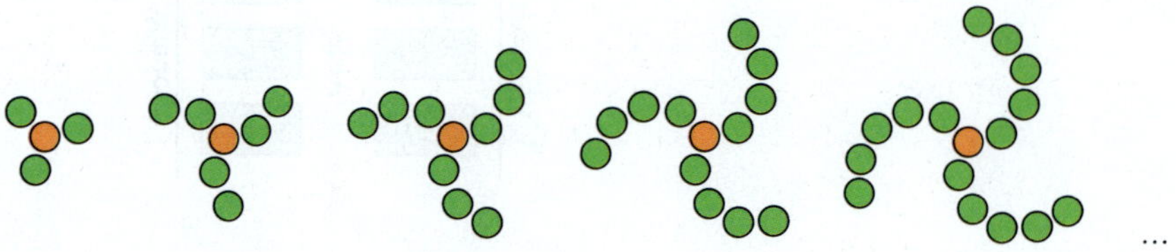

1. In the table below, write the number of small circles that the previous figures are made of. Imagine that the sequence of figures continues forever, so that for each counting number N, there is an Nth figure. What is an expression for the number of small circles in the Nth figure? Add this information to your table.

Position of Figure	Number of Small Circles in Figure
1	
2	
3	
4	
5	
6	
⋮	⋮
N	

2. Relate the structure of the expression you found in part (a) to the structure of the figures and use the relationship to explain why your expression makes sense.

3. How many small circles will the 38th figure in the sequence be made of? How can you tell?

4. Will there be a figure in the sequence that is made of 100 small circles? If yes, which one? If no, why not? Answer these questions in two ways: with algebraic equations and in a way that a student in elementary school who has not yet studied algebraic equations might be able to understand.

5. Will there be a figure in the sequence that is made of 125 small circles? If yes, which one? If no, why not? Answer these questions in two ways: with algebraic equations and in a way that a student in elementary school who has not yet studied algebraic equations might be able to understand.

Class Activity 9P How Are Expressions for Arithmetic Sequences Related to the Way Sequences Start and Grow?

CCSS CCSS SMP7, 8.F.4

This activity will help you notice an interesting connection between the way an arithmetic sequence starts and grows and an expression for the entry in position x of the sequence. (In the next activity, you'll explain why arithmetic sequences must always have equations of a specific type.)

1. For each of the next arithmetic sequences, guess an expression for the entry in position x. Then check your guesses.

   First sequence, increasing by 4
   5, 9, 13, 17, . . . Entry in position x: ___________
   Second sequence, increasing by 4
   7, 11, 15, 19, . . . Entry in position x: ___________
   Third sequence, increasing by 5
   7, 12, 17, 22, . . . Entry in position x: ___________

2. For each sequence in part 1, compare the expression you guessed with the way the sequence increases. What relationship do you notice?

3. For each sequence in part 1, compare the expression you guessed with the first entry of the sequence. What relationship do you notice?

4. Based on your observations in parts 2 and 3, guess the expression for the entry in position x of the next sequences. Then check your guesses.

   Fourth sequence, increasing by 3
   1, 4, 7, 10, . . . Entry in position x: ___________
   Fifth sequence, decreasing by 4
   7, 3, −1, −5, . . . Entry in position x: ___________

Class Activity 9Q Explaining Equations for Arithmetic Sequences

CCSS CCSS SMP3, SMP7, 8.F.4

1. The table below shows some entries for an arithmetic sequence whose first entry is 5 and that increases by 3.

Position	Entry
1	5
2	8
3	11
4	14
5	17
X	

a. If there were an entry in position 0, what would it be? Put it in the table.

b. Fill in the blanks to describe how to get entries in the sequence by *starting from the entry in position 0*.

- To find the entry in position 1: Start at ____ and add ____ 1 time.
- To find the entry in position 2: Start at ____ and add ____ 2 times.
- To find the entry in position 3: Start at ____ and add ____ 3 times.
- To find the entry in position 4: Start at ____ and add ____ 4 times.
- To find the entry in position 5: Start at ____ and add ____ 5 times.
- To find the entry in position X: Start at ____ and add ____ X times.

c. For each bullet in part (b), write an expression (using addition and multiplication) that corresponds to the description for finding the entry in the sequence.

Position 1 entry =

Position 2 entry =

Position 3 entry =

Position 4 entry =

Position 5 entry =

d. Let Y be the entry in position X. Use your work in parts (b) and (c) to find an equation relating X and Y.

e. Plot as many of the points (position, entry) as you can. What do you notice about the arrangement of these points?

X Position	Y Entry
1	5
2	8
3	11
4	14
5	17
X	Y =

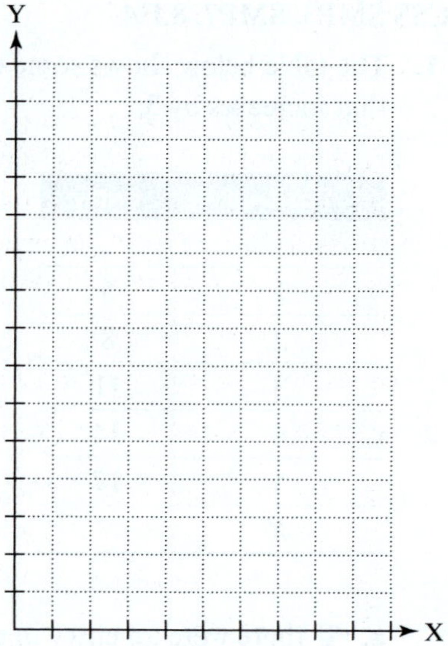

f. Discuss: How is part (b) related to going from point to point on the graph? How are the characteristics of your equation in part (d) related to your graph in part (e)?

2. Consider the arithmetic sequence whose first few entries are

$$2, 5, 8, 11, 14, \ldots$$

Let Y be the entry in position X. As in part 1, explain how to derive an equation relating X and Y. Graph some points and explain how characteristics of your equation are related to the graph and to the way the sequence grows.

9.6 Functions

Class Activity 9R What Does the Shape of a Graph Tell Us About a Function?

CCSS CCSS SMP2, SMP4, 8.F.2, 8.F.5

1. Items (a), (b), and (c) are hypothetical descriptions of a population of fish. Each description corresponds to a population function, for which the input is time elapsed since the fish population was first measured, and the output is the population of fish at that time. Match the descriptions of these population functions to the graphs and the tables. In each case, explain why the shape of the graph fits with the description of the function and the table for the function.

   a. The population of fish rose slowly at first, and then rose more and more rapidly.

   b. The population of fish rose rapidly at first, and then rose more and more slowly.

   c. The population of fish rose at a steady rate.

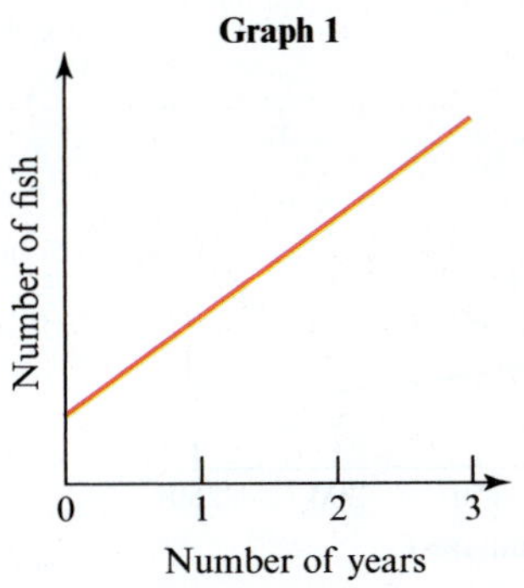

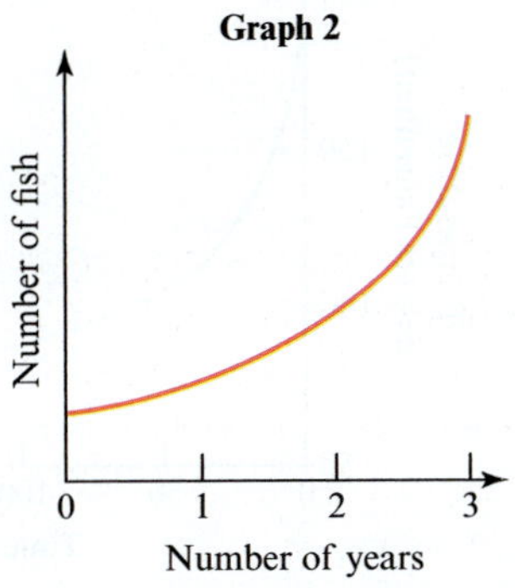

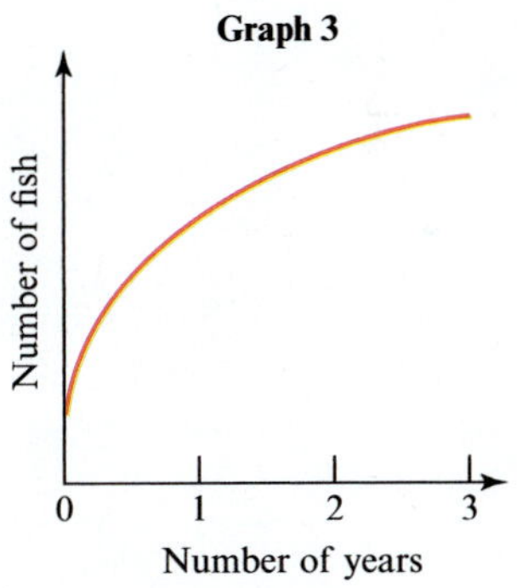

Table X

Years	Number of Fish
0	2000
1	8000
2	10,000
3	11,000

Table Y

Years	Number of Fish
0	2000
1	5000
2	8000
3	11,000

Table Z

Years	Number of Fish
0	2000
1	3000
2	5000
3	11,000

2. Hot water is poured into a mug and left to cool. This situation gives rise to a temperature function for which the input is the time elapsed since pouring the water into the mug and the output is the temperature of the water at that time. The graph of this function is one of the three graphs shown next. Which graph do you think it is, and why? For each graph, describe how water would cool according to that graph.

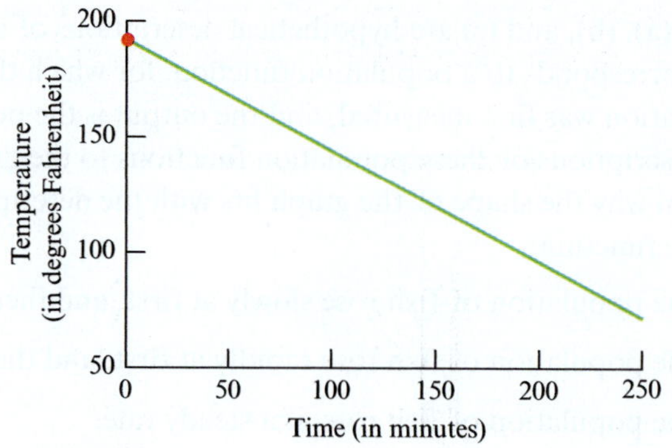

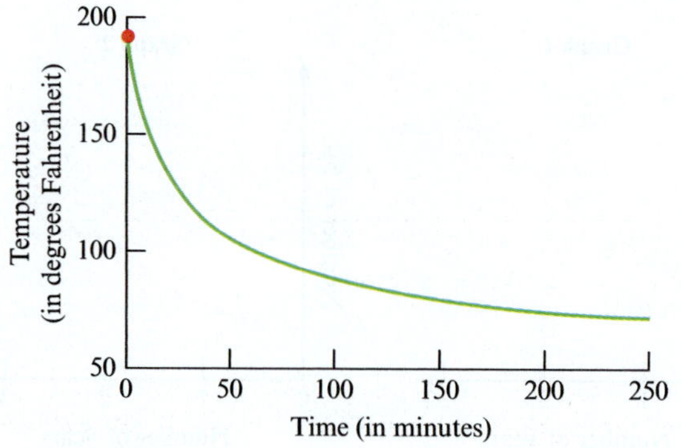

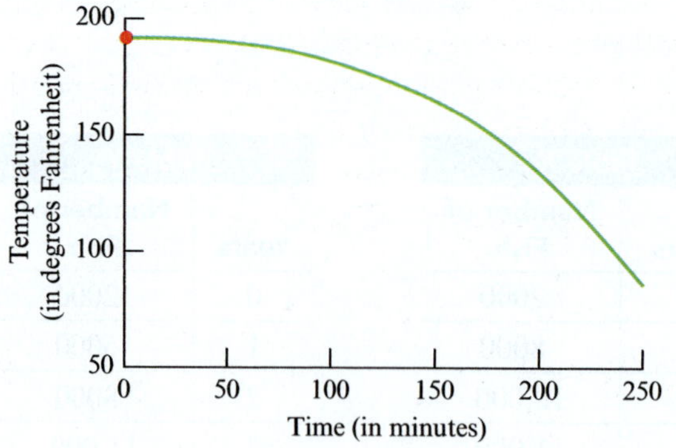

Which graph is the correct one for part 2? See page CA-186.

Class Activity 9S Graphs and Stories

CCSS CCSS SMP3, SMP4, 8.F.2, 8.F.5

1. A tagged manatee swims up a river, away from a dock. Meanwhile, the manatee's tag transmits its distance from the dock. This situation gives rise to a distance function for which the input is the time since the manatee first swam away from the dock and the output is the manatee's distance from the dock at that time. The graph of this distance function is shown below.

   Write a story about the manatee that fits with this graph. Explain how features of the graph fit with your story.

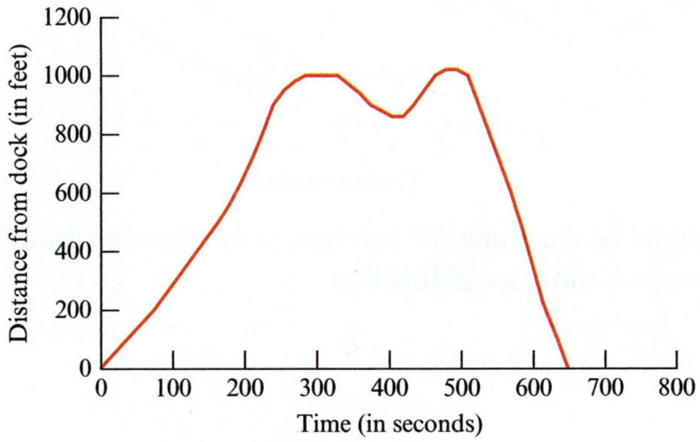

2. Carl started to drive from Providence to Boston, but after leaving he realized that he had forgotten something and drove back to Providence. Then Carl got back in his car and drove straight to Boston. This scenario gives rise to a distance function whose input is time elapsed since Carl first started to drive to Boston and whose output is Carl's distance from Providence. Could the next graph be the graph of the distance function described? Why or why not? If not, draw a different graph that could be the graph of the distance function. (Boston is 50 miles from Providence.)

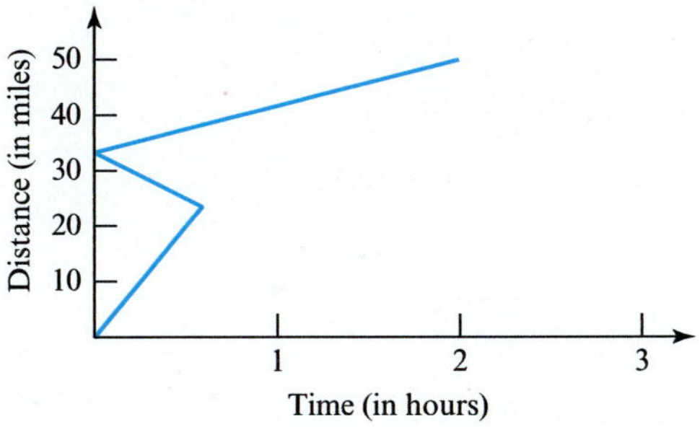

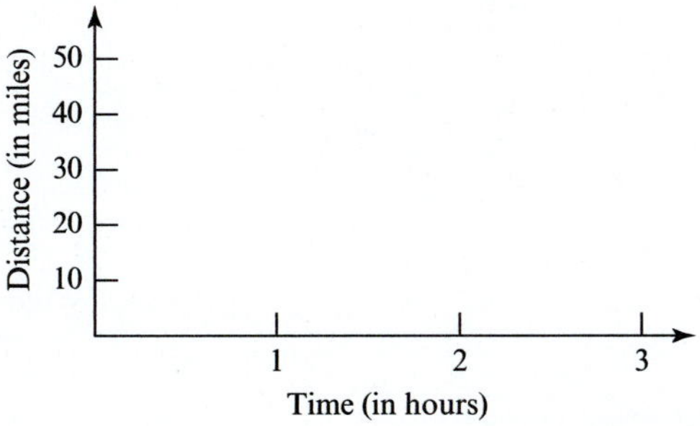

3. Here is what happened when Jenny ran a mile in 10 minutes. She got off to a good start, and ran faster and faster. Then all of a sudden, Jenny tripped. Once Jenny got back up, she started to run again, but at a slower pace. But near the end of her mile run, Jenny picked up some speed.

The next graph is supposed to fit with the story about Jenny's mile run. What is wrong with this graph? (Look for several errors.)

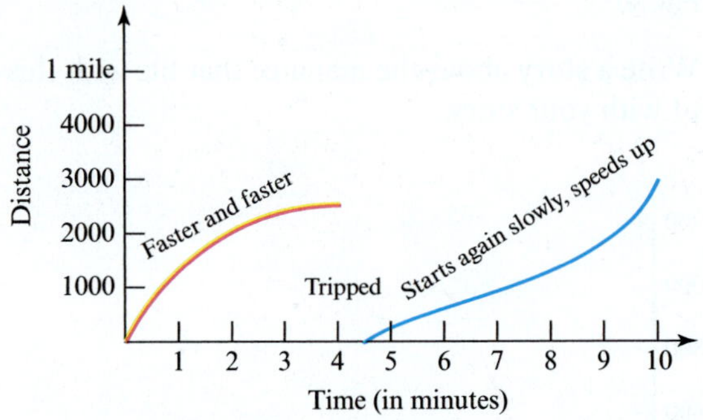

4. Sketch graphs that could be the graphs of two functions related to Jenny's mile run in part 3: a distance function and a speed function.

The correct graph on p. CA-184 is the second graph.

Class Activity 9T How Does Braking Distance Depend on Speed?

CCSS CCSS SMP4, SMP7, 8.F.2, 8.F.5

If you step on the brakes when you are driving, how far will you go until you come to a complete stop? The distance depends on the speed at which you are driving. Consider a *braking distance function*, which has independent variable s, the speed in miles per hour at which the car is traveling when the brakes are applied, and dependent variable d, the distance in feet which the car travels until it comes to a complete stop. (Note that we are not taking reaction time into account.)

1. Examine the graphs below and make an educated guess: Which of these graphs do you think a graph of the braking distance function will look like? Why? Which cannot possibly be the graph of the braking distance function? Why not?

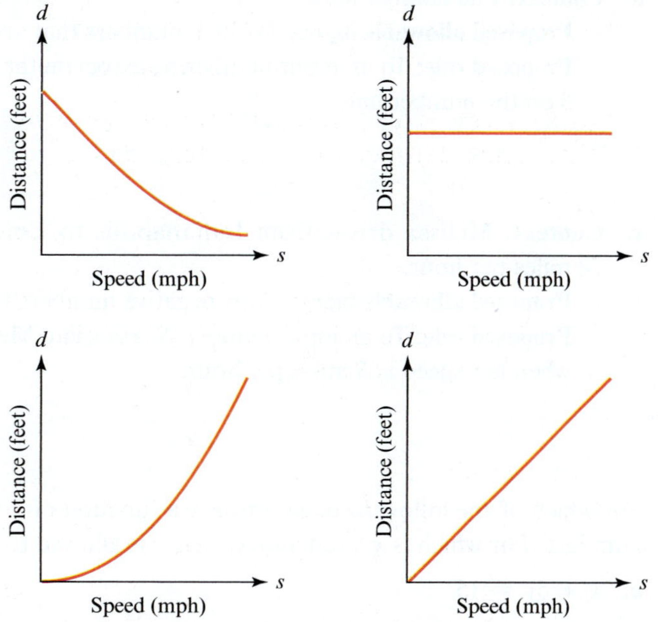

2. The braking distance function has the following property: if you double your speed, your braking distance will multiply by 4.

   a. Based on the property, which graph in part (a) represents the braking distance function? Explain how you can tell by reasoning about the graphs.

   b. Based on the property, which of the following equations could represent the braking distance function? Explain how you can tell by reasoning about the equations.

      i. $4d = 2s$

      ii. $d = 2s$

      iii. $d = \frac{1}{20}s^2$

Class Activity 9U Is It a Function?

CCSS CCSS SMP6, 8.F.1

Remember that a function is a rule that assigns one output to each allowable input.

1. Explain why each of the next proposed functions is *not* a function. Then modify the allowable inputs and rule so as to describe a function that is relevant to the given context.

   a. Context: Shopping at a local store.
   Proposed allowable inputs: Whole numbers between 0 and 100.
   Proposed rule: To an input number, N, associate the total cost of N items at the local store.

   b. Context: The number line.
   Proposed allowable inputs: Positive numbers that are less than 3.
   Proposed rule: To an input number, x, associate the number that is x units away from 5 on the number line.

   c. Context: Melissa drives from Indianapolis to Louisville. Her speed never exceeds 78 miles per hour.
   Proposed allowable inputs: Non-negative numbers that are less than 78.
   Proposed rule: To an input number, S, associate Melissa's distance from Indianapolis when her speed is S miles per hour.

2. For which of the following equations is y a function of x if x is allowed to vary over all real numbers? For which is x a function of y if y is allowed to vary over all real numbers?

   a. $x + 5y = 15$

   b. $x^2 + y = 100$

   c. $x^2 + y^2 = 100$

   d. $2 \cdot |x| = |y|$

3. For each example, determine if there could be a function that has those inputs and associated outputs. If so, describe a rule for such a function; if not, explain why not.

Example 1
Input: 1 → Output: 1
Input: 1.98 → Output: 1
Input: 2 → Output: 1
Input: 2.33 → Output: 1
Input: 2.7 → Output: 1
Input: 3.4 → Output: 1
Input: 3.6 → Output: 1

Example 2
Input: 1 → Output: 1
Input: 1.98 → Output: 2
Input: 2 → Output: 2
Input: 2.33 → Output: 2
Input: 2.7 → Output: 3
Input: 3.4 → Output: 3
Input: 3.6 → Output: 4

Example 3
Input: 1 → Output: 1
Input: 1.98 → Output: 1
Input: 2 → Output: 2
Input: 2.33 → Output: 2
Input: 2.7 → Output: 2
Input: 3.4 → Output: 3
Input: 3.6 → Output: 3

Example 4
Input: 1 → Output: 1
Input: 1 → Output: 1.98
Input: 2 → Output: 2
Input: 2 → Output: 2.33
Input: 2 → Output: 2.7
Input: 3 → Output: 3.4
Input: 3 → Output: 3.6

9.7 Linear and Other Relationships

Class Activity 9V Modeling Linear Relationships with Variables and Equations

CCSS CCSS SMP4, SMP8, 8.F.4

Remember these points when using variables:

- Variables stand for numbers; they are not labels.
- Define variables for quantities with care—for example, by saying "let x be the number of . . . ."

1. To make concrete, you need 3 times as much sand as cement. When Aaron was asked to formulate an equation about concrete, here's what he wrote:

   S = sand, C = cement

   $3S = C$

   Discuss Aaron's work. How could he revise it? Make a math drawing to help you explain.

2. For each of the following, make a table to show how two quantities in the situation vary together. Then define two variables and write and explain an equation to show how the variables are related.

   a. A mail order bead company sells beads for $10 per pound. The shipping is $7 for any amount of beads.

   b. The bead company earns $4 for each pound of beads it sells. The company has weekly expenses of $500 (no matter how many beads it sells).

3. The bead company got 150 pounds of turquoise beads and figures it will sell 8 pounds of turquoise beads every day.

Examine the two different ways that students made tables to relate quantities.

Work of Student 1	
Number of Days Elapsed	Number of Pounds Left
1	$150 - 8 = 142$
2	$142 - 8 = 134$
3	$134 - 8 = 126$
4	$126 - 8 = 118$
5	$118 - 8 = 110$

Work of Student 2	
Number of Days Elapsed	Number of Pounds Left
1	$150 - 8$
2	$150 - 2 \cdot 8$
3	$150 - 3 \cdot 8$
4	$150 - 4 \cdot 8$
5	$150 - 5 \cdot 8$

Discuss the two students' work. Which will be more useful for formulating an equation in two variables? Why?

4. Another bead company sells beads for \$12 for every 5 pounds of beads. The shipping is \$4 for any amount of beads.

Make a table to show how two quantities in this situation vary together. Then define two variables and write and explain an equation to show how the variables are related.

Class Activity 9W Interpreting Equations for Linear Relationships

CCSS CCSS SMP2, SMP4, 8.F.4

1. At the yogurt store, the cost of a cone of frozen yogurt depends on how much yogurt is in the cone. The cost and the amount of yogurt are related by the equation

$$C = 0.55Y + 1.15$$

where C is the cost in dollars and Y is the number of ounces of yogurt in the cone. Explain how to interpret 0.55 and 1.15 in terms of the situation.

2. A phone company charges \$0.50 for a 5-minute phone call plus an additional \$0.06 for each additional minute after that. Some students are trying to write equations to describe the relationship between the cost and the number of minutes of a phone call.

Discuss the following students' ideas and questions. Include clarifications or modifications as part of your discussion.

a. Niles writes the equation $C = 0.06(T - 5) + 0.50$.

b. Chad asks what the 0.06 multiplied by $T - 5$ stands for.

c. Francine wants to know what happens if T is 4. Does the equation work?

d. Anja writes the equation $C = 0.06T + 0.50$.

e. Quowanna wants to know how Niles and Anja defined their variables. She wonders if they defined T differently.

3. A grocer will spend \$150 on beans and tomatoes combined. Beans cost \$2 per pound and tomatoes cost \$3 per pound. Let B be a number of pounds of beans and T the corresponding number of pounds of tomatoes the grocer could buy.

Discuss and elaborate on the following students' ideas.

a. Mariah says she has a simple equation but it's not in the form $B =$ (some expression) or $T =$ (some expression).

b. Chris tries to formulate an equation in the form

$$T = 50 - (\text{something}) \cdot B$$

He found the 50 by reasoning that if the grocer buys no beans, he can buy 50 pounds of tomatoes.

c. DeShun says that for every additional 3 pounds of beans, the grocer must buy 2 fewer pounds of tomatoes, and she thinks this can help Chris find his equation.

4. Initially, a tank had 80 gallons of water in it when water started flowing out of it. Let W be the number of gallons of water left in the tank T seconds after water started flowing out of the tank. Suppose that W and T are related by the equation

$$W = 80 - T$$

Some students objected to this equation saying that the 80 is gallons and the T is seconds and it doesn't make sense to subtract seconds from gallons. Discuss!

Class Activity 9X Inversely Proportional Relationships Versus Decreasing Linear Relationships

CCSS CCSS SMP4, 8.F.3

Consider these two relationships:

Relationship 1: There are a number of hoses, all of the same size. Water flows out of all of these hoses at the same constant rate. Using 4 hoses, it takes 6 hours to fill an empty tub with water. Let x be the number of hoses used to fill the empty tub and let y be the number of hours it takes to fill the tub using that many hoses.

Relationship 2: Initially, there are 100 liters of water in a tub. Then water starts to flow out of the tub at a constant rate. After 20 minutes the tub is empty. Let x be the number of minutes since water starts flowing out of the tub and let y be the number of liters of water in the tub at that time.

1. In both relationship 1 and relationship 2, as x increases, what happens to y?

2. Make a table, draw a graph, and write an equation for Relationship 1 and for Relationship 2. What kinds of relationships are they and how can you tell?

Class Activity 9Y Is It Linear or Not?

CCSS CCSS SMP4, 8.F.3, 8.F.4

1. A group is throwing a benefit concert to raise money for a charity. Let x be the number of tickets the group might sell and let y be the net amount of money the group will raise if they sell that many tickets.

x	y
0	−500
20	−100
50	500
100	1500
200	3500

a. Could the relationship be linear or not? How can you tell?

b. Discuss the entries in the table that have a negative y-coordinate. Interpret these entries in terms of the scenario.

c. How much are tickets being sold for? How can you tell?

d. Where should the graph cross the x-axis and what is the significance of this?

2. A marble is dropped from the top of a tall building. Let x be the number of seconds elapsed since the marble was dropped and let y be the number of feet that the marble is above the ground at that time.

x	y
0	256
1	240
2	192
3	112
4	0

a. Could the relationship be linear or not? How can you tell?

b. How tall is the building? When does the marble hit the ground?

c. Describe how the marble falls.

d. What do you notice about how y changes as x increases by 1? Do you notice any pattern?

e. Find the changes in the changes in the y values in the table. What do you notice?

3. A bank account was opened and some money was put into it. Let x be the number of years since the money was put in the account and let y be the amount of money in the account.

x	y
0	100
5	200
10	400
15	800
20	1600

a. Could the relationship be linear or not? How can you tell?

b. How much money was put into the account initially?

c. Describe how the amount of money grows.

Class Activity 9Z What Kind of Relationship Is It?

CCSS CCSS SMP4

For each of the tables below, determine what kind of relationship the table exhibits and explain how you can tell. You do not need to find equations for the relationships.

Table A

x	y
1	3
2	8
3	15
4	24
5	35

Table B

x	y
1	7
3	10
7	16
13	25
21	37

Table C

x	y
1	3
2	6
4	12
8	24
16	48

Table D

x	y
1	3
2	6
3	12
4	24
5	48

Table E

x	y
1	25
2	20
3	15
4	10
5	5

Table F

x	y
1	60
2	30
3	20
4	15
5	12

Class Activity 9AA Doing Rocket Science by Reasoning About the Structure of Quadratic Equations

CCSS CCSS SMP7

1. Rocket scientists determined that t seconds after launch, a rocket's height, h, in feet above the ground will be given by the equation

$$h = 16(2 + t)(22 - t)$$

   a. Describe the structure of the expression on the right of the equal sign.

   b. Reason about the structure to determine which values for t will make the expression have a value of 0. To help your thinking: when you multiply numbers, how can the result be 0?

   c. Why would the rocket scientists want to know when the expression has a value of 0?

2. The rocket scientists determined that another equation for the height of the rocket is

$$h = 2304 - 16(t - 10)^2$$

a. Describe the structure of the expression on the right side of the equal sign.

b. Reason about the structure of $2304 - 16(t - 10)^2$ to determine the largest value it can have and to determine the value of t at which this occurs.

To help your thinking, consider these questions:

- Why can $16(t - 10)^2$ never be negative?
- How can you use the structure of $16(t - 10)^2$ to determine for which value of t it is 0?

c. Why would the rocket scientists want to know the largest value of the expression?

Class Activity 9BB Reasoning About the Structure of Quadratic Equations

CCSS CCSS SMP7

A company that sells Gizmos knows that its annual profit depends on the price it sells its Gizmos for. As the company changes the price of Gizmos, its profit changes.

The company has found three ways to write an equation for a profit function (by writing equivalent expressions). In each case, the independent variable, x, is the price of a Gizmo in cents and the dependent variable, y, is the company's annual profit in thousands of dollars.

1. Explain how to reason about the structure of Equation 1 to tell you when the company's profit is 0.

$$y = 2(x - 10)(80 - x) \qquad (1)$$

2. Explain how to reason about the structure of Equation 2 to tell you how the company can make its maximum profit.

$$y = 2450 - 2(x - 45)^2 \qquad (2)$$

3. Explain how to use Equation 3 to tell you what would happen if the company gave Gizmos away for free.

$$y = -2x^2 + 180x - 1600 \qquad (3)$$

4. What do parts 1, 2, and 3 tell you about the coordinates of points A, B, C, D, and E in the graph of the profit function?

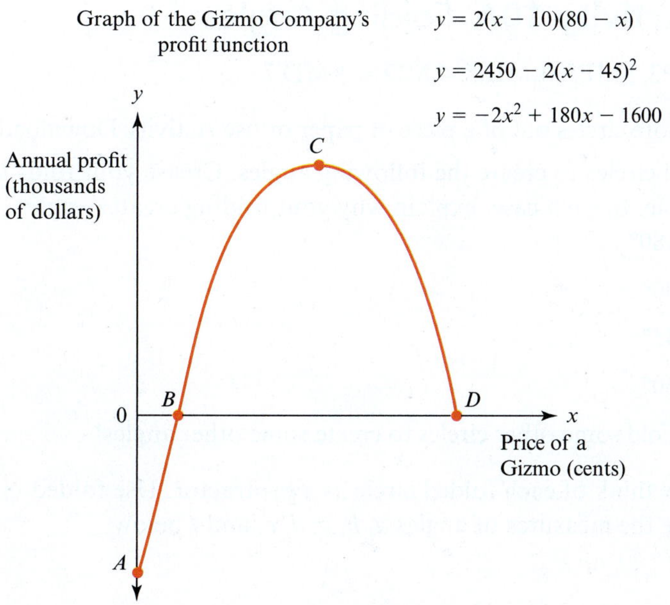

Geometry

10.1 Lines and Angles

Class Activity 10A Folding Angles

CCSS CCSS SMP3, SMP5, 4.MD.5, 4.MD.6, 4.MD.7

Cut 6 or more circles out of a piece of paper or use Activity Download 5.

1. Fold circles to create the following angles. Crease your folds to make the angles clearly visible. In each case, explain why your folding creates angles of that size.
   - **a.** 180°
   - **b.** 90°
   - **c.** 45°
   - **d.** 60°
   - **e.** Fold some other circles to create some other angles!

2. Now think of each folded circle as a protractor. Use folded-circle-protractors to determine the measures of angles *a*, *b*, *c*, *d*, *e*, and *f* below.

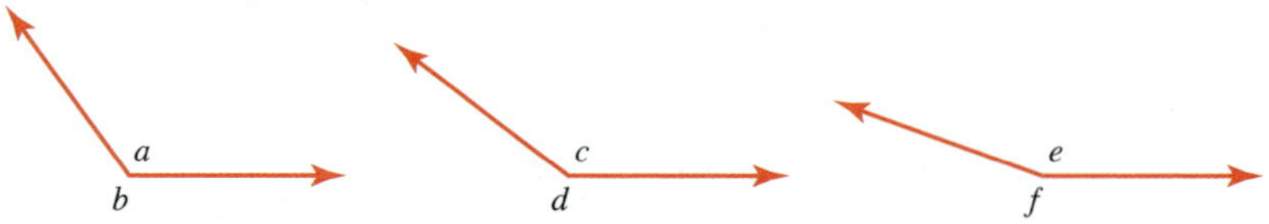

Class Activity 10B Angles Formed by Two Lines

CCSS CCSS SMP7, 7.G.5

1. The figure below shows 3 pairs of lines meeting (or you may wish to think of this as showing 1 pair of lines in three situations, when the lines are moved to different positions). In each case, how do opposite angles *a* and *c* appear to be related, and how do opposite angles *b* and *d* appear to be related?

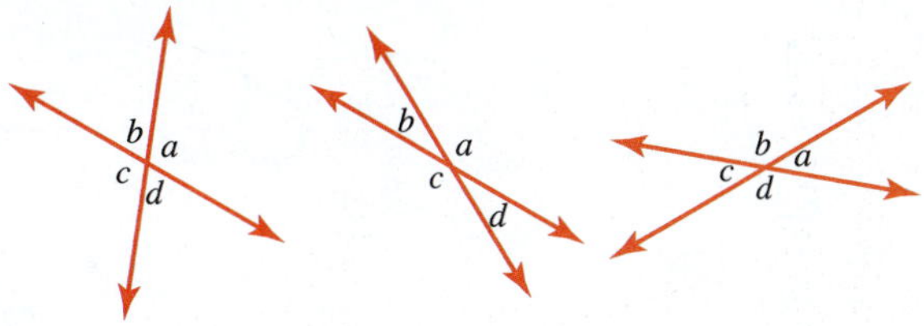

2. Do you think the same phenomenon you observed about opposite angles in part 1 will hold for *any* pair of lines meeting at a point? Do you have a convincing reason why or why not?

3. Explain *why* what you observed in part 1 about opposite angles must always be true by using the fact that an angle formed by a straight line is 180°. What does this fact tell you about several pairs of angles?

Class Activity 10C Angles Formed When a Line Crosses Two Parallel Lines

CCSS CCSS SMP3, 8.G.5

A pair of lines marked with arrows indicates that the lines are parallel.

Given 2 parallel lines and a transversal line that crosses the 2 parallel lines, as shown below, the Parallel Postulate says that $a = a'$, $b = b'$, $c = c'$, and $d = d'$.

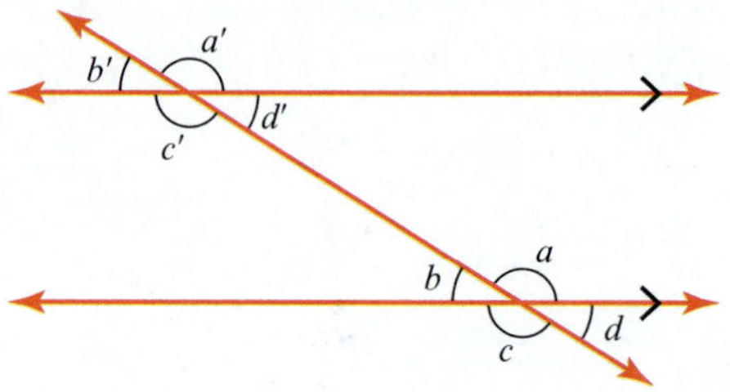

1. Given that lines m and n are parallel, use the Parallel Postulate and what we know about opposite angles to explain why $e = f$ (alternate interior angles are equal).

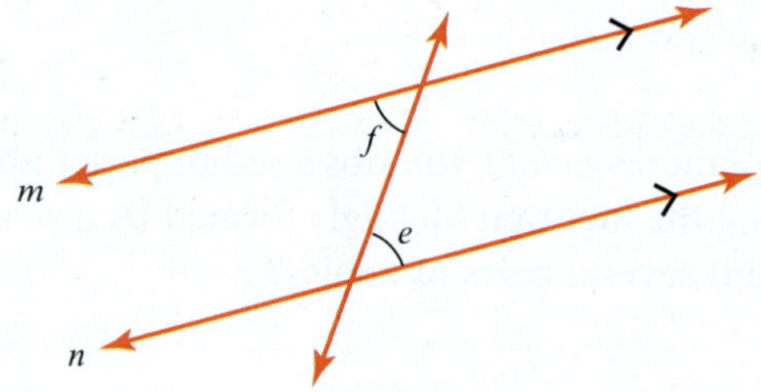

2. Given that lines p and q are parallel, use the Parallel Postulate and what we know about opposite angles to explain why $g + h = 180°$.

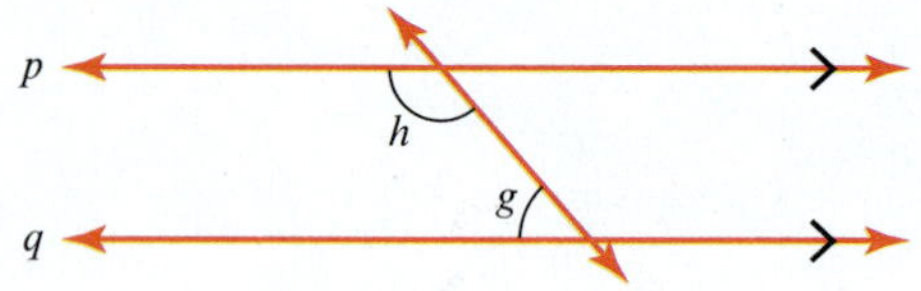

Class Activity 10D How Are the Angles in a Triangle Related?

CCSS CCSS SMP3, 8.G.5

You will need a ruler and scissors for this activity.

Work with a group of people. Each person in your group should do the following:

1. Using a ruler, draw a large triangle that looks different from the triangles of other group members. Cut out your triangle. Label the three angles a, b, and c.
2. Tear or cut all 3 corners off your triangle. Then put the angles together vertex to vertex, without overlaps or gaps. What do you notice? What does this show you about the angles in the triangle?

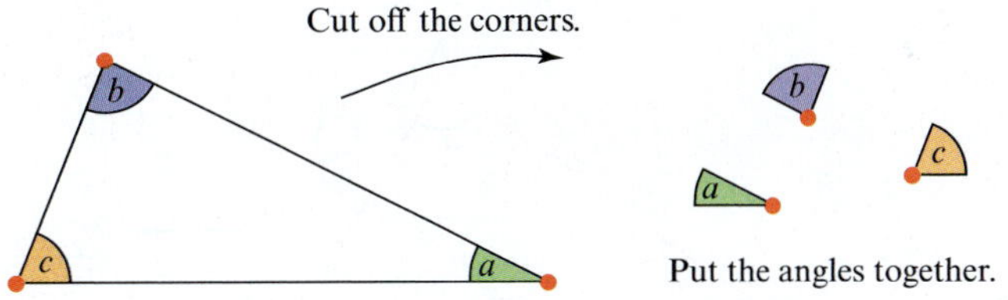

3. Discuss why the "putting the angles together" method of part 2 is not a proof. Consider these points:
   - Does the method show *exactly* how the angles are related?
   - Has every triangle been considered?

Class Activity 10E Drawing a Parallel Line to Prove That the Angles in a Triangle Add to 180°

CCSS CCSS SMP3, 8.G.5

This activity will show you a way to prove that the angles in a triangle add to 180°.

1. Given any triangle with (corner) points A, B, and C, let a, b, and c be the angles of the triangle at A, B, and C, respectively. Consider the line through A parallel to the side BC that is opposite A.

   What can you say about the 3 adjacent angles at A that are formed by the triangle and the line through A?

   What can you conclude about the sum of the angles in the triangle?

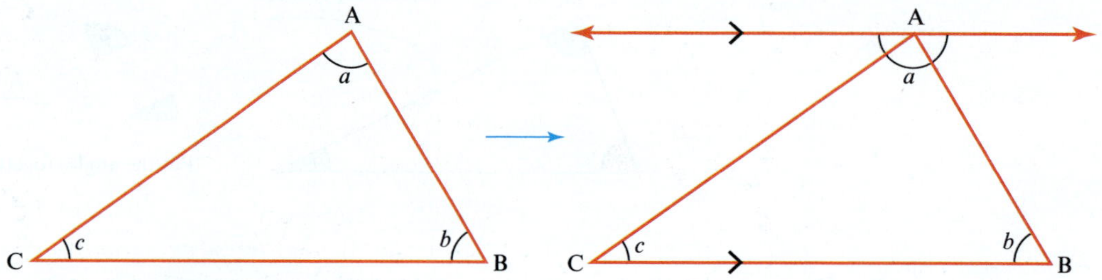

Draw a line parallel to BC through A.

2. What if you used a different triangle in part 1? Would you still reach the same conclusion?

Class Activity 10F Walking and Turning to Explain Relationships Among Exterior and Interior Angles of Triangles

CCSS CCSS SMP3, 8.G.5

You will need sticky notes and, if available, masking tape.

This activity will show you a way to understand why the angles in a triangle add to 180°. It is best done as a demonstration for the whole class.

1. Put 3 "dots" (sticky notes) labeled A, B, and C on the floor to create a triangle. If possible, connect them with masking tape. Label a point P on the line segment between A and B.
2. Choose two people: one to be a *walker* and one to be a *turner*. The rest are *observers*.

   The walker's job: Stand at point P, facing point A. Walk all the way around the triangle, returning to point P.

   The turner's job: Stand at one fixed spot, and face the same direction that the walker faces at all times. This means that when the walker turns at a corner, you should turn in the same way.

   The observers' job: Observe the walker and the turner, and make sure that they really are facing the same direction at all times.

   Repeat the walking and turning described above until everyone can confidently answer the following questions:

   a. Let's say that the walker and turner were facing north when the walker began walking around the triangle. Which directions did the turner face during the experiment? Were any directions left out? Were any directions repeated?

   b. What was the full angle of rotation of the turner when the walker walked once all the way around the triangle, returning to point P?

3. On the drawing below, show which angles the walker turned through at the corners of the triangle. Label these angles *d*, *e*, and *f*.

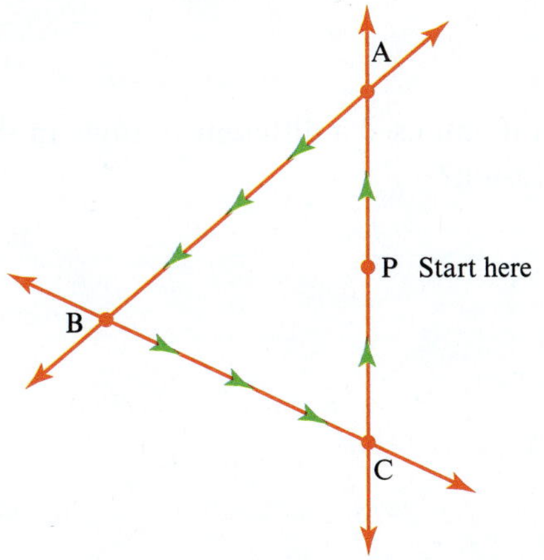

4. Based on your answer to part 2(b), what can you say about the value of $d + e + f$?

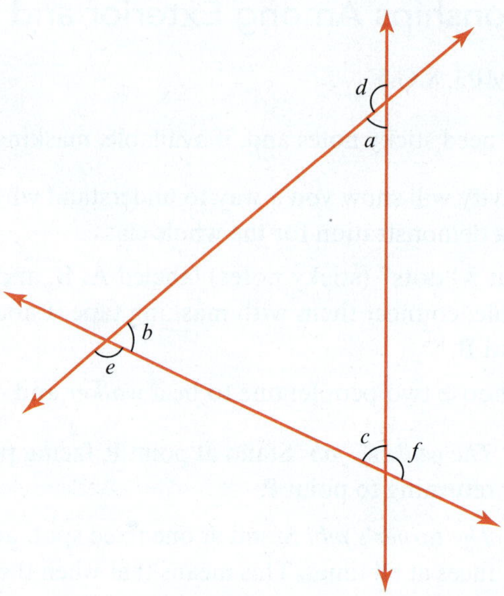

5. Check your answer to the previous part by using a protractor to find the angles d, e, f and adding them.

What if you used a different triangle? The values of d, e, and f might be different, but what about $d + e + f$?

6. You should have just found that the sum of the exterior angles of a triangle is 360°. In other words, $d + e + f = 360°$, where d, e, and f are the exterior angles, as shown in the figure above.

Use the formula $d + e + f = 360°$ to explain why the sum of the interior angles in a triangle is equal to 180°. In other words, show that $a + b + c = 180°$, where a, b, and c are the interior angles of a triangle, as shown in the figure.

Hint: What do you notice about $a + d$, $b + e$, and $c + f$? Can you use this somehow?

7. What if you used a different triangle in this activity? Would you still reach the same conclusion?

Class Activity 10G Angle Problems

CCSS CCSS SMP1

In some (but not all) of these problems, it will be helpful to add or to extend one or more lines or line segments.

1. Determine a formula for angle x in terms of angles a and b. Explain why your formula is valid (without measuring any angles).

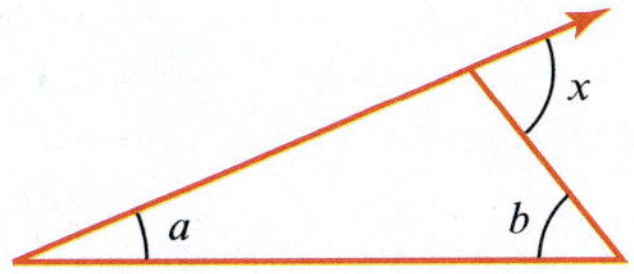

2. Given that the lines marked with arrows are parallel, determine the sum of the angles $a + b + c + d$. Explain why your answer is valid (without measuring any angles). See if you can find more than one explanation!

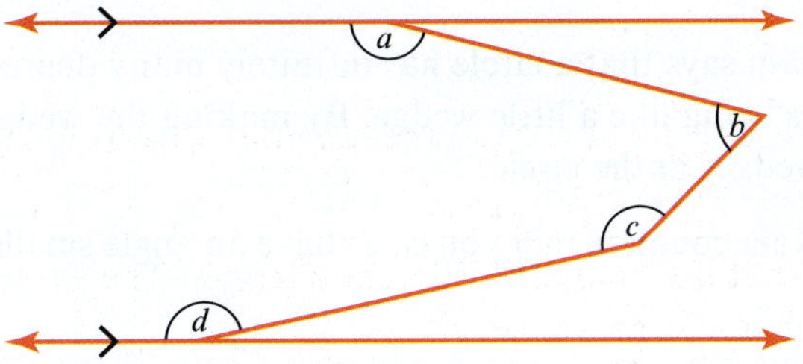

3. Determine the sum of the angles, $a + b + c + d + e$. Explain why your answer is valid (without measuring any angles). See if you can find two explanations!

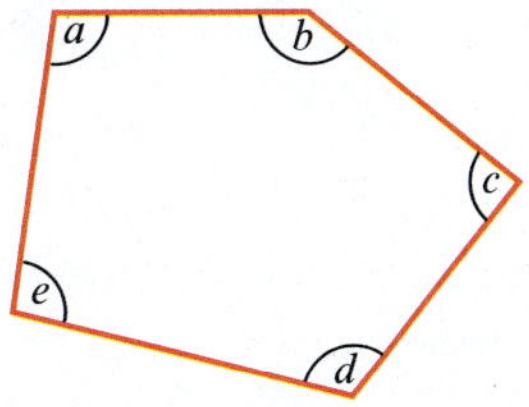

4. In part 3, what if the shape had N angles instead of 5 angles? Find a formula for the sum of the angles in terms of N and explain your reasoning.

Class Activity 10H Students' Ideas and Questions About Angles

CCSS CCSS SMP3

Discuss the following ideas and questions that students had about angles.

1. Harry has learned that the angles in a triangle add to 180° and that there are 360° in a circle. Harry wonders about the drawing below: "Shouldn't the circle be less than 180° because it is inside the triangle?"

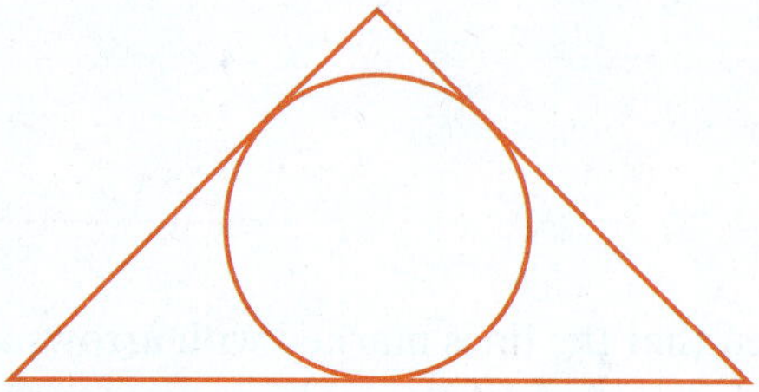

2. Sam says that a circle has infinitely many degrees because you can think of a degree as being like a little wedge. By making the wedges smaller, you can fit more and more wedges in the circle.

   Kaia counters that you can't have an angle smaller than 1°.

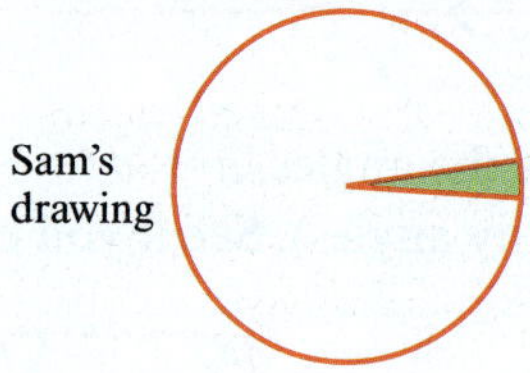

10.2 Angles and Phenomena in the World

Class Activity 10I Eratosthenes's Method for Determining the Circumference of the Earth

CCSS CCSS SMP4

The figure below shows a cross-section of the earth. At noon on June 21, the sun is directly overhead at location A, so that the sun's rays are perfectly vertical there. At the same time, 500 miles away at location B, the sun's rays make a 7.2° angle with the tip of a vertical pole (shown not to scale), which was determined by considering the shadow that the pole casts. Because the sun is far away, sun rays at the earth are (approximately) parallel. Use this information to determine the circumference of the earth, explaining your reasoning.

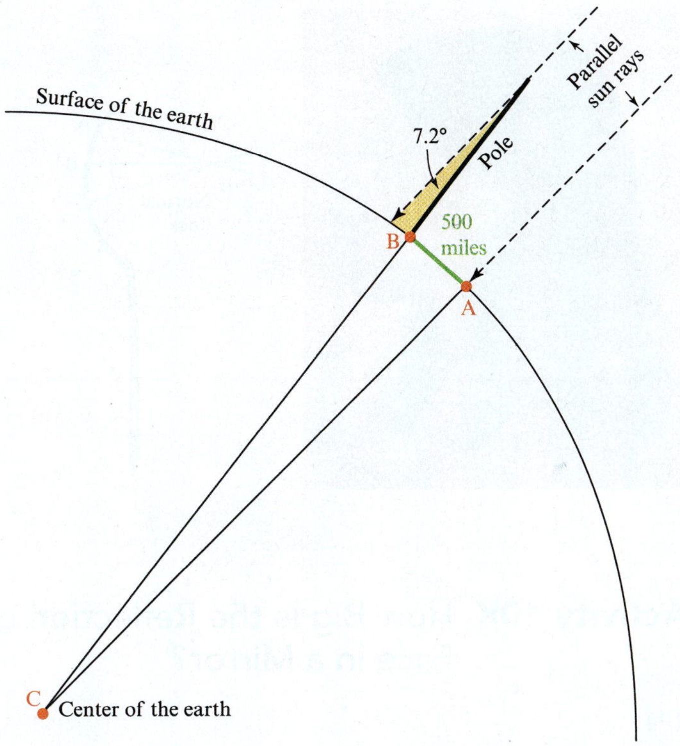

Class Activity 10J Why Do Spoons Reflect Upside Down?

CCSS CCSS SMP4

A large, reflective spoon would be helpful for this activity.

When you look at your reflection in the bowl of a spoon, you will notice that (in addition to looking quite distorted) your image will appear upside down. Have you ever wondered why? We will explore this now.

The figure below shows a person looking into the bowl of a (very large) spoon. Only a cross-section of the spoon is shown. The lines shown are the *normal lines* to the spoon at the points A, B, and C. Use these normal lines and the laws of reflection to determine what the person sees when she looks at points A, B, and C. (Assume that the person sees light that enters the center of her eye.)

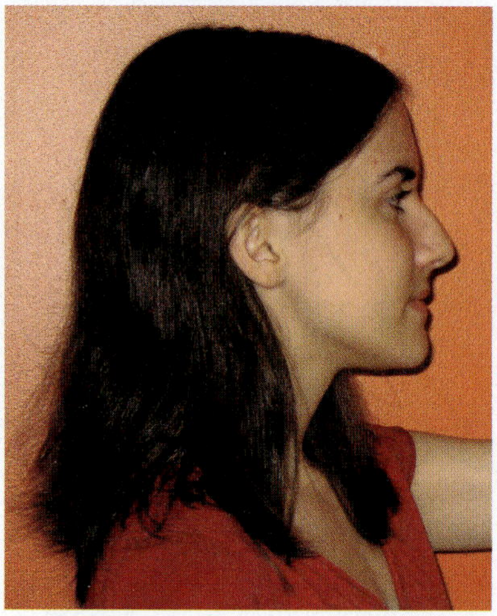

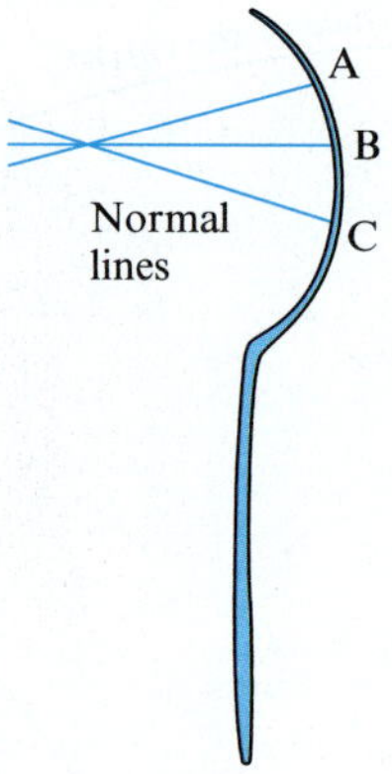

Class Activity 10K How Big Is the Reflection of Your Face in a Mirror?

CCSS CCSS SMP4

You will need a mirror and a ruler for the first part of this activity.

How big is the reflection of your face in a mirror? Is it the same size as your face, or is it larger or smaller, and if so, how much? The answers to these questions may surprise you.

1. Use a ruler to measure the length of your face from the top of your forehead to your chin. Now hold a mirror *parallel* to your face and measure the length of your face's reflection in the mirror, from the top of the forehead to the chin. Measure carefully!

   Hold the mirror closer to your face or farther away (but always *parallel* to your face), and repeat the measuring processes just described. The *position* of your reflection will probably change, but does the *size* of your reflection change or not? The answer may surprise you.

   Compare the length of your face and the length of your reflected face in the mirror. How do these lengths appear to be related?

2. The figures below show a side view of a person looking into a mirror. Using the laws of reflection, determine what the person will see at each of the points A, B, and C. We see objects by seeing the light that travels from the object to our eyes. So a person looking at a particular point sees the light that travels in a straight line from that point to the person's eye. To determine what a person sees at a point in a mirror, you must determine where the light at that point came from. For this you will need the laws of reflection.

3. Use the laws of reflection to show where the person looking into the mirror will see the top of her forehead and where she will see the bottom of her chin. Measure the length of the person's face and the length of her reflected face in the mirror. How do these lengths appear to be related? Your result should fit with what you discovered in part 1.

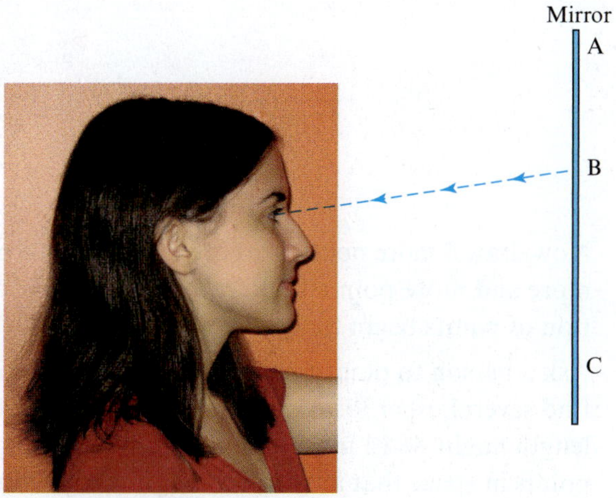

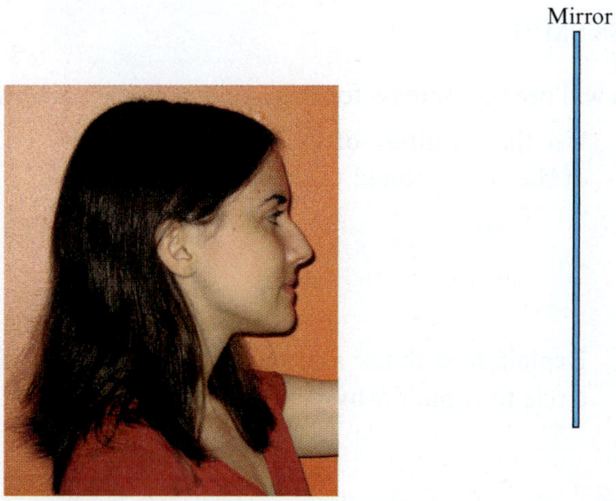

10.3 Circles and Spheres

Class Activity 10L Points That Are a Fixed Distance from a Given Point

CCSS CCSS SMP2

You will need a ruler for this activity.

1. Use a ruler to draw 5 different points that are 1 inch away from the point P:

   Now draw 5 more points that are 1 inch away from the point P. If you could keep drawing more and more points that are 1 inch away from point P, what shape would this collection of points begin to look like?

2. Ask a person to point to a particular point in space. Call that point P. Using a ruler, find several other locations in space that are 1 ruler-length away from point P. (A ruler-length might be 12 inches or 6 inches, depending on your ruler.) Try to visualize all the points in space that are 1 ruler-length away from your point P. What shape do you see?

Class Activity 10M Using Circles

CCSS CCSS SMP4

You will need a compass for this activity. Some string would also be useful.

1. Use the definition of a circle to explain why a compass draws a circle. How is the radius of the circle related to the compass?

2. Explain how to use a piece of string and a pencil to draw a circle. Use the definition of a circle to explain why your method will draw a circle.

3. Suppose that, 1 hour and 45 minutes ago, a prisoner escaped from a prison located at point P, shown on the map below. Due to the terrain and the fact that no vehicles have left the area, police estimate that the prisoner can go no faster than 4 miles per hour. Show all places on the map where the prisoner might be at this moment. Explain.

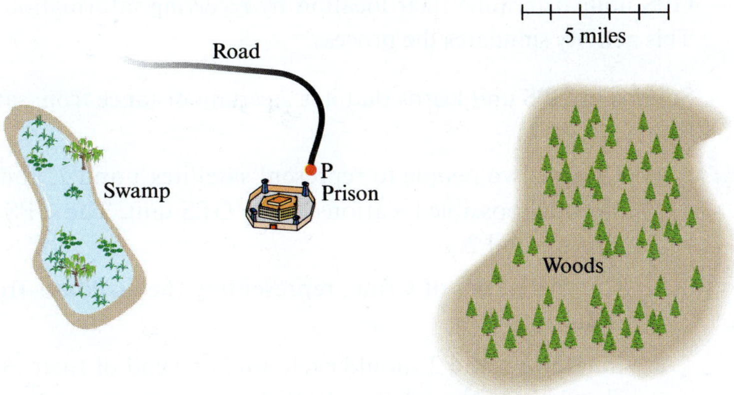

4. Some treasure is described as buried under a spot that is 30 feet from a spot marked X and 50 feet from a spot marked O. Use this information to help you show where the treasure might be buried on the map below. Is the information enough to tell you *exactly* where the treasure is buried? Explain.

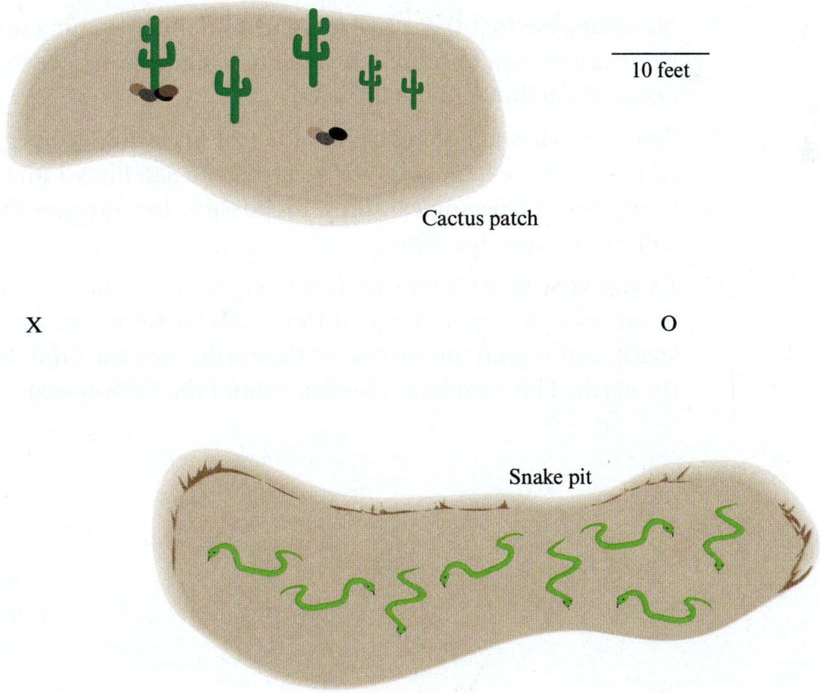

Class Activity 10N The Global Positioning System (GPS)

CCSS CCSS SMP4

You will need string for this activity. The activity is best as a whole-class demonstration.

GPS units determine their location by receiving information from satellites that orbit the earth. This activity simulates the process.

Suppose a GPS unit learns that it is a certain distance from satellite 1 and another certain distance from satellite 2.

1. Choose two people to represent satellites 1 and 2, and choose a third person who will show all possible locations of the GPS unit. The GPS person should stand between satellites 1 and 2.
2. Cut two pieces of string, representing the distances from satellites 1 and 2 to the GPS unit.
3. Satellites 1 and 2 should each hold one end of their piece of string, and the GPS person should hold the other ends of the two pieces of string in one hand, pulling the strings tight. (Everyone may have to adjust positions so that it is possible to do this. Once suitable positions are found, everyone should stay fixed in his or her position.)
4. The designated GPS person will now be able to show all possible locations of the GPS unit by moving the strings, while keeping them pulled tight and held in one hand (and while satellites 1 and 2 remain fixed in their positions).
5. Describe the shape of all possible locations of the GPS unit. How is this related to the intersection of two spheres? Are two satellites enough to determine a location?
6. Now suppose that there is also a third satellite beaming information to the GPS unit. Choose a person to represent satellite 3, and cut a piece of string to represent the distance of satellite 3 to the GPS unit.
7. Satellite 3 should hold one end of her or his string while the GPS person holds the other end in the same hand with the strings from satellites 1 and 2. By pulling all three strings tight, the GPS person can show all possible locations of the GPS unit. In general, there will be two such locations.

   As you've seen, with information from three satellites, a GPS unit can narrow its location to one of two points. If one of those two locations can be recognized as being in outer space, and not on the surface of the earth, then the GPS unit can report its location on the earth. This simulates the idea behind the GPS system.

Class Activity 10O Circle Designs

CCSS CCSS SMP5, SMP7

You will need various coins for part 1 and a compass for part 2.

1. Although the accompanying design looks complex, it is surprisingly easy to make. Use a compass (or a paper clip) to draw a design like the one below. Say briefly how to draw the design.

2. The next design was made by drawing part of the design of part 1. Use a compass (or a paper clip) to draw a design like this one. (Color it later if you like.)

3. To draw the designs below, use graph paper or draw square grids first. Then use a compass or paperclip. (Color the designs later if you like.)

10.4 Triangles, Quadrilaterals, and Other Polygons

Class Activity 10P What Shape Is It?

CCSS CCSS SMP6, K.G.2

Is it always obvious what type a shape is? Examine and discuss the examples below.

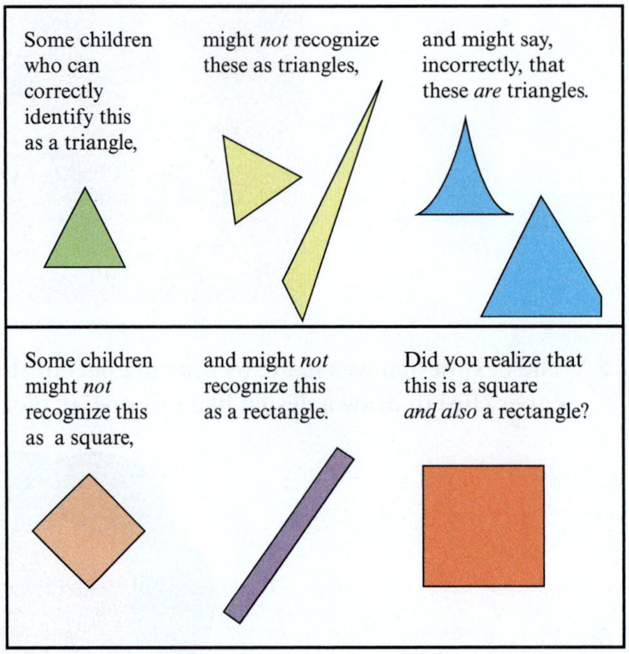

Now ponder this question: What would be a more reliable way to determine if a shape is or isn't of a specific type than simply observing what the shape looks like overall?

Class Activity 10Q What Properties Do These Shapes Have?

CCSS CCSS SMP7, 1.G.1

1. What properties or attributes do the shapes below have (or appear to have)? What do you notice about their sides? What do you notice about their angles? What other properties or attributes do you notice? List as many as you can find!

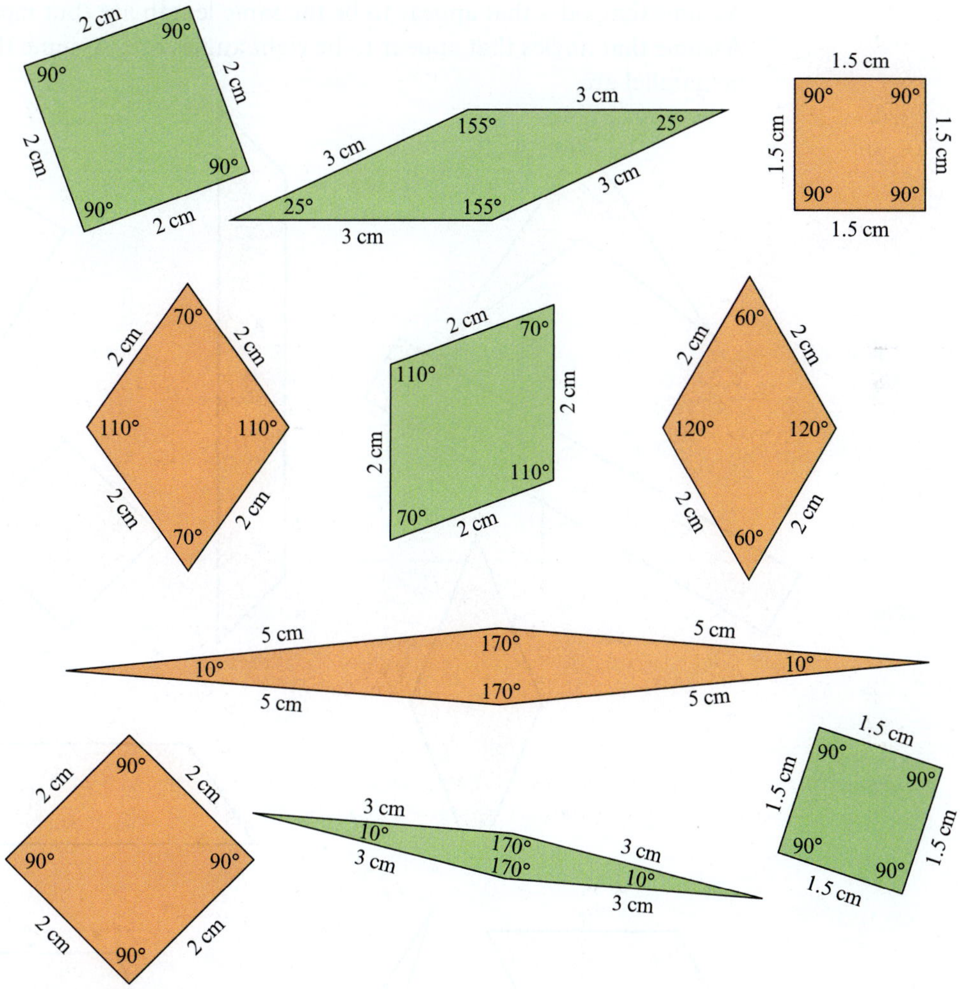

2. Which of the properties or attributes that you noticed in part 1 might be useful for deciding what type of shape they are? Which properties are *not* relevant for deciding what type of shape they are?

Class Activity 10R How Can We Classify Shapes into Categories Based on Their Properties?

CCSS SMP6, 2.G.1, 3.G.1, 4.G.2, 5.G.3, 5.G.4

1. For each category on the next page, identify *all* those shapes on this page that have *all* the attributes listed. Note that some shapes fit into more than one category and some shapes may not fit in any category.

   Assume that sides that appear to be the same length are (but measure if you aren't sure). Assume that angles that appear to be right angles are. Assume that sides that appear to be parallel are.

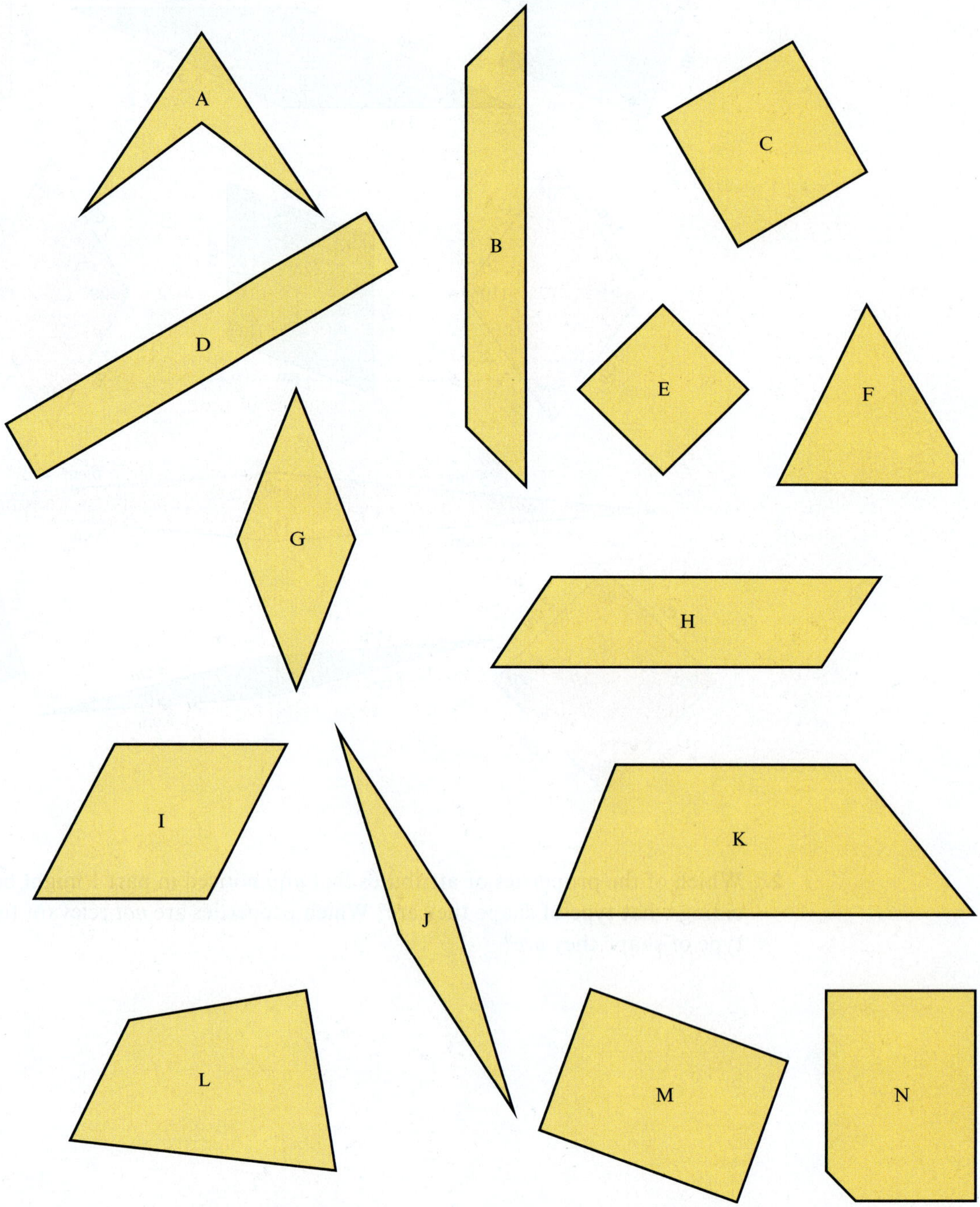

Category 1: 4 sides, 4 right angles, all 4 sides the same length

Category 2: 4 sides, 4 right angles	**Category 3:** 4 sides, 4 right angles, opposite sides parallel	**Category 4:** 4 sides, 4 right angles, opposite sides same length
Category 5: 4 sides, all the same length	**Category 6:** 4 sides, all the same length, opposite sides parallel	**Category 7:** 4 sides, all the same length, opposite angles same size

Category 8: 4 sides, opposite sides parallel	**Category 9:** 4 sides, opposite sides same length
Category 10: 4 sides, *at least* one pair parallel	**Category 11:** 4 sides, *exactly* one pair parallel

Category 12: 4 sides

2. Which categories in part 1 have exactly the same shapes? Try to draw a shape (using a ruler and protractor or geometry software) that will fit in one of those categories but not the other. Do you think it is possible?

For each category, provide its common name if you know it.

3. Which categories are a subcategory of another category? How can you tell that from lists of properties? Describe or draw a diagram to show how the categories in part 1 are related to each other.

Class Activity 10S How Can We Classify Triangles Based on Their Properties?

CCSS CCSS SMP5, SMP6, 4.G.2, 5.G.3, 5.G.4

1. For each category of triangles below, find all the triangles on the next page that have the given attribute.

Category 1: All 3 sides same length (equilateral triangles)	**Category 2:** All 3 angles same size

Category 3: At least 2 sides same length (isosceles triangles)	**Category 4:** At least 2 angles same size

Category 5: All angles are smaller than a right angle (acute triangles)

Category 6: Has a right angle (right triangles)

Category 7: Has an angle greater than a right angle (obtuse triangles)

2. Which categories in part 1 have exactly the same triangles? Try to draw (using a ruler and protractor or geometry software) a triangle that will fit in one of those categories but not the other. Do you think it is possible?

3. How are categories 1 and 3 related? Why?

4. How are categories 5, 6, and 7 related and how are these three categories related to the category of all triangles? Why?

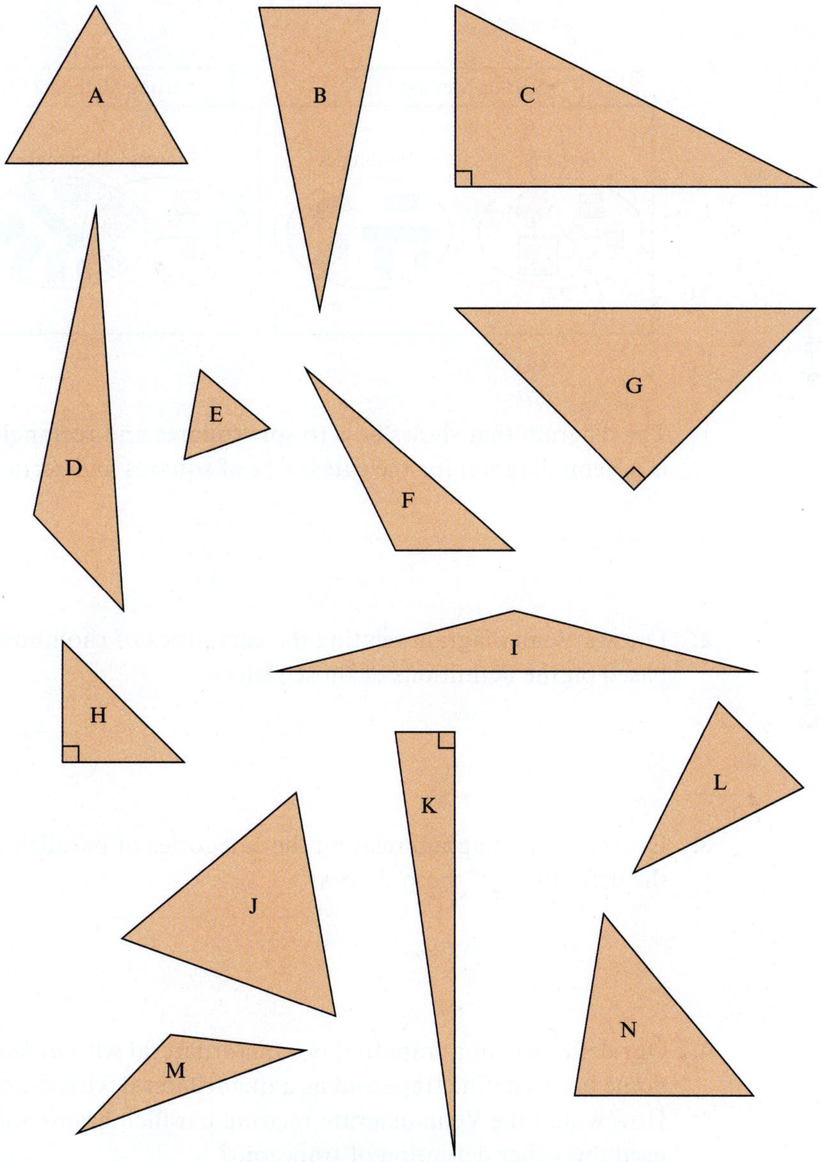

Note: Assume that sides that appear to be the same length are the same length.

Class Activity 10T Using Venn Diagrams to Relate Categories of Quadrilaterals

CCSS CCSS SMP6, SMP7, 5.G.4

1. Imagine that you have a large collection of plastic squares and rectangles in different sizes. You might sort those squares and rectangles in different ways.

   Which of the three ways of sorting the squares and rectangles shown below fits with the *definitions* of squares and rectangles? Explain!

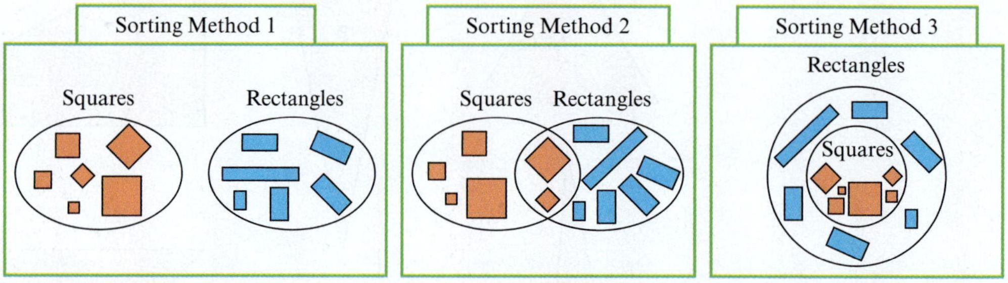

   The diagram that shows how to sort squares and rectangles *according to their definitions* is a Venn diagram for the categories of squares and rectangles.

2. Draw a Venn diagram relating the caregories of rhombuses, squares, and rectangles based on the definitions of those shapes.

3. Draw a Venn diagram relating the categories of parallelograms and trapezoids based on the definitions of those shapes.

4. Our definition of a trapezoid is a quadrilateral with *at least one* pair of parallel sides. Some books define trapezoid as a quadrilateral with *exactly one* pair of parallel sides. How would the Venn diagram relating parallelograms and trapezoids be different if we used this other definition of trapezoid?

Class Activity 10U Using a Compass to Construct Triangles and Quadrilaterals

CCSS CCSS SMP3, SMP5, 7.G.2

You will need a compass (for drawing circles) and ruler for this activity. Focus on the definition of a circle throughout the activity.

1. Use a compass to help you draw an isosceles triangle. Without measuring any side lengths, explain why your triangle must be isosceles.

2. Try to draw an equilateral triangle by using only a ruler and pencil, no compass. Why does this not work so well?

3. To make an equilateral triangle, follow the steps outlined in the figure below. Notice that you really need to draw only the top portions of the circles. Use this method to make examples of several different equilateral triangles.

   Explain *why* this method must always produce an equilateral triangle.

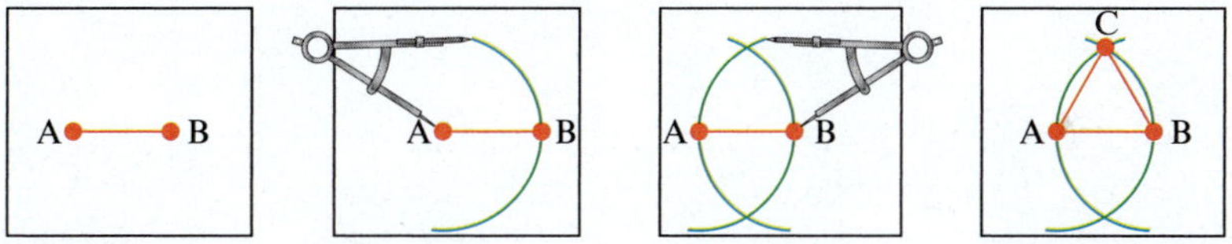

Step 1: Start with any line segment AB.

Step 2: Draw a circle centered at A, passing through B.

Step 3: Draw a circle centered at B, passing through A.

Step 4: Label one of the two points where the circles meet C. Connect A, B, and C with line segments.

4. In the figure below, the line segments AB and AC have the same length. Use a ruler and compass to draw a rhombus that has AB and AC as two of its sides. (You may wish to try drawing the rhombus without a compass first. Notice that it is difficult and clumsy to do so.)

Hint: To create the rhombus, you will need to construct a fourth point, D, such that the distance from D to B is equal to the distance from D to C, and such that these two distances are also equal to the common distance from A to B and A to C. Think about the *definition of circles* to help you figure out how to construct the point D.

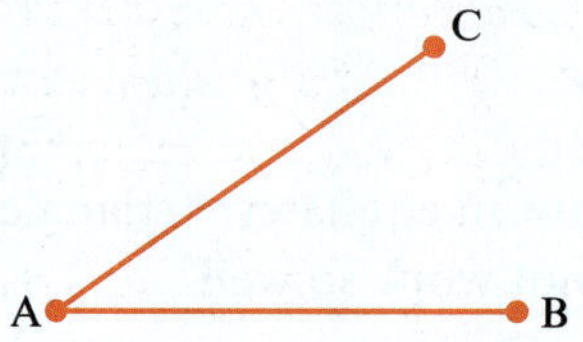

5. The line segment AB shown in the figure below is 4 inches long. Use a ruler and compass to construct a triangle that has AB as one of its sides, has a side that is 3 inches long, and has another side that is 2 inches long. Describe how you constructed your triangle, and explain why your construction must produce the desired triangle. *Hint*: Modify the construction of an equilateral triangle shown in part 3 by drawing circles of different radii.

6. Take a blank piece of paper. Use a ruler and compass to construct a triangle that has one side of length 6 inches, one side of length 5 inches, and one side of length 3 inches.

7. Is it possible to make a triangle that has one side of length 6 inches, one side of length 3 inches, and one side of length 2 inches? Explain.

Class Activity 10V Making Shapes by Folding Paper

CCSS CCSS SMP3, SMP5, 7.G.2

You will need paper, scissors, and a ruler for this activity.

1. To create an isosceles triangle, follow the next set of instructions.

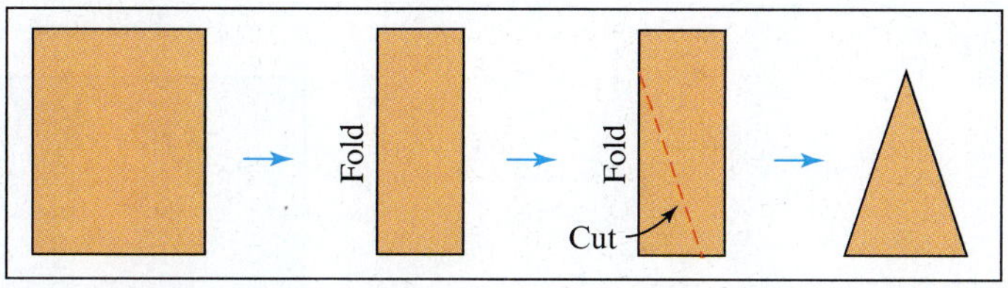

Fold a rectangular piece of paper in half. Then draw a line from the bottom edge to the folded side. Cut along this line. When you unfold, you will have an isosceles triangle.

   a. By referring to the definition of isosceles triangle, explain *why* the method described must always create an isosceles triangle.

   b. What properties does your isosceles triangle have? Find as many as you can. Explain if you can.

2. To create a rhombus, follow the next set of instructions.

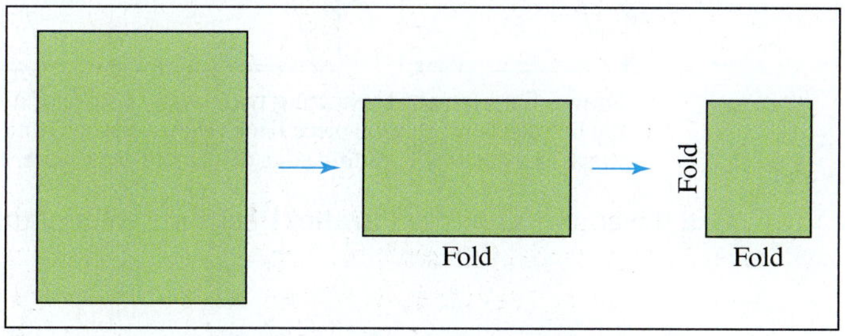

Step 1: Fold a rectangular piece of paper in half and then in half again, creating perpendicular fold lines.

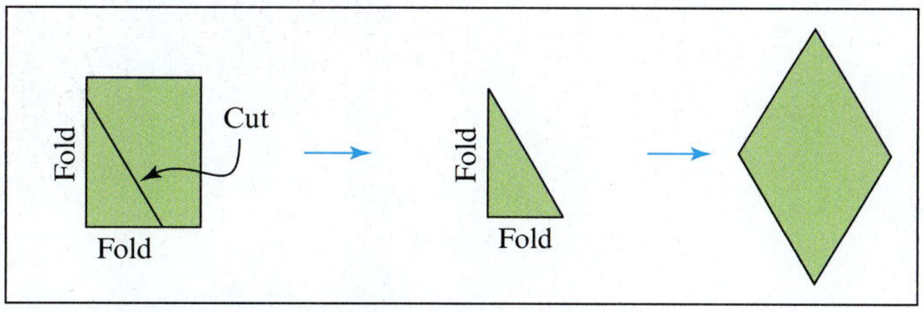

Step 2: Draw a line from anywhere on one fold to anywhere on the perpendicular fold. Cut along the line you drew. When you unfold, you will have a rhombus.

   a. By referring to the definition of rhombus, explain *why* the method described must always create a rhombus.

   b. What properties does your rhombus have? Find as many as you can. Explain if you can.

3. An ordinary rectangular piece of paper is one example of a rectangle. You can create other rectangles out of an ordinary piece of paper as follows: Fold the paper so that one edge of the paper folds directly onto itself. The opposite edge will automatically fold onto itself as well. Now unfold the paper and fold the paper again, this time so that the other two edges fold onto themselves. When you unfold, you can cut along the fold lines to create 4 rectangles.

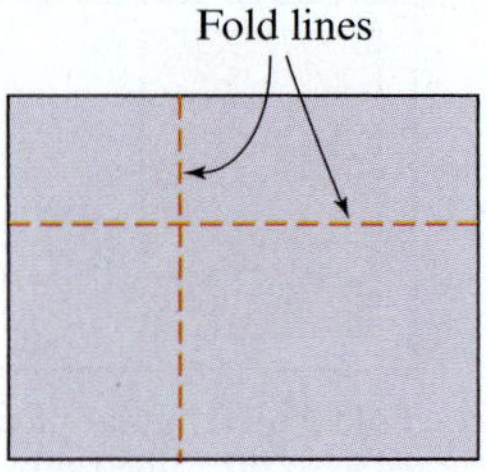

By referring to the definition of a rectangle, explain why the method just described must always create rectangles.

4. To create a parallelogram, start with a rectangular piece of paper of any size. Follow the next set of steps.

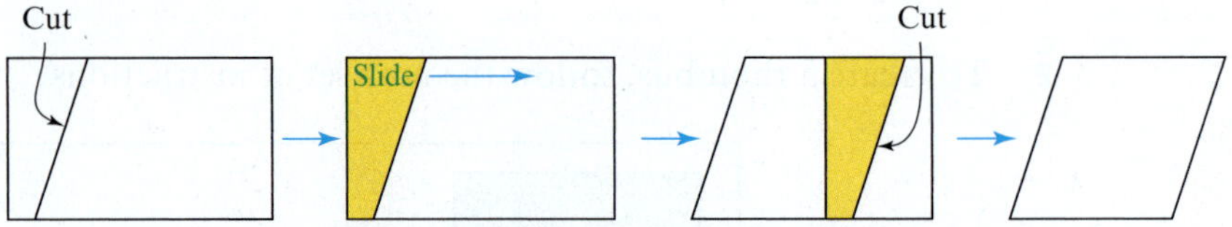

Draw a line segment connecting two opposite sides of a rectangle. Cut along the line segment. Put the piece back (shown shaded) and slide it over as shown. Mark and cut at the leading edge of the slid-over piece.

Use the converse of the Parallel Postulate to explain why the newly cut sides are parallel.

Class Activity 10W Making Shapes by Walking and Turning Along Routes

CCSS CCSS SMP1, 7.G.2, 7.G.5

1. The polygons below are regular polygons. For each polygon, give Robot Robby instructions for how to move and turn so that his path makes the polygon. In each case, Robby should turn back to face the way he started at the end.

   To help you determine Robot Robby's angles of turning, consider his total amount of turning as he goes all the way around the polygon. Which compass directions does he face? Are any repeated or omitted?

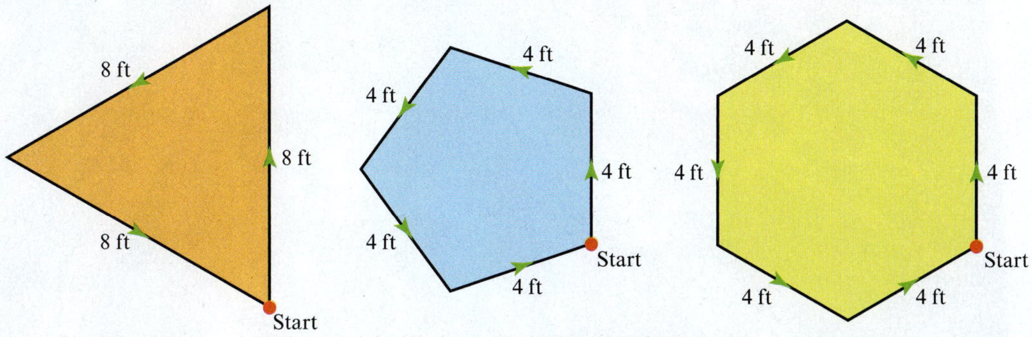

2. Give Automaton Amy instructions for how to move and turn so that her path makes a parallelogram that has a 3-meter side, a 5-meter side, and an angle of 60°. Explain how to determine the instructions.

CLASS ACTIVITIES

11

Measurement

11.1 Concepts of Measurement

Class Activity 11A The Biggest Tree in the World

CCSS CCSS SMP2

Each of the trees described below could perhaps qualify as the biggest tree in the world. Compare these trees. Why can reasonable people differ about which tree is biggest?

Tree 1: General Sherman is a giant sequoia in Sequoia National Park in California. It is 275 feet tall and has a circumference (at its base) of 103 feet and a volume of 52,500 cubic feet.

Tree 2: General Grant is a giant sequoia in Sequoia National Park in California. It is 268 feet tall and has a circumference (at its base) of 108 feet and a volume of 46,600 cubic feet.

Tree 3: Mendocino tree is a redwood tree near Ukiah, California. It is 368 feet tall and has a diameter of 10.4 feet, which means that its circumference should be about 33 feet.

Tree 4: A Banyan tree in Kolkata, India, has a circumference of 1350 feet (meaning the circumference of the whole tree, not just the trunk) and covers three acres.

Tree 5: A tree in Santa Maria del Tule near Oaxaca, Mexico, is 130 feet tall and is described as requiring 40 people holding hands to encircle it.

Class Activity 11B What Concepts Underlie the Process of Length Measurement?

CCSS CCSS SMP3, SMP5, 1.MD.2

Some students were asked to measure the length of a leaf using different objects. For each piece of (hypothetical) student work below, discuss ideas about length measurement that the student may not yet understand.

Student 1:

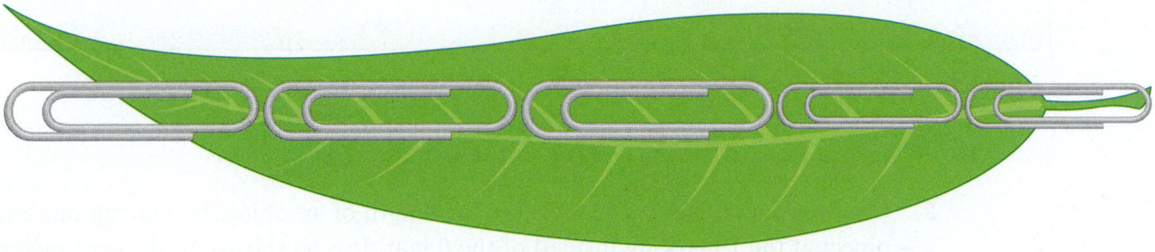

Student 2:

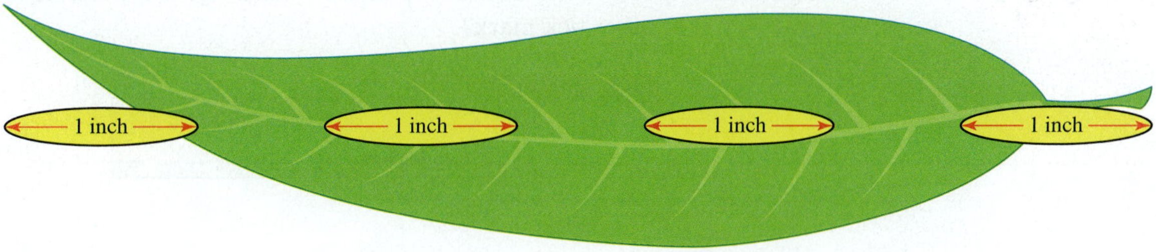

Student 3:

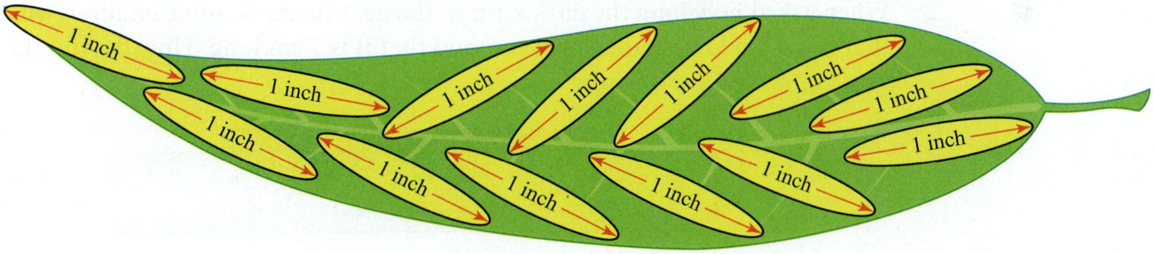

Student 4:

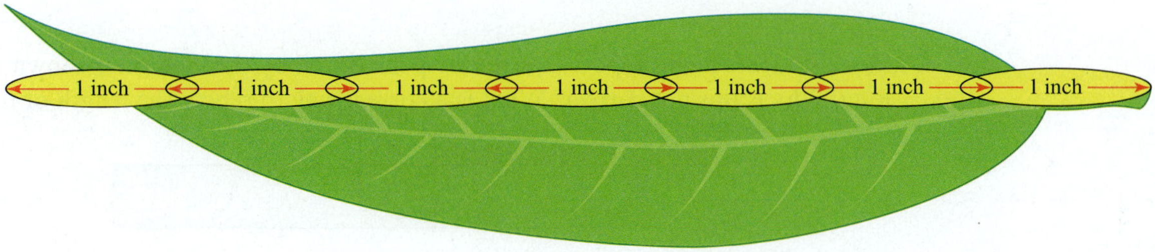

Class Activity 11C Making and Using a Ruler

CCSS CCSS SMP3, SMP5, 1.MD.2, 2.MD.1

1. Show and discuss how children could make their own inch-ruler using an inch-tile and a cardboard strip like the ones shown below. What do the tick marks and numbers on the ruler mean?

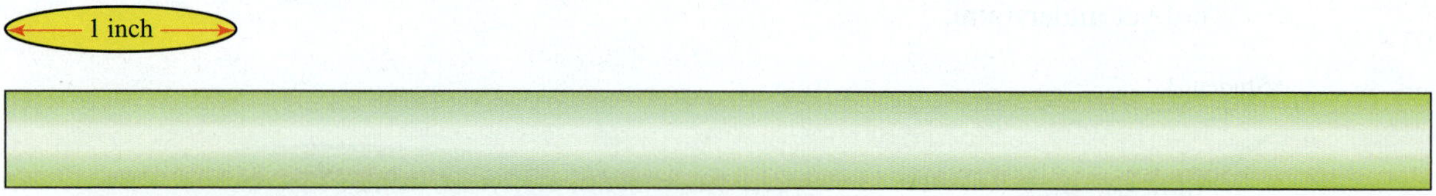

2. Children sometimes try to measure the length of an object by placing one end of the object at the 1 marking instead of the 0 marking, as shown on the centimeter ruler in the figure. Why is the strip below not 5 cm long, even though the end of the strip is at 5? Why might a child put one end of the strip at the 1 marking? Is it possible to measure by starting at 1 or at another tick mark?

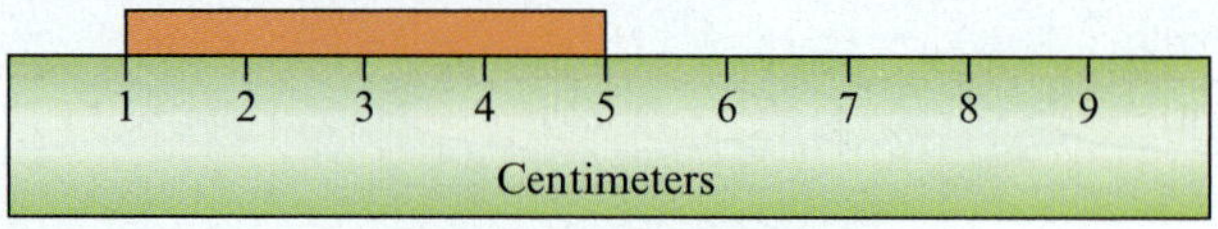

3. When asked how long the dark strip in the next figure is, some children will respond that it is 8 cm long. Others will respond that it is 7 cm long. How do you think children come up with these answers?

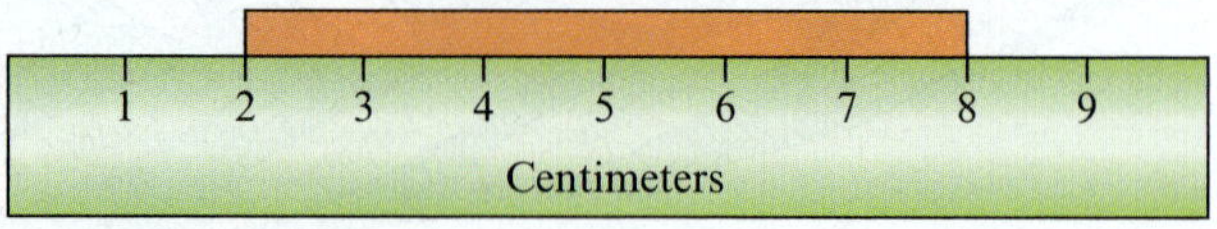

4. Some students might report that the strip measured by the inch ruler shown is 2.3 inches long. Why is this not correct? What is a correct way to report the length of the strip?

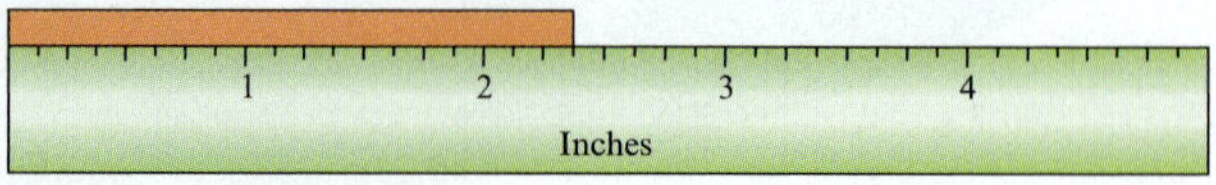

Class Activity 11D What Does "6 Square Inches" Mean?

CCSS CCSS SMP6, 3.MD.5, 3.MD.6

1. Discuss the following as clearly and concretely as you can, illustrating with 1-inch-by-1-inch tiles if available:

   What does it mean to say that a shape has an area of 6 square inches?

   Why is it easy to think that a 6-inch-by-6-inch square has area 6 square inches, and why is this not correct?

2. Which of the following describe the same area? Why?
   a. 4 square inches
   b. A 4-inch-by-4-inch square
   c. 4 in^2
   d. 4 in. $\times$ 4 in.
   e. A 2-inch-by-2-inch square
   f. 2 in^2

3. People sometimes say, "Area is length times width." Why is it not correct to characterize area this way?

11.2 Length, Area, Volume, and Dimension

Class Activity 11E Dimension and Size

CCSS CCSS SMP4

1. Imagine a lake. Describe one-dimensional, two-dimensional, and three-dimensional parts or aspects of the lake. In each case, state how you would measure the size of that part or aspect of the lake—by length, by area, or by volume—and name an appropriate U.S. customary unit and an appropriate metric unit for measuring or describing the size of that part or aspect of the lake. What are practical reasons for wanting to know the sizes of these parts or aspects of the lake?

2. Imagine a house. Describe one-dimensional, two-dimensional, and three-dimensional parts or aspects of the house. In each case, state how you would measure the size of that part or aspect of the house—by length, by area, or by volume—and name an appropriate U.S. customary unit and an appropriate metric unit for measuring or describing the size of that part or aspect of the house. What are practical reasons for wanting to know the sizes of these parts or aspects of the house?

11.3 Error and Precision in Measurements

Class Activity 11F Reporting and Interpreting Measurements

CCSS CCSS SMP6

1. **a.** Does a food label that says "0 grams trans fat in 1 serving" mean that the food contains no trans fat? If not, what does it mean?

   b. If a food label said "0.0 grams trans fat" would that mean there is no trans fat in the food?

2. One source says that the average distance from the earth to the sun is 93,000,000 miles, and another source says that the average distance from the earth to the sun is 92,960,000 miles. Can both of these descriptions be correct, or must at least one of them be wrong? Explain.

11.4 Converting from One Unit of Measurement to Another

Class Activity 11G Conversions: When Do We Multiply? When Do We Divide?

CCSS CCSS SMP3, 4.MD.1, 5.MD.1

1. Julie is confused about why we *multiply* by 3 to convert 6 yards to feet. She thinks we should *divide* by 3 because feet are smaller than yards.

   a. Make a math drawing to show how yards and feet are related. Take care that your drawing accurately portrays length as a one-dimensional, not a two-dimensional, attribute. Use your drawing and what multiplication means to explain why we multiply by 3 to convert 6 yards to feet.

   b. Discuss the relationship between the *size* of a unit and the *number* of units it takes to describe the length of an object.

   c. Try to think of other ways to discuss conversions. What problems or questions could you pose?

2. Nate is confused about why we *divide* by 100 to convert 200 centimeters to meters. He thinks we should *multiply* by 100 because meters are bigger than centimeters. Use several approaches to discuss converting 200 cm to meters.

Class Activity 11H Conversion Problems

CCSS CCSS SMP1, 5.MD.1, 6.RP.3d

1. Shaquila is 57 inches tall. How tall is Shaquila in feet?

   Should you multiply or divide to solve this problem? Explain. Describe a number of different correct ways to write the answer to the conversion problem. Explain briefly why these different ways of writing the answer mean the same thing.

2. Carlton used identical paper clips to measure the length of a piece of wood. He found that the wood is 35 paper clips long. Next, Carlton measured the length of the wood, using identical rods. He found that 2 rods are as long as 5 paper clips. How many rods long is the wood? Explain your reasoning.

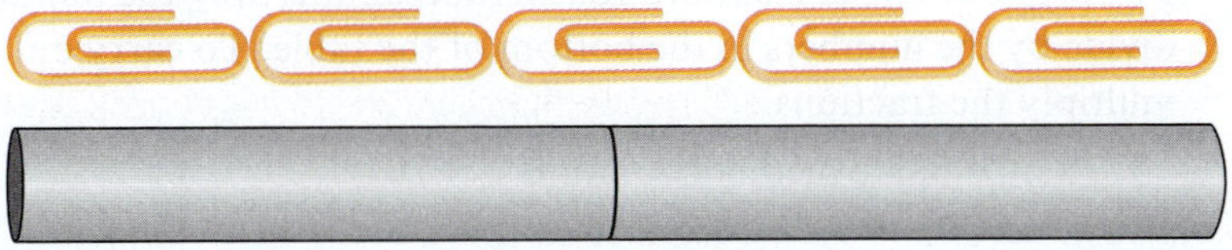

3. Suppose that the students in your class want to have a party at which they will serve punch to drink. The punch that the children want to serve is sold in half-gallon containers. If 25 children attend the party and if each drinks 8 fluid ounces of punch, then how many containers of punch will you need? Describe several different ways that students could correctly solve this problem. For each method of solving the problem, explain simply and clearly why the method makes sense.

Class Activity 11I Using Dimensional Analysis to Convert Measurements

CCSS CCSS SMP2

Methods A and B provide two ways of writing the steps for converting 25 meters to yards with dimensional analysis, using the fact that 1 in. = 2.54 cm.

Method A

25 m	100 cm	1 in.	1 ft	1 yd	
	1 m	2.54 cm	12 in.	3 ft	= 27.3 yd (approx.)

Method B

$$25 \text{ m} \times \frac{100 \text{ cm}}{1 \text{ m}} \times \frac{1 \text{ in.}}{2.54 \text{ cm}} \times \frac{1 \text{ ft}}{12 \text{ in.}} \times \frac{1 \text{ yd}}{3 \text{ ft}} = 27.3 \text{ yd (approx.)}$$

To carry out the calculations for method A, multiply the numbers in the top of the table and divide by the numbers in the bottom of the table. To carry out the calculations for method B, multiply the fractions.

1. **a.** Compare methods A and B.

   b. Method B works by starting with a measurement (such as 25 m) and repeatedly multiplying by certain fractions. Discuss the fractions that you multiply by. How are they chosen?

   c. Explain why method B works; in other words, explain why 25 meters really must be equal to 27.3 . . . yards. What is special about the fractions you multiply by that allows you to deduce this?

   d. If you converted 25 meters to yards by applying what multiplication and division mean, how would your calculations compare with the calculations you do with dimensional analysis?

2. Use dimensional analysis to convert 1 mile to kilometers, using the fact that 1 in. = 2.54 cm.

Class Activity 11J Area and Volume Conversions

CCSS CCSS SMP2

1. If 1 yard is equal to 3 feet, does this mean that 1 square yard is 3 square feet? Make a drawing to show how many square feet are in a square yard.

2. A rug is 5 yards long and 4 yards wide. What is the area of the rug in square yards? What is the area of the rug in square feet? Show two different ways to solve this problem. Explain each case.

3. A room has a floor area of 35 square yards. What is the floor area of the room in square feet? Explain your answer.

4. A compost pile is 2 yards high, 2 yards long, and 2 yards wide. Does this mean that the compost pile has a volume of 2 cubic yards? Explain.

5. Determine the volume in cubic feet of the compost pile described in the previous question in two different ways. Explain each case.

Class Activity 11K Area and Volume Conversions: Which Are Correct and Which Are Not?

CCSS CCSS SMP3

1. Analyze the calculations that follow, which are intended to convert 25 square meters to square feet. Which use legitimate methods and are correct, and which are not? Explain.

a. $25 \text{ m}^2 = 25 \text{ m} \times \dfrac{100 \text{ cm}}{1 \text{ m}} \times \dfrac{1 \text{ in.}}{2.54 \text{ cm}} \times \dfrac{1 \text{ ft}}{12 \text{ in.}} = 82 \text{ ft}^2$

b. $25 \text{ m}^2 = 25 \text{ m}^2 \times \dfrac{100 \times 100 \text{ cm}^2}{1 \text{ m}^2} \times \dfrac{1 \text{ in.}^2}{2.54 \times 2.54 \text{ cm}^2} \times \dfrac{1 \text{ ft}^2}{12 \times 12 \text{ in.}^2} = 269 \text{ ft}^2$

c. $25 \text{ m} = 25 \text{ m} \times \dfrac{100 \text{ cm}}{1 \text{ m}} \times \dfrac{1 \text{ in.}}{2.54 \text{ cm}} \times \dfrac{1 \text{ ft}}{12 \text{ in.}} = 82 \text{ ft}$

Therefore,

$$25 \text{ m}^2 = 82^2 \text{ ft}^2 = 6727 \text{ ft}^2$$

d. 25 square meters is the area of a square that is 5 meters wide and 5 meters long, so

$$5 \text{ m} = 5 \text{ m} \times \frac{100 \text{ cm}}{1 \text{ m}} \times \frac{1 \text{ in.}}{2.54 \text{ cm}} \times \frac{1 \text{ ft}}{12 \text{ in.}} = 16.404 \text{ ft}$$

Therefore,

$$25 \text{ m}^2 = 16.404 \times 16.404 \text{ ft}^2 = 269 \text{ ft}^2$$

2. Use the fact that 1 in. = 2.54 cm to convert 27 cubic feet to cubic meters in at least two different ways.

Class Activity 11L Model with Conversions

CCSS CCSS SMP4

1. How much water would you expect the people in a city with a population of 100,000 to use in 1 day? What size container would hold this amount of water? Compare this size container with something familiar.

   To answer these questions, first estimate how much water each person might use per day. Then do some calculations. Consider working with metric measurements.

   Note that many showers have a water flow of 10 liters per minute and many toilets use about 5 liters of water per flush.

2. How many square miles of land would you need to plant 1 million trees? 1 billion trees? 1 trillion trees? Compare these land areas to familiar areas. You may wish to assume you can plant trees in rows that are 10 feet apart and that the trees are 10 feet apart in each row.

CLASS ACTIVITIES 12

Area of Shapes

12.1 Areas of Rectangles Revisited

Class Activity 12A Units of Length and Area in the Area Formula for Rectangles

CCSS CCSS SMP2, 3.MD.5, 3.MD.6, 3.MD.7, 5.NF.4b

If available, square centimeter tiles would be helpful.

1. What does it mean to say that a shape has an area of 15 square centimeters?

2. The large rectangle shown here is 3 cm by 5 cm. What is a primitive way to determine the area of the rectangle in square centimeters by relying directly on the meaning of area?

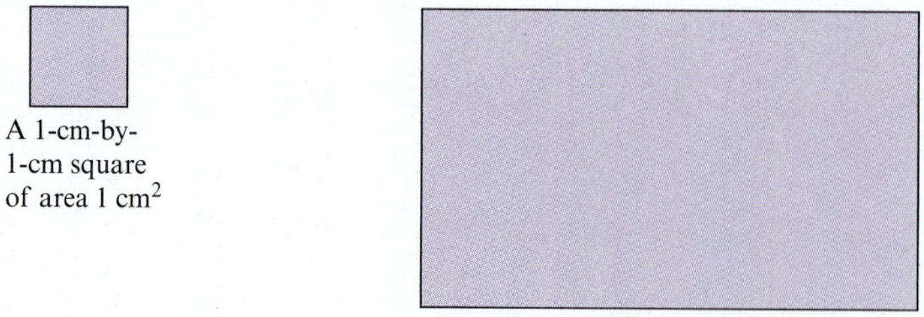

3. In Chapter 4, we defined multiplication in terms of equal groups. According to our definition, $3 \cdot 5$ means the number of units in 3 groups of 5 units each. Using our definition of multiplication, explain why the area of the large rectangle in part 2 is $3 \cdot 5 \text{ cm}^2$.

4. The *length times width* area formula for rectangles involves lengths, but doesn't explicitly involve equal groups. Discuss:

How are the *lengths* 3 cm and 5 cm *linked to* yet *different from* the numbers of groups and numbers of things in each group in your explanation for part 3?

5. **a.** Given that the line segment shown is 1 unit long, use the grid to lightly shade a rectangle that is $\frac{7}{10}$ units by $\frac{9}{10}$ units.

b. Apply the *length times width* formula to find the area of the shaded rectangle in part (a) and verify that the formula gives you the correct area for your rectangle. Attend carefully to units of area.

c. Show and describe the $\frac{7}{10}$- and $\frac{9}{10}$-unit lengths on your rectangle, keeping in mind that length is a one-dimentional attribute. Explain why it would be confusing to say that the small squares represent these lengths.

12.2 Moving and Additivity Principles About Area

Class Activity 12B Using the Moving and Additivity Principles

CCSS SMP3, 3.MD.7d

1. Use the moving and additivity principles (one or both) to determine the area of the L-shaped region in several different ways. In each case, explain your reasoning and write and evaluate an algebraic expression that fits with the strategy. Try to find strategies of the following types:
   - A simple subdividing strategy
   - A takeaway strategy
   - A move and reattach strategy
   - A combine two copies and take half strategy

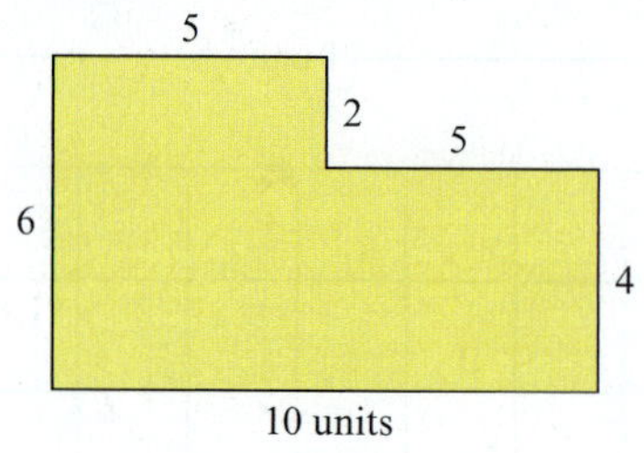

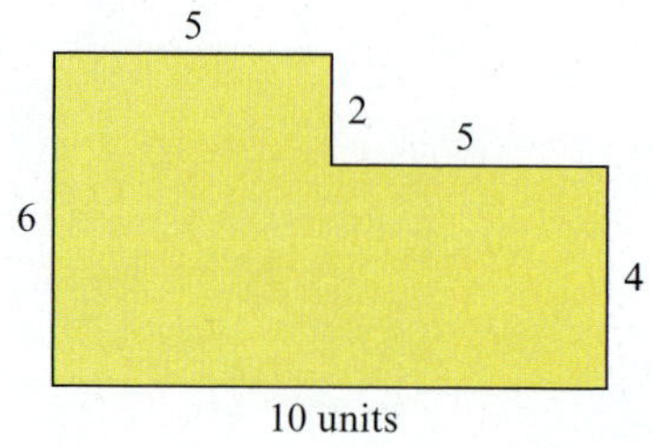

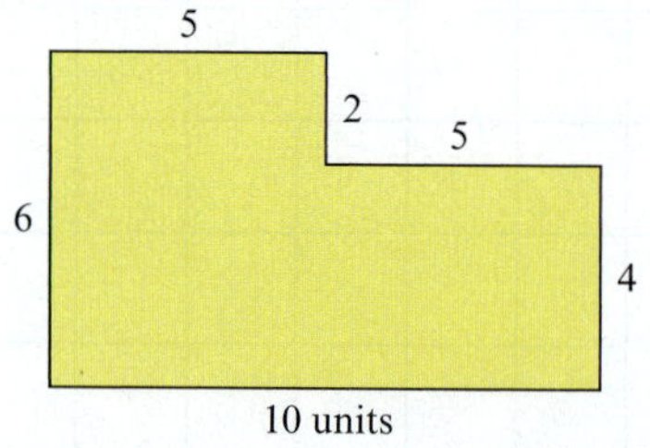

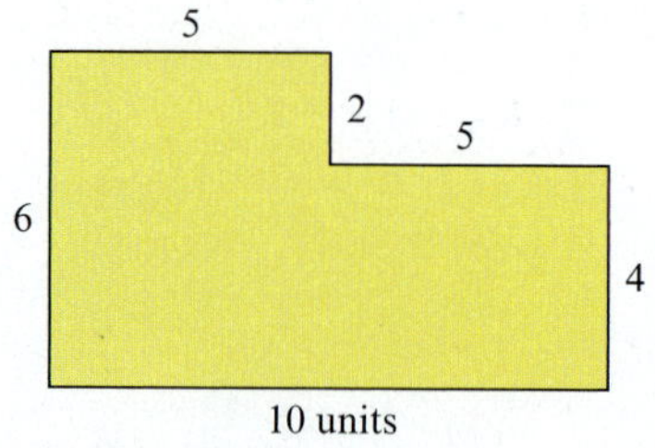

2. Use the moving and additivity principles (one or both) to determine the area of the shaded square, which is inside the 2-unit-by-2-unit square, in several different ways. In each case, explain your reasoning.

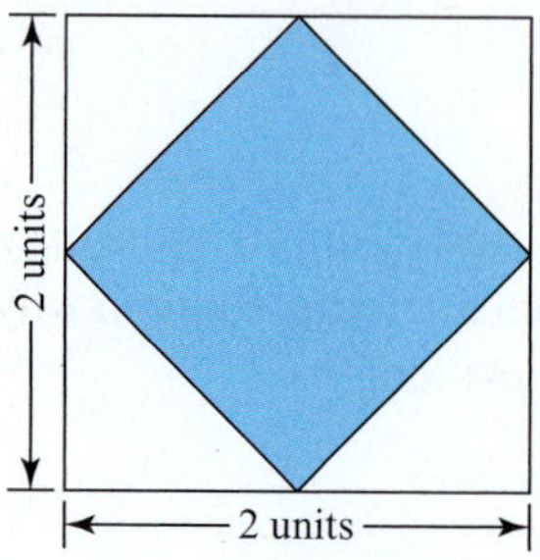

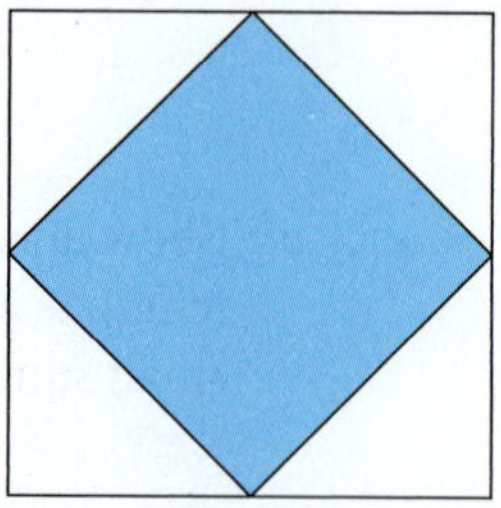

12.3 Areas of Triangles

Class Activity 12C Determining Areas of Triangles in Progressively Sophisticated Ways

CCSS CCSS SMP2, 6.G.1

1. Use the moving and additivity principles to determine the area of the next two triangles in three different ways: (a) by moving small pieces and relying directly on the definition of area, (b) by moving bigger chunks to create a rectangle, and (c) by viewing the triangle as part of a larger rectangle.

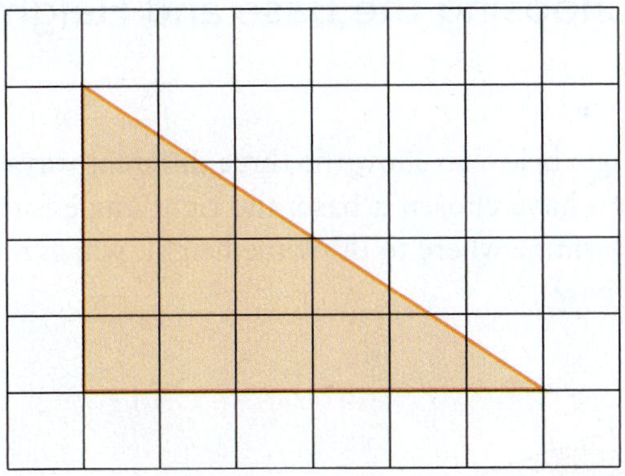
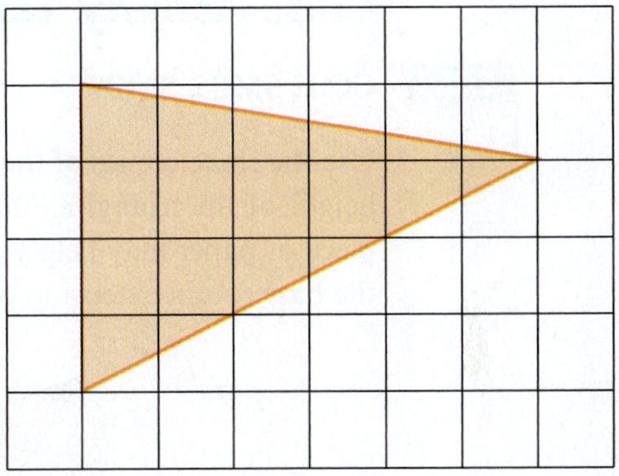

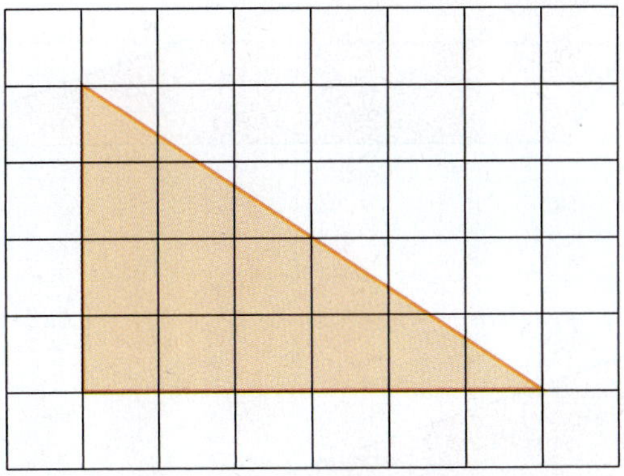
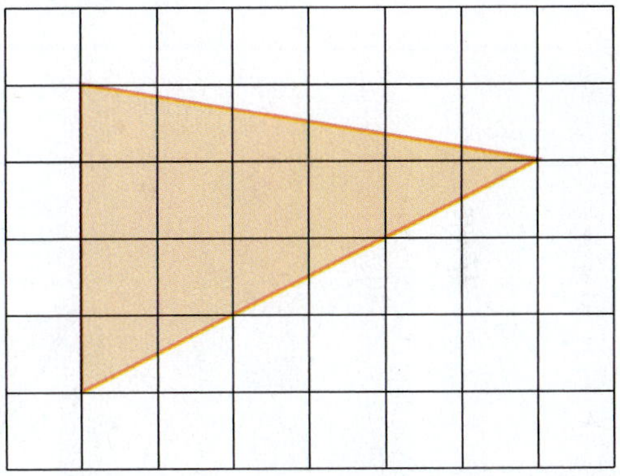

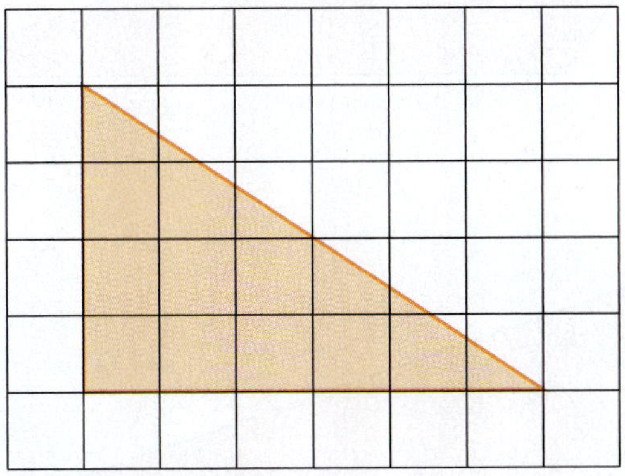
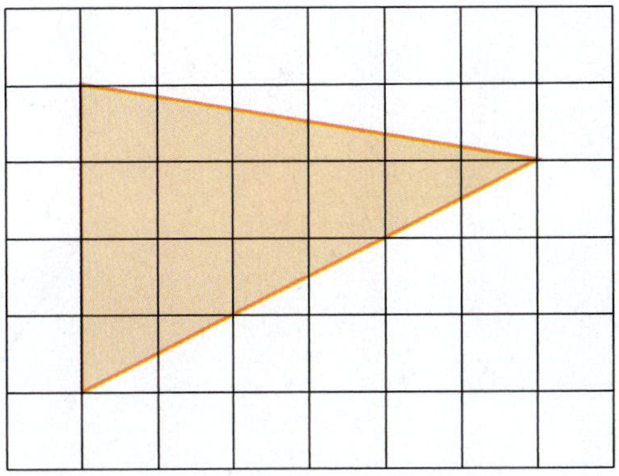

2. Think about some of the methods you used to determine the areas of the triangles in this activity: moving chunks, putting inside a rectangle and taking half, and putting inside a rectangle and taking away. Find arithmetic problems that can be made easy to solve by using numerical strategies that are similar in spirit to these geometric strategies. Describe how to solve these arithmetic problems and say briefly how the solution methods are roughly similar to the geometric methods.

Class Activity 12D Choosing the Base and Height of Triangles

CCSS CCSS SMP5, SMP7

Use the three copies of the triangle below to show the three different ways to choose the base and height of the triangles. Once you have chosen a base, the right angle formed by the corner of a piece of paper may help you determine where to draw the height, which must be perpendicular to the base (or an extension of the base).

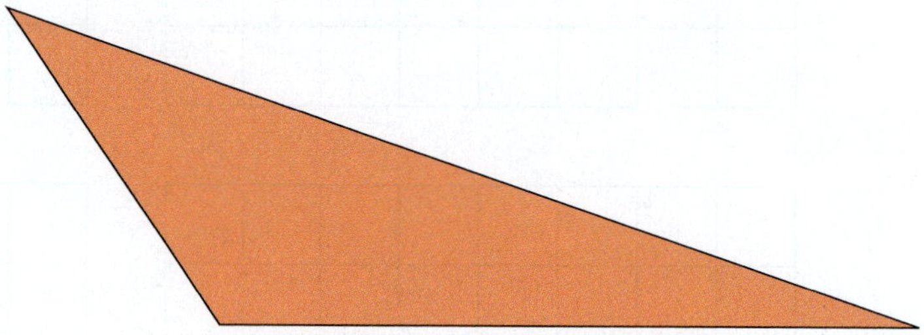

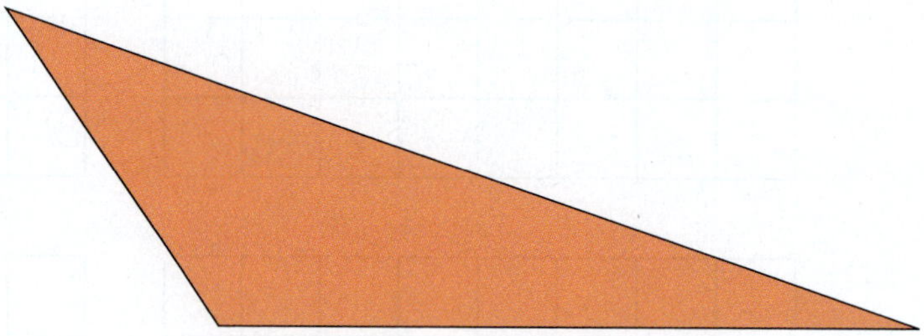

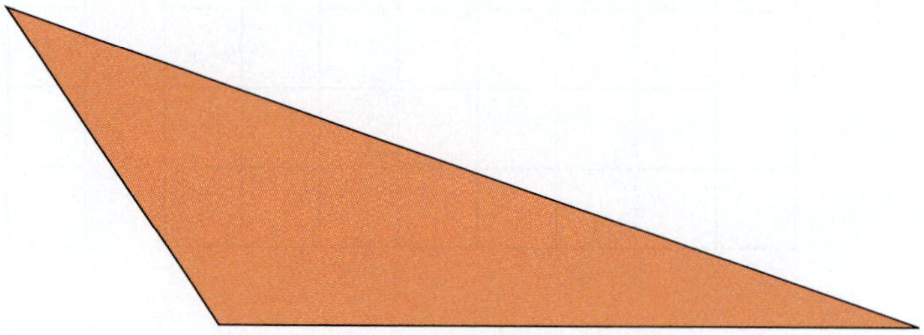

Class Activity 12E Explaining Why the Area Formula for Triangles Is Valid

CCSS CCSS SMP3, 6.G.1

1. Use the moving and additivity principles to explain in three ways why the next triangle has area $\frac{1}{2}(b \cdot h)$ square units for the given choices of b and h. One explanation should fit naturally with the expression $\frac{1}{2}(b \cdot h)$, another explanation should fit naturally with the expression $(\frac{1}{2}b) \cdot h$, and a third explanation should fit naturally with the expression $b \cdot (\frac{1}{2}h)$. Why is it valid to describe the area with any one of these three expressions?

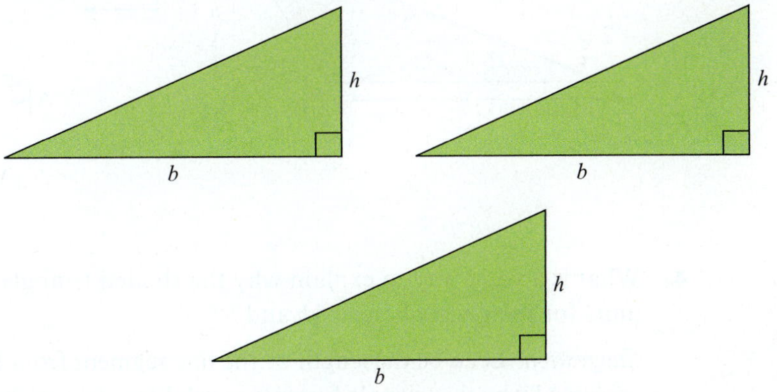

2. Use the moving and additivity principles to explain in two ways why the next triangle has area $\frac{1}{2}(b \cdot h)$ square units for the given choices of b and h. One explanation should fit naturally with the expression $b \cdot (\frac{1}{2}h)$ and the other should fit naturally with the expression $\frac{1}{2}(b \cdot h)$.

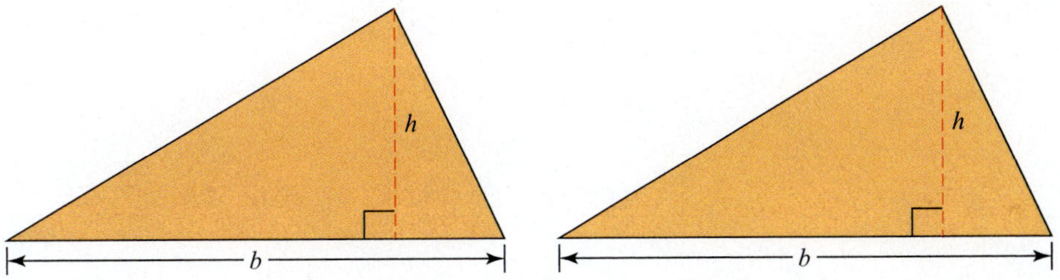

3. What is wrong with the following reasoning that claims to show that the area of the triangle ABC below is $\frac{1}{2}(b \cdot h)$ square units?

Draw a rectangle around the triangle ABC, as shown on the right in the figure below. The area of this rectangle is $b \cdot h$ square units. The line AC cuts the rectangle in half, so the area of the triangle ABC is half of $b \cdot h$ square units—in other words, $\frac{1}{2}(b \cdot h)$ square units.

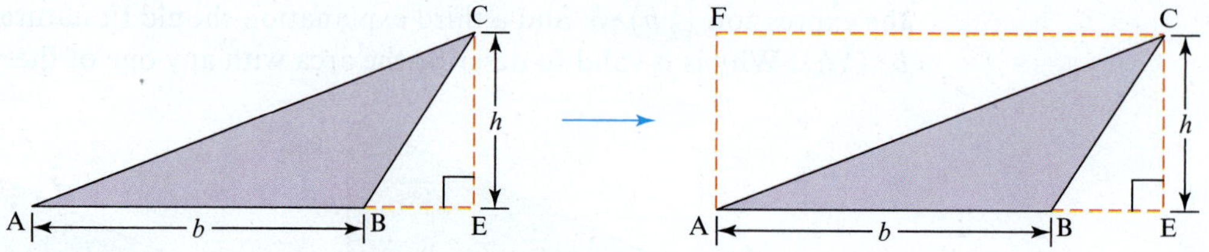

4. What is a valid way to explain why the shaded triangle in part 3 has area $\frac{1}{2}(b \cdot h)$ square units for the given choice of b and h?

Suggestion: Let a be the length of the line segment from B to E. Because we have already explained why the triangle formula is valid for *right triangles,* you may apply it to right triangles in the figure.

Class Activity 12F Area Problem Solving

CCSS CCSS SMP1, SMP3

1. Determine the area of the shaded triangle that is inside the rectangle. Explain your reasoning.

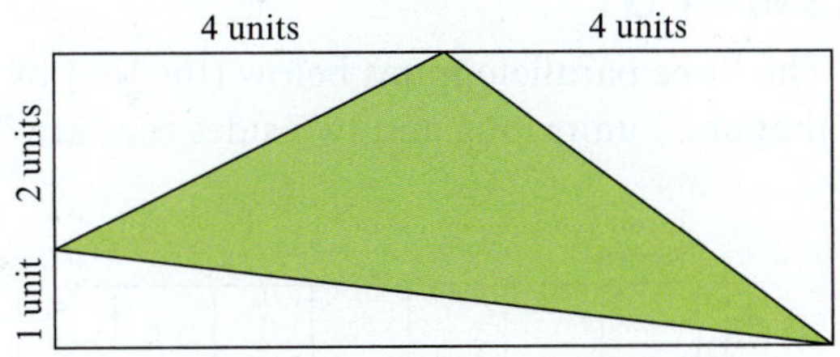

2. Determine the area of the shaded triangle that is in the rectangle in *two different ways.* Explain your reasoning in each case.

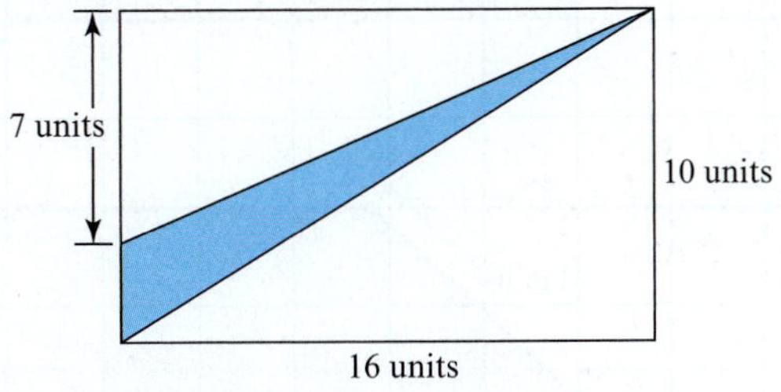

3. Determine the area of the shaded triangle in the figure below in *two different ways.* Explain your reasoning in each case.

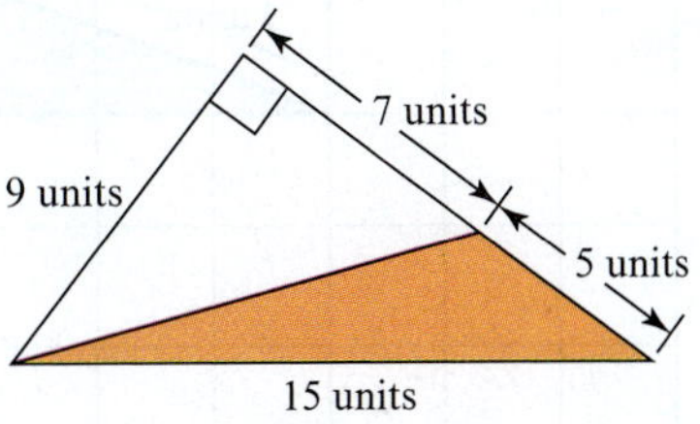

4. Determine the area of the shaded shape below. Explain your reasoning.

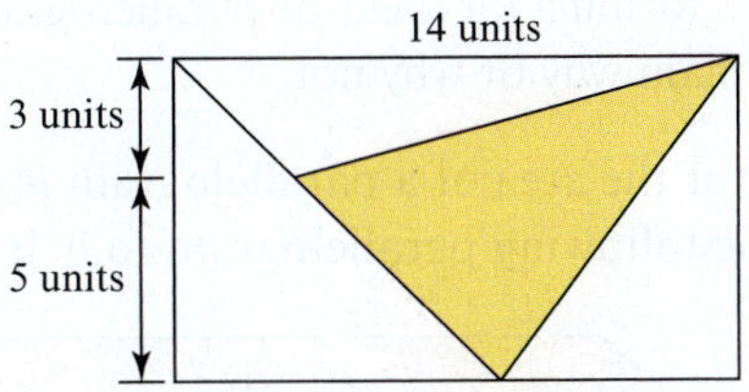

12.4 Areas of Parallelograms and Other Polygons

Class Activity 12G Do Side Lengths Determine the Area of a Parallelogram?

CCSS CCSS SMP8, 6.G.1

1. The three parallelograms below (the first of which is also a rectangle) all have two sides that are 3 units long and two sides that are 7 units long.

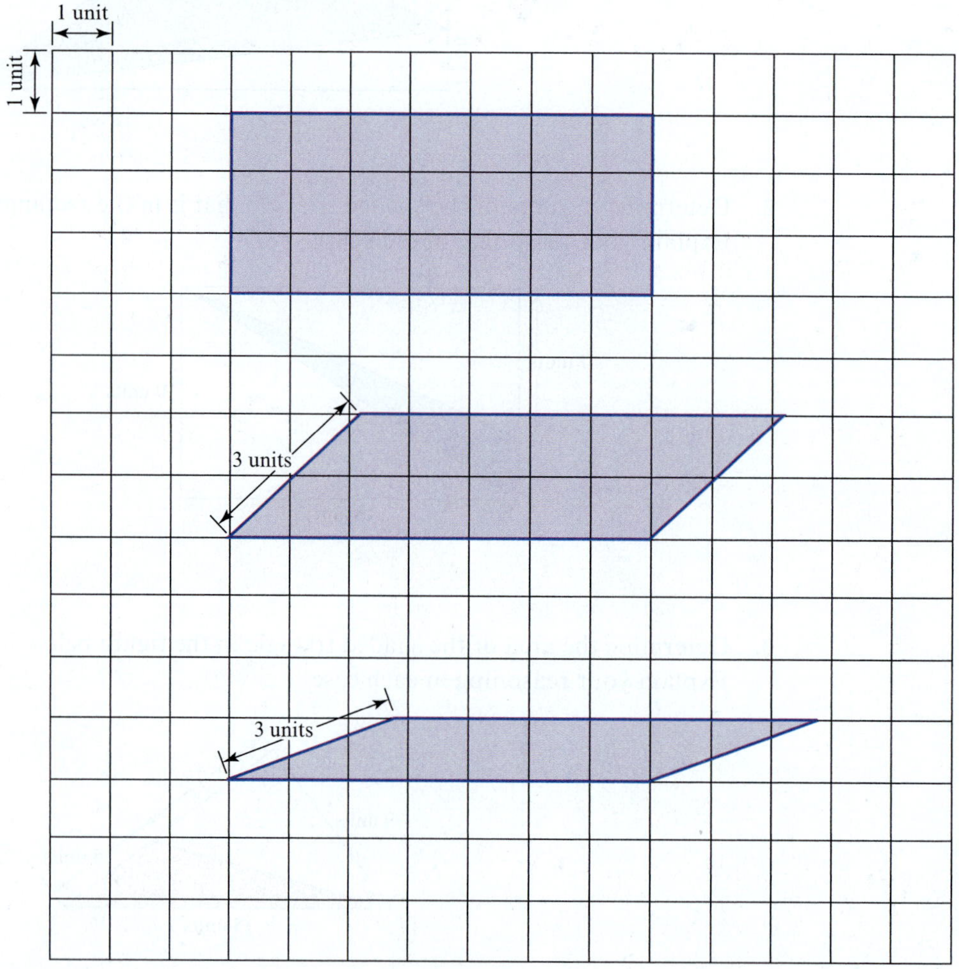

   a. Use the moving and additivity principles to determine the areas of the three parallelograms.

   b. Can there be a formula for areas of parallelograms that is only in terms of the lengths of the sides? Explain why or why not.

2. Find a formula for the area of a parallelogram *in terms of lengths of parts of the parallelogram.* Use the following parallelogram to help you describe your formula:

Class Activity 12H Explaining Why the Area Formula for Parallelograms Is Valid

CCSS CCSS SMP3, 6.G.1

1. Show how to subdivide and recombine the parallelogram below to form a b by h rectangle, thereby explaining why the area of the parallelogram is $b \cdot h$.

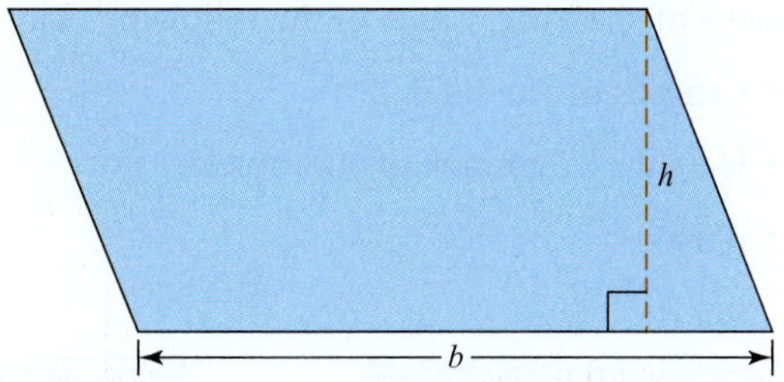

2. Explain why the area of the shaded parallelogram below is $b \cdot h$. To do so, consider using one of these approaches: (a) enclose the parallelogram in a rectangle, or (b) subdivide the parallelogram into two triangles.

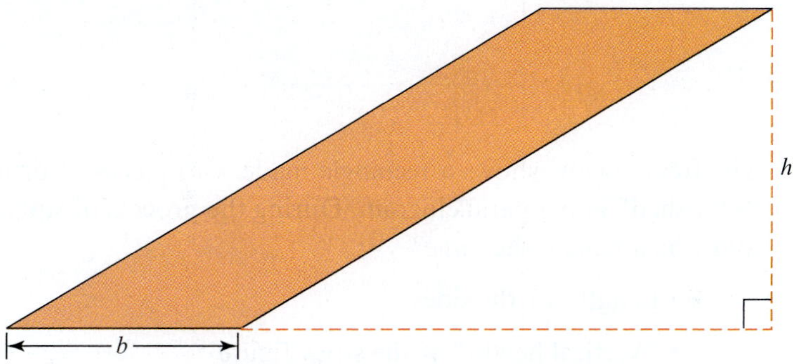

Class Activity 12I Finding and Explaining a Trapezoid Area Formula

CCSS CCSS SMP3, 6.G.1

Use several different methods to find and explain a formula for the area of a trapezoid that has parallel sides of length a and b and height h. Consider these ideas as well as others: (a) Combine two copies of the trapezoid to make a parallelogram. (b) Divide the trapezoid into *two* triangles, one with base b and one with base a.

If some of your formulas look different, explain why they are equivalent.

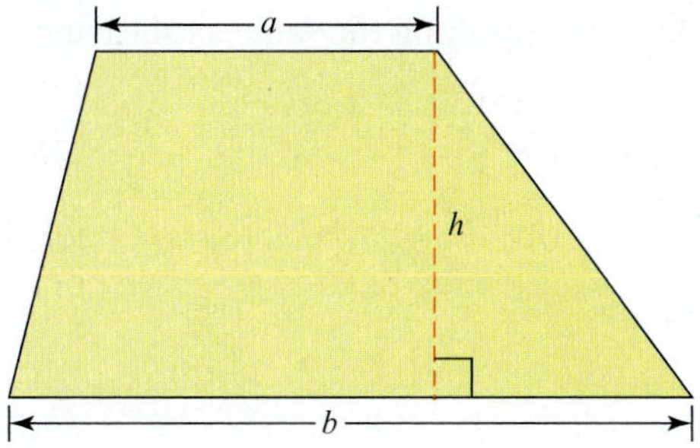

12.5 Shearing: Changing Shapes Without Changing Area

Class Activity 12J Is This Shearing?

CCSS CCSS SMP2

The figure below shows a rectangle made of toothpicks being sheared into a parallelogram. During the process of shearing, which of the following change and which remain the same?

- Lengths of the sides
- Height of the stack of toothpicks
- Area

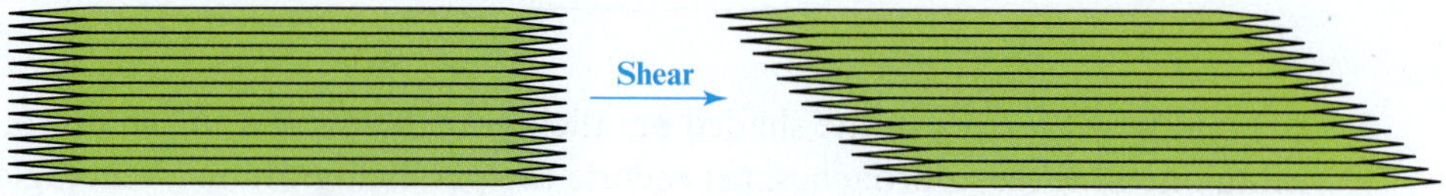

The figure below shows a rectangle made with pieces of drinking straws tied with a string being "squashed" into a parallelogram. During the process of squashing, which of the following change and which remain the same?

- Lengths of the sides
- "Vertical height" of the straw figure
- Area

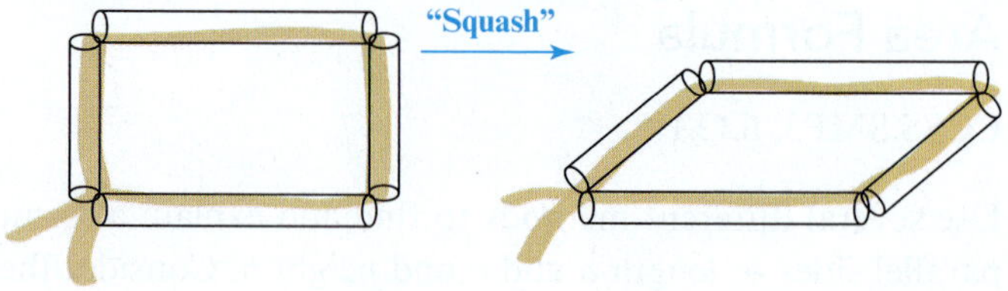

Is the process of squashing the same as shearing? Why or why not?

Class Activity 12K Solving Problems by Shearing

Solve the following problems by finding and shearing appropriate triangles. To shear your triangles, identify a base and shear *parallel to the base*.

CCSS CCSS SMP1

1. Triangle ABC is divided into a triangle, which has area T, and another region, which has area R. Show and explain how to use shearing to divide triangle ABC into *two* triangles, one of area T and one of area R.

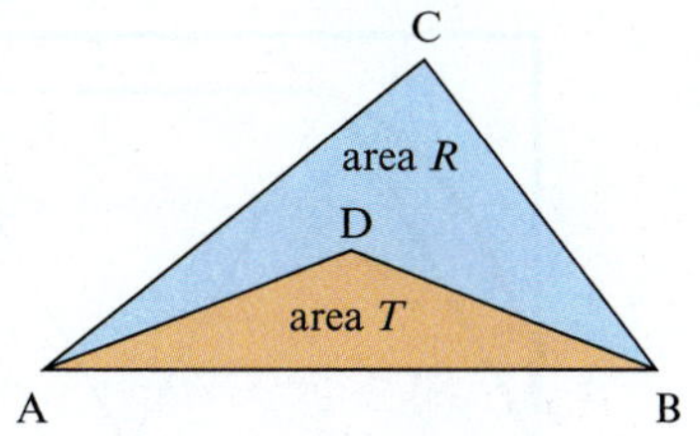

2. Show and explain how to use shearing to create a triangle that has the same area as the shaded region EFGH below.

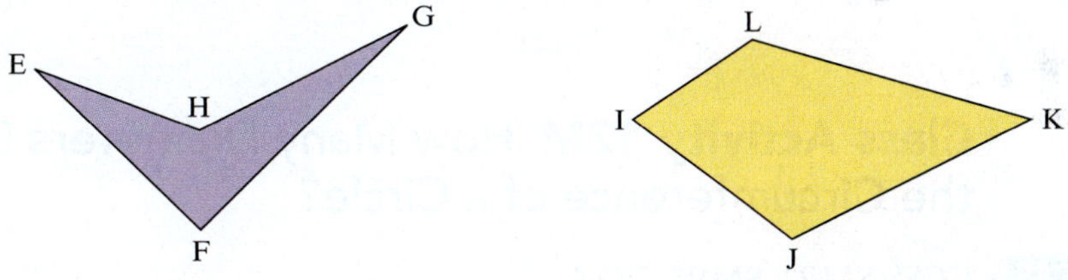

3. Show and explain how to use shearing to create a triangle that has the same area as the shaded region IJKL above.

12.6 Area and Circumference of Circles and the Number Pi

Class Activity 12L How Are the Circumference and Diameter of a Circle Related, Approximately?

CCSS CCSS SMP2

The circle below has diameter 1 unit. Notice that 6 equilateral triangles fit inside the circle and a square surrounds the circle. What does that suggest about the circumference of the circle? In particular, what are two numbers that the circumference lies between? Which of those two numbers does the circumference seem to be closer to?

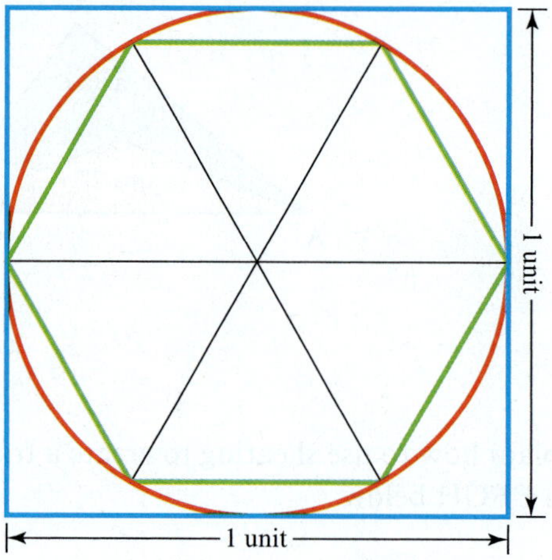

Class Activity 12M How Many Diameters Does it Take to Make the Circumference of a Circle?

CCSS CCSS SMP2, SMP5, 7.G.4

You will need Activity Download 6 and scissors for this activity.

1. Cut out the circumference strips from the Activity Download. By wrapping the edge of each circumference strip around its circle, verify that the length of each circumference strip really is the circumference of its circle.
2. Use Circle 1's diameter to measure the circumference strip for Circle 1. How many diameters does it take to make the circumference? Then use Circle 2's diameter to measure Circle 2's circumference strip and use Circle 3's diameter to measure Circle 3's circumference strip. What do you notice about all these measurements?
3. Based on this experiment, if the diameter of a circle is D centimeters and its circumference is C centimeters, how do you expect D and C to be related? Explain!

Class Activity 12N Where Does the Area Formula for Circles Come From?

CCSS CCSS SMP3, 7.G.4

You will need scissors for this activity.

1. On separate paper, draw a large circle, shade or color half of it, cut it into 8 "pie pieces," and arrange the pieces as shown below.

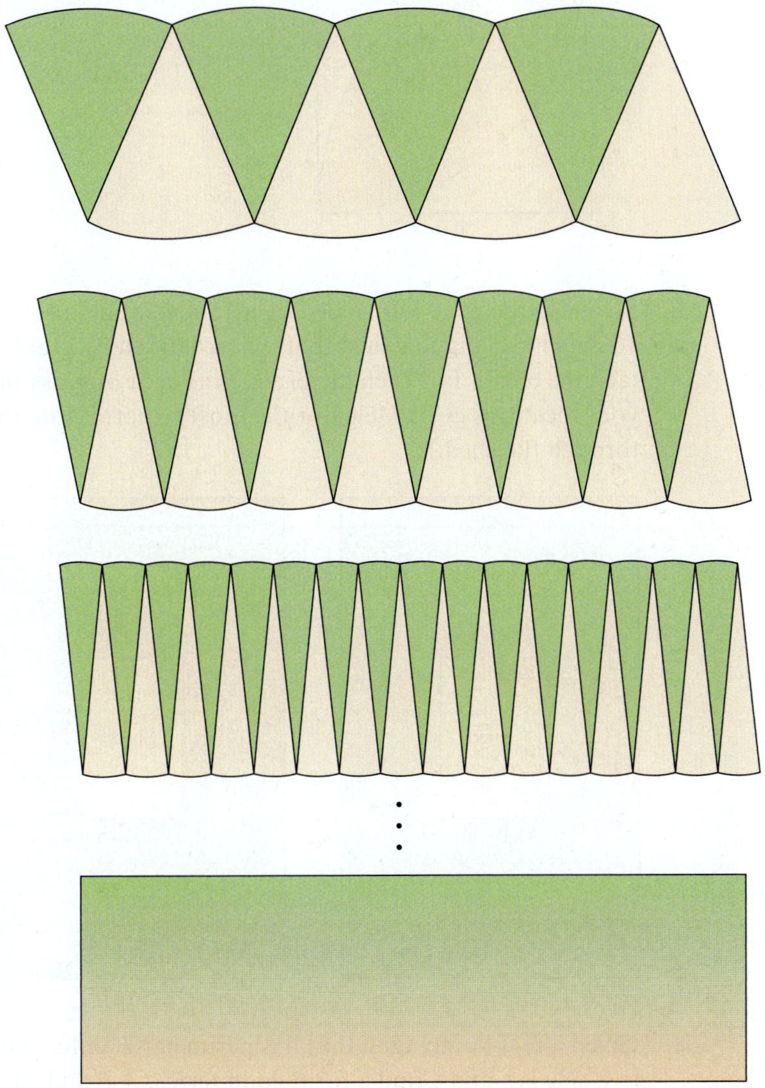

2. Now cut your circle into 16 "pie pieces." Arrange the 16 pieces as shown above.
3. Imagine cutting a circle into more and more smaller and smaller pie pieces and rearranging them as above.
   - What shape would your rearranged circle become more and more like?
   - What would the lengths of the sides of this shape be?
   - What would the area of this shape be?
4. Using your answers to part 3, explain why it makes sense that a circle of radius r units has area πr^2 square units, given that the circumference of a circle of radius r is $2\pi r$.

Class Activity 12O Area Problems

CCSS CCSS SMP1, 7.G.4

1. A reflecting pool will be made in the shape of the shaded region shown in the figure below. The arcs shown are from quarter-circles. What is the area of the surface of the pool? Explain your reasoning.

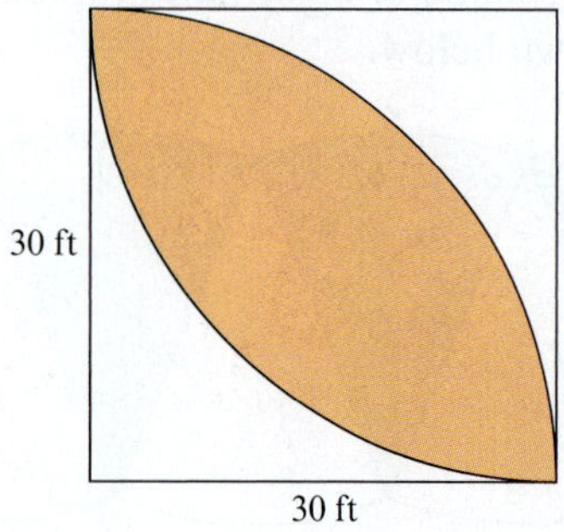

2. Melchior the goat will be tied by a 15 meter chain to the outside of a 10-meter-by-16-meter rectangular shed that is surrounded by grass. Try three different places to attach the chain. For each, determine the area of grass that Melchior could eat. Which of your locations gives Melchior the most to eat? (Note that Melchior cannot walk into or through the shed.)

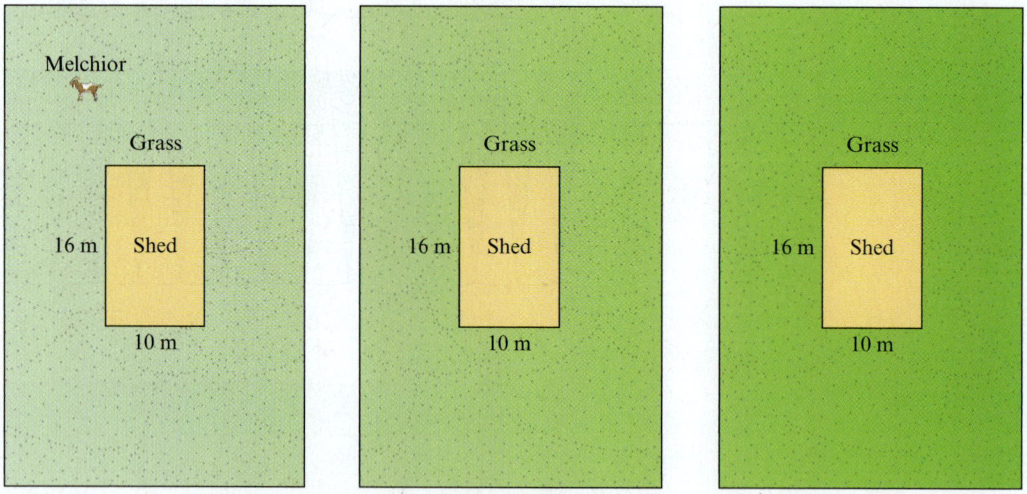

3. Given a regular octagon that has perimeter P units and whose distance from the center to a side is r units, find a formula in terms of P and r for the area of the octagon. Explain your reasoning. Is this similar to the case of a circle?

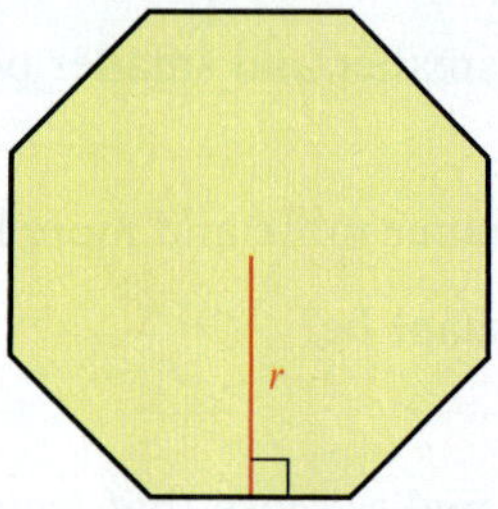

12.7 Approximating Areas of Irregular Shapes

Class Activity 12P Determining the Area of an Irregular Shape

CCSS CCSS SMP5

1. Think about several different ways that you might determine approximately the area of the surface of Lake Lalovely shown on the map below. Suppose that you have the following items on hand:

   - Many 1-inch-by-1-inch plastic squares
   - Graph paper (adjacent lines separated by $\frac{1}{4}$ inch)
   - A scale for weighing (such as one used to determine postage)
   - String
   - Modeling dough

   Which of these items could help you to determine approximately the area of the surface of the lake? How?

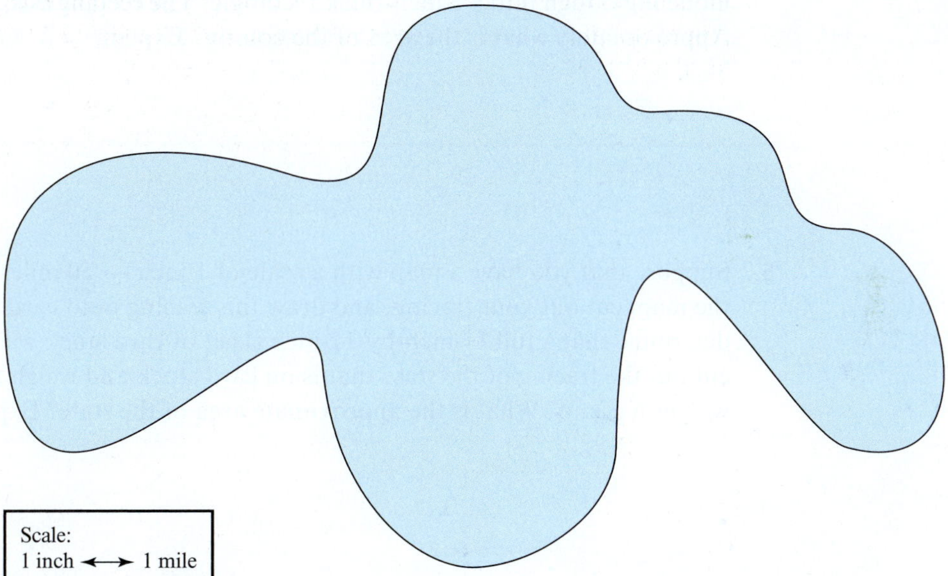

Scale:
1 inch ⟷ 1 mile
on map actual

2. Suppose that you have a map with a scale of 1 inch ↔ 100 miles. You trace a state on the map onto $\frac{1}{4}$-inch graph paper. (The grid lines are spaced $\frac{1}{4}$ inch apart.) You find that the state takes up about 108 squares of graph paper. Approximately what is the area of the state? Explain.

3. Some students were working on the problem in part 2 and made the initial calculations shown below. For each of these initial calculations, either explain the ideas that could be behind the calculations and use them to finish determining the area of the state, or explain how the calculations might lead to an incorrect area for the state.

a. $100 \div 4 = 25$ $\quad 25 \cdot 25 = 625$

b. $108 \div 4 = 27$ $\quad 27 \cdot 100 = 2700$

c. $4 \cdot 4 = 16$ $\quad 108 \div 16 = 6.75$

d. $\frac{1}{4} \cdot \frac{1}{4} = \frac{1}{16}$ $\quad 108 \cdot \frac{1}{16} =$

Now write a numerical expression for the area of the state and relate it to the correct calculation methods.

4. Suppose that you have a map with a scale of 1 inch ↔ 15 miles. You cover a county on the map with a $\frac{1}{8}$-inch-thick layer of modeling dough. Then you re-form this piece of modeling dough into a $\frac{1}{8}$-inch-thick rectangle. The rectangle is $2\frac{1}{2}$ inches by $3\frac{3}{4}$ inches. Approximately what is the area of the county? Explain.

5. Suppose that you have a map with a scale of 1 inch ↔ 50 miles. You trace a state on the map, cut out your tracing, and draw this tracing onto card stock. Using a scale, you determine that a full $8\frac{1}{2}$-inch-by-11-inch sheet of card stock weighs 10 grams. Then you cut out the tracing of the state that is on card stock and weigh this card-stock tracing. It weighs 6 grams. What is the approximate area of the state? Explain.

12.8 Contrasting and Relating the Perimeter and Area of Shapes

Class Activity 12Q Critique Reasoning about Perimeter

CCSS CCSS SMP3, 3.MD.8

You will need a centimeter ruler for part 2.

1. Johnny's method for calculating the perimeter of the shaded shape in Figure (a) is to shade the squares along the border of the shape, as shown in Figure (b), and to count these border squares. Therefore, Johnny says the perimeter of the shape is 24 units. Is Johnny's method valid? If not, why not?

(a) 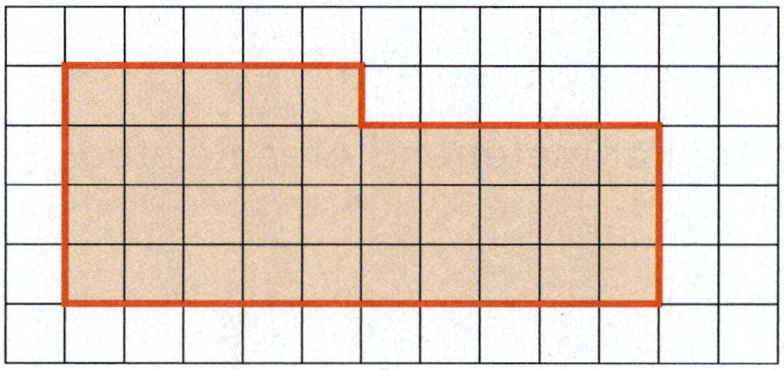

(b) 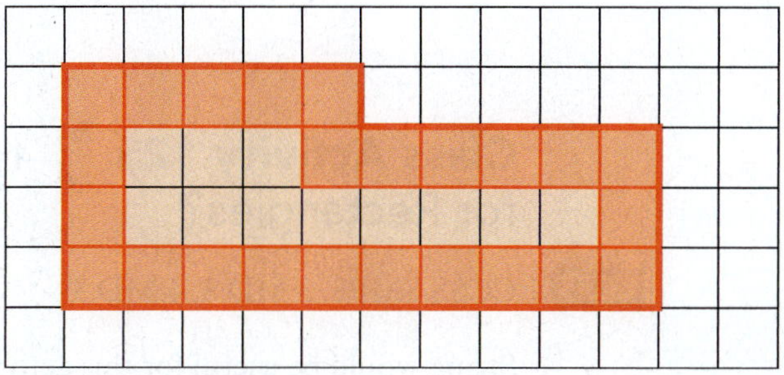

2. When Susie was asked to draw a shape with perimeter 15 cm, she drew a shape like the shaded one shown in the figure below on centimeter grid paper.

   a. Carefully measure the diagonal line segment in the shaded shape with a centimeter ruler. Then explain why the shape does not have perimeter 15 cm.

   b. Draw a shape that has perimeter 15 cm on the same graph paper. (The corners of your shape do not have to be located where grid lines meet.) Explain how you figured out how to make your shape.

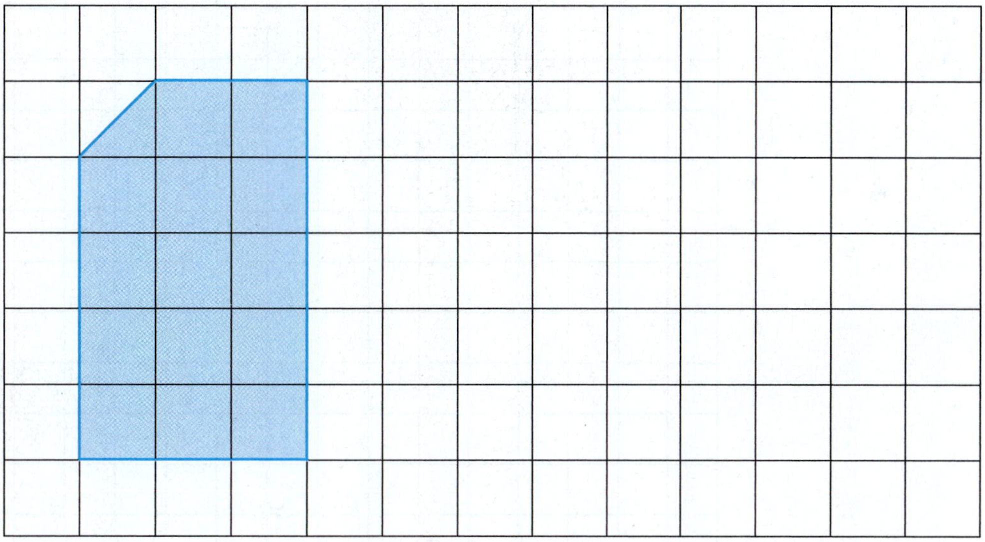

Class Activity 12R Find and Explain Perimeter Formulas for Rectangles

CCSS CCSS SMP7, 4.MD.3

Describe several different methods for determining the perimeter of a rectangle. For each method, write the corresponding formula for the perimeter P of an A-unit-by-B-unit rectangle.

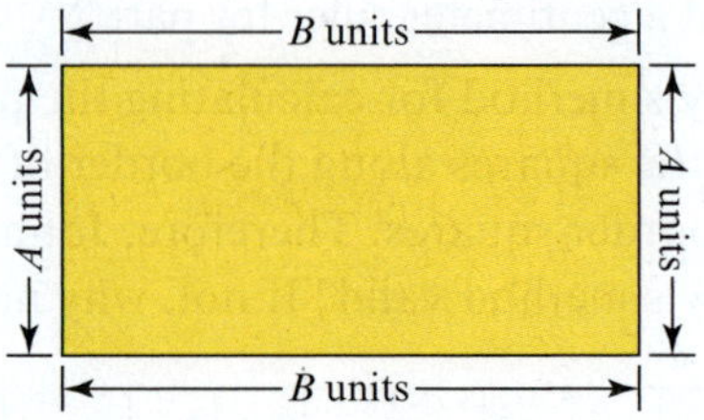

Class Activity 12S How Are Perimeter and Area Related for Rectangles?

CCSS CCSS SMP8, 3.MD.8, 4.MD.3

String would be useful for this activity.

1. If a rectangle has perimeter 20 units, then what could its area be? Draw at least 5 different rectangles of perimeter 20 units on the graph paper below. Include some rectangles that have sides whose lengths *aren't whole numbers of units.* Describe a strategy for finding rectangles whose perimeter is 20 units.

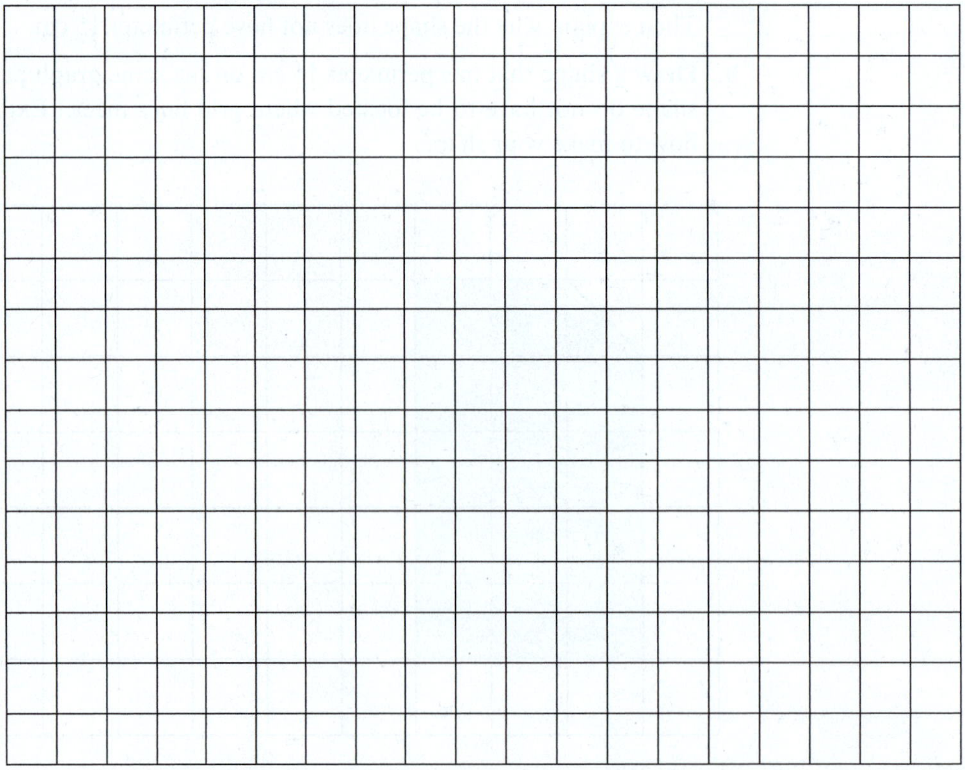

2. Find the area of each of your rectangles in part 1. In the table below, list the areas of your rectangles from part 1 in decreasing order. Below each area, draw a small sketch showing the approximate shape of the corresponding rectangle.

How are the larger-area rectangles qualitatively different in shape from the smaller-area ones?

Greater area ← → Smaller area

Area								
Shape								

3. Show how two people can use a loop of string and 4 fingers to represent all rectangles of a certain fixed perimeter.

Now consider all the rectangles of perimeter 20 units, *including those whose side lengths aren't whole numbers of units.* What are all the theoretical possibilities for the areas of those rectangles? What is the largest possible area, and what is the smallest possible area (if there is one)?

Class Activity 12T How Are Perimeter and Area Related for All Shapes?

CCSS CCSS SMP3, SMP5

String would be helpful for this activity.

1. Nick wants to find the area of the irregular shape below. He cuts a piece of string so that it goes all the way around the outside of the shape and then forms his piece of string into a square on top of graph paper. Using the graph paper, Nick gets a good estimate for the area of his string square and then uses the square's area as his estimate for the area of the original irregular shape.

   Discuss whether Nick's method is a legitimate way to estimate areas of irregular shapes.

2. Suppose that a forest on flat terrain has perimeter 200 kilometers, but there is no information on the shape of the forest. What can you say about the forest's area? Use a loop of string to help you think about this question. Then consider these questions:

   • What shape do you think would give the largest possible area for the forest? What is this area?

   • What range of areas do you think are possible for the forest?

12.9 Using the Moving and Additivity Principles to Prove the Pythagorean Theorem

Class Activity 12U Side Lengths of Squares Inside Squares

CCSS CCSS SMP7, 8.EE.2

Throughout this activity, assume that you don't yet know the Pythagorean theorem.

1. Use the moving and additivity principles (one or both) to determine the area of the shaded "tilted" square inside square (a). Then use the area of the tilted square to determine its side lengths. Repeat with square (b).

2. Draw your own "tilted squares" inside squares (c) and (d) and repeat the instructions for part 1.

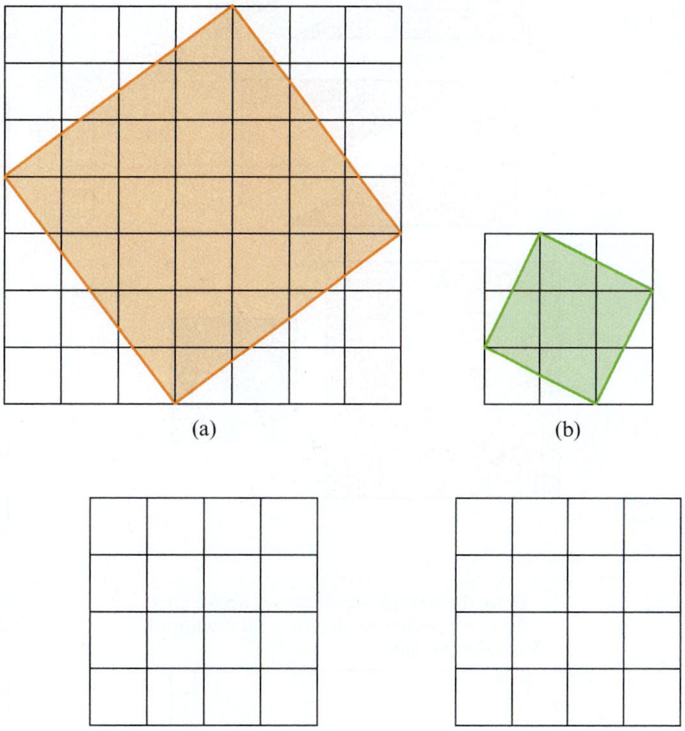

(a) (b) (c) (d)

Class Activity 12V A Proof of the Pythagorean Theorem

CCSS CCSS SMP3, 8.G.6

Scissors and Activity Download 7 might be helpful for this activity. Starting with any arbitrary right triangle, like the one below, we must explain why $a^2 + b^2 = c^2$ is true.

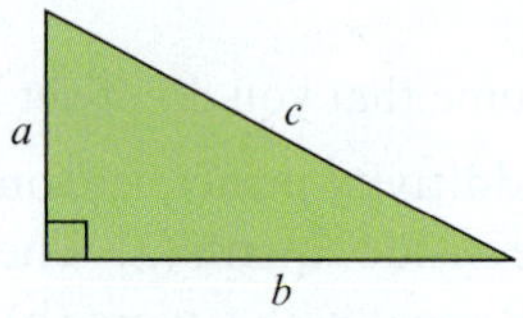

1. Follow the instructions below to show two different ways of filling a square that has sides of length $a + b$ with triangles and squares without gaps or overlaps. You may wish to cut out the squares and triangles on Activity Download 7.

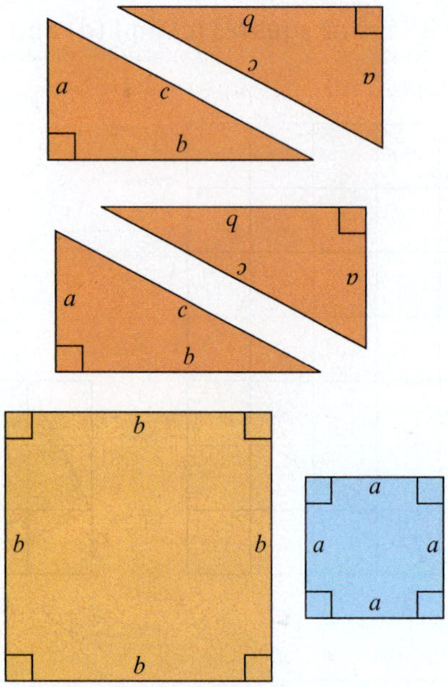

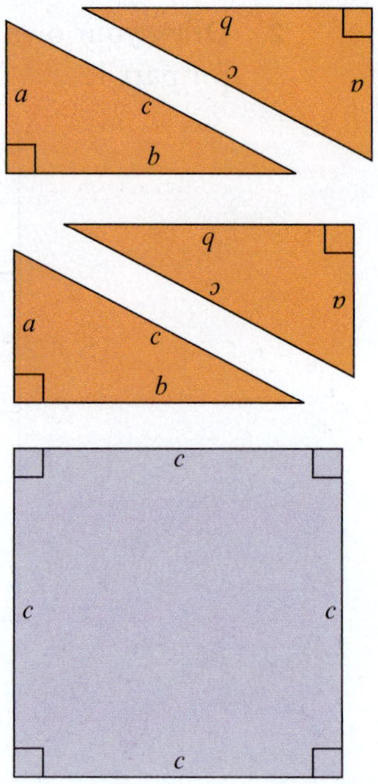

Draw the triangles and squares above into the square below so that they fill it without gaps or overlaps.

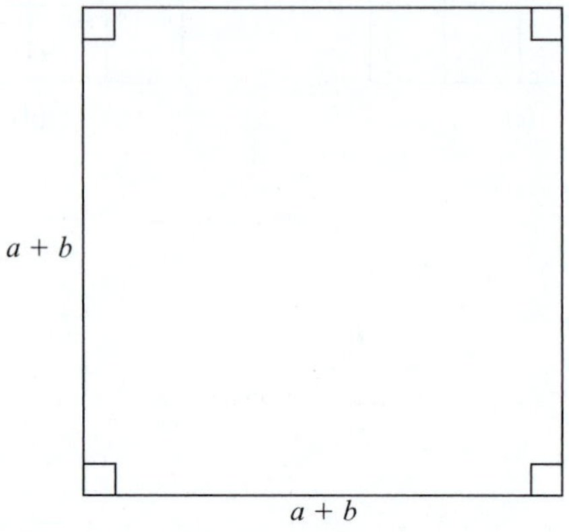

Draw the triangles and square above into the square below so that they fill it without gaps or overlaps.

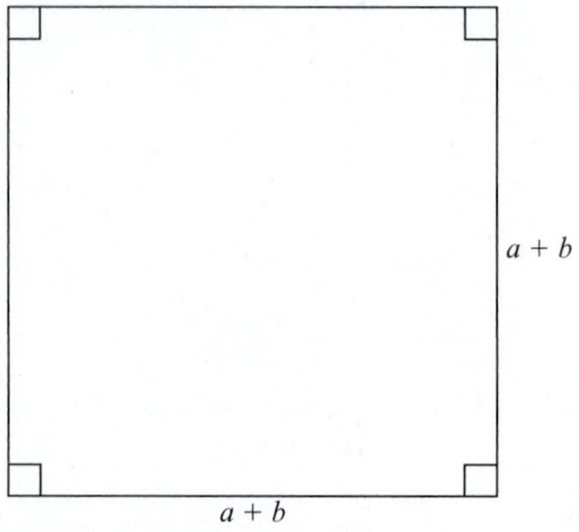

2. Now use part 1 and the moving and additivity principles about areas to explain why $a^2 + b^2 = c^2$. There are several different ways to do this.

Hint: Notice that *both* of the two ways of filling a square of side length $a + b$ use 4 copies of the original right triangle.

Summarize your proof of Pythagoras's theorem.

3. Here is a subtle point in the proof of the Pythagorean theorem: In both of the squares of side length $a + b$ on the previous page, there are places along the edges where 3 figures (2 triangles and 1 square) meet at a point. How do we know that the edge formed there really is a straight line and doesn't actually have a small bend in it, such as pictured in the figure below? We need to know that the edge there really is straight in order to know that the assembled shapes really do create large *squares* and not *octagons.* Explain why these edges really are straight. (*Hint:* Consider the angles at the points where a square and two triangles meet.)

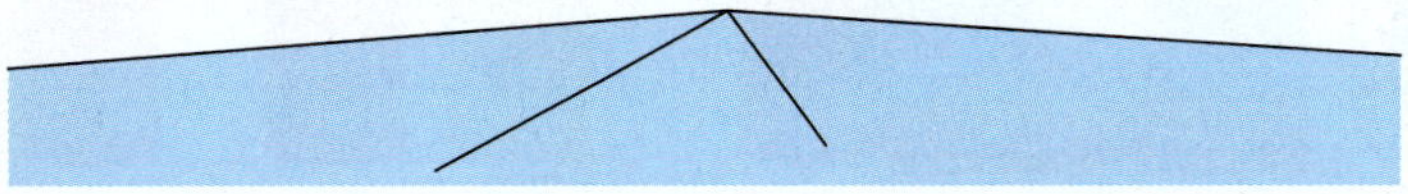

4. Here is another subtle point in the proof of the Pythagorean theorem. Why does the proof tell us about *all* right triangles? After all, we used one specific right triangle to explain the proof.

Solid Shapes and Their Volume and Surface Area

13.1 Polyhedra and Other Solid Shapes

Class Activity 13A Making Prisms and Pyramids

CCSS CCSS SMP7, K.G.5

You will need toothpicks or straws and gumdrops or modeling clay for this activity. If it is available, you could also use liquid soap.

Make the shapes listed below. In each case, visualize the shape first and predict how many toothpicks and clay balls you will need to make it.

1. Rectangular prism
2. Triangular prism
3. Pyramid with a triangle base
4. Pyramid with a square base

Some teachers like to dip these shapes into a liquid soap solution to show the faces of the shapes.

Class Activity 13B Analyzing Prisms and Pyramids

CCSS CCSS SMP7, SMP8

1. Answer the questions below *without* using a model. Use your visualization skills and look back at other models. (When you are done, verify with a model if one is available.)

- How many faces does a pentagonal prism have? Why? Describe the faces of a pentagonal prism. What kinds of shapes are they? How many of each kind of shape are there?
- How many edges and how many vertices does a pentagonal prism have? Explain.

2. Answer the questions below *without* using a model. Use your visualization skills and look back at other models. (When you are done, verify with a model if one is available.)

- How many faces does a pyramid with a hexagonal base have? Why? What kinds of shapes are they? How many of each kind of shape are there?
- How many edges and how many vertices does a pyramid with a hexagonal base have? Explain.

3. Discuss the difference between a rectangle and a rectangular prism.

Class Activity 13C What's Inside the Magic 8 Ball?

CCSS CCSS SMP7

You will need a Magic 8 Ball for parts 1 and 3 of this activity. (One or more can be shared by a class.) Scissors, tape, and Activity Downloads 8, 9, and 10 or snap-together plastic polygons would also be helpful.

1. There is a polyhedron inside the Magic 8 Ball. What can you tell about this polyhedron without breaking open the Magic 8 Ball?

2. Let's say someone wants to make a polyhedron to put inside a handmade Magic 8 Ball. This polyhedron would probably be very uniform and regular all the way around, so that all the answers would be equally likely to appear. Such a shape might have the following properties:

   - The faces are identical copies of one regular polygon; so all faces are equilateral triangles, or all faces are squares, or all faces are regular pentagons, and so forth.
   - The shape has no indentations or protrusions.
   - All the vertices are identical, so the same number of faces meet at each vertex.

   Make a guess: How many such shapes do you think there can be?

   Try to make some shapes that have the properties described above. Tape paper polygons together (use Activity Downloads 8, 9, and 10) or use snap-together plastic polygons. How many shapes can you find?

3. The actual Magic 8 Ball should contain one of the shapes that has the properties described in part 2. Which one must it be?

Class Activity 13D Making Platonic Solids

CCSS CCSS SMP1

You will need toothpicks and modeling clay (or gumdrops) for this activity. Scissors, tape, and Activity Downloads 8, 9, and 10 or snap-together plastic polygons would also be helpful.

Make all five Platonic solids by sticking toothpicks into small balls of clay. (Your dodecahedron may sag a little, but the others will be more stable.) Or make the Platonic solids by putting plastic polygons together or by using Activity Downloads 8, 9, and 10. Refer to the following descriptions of the Platonic solids:

Tetrahedron has 4 equilateral triangle faces, with 3 triangles coming together at each vertex.

Cube has 6 square faces, with 3 squares coming together at each vertex.

Octahedron has 8 equilateral triangle faces, with 4 triangles coming together at each vertex.

Dodecahedron has 12 regular pentagon faces, with 3 pentagons coming together at each vertex.

Icosahedron has 20 equilateral triangle faces, with 5 triangles coming together at each vertex.

Use your models to fill in the following table:

Shape	Number and Type of Faces	Number of Edges	Number of Vertices
Tetrahedron	4 equilateral triangles		
Cube	6 squares		
Octahedron	8 equilateral triangles		
Dodecahedron	12 regular pentagons		
Icosahedron	20 equilateral triangles		

Do you notice any relationships between the numbers?

13.2 Patterns and Surface Area

Class Activity 13E What Shapes Do These Patterns Make?

CCSS CCSS SMP7, SMP8, 6.G.4

Scissors, tape, and Activity Downloads 11, 12, and 13 would be helpful for this activity.

Small patterns for solid shapes are shown below. *Visualize* the shapes that these patterns will make. What shapes are they?

Compare the patterns and note similarities. Which shapes will be related or alike, and how will these shapes be related or alike?

Use Activity Downloads 11, 12, and 13 to make the shapes and check your answers.

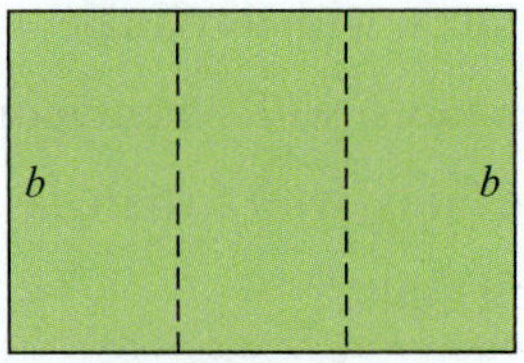

Pattern 1

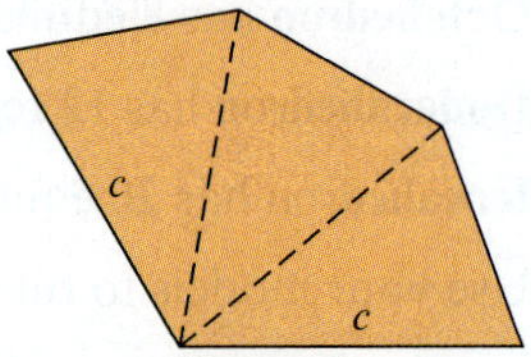

Pattern 2

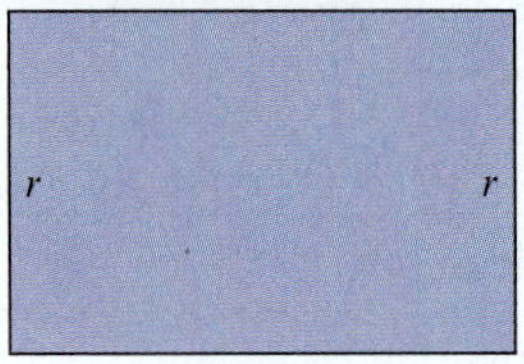

Pattern 3

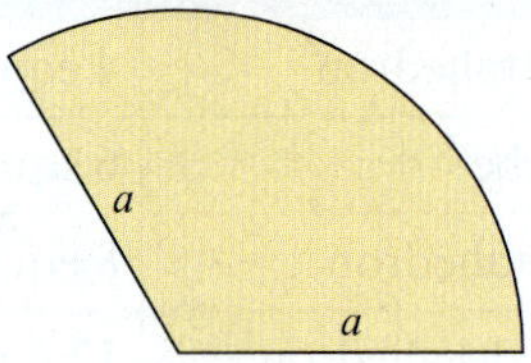

Pattern 4

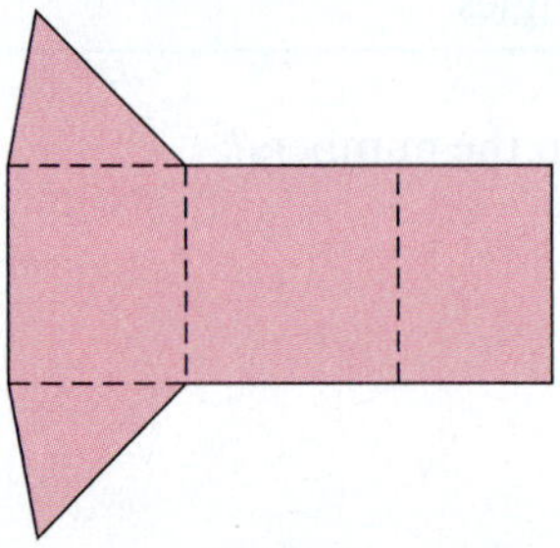

Pattern 5

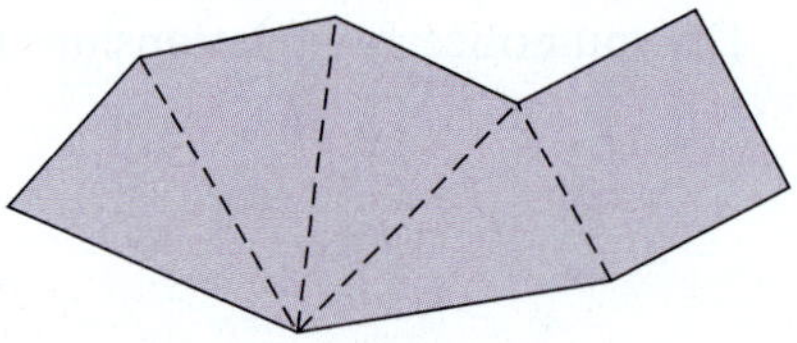

Pattern 6

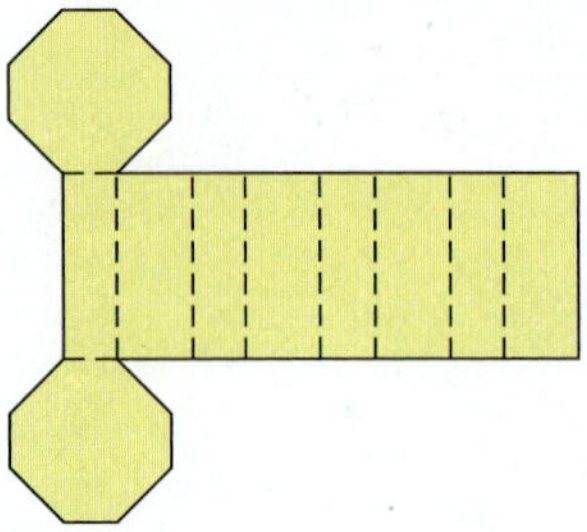

Pattern 7

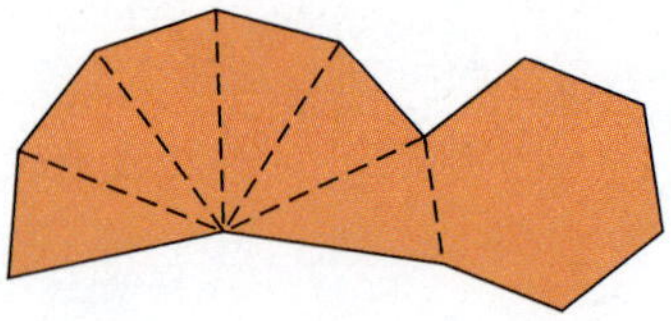

Pattern 8

Class Activity 13F Patterns and Surface Area for Prisms and Pyramids

CCSS CCSS SMP7, 6.G.4, 7.G.6, 8.G.7

You will need scissors and centimeter graph paper (use Activity Download 14) for this activity.

1. Use graph paper to make a pattern for a prism whose bases are a rectangle that is not a square. Include the bases in your pattern.

   Let's say that the width, length, and height of your prism are W cm, L cm, and H cm respectively (it's up to you which edge length you call which). Show which lengths on your pattern are W, L, and H. Then find a formula in terms of W, L, and H for the total surface area of a prism and explain your reasoning.

   Cut out your pattern and see if it works.

2. On centimeter graph paper (use Activity Download 14), make a pattern for a pyramid with a 6-cm-by-6 cm square base such that the apex will be 4 cm above the center of the base. Then find the surface area of the pyramid.

Class Activity 13G Patterns and Surface Area for Cylinders

CCSS CCSS SMP1, 7.G.6

You will need blank paper, scissors, and tape for this activity.

1. Take a standard 8.5-inch-by-11-inch piece of paper, roll it up, and tape it, without overlapping the paper, to make a cylinder without bases, as shown in the figure below. What is the area of the surface of the cylinder not including the bases? Why?

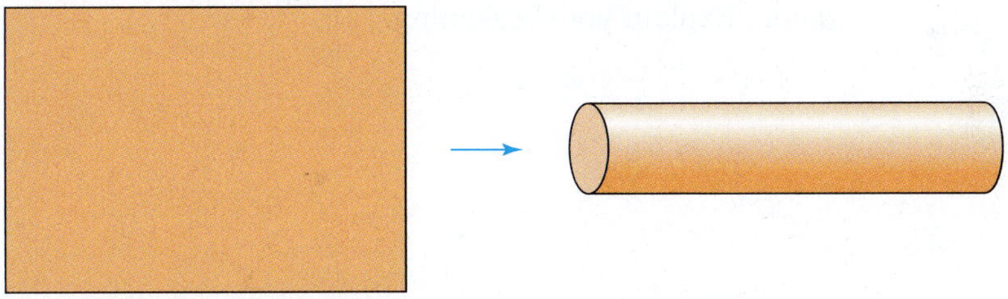

   What is the surface area of the cylinder (including the bases)? Explain.

2. A company wants to manufacture tin cans that are 3 inches in diameter and 4 inches tall. Describe the shape and dimensions of the paper label the company will need to wrap around the side of each can. Explain your reasoning.

Class Activity 13H Patterns and Surface Area for Cones

CCSS CCSS SMP1, SMP7, 7.G.4

You will need blank paper, scissors, and a compass for this activity.

1. How can you make a pattern for the lateral portion (i.e., the side portion, which does not include the base) of a cone? The pattern should not require the paper to overlap. Describe which edges would be joined to make the cone from the pattern.

2. On a piece of paper, draw half of a circle that has radius 5 inches. (*Suggestion:* Put the point of your compass in the middle of the long edge of the piece of paper.) Cut out the half-circle, and attach the two radii on the straight edge (the diameter) to form a cone without a base. Calculate the radius of the circle that will form a base for your cone. Explain your reasoning. Then draw this circle, and verify that it does form the base of the cone.

3. Make a pattern for a cone such that the base is a circle of radius 2 inches and the cone without the base is made from a half-circle. Determine the total surface area of your cone. Explain your reasoning.

4. Make a pattern for a cone such that the base is a circle of radius 2 inches and the lateral portion of the cone is made from part of a circle of radius 6 inches. What fraction of the 6-inch circle will you need to use? Determine the total surface area of your cone. Explain your reasoning.

Class Activity 13I Cross-Sections of a Pyramid

CCSS CCSS SMP7, 7.G.3

Modeling clay and dental floss would be helpful for part 2.

The following picture shows a pyramid with a square base like an Egyptian pyramid:

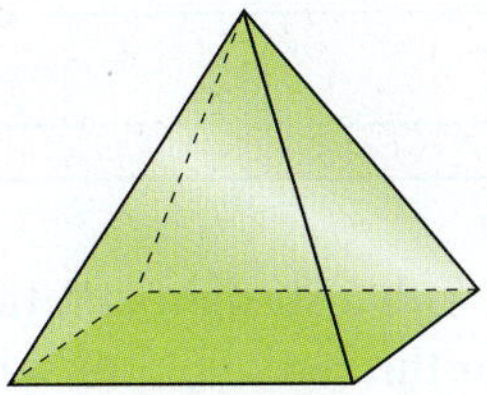

1. Visualize a plane slicing through the pyramid. The places where the plane meets the pyramid form a shape in the plane. Which shapes in the plane can be made this way, as a cross-section of the pyramid?

   List some plane shapes that you think *can* occur as a cross-section of the pyramid.

   List some plane shapes that you think *cannot* occur as a cross-section of the pyramid.

2. If available, use modeling clay to make a pyramid. Slice straight through the pyramid with dental floss, as if you were slicing the pyramid with a plane. Observe the cross-section that you create this way (i.e., the plane shape where the pyramid was cut). Put the pyramid back together and slice it in a different way.

   List the cross-sections you found by slicing the pyramid in various ways.

3. Think more about which plane shapes can and cannot occur as a cross-section of the pyramid. For each of the following shapes, either explain how the shape can occur as a cross-section of the pyramid or explain why it cannot occur: a trapezoid, a pentagon, a hexagon.

Class Activity 13J Cross-Sections of a Long Rectangular Prism

CCSS CCSS SMP7, 7.G.3

Modeling clay and dental floss would be helpful for this activity.

Suppose you have a very long 2-inch-by-4-inch board in the shape of a rectangular prism as pictured here:

If you saw through the board with a straight cut, the place where the board was cut makes a shape in a plane (a cross-section). In this activity, consider only cuts that go through the middle of the board, *not through the ends.*

1. Is it possible to saw the board with a straight cut in such a way that the shape formed by the cut is a square? Try to visualize if this is or is not possible.

2. Is it possible to saw the board with a straight cut in such a way that the shape formed by the cut is *not* a rectangle? (A square *is* a kind of rectangle.) Try to visualize if this is or is not possible. If the answer is yes, what kind of shape other than a rectangle can you get?

3. Use modeling dough to make a model of a long 2-inch-by-4-inch board. Make a straight slice through your model with dental floss. Describe the plane shape that the cut made. Now restore your model of the board to its original shape and make a different slice through your model. Keep trying different ways to slice your model. Describe all the different shapes you get where the model was cut.

4. Are there any plane shapes that can't arise from slicing the board? Give some examples. How do you know they can't arise? In particular, are these shapes possible: a pentagon? a quadrilateral that is not a parallelogram?

13.3 Volumes of Solid Shapes

Class Activity 13K Why the Volume Formula for Prisms and Cylinders Makes Sense

CCSS CCSS SMP3, SMP7, 5.MD.5a

Cubic-inch blocks would be helpful for this activity.

1. Build prisms of the indicated heights on top of the bases shown below.

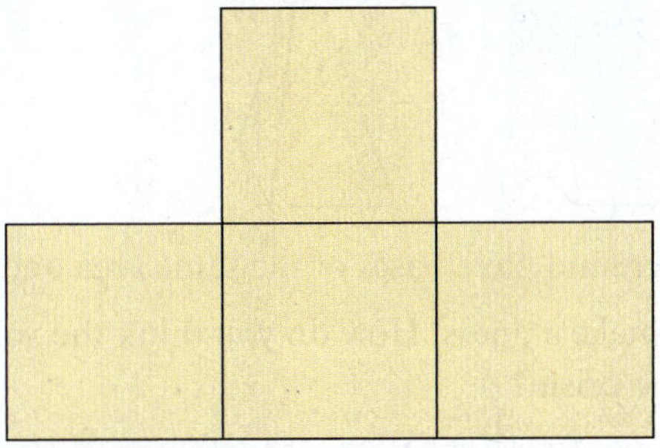

Build a 3-inch-tall prism that has this base.

Number of layers: ____________

Number of cubes in each layer: ____________

Expression for total number of cubes: ________

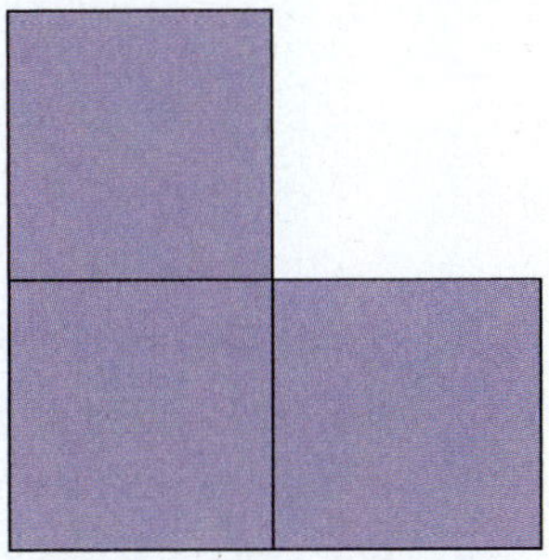

Build a 4-inch-tall prism that has this base.

Number of layers: ____________

Number of cubes in each layer: ____________

Expression for total number of cubes: ________

Then explain why the following volume formula for right prisms and cylinders makes sense:

$$\text{volume} = (\text{number of layers}) \cdot (\text{number of cubes in each layer})$$

2. Use the formula in part 1 to obtain this volume formula for right prisms and cylinders:

$$\text{volume} = (\text{height}) \cdot (\text{area of the base})$$

In this formula, the height is a *length* and the area of the base is an *area.* How are this length and area *linked to yet different from* the number of layers and the number of cubes in each layer in the formula in part 1?

3. Use the result of part 2 and Cavalieri's principle to explain why the formula

$$\text{volume} = (\text{height}) \cdot (\text{area of base})$$

gives the correct volume for an *oblique* prism or cylinder. Explain why the height should be measured perpendicular to the bases, and not "on the slant."

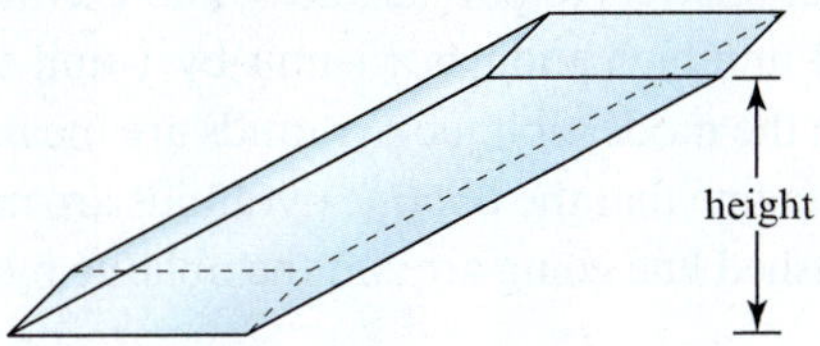

Class Activity 13L Comparing the Volume of a Pyramid with the Volume of a Rectangular Prism

CCSS CCSS SMP2, 8.G.9

You will need Activity Downloads 15 and 16 and scissors, tape, and dry beans or rice for this activity.

1. Cut out, fold, and tape the patterns on Activity Downloads 15 and 16 to make an open rectangular prism and an open pyramid with a square base.

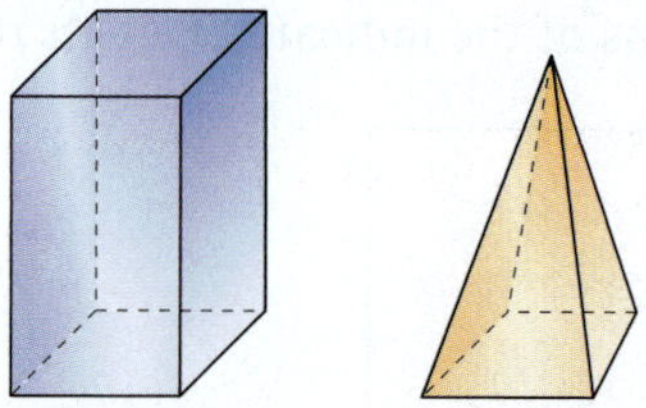

2. Verify that the prism and the pyramid have bases of the same area and have equal heights.

   Just by looking at your shapes, make a guess: How do you think the volume of the pyramid compares with the volume of the prism?

3. Now fill the pyramid with beans, and pour the beans into the prism. Keep filling and pouring until the prism is full. Based on your results, fill in the blanks in the equations that follow:

$$\text{volume of prism} = \_\_ \cdot \text{volume of pyramid}$$
$$\text{volume of pyramid} = \_\_ \cdot \text{volume of prism}$$

Class Activity 13M The $\frac{1}{3}$ in the Volume Formula for Pyramids and Cones

CCSS CCSS SMP2, 8.G.9

You will need Activity Download 17, scissors, and tape for this activity.

This class activity will help you see where the $\frac{1}{3}$ in the volume formula for prisms comes from.

1. Cut out three of the four patterns on Activity Download 17. The fourth pattern is a spare. Fold these patterns along the *undashed* line segments, and glue or tape them to make three *oblique* pyramids. Make sure the dashed lines appear on the outside of each oblique pyramid.

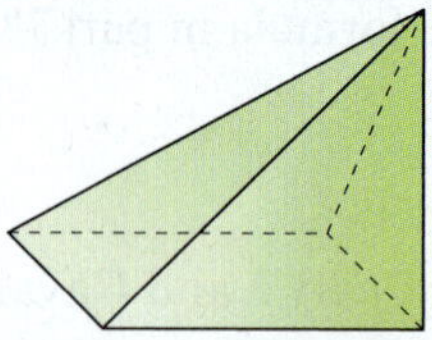

2. Fit the three oblique pyramids together to make a familiar shape. What shape is it? What is the volume of the shape formed from the three oblique pyramids? Therefore, what is the volume of one of the oblique pyramids?

3. Use your answers to parts 2 and 3, and Cavalieri's principle, to explain why a *right* pyramid that is 1 unit high and has a 1-unit-by-1-unit square base has volume $\frac{1}{3}$ cubic units. (The dashed lines on the model oblique pyramids are meant to help you see how to shear the oblique pyramid. Imagine that the oblique pyramids are made out of a stack of small pieces of paper, where each dashed line going around the oblique pyramid represents a piece of paper.)

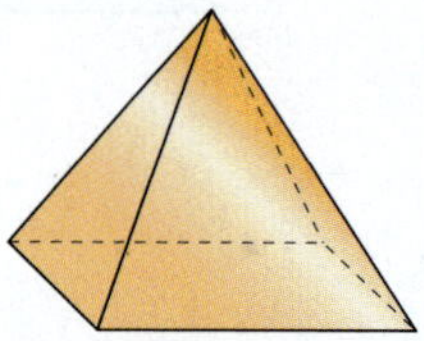

Class Activity 13N Volume Problem Solving

CCSS CCSS SMP1, 7.G.6, 8.G.9

1. A cone without a base is made from a half-circle of radius 10 cm. Determine the volume of the cone. Explain your reasoning.

2. Consider the block shown below.

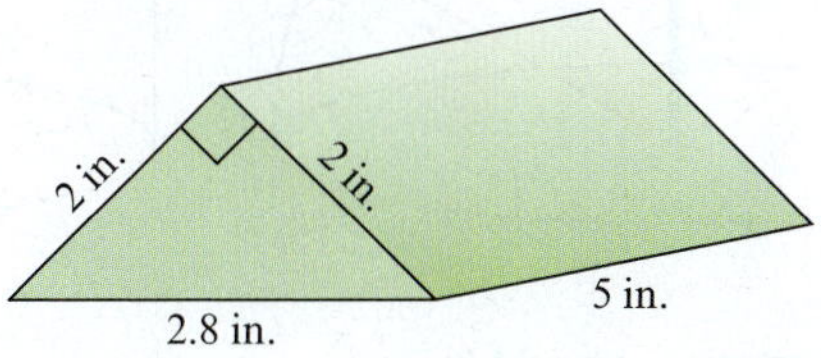

   a. Which of the following can be applied to determine the volume of the block: the pyramid volume formula, the prism volume formula, both, or neither? Explain, and determine the volume of the block with a formula if it is possible to do so.

   b. Determine the volume of the block in another way and explain your reasoning.

3. Determine the volume of the staircase shown below in two different ways.

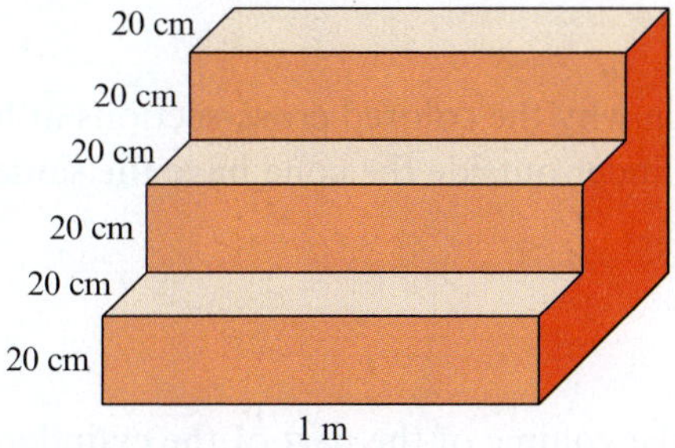

Class Activity 13O Deriving the Volume of a Sphere

CCSS CCSS SMP3

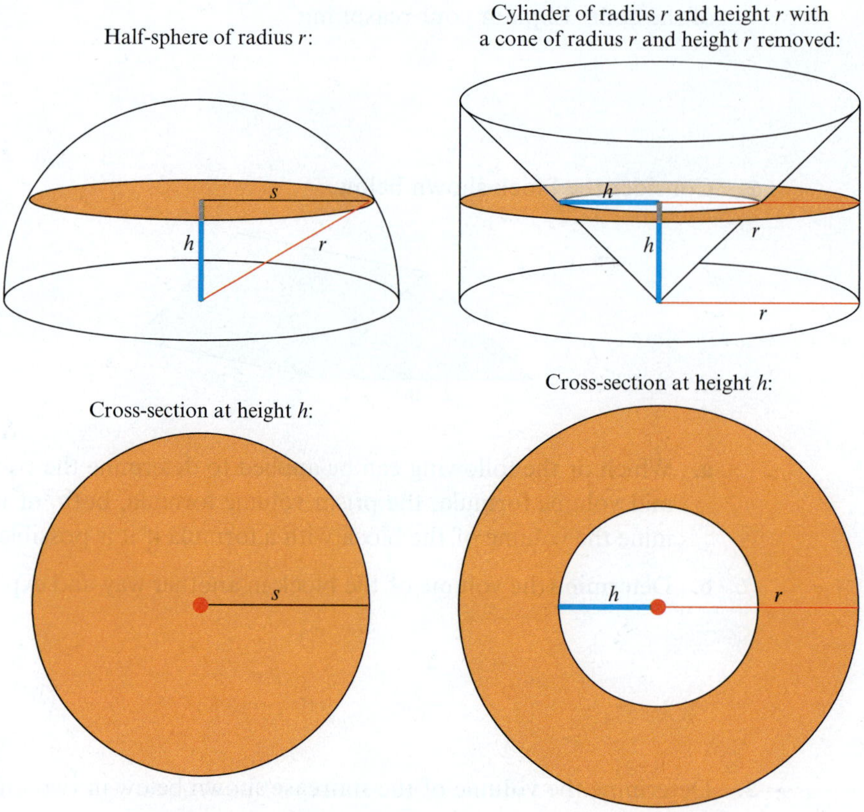

1. Explain why the colored cross-sections at height h of the half-sphere and of the part of the cylinder outside the cone have the same area.

2. Find the volume of the part of the cylinder that is outside the cone. By Cavalieri's principle and part 1, this volume is the same as the volume of the hemisphere. Use this to verify the volume formula for a sphere of radius r units.

Class Activity 13P Volume Versus Surface Area and Height

CCSS CCSS SMP1, SMP2

You will need several blank pieces of paper (including graph paper if available), a ruler, scissors, and tape for parts of this activity.

1. Young children sometimes think that taller containers necessarily hold more than shorter ones. Make or describe two open-top boxes such that the taller box has a smaller volume than the shorter box.

2. Students sometimes get confused between the volume and surface area of a solid shape and about how to calculate volume and surface area.

   a. Discuss the distinction between volume and surface area.

   b. Discuss the differences and similarities in the way we calculate the volume and surface area of a rectangular prism.

3. *A cylinder volume contest:* Each team starts with a single rectangular piece of paper that is A cm by $2A$ cm, so it is twice as long as it is wide. The team cuts their paper apart and tapes it together to make a cylinder with no bases. Each team must write a formula in terms of A for the volume of their cylinder. The team that makes the cylinder of largest volume wins the contest.

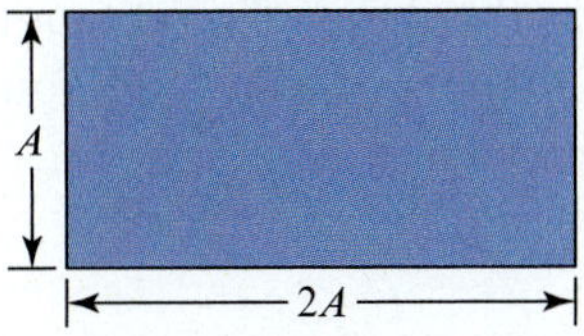

   a. Would it be possible to make a cylinder of even larger volume than the winning team's cylinder?

   b. What is the area of the lateral surface of the winning cylinder (not including the bases)? How does it compare to the other teams' cylinders?

   c. What is the height of the winning cylinder? How does it compare to the other teams' cylinders?

13.4 Volume of Submersed Objects Versus Weight of Floating Objects

Class Activity 13Q Underwater Volume Problems

CCSS CCSS SMP1, 6.G.2

1. A container can hold 2 liters. Initially, the container is $\frac{1}{2}$ full of water. When an object is placed in the container, the object sinks to the bottom and the container becomes $\frac{2}{3}$ full. What is the volume of the object in cubic centimeters? Explain.

2. A tank in the shape of a rectangular prism is 50 cm tall, 80 cm long, and 30 cm wide. First, some rocks are placed at the bottom of the tank. Then 80 liters of water are poured into the tank. At that point, the tank is $\frac{3}{4}$ full. What is the total volume of the rocks in cubic meters? Explain.

3. A fish tank in the shape of a rectangular prism is 1 m long and 30 cm wide. The tank is $\frac{1}{2}$ full. Then 30 liters of water are poured in and the tank becomes $\frac{2}{3}$ full. How tall is the tank? Explain your reasoning.

Class Activity 13R Floating Versus Sinking: Archimedes's Principle

CCSS CCSS SMP2, SMP4

For this activity you will need a milliliter measuring cup, water, and modeling clay that does not dissolve in water. In part 3 you will need a scale for weighing. In part 4 you will need a small paper cup and several coins or other small, heavy objects that fit inside the cup.

1. Pour water into your measuring cup and note the volume of water in milliliters. Form your clay into a "boat" that will float. By how much does the water level rise when you float your clay boat? Does this increase in water level tell you the volume of the clay?

2. Predict what will happen to the water level when you sink your clay boat: Will the water level go up or will the water level go down? Sink your boat and see if your prediction was correct.

   Did the increase in water level in part 1 tell you the volume of the clay, or not? Explain.

3. Exactly how much water does a floating object displace? Do the following to find out:
   a. Weigh your clay boat from part 1.
   b. Weigh an amount of water that weighs as much as your clay boat.
   c. Float your clay boat in the measuring cup and record the water level.
   d. Remove your clay boat from the water and pour the water you measured in part (b) into the measuring cup. Compare the water level now to the water level in part (c). If you did everything correctly, these two water levels should be the same.

      This experiment illustrates **Archimedes's principle** that a floating object displaces an amount of water that weighs as much as the object.

4. Determine approximately how much a quarter weighs in grams by using Archimedes's principle. Observe how much the water level rises when you put several quarters in a small cup floating in water in a measuring cup.

CLASS ACTIVITIES 14

Geometry of Motion and Change

14.1 Reflections, Translations, and Rotations

Class Activity 14A Exploring Reflections, Rotations, and Translations with Transparencies

CCSS CCSS SMP5, 8.G.1, 8.G.3

You will need a transparency, transparency markers, a paperclip, and a coordinate grid, such as Activity Download 18.

1. Draw a trapezoid in the top right portion of your coordinate grid. Label 4 points on your trapezoid A, B, C, and D. Put the transparency on top of your coordinate grid, draw your trapezoid onto the transparency, and label the corresponding points as A', B', C', and D'. Draw the x and y-axes on your transparency.

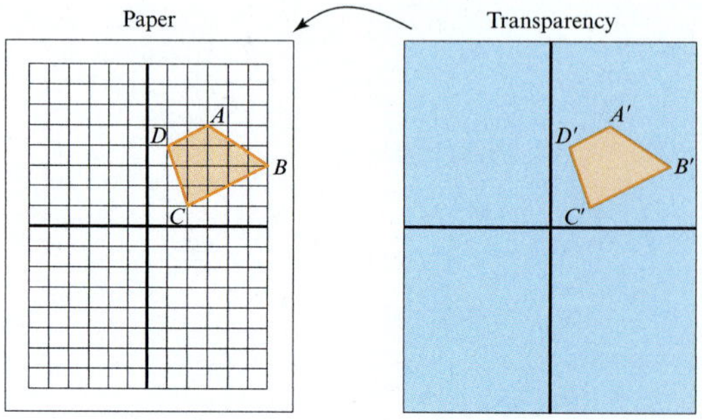

2. Perform each of the transformations in (a)–(f). Start with the transparency on top of the paper coordinate grid with the trapezoids aligned. Leave the paper fixed and move your transparency as described. In each case, answer the following: (i) is the image of a line segment a line segment of the same length? (ii) is the image of an angle an angle of the same size? (iii) are the images of parallel lines parallel lines?

a. Reflect across the x-axis by twirling the transparency around the location of the x-axis so as to flip the transparency upside down.

b. Reflect across the y-axis by twirling the transparency around the location of the y-axis so as to flip the transparency upside down.

c. Rotate 180° around the origin by pressing the tip of an unbent paperclip onto the transparency over the origin and rotating the transparency.

d. Rotate 90° counterclockwise around the origin using a paperclip as before.

e. Rotate 90° clockwise around the origin using a paperclip as before.

f. Translate by an arrow whose tail is at the origin and whose head is at the point (____, ____) (your choice).

3. Perform the transformations in parts (a)–(f) again. This time record the coordinates of A, B, C, D and their images. Describe how the coordinates of each point and its image are related.

Class Activity 14B Reflections, Rotations, and Translations in a Coordinate Plane

CCSS CCSS SMP7, SMP8, 8.G.3

As you draw the result of translating, reflecting, or rotating a shape, it will help you to consider where the vertices of the shape will go. It may also help you to consider the relative locations of vertices within the shape.

1. **a.** Draw the result of translating the shaded shapes in the coordinate plane below according to the direction and the distance given by the arrow. Explain how you know where to draw your translated shapes.

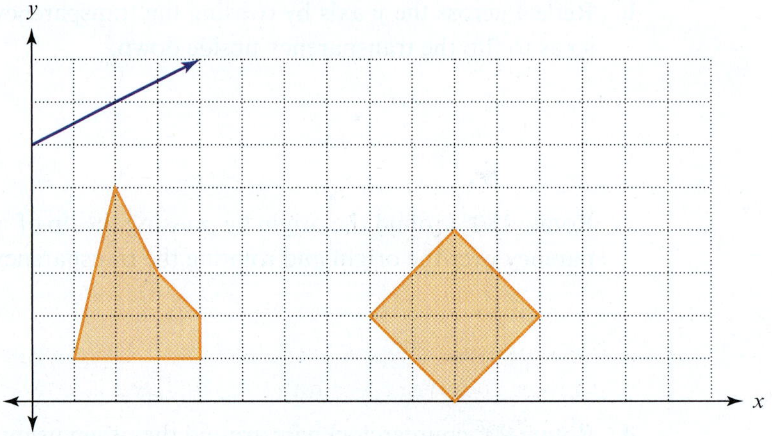

b. For the translation in part (a), what is the image of a point (a, b)?

2. **a.** Draw the result of reflecting the shaded shapes in the coordinate planes below across the y-axis. Then reflect the original shapes across the x-axis. Explain how you know where to draw your reflected shapes.

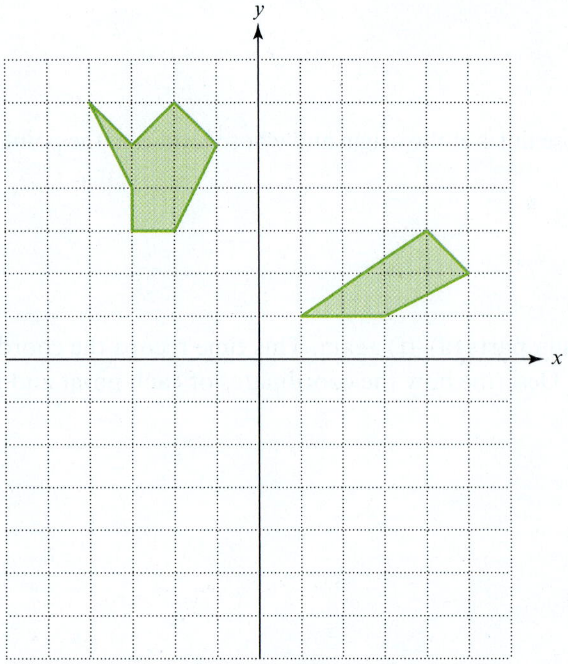

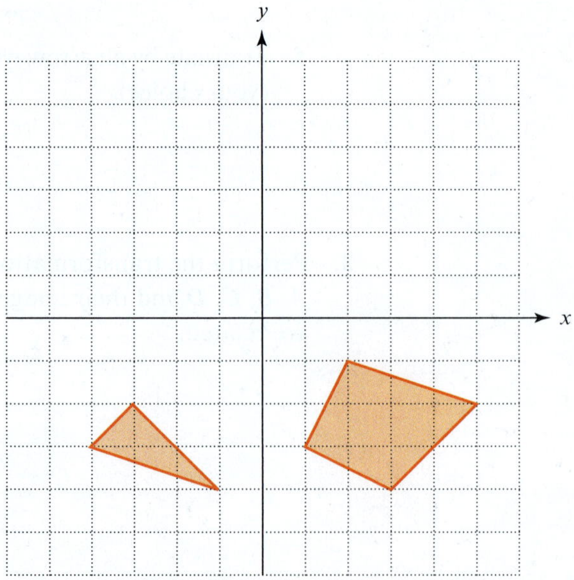

b. What is the image of a point (a, b) when it is reflected across the y-axis?

3. **a.** Draw the result of rotating the shaded shapes in the coordinate planes below by 180° around the origin (where the x- and y-axes meet). Explain how you know where to draw your rotated shapes.

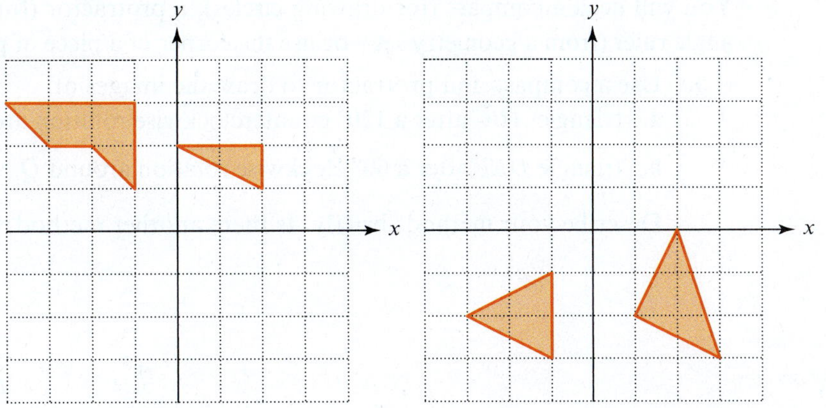

b. What is the image of a point (a, b) when it is rotated 180° around the origin?

4. **a.** Draw the result of rotating the shaded shapes in the coordinate planes below by 90° counterclockwise around the origin. Explain how you know where to draw your rotated shapes.

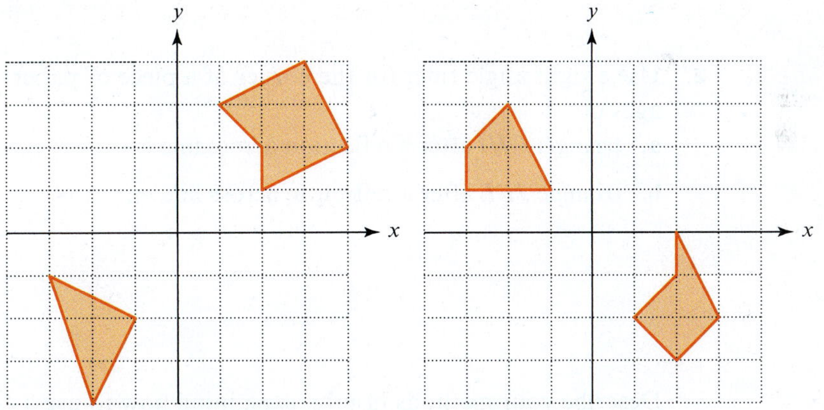

b. What is the image of a point (a, b) when it is rotated 90° counterclockwise around the origin?

Class Activity 14C Rotating and Reflecting with Geometry Tools

CCSS CCSS SMP5

You will need a compass (for drawing circles), a protractor (for measuring angles), and a right angle ruler (from a geometry set—or use the corner of a piece of paper).

1. Use a compass and protractor to draw the images of:
   a. triangle ABC after a 120° counterclockwise rotation around P;
   b. triangle DEF after a 60° clockwise rotation around Q.

   Describe your methods briefly. Is there another method?

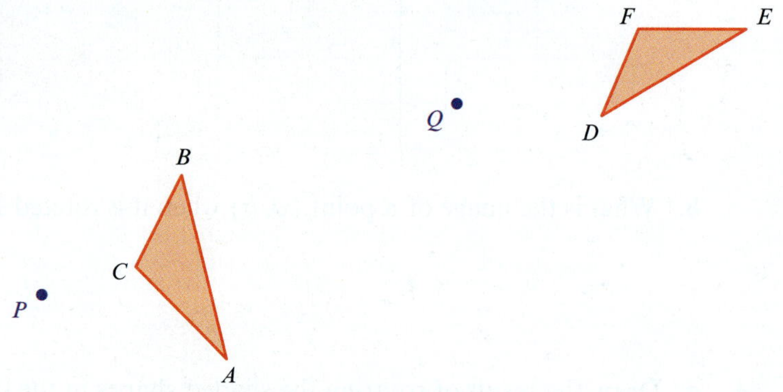

2. Use a right angle ruler (or the corner of a piece of paper) and compass to draw the images of:
   a. triangle GHI after a reflection across line ℓ;
   b. triangle JKL after a reflection across line m.

   Describe your methods briefly, explaining how to use a compass instead of a ruler to create desired distances. Is there another method?

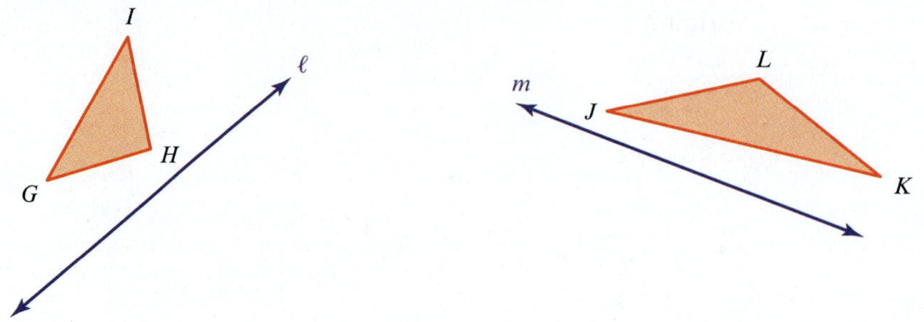

Class Activity 14D Which Transformation Is It?

CCSS CCSS SMP1, SMP7

1. The figures below show two transformations taking points A, B, C, D to points A', B', C', D'. What kind of transformations are they? How can you tell?

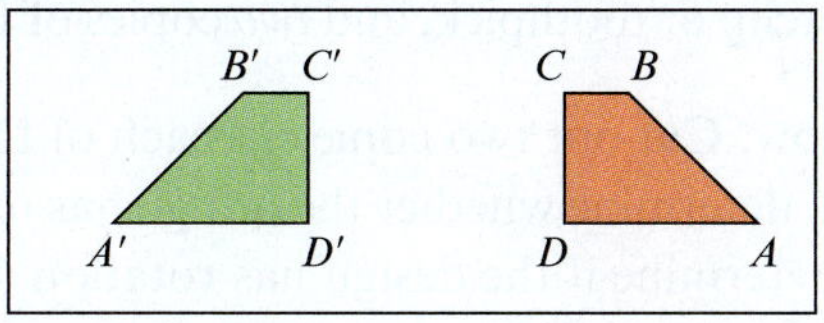

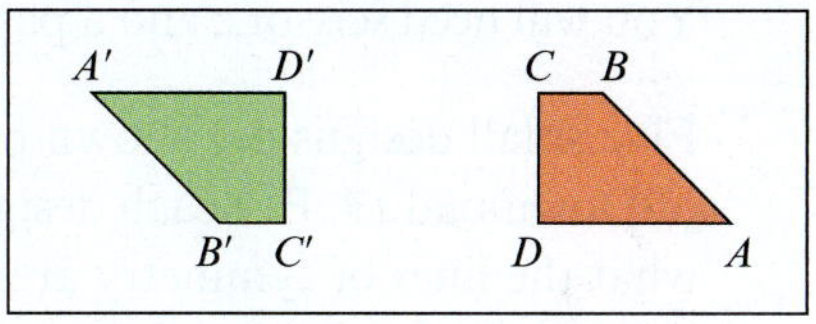

2. The figures below show two transformations taking points A, B, C to points A', B', C'. What kind of transformations are they? How can you tell?

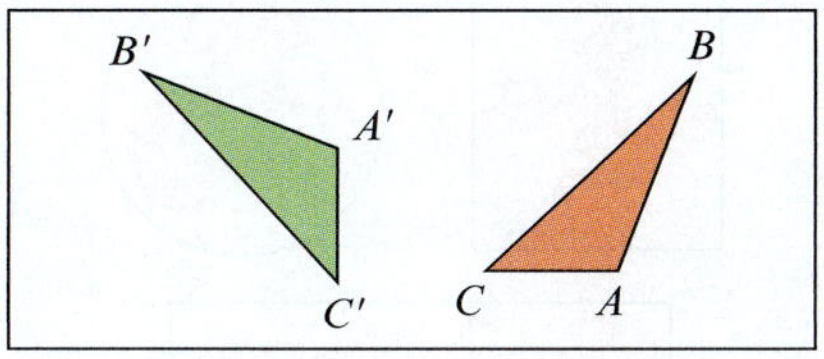

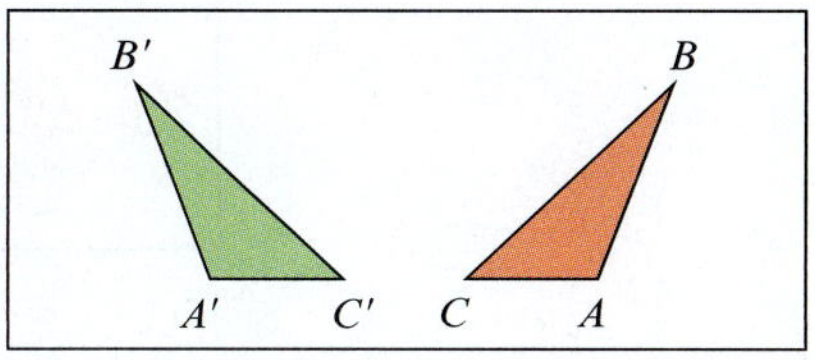

3. Does the figure show translation by the arrow that is shown? Why or why not?

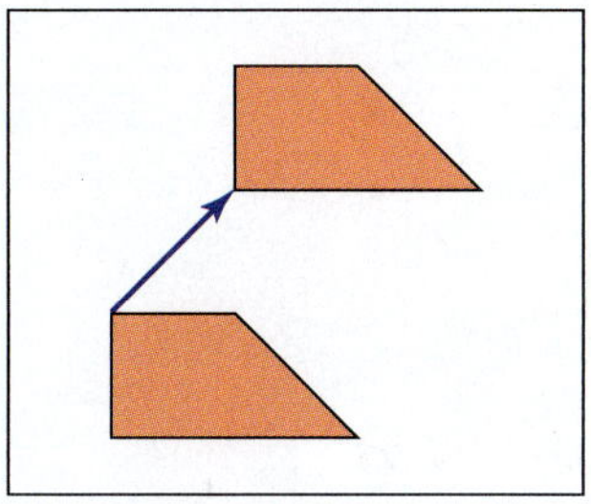

14.2 Symmetry

Class Activity 14E Checking for Symmetry

CCSS CCSS SMP7, 4.G.3

You will need scissors, and a paperclip or toothpick, and *two* copies of Download 19 for this activity.

Five small designs are shown below. Cut out two copies of each of these designs from two copies of Download 19. For each design, determine whether the design has reflection symmetry, and if so, what the lines of symmetry are; determine if the design has rotation symmetry, and if so, whether it has 2-fold, 3-fold, 4-fold, or other rotation symmetry; and determine whether the design has translation symmetry. Note that true translation symmetry can occur only with designs that are infinitely long, so you may need to imagine a design continuing on forever.

To check for symmetry you may want to put one copy of a design over the other, hold them up to a light, and move one copy while you keep the other fixed.

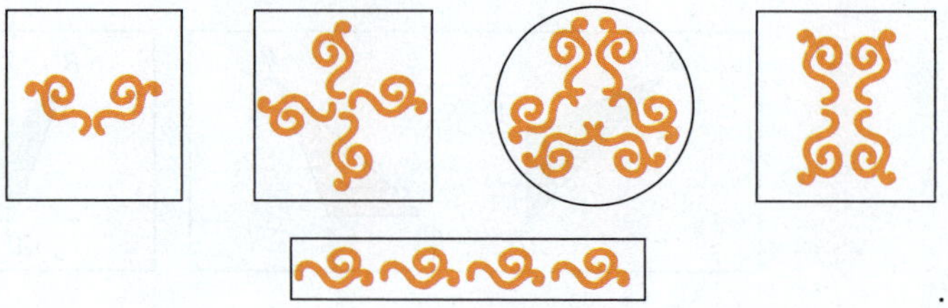

14.3 Congruence

Class Activity 14F Motivating a Definition of Congruence

CCSS CCSS SMP6, 8.G.2

You will need scissors and Download 20 (this one download is enough for four people).

Cut out one copy of Triangle 1 and one copy of Triangle 2 from Download 20 and share the rest of the triangles with your classmates. Put the two triangles on the desk in front of you and move them on the desk until you can determine if they are identical copies of each other. You may also flip the triangles upside down.

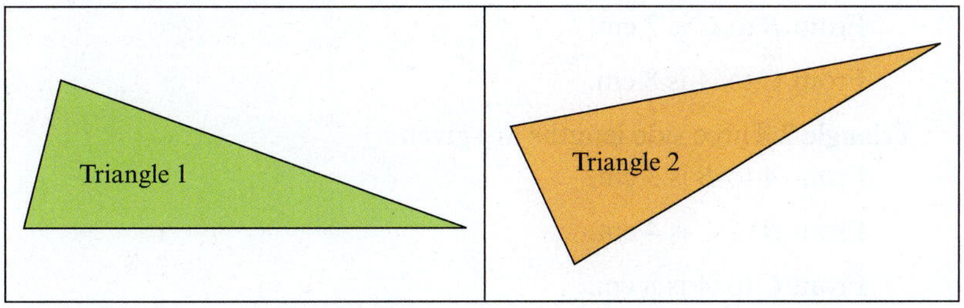

Class Activity 14G Triangles and Quadrilaterals of Specified Side Lengths

CCSS CCSS SMP7

You will need scissors, several straws, and some string for this activity.

1. Cut a 3-inch, a 4-inch, and a 5-inch piece of straw, and thread all three straw pieces onto a piece of string. Tie a knot so as to form a triangle from the three pieces of straw.
2. Now cut two 3-inch pieces of straw and two 4-inch pieces of straw, and thread all four straw pieces onto another piece of string in the following order: 3-inch, 4-inch, 3-inch, 4-inch. Tie a knot so as to form a quadrilateral from the four pieces of straw.

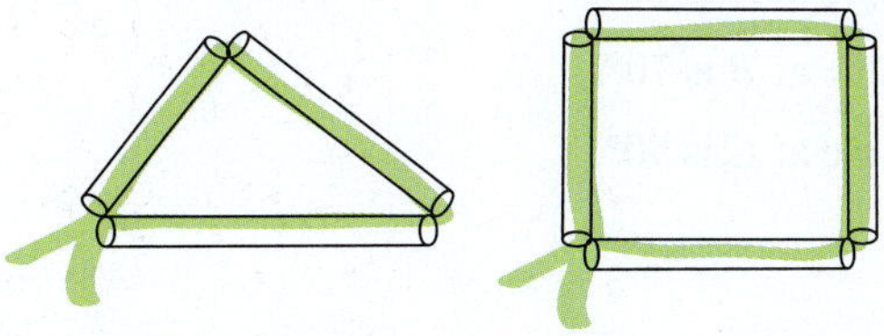

3. Compare your straw triangle and your straw quadrilateral. What is an obvious difference between them (other than the fact that the triangle is made of three pieces and the quadrilateral is made of four)?
4. When you made your triangle, if you had strung your three pieces of straw in a different order, would your triangle be different or not?

Class Activity 14H What Information Specifies a Triangle?

CCSS SMP5, 7.G.2

You will need a compass, protractor, ruler, and paper for this activity.

For each of the following, try to draw a triangle that has vertices A, B, and C and has the given specifications. Is there such a triangle? If so, think about whether any other such triangle will necessarily be congruent to yours or not. Then compare your triangle to a neighbor's. Are they congruent or not?

Triangle 1 Three side lengths are given:

From A to B is 6 cm.

From B to C is 7 cm.

From C to A is 8 cm.

Triangle 2 Three side lengths are given.

From A to B is 3 cm.

From B to C is 4 cm.

From C to A is 8 cm.

Triangle 3 Two side lengths and the angle between them are given:

From A to B is 5 cm.

The angle at A is 40°.

From A to C is 7 cm.

Triangle 4 A side length and the angle at both ends are given:

The angle at A is 30°.

From A to B is 8 cm.

The angle at B is 45°.

Triangle 5 Two side lengths and an angle that is not between them are given:

The angle at A is 20°.

From A to B is 8 cm.

From B to C is 4 cm.

Triangle 6 All three angles are given:

The angle at A is 20°.

The angle at B is 70°.

The angle at C is 90°.

Class Activity 14I Why Do Isosceles Triangles and Rhombuses Decompose into Congruent Right Triangles?

CCSS CCSS SMP3

1. Given that in triangle *ABC*, sides *AB* and *AC* have the same length, explain why the triangle decomposes into two congruent right triangles. To start your explanation, assume that we can do either of the following:
   - Add a line segment from *A* to the midpoint *M* of side *BC* (so that *BM* and *MC* are the same length).
   - Add a line segment from *A* to side *BC* that divides the angle at *A* in half.

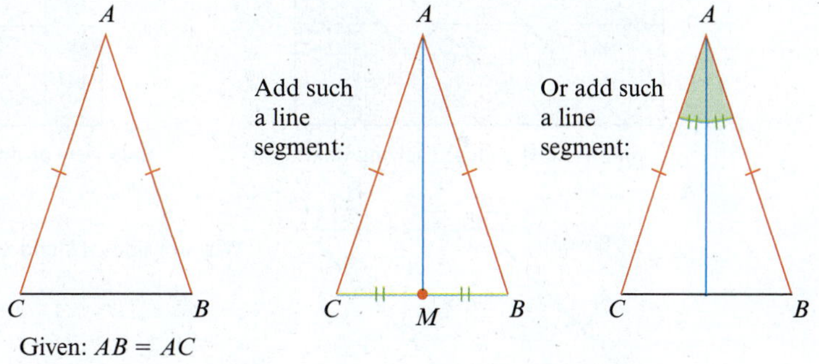

2. What is wrong with this proposed explanation for part 1:

   Add a line segment from *A* to the midpoint *M* of side *BC* that is perpendicular to side *BC* and divides the angle at *A* in half. Then because angles *BAM* and *CAM* are the same size and angles *BMA* and *CMA* are both right angles and therefore the same, it follows that angles *MBA* and *MCA* are also the same size (because the sum of the angles in a triangle is 180°). So all angles and all side lengths are the same in triangles *BAM* and *CAM*, and so those triangles are congruent right triangles.

3. Use the result of part 1 to explain why a rhombus decomposes into four congruent right triangles.

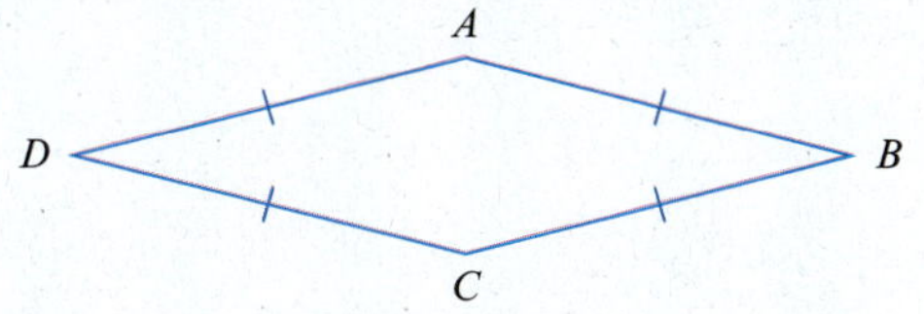

Class Activity 14J Sewing Boxes and Congruence

CCSS CCSS SMP3, SMP4

Sewing boxes and tool boxes are sometimes constructed as shown in the side view below so that *AD* and *BC* are the same length and *AB* and *CD* are the same length. Explain why sides *AB* and *CD* are guaranteed to remain parallel (which ensures that the contents of the top drawer won't spill out when opening the box).

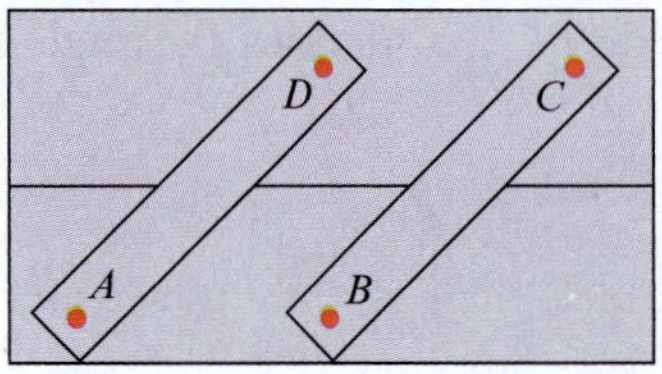

Side view of a closed sewing box.

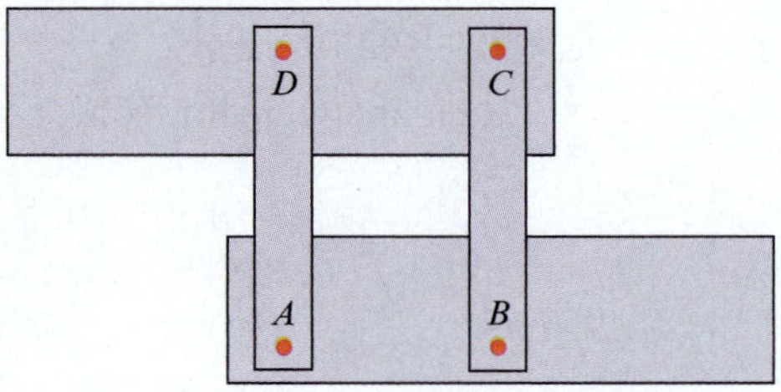

Side view of an open sewing box.

Given:

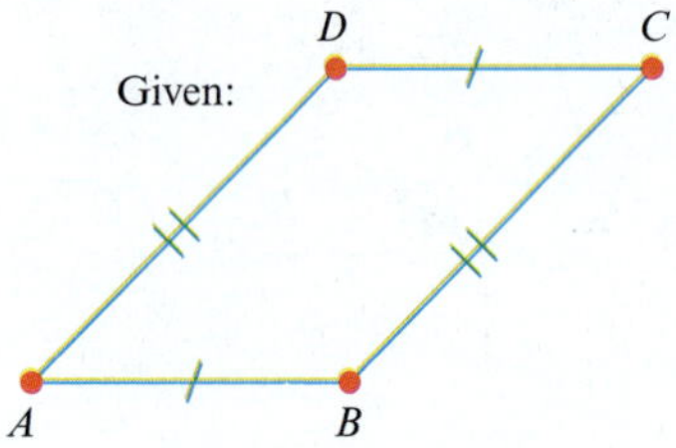

Why are sides *AB* and *CD* guaranteed to be parallel?

Class Activity 14K What Was the Robot's Path?

CCSS CCSS SMP1, SMP3, SMP4

A robot moves along a (flat) factory floor as follows:

1. Starting at point *A*, the robot moves straight to point *B*.
2. At point *B* the robot turns 110° counterclockwise.
3. The robot moves straight to point *C*.
4. At point *C* the robot turns 70° counterclockwise.
5. The robot moves straight to point *D*.
6. At point *D* the robot turns 110° counterclockwise.
7. The robot moves straight and stops when it gets back to point *A*.

What can you say about the distances that the robot traveled? Must any of them be related? Explain!

14.4 Constructions with Straightedge and Compass

Class Activity 14L How Are Constructions Related to Properties of Rhombuses?

CCSS CCSS SMP7

1. Below you see the result of using a straightedge and compass construction. There are two circles of the same radius, one centered at A and one centered at B. Draw the rhombus that arises naturally from this construction and that has A and B as vertices. Use the definition of rhombus to explain *why* the quadrilateral that you identify as a rhombus really must be a rhombus.

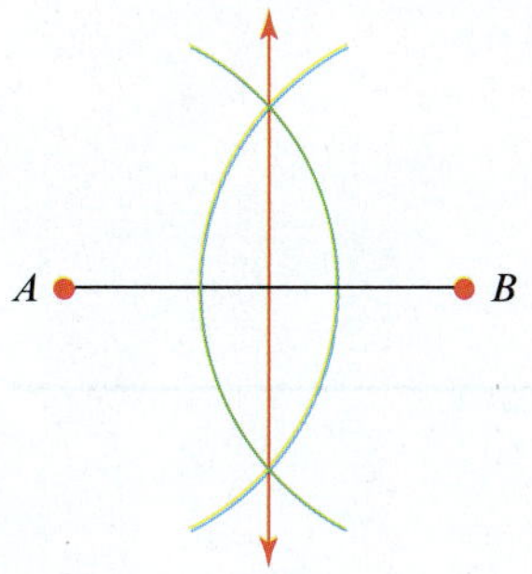

2. Which special properties of rhombuses explain why the line produced in the construction really is perpendicular to the line segment AB and divides the line segment AB in half? Explain.

3. Below you see the result of using a straightedge and compass construction. There are three circles of the same radius, one centered at P, one centered at Q, and one centered at R. Draw the rhombus that arises naturally from this construction. Use the definition of rhombus to explain *why* the quadrilateral that you identify as a rhombus really must be a rhombus.

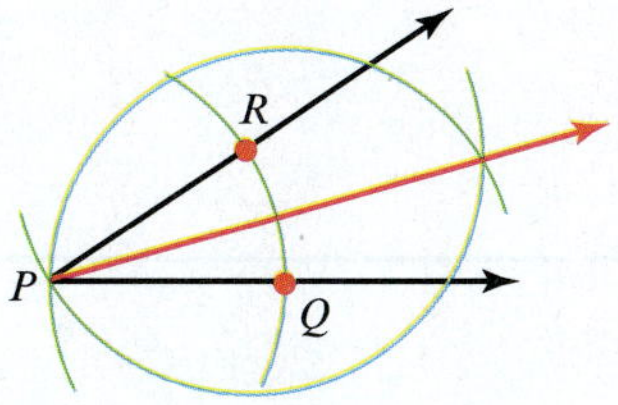

4. Which special properties of rhombuses explain why the construction divides the angle QPR in half? Explain.

Class Activity 14M Construct Parallel and Perpendicular Lines by Constructing Rhombuses

CCSS CCSS SMP1, SMP5

Use a straightedge and compass to do the following construction:

1. Construct a line that is parallel to the given line and passes through the point P by constructing a suitable rhombus. Briefly describe your method and explain what property of rhombuses you are using.

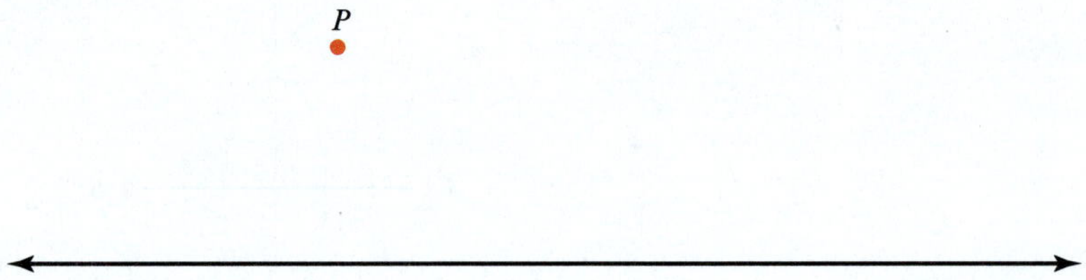

2. Construct a line that is perpendicular to the given line and passes through the point Q by constructing a suitable rhombus. Briefly describe your method and explain what property of rhombuses you are using.

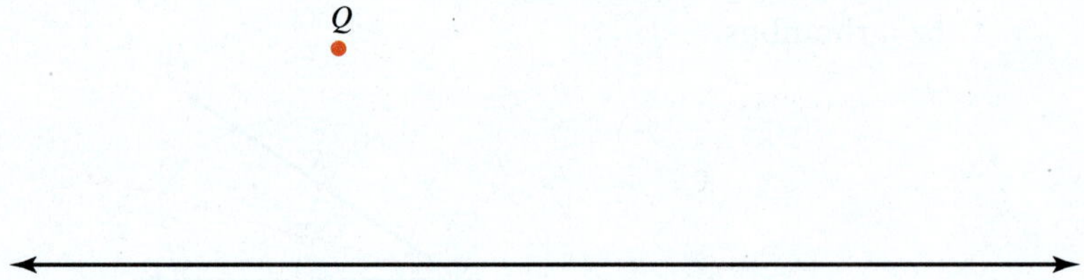

Class Activity 14N How Can We Construct Shapes with a Straightedge and Compass?

CCSS CCSS SMP1, SMP5

Use a straightedge and compass to do the following constructions:

1. Using a straightedge and compass, construct a line that is perpendicular to the line segment AB shown in the figure below and that passes through point A (*not* through the midpoint of line segment AB!). *Hint:* First extend the line segment AB.

   Now use a straightedge and compass to construct a square that has line segment AB as one side.

2. Using only a straightedge and compass (no protractor for measuring angles!), construct a triangle whose angles are 45°, 45°, 90°. Explain why your method produces the desired angles.

3. Using only a straightedge and compass (no protractor for measuring angles!), construct a triangle whose angles are 30°, 60°, 90° by first constructing an equilateral triangle. Explain why your method produces the desired angles.

4. Using a straightedge and compass, construct an octagon whose vertices all lie on the circle in the next figure and whose sides all have the same length. Explain your method.

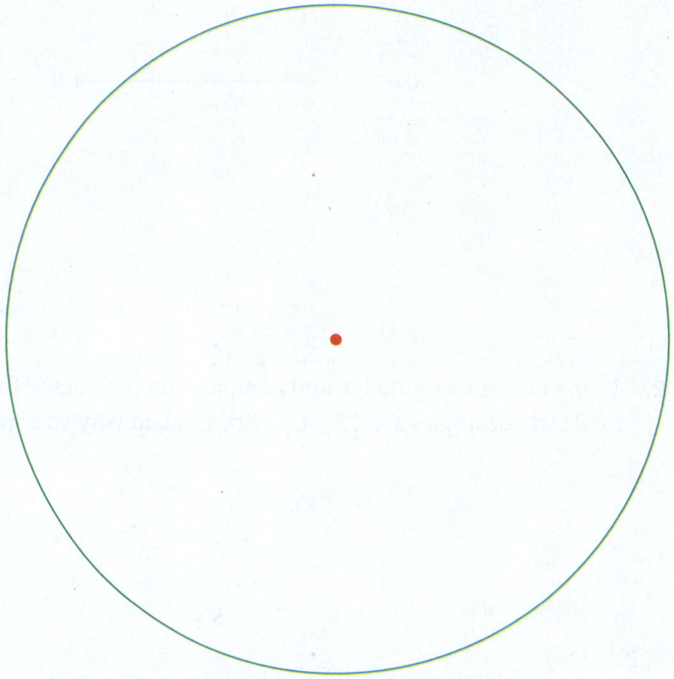

14.5 Similarity

Class Activity 14O Mathematical Similarity Versus Similarity in Everyday Language

CCSS CCSS SMP6

In mathematics, the terms *similar* and *similarity* have a much more specific meaning than they do in everyday language.

Examine the shapes below. In everyday language, we might say that all the shapes are similar. But mathematically, only the top two shapes on the right are similar to the original shape. How is the relationship of those two shapes to the original shape different from the relationship of the other two shapes to the original shape?

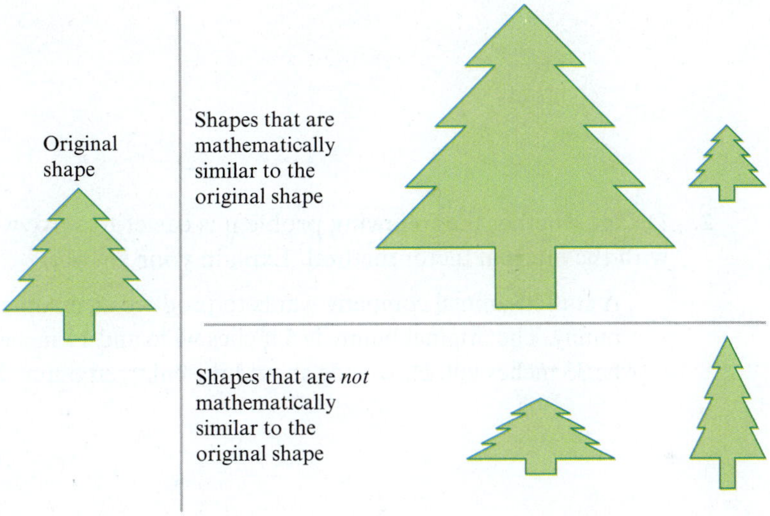

Class Activity 14P What Ways Can We Find to Solve Similarity Problems?

CCSS CCSS SMP1, 7.G.1

You have a poster that is 2 feet wide and 4 feet long. The poster has a simple design on it that you would like to scale up and draw onto a larger poster. The larger poster is to be 6 feet wide. How long should the poster be?

Find as many different ways as you can to solve this poster-scaling problem. In each case, explain your reasoning.

Class Activity 14Q Reasoning with the Scale Factor and Internal Factor Methods

CCSS CCSS SMP1, 7.G.1

1. Suppose that you have a postcard with an attractive picture on it and that you would like to scale up this picture and draw it onto paper that you can cut from a roll. The roll of paper is 20 inches wide, and you can cut the paper to virtually any length. If the postcard picture is 4 inches wide and 6 inches long, then how long should you cut the 20-inch-wide paper? Assume that the 4-inch side will become 20 inches long.

   Use two different methods to solve the postcard problem: the *scale factor* method and the *internal factor* method, and link each to a proportion you set up. In each case, explain why the method makes sense in as concrete a way as you can.

2. Decide whether the following problem is easier to solve with the scale factor method or with the internal factor method. Explain your answer.

   A stuffed-animal company wants to produce an enlarged version of a popular stuffed bunny. The original bunny is 6 inches wide and 11 inches tall. The enlarged bunny is to be 33 inches tall. How wide should the enlarged bunny be?

3. Decide whether the following problem is easier to solve with the scale factor method or with the internal factor method. Explain your answer.

   A toy company wants to produce a scale model of a car. The actual car is 6 feet wide and 12 feet long. The scale model of the car is to be $2\frac{1}{2}$ inches wide. How long should the scale model of the car be?

Class Activity 14R Critique Reasoning About Similarity

CCSS CCSS SMP3, 7.G.1

You will need scissors and Download 21 for part 3.

1. *Problem:* The picture on a poster that is 4 feet wide and 6 feet long is to be scaled down and drawn onto a small poster that is 1 foot wide. How long should the small poster be?

   Johnny solves the problem this way:

   > One foot is 3 feet less than 4 feet, so the length of the small poster should also be 3 feet less than the length of the big poster. This means the small poster should be $6 - 3 = 3$ feet long.

   Is Johnny's reasoning valid? Why or why not?

2. On graph paper, plot the widths and lengths of posters that are similar to a 4-foot-wide, 6-foot-long one. What do you notice about the points? Now plot Johnny's proposed width and length from part 1. What do you notice?

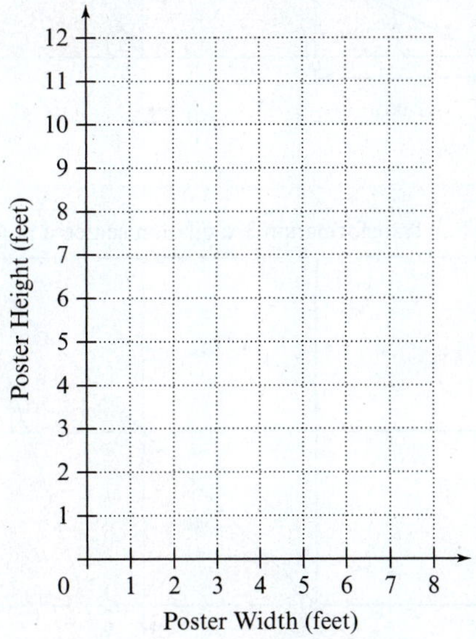

3. Using Download 21, cut out the following rectangles: a 4-unit-by-6-unit rectangle, several rectangles that are similar to it, and another rectangle that is not similar to it (label this one clearly to distinguish it from the others).

   Now try this "sighting" experiment. Hold a smaller rectangle in front of you and a larger similar one behind it. Then close one eye and adjust the two rectangles until they align exactly in your line of sight in both width and length. Try to do the same with a pair of rectangles that are not similar and note the difference.

14.6 Dilations and Similarity

Class Activity 14S Dilations Versus Other Transformations: Lengths, Angles, and Parallel Lines

CCSS CCSS SMP7

You might like to use Downloads 22 and 23 for this activity.

Examine the results of three different transformations below and create your own examples on the next page and on the Downloads. Then discuss what the different transformations do to the angles, lines, and lengths. Do angles remain the same size? Is a transformed line parallel to the original? Do parallel lines remain parallel? How are lengths in the original related to lengths in the transformed figure? How are the *dilations* different from the other transformations?

Original

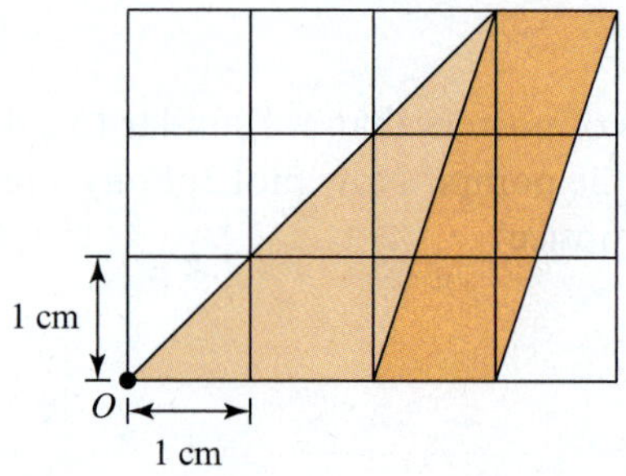

After Transformation 1 (horizontal scaling only)

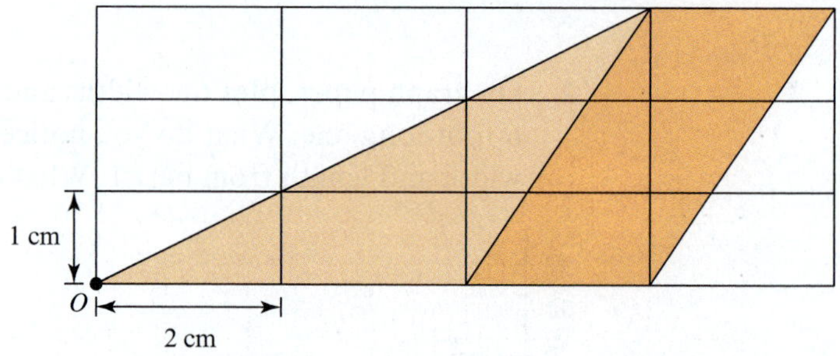

After Transformation 2
(vertical scaling only)

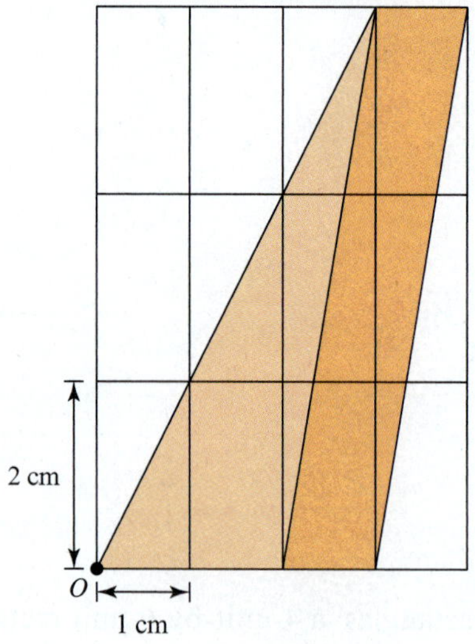

After Transformation 3, a dilation centered at O with scale factor 2

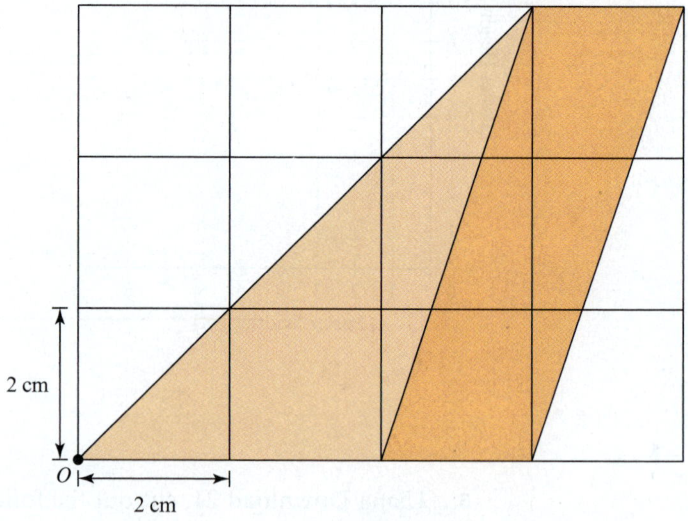

Original

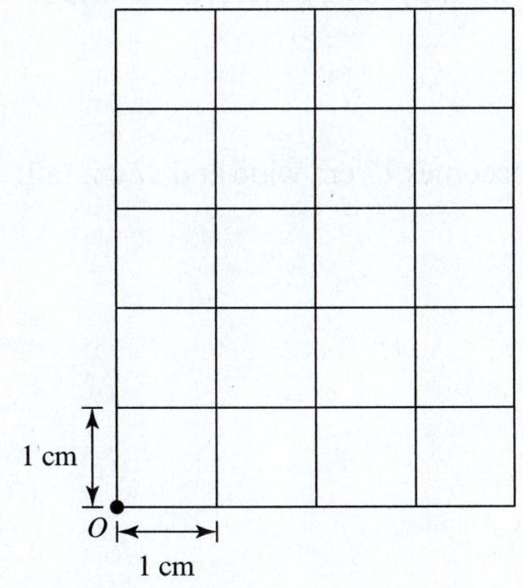

After Transformation 4 (horizontal scaling only)

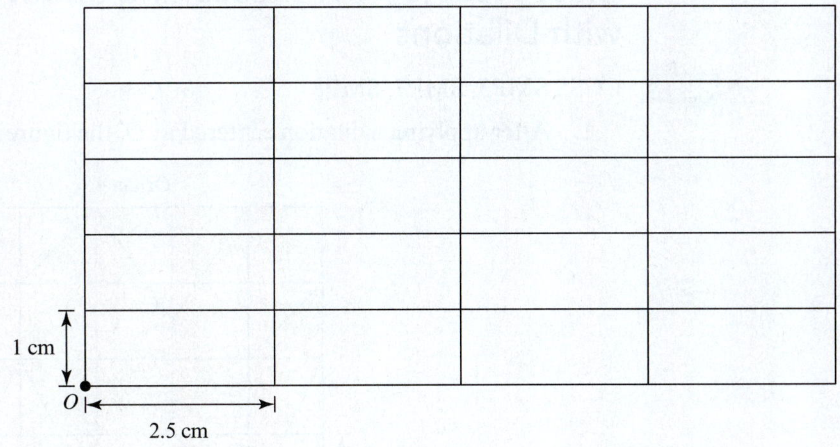

After Transformation 5
(vertical scaling only)

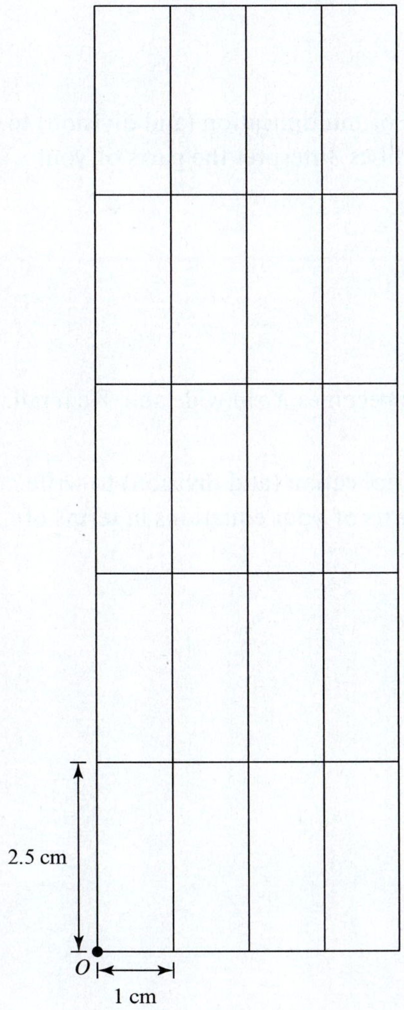

After Transformation 6, a dilation centered at O with scale factor 2.5

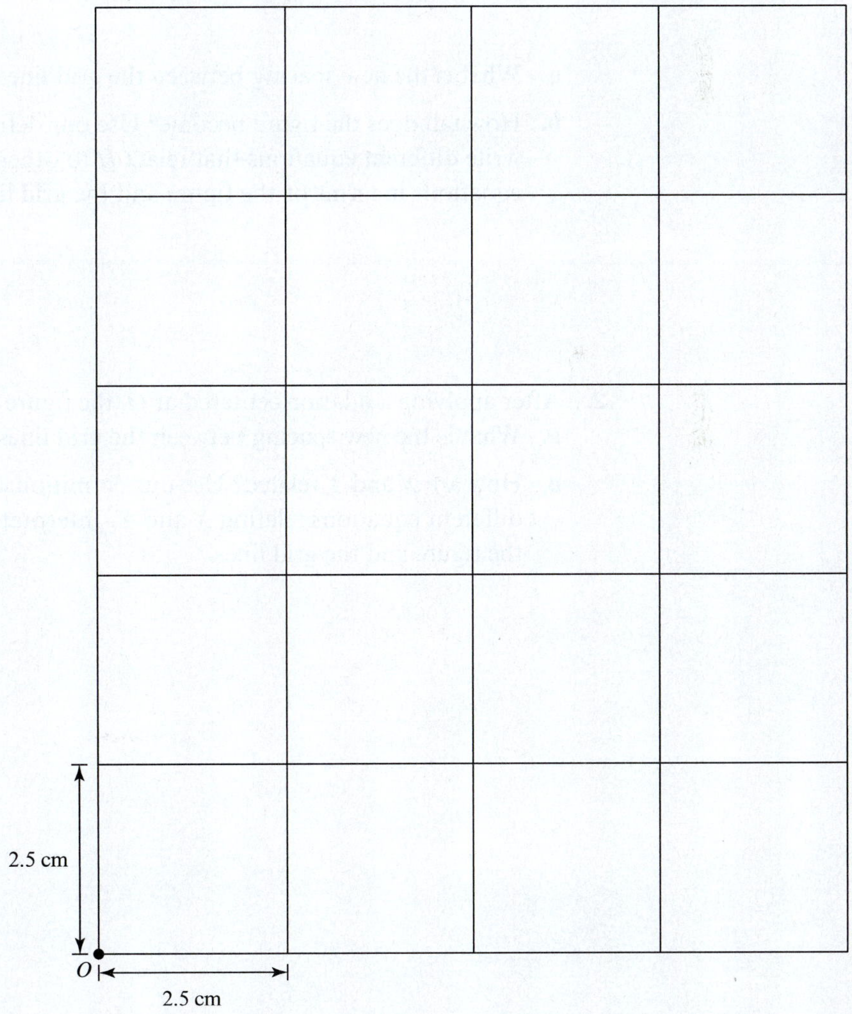

Class Activity 14T Reasoning about Proportional Relationships with Dilations

CCSS CCSS SMP3, SMP7, SMP8

1. After applying a dilation centered at O, the figure below becomes 17 cm wide and H cm tall.

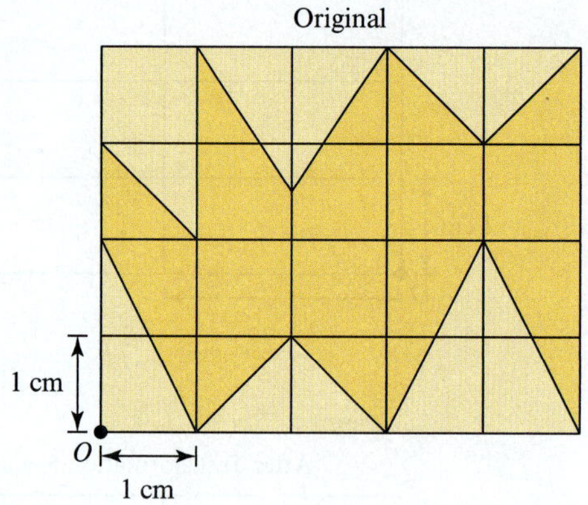

   a. What is the new spacing between the grid lines?

   b. How tall does the figure become? Use our definition of multiplication (and division) to write different equations that relate H to other quantities. Interpret the parts of your equations in terms of the figure and the grid lines.

2. After applying a dilation centered at O, the figure above becomes X cm wide and Y cm tall.

   a. What is the new spacing between the grid lines?

   b. How are X and Y related? Use our definition of multiplication (and division) to write different equations relating X and Y. Interpret the parts of your equations in terms of the figure and the grid lines.

Class Activity 14U How Can We Measure Distances by "Sighting"?

CCSS CCSS SMP1, SMP4

To do this activity, you will need your own ruler and one or more yardsticks and tape measures that can be shared by the class.

This activity will help you understand how the theory of similar triangles is used in finding distances by surveying.

1. Stand a yardstick on end on the edge of a chalkboard, or tape the yardstick vertically to the wall.

2. Stand back, away from the yardstick, in a location where you can see the yardstick. Your goal is to find your distance to the yardstick. Guess or estimate this distance before you continue.

   Record your guess of how far away the yardstick is here:

3. Hold your ruler in front of you with an outstretched arm. Make the ruler vertical, so that it is parallel to the yardstick. Close one eye, and with your open eye, "sight" from the ruler to the yardstick. Use the ruler to determine how big the yardstick appears to be from your location.

   Record the apparent size of the yardstick here:

4. With your arm still stretched out in front of you, have a classmate measure the distance from your sighting eye to the ruler.

   Record the distance from your sighting eye to the ruler here:

5. Use the theory of similar triangles to determine your distance to the yardstick. Sketch your eye, the ruler, and the yardstick, showing the relevant similar triangles. (Your sketch does not need to be to scale.) Explain why the triangles are similar.

   Is your calculated distance close to your estimated distance? If not, which one seems to be faulty, and why?

6. Now move to a new location, and find your distance to the yardstick again with the same technique.

Class Activity 14V How Can We Use a Shadow or a Mirror to Determine the Height of a Tree?

CCSS CCSS SMP1, SMP4

This activity requires several tape measures that can be shared by the class. Part 2 requires a mirror.

How could you find the height of a tree, for example, without measuring it directly? This class activity provides two ways to do this with *similar triangles*. Go outside and find a tree or pole on level ground whose height you will determine. Before you continue, guess or estimate the height of the tree.

First Method

This method will work only if your tree or pole casts a fully visible shadow. (So you need a sunny day.)

1. Measure the length of the shadow of the tree (from the base of the tree to the shadow of the tip of the tree).

2. Measure the height of a classmate, and measure the length of that person's shadow.

3. Sketch the two similar triangles in this situation. Explain why the triangles are similar. Use your similar triangles to find the height of the tree.

   Is your calculated height fairly close to your estimated height of the tree? If not, which one do you think is faulty? Why?

Second Method

1. Put a mirror flat on horizontal ground away from the tree with the reflective side facing up. Stand back from the mirror at a location where you can look into the middle of the mirror and see the top of the tree (you will probably need to move around to find this location).

2. Record the following measurements:
   a. The distance from your feet to the middle of the mirror when you are looking into the middle of the mirror and can see the top of the tree.
   b. The distance from the ground to your eyes.
   c. The distance from the middle of the mirror to the base of the tree.

3. Sketch the two similar triangles in this situation. Explain why the triangles are similar. Use your similar triangles to find the height of the tree.

   Is your calculated height fairly close to your estimated height of the tree? If not, which one do you think is faulty? Why?

14.7 Areas, Volumes, and Similarity

Class Activity 14W How Are Surface Areas and Volumes of Similar Boxes Related?

CCSS CCSS SMP2, SMP7, SMP8

You will need scissors, tape, and Downloads 24, 25 for this activity.

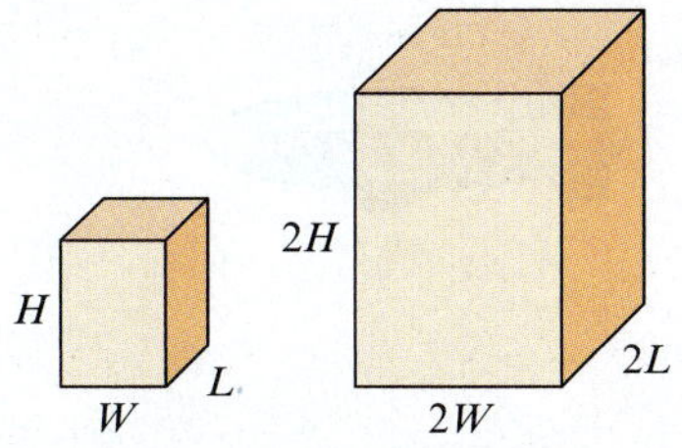

1. Cut out one box pattern from Download 24 (the other three are to share with classmates). Cut out the box pattern on Download 25. Fold these patterns to make two boxes (rectangular prisms), but leave them untaped, so that you can still unfold them. One box will have width W, length L, and height H; and one box will have width $2W$, length $2L$, and height $2H$. So the big box is twice as wide, twice as long, and twice as high as the small box.

2. By working with the patterns of the two boxes, determine how the surface area of the big box compares with the surface area of the small box. Is the surface area of the big box twice as large, three times as large, and so on, as the surface area of the small box? Look and think carefully—the answer may not be what you first think it is. Explain clearly why your answer is correct.

3. Now tape up the small box. Tape up most of the large box, but leave an opening so that you can put the small box inside it. Determine how the volume of the big box compares with the volume of the small box.

4. What if there were an even bigger box whose width, length, and height were each three times the respective width, length, and height of the small box? How would the surface area of the bigger box compare with the surface area of the small box?

   How would the volume of the bigger box compare with the volume of the small box?

5. Now imagine a variety of bigger boxes. Fill in the table below with your previous results and by extrapolating from your results.

Size of Big Box Compared with Small Box					
Length, Width, Height	2 times	3 times	5 times	2.7 times	k times
Surface area					
Volume					

Class Activity 14X How Can We Determine Surface Areas and Volumes of Similar Objects?

CCSS CCSS SMP1, SMP3, SMP4

1. If someone made a Goodyear blimp that was 1.5 times as wide, 1.5 times as long, and 1.5 times as high as the current one, how much material would it take to make the larger blimp compared with the current blimp? How much more gas would it take to fill this bigger Goodyear blimp, compared with the current one (at the same pressure)?

2. An adult alligator can be 15 feet long and weigh 475 pounds. Suppose that some excavated dinosaur bones indicate that the dinosaur was 30 feet long and was shaped roughly like an alligator. How much would you expect the dinosaur to have weighed?

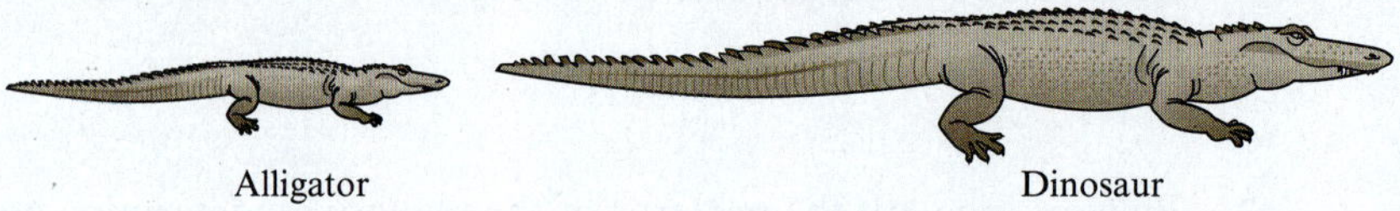

3. Explain why we *can't* reason in either of the following two ways to solve part 2 of this activity:
   - Each foot of the alligator weighs $475 \div 15 = 31.6\ldots$ pounds. So multiply that result by 30 to get the weight of the dinosaur as 950 pounds.
   - The dinosaur is twice as long as the alligator, so it should weigh twice as much, which is 950 pounds.

   What is wrong with those two ways of reasoning?

Class Activity 14Y How Can We Prove the Pythagorean Theorem with Similarity?

CCSS CCSS SMP3, 8.G.6

This activity will help you use similar shapes to prove the Pythagorean theorem.

Remember that the Pythagorean theorem says that for any right triangle with short sides of length a and b, and hypotenuse of length c,

$$a^2 + b^2 = c^2$$

1. Given any right triangle, such as the triangle on the left below, drop the perpendicular to the hypotenuse, as shown on the right.

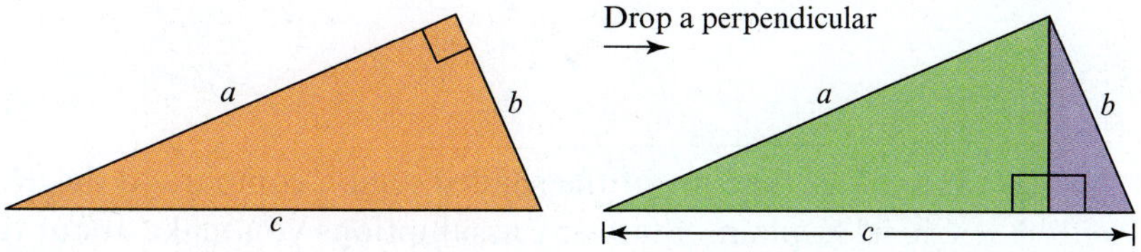

Use angles to explain why the two smaller right triangles on the right are similar to the original right triangle. (Do not use any actual measurements of angles, because the proof must be general—it must work for *any* initial right triangle.)

2. Now flip each of the three right triangles of part 1 over its hypotenuse, as shown below. View each of the three right triangles as taking up a percentage of the area of the square formed on its hypotenuse. Why must each triangle take up the same percentage of its square?

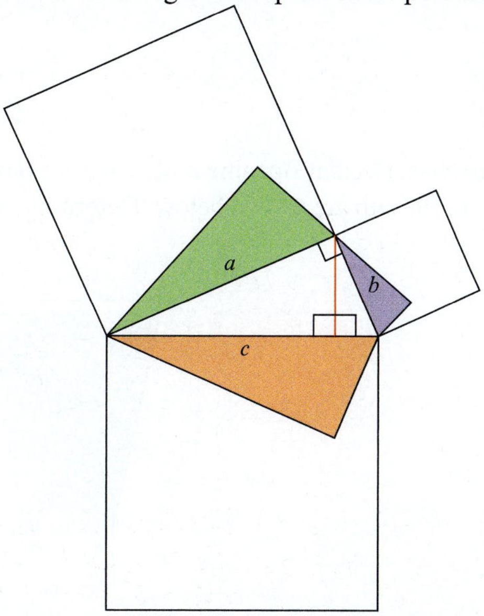

3. Let $P\%$ be the percentage of part 2. Express the areas of the three triangles in terms of $P\%$, and then explain why

$$P\% \cdot a^2 + P\% \cdot b^2 = P\% \cdot c^2$$

4. Use part 3 to explain why

$$a^2 + b^2 = c^2$$

thus proving the Pythagorean theorem.

Class Activity 14Z Area and Volume Problem Solving

CCSS CCSS SMP1, SMP4

1. Explain in at least two different ways why the area of the shaded region below is 3 times the area of the unshaded square inside it. (State any assumptions you make about the shapes.)

2. How does the area of the shaded region compare to the area of the unshaded region inside it? Explain. (State any assumptions you make about the shapes.)

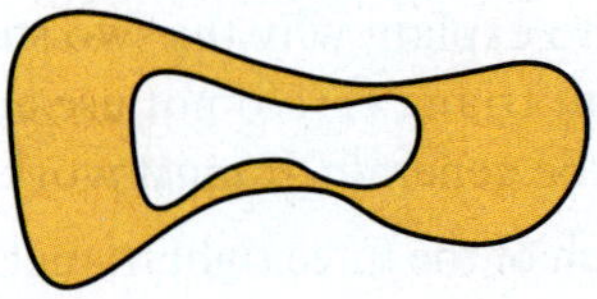

3. A cup has a circular opening and a circular base. A cross-section of the cup and the dimensions of the cup are shown below. Determine the volume of the cup. Explain your reasoning.

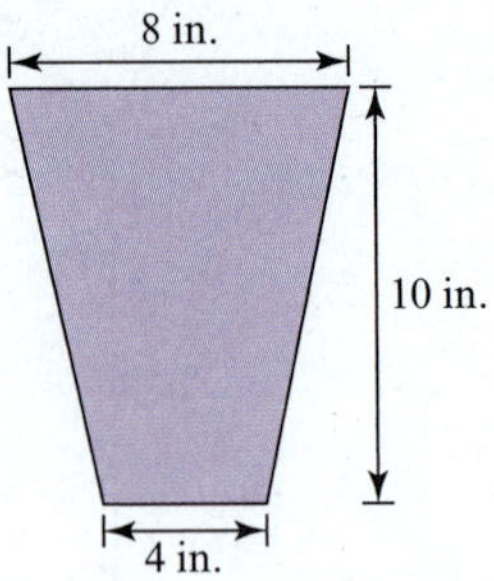

CLASS ACTIVITIES

15

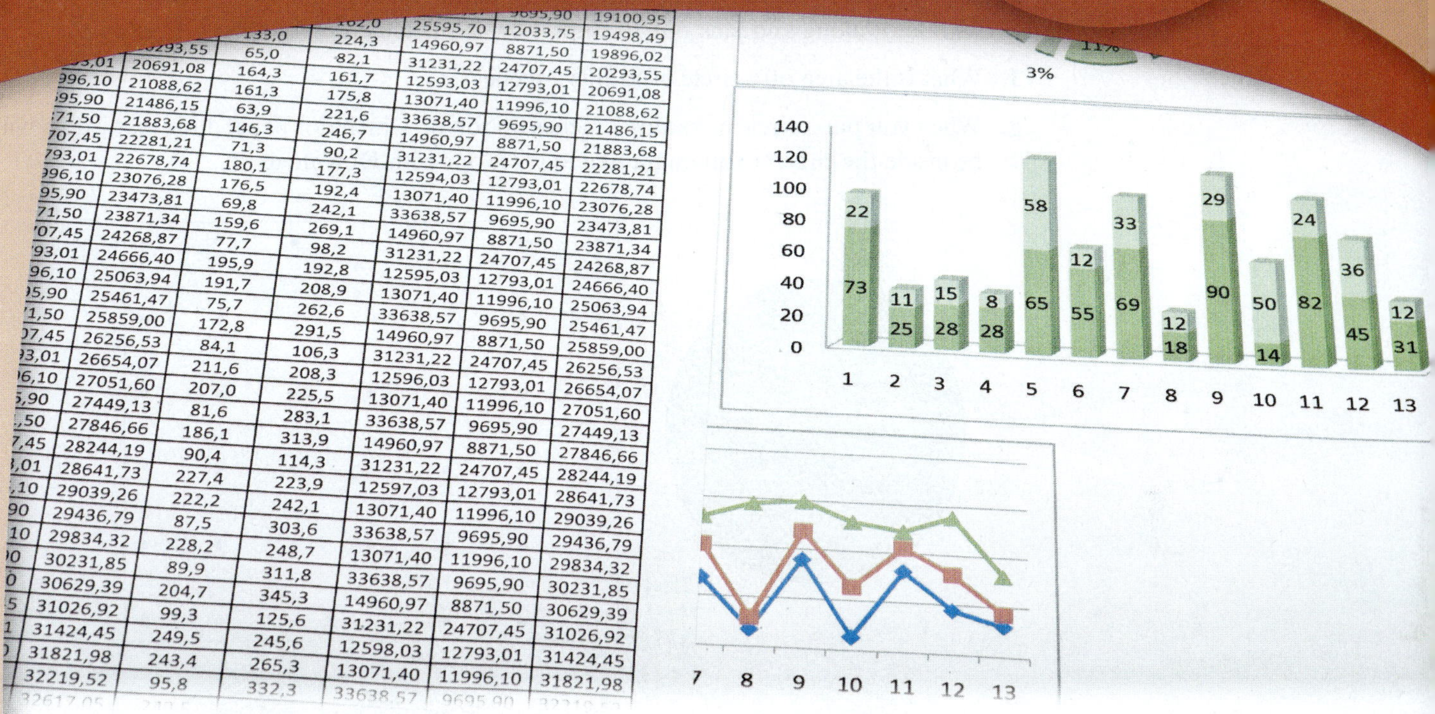

Statistics

15.1 Formulating Statistical Questions, Gathering Data, and Using Samples

Class Activity 15A Statistical Questions Versus Other Questions

CCSS CCSS SMP2, 6.SP.1

You may wish to use Download 26.

1. Which of the following questions are *statistical* questions? What makes them statistical?
   a. How heavy is Aneeth's backpack?
   b. How heavy are the backpacks of students in this class?

c. How much time did you spend doing homework yesterday?

d. How much time do students at this school spend doing homework?

e. A bag contains 50 red poker chips and 50 white poker chips. If you reach into the bag without looking and pick out 10 chips, how many red chips will you get?

f. What is the area of a circle of radius 2 inches?

g. When you put a circle of radius 2 inches on top of 1-inch dot paper, how many dots will be inside the circle? (You might like to try it out with Download 26.

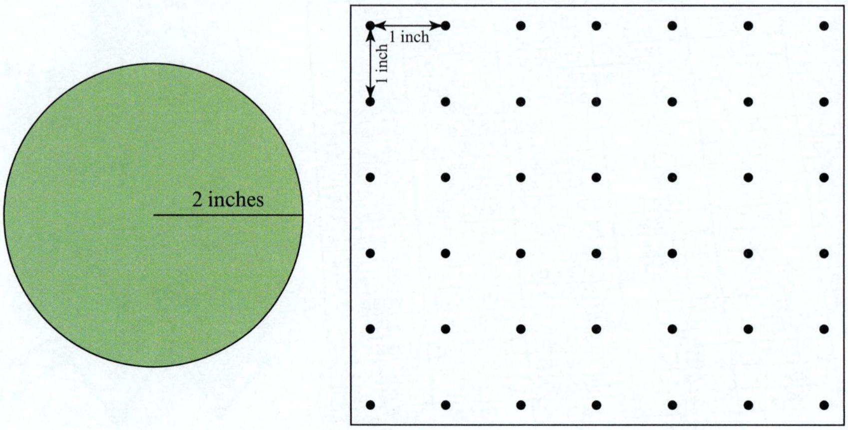

2. Write your own pair of contrasting questions, one that is statistical and one that is not. What makes one statistical and the other not?

Class Activity 15B Choosing a Sample

CCSS CCSS SMP2, 7.SP.2

A college newspaper wants to find out how the students at the college would answer a specific question of importance to the student body. There are too many students for the newspaper staff to ask them all. So the staff decides to choose a sample of students to ask. For each of the following ways that the newspaper staff could select a sample, discuss whether the sample is likely to be representative of the full student body or if there are reasons why the sample may not be representative.

a. Ask their friends.

b. Ask as many of their classmates as they can.

c. Stand outside the buildings their classes are in and ask as many people as they can who come by.

d. Stand outside the student union or other common meeting area, and try to pick people who they think are representative of the students at their institution to ask the question.

e. Generate a list of random numbers between 1 and the number of students at the college. (Many calculators can generate random numbers; random numbers can also be generated on the Internet.) Pick names out of the student phone book corresponding to the random numbers (e.g., for 123, pick the 123rd name), and contact that person by phone or by e-mail.

Class Activity 15C How Can We Use Random Samples to Draw Inferences About a Population?

CCSS CCSS SMP2, 7.SP.2

You may wish to use cups, plastic baggies, and beads, tiles, or other small objects in two different colors to simulate part 1.

1. Suppose you have a bowl filled with 250 beads, some red, the rest blue. You take a random sample of 10 beads and find that 7 are red and 3 are blue.

   Based on the random sample, how many red and blue beads should be in the full population of beads? To think about this question, imagine organizing the population of 250 beads in the following two ways:

   - **Baggies of 10 beads each:** Imagine a bunch of baggies each containing 7 red beads and 3 blue beads, just like the random sample.
   - **10 cups filled with beads:** Imagine distributing the random sample among 10 cups, so 7 cups have a red bead and 3 cups have a blue bead. Then imagine distributing all the beads among the cups.

   Explain how to reason in several different ways to infer the numbers of red and blue beads in the full population based on the random sample. In each case, support your reasoning with a math drawing and with expressions or equations involving 250, 10, 7, and 3.

2. Suppose you have a bowl filled with 650 beads, some green, the rest yellow. You take a random sample of 25 beads and find that 11 are green and 14 are yellow. Based on the random sample, how many green and yellow beads should there be? Explain how to reason in several different ways to infer those numbers. In each case, write and explain expressions for the numbers of green and yellow beads.

Class Activity 15D Using Random Samples to Estimate Population Size by Marking (Capture–Recapture)

CCSS CCSS SMP2, 7.SP.2

You will need a bag filled with a large number (at least 100) of small, identical beans or other small objects that can be marked (such as small paper strips or beads that can be colored with a marker) or swapped for items of another color.

Pretend that the beans are fish in a lake. You will estimate the number of fish in the lake without counting them all by using a method called *capture–recapture.*

1. Go "fishing:" Pick 60 "fish" out of your bag. Count the number of fish you caught, and label each fish with a distinctive mark (or swap it with another color). Then throw your fish back in the lake (the bag) and mix them thoroughly.
2. Go fishing again: Randomly pick 40 fish out of your bag. Count the total number of fish you caught this time, and count how many of the fish are marked.
3. Use your counts from parts 1 and 2 to estimate the number of fish in your bag. Explain your reasoning.

4. When Ms. Wade used the method described in parts 1 through 3, she picked 30 fish at first, marked them, and put them back in the bag. Ms. Wade thoroughly mixed the fish in the bag and randomly picked out 40 fish. Of these 40 fish, 5 were marked. Explain how to reason in several different ways to determine approximately how many fish were in the bag.

15.2 Displaying Data and Interpreting Data Displays

Class Activity 15E Critique Data Displays or Their Interpretation

CCSS CCSS SMP3

1. Ryan scooped some small plastic animals out of a tub, sorted them, and placed the animals to make a graph that looked like the following:

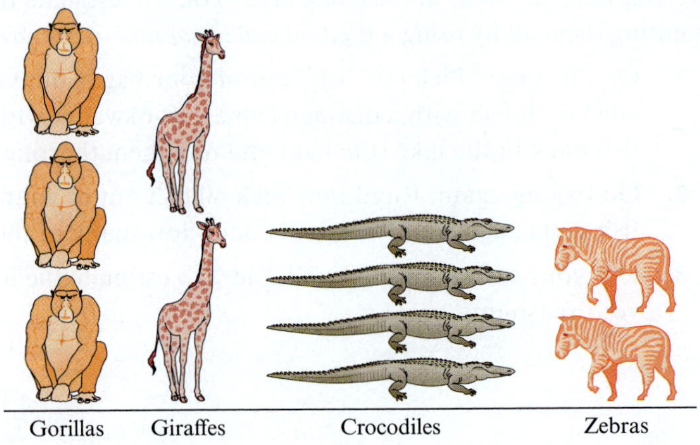

What might be a problem with Ryan's graph?

2. Critique the data display below. Would you recommend a different display?

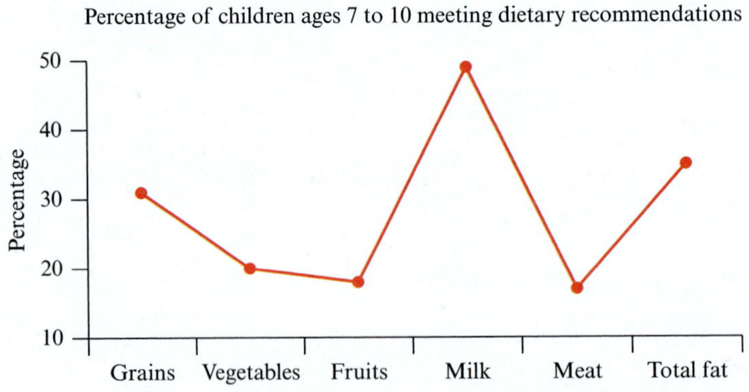

3. Critique the display below, which shows annual per capita carbon dioxide emissions in various countries, in light of Section 14.7 on how scaling affects volume.

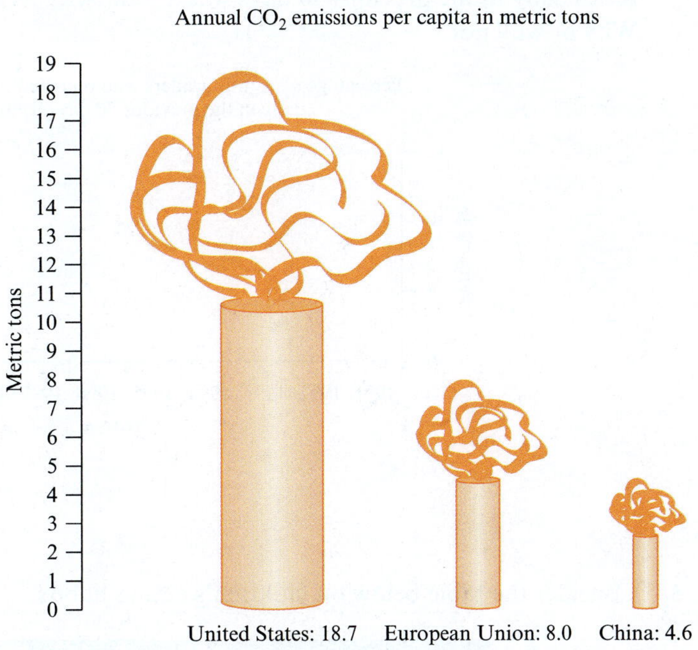

4. Students scooped dried beans out of a bag and counted the number of beans in the scoop. Each time, the number of beans in the scoop was recorded in the dot plot below.

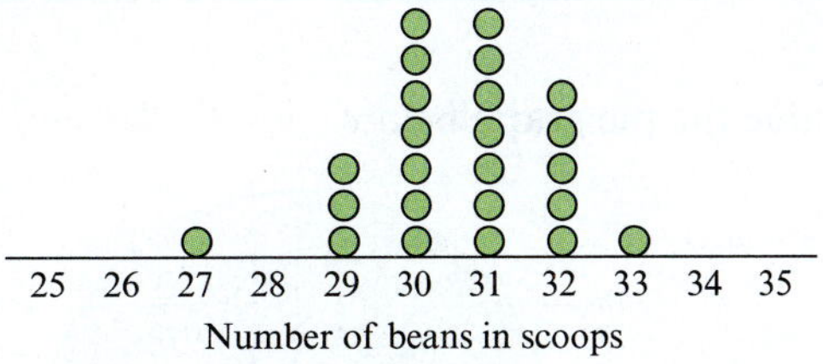

When a student was asked to make a list of the data displayed in the dot plot, the student responded thus:

1, 3, 7, 7, 5, 1

Critique the student's response.

5. Consider the line graph below about adolescents' smoking. Based on this display, would it be correct to say that the percentage of eighth-graders who reported smoking cigarettes daily in the previous 30 days was about twice as high in 1996 as it was in 1993? Why or why not?

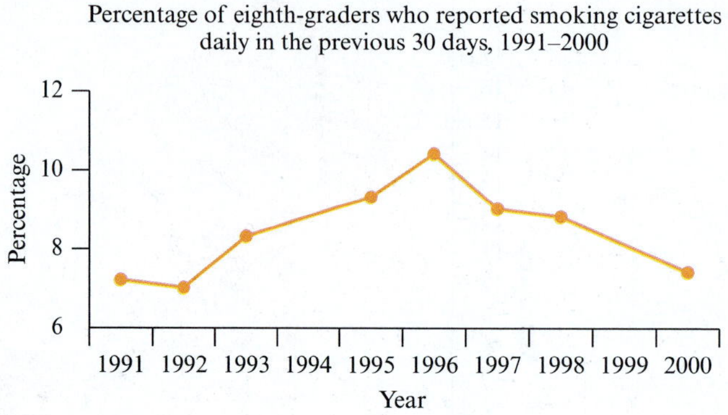

6. Consider the table below on children's eating habits.

Food	Percentage of 4–6-Year-Olds Meeting the Dietary Recommendation for a Food Group
Grains	27
Vegetables	16
Fruits	29
Saturated fat	28

Critique the pie graph below.

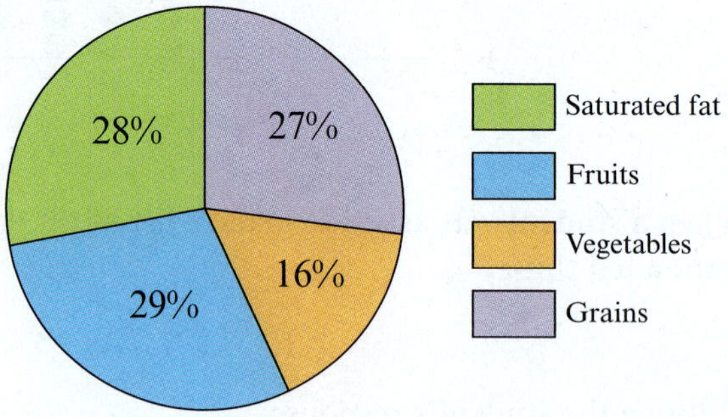

Class Activity 15F Display and Ask Questions About Graphs of Random Samples

CCSS CCSS SMP4, SMP8, 7.SP.2

For part 1 each pair or small group will need 20 sticky notes and a bag containing at least 30 objects that are identical except that 30% are one color and 70% are a second color.

Recall that the three levels of graph comprehension discussed in the text are as follows:

- **Reading the data,** which only requires a literal reading of the graph and does not require further interpretation.
- **Reading between the data,** which requires the ability to compare quantities (e.g., greater than, tallest, smallest) or the use of other mathematical concepts and skills (e.g., addition, subtraction, multiplication, division).
- **Reading beyond the data,** which requires the student to predict or infer from the data.

1. **a.** Randomly select 10 items from the bag and write how many items of the first color you selected on a sticky note. Put the items back in the bag and repeat the process until you have 20 sticky notes.

   b. Join with another group and find a good way to organize and display your 40 sticky notes. Make a drawing of your display.

   c. Ask and answer at least two questions about your graph for each of the three graph-reading levels.

2. In your classroom you have a box with 100 small square tiles in it. The tiles are identical except that some are yellow and the rest are blue. Your students take turns picking 10 tiles out of the box without looking. Then they record the number of yellow tiles (out of the 10) they picked on a sticky note and put the tiles back in the box. The class uses the sticky notes to make a dot plot on the chalkboard. It winds up looking like the one shown below (where each dot represents a sticky note).

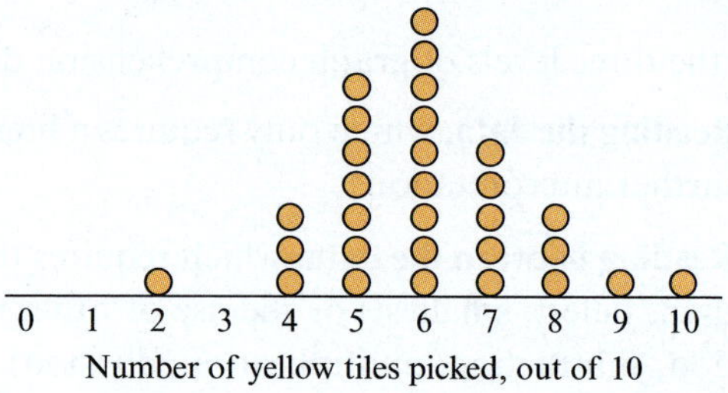

a. Write two "read the data" questions for the dot plot. Answer your questions.

b. Write two "read between the data" questions for the dot plot. Answer your questions.

c. Write two "read beyond the data" questions for the dot plot. Answer your questions (to the extent possible).

Class Activity 15G Displaying Data About Pets

CCSS CCSS SMP4, 3.MD.3

A class collected information about the pets they have at home, as shown in the table below.

Name	Pets at Home
Michelle	1 dog, 2 cats
Tyler	1 salamander, 2 snakes, 3 dogs
Antrice	1 hamster
Yoon-He	1 cat
Anne	none
Peter	2 dogs
Brandon	1 guinea pig
Brittany	1 dog, 1 cat
Orlando	none
Chelsey	2 dogs, 10 fish
Sarah	1 rabbit
Adam	none
Lauren	2 dogs
Letitia	3 cats
Jarvis	1 dog

1. Consider the following questions about pets:

   a. Are dogs the most popular pet?

   b. How many pets do most students have?

   c. How many students have more than one pet?

   d. Are most of the pets mammals?

   e. Write some other questions about pets that may be of interest to students and that could be addressed by the data that were collected.

2. Make each of the following data displays and use them to answer the questions from part 1. Observe that different graphs will be helpful for answering different questions.

   a. A display that shows how many students have 0 pets, 1 pet, 2 pets, 3 pets, and so on

   b. A display that shows how the *students* in the class fall into categories depending on what kind of pet they have

   c. Another display like the one in part (b), except pick the categories in a different way this time

   d. A display that shows how the *pets* of students in the class fall into categories

Class Activity 15H Investigating Small Bags of Candies

CCSS CCSS SMP4, 2.MD.10, 3.MD.3

For this activity, each person, pair, or small group in the class needs a small bag of multicolored candies. All bags should be of the same size and consist of the same type of candy. Bags should not be opened until after completion of the first part of this activity.

1. Do not open your bag of candy yet! Write a list of questions that the class as a whole could investigate by gathering and displaying data about the candies.

2. Open your bag of candy (but do not eat it yet!) and display data about your candies in two significantly different ways. For each display, write and answer questions at the three different graph-reading levels.

3. Together with the whole class, collect and display data about the bags of candies in order to answer some of the questions the class posed in part 1.

Class Activity 15I The Length of a Pendulum and the Time It Takes to Swing

CCSS CCSS SMP4, 5.G.2, 8.SP.1, 8.SP.2

A fifth-grader's science fair project[1] investigated the relationship between the length of a pendulum and the time it takes the pendulum to swing back and forth. The student made a pendulum by tying a heavy washer to a string and attaching the string to the top of a triangular frame, as shown in the diagram below.

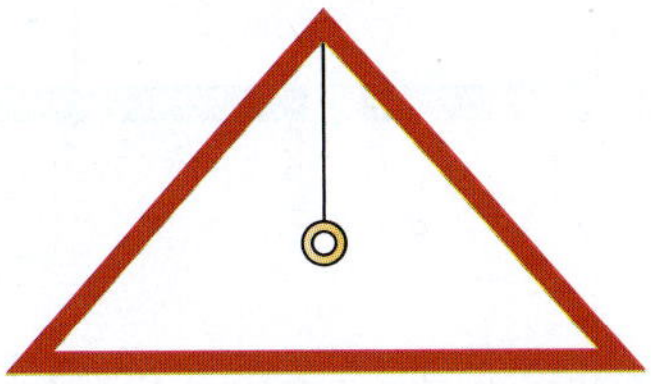

The length of the string could be varied. The table and scatterplot below show how long it took the pendulum to swing back and forth 10 times for various lengths of the string. (Several measurements were taken and averaged.)

Length of String in Inches	Time of 10 Swings in Seconds
1	2.61
2	2.97
3	3.04
4	3.41
5	3.96
6	4.13
7	4.22
8	4.5
9	4.64
10	5.13
11	5.32
12	5.56
13	5.62
14	5.87

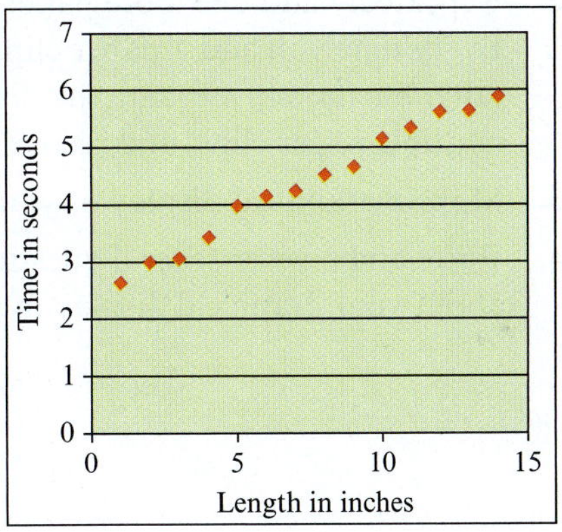

Write two or more questions about the scatterplot for each of the three graph-reading levels. Answer each question (to the extent possible).

[1]Thanks to Arianna Kazez for the data and information about the project.

Class Activity 15J Balancing a Mobile

CCSS CCSS SMP4, 5.G.2, 8.SP.1

For this activity, each person, pair, or small group in the class needs a drinking straw, string, tape, at least 7 paper clips of the same size, a ruler, and graph paper. You will use the straw, string, tape, and paper clips to make a simple mobile.

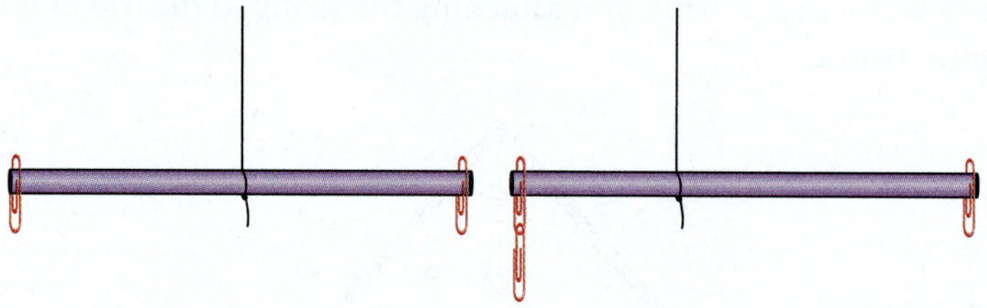

1. Tie one end of the string snugly around the straw. Tape one paper clip to each end of the straw. Hold the other end of the string so that your mobile hangs freely. Adjust the location of the string along the straw so that the straw balances horizontally. The string should now be centered on the straw, as in the picture on the left. Measure the distance on the straw from the string to each end.
2. Repeatedly add one more paper clip to one side of the straw (but not to the other side). Every time you add a paper clip, adjust the string so that the straw balances horizontally. Each time, measure the distance on the straw from the string to the end that has multiple paper clips, and record your data.
3. Make a graphical display of your data from part 2. (Use graph paper.)
4. Write and answer several questions at each of the three different graph-reading levels about your graphical display in part 3.

15.3 The Center of Data: Mean, Median, and Mode

Class Activity 15K The Mean as "Making Even" or "Leveling Out"

CCSS CCSS SMP2, 5.MD.2, 6.SP.3

You will need a collection of 16 small objects such as snap-together cubes or blocks for this activity.

1. Using blocks, snap cubes, or other small objects, make groups or towers with the following number of objects in the towers:

2, 5, 4, 1

Now "level out" or make the block towers even. That is, redistribute the blocks among your block towers until all 4 towers have the same number of blocks in them. How does this common number of blocks in each of the 4 towers compare with the mean of the list 2, 5, 4, 1?

2. For each list below, make block towers. Then level out the block towers and compare the common number of blocks in each tower with the mean of the list. In some cases you may have to imagine cutting your blocks into smaller pieces.

List 1:	1,	3,	3,	2,	1
List 2:	6,	3,	2,	5	
List 3:	2,	3,	4,	3,	4
List 4:	2,	3,	1,	5	

3. To calculate the mean of a list of numbers *numerically,* we add the numbers and divide the sum by the number of numbers in the list. So, to calculate the mean of the list 2, 5, 4, 1, we calculate

$$(2 + 5 + 4 + 1) \div 4$$

Interpret the *numerical* process for calculating a mean in terms of leveling out 4 block towers.

- When we add the numbers, what does that correspond to with the blocks?
- When we divide by 4, what does that correspond to with the blocks?

Class Activity 15L Solving Problems About the Mean

CCSS CCSS SMP1, SMP4

1. Suppose you have made 3 block towers: one 3 blocks tall, one 6 blocks tall, and one 2 blocks tall. Describe some ways to make 2 more towers so that there is an average of 4 blocks in all 5 towers. Explain your reasoning.

2. If you run 3 miles every day for 5 days, how many miles will you need to run on the sixth day in order to have run an average of 4 miles per day over the 6 days? Solve this problem in two different ways, and explain your solutions.

3. The mean of 3 numbers is 37. A fourth number, 41, is included in the list. What is the mean of the 4 numbers? Explain your reasoning.

4. Explain how you can quickly calculate the average of the following list of test scores without adding the numbers:

   81, 78, 79, 82

5. If you run an average of 3 miles a day over 1 week and an average of 4 miles a day over the next 2 weeks, what is your average daily run distance over that 3-week period?

   Before you solve this problem, explain why it makes sense that your average daily run distance over the 3-week period is *not* just the average of 3 and 4—namely, 3.5. Should your average daily run distance over the 3 weeks be greater than 3.5 or less than 3.5? Explain how to answer this without a precise calculation. Now determine the exact average daily run distance over the 3-week period. Explain your solution.

Class Activity 15M The Mean as "Balance Point"

CCSS SMP2, 6.SP.2, 6.SP.3

1. For each of the data sets below:
   - Make a dot plot of the data on the given axis.
   - Calculate the mean of the data.
   - Verify that the mean agrees with the location of the given fulcrum.
   - Answer this question: Does the dot plot look like it would balance at the fulcrum (assuming the axis on which the data is plotted is weightless)?

Data set 1:
3, 4, 4, 5, 5, 5, 6, 6, 7

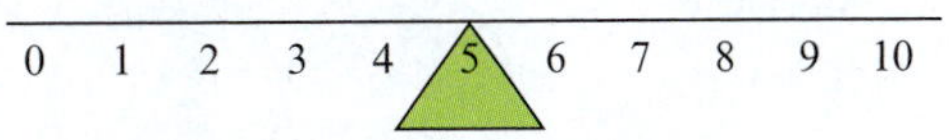

Data set 2:
4, 7, 7

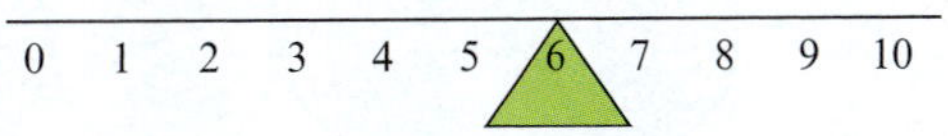

Data set 3:
2, 8, 8

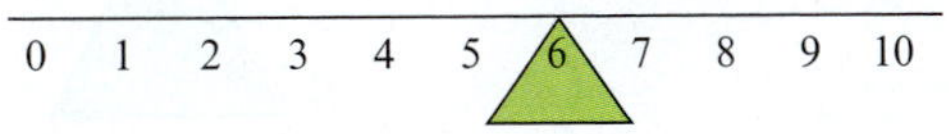

Data set 4:
1, 1, 1, 3, 3, 9

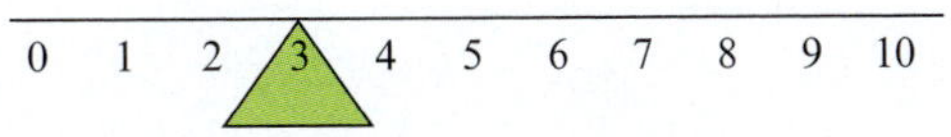

2. For each of the dot plots below, guess the approximate location of the mean by thinking about where the balance point for the data would be. Then check how close your guess was by calculating the mean.

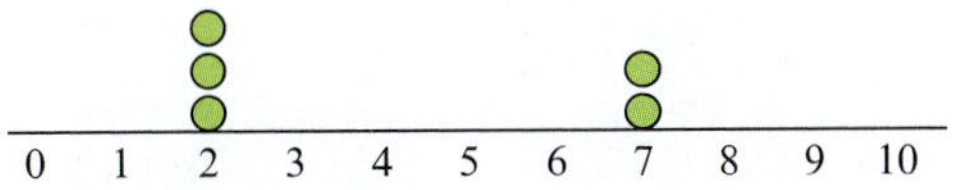

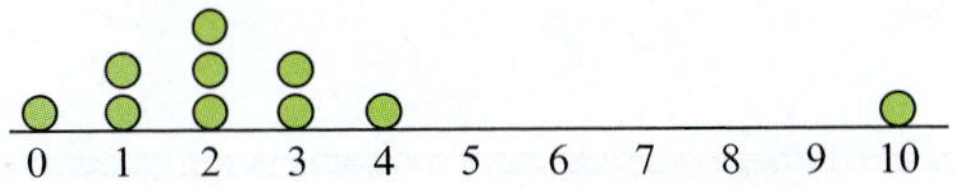

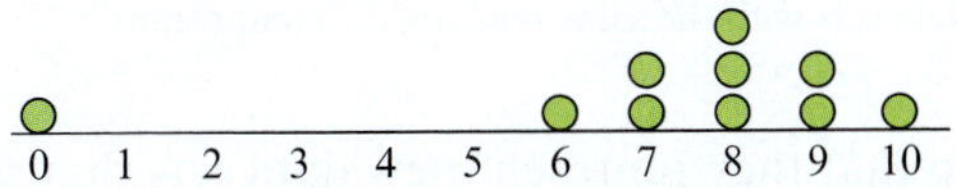

Class Activity 15N Same Median, Different Mean

CCSS CCSS SMP2, 6.SP.5d

You will need 9 pennies or other small objects for this activity.

The following data set is represented on the dot plot below with pennies:

4, 5, 5, 6, 6, 6, 7, 7, 8

Arrange real pennies (or other small objects) along the number line below to represent the same data set.

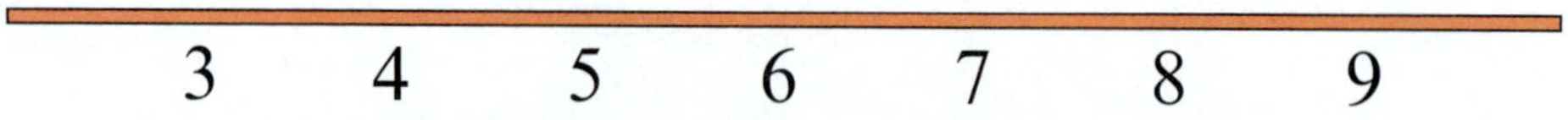

Show data sets with the same median, different means.

1. Rearrange your pennies so that they represent new data sets that still have median 6 but have means *less than* 6. To help you do this, think about the mean as the balance point. List your new data sets.
2. Rearrange your pennies so that they represent new data sets that still have median 6 but have means *greater than* 6. To help you do this, think about the mean as the balance point. List your new data sets.

Class Activity 15O Can More Than Half Be Above Average?

CCSS CCSS SMP2, 6.SP.5d

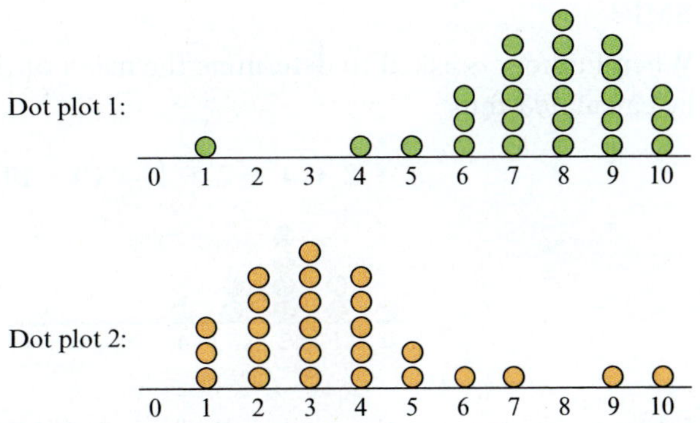

1. For each of the dot plots shown, decide which is greater: the median or the mean of the data. Explain how you can tell without calculating the mean.

2. A teacher gives a test to a class of 20 students.

   a. Is it possible that 90% of the class scores above average? If so, give an example of test scores for which this is the case. If not, explain why not.

   b. Is it possible that 90% of the class scores below average? If so, give an example of test scores for which this is the case. If not, explain why not.

3. A radio program describes a fictional town in which "all the children are above average." In what sense is it possible that all the children are above average? In what sense is it not possible that all the children are above average?

Class Activity 15P Critique Reasoning About the Mean and the Median

CCSS CCSS SMP3

1. When Eddie was asked to determine the mean of the data shown in the dot plot below, he calculated thus:

$$1 + 2 + 4 + 2 + 1 = 10, \quad 10 \div 5 = 2$$

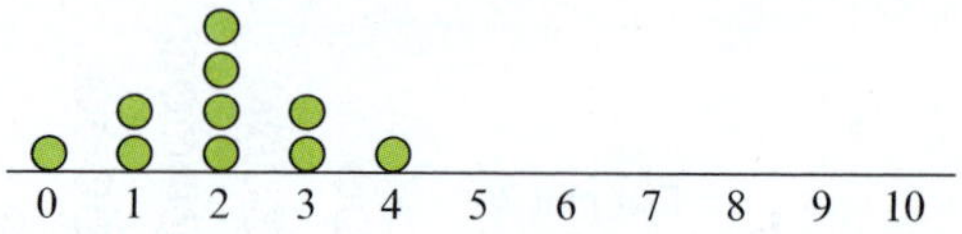

Eddie concluded that the mean is 2. Critique Eddie's reasoning.

2. Brittany said that the mean of the test data in the dot plot below is 3 because "all the students got the same score; all the scores are level of 3." Discuss!

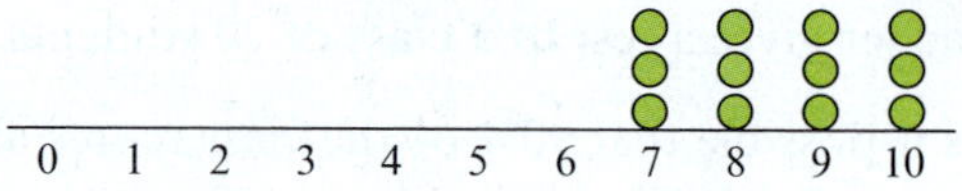

Student scores on a 10-point quiz

3. Critique the work of Student 1, Student 2, and Student 3 below.

Student 1: 5, 5, 6, 6, 6, 3, 4, 5, 4, 5, 5, 6, 4, 7, 5, 5, 5, 4, 6, 5
↑
Median: 5

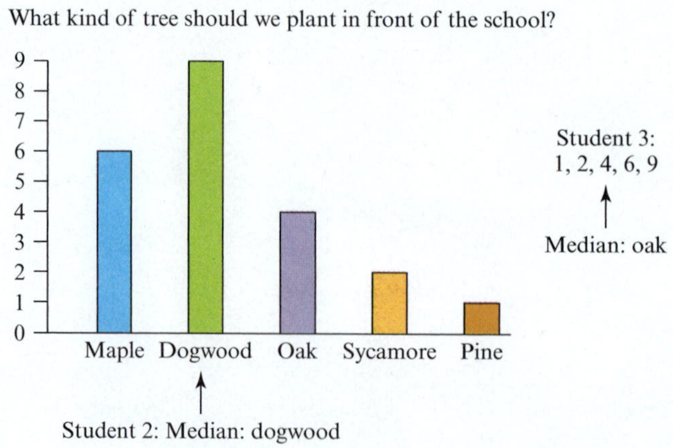

15.4 Summarizing, Describing, and Comparing Data Distributions

Class Activity 15Q What Does the Shape of a Data Distribution Tell About the Data?

CCSS CCSS SMP4, 6.SP.2

Examine histograms 1, 2, and 3 below and observe the different shapes these distributions take.

- The shape of histogram 1 is called *skewed to the right* because the histogram has a long tail extending to the right.
- The shape of histogram 2 is called *bimodal* because the histogram has two peaks.
- The shape of histogram 3 is called *symmetric* because the histogram is approximately symmetrical.

Histogram 1 Distribution of household income in the United States in 2007 (approximate)

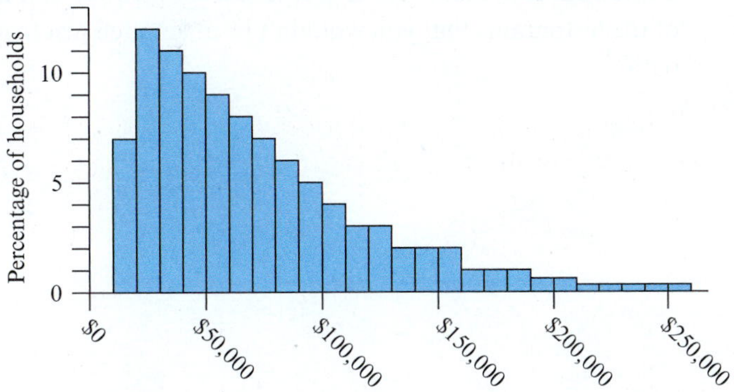

Histogram 2 Distribution of household income in hypothetical country A

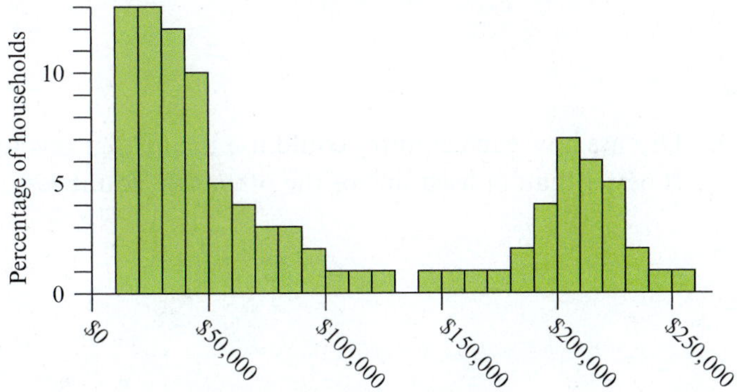

Histogram 3 Distribution of household income in hypothetical country B

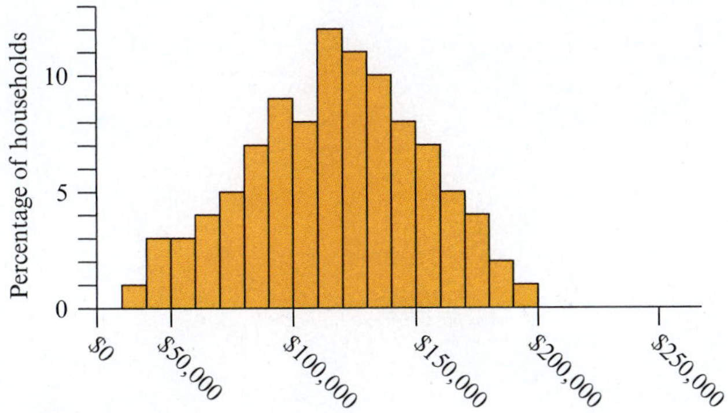

1. Write at least three questions about the graphs, including at least one question at each of the three levels of graph comprehension discussed in Section 15.2. Answer your questions.

2. Discuss what the shapes of histograms 1, 2, and 3 tell you about household income in the United States versus in the hypothetical countries A and B. What do you learn from the shape of the histograms that you wouldn't be able to tell just from the medians and means of the data?

3. Discuss how each country could use the histograms to argue that its economic situation is better than at least one of the other two countries.

Class Activity 15R Distributions of Random Samples

CCSS CCSS SMP4, 7.SP.2

For this activity, you will need a large collection of small objects (at least 50) in a bag. The objects should be identical, except that they should come in two different colors: 40% in one color and the remaining 60% in another color. Think of the objects as representing a group of voters. The 40% in one color represent "yes" votes and the 60% in the other color represent "no" votes.

1. You will be picking random samples of 10 from the bag and plotting the percentage of "yes" votes on a dot plot. Before you start picking these random samples, make a dot plot below to predict what your actual dot plot will look like approximately. Assume that you will plot about 30 dots.
   - Why do you think your dot plot might turn out that way?
   - How do you think the fact that 40% of the votes in the bag are "yes" votes might be reflected in the dot plot?
   - What kind of shape do you predict your dot plot will have?

Predicted

0 10 20 30 40 50 60 70 80 90 100

Percentage of "yes" votes in random samples of 10 objects

2. Now pick about 30 random samples of 10 objects from the bag. Each time you pick a random sample of 10, determine the percentage of "yes" votes and plot this percentage in a dot plot on the next page. Then return your sample to the bag.

   Compare your results with your predictions in part 1. Do you see the fact that 40% of the votes in the bag are "yes" votes reflected in the dot plot? If so, how? What kind of shape does the dot plot have?

For part 2: ____

0 10 20 30 40 50 60 70 80 90 100

Percentage of "yes" votes in random samples of 10 objects

For part 3: ____

0 10 20 30 40 50 60 70 80 90 100

Percentage of "yes" votes in random samples of 20 objects, rounded to the nearest ten

3. Next, you will repeat part 2, but this time you will pick random samples of 20 objects from the bag and you will round your percentages to the nearest ten. Before you start, how do you think this new dot plot will compare with your dot plot for part 2? After you are done, compare the two dot plots.

4. Your bag contains 40% "yes" votes. What range of percentages would you consider to be "pretty good predictions" of this 40%? For this range:

   a. What percent of the dots in your dot plot for part 2 are "pretty good predictions"?

   b. What percent of the dots in your dot plot for part 3 are "pretty good predictions"?

   c. If you were to pick random samples of 25 objects from the bag, how do you think the percentage of "pretty good predictions" would compare with (a) and (b)?

5. Compare the two histograms below. The first histogram shows the percentage of "yes" votes in 200 samples of 100 taken from a population of 1,000,000 in which 40% of the population votes "yes." The second histogram shows the percentage of "yes" votes in 200 samples of 1000 taken from the same population.

a. Compare the way the data are distributed in each of these histograms and compare these histograms with your dot plots in parts 2 and 3.

b. How is the fact that 40% of the population votes "yes" reflected in these histograms?

c. What do the histograms indicate about using samples of 100 versus samples of 1000 to predict the outcome of an election?

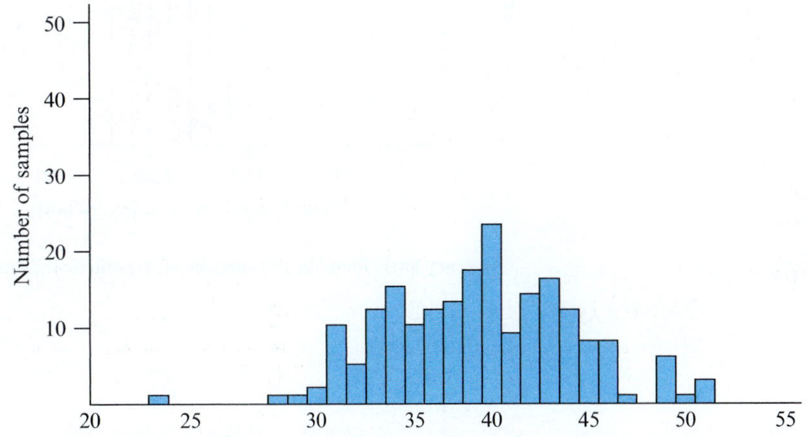

Percentage of "yes" votes in **samples of 100** taken from 1,000,000 voters in which 40% vote "yes."

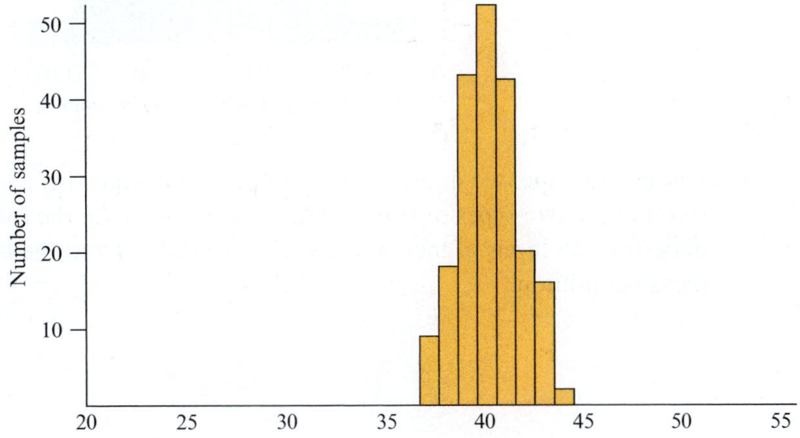

Percentage of "yes" votes in **samples of 1000** taken from 1,000,000 voters in which 40% vote "yes."

6. What if we made a histogram like the ones above by using the same population, but by picking 200 samples of 500 (instead of 200 samples of 100 or 1000)? How do you think this histogram would compare with the ones above? What if samples of 2000 were used?

Class Activity 15S Comparing Distributions: Mercury in Fish

CCSS CCSS SMP4, 6.SP.2

The two histograms below display hypothetical data about amounts of mercury found in 100 samples of each of two different types of fish. Mercury levels above 1.00 parts per million are considered hazardous.

Mercury levels found in 100 samples of hypothetical fish A

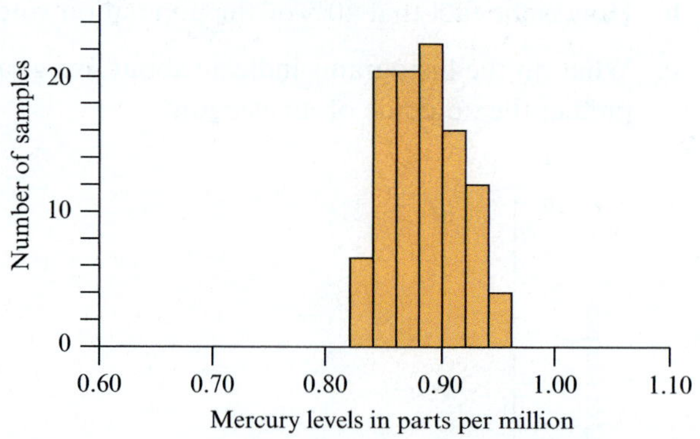

Mercury levels found in 100 samples of hypothetical fish B

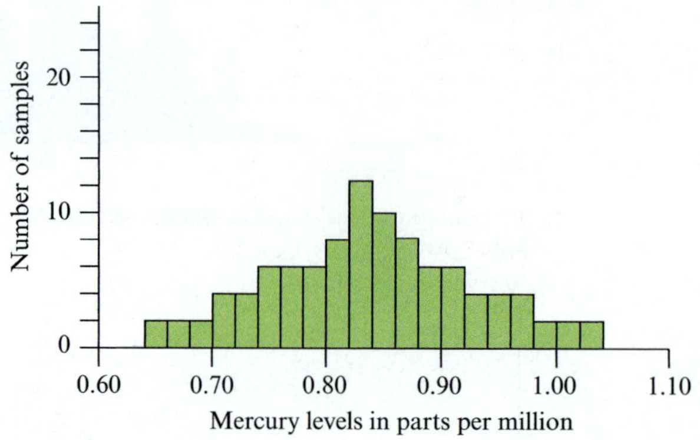

1. Discuss how the two distributions compare and what this tells you about the mercury levels in the two types of fish. In your discussion, take the following into account: median or mean levels of mercury in each type of fish, and the hazardous level of 1.00 parts per million.

2. In comparing the two types of fish, if you hadn't been given the histograms, would it be adequate just to have the medians or means of the amount of mercury in the samples, or is it useful to know additional information about the data?

Class Activity 15T Using Medians and Interquartile Ranges to Compare Data

CCSS CCSS SMP4, 6.SP.3, 6.SP.5

1. Determine the median, the first and third quartiles, and the interquartile range for each of the hypothetical taste-tester data shown in the following dot plots:

Taste-tester ratings of Coffee A:

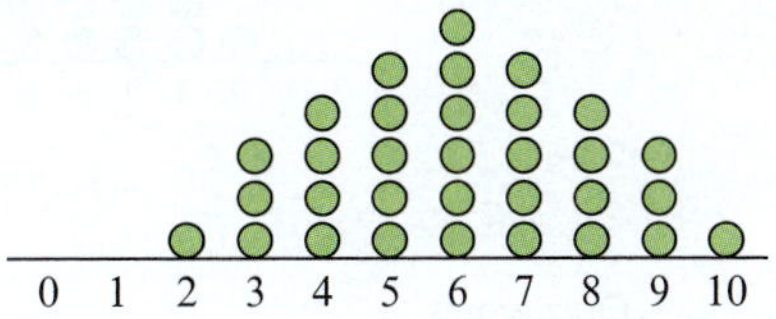

Taste-tester ratings of Coffee B:

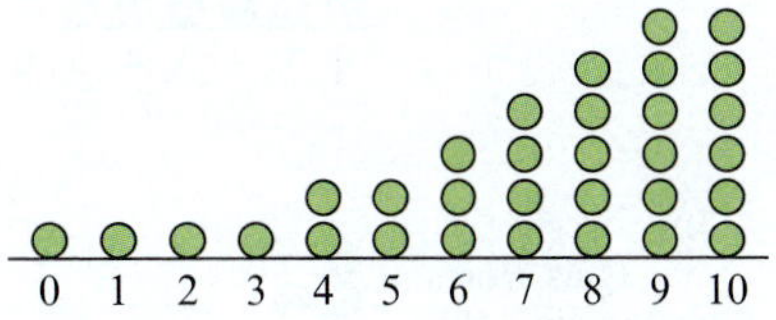

Taste-tester ratings of Coffee C:

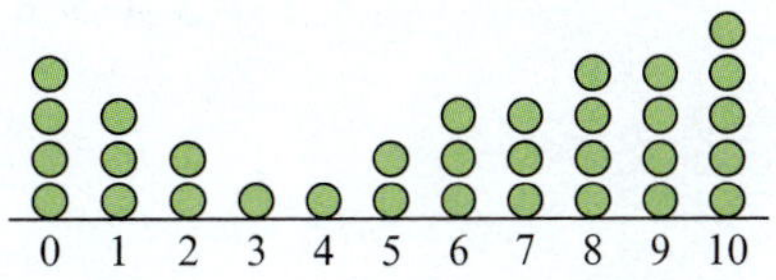

2. Discuss and compare the ratings of the three coffees using the medians and interquartile ranges from part 1. Which coffee seems to taste best? Which coffee seems to be most consistent?

Class Activity 15U Using Box Plots to Compare Data

CCSS CCSS SMP4, 6.SP.4

1. Make box plots for the 3 dot plots of hypothetical quiz scores in 3 classes that follow.

Quiz scores in class A:

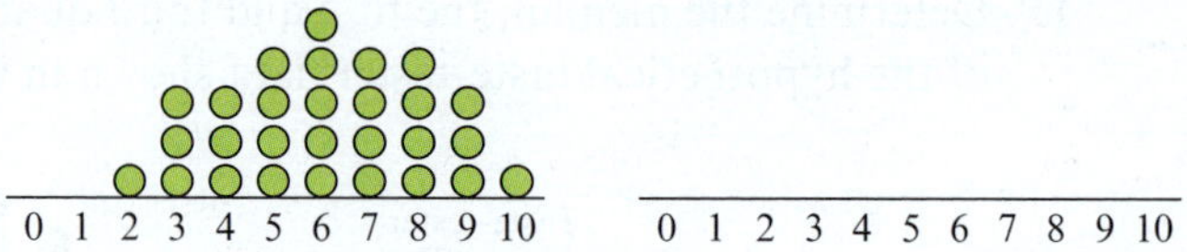

Quiz scores in class B:

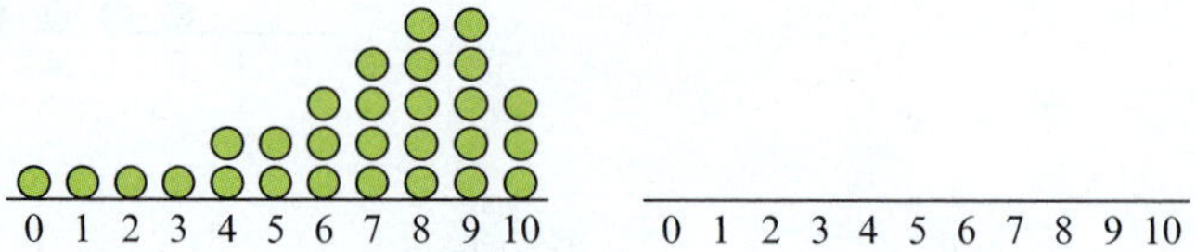

Quiz scores in class C:

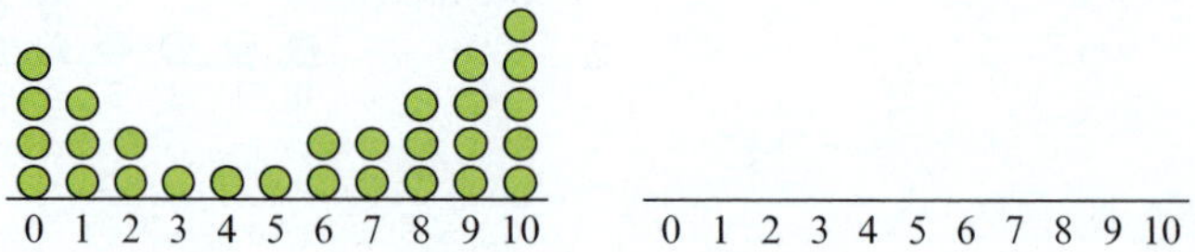

2. Suppose you had the box plots from part 1, but you didn't have the dot plots. Discuss what you could tell about how the 3 data sets are distributed. Compare how the three classes did on the quiz.

Class Activity 15V Percentiles Versus Percent Correct

CCSS CCSS SMP3

1. Determine the 75th percentile for each set of hypothetical test data in the 3 dot plots below.

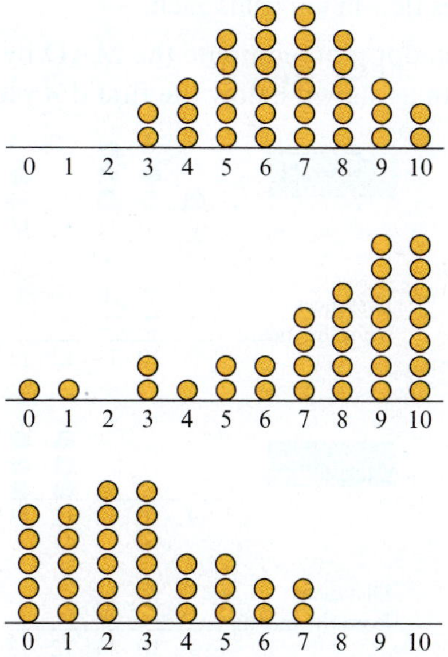

2. Discuss: On a test, is the 75th percentile the same as 75% correct?

3. Mrs. Smith makes an appointment to talk to her son Johnny's teacher. Johnny has been getting As in math, but on the standardized test he took, he was at the 80th percentile. Mrs. Smith is concerned that this means Johnny is really doing B work in math, not A work. If you were Johnny's teacher, what could you tell Mrs. Smith?

Class Activity 15W Comparing Paper Airplanes

CCSS CCSS SMP4, 6.SP.3, 6.SP.5c

The dot plots below show the distances, rounded to the nearest foot, that three different models of paper airplanes flew in ten trials each.

1. For each dot plot, compute the MAD by first replacing each dot with its distance from the mean as shown below the first dot plot.

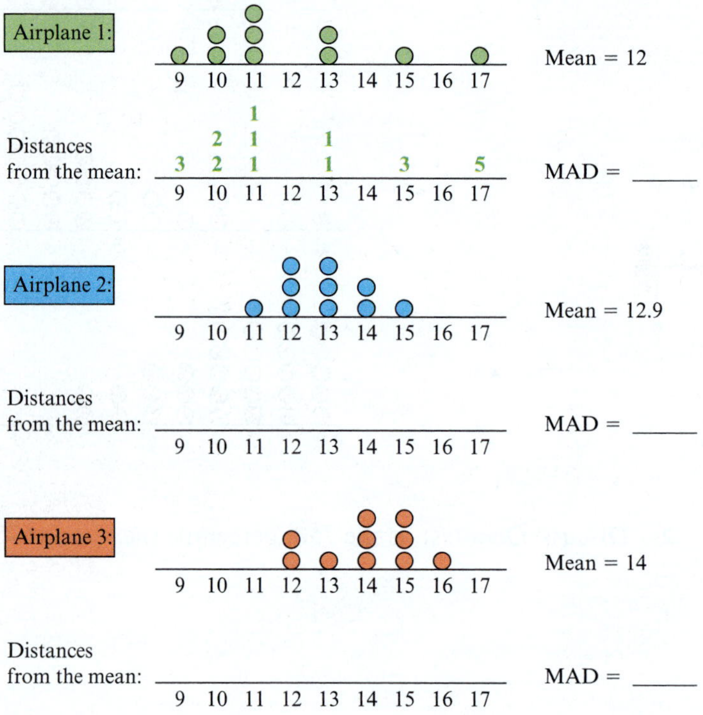

2. Use the means and the MADs from part 1 to discuss how the three paper airplanes compare. What do the means tell you? What do the MADs tell you? How confident can you be in saying that one airplane flies farther than another?

3. Suppose there is a fourth paper airplane for which the mean is 13.9 and the MAD is 1.1. Discuss how this fourth airplane compares with airplanes 2 and 3. Discuss how confident you can be in saying that one airplane flies farther than another.

Class Activity 15X Why Does Variability Matter When We Compare Groups?

CCSS CCSS SMP4, 7.SP.3, 7.SP.4

Some researchers are doing an experiment to see if a treatment is effective for an illness. They randomly assign patients to two groups: a treatment group and a control group. They then determine how long it takes each patient to recover from the illness.

In this activity you will consider two hypothetical scenarios for the experiment and think about why we should take measures of center *and* of variability into account when comparing the treatment and control groups.

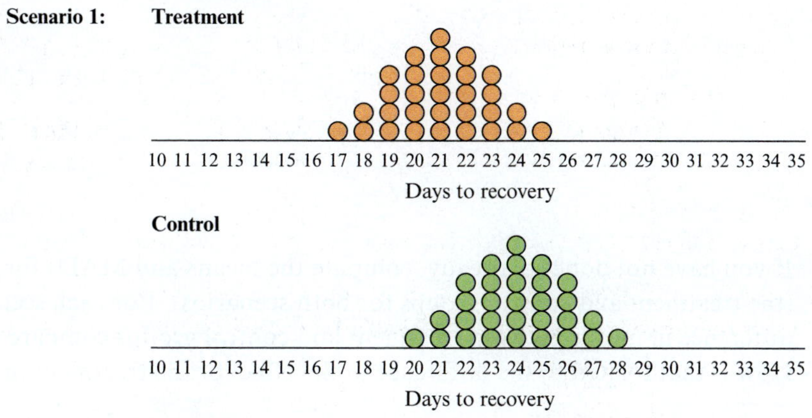

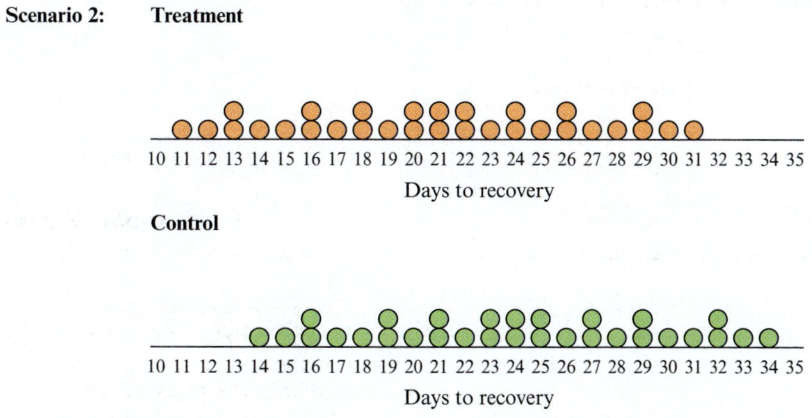

1. Compare the two hypothetical scenarios. How are they similar? How are they different?

2. If you have not done so already, discuss the following questions.
   a. How much better does the treatment group do than the control group? Can you quantify that?
   b. Does the treatment seem to have more of an effect in one scenario than in the other? How so?

3. If you have not done so already, compute the means and MADs for all four data sets (the treatment and control groups for both scenarios). For each scenario, how does the difference in means between treatment and control groups compare with the MAD? How is that related to the difference in the effect of the treatment in the two scenarios?

Class Activity 15Y How Well Does Dot Paper Estimate Area?

CCSS CCSS SMP4, 7.SP.3, 7.SP.4

You will need scissors and Downloads 27 and 28 for this activity.

Biologists sometimes use dot paper to estimate areas of leaves. If the dots are on a 1-inch grid, as in Download 27, then the number of dots the leaf covers gives an estimate for the area of the leaf in square inches.

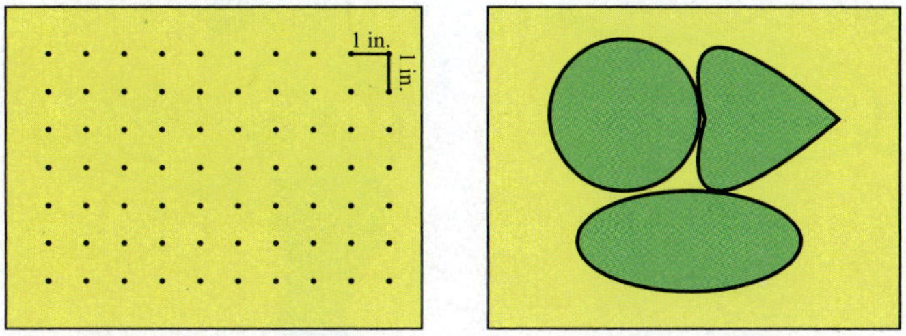

1. Cut out the circle, the oval, and the leaf-shape on Download 28. For each of those three shapes:
   - make a dot plot to show numbers of dots the shape covers as the shape is placed randomly on the dot paper (count only dots that are at least half-covered by the shape);
   - compute the mean number of dots the shape covers;
   - compute the MAD of the number of dots the shape covers.

2. Use your mean numbers of dots from part 1 as estimates for the areas (in square inches) of the three shapes. Based on these estimates, which shape has the least area and which has the greatest?

3. Discuss and use the variability you found in your data in part 1.
   - Which shape exhibited the least variability? Are you surprised?
   - Use the variability to discuss how close your estimated areas from part 2 are likely to be to the actual areas and how confident you are in identifying which shape has the least area and which has the greatest.

4. Compare your results from parts 2 and 3 with this information: the circle has radius 2 inches and the oval has the same area as the circle.

CLASS ACTIVITIES 16

Probability

16.1 Basic Principles of Probability

Class Activity 16A Probabilities with Spinners

CCSS CCSS SMP3, 7.SP.7a

1. Many children's games use "spinners." You can make a simple spinner by placing the tip of a pencil through a paper clip and holding the pencil so that its tip is at the center of the circle as shown. The paper clip should spin freely around the pencil tip.

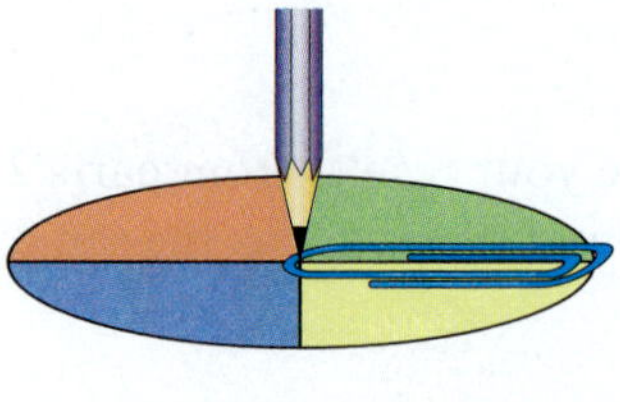

Compare the two spinners shown below. For which spinner is a paper clip most likely to point into a shaded region? Explain your answer.

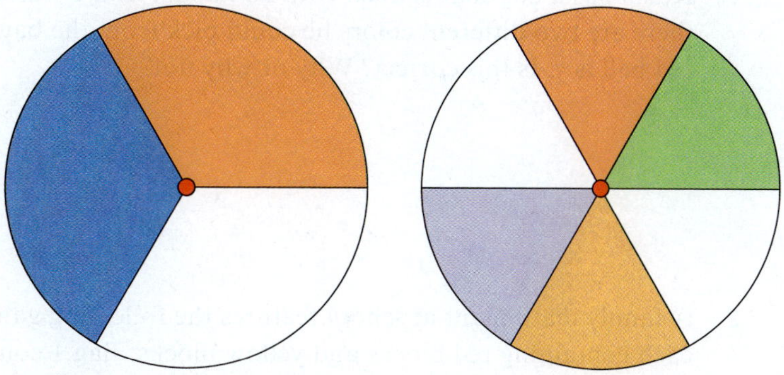

2. Compare the two spinners below. For which spinner is the paper clip that spins around a pencil point held at the indicated center point most likely to land in a shaded region? Explain your answer.

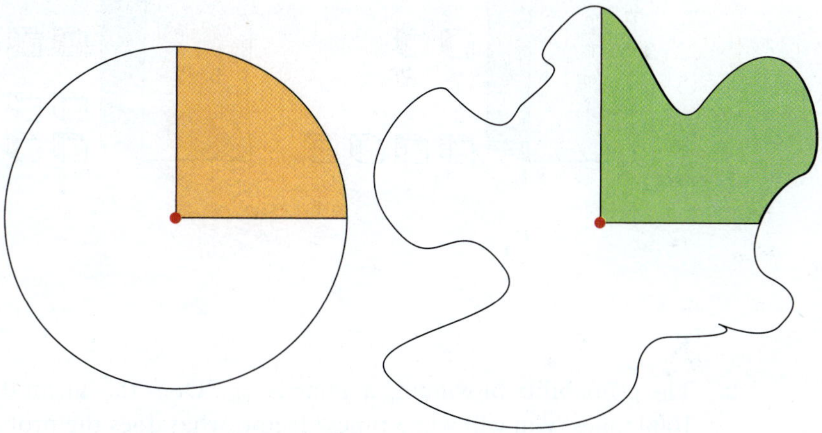

3. Draw a 4-color spinner (red, green, yellow, blue) such that

- landing on green is twice as likely as landing on red;
- landing on yellow is equally likely as landing on green;
- landing on blue is more likely than landing on yellow.

Determine the probabilities of landing on each of the colors on your spinner and explain your reasoning.

Could someone else make a different spinner that has the same properties above but has different probabilities than yours?

Class Activity 16B Critique Probability Reasoning

CCSS CCSS SMP3, 7.SP.6, 7.SP.7a

1. Kevin has a bag that is filled with 2 red balls and 1 white ball. Kevin says that because there are two different colors he could pick from the bag, the probability of picking the red ball is $\frac{1}{2}$. Is this correct? Why or why not?

2. A family math night at school features the following game. There are two opaque bags, each containing red blocks and yellow blocks. Bag 1 contains 2 red blocks and 4 yellow blocks. Bag 2 contains 4 red blocks and 16 yellow blocks. To play the game, you pick a bag and then you pick a block out of the bag without looking. You win a prize if you pick a red block. Eva thinks she should pick from bag 2 because it has more red blocks than bag 1. Is Eva more likely to pick a red block if she picks from bag 2 than from bag 1? Why or why not?

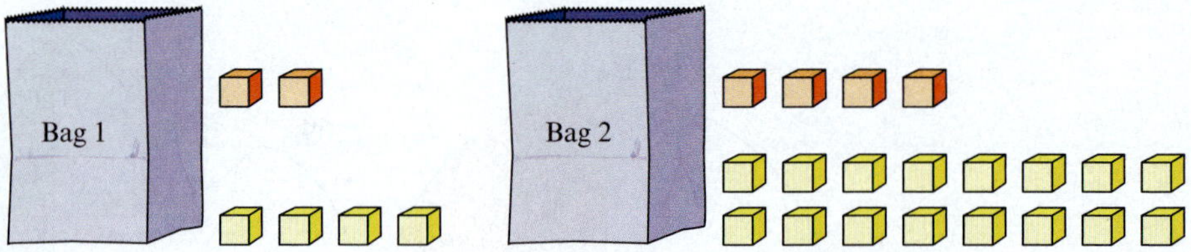

3. The probability of winning a game is $\frac{3}{1000}$. Does this mean that if you play the game 1000 times, you will win 3 times? If not, what does the probability of $\frac{3}{1000}$ stand for?

Class Activity 16C Empirical Versus Theoretical Probability: Picking Cubes from a Bag

CCSS CCSS SMP4, 7.SP.6

Each person (or small group) will need an opaque bag, 3 red cubes, 7 blue cubes, and a sticky note.

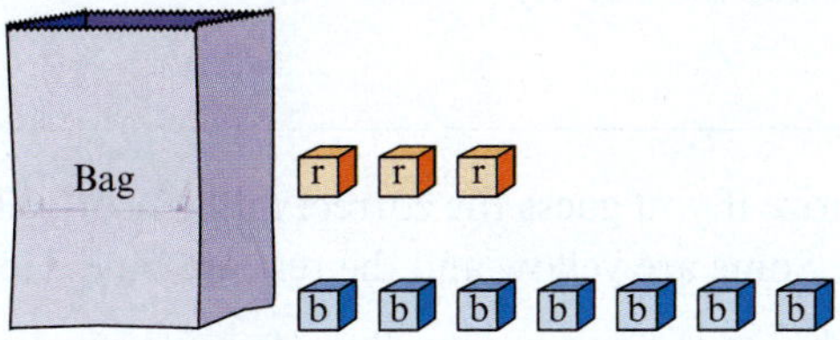

1. Put the 10 cubes in the bag, mix them up, and randomly pick a cube from the bag without looking. Record the color of the cube, and put the cube back in the bag. Repeat until you have picked 10 cubes. Record the number of red cubes you picked on your sticky note. Calculate the empirical probability of picking a red cube based on your 10 picks. Is it the same as the theoretical probability of picking red?

2. Use everyone's sticky notes from part 1 to create a class dot plot. How is the fact that there are 3 red cubes and 7 blue cubes in the bag reflected in the dot plot?

3. Use the class dot plot from part 2 to determine an empirical probability of picking a red cube from a bag containing 3 red cubes and 7 blue cubes. Compare this empirical probability with the theoretical probability of picking red.

Class Activity 16D Using Empirical Probability to Make Predictions

CCSS CCSS SMP4, 7.SP.6

A family math night at school includes the following activity. A bag is filled with 10 small counting bears that are identical except that 4 are yellow and the rest are blue. A sign next to the bag gives instructions for the activity:

> Win a prize if you guess the correct number of yellow bears in the bag! There are 10 bears in the bag. Some are yellow and the rest are blue. Here's what you do:
>
> - Reach into the bag, mix well, and pick out a bear.
> - Get a sticky note that is the same color as your bear, write your name and your guess on the note and add your note to the others of the same color.
> - Put your bear back in the bag, and mix well.

The sticky notes will be organized into columns of 10, so it will be easy to count up how many of each there are.

1. How will students be able to use the results of this activity to estimate the number of yellow bears in the bag?

2. What do you expect will happen as the night goes on and more and more bears are picked?

3. Discuss any additions or modifications you would like to make to the activity if you were going to use it for math night at your school.

Class Activity 16E If You Flip 10 Pennies, Should Half Come Up Heads?

CCSS CCSS SMP4, 7.SP.7b

You will need a bag, 10 pennies or 2-color counters, and some sticky notes for this activity.

1. Make a guess: What do you think the probability is of getting exactly 5 heads on 10 pennies when you dump the 10 pennies out of the bag?

2. Put the 10 pennies in the bag, shake them up, and dump them out. Record the number of heads on a sticky note. Repeat this for a total of 10 times, using a new sticky note each time. Out of these 10 tries, how many times did you get 5 heads? Therefore, what is the experimental probability of getting 5 heads based on your 10 trials?

3. Now work with a large group (e.g., the whole class). Collect the whole group's data on the sticky notes from part 2. Find a way to display these data so that you can see how often the whole group got 5 heads and other numbers of heads.

4. Is the probability of getting exactly 5 heads from 10 coins 50%? What does your data display from part 3 suggest?

16.2 Counting the Number of Outcomes

Class Activity 16F How Many Keys Are There?

CCSS CCSS SMP4, 7.SP.8b

Have you ever wondered: how can millions of *different* car keys be produced, even though car keys are not very big? Keys are manufactured to be distinct from one another by the way they are notched. Car keys have intricate notching. For simplicity, in this activity let's consider only simple keys that are notched on one side.

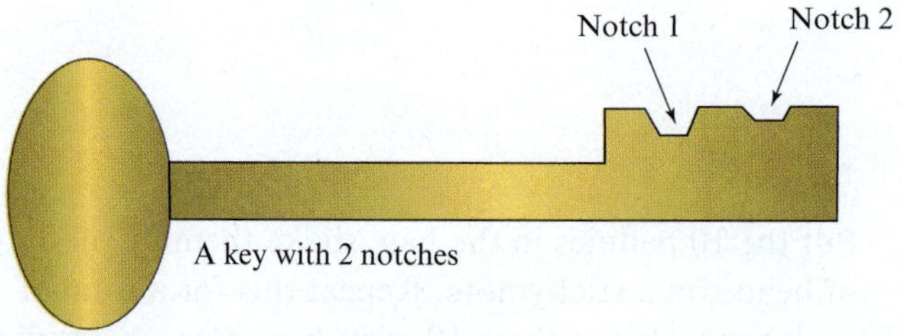

1. Suppose a simple type of key is to be made with 2 notches, and that each notch can be one of 3 depths: deep, medium, or shallow. How many different keys can be made this way? Explain.

2. Explain how to use multiplication to solve the problem in part 1 if you haven't already.

3. Now suppose the key is to be made with 4 notches, and each notch can be one of 3 depths: deep, medium, or shallow. How many keys can be made this way? Explain.

4. Now suppose the key is to be made with 10 notches, and each notch can be one of 5 depths. How many keys can be made this way? Explain.

Class Activity 16G Counting Outcomes: Independent Versus Dependent

CCSS CCSS SMP4, 7.SP.8b

1. How many 3-letter security codes can be made from the 4 letters A, B, C, D? For example, BAB and ABB are two such codes, and DAC is another. Explain.

2. How many 3-letter security codes can be made from the 4 letters A, B, C, D without using a letter twice? For example, BAC and ADB are two such codes. Explain.

3. Explain how to use multiplication to solve the problem in part 2 if you haven't already.

4. Contrast how you use multiplication to solve the problems in parts 1 and 2 and explain the distinction.

16.3 Calculating Probabilities of Compound Events

Class Activity 16H Number Cube Rolling Game

CCSS CCSS SMP4, 7.SP.8

Maya, James, Kaitlyn, and Juan are playing a game in which they take turns rolling a pair of number cubes. Each child has chosen a "special number" between 2 and 12, and each student receives 8 points whenever the total number of dots on the two number cubes is their special number. (They receive their points regardless of who rolled the number cubes. Their teacher picked 8 points so that the students would practice counting by 8s.)

- Maya's special number is 7.
- James's special number is 10.
- Kaitlyn's special number is 12.
- Juan's special number is 4.

The first person to get to 100 points or more wins. The students have played several times, each time using the same special numbers. They notice that Maya wins most of the time. They are wondering why.

1. Roll a pair of number cubes many times, and record the total number of dots each time. Display your data so that you can compare how many times each possible number between 2 and 12 has occurred. What do you notice?

2. Draw an array showing all possible outcomes on each number cube when a pair of number cubes are rolled. (Think of the pair as *number cube 1* and *number cube 2*.)
   a. For which outcomes is the total number of dots 7? 10? 12? 4?
   b. What is the probability of getting 7 total dots on a roll of two number cubes? What is the probability of getting 10 total dots on a roll of two number cubes? What about for 12 and 4?
   c. Is it surprising that Maya kept winning?

Class Activity 16I Picking Two Marbles from a Bag of 1 Black and 3 Red Marbles

CCSS CCSS SMP4, 7.SP.8

You will need an opaque bag, 3 red marbles, and 1 black marble for this activity. Put the marbles in the bag.

If you reach in without looking and randomly pick out 2 marbles at once, what is the probability that 1 of the 2 marbles you pick is black? You will study this question in this activity.

1. Before you continue, make a guess: What do you think the probability of picking the black marble is when you randomly pick 2 marbles out of the 4 marbles (3 red, 1 black) in the bag?

2. Pick 2 marbles out of the bag at once. Repeat this many times, recording what you pick each time. What fraction of the times did you pick the black marble?

3. Now calculate the probability theoretically, using a tree diagram. For the purpose of computing the probability, think of first picking one marble, then (without putting this marble back in the bag) picking a second marble. From this point of view, draw a tree diagram that will show all possible outcomes for picking the two marbles. But draw this tree diagram in a special way, *so that all outcomes shown by your tree diagram are equally likely.*

   Hints: The first stage of the tree should show all possible outcomes for your first pick. Remember that all branches you show should be equally likely. In the second stage, the branches you draw should depend on what happened in the first stage. For instance, if the first pick was the black marble, then the second pick must be one of the three red marbles.

   a. How many total outcomes for picking 2 marbles, 1 at a time, out of the bag of 4 (3 red, 1 black) does your tree diagram show?

   b. In how many outcomes is the black marble picked (on 1 of the 2 picks)?

   c. Use your answers to parts 3 (a) and (b) and the basic principles of probability to calculate the probability of picking the black marble when you pick 2 marbles out of a bag filled with 1 black and 3 red marbles.

4. Why was it important to draw the tree diagram so that all outcomes were equally likely?

5. Here's another method for calculating the probability of picking the black marble when you pick 2 marbles out of a bag filled with 1 black and 3 red marbles:

a. How many unordered pairs of marbles can be made from the 4 marbles in the bag?

b. How many of those pairs of marbles in part (a) contain the black marble? (Use your common sense.)

c. Use parts (a) and (b) and basic principles of probability to determine the probability of picking the black marble when you pick 2 marbles out of a bag containing 1 black and 3 red marbles.

6. Compare your answers to parts 3(a) and 5(a), and compare your answers to 3(b) and 5(b). How and why are they different?

Class Activity 16J Critique Probability Reasoning About Compound Events

CCSS CCSS SMP3, 7.SP.8

1. Simone has been flipping a coin and has just flipped 5 heads in a row. She says that because she has just gotten so many heads, she is more likely to get tails than heads the next time she flips. Is Simone correct? What is the probability that her next flip will be a tail? Does the answer depend on what the previous flips were?

2. Let's say you flip 2 coins simultaneously. There are 3 possible outcomes: Both are heads, both are tails, or one is heads and the other is tails. Does this mean that the probability of getting one head and one tail is $\frac{1}{3}$?

Class Activity 16K Expected Earnings from the Fall Festival

CCSS CCSS SMP1, 7.SP.7a

Ms. Wilkins is planning a game for her school's fall festival. She will put 2 red, 3 yellow, and 10 green plastic bears in an opaque bag. (The bears are identical except for their color.) To play the game, a contestant will pick 2 bears from the bag, one at a time, without putting the first bear back before picking the second bear. Contestants will not be able to see into the bag, so their choices are random. To win a prize, the contestant must pick a green bear first and then a red bear. The school is expecting about 300 people to play the game. Each person will pay 50 cents to play the game. Winners receive a prize that costs the school $2.

1. How many prizes should Ms. Wilkins expect to give out? Explain.

2. How much money (net) should the school expect to make from Ms. Wilkins's game? Explain.

16.4 Using Fraction Arithmetic to Calculate Probabilities

Class Activity 16L Using the Meaning of Fraction Multiplication to Calculate a Probability

CCSS CCSS SMP4

Use the circle below, a pencil, and a paper clip to make a spinner as follows: Put the pencil through the paper clip, and put the point of the pencil on the center of the circle. The paper clip will now be able to spin freely around the circle.

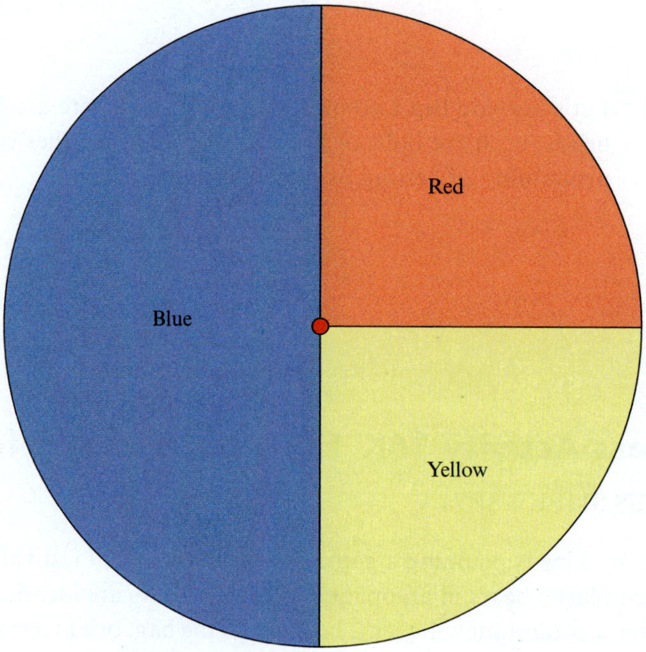

To win a game, Jill needs to spin a blue followed by a red in her next 2 spins.

1. What do you think Jill's probability of winning is? (Make a guess.)

2. Carry out the experiment of spinning the spinner twice in a row 20 times. (In other words, spin the spinner 40 times, but each experiment consists of 2 spins.) Out of those 20 times, how often does Jill win? What fraction of 20 does this represent? Is this close to your guess in part 1?

3. Calculate Jill's probability of winning theoretically as follows: Imagine that Jill carries out the experiment of spinning the spinner twice in a row many times. In the ideal, what fraction of those times should the first spin be blue? ______________ Show this by shading the rectangle below.

In the ideal, what fraction of those times when the first spin is blue should the second spin be red? ____________ Show this by further shading the rectangle below.

In the ideal, what fraction of pairs of spins should Jill spin first a blue and then a red? Therefore, what is Jill's probability of winning? ________ Explain how you can determine this fraction from the shading of the rectangle and from the meaning of fraction multiplication. Compare your answer with parts 1 and 2.

A rectangle representing many pairs of spins.

Class Activity 16M Using Fraction Multiplication and Addition to Calculate a Probability

CCSS CCSS SMP4

A paper clip, an opaque bag, and blue, red, and green tiles would be helpful.

A game consists of spinning the spinner in Class Activity 16L and then picking a small tile from a bag containing 1 blue tile, 3 red tiles, and 1 green tile. (All tiles are identical except for color, and the person picking a tile cannot see into the bag, so the choice of a tile is random.) To win the game, a contestant must pick the same color tile that the spinner landed on. So a contestant wins from either a blue spin followed by a blue tile or a red spin followed by a red tile.

1. Make a guess: What do you think the probability of winning the game is?
2. If the materials are available, play the game a number of times. Record the number of times you play the game (each game consists of both a spin *and* a pick from the bag), and record the number of times you win. What fraction of the time did you win? How does this compare with your guess in part 1?
3. To calculate the (theoretical) probability of winning the game, imagine playing the game many times. Answer the questions below in order to determine the probability of winning the game.

   a. In the ideal, what fraction of the time should the spin be blue? ________ Show this by shading the rectangle below.

   In the ideal, what fraction of those times when the spin is blue should the tile that is chosen be blue? ________ Show this by further shading the rectangle below.

   Therefore, in the ideal, what fraction of the time is the spin blue and the tile blue? ________ Explain how you can determine this fraction from the shading of the rectangle and from the meaning of fraction multiplication.

   b. In the ideal, what fraction of the time should the spin be red? ________ Show this by shading the rectangle below.

   In the ideal, what fraction of those times when the spin is red should the tile that is chosen be red? ________ Show this by further shading the rectangle below.

   Therefore, in the ideal, what fraction of the time is the spin red and the tile red? ________ Explain how you can determine this fraction from the shading of the rectangle and from the meaning of fraction multiplication.

   c. In the ideal, what fraction of the time should you win the game, and therefore, what is the probability of winning the game? Explain why you can calculate this answer by multiplying and adding fractions. Compare your answer with parts 1 and 2.

A rectangle representing playing the game many times.

ACTIVITY DOWNLOADS

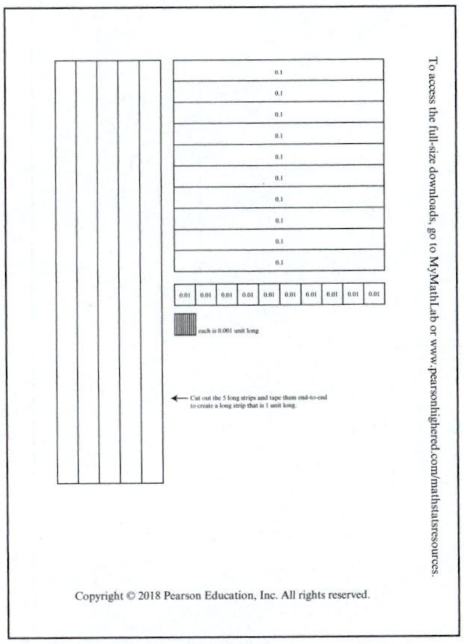

Activity Download 1

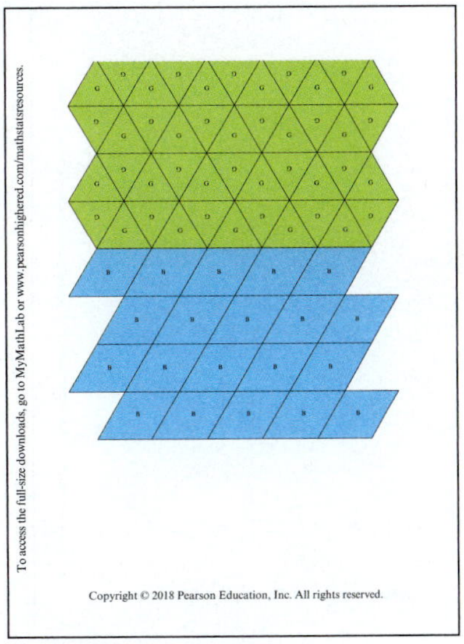

Activity Download 2

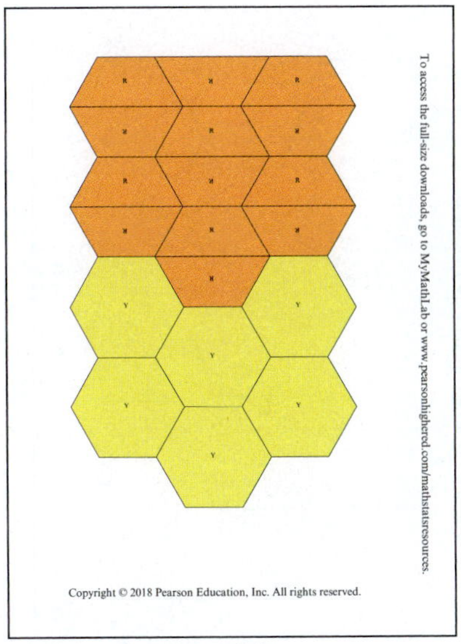

Activity Download 3

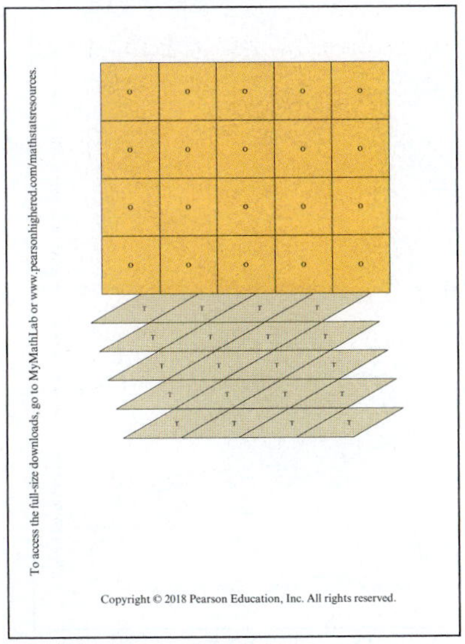

Activity Download 4

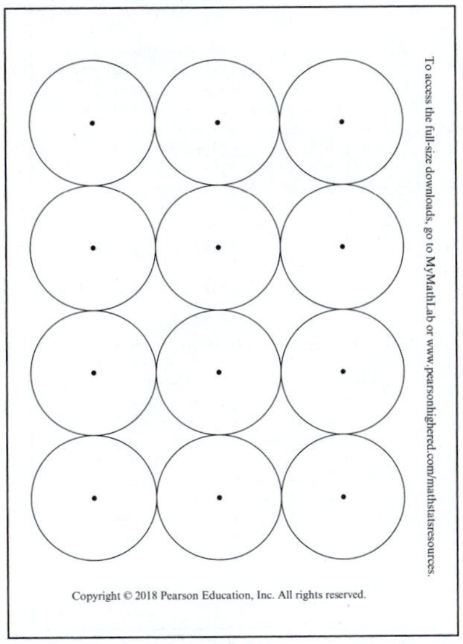

Activity Download 5

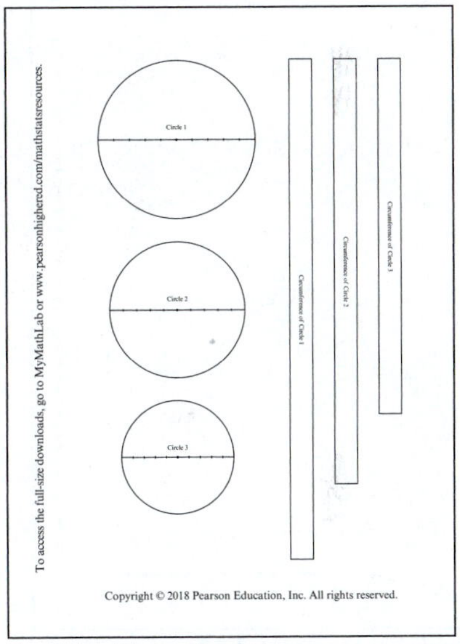

Activity Download 6

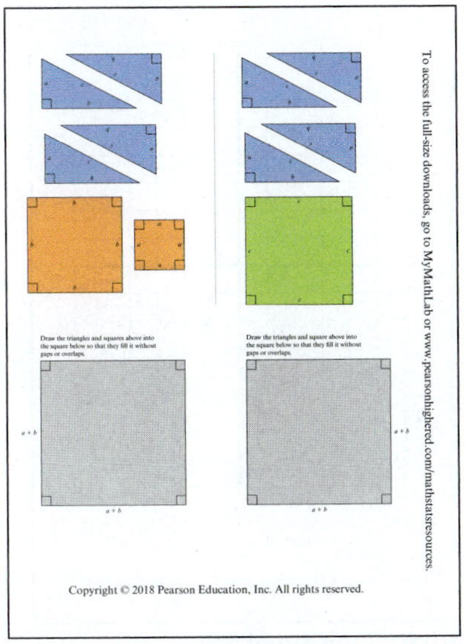

Activity Download 7

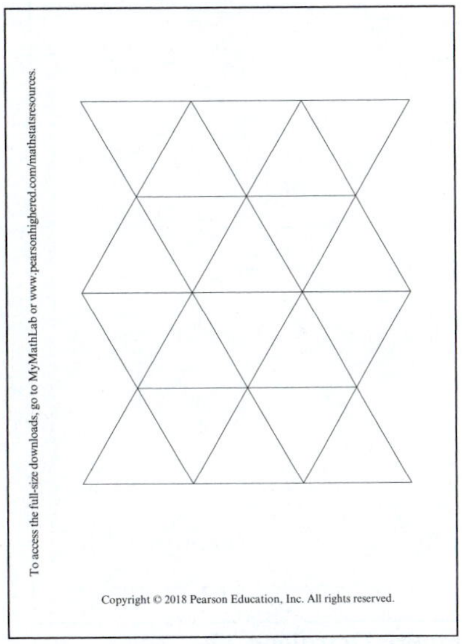

Activity Download 8

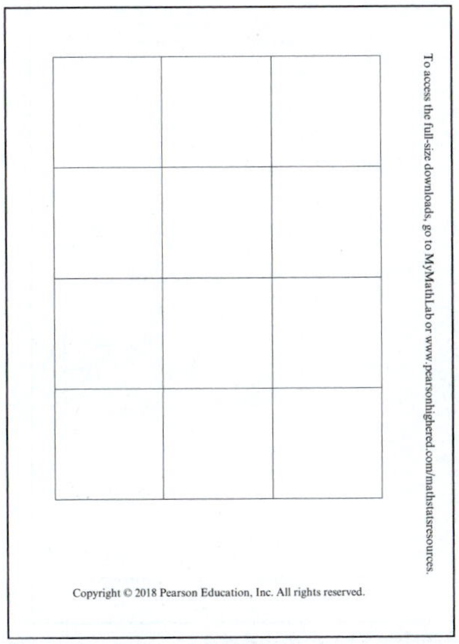

Activity Download 9

To access the full-size downloads, go to MyMathLab or www.pearsonhighered.com/mathstatsresources.

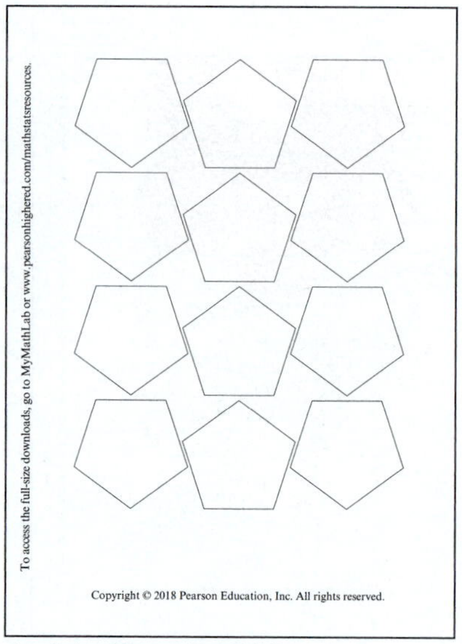

Activity Download 10

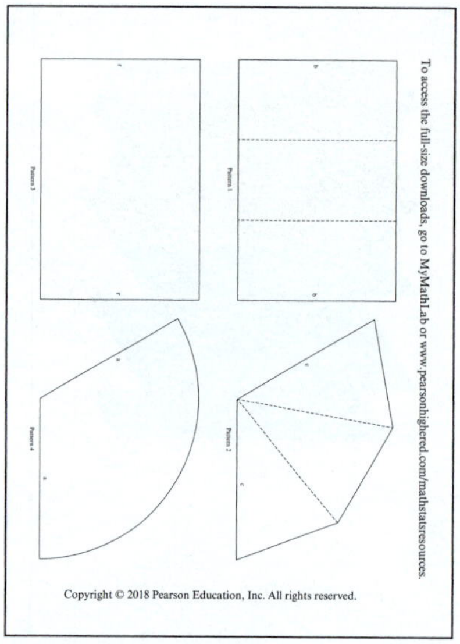

Activity Download 11

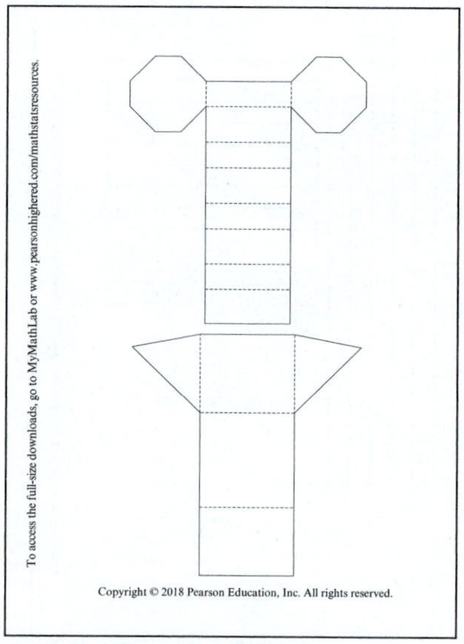

Activity Download 12

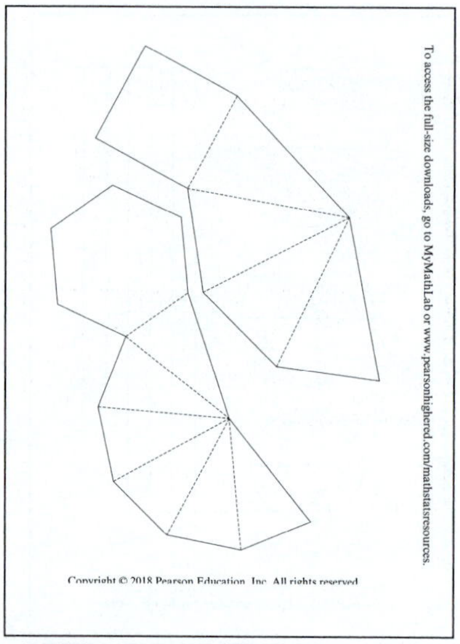

Activity Download 13

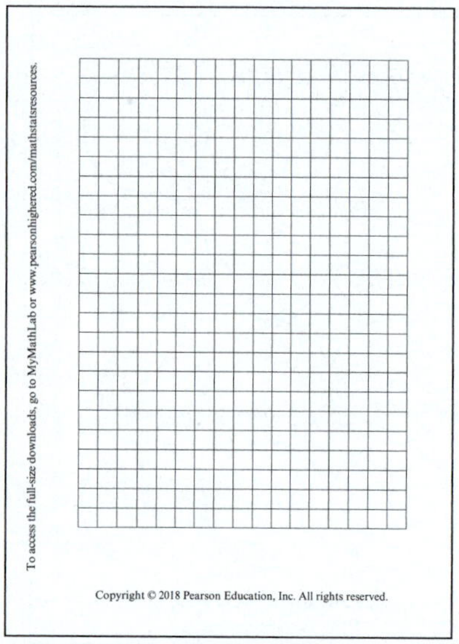

Activity Download 14

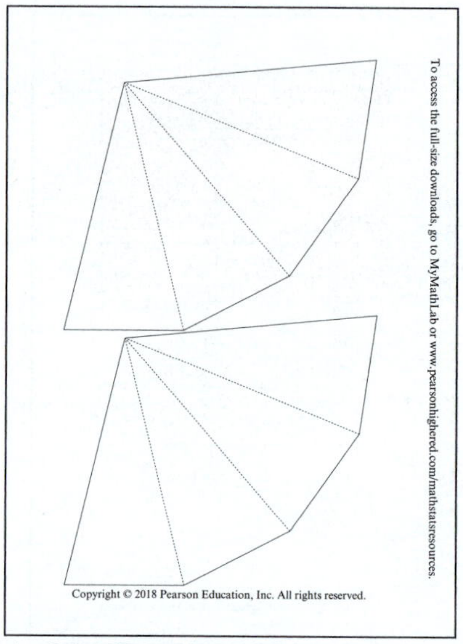

Activity Download 15

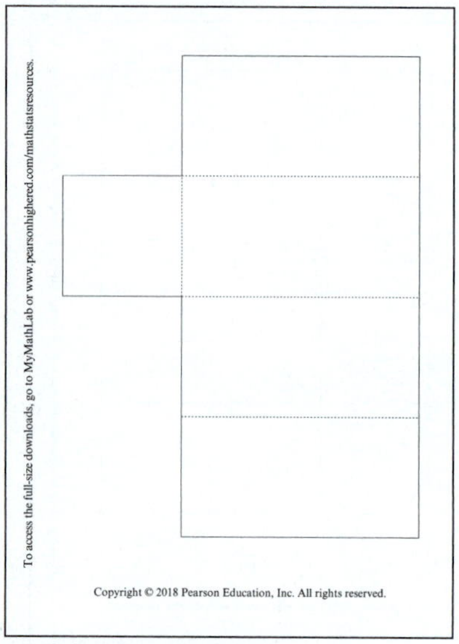

Activity Download 16

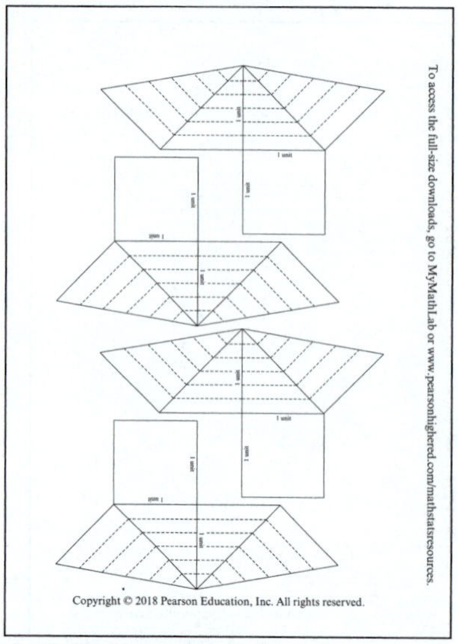

Activity Download 17

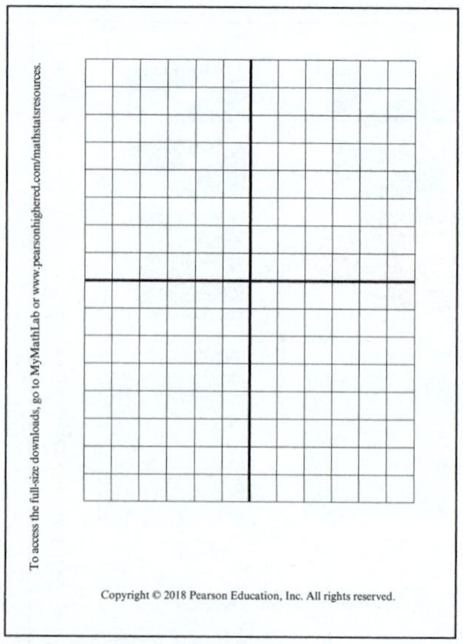

Activity Download 18

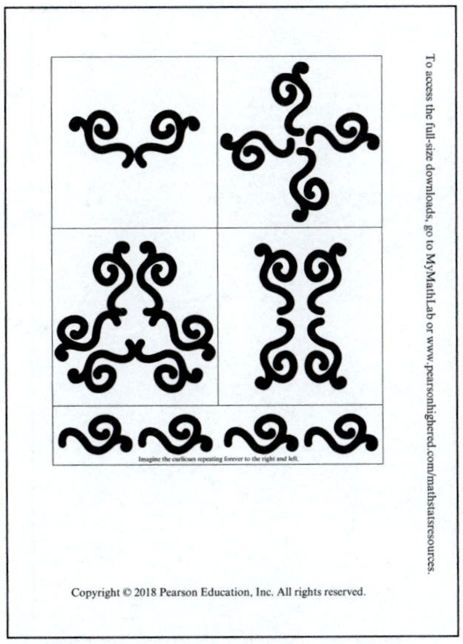

Activity Download 19

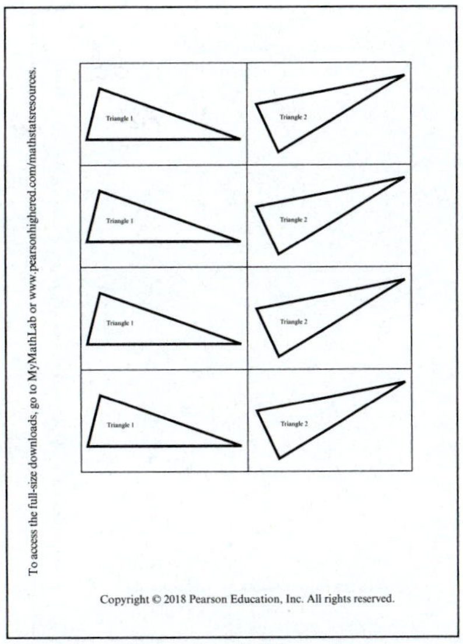

Activity Download 20

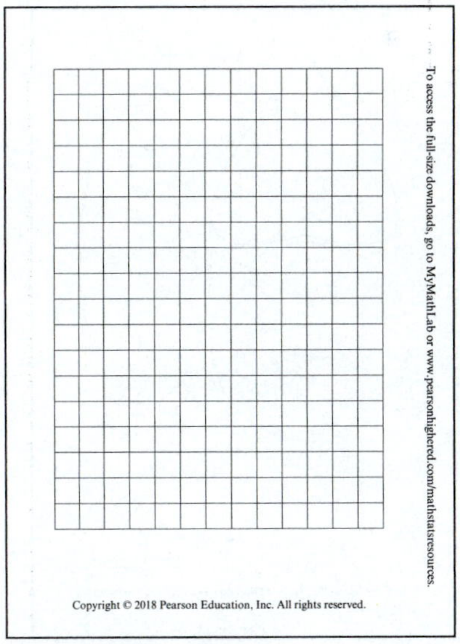

Activity Download 21

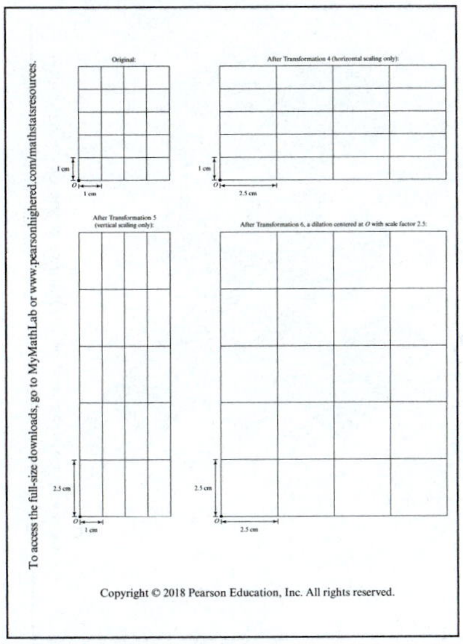

Activity Download 22

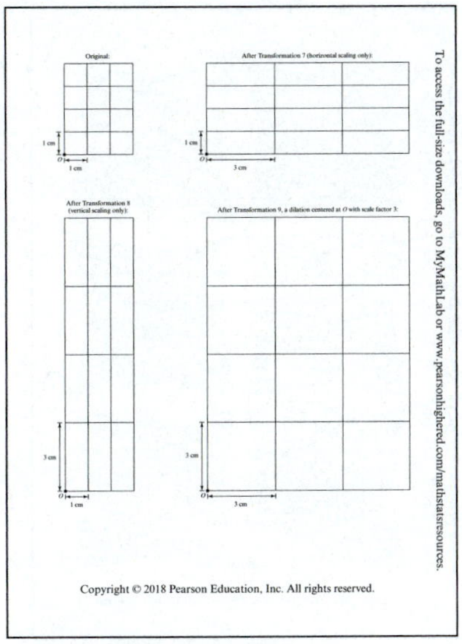

Activity Download 23

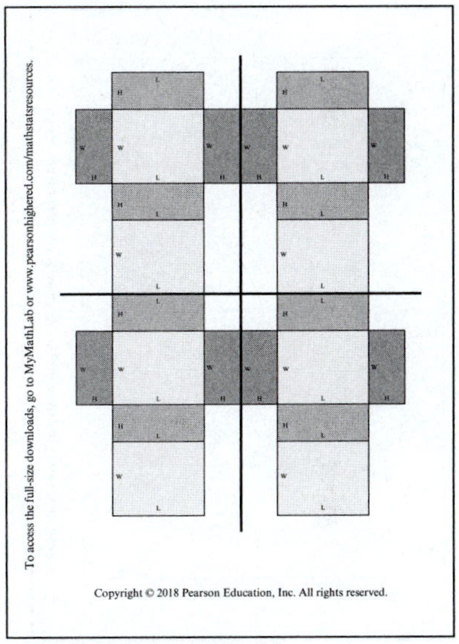

Activity Download 24

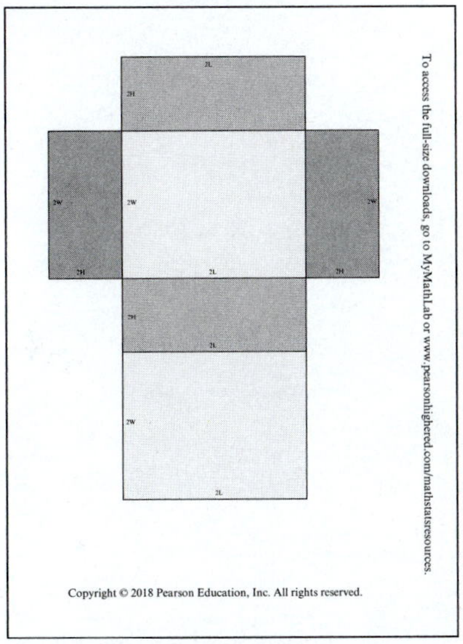

Activity Download 25

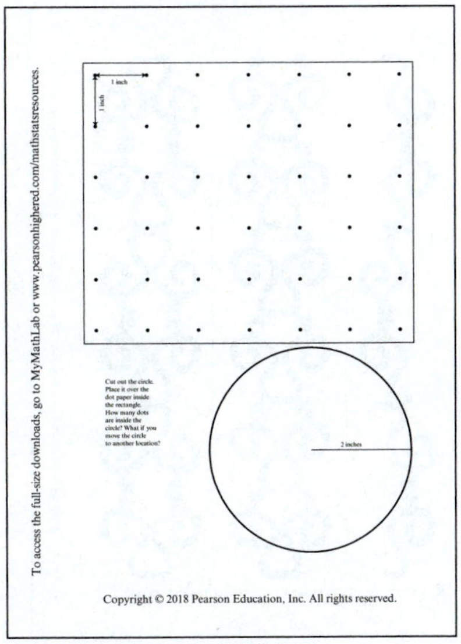

Activity Download 26

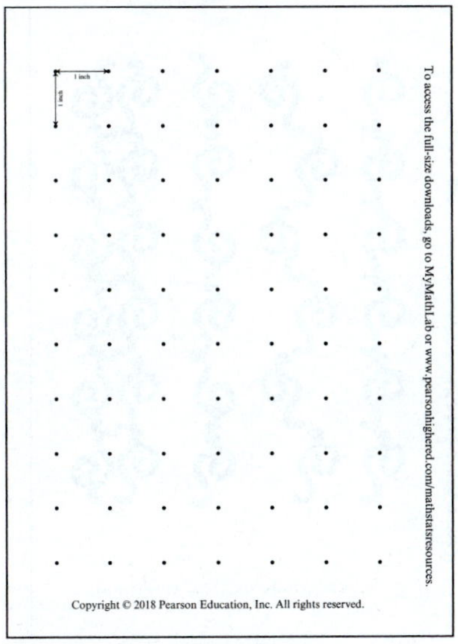

Activity Download 27

To access the full-size downloads, go to MyMathLab or www.pearsonhighered.com/mathstatsresources.

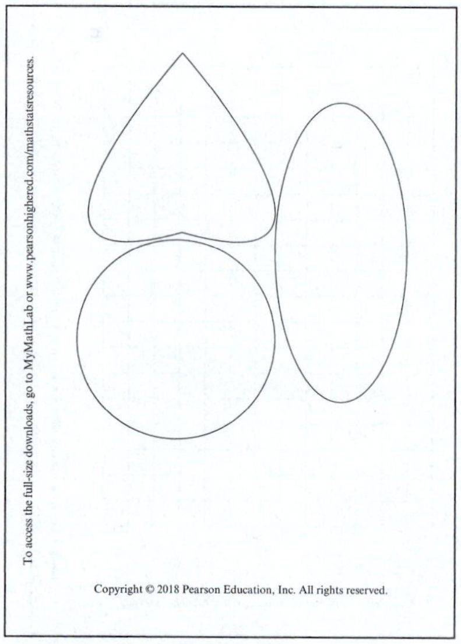

Activity Download 28

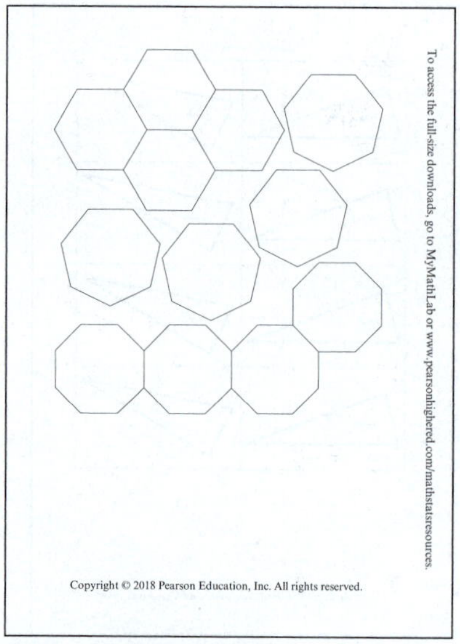

Activity Download 29

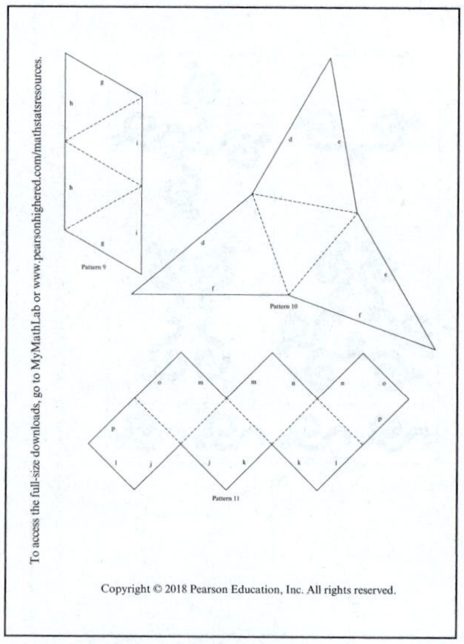

Activity Download 30

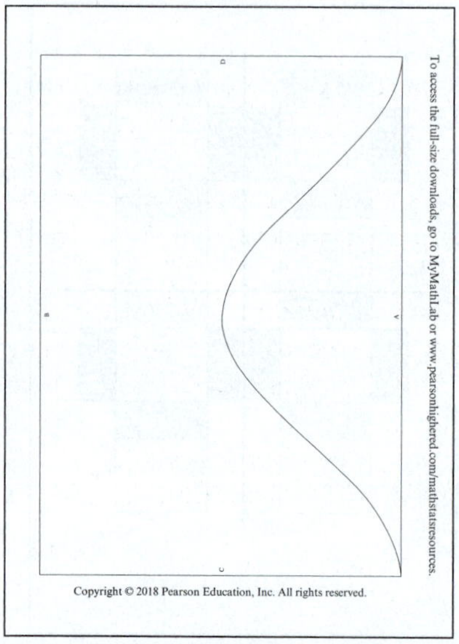

Activity Download 31

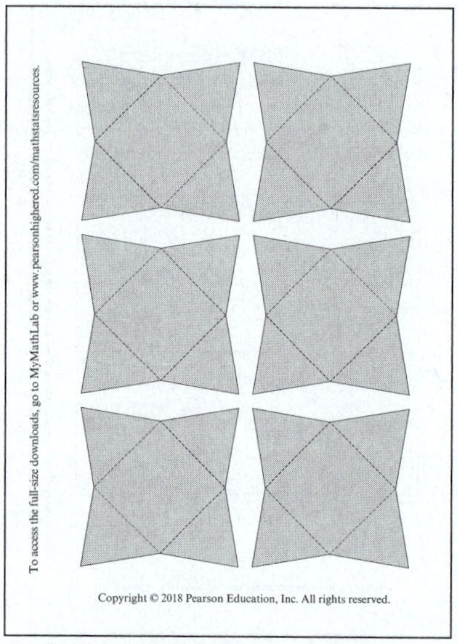

Activity Download 32

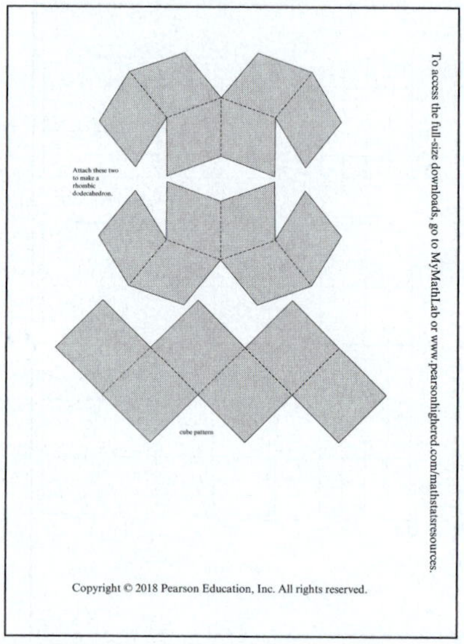

Activity Download 33

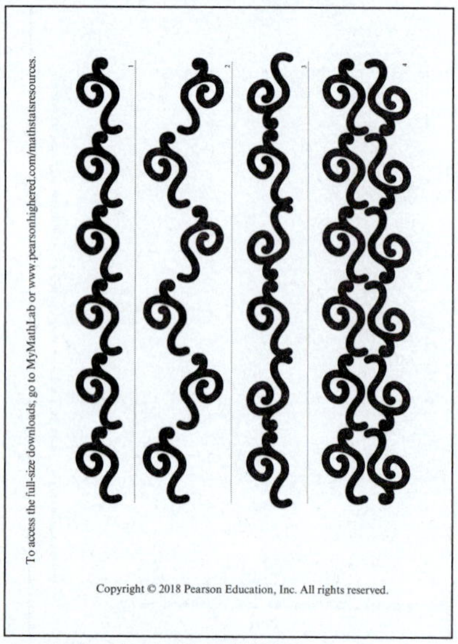

Activity Download 34

Activity Download 35

BIBLIOGRAPHY

[1] George W. Bright. Helping elementary- and middle-grades preservice teachers understand and develop mathematical reasoning. In *Developing Mathematical Reasoning in Grades K–12*, pages 256–269. National Council of Teachers of Mathematics, 1999.

[2] T. P. Carpenter, E. Fennema, M. L. Franke, S. B. Empson, and L. W. Levi. *Children's Mathematics: Cognitively Guided Instruction*. Heinemann, Portsmouth, NH, 1999.

[3] Frances Curcio. *Developing Data-Graph Comprehension in Grades K–8*. National Council of Teachers of Mathematics, 2d ed., 2001.

[4] K. C. Fuson. Developing mathematical power in whole number operations. In J. Kilpatrick, W. G. Martin, and D. Schifter, eds. *A Research Companion to Principles and Standards for School Mathematics*. Reston, VA, NCTM, 2003.

[5] K. C. Fuson, S. T. Smith, and A. M. Lo Cicero. Supporting First Graders' Ten-Structured Thinking in Urban Classrooms. *Journal for Research in Mathematics Education*, vol. 28, Issue 6, pages 738–766, 1997.

[6] Graham Jones and Carol Thornton. *Data, Chance, and Probability, Grades 1–3 Activity Book*. Learning Resources, 1992.

[7] Edward Manfre, James Moser, Joanne Lobato, and Lorna Morrow. *Heath Mathematics Connections*. D. C. Heath and Company, 1994.

[8] National Council of Teachers of Mathematics. *Navigating through Algebra in Pre-Kindergarten–Grade 2*. National Council of Teachers of Mathematics, 2001.

[9] National Research Council. *Adding It Up: Helping Children Learn Mathematics*. J. Kilpatrick, J. Swafford, and B. Findell, eds. Mathematics Learning Study Committee, Center for Education, Division of Behavioral and Social Sciences and Education. National Academy Press, Washington, DC, 2001.

[10] A. M. O'Reilley. Understanding teaching/teaching for understanding. In Deborah Schifter, ed., *What's Happening in Math Class?*, vol. 2: Reconstructing Professional Identities, pages 65–73. Teachers College Press, 1996.

[11] Dav Pilkey. *Captain Underpants and the Attack of the Talking Toilets*. Scholastic, 1999.

[12] Singapore Curriculum Planning and Development Division, Ministry of Education. *Primary Mathematics Workbook*, vol. 1A–6B. Times Media Private Limited, Singapore, 3d ed., 2000. Available at *http://www.singapore-math.com*

[13] K. Stacey. Traveling the Road to Expertise: A Longitudinal Study of Learning. In Chick, H. L. and Vincent, J. L. eds. *Proceedings of the 29th Conference of the International Group for the Psychology of Mathematics Education*, vol. 1, pp. 19–36. PME, Melbourne, 2005.

[14] V. Steinle, K. Stacey, and D. Chambers. *Teaching and Learning about Decimals* [CD-ROM]: Department of Science and Mathematics Education, The University of Melbourne, 2002. Online sample at *http://extranet.edfac.unimelb.edu.au/DSME/decimals/*

INDEX

CREDITS

Frontmattter
p. xi, xix, Sybilla Beckmann; p. xx (NSF logo), National Science Foundation.

Chapter 1
p. 1, Olena Bloshchynska/Shutterstock; p. 18, Sybilla Beckmann

Chapter 2
p. 41, Elisanth/Fotolia; p. 46 (daffodils), Image Source/Getty Images; p. 46 (boy doing yard work), Comstock/Stockbyte/Getty Images; p. 46 (children playing in water), Imagevixens/123RF; p. 46 (child building snowman), Tomsickova/Fotolia

Chapter 3
p. 92, Jeremy Richards/Fotolia

Chapter 4
p. 142, Ivonne Wierink/Fotolia; p. 164, Tatiana Popova/Shutterstock

Chapter 5
p. 196, Africa Studio/Fotolia

Chapter 6
p. 222, Coleman Yuen/Pearson Education Asia Ltd

Chapter 7
p. 281, Okea/Fotolia

Chapter 8
p. 336, Jenny Zhang/123RF

Chapter 9
p. 378, Kseniya Ganz/Fotolia

Chapter 10
p. 451, Sergii Mostovyi/Fotolia

Chapter 11
p. 492, Felix Lipov/Shutterstock

Chapter 12
p. 525, Galyna Andrushko/Shutterstock; p. 561, Janelle Lugge/Shutterstock

Chapter 13
p. 580, John Kasawa/Shutterstock; p. 584, Sybilla Beckmann; p. 585, Sybilla Beckmann

Chapter 14
p. 612, Lyusya_k/Fotolia; p. 623, Sybilla Beckmann; p. 628, Sybilla Beckmann; p. 629 (house construction), Sue Ashe/Shutterstock; p. 629 (folding laundry rack), Sybilla Beckmann; p. 629 (laundry rack), Sybilla Beckmann; p. 629 (skyscraper with crane), David Franklin/Fotolia; p. 633, Sybilla Beckmann; p. 636, Klubovy/E+/Getty Images; p. 637, Sybilla Beckmann; p. 658, Sybilla Beckmann

Chapter 15
p. 673, Zoonar/Sergii Dashkevych/Zoonar GmbH/Alamy Stock Photo

Chapter 16
p. 722, Yommy/Fotolia

Class Activities 1
p. CA-1, Olena Bloshchynska/Shutterstock

Class Activities 2
p. CA-21, Elisanth/Fotolia; CA-210, Sybilla Beckmann; CA-211, Sybilla Beckmann

Class Activities 3
p. CA-45, Jeremy Richards/Fotolia; p. CA-58, excerpt, Class Activity 3L, From Math 103 - Section 3.3, by Ashley Whitehead. Published by Prezi Inc., © 2013.

Class Activities 4
p. CA-67, Ivonne Wierink/Fotolia

Class Activities 5
p. CA-86, Africa Studio/Fotolia

Class Activities 6
p. CA-100, Coleman Yuen/Pearson Education Asia Ltd

Class Activities 7
p. CA-118, Okea/Fotolia

Class Activities 8
p. CA-140, Jenny Zhang/123RF

Class Activities 9
p. CA-158, Kseniya Ganz/Fotolia

Class Activities 10
p. CA-200, Sergii Mostovyi/Fotolia

Class Activities 11
p. CA-228, Felix Lipov/Shutterstock

Class Activities 12
p. CA-240, Galyna Andrushko/Shutterstock

Class Activities 13
p. CA-264, John Kasawa/Shutterstock

Class Activities 14
p. CA-280, Lyusya_k/Fotolia

Class Activities 15
p. CA-302, Zoonar/Sergii Dashkevych/Zoonar GmbH/Alamy Stock Photo

Class Activities 16
p. CA-340, Yommy/Fotolia